WIND ENGINEERING INTO THE 21ST CENTURY
VOLUME 3

PROCEEDINGS OF THE TENTH INTERNATIONAL CONFERENCE ON WIND ENGINEERING
COPENHAGEN / DENMARK / 21-24 JUNE 1999

Wind Engineering into the 21st Century

Edited by

A. Larsen
COWI Consulting Engineers and Planners, Lyngby, Denmark

G. L. Larose & F. M. Livesey
Danish Maritime Institute, Lyngby, Denmark

VOLUME 3

A.A. BALKEMA / ROTTERDAM / BROOKFIELD / 1999

The 10th International Conference on Wind Engineering has been organised by the Danish Wind Engineering Society

Cover photo:

The 10th ICWE logo is centred around a light blue cyclone, the logo of the Danish Wind Engineering Society.

The texts of the various papers in this volume were set individually by typists under the supervision of each of the authors concerned.

Published by
A.A. Balkema, P.O. Box 1675, 3000 BR Rotterdam, Netherlands
Fax: +31.10.413.5947; E-mail: balkema@balkema.nl; Internet site: www.balkema.nl
A.A. Balkema Publishers, Old Post Road, Brookfield, VT 05036-9704, USA
Fax: 802.276.3837; E-mail: info@ashgate.com

For the complete set of three volumes, ISBN 90 5809 059 0
For Volume 1, ISBN 90 5809 060 4
For Volume 2, ISBN 90 5809 061 2
For Volume 3, ISBN 90 5809 062 0

Printed in the Netherlands

Wind Engineering into the 21st Century, Larsen, Larose & Livesey (eds) © 1999 Balkema, Rotterdam, ISBN 90 5809 059 0

Table of contents

Buildings – General

Bluff body aerodynamics

Low-rise buildings

Wind tunnel techniques

Buildings – General

Wind Engineering into the 21st Century, Larsen, Larose & Livesey (eds) © 1999 Balkema, Rotterdam, ISBN 90 5809 059 0

Double modal transformation in continuous modelling

L.Carassale, G.Piccardo & G.Solari
Department of Structural and Geotechnical Engineering, University of Genoa, Italy

ABSTRACT: The modal transformation techniques, usually adopted in dynamic analyses, are based on the search for a space in which there are advantages in the solution or interpretation of the problem. Analogous transformation techniques have been recently developed with the aim of defining a space where a multi-dimensional random process is expressed by a linear combination of one-dimensional uncorrelated processes. This paper proposes a method, called Double Modal Transformation, by which the dynamic analysis of a linear structure is carried out through the simultaneous transformation of the equations of motion and the loading process. This technique is developed by a continuous approach and is applied to the evaluation of the along-wind response of slender structures. The solution is provided in closed form.

1 INTRODUCTION

Structural dynamic analyses are usually carried out by classical modal analysis. It is based on the search for a space in which the equations of motion result decoupled (e.g. Hurty & Rubinstein 1964). They are called principal equations and their unknown functions are the principal coordinates. Since, in general, only a limited number of principal coordinates is necessary for the description of the structural response, dynamic analyses require the solution of few decoupled equations. Therefore, the structural response is expressed as a series in which only a limited number of terms may be retained. The loading terms of the principal equations are evaluated by projecting the loading function into the space of the principal coordinates. If the load is expressed by a random field defined through its cross power spectral density function (cpsdf), this operation is carried out by calculating a frequency-dependent $2N$-dimensional integral, where N is the dimension of the domain. Such operation constitutes the most burdensome step in the dynamic analysis.

A recent development in the process theory is the application to a random field of transformation rules analogous to those usually adopted in structural dynamics. These techniques, called proper transformations, are based on the search for a space in which the random field is represented by a series of mono-dimensional uncorrelated processes (Armitt 1968, Spanos & Ghanem 1989, Bienkiewicz et al. 1992, Kikuchi et al. 1997, Di Paola 1998). It is interesting to note that a small number of mono-dimensional processes is generally sufficient to represent the random field with a good approximation; thus, only few terms may be retained in the series expansion.

The simultaneous application of classic modal analysis and suitable proper transformations of external loads leads to the Double Modal Transformation technique (DMT). Adopting this approach, the structural response is obtained as a double series in which the contributions of structural and loading modes are superimposed.

The formalisation and the application of this technique may be developed both with a discrete and a continuous approach. The discrete treatment is discussed by Carassale et al. (1999).

This paper develops a continuous approach which is applied, as an example, to the evaluation of the alongwind response of a slender structure. The solution is given in closed form.

2 PROPER COVARIANCE TRANSFORMATION

Let $v(z;t)$ be a Gaussian weakly-stationary nil mean random field, defined on the mono-dimensional domain $\mathcal{D}$, being t the time and z the abscissa of $\mathcal{D}$. Let $C_v(z;z')=E[v(z;t)v(z';t)]$ be the covariance function of the field v at the zero time lag, being E[.] the average operator. The generalization to a multi-dimensional domain does not introduce conceptual differences, but only formal complications.

Let λ_k be the eigenvalues of C_v, called covariance eigenvalues; $\phi_k(z)$ are the covariance eigenfunctions. They are the non-trivial solution of the integral equation:

$$\phi_k(z) = \frac{1}{\lambda_k} \int_{\mathscr{D}} C_v(z,z')\phi_k(z')\mathrm{d}z' \tag{1}$$

which represents a homogeneous Fredholm integral equation of the second kind; its kernel C_v is bounded, symmetric, real and positive definite, thus it has at least one eigenvalue; all its eigenvalues are real and positive, its eigenfunctions are real and orthonormal in the following sense (Kanval 1971):

$$\begin{aligned} &\int_{\mathscr{D}} \phi_h(z)\phi_k(z)\mathrm{d}z = \delta_{hk} \\ &\int_{\mathscr{D}}\int_{\mathscr{D}} C_v(z,z')\phi_h(z)\phi_k(z')\mathrm{d}z\mathrm{d}z' = \lambda_k\delta_{hk} \end{aligned} \tag{2}$$

where δ_{hk} is Kronecker's delta. Thanks to Eqs. (2), the covariance function can be expressed through the Fourier series:

$$C_v(z,z') = \sum_k \lambda_k\phi_k(z)\phi_k(z') \tag{3}$$

3 PROPER SPECTRAL TRANSFORMATION

Let $S_v(z,z';n)$ be the cpsdf of $v(z;t)$, being n the frequency. It is normalized by the relationship:

$$C_v(z,z') = \int_0^\infty \mathrm{Re}[S_v(z,z';n)]\mathrm{d}n \tag{4}$$

where Re[.] denotes the real part. Let $\gamma_k(n)$ be the eigenvalues of S_v, called spectral eigenvalues; $\theta_k(z,n)$ are the spectral eigenvectors. They are the non-trivial solution of the integral equation:

$$\theta_k(z;n) = \frac{1}{\gamma_k(n)} \int_{\mathscr{D}} S_v(z,z';n)\theta_k(z';n)\mathrm{d}z' \tag{5}$$

Since S_v is bounded, Hermitian and semi-positive definite, it has at least one eigenvalue; all its eigenvalues are real and positive and its eigenfunctions, in general complex, are orthonormal in the following sense:

$$\begin{aligned} &\int_{\mathscr{D}} \theta_h(z;n)\theta_k^*(z;n)\mathrm{d}z = \delta_{hk} \\ &\int_{\mathscr{D}}\int_{\mathscr{D}} S_v(z,z';n)\theta_h(z;n)\theta_k^*(z';n)\mathrm{d}z\mathrm{d}z' = \gamma_k(n)\delta_{hk} \end{aligned} \tag{6}$$

where θ_k^* is the complex conjugate of θ_k. Thanks to Eqs. (6), S_v can be expressed through the Fourier series:

$$S_v(z,z';n) = \sum_k \gamma_k(n)\theta_k(z;n)\theta_k^*(z';n) \tag{7}$$

Replacing Eqs. (3) and (7) into Eq. (4) provides the expression which links the eigensolutions of the covariance function and of cpsdf:

$$\begin{aligned} &\sum_k \lambda_k\phi_k(z)\phi_k(z') = \\ &= \sum_k \int_0^\infty \gamma_k(n)\mathrm{Re}[\theta_k(z;n)\theta_k^*(z';n)]\mathrm{d}n \end{aligned} \tag{8}$$

4 DYNAMIC RESPONSE BY DMT

Let us consider a linear visco-elastic structure geometrically corresponding to the mono-dimensional domain $\mathscr{D}$. The motion of the structure, in the Lagrangian space, is governed by a partial derivative problem in terms of the displacement function $q(z;t)$:

$$\mathscr{M}\ddot{q}(z;t) + \mathscr{C}\dot{q}(z;t) + \mathscr{K}q(z;t) = f(z;t) \tag{9}$$

where $\mathscr{M}$, $\mathscr{C}$, $\mathscr{K}$ are the inertia, damping and stiffness operators; $f(z;t)$ is the external force defined, with wide generality, through the formula:

$$f(z;t) = a(z)v(z;t) \tag{10}$$

where $a(z)$ is a deterministic function. Eq. (10) is usually solved by expressing $q(z;t)$ through the series:

$$q(z;t) = \sum_j \psi_j(z)p_j(t) \tag{11}$$

in which $\psi_j(z)$ are the eigenfunctions of the structure. The coefficients of the series $p_j(t)$ are called principal coordinates and constitute an infinite-dimensional Cartesian space called, principal space. By assuming that the structure has classic vibration modes, the motion is ruled by an infinite number of decoupled ordinary differential equations in the principal space:

$$\ddot{p}_j + 4\pi\xi_j n_j\dot{p}_j + (2\pi n_j)^2 p_j = \int_{\mathscr{D}} \psi_j(z)f(z;t)\mathrm{d}z \tag{12}$$

where n_j and ξ_j are the natural frequency and the damping ratio of the j-th mode, respectively.

By sorting the natural frequencies in increasing order, only a limited number of principal coordinates is significant for the evaluation of the structural response (Hurty & Rubinstein 1964).

The solution of Eqs. (11) and (12) can be carried out both in the time and in the frequency domain. The application of the DMT in the time domain is widely discussed by Carassale & Solari (1999); by adopting the frequency domain method, the cpsdf of the principal coordinates is given by:

$$\begin{aligned} &S_{p_{ij}}(n) = H_{p_i}^*(n)H_{p_j}(n)\cdot \\ &\cdot\int_{\mathscr{D}}\int_{\mathscr{D}} a(z)a(z')S_v(z,z';n)\psi_i(z)\psi_j(z')\mathrm{d}z\mathrm{d}z' \end{aligned} \tag{13}$$

where H_{p_j} is the complex frequency response function associated to the j-th principal coordinate and $H_{p_i}^*$ is the complex conjugate of H_{p_i}. By applying Eq. (7), the double integral in Eq. (13) can be decomposed in the following series:

$$S_{p_{ij}}(n) = H^*_{p_i}(n) H_{p_j}(n) \sum_k G_{jk}(n) G^*_{ik}(n) \gamma_k(n) \quad (14)$$

where G_{jk} are the cross-modal participation coefficients defined by:

$$G_{jk}(n) = \int_{\mathcal{D}} a(z) \theta_k(z;n) \psi_j(z) \mathrm{d}z \quad (15)$$

It is demonstrated that the covariance of different principal coordinates is almost zero if the structure is lightly damped and its natural frequencies are well separated (Elishakoff 1983). In this classic case, the variance of the displacement is given by:

$$\sigma_q^2(z) = \sum_j \psi_j^2(z) \sigma_{p_j}^2 \quad (16)$$

where $\sigma^2_{p_j}$ is the variance of the j-th principal coordinate:

$$\sigma_{p_j}^2 = \int_0^\infty H_{p_j}^2(n) \sum_k \left|G_{jk}(n)\right|^2 \gamma_k(n) \mathrm{d}n \quad (17)$$

Actually, several coefficients G_{jk} may be not significant, negligible or, even more, rigorously null. First of all this occurs when, sorting the natural frequencies in increasing order, only the first M_t principal coordinates contribute to the response and the terms G_{jk} with $j>M_t$ are therefore not significant. Analogously, by sorting the loading spectral eigenvalues in decreasing order, only the first N_s spectral principal components contribute to the excitation; it follows that also the terms G_{jk} with $k>N_s$ are not significant. Finally, due to the reciprocal shape of structural and spectral eigenfunctions, it may happen that $G_{jk} \simeq 0$; in such a case, the j-th structural mode is called quasi-orthogonal to the k-th loading spectral mode with respect to $a(z)$ (Eq. 15); ψ_j is called orthogonal to θ_k with respect to $a(z)$ when $G_{jk}=0$.

It follows that the dynamic response of a structure to a multi-dimensional loading field can be generally expressed by a double summation (Eqs. 16,17) of few structural modes and few loading spectral modes.

Under the condition in which $G_{jk}=0$ for every $j \neq k$, each structural mode is excited only by the corresponding loading mode:

$$\sigma_{p_j}^2 = \int_0^\infty H_{p_j}^2(n) \left|G_{jj}(n)\right|^2 \gamma_j(n) \mathrm{d}n \quad (18)$$

In such a situation the joint sorting of the natural frequencies in increasing order and of the spectral eigenvalues in decreasing order makes M_t principal coordinates enough to suitably represent the structural response. The case in which $M_t=1$ is not so unusual.

The evaluation of the probability distribution of the maximum response involves the knowledge of its spectral moments (Vanmarcke 1972). The application of the previous methods in this context does not involve any new remarkable concept.

A relevant particular case is when the loading process has harmonic content quickly decreasing on increasing the frequency. In the wind engineering field this is the typical case of the atmospheric turbulence excitation. Assuming again that the structure is lightly damped and its natural frequencies are well separated, Eq. (17) can be suitably approximated by (Simiu & Scanlan 1996):

$$\sigma_{p_j}^2 = \frac{1}{(2\pi n_j)^4} \sum_k \int_0^{+\infty} \left|G_{jk}(n)\right|^2 \gamma_k(n) \mathrm{d}n + \frac{1}{8(2\pi n_j)^3 \xi_j} \sum_k \left|G_{jk}(n_j)\right|^2 \gamma_k(n_j) \quad (19)$$

where the first and the second terms on the right hand side represent, respectively, the background and the resonant part of the response. Replacing Eq. (8) into Eq. (19), it follows:

$$\sigma_{p_j}^2 = \frac{1}{(2\pi n_j)^4} \sum_k B_{jk}^2 \lambda_k + \frac{1}{8(2\pi n_j)^3 \xi_j} \sum_k R_{jk}^2 \gamma_k(n_j) \quad (20)$$

where B_{jk} and R_{jk} are referred to as the cross-modal-participation coefficients of the background and resonant part of the response:

$$B_{jk} = \int_{\mathcal{D}} a(z) \phi_k(z) \psi_j(z) \mathrm{d}z \qquad R_{jk} = \int_{\mathcal{D}} a(z) \left|\theta_k(z;n_j)\right| \psi_j(z) \mathrm{d}z \quad (21)$$

It is relevant to observe that Eq. (20) provides the variance of the principal coordinates through an algebraic formula involving only eigenvalues and eigenfunctions of structure and loading processes. Also note that all previous remarks for the coefficients G_{jk} and the cross-modal orthogonality property can now be applied, separately, to the coefficients B_{jk} and R_{jk}.

5 EIGENQUANTITIES OF THE LONGITUDINAL TURBULENCE

Let us consider a slender structure in the atmospheric boundary layer. The mono-dimensional domain $\mathcal{D}$ is represented with the interval $[0, \ell]$, ℓ being the length of the structure. Let z be the abscissa on $\mathcal{D}$ and Z the height above the ground. Let $v(z;t)$ be a Gaussian random field constituted by the longitudinal component of the reduced atmospheric turbulence (with unit variance). The cpsdf $S_v(z,z';t)$ of $v(z;t)$, assumed to be real, is given by the equations

(Piccardo & Solari 1998a):

$$S_v(z,z';n) = \check{S}_v(\overline{Z},n)\exp\left[-\frac{nC_z|z-z'|}{U(\overline{Z})}\right]$$
$$\check{S}_v(Z,n) = \frac{6.868\,L_v(Z)/U(Z)}{[1+10.302\,n\,L_v(Z)/U(Z)]^{5/3}} \quad (22)$$

where $\check{S}_v(Z,n)$ is the power spectral density function (psdf) of the reduced turbulence, U is the mean wind velocity, $\overline{Z}$ is the reference height (Piccardo & Solari 1998a), C_z is the exponential decay factor along the structure and L_v is the integral length scale of turbulence.

S_v does not depend on z and z' separately, but only on the difference $(z-z')$; under this condition the random field $v(z;t)$ is said homogenous. Thanks to this property, the covariance function, defined by Eq. (4), can be approximated by the formula (Fig. 1):

$$C_v(z,z') \cong e^{-\beta\sqrt{\delta|z-z'|/\ell}}, \qquad \delta = \frac{\ell C_z}{L_v(\overline{Z})} \quad (23)$$

where β is a suitable parameter dependent on $L_v(\overline{Z})/U(\overline{Z})$, $\beta \simeq 0.38[L_v(\overline{Z})/U(\overline{Z})]^{-0.52}$

5.1 *Eigensolutions of the cpsdf*

Let us consider the integral equation (5) with the kernel expressed by Eq.(22a):

$$\theta(z;n) = \frac{\check{S}_v(\overline{Z};n)}{\gamma(n)}\int_0^{\ell} e^{-nC_z|z-z'|/U(\overline{Z})}\theta(z';n)\mathrm{d}z' \quad (24)$$

By virtue of the symmetry of the kernel, the eigensolutions can be expressed in closed form (Kanwal 1971, Carassale 1997):

$$\theta_k(z;n) = \sin(\mu_k z/\ell) + \frac{\mu_k}{\alpha}\cos(\mu_k z/\ell)$$
$$\alpha = C_z n\ell/U(\overline{Z}) \quad (25)$$

$$\gamma_k(n) = \check{S}_v(\overline{Z};n)\chi_k(n), \quad \chi_k(n) = 2\alpha\ell/(\alpha^2+\mu_k^2) \quad (26)$$

where μ_k is the k-th of the infinite roots of the transcendental equation:

$$2\cot(\mu_k) + \alpha/\mu_k - \mu_k/\alpha = 0 \quad (27)$$

Figure 2 shows the first three eigenfunctions, in the frequency range 0.01-1.00 *Hz*.

It is convenient to introduce the new parameter:

$$\varepsilon_k(\alpha) = \mathrm{arctg}[\mu_k(\alpha)/\alpha], \qquad k=1,2,\ldots \quad (28)$$

and express the eigenvalues and eigenfunctions through the relationships:

$$\chi_k(n) = 2U(\overline{Z})/nC_z\cos^2(\varepsilon_k), \quad k=1,2,\ldots \quad (29)$$

$$\theta_k(z;n) = D_k\sin\left(\frac{k\pi-2\varepsilon_k}{\ell}z+\varepsilon_k\right) \quad (30)$$

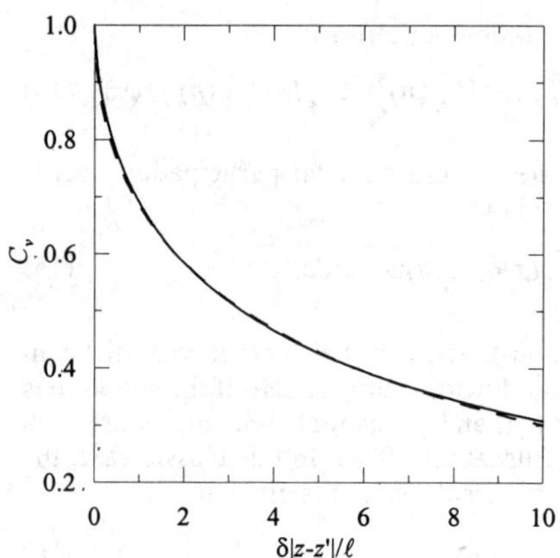

Figure 1. Covariance function C_v: Actual (solid line), approximated (dashed line). $L_v(\overline{Z})/U(\overline{Z})=1$.

$$D_k = \sqrt{\frac{2(k\pi-2\varepsilon_k)}{\ell[\sin(\varepsilon_k)\cos(\varepsilon_k)+k\pi-2\varepsilon_k]}} \quad (31)$$

Note that the parameter ε_k represents the phase of the modal shapes and its inverse, $g_k(\alpha)=1/\varepsilon_k(\alpha)$, is quasi-linear (Fig. 3). Thanks to this property $g_k(\alpha)$ can be expressed by:

$$g_k(\alpha) = 1/\varepsilon_k(\alpha) = 0.55 + 0.1k^{-2} + 0.318k^{-1.1}\alpha \quad (32)$$

Figure 4 shows the eigenvalues of S_v; the actual functions are compared with the approximate ones. It is possible to observe that the approximation is very good if $\alpha>3$. For ordinary structures, α is greater than 10.

5.2 *Eigensolutions of the Covariance Function*

Let us consider the integral equation (1) with the kernel C_v defined by Eq. (23):

$$\phi(z) = \frac{1}{\lambda}\int_0^{\ell} e^{-\beta\sqrt{\delta|z-z'|/\ell}}\phi(z')dz' \quad (33)$$

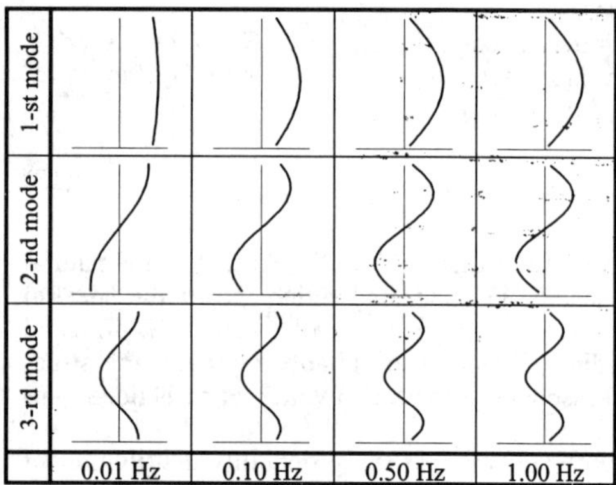

Figure 2. Eigenfunctions of $S_v(z,z';n)$.

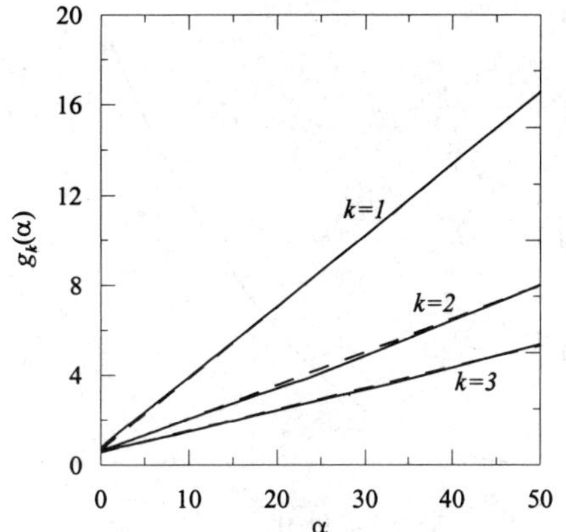

Figure 3. Comparison between the actual functions g_k (solid line) and the approximated Eq. (32) (dashed line).

It is a convolution type equation and admits a closed form solution if the kernel is an ℓ-periodic even function (Krasnov et al. 1971).

This property does not apply in the present case nevertheless the numerical testing demonstrated the effectiveness of such assumption. It follows that the eigensolutions of Eq. (33) are furnished by:

$$\lambda_k = \int_0^1 2e^{-\beta\sqrt{\delta\zeta/2}} \cos[(k-1)\pi\zeta]\, d\zeta, \quad k = 1,2,.. \tag{34}$$

$$\begin{aligned} \phi_{k,1}(z) &= \sin[2(k-1)\pi z/\ell] \\ \phi_{k,2}(z) &= \cos[2(k-1)\pi z/\ell] \end{aligned} \tag{35}$$

Each eigenvalue λ_k has multiplicity 2, except the first, that has a constant as the unique eigenfunction:

$$\phi_1 = 1/\sqrt{2} \tag{36}$$

The first eigenvalue is obtained in closed form; the second can be approximated as follows:

$$\lambda_1 = 4/\eta^2\left(1 - e^{-\eta} - \eta e^{-\eta}\right), \qquad \eta = \beta\frac{\sqrt{2}}{2}\sqrt{\delta} \tag{37}$$

$$\lambda_2 \simeq \eta/2\exp(-\eta^{0.7}) + 0.007 \tag{38}$$

Figure 5 compares the exact value of λ_2, obtained by Eq. (34), with the approximated Eq. (38), for usual values of η.

6 DYNAMIC ALONGWIND RESPONSE

To illustrate the reliability of the DMT in continuous modeling, two slender prismatic structures are considered (Piccardo & Solari 1998b). Both models are immersed in a wind field characterized by a logarithmic mean velocity profile, with a roughness coefficient z_0=0.7 m. It is supposed that their dynamic response can be represented by the contribution of the fundamental mode only.

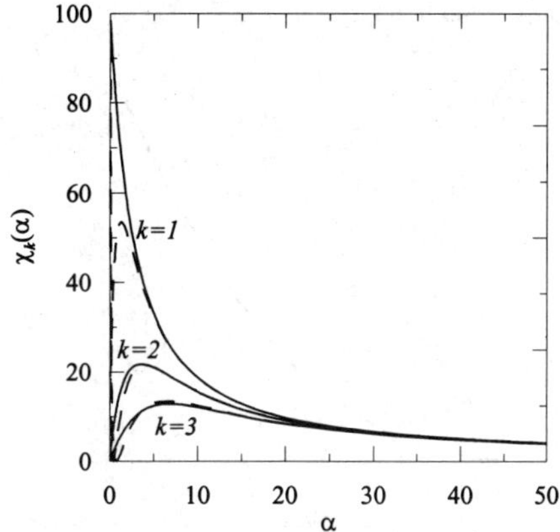

Figure 4. Eigenvalues of the cpsdf: actual (solid line), approximated (dashed line).

Model 1 simulates a reinforced-concrete chimney; it is ℓ=180 m high, with diameter b=5.6 m. The alongwind fundamental frequency is n_1=0.26 Hz and the relative modal shape is $\psi_1(z)=(z/\ell)^{2.15}$. The non-dimensional parameters introduced in the previous solutions are δ=10.7 and α=14.2.

Model 2 represents a double-hinged horizontal square section steel element of an industrial structure. It lies 20 m over the ground; it is ℓ=30 m long; its side is b=0.4 m. The alongwind fundamental frequency is n_1=1.5 Hz and the relative modal shape is $\psi_1(z)=\sin(\pi z/\ell)$; δ=3.7 and α=20.6.

Figures 6 and 7 show the rms of the alongwind response versus the shear velocity. The solid lines represent the actual solutions, obtained by numerically solving Eqs. (13), (16). The other lines correspond to different approximations obtained by using Eqs. (20) together with the eigensolutions provided by Eqs. (26,29-32) and (35-38).

For model 1 (Fig. 6), the dotted line represents the solution obtained by retaining only the first mode for both the background and the resonant

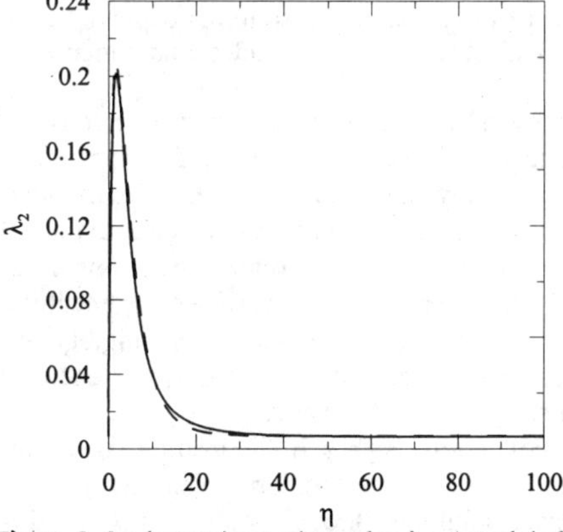

Figure 5. 2-nd covariance eigenvalue λ_2: Actual (solid line), approximated (dashed line).

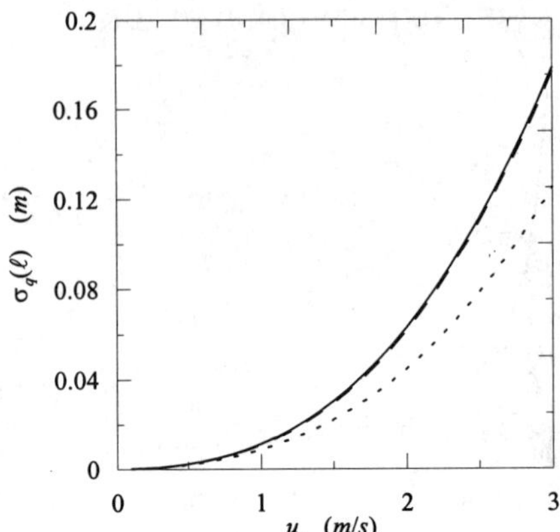

Figure 6. Rms of the displacement. Model 1.

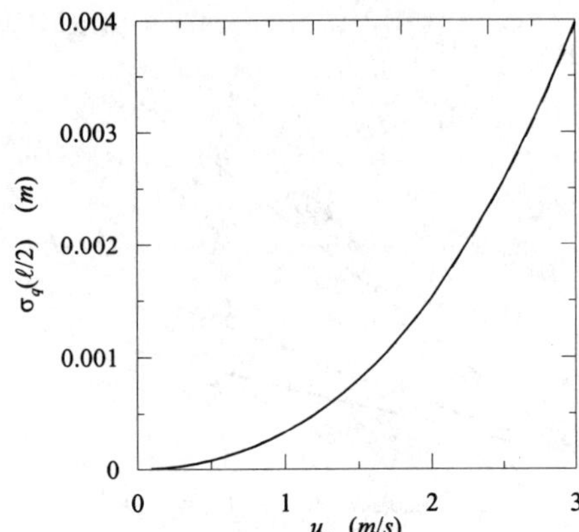

Figure 7. Rms of the displacement. Model 2.

components; the dashed line is obtained by using 2 and 5 modes for the background and resonant components, respectively, showing a very good approximation.

For model 2 (Fig. 7), the exact solution is obtained by retaining only the first mode for both the background and the resonant part. It is due to the similarity between the eigenfunctions of the structure and of the cpsdf (i.e. quasi-orthogonal property), that makes negligible the coefficients R_{jk} when $j \neq k$ (Eq. 21b).

7 CONCLUSIONS

The present paper applies modal transformation laws similar to those usually employed in structural analyses. Using these transformations a multi-dimensional loading process is represented by the superimposition of deterministic loading distributions modulated by uncorrelated random processes. The joint application of this technique with the classic modal analysis is called Double Modal Transformation (DMT). It expresses the response by a double series expansion whose terms depend on the loading and structural eigenvalues and eigenfunctions.

The alongwind response of slender structures to atmospheric turbulence can be correctly represented by retaining only few loading modes. In particular, the background contribution can be approximated through only few terms by virtue of the strongly decreasing character of the covariance eigenvalues. The number of necessary modes for suitably approximating the resonant term mainly depends on the first structural modal shape.

Using a continuous approach, the loading eigensolutions are obtained in closed form. This involves almost exact solutions through very expressive formulae and little computational burden.

REFERENCES

Armitt, J. 1968. Eigenvector analysis of pressure fluctuations on the West Burton instrumented cooling tower. *Central Electricity Res. Lab.* (UK), Internal Report RD/L/M 114/68.

Bienkiewicz, B.,H.J. Ham, Y. Sun 1992. Proper Orthogonal Decomposition. *J. Wind Eng. Ind. Aerodyn.* 50: 193-202

Carassale, L. 1997. The dynamic response of structures to wind action by DMT. Degree thesis (in Italian).

Carassale, L., G. Piccardo & G. Solari 1999. Double modal transformation and wind engineering applications. Submitted to *J. Eng. Mech. ASCE.*

Carassale, L. & G. Solari, in prep. Dynamic response of structure by discrete double modal transformation.

Davenport, A.G. 1962. The response of slender, line-like structure to a gusty wind. *Proc. Instn. Civ. Engrs.* 23: 389-408.

Di Paola, M. 1998. Digital simulation of wind field velocity. *J. Wind Eng. Ind. Aerodyn.* 74-76: 91-109.

Elishakoff, I. 1983. *Probabilistic methods in the theory of structures.* New York: J. Wiley & Sons.

Hurty, W.C. & M.F. Rubinstein 1964. *Dynamics of Structures.* Englewood Cliffs (New Jersey): Prentice-Hall Inc.

Kanwal, P. 1971. *Linear Integral Equation.* New York: Academic Press.

Kikuchi, H., Y. Tamura, H. Ueda & K. Hibi 1997. Dynamic wind pressures acting on a tall building model - Proper orthogonal decomposition. *J. Wind Eng. Ind. Aerodyn.* 69-71: 631-646.

Krasnov, M., A. Kiselev & G. Makarenko 1971. *Problems and exercises in integral equations*. Moscow: Mir Publishers.

Simiu, E. & R.H. Scanlan 1996. *Wind effects on structures*, 3rd Ed. New York: J. Wiley & Sons.

Solari, G. 1993. Gust buffeting. I: Peak wind velocity and equivalent pressure. *J. Struct. Eng. ASCE.* 119: 365-382.

Spanos, P.D. & R. Ghanem 1989. Stochastic finite element expansion. *J. Eng. Mech. ASCE.* 115: 1035-1053.

Piccardo, G. & G. Solari 1998a. Closed Form Prediction of 3-D Wind-Excited Response of Slender Structures. *J. Wind Eng. Ind. Aerodyn.* 74-76: 697-708.

Piccardo, G. & G. Solari 1998b. Generalized equivalent spectrum technique. *Wind & Structures* 1: 161-174.

Vanmarke, E.H. 1972. Properties of spectral moments with applications to random vibration. *J. Eng. Mech. Div. ASCE.* 2: 425-446.

Wind Engineering into the 21st Century, Larsen, Larose & Livesey (eds) © 1999 Balkema, Rotterdam, ISBN 90 5809 059 0

Wind pressure field simulation and comparison with experimental results

M.Gioffrè & V.Gusella
Department of Water and Structures Engineering, University of Perugia, Italy

M.Grigoriu
School of Civil and Environmental Engineering, Cornell University, Ithaca, N.Y., USA

ABSTRACT: The wind pressure field on tall buildings is numerically generated using multivariate translation processes. The theory of vector translation processes is briefly reviewed. Conditions on the matrix of cross-covariance functions are given to assure the applicability of the model. A simulation algorithm is proposed. The model is calibrated on experimental results obtained from wind tunnel tests on the CAARC standard tall building. The efficiency of the proposed methodology for reproducing the non-Gaussian nature on separated flow region is demonstrated.

1 INTRODUCTION

A correct modelling of aerodynamic loads acting on structures is of interest for two main reasons. The estimate of maximum wind load and wind load effects and the fatigue damage estimation. To aim this goal, the Gaussian model appears to be inadequate to describe the wind pressure nature. Experimental research has shown that the wind-structure interactions are responsible for the non-Gaussian features of the pressure time series acting on building surfaces (Peterka & Cermak 1975, Stathopoulos 1980).

Dealing with bluff bodies the complexity of wind structure interactions makes the use of theoretical methods possible only in particular cases. To overcome these difficulties the combination of experimental tests and simulation procedures can be a very useful tool. This approach has been followed by several researchers in the last years (Gurley et al. 1997, Kumar & Stathopoulos 1997). Nevertheless, the accuracy and efficiency in the generation of non-Gaussian wind pressure time histories is still being pursued.

Within this context an algorithm for generating multivariate non-Gaussian processes is proposed in this paper. The procedure uses the correlation-distortion method based on translation vector processes. This method does not require iterative procedures and is well suited when experimental data are available. In particular it requires cross-covariance functions and marginal distributions that can be directly estimated from data.

The proposed algorithm is applied to the generation of wind pressure time series in the separated flow regions of a prismatic tall building. The experimental data used to calibrate the model are constituted of the wind pressure fluctuations recorded in wind tunnel tests (Gioffrè et al. 1998).

2 MULTI-VARIATE SIMULATION

2.1 Model definition

The wind pressure fluctuations acting on the surfaces of a building can be modeled by a stochastic time dependent field. The coefficient of pressure $C_p = C_p(\zeta_1, \zeta_2, \zeta_3; t)$ can be used to describe the field where $\boldsymbol{\zeta} = \{\zeta_1, \zeta_2, \zeta_3\}$ are the coordinates of the considered point in space, and t is the time. It follows that $C_p(\boldsymbol{\zeta}; t)$ can be modeled by a so called one-variate four dimensional (*1-V*, *4-D*) stochastic non-Gaussian field. Nevertheless, information on the pressure field is generally available in a discrete number of points, e.g. the m tap locations where the pressure is measured in experimental tests.

The collection of the coefficient of pressure fluctuations $C_{p_1}(t), \ldots, C_{p_n}(t)$ into a vector allows the description of the pressure fluctuation by a discrete stochastic field represented by an m-variate one dimensional (m-*V*, *1-D*) stochastic vector process. Thus, the discretization transforms the *1-V*, *4-D* stochastic field into an m-*V*, *1-D* stochastic vector process.

The non-Gaussian nature of the wind pressure vector process on tall buildings can be successfully described, for practical applications, by a translation model (Gioffrè 1998). Let $\boldsymbol{Y}(t)$, $t \in \mathcal{R}$, be a stationary m-dimensional Gaussian vector process with zero mean, covariance functions $\rho_{kl}(\tau) =$

$E[Y_k(t)Y_l(t+\tau)]$, $k,l = 1,\ldots,m$, and marginal distributions equal to the standard Gaussian distribution $\Phi_{Y_k}(y_k) = P[Y_k(t) \leq y_k] = \Phi_Y(y_k)$, $k = 1,\ldots,m$. The translation vector process $\boldsymbol{X}(t)$ is defined by the non-linear memoryless transformation $\boldsymbol{g}: \mathcal{R}^m \rightarrow \mathcal{R}^m$ with monotonic components $g_k[Y_k(t)] = F_{X_k}^{-1}[\Phi_Y(Y_k(t))]$ (Grigoriu 1995) so that

$$X_k(t) = g_k[Y_k(t)] = F_{X_k}^{-1}[\Phi_Y(Y_k(t))] \tag{1}$$

$k = 1,\ldots,m$, is the kth component of $\boldsymbol{X}(t)$.

The vector translation process $\boldsymbol{X}(t)$ is completely characterized by the mapping $\boldsymbol{g}(\cdot)$ and the covariance functions $\rho_{kl}(\tau)$, $k,l = 1,\ldots,m$, because $\boldsymbol{Y}(t)$ is a Gaussian vector process.

The vector mean $\boldsymbol{\mu_X}$, and covariance functions $C_{X_kX_l}(\tau) = E[X_k(t)X_l(t+\tau)] - \mu_{X_k}\mu_{X_l}$ of $\boldsymbol{X}(t)$ can be directly calculated using the definitions in (1). The scaled covariance functions of $\boldsymbol{X}(t)$ are defined as

$$\xi_{kl}(\tau) = \frac{C_{X_kX_l}(\tau)}{\sigma_{X_k}\sigma_{X_l}} \tag{2}$$

$k,l = 1,\ldots,m$, and represent the covariance functions of the vector process with components $\tilde{X}_k(t) = [X_k(t) - \mu_{X_k}]/\sigma_{X_k}$, $k = 1,\ldots,m$.

2.2 Covariance functions

The covariance functions $\xi_{kl}(\tau)$ and $\rho_{kl}(\tau)$ are related by the equation

$$\sigma_{X_k}\sigma_{X_l}\,\xi_{kl}(\tau) = \int_{-\infty}^{\infty}\int_{-\infty}^{\infty} [g_k(y_{k1}) - \mu_{x_k}]$$

$$[g_l(y_{l2}) - \mu_{x_l}]\,\phi_{Y_{k1}Y_{l2}}[y_{k1}, y_{l2}; \rho_{kl}(\tau)]\,dy_{k1}dy_{l2} \tag{3}$$

$k,l = 1,\ldots,m$, where $\phi_{Y_{k1}Y_{l2}}[y_{k1}, y_{l2}; \rho_{kl}(\tau)]$ is the joint density of the dependent Gaussian variables $Y_k(t)$ and $Y_l(t+\tau)$ with correlation coefficient $\rho_{kl}(\tau)$. Equation (3) can be used to calculate the values of $\xi_{kl}(\tau)$ corresponding to $\rho_{kl}(\tau)$ for an arbitrary but fixed τ. Furthermore, the following inequality holds (Lancaster 1957)

$$|\xi_{kl}(\tau)| \leq |\rho_{kl}(\tau)| \qquad k,l = 1,\ldots,m \tag{4}$$

It is possible to verify that arbitrarily defined scaled covariance functions and marginal distributions are not always compatible with a description of the non-Gaussian vector process $\boldsymbol{X}(t)$ by means of a translation model. Considering the auto-covariance functions one has that, if $\rho_{kk}(\tau)$, $k = 1,\ldots,m$, is equal to 1, 0, or -1, the corresponding values of $\xi_{kk}(\tau)$ are 1, 0, or the lower bound (Grigoriu 1995)

$$\xi_{kk}^{*l} = \frac{E[g_k(Y_k)g_k(-Y_k)] - \{E[g_k(Y_k)]\}^2}{E[g_k(Y_k)^2] - \{E[g_k(Y_k)]\}^2} \tag{5}$$

respectively. If one considers the cross-covariance functions it happens that even the limit $\rho_{kl}(\tau) = 1$ does not always implies that $\xi_{kl}(\tau) = 1$. For example, assume the two-dimensional Gaussian vector with components $Y_1 = Y_2 = Y$ with mean zero and unit variance. Define the translation vector components as $X_1 = g_1(Y_1) = Y$ and $X_2 = g_2(Y_2) = Y^2$. The variables Y_1 and Y_2 are perfectly correlated but the expectation $E[X_1X_2]$ is equal to $E[Y^3] = 0$. It follows that for general translation vectors with target covariance functions $\xi_{kl}^0(\tau)$ and marginal distributions $F_{X_k}^0(x_k)$, Equation (3) has solutions $\rho_{kl}^0(\tau)$ if $\xi_{kl}^0(\tau)$ is in the interval $[\xi_{kl}^{*l}, \xi_{kl}^{*u}]$, $k,l = 1,\ldots,m$, for every τ. The lower bound ξ_{kl}^{*l} and the upper bound ξ_{kl}^{*u} are calculated corresponding to $\rho_{kl}(\tau) = -1$ and $\rho_{kl}(\tau) = 1$, respectively. It is to note that $\xi_{kk}^{*u} = 1$, $k = 1,\ldots,m$, for definition.

If solutions $\rho_{kl}^0(\tau)$ to Equation (3) exist, one has to verify the functions $\rho_{kk}^0(\tau)$ to be non-negative definite, i.e.

$$\boldsymbol{a}^T \boldsymbol{\rho}_{0kk}^{(n)} \boldsymbol{a} \geq 0 \qquad k = 1,\ldots,m \tag{6}$$

where n is a positive integer, $\boldsymbol{a} \in \mathcal{R}^n$ denotes an arbitrary vector, and $\boldsymbol{\rho}_{0kk}^{(n)} = \{\rho_{kk}^0(t_i - t_j)\}$, $i,j = 1,\ldots,n$, denotes a matrix consisting of the covariances of the components $Y_k(t_i)$ and $Y_k(t_j)$ of the random vector $\boldsymbol{Y}_k = \{Y_k(t_1),\ldots,Y_k(t_n)\}$. Alternative ways of checking whether or not $\boldsymbol{\rho}_{0kk}^{(n)}$ is nonnegative definite are

$$\lambda_{\min}\left[\boldsymbol{\rho}_{0kk}^{(n)}\right] \geq 0 \tag{7a}$$

$$\left|\boldsymbol{\rho}_{0kk}^{(n)}\right| \geq 0 \tag{7b}$$

where $\lambda_{\min}\left[\boldsymbol{\rho}_{0kk}^{(n)}\right]$ and $\left|\boldsymbol{\rho}_{0kk}^{(n)}\right|$ denote the smallest eigenvalue and the determinant of $\boldsymbol{\rho}_{0kk}^{(n)}$, respectively. Furthermore, it has to be verified if the matrix

$$\boldsymbol{\rho}^0(\tau) = \begin{pmatrix} \rho_{11}^0(\tau) & \cdots & \rho_{1m}^0(\tau) \\ \vdots & \ddots & \vdots \\ \rho_{m1}^0(\tau) & \cdots & \rho_{mm}^0(\tau) \end{pmatrix} \tag{8}$$

is non-negative definite for every time lag τ.

If at least one of the $\rho_{kk}^0(\tau)$ and/or $\boldsymbol{\rho}^0(\tau)$ are negative definite, then they are not covariances and it is not possible to represent the target non-Gaussian process as a translation vector process with scaled covariances $\xi_{kl}^0(\tau)$ and marginal distributions $F_{X_k}^0(x_k)$.

2.3 Simulation algorithm

The objective of the simulation algorithm is the generation of samples of the m-dimensional non-

Gaussian vector process $\boldsymbol{X}(t)$ with the target matrix of the $m \times m$ scaled covariance functions

$$\boldsymbol{\xi}^0(\tau) = \begin{pmatrix} \xi_{11}^0(\tau) & \cdots & \xi_{1m}^0(\tau) \\ \vdots & \ddots & \vdots \\ \xi_{m1}^0(\tau) & \cdots & \xi_{mm}^0(\tau) \end{pmatrix} \tag{9}$$

and the target marginal distributions $F_{X_k}^0(x_k)$, $k = 1, \ldots, m$ of the kth component of $\boldsymbol{X}(t)$.

The proposed simulation algorithm for non-Gaussian stochastic vector processes presents three main steps: model calibration, sample generation, and mapping. The model calibration consist in determining the the second-moment characteristics of the Gaussian image $\boldsymbol{Y}(t)$ of $\boldsymbol{X}(t)$ and the m memoryless non linear transformations $g_k(\cdot)$ of (1) such that the target marginal covariance matrix $\boldsymbol{\xi}^0(\tau)$ and marginal distributions $F_{X_k}^0(x_k)$ can be matched. Taking $F_{X_k}(x_k) = F_{X_k}^0(x_k)$ in (1) the vector non-linear transformation $\boldsymbol{g}(\cdot)$ is completely determined and the target marginal distributions are matched. It is worthwhile to note that this method is particularly attractive for pressure simulation since it is possible to match exactly the marginal distributions which can be very different from one tap location to another depending on the complexity of flow-structure interactions.

Once the transformation is set, the conditions on the covariances has to be verified. First, the upper and lower bounds, ξ_{kl}^{*u} and ξ_{kl}^{*l}, $k, l = 1, \ldots, m$, have to be determined using (3) to check if $\xi_{kl}^0(\tau) \in [\xi_{kl}^{*l}, \xi_{kl}^{*u}]$ for every τ. If this condition is verified then the functions $\rho_{kl}^0(\tau)$ can be calculated by using (3). The successive verification is on the non-negative definiteness of the functions $\rho_{kk}^0(\tau)$, $k = 1, \ldots, m$, and the matrix $\boldsymbol{\rho}^0(\tau)$.

The non-Gaussian vector process $\boldsymbol{X}(t)$ can be modeled by a vector translation process, in matching $\boldsymbol{\xi}^0(\tau)$ and $F_{X_k}^0(x_k)$, only if all these conditions are verified. Otherwise, two main strategies can be adopted. The first one is considering more general classes of non-Gaussian models that can better represent the desired statistics (Grigoriu 1995). The second is approximating $\boldsymbol{\xi}^0(\tau)$ and/or $F_{X_k}^0(x_k)$. The particular approximation to consider depends on the problem. For example, if one is more interested in calculating extremes, $\boldsymbol{\xi}^0(\tau)$ can be approximated and $F_{X_k}^0(x_k)$ can be matched exactly. The opposite choice can be adopted if one is interested in mean crossing rates.

Once all the conditions are verified, the second step of the simulation algorithm consist of generating realizations of the m-variate Gaussian process $\boldsymbol{Y}(t)$ with covariance functions $\rho_{kl}^0(\tau)$, $k, l = 1, \ldots, m$. Either one of the well established methods reported in literature can be used. In the present work a method based on the sampling theorem is preferred because it appears to be the most efficient since it only requires the knowledge of the covariance functions at discrete times, which can be directly estimated from data. This method utilize a local parametric representation of the components $Y_k(t)$ of $\boldsymbol{Y}(t)$, with frequency bands $[0, \omega_k^*]$. The representation is given by

$$Y_{k,w_k}(t) = \sum_{u=-w_{k,t}-w_k}^{w_{k,t}+w_k+1} Y_{k,u}\, \alpha_u(t; T_k)$$

$$w_{k,t} T_k \leq t \leq (w_{k,t}+1) T_k \tag{10}$$

$k = 1, \ldots, m$, where $T_k = \pi/\omega_k^*$, w_k is a positive integer determining the size of the window used in the representation, $w_{k,t}$ is the largest integer smaller than t/T_k, $Y_{k,u} = Y_k(uT_k)$ are values of component k of $\boldsymbol{Y}(t)$ at equally spaced times uT_k, called nodal points, and

$$\alpha_u(t; T_k) = \frac{\sin[\pi(t - uT_k)/T_k]}{\pi(t - uT_k)/T_k} \tag{11}$$

Background on this model are beyond the scope of this paper and can be found in (Grigoriu & Balopoulou 1993). Samples of $\boldsymbol{Y}(t)$ are generated using the numerical procedure based on this model and proposed in the work of Gioffrè (1998).

In the third and last step of the algorithm the generated samples of $\boldsymbol{Y}(t)$ are transformed into samples of $\boldsymbol{X}(t)$. Each component $y_k(t)$ is mapped to the corresponding $x_k(t)$ using Equations (1).

The main characteristics of the proposed algorithm can be summarized as follow:

1. Exact determination of $\boldsymbol{\rho}^0(\tau)$ by numerical or analytical tools once its existence is verified;

2. No need of iterative procedures because $\boldsymbol{\rho}^0(\tau)$ is calculated only once prior to the generation of the Gaussian samples;

3. The efficiency of the algorithm is only related on the effectiveness of the Gaussian simulation.

It is worth to note that the first two characteristics represent also the main advantages as opposed to other methods based on the correlation-distortion method reported in literature. In particular the methods reported in (Gurley et al. 1996, Gurley & Kareem 1998) use approximate relations between $\xi_{kl}(\tau)$ and $\rho_{kl}(\tau)$ (or, equivalently, between their Fourier transforms) based on Hermite polynomials and a limited number of statistical moments. Furthermore, the methods reported in (Popescu et al. 1998, Yamazaki & Shinozuka 1988) require several iterative generation of $\boldsymbol{Y}(t)$ to match the target correlation functions.

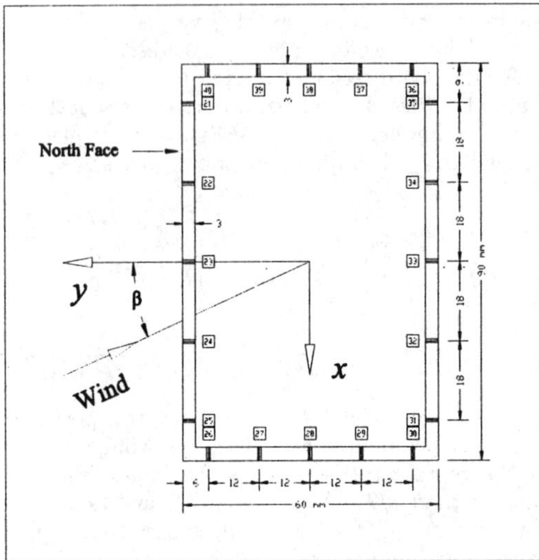

Figure 1. Cross-section of the model. Taps numbering at 2/3*H*.

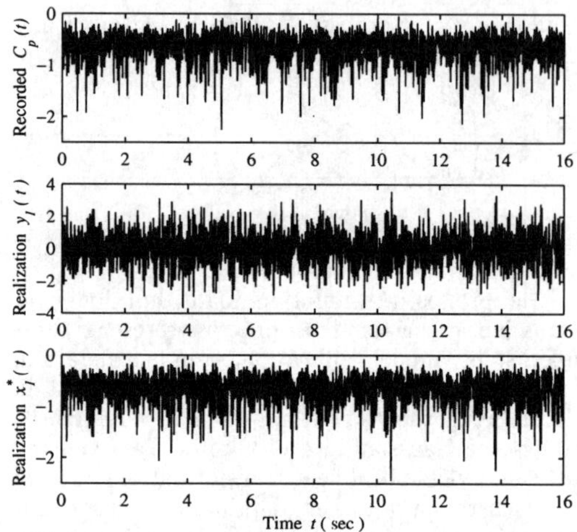

Figure 2. Recorded time series at Tap # 26, and samples $y_1(t)$ and $x_1^*(t)$ of the Gaussian and Lognormal process respectively.

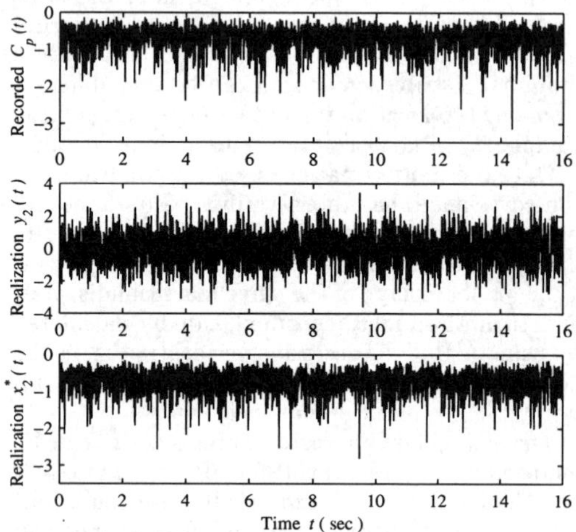

Figure 3. Recorded time series at Tap # 28, and samples $y_2(t)$ and $x_2^*(t)$ of the Gaussian and Lognormal process respectively.

3 WIND PRESSURE VECTOR PROCESS

3.1 Experimental data

Figure 1 shows a section of the CAARC standard tall building model utilized in the wind tunnel experimental tests (Gioffrè 1998). The scale of the model is 1:500 and the section is at 2/3 the height of the model (H = 366 mm). Three time series of coefficient of pressure fluctuations, $C_p(t)$, are used in this paper to apply the proposed algorithm. These time series were recorded in the separated flow region at taps #26, #28, and #30, with incoming wind normal to the north face ($\beta = 0°$). Plots of the recorded time histories are shown in Figures 2-4.

3.2 Model calibration

The $C_{p,j}(t)$, $j = 26, 28, 30$ are assumed to be the components X_k^*, $k = 1, 2, 3$, of an ergodic vector process $\boldsymbol{X}^*(t)$. The mean $\mu_{X_k^*}$, standard deviation $\sigma_{X_k^*}$, skewness coefficient $\gamma_{3,k}$, and kurtosis coefficient $\gamma_{4,k}$ of the components $X_k^*(t)$ are reported in Table 1.

Let $\tilde{X}_k(t) = (X_k^*(t) - \mu_{X_k^*})/\sigma_{X_k^*}$, $k = 1, 2, 3$, be the three standardized components of $\tilde{\boldsymbol{X}}(t)$. For simplicity of notation the 3-dimensional vector process $\boldsymbol{X}(t) = -\tilde{\boldsymbol{X}}(t)$ is considered in the following.

The normalized time series $X_k(t)$ are used to estimate the matrix of target scaled covariances $\boldsymbol{\xi}^0(\tau)$ and the marginal distributions $F^0_{X_k}(x_k)$, $k = 1, 2, 3$ of the translation model. The covariance functions are shown in Figure 6 (continuous line).

Table 1. Mean, standard deviation, skewness and kurtosis coefficients.

	μ	σ	γ_3	γ_4
$X_1^*(t)$	-0.6637	0.2828	-1.1281	5.0811
$X_1(t)$	0	1	1.1281	5.0811
Model L_1	0	1	1.1281	5.3452
$X_2^*(t)$	-0.7285	0.3313	-1.2736	6.0423
$X_2(t)$	0	1	1.2736	6.0423
Model L_2	0	1	1.2736	6.0162
$X_3^*(t)$	-0.7371	0.3937	-0.9572	4.5396
$X_3(t)$	0	1	0.9572	4.5396
Model L_3	0	1	0.9572	4.6724

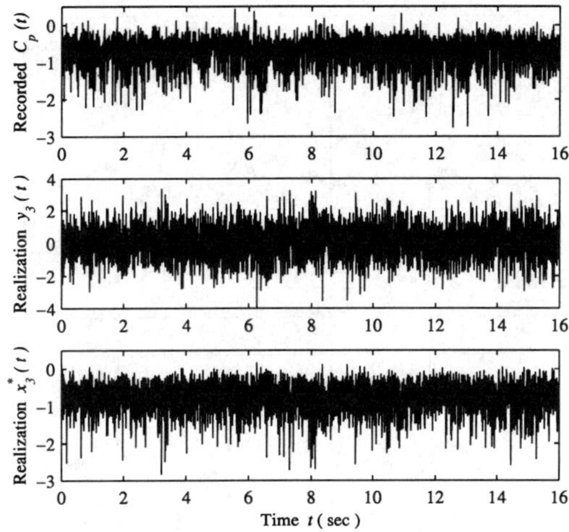

Figure 4. Recorded time series at Tap # 30, and samples $y_3(t)$ and $x_3^*(t)$ of the Gaussian and Lognormal process respectively.

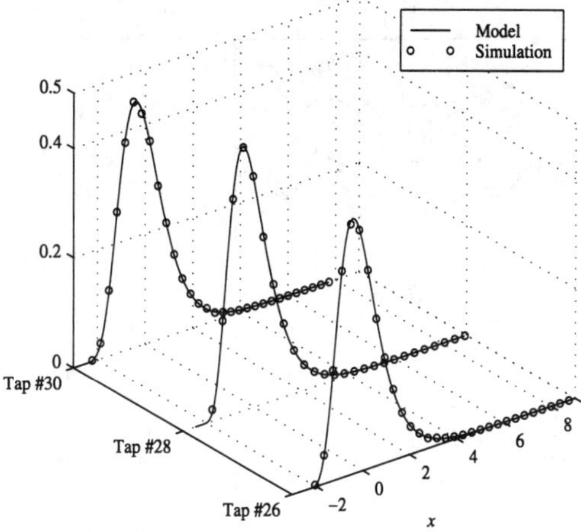

Figure 5. Estimated marginal densities from simulation and marginal densities of the Lognormal model.

The shifted Lognormal translation process

$$X_k(t) = a_k + \exp[\mu_{Y_k^*} + \sigma_{Y_k^*} Y(t)] \quad (12)$$

where a_k, $\mu_{Y_k^*}$, and $\sigma_{Y_k^*}$ are the parameters of the model, is found to be the better fit with the assumed time series (Gioffrè 1998). The first four moments of the assumed marginal distributions (indicated as Model L_k, $k = 1, 2, 3$) are reported in Table 1. The shape of the three marginal densities is shown in Figure 5 (continuous lines).

Once the non-linear transformations in (1) are determined, it is possible to calculate the lower and upper bounds of the covariances. It is found that $\xi_{kl}^0(\tau) \in [\xi_{kl}^{*l}, \xi_{kl}^{*u}]$, $k, l = 1, \ldots, m$, for every τ. It follows that an unique solution $\boldsymbol{\rho}^0(\tau)$ exist and can be calculated using Equation (3). The functions $\rho_{kk}^0(\tau)$, $k = 1, 2, 3$, and the complete matrix $\boldsymbol{\rho}^0(\tau)$ are found to be positive definite. All the required conditions are therefore satisfied and it is possible to represent the vector process $\boldsymbol{X}(t)$ by means of a translation model to match exactly the target marginal distributions $F_{X_k}^0(x_k)$, $k = 1, 2, 3$, and covariance matrix $\boldsymbol{\xi}^0(\tau)$.

3.3 Numerical results

Samples $\boldsymbol{y}(t)$ of the 3-variate Gaussian process $\boldsymbol{Y}(t)$ with zero mean and covariance matrix $\boldsymbol{\rho}^0(\tau)$ are generated using the method based on the parametric representation in (10). It is assumed that $w_k = w = 3$, and $T_k = T = 0.004$ s, $k = 1, 2, 3$. The obtained samples are mapped into samples $\boldsymbol{x}(t)$ of $\boldsymbol{X}(t)$ by the transformations (12).

Figures 2-4 show realizations of the components of the Gaussian vector $\boldsymbol{Y}(t)$ and the Lognormal vector $\boldsymbol{X}^*(t)$ after mapping. The overall comparison between the behaviour of the simulated time series and the recorded data from which the model was calibrated is satisfactory.

Figure 5 compares the marginal density functions of the Lognormal model and the estimation of the marginal histograms of the components $X_k(t)$ of $\boldsymbol{X}(t)$ based on 100,000 realizations of this vector process. Plotting the densities of each single tap in logarithmic scale it is evident that the estimated density from simulation is in agreement with the model also in the tail regions. These plots are not reported here because of the limited space available.

In Figure 6 the circles represent the estimated covariance functions based on 10,000 realizations of the vector $\boldsymbol{X}(t)$, which perfectly matches $\boldsymbol{\xi}^0(\tau)$.

4 CONCLUSIONS

A procedure for generating multivariate wind pressure time histories has been presented. The mathematical background and the conditions for the applicability of the theory of translation vector processes have been reported. Once these conditions are verified the generation of non-Gaussian time series does not require any iterative procedure. Furthermore, if the conditions are not verified, it is always possible to control the accuracy in the description of the second-moment characteristics and/or the marginal distribution of the target non-Gaussian vector. Depending on the particular problem study considered one can decide to approximate one or both $\boldsymbol{\xi}^0(\tau)$ and $F_{X_k}^0(x_k)$,

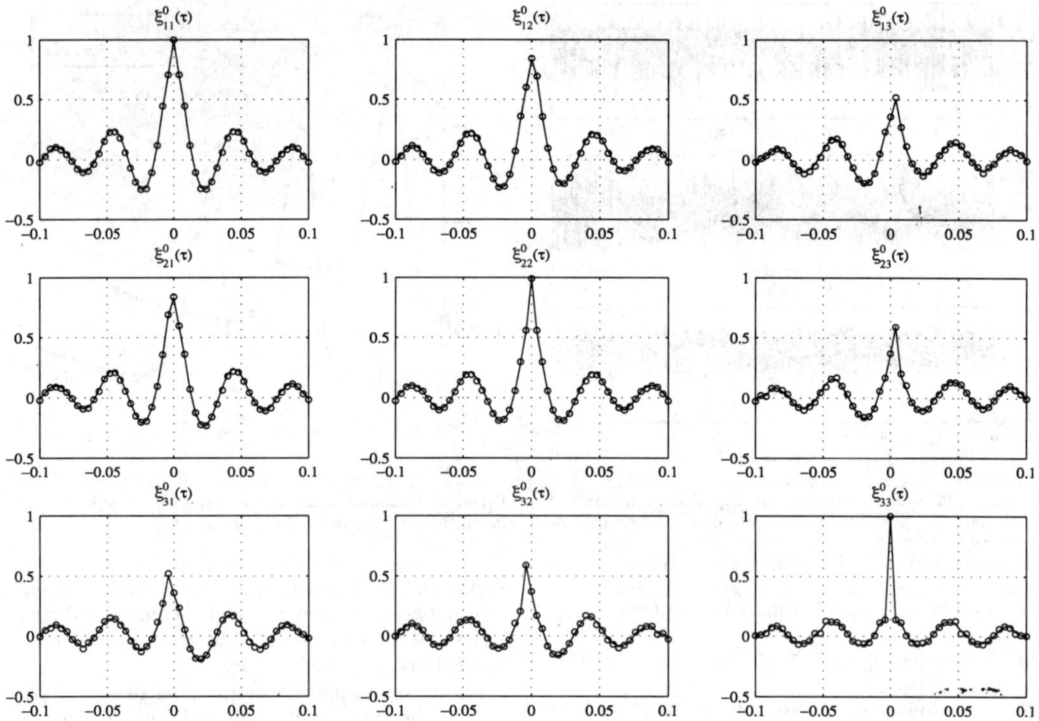

Figure 6. Target and simulated covariance functions.

$k = 1, \ldots, m$. The efficiency of the overall algorithm is strictly related to the generation of the parent Gaussian vector process. In this paper a method based on the sampling theorem is used because it only requires the knowledge of the covariance function at discrete times, which are directly estimated from experimental data.

The proposed procedure has been successfully applied to the generation of the wind pressure time series in the separated flow region of a tall building. The non-Gaussian features of interest in the pressure fluctuations can be described adopting the translation model because the required conditions are satisfied. The accuracy and the efficiency of the algorithm has been very satisfactory.

REFERENCES

Gioffrè, M. (1998). *Analysis and Simulation of non-Gaussian Processes with Application to Wind Engineering and Reliability*. Ph. D. thesis, University of Florence, Florence, Italy.

Gioffrè, M., V. Gusella, & M. Grigoriu (1998). Analysis of non-gaussian stochastic field of wind pressure. In *Proceedings, Fourth International Conference on Stochastic Structural Dynamics*, Notre Dame, Indiana.

Grigoriu, M. (1995). *Applied non-Gaussian Processes: Examples, Theory, Simulation, Linear Random Vibration and MATLAB Solutions*. New York:Prentice–Hall: Englewood Cliffs.

Grigoriu, M. & S. Balopoulou (1993). A simulation method for stationary gaussian random functions based on the sampling theorem. Probabilistic Engineering Mechanics 8(3-4), 239–254.

Gurley, K. R. & A. Kareem (1998). Simulation of correlated non-gaussian pressure fields. MECCANICA 33(3), 309–317.

Gurley, K. R., A. Kareem, & M. A. Tognarelli (1996). Simulation of a class of non-normal random processes. Journal of Non-Linear Mechanics 31(5), 601–617.

Gurley, K. R., M. A. Tognarelli, & A. Kareem (1997). Analysis and simulation tools for wind engineering. Probabilistic Engineering Mechanics 12(1), 9–31.

Kumar, K. S. & T. Stathopoulos (1997). Computer simulation of fluctuating wind pressures on low building roofs. Journal of Wind Engineering and Industrial Aerodynamics 71, 485–495.

Lancaster, H. O. (1957). Some properties of the bivariate normal distribution consider in the form of a contingency table. Biometrika 44, 289–292.

Peterka, J. A. & J. E. Cermak (1975). Wind pressure on buildings - probability densities. Journal of the Structural Division 101(ST6), 1255–1267.

Popescu, R., G. Deodatis, & J. H. Prevost (1998). Simulation of homogeneous nongaussian stochastic vector fields. Probabilistic Engineering Mechanics 13(1), 1–13.

Stathopoulos, T. (1980). Pdf of wind pressures on low-rise buildings. Journal of the Structural Division 106(ST5), 973–990.

Yamazaki, F. & M. Shinozuka (1988). Digital generation of non-gaussian stochastic fields. Journal of Engineering Mechanics, ASCE 114(7), 1183–1197.

Wind Engineering into the 21st Century, Larsen, Larose & Livesey (eds) © 1999 Balkema, Rotterdam, ISBN 90 5809 059 0

Application of cascade-correlation learning network for determination of wind pressure distribution in buildings

T. Girma
University of Roorkee, India (Deputed from Nazareth Technical College, Ethiopia)

P.N. Godbole
Civil Engineering Department, University of Roorkee, India

ABSTRACT: Usually the evaluation of wind load on buildings is carried out using codes and standards, whose specifications are generally based on wind tunnel tests performed on selected building structures with common shapes. For example IS:875 (part-3) 1987 gives specification for rectangular, square, cylindrical, and uniform shapes. Wind pressure determination in buildings having shapes different from those specified on codes and standards requires wind tunnel studies. However, wind tunnel studies can be expensive in terms of resources and time. This paper suggests the use of artificial neural network (ANN) trained with limited wind tunnel data for pressure distribution determination in buildings. ANN has been chosen because of its good generalization capability when trained with limited data. In addition solutions from ANN models are obtained easily and fast.

1 LITERATURE REVIEW

ANN have been successfully applied to many applications in civil engineering (Karuanithi et al. 1994, Flood et al. 1994, Kamarithi et al. 1992, Garrett et al. 1994, Attoh-okine et al. 1996, Khaundri et al 1995 and so many others). Khanduri et al. suggest a neural network approach for assessment of wind-induced interference effects on design loads for buildings. To test the applicability of the neural network methodology they have developed a back-propagation neural network example to predict the mean and dynamic along wind and across-wind interference factors for overturning moment when two identical square buildings interact. In their work the ANN models were trained to a mean square error threshold of 0.0001.

In the present study an ANN program has been developed based on cascade-correlation learning network (CCLN) using C++ language. Unlike back-propagation CCLN can determine its own size, depth and topology. It learns fast and retains the structure it has built. Moreover it is an incremental learning technique. Hence it is advantageous than back-propagation.

The ANN program with CCLN has been used to determine pressure distribution for a large multistoreyed office building for which limited wind tunnel test data was available.

2 CASCADE-CORRELATION LEARNING NETWORK (CCLN)

The cascade-correlation architecture (Figure.1) is a dynamic unit-growing learning network, which builds its hidden unit as well as hidden layer depth systematically (Fhalman 1990). It begins with some input and one or more outputs, but with no hidden units. The number of inputs and outputs dictated by the problem and by the I/O representation of the experimenter has chosen. Every input is connected to every output unit by a connection with an adjustable weight. There is also a bias input, permanently set to +1.

The output units may just produce a linear sum of their weighted inputs , or they may employ some non-linear activation function. For the present problem sigmoidal activation function whose output range can be changed as required has been used.

We add hidden units to the network one by one from a pool of candidate units. Each new hidden unit receives a connection from each of the network's original inputs and also from every pre-existing hidden unit. The hidden unit's input weights are frozen at the time the unit is added to the net; only the output connections are trained repeatedly. Each new unit therefore adds a new one unit "layer" to the network.

The CC architecture can be trained with a number learning algorithm. For present work, the quickprop algorithm developed by fahlman has been used [5]. The quickprop weight correction algorithm is a second-order method based loosely on Newton's method, but it is more heuristic than formal

$$\Delta W_{ji}(t) = \begin{cases} \varepsilon S_{ji}(t) & \text{if } \Delta W_{ji}(t-1) = 0 \\ \frac{S_{ji}(t)}{S_{ji}(t-1) - S_{ji}(t)} \Delta W_{ji}(t-1) & \text{if } \Delta W_{ji}(t-1) \neq 0 \text{ and } \frac{S_{ji}(t)}{S_{ji}(t-1) - S_{ji}(t)} < \mu \\ \mu \Delta W_{ji}(t-1) & \text{otherwise} \end{cases} \quad \ldots.(1)$$

where $S_{ji} = \frac{\partial E}{\partial W_{ji}}$ (slope)

ε = learning rate
μ = maximum growth factor

Several aspects of Eq.(1) are noteworthy. First, the difference of the first partial derivatives is used. This is an attempt to approximate the second partial derivative $\partial^2 E / \partial W^2_{ji}$. Second, $|\Delta W_{ji}(t)|$ cannot exceed $\mu|\Delta W_{ji}(t-1)|$; thus prevents single, large weight changes, which could negate many epochs of previous learning.

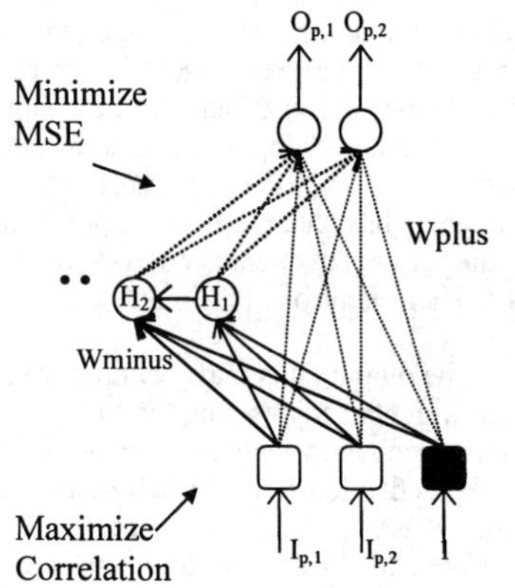

Figure 1 The structure of a cascade-correlation learning network. Solid connections are frozen (Wminus). Dotted connections (Wplus) are trained repeatedly.

At any step the network is trained until either a maximum number of iterations is reached, or a *patience* criterion is met. Patience rules are of the form *IF the error has not changed by at least p% over the last n epochs, THEN stop training.*

i) To create a new hidden unit, we begin with a candidate unit that receives trainable input connections from all of the network's external inputs and from all pre existing hidden units. The output of this candidate unit is not yet connected to the active network. We run a number of passes over the examples of the training set, adjusting the candidate unit's input weights after each pass. The goal of this adjustments is to maximize R, the sum overall output units O of the magnitude of the correlation between H, the candidate unit's value and E_o, the residual output error observed at unit O. we define R as

$$R = \sum_o \left| \sum_p (H_p - \overline{H})(E_{p,o} - \overline{E_o}) \right|$$

where O is the network output which the error is measured and p is the training pattern. The quantities $\overline{H}$ and $\overline{E_o}$ averaged overall patterns.
In order to maximize R, $\partial R/\partial W_i$ must be computed. we can expand and differentiate the formula for R to get

$$S = \partial R / \partial W_i = \sum_{p,o} \sigma_o (E_{p,o} - \overline{E_o}) f'_p I_{p,i}$$

where σ_o is the sign of the correlation between the candidate's value and output O, f'_p is the derivative for pattern p of the candidate unit's activation function with respect to the sum of its inputs, and $I_{p,i}$ is the input the candidate unit receives from unit i for pattern p.

Instead of a single candidate unit, it is possible to use a pool of candidate units, each with a different set of random initial weights. All receive the same input signals and see the same residual error for each training pattern. Whenever we decide that no further progress is being made, we install the candidate whose correlation score is best.

The use of this pool of candidates is beneficial in two ways [5]: it greatly reduces the chance that a useless unit will be permanently installed because an individual candidate got stuck during training, and (on a parallel machine) it can speed up the training because many parts of weight-space can be explored simultaneously. Small pools with four to eight candidate units can be taken. In present work, eight candidates has been taken for Scope Twin Tower problem and six candidates has been taken for all other problems.

ii) The connections to output units are trained to minimize the familiar sum square error

$$E = \frac{1}{2}\sum_{o,p} (T_{op} - O_{op})^2$$

where O_{op} is the net's output at unit o for pattern p and T_{op} is the target output.

E is minimized using gradient descent method:

$$S_{oi} = \frac{\partial E}{\partial W_{oi}} = \sum_{p} \delta_{op} O_{ip}$$

where $\delta_{op} = (T_{op} - I_{op}) f_p'$

and f_p' is the derivative of the activation function of the output units for pattern p, O_{ip} is the activation of input unit i (this unit could be a hidden unit also) and W output connection weights.

3 WIND PRESSURE DISTRIBUTION FOR MULTI STOREY BUILDING.

Wind tunnel study for a Twin Tower Office Building, from Wind Engineering center, Civil Engineering Department, University of Roorkee, has been used in the present study [9]. CCLN has been used to model the wind pressure distribution in the building to predict some pressure coefficients. The pressure coefficients to be predicted were samples intentionally omitted from the training data so that they can be used to test the validity of the results the network generates.

The Twin Tower complex is built on a circular base. It has two basements and a three storey high Podium with two towers rising above the podium. The North tower (A) is 17 storey high while the South tower (B) is 12 storey high. The towers are C shaped and face each other, each subtending an angle of 140^0 at the center. The total height from ground level upto the roof of the tower is 59.14 m for the South tower while or the North tower it is 76.43 m.

The Podium structure and the two towers are protected on their faces by curtain wall using Aluminum Alloy frames fixed to the RCC structure and special tempered glass being fixed to these frames. The building is designed to be fully air-conditioned.

One of the important loads in the design of such buildings is the wind load, and hence a knowledge of wind pressure on the various portions of the building is essential. While for cylindrical structure, the pressure coefficients are specified in the Indian Code IS : 875, the Scope Tower is not exactly a cylinder. Further, it is surrounded by several buildings of varying heights on all sides. In order to assess the pressure distribution on the building, therefore, a model study was made for different wind directions. The most prevalent directions were selected from the Wind Rose for the area. The predominant directions are SW-NNW or SE - ENE

3.1 Wind Pressure Measurement

For measurement of pressure, the 1:300 scale model prepared was provided with a total of 169 pressure points. The North tower was provided with 7 rows of 5 pressure points each on the outer and 6 rows in the inner face; the sides were provided with 5 rows of 3 pressure points each and 1 row of 2 pressure points each. For the south tower on the other hand, the number of rows was reduced to 4 while keeping the number of pressure points in each row the same as for the North tower.

After noting the model causes negligible blockage and the model boundary layer is turbulent, the 1:300 model was mounted on the turn table along with the models of the surrounding buildings at appropriate locations and pressures measured at all the 169 pressure points at wind velocity of 12 m/s. The data was collected and processed using the Scanivalve in conjunction with the

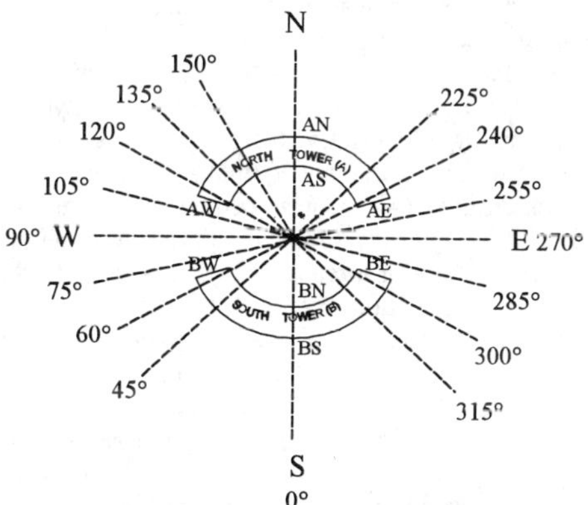

Figure 2 - Definition sketch for wind direction

Data Acquisition System and a software developed for data processing. The test was made for 15 different wind directions(i.e., 0^0, 45^0, 60^0, 75^0, 90^0, 105^0, 120^0, 135^0, 150^0, 180^0, 225^0, 250^0, 270^0, 285^0, 300^0, 315^0).

The pressure data can best be represented in the form of dimensionless pressure coefficient C_p, given by

$$C_P = \frac{\Delta P}{\rho V_o^2 / 2}$$

where $\Delta P = P - P_o$, P being the pressure at any point, p_0 the reference pressure, ρ is density of air and V_o is the reference velocity, while V_o is easily defined in a uniform velocity field, it is not so simple in a gradient velocity field such as a boundary layer flow. One could thus choose V_o either at a chosen reference height or the velocity at the level at which the pressure is measured. It is also worth pointing out that in field, P_o is the atmospheric pressure and thus ΔP yields the gauge pressure at any point. In the study, all C_p values have been worked out based on the velocity at the level of the top of the North tower as this is useful in identifying the point where the pressure will correspond to the maximum C_p. These values of C_p will not change so long as the velocity distribution remains the same as that of the test condition.

3.2 Preparation of Training and Testing data.

In preparing the training data different conditions have to be considered. Increasing the number of training patterns increases the potential level of accuracy that can be achieved by the network. A large training-pattern set, however, can sometimes overwhelm certain training algorithms, there by increasing the likelihood of an algorithm becoming stuck in a local error minimum [4]. Consequently, there is no guarantee that adding more training patterns leads to improved solutions. Moreover, there is a limit to the amount of information that can be modeled by a network that comprises a fixed number of hidden neurons; thus the gains that can be made by adding more training patterns quickly decrease.

Different pilot experiments were performed. For simplicity side BE which has got the least pressure points was chosen for pilot study. Of course the output of the network will be the pressure coefficient at each pressure point or any other intermediate point. Moreover, the input to the network is expected to include wind direction, and geometrical parameters (for example the corresponding height of the pressure points). Concerning target output, out of the 15 wind directions, pressure coefficient corresponding to 60° wind direction was randomly selected for testing purpose and it was decided to use all the other pressure coefficients corresponding to 14 different wind directions for training the network. This decision was made after observing the highly distributed nature of the pressure coefficients. In selecting specifically which parameters to be included in the input vector different options were tried. First, It was thought, all the twelve pressure points in face BE were to be used at a time for training the network so that it will be able to model pressure coefficients at any point in face BE. The input to the network must contain all necessary parameters that will enable the network differentiate between each pressure points. For this purpose, it will be reasonable to include horizontal and vertical coordinates of the pressure points and the wind direction in the input vector.

From the above discussions it would be easy to determine the exact size of the input and target output vector. The target out put vector will be a 196 (12x14) by 1 vector. And the input will be a 196 by 3 vector. The wind directions were normalized between 0 and 1.

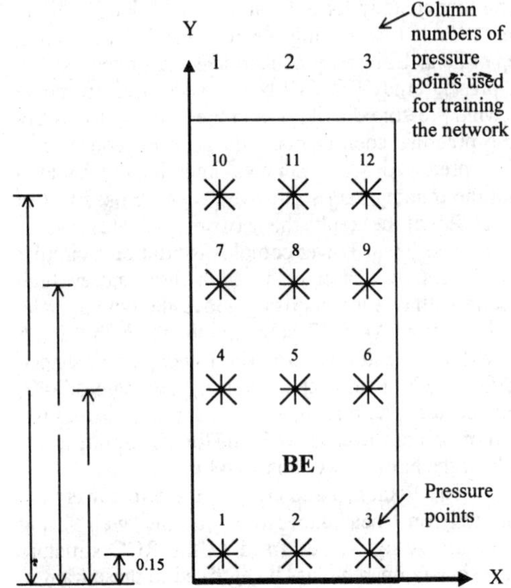

Figure 3 Vertical distance measurements

And the horizontal and vertical coordinates of pressure points were taken in mm. The network was able to be trained to threshold mean square error of 0.0001 with 15 minutes.

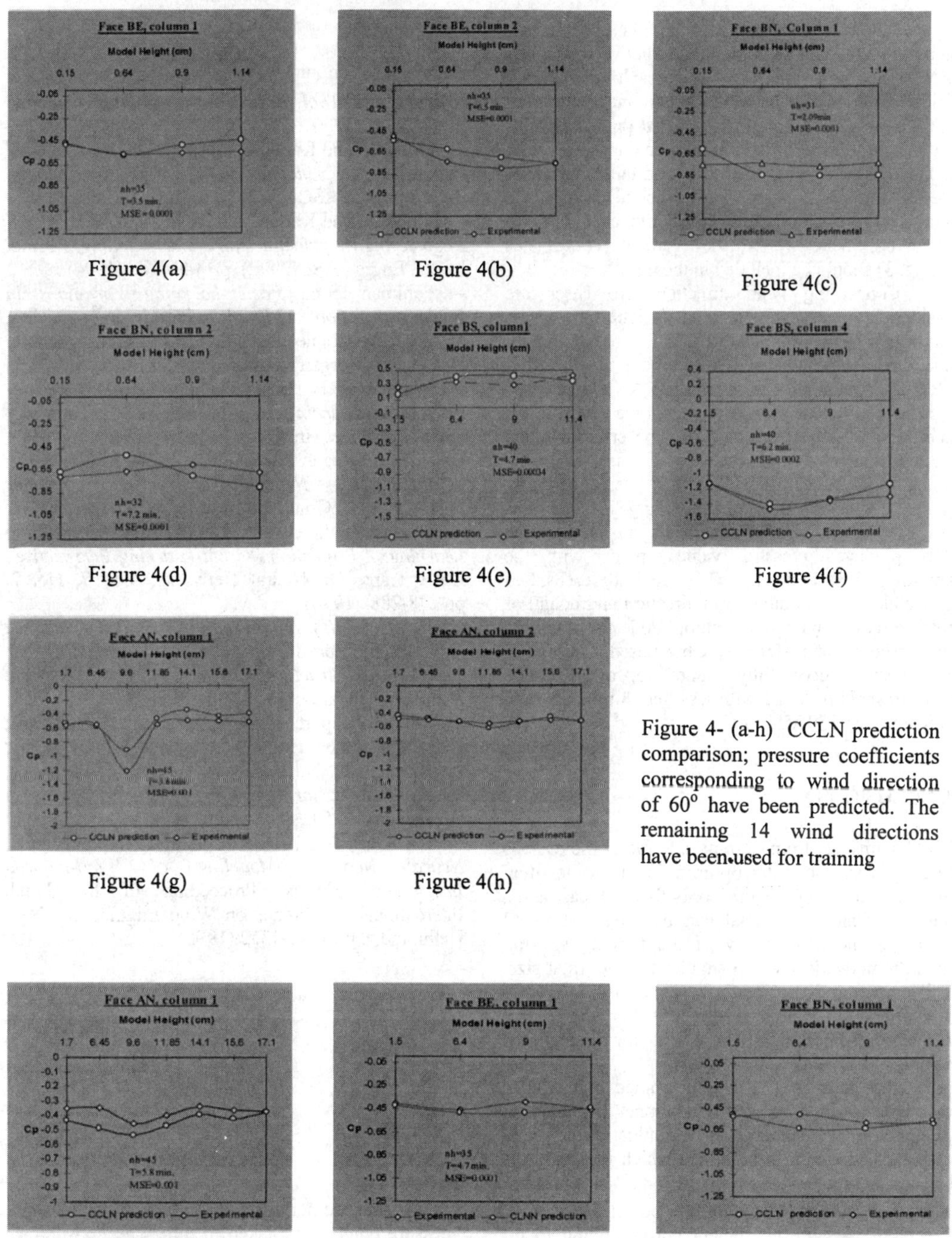

Figure 4(a) Figure 4(b) Figure 4(c)

Figure 4(d) Figure 4(e) Figure 4(f)

Figure 4(g) Figure 4(h)

Figure 4- (a-h) CCLN prediction comparison; pressure coefficients corresponding to wind direction of 60° have been predicted. The remaining 14 wind directions have been used for training

Figure 5(a) Figure 5(b) Figure 5(c)

Figure 5 (a-c) CCLN prediction comparison; pressure coefficients corresponding to wind direction of 90° have been predicted, here 12 wind direction have been used for training excluding 90°, 135°, 285° randomly from training set.

However if we consider face AN The size of the input vector will be 490x3 and target vector will be 490x1, and these are large vectors. It was finally decided to train the network taking one column of pressure points at a time instead of all pressure points in a face at a time. Which will reduce the input vector for face BE into 56 (4x14) by 2, and target vector or face BE into 56X1 array. In the second case only a column of pressure points at a time are going to considered, hence only the vertical coordinates (Figure. 3) should be included in the input vector.

Concerning wind direction two cases are considered. First, a single wind direction has been randomly selected for testing and the remaining 14 wind directions were used for training the network. After the training was over the ANN predicted pressure coefficient values were compared with the experimental values (testing set). The error obtained was less than 15 % in most cases (Figure. 4). In the second case three wind directions were randomly selected for testing and the remaining 12 directions were used for training. Again in the second case the ANN predicted pressure values match with the experimental values (Figure. 5). In both cases two input nodes for encoding wind direction and height of pressure points and a single output node for obtaining pressure coefficient values have been used. The CCLN models were successfully trained for mean square error threshold of 0.001 with less than 8 minutes time using Pentium 150 MHz. PC.

CONCLUSIONS

CCLN learns fast from examples. Some of the reasons are it uses second order optimization in its learning algorithm quickprop, requires only forward pass and a single residual error signal can be broadcast to all candidates moreover many of the training are run when the network is much smaller than its final size. Each unit in a CCLN performs as a specific feature detector and its input weights are not altered once they are frozen. This feature is useful for incremental learning.

The ability of CCLN to be trained to generalize when presented with limited wind tunnel data makes it an attractive application for determination of wind pressure distribution in buildings which in turn saves time and resources.

REFERENCES

– Eberhart, R.C., and Dobbins, R.W., *Neural Network P.C. Tools : A Practical Guide*, Academic Press, USA. (1990)

– Flood, I., and Kartam, N., *Neural Network in Civil Engineering I : principles and understanding.* J.comp, in Civ. Engrg., ASCE, 8(2), pp 131-149. (1994)

– Flood, I., and Kartam, N., *Neural Network in Civil Engineering II : principles and understanding.* J.comp, in Civ. Engrg., ASCE, 8(2), pp 149-162. (1994)

– Fahlman, S.E., *Faster-Learning Variations in Back-propagation: An Empirical Study*, In proceeding of the 1988 Connectionist Models Summer School. San Mateo. Morgan Kaufmann. (1988)

– Fahlman, S.E., and Lebiere, C., *The Cascade-Correlation Learning Architecture.*, In D.S. Touretzky (eds), Advances in Neural Information Processing Systems II, Morgan Kaufmann, pp 524-532. (1990)

– Fu, L.M., *Neural Network in computer intelligence*, McGraw-Hill Inc..(1994)

– Hwang, J.N. et al., *The Cascade-Correlation Learning: A Projection Pursuit Learning Perspective.*, IEEE Trans. On Neural Networks, Vol. 7, No. 2, pp.278-288. (1996)

– IS:875 (1987), *Indian Standard Code of Practice for Design Loads (other than Earthquake) for Buildings and Structures, part-3, Wind Loads:*, Bureau of Indian Standards, New Delhi, India.

– Internal Report, Civil Engineering Department, University of Roorkee.

– Karunanithi, N., Grenney, W.J., and Whitley, D., *Neural Networks for River Flow Prediction,* J.comp, in Civ. Engrg., ASCE, 8(2), pp 201-219.(1994)

– Khanduri, A.C., Bedard, C., and Stathopoulos,T., *Neural Network Modeling of Wind-induced Interference Effects*, Proceeding of the Ninth International Conference on Wind Engineering, New Delhi, India, pp 1341-1352.(1995)

Wind Engineering into the 21st Century, Larsen, Larose & Livesey (eds) © 1999 Balkema, Rotterdam, ISBN 90 5809 059 0

Dynamic responses and equivalent wind loads of the Jin Mao Building in Shanghai

M.Gu, Y.Zhou, F.Zhang & H.F.Xiang
State Key Laboratory for Disaster Reduction in Civil Engineering, Tongji University, Shanghai, People's Republic of China

ABSTRACT: Formulas for static equivalent wind loads on tall buildings are presented in this paper. The resonant component of the equivalent wind load is represented by the inertial wind force due to the structural vibration, while the background component is derived based on the "load response correlation" method. A high frequency force balance test and an aeroelastic full structural model test on the Jin Mao Building with a height of 420.5 m located in Shanghai where typhoon often attacks have been carried out in the TJ-2 Boundary Layer Wind Tunnel in Tongji University. Dynamic responses of the Jin Mao Building and the equivalent wind loads on the building are finally given.

1 INTRODUCTION

In recent years, a great number of super-tall buildings have been constructed in China. Among these buildings the Jin Mao Building with a height of 420 m is the tallest one. For such a super-tall building located in Shanghai where typhoon often attacks, a wind-resistant study is necessary to be made.

Wind-induced vibration of super high-rise buildings is the critical factor in the structural design procedure. From the practical viewpoint, wind-induced vibration of buildings is classified as along-wind vibration and cross-wind vibration. The former is well known due primarily to the work of Davenport and other researchers in the early 1960s (Davenport 1961, 1967). Random vibration theory can be applied to analyze the responses. The mechanism of latter is much more complex than that of the former. Random vibration theory is difficult to be directly used in the analysis of cross-wind vibration of high-rise buildings, while model tests in wind tunnels play the main tool in this study. The relevant codes of few countries have dealt with the cross-wind responses of high-rise buildings.

High-frequency force balance model test is presently wide used for wind-resistant design of high-rise buildings for its low cost, low time consuming and convenience for use (Cook 1983, Tschanz 1983). However, it is well known that the effects of non-ideal model shape and aerodynamic damping are still problems in the use of this technology. Vickery (1970) pointed out that "even for large deviations of the actual mode shape, the errors are roughly 1-3%, namely the effects of moderate deviations from a straight line mode-shape insignificant". This conclusion has been applied to high-frequency force balance and other techniques. Presently, Zhou, Gu and Xiang (1998) indicated that the mode shape correction factor proposed by Vickery is essentially the first-mode displacement correction factor and can be neglected in the analysis on wind-induced vibration displacement of high-rise buildings; but for some other responses, such as the base shearing force, a proper correction factor should be taken into account.

As for the aerodynamic damping, Davenport (1979) gave formulas for along-wind and cross-wind aerodynamic damping ratios based on quasi-steady theory. Generally speaking, the effects of cross-wind aerodynamic damping ratio on structural vibration are much greater than those of along-wind one. A study on the Jin Mao Building (Ho et al. 1994) indicated that the structural loads in cross-wind direction increases by about 36% due to aerodynamic damping.

An aeroelastic building model is considered to be a most effective model which is able to overall reflect the important factors, such as aerodynamic damping, non-ideal mode and multi-mode coupling, etc., which affect the vibration of high-rise buildings due to wind, (Isyumov 1983). Even then, it has been seldom used to study on wind-induced vibration of high-rise buildings because the model test is time- and cost-consuming.

Formulas for calculation of equivalent dynamic loads are firstly given in this paper. A high-frequency force balance test on the rigid model and an aeroelastic full model test of the Jin Mao Building as well as the results are then presented. Finally some of these results are compared with each other and with those by Ho et al. (1994).

2 THEORETICAL FORMULAS

The equivalent wind load on a tall building is separately evaluated as mean, background and resonant components. The formulas for the mean wind force are not shown here. Different from the "gust loading factor" method, the background equivalent wind load component is derived based on the "load response correlation" method, which was originally introduced by Kasperski et al. (1992). The resonant component is represented by the inertial wind force.

The basic assumptions for deriving these formulas are given as,

1. The first mode shape function of the building:

$$\varphi_1(z) = c(z/H)^{\beta} \tag{1}$$

where c is a constant; H is the height of the building; β is the exponent of the first mode shape function.

2. The fluctuating wind force acting on the building per unit length:

$$P(z) = P_H(z/H)^{\gamma} \tag{2}$$

where P_H is the fluctuating force at the top of the building per unit length; γ equals to α (the exponent of the mean wind profile) for the along-wind vibration of buildings.

3. The influence function for the responses of buildings under the action of equivalent wind loads:

$$i(z) = i_c(z/H)^{\beta_0} \tag{3}$$

where i_c is a constant; β_0 is the exponent of the influence function. The Equation (3) is for base shearing force when i_c=1 and β_0=0; and is for base bending moment when i_c=H and β_0=1; etc.

Based on the above assumptions, the RMS value of first mode resonance displacement can be calculated using the following equation:

$$\sigma_{DR} = \frac{P_H(1+2\beta)}{m_0 H\omega_1^2}\sqrt{\frac{\pi f_1 S_P(f_1)}{4\zeta_1}}\left|J(\gamma,\beta,f_1)\right|^2 \left(\frac{z}{H}\right)^{\beta} \tag{4}$$

where m_0 is the mass per unit length of the building; ω_1 is the first circular frequency of the building; and

$$\left|J(\gamma,\beta,f_1)\right|^2 = \int_0^H\int_0^H \left(\frac{z_1}{H}\right)^{\gamma+\beta}\left(\frac{z_2}{H}\right)^{\gamma+\beta} R(z_1,z_2,f)dz_1dz_2 \tag{5}$$

where $R(z_1,z_2,f)$ is the correlation function of the fluctuating wind force. The RMS resonance equivalent wind load equals to the inertial force of the structure, that is

$$\sigma_{FR} = m(z)\omega_1^2\sigma_{DR} \tag{6}$$

The background responses at frequencies below the natural frequency are essentially quasi-static and the structural responses are not affected by the structural dynamic behavior. The determination of the static equivalent wind load associated with the background response is somewhat difficult because of the correlation of the applied wind pressure. The "load response correlation" method is used here to derive the relevant formulas. The RMS background equivalent wind load is

$$\sigma_{FB}=\frac{P_H(2+2\gamma)}{H}\left(\int_0^{\infty}S_P(f)\left|J(2\gamma,1,f)\right|^2 df\right)^{1/2}\left(\frac{z}{H}\right)^{2\gamma} \quad (7)$$

The background responses is shown as

$$\sigma_{rB}=\frac{P_H i_c(2+2\gamma)}{(1+2\gamma+\beta_0)}\left(\int_0^{\infty}S_P(f)\left|J(2\gamma,1,f)\right|^2 df\right)^{1/2} \quad (8)$$

3 HIGH FREQUENCY FORCE BALANCE (HFFB) TEST

Figure 1 shows the elevation of the Jin Mao Building. To understand the wind loads on this super-tall building and evaluate its dynamic responses, a high frequency balance model test was carried out in the TJ-2 Boundary Layer Wind Tunnel in Tongji University, which has a working section with a width of 2.5 m and a height of 3 m. A rigid model of the tower of the Jin Mao Building was built to a scale of 1:500 according to the height of the real building and the size of the working section of the wind tunnel.

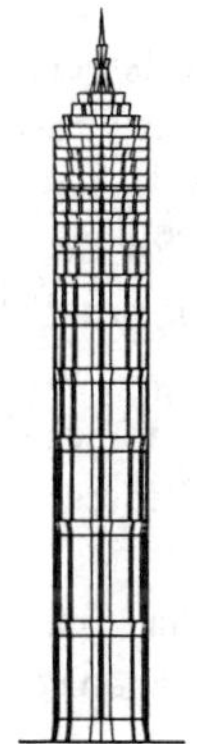

Figure 1 Elevation of the Jin Mao Building

3.1 Simulation of wind characteristics

According to the Chinese Code for Wind Loads on Buildings, the basic wind pressure in Shanghai area is 0.55 kN/m^2, corresponding to the basic wind speed of 29.7 m/s. Further the wind speeds at the gradient height in the B category of terrain for some return periods are shown in Table 1

Table 1 Wind speeds at gradient heights

Return period(year)	10	30	50	100
Wind speed at G.H.(m/s)	46.2	52.5	55.7	59.9

G.H.: means gradient height

The surroundings of the Jin Mao Building are roughly divided into three sectors with similar wind characteristics, corresponding to three categories of terrains, that is B, C and D, representing open terrain, suburb terrain and center of large city, respectively. The component α of the mean wind profile are 0.16, 0.22 and 0.30 for the B, C and D categories of terrains, respectively; and the corresponding gradient heights are 350, 400 and 450m.

Wind characteristics simulation was also performed at a length scale of 1:500. This simulation for the building site was achieved by a combination of turbulence generating spires and a barrier at the entrance of the wind tunnel, roughness elements along the wind tunnel floor upstream of the model. Figure 2 shows the simulated mean wind speed

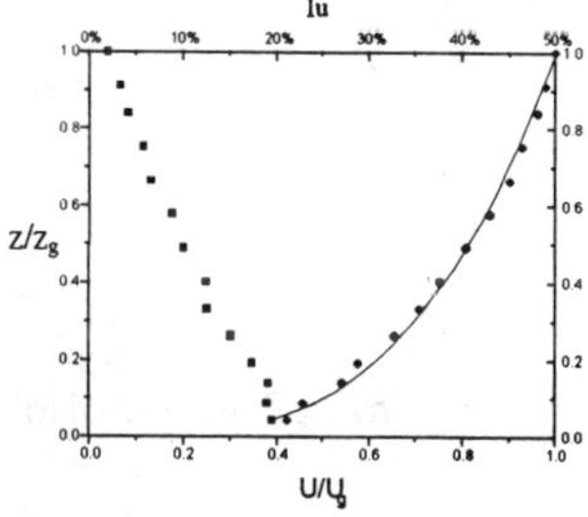

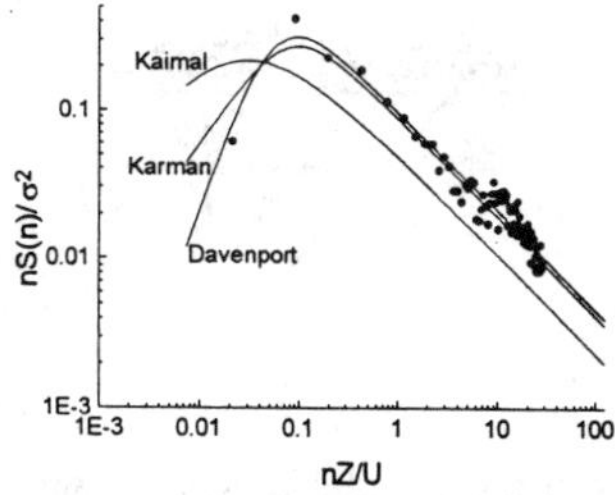

Figure 2 Simulated results of Terrain D

profile, turbulent intensity and power spectrum of terrain D. The turbulence scales are about 0.38m and 0.36 m for Terrain B and D, respectively.

The immediate surroundings of the Jin Mao Building, representing both the primary condition and future fully developed condition were also separately represented in the wind tunnel. For the so-called primary condition, no neighbor building is considered; the developed condition means the scaled replication of the immediate surroundings of the building within a full scale radius of about 600 m (see Fig.3) in the wind tunnel. It should be stressed that there is another super-tall building named Huanqiu Finance Building with a height of 460m near the Jin Mao Building. The effects of this super-tall building on the vibration of the Jin Mao Building are seen significant as indicated hereafter.

The definitions of the model's coordinate system and the wind direction are shown in Figure 4.

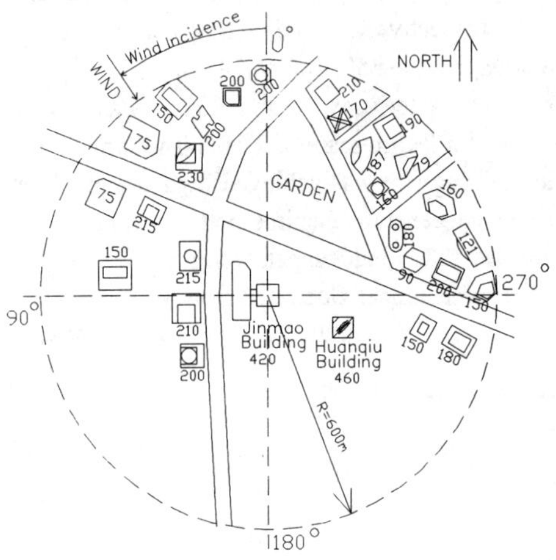

Figure 3 Immediate surroundings of the Jin Mao Building

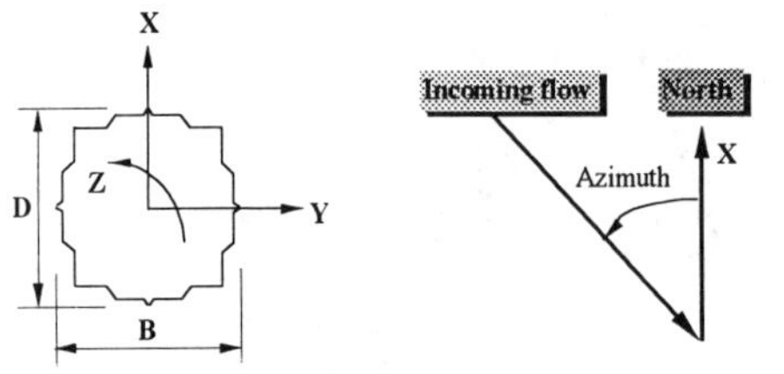

Figure 4 Definitions of the coordinate system and wind direction

3.2 System of balance and building model

A strain gauge balance for five components of aerodynamic forces was designed and manufactured by the authors and the co-researchers. The natural frequencies of the balance without load (model) on it are 188, 208, 273, 768 and 843 Hz for the two shearing force, a torsion and two bending moment components, respectively.

The rigid foam model of the building is about 0.8 m in height, and weighs 650 grams. The frequencies of the system of the balance-model system are 117, 123, 120, 36 and 38 Hz for the five components, respectively. The concerned maximum frequency of the aerodynamic forces acting on the model is about 18 Hz for the testing wind speed of 12m/s. It seems that the latter two frequencies of the balance-model system are somewhat low. To improve the frequency response characteristics of the balance-model system, low-pass filters were used and the filtering frequencies were set to be 25 Hz.

3.3 Main testing results

Figure 5 shows the base bending moment coefficients in x and y directions versus wind attack angles. From figure 5(a), it is seen that the absolute values of $C_{\overline{M}x}$ for the wind attack angles of 90° and 270°, namely the along-wind mean base bending moment coefficients, are larger than those for the other wind attack angles. On the other hand, the values of $C_{\tilde{M}y}$ for the wind attack angles of 90° and 270°, namely the RMS values of cross-wind base bending moment coefficients, are larger than those for the other wind attack angles (Fig.5(b)).

The comparison between the coefficients for the primary condition in B and D categories of wind field conditions indicates that the increase of ground roughness decreases the mean aerodynamic forces while increases fluctuating aerodynamic forces acting on the Jin Mao Building.

It can also be seen from this figure that the aerodynamic force coefficients are almost symmetric about the wind attack angle of 180° for the primary condition. The Huanqiu Finance Building is in the direct front of the Jin Mao Buildings when wind attack angle is 247°. Figure 5 shows that for the wind attack angle between about 203° to 280° the RMS force coefficients become larger but the absolute values of the mean force coefficients become smaller for the developed condition than the corresponding values for the primary condition. This means that the disturbance on the wind-induced vibration of the Jin Mao Building mainly come from the Huanqiu Building with a height of 460 m but from other buildings which have about half height of Jin Mao Building. It is obvious that the disturbance of the Huanqiu Building reduces mean wind loads while increases fluctuating wind loads acting on the Jin Mao Building.

The maximum value of the base bending moment coefficient is about 0.06 for wind attack angle of about 270°, which is the cross-wind moment coefficient.

Other parameters, such as the base shearing force and torsion moment coefficients for different conditions have the same tendencies with the base bending moment coefficients as mentioned above.

Figure 6 shows the power spectrum density functions (PSDF) of the base bending moment coefficients in x and y directions for the developed condition, D category of wind field and wind attack angle of 270°. The vortex-shedding phenomenon can be clearly seen from the PSDF of the coefficient of the cross-wind base bending moment. The energies of the cross-wind base bending moment in the interested frequency region are much greater than those of the along-wind base bending moment.

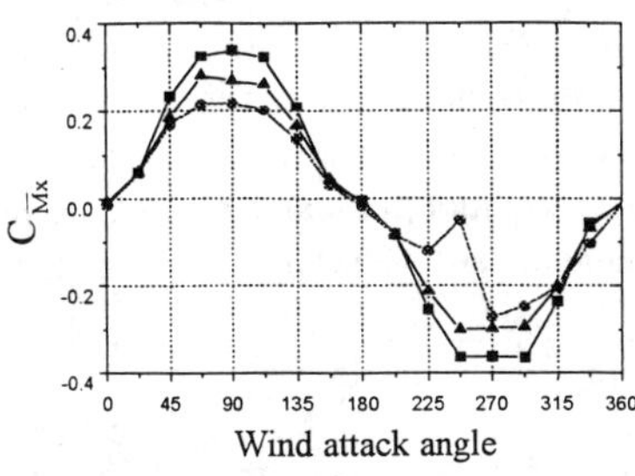

(a) ●D. condition, Terrain D;
▲ P. condition, Terrain D; ■ P. condition, Terrain B

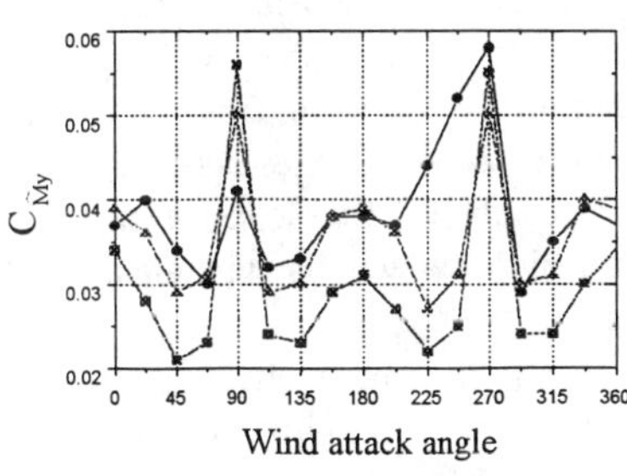

(b) ●D. condition, Terrain D;
▲ P. condition, Terrain D; ■ P. condition, Terrain B

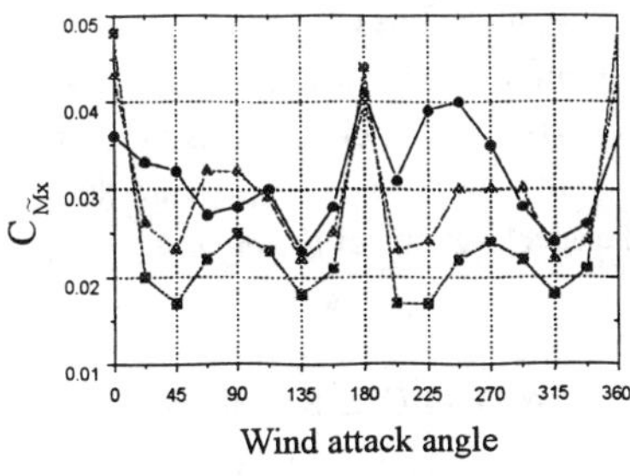

(c) ●D. condition, Terrain D;
▲ P. condition, Terrain D; ■ P. condition, Terrain B

Figure 5 Base bending moment coefficients versus wind attack angles for different terrain conditions

4 AEROELASTIC MODEL TEST

Length scale of the aeorelastic full model of the building was also 1:500 as the model for the force balance test, and the wind speed scale was selected as 1:6. Other scales, such as the stiffness scales and

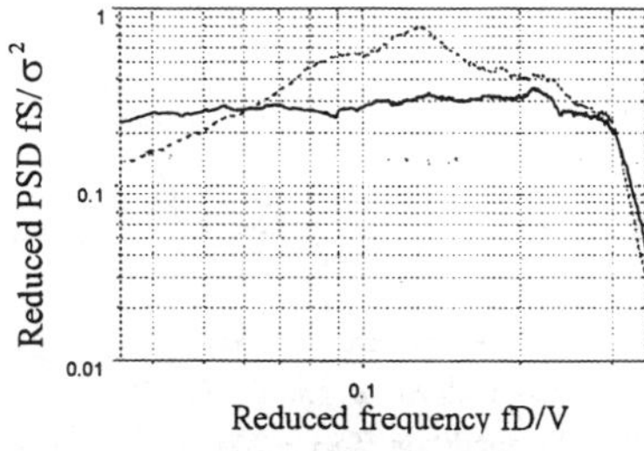

Figure 6 Power spectrum density functions of the base moment coefficient (----My, ——Mx)

mass scale, etc., for the model design can be deduced based on these two scales.

The model was composed of a frame of six stories made of aluminum alloy as a skeleton for the simulation of structural stiffness and a "clothing" for simulating the configuration and the mass distribution of the building. The main frequencies of the model are listed in Table 2 with the computed ones in it. The measured first bending and torsion frequencies converted from the model to the prototype are 0.162Hz and 0.59Hz, respectively; while the corresponding computed results are 0.165 and 0.64 Hz. The errors between the frequencies of model and the expected ones of the first bending and torsion modes are acceptable. From this table, it can also be seen that the second mode bending frequency of the model is much smaller than the calculated one. The damping ratios of the model for the three modes are all about 1.8%.

Table 2 Main frequencies of aeroelastic model

	Frequency		
	Model	Calculation	Error
1st bending mode	0.162	0.1615	0.3%
2nd bending mode	0.3	0.446	33%
Torsional mode	0.59	0.64	8%

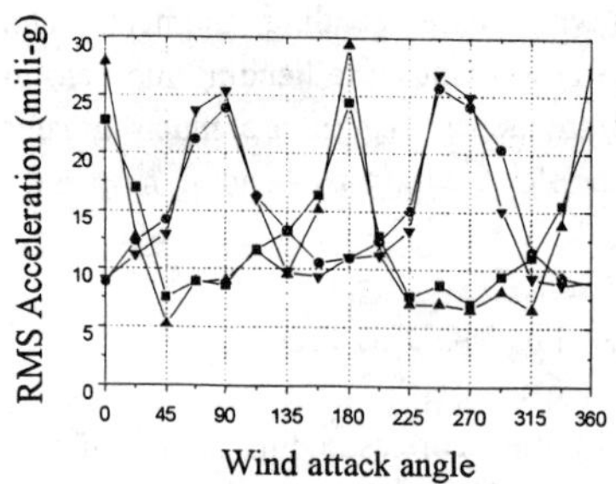

(a) ●X, ■ Y (P. condition, Terrain B)
▼X, ▲ Y (P. condition, Terrain D)

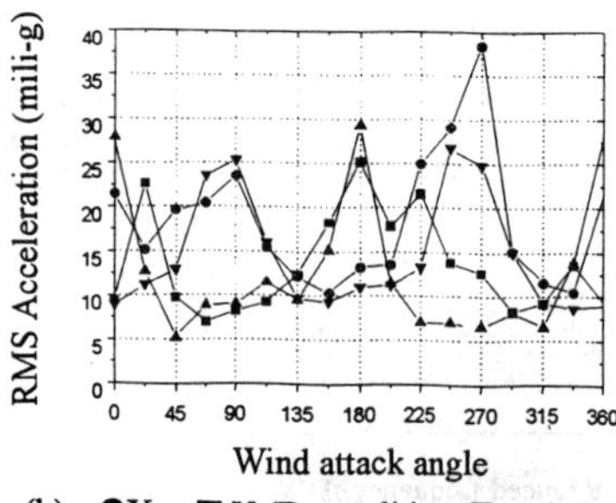

(b) ●X, ■ Y (D. condition, Terrain D)
▼X, ▲ Y (P. condition, Terrain D)

Figure 7 Acceleration versus wind attack angles

Four mini-accelerometers were mounted on the sixth floors of the model, corresponding to the height of 334 m of the actual building. The vibration signals in both x and y directions were recorded with these accelerometers and a computer. The testing wind field conditions for this test are the same as those for the HFFB test. The testing wind speeds are from 5 to 12m/s with step of 1m/s.

Figure 7 shows the RMS values of accelerations of the Jin Mao Building for the primary and developed conditions at 10 m/s testing wind speed versus different wind attack angles. For the primary condition, the acceleration responses of the building are almost symmetric about the wind attack angle of 180°, which are almost the same as the base bending moment coefficients from the high-frequency balance tests as mentioned before. Figure 7 (b) also clearly shows that the disturbance from the Huanqiu Finance Building has great effects on the wind-induced vibration of the Jin Mao Building. For developed condition, the maximum cross-wind RMS acceleration value is about 1.3 times as larger as that for the primary condition. But it seems that this disturbance has little effect on the along-wind response of the Jin Mao Building.

The comparison between the acceleration responses on the primary condition for B and D categories of wind field conditions indicates that the increase of ground roughness increases acceleration responses of the Jin Mao Building.

In addition, it can be known that the cross- and along-wind vibration responses are both mainly contributed from its first bending modes in x and y directions according to the power spectrum density functions of the accelerations of the building.

5 EQUIVALENT WIND LOADS AND DYNAMIC RESPONSES

5.1 Equivalent wind loads

The equivalent wind load was estimated as the combination of mean, background and resonant

combination of mean, background and resonant components based on the results from high-frequency force balance tests and the analysis method presented in the Section 2 in this paper. Dynamic responses, such as vibration displacement, acceleration and base bending moment etc., of this building were also evaluated. Figure 8 shows the background and resonance equivalent wind loads for a 50-year return period, terrain D and wind attack angle of 270°. The structural damping ratio is assumed to be 2.5%.

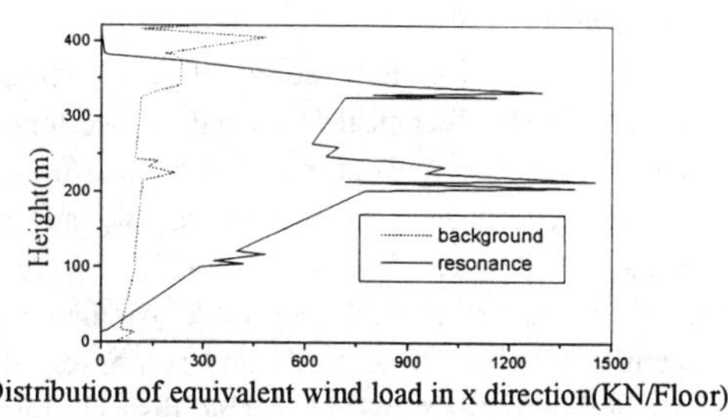

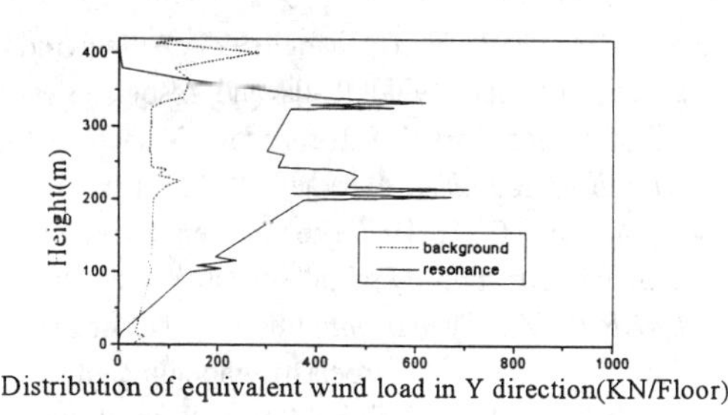

Figure 8 Equivalent wind loads

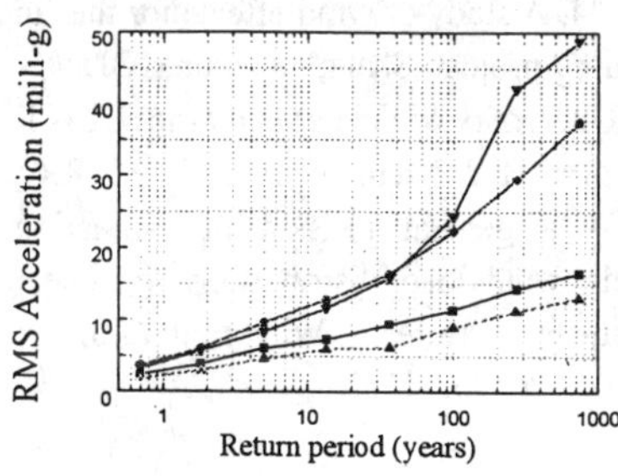

Figure 9 Comparison of accelerations

5.2 Acceleration and comparison with HFFB test

The wind-induced accelerations of the building were obtained by theoretical analysis based on the data from the force balance test. These results are then compared with the corresponding ones from the aeroelastic model test. For this comparison, the structural damping ratio is assumed to be 0.18 in the calculation procedure of the acceleration response based on the HFFB test, which is the same as that of the aeroelastic model. The acceleration results can be used to judge occupant comfort.

Figure 9 presents the RMS acceleration values at the height of 334 m of the Jin Mao Building from the two kinds of tests for the primary condition versus different-year return period. For the testing wind speeds lower than the wind speed of 100-year return period, the cross-wind acceleration responses from the two kinds of tests are shown to be almost the same. It seems that, more precisely, the along-wind acceleration obtained from the high-frequency balance test is somewhat larger than that from the aeroelastic model test. This may because the effects of the along-wind aerodynamic damping which appears positive value. For the wind speeds higher than the wind speed of 100-year return period, the cross-wind acceleration responses from the aeroelastic model test are gradually much larger than those from HFFB test.

From Figure 9 it can also be known that The maximum RMS values of the cross-wind acceleration and the along-wind acceleration for 10-year return period according to Chinese code are about 12 mili-g and 7 mili-g, respectively. It can be concluded that the total peak value comprised of cross- and along-wind peak accelerations is larger than the allowable value for human comfort.

5.3 Comparison with other testing results

Some main results from our research and the similar research carried out by BLWT Laboratory in West Ontario, Canada, have been compared and they show in agreement to a great extent. Table 3 shows the comparison between the base bending moments from the author and Ho et al.(1994) , respectively. The actual wind speeds are both 40m/s, It is seen that the results agree well.

Table 3 Comparison of base bending moments

Peak base bending moment (kN*m) (damping ratio 2..5%)	Primary condition		Developed condition	
	X	Y	X	Y
Authors' tests	5.12×10^6	5.12×10^6	6.98×10^6	5.99×10^6
Ho et al.'s tests	5.37×10^6	5.09×10^6	6.66×10^6	6.78×10^6

(The testing wind speeds are both 40m/s; the damping ratios are both 2.5%)

6 CONCLUDING REMARKS

This paper presents formulas for dynamic equivalent wind load on high-rise buildings. The dynamic equivalent wind load is separately evaluated as background and resonant components. Different from the "gust loading factor" method, the background equivalent wind load component is derived based on the "load response correlation" method. The resonance equivalent wind load component is represented by the inertial wind force. A high frequency force balance test and an aeroelastic full structural model test on the Jin Mao Building were carried out. Dynamic responses of the Jin Mao Building and the equivalent wind loads on the building are given. The results from both the HFFB test and the aeroelastic model test agree to some extent, and they are also in agreement with the testing results carried out in the University of Western Ontario, Canada. The acceleration response of the building for 10-year return period indicates that the acceleration response is larger than the allowable value for human comfort.

REFERENCES

Davenport, A. G. 1961. The application of statistical concepts to the wind loading of structures. Proc. Inst. Civ. Eng. 19: 449-472

Davenport, A. G. 1967. Gust loading factors. J. Struct. Div. ASCE. 93(3):11-34

Cook, N. J. 1983. A sensitive 6-component high-frequency-range balance for building aerodynamics. *Phys. E. Sci. Instrum.* 116: 390-393

Tschanz, T. & A. G. Davenport 1983. The base balance technique for the determination of dynamic wind loads. *J. Wind Eng. & Indust. Aerodyn.* 13: 429-439

Vickery, B. J 1970. On the reliability of gust loading factor. Proc. Technical Meeting concerning Wind Loads on Buildings and Structures, Building Science Series 30, National Bureau of Standards: 296-312. Washington DC.

Zhou, Y., M. Gu & H. F. Xiang 1998. Alongwind static equivalent wind loads and responses of tall buildings. Part I: Unfavorable distributions of static equivalent wind loads. *J. Wind Eng. & Indust. Aerodyn.* 79: 135-150

Zhou, Y., M. Gu & H. F. Xiang 1998. Alongwind static equivalent wind loads and responses of tall buildings. Part II: Effects of mode shapes. *J. Wind Eng. & Indust. Aerodyn.* 79: 151-158

Davenport, A. G. 1979. The influence on the aeroelastic responses of tall structures to wind. *LAHR/IUTAM Symposium*: 681-695. Karlsruhe.

Isyumov, N. 1982. The aeroelastic modeling of tall building. In T. A. Reinhold(ed.), Wind Tunnel Modeling for Civil Engineering Applications: 373-407

Ho, T.C.E., N. Isyumov, M. J. Mikitiuk & G. J. Crooks 1994. A study of wind effects for the Jin Mao Building project, Shanghai, China. BLWT-SS1. The University of Western Ontario: BLWT

Kasperski, M. And H. J. Niemann 1992. The L.R.C. method — a general method of estimating unfavorable wind load distributions for linear and non-linear structures. J. Wind Eng. & Indust. Aerodyn. 41-44:1753-1763

Wind Engineering into the 21st Century, Larsen, Larose & Livesey (eds) © 1999 Balkema, Rotterdam, ISBN 90 5809 059 0

Structural loads and responses from the integration of instantaneous pressures

T.C.E. Ho, G.R. Lythe & N. Isyumov
Boundary Layer Wind Tunnel Laboratory, The University of Western Ontario, London, Ont., Canada

ABSTRACT: This paper examines the effectiveness of using simultaneous pressure data, taken at discrete measurement locations, to determine the overall structural wind loads. This technique is compared to the traditional force balance technique, where generalized forces are estimated from the measured base moments. The inherent assumptions of the force balance technique and its applicability are discussed and compared with the pressure integration method. The basic model and test requirements for the point pressure data, needed to effectively determine the structural loads in terms of pressure tap coverage and frequency resolution, are also examined.

The structural loads obtained using point pressure data are found to be comparable to those from the force balance technique. Furthermore, this approach provides valuable additional information on the overall wind loads and eliminates a number of uncertainties in the estimation of the generalized forces. This results in better estimates of the generalized forces and subsequently better estimates of structural wind loads and wind-induced responses.

1 INTRODUCTION

The development of solid state pressure scanners permits simultaneous measurement of pressures at many pressure taps located on exterior surfaces of a building or structure. This permits the determination of instantaneous overall wind forces on most buildings and structures with good spatial and frequency resolution and offers important opportunities for improving the determination of structural wind loads for those buildings and structures where wind-induced resonant vibrations are significant. Commonly, such information is now obtained from force balance model tests (Tschanz 1983). These are not without difficulties for structures with unusual mode shapes. Aeroelastic studies, which can provide accurate information in all situations, are substantially more expensive and tend to be justified only in situations where motion-induced or "aeroelastic" effects are significant (Isyumov 1982).

There are a number of advantages in determining the structural wind loads from the spatial integration of simultaneously measured instantaneous pressures. The foremost is that generalized wind forces for particular modes of vibration can be determined without the limiting assumptions of conventional force balance model tests. For example, actual modal displacements can be used to instantaneously weight the external pressures and to arrive at the corresponding generalized forces. No corrections are needed when mode shapes depart from linear variations with height, as are necessary for conventional force balance model tests in which the generalized forces are estimated from the base moments. In particular, the generalized torque is obtained directly rather than by inference. Also, it is possible to examine higher modes of vibration. While not important for tall buildings, higher modes can become important for towers and other more flexible structures. This technique has been effectively used for estimating the wind-induced response of the projected 558m Observation/Telecommunication Tower in Jakarta, Indonesia. An important advantage of the pressure integration technique is that it allows the assembly of generalized forces for three-dimensional mode shapes without having to establish the correlation between the translational and rotational force components. These are difficult to ascertain and conventional force balance model techniques therefore primarily rely on the direct measurements of x, y and torsional base moments, neglecting the various cross terms.

Another major advantage of this technique is that it allows a detailed description of the background or "non-resonant" component of the time-varying wind loads. This is important for buildings of low and intermediate height, where resonance may not dominate the wind-induced dynamic response.

Initial information on the technique of integrating simultaneously measured pressures has been provided by Ho (1997). Substantially greater experience has now been gained and the purpose of this paper is to review the advantages, as well as the limitations of this approach. Results are presented to compare structural loads and responses in situations where mode shapes depart from a linear variation with height and have varying degrees of participation of sway in orthogonal x, y directions and rotation. Also examined are questions of the minimum spatial and frequency resolution needed to achieve acceptable accuracy.

2 NON-LINEAR MODE SHAPE

The high frequency force balance technique works well when the mode shapes are linear or nearly linear. The generalized force for a translational mode is written as

$$F^*(t) = \int_0^H F(z,t)\phi(z)dz \qquad (1)$$

where $F(z,t)$ is the instantaneous force per unit height at location z and time t and $\phi(z)$ is modal displacement at z. For a linear mode shape, $\phi(z) = z/H$, an exact relationship between the generalized force, F^*, and base overturning moment, BM, is obtained.

$$F^*(t) = \frac{1}{H}\int_0^H F(z,t)zdz = \frac{1}{H}BM(t) \qquad (2)$$

In practice, many buildings respond in a combination of flexure and shear, resulting in nearly linear mode shapes. This is not necessarily true for very slender buildings, where the flexural action may dominate.

Load distributions with height are readily available from the integration of pressure data and instantaneous modal forces can be assembled using equation 1 for all wind direction, using actual mode shapes and the measured wind forces. This represents an accurate description of the modal forces in all situations. On the other hand, the use of equation 2 is an approximation, unless $\phi(z)$ does equal z/H. Errors which result from using the simple relationship of equation 2, depend on the degree of nonlinearity of the actual mode shape and the variation of the instantaneous wind forces over the height of the building. Universal corrections are not possible because of the large variations in the load distributions for different wind angles, due to effects of the topography and surroundings, as well as the spatial and temporal variations in the natural wind.

Figure 1 shows the ratios of the mean generalized forces calculated using pressure integration and the force balance method based on equation 1. In these comparisons, the pressure data have also been integrated to emulate the base moments from a force balance test. With the pressure integration method being a more exact calculation, these ratios can be viewed as the mode shape correction required for the generalized forces determined from the force balance method. Figure 1 shows two cases. Case 1 is for an uncoupled mode with nearly linear mode shape and Case 2 is for a sway mode, lightly coupled with torsion. The generalized forces from the force balance technique are assembled using the measured base moment scaled by the modal contribution from each directional component. No mode shape correction was included. It should be noted that this comparison has included both the effect of mode shape and the variation of loading shape over the height of this building in an urban environment. As mentioned above, no consistent correction factor is expected. Similar results are found for the background dynamic measurements.

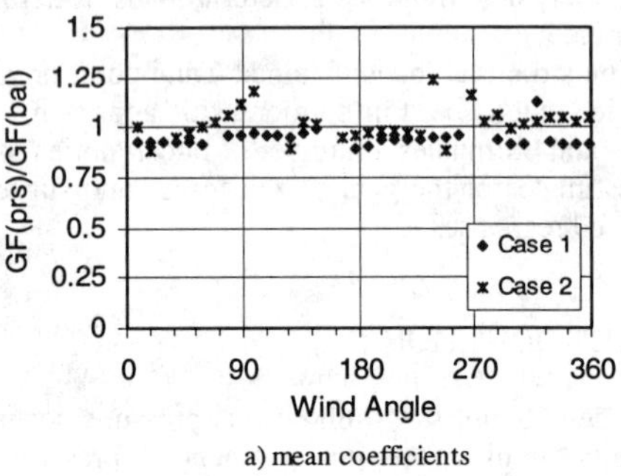

a) mean coefficients

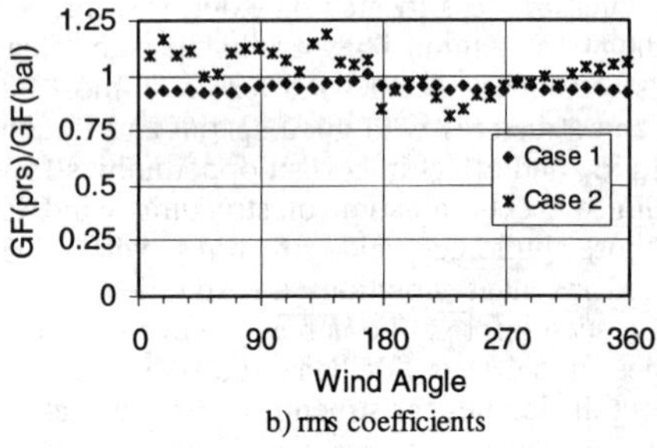

b) rms coefficients

Figure 1 Ratios of Mean Generalized Forces

The resonant component of the building response is determined from the spectra at the reduced frequency, calculated using the natural frequency of the building and the design wind speed. Vickery et. al. (1985) have suggested that the ratio of the spectral coordinates at the design reduced frequency to be used as mode shape correction.

$$\chi(f_o,\phi) = \sqrt{\frac{S_{F^*}(f_o,\phi)}{S_{F^*}(f_o,z/H)}} \qquad (3)$$

This is used to address responses dominated by the resonant dynamic component, such as the accelerations.

In practice, the details of the wind-induced forces, $F(z,t)$, are not available and the mode shape correction factor is pre-determined with an assumed alongwind force spectrum. These mode shapes corrections are applied to the calculated acceleration of the building which is in turn used to determine the inertial loads due to resonance. These are combined with mean and background dynamic base moments, from direct measurements in the wind tunnel tests, to obtain wind-induced peak loads. Furthermore, these mode shape corrections are applied equally to all wind directions, as it was observed by Vickery et. al. (1985) that the calculated corrections, based on the drag direction, appear to be also representative for the cross-wind response.

The availability of the point pressure data allows further investigation into these applications. Figure 2 shows an example of the mode shape correction determined from the ratio of the generalized force spectra from the pressure integration method and the force balance method.

$$\chi(f,\phi) = \sqrt{\frac{S_{F^*}(f,\phi)_{pressure}}{S_{F^*}(f,z/H)_{balance}}} \quad (4)$$

Also included in the figure is the empirical mode shape correction determined from equation 3 using the assumed force spectra. It can be seen that this is representative of the actual correction required. Note that this would have included the difference found in the background dynamic component and only a single point is applicable for each case with a unique building frequency and design wind speed.

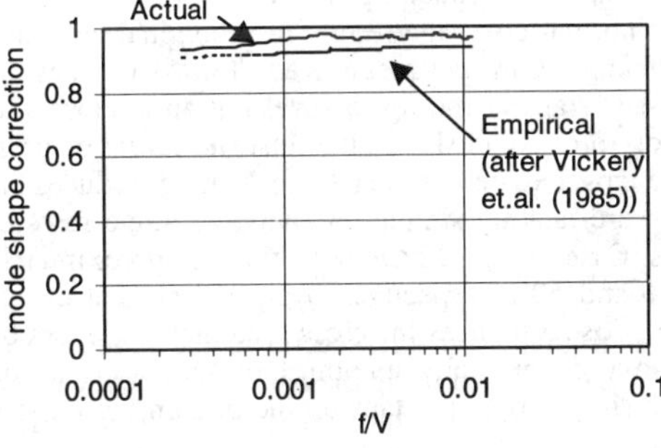

Figure 2 Mode Shape Correction - Frequency Domain

3 OTHER APPROACHES

Other methods of estimating the generalized forces are also available. For example, Xie and Irwin (1997) used the measured overturning moments and shears to fit an assumed force distribution varying linearly with height in the time domain. This distribution is then used to calculate the generalized forces. This can provide better estimate than the analytical method suggested by Vickery et. al. (1985). Yip (1995) has suggested to use the shear, moment and cross spectra to arrive at an estimated forcing function in the frequency domain for the derivation of the spectra of modal forces. The practicality of this method will need to be investigated. Fortunately, all of these difficulties are avoided with the use of pressure point data.

4 GENERALIZED TORQUE

The base torque, *BT*, measured with the force balance provides only an approximate estimate of the generalized torque and substantial adjustment is needed, even when the mode shape of the fundamental torsional mode is linear. By definition the generalized torque is

$$F_\theta^*(t) = \int_0^H T(z,t)\phi(z)dt \quad (5)$$

It is common to use

$$F_\theta^*(t) \sim C\int_0^H T(z,t)dt \approx C \times BT(t) \quad (6)$$

A value of C=0.7 is generally used. This is derived from a consideration of a range of loading and mode shapes. With the pressure test data, the rotational component of the generalized forces can be determined directly without any assumptions. Figure 3 shows the values of C for an actual building. It shows that the values are close to 0.7 in most cases but some are higher for particular wind angles. In this case, partial shielding of the lower part of the building by neighbouring buildings for wind angles of 110° and 210° may have provided larger contribution to the torque from the top part of the building. With an approximately linear mode shape, this would produce a generalized torque larger than the base torque, resulting in a relatively large mode shape correction. On balance, C=0.7 has been shown to provide a good correction to apply to the base torque. Note that $C \approx 0.7$ for wind directions where the wind induced torque is significant, see Figure 3. Nevertheless, there are difficulties which cannot be accounted for by a nominal mode shape correction applied to a base torque. The pressure integration eliminates this uncertainty.

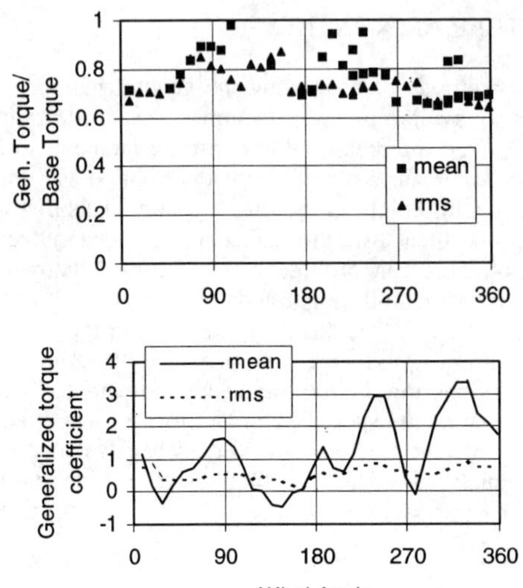

Figure 3 Correction For Generalized Torque

5 COUPLING OF COORDINATES (DEGREES OF FREEDOM)

In force balance tests, the base moments are measured in the two sway directions and in torsion independently. For situations with coupled degrees of freedom, the assembly of the modal forces requires the summation of the directional components.

$$F^*(t) = F_x^*(t) + F_y^*(t) + F_\theta^*(t) \tag{7}$$

where F_x^*, F_y^* and F_θ^* are the directional components of the generalized forces for that mode. The determination of the generalized forces from the directional components in the time domain is straightforward; however, their determination from the statistics of the components requires their correlations to be known. This can be accomplished with additional measurements of the various cross-spectra. However, cross spectra are still not yet routinely measured in force balance tests. A major computational effort is needed if mode shape corrections are to be included. For structures with strong coupling and highly correlated aerodynamic behaviour, this can be important.

In using pressure integration, the assembly of the generalized forces is carried out instantaneously and the correlation among components are inherently included. This is a much more robust approach and has attractive consequences on the overall reliability of the estimates.

6 RESOLUTION

The measurement of pressures at discrete points represents an approximation in the determination of wind loads. For a simple geometry, the layout of the pressure taps can be arranged to capture the loading pattern reasonably well. This is aided by the relatively gradual variation in the force distribution with height. For more complex geometry with many cutbacks, a large number of pressure taps may be required for a complete coverage. This is seldom a problem for state-of-the-art pressure scanning systems, capable of simultaneous measurements at some 512 to 1024 locations. Nevertheless, there is a practical limit for designs with highly complex architectural details where dense pressure tapping can become impractical. In these cases, the force balance technique may still offer a better way to measure the overall modal forces.

For most buildings, integration of forces from pressure data is an effective way to obtain the overall load distribution. "Rings" of pressure taps can usually be used effectively to describe loads at various heights along the building. With high frequency wind force fluctuation typically associated with wave lengths of the order of 40 to 60 m, the normal resolution of pressure taps spaced about 10 storeys apart or less is sufficient for obtaining the overall structural loads. The variation of pressures across the width of the building is typically higher and pressure taps are usually spaced more densely along particular rings.

Results from a typical building have been analyzed with different tap densities with the purpose of assessing a minimum spatial tap resolution requirement for a relatively simple building. Preliminary results indicate that typical pressure tap layouts used for local pressure tests also provide good structural load information. Figures 4 and 5 show the generalized forces from a typical pressure tap layout and three cases of reduced pressure taps on a building. The normal coverage results in a nominal coverage of about 240 m^2 per pressure tap. Case 1 reduces the pressure taps at the lower level but an attempt was made to consciously include just the levels of pressure taps that are required. Case 2 further reduces the pressure taps by simply removing alternate rows of taps. Cases 1 and 2 have reduction of approximately 40% and 60% respectively from the normal coverage. Case 3 reduces the cross sectional tap layout by removing nominally one-third of the locations on each ring of pressure taps on the building cross section.

The sampling requirement for the pressure tests should be viewed through the context that the data will be used both for the local pressure loads and structural loads determinations. For local pressures, a full scale sampling period of about an hour, with a cutoff frequency at an equivalent full scale frequency

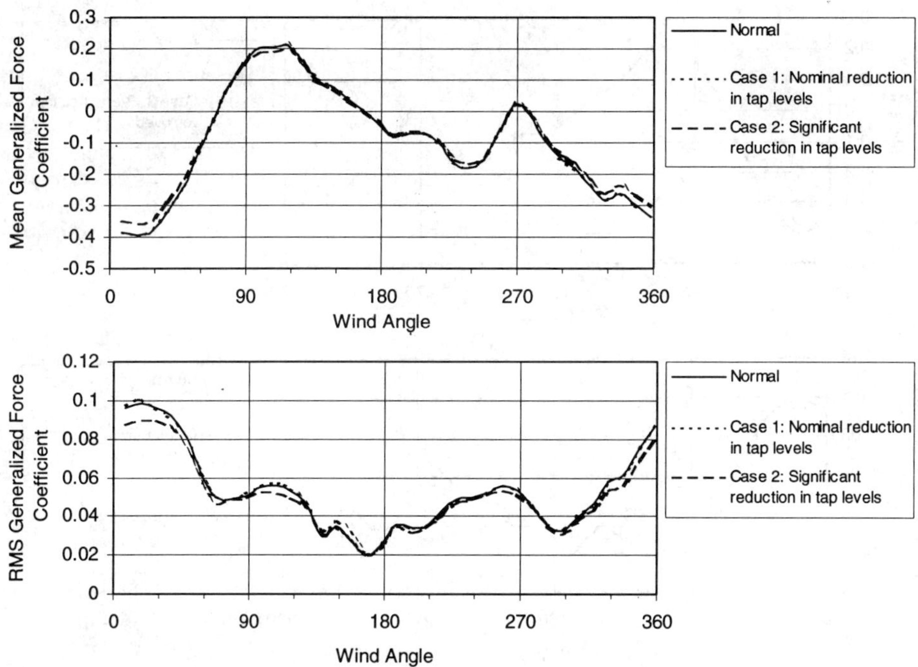

Figure 4 Effect of Reduced Tap Levels

of about 2 to 3 Hz will suffice. For structural loads, it is necessary to measure the spectral densities. The record length has to allow for good frequency resolution near the natural frequency of the building, and for an acceptable coefficient of variation of the spectral estimates. The number of spectral blocks must be 50 to 100 for coefficients of variation of 0.14 and 0.1 respectively. The upper limit frequency requirement for structural loads is usually lower than for local pressure requirements, with fundamental natural frequencies of buildings usually falling between 0.15 to 0.4 Hz. For a typical building tested at a length scale of 1:400 and wind tunnel free stream velocities of about 15 m/s, the sampling requirement will be in the range of 400 to 500 Hz for 1 to 2 minutes.

7 CONCLUSIONS

The pressure integration method has been shown to be an effective method for determining structural loads and responses. Its capability of providing the details of wind loads with height eliminates some of the uncertainties associated with estimates of generalized forces from base moments and base torque, used by the force balance technique, and allows better estimates of the generalized forces. The direct assembly of instantaneously acting loads on different parts of the structure at any instant of time without the complicating requirement of knowing their correlation is a major simplification in the analysis effort, and one which improves the reliability of the results. The availability of the details of wind loads also allows further investigations of the action of wind on particular structural elements which experience the combined action of wind forces in the sway and torsional directions.

The requirement on pressure tap coverage is shown to be well within the current practice. Sampling requirements are also within the capability of current data acquisition systems. Furthermore, standard pressure testing is more robust and is carried out routinely in most test facilities. The faster test wind speeds commonly used in pressure tests, also carry the advantage of avoiding borderline *Re* dependence for relatively small model dimensions and lower wind speed used in the force balance tests.

In cases where timing for design data is an issue, the force balance test is still the experimental method of choice as it can provide a good estimate of the structural loads and responses in a timely manner. When timing is less pressing, a pressure study is capable of providing both local loads for the design of cladding and improved overall wind-induced loads and structural responses for most buildings and structures.

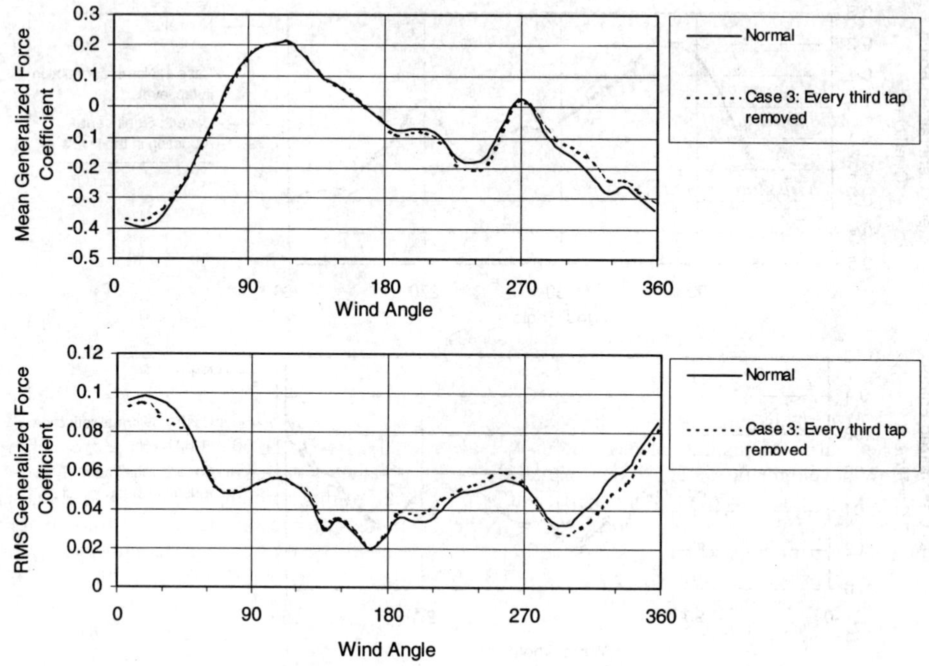

Figure 5 Effect of Reduced Taps Within Tap Levels

REFERENCES

Ho, T.C.E. 1997. The determination of overall structural loads and responses with high-speed scanned wind tunnel pressure data. *Proceedings of the Eighth U.S. National Conference on Wind Engineering*. Baltimore, Maryland, USA. June 5-7. On CD-ROM.

Isyumov, N. 1982. The aeroelastic modelling of tall buildings. *Proceedings of the International Workshop on Wind Tunnel Modelling Criteria and Techniques.* National Bureau of Standards. Gaithersburg, Maryland. April 14-16.

Tschanz, T. 1983. *The Base Balance Measurement Technique and Applications To Dynamic Wind Loading of Structures.* Ph.D. Thesis. The University of Western Ontario.

Vickery, P.J., A. Steckley, N. Isyumov & B.J. Vickery 1985. The effect of mode shape on the wind-induced response of tall buildings. *Proceedings of the Fifth National Conference on Wind Engineering.* Lubbock, Texas, USA. November 6-8.

Xie, J. & P.A. Irwin 1997. Application of the Force Balance Technique to a Building Complex. *Proceedings of the Eighth U.S. National Conference on Wind Engineering*. Baltimore, Maryland, USA. June 5-7.

Yip, D.Y.N. 1995. *Wind-Induced Dynamic Response of Tall Buildings With Coupled 3D Modes of Vibration.* Ph.D. Thesis. University of Auckland.

Wind Engineering into the 21st Century, Larsen, Larose & Livesey (eds) © 1999 Balkema, Rotterdam, ISBN 90 5809 059 0

Wind tunnel tests of a cable supported roof of a stadium

R.Jóźwiak & J.Kacprzyk
Institute of Aviation, Warsaw, Poland

J.A.Żurański
Building Research Institute, Warsaw, Poland

ABSTRACT: This paper presents the results of wind tunnel tests of Silesian Stadium in Chorzów. Tests were carried out at the Institute of Aviation in Warsaw. Model, made to 1:150 scale, was set up behind a turbulent flow development system. Experimental data were obtained by measuring the values of average and instantaneous fluctuations of pressure coefficient C_p on the roof of this stadium. It has been found that for the same directions of wind maximum local values of the mean pressure coefficient C_p are approximately –2.0. Instantaneous fluctuations of the pressure coefficient may be about 3.5 times higher than the average pressure values. The intensity of the pressure coefficient fluctuations falls mainly in the range from 0.25 to 0.4, except in some points, where the average values of the pressure coefficient are very small. Autocorrelation functions and spectra were also obtained.

1 INRODUCTION

A large cable supported roof is being currently designed for the Silesian Stadium in Chorzów, Upper Silesia, Poland[*)]. The roof, with its total surface area of 47,700 sq.m. shall top up a stadium designed for a 60,000-strong audience. Its external ridge will be of almost circular form, about 350 m in diameter, while the internal ridge will be elliptical. The steel trusses are designed supported by back-stays from 40 towers and tensioned by the inner and outer rig cables.

Model tests were carried out in two wind tunnels of the Institute of Aviation in Warsaw. The test programme used rigid models scaled 1:150, upon which the values of average pressure and pressure fluctuations were measured.

The preliminary tests were made in a wind tunnel with a test section of 1.5 m. in diameter, on two-dimensional model of stadium sections, delimited by boundary boards, with two types of roof shape. The objectives of these tests were: to select the methodology and technique of measurements, to optimise the location of points of measurement over the surface of the roof, to find possible simplified models of the whole stadium, to determine the impact of the inclination angle of front and rear slopes of the roof and the impact of the shape of the roof on local distribution of pressure and loading of a given form of the roof.

Conclusions from preliminary aerodynamic tests provided the basis for selection of target shape of the roof with defined angles of inclination of front and rear roof slopes.

The tests of suitably projected model of the whole stadium (including that scaled 1:150) were carried out in a large wind tunnel with an open test section measuring 5m diameter. These tests were made with simulation of actual conditions occurring upon the area, where the stadium is located.

Experimental data were obtained from measurements of mean and fluctuation pressures on the roof of this stadium.

2 RESEARCH CONDITIONS

The main tests were conducted in the Institute of Aviation in Warsaw, in a wind tunnel with an open test section, 5m diameter and range of airspeed from 1 to 57 m/s.

The tunnel is equipped with a simulating plate, sized 5.25 x 7.55 m. The boundary layer was simulated using various perturbation elements, as spires and components to simulate roughness of ground, in the form of wooden blocks. It was described in Preprints EECWE'94 [6].

The model of the whole stadium was installed on the floor plate, upon a rotated table, 3.3 m in diameter, allowing for control of wind direction. The

[*)] Design Office (Zakład Projektowania i Wdrożeń TB, Katowice) : architecture-Z.Pelczarski, steel structures-J.Mach, RC structures-T.Badora.

model, scaled down to 1/150 of its actual size, was placed behind a turbulent flow development system, with length of approximately 6m. The measurements were made over the whole range ($0^0 \div 360^0$) of wind directions.

The boundary layer, modelled for suburban environment, was characterised by the exponent of velocity profile, $\alpha = 0.19$ according to Polish Standard. The resulting profiles of velocity are shown in Figure 1.

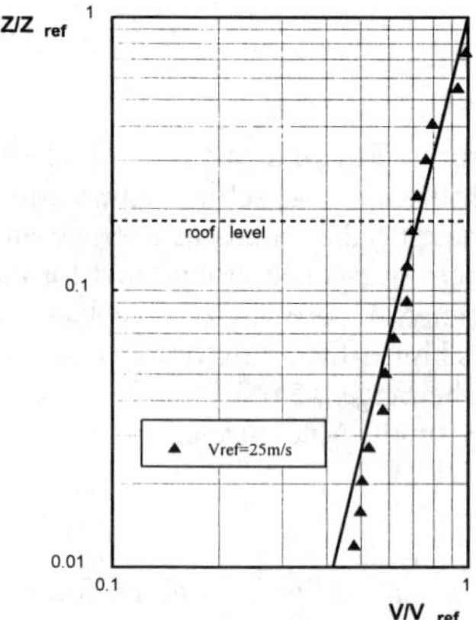

Figure 1. Variation of velocity profile with height

Measurements of the turbulence intensity profile in the boundary layer were made using a DISA 55m, hot wire anemometer. The intensity of turbulence was about 10% at the roof level of the model, corresponding to the height of 40m in the full scale. The turbulence intensity profile of the model of the boundary layer is shown in Figure 2. According to ISO standard, the value of turbulence intensity should be about 11%.

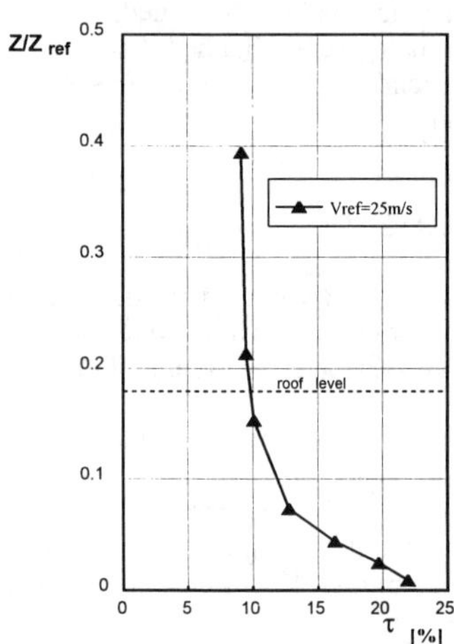

Figure 2. Variation of turbulence intensity profile with height

Dynamic pressure of velocity in a boundary layer and mean values of pressure for the roof surfaces were recorded using a Scanivalve sampling scanner equipped in PDCR22 pressure transducers. Instantaneous fluctuations of pressure coefficient C_p were recorded using Kulite pressure transducers type LQ-125-106.

3 MODEL OF THE STADIUM

The model of the stadium, made to 1:150 scale, was shown in Figures 3, 4 and 5. The whole model of the stadium measures ca 3.0 m in diameter, while the roof diameter is ca 2.33 m (which corresponds to 350 m in the real object). The model of the stadium includes audience tribunes projected in the stadium cavity and additional tribunes with modelled steps and the external stadium embankment and surrounding buildings. Also the roof geometry was modelled in detail, together with the trusses (upon which the roof is supported) and pillars, on which the roof structure with trusses is suspended. Additionally, the internal and external rings were modelled. The stadium is divided into 40 sectors with tribunes for the audience. This division is reflected in construction of the stadium roof, which also consists of 40 segments, corresponding to each individual stadium sector. The roof is formed as a pyramid. Each roof segment consists of 4 or 3 pyramids, depending on the length (chords) of its segments, which vary along the stadium's perimeter, depending on the height of tribunes. The roof segments numbered from 4 to 16 and from 24 to 36 are formed of 4 pyramids, while segments 1-3, 17-23 and 37-40 are formed of 3 pyramids. The principle adopted in roof design is such that all tops of pyramids are on the same height over the whole perimeter of the stadium. The front roof slope is inclined at 5%, and the rear slope – at 6%.

The points on the roof surface, where the average pressure is measured, are located in 11 segments of one, symmetrical half of the roof, composed of 20 segments. As a result, 160 measurement points were located upon one half of the roof surface, with diameter of 0.6 mm. Additionally, measurement of instantaneous pressure values were made in 16 selected points of the roof surface.

A sketch of the stadium's cross-section taken across sectors 10 and 30 was shown in fig. 3. Fig. 4 shows the sketch of the stadium cavity with additional tribunes and the roof. A photograph of the examined model of the stadium was shown in fig. 5.

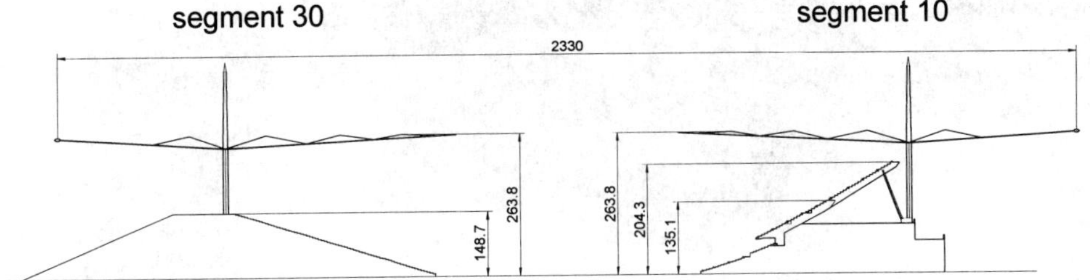

Figure 3. A sketch of the section of the stadium through segments 10 and 30

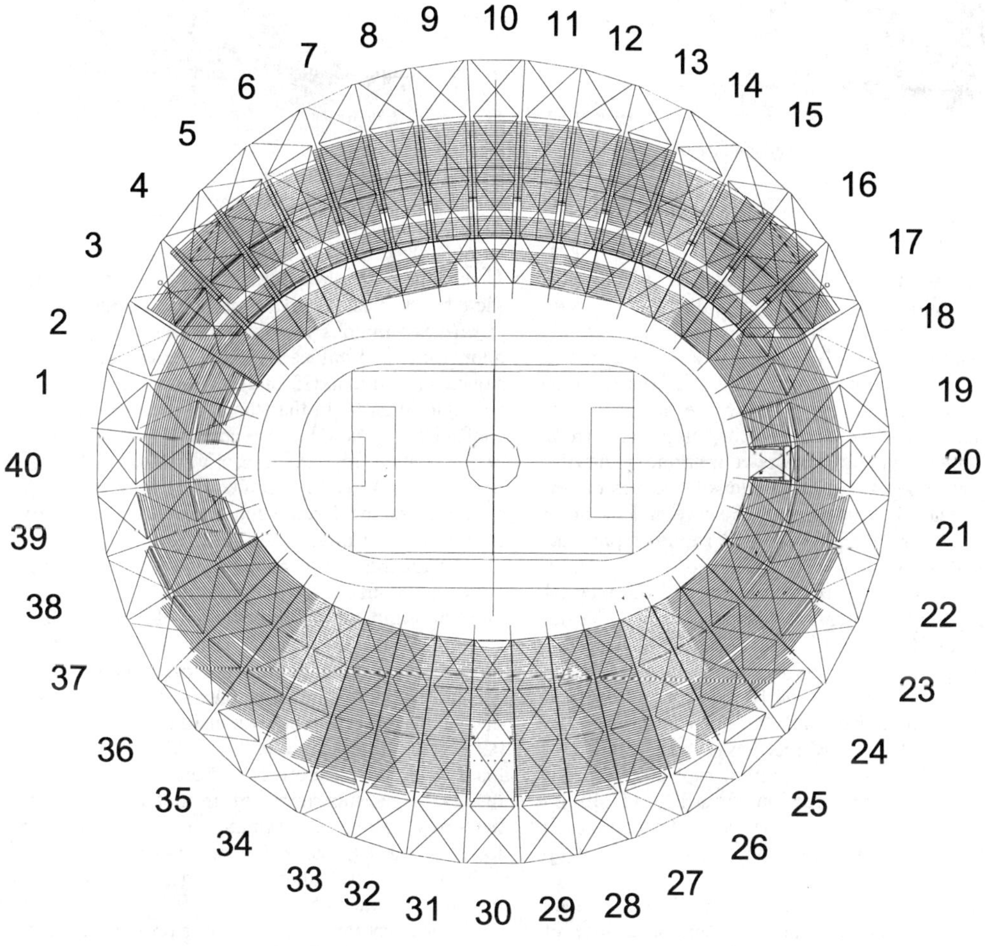

Figure 4. Rostra of the stadium and roof – 40 segments

Figure 5. A photograpn oi the stadium model.

4 RESULTS

Figure 6 presents example results of average pressure values on 11 segments of the stadium roof for the wind directions $\beta=0°$. Average pressure coefficient values $C_p = (p_i - p_o)/q$, where q is a dynamic pressure of wind flow on the height of the stadium roof. These graphs show pressure coefficient C_p distributions along the chord of the segments (roof width) on the upper and lower roof surfaces and distributions of roof load (as a sum of coefficients on the upper and lower surface). Positive and negative C_p values represent positive and negative pressure values, respectively. The presented results show high loads in windward roof segments, on external roof slopes of segments 28-32 and on the internal roof slopes of segments 5-15. The first is caused by the upward flow due to the external grand stand slope, the second due to the flow generated by field of the stadium and rostra. After consideration of the whole range of wind directions we can state that maximum C_p values on external roof slopes occur within range of wind directions from 120^0 to 240^0 on segments 5-15 (in the area, where the tribunes reach their maximum height), which assume levels of up to +1.2.

Analysis of results showed that loads on each individual segment of the roof and on surface of each individual pyramid of these segments change dramatically depending on direction of wind. The distribution of maximum and minimum loads (which represent the totals of pressure coefficients on upper and lower roof surfaces) on each individual pyramid of 11 roof segments, within the full range of wind directions from 0^0 to 360^0 was shown in figure 7. These loads were illustrated on a bar chart, in which each bar represents loads of one pyramid upon a given segment. As is seen from this figure, maximum roof load values were registered on rear roof slope, on segments 13 and 15, for wind blowing at an angle of 210^0. In that case the values of pressure coefficient C_p, which represent unit loads of surface of these pyramids, are respectively $C_{p13} = 1.82$ and $C_{p15} = 1.85$. A negative value of C_p, according to the assumed frame of reference, indicates that the roof loading force is directed upwards.

As regards the front slope of the roof, the highest load values were observed upon the first pyramid, on segment 13, for winds blowing at angle $\beta = 60^0$, where $C_p = -1.05$. Minimum loads are observed on central pyramids of segments 17, 20, 22 and on third pyramid of segment 24. In these cases the values of C_p are positive and oscillate around 0.3. A positive value of C_p means that the roof loading force is directed downwards. Figure 8 shows results of measurements of instantaneous fluctuations of C_p along the segment chord, as an example for segment 10, and wind blowing at angle of 0^0. This figure also shows average, minimum and maximum values of $C_p = C_{pupper} - C_{plower}$. These results show greatest fluctuation of pressure upon the front surfaces of the segment, from the windward side. In that case maximum instantaneous C_p values are approximately 3.5 times in excess of the average value of this coefficient.

On the basis of measurements made over the full range of wind directions it was shown that the maximum instantaneous value of C_p obtained for

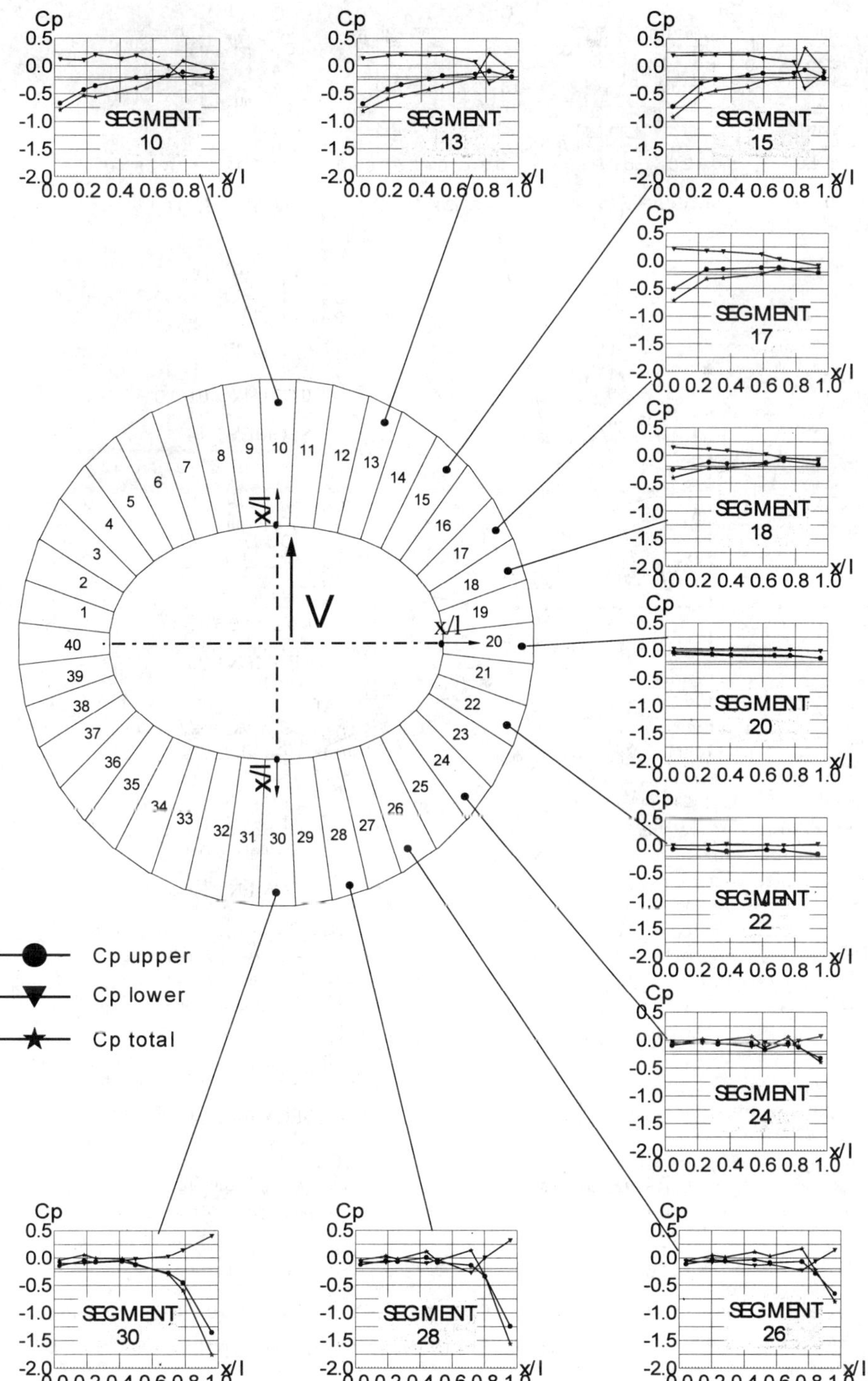

Figure 6. Coefficient pressure distribution along segment chord, $\beta=0°$

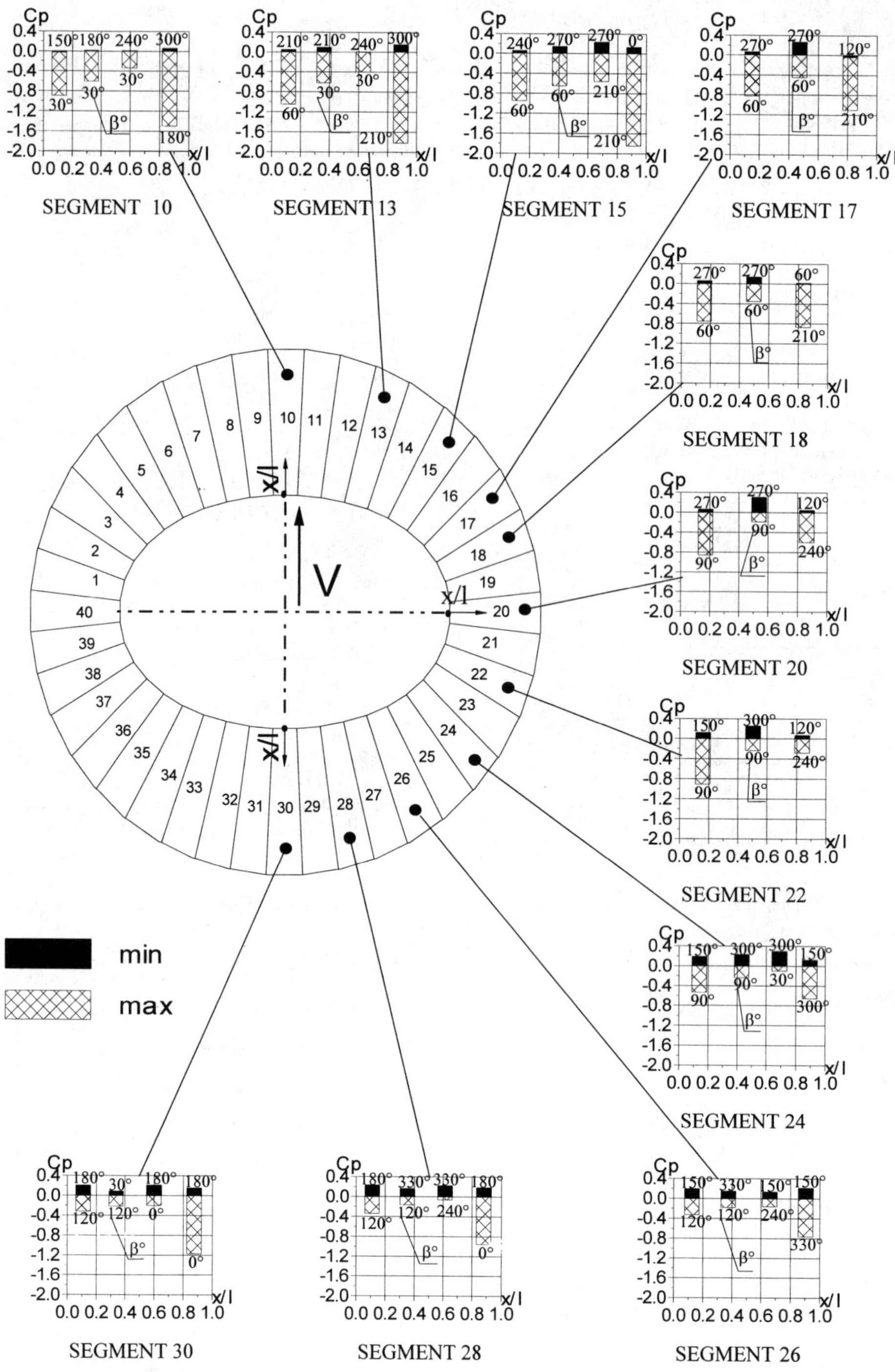

Figure 7. Distribution of maximum and minimum loading on each pyramid in range of directions of wind from 0° to 360°

wind direction of $\beta = 240^0$ upon roof segment no. 15 is –4.4.

The measurements showed that fluctuations of pressure upon the upper roof surfaces are 2-4 times greater that upon lower surfaces. Furthermore, smaller fluctuations of pressure are observed upon that part of the roof, which is shadowed by the opposite part.

The intensity of C_p fluctuation, σ_{Cp}/C_p falls mainly within the range 0.25 to 0.4, with the exception of some points with values of 0.5 to 0.7, and exceptionally 1.0 and 2.6, where average C_p values are very small.

Furthermore, autocorrelation functions and spectra were obtained. Spectrum analysis shows that the main spectrum occurs in general below 25 Hz, and the amplitude of pressure fluctuations reaches its maximum for frequencies ranging from 0 to 10 Hz. In real world this corresponds to frequencies from 0 to 0.8 Hz.

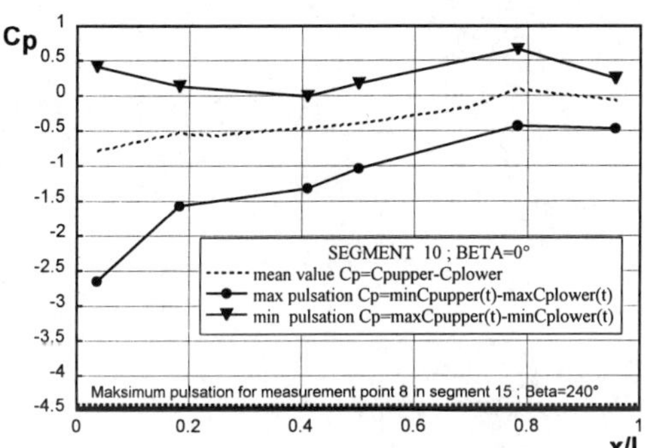

Figure 8. Maksimum and minimum value of sum of pulsation for upper and lower side of roof

5 CONCLUSION

In conclusion, we can state that the complicated structure of the stadium, significantly varying height of tribunes, buildings, changed width of the roof and elliptical shape of the roof's inner edge generate strong asymmetry in the air flow over the roof together with high intensity of turbulence of the air stream, and consequently lead to very strong variation of loads on individual roof slopes.

The greatest loads occur in zones of the roof, from the windward side. These are also the zones, in which the highest pressure fluctuations are observed.

In general, we can state that maximum instantaneous roof loads are much in excess of average values. In some zones of several roof segments, particular directions of wind generate maximum loads even 3.5 times higher than the average values.

REFERENCES

Borri, C., M. Majowiecki, & P. Spinelli, 1992. *Wind response of a large tensile structure: the new roof of the Olympic Stadium in Rome*, Journal of Wind Engineering and Industrial Aerodynamics, 41-44 , 1435-1446.

Holmes, J.D. 1995. *Methods of fluctuating pressure measurements in wind engineering*, Ninth International Conference on Wind Engineering, ed. Wiley.

Höffer, R., J.D. Holmes, M. Kasperski, & A.J. Niemann, 1996. *Wind tunnel tests in building aerodynamics*, Dynamic of Civil Engineering Structures, Ed. W.B. Krätzig, H.J. Niemann.

Jóźwiak, R. 1998. *Wind tunnel tests of roof segments of the Silesian Stadium in Chorzów,* Int. Rep.146/BA/98/D.

Jóźwiak, R., & J. Kacprzyk, 1998. *Wind tunnel investigation of a cable supported stadium roof of the Silesian Stadium in Chorzów,* Int. Rep.147/BA/98/D.

Kacprzyk, J., R. Jóźwiak, J.A. Żurański, 1994. *Simulation of natural wind in a wind tunnel with an open, short working section,* Preprints EECWE'94, part 2

Niemann, H.J. 1987. *Loads and response of a cantilevered roof due to wind turbulence*, Seventh International, Conference on Wind Engineering, Preprints, vol. 3.

Vickery, B.J. & M. Majowiecki 1992. *Wind induced response of a cable supported stadium roof,* Journal of Wind Engineering and Industrial Aerodynamics, 41-44, 1447-1448.

Wind Engineering into the 21st Century, Larsen, Larose & Livesey (eds) © 1999 Balkema, Rotterdam, ISBN 90 5809 059 0

Study on a method to identify external forces and damping parameters of structure from random response values

M. Kanda, E. Maruta & D. Inaba
Department of Architecture and Architectural Engineering, College of Industrial Technology, Nihon University, Narashino, Chiba, Japan

T. Matsuyama
Kumagaigumi Research Institute, Tsukuba, Ibaragi, Japan

ABSTRACT: Proposed in this paper is a method of identifying external forces and damping parameters of structure from random response values. This method estimates external forces acting on structure from response values by means of the inverse procedure of response analysis in the frequency range. An appropriate value of the damping parameter is identified by investigating the spectrum of external force for various damping parameters. This is very effective in cases of involving analyzing aerodynamic vibration tests in wind tunnels and field measurements of real structure vibration. The advantage of this method over other methods is that it can more accurately identify damping parameters even with less data.

1. INTRODUCTION

In order to accumulate basic data to improve structural design technique, many experiments and field measurements to identify damping parameters and external forces of structure are being carried out. Various methods of identifying damping parameters from field measurements were proposed, and got noteworthy results. Additionally apparent damping effects by external force such as aerodynamic damping were investigated by means of wind tunnel tests. Furthermore, acting forces on structures were also estimated from experiments and field measurements. However, these methods are necessary for measuring a large amount of data and for carrying out a large scale analyses when damping parameters are estimated. Moreover, the accuracy of identified values is limited. In particular, there might not be an appropriate method of identifying external wind force acting on structure. These facts prevent improvement in design technique of structure. Thus, the technique can be improved if the properties of the external forces and the damping parameters can be investigated easily.

Based on the above, the purpose of this paper is to propose a new method of identifying external forces and damping parameters from random response values of structure.

2. OUTLINE OF THIS METHOD

The basic concept underlying this method is shown in Figure1. This method estimates damping parameters and external forces by means of the inverse procedure of response analysis in the frequency range. The appropriate damping parameter is estimated from the spectra of external forces by trial and error. The basic principle and procedure are as follows:

If it is assumed that the orthogonality condition applies not only to the mass and stiffness but also the damping, it is convenient to invoke reduce down to a-single-degree-of-freedom. Based on the property, the equation of motion for a-single-degree-of-freedom system is expressed as follows:

$$\ddot{X} + 2\xi\omega_0\dot{X} + \omega_0^{\,2}X = f / M \tag{1}$$

where, M, ω_0 and ξ represent mass, natural circular frequency and damping parameter respectively. X, Xand X represent acceleration,

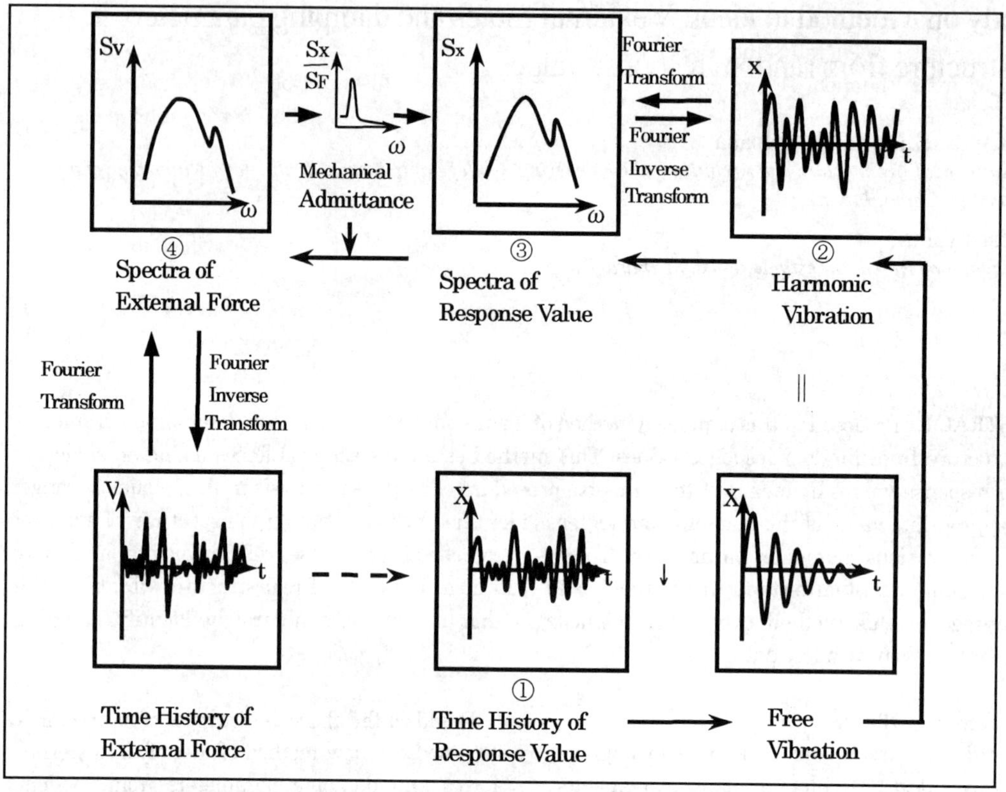

Sv
ω
④
Spectra of External Force

Sx / SF
ω
Mechanical Admittance

Sx
ω
③
Spectra of Response Value

Fourier Transform

Fourier Inverse Transform

x
t
②
Harmonic Vibration

Fourier Transform

Fourier Inverse Transform

V
t
Time History of External Force

X
t
①
Time History of Response Value

X
t
Free Vibration

Figure1 The Concept of This Method

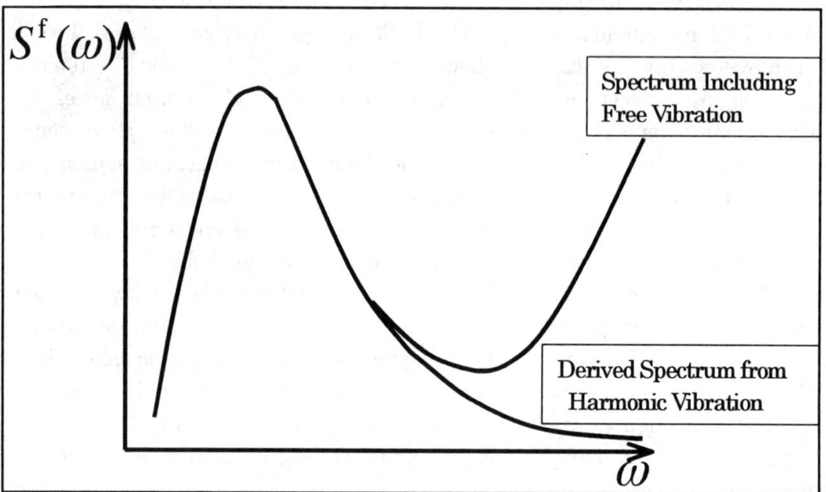

Figure2 The Differences between the Spectra of External Forces

velocity and displacement respectively. f represent external force. Generally, the response values derived from Equation. (1) are expressed as the sum of the harmonic vibration and the free vibration. With regard to the response values, the displacement is expressed as follows:

$$X = X_H + X_F \tag{2}$$

where, X_H represents the harmonic vibrations of displacement, and X_F represents the free vibration of displacement. It is assumed that external force f can be expressed by the summation of many sine waves. The amplitudes and phases of the sine waves of natural circular frequency ω_0 are expressed as α_k and ϕ_k respectively. External force f(t) at time t can be expressed as follows:

$$f(t) = \sum_{k=1}^{n} \alpha_k e^{i(\omega_k t + \varphi_k)} \tag{3}$$

When the force acts on a vibration system of mass M, natural circular ferquency ω_0 and damping parameter ξ respectively, $X_H(t)$ and $X_F(t)$ of the vibration system can be expressed as follows:

$$X_H(t) = \sum_{k=1}^{n} \beta_k e^{i(\omega_k t + \varphi_k)} \tag{4}$$

$$X_F(t) = A_F e^{-\xi\omega_0 t} \cos(\sqrt{1-\xi^2}\,\omega_0 t - \phi_F) \tag{5}$$

where,

$$\alpha_k = -M\omega_0^{\,2} \beta_k \gamma_k \tag{6}$$

$$\varphi_k = \phi_k + \Phi_k \tag{7}$$

$$\gamma_k = \sqrt{(1-(\omega_k/\omega_0)^2)^2 + 4\xi^2(\omega_k/\omega_0)^2} \tag{8}$$

$$\Phi_k = \tan^{-1} 2\xi(\omega_k/\omega_0)/(1-(\omega_k/\omega_0)^2) \tag{9}$$

A_F and ϕ_F are amplitude and phase of free vibration respectively. These are determined by the initial conditions. Real and imaginary number of Fourier spectra are represented by A_k and B_k respectively. Then, β_k and ϕ_k are expressed as follows:

$$\beta_k = \sqrt{A_k^{\,2} + B_k^{\,2}} \tag{10}$$

$$\phi_k = \tan^{-1}(B_k/A_k) \tag{11}$$

Based on the principle of vibration system and equations, the procedure for identifying damping parameters and external forces from displacement response values is as follows:

Ⅰ. Response values are obtained.

Ⅱ. The free vibration values which are generated from the assumed initial values and the damping parameter are taken from the response values to evaluate the harmonic vibration values.

Ⅲ. The harmonic vibration values obtained in Ⅱ are Fourier transformed and the Fourier spectrum of harmonic vibration values is estimated.

Ⅳ. The spectra of external force are derived from the spectra of harmonic vibration by means of the inverse transfer function of a-single-degree-of-freedom vibration system.

Ⅴ. The aptitude of free vibration values is determined from the identified spectra of external forces.

Ⅵ. Processes Ⅱ to Ⅴ are repeated until the appropriate free vibration values are determined. The damping parameter of the determined free vibration values becomes the appropriate damping parameter of the structure.

Ⅶ. The external force on time history is evaluated from the spectra of external force by means of Fourier inverse transformation. Then, if apparent damping effects are included, it is necessary to add the effect to the the value of evaluated external forces.

The concept for determing the suitable value for the damping parameter in Ⅴ is shown in Figure2 imitatively. The ordinary amplitude spectrum of external force doesn't have power in the high frequency range. On the other hand, the amplitude spectrum of external force, in the case when free vibration isn't reduced perfectly, increases as frequency increases. The free vibration value for which the increase in the high frequency range is smallest is considered suitable, and its damping parameter is considered suitable, too.

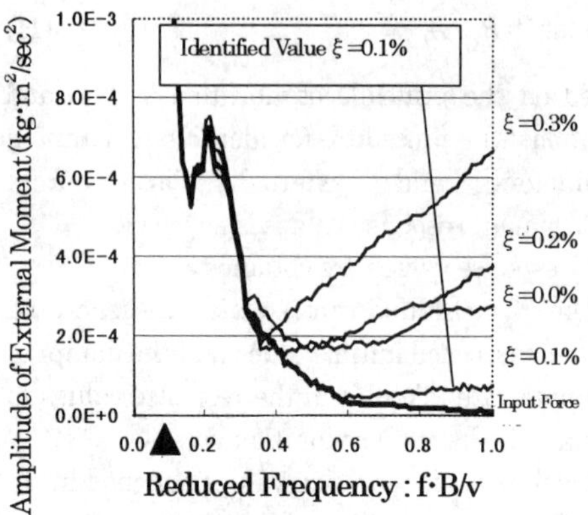

Figure 3-a Spectrum of External Force (: ξ =0.1%)

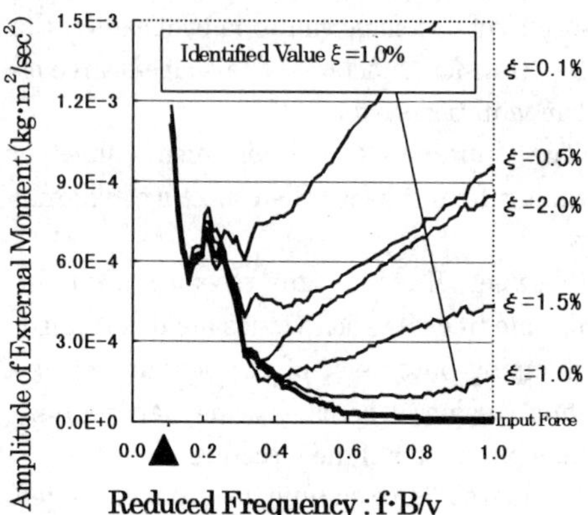

Figure 3-b Spectrum of External Force (: ξ =1.0%)

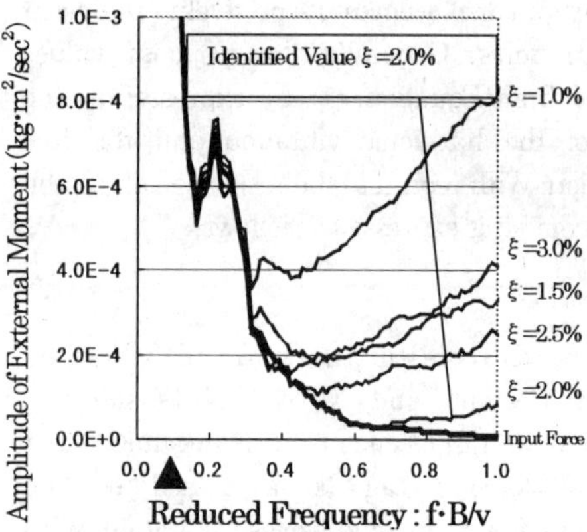

Figure 3-c Spectrum of External Force (: ξ =2.0%)

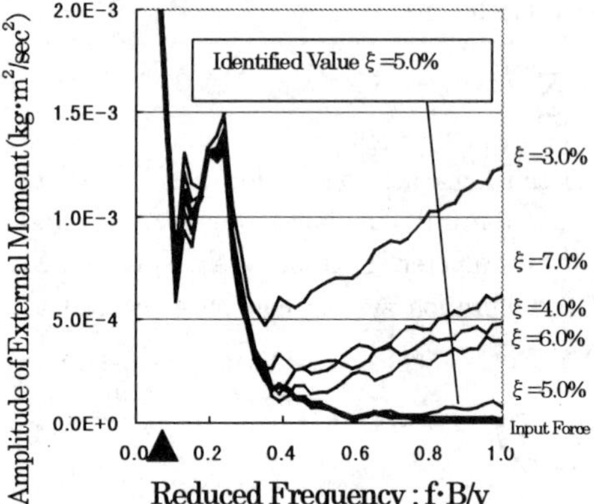

Figure 3-d Spectrum of External Force (: ξ =5.0%)

3.THEVALIDITYOFTHISMETHOD

In order to verify the validity of this method, the damping parameters and external forces are identified from random response values of a-single-degree-of-freedom system by means of computer simulation and a wind tunnel vibration test .

In the computer simulation, the natural circular frequency and the mass of this vibration system are ω_0=66.57rad/sec and M=0.219Kgcm^2 respectively. The damping parameters are ξ =0.1,1.0,2.0 and 5.0%. The number of sampling data is 1024 in the case of ξ =0.1,1.0 and 2.0%, and is 512 in the case of ξ =5.0%. The sampling time interval is 1000 μ sec.

The estimated spectrum and time histories of the external forces in the case of ξ =1.0 and 2.0% are shown in Figure3 and Figure4 respectively. Every graphs in Figure3 indecats that increase of spectra in the high frequency ranges are suppresed as the assumed ξ approaches to input value. From Figure4, the estimated values of external force on time history correspond to the input values of external force very well in the case of ξ =2.0%. However, the estimated values don't correspond to the input values very well in the case of ξ =1.0%. There are some differences between the estimated value

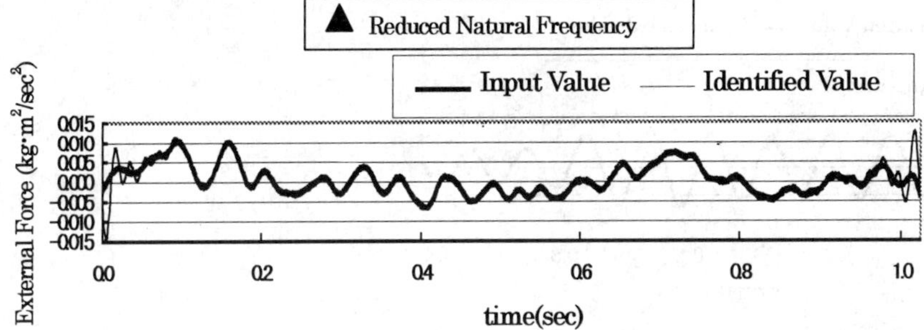

Figure4-a Comparison Identified Value with Input Value (ξ =1.0%)

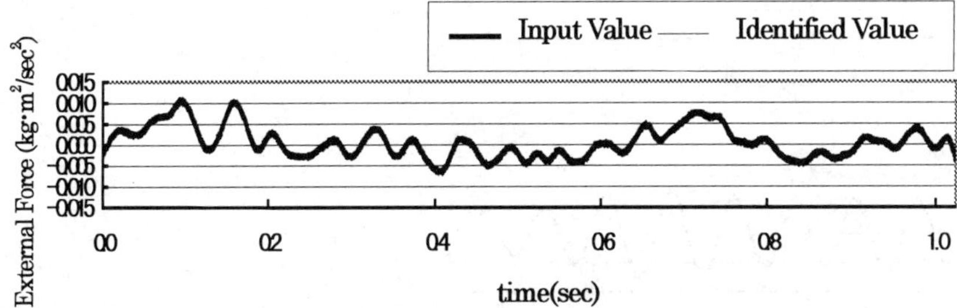

Figure4-b Comparison Identified Value with Input Value (ξ =2.0%)

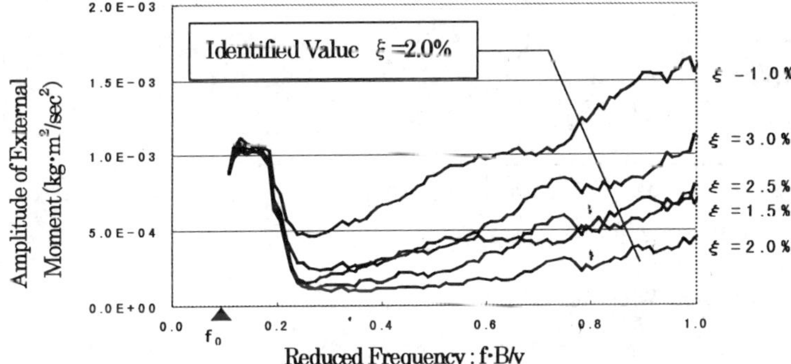

Figure 5 Spectrum of External Force

in the case of ξ =1.0% and 2.0%. The difference is especially remarkable at the beginning and end of the time histories.

The damping parameter and external forces are identified from the response values, which is obtained from the wind tunnel vibration test, in across wind direction. The natural circular frequency, mass, number of sampling and sampling time interval are similar to the previous case. ξ of the system is set at approximately 2%.

The identified spectra from the response values are shown in Figure5. If the assumed ξ approaches the input value, the increase of the spectrum in high frequency ranges is suppressed as frequency increases in the previous case. So, the identified value of the damping parameter is determined to be 2.0%. The time history of total response value, free vibration, harmonic vibration and external wind force when damping parameter and initial parameter are considered

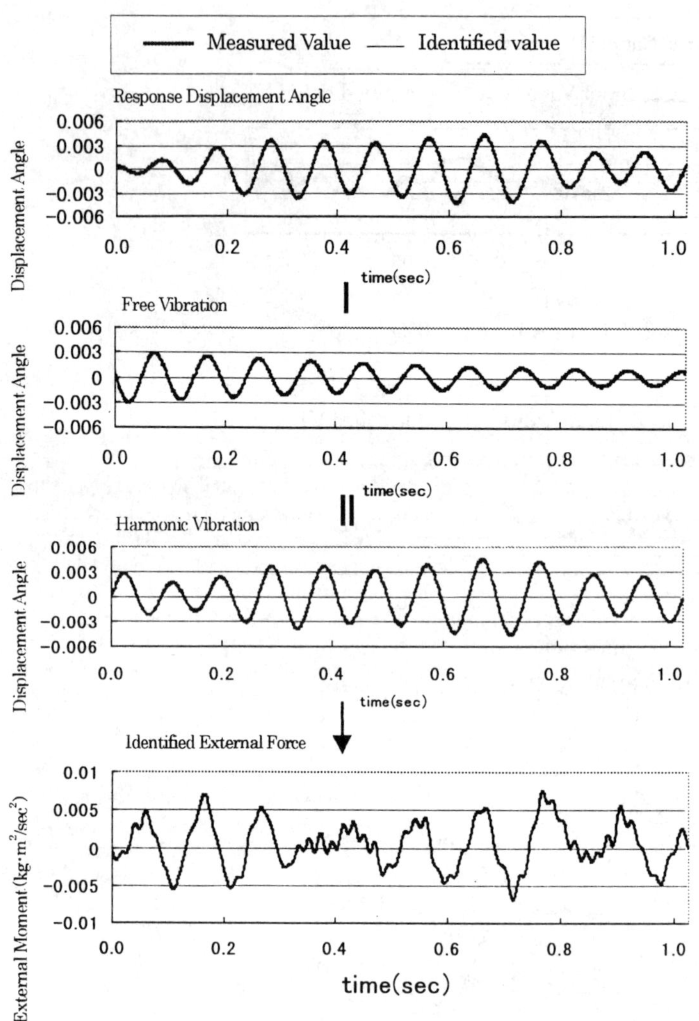

Figure.6 Identification of External Force from Wind Tunnel Test

as suitable is shown in Figure6. Especially, the time history of total response value is compared with the response value obtained from the estimeted external force by means of step-by-step integration analysis on time history. The total response value obtained from wind tunnel vibration tests corresponds very well to the response value obtained from the estimated external force.

4. CONCLUSION

The result obtained though this study can be summarized as follows:

1) A new method of identifying damping parameters and external forces is proposed.
2) The proposed method can identify the values accurately even if there is little data.
3) This method can be widely applied, for example, for analysis of the results of wind tunnel tests and field measurements of structure vibration.

5.REFERENCE

1) Y. TAMURA and ALAN JEARY: Meeting on Structual Damping International Wind Engineering Fourum and Additional Paper. Journal of Wind Engineering and Industrial Aerodynamics, Vol. 59,1996

Wind Engineering into the 21st Century, Larsen, Larose & Livesey (eds) © *1999 Balkema, Rotterdam, ISBN 90 5809 059 0*

Wind tunnel tests on a space grid roof

N. Lakshmanan, S. Selvi Rajan, S. Arunachalam & J. Shanmugasundaram
Structural Engineering Research Centre, Madras, India

ABSTRACT: This paper presents the results of a wind tunnel investigation on a 1:50 scale model of a space grid roof with a central column, conducted under open terrain conditions using the atmospheric boundary layer wind tunnel available at SERC, Madras. Special measuring devices were fabricated for evaluating the overall lift force, and moments about two orthogonal axes and the drag force, at varying angles of attack. The space grid roof had a natural frequency of 32 Hz on the starred cantilever support system. Using the inverse mechanical admittance function, the forces and moments developed on a rigid roof were evaluated. Assuming a linear variation of pressure profile from edge to edge, the maximum and minimum pressure coefficients were derived and compared with corresponding values given in IS:875- Part-3-(1987) on "Wind Loads" for mono-slope and duo-slope roofs. Recommendations are given for pressure and drag coefficients to be considered in analysis and design of similar space grid roofs.

1. INTRODUCTION

Space grids are becoming more and more popular for providing sport halls, stadia, assembly halls, open-air theatres, etc., Modular prefabrication and special jointing techniques have been developed at SERC, Madras for providing roofing over a given large area. Boundary layer wind tunnel studies and full-scale measurement studies for evaluation of pressure distribution and drag coefficients, on a space grid roof or on a similar structure are very scanty in literature. Kim and Mehta conducted full-scale measurements on evaluation of the average uplift force on a flat roof area (Kim & Mehta 1979). The wind induced response of a large cantilevered grandstand was studied in a wind tunnel (Melbourne & Cheung 1988) and it was shown, based on measurement of displacement of the leading edge of the roof, that such structures are highly wind sensitive. Detailed information on pressure distribution over the roof and drag coefficients on a space grid roof covering a large area and supported by widely spaced columns is seldom available for a designer, even in Codes of practice, for example, the Indian standards on wind loads (IS:875, Part 3, 1987, BIS, New Delhi). The present study attempts to evaluate these forces through a wind tunnel test. An experimental space grid roof covering an area of 15 m x 16 m is available at SERC, Madras. It essentially consists of pyramidal space grid modules, connected by a system of bolts and nuts. The prototype space grid has a slope of 1:20 and is made of angle sections of less than 75 mm width. The grid roof is supported on four H-columns made of reinforced concrete which are located perpendicular to the diagonals of the roof. The side elevation of the prototype structure is shown in Figure 1. Corrugated asbestos sheet roofing is provided over the roof.

2. WIND TUNNEL EXPERIMENTAL PROCEDURE

2.1 Model details

A model scale of 1:50 with zero roof slope was selected for the present study consistent with the size of the prototype space grid roof. While it is essential that the flow simulation has to be carried out exactly as per the requirements of an open terrain category with a flow scale of 1:50, it was felt that perfect matching and detailing of all the sections in the prototype is neither practical nor necessary, so far the overall drag forces and uplift forces are concerned. Welded wire mesh of 25 mm x 25 mm made with 2.4 mm diameter mild steel wires was used on the top and bottom layers. The bottom layer has 12 x 13 rods while the top layer 13 x 14 rods. The bottom layer was staggered by 12.5 mm in both the directions with the result that the nodal points on the top plane were exactly over

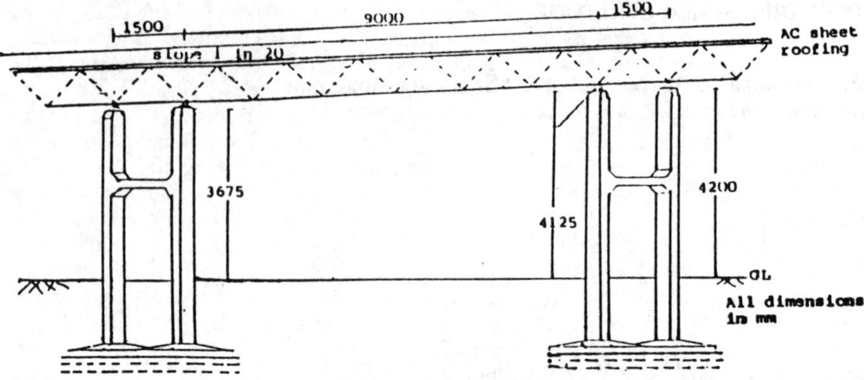

Figure 1. Side elevation of prototype space grid roof

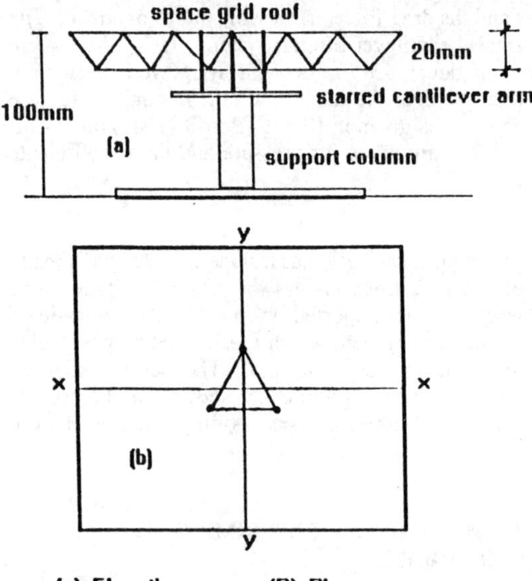

Figure 2. Combined load measurement system

the centre of the squares in the bottom plane. The overall depth, including the sheet on top was 20 mm. An innovative lacing pattern using 1.2 mm mild steel wires between the top and bottom nodes was adopted to complete the model. The model was thoroughly cleaned and electroplated. The total weight of the model space grid roof was 730 gm. The model was supported at a height of 90 mm above the floor of the tunnel as shown in Figure 2.

2.2 *Load measuring devices*

A special load measuring device with a starred cantilever arrangement as shown in Figure 3 was designed and fabricated for evaluating the overall uplift forces and moments about two orthogonal axes. The measurement scheme provided a statically determinate system for force evaluation at individual connecting points. Another load measuring device, as shown in Figure 4, was also specially designed and fabricated for determining the drag force. The system was calibrated using dead weight system by applying loading at centre of the model as also moments about x and y axes separately. Based on the calibration, an output voltage of 1V corresponded to an uplift force of 1.83 kg and a drag force of 1.67 kg (0 degree angle of attack).

2.3 *Test results*

The mean velocity at the level of the roof was 9.75 m/s with a corresponding mean dynamic pressure equal to 5.7 kg/m^2. The measurements were taken at various angles of attack between 0° and 180° in steps of 30°. The natural frequency of the starred cantilever being 32 Hz (with the velocity ratio of unity), the model response corresponds to that of a dynamically sensitive roof with a natural frequency of 0.64 Hz. Figure 5 illustrates spectra of forces on space grid roof for 0° and 90° angles of attack. Similar spectra were also obtained for other angles of attack.

3. ANALYSIS OF TEST DATA AND DISCUSSIONS

It can be clearly seen from Figure 5 that the resonant peak occurs at 32 Hz as expected. From the measured time history records of forces and moments, which corresponded to the flexible model, the respective spectra were deduced using the "global software" and presented in Figure 5. An analytical procedure using the inverse mechanical admittance function was employed to derive the spectral values corresponding to an equivalent rigid model. Such a procedure is analogous to the use of frequency response function or transfer function for applying correction to the tubing response in the conventional study of pressure measurements in wind tunnels (Cook, 1990). The statistics of uplift forces and moments for the case of flexible and rigid roofs for different angles of attack are shown in Tables 1 & 2. These values clearly indicate that the fluctuating components are significantly higher than the respective mean values and this can be attributed to the turbulence in the approach flow and its interaction at leading edges in addition to the dynamic sensitivity of the flexible roof.

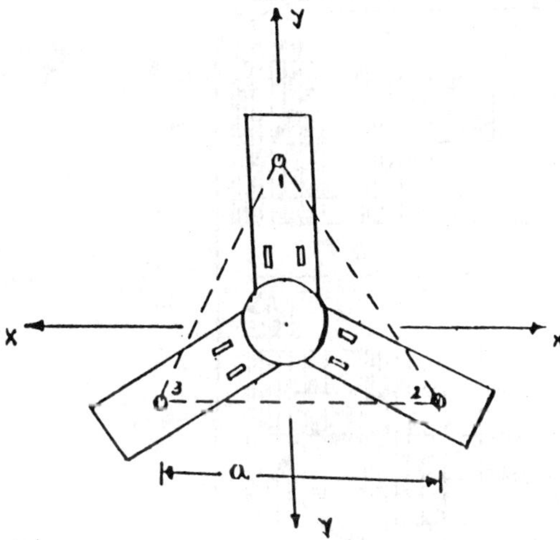

Figure 3. Starred cantilever arm

Based on the results of force coefficients corresponding to the rigid model and with the assumption of quasi-steady aerodynamics, the maximum positive and negative pressure coefficients have been computed as equal to 0.20 and -0.34 respectively. The measured data clearly showed that the pressure loading was not uniform across the roof and hence by assuming a linear variation of pressure, the overall net pressure coefficients duly accounting for the differential bending moment across the roof, were computed. These pressure coefficients have been compared with those given in the Indian Standards on wind loads (IS:875-Part 3-1987) for the case of monopitch and duopitch roofs with zero degree roof angle, and they are presented in Table 3. A good correlation can be seen between the experimental and Codal values. The coefficients for pressures and moments for the flexible case were seen to be relatively more than those for rigid case, as one would expect. For example, the average standard deviation of force coefficients for the flexible roof was seen to be nearly 1. 5 times the corresponding value in the case of rigid roof, based on test data.

$$F = F_1 + F_2 + F_3$$

$$M_x = (2F_1 - F_2 - F_3)(h/3)$$

$$M_y = (F_2 - F_3)\,a$$

$$h = 0.8667\,a$$

$$a = 62.50\ \text{mm}$$

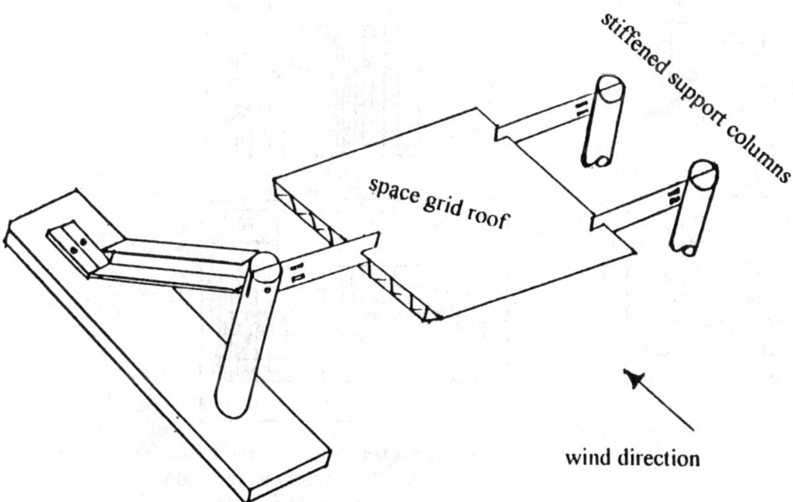

Figure 4. Instrumentation for measurement of drag force

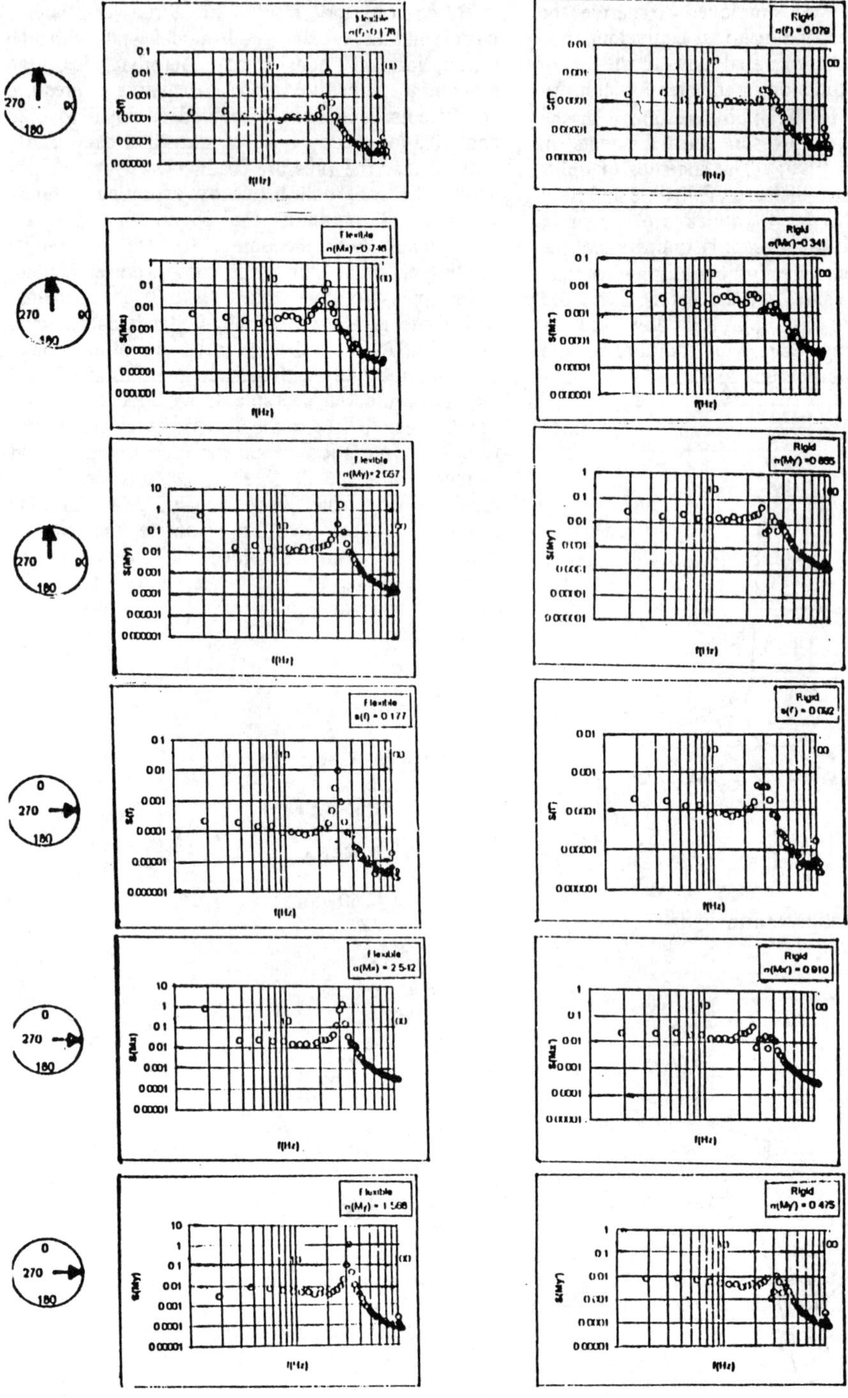

Figure 5. Spectra of forces in space grid roof for 0° and 90° wind angles

Table 1. Statistics of forces on flexible roof.

Force/moment	Wind angle (deg)	Mean	Max.	Min.	Std. dev.
Force (kg)	0	-0.084	0.333	-0.480	0.129
	30	-0.051	0.510	-0.500	0.132
	60	0.160	0.570	-0.270	0.126
	90	0.020	0.430	-0.480	0.129
	120	0.015	0.520	-0.525	0.132
	150	0.010	0.400	-0.360	0.126
	180	0.045	0.390	-0.540	0.120

Force/moment	Wind angle (deg)	Mean	Max.	Min.	Std. dev.
Mx (kg-cm)	0	0.082	3.990	-3.770	1.170
	30	-0.556	5.937	-6.166	1.460
	60	-1.020	4.220	-6.380	1.637
	90	-1.246	5.157	-8.240	1.615
	120	-1.091	4.924	-7.592	1.660
	150	-0.590	3.484	-4.668	1.070
	180	0.007	3.995	-3.923	1.110

Force/moment	Wind angle (deg)	Mean	Max.	Min.	Std. dev.
My (kg-cm)	0	1.070	6.643	-4.340	1.69
	30	1.015	7.045	-4.680	1.60
	60	0.926	5.020	-3.100	1.183
	90	-0.048	3.626	-3.835	1.154
	120	-0.490	4.437	-5.265	1.213
	150	-0.711	4.517	-6.915	1.686
	180	-0.803	5.000	-6.890	1.58

Table 2. Statistics of forces on rigid roof.

Force/moment	Wind angle (deg)	Mean	Max.	Min.	Std. dev.
Force (kg)	0	-0.084	0.257	-0.425	0.105
	30	-0.051	0.300	-0.388	0.090
	60	0.160	0.433	-0.113	0.084
	90	0.020	0.283	-0.280	0.075
	120	0.015	0.310	-0.310	0.078
	150	0.010	0.230	-0.210	0.072
	180	0.045	0.240	-0.271	0.066

Force/moment	Wind angle (deg)	Mean	Max.	Min.	Std. dev.
Mx (kg-cm)	0	0.082	1.344	-1.162	0.378
	30	-0.556	2.735	-3.399	0.740
	60	-1.020	1.970	-4.078	0.934
	90	-1.246	2.362	-5.190	0.910
	120	-1.091	2.438	-4.905	0.974
	150	-0.590	1.873	-3.056	0.647
	180	0.007	1.958	-1.878	0.543

Force/moment	Wind angle (deg)	Mean	Max.	Min.	Std. dev.
My (kg-cm)	0	1.070	3.889	-1.667	0.855
	30	1.015	4.200	-1.867	0.810
	60	0.926	3.213	-1.317	0.660
	90	-0.048	1.464	-1.604	0.475
	120	-0.490	2.016	-2.920	0.617
	150	-0.711	1.636	-3.495	0.757
	180	-0.803	2.170	-3.925	0.810

In addition to overall net pressure coefficients, the Code also recommends local pressure coefficients for mono-slope and duo-slope roof cases. The corresponding largest positive and negative pressure coefficients, based on test data, both for the alongwind and acrosswind directions have been computed. In both cases the edge distance has been assumed to be equal to 0.10 w (or 0.1L) as specified in the Code. Further, as can be seen from Table 1, when maximum moment occurs in one direction, there exists also some amount of moment in the other direction, the effect of which is also included in computing local pressure coefficients. There is a good agreement observed between the Codal and experimental values, in the case of alongwind direction, as shown in Table 3. However in the case of acrosswind direction compared to the test values of + 0.75 and -0.75, distances wherein the local pressures are maximum is taken as 0.05L, then the pressure coefficients will be ± 1.21 which are far closer to the Codal values.

The measured values of drag force are given in Table 4. The maximum value of mean drag force was found to be 0.117 kg along the diagonal direction. The drag force was computed using

Table 3. Comparison of experimental results with values derived using IS:875 (Part 3)-1987

Pressure Coefficient	Expt.	Mono slope Value	Duo pitch Value
Max – ve			
i) overall	-0.970	-1.10	-0.825
ii) local	-1.380	-1.40	-1.500
Max + ve			
i) overall	0.51	0.44	0.525
ii) local	1.02	1.10	1.30

multiple frame analogy with a frame spacing ratio of 0.5 and effective solidity ratio of 0.25. A comparison with the Indian Code, IS: 875, shows that the Codal value was over estimated and it is suggested that a shielding factor of 0.5 seems more reasonable.

Based on the measured data shown Table 1, it can be seen that except for the along wind moment, M_x, the mean values for drag force and across wind bending moments can be reasonably assumed to be equal to zero. With this assumption the variance of the response of the flexible space grid roof can be computed from the variance of the response of a

Table 4. Drag forces and drag moments.

Angle of attack	Forces (kg & kgcm)	Mean	Max.	Min.	Std. Dev.
0°	F_D	-0.050	0.015	-0.130	0.018
	M_x	-	-	-	-
	M_y	-0.18	0.054	-0.468	0.066
30°	F_D	-0.082	0.056	-0.268	0.033
	M_x	-0.148	0.101	-0.663	0.060
	M_y	-0.266	0.175	-1.147	0.102
45°	F_D	-0.094	0.117	-0.260	0.047
	M_x	-0.234	0.295	-0.662	0.120
	M_y	-0.234	0.295	-0.662	0.120
60°	F_D	-0.082	0.056	-0.268	0.033
	M_x	-0.266	0.175	-1.147	0.102
	M_y	-0.148	0.101	-0.663	0.060
90°	F_D	-0.050	0.015	-0.130	0.018
	M_x	-0.180	0.054	-0.468	0.066
	M_y	-	-	-	-

Table 5. Comparison of computed and experimental values for forces

Parameter			Rigid	Flexible
Force (kg)	Expt. (Ave.)	Mean	0	0
		Max.	0.286	0.450
		Min.	-0.286	-0.450
	Computed (Ave.)	Mean	0	0
		Max.	0.280	0.450
		Min	-0.280	-0.450
M_x (kgcm)	Expt. (Ave.)	Mean	-1.04	-1.04
		Max.	-4.240	-7.070
		Min	2.016	4.690
	Computed (Ave.)	Mean	-1.040	-1.040
		Max.	-3.975	-6.970
		Min	2.935	4.890
M_y (kgcm)	Expt. (Ave.)	Mean	0	0
		Max.	-1.570	-3.860
		Min	1.570	3.860
	Computed (Ave.)	Mean	0	0
		Max.	-1.575	-4.025
		Min	1.575	4.025

rigid roof using the following expression:

Variance (Flex-roof)= Variance (rigid) + Variance (resonance).

In the above, the value of peak factor used for force and across wind moment is taken as 3.5, whereas, for along wind moment, the value is taken as 3.25 for σ_{rigid} and 3.75 for $\sigma_{resonance}$. Table 5 shows a very good agreement between measured values of peak forces, and moments and those values computed using the above suggested peak factors.

4. CONCLUSIONS

A detailed wind tunnel experimental programme has been conducted to measure the drag, up lift force and moments developed due to flow of wind over a 1:50 model space grid roof. Based on this study the following conclusions are arrived at:

i) The instrumentation systems developed for determining overall uplift and drag forces have given excellent performance leading to development of simplified methodologies of wind tunnel testing for overall design of structures.

ii) The inverse mechanical admittance function approach used to obtain the spectrum of response for rigid roof from that of a flexible roof has given excellent results.

iii) The statistics of forces and moments of the rigid model derived from the flexible model have been used to deduce the pressure coefficients for the rigid roof. These values are in very good agreement with corresponding values recommended in IS: 875 (Part 3) 1987, for mono, and duo pitch free roofs. The local pressure coefficients derived using alongwind and acrosswind moments also match well with the Codal values. However, in the acrosswind moments, the edge distance over which these high pressures act is estimated to be only 0.05 times the length as against 0.1 times the length as given in the Code.

iv) The variance of response of a flexible roof can be computed from the variance of a rigid roof by assuming suitable values for peak factors as illustrated in this paper.

v) The test results indicate that the drag forces are over estimated when computed using clause: 6.3.3.3. and 6.3.3.4 of IS:875(Part 3)-1987. A shielding factor of 0.50 seems to be more appropriate. Also the top and bottom layers of the grid roof are subjected to different velocities, which is to be suitably accounted for.

5. REFERENCES

Cook, N.J. 1990 The designer's guide to wind loading of building structures, Part 2, static structures, London: Buuterworths.

IS:875(Part 3) 1987 Indian standard Code of practice for design loads (other than earthquakes) for buildings and structures, wind loads, Bureau of Indian Standards, New Delhi.

Kim, S. & K.C.Mehta, 1979, Proc. 5th Intl. Conf. Wind Engineering, Colorado, July 1979: 357-367, New York; Pergoman.

Melbourne, W.H. & J.C.K. Cheung 1988. Reducing the wind loading on large cantilevered roofs, J.Wind Engg. and Ind. Aerodynamics, 28, 401-410.

Wind Engineering into the 21st Century, Larsen, Larose & Livesey (eds) © 1999 Balkema, Rotterdam, ISBN 90 5809 059 0

Resonant and background crosswind response components in turbulent flow

W.H.Melbourne & J.C.K.Cheung
Department of Mechanical Engineering, Monash University, Clayton, Vic., Australia

ABSTRACT: Examples have been given of the relative contributions of the background and resonant response components to the total cross-wind response for square, circular and rectangular shaped buildings. The relative effects of cross-section shape and damping were shown to be very complex. The importance of separating out the resonant component of the cross-wind response to use in the determination of building acceleration for human comfort considerations has been illustrated.

1 INTRODUCTION

The majority of the cross-wind response of tall buildings and structures, under the action of turbulent wind flow, can be identified with a background component containing a broad range of frequencies associated with the flow turbulence spectrum and resonant components at discrete structural frequencies. The latter are normally associated with narrow band energy from either direct turbulence input or from wake excitation. The analytical and experimental identification and quantification of these two components of response is important in the evaluation of acceleration levels, for human occupancy comfort, and fatigue estimates, both of which depend primarily on the resonant components.

For along-wind response it is relatively easy, both analytically and experimentally, to separate the resonant component from the mean and background components using quasi-steady assumptions and the gust factor approach pioneered by Davenport and Vickery. However for the cross-wind response, which tends to be more important for both acceleration and fatigue considerations, it is not so easy to separate out the resonant component analytically, because its magnitude depends much more critically on reduced velocity, the shape of the cross-wind force spectrum (which in turn depends on turbulence and body shape, particularly small changes to edge conditions) and the structural and aerodynamic damping. Similarly the background component of response, apart from its direct dependence on the cross component of turbulence, is also quite dependent on reduced velocity, and other turbulence effects and body shape. Hence the relative contributions of background and resonant components can vary considerably.

For a majority of buildings and structures the effects of fatigue and acceleration due to wind excitation are not very critical to design. However in some cases, and they tend to be for relatively major structures, where the design is being affected by fatigue or acceleration considerations, it is much more important to be able to quantify the resonant component, or components, primarily responsible. This paper will use wind tunnel measurements on model buildings to explore some of the above influences on the relative magnitudes of the resonant and background components of cross-wind response of buildings and structures in turbulent flow, and to discuss the mechanisms involved.

2 MODEL MEASUREMENTS OF REPONSE OF A SQUARE AND A CIRCULAR BUILDING

For most buildings with design significant cross-wind response the operating reduced velocities are relatively high where the cross-wind response tends to be dominated by the first cantilever modes of resonant response. Hence using the wind tunnel model moment measurements of the total response

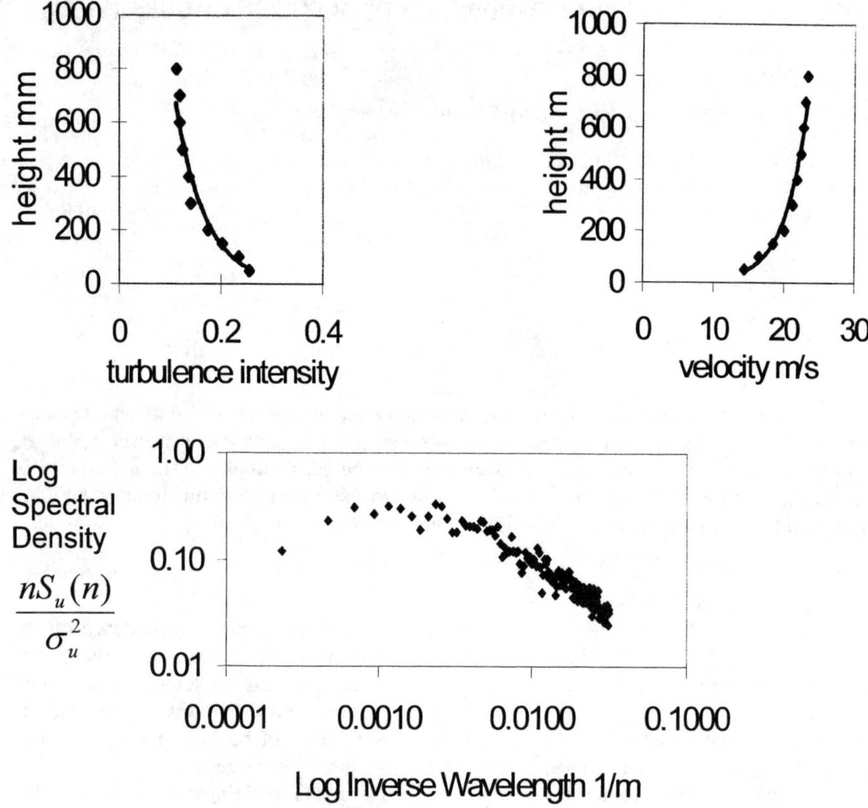

Figure 1. Velocity and turbulence intensity profiles and spectral density of the longitudinal component for the 1/400 scale natural wind boundary layer model in full scale dimensions.

was not seen as overly conservative when predicting acceleration conditions or fatigue loading histories. However as data were needed for higher damping values and structures undergoing aerodynamic interference it was found, when filtering the response signals into a dominant resonant component and background component by high pass and low pass filtering, that no clear picture emerged as to the relationship between the two major components. As it was just as important to know when the resonant component was dominant as well as knowing when some reductions could be expected it was decided to undertake a small study to quantify these effects on a square and circular sectioned tall building. To this end measurements were made for two buildings with the following characteristics:

height, h	=	240m
planform	=	40x40m square and 40m diameter
height/width ratio	=	6:1
1st and 2nd cantilever mode frequencies	=	0.19
building density	=	250 kg/m^3

These buildings were tested at a model scale of 1/400 in a natural wind boundary layer model with the characteristics shown in Figure 1. These were nominally the characteristics of flow over suburban terrain.

2.1 *Data for the square and circular buildings*

Measurements of base bending moment were made using linear mode aeroelastic models, that is, models mounted on a balance which oscillate only in a linear mode about two orthogonal axes at the base. These axes were aligned to measure along-wind and cross-wind components and in the case of the square sectioned building the wind direction was normal to one face. For this study only the cross-wind standard deviation base moment will be used and these have been normalised as a moment coefficient and plotted against reduced velocity defined as follows:

$$C_{\sigma_{M_y}} = \frac{\sigma_{M_y}}{\frac{1}{2}\rho \overline{V}_h b h^2} \quad (1)$$

$$V_r = \frac{\overline{V}_h}{nb} \quad (2)$$

where

σ_{M_y} = cross-wind standard deviation base moment
$\overline{V}_h$ = mean wind speed at top of building height
b = building upstream face width or diameter
h = building height
n = 1st mode frequency of building
ρ = air density.

In Figures 2 to 5 the cross-wind standard deviation base moment coefficients are given for the square and circular tall buildings for critical damping values of $\zeta = 0.01$ and 0.05. These moment coefficients are given as the total standard deviation and then the high pass and low pass filtered components to quantify that part of the total standard deviation response which is attributable to the resonant component and that part which is attributable to background lower frequencies. To visualise these components an example spectrum of the total cross-wind response moment of the square section building is shown in Figure 6 for a critical damping ratio of $\zeta = 0.05$, and reduced velocity $V_r = 5$. The background excitation is due to the lateral component of turbulence and the broad band wake excitation. The high pass filter was set to include all the resonant component, i.e. that component centred around the 1st mode building frequency, and to cut off the frequency contributions below the resonant band. As the operating reduced velocity approaches the peak of the wake (vortex) excitation, ($V_r \approx 10$ and 6 for the square and circular cross-section buildings respectively), the total response increases and that part of the response due to the wake excitation moves away from the direct turbulence excited background response and merges with the resonant component.

From these measurements it can be seen that for the square building, with $\zeta = 0.001$, as the operating reduced velocity increases from about 2 to 10, where the frequency of the peak of the wake (vortex) excitation coincides with the 1st mode building frequency, the contribution of the resonant component to the total response varies from about 80% to 98%. For the circular building with $\zeta = 0.01$, it can be seen that the contribution of the resonant component to the total response is almost constant with reduced velocity at about 90% and only approaching a lower value of about 85% at the lowest reduced velocity measured. The effect of increasing the structural damping to $\zeta = 0.05$ has reduced the relative contribution of the resonant component by about a further 5% in both cases. This small additional reduction of the ratio of resonant component to total response with increased damping was small compared with some earlier indications that the background component might be relatively less affected by increased damping. However these results do not provide any conclusive evidence in this respect because the resonant component here is always a large part of the total response. In some other results to be given later, where the contribution of the resonant component is relatively less, the background component is shown to be significantly less effected by increased damping than the resonant component.

3 ACCELERATION EVALUATION FOR THE SQUARE BUILDING

Acceleration is a function of frequency squared and except for very tall buildings, the majority of the background is at frequencies around a fifth to a tenth that of the 1st mode resonant frequency. Accordingly, with the squared effect it can be seen that the resonant component will tend to dominate the acceleration levels felt by upper level building occupants. An illustration of an acceleration calculation based on the resonant and total components for the square tower for a damping ratio of $\zeta = 0.01$ in suburban terrain is given as follows:

h = 240m, b = 40m, n = 0.19 Hz.

For an annual maximum mean wind speed at 240m of $\overline{V} = 22\ \mathrm{ms}^{-1}$ $\quad V_r = \frac{20}{0.19 \bullet 40} = 2.6$, and from Figure 2 $C_{\sigma_{M_{total}}} = 0.50$ and $C_{\sigma_{M_{resonant}}} = 0.39$.

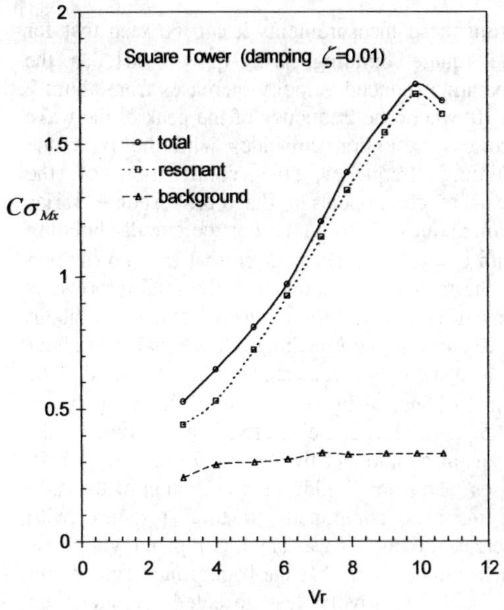

Figure 2. Cross-wind standard deviation base moment as a function of reduced velocity for the square building with damping ratio 0.01, for wind normal to a face.

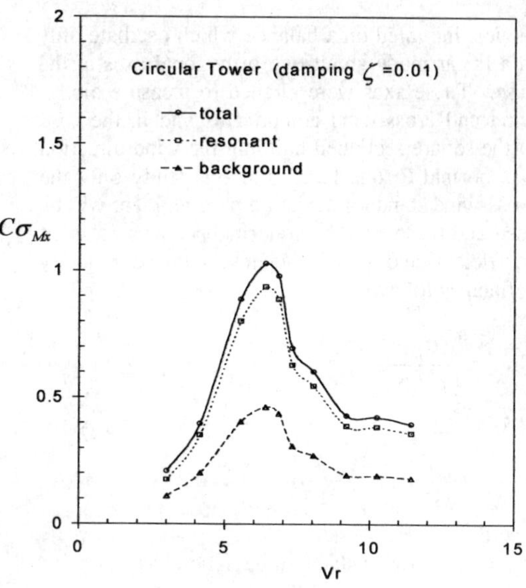

Figure 4. Cross-wind standard deviation base moment as a function of reduced velocity for the circular building with damping ratio 0.01.

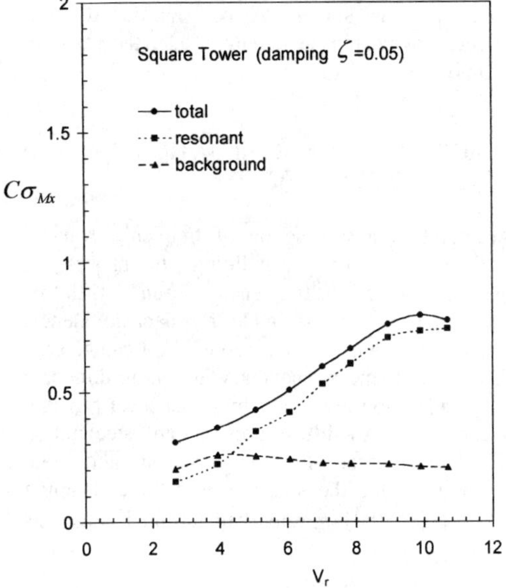

Figure 3. Cross-wind standard deviation base moment as a function of reduced velocity for the square building with damping ratio 0.05, for wind normal to a face.

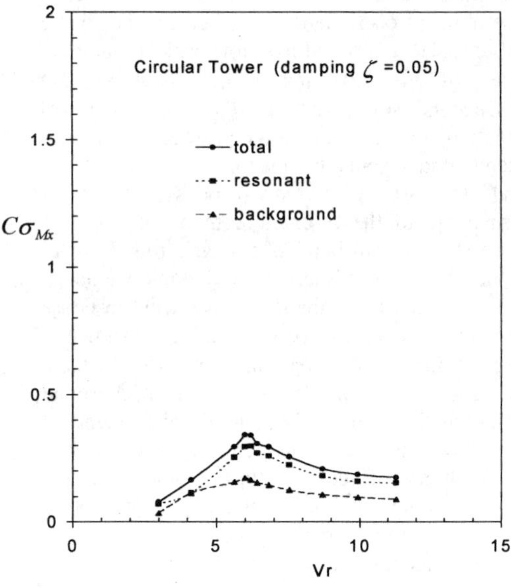

Figure 5. Cross-wind standard deviation base moment as a function of reduced velocity for the circular building with damping ratio 0.05.

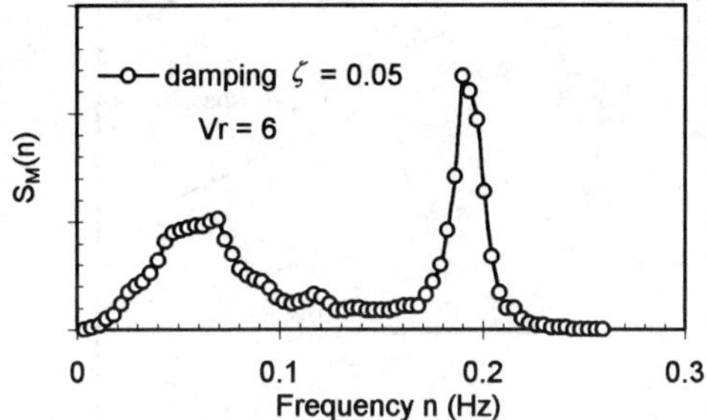

Figure 6. Spectrum of the cross-wind base moment for the square building for damping ratio 0.05, and reduced velocity 6.0.

For a square building with constant density of 250 kgm^{-3} and a linear mode, the inertial base moment for unit displacement (of 1m) at the top is given by

$$
\begin{aligned}
M_{inertia} &= \tfrac{1}{3} \bullet b^2 h \rho_{bld} (2\pi n)^2 h \\
&= \tfrac{1}{3} 40^2 \bullet 240 \bullet 250 (2\pi \bullet 0.19)^2 240 \qquad (3) \\
&= 10.9 \times 10^9 \text{ Nm}
\end{aligned}
$$

Evaluating the peak moment for a 10 minute maximum period from

$$M = gC_{\sigma_M} \tfrac{1}{2} \rho \bar{V}_h^2 bh^2 \quad \text{for} \quad g = \sqrt{2\ell n Tn} \qquad (4)$$

gives

$$\hat{M}_{total} = 1.030 \times 10^9 \text{ Nm}$$

and

$$\hat{M}_{resonant} = 0.803 \times 10^9 \text{ Nm}$$

These peak moments then give peak displacements of

$$\hat{y}_{total} = \frac{1.030}{10.9} = 0.945\text{m}$$

and

$$\hat{y}_{resonant} = \frac{0.803}{10.9} = 0.0737\text{m}$$

and peak accelerations from $\ddot{y} = y(2\pi n)^2$ of

$$
\begin{aligned}
\hat{\ddot{y}}_{total} &= 0.135 \text{ ms}^{-2} \\
&= 13.7 \text{ mill} - g
\end{aligned}
$$

and

$$
\begin{aligned}
\hat{\ddot{y}}_{resonant} &= 0.105 \text{ ms}^{-2} \\
&= 10.7 \text{ mill} - g
\end{aligned}
$$

These accelerations compare with the 1 year return period peak acceleration criterion from Melbourne and Cheung (1988) of

$$
\begin{aligned}
\hat{\ddot{y}} &= \sqrt{2\ell nTn}\,(0.68 + \frac{\ell nR}{5})\exp-(-3.65 - 0.41\,\ell n\, n) \qquad (5) \\
&= 0.108 \text{ ms}^{-2} = 11.2 \text{ milli} - g
\end{aligned}
$$

In this case the difference between using the total moment response versus the resonant component only for the acceleration calculation is one showing an exceedance of the acceleration criterion versus the other showing the criterion is achieved.

It should also be noted that when determining moment and shear force distributions up the building from base moment data, of the type measured here, that the inertial distribution of load should be based on the resonant component of the standard deviation moment. However these calculations are normally based on ultimate limit state design conditions, which are for higher reduced velocities, where the difference between the total moment response and the resonant component is much less significant.

4 MODEL MEASUREMENTS OF RESPONSE FOR A RECTANGULAR BUILDING

It is of interest to carry this discussion to a rectangular building with the wind blowing normal to the smaller face. For this type of planform the cross-wind lift (sideforce) curve slope is relatively higher and if the leading edge flow separations re-attach the wake energy (vortex excited) is less which leads to a situation where the background component increases and the resonant component decreases. An example of this can be given for a 2:1 rectangular building with the following characteristics:

height, h	=	100m
windward face width, b	=	15m
streamwise face width, d	=	30m
1st and 2nd mode frequencies, $n_{1,2}$	=	0.5 Hz
building density, ρ_s	=	270kgm^{-3}
turbulence intensity, $(\frac{\sigma_v}{V})_{100m}$	=	0.20

In Figure 7 the cross-wind standard deviation base moment components for this rectangular building are given as a function of reduced velocity for two damping ratios ζ = 0.01 and 0.03. A response spectrum is also shown inserted. In Figure 8 the

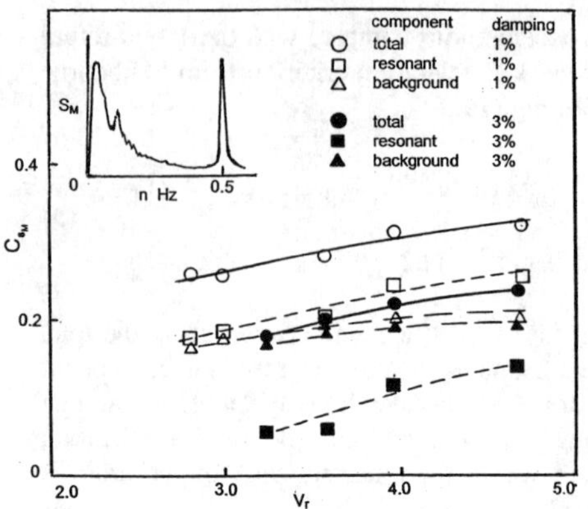

Figure 7. Standard deviation cross-wind base bending moment coefficients as a function of reduced velocity and damping with resonant and background components obtained by high-pass and low-pass filtering, for the 2:1 rectangular building with wind normal to the small face.

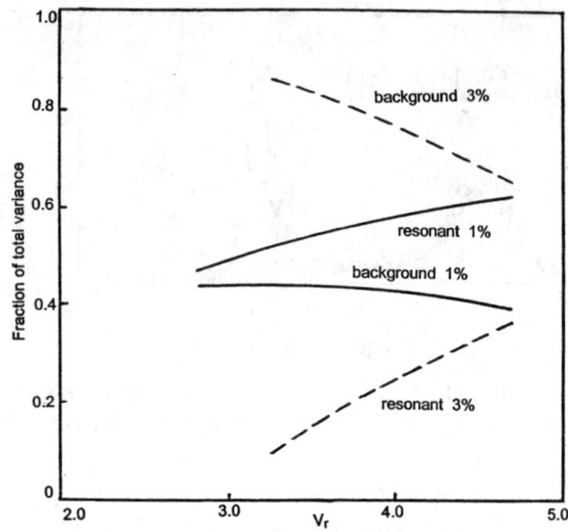

Figure 8. Fraction of total variance of the cross-wind resonant and background components as a function of reduced velocity and damping from Figure 7.

resonant and background components are plotted as a fraction of the standard deviation squared (variance).

It can be seen that for a damping ration of 1%, at the lowest reduced velocities, the resonant and background components contribute almost equally whilst at a damping ratio of 3% the resonant component is significantly damped and the response is dominated by the background component. At the higher reduced velocities shown, (still 50% below the peak of the cross-wind force spectrum), it can be seen that the resonant component (as a square of the standard deviation) is rising to 60% of the total and this percentage would continue to rise as the reduced velocity approached the peak of the cross-wind force spectrum. This example has been included to show just how complex the relative cross-wind exciting and damping processes can be.

5 CONCLUSIONS

Examples have been given of the relative contributions of the resonant and background components to the total cross-wind response for a square and a circular sectioned tall building. For the square building, with damping $\zeta = 0.01$ as the reduced velocity increases from about 2 to 10 the contribution of the resonant component to the total response rose from about 80% to 98%. For the circular building the contribution of the resonant component to the total response was more constant rising only from about 85% to 90%. The effect of increasing damping to $\zeta = 0.05$ was shown to reduce the resonant component contribution by about a further 5% for both sections over the broad range of reduced velocities tested. By contrast the example for a 2:1 rectangular building showed that for damping $\zeta = 0.01$ the contribution of the resonant and background components were almost equal, and for higher damping the background component became dominant.

An example calculation of the top of building acceleration for the square sectioned building illustrated the overestimation which would occur if the total standard deviation moment was used relative to the resonant component which is primarily responsible for the acceleration.

The relative contributions of the background and resonant components and the effect of damping were shown to be very complex.

ACKNOWLEDGEMENTS

The support of the Australian Research Council for this work is gratefully acknowledged as is also the assistance of Mr H Fricke and General Staff at Monash University for their contributions.

REFERENCES

Melbourne WH and Cheung JCK, 1988. Designing for serviceable accelerations in tall buildings, *4th Intl Conf on Tall Buildings*, Hong Kong and Shanghai, pp 148-155.

Wind Engineering into the 21st Century, Larsen, Larose & Livesey (eds) © 1999 Balkema, Rotterdam, ISBN 90 5809 059 0

Full-scale measurements of fluctuating wind-induced internal pressures in a high-rise building

Takeshi Ohkuma
Department of Architecture and Building Engineering, Kanagawa University, Yokohama, Japan

Nobuo Kato, Yoshinori Niihori & Tsuyoshi Kurita
Institute of Technology, Tokyu Construction Company Limited, Sagamihara, Japan

ABSTRACT: This paper investigates the characteristics of wind-induced internal pressure in a high-rise building. Full-scale measurements were taken using absolute pressure meters, which allowed high-resolution measurements. The authors have reported the characteristics of mean wind-induced internal pressures obtained from full-scale measurements in a previous paper (Kato 1997). The present paper describes the characteristics of fluctuating wind-induced internal pressure obtained from full-scale measurements.

1. INTRODUCTION

Wind forces are usually predominant in the design of claddings (curtain walls, glass, etc.) for high-rise buildings, so it is important to develop a wind resistant cladding design. The wind forces acting on claddings are estimated as the differential pressure between the external wind pressure and the internal pressure. The characteristics of the wind forces acting on an actual building have to be measured to accurately estimate the wind forces and to thoroughly understand their real phenomena. However, there are very few examples of full-scale measurements (Fujii 1994), (Kanda 1988) because of the many problems such as cost and client approval. In particular, since the internal pressure is lower than atmospheric pressure, absolute pressure transducers with high resolution must be used for direct measurements of internal pressure. However, most full-scale measurements have been made by means of differential pressure transducers, because no highly reliable absolute pressure transducer is available. Furthermore, due to uncertainties in internal pressure, large ambiguities remain when measuring the aerodynamic coefficient during a wind pressure test.

The authors have carried out full-scale measurements of a 120-meter-high high-rise building to obtain sufficient data for wind resistant design of its claddings. They have reported the characteristics of mean wind-induced internal pressures in previous papers (Kato 1997), (Kato 1997). The present paper describes the fluctuating wind-induced internal pressures measured on 17th Sep. 1995 and 22nd Sep.1996 when typhoons hit the Kanto area.

Photo 1 View of the subject building

Photo 2 Two installed anemometers

2. BUILDING OUTLINE

The Setagaya Business Square, which is the subject of these full-scale measurements, is an office building with 29 stories above ground (see photo 1). The surrounding district has private house of about 2 stories standing close together, and is studded with 4-8 story buildings. The west and east sides of the building are set back at the 11th and 26th levels. The plan shape is 45.9m x 45.9m for the 1st - 10th floors, 39.5m x 45.9m for the 11th - 25th floors, and 33.1m x 45.9m for the 26th -29th floors. The corners of each level are chamfered by 3.2m x 3.2m. The main structural parts are of steel, with an eaves height of 120m. A tower about 30m tall has been erected on the rooftop.

3. MEASURING METHOD

3.1 Wind Direction and Wind Velocity

Ultrasonic anemometers were used for the measurements of wind direction and wind velocity. Two transducers were installed on the tower (about 140m above the ground) to measure the along-wind(u), across-wind(v) and vertical-wind(w) components of wind velocity (see photo 2).

3.2 External Pressure

Fifty-one openings for external pressure measurements were made in the outer wall surface of the building. Internal pressure differences were measured by conducting the pressure acting on the external wall into a differential pressure gauge through an aluminum pipe and a vinyl tube (with an inside diameter of 10mm and a total length of about 3m).

3.3 Internal pressure

Highly precise absolute pressure transducers with a crystal oscillator were used to measure the internal pressure(see photo 3). Five transducers were installed in the building and one was placed outside to measure the atmospheric pressure at each point (see Fig. 1). The measurement ranges of the absolute pressure transducers used for the full-scale measurements were 800-1100 hPa. The resolution was 5/1,000,000 for the range, and measurements could be made in approximately 0.15Pa steps. Absolute pressure transducers were installed in the elevator halls located at the west corners of the 7th floor (25m above ground), 13th floor (49m above ground), 18th floor (69m above ground) and the 24th floor (97m above ground), and in the office on the 13th floor. The office, with a floor area of about 1,075m^2, was not used during the measurements and there were no partitions or desks. The exterior absolute pressure transducer was installed 1.5m above the ground in an instrument shelter about 20m east of the

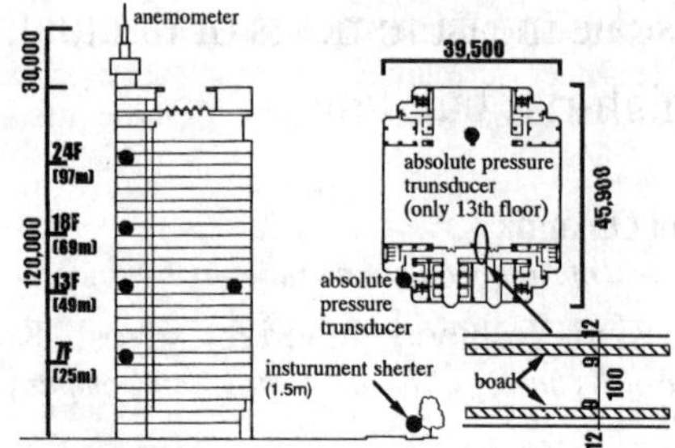

Figure 1 Installation points for absolute pressure transducers

Photo 3 Installed absolute pressure transducer and personal computer

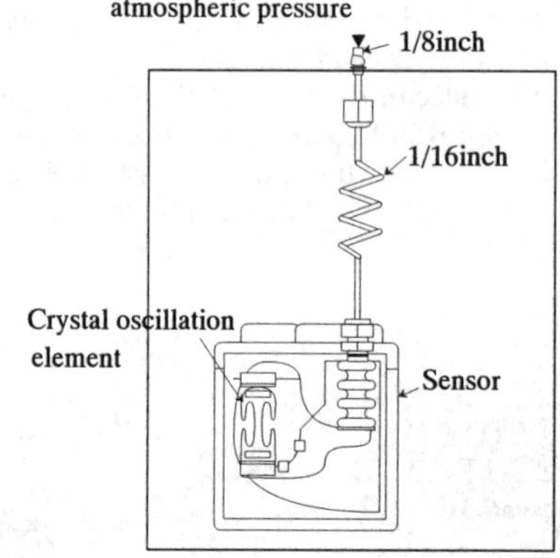

Figure 2 Absolute pressure transducer

Table 1 Specifications of absolute pressure transducer

Sensor	Crystal oscillation element
Range	800hPa～1100hPa
Resolution	0.15Pa
Response time	0.25sec

building subject to full-scale measurements. The instrument shelter, surrounded by 2m-high shrubs and a hedge, was placed where wind seldom occurs(see photo 4).

Photo 4 Installation of meteorological instrument shelter

3.4 *Data collection*

Simultaneous sampling (at 25Hz) was carried out of wind direction, wind velocity, wind pressure and acceleration for a total of 64 channels, using a personal computer. The atmospheric pressure was recorded each second. The mean value, standard deviation, maximum and minimum calculated over each ten-minute period were recorded on an internal hard disk with a personal computer located near each of the absolute pressure transducers (see photo 3).

4 FULL-SCALE MEASUREMENT RESULTS

4.1 Results for September 22 1996 from 12:00 to 18:00

The variation in wind velocity, wind direction, atmospheric pressure and internal pressure of the 13th and 24th floors measured on September 22, 1996 from 12:00 to 18:00 are shown in Figure 3.

Wind direction was approximately 330 degrees at about noon, but gradually changed to 300 degrees by 6 p.m., that is, from north to north-northwest. The maximum mean wind velocity for ten minutes was 29.4m/s between 14:40 and 14:50. The maximum instantaneous wind velocity during that period was 38.2m/s. The atmospheric pressure above the ground surface showed it's lowest value of about 976hPa at about 3 p.m. The internal pressures on the 13th and 24th floors changed to about atmospheric pressure and showed their lowest values of about 973hPa and 966hPa respectively, at about 3 p.m.

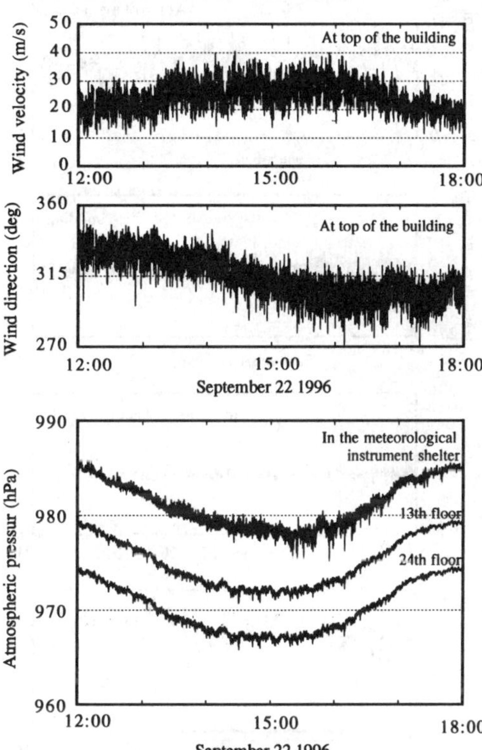

Figure 3 Simultaneous time traces of wind velocity, wind direction and atmospheric pressure

4.2 Noise reduction of atmospheric pressure

In this paper, the internal pressure(Piz) is given by Eq.(1).

$$Piz = Pz - Prefz \qquad \text{Eq.(1)}$$

$$Prefz = 1 / 10^{k} \times P_0 \qquad \text{Eq.(2)}$$

$$k = z / 18400\{(1 + temp / 273)\}$$

$$z = 18400(1 + temp / 273) \times \log_{10}(P_0 / Pzref)$$

barometric height formula

where Pz: the internal atmospheric pressure measured at each point in the building (hPa), Prefz: the outside atmospheric pressure given by Eq.(2) led from barometric height formula (hPa), P_0: atmospheric pressure measured at ground level outside the building (hPa), z: the measuring height of internal pressure (m), t: temperature (℃)

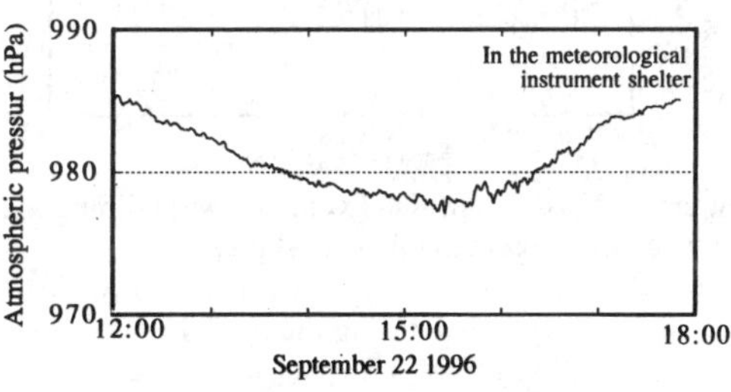

Figure 4 Simultaneous time trace of atmospheric pressure after moving average

Table 2 The values of mean wind direction, mean wind velocity and turbulence intensity for wind velocity obtained every ten minutes from 14:20 to 15:10

time	10 minitue wind velocity				10 minitue wind direction	
	mean (m/s)	gust factor	turbulence intencity	turbulence scale (m)	mean (deg)	rms (deg)
14:20~14:30	28.8	1.54	0.14	208	315	5.0
14:30~14:40	27.3	1.34	0.14	128	314	5.8
14:40~14:50	29.4	1.30	0.13	200	313	5.2
14:50~15:00	25.8	1.43	0.15	191	311	5.3
15:00~15:10	26.2	1.40	0.14	280	308	5.3

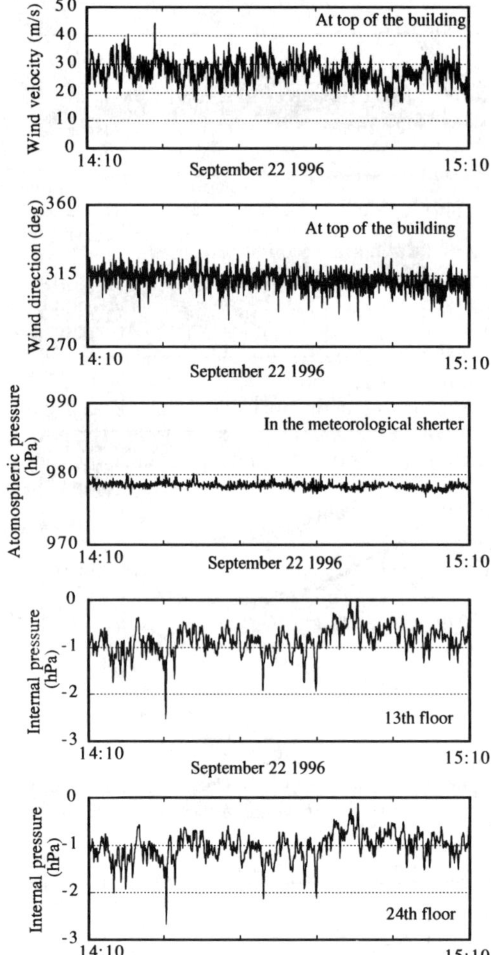

Figure 5 Variation of wind velocity, wind direction , atmospheric pressure and internal pressure

The atmospheric pressure above the ground surface, shown in Figure 3, contains high frequency fluctuations, but it take into account that atmospheric pressure contains high frequency fluctuations. These fluctuations are considered to cause disturbances of wind velocity fluctuation. Thus, in this paper, the atmospheric pressure above the ground surface from which high frequency fluctuations have been smoothed by a moving average is defined as the reference atmospheric pressure above the ground surface. The smoothed atmospheric pressure above the ground surface is shown in Figure 4.

4.3 Results on September 22 1996 from 14:20 to 15:10

The values of mean wind direction, mean wind velocity and turbulence intensity for wind velocity obtained every ten minutes from 14:20 to 15:10 are shown in Table 2. The mean wind velocites were between 25.8m/s and 29.4m/s. The turbulence intensity and gust factor were almost equable, and were about 0.14 and 1.4, respectively. The mean wind direction changed by only 7 degrees from 315 degrees to 308 degrees. The standard deviation of fluctuating wind direction was about 5 to 6 degrees. However, the turbulence scale values were uneven, from 128m ~ 280m.

The variation in wind direction, wind velocity, atmospheric pressure above ground surface and internal pressure on the 13th and 24th floors are shown in Figure 5. The atmospheric pressure above the ground surface was about 978hPa. Internal pressures on the 13th and 24th floors had spike-like fluctuations that frequently occur in external negative pressures. These spikes did not necessarily coincide with the external pressures.

4.4 Variation of the Avaraged time

The variation in the coefficient of fluctuating internal pressure and minimum internal pressure whose averaged time changed from 1 second to 600 seconds are shown in Figure 6. The fluctuating internal pressure coefficient and the minimum internal pressure coefficient were almost equal until the average time

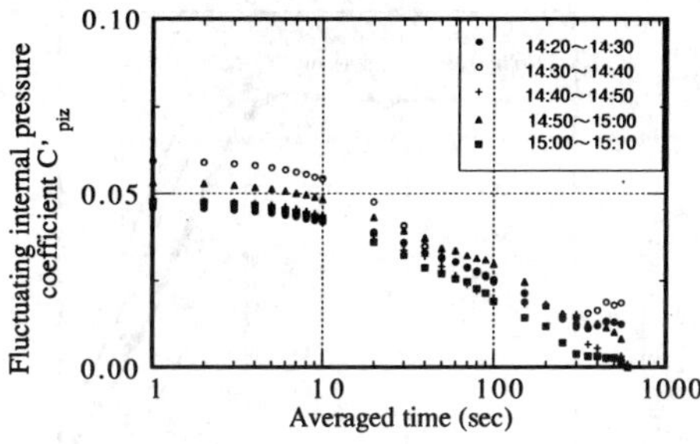

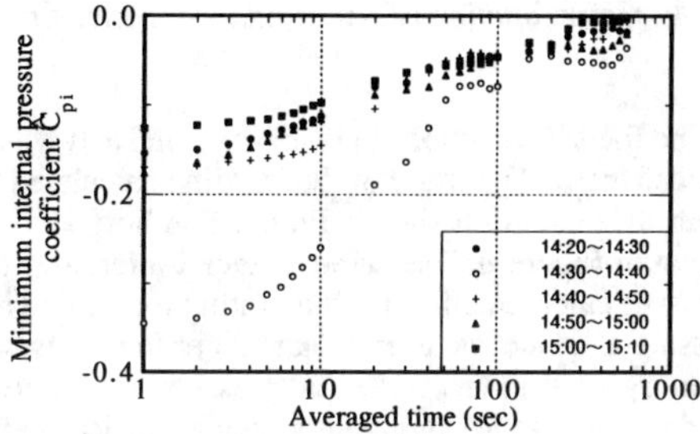

Figure 6 The coefficient of the fluctuating wind-induced internal pressures

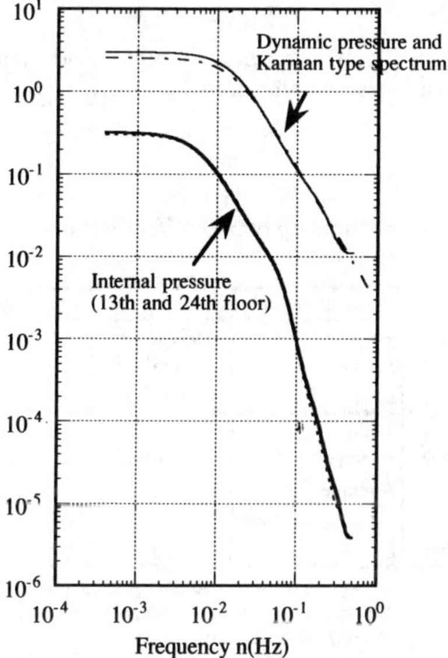

Figure 7-1 Power spectral density of dynamic pressure and wind-induced internal pressures (vertical axis: Sp(n)/qref²)

reached about 10 seconds. It is considered that the power of fluctuation over 10 second-period caused by leakage pass was low.

4.5 Power spectral density

The power spectral density of wind velocity and that of the wind-induced internal pressures for 50 minutes are shown in Figure 7. It also shows Karman type spectrum. The vertical axis in Figure 7 indicats $Sp(n)/qref^2$, $nSp(n)/\sigma^2$, $nSp(n)/qref^2$. Where qref is the dynamic pressure at top of the building.

There was a peak at a lower frequency in the power spectral density of the fluctuating wind-induced internal pressures than in the wind pressures. The power spectral density of wind-induced internal pressures decreased sharply beyond 0.06Hz because of the filter effects of the leakage pass.

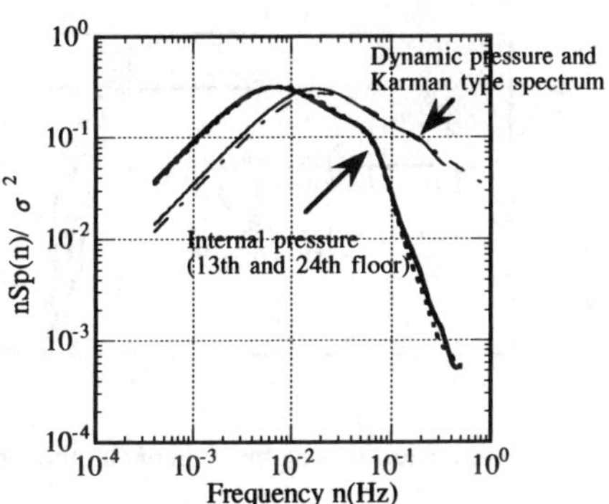

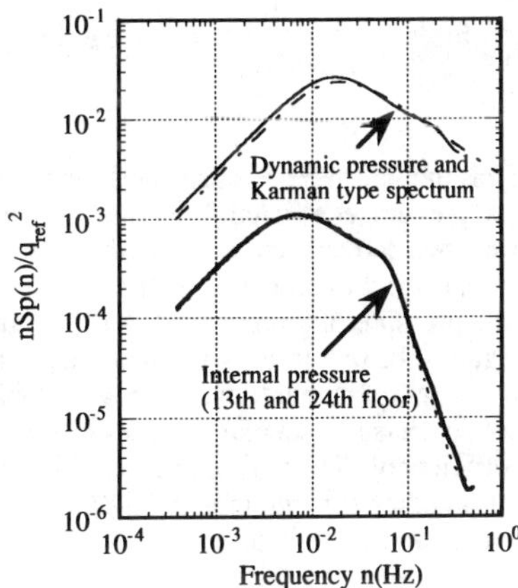

Figure 7-2 Power spectral density of dynamic pressure and wind-induced internal pressures (vertical axis: $nSp(n)/\sigma^2$ and $nSp(n)/qref^2$)

4.6 Transfer function of wind-induced internal pressure

The transfer functions between external and wind-induced internal pressure of the building calculated by Saathoff's equation shown Eq.(3)(Saathoff 1981)is shown in Figure 8. The wind-induced internal pressure was calculated by the fluctuating external wind pressure as a sine wave, and the transfer functions was given by wind-induced internal pressure and external wind pressure ratio. Parameters such as orifice coefficient suggested by Liu et al.(Liu 1990)were used. It was found that the larger the opening ratio of a gap(a/V_0), the higher the frequency at which the transfer functions start to decrease. In the measurement, this is the frequency at which power spectral density starts to decrease steeply when a/V_0 equals approximately 5×10^{-5}. The grade of ventilation of the measured building corresponds to the average by comparing the area of the measured openings in the existing buildings shown in Table 3 and was considered to be a reasonable value.

The fluctuating wind-induced internal pressure obtained from Eq.(3) where there is a small opening might be inaccurate because the viscous effects were larger than the inertia effects. It was, however, considered that the fluctuating wind-induced internal pressure can be obtained from the leakage pass phenomenon by determining the appropriate value of the parameters fixed by the opening ratio of the gap.

The resonant frequency of the fluctuating wind-induced internal pressure was approximately 0.27Hz as obtained from Eq.(5), 5×10^{-5} as an opening ratio of the gap, and the power level at that frequency can be considered negligible by comparing the power spectral density shown in Figure 7.

$$\frac{\rho Le\, V_0}{k\, a\, n\, P_0}\ddot{C}_{pi} + \frac{\rho\, q\, V_0^{\,2}}{2(k\, n\, a\, P_0)^2}\left|\dot{C}_{pi}\right|\dot{C}_{pi} + C_{pi} = C_{pe} \qquad \text{Eq.(3)}$$

where Cpi; the internal pressure coefficients, Cpe; the external pressure coefficients; dot and double dots on represent, respectively, the first and the second derivatives with respect to time t, ρ is the density of air and P_0 is the pressure of ambient air; a is the area of the opening; n is the polytropic exponent (equal to 1.4 for adiabatic air); V_0 is the internal volume of room; q is the dynamic pressure (stagnation pressure); k is the orifice coefficient of the opening and Le is the length of the air plug at the opening(equal to 0.88). Helmholtz oscillator without a neck it is :

$$Le = \sqrt{\frac{\pi A}{4}} \qquad \text{Eq.(4)}$$

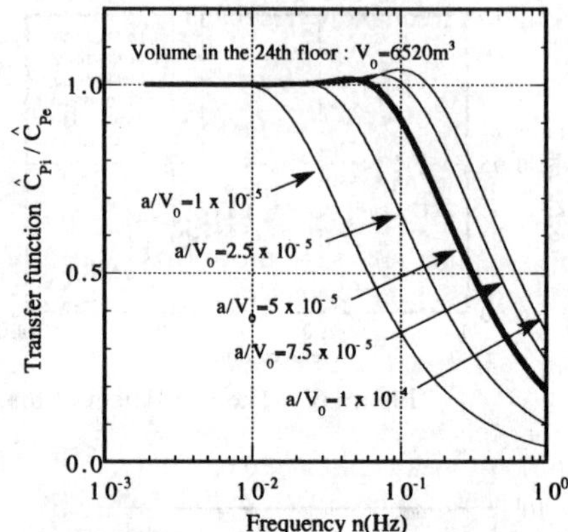

Figure 8 Transfer function between wind-induced internal pressure and external pressure

Table 3 Relationship between Ventilation grade and opening ratio

	Ventilation grade	Opening ratio	Reference
Office	Tight	0.5cm²/m²	Tamura 1976
Office	Average	1.0cm²/m²	
Office	Loose	2.0cm²/m²	
Office	Tight	0.5cm²/m²	Murakami 1983
Office	Average	0.8cm²/m²	
Office	Average ~ loose	1.5cm²/m²	
Office	Loose	2.8cm²/m²	

These values are shown opening area / floor area. So opening ratio (a/V_0) using this investigation are $\Phi_0 = a/(B/4h)$.
where B: width of building(=39.5m), h: height of one floor(h=4m)

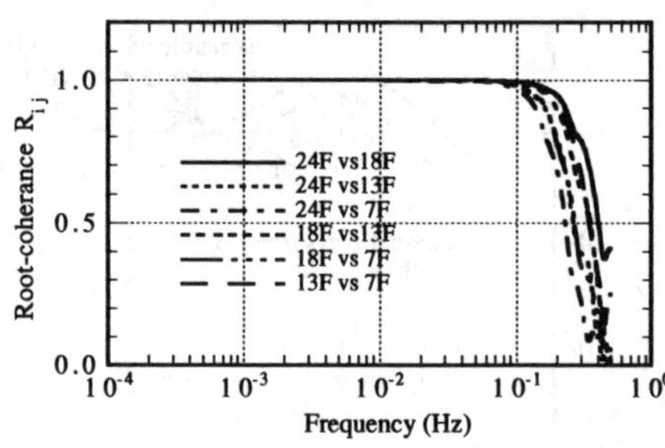

Figure 9 Root-coherence beween each measuring points

The undamped resonant frequency is :

$$n_0 = \frac{\omega_0}{2\pi} = \frac{1}{2\pi}\sqrt{\frac{k\,a\,n\,P_0}{\rho\,Le\,V_0}} \qquad \text{Eq.(5)}$$

This equation is a well-known formula for the Helmholtz resonance frequency.

4.7 Root-coherence between measurement points

The root-coherence (r_{ij}) between the 24th, 18th, 13th and 7th floors measured on September 17 1995 8:00 ~9:20 are shown in Figure 9. The root-coherence values lower than 0.1Hz were 1.0 for perfect correlation, because the fluctuating wind-induced internal pressure in the building occurred at the same time. The fluctuating wind-induced internal pressure can simulate Eq.(3), based on the assumption of the model of one node in the building.

4.8 Values of fluctuating and minimum wind-induced internal pressures

The variation of fluctuating wind-induced internal pressure coefficients (C'pi) given by Eq.(6.2) for changing wind velocity are shown in Figure 10. The values of C'pi were approximately 0.05 higher than the mean wind velocity of 25m/s.

The variation of mean wind-induced internal pressure coefficients (Cpi) given by Eq.(6.1) and minimum wind- induced internal pressure coefficients ($\hat{C}_{pi}$) given by Eq.(6.3) for changing dynamic pressure at top of the building are shown in Figure 11. The values of Cpi were approximately -0.21. These values agreed with a value of correlation coefficient between mean dynamic pressure at top of the building and internal pressure shown in Figure 12. The minimum wind-induced internal pressure coefficients $\hat{C}_{pi}$ higher than the mean wind velocity of 25m/s were approximately -0.4.

$$C_{pi} = \frac{P_i}{q_{ref}} \qquad \text{(6.1) mean wind-induced internal pressure coefficient}$$

$$C'_{pi} = \frac{\sigma_{pi}}{q_{ref}} \qquad \text{(6.2) fluctuating wind-induced internal pressure coefficient}$$

$$\hat{C}_{pi} = \frac{\hat{P}_i}{q_{ref}} \qquad \text{(6.3) minimum wind-indued internal pressure coefficent}$$

where P_i is the mean of internal pressure, σ_{pi} is the rms of internal pressure, $\hat{P}_i$ is the minimum of internal pressure, q_{ref} is the dynamic pressure at top of the building

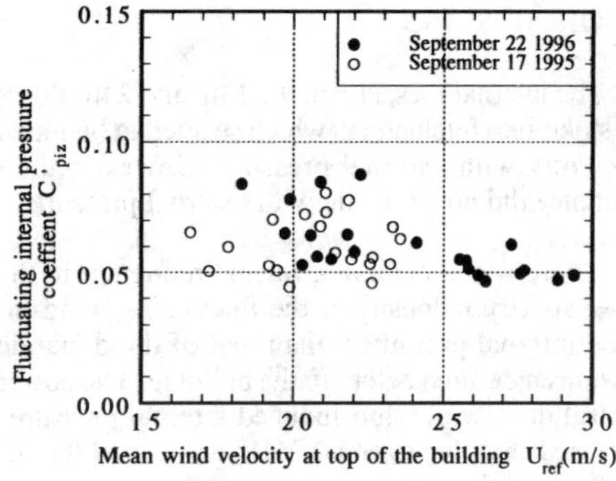

Figure 10 The variation of fluctuating wind- induced internal pressure coefficients for changing wind velocity

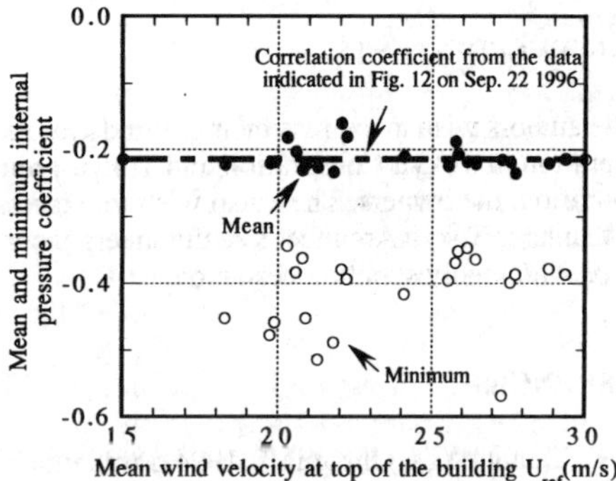

Figure 11 The variation of mean and minimum wind-induced internal pressure coefficients for changing wind velocity

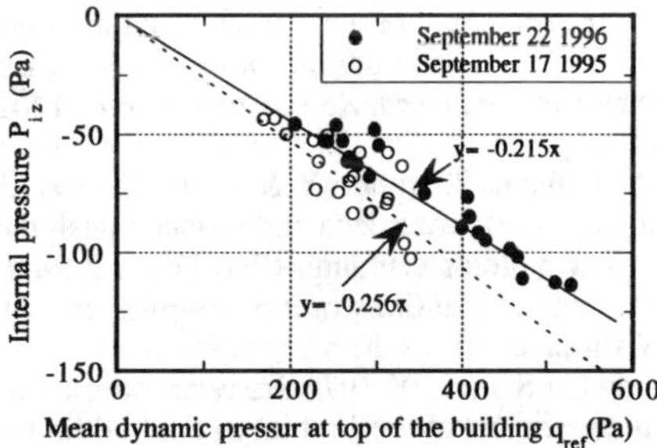

Figure 12 Correlation between mean dynamic pressure at top of the building and internal pressure corrected for barometric heigt pressure

5. CONCLUSIONS

1. The internal pressure on the 13th and 24th floors had spike-like fluctuations which seemed to be found frequently with external pressure. However, their occurence did not coincide with external pressure.

2. There was a peak at a lower frequency in the power spectrum density of the fluctuating wind-induced internal pressures than that of the dynamic pressures measured at top of the building. The power spectral density of wind-induced internal pressures decreased sharply beyond 0.06Hz because of the filter effects of the leakage pass.

3. The coefficients of fluctuating wind-induced internal pressures were about 0.05 and their peak coefficients were about -0.4. These values were smaller than the A.I.J recommended ones(A.I.J. 1993).

ACKNOWLEDGMENTS

The authors wish to express their gratitude for the cooperation of Tokyu Corporation and Tokyu Land Corporation, the owners. They also wish to express their thanks to Tokyu Architect's & Engineers Incorporated who kindly supplied design data.

REFERENCES

Fujii K. Tamura Y. & Ohkuma T. 1994. Problems in the structural engineering division found through full-scale measurements. *Journal of Wind Engineering of Japan.* 59: 61-66. (in Japanese)

Kanda J. & Ohkuma T. 1988. Recent developments in full-scale wind pressure measurements in Japan. *Journal of Wind Engineering of Japan.* 37: 163-172.

Kato N. Niihori Y. Kurita T. & Ohkuma T. 1997. Full-scale measurement of wind-induced internal pressure in a high-rise building. *Journal of Wind Engineering and Industrial Aerodynamics*. 69-71: 619 - 630.

Kato N. Ohkuma T. Niihori Y. & Kurita T. 1997. A study on wind forces for wind resistant design of high-rise building claddings (Part 1). *A.I.J. Journal of Structural and Construction Engineering*. 500: 1-6. (in Japanese)

Tamura G.T. Shaw C.Y. 1976 Study on exterior wall air tightness and air infiltration of tall building. *ASHRAE Transaction.*

Murakami S. & Yoshino H. 1983. Investigation of airtightness of houses. *A.I.J. Journal of Structural and Construction Engineering*. 325:104-115. (in Japanese)

Saathoff P.J. & Liu H. 1981. Building internal pressure : sudden change. *A.S.C.E. Journal of the Engineering Mechanics Division*. 107: 309-321.

Liu H. & Rhee K. H. 1990. Helmholtz oscillation in building models. *Journal of Wind Engineering and Industrial Aerodynamics*. 30: 95 -115.

Archtectual Insutitute of Japan(AIJ). 1993. Recommendation for Loads on Building.

Wind Engineering into the 21st Century, Larsen, Larose & Livesey (eds) © 1999 Balkema, Rotterdam, ISBN 90 5809 059 0

Experimental study of wind effects on flat porous fabric roofs

A. M. Vitale & C.W. Letchford
Department of Civil Engineering, The University of Queensland, Brisbane, Qld, Australia

ABSTRACT: This paper presents the results of a parametric wind tunnel study of wind effects on flat fabric roofs in uniform flow. Parameters investigated include the fabric mass, porosity and pretension. Fabric mass is found to influence tensions induced during flapping. Fabric porosity suppresses the static response and destroys the periodicity of the dynamic response. Fabric pretension suppresses the dynamic response of impervious fabrics.

1 INRODUCTION

Porous fabric structures are a cost-effective option for covering outdoor areas. They are used for the provision of sun protection, hail protection and wind protection. Australia has the highest incidence of skin cancer in the world due to the hot climate and outdoor-orientated lifestyle. It is not surprising then that the supply and erection of porous fabric structures has become a recent boom industry in this country. Shade structures for spectator comfort can now be seen at prominent sporting stadiums and will be on display in front of a world audience at the Sydney 2000 Olympic Games.

There are two broad classes of porous fabric roof. The first class consists of small, single span (10-20 m) pitched canopies used for coverage of outdoor public areas such as sports grounds, playgrounds, schools, childcare centres, shopping centres and swimming pools. The second class of structure is the flat profile, continuous span structure used for covering large areas such as livestock feedlots, fruit and vegetable growing plantations, plant nurseries and car storage yards.

This second class of structure has been observed to be particularly sensitive to dynamic wind effects due to the large spans and flat profiles involved. Moderate to high velocity winds can induce travelling and standing waves in the fabric roof. This can cause high stress concentrations in the fabric near rigid supporting members, which may lead to failure of the structure. In October 1998, several hectares of porous fabric roofing covering a car storage site in Brisbane were destroyed by high velocity winds generated during a severe tropical thunderstorm. Oscillations in the fabric roof were reported as being as large as 2 metres in amplitude prior to failure of the structure.

In this paper, results are presented of the wind tunnel testing of sectional fabric roof models in uniform flow. The influence of fabric mass, porosity and pretension are examined. Minami et al. (1993, 1994) investigated the dynamic response of a membrane suspended in a uniform air flow using similar wind tunnel techniques, however these tests were limited to impermeable membranes.

2 EXPERIMENTAL APPARATUS

2.1 *Wind tunnel*

Wind tunnel testing was carried out in the Number 3 Aerodynamic Wind Tunnel at the University of Queensland. The working section has nominal dimensions of 1.215 m width and 0.815 m height with 0.18 m x 0.27 m corner fillets. Two variable speed DC motors enable wind speeds of up to 30 m/s to be produced with turbulence intensities less than 0.2%.

2.2 *Model arrangement*

The fabric specimens were suspended at mid-tunnel height by two 25 mm diameter hollow circular section supports. Each end of the fabric was wrapped several times around the support. This support condition enabled the sag of the fabric to be easily adjusted by winding the fabric off or onto the support bars. The span between support bars, L, was 1.095 m. The width, W, of the fabric models was 0.65 m. The sag ratio, S_r (i.e. mid-span sag divided by the span), of the model was used as a measure of the

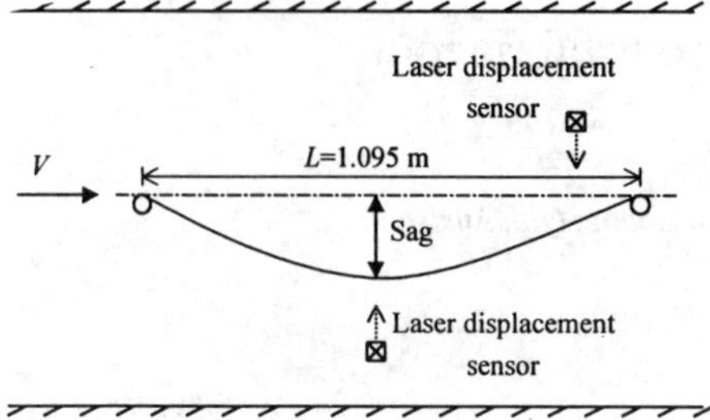

Figure 1. Arrangement of model in wind tunnel.

level of pretension in the fabric. The model arrangement is shown in Figure 1.

2.3 *Instrumentation*

The uniform wind velocity was measured using a pitot-static tube placed 280 mm upstream of the windward support. Static and dynamic deformations of the fabric were measured using two laser displacement sensors. One sensor was located at mid-span and the other sensor was positioned 180 mm upstream of the leeward support. Spectral analysis of the laser sensor output signals enabled the frequency content of the dynamic response of the fabric to be estimated. A visual record of the fabric response was taken with a video camera.

Each end of the support bars was connected to a two-component force balance outside of the wind tunnel side walls. The force balances consisted of two strain-gauged cantilever transducers arranged at right angles to each other to measure the horizontal and vertical forces transmitted to the ends of the support bars.

2.4 *Data Acquisition*

30 second duration records of the laser displacement sensor and force balance signals were acquired at a sampling frequency of 100 Hz. The first symmetric natural frequency of the support bars was approximately 25 Hz. The force balance signals were digitally low-pass filtered at 20 Hz to eliminate noise generated at this natural frequency. At several of the highest velocities tested, the oscillating frequency of the fabric approached the 25 Hz resonant frequency of the supports. It was then impossible to separate the required force signal from the spurious resonant response of the force balance.

After filtering, each force balance signal was added to its' counterpart at the opposite end of the support. The mean aerodynamic drag forces acting on the bluff-body support bars were subtracted from the total horizontal forces. These drag forces were estimated from a wind tunnel run with no fabric suspended between the supports. Thus, the final reduced forces represented the total horizontal and vertical components of the wind-induced fabric tension acting on the windward and leeward supports.

2.5 *Fabrics*

A range of solid and porous fabrics was tested in the wind tunnel. Since the mass of the fabrics was expected to have a strong influence on the dynamic response, mass scaling of the fabrics had to be considered. While there was no particular prototype structure in mind, the model span was considered to be 10 to 20 times smaller than that of typical full-scale structures. This means that the mass per unit area of the model fabrics was required to be 10 to 20 times less than that of shadecloth (but with the same porosity). This was not an easy requirement to achieve and model fabrics with mass scales of only 2 to 8 were tested.

The static pressure loss coefficient, K, was used as a measure of the fabric porosity. This parameter is easily estimated by measuring the static pressure drop across the fabric in a wind tunnel. The coefficient indicates a fabric's resistance to through-flow. It is defined as:

$$K = \frac{\Delta p}{1/2\,\rho V^2} \qquad (1)$$

where Δp is the static pressure drop across the fabric, ρ is the air density and V is the wind velocity upstream of the fabric.

Figure 2 gives the mass and porosity characteristics of the model fabrics. The model fabrics were selected to have a similar porosity to common grades of shadecloth.

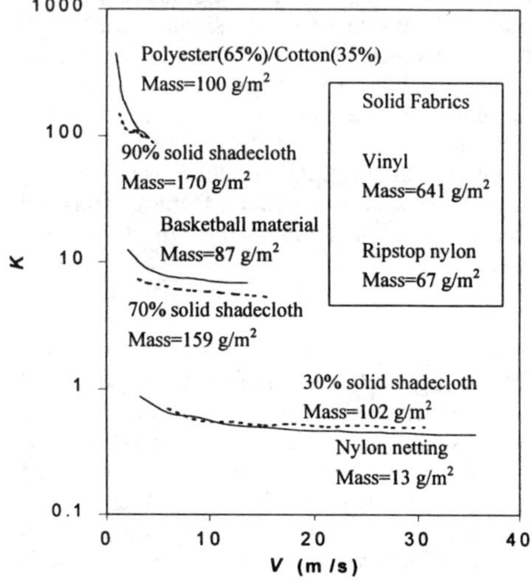

Figure 2. Mass and porosity characteristics of fabrics.

3 RESULTS

3.1 *Response configurations*

During the course of the wind tunnel testing there were three different stable response configurations observed. At zero wind speed, the fabric hangs in a catenary between the two supports. In the first stable configuration, the windward half of the fabric is lifted slightly upwards while the leeward half is pushed significantly downwards. There is a negative (i.e. downward) net lift force acting on the fabric. This downward deflected configuration is shown in Figure 3a and will be denoted by the symbol 'D'.

The second stable configuration involves the leeward half of the fabric being pushed upwards. There is a positive (i.e. upward) net lift force acting on the fabric. This configuration will be denoted as 'U' and is shown in Figure 3b. The U configuration was generally obtainable by manually forcing the fabric in the D configuration upwards, causing the fabric to snap through. These two stable configurations are static responses.

The third stable configuration was a dynamic response. This response took the form of a travelling wave located mostly in the leeward half-span of the fabric. The windward half of the fabric was generally near to horizontal. The mean net lift force acting on the fabric is close to zero as the fabric oscillates about a line joining the two supports. Minami et al. (1993, 1994) referred to this dynamic response as 'flutter'. In this paper the response will be termed 'flapping' so as not to confuse it with the term 'flutter' reserved for the coupled degree of freedom instability of bridge decks and aerofoils. Figure 3c shows the flapping configuration which will be denoted as 'F'.

The particular configuration that the fabric adopted depended on the wind velocity, the sag ratio, the fabric mass and porosity. Under certain conditions there was more than one stable configuration possible. Table 1 lists the configurations adopted by each different fabric. The fabrics will be abbreviated as 'RN' for ripstop nylon, 'V' for vinyl, 'PC' for polyester/cotton, 'B' for basketball fabric and 'NN' for nylon netting. The velocity, V, has been non-dimensionalised using the Froude Number, defined as $Fr=V/(gL)^{0.5}$, where L is the span.

The solid fabrics were only able to flap at the largest sag ratio. The porous fabrics were able to flap at all sag ratios (except PC at S_r=0.016).

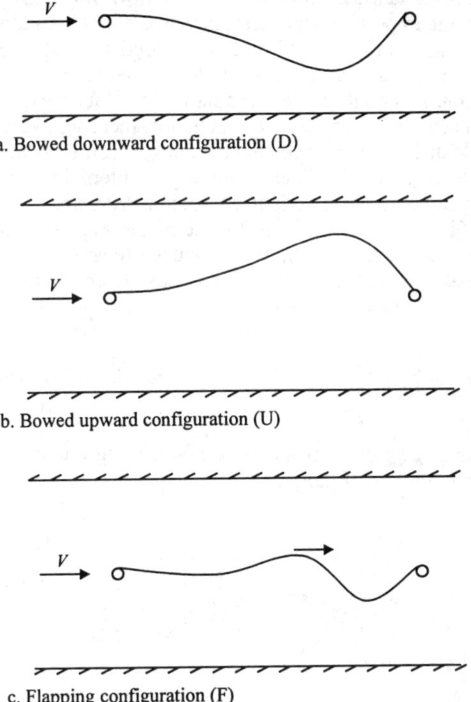

a. Bowed downward configuration (D)

b. Bowed upward configuration (U)

c. Flapping configuration (F)

Figure 3. Fabric response configurations.

Table 1. Response configurations for wind tunnel tests.

Fabric	Config.	S_r=0.177	S_r=0.122	S_r=0.087	S_r=0.043	S_r=0.016
RN	D	$Fr<3.87$	All Fr	All Fr	All Fr	All Fr
RN	U	-	All Fr	All Fr	All Fr	All Fr
RN	F	$Fr\geq3.87$	-	-	-	-
V	D	$Fr<3.08$	All Fr	All Fr	All Fr	$Fr>6.53$
V	U	-	$Fr>3.34$	$Fr>2.85$	$Fr>2.62$	All Fr
V	F	$Fr\geq3.08$	-	-	-	-
PC	D	$Fr\leq0.74$	$Fr\leq0.86$	$Fr\leq1.95$	$Fr\leq1.73$	-
PC	U	-	-	-	$Fr=2.43$	All Fr
PC	F	$Fr\geq0.79$	$Fr\geq1.31$	$Fr\geq2.51$	$1.94\leq Fr\leq2.21$ & $Fr\geq2.87$	-
B	D	-	-	-	-	-
B	U	-	-	-	-	-
B	F	All Fr	All Fr	All Fr	All Fr	All Fr
NN	D	-	-	-	-	-
NN	U	-	-	-	-	-
NN	F	All Fr	All Fr	All Fr	All Fr	All Fr

3.2 *Static response*

The static response was mainly exhibited by the impervious fabrics. Since no through flow was possible, the impervious fabric surface was able to sustain the large pressure differential required to push the fabric upwards or downwards to the stable configuration. Although the leeward half of the fabric experienced the larger aerodynamic loads, there was very little difference between the tensions measured at the ends of the fabric. This smoothing of internal forces is a common characteristic of tension structures.

Figures 4 shows the influence of the sag ratio on the fabric tension. The wind-induced tension is expressed in terms of the mean tension coefficient, $C_{T,mean}$ defined as:

$$C_{T,mean} = \frac{T_{mean}}{1/2\,\rho V^2 L} \tag{2}$$

where T_{mean} is the mean wind-induced fabric tension per unit width of fabric.

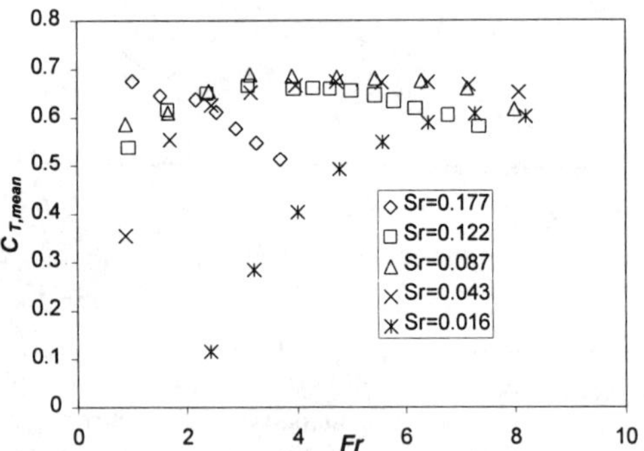

Figure 4. Effect of sag ratio on fabric tension (fabric RN, windward tension, configuration D).

Surprisingly, the level of pretension in the fabric has little effect on the wind-induced tensions in the static configuration. $C_{T,mean}$ generally lies between 0.6 and 0.7. Similar tensions were measured while the fabric was in the U configuration. Values of $C_{T,mean}$ measured for the vinyl fabric were also between 0.6 and 0.7. The fabric mass does not seem to strongly influence the response of the fabric in the static configuration.

Minami et al. (1994) measured the horizontal force acting on the windward support bar for three aramid fibre fabrics FAB47, FAB140, FAB251. The numbers in the symbols represent the mass (g/m^2) of the fabrics. Figure 5 compares these results with the horizontal component of fabric tension measured in the current tests. The mean horizontal tension coefficient, $C_{TH,mean}$, is defined in the same way as for $C_{T,mean}$.

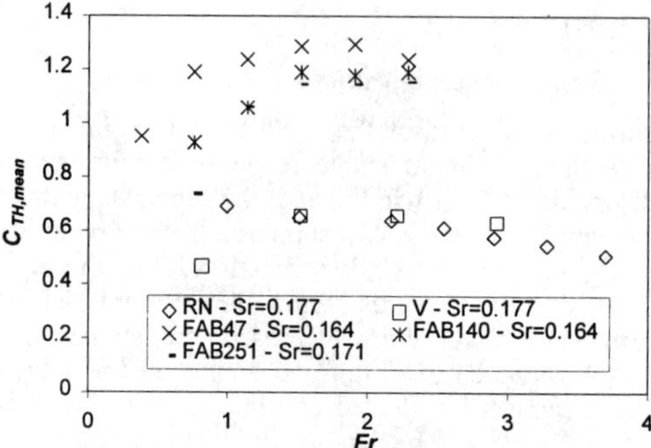

Figure 5. Horizontal tension components (configuration D).

In the study by Minami et al. (1994), $C_{T,mean}$ took values between 1.1 and 1.3. These are approximately twice the magnitude of the values determined from the present testing. The results do confirm that the fabric mass is not an important parameter for the static response. It is strongly suspected that the aerodynamic drag caused by the windward support bar has not been subtracted from the measured horizontal forces of Minami et al. The dimensions of this bar are not given. In addition, these tests probably achieved a more accurate two-dimensional flow condition since the fabric occupied the entire tunnel width. In the present study the model occupied only half the tunnel width and may have experienced end-effects.

3.3 *Dynamic response*

3.3.1 *General*

The flapping configuration was typically the result of a sudden transition from the D configuration at a critical flow velocity. As the wind speed is increased in the D configuration, the windward half of the fabric is lifted upward, while the position of maximum sag moves leeward. At the transition from D to F, the hump in the leeward fabric becomes unstable. It is lifted upward and starts to oscillate, but does not fully snap through to the U configuration. It is emphasized that this is a rapid transition. The measured net lift forces acting on the solid fabrics are still strongly negative immediately prior to transition. The mean net lift forces in the F configuration are approximately zero.

Once transition has been initiated, the F configuration can be sustained at lower velocities than the critical starting velocity. The introduction of porosity into the fabric causes the F configuration to be initiated at lower velocities. Flow through the fabric surface tends to equalize the pressure on each side and this makes it difficult for the static configuration to be sustained.

At high values of S_r, the F configuration takes the form of a travelling wave starting near mid-span. The travelling wave is approximately sinusoidal in shape with a single wavelength being formed. As the initial sag becomes smaller, the position of wave initiation moves downstream and the sinusoidal waveform becomes increasingly distorted, possibly due to the superposition of higher order waves. This phenomenon was similarly observed in the tests of Minami et al. (1993, 1994). As the fabric porosity increases, the flapping also becomes more distorted and relocates closer to the leeward support.

3.3.2 *Flapping frequency*

In the F configuration, the impervious fabrics and fabric PC flapped with a dominant discrete frequency (i.e. the frequency of the travelling wave). The higher porosity B fabric exhibited this discrete flapping at low flow velocities only. At higher velocities, the flapping became the result of a broad band of frequency components. The frequency content of the highest porosity NN fabric flapping was broadband at all flow velocities. Flow through the porous fabrics has the effect of disturbing the two-dimensional nature of the travelling wave (i.e. flapping is poorly correlated across the width of the fabric).

In Figure 6, the measured flapping frequencies are shown for the maximum sag ratio. The frequency is expressed in the form of the Strouhal Number, St, defined as.

$$St = \frac{fL}{V} \tag{3}$$

where f is the flapping frequency of the fabric. The results of the study by Minami et al. (1994) are also shown in Figure 6. These include the aramid fibre fabrics (FAB47, FAB140, FAB251) as well as a teflon coated glass fibre fabric (CFAB1268). Again, the number in the symbol represents the mass (g/m^2) of the fabric.

At the larger velocities, the values of St were reasonably constant. This means that the flapping frequency increased linearly with the flow velocity. The fabrics with a mass in the order of 100 g/m^2 (e.g. RN, PC, FAB47, FAB 140) flapped with a Strouhal Number of approximately unity. The fabrics with a mass in the order of 1000 g/m^2 (e.g. V, CFAB1268) flapped with a Strouhal Number of approximately 0.1. This suggests that flapping frequency may be roughly inversely proportional to the fabric mass. The sag ratio was found not to have a significant effect on the flapping frequency. The Strouhal Numbers measured in the current study compare favourably with those measured by Minami et al. (1994).

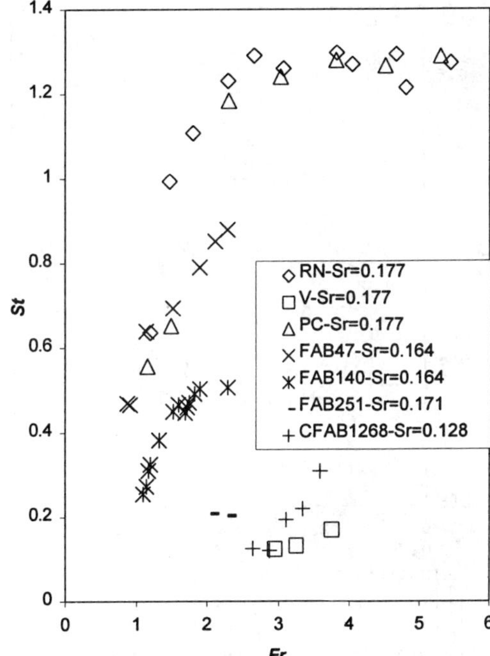

Figure 6. Flapping frequencies of fabrics.

3.3.3 *Flapping tensions*

In the F configuration, the tension in the fabric fluctuates. Figure 7 presents the mean windward tension coefficients for the largest sag ratio. Mean values were calculated over the 30 second duration of the acquired samples.

The mean flapping tensions are significantly smaller than the tensions for the static configuration. The mean tension coefficient for the lighter fabrics tested (all except fabric V) is generally between 0.03 and 0.05. The higher porosity fabrics (B, NN) have $C_{T,mean}$ values closer to 0.03 while the more impervious PC and RN are approximately 0.05. The significantly heavier V fabric has larger tension coefficients of 0.2 to 0.25. The sag ratio did not have a significant influence on the mean tensions.

The mean windward horizontal tension coefficients measured by Minami et al. (1994) are also shown in Figure 7. The slope of the fabric at the leading edge during flapping is very close to horizontal and so the horizontal component of fabric tension is effectively the same as the absolute tension. For the lighter fabrics, the mean tensions measured by Minami et al. are significantly larger than those of the current study. Again, it is suspected that this discrepancy might be explained by the influence of the windward support bar and the better achievement of two-dimensional flow conditions in

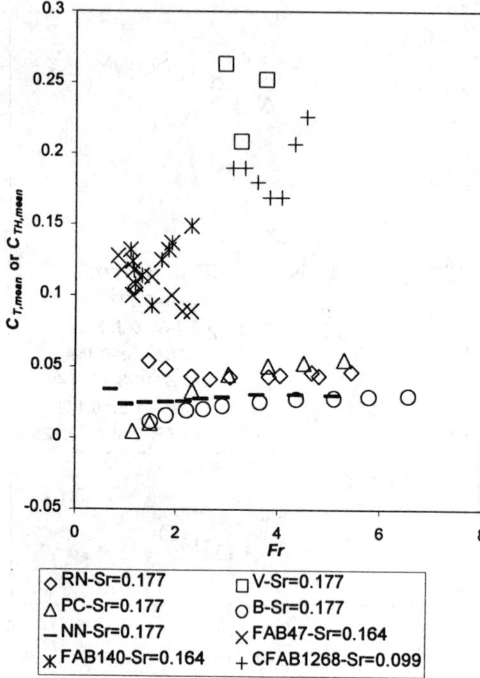

Figure 7. Mean windward tension coefficients during flapping.

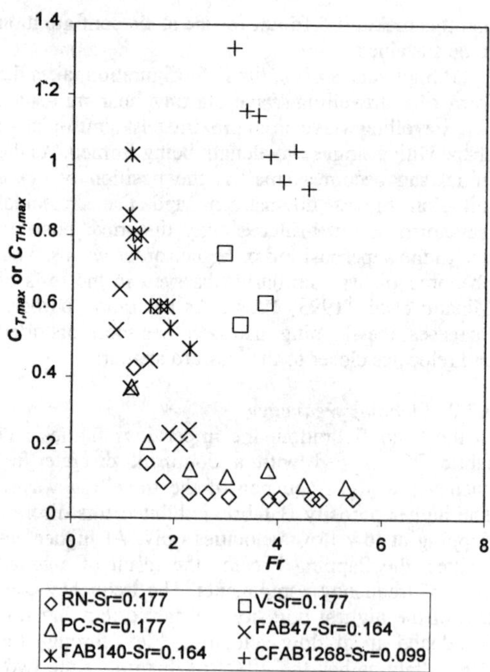

Figure 8. Maximum windward tension coefficients during flapping.

the previous study. The mean tensions for the heavy fabrics (V, CFAB1268) compare quite well.

The maximum windward tensions recorded during the 30 second sample durations are shown in Figure 8. The maximum tension coefficient, $C_{T,max}$, is defined in the same way as $C_{T,mean}$, but using the maximum tension instead of the mean tension. The results for the higher porosity B and NN fabrics have been omitted because the signal-to-noise ratio of the force balance measuring system was too low to allow maximum tensions to be confidently predicted for these fabrics. Maximum horizontal windward tensions obtained by Minami et al. (1994) are also shown in Figure 8 (these tensions were measured over a 10 second duration sample of flapping).

The maximum tensions increase as the fabric mass increases. The maximum tension occurs when the crest of the travelling wave approaches the rigid leeward support bar. The motion of the oscillating fabric is forced to come to a sudden stop at the support. This is achieved by an increase in fabric tension. Heavier fabrics will have a larger inertia associated with their oscillation and, accordingly, a larger tension will be required to rapidly halt this inertia.

A consequence of this mechanism is that the maximum tensions measured for the lighter fabrics are not significantly larger than the mean tensions. However, the maximum flapping tensions of the heavier fabrics are well in excess of the mean tensions and approach magnitudes comparable with the tensions induced in the static configurations.

4 CONCLUSIONS

A flat fabric roof suspended from its' ends can take three stable configurations in the wind: bowed upwards and downwards (static configurations) and flapping (dynamic configuration). The flapping configuration consists of a travelling wave in the leeward half-span of the fabric.

In the static configuration, the fabric pretension and mass does not significantly influence the wind-induced tensions in the fabric. Increasing fabric porosity causes suppression of the static configurations in preference for the dynamic configuration.

Fabrics with a porosity of 10% or less flap with a dominant frequency that increases linearly with the flow velocity. This frequency decreases with increasing mass but is relatively insensitive to the fabric pretension. At higher porosities, flow through the fabric destroys the two-dimensional nature of the flapping. Flapping then occurs with a wide range of frequencies.

Mean tensions induced in the fabric during flapping are significantly lower than in the static configuration. Peak tensions during flapping are caused by the abrupt retardation of the travelling wave as it approaches the leeward support. Peak tensions increase for fabrics with larger mass and can equal those of the static configurations. Increasing fabric porosity does not significantly lower the flapping tensions. Fabric pretension is effective in suppressing flapping for impervious fabrics.

REFERENCES

Minami, H., Okuda Y. & Kawamura S. 1993. Experimental studies on the flutter behaviour of membranes in a wind tunnel. *Space Structures 4* Vol. 1: 935-945. London: Thomas Telford.

Minami, H., Okuda Y. & Kawamura, S. 1994. Experimental study and analysis on the fluttering of a membrane suspended in a uniform air flow. Part 1: Experimental investigation on the behaviour and the bearing force of membrane fluttering. *Mem. Fac. Eng., Osaka City Univ.* Vol. 34: 65-76.

Wind Engineering into the 21st Century, Larsen, Larose & Livesey (eds) © *1999 Balkema, Rotterdam, ISBN 90 5809 059 0*

Wind-induced response control of a slender tall building with active-passive mass damper

M.Watakabe, O.Chiba & M.Tohdo
Technical Research Institute, Toda Corporation, Japan

T.Fujita
Institute of Industrial Science, University of Tokyo, Japan

ABSTRACT: The authors have executed research and development of an active-passive mass damper (hereinafter referred to as APMD) since 1990(Fujita *et al.* 1992). This mass damper has been installed in a slender tall building with an aspect ratio of eight since November 1997. The mass damper system can function mainly as an active mass damper to effectively control moderate vibrations of the building caused by seasonal winds or weak earthquakes and also can function as a passive mass damper against strong winds or earthquakes. Simulation analysis of wind-induced response, forced vibration tests and observation of behaviors in moderate winds were carried out to investigate the response control performance of the mass damper system. Through the results obtained by these investigations, it was concluded that the mass damper system was quite effective for the response control of the slender tall building against wind loads.

1 INRODUCTION

In recent years, the increasing number of high-rise buildings, sightseeing towers and other flexible and dynamically sensitive structures in Japan has generated more interest in the field of structural vibration control. This interest and awareness have manifested themselves in the construction of many structures equipped with vibration control systems to improve habitability. The authors have executed research and development of a new type of hybrid mass damper using hydraulic actuators for response control of tall buildings since 1990, and have already reported on response control performance of the APMD installed on a large-scale test building(Watakabe *et al.* 1994). When moderate vibrations occur, the APMD system works as an active mass damper which can effectively control the vibrations. Furthermore, the APMD automatically changes to a passive mode which makes the mass damper work as a passive one when large vibrations of the building are induced. In the initial design process of a slender tall building, there was some anxiety for vibration of the building during seasonal winds because of its slenderness and its small structural damping. Based on this recognition, installation of the APMD system was decided to reduce vibrations and increase serviceability. The slender tall building equipped with the APMD system was constructed in February 1998.

This paper presents the mass damper system and its control performance obtained by simulation analysis, forced vibration tests and observation of behaviors in moderate winds.

2 OUTLINE OF BUILDING AND APMD SYSTEM

A photo of the building concerned and outline of the observation system are shown in Figure 1. The building on which the APMD system was installed is a steel truss tube structure with a 154m height, a 6,877 tons total weight, and a typical floor area of 19m × 19m. The first natural frequencies of the building in the X- and Y-directions are 0.35 and 0.41 Hz, respectively. Earthquake and wind-induced vibrations have been observed to investigate the vibration control performance of the APMD system. The responses of the building were recorded by

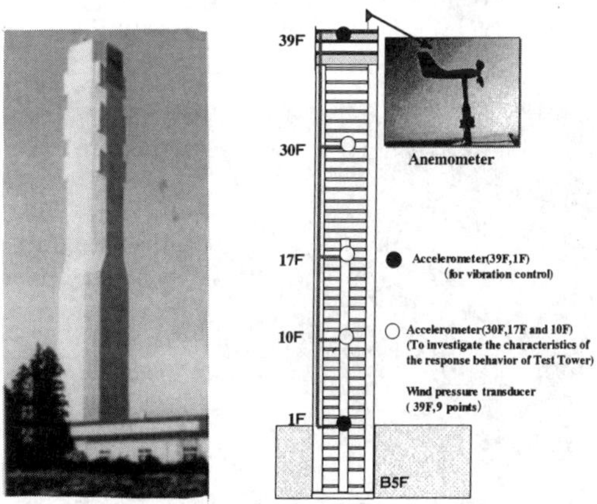

Figure 1. Full view of the building concerned locations and of measuring devices

accelerometers which were set on the 1st,10th,17th,30th and 39th floors (5 sets,10 components). Velocity and displacement transducers (2 sets, 4 components) were set on the damper. The wind velocity and wind direction were measured by a windmill-type anemometer on the top of the building. The wind pressures were measured by using wind pressure transducers at 9 points on the façade at the 148m above the ground level. The data were recorded by digital data recorders.

Table 1 summarizes the mass damper parameters. Figure 2 shows a photo of the APMD installed on the 39th floor of the building. Figure 3 shows the diagram of the hydraulic circuit. The moving mass has an outer size of $5 \times 5 \times 2.48^H$ m. The moving mass weights of the APMD are 34.5 tons and 26.8 tons in the X- and Y-directions, respectively. The main mass is supported by a XY-motion mechanism which provides free parallel movements in the horizontal directions. The main mass and intermediate frame form a moving mass in the X-direction, and only the main mass in the Y-direction. The switching rule to either active or passive mode is determined by the system pressure and the mass displacement. The hydraulic actuator is mounted in each direction. The control system for the 1st lateral vibration mode in two directions were designed by the Linear Quadratic (LQ) optimal control theory. The APMD system has two operational modes, that is, the active mode and the passive mode. The mass damper works actively to control the moderate

Table 1. Specifications of the APMD

	X-direction	Y-direction
Out size	$5.0 \times 5.0 \times 2.48^H$ m	
Moving mass weight	34.5 tons	26.8 tons
Effective mass ratio	1.73 %	1.41 %
Natural frequency	0.35 Hz	0.39 Hz
Maximum displacement	±40.0 cm	
Friction coefficient	3.5/1000	3.8/1000
Pump unit	11.0 kW	
Accumulator	60.0 l	

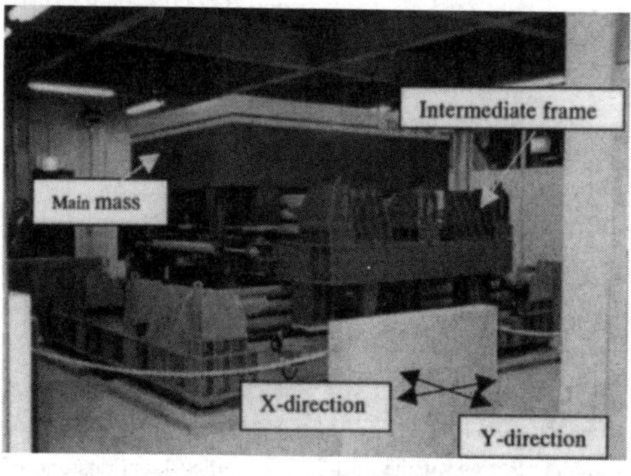

Figure 2. View of the APMD

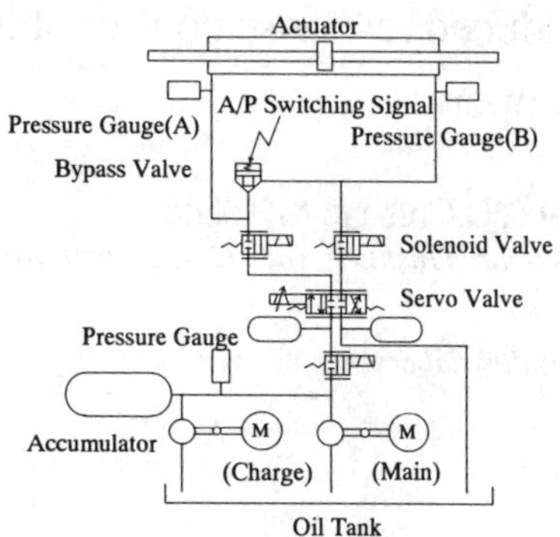

Figure 3. Diagram of the hydraulic circuit

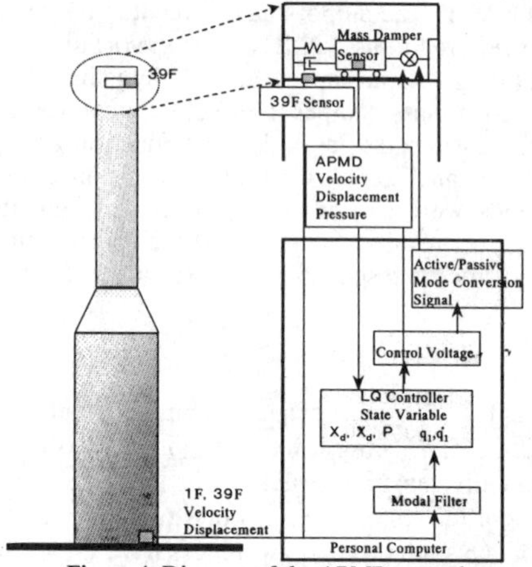

Figure 4. Diagram of the APMD control system

vibrations caused by seasonal winds. On the other hand, it works passively to control the large vibrations caused by strong winds or earthquakes.

Figure 4 shows a block diagram of the overall control system, which is actually comprised of a pair of the control systems that operate in independent of the X- and Y- responses. The trigger signal and the hydraulic control ON/OFF signal, however, are in-common. This system monitors the absolute velocity of the 39th floor of the building; and if this measured value exceeds a threshold one, the trigger signal is generated to start a control response. During the control operating, the system inputs the velocity and displacement of the mass damper, the actuator pressure, and the velocity and displacement of the 39th and the 1st floor into the personal computer via the A/D converter. The control voltage and the bypass opening signal are then computed and output to the hydraulic control panel through D/A conversion. The hydraulic control panel turns the hydraulic pump ON and OFF, opens and closes various valves, and detects some system anomalies.

The purpose of the APMD system was to reduce RMS values of the response accelerations by about 1/3～1/2 during moderate seasonal winds under active control, and by about 1/2～2/3 during strong winds under passive control in comparison with those when non-controlled.

3 ANALYTICAL MODELS

3.1 *Analytical model of building and mass damper system*

Figure 5 shows an analytical model of the building where the mass damper is installed on the 39th floor. The X- and Y- directions are independently treated with a one-directional analysis model. The equations of the motion are represented as

$$[M]\{\ddot{X}\}+[C]\{\dot{X}\}+[K]\{X\}= \{s\}\{c_d\dot{x}_d+k_dx_d+\mu m_d g\,\mathrm{sgn}(\dot{x}_d)-F\}-[M][1]\ddot{z}+\{W\} \quad (1)$$

$$m_d(\ddot{x}_d+\ddot{x}_s+\ddot{z})+c_d\dot{x}_d+k_dx_d+\mu m_d g\,\mathrm{sgn}(\dot{x}_d)=F \quad (2)$$

where $[M]$,$[C]$,$[K]$(n×n) are respectively the building's mass, damping and rigidity matrices;$\{x\}$(n×1) is the state vector and x_s is the relative displacement vector of the floor installed the mass damper to the ground surface. $\ddot{z}$ is the ground surface acceleration due to seismic motion; $\{s_i\}$=(0···010···0)is a vector representing the floor installed of the mass damper; $\{1\}=(1\cdots1)^T$ (n×1); $\{w\}=(w_1w_2\cdots w_n)^T$ (n×1) is the external force vector due to wind; x_d, m_d, k_d, and η are respectively the relative displacement, mass, spring constant, and dynamic friction coefficient of the mass damper; and F is the control force produced by the hydraulic actuator.

3.2 *Analysis model of hydraulic actuator*

The hydraulic actuator's characteristic equation is represented as

$$a\dot{x}_d+\gamma p_L=b(u-k_{fd}x_d)\sqrt{1-\mathrm{sgn}(u-k_{fd}x_d)p_L/p_s} \quad (3)$$

where a is the piston's sectional area, γ a constant indicating the rigidity of the cylinder; b is the flow-rate gain of the servo valve and u its input voltage; p_L the load pressure of the cylinder, p_s system pressure; and k_{fd} is the feedback gain to x_d. The dynamic characteristic equation of the system pressure is represented as

$$(a|\dot{x}_d|-Q_l)\Delta t=V_l\frac{p_g}{p_{s2}}\eta\left\{(p_{s1}/p_{s2})^{\frac{1}{n}}-1\right\}\Big/\left\{(p_{s1}/p_{s2})^{\frac{1}{n}}\right\} \quad (4)$$

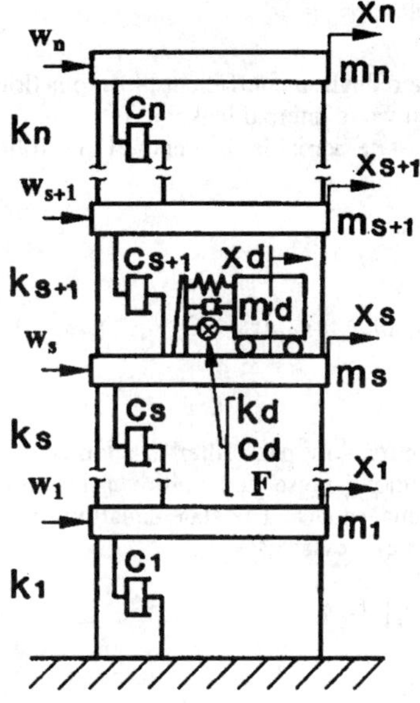

Figure 5. Analytical model

$$(Q_l-a|\dot{x}_d|)\Delta t=V_l\frac{p_g}{p_{s1}}\eta\left\{(p_{s2}/p_{s1})^{\frac{1}{n}}-1\right\}\Big/\left\{(p_{s2}/p_{s1})^{\frac{1}{n}}\right\} \quad (5)$$

where p_{s1} and p_{s2} are respectively the system pressure at t and $t+\Delta t$ time; Q_l is input flow-rate from pump unit; V_l is the accumulator volume, p_g nitrogen gas closed input pressure, η effect ratio of the accumulator, and n polytropic number of nitrogen gas.

4 DESIGN OF CONTROL SYSTEM

4.1 *LQ optimal regulator design*

Controller is designed for X- and Y- translation of the building independently. Using the modal matrix$[T]$, a physical coordinate system {x} was converted to a mode coordinate system {q}. When the 1st modes are adopted, and mass damper friction is neglected, the following reduced-order are obtained from Eqs. (1) and (2).

$$m_1^*\ddot{q}_1+c_1^*\dot{q}_1+k_1^*q_1=t_{s1}(c_d\dot{x}_d+k_dx_d-ap_L) \quad (6)$$

$$m_d(\ddot{x}_d+t_{s1}\ddot{q}_1)+c_d\dot{x}_d+k_dx_d=ap_L \quad (7)$$

where m^*_1, c^*_1 and k^*_1 are respectively the 1st mode generalized mass, damping and stiffness; t_{s1} represents the 1st -component of the modal matrix; index s represents floor installed the mass damper.

The hydraulic actuator's characteristic equation was linearized as following

$$a\dot{x}_d + \gamma \dot{p}_L = b\left(u - k_{fd}x_d\right) - lp_L \tag{8}$$

where l is a reduction coefficient of output flow-rate due to servo valve internal leakage.

If the state variable is denoted by following equation

$$\{X\} = \{\dot{q}_1 \; q_1 \; \dot{x}_d \; x_d \; p_L\}^T \tag{9}$$

the state equation is expressed as

$$\dot{X} = AX + Bu \tag{10}$$

Furthermore, low pass filter is adopted for u in order to remove noise and spillover instability of higher residual modes. The state equation of the low pass filter is expressed as

$$\{\dot{X}_F\} = [A_F]\{X_F\} + [B_F]u$$

$$y_F = [C_F]\{X_F\} \tag{11}$$

combining the matrix equations Eqs.(10) and (11), the extended system can be written as

$$\begin{Bmatrix} \dot{X} \\ \dot{X}_F \end{Bmatrix} = \begin{bmatrix} A & BC_F \\ 0 & A_F \end{bmatrix} \begin{Bmatrix} X \\ X_F \end{Bmatrix} + \begin{Bmatrix} 0 \\ B_F \end{Bmatrix} u \tag{12}$$

The feedback gain matrix is determined by the LQ optimal control theory under the extended system Eqs.(12) to decrease the relative velocities and displacements of the 1st mode of the building.

The state variables which can be directly measured are the damper's relative displacement x_d and relative velocity x_d and the hydraulic-actuator internal load pressure p_L. Since it is impossible to directly obtain q_l and q_1, they are calculated via a modal filter, using the building's absolute velocity and displacement.

4.2 *Active-passive mode switching rules*

For active-passive switching, operational limits of this mass damper are mainly the actuator's system pressure and the displacement of the moving mass. The pressure mater of the accumulator searches the system pressure. To avoid these operational limits, the active-passive mode switching is applied. The mode switching rules are expressed as follows.

if ($p_s<p_{sP}$) or ($x_d>x_{dP}$) then Active to Passive. (13)

if ($p_s>p_{sA}$) and ($x_d<x_{dA}$ for 15sec) then Passive to Active. (14)

where p_s and x_d are respectively the system pressure the displacement of the moving mass; p_{sP} and x_{dP} are the switching values to the passive mode for the mass damper system pressure and the displacement respectively, and p_{sA} , and x_{dA} are the switching values to the active mode.

The limits values of p_{sP}, p_{sA}, x_{dP} and x_{dA} take respectively 1.2×10^7Pa, 1.5×10^7Pa, 0.35m and 0.05m.

5 RESPONSE ANALYSIS OF WIND-INDUCED VIBRATION FOR RESPONSE CONTROL PERFORMANCE

Wind responses of the building with and without the APMD were analyzed to confirm the effect of the APMD system. The damping ratios used in the analyses were assumed to be 0.5 % when non-controlled and 8 % when controlled. The response analyses of wind-induced vibration were carried out for seasonal, 1-year and 5-year return period wind excitations based on pressure characteristics by wind tunnel tests (Ebisawa *et al.* 1997). Figure 6 shows examples of time histories of the building responses in the across-wind direction obtained by the response analysis for a seasonal wind excitation. In order to investigate the control effect of the APMD system, the power spectra were calculated using the response acceleration shown in Figure 6. The results were shown in Figure 7. Comparing with the responses when non-controlled, the time history and the spectrum when controlled clearly show increase of

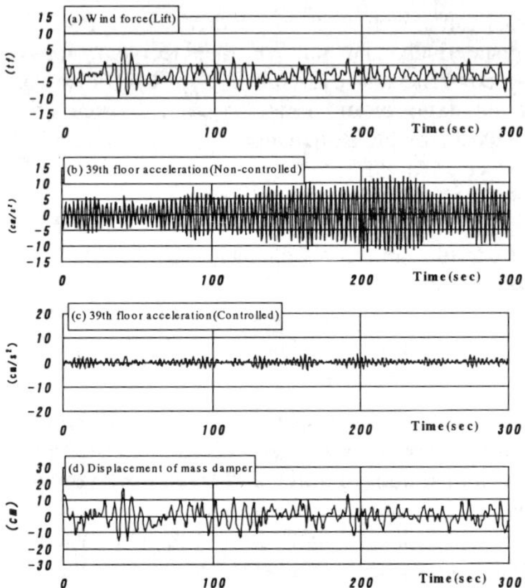

Figure 6. Comparison of time histories of responses in the across-wind direction when non-controlled and controlled (Seasonal wind, Mean wind velocity at 158m above the ground:14m/sec)

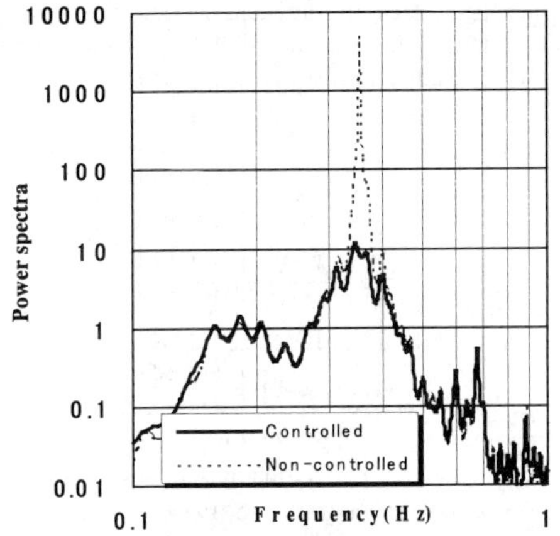

Figure 7. Power spectra of the 39th floor accelerations in the across-wind direction (Seasonal wind)

Table 2. Results of the response analyses for simulated winds

Direction of Wind-induced vibration	Simulated winds (Return period)	Accelerations				Ratio of the controlled to the non-controlled	
		Non-controlled		Controlled			
		Max. (cm/s^2)	RMS (cm/s^2)	Max. (cm/s^2	RMS (cm/s^2)	Max.	RMS
Across-wind direction	Seasonal	12.4	4.65	3.2	0.95	0.26	0.21
	1- year	28.6	10.72	7.5	2.26	0.26	0.21
	5- year	40.1	15.02	10.5	3.27	0.26	0.22
Along-wind direction	Seasonal	6.4	2.20	2.1	0.63	0.33	0.29
	1- year	14.7	5.07	5.8	1.56	0.39	0.31
	5- year	20.7	7.11	7.1	2.21	0.34	0.31

damping effects by the APMD. Table 2 shows the results of the analyses for seasonal wind, 1-year and5-year return period wind excitations. Through the installation of the APMD, the peak and the R.M.S. values of the accelerations in across-wind under the active mode were reduced to 1/5~1/4 of those when non-controlled. The damping effect in the across-wind direction was more effective than that in the along-wind direction.

6 FORCED VIBRATION TESTS AND VERIFICATION OF CONTROL EFFECT

Forced vibration tests were carried out by performing sinusoidal wave excitation using the APMD installed on the 39th floor of the building. An example of the response control of the APMD in the tests was shown in Figure 8. The response velocity obtained on the 39th floor and the control voltage to the actuator are shown in Figure 8(a), (b) , respectively. Figure 8 (a) shows the 39th floor velocity when excited by the APMD in the first 70 seconds period and then controlled by the APMD after the period. Figure 8(c) is the enlarged time history of the 39th floor velocity when controlled and shows that the damping for the 1st mode increased by 8~11%.

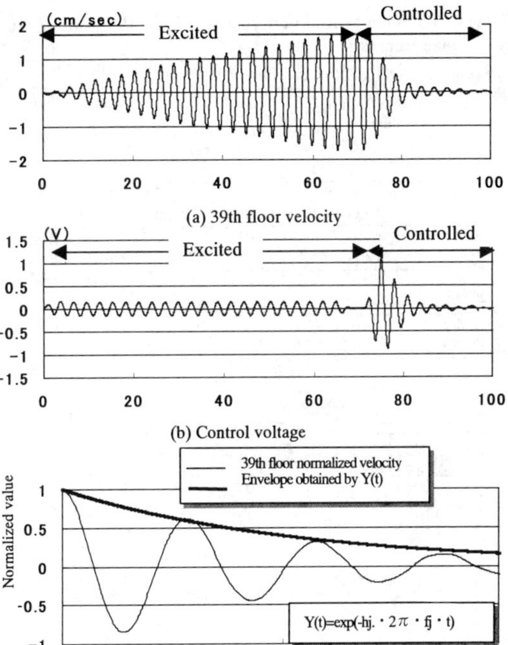

Figure 8. Response control performance of the APMD in the forced vibration test

7 CONTROL PERFORMANCE IN ACTUAL WINDS

Observation of the behaviors in the event of moderate seasonal wind excitations has been carried out to investigate the performance of the APMD system since February 1998. Wind records measured during moderate seasonal winds are listed in Table 3. The performance of the APMD was investigated by comparing the responses when controlled and non-controlled during wind excitations. The results are summarized in Table 4. The wind-induced peak accelerations were substantially reduced by the APMD as shown in Table 4 for the along-wind and across-wind responses, respectively. The peak and

Table 3. Strong wind observation records

Date	Max. wind velocity(m/s)	Mean wind velocity(m/s)	Mean wind direction
Mar. 7, '98	29.4	8.7	WNW
Mar. 15, '98	31.4	14.2	WNW
Mar. 20, '98	33.7	16.6	SSW
Sep. 15, '98	52.7	27.6	SSW

Table 4. Vibration control performance of APMD in strong wind observation records

Date	Direction of wind-induced vibration	Accelerations Non-controlled Max. (cm/s²)	RMS (cm/s²)	Controlled Max. (cm/s²)	RMS (cm/s²)	Ratio of the controlled to the non-controlled Max.	RMS
Mar. 7, '98	Across-	12.8	3.28	2.8	0.89	0.22	0.27
	Along-	5.8	1.63	2.0	0.58	0.35	0.35
Mar. 15, '98	Across-	13.5	3.04	3.1	0.80	0.23	0.26
	Along-	5.1	1.13	2.5	0.51	0.49	0.45
Mar. 20, '98	Across-	6.5	1.91	2.5	0.56	0.38	0.29
	Along-	7.2	2.11	2.7	0.60	0.38	0.28
Sep. 15, '98	Across-	44.8	18.1	10.0	3.61	0.22	0.20
	Along-	25.8	8.38	5.7	2.51	0.22	0.29

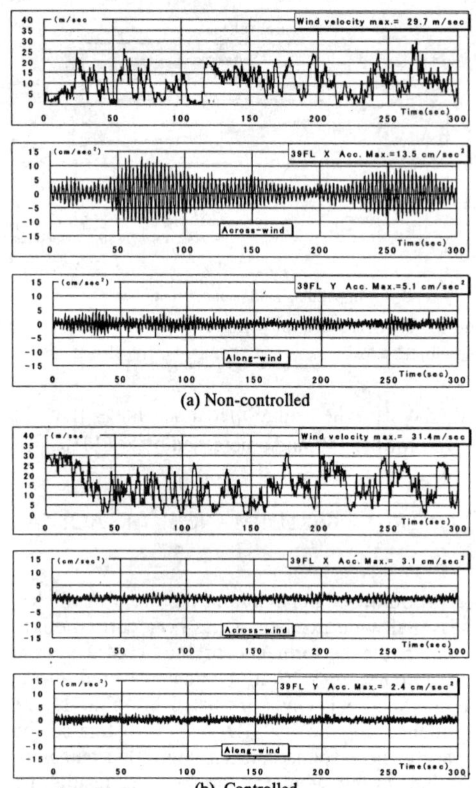

Figure 9. Comparison of response when non-controlled and controlled observed during a seasonal wind on March 15, 1998

the RMS values of the accelerations in the active mode were reduced to about 1/4.5～1/2.0(average 1/3) of those when non-controlled. Examples of the acceleration records obtained on the 39th floor of the building are shown in Figure 9. Figure 9(a) shows the time histories of the wind velocity and the 39th floor responses of the building when non-controlled. Figure 9 (b) shows the time histories of those when controlled. From these results, the peak acceleration

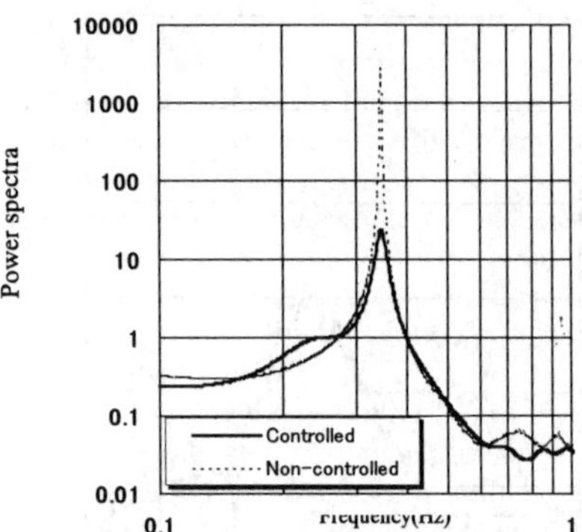

Figure 10. Power spectra of the 39th floor acceleration in the across-wind direction (March 15, 1998)

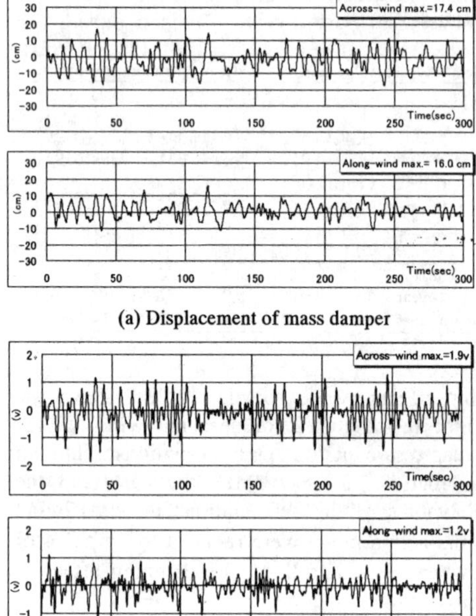

Figure 11. Displacement and control voltage of hybrid mass damper (March 15, 1998)

values of the39th floor when non-controlled and controlled were 13.5 cm/s^2 and 3.1 cm/s^2 , respectively. The peak acceleration when controlled was reduced to about 1/4.4 of that when non-controlled. Furthermore, in order to investigate the control effect, the power spectra were calculated from the 39th floor acceleration records in Figure 9(a), (b), as shown in Figure 10. From the comparison, it was verified that the damping of the building was obviously increased by installing the APMD and the performance during seasonal prevailing winds satisfied client's requirement. Figure 11 indicates the control voltage and mass damper response in the X- and Y-directions for the active control. When the mode is active control, the control voltage and mass damper response increase and the response of the building becomes small. The Lissajous diagram of the 39th floor acceleration records in the both along- and across-wind directions when controlled is compared with that when non-controlled as shown in Figure 12. It was confirmed that, in the case when non-controlled, the across-wind response grew larger than the along-wind response. This property was similar to the results obtained through the aerodynamic vibration experiment in the wind tunnel tests. The Lissajous diagram of the along-wind and across-wind responses when controlled showed the behavior like a circle, and the control performance was confirmed after the installation of the APMD.

Moreover, the performance obtained during the actual wind was almost identical with that analyzed for the simulated wind force. From the results, it was confirmed that the response analysis of wind-induced vibration was available to estimate the control performance of the APMD for the building. No uncomfortable vibration was observed under the controlled state, although there were many noticeable vibrations in the case when non-controlled. From the wind response analysis for seasonal wind excitations, it was confirmed that the peak acceleration values were much below the recommended level suggested by the guideline of A.I.J.(A.I.J. guideline 1994) for the evaluation of habitability against building vibrations. Likewise, the values of the maximum response acceleration obtained by the observation record under seasonal wind excitations with a occurrence probability of two or three times in a year shown in Figure 13 were also less than commonly accepted limits recommended by the A.I.J. guideline. Accordingly, based on the A.I.J. guideline shown in Table 5, the rank of the habitability was improved from the state beyond the rank Ⅲ to the rank Ⅱ by the APMD.

And next, the response control performance presents measurements performed on the building under Typhoon. Figure 14 plots the maximum acceleration of the control response and non controlled response at the 39th level of the building against the mean wind speed. It is confirmed from Figure 14 that the control response is clearly reduced to about 1/4～1/3 of those when non-controlled. Figure 15 compares building responses for active and passive modes and non-controlled in the 1st mode, where the results are shown as 39th floor's acceleration responses. In the 1st mode with active control, the responses was reduced to about 1/4 of those non-controlled, where as the corresponding reduction in passive control was about 1/2. When the mode is switched from passive

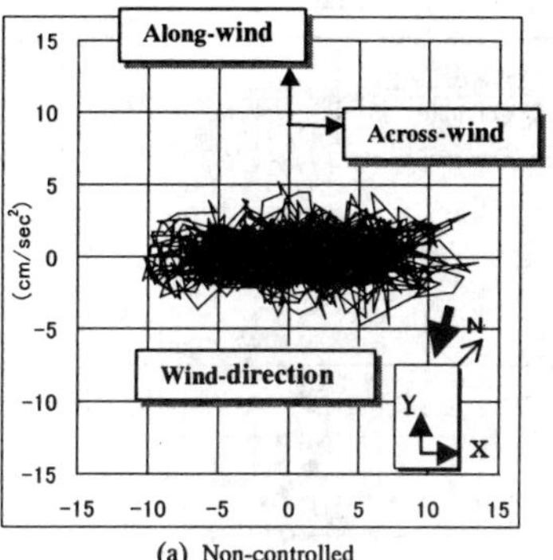

(a) Non-controlled

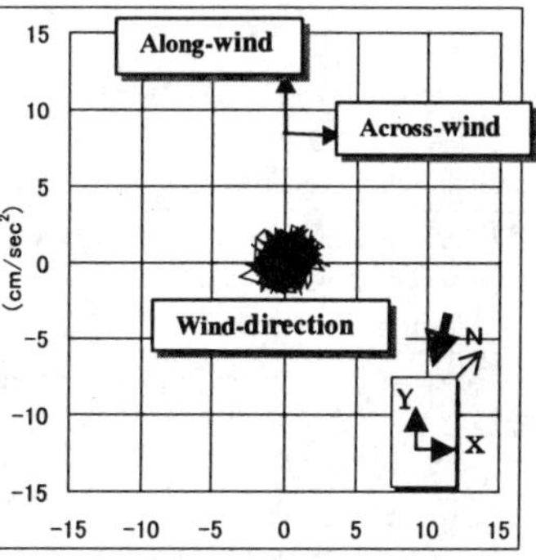

(b) Controlled

Figure 12. Comparison of Lissajous diagrams of the 39th floor acceleration when non-controlled and controlled (March 15, 1998)

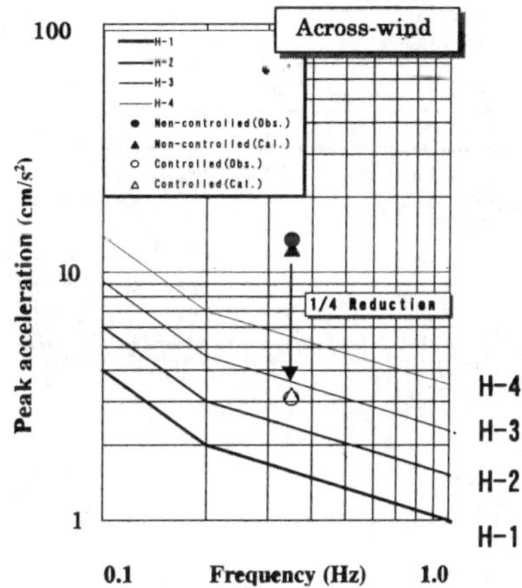

Figure 13. Estimation of habitability of the building

Table 5. Evaluation of rank of A.I.J. guideline

	Rank Ⅰ	Rank Ⅱ	Rank Ⅲ
Office	H- 2	H- 3	H- 4

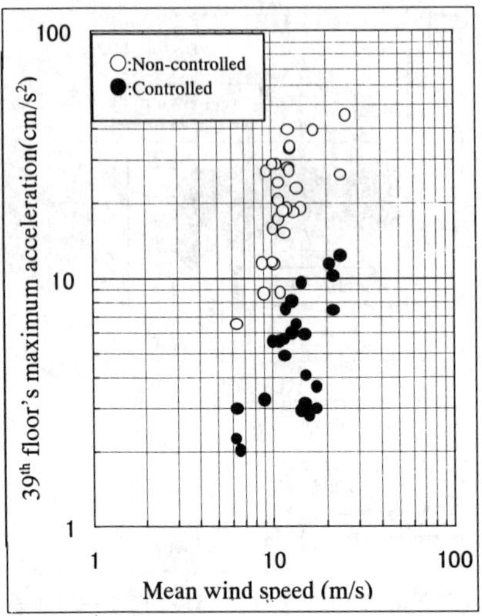

Figure 14. 39th floor's maximum acceleration of the building versus mean wind speed (Typhoon 9805)

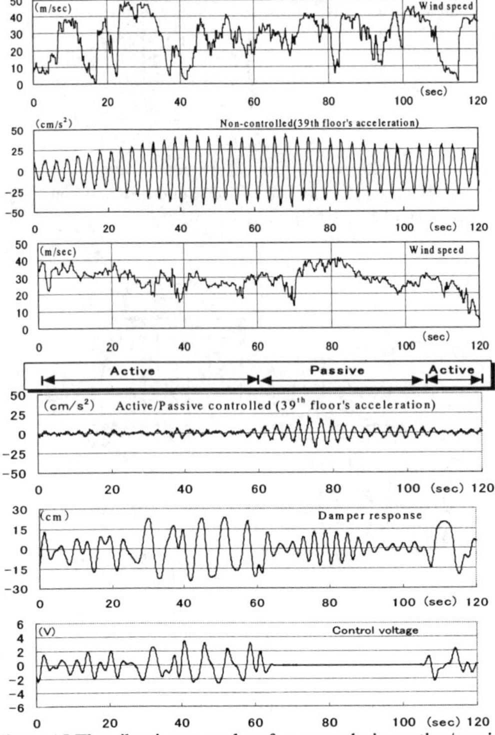

Figure 15.The vibration control performance during active/passive switching

to active, the control voltage and mass damper response sharply increase and the response of the building becomes small instantaneously. Then, it is confirmed that the active/passive mode is smoothly switched from active to passive by continuously opening/closing the bypass valves. The mode switching does not produce impulsive response of the building.

8 CONCLUSIONS

The forced vibration tests and the observation of the behaviors in the moderate wind excitations were carried out to investigate the response control performance of the APMD applied to the slender tall building. The conclusions are obtained as follows;

(1) The forced vibration tests showed that the damping for the 1st mode of the building increased by 8～11%, and the spillover instability did not occur.

(2) Through the observation during seasonal winds, it was confirmed that the peak and the RMS values of the accelerations under the active mode were reduced to about 1/4.5～1/2.0 (average 1/3) of those when non-controlled. This result was almost the same as the design analysis for simulated wind excitations.

(3) Based on the A.I.J. guideline for the evaluation of habitability against building vibrations, the rank of the habitability was improved from the state beyond the rank Ⅲ to the rank Ⅱ by the APMD.

(4) It was confirmed that the active/passive mode switching to avoid the operational limits was effective against strong winds.

ACKNOWLEDGEMENTS

Observation of the response control performance of the installed mass damper described in this paper was supported by Nippon OTIS Elevator Company. This support is gratefully acknowledged.

REFERENCES

Fujita, T.,Takanashi, S.,Matsumoto, K., Kinoshita, M. and Miyano, H. (1992) Mass damper with convertible active and passive modes for response control of buildings, *Proceeding of 10th World conference on earthquake engineering, Madrid*, vol. 4 pp2079-2083.

Watakabe, M.,Chiba, O.,Izumi, N., Okamoto, M. and Fujita, T. (1994) Wind response measurements for hybrid mass damper with convertible active and passive modes, *Proceeding of 13th National symposium on wind engineering, Tokyo*, pp467-472, (in Japanese)

Ebisawa, H., Watakabe, M.,Chiba, O.,Izumi, N. and Iwashita, S. (1997) Study on Response Control System with Hybrid Mass Damper Applied to A Slender Building (Part-1) Outline and results of wind tunnel test. , *Preprints of Annual Meeting of Architectural Institute of Japan* (in Japanese)

Architectural Institute of Japan, (1994) *Guidelines for the Evaluation of Habitability to Building Vibration,* (in Japanese), *Maruzen.*

Wind Engineering into the 21st Century, Larsen, Larose & Livesey (eds) © 1999 Balkema, Rotterdam, ISBN 90 5809 059 0

Wind action on the new Congress Centre, Lucerne, Switzerland

M.Wieland
Electrowatt Engineering Limited, Zürich, Switzerland

N.Isyumov & R.T.Edey
Boundary Layer Wind Tunnel Laboratory, University of Western Ontario, London, Ont., Canada

ABSTRACT: Wind tunnel model tests of the new Lucerne Congress Centre were made at a geometric scale of 1:200 and included simultaneous measurements of the instantaneous pressures at 256 locations on its exterior surfaces. These data were used to provide information on local peak pressures and suctions acting on components of the building envelope, as well as the overall wind loads on the primary structure. The emphasis of this paper is to describe the overall structural wind loads and responses and to examine the additional effects of wind-induced resonant vibrations. The additional inertia forces due to resonant vibrations were found to increase the peak overall structural loads and responses by some 15%.

The Congress Centre has been instrumented with accelerometers and strain gauges to obtain feedback on its dynamic properties and to monitor its performance under the action of wind and other loads. These measurements are continuing.

1. BACKGROUND

The new Lucerne Congress Centre is located on the shore of Lake Lucerne (Vierwaldstaettersee) adjacent to the main railway station and main public square. Wind engineering studies were carried out at the Boundary Layer Wind Tunnel Laboratory (BLWTL) of The University of Western Ontario in order to assist Electrowatt Engineering Ltd. (EWE) with wind-related design considerations, Harris et al. (1994). The massing of the Congress Centre is substantially greater than that of its immediate surroundings and there were concerns for the effects of wind on the design and performance of the Congress Centre itself and its potential impact on the wind environment in surrounding areas. A particularly important question was the wind loading and the response of the cantilevered portion of the main roof. At the north-east corner, the roof has an unsupported overhang of about 35m and 24m in the north and the east directions respectively. The wind tunnel study provided information on the wind-induced loads on the main structural system of the roof and its vertical deflections. The wind tunnel model study also provided information on local peak wind-induced pressures and suctions on components of the building envelope, as well as wind conditions in surrounding public areas. The emphasis of this paper is on the structural wind-induced loads and responses, and in particular the estimate of additional loads due to wind-induced resonant vibrations of the roof.

Since the construction of the Congress Centre, the roof has been instrumented with accelerometers and strain gauges to obtain feedback on its dynamic properties and to monitor its wind-induced motions and potential fatigue damage to critical joints and connections of structural members.

2. WIND TUNNEL MODEL STUDIES

Wind tunnel model tests were carried out using a 1:200 geometric scale pressure model of the Congress Centre and followed currently accepted practice ASCE (1987). Significant buildings within a full-scale radius of about 240m were included in the model. Wind tunnel tests were carried out for a full range of wind directions and for upstream exposures which simulated the terrain roughness of the surrounding suburban and urban areas and fetches over Lake Lucerne. Views of the model in the wind tunnel are shown in Figure 1. Figure 1a is the view looking towards Lake Lucerne. Figure 1c shows the prominent two-way cantilever portion of the roof at the north-east corner.

In the wind tunnel experiments, instantaneous pressures were measured at up to 256 locations distributed over roof and facade surfaces of the Centre. Simultaneous time histories of these pressures were recorded at a model sampling rate of 400 Hz. This sampling rate permitted pressure fluctuations to be measured up to a maximum full-scale frequency of

a/ View of Model in Wind Tunnel Looking Towards Lake Lucerne

b/ Close-up of Model Showing 2-way Cantilevered NE Corner of Roof

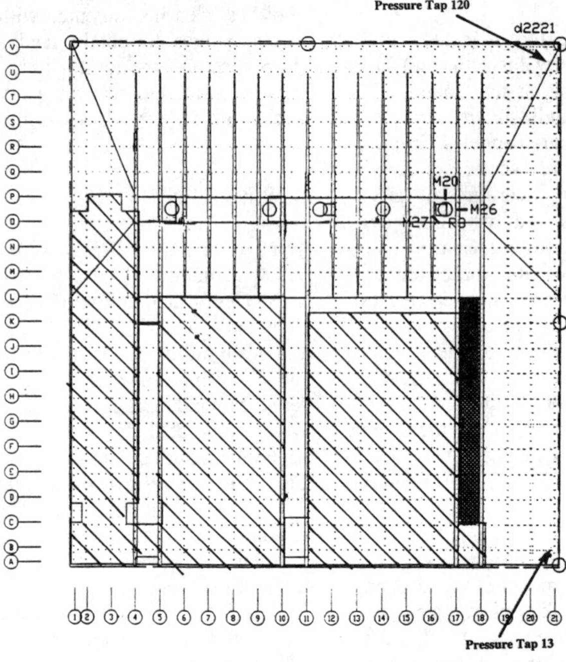

c/ Tap layout, Structural Quantities Measured

Figure 1 **Views of 1:200 scale Wind Tunnel Model of Lucerne Congress Center**

3.3 Hz at the 30-year return period wind speed. Experience with wind action for other large span roofs has indicated that resonant effects are generally minimal if the frequency of vibration is 1 Hz and higher. Some building codes, ASCE (1995), use this 1 Hz as a nominal divide between buildings and structures with and without significant wind-induced resonant vibrations, which subject the structure to additional inertia forces. It is this particular aspect of the study which is highlighted in this paper.

A dynamic analysis of the roof was carried out by EWE. The first mode of vibration was found to largely comprise a "flapping" of the cantilevered north-east portion of the roof. The natural frequency for this mode was predicted as 1.33 Hz. Wind-induced resonant vibrations for higher modes appeared to be unlikely with the exception of Mode 6, which comprised an asymmetric motion of the overhang along the north edge of the roof, with an estimated frequency of 2.2 Hz. Of these, the first mode was judged to be potentially more sensitive to wind-induced excitation as it involved a much smaller portion of the roof and could therefore develop more significant wind-induced generalized forces.

3. TYPICAL RESULTS

Local pressures

A sample time history of the local wind-induced pressure is shown in Figure 2. This is the differential pressure at the north-east corner of the cantilevered roof. The differential pressure is negative (uplift) and its time signature is presented for $\alpha = 45^{\circ}$, which is approximately a cornering wind and for a gradient wind speed of 46 m/s, which is estimated to have an average recurrence interval of 30 years. Both the pressure, as well as time are given in full scale units. The signature is a typical one, showing that peak local suctions tend to be of an intermittent nature and of very short duration. The bulk of the variance of the pressure fluctuations comes from lower frequencies, as can be seen from the power spectrum, which is also included in Figure 2. The power spectral density is in a normalized form; however, the frequency is shown as a full scale wave number. The corresponding wave numbers for the natural frequencies of the first and sixth modes of vibration are approximately 0.029/m and 0.048/m. Clearly, there is energy in the pressure fluctuations at these frequencies. A power spectrum for a pressure tap near the south-east corner of the roof for the same wind direction and wind speed shows that significant contributions to the variance of the pressure fluctuations can come from higher frequencies.. It must be concluded therefore that wind-induced resonant vibrations of these two modes cannot be ruled out without a thorough analysis.

Structural Loads From Spatially Averaged Pressures

Simultaneous time histories of the instantaneous pressures at all pressure taps on exterior surfaces of the Congress Centre were recorded. A subsequent analysis was carried out to determine various structural loads and responses. This can be done by assigning a surface area and influence coefficient to each pressure tap and computing their effects on the structure at every instant of time. Figure 3 shows time histories of various structural actions obtained for the roof of the Congress Centre for a wind direction of $\alpha = 45^{\circ}$ and a gradient wind speed of 46 m/s. The quantities shown are the vertical deflection at the north-east corner, a bending moment in one of the main north-south girders supporting the cantilevered north-east portion and the reaction in the pier supporting this roof girder. All quantities are shown in full-scale units. The sign convention for deflections is positive upwards and positive reactions are downward. Clearly, this shows the uplift action of the wind on this portion of the roof.

Information, such as shown in Figure 3, requires a full structural analysis to determine the influence coefficients which must be applied to the wind forces acting at the individual pressure taps. No assumptions about correlations or aerodynamic admittances are needed, as long as the forces or structural actions at particular instants of time are assembled from the corresponding simultaneously measured component pressures. The opportunities of this technique for tracking the behaviour of structures under wind action are virtually unlimited, providing that there is no appreciable amplification of the action of the dynamic wind forces by the structural system. In short, providing that the wind-induced resonant vibrations are minimal. As a result, the data shown in Figure 3 are incomplete estimates of the wind action in situations where the effects of additional resonance are significant. In most situations, including the Congress Centre, it is reasonable to proceed with the assumption that the response of a roof structure to the instantaneous action of wind forces can be taken to be approximately static (without mechanical amplification) and to allow for additional resonant effects as an upward correction. This has been the approach taken in this study.

4. EFFECTS OF WIND-INDUCED RESONANT VIBRATIONS

The simultaneous pressure data were used to determine the generalized forces for vibration modes 1 and 6, which represent the lowest two modes with significant movement of the overhanging portion of the roof. The generalized force for mode "j" is defined as:

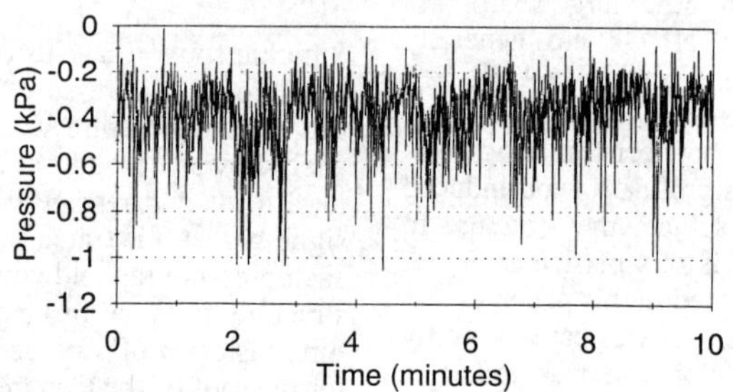

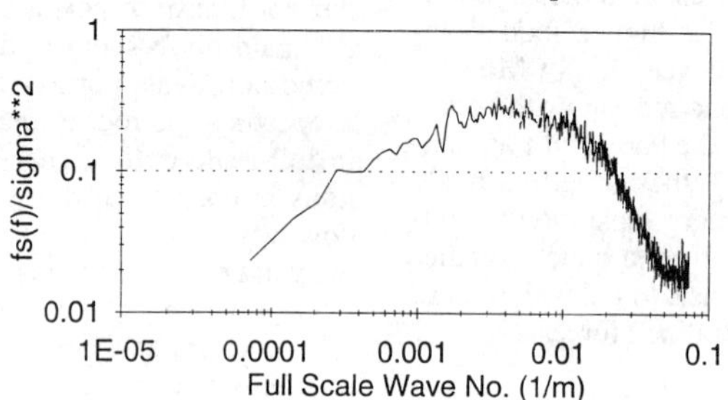

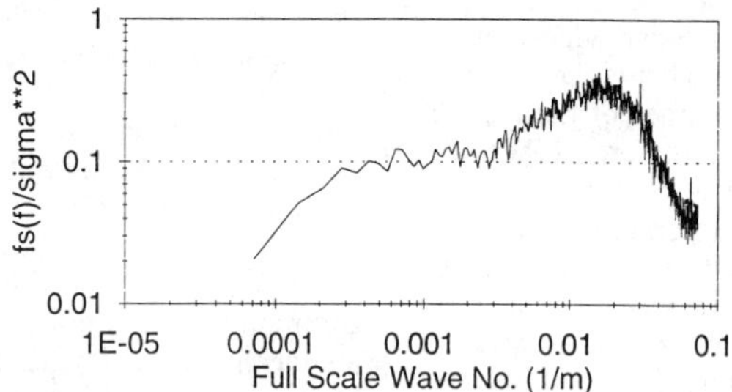

Figure 2 **Time Histories and Power Spectra of Different Pressures as Tap 120 (NE corner of roof) and Power Spectrum of Differential Pressure at Tap 13 (SE corner of roof) for winds of the Lake α=45° V_g=46 m/s**

$$Q^*(t)_j = \sum_{i=1}^{n} a_i \, p_i(t)\phi_{i,j} \tag{1}$$

where α_i = tributary area associated with pressure tap i; $p_i(t)$ = differential wind pressure at tap i at time t; $\phi_{i,j}$ = model amplitude of mode "j" at tap "i". The modal amplitudes are normalized so that their maximum value for each mode is 1.0.

The mode shapes for modes 1 and 6, namely $\phi_{i,1}$ and $\phi_{i,6}$, were calculated by EWE. Power spectra of $Q^*_{j=1 \text{ and } 6}$ are shown in Figure 4 for $\alpha = 45°$ and $\overline{V}_g$ = 46 m/s. The spectral densities, shown in normalized form, are small but not negligible at the full scale wave numbers of $f_{o_1}/\overline{V}_g$ = 0.029/m and $f_{o_6}/\overline{V}_g$ = 0.048/m.

The maximum effective value of the generalized force for a particular mode of vibration of the roof can be expressed as:

$$\hat{Q}^* \approx \overline{Q}^* + g\sqrt{\sigma^2_{Q^*} + \frac{\pi}{4}\frac{f_o\, S_{Q^*}(f_o)}{\varsigma_{Total}}} \tag{2}$$

where $\overline{Q}^*$ and σ_{Q^*} are the mean and quasi-static or background RMS values of the generalized force Q^* and $f_o\, S_{Q^*}(f_o)/\sigma^2_{Q^*}$ is the normalized spectral density of Q^* at f_o, the natural frequency of vibration of this mode. All three quantities were measured in the wind tunnel study for modes 1 and 6.

The peak factor g is typically in the range of 3.5 to 4.0. The total damping is $\varsigma_{Total} = \varsigma_s + \varsigma_a$. The aerodynamic damping ς_a is not exactly known, however experience with other large roofs indicates that $\varsigma_a > 0$. Therefore $\varsigma_{Total} = \varsigma_s$ is expected to be conservative. A value of $\varsigma_{Total} = 0.01$ has been used in the analysis. This is regarded as a good nominal value for a steel structure.

For convenience, equation 2 for the maximum generalized force can be rewritten to be in the form

$$\hat{Q}^* = \overline{Q}^* + \sqrt{\tilde{Q}_B^{*2} + \tilde{Q}_R^{*2}} \tag{3}$$

where $\tilde{Q}_B^*$ and $\tilde{Q}_R^*$ are the background (quasi-static) and additional resonant components of the dynamic portions of Q^* respectively. Namely

$$\tilde{Q}_B^* = g\sigma_{Q^*} \tag{4}$$

$$\tilde{Q}_R^* \cong g\sigma_{Q^*}\sqrt{\frac{\pi}{4}\frac{f_o\, S_{Q^*}(f_o)}{\sigma^2_{Q^*}\,\varsigma_{Total}}} \tag{5}$$

The mean value $\overline{Q}^*$ and the quasi-static dynamic value $\tilde{Q}_B^*$ are already accounted for in the measured peak wind-induced pressures and structural loads. What is unaccounted for is $\tilde{Q}_R^*$, which represents the effects of the additional "inertia forces", which result due to wind-induced resonant vibrations. The maximum value of the corresponding additional pressure experienced by the roof at location "i" can be expressed as follows:

$$\Delta p_{e_i} = m_i\, \hat{\ddot{y}}_i / a_i = \mu_i\, \hat{\ddot{y}}_i \tag{6}$$

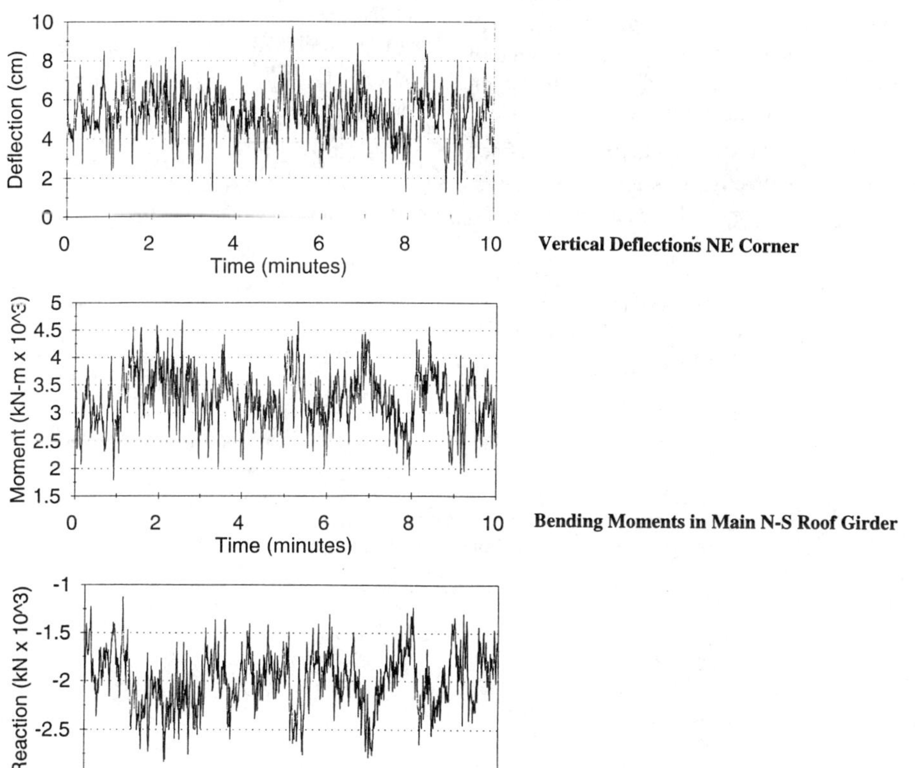

Figure 3 **Time Histories of Structural Actions for Wind of the Lake α=45° V_g=46 m/s Excluding Additional Wind Induced Resonance Effects**

where $\hat{\ddot{y}}$ is the maximum acceleration at location "i", due to resonant vibrations in the mode considered, and m_i and a_i are the mass and areas of the roof associated with location "i". Therefore, $\mu_i = m_i / a_i$ is the corresponding mass of the roof per unit area.

The value of the peak acceleration can be written as

$$\hat{\ddot{y}} \approx (2\pi f_o)^2 \frac{\tilde{Q}_R^*}{K^*} \tag{7}$$

where K^* is the generalized stiffness.

Since $K^* = (2\pi f_o)^2 M^*$, where M^* is the generalized mass, the effective additional pressure due to resonant vibration at location "i" becomes

$$\Delta p_{e_i} \approx \frac{\mu_i}{M^*} \phi_i \, g \, \sigma_{Q^*} \sqrt{\frac{\pi}{4} \frac{f_o S_{Q^*}(f_o)}{\sigma_{Q^*}^2 \varsigma_{Total}}} \tag{8}$$

Equation 8 was used to estimate this additional pressure Δp_e for modes 1 and 6 of the Final Configuration of the Congress Centre for selected wind directions and for a wind speed corresponding to a return period of 30 years. EWE provided information on $\mu_i, \phi_{i,j}, M_j^*$ and f_o for this analysis.

To simplify the analysis further, Δp_e^* was defined as the value of the effective pressure at the location where $\phi_i = \phi_{max} = 1$. For mode 1, this occurs at the north-east corner of the roof. The mass per unit area at this location is $\mu = \mu_i$ = 200 kg/m^2. For mode 6, ϕ_{max} occurs along the north edge around beam line 8. The mass per unit area at this location is $\mu = \mu_i$ = 140 kg/m^2.

$$\therefore \Delta p_e^* = \frac{\mu}{M^*} g \, \sigma_{Q^*} \sqrt{\frac{\pi}{4} \frac{f_o S_{Q^*}(f_o)}{\sigma_{Q^*}^2 \varsigma_{Total}}} \tag{9}$$

The largest values of Δp_e^*, estimated for modes 1 and 6, for the 30-year return period wind speed, are shown in Table 1. These values are based on ς_{Total} = 0.01 and a peak factor of g = 3.8. The values of Δp_e^* in Table 1 are for wind directions from the important northeast wind quadrant. Lower values are obtained for other parts of the compass, due to increased shelter. As a result, it is reasonable to reduce the values of Δp_e^* given in Table 1 by a directional factor of 0.7.

The additional effective pressure Δp_e^* or $0.7\Delta p_e^*$ and its effects on various structural load quantities must be combined in the same manner as the addition of $\bar{Q}^*$, $\tilde{Q}_B^*$ and $\tilde{Q}_R^*$. In other words, the effective local pressure becomes

$$\hat{p}' = \bar{p} + \sqrt{\tilde{p}^2 + \Delta p_{e_i}^2} \tag{10}$$

where $\hat{p}'$ becomes the new peak local pressure, in-

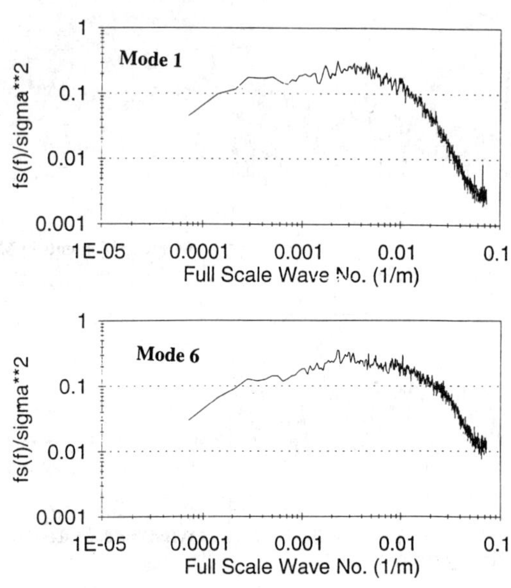

Figure 4 **Spectra of the Generalized Wind Induced Force of Mode 1 (fo = 1.33 Hz) and Mode 6 (fo = 2.20 Hz) for winds of the Lake α=45° V_g=46 m/s**

Table 1. Largest Estimated Value of Additional Effective Pressure Due to Wind-Induced Resonant Roof Vibrations at the 30-Year Return Period Wind Speed

Mode	f_o (Hz)	M^* (Kg)	μ kg/m^2	Δp_e^* (kPa)	$0.7\Delta p_e^*$ (kPa)
1	1.33	25,500	200	1.31	0.92
6	2.2	46,300	140	0.24	0.17

Table 2. Summary of the Estimated Increases in Structural Actions Due to the Effects of Resonant Vibrations.

Quantities	Increase Due to Resonance $\frac{(Peak\ Value)_{with\ res.}}{(Peak\ Value)_{without\ res.}}$			
	Least Affected		Most Affected	
	Positive	Negative	Positive	Negative
Vertical Deflections of Roof	1.0	1.0	1.14	1.16
Reactions Under Main Roof Girders	1.0	1.0	1.56	1.09
Bending Moments in Main Roof Girders	1.0	1.0	1.08	1.29

cluding the effects of resonance, and $\overline{p}$ and $\widetilde{p}_B$ are mean and quasi-static fluctuating components, as directly measured in the wind tunnel. The increase in the peak pressure, due to the inclusion of resonant effects, becomes

$$\eta_p = \frac{\hat{p}'}{\hat{p}} \tag{11}$$

where $\hat{p}$ is the peak pressure, without resonant effects, namely $\hat{p} = \overline{p} + \widetilde{p}_B$.

Similarly, the peak value of any formed quantity X becomes

$$\hat{X}' = \overline{X} + \sqrt{\widetilde{X}_B^2 + \widetilde{X}_R^2} \tag{12}$$

where $\overline{X}$ and $\widetilde{X}_B$ are the mean and quasi-static fluctuating components of the structural quantity X, and $\widetilde{X}_R$ is the additional component due to resonance, obtained by applying $(\Delta p_e)_i$ to the structure. The increase due to resonance becomes

$$\eta_x = \hat{X}'/\hat{X} \tag{13}$$

For example, the increase in one of the deflections, becomes

$$\eta_\Delta = \frac{\hat{\Delta}'}{\hat{\Delta}} \tag{14}$$

where

$\hat{\Delta}' = \overline{\Delta} + \sqrt{\widetilde{\Delta}_B^2 + \widetilde{\Delta}_R^2}$ and where and $\widetilde{\Delta}_B = g\sigma_{\Delta_B}$ and $\widetilde{\Delta}_R = g\sigma_{\Delta_R}$.

Since the various dynamic effects are added as the sum of squares of independent quantities, the effects of mode 6 become negligible. As seen from Table 1, the further increase due to including the effects of mode 6 in addition to mode 1 would be approximately $\sqrt{1^2 + 0.24^2/1.31^2} = 1.017$ or less than 2% of a relatively small quantity. As a result, only the additional effects of resonant vibrations in mode 1 were considered.

A summary of the effects of additional resonant vibrations on roof deflections, wind-induced reactions and wind-induced bending moments in the main roof girders is provided in Table 2. For each category of structural action, increases are shown for the negative and positive values of the least and most effected quantities.

The important deflection at the north-east corner of the roof is increased by approximately 15%. Increases of the reactions and bending moments, were found to depend on their magnitudes and signs. Typically, the numerically larger of the positive and negative peak values is increased by up to about 15%. The substantially larger increases were found to occur for numerically smaller and therefore less significant quantities.

5. PERFORMANCE TO-DATE

A programme of monitoring the full-scale performance of the Congress Centre roof has been initiated using an instrumentation system developed by EWE. This includes accelerometers at the north-east corner and at the centre of the cantilevered north edge of the roof. Also included are strain gauges on some of the key structural members. An anemometer has been positioned above the roof. Its present location is not a satisfactory one as it appears to be in a region where the flow is influenced by the presence of the roof. So far, it has not been possible to correlate these anemometer measurements with data from other weather stations in the Lucerne area.

Spectral analysis of the measured vertical accelerations have shown that the fundamental frequency of the roof is approximately 1.3 Hz. This is consistent with the first mode frequency of 1.33 Hz, calculated by EWE. A full-scale database is only –now emerging and it appears that the damping of the first mode of vibration may be comparable or smaller than 1% of critical, as assumed in the analysis. It is premature to draw specific conclusions on the effective damping at this time.

The maximum dynamic vertical deflection measured so far is 5.5mm obtained during a weather event with a maximum wind speed of 56 kpm from the north. The maximum wind event recorded so far has been a wind speed of 72 kpm from a wind direction of 312°. The corresponding maximum dynamic deflection at the north-east corner was measured to be 1.7mm. Data from the 9 strain gauges installed in one of the primary steel girders of the roof have not yet been fully examined. These are only initial data and it is premature to draw specific conclusions, particularly as the recorded wind speeds appear to be biased by the aerodynamic influence of the roof.

6. CONCLUSIONS

This paper has presented the main findings of a wind engineering study for the new Congress Centre in Lucerne, Switzerland. The particular emphasis in this paper has been the use of simultaneously measured pressures on exterior surfaces of the Congress Centre to arrive at wind loads and responses of the primary structural system of the roof. In particular, the emphasis has been to quantify the effects of additional inertia forces which act on the structure in the presence of wind-induced resonant vibrations. Theoretical procedures for this analysis have been presented in the paper.

The special analysis which has been carried out indicates that the effects of resonant vibrations are mainly limited to the overhanging north-east portion of the roof and are predominantly due to wind-induced vibrations in the first mode. The effects of resonant vibrations on the actual local peak pressures and suctions are minimal. For example the largest increase of the local peak suctions was found to be approximately 2%. This is within the accuracy of the overall study and does not justify any special further attention. On the other hand, wind-induced resonant vibrations were found to have an appreciable effect on the wind-induced loads on the primary structure of the roof. The peak deflections at the north-east corner of the roof were found to increase by approximately 15%. Increases in the reactions and bending moments were found to depend on their actual magnitudes and sign. Typically, the numerically larger of the positive and negative peak values increased up to about 15%. On the other hand, substantially larger increases were predicted for structural actions not particularly affected by the mean and background or non-resonant components of the wind load. For example, downward acting wind forces at reaction points of the cantilevered portions of the roof were generally small as the primary wind action is uplift. In this situation, resonant vibrations, which has similar positive and negative amplitudes, resulted in a substantial relative increase of the downward wind force.

Full-scale monitoring of the Congress Centre has been prompted both by scientific interest and the desirability to confirm the predicted performance of the roof. Measurements to-date have indicated a fundamental mode frequency of the roof of 1.3 Hz. This is in good agreement with the theoretical estimate of 1.33 Hz made by EWE. The monitoring programme has just recently become fully operational and further opportunities for comparison are expected.

REFERENCES

Harris, C., Edey, R.T. & Isyumov, N. 1994. A study of wind effects for the Congress Centre, Lucerne, Switzerland, *The University of Western Ontario* Report, BLWT-SS19-1994.

ASCE. 1987. Wind tunnel studies of buildings and structures, *American Society of Civil Engineers Manual of Pracatice*, No. 67.

ASCE Standard 7-95. 1995. Minimum design loads for buildings and other structures, *American Society of Civil Engineers*, New York, N.Y., U.S.A.

Wind Engineering into the 21st Century, Larsen, Larose & Livesey (eds) © 1999 Balkema, Rotterdam, ISBN 90 5809 059 0

Modelling tuned liquid dampers as sloshing-slamming dampers

Swaroop Yalla & Ahsan Kareem
Department of Civil Engineering and Geological Sciences, University of Notre Dame, Ind., USA

ABSTRACT: Tuned Liquid dampers (TLDs) have been implemented to suppress the response of structures to environmental loading, particularly wind induced vibrations, for sometime now. Although the installation, operation and maintenance of these dampers is relatively simple, theoretically very few models exist which model their dynamic features. The response of TLDs is non-linear and is characterized by an amplitude dependent frequency response function. In this study, the non-linear TLD characteristics are modeled as a sloshing-slamming damper. This mechanical model is based on the experimental data available and incorporates the impact phenomenon with liquid sloshing at higher amplitudes of excitation.

1.0 INTRODUCTION

A Tuned Liquid Damper (TLD)/Tuned Sloshing Damper (TSD) consists of a container partially filled with a liquid, usually water, placed on top of a structure. Like a Tuned Mass Damper (TMD), it imparts indirect damping to the primary system thereby reducing response. The energy dissipation occurs through various mechanisms : viscous action of the liquid, wave breaking, contamination of the free surface, screens and baffles, and container geometry. The damping characteristics of a TLD are amplitude dependent. The TLDs can be broadly classified into two categories : shallow-water and deep-water dampers. This classification is based on the ratio of the water depth to the water surface length in the direction of the motion. In the shallow water case, the TLD damping results primarily from energy dissipation through the action of the liquid's internal viscous forces and from wave breaking. For the deep-water damper, baffles or screens are needed to enhance damping. One drawback of the deep-water damper is the fact that a large portion of water does not participate in sloshing and adds to the dead weight.

The initial proposal to apply liquid dampers to building structures for reduction of wind-induced motion was reported by Modi and Welt, 1987; Kareem and Sun, 1987. Implementation of TLDs on actual structures was first done in Japan, examples include Nagasaki airport tower, Yokohoma marine tower, Shin Yokohoma Prince hotel and Tokyo international airport tower (Tamura, et al., 1995).

The motion of liquids in rigid containers has been the subject of many studies in the past few decades due to its frequent application in several engineering disciplines. A large number of the present and past studies have been based on linear and/or nonlinear potential flow theory. The numerical model based on the fully non-linear, non-dispersive shallow water theory may predict some important aspects of TLDs such as the hardening effect; however, it requires lengthy computations which makes a fully coupled (liquid-structure) analysis quite cumbersome. Mechanical modeling of sloshing dampers has been done by other researchers as well. An equivalent mechanical model for the sloshing of liquids in rectangular tanks has been proposed by Abramson, 1966. This model is based on the linear wave theory and assumes that no wave breaking takes place. Several researchers have conducted experimental studies on these type of dampers to model the non-linear characteristics with empirical equations (Fujino, et al., 1992; Reed, et al., 1996). Sun, et al., (1995) have presented a TMD analogy for the non-linear sloshing in TLDs. The interface force between the damper and the structure is conceptualized as a force from a virtual mass and dashpot. These analytical values were compared to a series of experiments where good fits were obtained in most cases. In this paper, the sloshing phenomenon for shallow and intermediate depths has been modeled as a sloshing-slamming damper. This model is appropriate in light of the existing experimental evidence and includes the impact phenomenon with the liquid sloshing at higher amplitudes of excitation.

2.0 BACKGROUND

The proposed sloshing-slamming model is a combination of two models, namely the linear sloshing model and the impact damper.

2.1 Linear sloshing mechanical model

A simplified mechanical model of sloshing in rectangular tanks is based on linear potential flow theory (Abramson, 1966). The model accounts for all the essential features of the influence of tanks on the response of the primary system without resorting to a complicated finite element model. The following assumptions are made: (a) rigid rectangular tank (b) homogenous liquid (c) incompressible and non-viscous liquid (d) no sources or sinks (e) small wave slopes, and (f) irrotational flow. The schematic details of the equivalent mechanical model are shown in Fig. 1(a). It can be shown on the basis of linear wave theory that,

$$m_n = M_l\left(\frac{8\tanh\{(2n-1)\pi r\}}{\pi^3 r(2n-1)^3}\right); \quad n=1, 2........ \tag{1}$$

$$\omega_n^2 = g(2n-1)\pi\tanh\{(2n-1)\pi r\}/a; \quad n=1, 2...... \tag{2}$$

where n is the sloshing mode, m_n is the mass of liquid acting in that mode, ω_n is the frequency of sloshing, $r = h/a$ where h is the height of water in the tank and a is the length of the tank in the direction of excitation, M_l is the mass of the total water in the tank and m_o is the 'dead mass' which does not participate in sloshing, and is given by,

$$m_0 = M_l - \sum_{n=1}^{\infty} m_n \tag{3}$$

Usually, the fundamental mode of liquid sloshing, $n = 1$, is used for analysis. The dynamic response using this model is similar to that of a TMD with double peaks of equal height at optimal damping which change to a single peak as the damping increases. However, this model works well for small amplitude excitations, where wave breaking and nonlinearities are not introduced into the system.

2.2 Impact damper

An impact damper is a small rigid mass placed in a container which is firmly attached to the primary system (Fig. 1(b)) (Lieber and Jensen, 1945). A clearance exists between the container and the impact damper so that collisions take place intermittently as soon as the displacement of the primary system exceeds the clearance. These collisions produce energy dissipation and exchange of momentum. The primary source of attenuation of motion in the primary system is due to this exchange of momentum. This momentum exchange reverses the direction of motion of the impact damper.

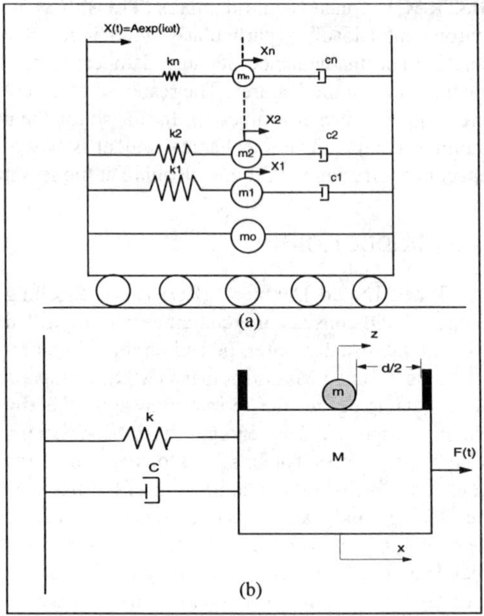

FIGURE 1. (a) Mechanical model of sloshing liquid (b) Impact damper model

The equations of motion are written as,

$$\begin{aligned} M\ddot{x} + C\dot{x} + Kx &= F(t) \\ m\ddot{z} &= 0 \end{aligned} \tag{4}$$

and the velocity of the primary system after collision is given as,

$$\dot{x}_{ac} = \frac{(1-\mu e)}{(1+\mu)}\dot{x}_{bc} + \frac{\mu(1+e)}{(1+\mu)}\dot{z}_{bc} \tag{5}$$

where e is the coefficient of restitution of the materials involved in the collision and μ is the mass ratio =m/M. The subscripts ac and bc refer to after-collision and before-collision states, respectively. The velocity of the impact mass is reversed after each collision. The numerical simulation aspects of this problem are discussed in the next section.

3.0 SLOSHING-SLAMMING MODEL

In the experiments conducted by Sun, et al., 1995, the TLD was excited by a shaking table. The results are shown in Fig. 2, where the jump frequency and the damping ratio are plotted against the amplitude of base excitation. Note that these experiments were conducted on an uncoupled TLD excited at the base. Figure 2 (a) shows that there is a hardening effect at higher amplitudes of excitation. This type of nonlinear hardening occurs at frequency ratios ($\beta = \omega_e/\omega_s$) > 1.00, where ω_e is the frequency of excitation and ω_s is the linear sloshing frequency, and is dependent on the amplitude of excitation. It has been observed that as the amplitude of excitation increases, the energy dissipation occurs over a broader range of frequencies, and therefore, the TLD behaves as a more robust system. Modeling of the uncoupled TLD system will be discussed later in the context of the base shear force.

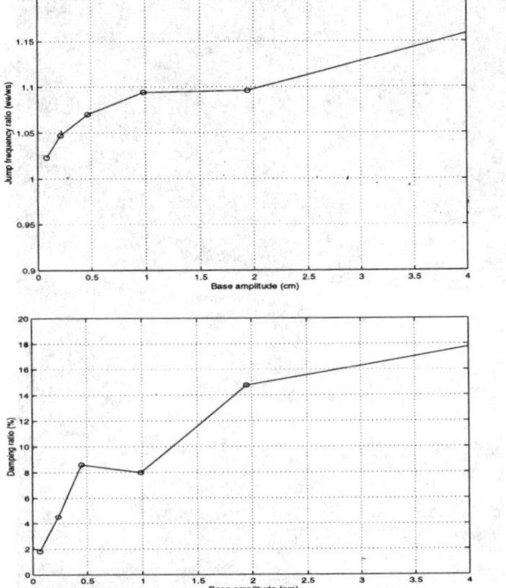

FIGURE 2. Variation of (a) jump frequency and (b) damping ratio of the sloshing damper with the base amplitude (*Experimental results taken from Sun, et al., 1995*).

Here attention is focussed on the coupled TLD-structure system which exhibits certain prominent characteristics as the amplitude of excitation increases. Experimental studies suggest that the frequency response of TLD's, unlike a TMD's, is excitation amplitude dependent. It has been observed that due to the increased damping (brought about by wave breaking and other associated phenomenon), the frequency response function changes from a double peak to single peak curve.

Based on the visual evidence provided by videos of sloshing experiments (Fig. 3), the authors observed that at higher amplitudes the sloshing phenomenon more closely resembles an impact or slamming phenomenon. The liquid mass rolls and slams against the container walls periodically. Similar observations have been made regarding the impact of breaking waves on bulkheads in coastal engineering applications. Most numerical models do not account for this impinging force on the walls of the container.

The proposed model includes the impact phenomenon in the overall model, and is shown schematically in Fig. 4. The model proposes a mass exchange taking place between the two systems; i.e., at higher amplitudes, some portion of the linear sloshing action is transferred to the impact damper mechanism. The model attempts to introduce amplitude-dependent mass convection effect known to exist in nonlinear sloshing.

The amount of transfer can be related to the amplitude of excitation. We introduce a parameter γ, which is an indicator of the portion of linear mass m_1 acting in the linear mode. Since the total mass is conserved, this implies that the rest of the mass is acting in the impact system. For example, $\gamma = 1.0$ means all of the mass m_1 is acting in the linear sloshing system. After the mass exchange has taken place, the new masses $\underset{\sim}{m}_1$ and $\underset{\sim}{m}_2$ in the linear sloshing system and the impact system, respectively, are given by,

$$\underset{\sim}{m}_2 = m_2 + (1-\gamma)m_1 \tag{6}$$

$$\underset{\sim}{m}_1 = \gamma m_1 \tag{7}$$

At low amplitudes, there is no mass exchange and the linear theory holds, but as the amplitude increases, γ decreases and the impact mass increases. Moreover, since m_1 is decreasing, the frequency of sloshing increases, which explains the hardening effect. Now, to relate the variation of the parameter γ with the amplitude of excitation, one uses the relation (which for simplicity can be assumed to be linear) between base-amplitude of excitation and the jump frequency ratio, given in Fig. 2(a),

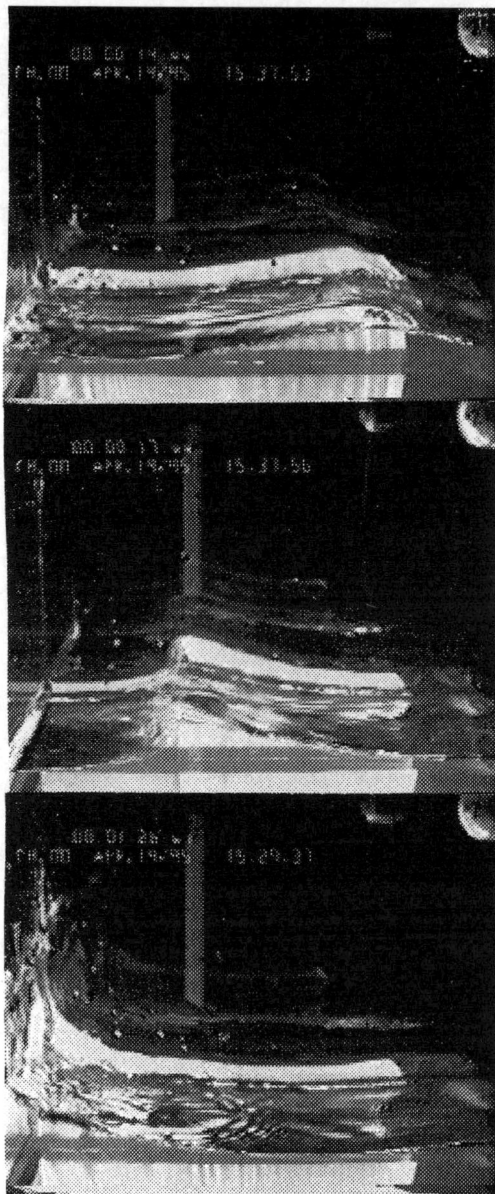

FIGURE 3. Frames from the sloshing experiments video at high amplitudes. (*Video of TLD experiments, Courtesy, Dr. D.A. Reed*)

$$\beta = \kappa A + 1 \tag{8}$$

where A is the base amplitude of excitation and κ is the slope of the linear relationship obtained by curve-fitting through the data in Fig. 2(a).

Moreover, note that: $\omega_1^2 = \frac{k_1}{m_1} = \frac{\omega_1^2 m_1}{m_1}$, which from Eq. 7 implies $(\beta)^2 = \frac{1}{\gamma}$. Substituting Eq. 8 in this relation gives,

$$\gamma = \frac{1}{(\kappa A + 1)^2} \tag{9}$$

Equation 9 holds for the range in which the linear relationship given in Fig. 2(a) is valid. Note that this is a general formulation and any relation (linear, bi-linear or non-linear) can be introduced from which the mass exchange parameter (γ) can be related to the amplitude of excitation (A). The system of equations for the system shown in Fig. 4 can be written as:

$$\begin{aligned} M\ddot{X} + (C + c_1)\dot{X} + (K + k_1)X - c_1\dot{x}_1 - k_1 x_1 &= F(t) \\ m_1\ddot{x}_1 + c_1\dot{x}_1 + k_1 x_1 - c_1\dot{X} - k_1 X &= 0 \\ m_2\ddot{z} &= 0 \end{aligned} \tag{10}$$

These are the three governing equations. After each collision the velocity of the secondary system, which is coupled to the impact system, changes in accordance with Eq. 5. A collision is numerically simulated at the time when the relative displacement between m_1 and m_2 is close to $d/2$ within a prescribed tolerance. The relative tolerance has been taken as $1x10^{-6}$. Since the relative displacements have to be checked at each instant of time, time domain numerical integration methods have to be used to solve the system of equations. In order to construct the frequency response curves, the maximum steady state response was read off at each excitation frequency and was plotted for a range of frequencies.

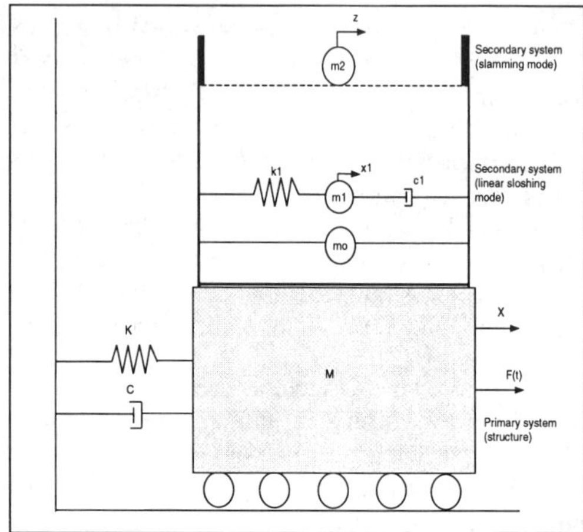

FIGURE 4. Proposed sloshing-slamming model

4.0 NUMERICAL SIMULATION

A numerical study was conducted using the experimental model parameters in Fujino, et al., 1992 which are listed in Table 1. Figure 5 shows the transition of the frequency response curves as the mass exchange parameter is varied. This can also be seen as the amplitude dependent variation of the frequency response function.

Note that the frequency response function changes from a double peak to a single peak curve corresponding to non-linear slamming phenomenon at higher amplitudes. It can also be observed that the linear model and the sloshing-slamming model give similar frequency response curves at low amplitudes of excitation. This means that the linear theory holds well for low amplitudes of excitation.

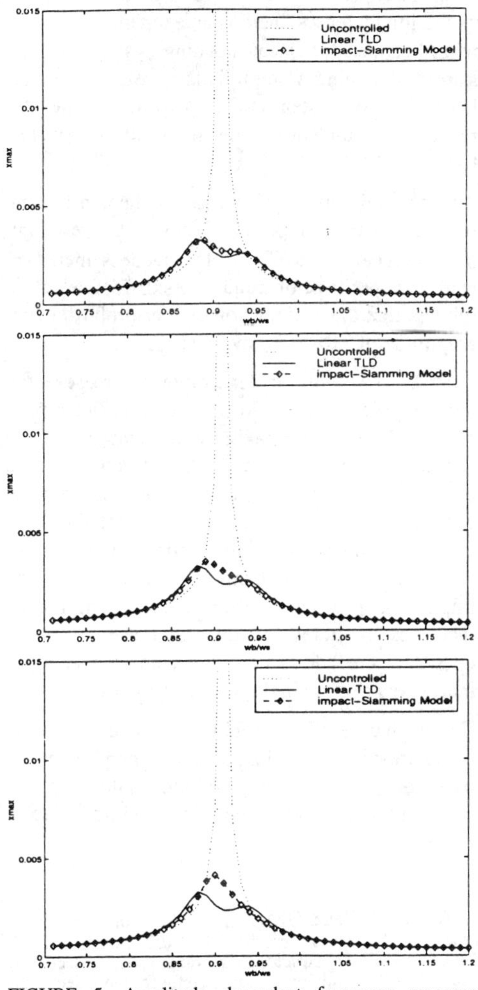

FIGURE 5. Amplitude dependent frequency response function (a) $\gamma = 1.00$ (b) $\gamma = 0.99$ (c) $\gamma = 0.97$

TABLE 1. Parameters for Simulation

Primary system mass, m (Kg)	168 Kg
Primary system damping	0.32 %
Natural freq. of primary system	5.6 rad/s
length of tank, a	25 cm
breadth of tank, b	32 cm
height of liquid, h	2.1 cm
coefficient of restitution, e	0.4
clearance for impact do/(F/k)	20

5.0 BASE SHEAR FORCE

The non-dimensionalized base shear is plotted for various amplitudes based on experimental data from Fujino, et al., 1992 in Fig. 6 (a). One can note clearly the presence of the jump and hardening phenomenon. Furthermore, the range of frequencies over which the TLD is effective increases as the base amplitude increases. In order to model this type of non-linearity in the spring, let us consider two models based on a nonlinear oscillator. Both simulate the non-linearity in the base shear force effectively, but one of the models explains the physical phenomenon in a more intuitive way, in accordance with the sloshing-slamming model discussed earlier. Consider an oscillator with non-linear cubic stiffness and damping term, whose equation of motion is given as,

$$m\ddot{x} + c\dot{x} + kx + \lambda\dot{x}^3 + \Gamma x^3 = F_o \sin(\omega_e t) \tag{11}$$

where λ and Γ are the non-linear cubic damping and stiffness constants, respectively. The maximum base shear or the maximum force exerted on the wall of the container by the sloshing liquid is given as,

$$F_b = max\{kx + c\dot{x} + \lambda\dot{x}^3 + \Gamma x^3\} \tag{12}$$

and the normalized maximum base shear force is given as,

$$F_b' = F_b/(m\omega_e^2 A) \tag{13}$$

The non-dimensionalized base shear force defined in Eq. 13 is plotted in Fig. 6(b) as the cubic stiffness and damping terms in Eq. 11 and the amplitude of excitation are increased (model 1). It is noted that these plots are quite close to the experimental results given in Fig. 6(a). However, this model does not give us any physical insight into the problem. On the other hand, if a small (but constant) cubic stiffness and damping terms are retained, and concomitantly the mass is reduced and the linear mode damping is increased, the resulting non-dimensionalized base

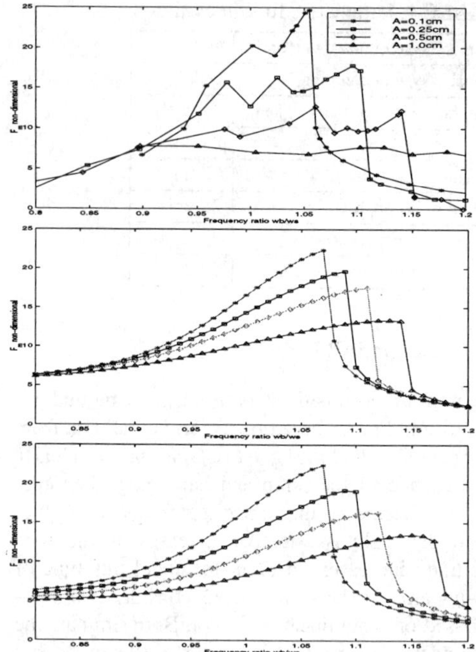

FIGURE 6. (a) Variation of the non-dimensionalized base shear force with the frequency ratio. (*experimental results taken from Fujino, et al. 1992*); Numerical simulations using duffing models (b) model 1 (c) model 2.

shear force is shown in Fig. 6 (c). These plots (model 2) are similar to the results in Fig. 6 (b). Model 2 exhibits a clear reproduction of the measured data as well as reflects the physics introduced by the change in the role of the mass term and added damping due to the wave impact.

Some salient points to be noted are:

1. Introducing a non-linear cubic stiffness and damping term is justified. From experimental data it is evident that even at low amplitudes there is a presence, albeit small, of non-linearity.
2. The decreasing mass in the primary sloshing system can be explained from our sloshing-slamming model. Some of the mass transfers to the impact sub-system via the mass transfer as described in the earlier section.
3. The increase in damping is due to the impact damper action and the formation of breaking waves. This has been verified experimentally. (Fig. 2(b)).

6.0 CONCLUSIONS

In this paper a new mechanical model of tuned sloshing dampers is presented. It has been noted that previous models, both mechanical and numerical, have certain shortcomings in explaining some features of TLDs. It has been seen that the sloshing motion of liquid at high amplitude excitations resembles more closely an impact or slamming phenomenon. Water convects as a lumped mass and strikes the sides of the wall producing a change of momentum. The proposed sloshing-slamming model builds on the experimental evidence available and incorporates the impact phenomenon in the overall model. The following specific conclusions are drawn from our studies:

1. The linear model performs satisfactorily at low amplitudes of excitation. As the amplitude of excitation increases, there is an exchange of liquid mass between the linear sloshing system and the impact slamming system. This is responsible for hardening the system and making the frequency response dependent on the amplitude of excitation.
2. At higher amplitudes of excitation, the non-linearity reduces the damper efficiency at the resonant peak. However, the TLD effectiveness increases over a wider range of frequencies. Effectively, this implies that the TLD becomes more robust as the amplitude of excitation increases.
3. As the amplitude of excitation increases, the damping increases due to complex non-linear phenomena like wave-breaking and slamming which cannot be quantified easily. An attractive feature of the proposed model is that one does not have to explicitly specify how the damping of the TLD varies with amplitude. This model implicitly provides the additional damping in the form of an impact damper. The fact that the damping indeed increases is noted in the frequency response curves where added damping (higher than optimal) gives rise to single peak curves (Fig. 5).
4. The exchange of mass taking place between the linear mode and the impact mode can be related by empirical relationships which relate the variation in jump frequency and the amplitude of excitation.

7.0 ACKNOWLEDGMENTS: The authors gratefully acknowledge the support provided by NSF Grant CMS-95-03779 under the NSF Structural Control Initiative directed by Drs. S. C. Liu. The authors are also grateful for the video of the sloshing experiments provided by Dr. D.A.Reed.

8.0 BIBLIOGRAPHY

1. Abramson, H.N., (ed.) (1966), "The Dynamic Behavior of Liquids in Moving Containers", NASA, SP-106.

2. Fujino,Y., Sun, L.M., Paceno,B., and Chaiseri, P., (1992), "Tuned Liquid Damper (TLD) for Suppressing Horizontal Motion of Structures," J. Engg. Mech., Vol.118, 2017-2030.

3. Kareem, A. and Sun, W.J., (1987), "Stochastic Response of Structures with Fluid-containing Appendages," J. of Sound and Vibration, 119(3), 389-408.

4. Reed, D., Yu, J., Yeh, H., and Gardarsson, S., (1996), " Investigation of Tuned Liquid Dampers under Large Amplitude Excitation," J. Engg. Mech., 405-413.

5. Sun, L. M., Fujino,Y., Paceno, B., and Chaiseri, P., (1995), "The Properties of Tuned Liquid Dampers using a TMD Analogy," Earthquake Engg. Struc. Dyn., Vol. 24, 967-976.

6. Tamura, Y., Fujii, K., Ohtsuki, T., Wakahara, T., Kohsaka, R. (1995), "Effectiveness of Tuned Liquid Dampers under Wind Excitation," Engineering Structures, 17 (9), 609-621.

7. Modi, V.J. and Welt, F. (1987), "Vibration control using nutation dampers", International conference on Flow induced Vibrations, R. King (ed.), BHRA, London, 369-376.

8. Lieber, P. and Jensen, D.P., (1945), "An Acceleration Damper: Development, Design and Some Applications," Trans. ASME, Vol. 67, 523-530.

Wind Engineering into the 21st Century, Larsen, Larose & Livesey (eds) © 1999 Balkema, Rotterdam, ISBN 90 5809 059 0

Wind tunnel study of a building structure incorporating viscous-damping walls

Ngai Yeung & Austin D.E. Pan
Department of Civil Engineering, The University of Hong Kong, People's Republic of China

Abstract: This paper presents the results of an experimental study on the effectiveness of VD walls for controlling wind-induced vibrations in building structures. In a wind tunnel study, a simple four-story building, square in plan, was constructed and instrumented. A series of wind tunnel tests of the model with and without viscous-damping walls were conducted. The effectiveness of VD walls under different fluid levels and wind speeds and attack angles was investigated. The experimental results showed that the viscous-damping wall is highly effective in most cases in reducing wind-induced vibrations. The design and benefits of viscous-damping walls for building construction are discussed.

1 INTRODUCTION

The trend towards higher and more flexible buildings has produced structures that are more sensitive to wind excitations. An essential way to controlling the wind-induced vibrations is to achieve a significant level of damping for structures. Passive control devices are most commonly used for this purpose. In this paper a passive control device referred to as viscous-damping (VD) wall is presented and its behavior and effectiveness under wind are investigated.

Viscous-damping (VD) wall was first proposed by a Japanese engineer, Mitsuo Miyazaki, in the 1980's as a form of passive vibration control device that would substantially increase the damping values of buildings and thus reduce the dynamic responses (Miyazaki *et al.* 1986).

A VD wall unit is illustrated in Figure 1. Viscous fluid is sandwiched by three wall plates. During construction, the inner plate is attached to the upper floor of a building story while the two outer plates are mounted on the lower floor of the same story. When a building is subjected to dynamic forces induced by wind or earthquake, each story will experience relative horizontal displacements between the upper and lower floors (Figure 2). This inter-story drift causes the inner plate to move relative to the outer plates. This action, due to the presence of the viscous fluid between the plates, generates viscous-damping and restoring forces to oppose the inter-story drift. The overall dynamic response of the building is thereby reduced.

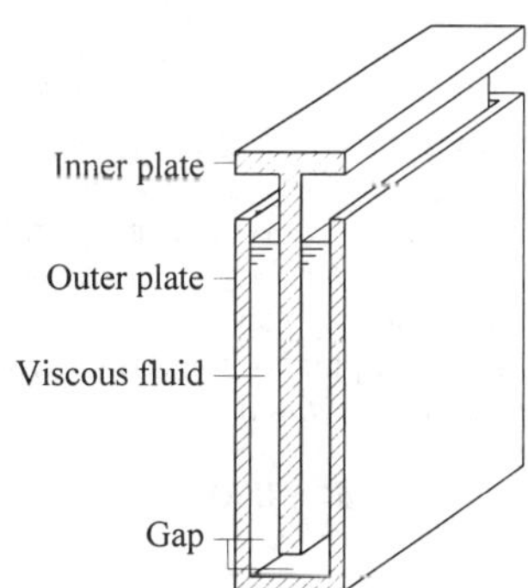

Figure 1. Viscous-damping wall unit

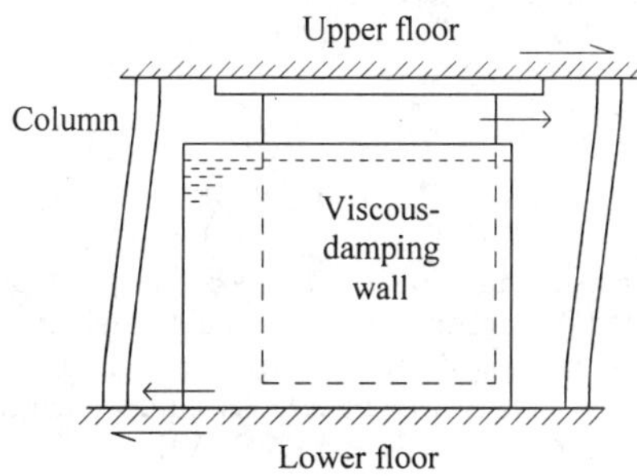

Figure 2. Inter-story drift and relative displacement

The test results of an earlier model incorporated with VD walls have been reported by Yeung and Pan (1997 and 1998). This paper presents the results of a new model in which the scaling has been selected to produce a larger vibrational response due to vortex shedding. Five series of wind tunnel test were conducted for five kinds of models referred to as the model without fluid and with 1/4, 1/2, 3/4 and full of fluid. In each series, the mean wind speed was changed at five levels from 6 *m/s* to 14 *m/s* at 2 *m/s* intervals. The angle of wind attack varied from 0 degree to 90 degree at 15 degree increments.

2 DAMPING MECHANISM

The viscous-resistant force (Q_w) generated in the wall can be expressed by the Voight-Kelvin model as follows:

$$Q_w = Q_c + Q_k \tag{1}$$

where Q_c is the damping force due to velocity gradient in the fluid and Q_k is the restoring force due to relative displacement.

The formula for the damping force Q_c is expressed as:

$$Q_c = \mu A \left(\frac{dv}{dy} \right)^{\alpha} \tag{2}$$

where A is the area of the plate, μ is the fluid viscosity, dv is the relative velocity, and dy is the gap distance between the two adjacent plates. Since the viscous fluid used in VD walls is not a perfect Newtonian fluid, an exponent factor α obtained from experiments is applied to the term dv/dy.

The restoring force, which is dependent on the relative displacement between plates, may be expressed as (Arima *et al.* 1988):

$$Q_k = \mu A \frac{\delta^{\beta}}{dy^2} \tag{3}$$

where δ is the relative displacement and the exponent factor β is again obtained from experiments.

3 DESCRIPTION OF THE BUILDING MODEL

An aeroelastic building model was designed and constructed for testing in the wind tunnel at the University of Hong Kong. Testing of the model was conducted with and without viscous fluid so that the behavior and effectiveness of VD walls in controlling the wind-induced vibrations could be compared.

3.1 *Aeroelastic scaling of the building model*

A four-story aeroelastic building model was designed and constructed using the scaling factors shown in Table 1.

Table 1. Aeroelastic scaling

Property	Modeling parameter	Value
Length	L_m/L_p	1/12
Velocity	V_m/V_p	1/5
Time	$T_m/T_p=V_pL_m/V_mL_p$	5/12
Acceleration	$a_m/a_p=V_m^2L_p/V_p^2L_m$	12/25
Material density	ρ_m/ρ_p	1
Air density	$\rho_{air\ m}/\rho_{air\ p}$	1
Mass	$m_m/m_p=\rho_mL_m^3/\rho_pL_p^3$	1/1728
Frequency	$N_m/N_p=T_p/T_m$	2.4
Mass moment of Inertia	$J_m/J_p=\rho_mL_m^5/\rho_pL_p^5$	4.02E-06
Stiffness	$K_m/K_p=\rho_mL_m^3T_p^2/\rho_pL_p^3T_m^2$	1/300
Moment of inertial	$I_m/I_p=K_mL_m^3/K_pL_p^3$	1.93E-06
Force	$F_m/F_p=\rho_mL_m^4T_p^2/\rho_pL_p^4T_m^2$	2.78E-04
Moment	$M_m/M_p=\rho_mL_m^5T_p^2/\rho_pL_p^5T_m^2$	2.31E-05
Damping	ζ_m/ζ_p	1

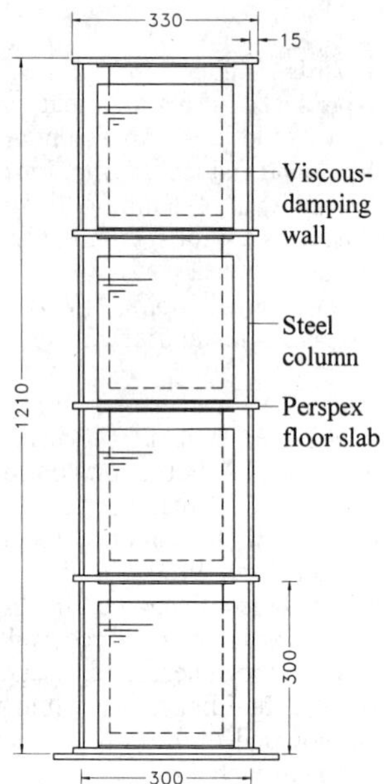

Figure 3. Dimensions of the building model

3.2 *Building model*

An elevation view of the finished four-story building model with VD walls installed in each story can be seen in Figure 3 (with exterior cladding removed). Overall dimensions of the test model are

330-*mm* by 330-*mm* in plan and 300-*mm* story heights. The floor slabs and wall plates were made of perspex and the columns of stainless steel. The diameter of the column is 8 *mm*. The thickness of walls and floor slabs are 6 *mm* and 10 *mm*. A single viscous-damping wall unit was installed at the center of each floor.

3.3 *Viscous-damping wall unit*

The viscous-damping wall unit used in the model is shown in Figures 4 and 5. It has overall dimensions of 240-*mm* in width and 290-*mm* in height. The size of the gap between the inner plate and outer plates is 4 *mm*.

Silicone fluid (Jaeger 1408s) with a viscosity of 12,000 *centistokes* and density of 974 kg/m^3 at 25 °C was used to fill into the VD wall units during testing.

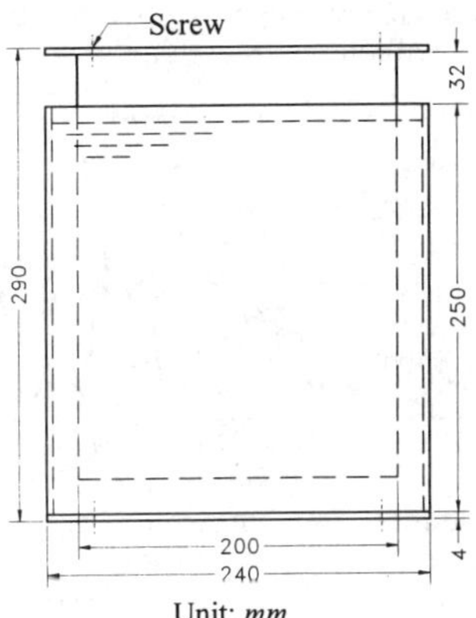

Figure 4. Dimensions of the VD wall unit

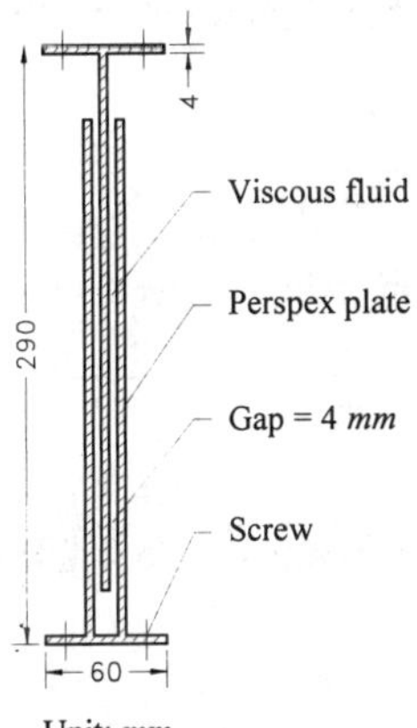

Figure 5. Cross-section of the VD wall unit

4 EXPERIMENTAL SET-UP

4.1 *Wind tunnel*

Wind tunnel testing was carried out in the Environmental Wind Tunnel at the Department of Civil Engineering, the University of Hong Kong. The working section of the wind tunnel is 3-*m* wide and 1.8-*m* high and is approximately 18 *m* long. The 2-*m* diameter wind fan can produce a maximum air-stream speed of 18 *m/s*. The temperature in the wind tunnel is controlled at a constant 25 °C.

4.2 *Instrumentation of the building model*

To measure the dynamic response, the model was instrumented with piezoelectric accelerometers placed in both the along-wind and cross-wind directions on each floor level and the roof. Pressure gauges were placed on the exterior cladding of the model to measure the wind pressure. Two displacement meters (LVDT) were installed at the roof level to measure the maximum horizontal drift. Force transducers were installed at the base of each column to measure the axial forces during vibration.

4.3 *Experimental program*

Static tests were conducted to determine the stiffness of the model. Free vibration tests were performed by giving an initial horizontal displacement at the roof level of the model and then releasing. These tests were conducted for cases without viscous fluid and with different height level of viscous fluid. Based on these results, the damping ratios were computed.

A series of tests, of the building model with and without viscous fluid, were conducted to investigate the behavior and effectiveness of VD walls in controlling wind-induced vibrations for flexible, multi-story building structures.

To measure the dynamic responses of the model with and without viscous fluid, a series of wind tunnel tests were conducted for a total duration of 9 minutes and at 5 levels of wind speed. Starting off at 6 *m/s*, the mean wind speed was increased to 8 *m/s*, then back down to 6 *m/s*, and in the final stage of testing the mean wind speed was turned up to 14 *m/s*.

The dynamic responses of the building model in both along- and cross-wind directions were measured by the accelerometers and transferred to the computer for processing in real-time. On each floor, two accelerometers manufactured by PCB Piezotronics, Inc. (Model 352B68) were mounted in the center to measure the two translational accelerations. Another accelerometer was on the edge of the floor plan to measure the torsional acceleration. Four accelerometers (Type B&K 4370 with Amplifier B&K 2635) were installed in the center of the four floors which were used to measure the along-wind displacement by integrating the acceleration signal with a built-in circuit. The displacements thus measured were compared with those recorded by the LVDTs.

The wind pressure on the cladding of the model was recorded by the pressure gauges; and the axial forces inside columns were measured by the force transducers. The mean wind speed was measured using a pitot tube in the test section of the wind tunnel. The accelerometers, force transducers and LVDTs and their corresponding signal conditioners were calibrated prior to the experiments.

All signals from the various sensors were filtered and amplified before being transferred to the computer for display and storage. The feedback of the measured data was used to adjust wind tunnel operating and testing process. The signals were sampled and analyzed using the Global Lab software package with an A/D DT 709Y & DT 752 data acquisition board.

5 EXPERIMENTAL RESULTS

The degree of reduction in real-time acceleration and displacement histories, RMS accelerations, and acceleration response spectra were used as a measure of the effectiveness of the VD walls. The effects of different wind attack angles were also tested and the results presented.

5.1 *Free vibration test*

Based on the results of the free vibration tests, the damping ratios were computed to be approximately 4% for the model without viscous damping and 20% for the model with viscous fluid half-filled. Damping values in this range have been reported in actual buildings where full-scale VD walls have been installed (Miyazaki 1992). The fundamental natural frequency of the test model without viscous fluid was computed to be 2.15 *Hz*. The frequencies of other modes are 7.82 *Hz*, 16.24 *Hz* and 24.45 *Hz*, respectively. The fundamental natural frequency increased to 2.35 *Hz* for the model with viscous fluid half-filled and 2.74 *Hz* with viscous fluid fully filled. The increase in its fundamental frequency was due to the added stiffness from the VD walls. The other modes of vibration could hardly be detected in the model with viscous fluid because the model was highly damped.

5.2 *Along-wind response*

During the wind tunnel testing, the model without viscous fluid deflected rapidly when the wind speed was changed and, under the influence of wind turbulence, it vibrated strongly in the translational and torsional modes. A remarkable improvement in the response of the model was observed after viscous fluid was added. With change in wind speed, the VD model deflected gradually toward a new equilibrium position, and vibrations in both the translational and torsional modes were hardly perceptible.

A wind pressure time history is shown in Figure 6, with the mean wind speed varying from 6 *m/s* to 14 *m/s*. Under these levels of wind-excitation, the acceleration of the model structure varied accordingly. Figures 7 and 8 show the along-wind acceleration time histories of the roof of the model. When the wind speed was under 8 *m/s*, both models with and without viscous fluid had no obvious vibrations that could be observed by sight. After the wind speed was increased, the model without viscous fluid experienced severe vibrations, but the vibrations of the model with viscous fluid remained at low levels, in the range of 40% to 60% of the accelerations without viscous fluid.

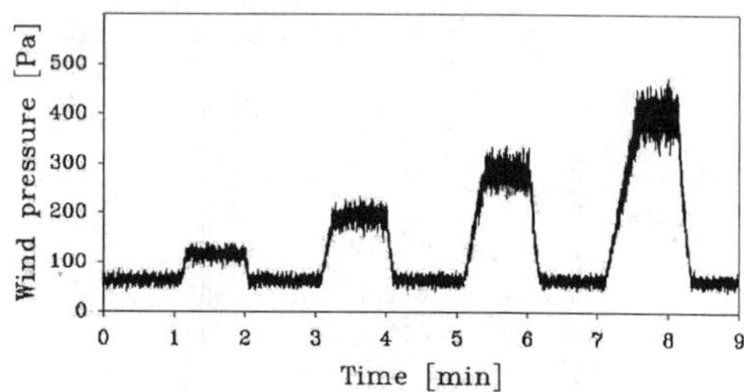

Figure 6. Wind pressure history

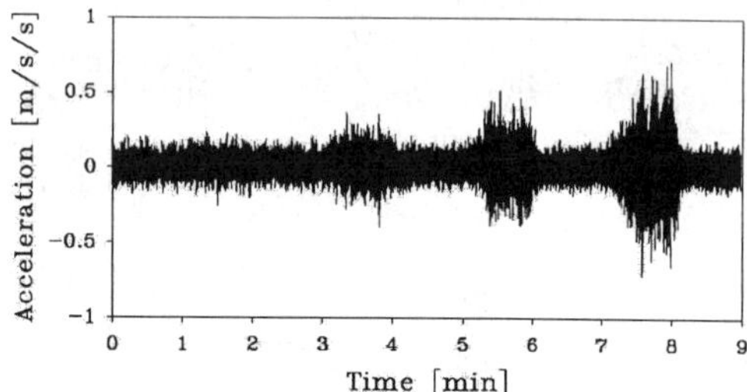

Figure 7. Along-wind acceleration without fluid

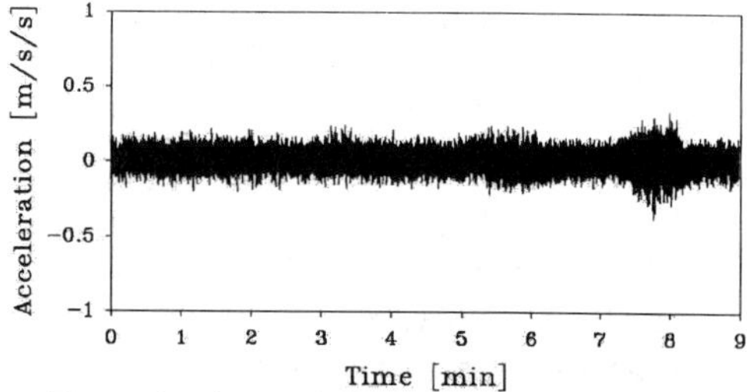

Figure 8. Along-wind acceleration with fluid

The reduction of the vibration induced after adding viscous fluid could also be seen from the measured displacements. Figures 9 and 10 show the displacement time histories recorded on the roofs. A comparison of these test results shows that the reduction in displacement of the models is on the order 60% to 80% after viscous fluid is added.

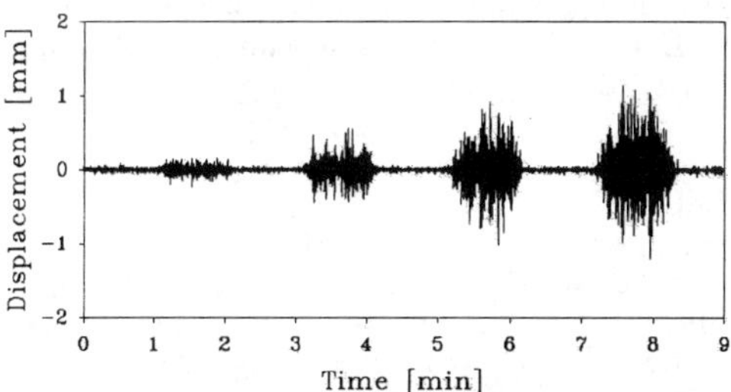

Figure 9. Drift of the roof without viscous fluid

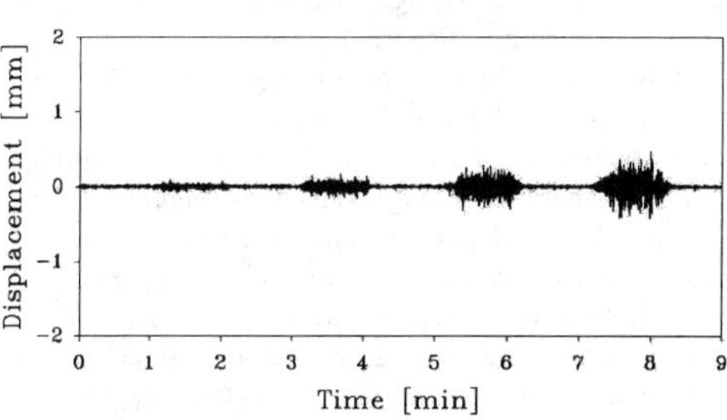

Figure 10. Drift of the roof with viscous fluid

5.3 *Cross-wind response*

Even though the VD wall units were aligned in the along-wind direction, all the wind tunnel tests showed that the effectiveness of viscous damping extended to the response of the model in the cross-wind direction. In both the along-wind and cross-wind directions, VD walls helped reduce the lateral displacement by at least 40%.

5.4 *Acceleration response spectra*

The response spectra of accelerations at the roof of the building model with and without viscous fluid are shown in Figures 11 and 12. They were obtained by Fast Fourier Transform of the recorded responses of the accelerations at the roof level.

Both the along-wind spectra and the cross-wind spectra indicate that the vibration is dissipated by the addition of viscous fluid into the VD wall units. In both figures, the first mode of vibration is shifted to a higher frequency compared to that of the model without viscous fluid. As mentioned in the free vibration tests, this is due to the added stiffness from the VD walls.

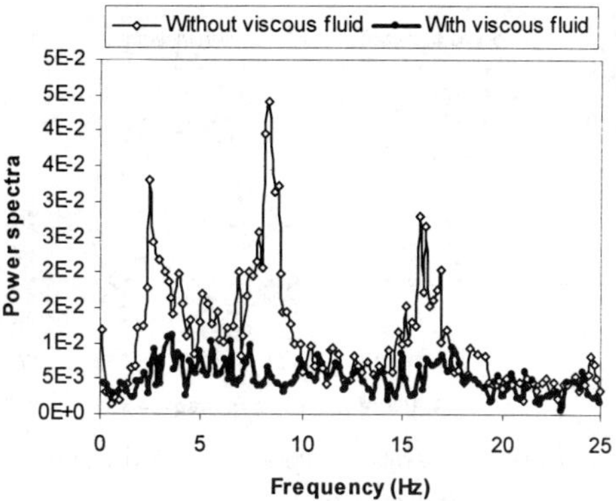

Figure 11. Along-wind response spectra

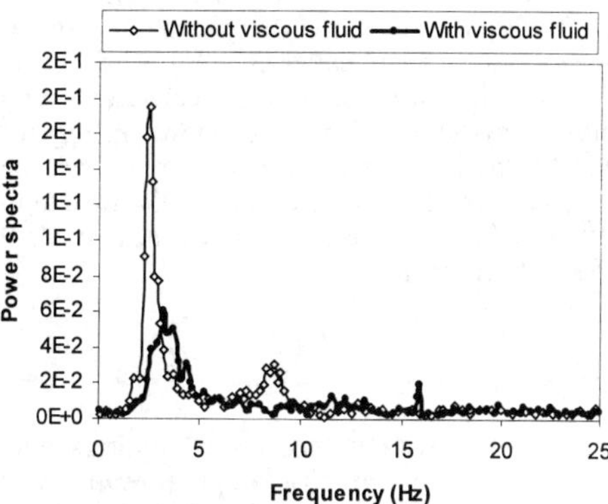

Figure 12. Cross-wind response spectra

5.5 *Effect of wind attack angles*

In real situations the wind approaches the building structure from all possible directions. The test results of the effect of wind attack angles on the dynamic response in both the along-wind and cross-wind directions are shown in Figures 13 and 14. The root-mean-square (RMS) of acceleration response is plotted for 15 degree increments.

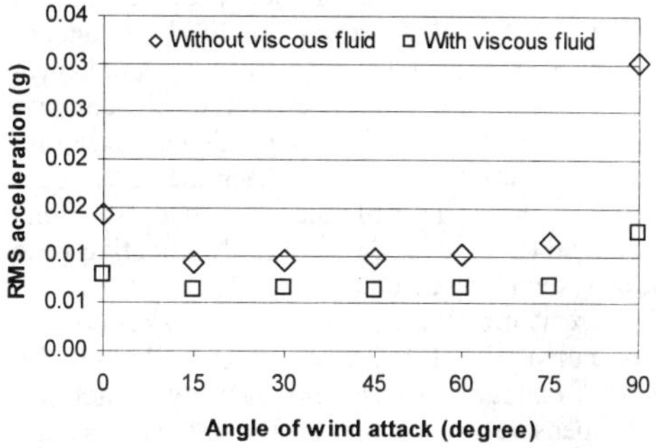

Figure 13. Along-wind response under different attack angles

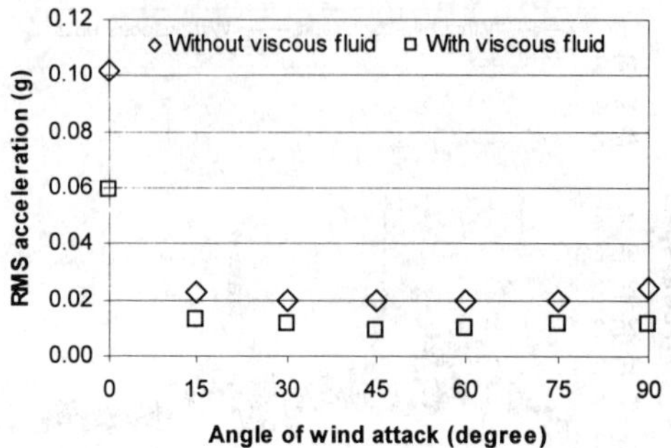

Figure 14. Cross-wind response under different attack angles

With change in wind speed, the results of the dynamic response under different attack angles maintained a similar pattern. At every angle concerned, the wind-induced acceleration was suppressed by 40% to 70%. As shown in Figures 13 and 14, the largest responses occurred when the wind acted normal to the face of the building due to vortex shedding (i.e., angle of wind attack equals 0 degree or 90 degree).

6 CONCLUSIONS

Wind tunnel testing of an aeroelastic building model with and without viscous-damping walls was conducted to investigate the behavior and effectiveness of VD walls in controlling wind-induced vibrations in building structures. The scaling of the model was selected to induce large vibrations due to vortex shedding. Free vibration tests of the model structure with and without VD walls added showed that VD walls could significantly increase the structural damping, and consequently reduce the vibration.

With and without viscous fluid added in the wall units, the model structure was tested under different intensities of wind. The results show that the dynamic response of the model without viscous fluid was substantial. After viscous fluid was added, vibrations were reduced significantly. This indicates that VD walls are highly effective in reducing along-wind vibrations. The results also indicate that VD walls are effective in controlling the cross-wind vibrations. With VD walls placed in two directions perpendicular to each other, a highly effective pass control system may be created.

For resisting wind- and earthquake-induced forces, conventional building design relies on the strategies of high stiffness and ductility, which are often expensive and unreliable. In contrast, viscous-damping walls may offer a new design strategy of high damping that is simple, functional and cost-effective. Already in Japan, there are a number of new buildings that have incorporated viscous-damping walls (PSRG 1997).

ACKNOWLEDGEMENTS

This research was supported by the Committee on Research and Conference Grants of the University of Hong Kong. This support is gratefully acknowledged.

REFERENCES

Arima, F., M. Miyazaki, H. Tanaka & Y. Yamazaki 1988. A study on buildings with large damping using viscous damping walls. *Proceedings of Ninth World Conference on Earthquake Engineering*, Vol. V: 821-826.

Miyazaki, M. 1992. Design of a building with 20% or greater damping. *Proceedings of the Tenth World Conference on Earthquake Engineering*: 4143-4148.

Miyazaki, M., Y. Kitada, F. Arima & I. Hristov 1986. Earthquake response control design of buildings using viscous damping walls. *Proceedings of the First East Asian Conference on Structural Engineering and Construction*: 1882-1891.

PSRG (Protective Systems Research Group of the Earthquake Engineering Research Center at the University of California at Berkeley). 1997. Buildings in Japan incorporating passive energy dissipation systems. *URL: <http://www.eerc.berkeley.edu/prosys/japanpedbldgs.html>*

Yeung, N. & Pan, A.D.E. 1997. Viscous-damping wall and cladding as passive control for high-rise buildings. *Proceedings of the 2nd European & African Conference on Wind Engineering*, Genova, Italy, Vol. II: pp. 1643-1650.

Yeung, N. & Pan, A.D.E. 1998. The effectiveness of viscous-damping walls for controlling wind vibrations in multi-story buildings. *Journal of Wind Engineering and Industrial Aerodynamics*, 77&78, pp. 337-348.

Wind Engineering into the 21st Century, Larsen, Larose & Livesey (eds) © *1999 Balkema, Rotterdam, ISBN 90 5809 059 0*

Wind pressure distribution on a stadium roof by wind tunnel model test

L.D.Zhu, W.Chen, Z.C.Shi & F.Zhang
State Key Laboratory for Disaster Reduction in Civil Engineering, Tongji University, Shanghai, People's Republic of China

ABSTRACT: It is necessary to use wind tunnel model test to determine the wind pressure on the roof of a low-rise stadium with distinctive exterior shape. This kind of test is similar to that one for determining the wind pressure on an ordinary building, but it has its own characteristics. The main one is that, because the roof of a stadium is usually opened, both the up and down surfaces is subjected to wind action, hence, wind pressures on both surfaces should be measured. In this paper, taking Hongkou Stadium in Shanghai, China, as an engineering background, a simplified data processing method is presented at first to reduce the requirements for measuring technique and instruments. Then, the corresponding wind tunnel model test is discussed for measuring wind pressure on the stadium roof. Finally, some conclusions are made for the wind pressure distribution on the roof of the stadium like Hongkou Stadium.

1 INRODUCTION

Athletic sports and games, such as football game, become more and more popular in China. To meet this situation, several large stadiums have been built (e.g. Shanghai Stadium in Shanghai possessing about 80,000 seats etc.), or are under construction (e.g. Yellow Dragon Stadium in HangZhou – the Capital of Zhejiang Province, which can contain more than 50,000 spectators; Hongkou Football Stadium in Shanghai containing more than 30,000 spectator seats, etc.), and some else in several big cities are under designing or planing (e.g. Wuhan – the Capital of Hubei Province; Nanjing – the Capital of Jiangsu Province; Nanning – the Capital of Yunnan Province; and Qingdao in Shandong Province, etc.).

The wind pressure acting on the roof of a stadium is one of the basic parameters for structural design. However, it is difficult, even impossible to determine the wind load distribution on a stadium roof only by means of the current load code, because each stadium has its own distinctive exterior shape, which, in general, is quite complex. So, the best way to meet this task is the full model wind tunnel test.

There are two main types of full model wind tunnel test, called aeroelastic full model test and rigid full model test respectively. The first one considers the aeroelastic effect between vibrating structure and incident wind flow, and is used to measure the dynamic response of the structure in the boundary layer wind field. It requires higher technique, and is more complicated and expensive than the second one. In the case of the second one, the aeroelastic effect is neglected, therefore, it is mainly used to determine the distributions of mean and fluctuated wind pressures acting on the structure.

For those nearly rigid structures, the amplitude of structural vibration is very small, and the aeroelastic effect and the inertial force can be neglected totally, thus, the wind load can be determined only through rigid full model wind tunnel test. For the flexible structure, beside the rigid full model test, it is necessary to carry out the aeroelastic full model test to determine the acceleration response and inertial force because the structural vibration is remarkable and there is strong interaction between structure and incident wind flow. For those structures between flexibility and rigidity, on the one hand, the amplitude of its vibration is not large, and the aeroelastic effect can be neglected; but on the other hand, the structural inertial force is not small enough to be neglected, thereby it should be taken into account. In this case, the mean and fluctuated wind pressures on a structure can be determined trough rigid full model test, and the structural dynamic response and inertial force can be calculated approximately (Simiu & Scanlan 1996, Zhang 1990) based on the results of the rigid full model test.

The roof truss system of Hongkou Stadium has medium stiffness. Therefore, the way of rigid full model test and approximate analysis was chosen to determine the wind load on its roof.

In this paper, taking Hongkou Stadium as an engineering background, a simplified data processing method is presented at first, to reduce the requirements for measuring technique and instruments. Then, the corresponding rigid full model wind tunnel test is discussed for measuring wind pressure on the stadium with a large span cantilevered roof. Finally, according to the test results, several conclusions are made for the wind pressure distribution on the roof of the stadium like Hongkou Stadium.

The further approximate analysis for dynamic response of this Stadium was conducted by Professor B.C.Huang et al, Tongji University, therefore it will not be introduced in this paper.

2 CONCISE DESCRIPTION OF HONGKOU STADIUM

Hongkou Stadium was designed by Shanghai Institute of Architectural Design & Research of China. Its roof is a hoop spatial structure with the shape of a saddle. It consists of a complex spatial steel truss system and special thin film tautly mounted under the truss system. The whole steel truss system is comprised of 22 radial spatial sub-trusses and a number of spatial sub-trusses in hoop direction. The film is divided into 22 blocks by the radial sub-trusses. The overall size of the roof is about 214m along east-west direction and 205m along north-south direction, and in the center of the roof, i.e. above the football field, there is an elliptic opening of about 98m and 130m in length along east-west and north-south direction respectively. The highest point of the roof is about 60m above the ground. The whole roof is supported on 22 columns, and there is almost no wall between the roof and the bleachers, therefore wind can blow through the stadium. Obviously, the flow around the stadium is disturbed strongly by the exposed truss system, then, becomes very complicated.

3 CHARACTERISTICS OF RIGID FULL MODEL WIND TUNNEL TEST FOR WIND PRESSEURE ON STADIUM ROOF

Wind pressure measurement between a stadium and an ordinary building has much more similarities. For example, some principles of similarity, such as geometrical similarity, must be satisfied; simulation of the main characteristics of incident wind field, such as profile of mean wind velocity, profile of turbulent intensity, power spectrum, etc., should be taken into account (Simiu & Scanlan 1996). And the surroundings within appropriate area around the main structure should be simulated; the arrangement of pressure taps, reference point of wind pressure, reasonably selecting sampling rate and sampling length of the testing signals, etc., will also be taken into consideration. In addition, the models should be with enough rigidity and strength to avoid the occurrence of significant vibration or damage. However, there are two main differences between these two types of structures as follows.

The roof of a stadium usually opened and both the up and down surfaces is subjected to wind action, but only exterior surface is blown by wind for an ordinary building due to its closed outer wall. Therefore, wind pressures on both surfaces should be measured for the roof of a stadium while only the external wind pressure is to be measured for the ordinary building.

It is well known that the existence of measuring equipment, such as pressure measuring pipes and scan valve etc., should not affect the wind filed about structures. For ordinary building, they can be easily placed in the inner space of the model, but it is more difficult and complicated for the stadium to arrange measuring equipment because of the its opened space and very thin roof.

Strictly speaking, for the roof of a stadium, wind pressures on both the up and down surfaces must be measured simultaneously, then the desired wind pressure is the difference of the time signals between these two surfaces. To achieve this, the highly valuable multi-channel synchronizing data acquisition equipment and a large amount of wind pressure sensors must be available. To reduce the requirements for measuring technique and instruments, a simplified data processing method is introduced in the next section.

4 SIMPLIFIED METHOD FOR DATA REDUCTION

Suppose $X(t)$ and $Y(t)$ represent, respectively, the time signals of wind pressures, or wind pressure coefficients, simultaneously measured on the up and down surfaces of a measuring point on the stadium roof, $Z(t)=Y(t)-X(t)$ is the desired net wind pressure, or wind pressure coefficient of time signals. They are all random processes, and generally are supposed to be ergodic. The mean and RMS values of the net wind pressures are

$$\mu_Z = EZ = E(Y - X) = \mu_Y - \mu_X \tag{1}$$

$$\begin{aligned}\sigma_Z &= \sqrt{E\left[(Z-\mu_Z)^2\right]} \\ &= \sqrt{\sigma_Y^2 + \sigma_X^2 - 2\left[E(XY) - \mu_X \mu_Y\right]}\end{aligned} \tag{2}$$

Where μ , σ and σ^2 represents mean value, root mean square(RMS) value and variance respectively; E is time expectation operator.

From the primary testing data of Hongkou Stadium, it is found that in the middle area and leeward edge area of the roof, which take the majority of measuring points, the values of $|E(XY)-\mu_X \mu_Y|$ are close to zero. Although for the windward edge area of the roof, the values of $|E(XY)-\mu_X \mu_Y|$ is a little larger than the former, it is still a small value relative to the corresponding values of $\sigma_X^2+\sigma_Y^2$. Therefore it can be supposed that

$$E(XY) \approx \mu_X \mu_Y \tag{3}$$

In fact, the incident wind flow is divided into two separated flows by the stadium roof plate. As described in section 2, the roof of Hongkou Stadium is composed of a complex spatial steel truss system and special thin film tautly mounted under the truss system. When the wind flow passes the stadium roof, the up flow is strongly disturbed by the truss system while the down flow is disturbed in a smaller degree. Therefore, the up and down flows are poorly correlated to each other in the middle area and the leeward edge area of the roof, i.e. formula (3) is well satisfied. For the windward area of the roof, the correlation between the up and down flows is a little better than that one for the middle and leeward areas because flows are not disturbed sufficiently, thereby, the deviation of formula (3) is a little lager in this area. A detailed research on the correlation between the up and down flows has been carrying out for the various types of the roof system.

Considering the approximate relation (3), the formula (2) yields:

$$\sigma_Z \approx \sqrt{\sigma_Y^2 + \sigma_X^2} \tag{4}$$

Formula (1) means that the mean value of the net wind pressure on the roof is equal to the difference between the mean wind pressures on down surface and up surface. Formula (4) indicates that the RMS value of the net pressure approximately equals the root square of the sum of the variances of wind pressures on up and down surfaces. Thus, provided that X, Y and Z are all ergodic, we can measure the wind pressures on the up and down surfaces separately at different time. And then the values of σ_X^2 and σ_Y^2 can be calculated based on these separately recorded samples of X and Y, and the RMS value of the desired net wind pressure on the roof can be determined in light of formula (4). This method makes the testing work doubled, but the requirements, for the measuring technique and equipment, are reduced and the results still keep reasonable.

Let' s denote $p_{um}(t)$ and $p_{dm}(t)$ as the time signals recorded on the up and down surfaces of one measuring point on the roof of the stadium model respectively. Then, the corresponding time signals of wind pressure coefficients on up and down surfaces, which are dimensionless and can be applied to both the model and prototype, are respectively,

$$C_{p_u}(t) = \left[p_{um}(t) - q_{\infty m}\right] / \left(q_{0m} - q_{\infty m}\right) \tag{5}$$

$$C_{p_d}(t) = \left[p_{dm}(t) - q_{\infty m}\right] / \left(q_{0m} - q_{\infty m}\right) \tag{6}$$

Where q_{0m} and $q_{\infty m}$ are, respectively, the local pressure (or overall pressure) and the ambient pressure (or static pressure) at reference point in wind tunnel, which are often measured with Pitot tube. $q_{rm}=q_{0m}$ $q_{\infty m}$ is the reference dynamic pressure for the model. The location of the reference point must be chosen carefully. The wind flow in vicinity to the reference point must not be disturbed. Then, we can easily calculate the mean values and variances of Cp_u of Cp_d, further, with formula (1) and (4), obtain the mean value(Cp_{mean}) and RMS value(Cp_{rms}) of the net wind pressures on the roof. From the viewpoint of probability, the maximum and minimum net coefficients of wind pressures can then be determined with the following formulas,

$$C_{p\max} = C_{pmean} + gC_{prms} \tag{7}$$

$$C_{p\min} = C_{pmean} - gC_{prms} \tag{8}$$

Where, the peak factor g is set to 2.5 based on Chinese Code on Wind Load(Chinese Architectural Industry Press 1991). Thus, under the assumption of normal distribution of wind pressure fluctuation, the probability that the instantaneous net wind pressure exceeds the range of $[Cp_{max}, Cp_{min}]$ is only about 1.24%.

Having got Cp_{max} and Cp_{min}, we can determine the maximum net wind pressure, p_{max}, and the minimum net wind pressure, p_{min}, on the roof of prototype as follows,

$$p_{\max} = C_{p\max}\, q_r \tag{9}$$

$$p_{\min} = C_{p\min}\, q_r \tag{10}$$

Where, q_r is the reference dynamic wind pressure for prototype in real wind field, determined according to Chinese Code on Wind Load(Chinese Architectural Industry Press 1991), and the height of reference point in real filed is determine by that one in wind tunnel and the adopted length scale.

5 BRIEF DESCRIPTION OF RIGID FULL MODEL WIND TUNNEL TEST FOR HONGKOU STADIUM

The rigid full model wind tunnel test for the Stadium was carry out in TJ-3 Wind Tunnel at Tongji University in Shanghai of China. It is a large boundary layer wind tunnel with a test section of 15m in width, 2m in height and 14m in length. Its wind velocity can be continuously adjustable from 0.2m/s to 17.6m/s. The test wind speed was chosen as 15m/s. The model length scale was 1:150. The up and down wind pressures, including both the mean and fluctuated pressures, at 545 valid measuring points on the

Figure 1. Models of the stadium and its surroundings

roof were measured in 24 incident wind directions around a circle with an interval of 15°, where 0° incident direction indicates that wind blows from due north. The mechanical scan valve system produced by Scannivalve Co., USA, was used to take measurement, and the measured data were processed with self-developed software. The sampling frequency was about 620Hz. The original records and processed data and results were stored in MO disks of 2.6GB capacity.

5.1 *The models*

The stadium model is near round, and the height and the diameter are about 40cm and 150cm respectively. The design and manufacture of the model strictly subjected to the principle of geometric similarity except that the thickness of the model roof plates (3.5mm) is somewhat thicker than it should be according to similarity principle, for the purpose of embedding the pressure pipelines in it. Computer controlled small millers were used for the model manufacture, especially for manufacture of roof plates and bleachers. All the 22 blocks of the roof plates were made of 3.5mm perspex plate, and were hot-pressed to the desired shape after printed-circuit-like narrow troughs were made on them. The inner and outer edges of the roof plate were thined and shaped to streamline form. The truss system was made of welded brass bars with the various diameters of 2.5mm, 2.2mm, 0.6mm, etc. The roof plates were fixed under the truss structure block by block, then the gaps between each two blocks were filled with epoxy resin to avoid leakage of wind. The other parts of the stadium model were made of perspex or ABS plastics. The bleachers of the model can be disassembled to facilitate the measuring equipment.

The surroundings within the round area of about 360m radius were also simulated including about 110 buildings, a section of railway viaduct, and trees in the Hongkou Park, etc. The models of those tall buildings and viaduct were made of ABS plastics while the models of those low-rise buildings were made of foam plastics. All the surrounding models together with the stadium model were placed on a rotatable round plank with a diameter of 4.8m, and the stadium model was located in the center of the plank (Fig. 1).

5.2 *Arrangement of pressure-measuring points and pressure-conveying pipelines*

There are 22~35 pressure measuring points distributed on each of the 22 roof blocks, 614 in all (Fig.2). A double mouth tap, a ϕ1.5mm cylindrical orifice penetrating through the perspex roof plate and

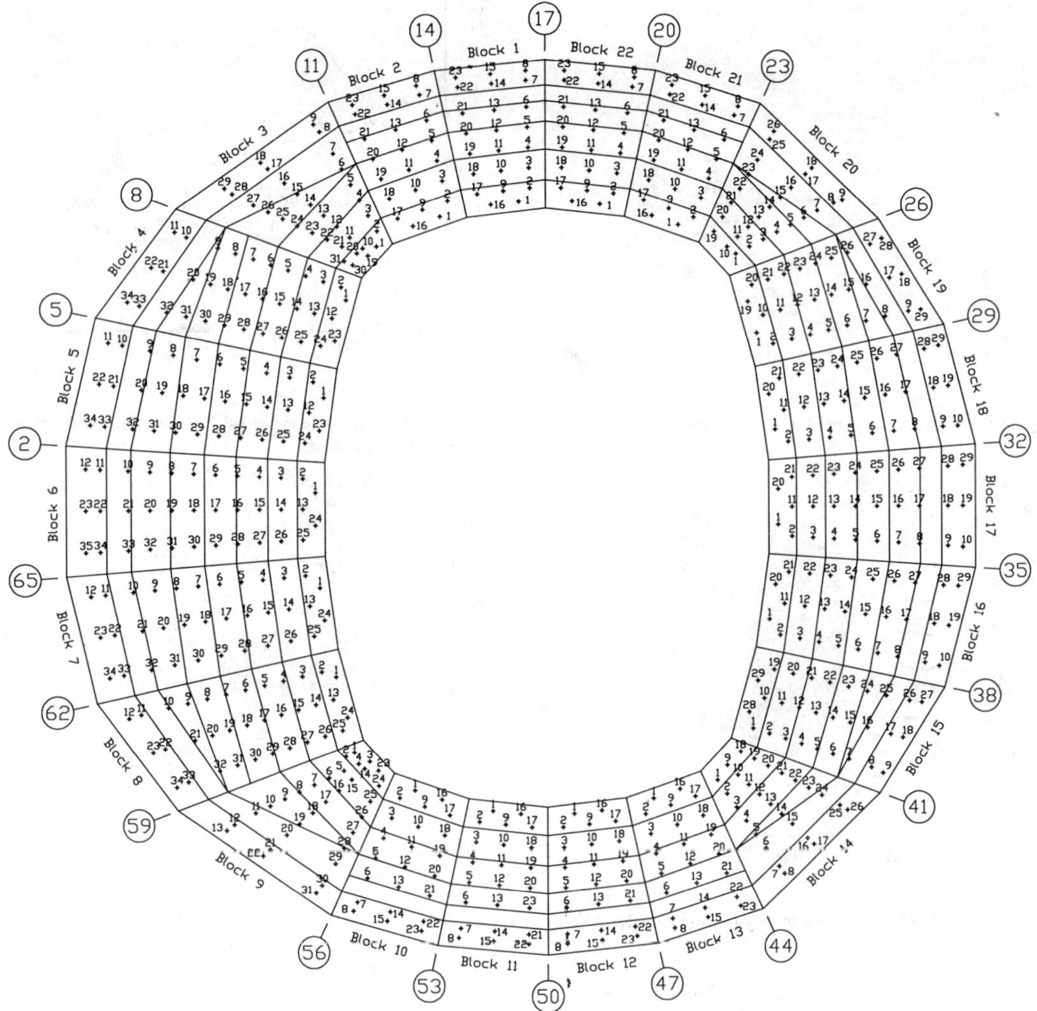

Figure 2. Schematic diagram of measuring point arrangement on the roof of Hongkou Stadium

perpendicular to both the up and down roof surfaces, was installed at each pressure-measuring point. Each tap was connected to pressure scan valve through a pressure-conveying pipeline. Each pipeline consisted of three parts. One part, liked to the tap, was a pliable copper pipe with the inner diameter of 1.1mm and the outer diameter of 1.6mm. To avoid disturbing the wind flow, it was embedded in a preset trough on roof plate and stretched down under the bleachers, nestled against the side surface of the column. The another part, connected to scan valve located under the bleachers, was a supple pipe of plastics with the inner diameter of 1.5mm and the outer diameter of 2.2mm. The two parts were connected through an elaborate short damping. The whole length of every pressure-conveying pipeline was limited within 80cm for obtaining the fluctuated pressure with enough accuracy. During pressure measurement on the up surface, the down mouths of the taps were sealed up with removable adhesive paper, and vice versa.

It is notable that, for each valid measuring point, the pressure-conveying system must be unimpeded, and be sealed air-tightly except for the opened one of its two tap mouths. So, it is very important to check every pressure-conveying pipeline carefully before the commencement of every time of pressure measurement. In this test, the method of applying a constant pressure on every tap was used to check the behaviour of pressure-conveying systems. It was found that the pressure-conveying systems of 69 measuring points were not perfect, either somewhat

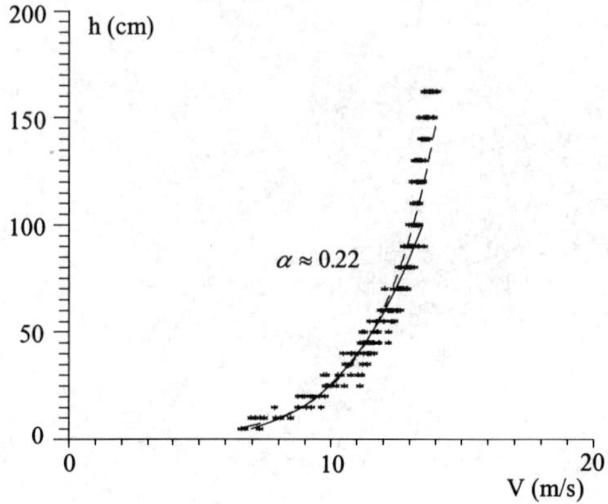

Figure 3. Profile of mean speed in wind tunnel

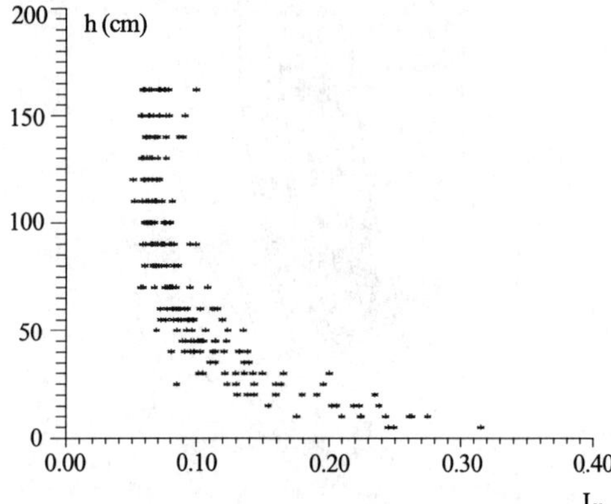

Figure 4. Profile of turbulence intensity in wind tunnel

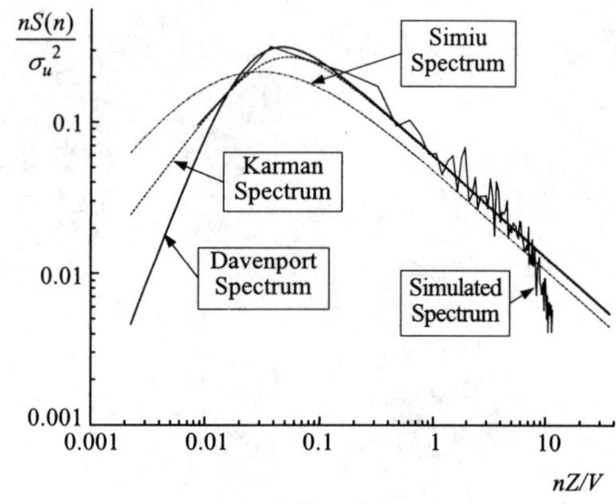

Figure 5. Spectrum of longitudinal velocity fluctuations at 30cm height in wind tunnel

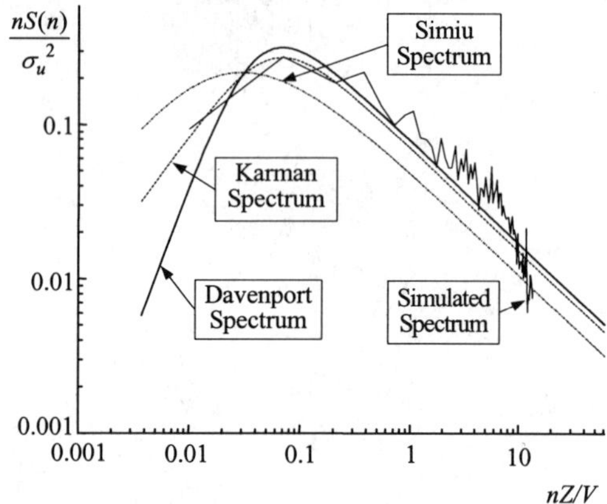

Figure 6. Spectrum of longitudinal velocity fluctuations at 50cm height in wind tunnel

blocked or somewhat leaky, and could not be repaired, then had to be forsaken. Finally, the total number of valid measuring points was 545. However, that is enough to determine the wind pressure distribution on the roof.

5.3 *Simulation of atmospheric boundary layer*

Houngkou Stadium is located in the northeast urban area of Shanghai. According to Chinese Code on Wind Load(Chinese Architectural Industry Press 1991), the local terrain belongs to type-C, therefore, the profile of ten-minute mean wind speed can be described approximately by the power law of 0.22 exponent, and the gradient height, i.e. the depth of the atmospheric boundary layer, is specified as 400m.

The simulation of atmospheric boundary layer in wind tunnel, TJ-3, was carried out in light of the chosen length scale (1:150) and the method given by Shi (1994), and was achieved by means of spires and roughness blocks.

Figure 3 and Figure 4 show, respectively, the mean speed profile and turbulence intensity profile of the simulated boundary layer wind field in the wind tunnel. The exponent of the fitted power law is close to the required value. The mean turbulence intensities are, respectively, about 11% and 20% with heights of 6.7cm and 40cm in the wind tunnel, corresponding to the prototype heights of 10m and 60m (the top level of the stadium).

Figure 5 and Figure 6 show, respectively, the spectra of longitudinal velocity fluctuations with heights of 30cm and 50cm in wind tunnel. The curves of the corresponding well-known empirical spectra suggested by Simiu, Karman and Davenport are drawn in the Figures as well (Simiu 1974, Simiu & Scanlan 1996, Davenport 1961, Zhang 1990). The comparison shows the good agreement between the

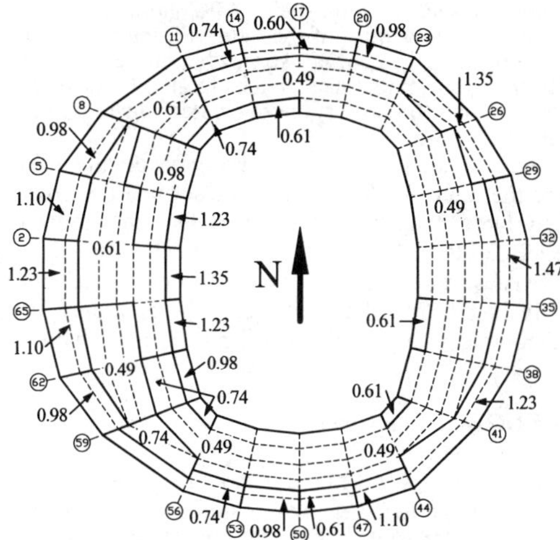

Figure 7. Zoned diagram of maximum coefficients of net wind pressure on the roof of Hongkou Stadium

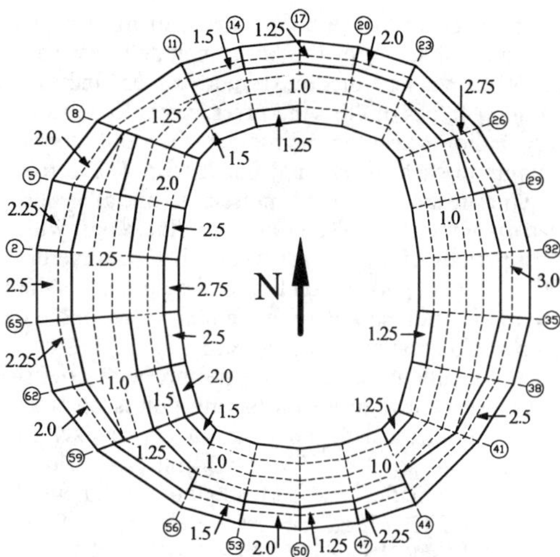

Figure 9. Zoned diagram of maximum wind pressure on the roof of Hongkou Stadium (100 year return period)

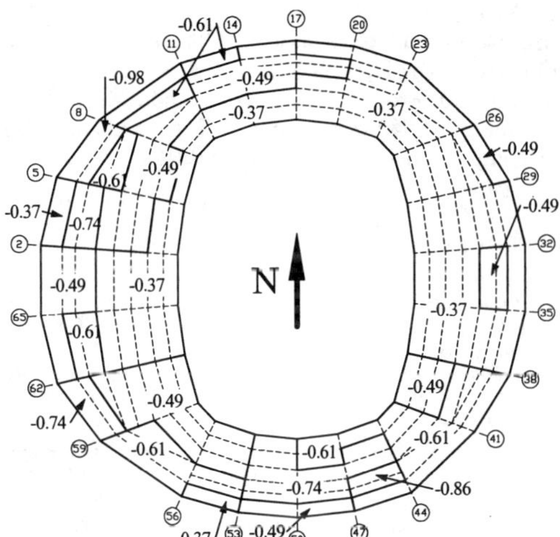

Figure 8. Zoned diagram of minimum coefficients of net wind pressure on the roof of Hongkou Stadium

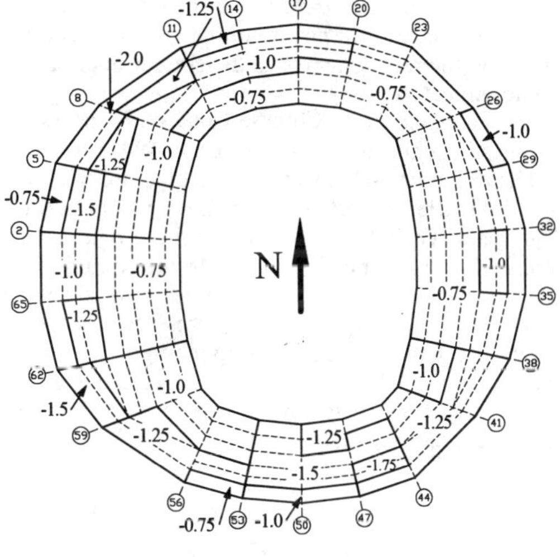

Figure 10. Zoned diagram of minimum wind pressure on the roof of Hongkou Stadium (100 year return period)

spectra of the simulated boundary wind filed and the empirical spectra.

5.4 *The reference point and reference dynamic pressure*

It is most convenient to choose the gradient height as the height of the reference point, but, in wind tunnel, the reference point was set in the height of 1.62m due to the limited height of the wind tunnel, corresponding to 243m in the real wind field. For the purpose of convenient use of the test results, the directly measured pressure coefficients based on the foregoing reference point at 243m were converted into those based on the reference point at 400m, the gradient height over type-C terrain. The converting factor was $f_C = (243/400)^{2\times0.22} = 0.8031$. All the pressure coefficients provided in this paper are converted with this factor. Thus, the reference wind pressure is taken as the dynamic pressure of gradient wind.

6 MAIN RESULTS OF TEST

The detailed test results are provided in the non-published laboratory report (Zhu & Chen 1998). The results, given in the report in the form of table or distribution diagram or the both, include mean values, RMS values, the maximum values and the minimum values of the wind pressure coefficients at every valid measuring points on the roof corresponding to each of 24 wind incidences. The corresponding results of wind pressures on all valid measuring points are also provided in the report with return periods of 30, 50, 100 years. The positive direction of wind pressure on the roof is defined as the upward direction normal to the roof surfaces, and the negative direction points downward.

The maximum value of the net pressure coefficient among those of all measuring points on west part of the roof and for all wind incidences is 1.30 and located at point 6-24(i.e. No.24 on block No 6, similarly hereinafter) and corresponding to the wind incidence of 45°. That one for the east part of the roof is 1.47 and located at point 18-19 and corresponding to the wind incidence of 105°. The minimum one is -0.98 and located at point 4-11 and corresponding to 0° wind incidence for the west roof, and is -0.85 and located at 13-13 and corresponding to 165° wind incidence for the east roof. Figure 7 is the zoned diagram of maximum net pressure coefficients on the roof. Figure 8 is the similar one for the minimum net value.

In accordance with Chinese Code on Wind Load (Chinese Architectural Industry Press 1991), V_g and q_r above the type-C terrain in Shanghai are 57.1m/s and 2.038kPa, respectively, for the return period of 100 years. Thus, the maximum positive pressure is 3.0kPa located at point 18-19 and corresponding to the wind incidence of 105°. The maximum absolute value of negative (point down) pressure is 2.0kPa located at point 4-11 and corresponding to the wind incidence of 0°. Figure 9 and Figure 10 are, respectively, the zoned diagram of maximum and minimum net pressures on the roof.

7 CONCLUSIONS

Wind pressures on the up and down surfaces of the stadium roof may be measured separately. Although the testing work is doubled, the requirement for measuring equipment, technique and the test costs can be reduced, while the results remain reasonable.

Non-simultaneous pressure measuring on both surfaces is based on some primary test investigating results from the stadium model. The degree of correlation between the up and down flows needs to be further studied in detail.

The net positive pressure (point up) is much larger than the net negative pressure (point down). The ratio between the former and the latter may reach 2~3.

Pressures measured on the region near both the inner and outer edges of the roof hoop are much larger than measured on the middle region. The former may be 2~3 times of the latter. Furthermore, the outer edges of the roof are somewhat up-bent (about 8°), therefore undertake much large net positive pressure than the corresponding inner edges in general.

Overall, the net positive pressures on the roof over the main platform (the highest part) are obviously larger than the pressures on the other 3 parts because of the "wind catching" effect.

Because the height of the roof hoop varies violently in the four diagonal areas, the pressures are rather larger than those on the relatively flat region.

ACKNOWLEDGMENTS

The project of wind tunnel rigid full model test of Hongkou Stadium was greatly supported by Shanghai Institute of Architectural Design & Research, especially in financial and technical support. We would like to express our heartfelt gratitude to Mr. X.M.Huang & Mr. Y.R.Lin, the senior engineers of the Institute for their kindly help, and Prof. H.F.Xiang for his support and technical guidance. Also, we would like to give our utmost thanks to our colleagues: Mr. H.Y.Mao, Ms X.Y.Wu, Ms H. Ha, Mr. R.Q.Du, Prof. Z.X.Lin, Ms L.Zhou Mr. H.J.Li, and the postgraduate students: Mr. Y.Quan, Mr. P.Huang, for their participating in the test work.

REFERENCES

Chinese Architectural Industry Press. 1991. *Overall of the current used code on architectural structures (I)* (in Chinese. Chinese Architectural Industry Press

Davenport, A. G. 1961. The application of statistical concepts to the wind loading of structures. *Proc. Of the Iinst. Of Civ. Engrs, London.* V19: 449-472.

Shi, Z.C. 1994., Adjusting method studies on neutral atmospheric boundary layer simulating tests. *J. of Tongji University*, V122, No. 4: 469-474.

Simiu,E. 1974. Wind spectra and dynamic alongwind response, *J. Struct. Dir.*, ASCE, V100, No. ST9, Proc. Paper 10815: 1897-1910

Simiu, E. & Scanlan R.H. 1996. *Wind effects on structures: Fundamentals and applications to design.* New York: John Wiley & Sons, INC.

Zhang, X.T. 1990. *Theory of wind loads on engineering structures and handbook for wind-resistant calculation* (in Chinese). Shanghai: Press of Tongji University.

Zhu, L.D. & Chen, W. 1998. Model test and research on wind load distribution and wind-resistant analysis for Shanghai Hongkou Stadium (II) — Wind tunnel model test and research on wind load distribution(in Chinese). Laboratory report of No. WT-9804, State Key Laboratory for Disaster Reduction in Civil Engineering, Tongji University, Shanghai, China.

Bluff body aerodynamics

Wind Engineering into the 21st Century, Larsen, Larose & Livesey (eds) © 1999 Balkema, Rotterdam, ISBN 90 5809 059 0

Wavelet analysis of pressure-velocity correlations on bluff bodies

R.E.Akins
Washington and Lee University, Lexington, Va., USA

T.A.Reinhold
Clemson University, S.C., USA

ABSTRACT: Greater understanding of the details of flow fields around bluff bodies and the separated regions associated with these flows is necessary to advance the knowledge of fluid-structure interactions. These mechanisms are particularly important in wind engineering in determining design pressures on structures. In almost all instances the maximum design pressure occurs in regions of separated flow. The details of the flow in these separated regions and mechanism which determines the interaction between velocity fluctuations and pressure fluctuations are not well understood. In addition, the effect of these phenomena on the criteria for physical or numerical modeling of flows in these regions are not well established. Measurements have been conduced using a laser Doppler velocimeter (LDV) to examine the structure of separated shear layers and the associated regions of separated flow. Pressure measurements have been taken concurrent with the velocity measurements. Wavelet analysis was used to correlate the pressures and velocities.

1. INTRODUCTION

Previous efforts to relate pressure and velocity fluctuations have used thermal anemometry located upwind of the bluff body (Hajj & Tieleman 1996). The results presented in this paper allow correlations between velocities both upwind of the structure and within separated regions to be used in conjunction with pressures on stagnation faces and in separated regions.

Wavelet analysis is employed to determine the relationship between velocity and pressure fluctuations as a function of both frequency and time. This technique will allow isolation of extreme events and will describe frequency ranges with substantial correlation.

Data have been collected for a single bluff body in a boundary-layer flow to determine if this technique can provide useful insight into the causes of extreme pressures. This method of collection and reduction of data will allow the isolation of mechanisms causing peak pressures and may provide a basis for a systematic investigation of methods of reduction of peak pressures on bluff bodies.

The results should be considered preliminary in that the measurement techniques are emphasized and no effort has been made to systematically vary incident flow, building geometry, nor incident wind direction.

2. BACKGROUND

Recent efforts to correlate incident flow with pressures in separated regions reported by Hajj & Tieleman (1996) and Hajj et al (1998) have used a reference wind speed measured several building heights upwind of a model or structure. These studies concentrated on peak pressures. This work attempts to look at velocities measured much closer to a model in a boundary-layer wind tunnel. Incident velocities were measured very close to the model and in separated regions both on the roof and near the separated regions on the sides of the model in order to correlate the velocities with pressures.

3. APPROACH

The LDV system allows measurement of reversing flows and locally large turbulent intensities. The probe is nonintrusive; the location of the measurement is not disturbed by the measurement. In order to use a LDV system in a boundary-layer wind tunnel, the flow must be seeded with particles. Akins & Reinhold (1998) summarized previous efforts at seeding and use of LDV systems in boundary-layer wind tunnels. In order to use the LDV system in a large cross-section tunnel, 6

m^2, a commercial theatrical smoke generator using alcohol-based smoke was placed at the entrance of the test section to seed the entire flow. Data rates of 2-3 kHz were consistently attained in the undisturbed flow of the boundary layer. Near the model and in separated region, data rates of 500 Hz were attained.

A single component of velocity was measured using a TSI Model 9253-350 probe mounted on a traverse. Data were collected for periods from 20 to 60 seconds depending on data rates to obtain a record length of 50,000 data points. Random sampling was used and pressures were measured as the same time as the velocities. The LDV system only samples data when particles are in the measurement volume. The data points are therefore unequally spaced and all analysis must be performed with specially written algorithms.

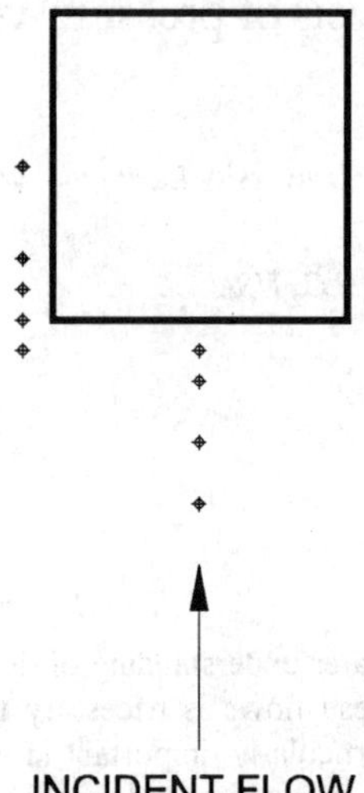

Figure 1. Velocity measurement locations.

4. BOUNDARY LAYER SIMULATION

These measurements were completed in the boundary-layer wind tunnel of the Wind Load Test Facility of the Department of Civil Engineering, Clemson University. The tunnel is open return with the fans in the upwind position. A settling chamber with a grill and screens provides a uniform incident flow. The test section has a cross section of 3m by 2m and a working length of 16m. The boundary layer is developed using a combination of trips, spires, and roughness elements. The flow used in this study was developed for use in simulations of flow around low-rise structures and is representative of flow over open terrain at a nominal scale of 1:50. The measurements reported in this paper are intended to demonstrate a measurement technique, not a specific flow condition. As such, the boundary layer should be considered representative of atmospheric flows in general, not a specific flow. The bondary layer had a power law exponent of 0.16, and a longitudinal turbulence intensity of 16% and a longitudinal integral scale of 0.64 m at the height of the model.

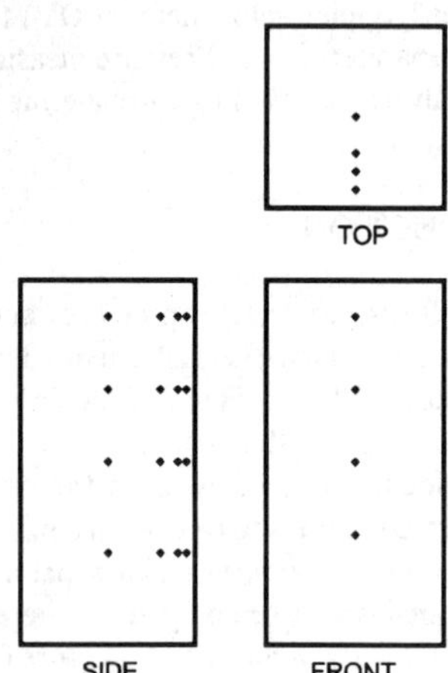

Figure 2. Pressure tap locations.

5. RESULTS

A single isolated rectangular prism (W=D= 0.13m, H= 0.25m) was mounted with the incident flow normal to the surface. Velocities were measured at four vertical locations, z/H= 0.3,0.5,0.7, and 0.9 for the locations shown in a top or plan view in Figure 1. Pressures were measured at four locations simultaneously on the front, top or sides of the model. The locations of the pressure taps are shown in Figure 2.

5.1 Wavelet analysis

A data reduction program was written to perform wavelet analysis on the unequally spaced data produced by the LDV system. A direct wavelet analysis was performed using a Daubechies D4 wavelet. Hubbard (1996) provides an excellent summary of wavelets.

Wavelet analysis uses a moving window and is able to isolate variations in a signal both as a function of frequency and time. It differs from traditional Fourier analysis in that the wavelet analysis is capable of looking at the frequency content of discrete regions of a time series instead of in essence averaging over the

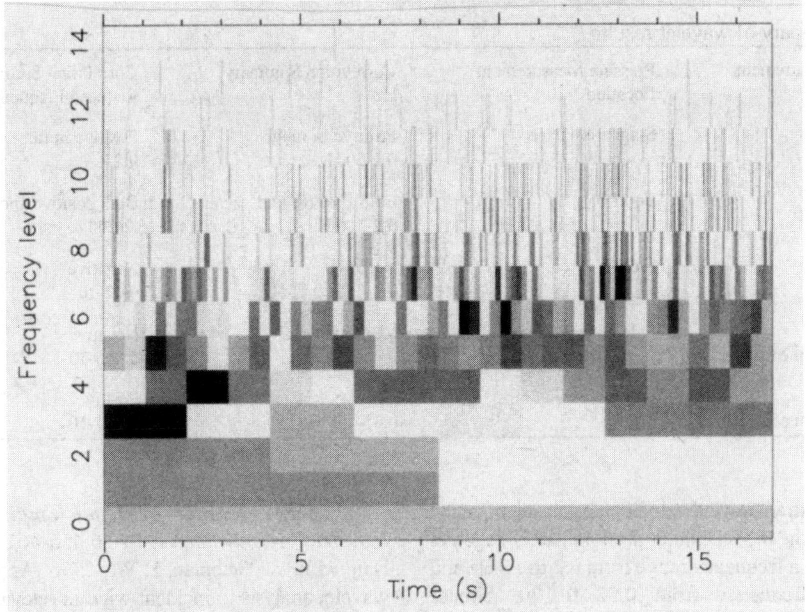

Figure 3. Typical scalogram of pressure/velocity wavelet analysis.

entire record as is done with Fourier analysis. Hajj et al (1998) used wavelet analysis to examine extreme peak pressures. The results presented in this paper examine all frequencies and magnitudes of fluctuations. The use of a cross-wavelet analysis of pressure and velocity corresponds to both a cross-spectral analysis in the frequency domain and a cross-correlation analysis in the time domain. A cross-wavelet is the analog of a mean-square map for two different signals (Newland 1993). It can be normalized to a range of +1 to -1 where +1 indicates perfect positive correlation between signals, and -1 indicates perfect negative correlation between signals. A value of the cross-wavelet of 0 indicates no correlation between the two signals. Mean-square maps of wavelets are also referred to as scalograms and are usually plotted in color. A typical result is shown in grey shades in Figure 3. The darker regions correspond to large absolute values of correlation and the light or white areas are regions of low or no correlation. The vertical scale, labeled as frequency level, corresponds to the range of frequencies examined by the wavelet. A frequency range of 0 corresponds to a frequency with a period of the length of the record analyzed. The higher ranges correspond to frequencies with a period of the record length divided by 2^N where N is the frequency level. It is not possible to determine the difference between positive and negative values in the grey shading of Figure 3.

5.2 Summary of findings

A qualitative description of the findings is included in Table 1. The cross-wavelet between particular velocity and pressure measurement locations can not be presented or discussed due to space limitations. The behavior of general cases, for example upstream velocity measurements and pressures on the upwind or stagnation face are summarized. In all of these cases the qualitiative observations are based on multiple data records, some with the velocity measured at the same elevation of the pressure measurement and some at different elevations.

For a velocity measurement upwind of the model a distance 0.3H and pressures on the stagnation face, a positive correlation exists intermittently as a function of time and frequency. The density of the grey areas in Figure 3 is representative of the regions which would have significant correlation. The frequency range of the regions with correlation was from .02-50 Hz in dimensional terms and from 0.02 to 4 in terms of a reduced frequency, nH/U, and a nominal mean wind speed at the height of the model of 3 ms^{-1}. Roof pressures for the same velocity measurement locations showed both positive and negative correlations in a frequency range of 0.2 to 500 Hz and a corresponding reduced frequency range of 0.02 to 200. The pressure measurement locations on the roof are in a separated region. Side pressures

Table 1 Summary of wavelet results

Velocity Measurement Location	Pressure Measurement Location	Correlation Summary Hz	Correlation Summary Reduced Frequency
Upwind 0.3H Centerline	Stagnation Face	Positive in range 0.2-50 Hz	Positive in range 0.2-4
Upwind 0.3H Centerline	Top/Roof	Both Positive and negative 0.2-500 Hz	Both positive and negative .2-200
Upwind 0.3H Centerline	Side	Negative 0.2-50 Hz	Negative 0.02-20
Upwind 0.1H Side 0.1H from edge	Side upwind taps	Negative 0.05-150 Hz	Negative 0.005-10
Upwind 0.1H Side 0.1H from edge	Side downwind taps	Positive 0.05-150 Hz	Positive 0.005-10

correlated with an upwind velocity measurement on the centerline of the model exhibit predominately negative correlation in a frequency range from 0.2 to 50 Hz and a reduced frequency from 0.02 to 20. Similar correlation patterns were observed for the three pressure measurement locations for velocities measured at upwind locations of 0.2H and 0.1H.

A second set of correlations were examined for a reference velocity location 0.1H upwind of the model and 0.1H from the edge of the model. This location is shown in Figure 1. For the upwind taps, those closer to the leading edge, the correlations are predominantly negative in a frequency range of 0.5 to 150 Hz and the corresponding range of reduced frequencies form 0.005 to 10. The correlations of velocity with pressure for the pressure tap locations nearer the center of the face are positive in the same frequency range.

6. CONCLUSIONS

The use of wavelet analysis techniques coupled with velocity measurement near a bluff body using a LDV system allows a better understanding of the interaction of the incident velocity field and the fluctuating pressures on the surface of the body. Intermittent correlation over a range of frequencies or wave numbers were observed in both stagnation and separated regions. The frequency ranges of significant correlation will provide guidance in developing guidelines for simulation of atmospheric flows.

7. REFERENCES

Akins, R.E. & Reinhold, T.A., 1998. Laser Doppler velocimeter measurements of separated shear layers on bluff bodies. *Journal of Wind Engineering and Industrial Aerodynamics*, 74-76:455-461.

Hajj, M.R. & Tieleman, H.W., 1996. Applications of wavelet analysis to incident wind in relevance to wind loads on structures. *ASME Journal of Fluids Engineering*, 118:874-876.

Hajj, M.R., Jordan, D.A,. & Tieleman, H.W., 1998. Analysis of atmospheric wind and pressures on low-rise building, *Journal of Fluids and Structures*, 12:537-547.

Hubbard, Barbara Burke, 1996. *The World According to Wavelets*. Wellesley, Mass, AK Peters.

Newland, D.E., 1993. *An Introduction to Random Vibrations, Spectral and Wavelet Analysis, Third Edition*. Essex, England, Longman.

8. ACKNOWLEDGEMENT

Mr. Nahan Hottle worked to help develop the wavelet analysis software for unequally spaced data while a student at Washington and Lee University.

Wind Engineering into the 21st Century, Larsen, Larose & Livesey (eds) © 1999 Balkema, Rotterdam, ISBN 90 5809 059 0

A wind tunnel study of turbulent flow behind a backward facing step with and without flow underneath the step

O.O. Badran
Department of Mechanical Engineering, Al-Balqa Applied University, Jordan
H.H. Bruun
Department of Mechanical Engineering, Bradford University, UK

ABSTRACT : The purpose of this paper is to present the Flying Hot – Wire (F.H.W.) measurements and to interpret these in terms of the flow mechanisms of the phenomena associated with two – dimensional separation of the turbulent boundary layer behind a backward – facing step (B.W.F.S.). The inlet conditions were also changed by carrying out tests both with (Case I) and without (Case II) flow underneath the step configuration. The Flying X-Hot – Wire probe was used to measure the U and V component of the flow field over the step, with particular emphasis on the flow inside and surrounding the region of recirculating flow and downstream the reattachment point. From the results of the two cases (Case I, and Case II), it can be concluded that the different inlet conditions have an effect on the turbulence intensity upstream of the step, and on the flow structure (e.g. turbulence normal and shear stresses) in the free shear layer downstream of the step, especially at axial positions closest to the step shoulder, in addition to the change in reattachment length. Also it was found that by blocking the flow channel underneath the step, the flow separated from the leading edge of the upstream plate resulting in an increase in the boundary layer thickness and a substantial increase in the magnitude of the turbulence intensity and its vertical distribution.

1 INTRODUCTION

Flow separation is important in many engineering applications (e.g. aerofoil, backward-facing step, bluff bodies, cylinders, etc.) In order to understand the main physical features of the flow field, there is a need for experimental data to identify the flow behavior and explain it. Although a large number of investigations of separated flow have been studied (i.e. Hsu (1950), Tani et al. (1966), Baker (1977), Durst & Tropea (1981), Adams & Johnston (1988)) the phenomena of turbulent separation and reattachment are still not very well understood. There is therefore a need for further data to clarify many of the ideas on separated flow and reattachment length. To study the effect of the inlet and exit conditions on the separated flow field, experiments were carried out both with flow underneath the step and with this flow passage blocked Also, for both flow conditions, tests were carried out with sharp leading edge, Badran (1993).

2 EXPERIMENTAL FACILITIES

The original mechanism of the present F.H.W. was designed and constructed by Jaju (1987) and it is similar to the one implemented at Imperial College, London and reported by Thomson & Whitelaw (1984).

A Stationary hot-wire probe is only applicable in low/moderate turbulence intensity flows duc to thc reverse flow ambiguity. This problem can be resolved by using the flying hot-wire technique, which enables accurate measurements in regions where flow reversal occurs or in regions of high turbulence intensity.

The basic principle of the flying X hot –wire (F.H.W.) technique can be explained with reference to Figure 1. Consider a surface with a (2-dimensional) separated flow region. A space –fixed co-ordinate system is introduced, in which the flow velocity vector $\vec{V}$ and its velocity components U and V are to be evaluated. Based on the geometry of the flying hot –wire mechanism the X probe used will follow a prescribed curve (a). At time t, the probe is assumed to be at a known position (X_p, Y_p) and move with a known probe velocity $\vec{V}_p$.

The moving hot-wire probe is exposed to the relative velocity $\vec{V}_r$, and this velocity vector is for an X probe normally evaluated in terms of the velocity components (U_r', V_r') in a probe-stem aligned co-ordinate system. Provided the orientation of the probe – stem relative to the space – fixed co-ordinate system is known, then the corresponding space – fixed velocity components $(\vec{U}_r, V_r)$ can be evaluated. Having measured $\vec{V}_p$ and $\vec{V}_r$ the flow vector $\vec{V}$ is obtained from $\vec{V} = \vec{V}_p + \vec{V}_r$ as illustrated in Figure 1. Provided that the magnitude of the probe velocity V_p is greater than the magnitude of the flow velocity V then the relative velocity Vector $\vec{V}_r$ will remain within the approach quadrant of the X probe, and the hot –wire signals can be interpreted uniquely (Bruun et al. 1991).

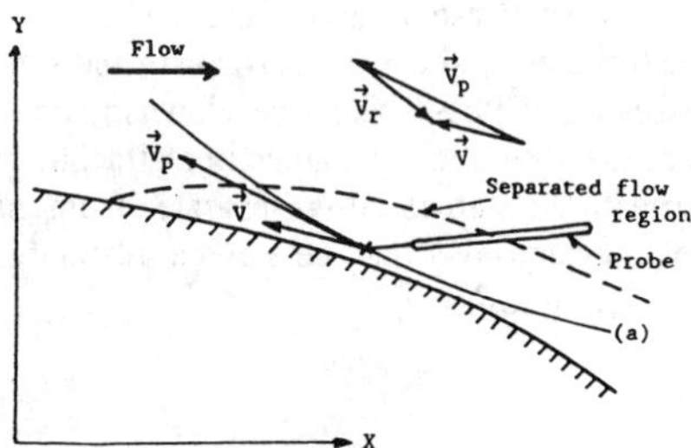

Figure 1. Principle of measurements with a F. H. W.

The development of a F.H.W. system involves the selection of a suitable probe path and a related drive mechanism. In this investigation a bean –shaped curve path was selected, and it was achieved using the four bar mechanism shown in Figure 2. Mechanically, the motion is implemented by rotating a flywheel linked to the arm OR. The advantage of this F.W.H. system is that only part of the arm and the probe needs to be placed inside the wind tunnel, thus minimizing the blockage effect.

An electric motor was used to drive the flywheel of the four bar mechanism. As shown in Figure 2, the probe starts from a rest position and a significant angular acceleration is required to achieve the required probe velocity $\vec{V}_p$ during a single sweep. Consequently, to avoid excessive torque requirements, the mass of rotating parts of the system was minimized, and a DC printed circuit motor of a reasonably small size (≅1kg) was selected. This type of motor could be overrun for a small period of time above its normal operating current, to achieve the high acceleration required during its first revolution. It could also be controlled easily for start/stop operations.

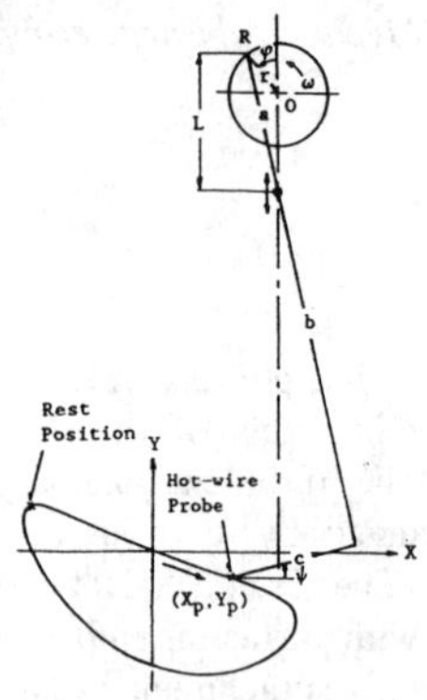

Figure 2. Four bar F.H.W. probe mechanism (r = 60mm , a = 160mm , b = 410 mm , C = 147mm).

The notations for the step and F.H.W. used in this study are shown in Figure 3.

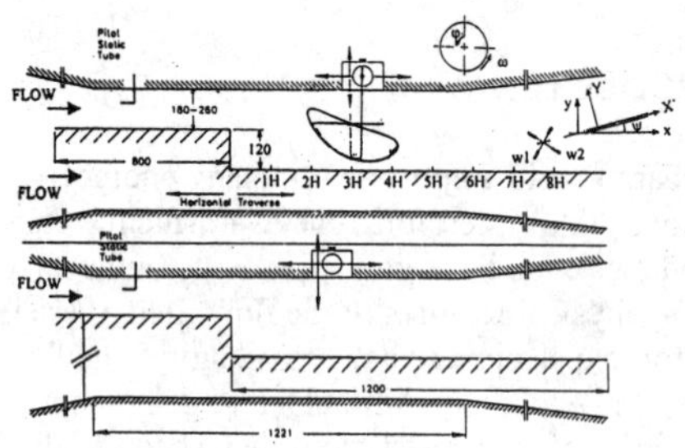

Figure 3 . Backward-facing step model.

3 RESULTS: -

This paper presents F.H.W. measurements of the mean velocity, normal and shear stresses in the separated flow region behind a backward –facing step. The free stream velocity used for all measurements was 10m/sec which corresponds to $Re_H=0.79 \times 10^5$.

3.1 *Case I*

This test (Case I Flow underneath the step configuration) was used as a reference, as this condition result in a nearly normal turbulent boundary layer on the upstream plate just before the step. Measurements were carried out upstream of the step at X/H= - 0.3 of the mean velocity $\overline{U}$ (Fig. 4a) and the fluctuation $\overline{u}^2$ (Fig. 4b), using a stationary hot-wire anemometer. From Figure 4a the boundary layer thickness, δ, was evaluated as δ/H= 0.8,with δ being the value of Y at which U=0.99U_∞.

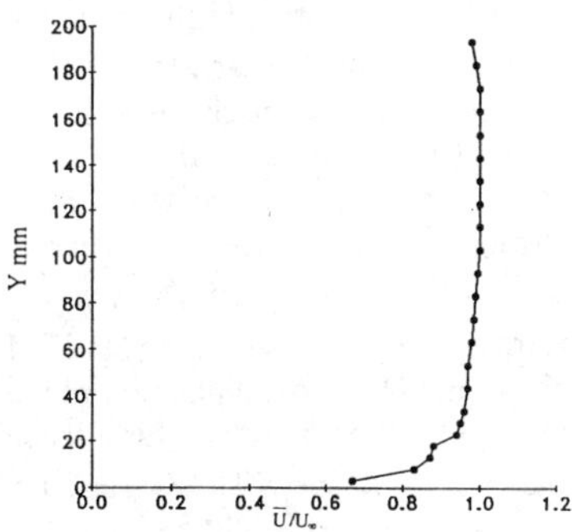

Figure 4a. Inlet velocity profile $\overline{U}/U_\infty$ (Case I).

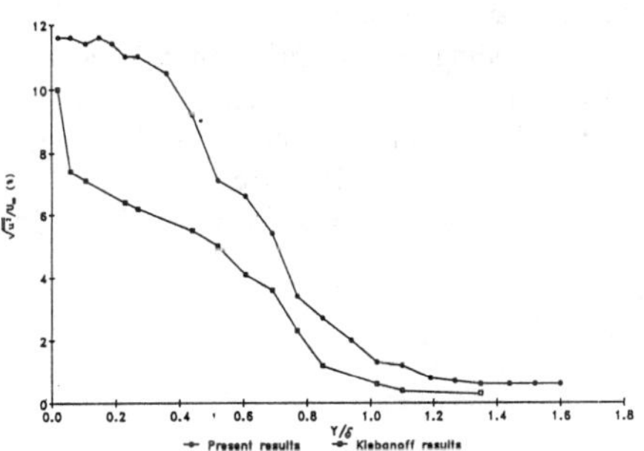

Figure 4b. Normal stress $\sqrt{\overline{u}^2}/U_\infty$ % profile (Case I).

In Figure 4b the turbulent profile $\sqrt{\overline{u}^2}/U_\infty$(%) is compared with the measurements of the longitudinal turbulent fluctuations in the boundary layer of a flat plate by Klebanoff (1955). He obtained high values of $\sqrt{\overline{u}^2}/U_\infty$ near the wall of the flat plat, which is similar to the measured data for Case I. However, the difference in the shape of turbulent profile above the wall may be due to the difference in the boundary layer thickness (δ).

3.1.1 *Mean Velocity Results*

The free stream Velocity U_∞ was used as a reference, and the velocity is presented and described in terms of the ratio of the local mean velocity to the free stream velocity. The velocity vector field for Case I is shown in Figure 5.

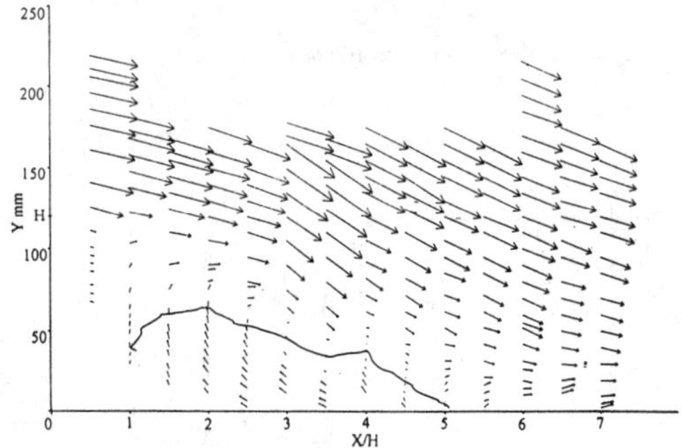

Figure 5. Measured mean velocity vectors.

Figure 5 also contains the position of the $\overline{U}$ =0. From Figure 5 several features can be identified. The separation occurs at the step shoulder and the flow is seen to reattach at (X_R/H ≅ 5). The initial free shear layer is seen to be deflected continuously downwards towards the reattachment point (X_R/H≅5). The $\overline{U}=0$ line separates the reverse flow (inner region) from the outer streamwise flow region, and the height of the $\overline{U}$ = 0 curve from the wall varies along the axial axis downstream of the step. The reverse flow region is seen to extend from X/H ~ 1 to 5, with a maximum vertical extent of y = 64mm occurring at the axial distance X/H = 2.

Beyond the reattachment point a new sub-boundary layer started to develop which will develop into a fully turbulent boundary layer condition. Due to the expansion over the step, the maximum value of the velocity vector in the potential flow, $\overline{U}/U_\infty$, was about 0.9. It is observed that the deflection of the outer flow is increased significantly at the axial distance X/H = 3, due to the effect of separation bubble underneath this region.

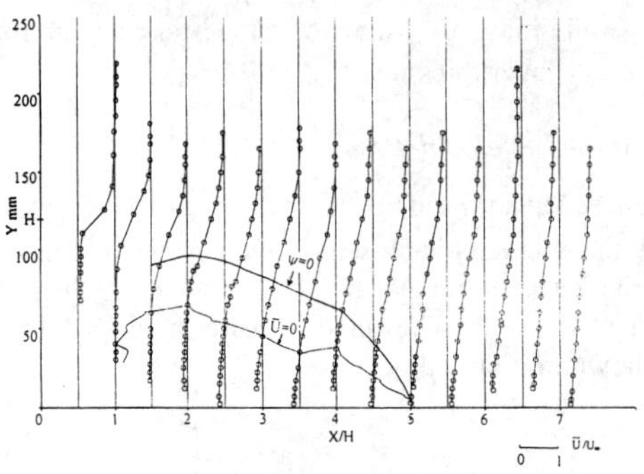

Figure 6a. Mean velocity profile of $\overline{U}/U_{\infty}$.

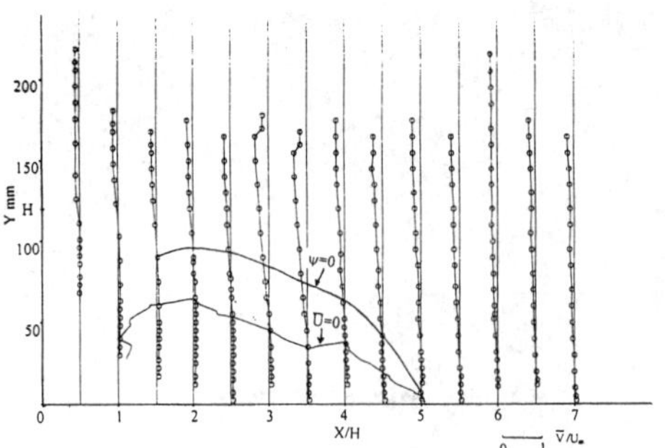

Figure 6b. Mean velocity profile $\overline{V}/U_{\infty}$

Figure 6.a shows the profiles of the $\overline{U}/U_{\infty}$ velocity component at various downstream locations from the step shoulder. The dividing streamline (ψ) has been drawn on these figures. At X/H = 0.5 and 1, the measured part of the mean velocity profile is seen to have positive values, at all measured positions. From X/H = 1.5 to 4.5 the $\overline{U}/U_{\infty}$ velocity profiles clearly identify the existence of the separation bubble. Inside the reverse flow region the magnitude of the maximum negative $\overline{U}$ velocity component ($\overline{U}/U_{\infty}$) is about − 0.17(-17%) and it occurred at X/H = 3.5. The velocity profiles beyound reattachment (X/H>5) show the initial development of the turbulent bundary layer towards that of a normal turbulent boundary layer. In the shear layer region surrounding the separation bubble the values of $\overline{U}$ increase gradually in the vertical direction towards the value of the free stream velocity in potential flow region.

Figure 6b shows the vertical mean velocity profiles (V-component) at various cross- sections downstream of the step. The scale used for $\overline{V}/U_{\infty}$ is the same as that for $\overline{U}/U_{\infty}$ in Figure 6a to allow direct comparison. As expected in the separated flow region, the vertical mean velocity component is observed to be positive, while outside the separation region high negative values occur, representing the deflection of the potential flow.

3.1.2 *Turbulent stresses*

Figures 7 - 9 show the distribution of the turbulent normal stresses $\overline{u}^2$, v^2 and shear stress uv , respectively, downstream of the backward –facing step for Case I (flow underneath the step configuration). The data are presented in the normalized form as, for example, $\overline{u}^2/U^2_{\infty}$ x 10^3 , $\overline{v}^2$ / U^2_{∞}x 10^3 , $-\overline{uv}$ / U^2_{∞}x 10^3, and the dividing streamline and $\overline{U}$ = 0 are identified on each of the diagrams. The vertical position of the maximum values for the normal turbulence intensities $\overline{u}^2$ and $\overline{v}^2$ between X/H = 0.5 and 1.5 occur at y ≅ H, and nearly coincide with the position of the dividing streamline. Tani et al. (1961) , Chandersuda (1975) and Baker (1977) also record this feature in their experimental results.

The corresponding vertical variation in $\overline{u}^2$ and $\overline{v}^2$ show low values inside the recirculation region, followed by large values in the shear flow region and a decrease to the free stream value in the potential flow above the shear layer flow. The

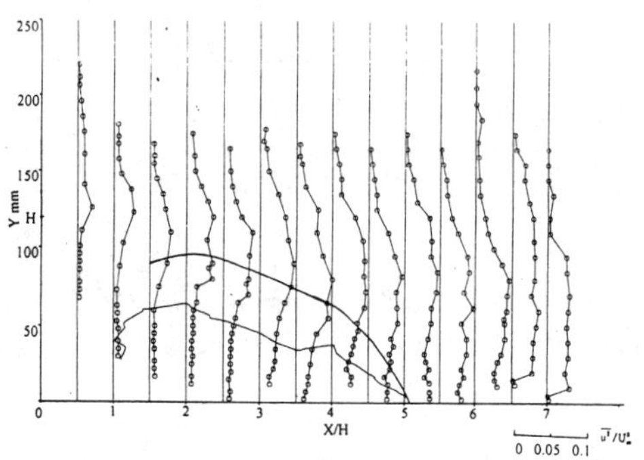

Figure 7 Normal stress $\overline{u}^2/U^2_{\infty}$ x 10^3

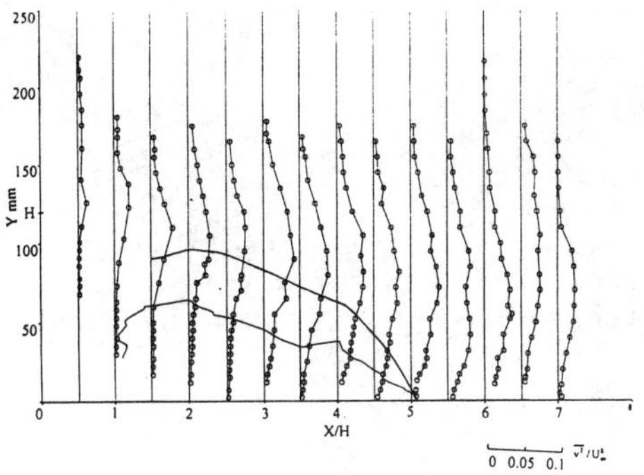

Figure 8 Normal stress $\overline{v}^2$ / U^2_∞ x 10^3

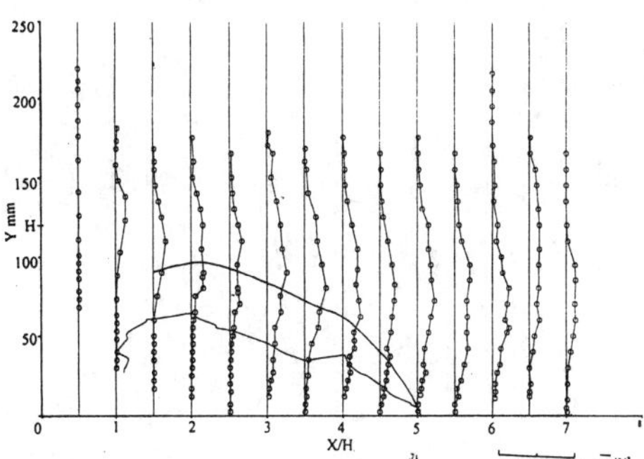

Figure 9 Shear stress, $-\overline{uv}$ / U^2_∞ x 10^3

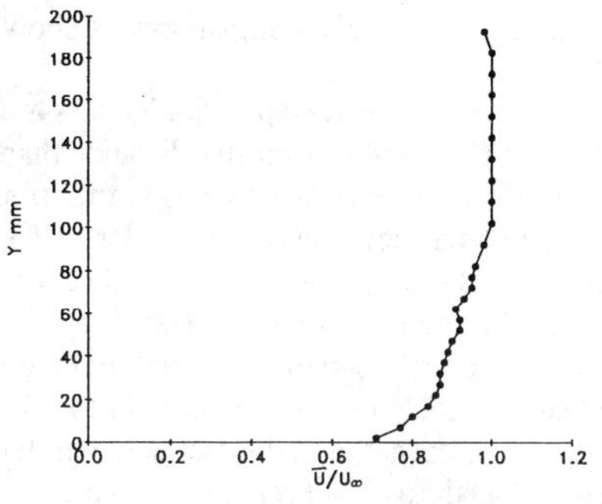

Figure 10a. Inlet velocity profile $\overline{U}/U_\infty$ (Case II)

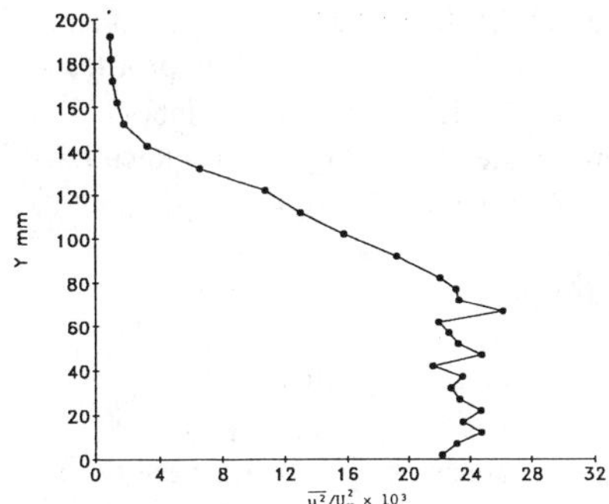

Figure 10b Normal stress $\overline{u}^2/U^2_\infty$ (Case II)

longitudinal stress values increased from $\overline{u}^2_{max}/U^2_\infty$=0.019 at X/H = 0.5 to a maximum value of $\overline{u}^2_{max}/U^2_\infty$= 0.05 at X/H = 3.5, followed by a region of constant values until X/H = 6 before decreasing relatively sharply.

3.2 *Case II*

This section highlights some further effects of changes to inlet conditions on the flow downstream of the step, using the same sharp edge section as for Case I , but with the flow channel underneath the step blocked with a board .

3.2.1 *Inlet profiles*

The inlet profile for the mean velocity $\overline{U}/U_\infty$ and turbulence quantity $\overline{u}^2/U^2_\infty$ are shown in Figure 10a and 10b, respectively.

From Figure 10a it is noticed that inlet turbulent boundary layer thickness has become consistently thicker (δ/H = 1.2) than in Case I (δ /H = 0.87), and the velocity profile displays lower values near the wall .

The profile of the normal stress (Fig. 10b), $\overline{u}^2/U^2_\infty$, shows a significant increase in its maximum value of $\overline{u}^2_{max}/U^2_\infty$= 26 x$10^{-3}$ near the wall, as compared with a value of $\overline{u}^2/U^2_\infty$ = 13.5 x 10^{-3} in Case I. Also, the turbulent intensity remained high and nearly constant for a significant vertical distance, up to Y/H ≅0.6, due to the separation formed at the leading edge, which greatly enhanced the turbulence in the boundary layer on the upstream plate .

3.2.2 *Mean Velocity Results*

The velocity vector field representation gives a clear indication of the differences in flow field between

Case I and Case II. This comparison is shown in Figure 11.

It is observed that the re-separation occurs at X/H =0.7 for Case II, which is a shorter distance than for Case I (X/H =1.2), reflecting changes in the strength of the corner eddies near the step shoulder. Also, the magnitude of the velocity vector in case II was higher near the step between X/H = 1 to 2. The vertical extent of the separation bubble region for Case II can be seen to be smaller than for Case I, except near the corner, and the reattachment length has been reduced to $X_R/H = 4.5$. In both cases the maximum vertical extent of the $\overline{U} = 0$ curves was Y/H = 0.5 located at X/H = 1 for Case II and X/H = 2 for Case I. In the potential flow region the deflection of the flow between X/H = 3 and 5.5 is seen to be greater in Case I. The profiles of the corresponding mean velocity component $\overline{U}$ are shown in Figure 12, giving a clear indication of the locations of the curves $\overline{U} = 0$ and $\psi = 0$.

3.2.3 *Turbulent Stresses*

The measured profiles of, $\overline{u}^2/U^2_\infty$, $\overline{v}^2/U^2_\infty$ $-\overline{uv}/U^2_\infty$ are shown in Figures 13 – 15. As expected, the general flow development is similar to the previous case with two distinct regions. Firstly, a shear layer region containing high turbulence levels can clearly be identified.

Beneath the inner region of the shear layer ($\overline{U} = 0$) is a region of the recirculating flow in which the turbulence levels are low, increasing axially towards the reattachment zone. However, it should also be noted that as the turbulent intensity in the inlet boundary layer has been increased, this is also reflected both inside the reveres flow region and outside the free shear layer. Downstream of the reattachment zone the turbulent intensities and shear stress continue to decline.

Inside the reverse flow region the value of the normal and shear stresses of $\overline{u}^2/U^2_\infty, \overline{v}^2/U^2_\infty$, $-\overline{uv}/U^2_\infty$ for Case II are nearly twice those for Case I, and the values increase axially from step shoulder to reattachment point. Also, in the shear layer region for Case II the maximum values of the normal and shear stresses between X/H = 0.5 ~2.5 are nearly 15% higher than for Case I. This is followed by a rapid decline in the maximum values of $\overline{u}^2, \overline{v}^2$, $-\overline{uv}$, to values 4% lower than for Case I at X/H > 5.

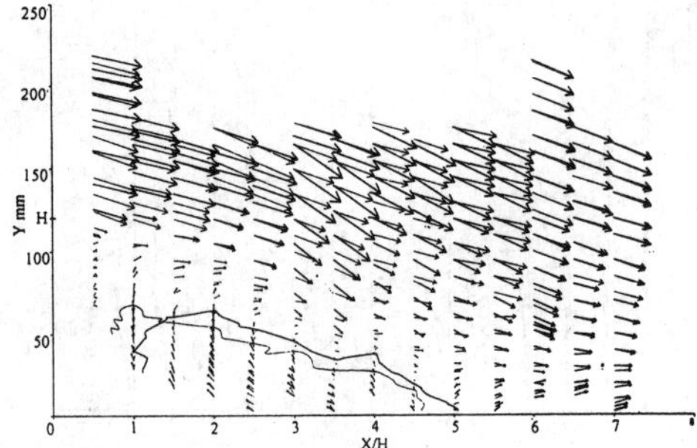

Figure 11. Measured mean velocity vectors .

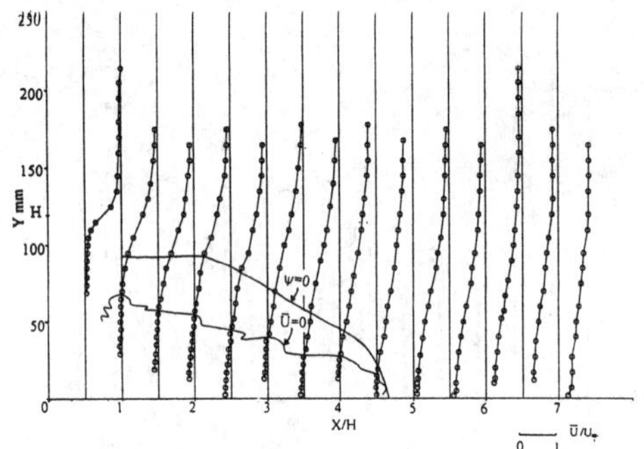

Figure 12 Mean velocity of $\overline{U}/U_\infty$

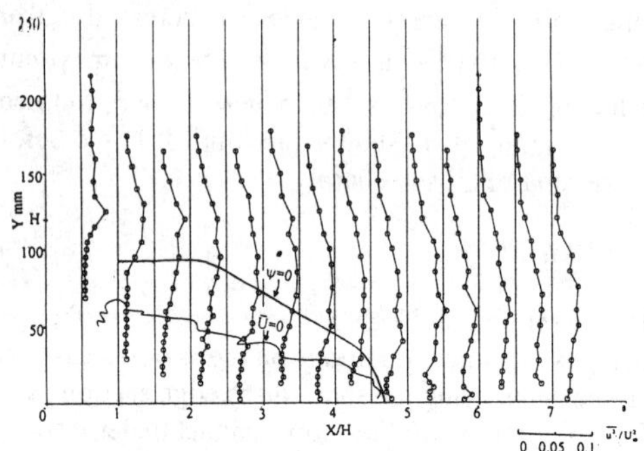

Figure 13. Normal stress $\overline{u}^2/U^2_\infty$ x 10^3

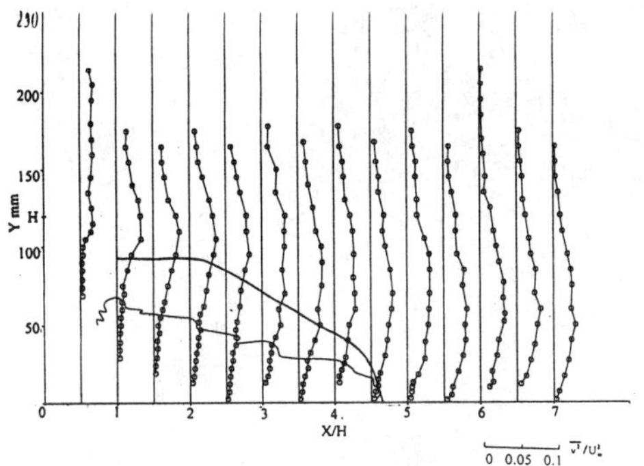

Figure 14. Normal stress $\overline{v}^2 / U^2_\infty$ x 10^3

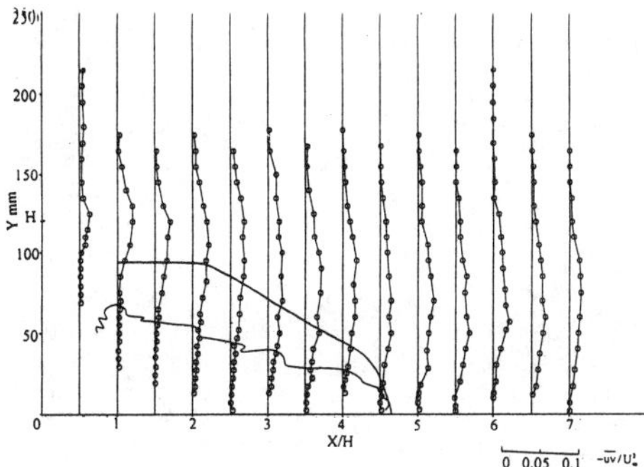

Figure 15. Shear stress $-\overline{uv} / U^2_\infty$ x 10^3

A change can be observed in the peak values and their locations between the two cases. For Case I the peak value of the normal stresses ($\overline{u}^2_{max} / U^2_\infty = 50$ x 10^{-3}, $\overline{v}^2_{max} / U^2_\infty$ = 42x 10^{-3}) and shear stress ($-\overline{uv}_{max} / U^2_\infty =$ 28 x 10^{-3}) occurs upstream of reattachment at X/H=3 ~3.5, and for Case II the peak value of the longitudinal stress ($\overline{u}^2_{max} / U^2_\infty =$ 55x10^{-3}) occurred half a step height downstream of the reattachment, while the vertical and shear stresses ($\overline{u}^2_{max} / U^2_\infty =$ 37x 10^{-3}, $-\overline{uv}_{max} / U^2_\infty = 21.2$ x10^{-3}) occurred at X/H= 2 .

4 CONCLUSIONS

From these results the following conclusions can be made: the blockage underneath the step (Case II) increased the turbulence intensity upstream of the step due to the separation occurring at the leading edge. This affected the flow structure downstream of the step, such as increasing the turbulent normal and shear stresses inside the separated flow and free shear layer regions, and reducing the reattachment length (e.g. for the no flow case (Case II) occurred at X_R /H = 4.5, while for the flow case (CaseI), occurred at X_R / H = 5).

F.H.W. anemometry seems to be one of the suitable instruments for investigation in the separated flow region.

REFERENCES

Adams, E.W. & J.P. Johnston 1988. Effects of the separating layer on the reattachment flow structure. Part 2: Reattachment length and wall shear stress. *Experiments in Fluids.* 6 : 493-499 .

Badran, 0.0 1993. A Flying Hot –Wire study of separated flows, Ph.D. Thesis Bradford University, UK.

Baker, S. 1977. Regions of re- circulating flow associated with two – dimensional steps. Ph.D. thesis, University of Surrey , UK.

Bruun, H.H., O.O Badran. & H.H Al- Kayiem, 1991. "A Flying hot –wire probe study of flow behind a backward – facing step" Proc of 2nd World Conf of Experimental Heat Transfer. Fluid Mech and Thermo, Dubrovnik, Yugoslavia.

Chandersuda, C. 1975 . A reattaching turbulent shear layer in incompressible flow Ph.D. Thesis, Imperial College of Science and Technology , UK.

Durst, F. & C. Tropea 1981. Turbulent backward – facing step flows in two –dimensional ducts and channels .*Proc. of 3rd symposium on Turbulent shear flows* . 1-6 (Davis CA, USA),1-6.

Hsu, H.C. 1950. Characteristics of mean flow and turbulence at an abrupt two-dimensional expansion. Ph.D. Thesis, Department of Mechanics and Hydraulics, state University of Iowa. USA.

Jaju, O. A. R 1987. Development of a Flying hot wire probe system . M.Phil Thesis, Bradford University , UK.

Klebanoff, D.S 1955. Characteristics of turbulence in a boundary layer with zero gradient. NACA , Report 1247

Tani, I, M. Iuchi, & H., Komeda, 1961. Experimental investigation of flow separation associated with a step or a groove. *Aeronautical Research Institute*, University of Tokyo, Report No.364, 27. No.4, 119-137.

Thompson, B.E. & J.H., Whitelaw 1984. Flying Hot-Wire Anemometry. *Experiments in Fluids.* 2:47-55.

Wind Engineering into the 21st Century, Larsen, Larose & Livesey (eds) © 1999 Balkema, Rotterdam, ISBN 90 5809 059 0

Quasi-steady models of aerodynamic loads on two circular cylinders in side-by-side arrangement

E. Błazik-Borowa
Department of Structural Mechanics, Lublin Technical University, Poland

ABSTRACT: The paper deals with the quasi-steady model of vibration of cylinders in a side-by-side arrangement. The cylinders are elastically supported and they have equal diameters. The model assumes that both cylinders participate in process of vibration and they have two degrees of freedom. The movement of the cylinders is written as a set of four nonlinear differential equations. On the basis of comparison numerical and research results the author checks the correctness of the quasi-steady model for this arrangement.

1 THE PRINCIPAL ASSUMPTIONS AND RELATIONS

On the basis of static and dynamic researches of two identical circular cylinders it was observed that there are some cylinders arrangements at which the bistable flows inducing cylinders vibration exist (comp. Zdravkovich and Pridden 1977, Zdravkovich 1987, Zdravkovich and Medeiros 1991, Zhang and Melbourne 1995). One of these arrangements is a cylinders arrangement so that the velocity direction is nearly perpendicular to the axis joining the midpoints of the cylinders (comp. Figure 1). Such an arrangement is called a side-by-side arrangement. When the distance between the centers of cylinders is smaller than about 3.0 diameters the flow passing through the gap is biased to one side. The biased flow is not stable but changes sides at irregular intervals of time.

In the paper the quasi-steady model of cylinders vibration in case of the side-by-side arrangement of these cylinders will be presented. This model has been previously prepared for interference galloping and it has been described in detail in the following papers: Błazik-Borowa and Flaga 1995, Błazik-Borowa and Flaga 1997, Błazik-Borowa and Flaga 1998. The quasi-steady model for the side-by-side arrangement is based on similar assumptions, thus:

1. Diameters of crosswise intersections of cylinders are equal.
2. Position of cylinder 2 is found within the region with the following limitations: $1.2<L_y/D<3$ and $-0.2<L_x/D<0.2$ (comp. Figure 1), that is in the region of bistable flows.
3. Relative angles $\breve{\mu}_r$ and μ_r are angles between the directions perpendicular to the relative wind directions and the axis joining the midpoints of the cylinders.
4. Cylinders are made of rigid material.
5. Flow is viscous and incompressible.
6. Cylinders are elastically supported and they have two degrees of freedom: along the average wind direction and in the direction perpendicular to it.
7. Boundary disturbance effects are neglected.
8. Problem is treated as flat.
9. Global coordinate system is linked to the midpoint of an first cylinder and x axis is parallel to the average wind direction.
10. Quantities such as: the instantaneous wind velocity $\mathbf{V}_s$, the components of the instantaneous wind velocity towards x and y axes $\mathbf{u}_s$ and $\mathbf{v}_s$, the instantaneous angle of wind attack γ_s, are the average values from the region of averaging $S=\Delta y x H$, where: $H=\kappa_1 D$; $\Delta y=\kappa_2 D$ - the across dimension of the wake formed behind the two cylinders $\kappa_1, \kappa_2 \in (5\div 10)$. The spatial region of averaging is located in front of a first cylinder in the distance at least D.
11. Aerodynamic load is constant on entire length of cylinders.
12. Determination of the physical model and then the mathematical one of a wind load for first and second cylinders is based on a dimensional analysis and quasi-steady theory in aerodynamics of slender structures.

Assuming appropriate trigonometric dependence (comp. Figure 1), the components of wind load towards the axes of a global coordinate system influencing the cylinders can be given by:

- for the first cylinder

$$\breve{\mathbf{W}}_x(\breve{\alpha}_r,\breve{\mu}_r)=$$

$$=\left[\breve{\mathbf{W}}_{x_3}\left(\breve{\mu}_r\right)\cos\breve{\alpha}_r-\breve{\mathbf{W}}_{y_3}\left(\breve{\mu}_r\right)\sin\breve{\alpha}_r\right] \quad (1)$$

$$\breve{\mathbf{W}}_y(\breve{\alpha}_r,\breve{\mu}_r)=$$

$$=\left[\breve{\mathbf{W}}_{x_3}\left(\breve{\mu}_r\right)\sin\breve{\alpha}_r+\breve{\mathbf{W}}_{y_3}\left(\breve{\mu}_r\right)\cos\breve{\alpha}_r\right] \quad (2)$$

- for the second cylinder

$$\mathbf{W}_x(\alpha_r,\mu_r)=$$

$$=\left[\mathbf{W}_{x_4}\left(\mu_r\right)\cos\alpha_r-\mathbf{W}_{y_4}\left(\mu_r\right)\sin\alpha_r\right] \quad (3)$$

$$\mathbf{W}_y(\alpha_r,\mu_r)=$$

$$=\left[\mathbf{W}_{x_4}\left(\mu_r\right)\sin\alpha_r+\mathbf{W}_{y_4}\left(\mu_r\right)\cos\alpha_r\right] \quad (4)$$

where: $\breve{\mathbf{W}}_{x3}$, $\breve{\mathbf{W}}_{y3}$ - components of interferential load on the length unit of the first cylinder along axes x_3 and y_3, respectively; $\mathbf{W}_{x4}$, $\mathbf{W}_{y4}$ - components of interferential load on the length unit of the second cylinder along axes x_4 and y_4, respectively. The axes x_3 and y_3 are directed along and across the direction of the relative velocity $\breve{\mathbf{V}}_r$ of the first cylinder, and the axes x_4 and y_4 are directed along and across the direction of the relative velocity $\mathbf{V}_r$ of the second one.

On the basis of dimensional analysis, the above components of aerodynamic load can be written as follows:

$$\breve{W}_i(\breve{\mu}_r)=\frac{1}{2}\rho\breve{V}_r D\breve{C}_i(\breve{\mu}_r),\ i=x_3\,,y_3 \quad (5)$$

$$W_i(\mu_r)=\frac{1}{2}\rho V_r D C_i(\mu_r),\ i=x_4\,,y_4 \quad (6)$$

The above aerodynamic loads of the cylinders are described as functions of static aerodynamic drag and lift coefficients:

$$\breve{C}_{x_3}\left(\breve{\mu}_r\right)=\breve{C}_x\left(\theta=\breve{\mu}\right),\ \breve{C}_{y_3}\left(\breve{\mu}_r\right)=\breve{C}_y\left(\theta=\breve{\mu}_r\right),$$

$$C_{x_4}\left(\mu_r\right)=C_x\left(\theta=\mu_r\right),\ C_{y_4}\left(\mu_r\right)=C_y\left(\theta=\mu_r\right) \quad (7)$$

They are obtained for a particular case that is when the cylinders do not vibrate so they do not change their locations and at the same time the velocity and the direction of the wind are constant. The arrangement of these cylinders correspond to the momentary arrangement of vibrating cylinders. The aerodynamic coefficients have been adopted the following the papers of Zdravkovich and Pridden 1977, ESDU 84015 1984.

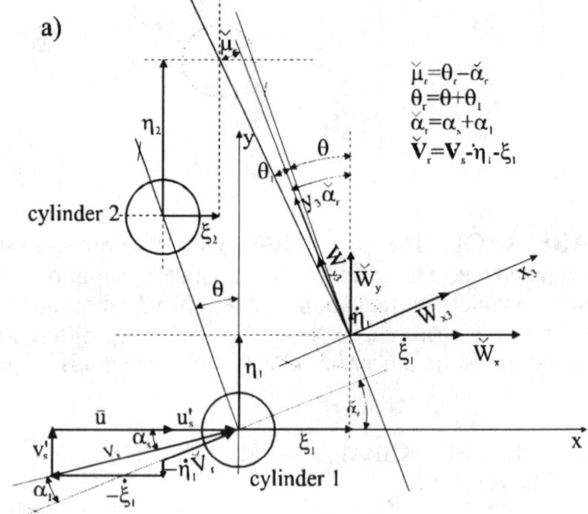

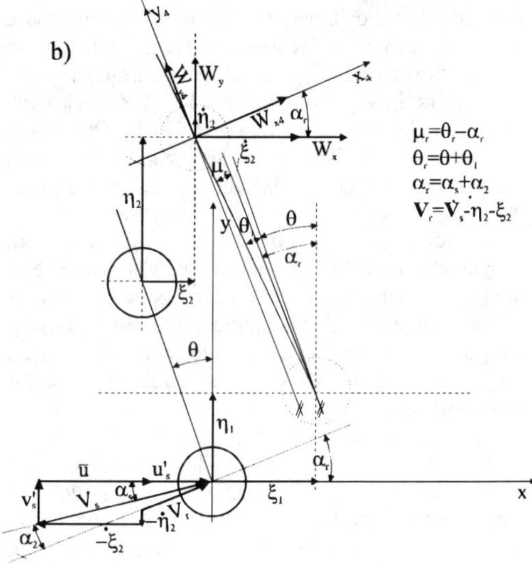

Figure 1. The arrangement of the cylinders and basic notations.

The relative angles of the wind attack are angles between the directions perpendicular to the direction of a respective relative velocity and a line joining centers of displaced cylinders (cylinders in a new location) and they are described by the equations (comp. Figure 1):

- for the first cylinders

$$\breve{\mu}_r=\breve{\mu}_r(z,t)=(\theta+\theta_1)-(\alpha_s+\alpha_1)=\theta_r-\breve{\alpha}_r \quad (8)$$

$$\breve{\mu}_r(t) = \arctan \frac{\left(L_y \tan\theta - \xi(t)\right)\left(u_s(t) - \dot{\xi}_1(t)\right) - (v'_s(t) - \dot{\eta}_1(t))\left(L_y + \eta(t)\right)}{\left(u_s(t) - \dot{\xi}_1(t)\right)\left(L_y - \eta(t)\right) + \left(L_y \tan\theta + \xi(t)\right)(v'_s(t) - \dot{\eta}_1(t))}$$

– for the second cylinders

$$\mu_r = \mu_r(z,t) = (\theta + \theta_1) - (\alpha_s + \alpha_2) = \theta_r - \alpha_r \quad (9)$$

$$\mu_r(t) = \arctan \frac{\left(L_y \tan\theta - \xi(t)\right)\left(u_s(t) - \dot{\xi}_2(t)\right) - (v'_s(t) - \dot{\eta}_2(t))\left(L_y + \eta(t)\right)}{\left(u_s(t) - \dot{\xi}_2(t)\right)\left(L_y - \eta(t)\right) + \left(L_y \tan\theta + \xi(t)\right)(v'_s(t) - \dot{\eta}_2(t))}$$

where u_s and v'_s - spatially averaged components of the wind velocity in the direction x and y; $\xi_1, \eta_1, \xi_2, \eta_2$ - components of displacement of the first and second cylinders towards axes x and y, respectively; $\xi = \xi_2 - \xi_1$; $\eta = \eta_2 - \eta_1$.

The vectors of the relative velocities are defined as (comp. Figure 1):

– for the first cylinder

$$\breve{\mathbf{V}}_r = \breve{\mathbf{V}}_r(z,t) = \mathbf{V}_s(z,t) - \dot{\eta}_1(t) - \dot{\xi}_1(t) \quad (10)$$

– for the second cylinder

$$\mathbf{V}_r = \mathbf{V}_r(z,t) = \mathbf{V}_s(z,t) - \dot{\eta}_2(t) - \dot{\xi}_2(t) \quad (11)$$

2 MOTION EQUATIONS

The equations describing the movement of these cylinders towards axes x (the average wind direction) and y (the direction perpendicular to the average wind direction) are formulated according to the assumption that each cylinder is a system with two degrees of freedom. The right sides of these equations are the sums of acceleration, damping and stiffness forces. The left sides are equal to the values of the wind loads. The motion equations can be written in the form of a set of four nonlinear differential equations:

– for the first cylinder

$$m\ddot{\xi}_1(t) + C_{\xi_1}\dot{\xi}_1(t) + K_{\xi_1}\xi_1(t) = \breve{W}_x(t) \quad (12)$$

$$m\ddot{\eta}_1(t) + C_{\eta_1}\dot{\eta}_1(t) + K_{\eta_1}\eta_1(t) = \breve{W}_y(t) \quad (13)$$

– for the second cylinder

$$m\ddot{\xi}_2(t) + C_{\xi_2}\dot{\xi}_2(t) + K_{\xi_2}\xi_2(t) = W_x(t) \quad (14)$$

$$m\ddot{\eta}_2(t) + C_{\eta_2}\dot{\eta}_2(t) + K_{\eta_2}\eta_2(t) = W_y(t) \quad (15)$$

These equations can be written in the more developed forms:

– for the first cylinder in the direction x

$$m\frac{2}{\rho D}\ddot{\xi}_1(t) + C_{\xi_1}\frac{2}{\rho D}\dot{\xi}_1(t) + K_{\xi_1}\frac{2}{\rho D}\xi_1(t) + \left(-\breve{C}_{x_3}(\breve{\mu}(t))u_s(t) + \breve{C}_{y_3}(\breve{\mu}(t))v'_s(t) + \breve{C}_{x_3}(\breve{\mu}(t))\dot{\xi}_1(t) - \breve{C}_{y_3}(\breve{\mu}(t))\dot{\eta}_1(t)\right) \times \sqrt{\left(u_s(t) - \dot{\xi}_1(t)\right)^2 + \left(v'_s(t) - \dot{\eta}_1(t)\right)^2} = 0 \quad (16)$$

– for the first cylinder in the direction y

$$m\frac{2}{\rho D}\ddot{\eta}_1(t) + C_{\eta_1}\frac{2}{\rho D}\dot{\eta}_1(t) + K_{\eta_1}\frac{2}{\rho D}\eta_1(t) + \left(-\breve{C}_{x_3}(\breve{\mu}(t))v'_s(t) - \breve{C}_{y_3}(\breve{\mu}(t))u_s(t) + \breve{C}_{y_3}(\breve{\mu}(t))\dot{\xi}_1(t) + \breve{C}_{x_3}(\breve{\mu}(t))\dot{\eta}_1(t)\right) \times \sqrt{\left(u_s(t) - \dot{\xi}_1(t)\right)^2 + \left(v'_s(t) - \dot{\eta}_1(t)\right)^2} = 0 \quad (17)$$

– for the second cylinder in the direction x

$$m\frac{2}{\rho D}\ddot{\xi}_2(t) + C_{\xi_2}\frac{2}{\rho D}\dot{\xi}_2(t) + K_{\xi_2}\frac{2}{\rho D}\xi_2(t) + \left(-\breve{C}_{x_4}(\breve{\mu}(t))u_s(t) + \breve{C}_{y_4}(\breve{\mu}(t))v'_s(t) + \breve{C}_{x_4}(\breve{\mu}(t))\dot{\xi}_2(t) - \breve{C}_{y_4}(\breve{\mu}(t))\dot{\eta}_2(t)\right) \times \sqrt{\left(u_s(t) - \dot{\xi}_2(t)\right)^2 + \left(v'_s(t) - \dot{\eta}_2(t)\right)^2} = 0 \quad (18)$$

– for the second cylinder in the direction y

$$m\frac{2}{\rho D}\ddot{\eta}_2(t) + C_{\eta_2}\frac{2}{\rho D}\dot{\eta}_2(t) + K_{\eta_2}\frac{2}{\rho D}\eta_2(t) + \left(-\breve{C}_{x_4}(\breve{\mu}(t))v'_s(t) - \breve{C}_{y_4}(\breve{\mu}(t))u_s(t) + \breve{C}_{y_4}(\breve{\mu}(t))\dot{\xi}_2(t) + \breve{C}_{x_4}(\breve{\mu}(t))\dot{\eta}_2(t)\right) \times \sqrt{\left(u_s(t) - \dot{\xi}_2(t)\right)^2 + \left(v'_s(t) - \dot{\eta}_2(t)\right)^2} = 0 \quad (19)$$

where ρ - air mass density; m - mass per length unit of the cylinder; D - cylinder diameter; $K_{\xi_1}, K_{\eta_1}, K_{\xi_2}, K_{\eta_2}, C_{\xi_1}, C_{\eta_1}, C_{\xi_2}, C_{\eta_2}$ - damping and stiffness coefficients of the first and second cylinder towards the axis x and y, respectively.

The above system of four differential equations of the second order has been replaced by the system of eight equations of the first order and next it has been solved numerically with Runge-Kutty's method.

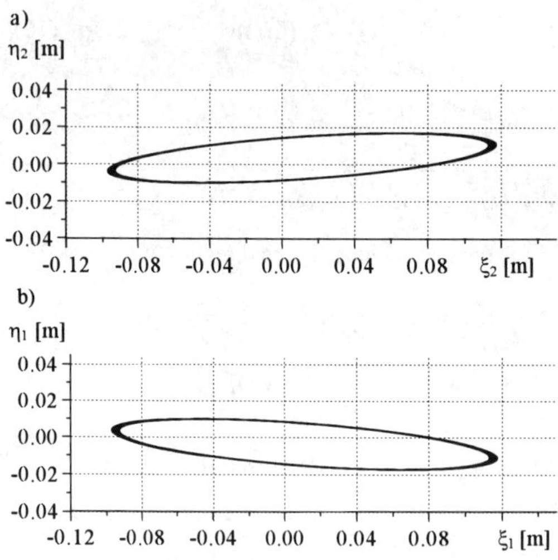

Figure 2. Trajectory forms of cylinders at the steady flow with $V_s=\bar{u}=16$ m/s and $L_y/D=1.2$, $\theta=0°$, $Sc=50$ (Sc - Scruton number).

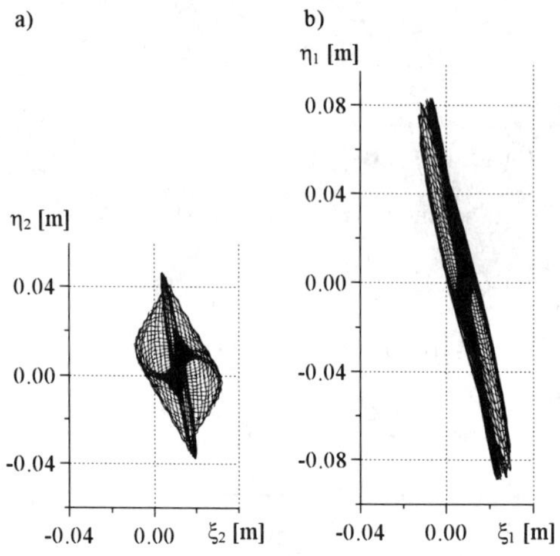

Figure 3. Trajectory forms of cylinders at the steady flow with $V_s=\bar{u}=16$ m/s and $L_y/D=1.2$, $\theta=5°$, $Sc=50$.

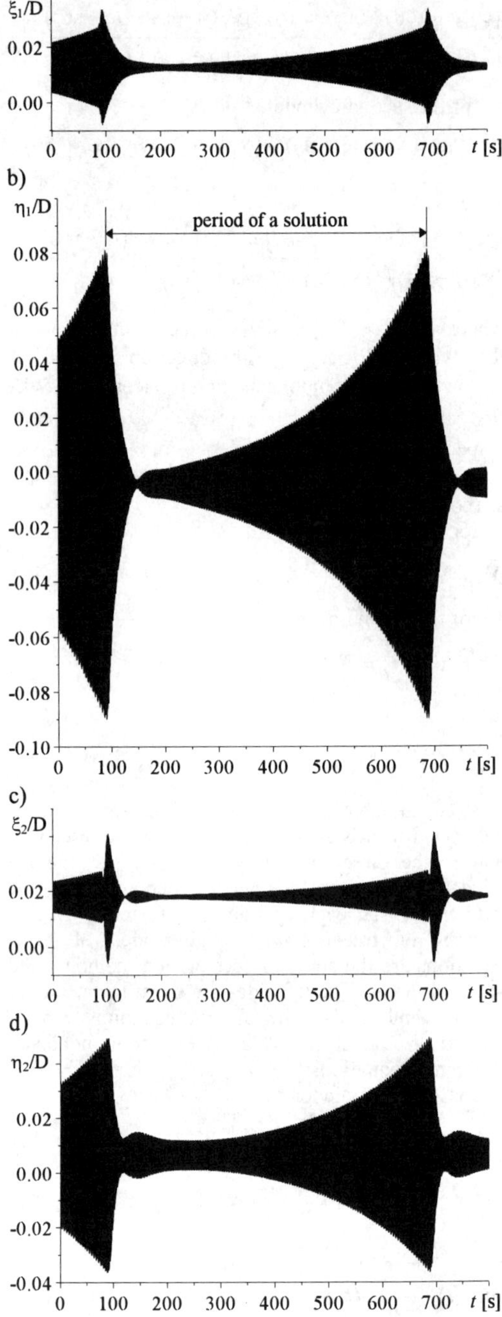

Figure 4. Relative displacements of cylinders as functions of time at the steady flow $V_s=\bar{u}=20$ m/s and $L_y/D=1.2$, $\theta=5°$, $Sc=50$.

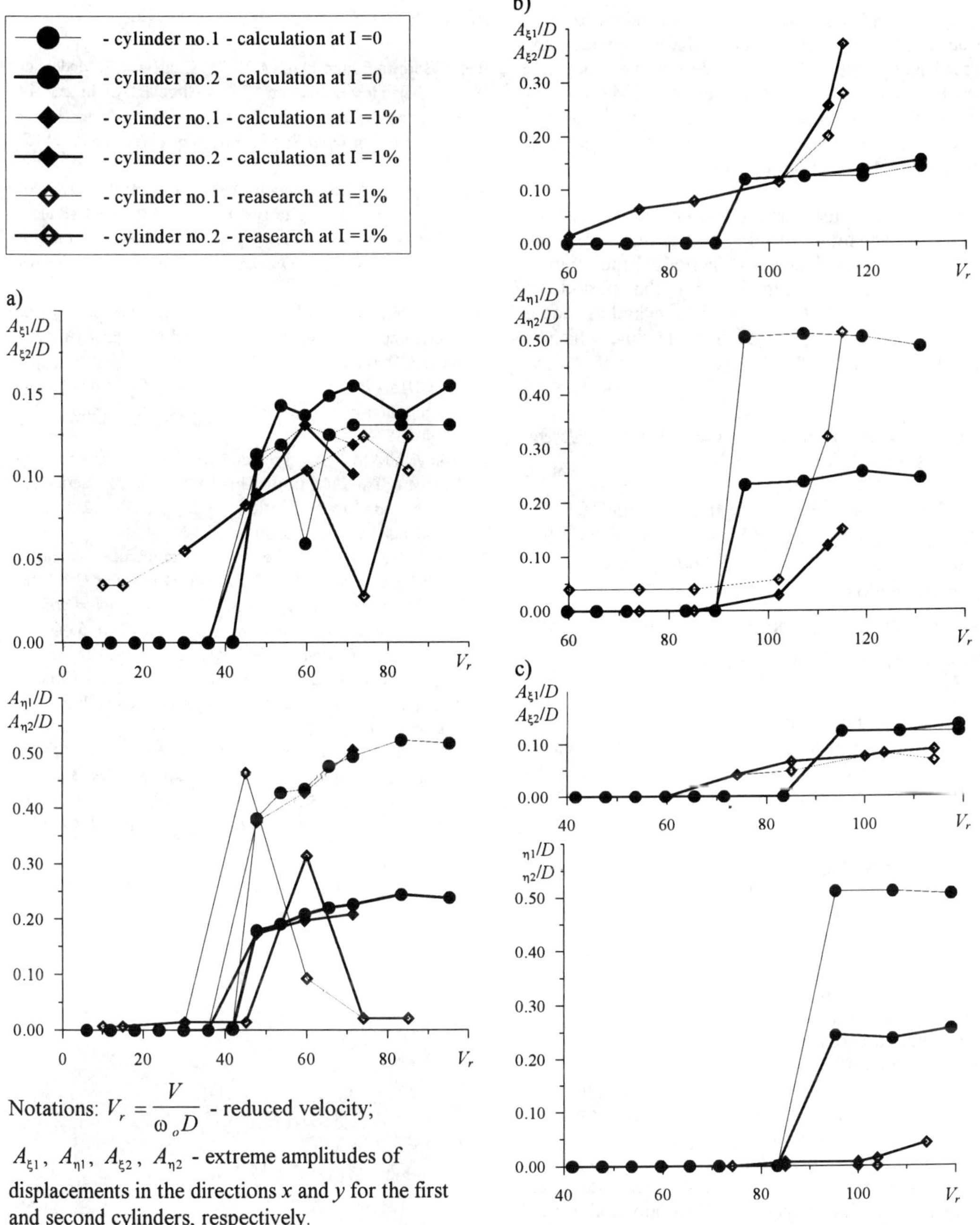

Notations: $V_r = \dfrac{V}{\omega_o D}$ - reduced velocity;

$A_{\xi 1}$, $A_{\eta 1}$, $A_{\xi 2}$, $A_{\eta 2}$ - extreme amplitudes of displacements in the directions x and y for the first and second cylinders, respectively.

Figure 5. The comparison of numerical and research results (Zdravkovich and Medeiros 1991) at L_y/D=1.2 and θ=5°; a) numerical analyses and researches are made at the flow with the turbulence intensity I_u=1% (where $I_u = 100\% \sigma/\overline{u}_s$ - turbulence intensity) for Sc=10; b) numerical analyses are made at the steady flow and researches are made at the flow with the turbulence intensity I_u=1%, the Scruton number is equal to 50 and is the same for all results; c) numerical analyses are made at the steady flow and researches are made at the flow with intensity turbulence I_u=1%, the Scruton number is equal to 100 and is the same for all results.

Figures 2 shows trajectories for cylinders located perpendicular to a wind direction that means they are located along axis y. Both cylinders move along the similar regular ellipses. In the case of cylinders which are situated in other way (for example $\theta=5^{\circ}$) trajectories of the cylinders become less regular and amplitudes of vibration considerably differ (comp. Figure 3).

It can be noted that solution of the set of differential equations (16-19) is periodic. We can differentiate two kinds of periods: the period between extreme amplitudes and the period of vibration shown in Figure 4. The cylinders come back on their initial paths after some time which is equal to the period.

3 THE VERIFICATION OF THE QUASI-STEADY MODEL

In order to check the quasi-steady model for the side-by-side arrangement the numerical analysis has been carried out for the following parameters: $L_y/D=1.2$ and $\theta=5^{\circ}$, $D=0.168$ m, $m=87.7$ kg/m, $\omega_O=6.41$ rad/s (where ω_O natural circular frequency of a cylinder), $K_{\xi_1}=K_{\eta_1}=K_{\xi_2}=K_{\eta_2}=3000\,\mathrm{N/m}$, $\rho=1.25$ kg/m^3.

Figure 5 shows the comparison of numerical results and the results at the flow with intensity turbulence equaling 1%, which are presented by Zdravkovich and Medeiros 1991.

4 CONCLUSIONS

Since the values of calculated and measured amplitudes are similar and the values of the reduced critical velocities are also similar, it can be concluded that the quasi-steady model describes the discussed phenomenon correctly.

In this moment it can be noted on the basis Figures 2-5 that:

1. The critical velocity of inducing vibration exists and it depends on Scruton number and the turbulence intensity of a wind velocity.
2. If cylinders are located perpendicularly to a wind direction then displacements of cylinders are symmetric with respect to a line along the wind direction. For other side-by-side arrangements cylinder trajectories become less regular and they are not symmetric.

Thus, in future, the described model will be applied to analyze vibrations of cylinders in a side-by-side arrangement.

REFERENCES

Błazik-Borowa,E. and Flaga,A. 1995. Quasi-steady Models of Aerodynamic Loads on Two Circular Cylinders in Staggered Arrangement. *The Proc. Ninth International Conference on Wind Engineering, Vol.1, New Delhi 1995*: 95-107.

Błazik-Borowa,E. and Flaga,A. 1997. Numerical analysis of interference galloping of two Identical Circular Cylinders at unsteady air onflow. *2nd European & African Conference on Wind Engineering „EACWE'97", Genova 1997*: 361-370.

Błazik-Borowa,E. and Flaga,A. 1998. Numerical analysis of interference galloping of two Identical Circular Cylinders. *Wind & Structures, Vol.1, No. 3*: 243-253.

ESDU 84015 1984. *Cylinders groups: mean forces on pairs of long circular cylinders, Engineering Sciences Data*, August 1984.

Zdravkovich,M.M. & Pridden,D.L. 1977. Interference Between Two Identical Circular Cylinders; Series of Unexpected Discontinuities. *Journal of Industrial Aerodynamics*: 255-270.

Zdravkovich,M.M. 1987. Review of Interference - Induced Oscillations in Flow Past Two Parallel Circular Cylinders in Various Arrangements. *Seventh International Conference on Wind Engineering, vol.2, Aachen*: 51-66.

Zdravkovich,M. & Medeiros,E. 1991. Effect of Damping on Interference-Induced Oscillations of Two Identical Circular Cylinders. *Journal of Wind Engineering and Industrial Aerodynamics*:197-211.

Zhang,H. & Melbourne,W.H. 1995. Effects of Free Stream Turbulence on Flow Interference Between Two Circular Cylinders in Side-By-Side Arrangement. *Ninth International Conf. on Wind Eng., New Delhi 1995*: 141-152.

Wind Engineering into the 21st Century, Larsen, Larose & Livesey (eds) © *1999 Balkema, Rotterdam, ISBN 90 5809 059 0*

Wind tunnel and numerical modelling of the wind flow over parallel 2D ridges

A.J. Bowen
University of Canterbury, Christchurch, New Zealand
S. Koren & E. Arad
RAFAEL, Haifa, Israel

ABSTRACT: Wind tunnel and numerically-simulated data are presented for strong-wind flows over several sets of 3 concatenated 2D parallel ridges of simple identical shape. The purpose of the study is to improve the understanding of strong neutrally-stable wind flows over complex terrain. A comparison with similar wind-tunnel measurements taken over isolated single ridges of the same proportions is used to identify some effects of multiple ridges. Numerical simulation is used to validate the accuracy of small-scale complex terrain modelling in wind tunnels and to investigate the effects of Reynolds number.

1 INTRODUCTION

The effects of complex terrain on the wind flow close to the ground is of great importance for the estimation of wind loads on buildings and structures which are situated in hilly terrain. Although there is a growing data base from reliable measurements over isolated hills such as the full-scale measurements (Taylor & Teunissen, 1987) and wind-tunnel model tests (Teunissen et al, 1987) over the Askervein Hill, there is very little information involving complex terrain. The dearth of practical, quantitative information on conditions in complex terrain for the practising engineer is evident in the various national standards for wind loads on structures such as NZ4203/4 (1992) and AS1170.2 (1989). These codes use crude wind-speed multiplying factors for topography which are based on limited experimental data taken from wind flows over simple, isolated hills. The effects on these multipliers from complex upstream terrain is neither addressed nor understood.

As a result of the lack of reliable prediction methods for flows over complex terrain, there has been a steady demand for wind-tunnel tests featuring sites in such surroundings. Establishing confidence in small-scale wind-tunnel tests is crucial and can be done through comparisons with reliable full-scale data. This has been achieved with single isolated hills such as the Askervein hill over a range of model scales down to 1:2500 (Teunissen et al, 1987). Some useful modelling techniques have been developed as a result. In contrast, no such guidelines are available for modellers using small-scale complex terrain models. There are however, several publications such as Britter (1982), Tieleman (1982) and Cermak (1984) concerning the theoretical issues involved in such simulations.

Furthermore, the effects of scale on the accuracy of the wind flow over a complex terrain model in the wind-tunnel are not fully understood. As the scale of the model is reduced below 1:1,000, the accuracy of the flow simulation becomes increasingly uncertain. Some reference to scaling problems has been reported by Zhao Y. Wang et al (1996) and Bowen (1998).

2 WIND TUNNEL MODELLING

2.1 *Wind tunnel*

The tests were conducted at the open-circuit wind tunnel at the Department of Mechanical Engineering, University of Canterbury, NZ. It has a 3:1 contraction leading into a 12.2m long by 1.22m square working section which is fitted with an adjustable roof. The free-stream turbulence properties were adjusted using a set of spires at the entrance to the working section. Two Woods 26kW contra-rotating fans with pneumatic blade-pitch control are mounted in series downstream of the working section and provide flow speeds in the test section of 13m/s.

2.2 *Topographic model*

2.2.1 *Hill model shapes*

The model hill shape was chosen to represent a generic 2D ridge which had previously received analytical (Jackson and Hunt, 1975) and experimental attention (Bowen, 1983). The cross-sectional shape of the two-dimensional ridge is defined by;

$$Z_H = \frac{H}{1+\left[\frac{X}{L}\right]^2}$$

where H is the hill height at ridge top, L the half-length of the hill, X the distance from the ridge top location and Z_H the local elevation of the hill.

The hill models were made from a dense polyurethane foam which was shaped using a numerically-controlled milling machine. Three sets of a steep (H/2L = 0.4) and a gradual version (H/2L = 0.2) of the same hill shape were made and investigated. The height of the hills (model H = 70mm, full-scale 175m) was chosen in an attempt to typify the hills that surround the Wellington City area of New Zealand. This area is regularly the subject of wind-speed investigations (Carpenter et al, 1993) due to its exposed nature. Spacing between the multiple ridges was chosen at 10 and 15 hill-heights which correspond to typical full-scale distances of 1.7km and 2.6km respectively. The surface roughness over the hills was kept the same as that of the tunnel floor upstream of the models in order to eliminate any unnecessary perturbations at the upwind edge of the models. A schematic view of the wind-tunnel set up is shown in Figure 1.

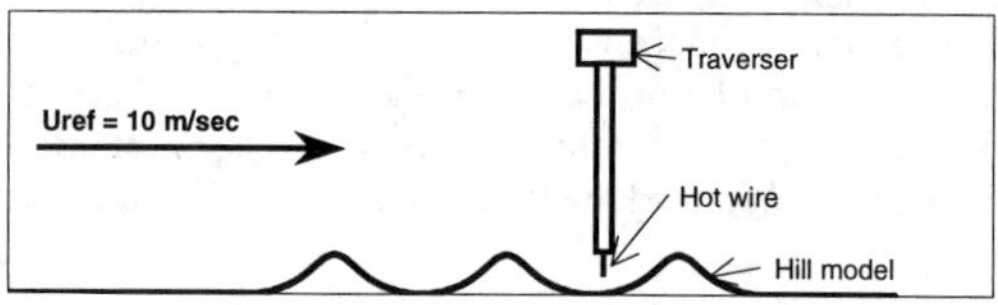

Figure 1. Schematic layout of the wind-tunnel tests.

2.2.2 *Geometric scale*
A model scale of 1:2,500 was chosen to avoid significant tunnel blockage effects and to provide a sufficiently long fetch over the upwind terrain. For isolated hills, the fetch length is simply determined by the length of roughened floor required to generate a correctly scaled boundary-layer with the required level of turbulence. However, for complex terrain models, an inner boundary layer develops over the topography upstream of the site in a unique way. Strataridakis & White (1999) suggests that in order to model the average flow speed accurately over complex terrain, a fetch equal to 15-20 times the required boundary-layer (BL) height is essential. For accurate turbulence intensities the fetch should be 50 BL heights, whereas a 100-500 BL height fetch is necessary for the turbulence energies. The use of spires as in the present tests may shorten the required fetch needed to satisfy all criteria to the order of 20 BL heights.

In order to minimise tunnel-blockage effects, the wind-tunnel model should be made sufficiently small so that the tunnel roof is above the maximum height of influence of the model. That way, the deflected stream-lines over the hill model are unaffected. The ratio of hill height to the height of the tunnel working section was 6%. Any blockage effect is exacerbated by the large along-wind extent of the 3 ridge model configuration (Goh 1981). However, these effects are assumed to be insignificant for all the current tests. Success in modelling at these small scales has been proven earlier with the Askervein Hill tests (Teunissen et al, 1987).

2.3 *Wind-tunnel flow characteristics*

2.3.1 *Reference profile*
Flow speeds in the free-stream were measured along the empty tunnel at a height of Z=570 mm above the floor and at the centre-line (Y=600 mm from the walls). The free-stream velocity along the tunnel varied by less than 1.5% and therefore may be assumed to be constant in the working section. This had been achieved by appropriate adjustment of the flexible roof over the test section, to compensate for the growth in the boundary layer on the tunnel floor, walls and ceiling. The reference position for the model tests was defined at X=6,800mm downstream of the beginning of the tunnel working-section. The test area for the hill models was located just downstream of this reference position.

Log-linear plots of the reference velocity profile enable the surface roughness length, z_0 to be found using the interception of the linear section of the profile (within approximately the lower 1/3rd of the boundary layer) with the height axis. At the reference position, the surface roughness-length of the model boundary-layer is estimated to be 0.017mm (4cm full scale). The gradient of a straight line fit to the log-log plot of the reference flow-speed profile provides the power-law exponent. The exponent is approximately constant with X and the value of α=0.16 is appropriate for the reference conditions. The repeatability of the measurements has been checked and is considered satisfactory.

The conditions in the wind tunnel were adjusted so that the modelled full-scale characteristics at the reference position, and listed in Table 1, represent the idealised situation of strong-wind, neutrally-stable conditions over flat, open terrain.

Table 1. Full-scale boundary-layer characteristics of wind-tunnel tests based on a model scale of 1:2,500.

Full-scale parameter	**Current Tests**
Boundary-layer depth, m	375
Surface Roughness, z_o cm	4
Turbulence intensity at 10m	12 %
Power law exponent, α	0.16
U_{10}/U_δ	0.55
u_*/U_δ	0.04
$K_{10} = (u_*/u_{10})^2$	0.005
$u_* z_o/\nu$	0.5

Standardised strong wind, neutrally stable conditions were first identified by Davenport (1960) and have since been confirmed by Plate (1971) and Cermak & Cochran (1992). Similar conditions are to be found for terrain category 2 in AS1170 (1989) and with the ESDU 8206 (1993) model (except for the boundary-layer height). The current model test set-up is almost identical to some of the earlier model Askervein hill tests reported by Teunissen et al (1987).

The thickness of the model boundary-layer at the reference position was 150mm (375m at full-scale). However, little emphasis should be placed on the depth of the boundary-layer as a scaling parameter due to its sensitivity to atmospheric stability (Bowen 1998). In trying to match model and full-scale values for the mean-flow parameters in Table 1, a wide range of geometric scales is possible. However, the choice of the geometric scale for such a generic model test as the current one does have considerable tolerance as the heights of the equivalent full-scale hills which these tests refer to also vary significantly. The surface roughness length was considered the best indicator here with due reference to the friction velocity and to the levels of turbulence. Reasonable agreement in these characteristics between the model reference boundary-layer and the full-scale equivalent is evident when a geometric model-scale of 1:2,500 is adopted.

2.3.2 *Aerodynamic roughness*

An indication of the level of aerodynamic roughness is given by the roughness Reynolds number, $u_* k_s/\nu > 70$ (Schlichting 1979). If a direct relationship between the sand roughness, k_s and the surface roughness length, z_o is assumed (say, $k_s = 25z_o$ Teunissen et al, 1987), the Reynolds number becomes $u_* z_o/\nu > 3$. The value for the present tests is about 0.5, which indicates that the boundary layer lies in the transition range ($0.2 < u_* z_o/\nu < 3$), where fully turbulent flow is likely but cannot be guaranteed. However the turbulence-intensity profiles indicate that turbulent flow was present, although the measured level of turbulence intensity at 4mm (10m full-scale) of 12% is low compared to the expected value of 18-20% which is typical for the rough terrain.

Schlichting (1979) also provides a limit for the independence of the surface skin friction coefficient from surface roughness effects using an analogy based on pipe flow criteria (Moody diagram) and the Reynolds number, $Re_x=U_0.X/\nu$ and X/k_s. In effect, this condition ensures aerodynamically rough flow by using physical rather than flow parameters such as u_*. For the current tests, $Re_x = 5 \times 10^6$ and $X/k_s = 1.7 \times 10^4$ which confirms that the flow lies in the transition zone and is therefore weakly dependent on the Reynolds number and surface roughness.

The onset of flow separation over the model is affected by the level of surface roughness (Britter, 1982), with rough surfaces promoting separation due to the relatively low levels of momentum close to the ground. This behaviour is in contrast with the flow around bluff bodies where rougher surfaces cause greater mixing close to the surface and the promotion of attached flows through higher levels of momentum.

Teunissen et al (1987) noted that in order to achieve a suitable roughness length for small-scale models, the model surface roughness had to be disproportionally large compared with its full-scale equivalent, which violated the geometric scaling. The small-scale features of the terrain help to provide an apparent surface roughness that promotes aerodynamically rough flow. The physical difference between surface-roughness elements and small-scale topographic features become more blurred as the model scale is reduced.

Ferreira & Veigas (1995) shows experimentally that the effects of Reynolds number on the accuracy of the model flow simulation over hills can be significant. For the current tests, the Reynolds number is 5×10^4, based on the free-stream flow-speed and the height of the model hills. Ferriera & Veigas (1995) reports a big change in flow conditions below $Re = 8 \times 10^4$ for their test set-up based on changes in the drag coefficient. As a result, a minimum value of 10^5 for wind tunnel tests was recommended. The effects of Reynolds number on the present results were investigated numerically and are discussed later.

2.3.3 *Flow turbulence*

The turbulence intensity in the upstream reference flow for the current tests varies from 12% at a height of 4mm to 0.5% in the free-stream flow above 150mm. The turbulent structure of the boundary-layer over complex terrain is unlikely to be in equilibrium with the underlying surface (Tieleman 1982). This is particularly true for relatively short models such as those in the current tests. It is therefore inappropriate to obtain roughness lengths and the friction velocities from locally observed mean-wind profiles. Therefore the flat-plate flow models in equilibrium that are widely used to explain the behaviour of turbulent flows over complex terrain should be interpreted with care.

No turbulence spectra were measured during the current tests, so scaling to the turbulence length scales could not be done. Reliance is made on earlier turbulence measurements reported by Stock & Bowen (1992) using the 1:2,500 Askervein model in the same wind tunnel and reference flow conditions. The dimensionless peak frequency for the longitudinal turbulence is reported to have been 0.02 at the equivalent full-scale height of 10m and is considered adequate for the current tests (Panofsky & Dutton, 1984). The boundary-layer thickness determines the largest length scale of turbulence present in the model boundary layer.

2.4 *Test procedures*

2.4.1 *Flow measurement*

A Thermo Systems Inc. (TSI) #1054b linearized anemometer was used for the flow measurements. It consisted of a horizontal 5μ tungsten, single-wire sensor mounted on a three-dimensional traverser. Cleaning with alcohol and calibration of the hot wire with a TSI calibrator were performed daily.

The tunnel flow speed was monitored continuously with a pitot-static tube which was mounted from the sidewall, well upstream of the model test area. The current tests were carried out at a tunnel flow speed of about 11m/s. Periodic readings from the hot wire at the upstream tunnel-monitoring and reference position were taken to check the calibration. If the differences exceeded 5%, a new calibration was conducted. The operational resistances of the hot-wire system were also checked periodically. Hot-wire measurements of flow speed were later normalised by a nominal free-stream reference speed (U_{ref}=10m/s) to correct for changes in the wind-tunnel speed during each profile traverse. The flow-speed readings were also corrected for deviations due to the calibration temperature and adjusted to standard atmospheric pressure. The first reading was repeated at the end of each profile measurement and the whole data set rejected if significant differences due to drifting of the calibration were evident at that same position.

A PC-486/33 computer controlled the movement of the traverser and measurement probe in the 3 orthogonal directions. Due to the small model scale, great care was given to the accurate positioning of the probe during the initial setting-up, in order to minimise position errors. Consequently, they are considered to be less than 1mm.

2.4.2 *Data acquisition and processing*

Using the same computer, the anemometer output was sampled at 500Hz over a 2sec record period using a Boston Technology AD board #PC30 DS with in-house software and filtered with a 250Hz low-pass filter.

At the geometric scale of 1:2,500, the 2sec record length is equivalent to a full-scale record of 83 minutes. This is well over the time considered necessary for an accurate time-average in the full-scale situation. The record length is more than adequate, especially as the wind-tunnel flow is comparatively well behaved and stationary compared to full-scale wind flows.

2.4.3 *Data corrections*

Three systematic corrections were applied to the raw mean flow-speed profile measurement data. They are required since the changes in flow speed must only be due to the hill model and not to other extraneous effects from the operation of the tunnel or from spatial variations throughout the flow. The speed-up data are the result of subtracting two relatively large but often similar flow speeds, so accuracy here is of particular importance.

a) Wind-tunnel speed fluctuations: This correction compensates for the variations in the speed of the wind-tunnel as it was difficult to keep the free-stream flow speed constant throughout the test programme. The measured local flow speeds were multiplied by the ratio of the nominal speed, 10 m/sec over the actual reference flow speed. The actual flow speed was recorded during the tests by a reference pitot-static tube situated upstream in the undisturbed flow above the boundary layer.

b) Zero drift: Drifting of the zero setting of the hot-wire anemometer was a spasmodic problem throughout the tests. To monitor this disturbance, the first reading of each profile set taken at 5mm height, was repeated at the completion of each profile. Any differences were considered to be solely due to the zero drift of the anemometer as any initial positioning errors would be present in both the readings. Most files exhibited changes of less than 1 or 2% with a further 8 below 5%. If there was excessive drift (≥ |5%|), then the data set was rejected. All the approved files containing acceptable drift were corrected assuming a linear variation during the profile measurement set.

c) Boundary-layer growth: Measurements taken at a constant height over a uniformly roughened floor without a model installed, would show a small but significant drop in flow velocity with distance downwind. This is due to the natural growth in the thickness of the boundary-layer along the tunnel. As this effect is extraneous to any topographic effects, a correction is needed if there is a significant downwind distance over a complex terrain model. A suitable correction has been used by AES Canada for the Askervein hill tests (Teunissen & Flay, 1981). It is based upon the common empirical prediction for the growth of the boundary-layer depth, δ in a turbulent boundary-layer,

$$\frac{\delta}{X} = 0.37 \, Re_X^{-\frac{1}{5}}$$

The boundary-layer profile is assumed to remain the same shape, defined by a power-law profile of power α, between the upstream reference position X_R and the local measurement position X_N. Rather than adjusting the height, a simpler approximate correction, C may be added to the uncorrected flow speed at position X_N for any height within the boundary layer, where,

$$C = U_R \cdot \left[1 - \left(\frac{X_R}{X_N} \right)^{0.8\alpha} \right]$$

The derivation of this correction is based on the flow over a flat surface without a hill model.

However, the presence of the model during the actual tests is assumed to have no effect on the rate of boundary-layer growth. Tests in the tunnel over a flat roughened floor have confirmed the efficacy of this correction method.

2.5 *Wind-tunnel results*

The variations of mean-velocity, turbulence intensity and fractional speed-up factors with height above 3 concatenated ridges are compared with those over an isolated single ridge of the same shape. Figure 2 shows the profiles over relatively steep ridges (H/2L=0.4) with a close ridge spacing (S/H=10), where H is the hill height (70mm), L is the half-length of hill and S is the ridge spacing. Figure 3 shows the profiles over more gradual ridges (H/2L=0.2) of similar shape but with a wider ridge spacing (S/H=15). The reference profile represents the unaffected flow over level ground.

The influence of the multiple-ridge hill models extends beyond the upper level of the measurements (400mm, Z/H≈6). The mean wind velocity-height profiles are nearly uniform over all the 3 ridges in both model sets. The velocities close to the ground are generally higher over the multiple-ridges than over the isolated ridges. A low-level jet is also evident in Figure 2a over the steeper ridges close to the ground.

Turbulence intensities are significantly higher over the 2nd and 3rd steep ridges (fig 2b) compared with the isolated ridge case. They are also greater with the shorter spacing between ridges due to the higher incidence of separated flow occurring in the valleys between them. Over the first steep hill and all the gradual hill tops (figs 2b, 3b), turbulence levels are lower or similar to the respective isolated hill case. In the wider-spaced configurations, S/H=15 (fig 3b), turbulence levels appear to reach a maximum at some distance above the ground. In the steep hill and close spacing case (fig 2b), the data are very scattered, especially over the second hill top.

The fractional speed-up is the percentage difference between the local wind speed and the reference undisturbed speed at the same height. These profiles are approximately linear when plotted on a semi-log height scale with values rising to a maximum of about 90% near the surface of both hills (figs 2c, 3c). These maxima are significantly higher than the 65-80% recorded over the isolated single ridges of the same shape.

The wind-tunnel data indicate that an internal boundary-layer develops over the multiple ridges. This layer contains higher levels of turbulence than the upstream reference flow. These conditions result in a modified speed-up behaviour compared to that over an isolated ridge. Such differences increase when separation occurs in the valleys between the ridge lines, as in the case of the steeper hills.

3 NUMERICAL MODELLING

3.1 *Numerical method*

The incompressible Navier-Stokes equations for the flow over the model hills were solved using Fluent 5 CFD code. Fully turbulent flow was modelled in the test area as Re_x at the reference position is about $5x10^6$ and the measured turbulence intensity is adequate. Turbulence mechanism was modeled using Spalart Allmaras one-equation model. This model was selected since it combines robustness and a fair predictive capability for separated flow. As the wind velocity did not exceed 15 m/sec through out the solution domain, the incompressible formulation is a proper representation for this flow case. The wall roughness is defined using the roughness height. A typical open-country surface roughness length is about 4cm. For the 1:2500 model it was converted to a sand roughness of k_s=0.4mm.

The solution domain extends to 2m above the hill models. The length of the computational domain is 5.4m, and the models are located in the centre. Turbulent inflow conditions were set at the domain entrance. The turbulent profile is the result of the numerical solution of the flow over a flat plate of the length of 6.8m, which matches the experimental setup. On the far-field boundary symmetry conditions were set, and outflow conditions were defined in the solution domain downstream boundary. The governing equations are discretized on a quadrilateral two-dimensional mesh. 61 mesh points were used in the vertical direction, with strong stretching towards the wall. As a result, Δy^+ on the wall is within the order of 1. About 200 mesh points were set in the streamwise direction. The mesh independence of the numerical solution was checked by doubling the mesh size, showing only marginal variation in the results.

A segregated implicit scheme is used for the solution. The momentum equations are solved by the QUICK scheme. A Poisson type equation is solved for the pressure correction, with the code standard pressure interpolation (pressure values at the cell faces are interpolated using the momentum equation coefficients). Pressure-velocity coupling are done using SIMPLEC method. The turbulence model equation is also solved by the QUICK method, using the standard model coefficients. The norms of the residual of the governing equations were reduced by more than 4 orders of magnitude within 600 iterations, for the above-described mesh.

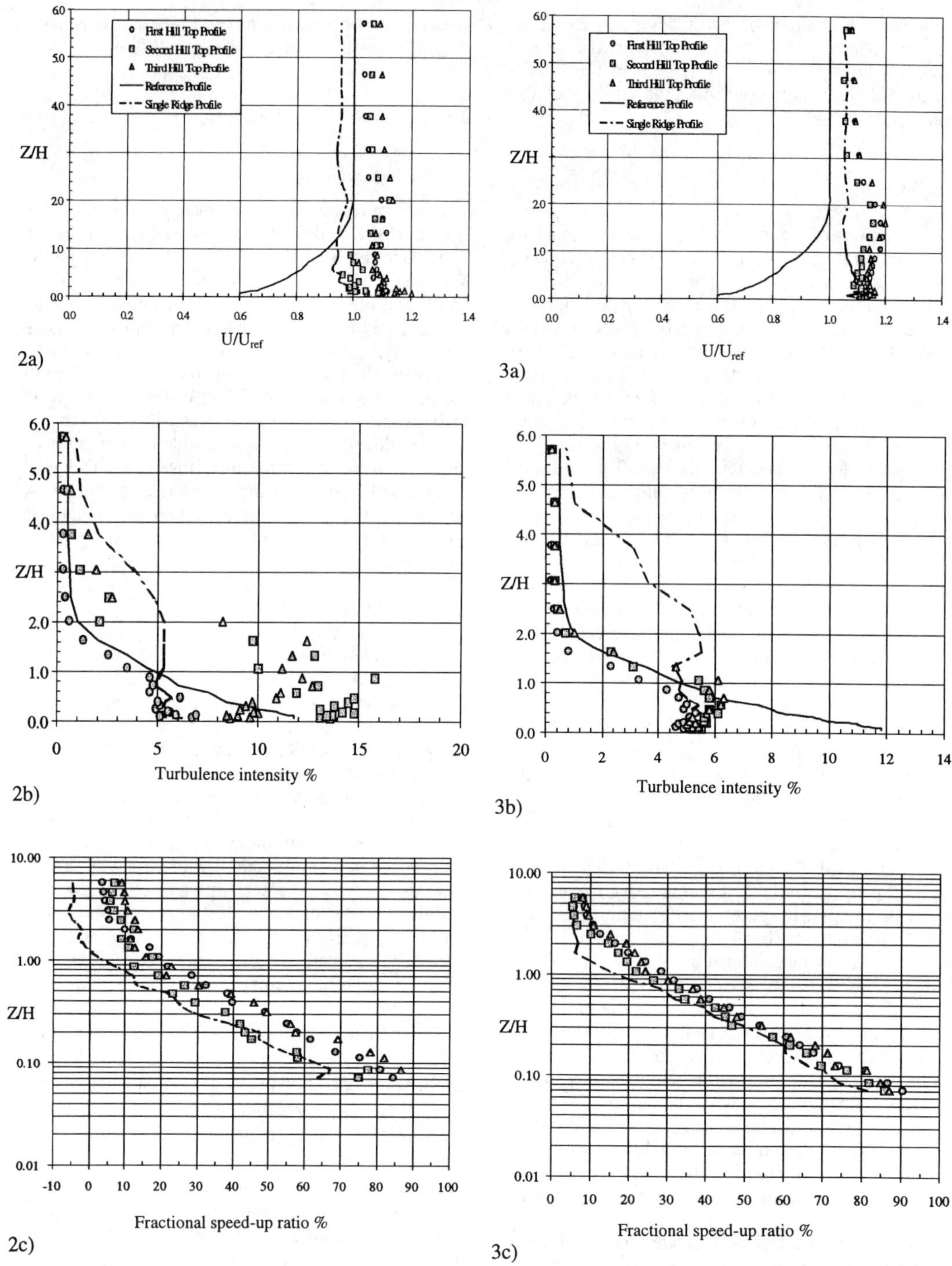

Figure 2. Wind flow characteristics over 3 steep ridges taken from wind-tunnel experiments. H/2L=0.4 with close spacing, S/H=10.

Figure 3. Wind flow characteristics over 3 gradual ridges taken from wind-tunnel experiments. H/2L=0.2 with wide spacing, S/H=15.

3.2 *Results and discussion*

The numerical results are validated by comparing the velocities over the isolated hilltops and valleys as well as on the concatenated cases. A representative result is shown in Figure 4, where the differences between the experimental data and the simulation do not exceed 7%. The computed thickness of the boundary layer is in good agreement with the experimental data. However, in the case of an isolated steep hill (H/2L=0.4) there is a shift in the velocity magnitude (but not in the boundary-layer thickness) between the methods. Since such a shift was not observed in any other case it could be attributed to a measurement error.

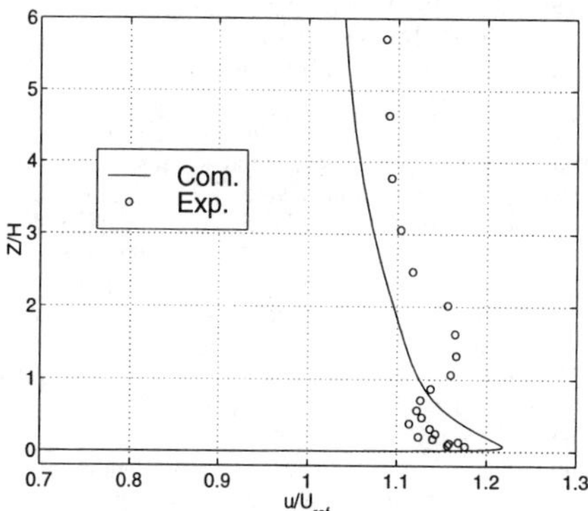

Figure 4. Comparison of flow velocities over first hill-top, H/2L=0.4, S/H=15 (Com.= computation, Exp.= experiment).

In Figures 5-7 the velocity profiles over the concatenated (S/H=10) steep hilltops are plotted. Over the first and second hilltops, a fair agreement in both velocity magnitude and boundary-layer thickness is obtained. However, over the third hill there is a peak in the measured velocity near the surface that was not reproduced by the computation. Comparable agreement was obtained in the case of larger distance between the hills (S/H=15). As expected, a better matching was achieved in the case of the single and concatenated gradual hills (H/2L=0.2).

The computational result allows an insight into the structure of the flow field, as shown in Figure 8. In the case of steep closely-spaced hills, the area of influence of the hills extends to several hill heights. Consequently, measurements that were assumed to be taken in an undisturbed flow field were actually within this domain. This is not the case for neither the isolated nor the gradual concatenated hills configurations.

The effect of Reynolds number (Re_x) is evaluated in Figure 9. The velocity profiles over the second hilltop in the steep closely-spaced case are compared for various Re_x. It appears that within the boundary layer, the velocity magnitude is proportional to Re_x. However, while Re_x varies by half an order of magnitude, the variation of the velocity is within the deviation from the experimental results. The major change in flow conditions reported by Ferreira & Veigas (1995) at $Re_x = 8 \times 10^4$ was not observed in this case.

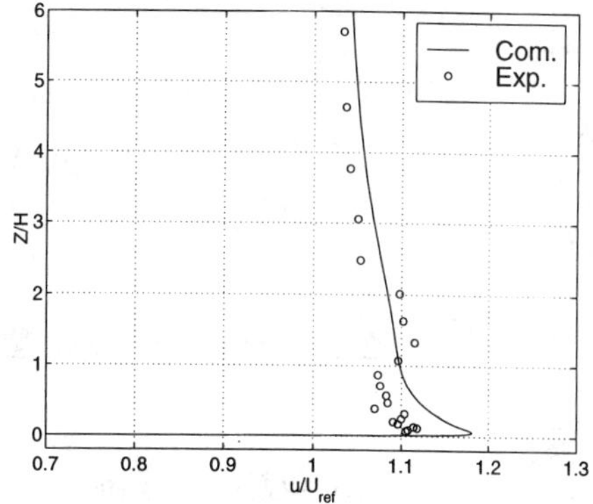

Figure 5. Comparison of flow velocities over first hill-top, H/2L=0.4, S/H=10.

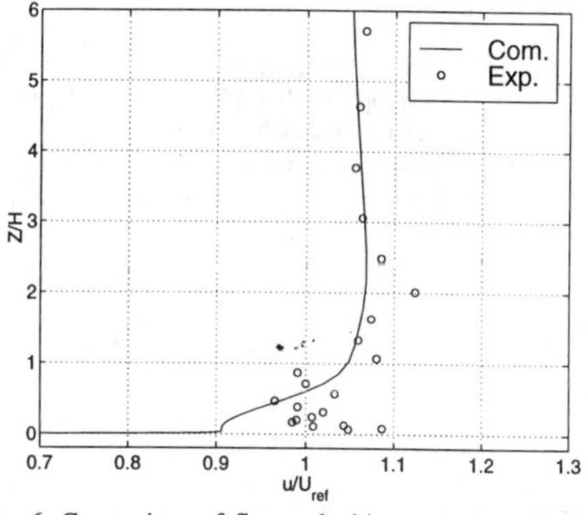

Figure 6. Comparison of flow velocities over second hill-top, H/2L=0.4, S/H=10.

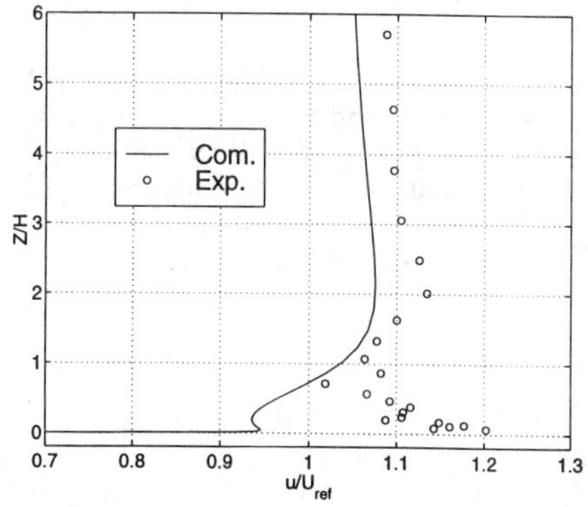

Figure 7. Comparison of flow velocities over third hill-top, H/2L=0.4, S/H=10.

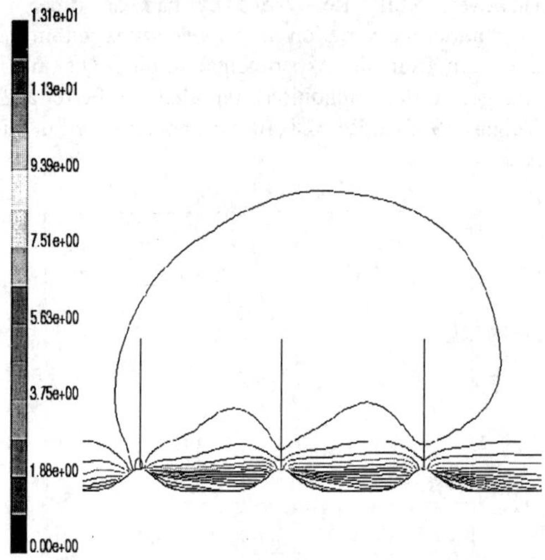

Figure 8. Contours of flow velocity over 3 hills, H/2L=0.4, S/H=10 (flow from left to right).

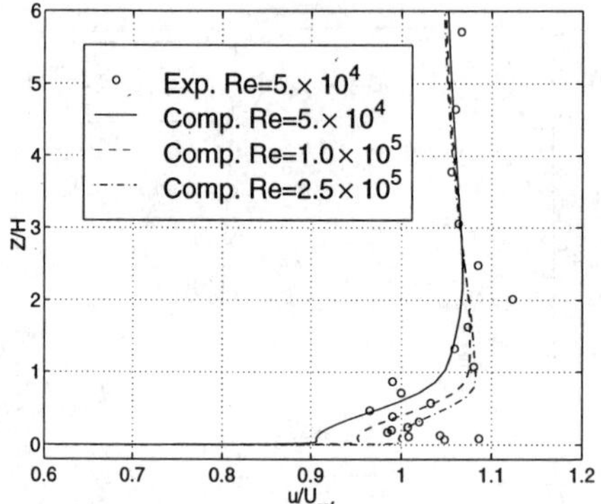

Figure 9. Effect of Reynolds number on the velocity over the second hill-top, H/2L=0.4, S/H=10.

4 CONCLUSIONS

This paper provides an insight into the wind-flow over parallel ridges and has identified some significant effects due to the upwind complex terrain. Encouraging agreement has been obtained with numerical simulations which has also confirmed that the flow is sensitive to Reynolds number.

5 REFERENCES

AS1170.2 1989. Australian Loading Code Part 2.

Bowen, A.J. 1998. Some complications from modelling the wind flow over complex terrain at small geometric scales. Proc. 13th Australasian Fluid Mechanics Conf., Monash University, Melbourne Australia, 13-18 December, 405-408.

Bowen A.J. 1983. The prediction of mean wind speeds above simple 2D hill shapes. Journal of Wind Engineering and Industrial Aerodynamics 15, 259-70.

Britter, R. 1982. Modeling flow over complex terrain and implications for determining the extent of adjacent terrain to be modelled. Proc. Int. Conf. on wind modeling criteria and techniques in civil engineering applications. Gaithersburg, USA, 186-196.

Carpenter, P., Jamieson, N.J., & Reid, S.J., 1993. Wind flow over complex topography and its prediction. Report 93-29135. Central Laboratories, PO Box 30-845, Lower Hutt, New Zealand, 28.

Cermak, J.E. 1984. Physical modelling of flow and dispersion over complex terrain. Boundary-layer Meteorology, 30, 261-292.

Cermak, J.E & Cochran, L.S. 1992. Physical modeling of the atmospheric surface layer. Jour. Wind Engineering and Industrial Aerodynamics, 41-44, 935-946.

Davenport, A.G. 1960. The relationship of wind structure to wind loading". National Physical Laboratory Symposium No 16. Wind effects on Buildings and Structures, HMSO London, 54-102.

ESDU 82026. 1993 Strong winds in the atmospheric boundary-layer. Part 1 Mean-hourly wind speeds. ESDU International plc. 27 Corsham St. London N1 6UA England.

Ferreira, A.D., & Viegas, D.X. 1995 Wind tunnel simulation of the adiabatic flow around a set of sinusoidal hills. Proc. 9th Int. Conf. on Wind Engineering, New Delhi, India. 311-21.

Jackson, P.S. & Hunt, J.C.R. 1975. Turbulent flow over a low hill. Quart. J. Roy. Meteorol. Soc. 101, 833-51.

Goh, C.B. 1981. Blockage removal in topographic modelling. PhD thesis, Department of Mechanical Engineering, University of Auckland, New Zealand.

NZ4203/4, 1992. New Zealand Standard. General Structural Design and Design Loadings for Buildings. Part 4 Wind Load Provisions. Standards Association of New Zealand, Private Bag Wellington.

Panofsky, H.A. & Dutton, J.A. 1984. Atmospheric Turbulence. Wiley Interscience, ISBN0-47105714-2.

Plate, E.J. 1971. Aerodynamic Characteristics of Atmospheric Boundary Layers. Office of Information Services, US Atomic Energy Commission,. NTIS TID-25465, 190.

Schlichting, H. 1979. Boundary-Layer Theory. McGraw Hill Book Co. 7th Edition.

Stock, D.E. & Bowen, A.J., 1992. Wind tunnel measurements of the turbulence structure of wind flow over Askervein Hill. Proc. American Meteorological Soc. Tenth Symposium on Turbulence and Diffusion, Portland Oregon,

Strataridakis C & White, B., 1999. Turbulence measurements for wind-tunnel siting on complex terrain. Proc. ASME Wind Energy Symposium, Reno USA, AIAA-99-0054.

Taylor, P.A., & Teunissen, H.W. 1987. The Askervein Hill Project: Overview and background data. Boundary-Layer Meteorology, 39, 15-39.

Teunissen, H.W. & Flay, R.G.J. 1981. Wind tunnel simulation of the planetary boundary layer over an isolated hill. Part 1 Rough model. Internal Report MSRB-81-3, Atmospheric Environment Service, Downsview Canada.

Teunissen, H.W., Shokr, M.E., Bowen, A.J., Wood, C.J. & Green, D.W.R. 1987. The Askervein Hill project: Wind tunnel simulations at three length scales. Boundary-layer Meteorology, 40, 1-29.

Tieleman, H.W. 1982. Simulation criteria based on meteorological or theoretical considerations. Proc. Int. Conf. on wind modeling criteria and techniques in civil engineering applications. Gaithersburg, USA, April, 186-196.

Zhao Yin Wang, Plate, E.J., Rau, M., & Keiser, R. 1996. Scale effects in wind tunnel modelling. Jour. Wind Engineering and Industrial Aerodynamics, 61, 113-130.

Wind Engineering into the 21st Century, Larsen, Larose & Livesey (eds) © 1999 Balkema, Rotterdam, ISBN 90 5809 059 0

An experimental study of the power imparted by wind to a twin cylinder bundle

D. Brika & A. Laneville
Département de génie mécanique, Université de Sherbrooke, Qué., Canada

ABSTRACT: This paper deals with an experimental study of the power imparted by wind to a twin cylinder bundle coupled only by flow. The cylinders are in tandem and staggered arrangements for separations typical of transmission lines. The results indicate that the presence of the upstream cylinder and its vibration have a great influence on the power imparted to the downstream cylinder. This power depends strongly on the separation between the cylinders and the wake vortex pattern of the upstream cylinder.

1. INTRODUCTION

This study deals with an extension of two previous ones [Brika & Laneville (1996), Brika & Laneville (1997)] in which it is observed that, in comparison with the results of an isolated cylinder, the presence of a fixed upstream cylinder influences the aeolian oscillations of the downstream flexible cylinder as follows:

- the power imparted by wind is greater;
- this power decreases with increasing value of the spacing ratio between the cylinders and with the bundle inclination angle, β (with respect to the flow direction);
- beyond β= 20°, the downstream cylinder behaves as an isolated one;
- the hysteresis phenomenon observed in the case of a single cylinder is absent when the downstream cylinder is immersed in the wake of the upstream cylinder.

Since real conductors are free-vibrating, the objective of the present study is to investigate experimentally the effect of the vibration of the upstream cylinder on the power imparted by wind to the downstream cylinder. The cylinders are coupled only in an aerodynamic way. To the authors' knowledge, similar studies have not been conducted previously.

2. EXPERIMENTAL SETUP

The experimental setup for each model is similar to the one used in the two previous studies [Brika & Laneville (1996), Brika & Laneville (1997)] and is described in detail in [Brika & Laneville (1993)]. The cylinders are identical and have a smooth surface, an external diameter D= 33.4 mm, a length L= 3.3 m and a natural frequency f_n= 17.58 Hz. Figure 1 shows a schematic view of the bundle with the different parameters studied, i.e. the separation L between the cylinders, the bundle inclination angle β with respect to the flow direction and the plane of vibration characterized by angle α with respect to the cross-flow direction. Six spacing ratios have been examined, i.e. L/D= 7, 10, 13, 16 and 25 when the cylinders are in tandem arrangement and four bundle inclinations β= 0, 5, 10 and 15° for a typical separation L/D= 13. The turbulence intensity is of the order of 0.1% and the Reynolds number varies between 2000 and 12000.

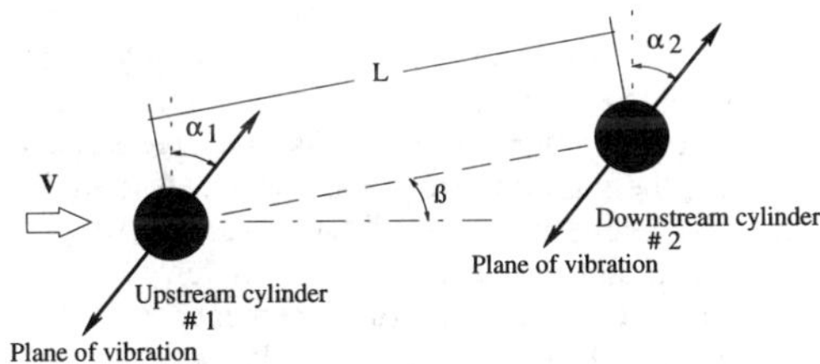

Figure. 1: A schematic view of a twin cylinder bundle.

The main difficulty encountered with the present experimental setup is to get two models with exactly the same natural frequencies; otherwise, a difference, even small, may result in an undesirable, although natural, beating phenomenon. This has been overcome by adding small rings, 1.6 mm thick (1.4 grams), at the ends of the model with higher frequency. The models' natural frequencies were matched within 1 mHz.

3. EXPERIMENTAL PROCEDURE

Since the spacing ratios examined herein are relatively high, the upstream cylinder seems to behave as an isolated one. The power imparted by wind to this cylinder is then considered equal to that of a single cylinder. The paper focuses on the power imparted to the downstream cylinder. Its evaluation is based on the logarithmic decrement method described in detail in [Brika & Laneville (1996)]. As the part of the model internal to the wind tunnel vibrates in a sinusoidal mode shape, the results obtained require no transformation to apply to real conductors in opposition to the results obtained with elastically-mounted models. For each flow velocity, the power input has been measured under the following conditions:

a) both cylinders are released simultaneously from rest (case R-R);
b) the downstream cylinder is released from rest while the upstream one is vibrating with its steady-state amplitude and its wake vortices are of the 2S type (case S-R);
c) the same as b) but the vortices are of the 2P type (case P-R).

It is worthwhile to remember here that in the 2S mode two vortices are shed from the cylinder per cycle of oscillation while in the 2P mode there are four vortices. The establishment of a vortex pattern in the wake of the upstream cylinder depends on the motion history and the initial conditions of excitation [Brika & Laneville (1993)].

4. DYNAMIC RESPONSE

Figure 2 shows an example of the dynamic response of both cylinders as a function of the reduced velocity $V_r = V/fD$, V is the free-stream velocity, f, the vibrating frequency ($f \approx f_n$), A_1 and A_2 are the single amplitudes of vibration respectively of the upstream and the downstream cylinder, Θ is the phase angle between the displacements of the cylinders and Φ is the phase angle between the vortex shedding and the displacement of the downstream cylinder. These results are obtained with a progressive increase or decrease of flow velocity while the cylinders are vibrating with their steady-state amplitudes of the previous velocity. The response of the upstream cylinder (A_1/D) is similar to that of an isolated cylinder [Brika & Laneville (1993)] and differs only by a slight shift to a higher velocity of the position of the lower critical velocity (LCV) where a discontinuity occurs. The disturbance of the wake of the windward cylinder

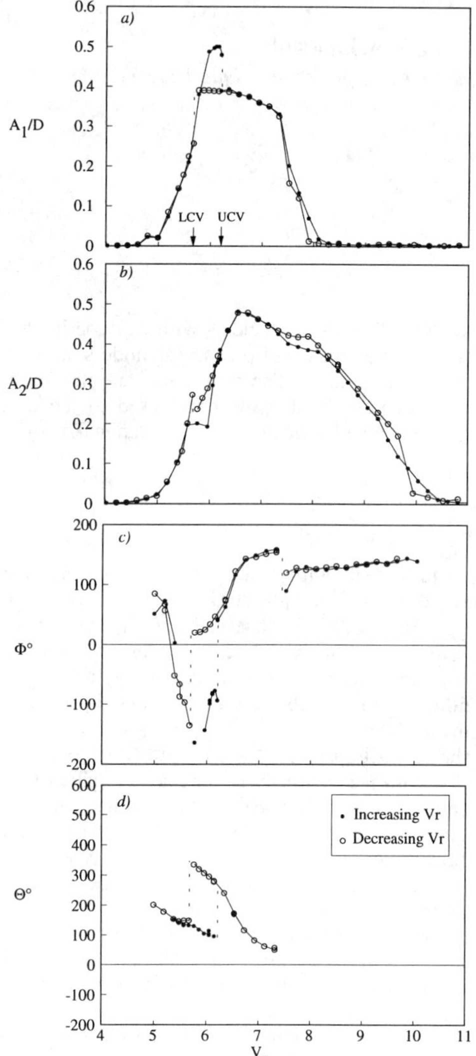

Figure. 2: Characteristics of the dynamic response of both cylinders ($L/D = 13$, $\beta = \alpha_1 = \alpha_2 = 0°$)

due to the motion of the downward cylinder may be the cause of this difference. This Figure also indicates the presence of two discontinuities in A_1/D, A_2/D, Θ and Φ as in the case of an isolated cylinder in addition to a third discontinuity in A_2/D and Φ coinciding with the end of synchronization of the upstream cylinder. A_1/D, A_2/D, Θ and Φ present in their variation with V_r a hysteresis phenomenon, one branch of which is obtained with a progressive increase of flow velocity and the other with a progressive decrease of V_r. Each branch is associated with a particular vortex pattern in the cylinder wake. The discontinuities observed in the above parameters are simultaneous.

5. POWER INPUT

The power imparted by wind to the downstream cylinder is evaluated in accordance with the procedure explained in section III. The power is estimated for different velocities, mainly around the resonance, and the maximum power presented herein is given by the envelope of the different curves. Figure 3 shows an example of the power input obtained under the three different conditions

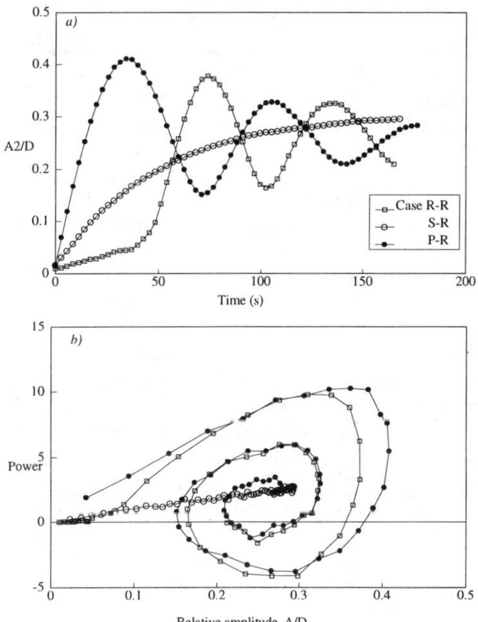

Figure. 3: The time-response and the corresponding power imparted to the downstream cylinder at V_r= 5.97 (L/D= 10, ß= α1= α2= 0°). a), time response; b), power input.

described previously. Are also presented in this Figure the corresponding time responses of the downstream cylinder which indicate that when the wake of the upstream cylinder is of the 2S type (case S-R), the time response of the rear cylinder is similar to the time response of an over-damped second-order system to a step input (constant flow velocity!) while its response resembles that of an under-damped second-order system submitted to the same input when the wake is of the 2P type (case P-R). The final steady-state amplitudes are, however, different for both cases. On the other hand, a previous study [Brika & Laneville (1993)] has shown that the wake vortex pattern of an isolated cylinder is of the 2P type when the model is released from rest. Then in the cases (R-R) and (P-R) the vortices shedding from the upstream cylinder are of the same type (2P mode). The response is, however, different in the early stages of motion as shown in Figure 3(a). This results in a different power at small amplitudes.

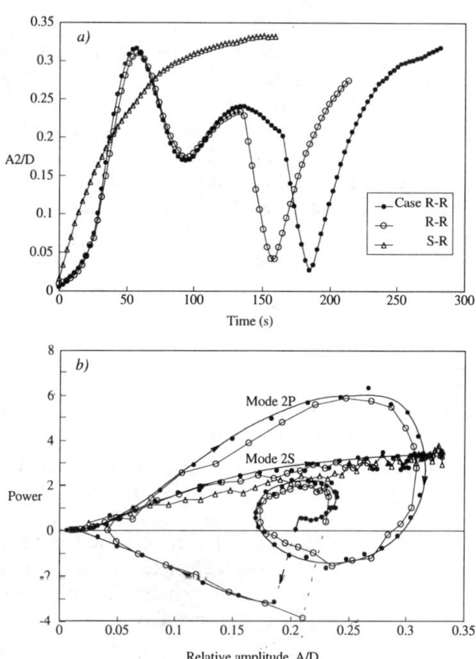

Figure. 4: The time-response and the corresponding power imparted to the downstream cylinder showing the bifurcation phenomenon at V_r= 5.7 (L/D= 10, ß= α_1= α_2= 0°). a), time-response; b), power input.

In the case (R-R), the transient response A_2/D presents an inflection point beyond which the response becomes quasi-similar to the case (P-R).

This indicates that P is also influenced by the amplitude of vibration of the upstream cylinder. The point of inflection which occurs around A_2/D= 0.05 marks the trigger of synchronization between the vortex shedding and the cylinder vibration. The occurrence of this point is delayed by an increase of the flow velocity and the curves of the power input for both cases (R-R) and (P-R) become more distinct. For a smaller velocity, another phenomenon has also been observed as shown in Figure 4. When both cylinders are released from rest, the upstream one undergoes a bifurcation in its displacement at a given amplitude as in the case of an isolated cylinder [Brika & Laneville (1993)]. This bifurcation is

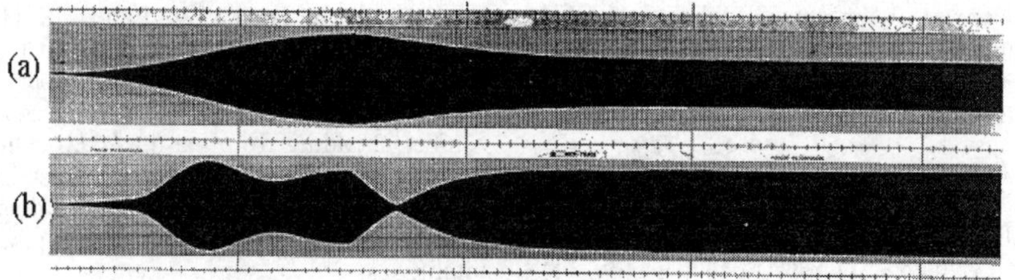

Figure 5. Traces of both cylinders showing the bifurcation phenomenon (L/D= 7, V_r= 5.8, ß= α_1= α_2= 0°)

accompanied by a similar one in the displacement of the downstream cylinder as illustrated clearly in the records of Figure 5. Results of flow visualization showed that the bifurcation is attributed to a sudden change of the vortex pattern from the 2P to the 2S mode. Beyond the bifurcation, $A_1(t)$ decreases progressively and tends to the steady-state amplitude of the 2S mode, while $A_2(t)$ decreases rapidly to almost zero before undergoing a progressive increase to another steady-state value (similar to the case (S-R)). Prior to the bifurcation, $A_2(t)$ presents a modulation similar to that of Figure 3(a) but not past. The power input *P* presents similar variation with A_2/D. Results of Figures 3 and 4 also indicate that, for the same flow velocity, the power imparted by wind to the downstream cylinder is greatly influenced by the wake flow pattern of the upstream cylinder. This power is much less important when the upstream cylinder wake is of the 2S type than when it is of the 2P type. This observation is very important since it has not been noticed before and may be the source of discrepancy between the results of previous studies [Brika & Laneville 1996].

5.1 *Effect of the Reduced Velocity*

Figure 6 shows the effect of the reduced velocity V_r on the power imparted by wind to the downstream cylinder when both cylinders are released simultaneously from rest (case R-R). The power input increases with increasing V_r, reaches a maximum value at V_r= 6.4 and then decreases with V_r until the velocity of the end of synchronization of the upstream cylinder is reached. Beyond this velocity, *P* increases again with V_r, reaches a second maximum value between V_r = 7.3 and 7.4. This last maximum power is the one evaluated in the previous study [Brika & Laneville (1997)] since the upstream cylinder comes to rest beyond the end of synchronization.

For the case (P-R), results roughly similar to those of Figure 6(a, b) have been obtained with a maximum occurring around V_r= 6.4. When the wake vortex mode of the upstream cylinder is of the 2S type (case S-R), the maximum power is, however, around V_r= 5.8 and is much smaller than the two previous cases in which the vortex mode is of the 2P type.

Figure 7 shows a comparison between the maximum wind power imparted to the downstream cylinder obtained with the three different cases (R-R, P-R and S-R) and the power imparted to an isolated cylinder for (L/D= 13, α_1= α_2= β= 0°). We notice first that in the case (R-R), two envelopes of the power input have been obtained. The lower envelope (triangle symbol) is obtained during the transient oscillations of the upstream cylinder, i.e. during early stages of the cylinders build-up from rest, which is referred to 'small A_1', while the upper envelope (empty circles) is obtained when A_1 approaches its steady-state value and A_2 begins its beating-like modulation (see Figure 6a). The difference between the values of the two envelopes shows again the effect of the vibration amplitude A_1 on the power input of the downstream cylinder. The upper envelope comes near the maximum power curve of the case (P-R) since as A_1 approaches its steady-state value, both cases become similar. Figure 7 also indicates that the maximum power is obtained in the case (P-R), i.e. when the downstream cylinder is released from rest while the upstream one is vibrating with its steady-state amplitude of the 2P mode. For an amplitude A_2/D= 0.25 and L/D= 13, this power input is 410% higher than in the case (S-R), 190% higher than the power imparted to a single cylinder and 40% higher than that of the case (R-R) in which both cylinders are released simultaneously from rest. This last difference is due to the contribution of the vibration amplitude A_1/D.

5.2 *Effect of the Spacing Ratio*

Figure 8 shows the effect of the spacing ratio L/D on the maximum power imparted by wind to the downstream cylinder for the tandem arrangement and the case (R-R). Beyond A_2/D= 0.15, the power input decreases with increasing value of L/D and tends progressively to that of an isolated cylinder for larger separations between the cylinders. This is attributed mainly to the fact that vorticity, and hence lift force, generated by the upstream cylinder decreases as it is transported downstream [Sarpkaya (1979)]. The additional excitation of the rear cylinder diminishes with L/D and this cylinder tends to behave as an isolated one in the case of larger separations.

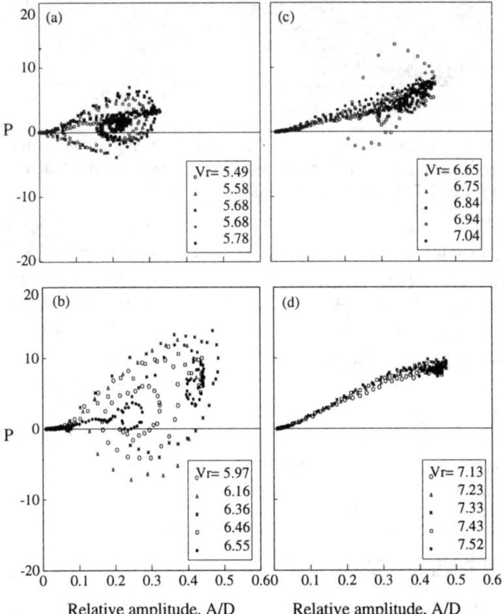

Figure. 6: Example of the effect of the reduced velocity on the power imparted to the downstream cylinder (case R-R, L/D= 10, ß= α_1= α_2= 0°). a), 5.4≤ V_r≤ 5.8; b),.5.8≤ V_r≤ 6.5; c), 6.5≤ V_r≤ 7.0.; d), 7.0≤ V_r≤ 7.5.

For the case (P-R), the effect of L/D is shown in Figure 9. This effect is more significant and the maximum power for each separation is greater than that of the previous case. When the wake vortex pattern of the upstream cylinder is of the 2S type (case S-R) the maximum power input is, however, much less important and decreases also with increasing value of L/D as illustrated in Figure 10.

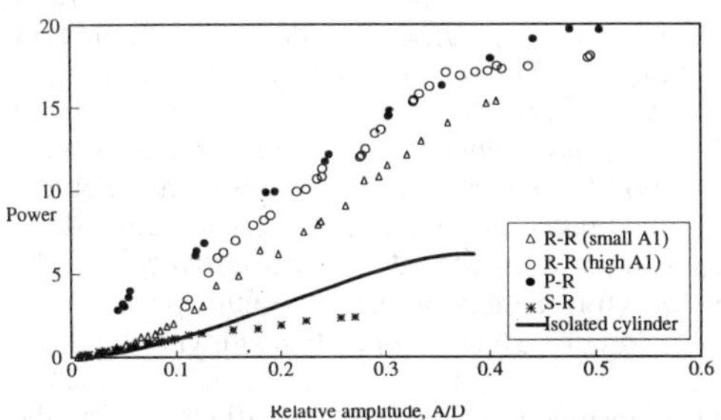

Figure.7: Comparison between the maximum power imparted to the downstream cylinder obtained under different conditions (L/D= 13, ß= α_1= α_2= 0°).

5.3 *Effect of the Bundle Orientation*

Figures 11 and 12 show the effect of the bundle orientation angle β with respect to the flow direction

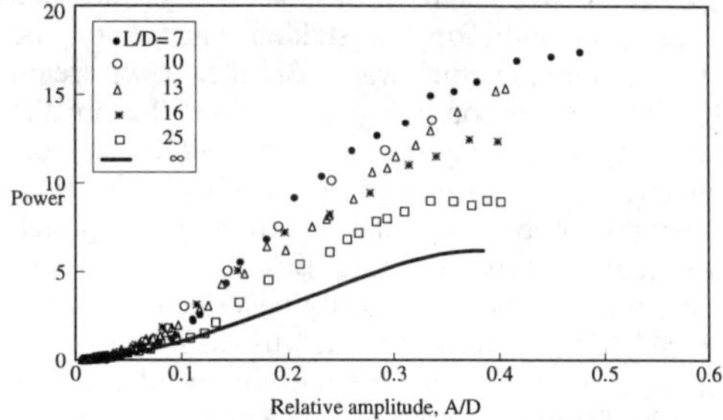

Figure. 8: Effect of L/D on the maximum power imparted to the downstream cylinder in the case R-R (ß= α_1= α_2= 0°).

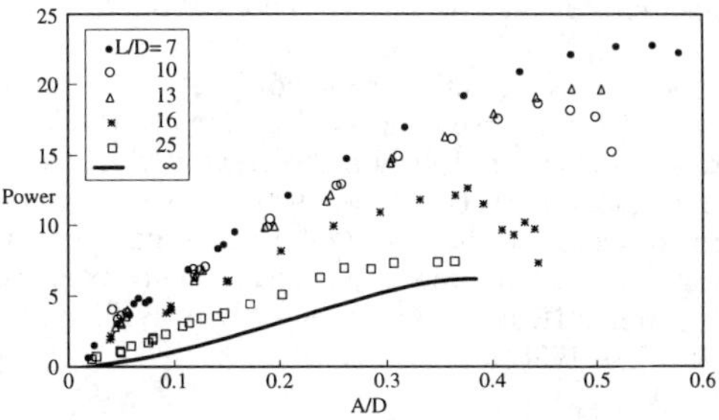

Figure. 9: Effect of L/D on the maximum power imparted to the downstream cylinder in the case P-R (ß= α_1= α_2= 0°).

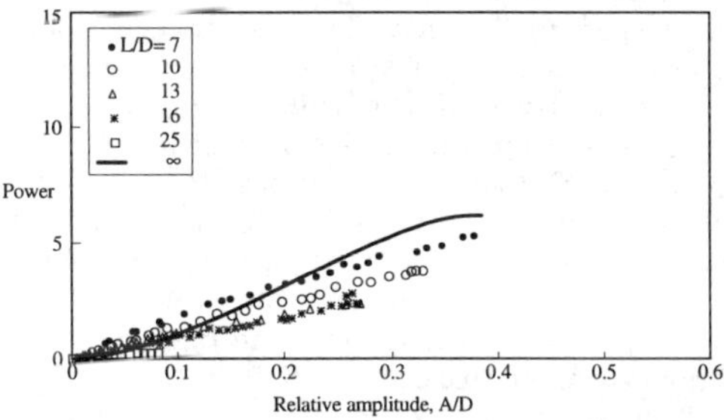

Figure. 10: Effect of L/D on the maximum power imparted to the downstream cylinder in the case S-R (ß= α_1= α_2= 0°).

respectively for the cases (P-R) and (S-R). In the case (P-R), the power input decreases with increasing value of β and tends progressively to the one of a single cylinder. Beyond β= 20° the downstream cylinder will be outside the wake of the upstream cylinder and will behave as an isolated one [Brika & Laneville (1997)]. When the wake vortex pattern of the upstream cylinder is of the 2S type, the power input, contrary to the previous case, increases now with increasing value of the bundle inclination angle for β< 15° as shown in Figure 12. For β= 15° a bifurcation is observed in the downstream cylinder displacement build-up from rest while the upstream cylinder is vibrating with its steady-state amplitude of the 2S mode. This bifurcation is accompanied with a sudden change of the phase angle Φ between the vortex shedding and this displacement indicating a sudden change of the vortex pattern in the wake of the downstream cylinder. The discontinuity at $A_2/D= 0.2$ in the power input of Figure 12 corresponds to this bifurcation. Prior to this phenomenon ($A_2/D< 0.2$), the maximum power as shown in the Figure is much lower than that of the case $\beta= 10°$ and doesn't respect the tendency previously described. This can be explained as follows: the downstream cylinder is initially outside the wake of the windward cylinder when the former is at rest or vibrating with a small amplitude, in this case it behaves as an isolated cylinder; if it continues to vibrate beyond a critical amplitude ($A_2/D= 0.2$ in the present experiments), then it crosses the boundary of the upstream cylinder wake which influences its own wake and its characteristics. Similar behavior has also been observed in the case of a fixed upstream cylinder [Brika & Laneville (1997)] the vortex of which is of the 2S mode type. However, such behavior has not been observed in the case (P-R) since the cylinder wake of the 2P mode is larger and expands downstream [Brika & Laneville (1996), Williamson & Roshko (1988)].

5.4 *Effect of the Plane of Vibration*

The plane of vibration, characterized by the angle α with respect to the cross-flow direction, is not of major concern in the present study since a previous investigation [Brika & Laneville (1997)] has shown that the power imparted to a cylinder vibrating in a direction α is roughly proportional to that of the orthogonal plane ($\alpha= 0°$) by $\cos^2\alpha$. In the present study only two different cases have been examined as illustrated in Figure 13. The results of the maximum power input are given in Figures 14 and 15 respectively for the cases (P-R)

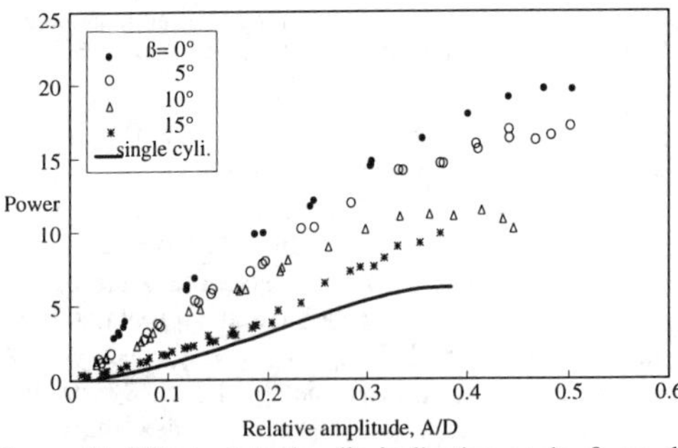

Figure. 11: Effect of the bundle inclination angle, ß, on the maximum power imparted to the downstream cylinder in the case P-R ($L/D= 13$, $\alpha_1= \alpha_2= 0°$).

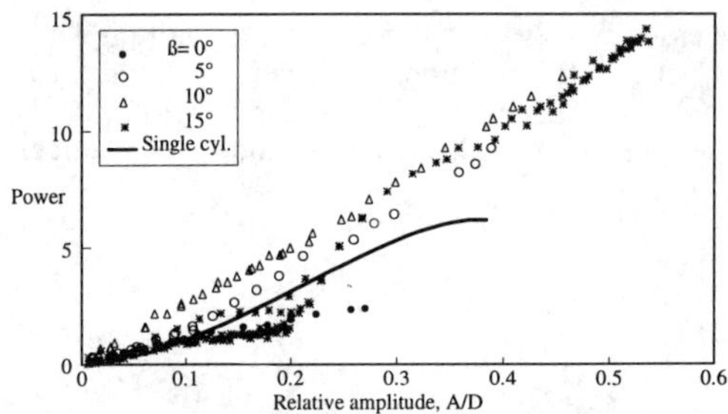

Figure. 12: Effect of the bundle inclination angle, ß, on the maximum power imparted to the downstream cylinder in the case S-R ($L/D= 13$, $\alpha_1= \alpha_2= 0°$).

and (S-R). For the cases (a) and (b), i.e. $\alpha_1= \alpha_2= 15°$ and $\alpha_1= -\alpha_2= -15°$, the power input seems to be not greatly affected by the orientation of the plane of vibration of the upstream cylinder. We notice, however, a slight difference for case (a) ($\alpha_1= 15°$) at larger amplitudes of oscillation. In comparison with the orthogonal vibration, a slight decrease of P with increasing α is observed in the case (P-R) while an increase of P is noticed at higher amplitudes in the case (S-R).

6. SUMMARY AND CONCLUSION

Figure 16 summarizes, for two different spacing ratios $L/D= 13$ and 25, the variation of the maximum power imparted by wind to a flexible cylinder with the three different configurations investigated up till now, i.e.:

- a single cylinder
- a cylinder in the wake of a fixed upstream cylinder
- a bundle of two uncoupled free-vibrating cylinders

The results of this Figure indicate that the presence of an upstream cylinder has a great influence on the power imparted by wind to the downstream cylinder. This influence is more important when the upstream cylinder is vibrating and the cylinders are close to each other. Thus for a typical separation $L/D= 13$ and an amplitude of oscillation $A_2/D= 0.3$, the specific power input P increases by about 50% of the power imparted to an isolated cylinder when the upstream cylinder is fixed and by 164% when this later is vibrating. For the largest separation examined ($L/D= 25$) these increases are only 14 and 35% respectively. For the free-vibrating cylinders, the maximum power input given in Figure 16 is the one obtained with the (P-R) case.

The results are also greatly influenced by the wake vortex pattern of the upstream cylinder. The power input is much less important when this pattern is of the 2S type than when it is of the 2P type. For the staggered arrangement, the power input decreases progressively with increasing value of the bundle inclination angle β when this pattern is of the 2P type and increases with β (β≤ 10°) when the pattern is of the 2S type. Beyond β= 20°, the downstream cylinder behaves as an isolated one. To simulate spacer dampers, a mechanical coupling must be examined and this will be the subject of a future work.

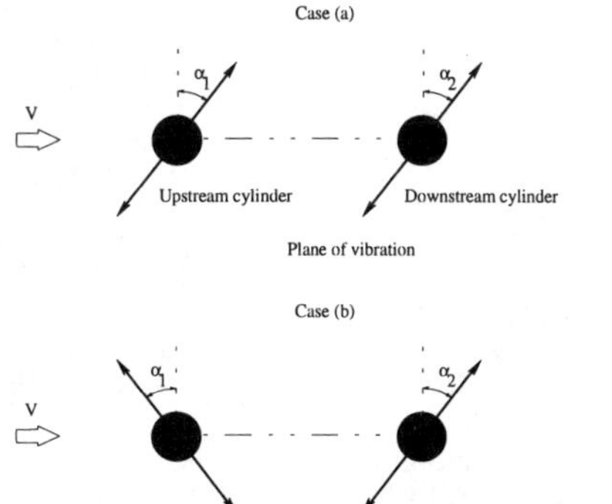

Figure. 13: The two cases of the non-orthogonal planes examined in the present study (L/D= 13, ß= 0°). a), $\alpha_1= \alpha_2=$ 15°; b), $-\alpha_1= \alpha_2=$ 15°

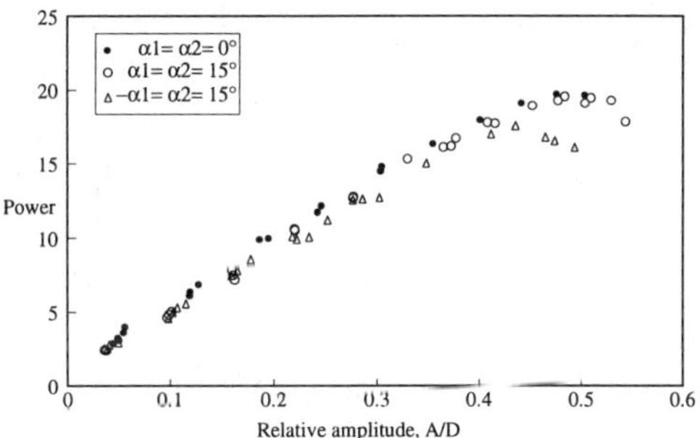

Figure. 14: Effect of the plane of vibration on the maximum power imparted to the downstream cylinder in the case P-R (L/D= 13, ß= 0°).

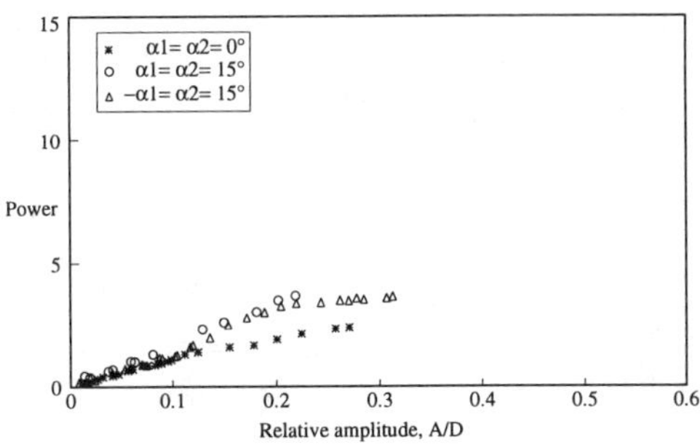

Figure. 15: Effect of the plane of vibration on the power imparted to the downstream cylinder in the case S-R (L/D= 13, ß= 0°).

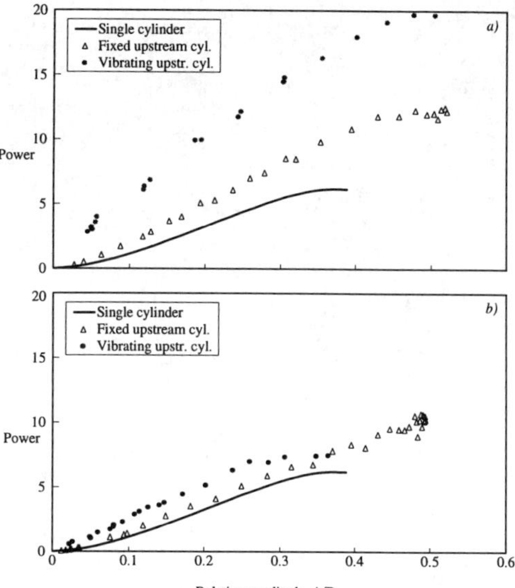

Figure. 16: Effect of the upstream cylinder on the maximum power imparted by wind to the downstream cylinder (ß= α_1= α_2= 0°). a), L/D= 13; b), L/D= 25.

ACKNOWLEDGEMENT

The authors gratefully acknowledge the financial support of Hydro-Québec and the Natural Sciences and Engineering Research Council of Canada.

REFERENCES

Brika, D. & A. Laneville, 1996 A Laboratory Investigation of the Aeolian Power Imparted to a Conductor Using a Flexible Circular Cylinder, *IEEE Trans. Power Delivery* 11, 1145-1152.

Brika, D. & A. Laneville, 1997 The Power Imparted by Wind to a Flexible Circular Cylinder in the Wake of Another Stationary Cylinder, *IEEE Trans. on Power Delivery* 12 , 398-405.

Brika, D. & A. Laneville, 1993 Vortex-Induced Vibrations of a Long Flexible Circular Cylinder, *Jour. Fluid Mechanics* 250, 481- 508.

Sarpkaya, T. 1979 Vortex-Induced Oscillations - A selective Review, Trans. ASME, *J. Applied Mechanics* 46, 241-258

Williamson, C.H.K. & A. Roshko, 1988 Vortex Formation in the Wake of an Oscillating Cylinder, *Jour. Fluids and Structures* 2, 355-381.

Wind Engineering into the 21st Century, Larsen, Larose & Livesey (eds) © 1999 Balkema, Rotterdam, ISBN 90 5809 059 0

Improved blockage corrections for bluff bodies in closed and open wind tunnels

K.R.Cooper
National Research Council of Canada, Ottawa, Ont., Canada

E.Mercker
German-Dutch Wind Tunnel, Emmeloord, Netherlands

J.Wiedemann
FKFS, University of Stuttgart, Germany

ABSTRACT: Recent developments in bluff-body blockage corrections for closed-jet and open-jet wind tunnels are presented, including an alternative development of Maskell's method for closed-wall wind tunnels. This derivation breaks the correction into dynamic pressure and incremental drag adjustments and is shown to remove a tendency for over-correction at large blockage, making the method more widely applicable than previously. A new development in open tunnel corrections has identified important effects at the tunnel nozzle and collector that may dominate the classical blockage effects. This approach brings measurements made using collector or plenum static pressure references into close agreement.

1 INRODUCTION

Bluff bodies impose a more serious constraint in a wind tunnel than do streamlined bodies. In particular, their large wakes produce blockage effects that dominate solid blockage effects.

Maskell 1965 developed a method for estimating wake blockage, based on momentum analysis (drag), that has been widely utilised. While it was a great improvement over the small-wake, classical analysis, it has been shown to over-estimate the drag correction at large blockage.

The open-jet wind tunnel community commonly assumed that the corrections were small and could be neglected. However, it was known that different results could be obtained, depending on the dynamic-pressure measurement technique employed. The resulting uncertainties were tolerated or ignored because they were thought to be small and because no other option was available.

Recent developments in correction technologies offer an opportunity to improvement these circumstances and form the subject matter of this paper.

2 CHARACTERISATION OF BLOCKAGE EFFECTS

The presence of a model in a bounded or a free jet causes significant flow distortions to occur. They are driven by a changed static pressure distribution in the wind tunnel over the vicinity of the model and appear at the model as a test speed change, as increments in forces due to velocity gradients and as flow angle distortions in the pitch and the yaw planes. The first two items, speed changes and gradient effects, are usually combined as 'blockage', while the angle distortions due to the cross-stream induced velocity distributions, are referred to as 'interference'. Only the former will be dealt with here.

The assumptions that are commonly made in the analysis of blockage require that a model be 'small in the test section', permitting linear theory to be employed. This size assumption also implies that the induced velocities at one point on the model, typically the centre of the model, represent the blockage effect over the whole model. This assumption is usually not satisfied although it turns out that the model only must be 'small enough', implying a size for which the correction procedure will work. One of the unresolved issues is what the upper limit on size may be.

2.1 *Solid and wake blockage*

The major blockage components result from:

1. velocity increments induced by the displacement effect of the model volume and its separation bubble - termed solid blockage;
2. velocity increments induced by the displacement effect of the wake - termed wake blockage;
3. incremental forces due to velocity gradients induced by the wake.

In general, these effects differ between closed and open tunnels in both sign and magnitude.

2.1.1 *The closed tunnel*
The model, any separation bubbles on it, and its wake, obstruct the flow in a closed tunnel, increasing the flow speed as a consequence of continuity, thereby producing a reduction in static pressure in the vicinity of the model.

The solid component of blockage produces a positive velocity increment at the model, but no gradient, whereas the wake blockage induces both a positive velocity increment and a velocity gradient at the model. The forces on the model are increased by the speed increase whereas the surface pressures are affected by the speed increase and by the reduction in reference static pressure.

This dual effect on pressures is important to recognise, as it has been improperly interpreted previously. As a consequence, negative pressures are more affected than positive pressures and no change to the stagnation pressure will occur in a constant-energy, incompressible wind tunnel flow.

2.1.2 *The open tunnel*
The blockage effects are generally reversed in the open tunnel, as the free jet over-expands and the local velocity diminishes. The velocity increments are between one-quarter and one-half of those in the closed tunnel and are negative.

Solid blockage produces a negative velocity increment at the model while the wake produces no velocity increment. Thus, the classical view of the open tunnel is that it has no wake blockage.

The open tunnel also has been found to have additional influences at the entry and the exit of the test section that produce effects like those in closed tunnels (Mercker & Wiedemann 1996; Mercker et al. 1997). These effects are not included in the classical open-tunnel literature and will be dealt with in the discussion of the new open-tunnel correction procedure.

2.2 *Correction methodology*

An adequate correction must address the components of blockage described above and do so in a physically correct fashion. The total blockage effect on velocity is a summation of positive or negative velocity increments at the model location, such that,

$$(q_c/q_u)=(1+\varepsilon_s+\varepsilon_w+\sum\varepsilon_i)^2=(1+\varepsilon_t)^2\approx(1+2\varepsilon_t) \quad (1)$$

where each $\varepsilon = (U_b-U_m)/U_m=\Delta u/U_m$ is an increment in non-dimensional velocity induced by the boundary constraint; U_b is the blocked velocity due to a particular effect; U_m is the upstream measured velocity far from the model; the subscripts 'u' and 'c' refer to uncorrected and corrected, respectively; ε_t is the total blockage that includes the solid blockage component, ε_s, the wake blockage component, ε_w and any additional blockage components, $\sum\varepsilon_i$, due to mounting system, measuring system and upstream and downstream effects that appear as velocity increments at the model.

The major gradient effect is on drag and is caused by the presence of the viscous wake, so the correction to drag appears as,

$$C_{Dc}=(C_{Du}+\Delta C_{Dw})/(q_c/q_u) \quad (2)$$

where ΔC_{Dw} is the drag increment that results from wake blockage. At all times the drag is that measured in wind axes, that is, in an axis system aligned with the mean tunnel flow.

The correction to pressure coefficient is made as,

$$(1-C_{pc})=(1-C_{pu})/(q_c/q_u) \quad (3)$$

This can be rewritten as,

$$C_{pc}=[1-(q_u/q_c)]+(C_{pu}/(q_c/q_u)) \quad (4)$$

where the first term in square parentheses on the right-hand-side of equation (4) is the correction for the static pressure offset and the second term is the correction for the flow speed increase. It is apparent from equation (3) that a pressure coefficient of unity is unchanged by blockage, as should be the case.

3 MEASURED BLOCKAGE BEHAVIOUR

A reasonable body of literature has been generated on blockage effects on two- and three-dimensional bodies in closed tunnels. The data base for open tunnel blockage is smaller. Examples of measurements in closed and open tunnels on a range of bluff models are presented in Figures 1 and 2, respectively.

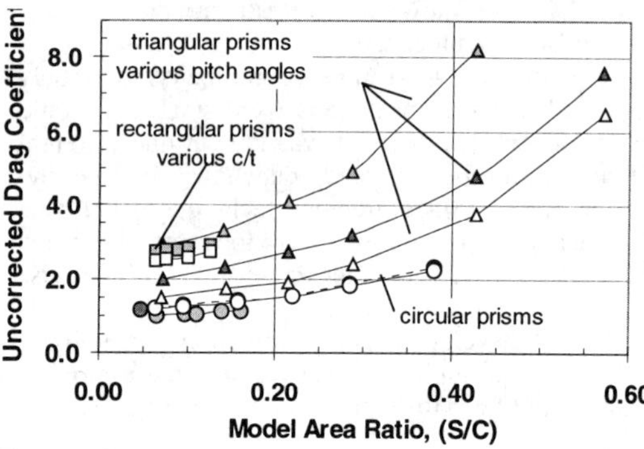

Figure 1. Summary of two-dimensional blockage behaviour in closed tunnels.

The drag data of Figure 1 show typical closed-tunnel blockage, having a monotonic increase in measured drag coefficient with increasing model-to-tunnel area ratio, (S/C) = (h/t). S is the model frontal area, C is the tunnel cross-sectional area, t is the model thickness and h is the test section height. No two-dimensional data are presented for open tunnels because none were available. The data are grouped by model type, primarily through drag level.

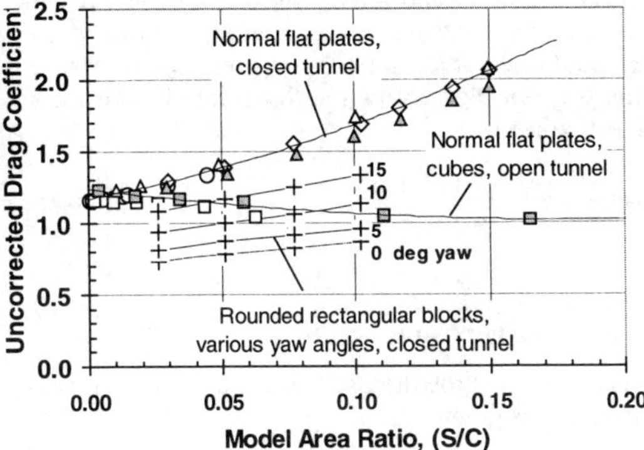

Figure 2. Summary of three-dimensional blockage behaviour in closed and open tunnels.

Figure 2 presents similar data for three-dimensional bodies in closed and open tunnels, which have the expected opposite blockage trends

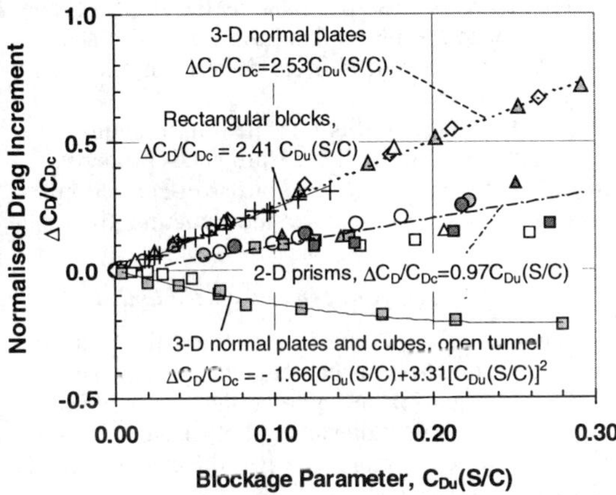

Figure 3. Summary of two-dimensional and three-dimensional blockage behaviour in closed and open tunnels.

The data from both figures may be collapsed as normalised increments in drag coefficient, $\Delta C_D/C_{Dc}=(C_{Du}-C_{Dc})/C_{Dc}$ plotted against the blockage parameter, $C_{Du}(S/C)$, as in Figure 3. Now the measurements form three distinct groups based on tunnel type and model type.

Regression lines through each provide an indication of the relative magnitude and sign of the blockage effect. The two-dimensional models in the closed tunnel have only 40 percent of the blockage effect of the three-dimensional models. The blockage effect for the three-dimensional models in the open tunnel is approximately 66 percent of that in the closed tunnel, based on the linear term in the regression of the open-tunnel results. This is a larger effect than that predicted by classical theory.

The constants in the fits of the closed-tunnel data should be the constants determined by Maskell in the development of his correction method.

4 CORRECTION METHODS

The presentations of the correction methods are separated by tunnel type. The methods chosen for discussion include analytical formulations based on geometry and measured model drag forces. The family of methods that have been developed for closed tunnels that are based on boundary measurement techniques are not included here (Hackett et al. 1979, Cooper et al. 1996, Ewald et al. 1998).

Cooper et al. 1996, offers a detailed guide to the application of two wall-pressure-measurement methods to automotive models. These methods also should be applicable to three-dimensional wind engineering situations, especially to model buildings in a cityscape where the total drag is not known.

4.1 *Concept of the duplex tunnel*

Most wall corrections consider models that are centrally mounted in the test section. When a model is floor mounted, as in the case of a building model, a cityscape or an automobile, the floor plane is taken as a plane of symmetry. The tunnel/model combination is reflected about this plane, producing an imaginary, duplex tunnel, with a now centrally-located model of double size in a doubled tunnel The sketch of Figure 4 shows the concept.

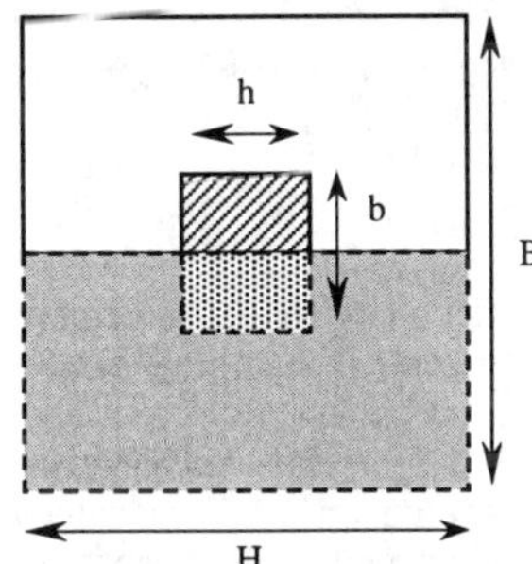

Figure 4. Sketch showing the reflected duplex tunnel cross section employed for floor-mounted models.

4.2 *The two-step derivation of Maskell's wake-blockage method for closed tunnels*

Maskell 1965 made a momentum-based analysis of the wake blockage of bodies with large separated wakes. The development utilised flat-plate data, for which the primary source of blockage was due to the wake, since the model volume was negligible. Maskell derived the correction by first assuming that the wake boundary was unaffected by wall constraint. It lead to a partial correction of the form,

$$\left(\frac{q_c}{q_u}\right)=\left(\frac{C_{Du}}{C_{Dc\infty}}\right)=1+\theta C_{Dc\infty}(S/C) \tag{5}$$

where θ is an empirical wake-blockage constant and $C_{Dc\infty}$ is the corrected drag coefficient, excluding the constraining effect of the walls on the wake, that must be determined by iteration in equation (5).

Inclusion of the distortion of the wake boundary by the tunnel walls provided an additional component of the blockage that was included in the correction to complete it as,

$$\left(\frac{q_c}{q_u}\right)_1=\left(\frac{C_{Du}}{C_{DcM1}}\right)=1+\theta C_{Dus}(S/C)\approx 1+2\varepsilon_{ws} \tag{6}$$

where ε_{ws} is the interference velocity due to the separated wake; C_{Dus} is the component of drag due to separated flow; and C_{DcM1} is the standard, single-step, Maskell wake-blockage corrected drag. The separated-flow drag is commonly taken as the measured drag with little error for flat plates or bodies with leading-edge separation. However, for bodies with separation at a bluff base, like automobiles, any drag due to vortex flows or due to skin friction should be removed. Such a drag separation is frequently difficult or impossible, and the full drag is usually used, producing an overestimate of the wake blockage.

If a drag-component separation is possible, the wake-blockage correction can be written to include a skin-friction term and a wake term as,

$$(q_c/q_u)=[1+\varepsilon_{wa}+\varepsilon_{ws}]^2\approx[1+2\varepsilon_{wa}+2\varepsilon_{ws}]$$

$$(q_c/q_u)=\left\{1+\left(\frac{S}{C}\right)\left[\frac{C_{Do}}{2}+\theta(C_{Du}-C_{Di}-C_{Do})\right]\right\} \tag{7}$$

where ε_{wa} is the interference velocity due to skin friction (the attached-flow, thin-wake component); C_{Do} is the skin friction coefficient; and C_{Di} is the induced drag due to either lift or side forces

The blockage constant, θ, can be approximated for a wide range of bodies, from those of infinite aspect ratio to those of finite aspect ratio, by,

$$\theta=0.96+1.94\exp[-0.06(h/b)] \tag{8}$$

where (h/b) is the model aspect ratio in front elevation, defined as the ratio of the duplex model height, h, over the duplex model span (breadth), b, where h is chosen as the larger of the two dimensions, resulting in (h/b)>1 always.

The wake constraint should appear at the model as a drag increment. Since Maskell included the drag increment in the dynamic-pressure correction, equation (5), it provides a correct adjustment to drag but is an over-correction of the test speed and the other force, moment and pressure coefficients.

Hackett 1996 and AIAA discussions 1998 recognised that the difference between equations (5) and (6) could be solved for the drag increment, permitting the blockage correction to applied in two parts. The two-step equation is,

$$C_{DcM2}=\frac{(C_{Du}+\Delta C_{DM})}{(q_c/q_u)_1}$$

$$=\frac{(C_{Du}+\Delta C_{DM})}{1+\theta(C_{DcM1}-\Delta C_{DM})(S/C)} \tag{9}$$

where ΔC_{DM} is the drag increment due to wake constraint. It is given by,

$$\Delta C_{DM}=\frac{C_{Dus}}{(1+\theta C_{Dus}(S/C))}+$$

$$\left[\frac{1}{2\theta(S/C)}\right]\left[1-\sqrt{1+4\theta C_{Dus}(S/C)}\right] \tag{10}$$

This correction is applicable as the full blockage adjustment to flat plates that have no solid blockage or to a model immersed in its wake from a leading-edge separation.

The advantage of this formulation, in addition to its ability to correct all measurements properly, is that it removes the tendency to over-correction to drag found in the original version, equation (6).

4.3 *Solid blockage for open and closed tunnels*

When the flow is attached to near the point of maximum thickness, as in a circular cylinder, or to a bluff base, as for an automobile, an additional blockage component due to solid blockage should be added to the correction. The two-dimensional solid blockage correction is,

$$\varepsilon_{2s}=\frac{k_{2s}}{\beta^3}\left(\frac{t}{h^*}\right)^2\lambda_2,\quad \lambda_2=\left(\frac{2A}{\pi tc}\right)\left[\left(\frac{c}{t}\right)+1.2\beta\right] \tag{11}$$

where h^*=h, the tunnel height, for a closed tunnel; $h^*=h/(1+\varepsilon_{qn})$ for an open tunnel; A is the cross-sectional area of the model; t is the model thickness; and c is the model chord. The open tunnel height is decreased to account for the extra expansion of the emerging jet caused by the presence of a model through the nozzle blockage factor, ε_{qn} (see Section 4.4.1 for the definition of ε_{qn}). The Mach number correction, $\beta=(1-M^2)^{1/2}$, is negligible for low-speed, boundary-layer tunnel studies but can be important for high-Reynolds-number testing of two-dimensional as the Mach number increases. The required blockage constants in equation (11) are,

$$k_{2s}^{closed} = \left(\frac{\pi^2}{12}\right), \; k_{2s}^{open} = \left(\frac{-\pi^2}{24}\right), \; \beta = \sqrt{1-M^2} \quad (13)$$

The open-tunnel correction is of opposite sign and one-half the magnitude of that for the closed tunnel.

The three-dimensional solid blockage correction is (Mercker & Wiedemann 1996),

$$\varepsilon_{3s} = \frac{k_{3s}}{\beta^3}\left(\frac{B}{H}+\frac{H}{B}\right)^n \left(\frac{V}{2L_m}\right)^{1/2} \left(\frac{S}{C^{3/2}}\right) \quad (14)$$

where B and H are as defined in Figure 4; L_m is the model length; C is the tunnel cross-sectional area; V is the model volume; n=1.00 and k_{3s}=0.41 for closed tunnels; and n=2.94 and k_{3s}=-0.032 for open tunnels. C and V are the single or duplex tunnel values.

Again, the open tunnel correction is of opposite sign to that for the closed tunnel and, for a test-section aspect ratio of (B/H)=1, the open tunnel magnitude is 0.3 of that for the closed tunnel.

4.4 *A recent development in open-tunnel corrections*

The classical, open-tunnel correction components summarised above have been found to be incomplete. There are important, model-induced effects at the open-jet nozzle and at the entrance to the collector that resemble closed-tunnel behaviour and that affect the flow at the model.

Mercker & Wiedemann 1996, Mercker et al. 1997 and Mercker & Wiedemann 1999 have identified an expanded set of blockage effects. They have developed a correction that contains the following components,

1. *Nozzle blockage, ε_{ij}* – The interference of a model on the nozzle changes the calibration of the dynamic-pressure measuring system from the empty-tunnel value. The distortion is measured differently by the nozzle (j=n) and the plenum (j=p) methods. The effect appears as a positive velocity increment at the model. The subscript i=2 or i=3 denotes a two-dimensional or a three-dimensional geometry
2. *Solid blockage and jet expansion, ε_s* - A free-jet flow over-expands at a model, reducing the velocity at the model to a value below that measured during the empty-tunnel calibration. Proximity of a model to the nozzle increases this effect, inducing an additional negative velocity increment at the model due to extra jet expansion produced by the nearness of the model.
3. *Collector/wake blockage, ε_c* - The solid-wall constraint on the wake at the collector produces a positive velocity increment at the model.

In addition, the effect on drag of the empty-tunnel pressure gradient must be removed. Mercker & Wiedemann 1997 propose a method for this correction.

4.4.1 *Nozzle blockage factors*

The effect of a model at the nozzle had not been fully understood before the work of Mercker & Wiedemann 1996. It had been known (Kuhn 1995) that the two methods of measuring dynamic pressure in open tunnels, using either a static pressure measurement in the nozzle or a static pressure measurement in the plenum surrounding the free jet produced significantly different drag coefficients. This occurs because each method produces a different dynamic pressure measurement when a model is present, even though they produced identical results when the empty test section was calibrated.

The reason for the differences can be seen in Figure 5 (Ewald et al. 1998), where the distortion produced by a model at the nozzle is related to the far-away, undisturbed flow case.

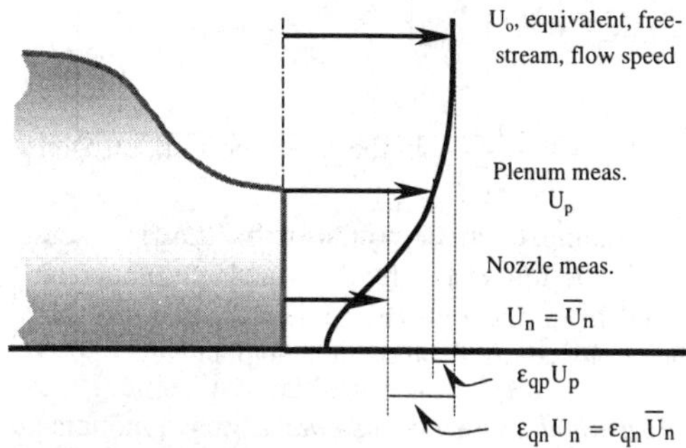

Figure 5. Illustration of the measurement of the approach flow-field in an open tunnel and its relationship to the free-stream equivalent.

The nozzle method measures the average velocity through the nozzle and, due to model effects on the nozzle flow, under-measures the equivalent, free-stream velocity. The plenum method measures the velocity at the jet boundary, and also under-measures the equivalent free-stream approach value, but by less than with the nozzle measurement. The relationship of these measurements to each other and to the equivalent, free-stream approach flow will be a function of the model geometry and of the proximity of the model to the nozzle.

A correction for this model influence can be obtained by representing the model by a suitably-sized line or point doublet for the two-dimensional and three-dimensional cases, respectively (Mercker & Wiedemann 1999). The velocity perturbation produced by the doublet at the nozzle exit plane gives the ratio of, the average nozzle flow or the velocity at the edge of the jet, to U_o, the equivalent onset flow. Momentum arguments were then used to ensure that both velocity measurements produced the same effect at the model. The nozzle blockage is,

$$\varepsilon_{in} = \varepsilon_{iqn}\chi_n, \quad \varepsilon_{iqn} = \frac{f_{in}}{(1-f_{in})} \approx f_{in} \tag{15}$$

where ε_{iqn} are the nozzle blockage factors at the nozzle plane; and χ_n is the relaxation factor that transfers the jet distortion from the nozzle plane to the model centre. The last approximation in equation (15) is valid for small blockage where $f_{ij} \ll 1$.

The plenum blockage is,

$$\varepsilon_{ip} = \varepsilon_{iqp} + (\varepsilon_{in} - \varepsilon_{iqn}) \tag{16}$$

where ε_{ipn} are the plenum blockage factors. The nozzle components in parentheses on the right-hand-side provide the momentum match just mentioned.

The parameters in equation (15) are:

Nozzle method, two-dimensional

$$f_{2n} = \left(\frac{R_m^2}{hx_m}\right)\tan^{-1}\left(\frac{h}{x_m}\right), \quad \chi_{2n} = \frac{h^2}{\left(x_m^2 + h^2\right)} \tag{17}$$

where $R_m = \frac{t}{(2\sqrt{\lambda_2})}$ is the effective radius of the approximating cylinder produced by the line doublet; h is the tunnel height; λ_2 is a shape parameter obtained from equation (11); and x_m is the distance of the model centre from the nozzle plane.

Nozzle method, three-dimensional

$$f_{3n} = \frac{R^3}{\left(x_m^2 + R_n^2\right)^{3/2}}, \quad \chi_{3n} = \frac{R_n^3}{\left(x_m^2 + R_n^2\right)^{3/2}},$$

$$R_n = \sqrt{\frac{C}{\pi}}, \quad R = \sqrt[3]{12\lambda_m}\sqrt{\frac{S}{2\pi}}, \quad \lambda_m = \left(\frac{1}{2}\right)\frac{L_m}{\sqrt{\frac{S}{\sqrt{2}}}}\left(\frac{V}{SL_m}\right) \tag{18}$$

where R is the radius of the approximating sphere produced by the point doublet; λ_m is a shape factor; S is the model reference area; C is the nozzle area; R_n is the nozzle hydraulic diameter; L_m is the model length; and V is the model volume. All model lateral dimensions, areas and volumes are duplexed for ground-based models, as is the tunnel area.

Plenum method

$$f_{2p} = R^2\left(\frac{h^2 - x_m^2}{\left(h^2 + x_m^2\right)^2}\right)$$

$$f_{3p} = \left(\frac{R^3}{\left(R_n^2 + x_m^2\right)^{3/2}}\right) - \left(\frac{(3/2)R^3R_n^2}{\left(R_n^2 + x_m^2\right)^{5/2}}\right) \tag{19}$$

Collector blockage

The collector blockage has the same form in two or three dimensions, with the appropriate constants.

$$\varepsilon_c = \varepsilon_{iw}\frac{h_c^2}{\left[h_c^2 + (L_{ts} - x_m)^2\right]} \tag{20}$$

where $\varepsilon_{2w} = \frac{\theta}{2}C_{Dus}\left(\frac{t}{h_c}\right)$, $\varepsilon_{3w} = \frac{\theta}{2}C_{Dus}\left(\frac{S}{C}\right)$, or

$$\varepsilon_{3w} = \left(0.41 + \frac{C_{Du}}{4}\right)\left(\frac{S}{C}\right) \tag{21}$$

The second expression for ε_{3w} was specifically derived for automobiles (Mercker 1986) and should also be applicable to other models with flow attached to a bluff base.

Application of the blockage components

The complete nozzle-method dynamic-pressure correction has the form,

$$\boxed{(q_c / q_u) = (1 + \varepsilon_s + \varepsilon_n + \varepsilon_c)} \tag{22}$$

The complete plenum-method dynamic-pressure correction has the form,

$$\boxed{(q_c / q_u) = (1 + \varepsilon_s + \varepsilon_n + \varepsilon_c + \varepsilon_{qp} - \varepsilon_{qn})} \tag{23}$$

4.5 *Tunnel pressure gradient*

If the open or closed tunnel has a longitudinal pressure gradient it will affect the wind-axis drag and must be corrected for. This correction is not blockage, per se, but is usually applied at the same time as the blockage corrections. The correction is in the form of a drag increment that is added to the model prior to the application of the blockage correction, as in equation (2).

5 EXPERIMENTAL VERIFICATION

The efficacy of the correction methods presented will now be addressed by their application to several data sets, including some of those in Figures 1 and 2. The usual procedure for investigating blockage is to make measurements on a set of increasingly larger models in a given test section, to make measurements on a model in several wind tunnels or to make measurements on one model in a wind tunnel that has a variable test section size. The intention in each case is to exclude all other effects than blockage which, as it turns out, is usually impossible. This is especially so for two-dimensional testing, where changing blockage by increasing model size also reduces aspect ratio. The result is an increasing contamination of the data at larger sizes with three-dimensional flow effects. This can be seen in the data of Figure 3. More variability is present in the two-dimensional data than in the three-dimensional data, as the model aspect ratios varied from of order 15 at low blockage to unity at high blockage.

5.1 *Closed tunnels*

5.1.1 *Three-dimensional data*

The improvements provided by the two-step Maskell method are readily quantified by data from three-dimensional flat plates, from parachutes and from two-dimensional prisms of several cross-sections.

The parachute example is particularly instructive. It involved the testing of three parachute in six closed wind tunnels (Macha & Buffington 1985). It was found that Maskell's equation over-corrected unless the blockage constant was reduced from 2.5 to 1.85. When the two-step method was applied, equations (6), (9), (10), with a blockage constant of 2.5, an excellent correction resulted, as shown by Figures 6 and 7.

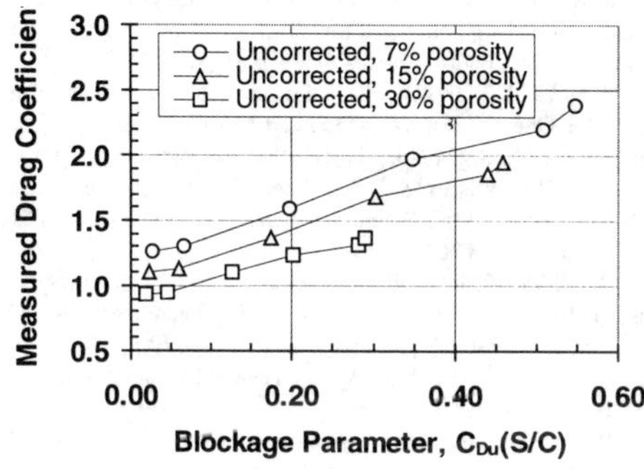

Figure 6. Measured parachute data (Macha & Buffington 1985).

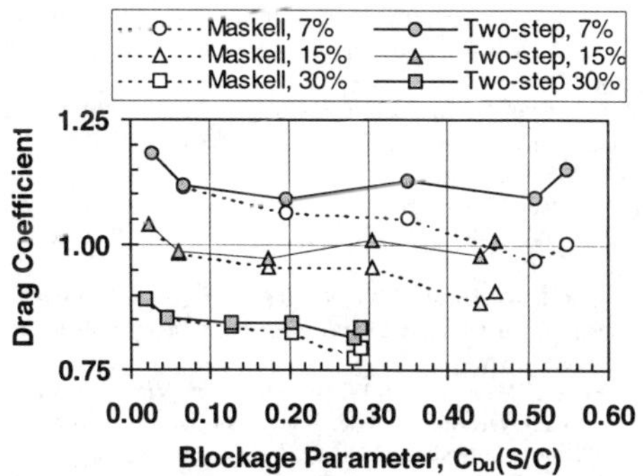

Figure 7. Application of the two-step correction to parachutes (Macha & Buffington 1985).

When the method is applied to flat plates (Gould 1970), a similar result is found, as shown in Figure 8. The drag increment becomes an increasingly large part of the correction at the largest blockages.

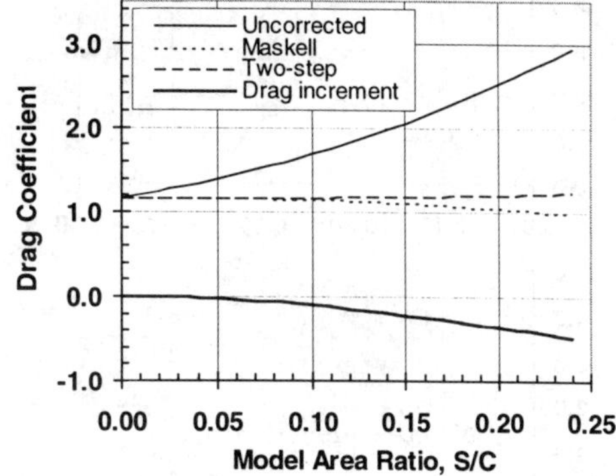

Figure 8. Application of the two-step correction to flat plates (Gould 1970).

5.1.2 *Two-dimensional data*

Several examples can be found in circular prisms (Farell et al. 1976), rectangular prisms (Courchesne & Lanneville 1979) and triangular prisms (Ramamurthy & Ng 1973).

The circle and rectangle data are presented in Figure 9. The circular prism is better corrected by the two-step method, while both over-correct the rectangular model data.

The final closed tunnel data set is presented in Figure 10 for a triangle at an angle of attack of 30 degrees, resulting in both lift and drag on the body. The drag is again well corrected by the two-step method, although the lift data is slightly under-corrected at high blockage.

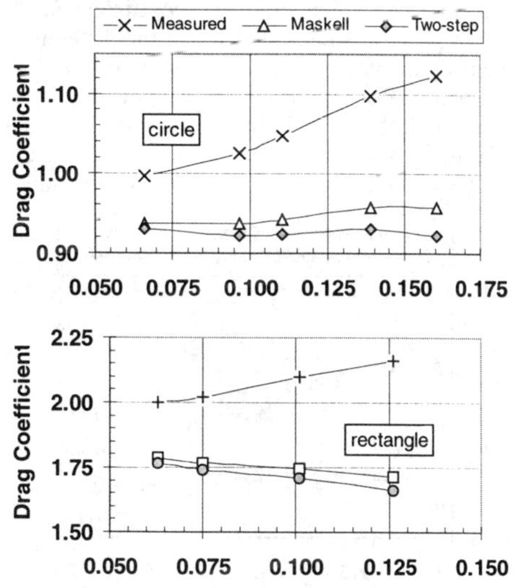

Figure 9. Application of the two-step method to rectangular and circular prisms.

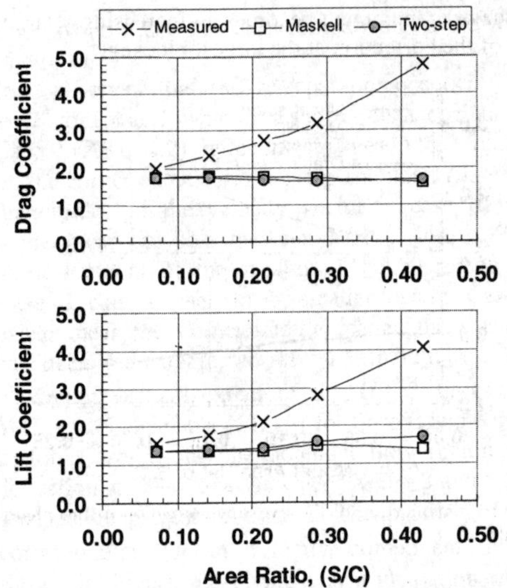

Figure 10. Application of the two-step correction to a triangular prism with lift.

5.2 *Open tunnels*

The open-tunnel correction is new and only one three-dimensional data set was available. These measurements were the drag coefficients of two full-scale automobiles with changes in axial position in the test section of an open tunnel (Schwartekopp 1999) for both dynamic-pressure measurement methods, Figure 11. The uncorrected data showed an increase in the drag for both models as they approached the nozzle, and showed a large difference between the nozzle- and plenum-based values.

After the correction, the drag coefficient variations with position were reduced and the differences between dynamic-pressure measurement methods were reduced. The sedan data were better corrected than the van data. It was believed that the static taps used for the nozzle measurements were too close to the nozzle exit and were interfered with by the van.

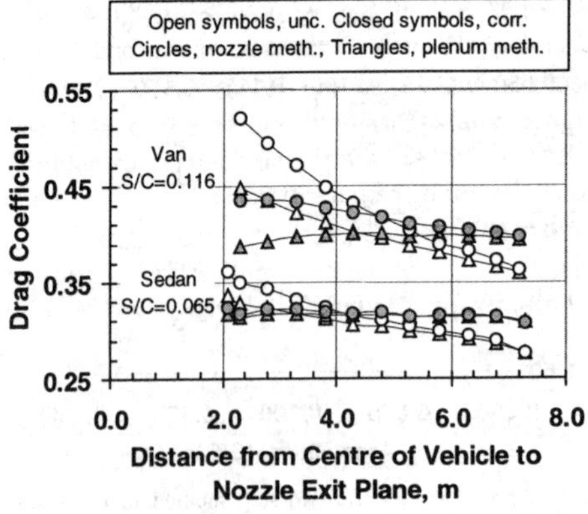

Figure 11. Application of open-tunnel correction method.

6 CONCLUDING REMARKS

New developments in bluff-body blockage corrections in closed and open tunnels have been summarised. An alternative derivation of Maskell's method has been shown to make its application much more general and more accurate in closed tunnels.

A new open-tunnel correction has provided valuable insight into open-tunnel blockage and an explanation of the differences in dynamic-pressure measured by the two primary techniques. The correction offers a useful data improvement, being able to reduce the effect of model position on the measured drag coefficient. This method is under active development and can be expected to improve further.

REFERENCES

Cooper K.R. (Ed) 1996. Closed-Test-Section Wind Tunnel Blockage Corrections for Road Vehicles. *Soc. of Auto. Engrs*, SAE SP-1176, Warrendale, PA, USA.

Courchesne J. & A. Laneville 1979. A Comparison of Correction Methods in the Evaluation of Drag Coefficient Measurements for Two-Dimensional Rectangular Cylinders. *JFE* 79WA/FE-3, 1979.

Ewald B. (Ed) 1998. Wind Tunnel Wall Corrections. Ch. 6, pp. Bluff-Body Blockage Corrections in Closed- and Open-Test-Section Wind Tunnels (K. Cooper). *AGARD*, AGARDograph 336.

Farell C. & O. Guven & S. Carrasquel & V.C. Patel 1977. Effect of Wind-Tunnel Walls on the Flow Past Circular Cylinders and Cooling Tower Models. *Am Soc of Mech Engrs*.

Gould R.W.F. 1970. Wake Blockage Corrections in a Closed Wind Tunnel for One or Two Wall-Mounted Models Subject to Separated Flow. *ARC* R&M No. 3649, HMSO, London, UK, 1970.

Hackett J.E. & D.J. Wilsden & D.E. Lilley 1979. Estimation of Tunnel Blockage from Wall Pressure Signatures: A Review and Data Correlation. *NASA* CR-152,241.

Hackett J.E. 1996. Tunnel-Induced Gradients and Their Effect on Drag. *AIAA 34th Aerospace Sciences Meeting*, Paper 96-0562, Reno, NV, Jan. 1996. AIAA Trans, Vol. 34, No. 12.

Hackett J.E. & K.R. Cooper & M. Mokry 1998. AIAA Journal discussions, Feb., 1998.

Kuhn A. 1995. Ermittlung der Ansrtomgeschwindigkeit und der effektiven Windgeschwindigkeit in ¾-offen Fahrzeugwindkanalen. *ATZ Automobiltechnische Zeitschrift* 97 (1995) 2, Germany.

Macha J.M. & R.J. Buffington 1985. An Experimental Investigation of Wall-Interference Effects for Parachutes in Closed Wind Tunnels. *Sandia National Laboratories*, Report SAND89-1485, Albuquerque, USA.

Maskell E.C. 1965. A Theory of the Blockage Effects on Bluff Bodies and Stalled Wings in a Closed Wind Tunnel. *ARC* R&M No. 3400, HMSO, London, UK.

Mercker E. 1986. A Blockage Correction for Automotive Testing in a Wind Tunnel with Closed Test Section. J. Wind Eng. & Ind. Aerod. 22, 1986.

Mercker E. & J. Wiedemann 1996. On the Correction of Interference Effects in Open Jet Wind Tunnels. *Soc Auto Engrs*, SAE 96061, Warrendale, USA.

Mercker E. & G. Wickern & J. Wiedemann 1997. Views on Nozzle Interference Effects in Open Jet Wind Tunnels. *Soc Auto Engrs*, SAE 970136, Warrendale, USA.

Mercker E. & J. Wiedemann 1999. Betrachtungen uber die korrektur von windkanalinterferenzen in offenen Meβstrecken. *Proc. 3 Stuttgarter Symp*, FKFS, Germany.

Ramamurthy A.S. & C.P. Ng 1973. Effect of Blockage on Steady Force Coefficients. *Proc. Am. Soc. Civil Engrs., J. Engng. Mech. Div.*, Vol. 99, No. EM4, pp. 755-772.

Schwartekopp B. 1999. Private comm., *Daimler-Benz*, Germany.

Wind Engineering into the 21st Century, Larsen, Larose & Livesey (eds) © 1999 Balkema, Rotterdam, ISBN 90 5809 059 0

A general method for the correction of end effects on the aerodynamic drag of bluff, two-dimensional bodies in wind tunnels

K.R.Cooper, S.J.Zan & N.G.Ball
National Research Council of Canada, Ottawa, Ont., Canada

ABSTRACT: A correction procedure has been developed at the National Research Council of Canada (NRCC) that permits a correction to drag measurements for the three-dimensional flow distortions produced by end effects on a bluff, two-dimensional model mounted in a wind tunnel. The procedure is general because, once derived for a given set of end conditions, it can be applied to geometrically similar bodies as a simple analytical adjustment that is a function of measured drag and model aspect ratio.

1 INRODUCTION

The fundamental aerodynamic characteristics of bluff shapes are commonly determined by two-dimensional testing of the sections. The assumption made is that the body suffers no flow distortions due to end effects because the model is reflected to infinity at the wind tunnel boundaries. The reality is that leakage and three-dimensional flow effects at the wind tunnel boundaries cause distortions of the flow over the ends of a model and consequent errors in the measured aerodynamic drag. The most common boundary effects are leakage at the tunnel boundary, where a model connects to a balance and mounting system, and three-dimensional flows, usually in the form of horse-shoe vortices, at the junctions between the ends of the model and the tunnel boundaries due to model/wall-boundary-layer interactions. Separate endplates and end seals can minimise these effects but cannot completely remove them, resulting in unknown, residual bias errors in all two-dimensional measurements.

Several studies at the National Research Council of Canada (NRCC) have shown that leakage effects into the wake separation bubble are the major source of error (Ball 1990, Cooper et al. 1997). Leakage may distort the base pressure distribution more than 10 diameters from the end of a model. Thus, a model with an aspect ratio of 20 would have no region of two-dimensional flow.

A method has been developed to assess and remove this error. It requires a set of calibration measurements on a body typical of that to be tested - a circular cylinder, for example, in the case of electrical power conductor drag measurements - that provide the input data for the correction procedure. It can also provide insight into the measurement errors induced by the boundary condition chosen.

The procedure is based on measurement of the base pressure distribution over the length of a typical model and its correlation with the drag of the model. The base pressure measurements can be made with flush taps or, in instances where pressure tapping is difficult, with a probe inserted into the wake separation bubble from downstream of the model. Once the calibration measurements have been completed, data on similar bodies can be corrected for the majority of the error that occurs.

2 TYPICAL LEAKAGE EFFECTS

The NRCC has developed a sealing system following the work of Kubo et al. 1989, utilising 150-mm diameter disks, attached to each end of a model, that are fitted flush with the surface of separate end plates. This disk is adequate for circular models up to a diameter of 50 mm. Larger models would require a proportionally larger sealing disk. The end plates remove the tunnel floor and ceiling boundary layers while the disks seal the wake separation bubble against leakage. This system is shown in Figure 1. A single sealing disk has a tare drag coefficient of C_D=0.0024 leading to only a small drag bias that can be readily removed.

As part of the development of this sealing system, the effect of leakage into the wake separation bubble was investigated by the introduction of increasing annular gaps at each end of a 41 mm diameter, 30-strand conductor model.

Figure 1. 150 mm sealing disk set flush with the local floor.

2.1 *Base pressure measurement techniques*

The base pressure distribution was initially measured by a 3.2 mm diameter, 760 mm long total pressure probe inserted into the wake of the model from downstream. This method was used because it avoided pressure-tapping stranded models and because it was an efficient technique with which to scan many models. The probe was attached to a traverse that travelled over the lower half of the model. Calibration measurements had shown that this technique gave an acceptable measure of base pressure.

However, the traversing system produced a small, repeatable, lateral movement of the probe tip during the survey. This movement resulted in the s-shaped pressure distributions seen in Figure 2 over the central part of the model. The unaffected part of the model was difficult to identify under this circumstance and was taken as that region over which the pressure distribution was the same at each leakage.

Later, base pressure measurements were made on a smooth circular model using flush taps on the rear face of the model to remove this unwanted variability and to provide a more detailed data set.

2.2 *Initial probe base pressure measurements*

The span-wise base-pressure distribution was measured with a total pressure probe to assess the extent of two-dimensional flow on a conductor model. The distributions showed that the flow was distorted for more than 10 model diameters from the tunnel floor at sub-critical Reynolds numbers. The model aspect ratio was 44, so nearly one-half the model length was affected at the largest leakage gaps.

This behaviour is seen in Figure 2, which presents base pressure distributions as a function of gap size at $Re=1.47x10^5$. Drag was not measured simultaneously because model/balance deflections were sufficiently large to cause fouling at the smaller gaps and to change the gap size at the larger gaps when force measurements could have been made. The model was locked in place to ensure a uniform gap during these measurements.

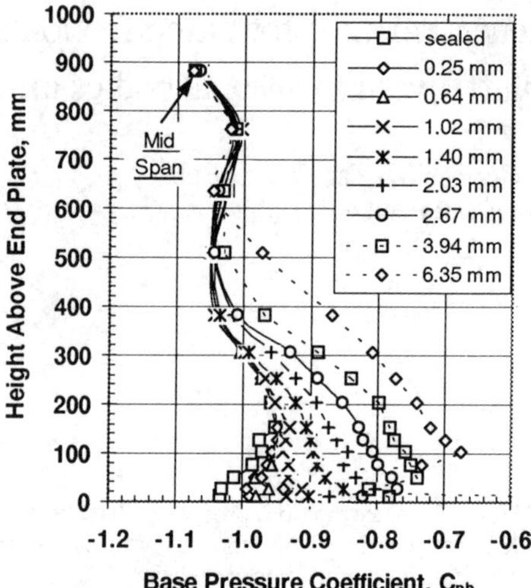

Figure 2. Base pressure distribution over one-half a stranded conductor model as a function of root leakage.

A set of simultaneous drag and base pressure measurements were then made at four Reynolds numbers with the 150 mm flush seals to quantify their behaviour. The data exhibited the linear correlation between drag coefficient and average base pressure coefficient shown in Figure 3.

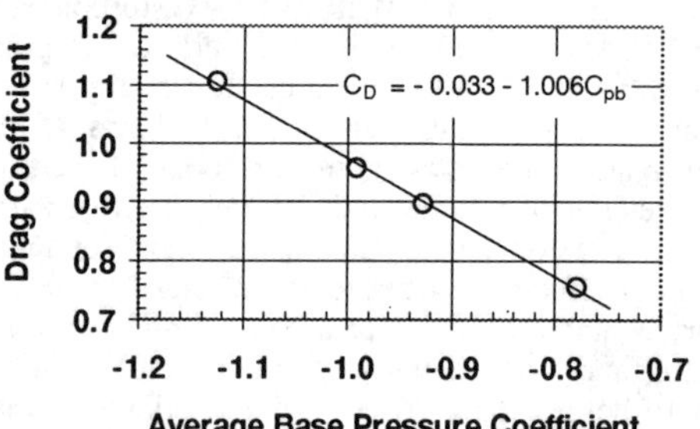

Figure 3. Correlation of average base pressure and drag coefficient for a model of a high-voltage transmission conductor.

This correlation suggested the possibility of estimating drag coefficient errors as a function of leakage from the base pressure data of Figure 2. It also suggested the possibility of a correction for three-dimensional flow effects using the measured base pressure distribution.

The drag deficits were obtained from the average pressures over the model ends, compared to the average pressure over the central portion of the model. The differences in average base pressure over these two regions, weighted by the appropriate model length ratios, were used with the slope of the correlation equation to compute the corresponding drag deficit on the model as,

$$\Delta C_D = -1.006\Delta\bar{C}_{pb} = -1.006(\bar{C}_{pbc} - \bar{C}_{pb})$$

$$= -1.006\left(\overline{C}_{pbc} - \overline{C}_{pbe}\right)\left(\frac{z_e}{L/2}\right) \quad (1)$$

where the overbar indicates a spatial average; C_{pb} is the base pressure coefficient; z is the length coordinate; L is the model length; and the additional subscripts 'e' and 'c' refer to the affected end region and the unaffected central region, respectively. The pressure over the central region was assumed to be the true, two-dimensional value.

Table 1 summarises the drag coefficients estimated from the average base pressures and the correlation equation in Figure 3. The table also presents estimates of the drag deficits (errors) due to end effects and of the corrected drag coefficients. The mean corrected drag coefficient was 1.010, with a standard deviation of 0.003. The largest measured drag error was 15.7% while complete sealing still left a drag deficit of 1.9% due to the root-junction vortices.

Table 1: Estimated Drag Coefficient Error Due to Root Leakage Using the Correlation Equation of Figure (3)

Gap (mm)	C_{pb}	C_D (Fig. 3 correlation)	ΔC_D	$C_{D,corr}$	% error
sealed	-1.014	0.987	0.019	1.006	1.9
0.00	-1.012	0.985	0.028	1.013	2.9
0.25	-1.011	0.984	0.025	1.009	2.5
0.64	-1.011	0.984	0.029	1.013	2.9
1.02	-1.004	0.977	0.033	1.010	3.4
1.40	-0.990	0.963	0.043	1.006	4.5
2.03	-0.977	0.950	0.061	1.011	6.4
2.67	-0.961	0.934	0.077	1.011	8.3
3.94	-0.937	0.910	0.096	1.006	10.6
6.35	-0.905	0.877	0.138	1.015	15.7

Base pressure and drag were measured simultaneously with the 150 mm sealing disks to quantify their performance. The uncorrected, measured drag coefficient was C_D=0.964 and the corrected value increased to C_D=1.002.

This preliminary study demonstrated the possibility of a correction to drag for end effects through the use of base pressure measurements.

3 DEVELOPMENT OF THE METHOD

It was expected that three-dimensional flow effects at the model ends would be non-dimensionally similar for a family of similar model with geometrically similar end conditions. Thus, a 25mm model with a 2mm annular clearance would be expected to have the same non-dimensional leakage effect as a 50mm model with a 4mm annular clearance. Once this leakage effect is determined for one model, it should enable drag measurements on both to be corrected. Even more useful would be a sealing condition that would allow many models to be tested without worrying about the root clearance or changes to it due to model and balance deflections. Such a seal is that supplied by the disk shown in Figure 1. As long as the separation bubble *closes on the disk*, the end condition seen by many models should be similar. If this is so, then base pressure measurements made with this sealing method on a typical model in a family should enable correction of the drag of all models in the family. The only limitations are that the model size not increase to the point that wake closure moves off the sealing plate and that the model aspect ratio not decrease to the point at which no central region of "two-dimensional flow" remains.

Figure 4 demonstrates the nearly identical non-dimensional base pressure distributions over the ends of conductors having diameters of 30 mm and 41 mm, supporting the previous assumption.

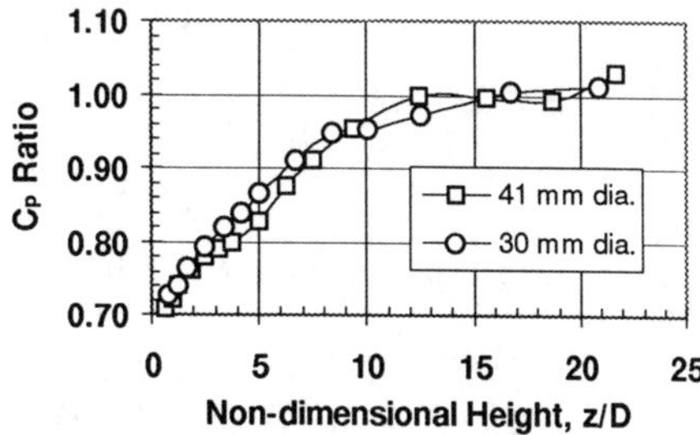

Figure 4. Comparison of pressure distributions on stranded conductor models with 150 mm flush sealing disks.

3.1 *Development of the correction equation*

The drag error due to three-dimensional flows induced by a set of end conditions is a function of many variables, including: the base pressure magnitude (drag), thickness of the end plate boundary layer and leakage at the root of the model.

For a given end condition that minimises wall boundary layer effects, the distortion due to the end conditions is primarily due to leakage flows into the wake of the model. Under these circumstances, it is expected that the magnitude of the distortion in the base-pressure field, and the resultant drag, would be a function of the magnitude of the base pressure in the central, unaffected region of the model.

Several assumptions must be satisfied for this to occur. These are:

1) That the sectional drag coefficient over any small element of the model span, $c_d(z)$, is a linear function of the local base pressure. That is,

$$c_d(z) = mC_{pb}(z) + b \quad (2)$$

where m and b are unknown coefficients that are universal for similar models and end conditions.

2) That each model has a sufficiently high aspect ratio for two-dimensional flow to exist over a cen-

tral region. A lower limit on the length of this region could be one-third of the model length.

3) That the distorted base pressure distributions over the end regions are functions of the average base pressure in the central zone only, for a given end condition.

4) That the end condition is geometrically scaled for each model, or is fixed and designed to provide the same leakage behaviour for each model.

3.1.1 *Analysis*

The average drag coefficient, C_{Dm}, measured on a conductor of finite length by a balance in a wind tunnel, can be written in terms of the sectional drag distribution, $c_d(\zeta)$, as,

$$C_D = \frac{1}{\eta}\int_0^{\eta} c_d(\zeta)d\zeta$$

$$= \frac{1}{\eta}\left[\int_0^{\eta_1} c_d(\zeta)d\zeta + \int_{\eta_1}^{\eta_2} c_d(\zeta)d\zeta + \int_{\eta_2}^{\eta} c_d(\zeta)d\zeta\right]$$

$$= \frac{1}{\eta}\left\{\int_0^{\eta_1}[C_{Dt} - \Delta c_d(\zeta)]d\zeta + \int_{\eta_1}^{\eta_2}[C_{Dt} - \Delta c_d(\zeta)]d\zeta + \int_{\eta_2}^{\eta}[C_{Dt} - \Delta c_d(\zeta)]d\zeta\right\}$$

$$= C_{Dt} - \left(\frac{1}{\eta}\right)\left\{\int_0^{\eta_1}\Delta c_d(\zeta)d\zeta + \int_{\eta_2}^{\eta}\Delta c_d(\zeta)d\zeta\right\}$$

$$= C_{Dt} - \left\{\left(\frac{\eta_1}{\eta}\right)\Delta C_{D1} + \left(\frac{\eta - \eta_2}{\eta}\right)\Delta C_{D2}\right\} \tag{3}$$

where C_{Dt} is the true, two-dimensional drag coefficient that exists over the central region of the model; Δc_d is the sectional drag deficit in the affected end regions; η = L/D is the non-dimensional length (aspect ratio) of the model, normalised by the model diameter, D; $\eta_1 = z_1/D$ is the length of one affected end of the model; $(\eta_2-\eta) = (z_2-L)/D$ is the length of the other affected end; and ΔC_{D1} and ΔC_{D2} represent the average drag errors over the ends of the model.

The limits of integration in equation (3) are not constrained to be exactly equal to the affected lengths, which are here treated as different, the most general case. Since the integrals are of the drag deficits over the ends of the model, which are zero outside the affected regions, they will not be in error if the integrations are performed over a distance longer than the affected lengths. There is a considerable simplification in the method if the two lengths are chosen to be the same and greater than the largest affected length under any expected test condition. Thus, no precise knowledge of affected length variation with base pressure magnitude is required for application of the method.

Without any error or loss of generality, the choice of affected end length can be made as $\eta_e/2 \equiv \eta_1 \equiv (\eta-\eta_2)$, where $\eta_e/2$ must be longer than the longest affected end length, and equation (3) reduces to,

$$C_{Dm} = C_{Dt} - \left(\frac{\eta_e}{2\eta}\right)(\Delta C_{D1} + \Delta C_{D2}) = C_{Dt} - \left(\frac{\eta_e}{2\eta}\right)\Delta C_D$$

which can be solved for the true drag coefficient, giving the correction equation,

$$\boxed{C_{Dt} = C_{Dm} + \left(\frac{\eta_e}{2\eta}\right)\Delta C_D} \tag{4}$$

A relationship is required between the true drag coefficient and the drag deficit to allow equation (4) to be applied. This relationship must be valid for a given set of end conditions and for a range of model sizes to be generally useful.

The drag deficit can be written in terms of the base pressure, through the drag/base-pressure correlation, as,

$$\Delta C_D = m[\Delta\bar{C}_{pb1} + \Delta\bar{C}_{pb2}]$$

$$= m[(\bar{C}_{pbc} - \bar{C}_{pb1}) + (\bar{C}_{pbc} - \bar{C}_{pb2})]$$

$$= m[2\bar{C}_{pbc} - (\bar{C}_{pb1} + \bar{C}_{pb2})] \tag{5}$$

The average pressures from the affected ends can be found from the average base pressure and the central average base pressure as follows

$$\bar{C}_{pb} = \frac{1}{\eta}\left[\int_0^{(\eta_e/2)} c_{pb}(\zeta)d\zeta + \int_{(\eta_e/2)}^{(\eta-\eta_e/2)} c_{pb}(\zeta)d\zeta + \int_{(\eta-\eta_e/2)}^{\eta} c_{pb}(\zeta)d\zeta\right]$$

$$= \left(\frac{\eta - \eta_e}{\eta}\right)\bar{C}_{pbc} + \left(\frac{\eta_e}{2\eta}\right)(\bar{C}_{pb1} + \bar{C}_{pb2}) \tag{6}$$

Using equation (6) in (5), and after some manipulation, the drag deficit becomes,

$$\Delta C_D = 2m\left(\frac{\eta}{\eta_e}\right)(\bar{C}_{pbc} - \bar{C}_{pb}) \tag{7}$$

where $\bar{C}_{pbc} = \left(\frac{1}{\eta - \eta_e}\right)\int_{(\eta_e/2)}^{(\eta-\eta_e/2)} c_{pb}(\zeta)d\zeta$

$$\bar{C}_{pb} = \left(\frac{1}{\eta}\right)\int_0^{\eta} c_{pb}(\zeta)d\zeta, \quad \bar{C}_{pb1} = \left(\frac{2}{\eta_e}\right)\int_0^{(\eta_e/2)} c_{pb}(\zeta)d\zeta,$$

$$\bar{C}_{pb2} = \left(\frac{2}{\eta_e}\right)\int_{(\eta-\eta_e/2)}^{\eta} c_{pb}(\zeta)d\zeta$$

3.1.2 *The drag correlation*

The correction procedure depends upon a linear relationship between drag coefficient and base pressure coefficient, as found in Figure 3. The final drag/base-pressure correlation was developed from the base pressure and drag data measured on both stranded and smooth circular cylinders. The drag measurements for these models are presented in Figure 5 and typical base pressure distributions for the smooth model can be seen in Figure 6. The pressure measurements were made with 55 flush taps on the model base of variable spacing to give higher resolution near the ends of the model.

A reduced critical Reynolds number on the smooth model was achieved by tripping the flow with thin strips of plastic tape at ± 45 degrees from the stagnation point. The drag curves show the expected behaviour, with the tripped model having a critical Reynolds number of 4.0×10^4.

The base-pressure distributions are more uniform than those measured with the probe and clearly show a pressure rise due to leakage and a negative pressure peak due to the root-junction vortex near the ends of the model. The distributions over both ends are different and the affected regions are smaller at the higher Reynolds numbers with the trip. Many of the tripped-model data points in the transitional Reynolds number region were unusable because no central region of constant base pressure was present due to varying flow transition over the length of the model.

The resulting drag/base-pressure correlation is shown in Figure 7. Data for the stranded and the smooth models collapse on the same curve, demonstrating the generality of the procedure.

The final step requires that the drag deficit be correlated with the true drag to permit evaluation of correction equation (4). The drag deficit was calculated from equation (7) while the true drag coefficient was obtained from the drag correlation equation of Figure 7 and the average central base pressure coefficient (defined just after equation (5)). The stranded and smooth model data align well, as seen in Figure 8, with the smooth model results concentrated at the subcritical and transcritical Reynolds numbers and the stranded model filling in between.

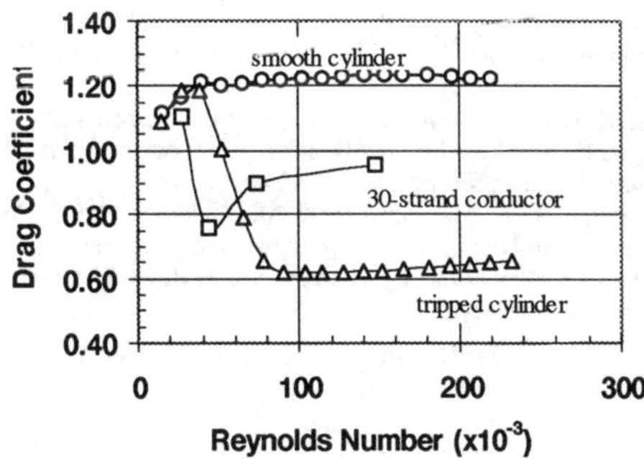

Figure 5. Drag behaviour with Reynolds number.

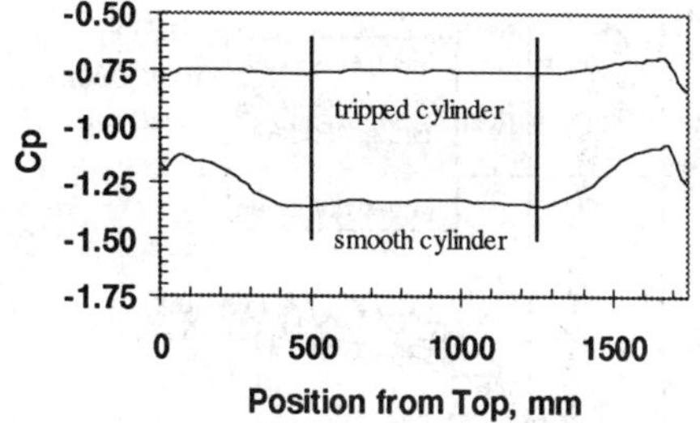

Figure 6. Typical base pressure distributions (Re = 2.1×10^5).

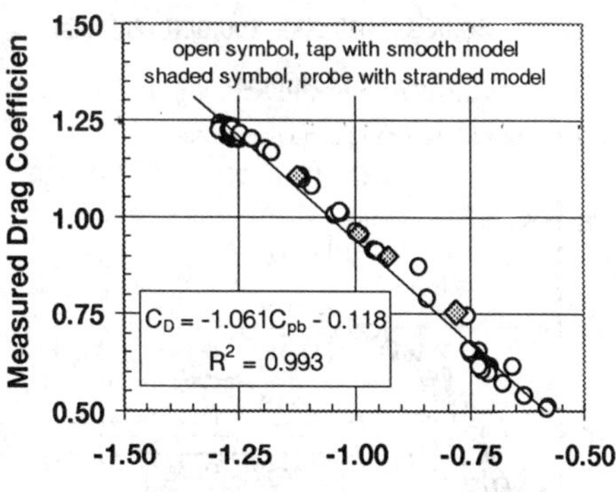

Figure 7. Combined drag/base-pressure correlation.

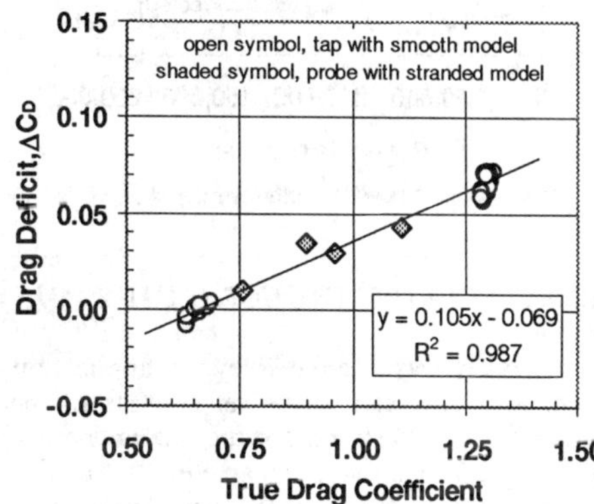

Figure 8. Correlation of the measured drag deficit with the true, two-dimensional drag coefficient.

When the correlation equation in Figure 8 is substituted into equation (4), the final correction equation is obtained as,

$$C_{Dt} = \left[\frac{C_{Dm} + b_\Delta \left(\eta_e / \eta \right)}{1 - m_\Delta \left(\eta_e / \eta \right)} \right] = \left[\frac{C_{Dm} - 0.069 \left(\eta_e / \eta \right)}{1 - 0.105 \left(\eta_e / \eta \right)} \right] \quad (8)$$

where the delta subscripts indicate constants from the drag deficit correlation of Figure 8.

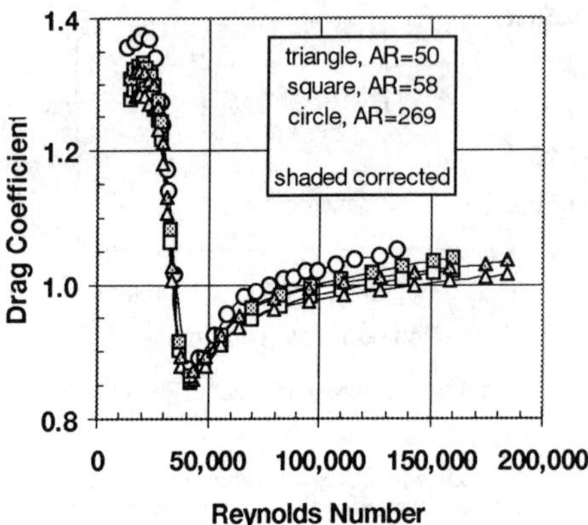

Figure 9. Corrected 24-strand conductor data

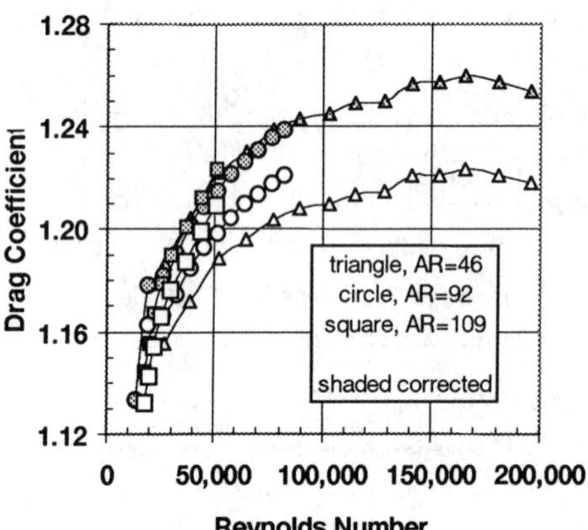

Figure 10. Corrected smooth circular cylinder data

4 APPLICATION OF THE CORRECTION

The correction was evaluated by its application to two available data sets. One set consists of measurements on two 24-strand conductors having aspect ratios of 57.7 and 49.6 diameter in the two-dimensional test rig in the 2m x 3m wind tunnel of the NRCC (Cooper et al. 1997). These measurements were compared with measurements on a 24-strand model having an aspect ratio of 269 in the 9m x 9m wind tunnel of the NRCC (Krishnasamy et al. 1991). The other set of data consists of measurements on smooth circular models having aspect ratios of 45.8, 91.8 and 109.1 in the two-dimensional test rig (Cooper et al. 1998).

The results are presented in Figures 9 and 10 for the stranded and the smooth models, respectively. Both graphs show an improvement when the correction is applied, in that the results at the lower aspect ratios in the 2m x 3m wind tunnel are brought into close agreement. The comparison of the 2m x 3m tunnel, stranded-cylinder results with the 9m x 9m tunnel result is still not perfect, although the correlation between the two tunnels is improved by the correction. Some of the residual differences may be related to model or tunnel influences other than end effects.

5 CLOSING REMARKS

A method has been described that corrects aerodynamic drag measurements for three-dimensional flow effects at the ends of a bluff two-dimensional model in a wind tunnel. The correction is in the form of a simple equation that is founded on the relationship between the sectional drag of a bluff model and the local base pressure. It can be generalised to families of similar models with a given type of end condition.

The procedure was applied to several trial data sets and was shown to be effective. The drag error obtained for bodies having drag coefficients near C_D=1.2 with one type of end conditions was found to be of order ΔC_D=0.035.

The calibration measurements were also useful in specifying the minimum model aspect ratio that would be permissible for application of the correction procedure. With the current test rig, this would be AR=39, assuming the minimum acceptable region of two-dimensional flow in the centre of the model would be one-third of the model length.

Finally, the correction equation predicts that a model aspect ratio of greater than 250 would be required with the 150 mm disk seals for the measured drag to be within 0.5% of the true value.

REFERENCES

Kubo, M., M. Miyazaki & K. Kato 1989. Effects of End Plates and Blockage of Structural Members on Drag Forces. *Journal of Wind Engineering and Industrial Aerodynamics*, Elsevier, the Netherlands.

Ball, N.G. 1990. *Determination of the Aerodynamic Drag of Overhead Electric Power Cables - Phase II*. LTR-LA-314, NRC, Ottawa, Canada. Controlled Distribution.

Cooper, K.R., S.J. Zan & N.G. Ball 1997. Wind Tunnel Measurements of Aerodynamic Forces on Single and Bundled Power Conductor Models, Including a Drag-Correction Technique for End Effects. LTR-A-008, NRC, Ottawa, Canada.

Cooper, K.R. & S.J. Zan 1999. Further measurements of the drag and base pressure distributions of smooth circular cylinders. To be published, NRC, Ottawa, Canada.

Krishnasamy, S.G., M. Tabatabi & M. Kastelein 1991. Response of Smooth Body, Trapezoidal Wire Compact Conductors to Wind Loading. Report No. B91-44-K, Ontario Hydro, Toronto, October 1991.

Wind Engineering into the 21st Century, Larsen, Larose & Livesey (eds) © 1999 Balkema, Rotterdam, ISBN 90 5809 059 0

Full scale and model studies of wind loads on a pair of circular cylinders

J.K.Galsworthy & B.J.Vickery
Boundary Layer Wind Tunnel Laboratory, Faculty of Engineering Science, University of Western Ontario, London, Ont., Canada

ABSTRACT: The paper describes a full scale investigation of the wind effects on a group of reinforced concrete chimneys conducted by the University of Western Ontario. Excessive across wind motions resulted in a failure in one of the stacks. The winds associated with the failure were in a direction that was inline with an upstream chimney that was roughly 25% larger in diameter with a spacing ratio of 5.1 diameters. Results suggested that the lift forces due to vortex shedding were amplified by a factor of 3.7 due to buffeting from the upstream structure. Subsequent to the full scale study, model tests were performed in the 1 MW wind tunnel at Monash University to further investigate the effect of buffeting. The maximum Reynolds number achieved was 10^6 compared to the full scale Reynolds of order 10^7. The experimental results confirmed, at least in a qualitative sense, that the shedding and buffeting peaks merge causing an increase in the total lift force.

1 INTRODUCTION

The present paper deals with wind effects on a pair of circular cylinders. Most data concerning flow interference between two cylinders was collected from wind tunnel experiments at subcritical Reynolds numbers and for identical diameters. The subject has been reviewed by Zdravkovich, 1977 and still attracts many researchers.

The results from such experiments are difficult to apply to full scale structures such as chimneys since Reynolds numbers can exceed 10^7. Also, the effects of surface roughness, turbulence and aspect ratio influence the loads. This emphasizes the value of full scale data. The data that exist are for the case of an isolated chimney free from interference effects and its application to grouped chimneys would be unwise.

The aim of the present paper is not to address the detailed fluid mechanics, but the wind engineering problem of the wind induced response of chimneys. Section 2 describes a full scale investigation of grouped reinforced concrete chimneys. A failure in one of the stacks occurred due to excessive across wind motions. The spacing ratio was large enough to suggest wake buffeting appeared to be the cause of the observed behaviour. In section 3, model studies of the buffeting effect are described. The results are presented for the case of a larger diameter cylinder upstream for Reynolds numbers in the upper supercritical range (Re = 10^6). Section 4 compares the results from both studies and shows that the model tests are consistent with the full scale observations.

2 FULL SCALE STUDY

The three chimneys studied were of the 'windshield' type where a reinforced concrete shell 'shields' the free standing unreinforced brick liner from the wind. The layout of the site is shown in Figure 1.

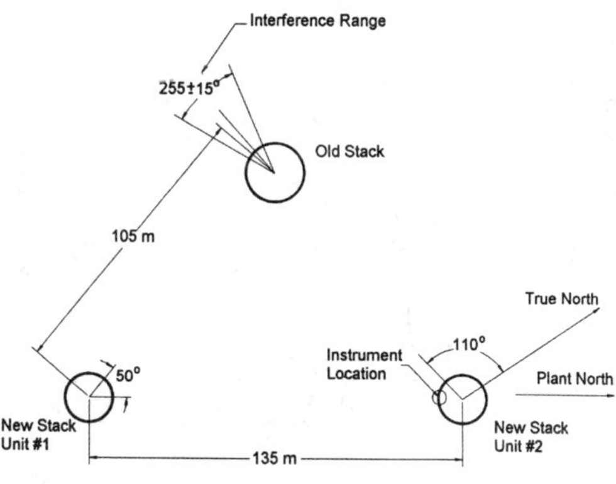

Figure 1. Site arrangement for full scale study.

Of the three chimneys, two were recently constructed (<1 year old at the time of study) and the third was taken out of service. The two new chimneys were identical and 250 m in height and the diameter at ¾ height was 16.5 m. The height of the older stack was 335 m and the diameter was roughly 25% larger at ¾ height of the new stack. The inline spacing ratio from the old to either of the new chimneys was 5.1 diameters and 8.2 diameters between the newer chimneys.

The study was initiated after a structural failure occurred in one of the new chimneys (Unit #2 in Figure 1). The failure consisted of an impact between the liner and shell at the top of the chimney. The impact was not 'hard' but steelwork and the chimney cap were both damaged. While the failure was not catastrophic, it did affect the serviceability of the chimney and the potential for a more severe impact between shell and liner was obvious. Evidence suggested that the failure resulted from across wind motions associated with a wind direction along the line from the old to the unit #2 chimneys.

Accelerometers were installed at a height of 170 m on the new chimneys on the liner and shell in both the North-South and the East-West axes, eight in total. The purpose of the instrumentation was twofold; first, to confirm and calibrate the frequencies and mode shapes and second, to determine displacements and predict the peak across-wind motions.

2.1 *Free Vibration Analysis*

An analysis of the undamped modes of vibration revealed that the motions of the liner and shell were coupled through a common foundation. In the first (liner) mode, the motions of the liner were in phase and exceeded those of the shell. In the second (shell) mode, the shell and liner motions are 180° out of phase with the shell motions exceeding those of the liner.

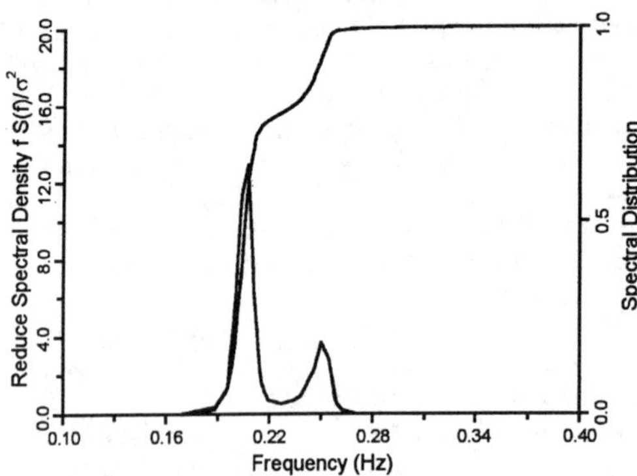

Figure 2. Spectrum of liner accelerations at 170 m.

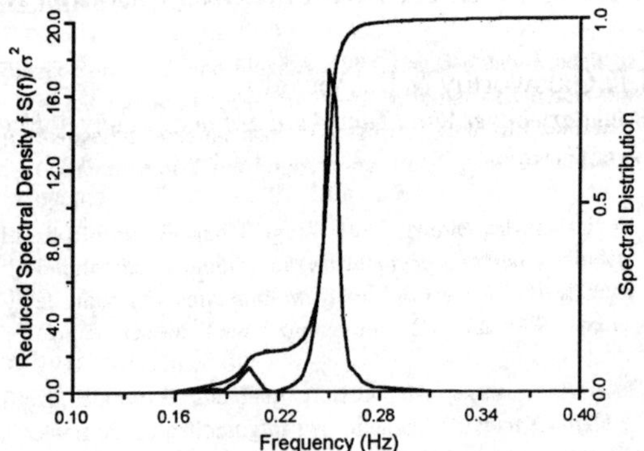

Figure 3. Spectrum of shell accelerations at 170 m.

Spectral analysis of the acceleration time series gave the frequencies and relative motions of the liner and shell in the two modes. Example spectra of the liner and shell accelerations are shown in Figures 2 and 3, respectively. The spectra reveal the two modes with frequencies equal to .205 Hz for the liner mode (f_{liner}) and .255 Hz for the shell mode (f_{shell}). Using the elastic moduli of the concrete and brick and the

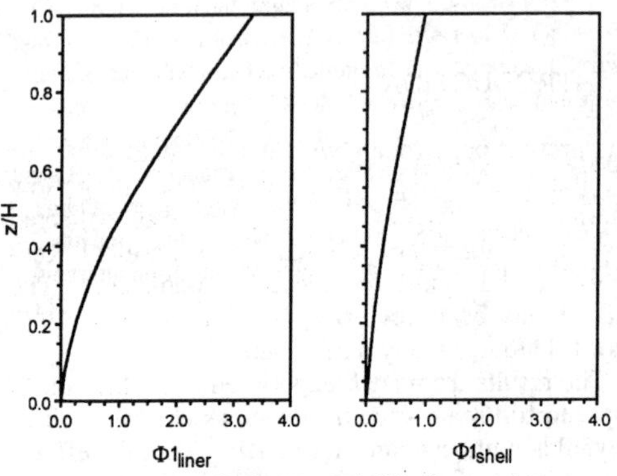

Figure 4. Mode shape for liner and shell in liner mode.

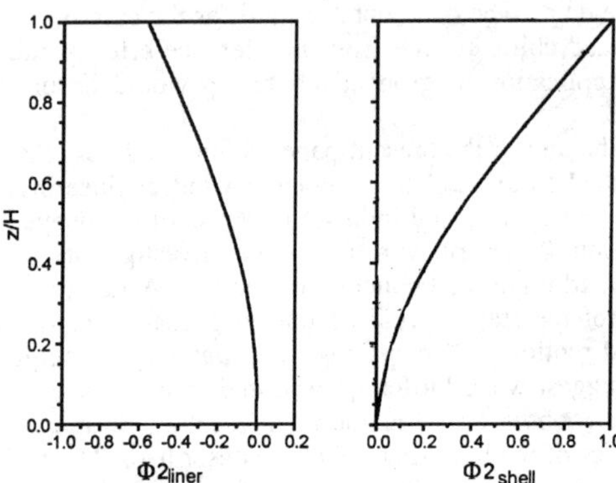

Figure 5. Mode shape for liner and shell in shell mode.

rotational stiffness of the foundation as variables, the results of the free vibration analysis were adjusted to match the measured relative motions and frequencies. The mode shapes are shown in Figure 4 for the liner mode and Figure 5 for the shell mode. Both mode shapes are normalized to unit tip amplitude for the shell. In the liner mode, the ratio of liner to shell tip deflections was 3.26 and in the shell mode the same ratio was 0.58. The shell mode dominated the response due to the large generalized mass of the liner mode.

2.2 *Prediction Model*

The results of the free vibration analysis were used in conjunction with the model of Vickery and Basu (1983) to predict the maximum expected tip amplitudes. Due to the short period of study (4 months), the wind speeds and directions associated with the maximum response were not observed. However, an effort was made to extrapolate to the maximum response from the observed events.

The variation of the root mean square (RMS) tip amplitude with wind speed is of the form:

$$\frac{\sigma_V}{\sigma_{V_{crit}}} = k^{\frac{3}{2}} \cdot \exp\left\{-\frac{1}{2}\left(\frac{1-k^{-1}}{B}\right)^2\right\} \tag{1}$$

where k is the ratio of the wind speed to the critical wind speed (V/V_{crit}). The bandwidth parameter, B, is a function of turbulence intensity and 0.35 was chosen as a representative value for the site (forested terrain). Figure 6 shows the behaviour predicted by the equation above where the velocity is normalized by the critical velocity of the shell mode defined by,

$$V_{crit} = \frac{f_{shell} D}{S_t} \tag{2}$$

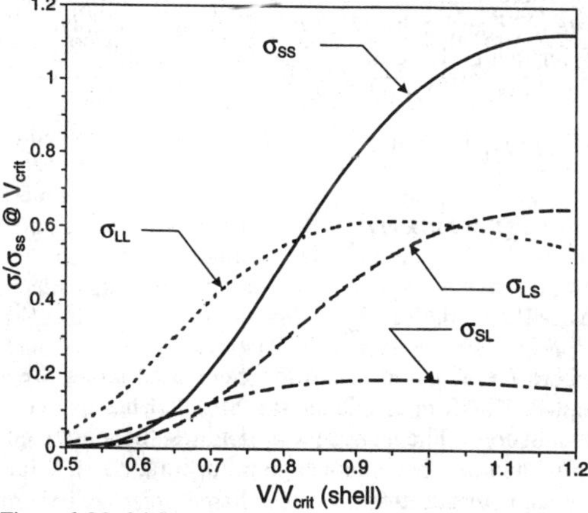

Figure 6. Model for extrapolating to maximum response.

where D is the diameter of the shell and S_t is the Strouhal number (S_t = 0.205), both at 5/6H. The amplitudes are normalized by the predicted shell deflection in the shell mode (σ_{ss}).

The values of σ_{SS}, σ_{LL}, σ_{SL}, and σ_{LS} were determined from the measured response spectra. These were used to define k as this ratio varies strongly with the ratio σ_{SS}/σ_{LL}. This enabled the estimation of the maximum response which occurs near k = 1.18.

2.3 *Results*

Over the period of study, eight usable (0.65< k < 0.80) records were obtained. Wind direction was estimated from an on-site anemometer (10 m) and the records from a nearby airport. The direction was also deduced from the response envelope (Figure 7), assuming the principal axes are aligned the along and across wind directions.

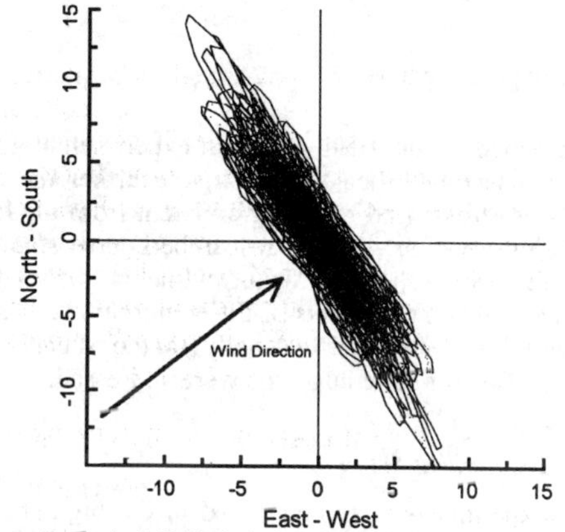

Figure 7. Typical X-Y plot of shell acceleration at the 170 m level

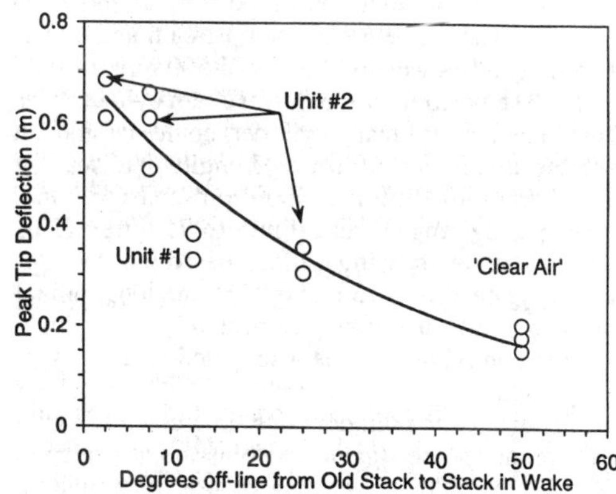

Figure 8. Variation of the predicted peak shell deflection with wind direction.

The predicted peak shell deflections are shown in Figure 8. 'Clear Air' refers to those observations where no wake interference occurred. The line shown in Figure 8 has no statistical significance and is meant only to show a trend.

The best estimate of the maximum shell deflection is 0.66 m for the wind direction inline with the old chimney. For wind directions where no significant interference is present, the best estimate is 0.18 m. This suggests that the amplification due to wake effects is equal to 3.7 with a probable range of ± 0.7. The relative motion between would be approximately 1 m, which would exceed the nominal clearance between the liner and shell, equal to 0.9 m.

A probability analysis of the records from the nearby airport suggested that winds causing these level of motions would be seen approximately once in five years. Therefore, under the most unfavourable conditions, damage to the steelwork and chimney cap would most likely be much worse than that experienced and a failure of the liner could be expected.

3 MODEL STUDY

Subsequent to the field study, an experimental program was established to investigate further the effects of across wind buffeting apparent from the full scale observations. Many researchers have studied interference between circular cylinders, but little data existed on the effects of different diameters and Reynolds numbers were usually in the subcritical range. Both of these concerns were addressed.

3.1 *Experiment Setup*

The experiments were conducted in the high speed working section of the 1 MW wind tunnel at Monash University. Dimensions were 5 m high by 4 m wide by 20 m long. Two circular cylinders spanning the width were placed in the flow as shown in Figure 9. The diameters of the upstream and downstream cylinders were 630 mm and 500 mm, respectively. The position of the downstream cylinder was fixed and the upstream cylinder could be moved with the use of a forklift truck. Longitudinal spacing ranged from 2000 mm to 3000 mm and the transverse spacing ranged from 0 mm to 750 mm. This gave maximum spacing ratios, based on the upstream cylinder, of 4.75 and 0.84 in the longitudinal and transverse directions, respectively. A single cylinder configuration was also tested as a basis of comparison.

The upstream cylinder was instrumented with pressure taps at the stagnation point and at angles of 90°, 180° and 270°. For the downstream cylinder, a ring of 32 taps at equal angles was installed on the centreline of the wind tunnel. Surface pressures were recorded for each tap at a sampling frequency of 200 Hz for 120 seconds.

Due to the size of the pressure scanning equipment, it was placed outside the wind tunnel and the required length of the tubes was 2.5 m. The resulting amplitude response was ±5% up to 20 Hz and rolled off to -40% at 40 Hz. The phase angle was linear in the range of interest (< 20 Hz). A digital filter was used to correct and low-pass (30 Hz) the time series prior to analysis.

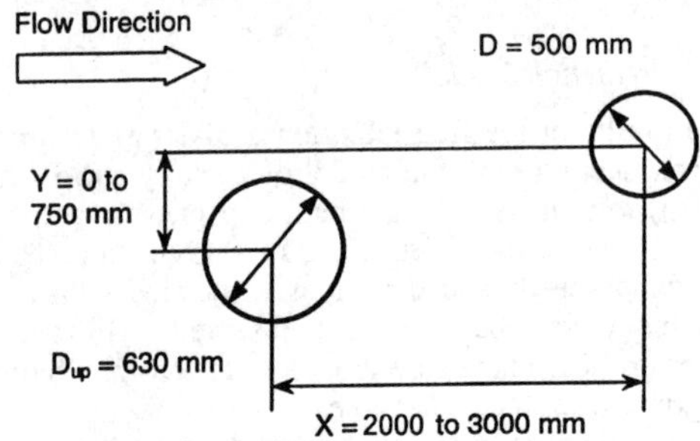

Figure 9. Setup for two cylinder configurations.

3.2 *Single Cylinder Results*

The Reynolds number for the single cylinder tests ranged from 2.2 x 10^5 to 10^6. The incident flow was uniform with a turbulence intensity of 5% and an integral length scale of 0.14 m. Measurements of the surface roughness indicated a mean height of approximately 2 μm giving a relative roughness of 4 x 10^{-6}.

Lift and drag coefficients were formed by integrating the pressure distribution around the ring for each time step. The resulting time series were analyzed and the mean drag and root mean square lift coefficients and the Strouhal numbers were calculated. The results were corrected for blockage effects using the empirical results of Modi and El-Sherbiny (1971). The correction formulae are:

$$C_{Dc} = C_D\left(1-1.404\cdot C_D\cdot\frac{S}{C}\right) \tag{3}$$

$$C'_{Lc} = C'_L\left(1-1.83\cdot C_D\cdot\frac{S}{C}\right) \tag{4}$$

$$S_{tc} = S_t\left(1-0.551\cdot C_D\cdot\frac{S}{C}\right) \tag{5}$$

where C_D, C'_L and S_t are the calculated mean drag and RMS lift coefficients and Strouhal number, respectively. The subscript c denotes the corrected value and S/C is the blockage ratio, equal to 10% for all test configurations.

Figure 10 shows the variation of the corrected coefficients with Reynolds Number. The drag coefficient increases with the Reynolds number, which is consistent with other measurements in the upper supercritical range. No discernable Strouhal number was observed below Re = 6.5 x 10^5, above this value, S_t is near constant and equal to 0.26. The maximum value of the RMS lift coefficient is near 0.3 at Re = 2.2 x 10^5 and approaches a constant value of 0.125 where strong vortex shedding returns.

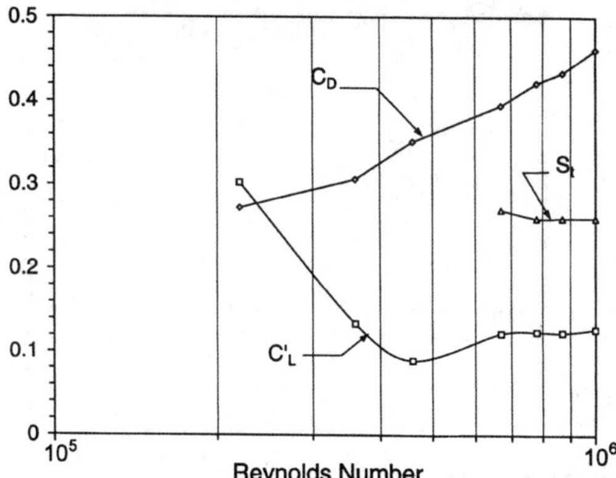

Figure 10. Variation of the mean drag and RMS lift coefficients and Strouhal number over the Reynolds number range tested.

The Strouhal number was determined from the computed spectra of the lift coefficients. The spectrum of the lift coefficient for Re = 10^6 is shown in Figure 11. The spectral peak was fitted to the equation,

$$\frac{fS(f)}{{C'_L}^2} = \frac{af}{\sqrt{\pi}Bf_s}\exp\left\{-\left(\frac{f_s - f}{Bf_s}\right)^2\right\} \tag{6}$$

where a is the fraction of the total variance under the peak, B is a bandwidth parameter and f_s is the central shedding frequency. The solid line shows the fit with the parameters shown in the legend. The abscissa is the reduced frequency based on the downstream diameter and the free stream velocity.

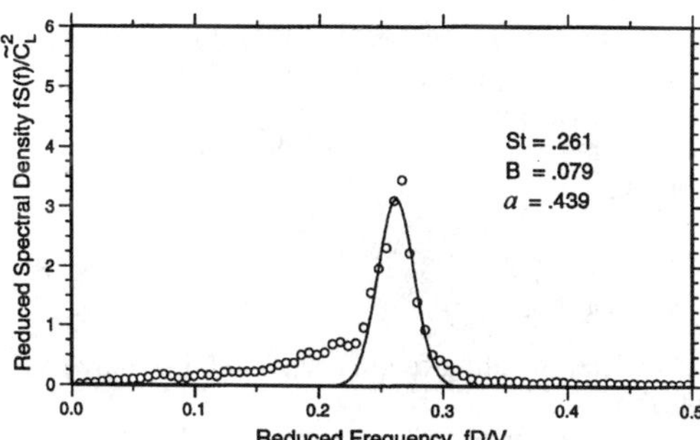

Figure 11. Spectrum of the lift coefficient for a single cylinder at Re = 10^6.

The spectrum reveals a significant amount of energy outside the peak. This is due to the lateral component of turbulence in the incident flow. A portion of the energy under the peak is also due to turbulence but no effort was made to distinguish between lift forces due to turbulence and those due to vortex shedding.

3.3 *Two Cylinder Results*

The results for the two cylinder configurations are presented for Re = 10^6, based on the downstream diameter. A total of sixteen cases were tested with the longitudinal spacing ratio, X/D_{up} between 3.2 and 4.75 and the transverse ratio, Y/D_{up} between 0 and 1.2.

The variation of the RMS lift coefficient is shown in Figure 12. For the tandem configurations, the increase in the lift coefficient due to wake turbulence is a maximum. This is expected since the lateral turbulence intensity in maximum on the centreline of the wake.

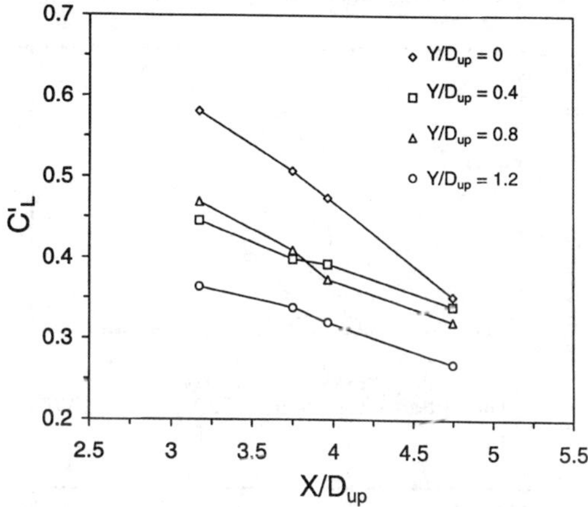

Figure 12. RMS Lift coefficient for two cylinder configurations.

The increase in the magnitude of the lift coefficient is significant for all cases. The maximum value is 0.58, which is close to 5 times the value measured for the single cylinder case. The minimum amplification measured is greater than 2. If the results are extrapolated to X/D_{up} = 5.1, the lift coefficient is roughly 0.31 which gives a multiplier of 2.5.

The variation of the Strouhal number for the same configurations is shown in Figure 13. Here, S_t is based on the downstream diameter and the free stream velocity as for the single cylinder case. For the tandem configurations, S_t is roughly 80% of the single cylinder value. The decrease is not as strong for the staggered configurations as might be expected. The behaviour around X/D_{up} equal to 3.8 to 4 could be due to a change in flow pattern but the exact cause is unclear.

The observed change in the Strouhal number can be explained by examining the spectra of the lift coefficient from the downstream cylinder and the spectra of the differential pressures at 90° and 270° from the upstream cylinder. The downstream lift

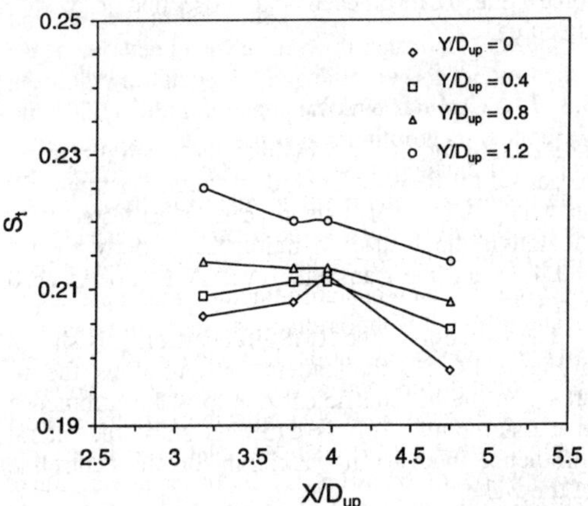

Figure 13. Strouhal numbers for two cylinder configurations.

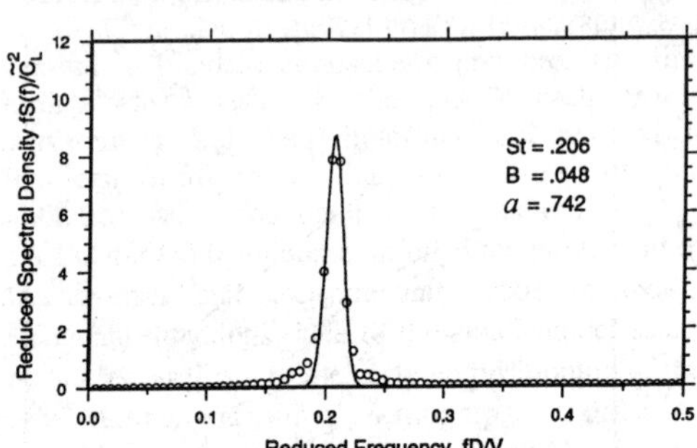

Figure 14. Downstream lift spectrum for X/D_{up} = 3.2 and Y/D_{up} = 0.

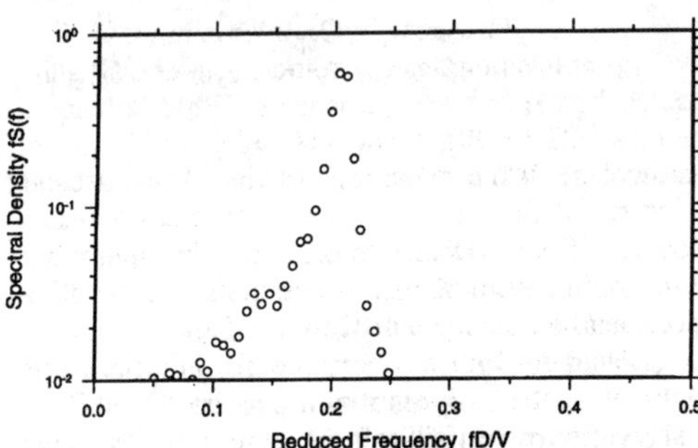

Figure 15.Differential pressure spectrum (90° and 270°) for the upstream cylinder for X/D_{up} = 3.2 and Y/D_{up} = 0.

and upstream differential pressure spectra for the tandem arrangement of X/D_{up} equal to 3.2 are shown in Figures 14 and 15, respectively. In both figures, the abscissa is the same as format as Figure 11.

The frequency of the downstream lift force is very narrowly banded about the Strouhal frequency of the upstream cylinder. This suggests a 'lock-in' type of behaviour where the two shedding frequencies coalesce.

Spectra for the tandem arrangement of X/Dup = 4.75 are shown in Figures 16 and 17. Here the same behaviour is observed with a slight broadening of the spectrum in Figure 16 compared to that in Figure 14.

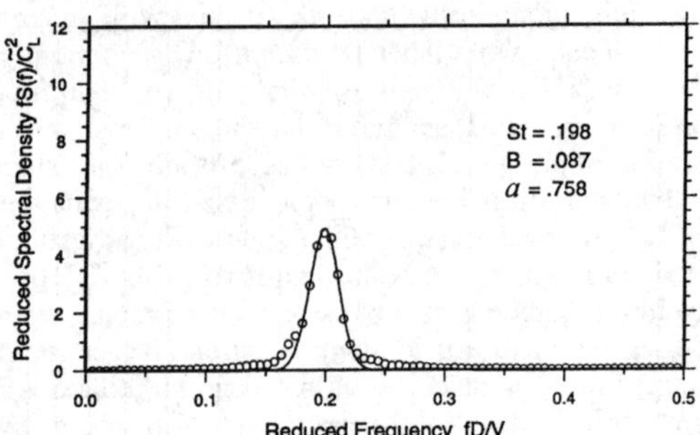

Figure 16. Downstream lift spectrum for X/D_{up} = 4.75 and Y/D_{up} = 0.

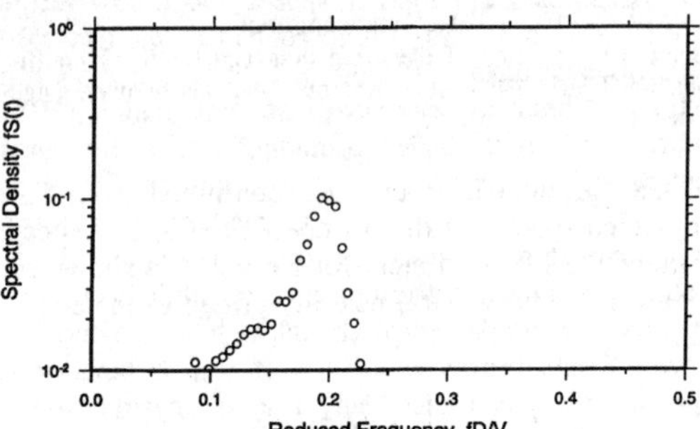

Figure 17. Differential pressure spectrum (90° and 270°) for the upstream cylinder for X/D_{up} = 4.75 and Y/D_{up} = 0.

It is unclear if the flow characteristics change between the two cases but an explanation of the latter at least is given. The diameter ratio,

$$\frac{D}{D_{up}} \cong 0.8 \tag{7}$$

suggests that, if the upstream cylinder is unaffected, the shedding frequency would be 80% of the single cylinder case. For spacing ratios greater than say 4, the expected mean velocity in the wake would be given as,

$$\frac{V_{wake}}{V_{freestream}} \cong 0.8 \tag{8}$$

and both the shedding frequencies would coalesce as the results show.

This explanation gives some insight into the cause of the full scale behaviour. The spacing ratio of 5.1 in full scale could not be duplicated due to space limitations but given the encouraging model study results, an attempt was made to compare the two studies.

4 COMPARISON OF FULL AND MODEL STUDIES

The across wind response of a chimney near the critical speed is of the form:

$$\frac{\sigma_{V_{crit}}}{D} = C\frac{C'_L}{S_t^2}\sqrt{\frac{\ell}{B}}\frac{1}{m_r}\frac{1}{\left(\zeta_s + \frac{Ka}{m_r}\right)^{1/2}} \tag{9}$$

where C is a constant for a given chimney and is a function of mode shape and geometry, ζ_s is the structural damping, m_r is the reduced mass, ℓ is a correlation parameter near 1.2 diameters and D is a characteristic diameter. The aerodynamic damping term, K_a is negative near the critical speed and for large reinforced concrete chimneys is equal to the value at small amplitudes of roughly -0.6.

A comparison can be made using the experimental data. Due to the lack of information to support otherwise, all variables in Equation 9 are treated as constants except for the RMS lift coefficient and the Strouhal number. The suggested amplification of the lift force is then given by the ratio,

$$\frac{\left(C'_L / S_t^2\right)_{\text{two cylinders}}}{\left(C'_L / S_t^2\right)_{\text{single cylinder}}} \tag{10}$$

where the lift coefficient for the single cylinder is taken from the model study. The total value is used as opposed to that predicted by the fitted curve in Figure 11.

The results of Equation 10 are shown in Figure 18. Also, included is the predicted ratio from the field study for the inline case.

Figure 18 shows that the amplification of the lift force decreases as the spacing between the cylinders increases. The magnitudes predicted appear to be somewhat higher than the full scale prediction but are not unreasonable given the uncertainties in both studies. If the curve for the inline model results were extrapolated to a spacing of 5.1 diameters, the amplification would be roughly 4.7. When compared to the range given in Section 2 (3.7 ± 0.7), this value lends some confidence to the model results.

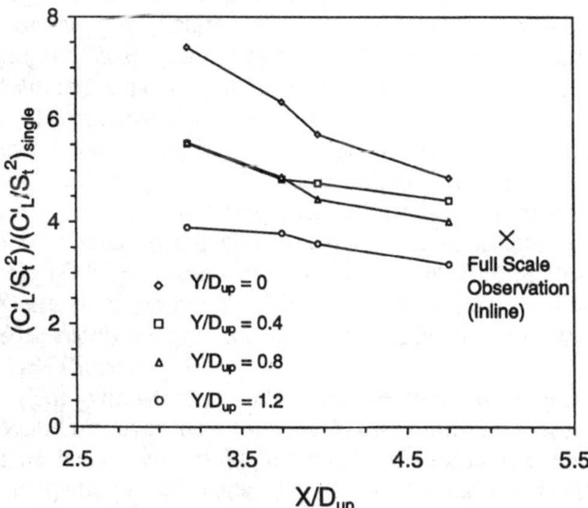

Figure 18. Predicted amplification of the lift force due buffeting forces.

5 CONCLUSIONS

The results from field and model studies of wind loads on a pair of circular cylinders are presented. These results show that lift forces due to wake turbulence amplify the lift forces associated with vortex shedding from an isolated cylinder. The field study suggested and the model study confirmed that the frequency of these forces coalesce. The magnitude of the force amplification from each study compared reasonably well considering the uncertainties.

These results stress the importance of including interference effects in assessing wind loads for grouped chimneys. The current state of knowledge is far from complete, especially at Re typical for full scale structure, and further field data are required.

ACKNOWLEDGEMENTS

A grant and postgraduate scholarship from the Natural Science and Engineering Research Council of Canada supported the work described herein.

REFERENCES

Modi, V.J. & S. El-Sherbiny 1971. Effect of wall confinement on aerodynamics of stationary circular cylinders. *Proc. of the 3rd Int. Conf. on Wind Effects on Buildings and Structures, Tokyo:* 365-375.

Vickery, B.J. 1995. A Field Investigation of the Wind-Induced Motions of Two 825 ft. Reinforced Concrete Chimneys Located Near Cheshire, Ohio. *Boundary Layer Wind Tunnel Report BLWT-SS20-1995.*

Vickery, B.J. & R.I. Basu 1983. Across-Wind Vibrations of Structures of Circular Cross-Section. Parts I and II. *J. of Wind Eng. and Industrial Aerodynamics.* 12:49-73.

Zdravkovich, M.M. 1977. Review of flow interference between two circular cylinders in various arrangements. *J. of Fluids Eng., Trans. ASME:* 618-633.

Wind Engineering into the 21st Century, Larsen, Larose & Livesey (eds) © 1999 Balkema, Rotterdam, ISBN 90 5809 059 0

Aspect ratio effects on the aerodynamics of short circular cylindrical structures

R. K. Garg
Structural Engineering Research Centre, Kamla Nehru Nagar, Ghaziabad, India
M. Kasperski & H.-J. Niemann
Building Aerodynamics Laboratory, Ruhr University Bochum, Germany
K. K. Chaudhry
Department of Applied Mechanics, Indian Institute of Technology, Delhi, New Delhi, India

ABSTRACT: Experiments have been conducted on models of finite height circular cylindrical bodies in simulated atmospheric boundary layers in wind tunnel. The effect of aspect ratio on various aerodynamic parameters such as drag, lift and cross-correlation properties has been studied. Although the results are presented for cylinders of aspect ratio up to 6, the study highlights distinct three-dimensional features such as the variation of aerodynamic parameters along the height of the cylinder. The results are not only useful for data base on aerodynamics of circular bodies but also for response studies of cylindrical structures.

1 INTRODUCTION

The aerodynamic behaviour of wind flow around circular cylinders in two-dimensional flow conditions has been extensively studied in the recent past. In reality, the wind flow around most civil engineering structures is of three-dimensional nature involving several features such as finite height, different end conditions, aspect ratio (AR), cross-sectional shape, non-prismatic shape (taper) besides the incoming flow conditions like uniform smooth flow or turbulent boundary layer flows. Limited results are available involving these features, but a comprehensive evaluation of structure flow interaction considering three dimensional phenomenon is lacking [see Basu (1986), Niemann & Hölscher (1988), Garg & Niemann (1995)].

Effect of free end on the aerodynamics of circular cylinders has been studied by Gould et al. (1968) on AR ≤ 12, by Okamoto & Yagita (1973) on $1 \leq$ AR ≤ 12.5, by Etzold & Fiedler (1976) on AR ≤ 8, by Farivar (1981) on AR ≤ 13, by Fox & West (1993) on AR ≤ 40, by Uematsu & Yamada (1994) on AR $\leq$ 5. The previous studies have revealed certain unique features of fluctuating pressure distribution and of vortex shedding (Strouhal number) variation for AR < 7. However, these results on vortex shedding pattern (needed for deducing lift force characteristics) were based on measurements of pressures along the generatrix or from wake flow measurements in low turbulence uniform flows which are approximate and are not based on actual fluctuating lift values.

The present study focuses on the influence of aspect ratio by carrying out experiments on circular cylindrical rigid models of different aspect ratios (AR = 1.71, 2.57, 4.28, 6.0 & 8.5) under simulated atmospheric wind environment in the boundary layer wind tunnel. An extensive study of the pressure field around the cantilevered cylinders of circular cross - section has been carried out. The pressures measured around the circumference of the cylinder and at various levels along the height of the cylinder are used to compute the mean, fluctuating and cross correlation of drag as well as lift forces. Effect of top end opening / closure of the cylinders has also been studied.

2 EXPERIMENTAL TECHNIQUE

The test model of circular cross-section was fabricated in modular form consisting of several segments. Each segment was made of wood having outer diameter as 70mm and of height as 60mm. By choosing an appropriate number of modules, it was possible to construct models of different aspect ratios viz. 1.71, 2.57, 4.28, 6.0 and 8.5, the results of cylinder AR 8.5 have been presented in Garg and Niemann (1995).

The experiments were carried out in a Wind Tunnel (WT) at Ruhr University, Bochum having test section 1.8mx1.6mx9.3m (breadth x height x length). The atmospheric boundary layers (ABL) were simulated by employing spires along with surface roughness on the floor of the wind tunnel. The power law index of simulated ABL was 0.17 and other details like turbulence intensity and length scale variation along the height can be seen in Garg et al (1997).

The fluctuating circumferential (at 12 points) pressures were measured on the model through piezoelectric type pressure transducers. The pneumatic tubes were optimized using techniques of Holmes and Lewis (1987) having admittance function nearly flat up to 100 Hz. The sampling rate was 1600 Hz with sampling time of 300 sec. The test Reynolds number lies in the sub-critical regime.

3 EFFECT OF DIFFERENT ASPECT RATIOS ON AERODYNAMIC PARAMETERS

Instantaneous drag and lift forces at a section on the cylinder have been deduced after integrating the pressure data measured at that section around the circumference with appropriate directional weighing functions. The aerodynamic coefficients have been normalised by local mean flow velocity. The Strouhal number, S(H) has been normalised by the mean velocity at the top of the model, $U_{(H)}$.

Results of studies for four aspect ratios namely 1.71, 2.57, 4.28 and 6 are being presented here.

3.1 *Mean drag characteristics*

The variation of mean drag coefficient, C_D based on local flow velocity along the height for cylinders of different aspect ratio is shown in Figure 1. The maximum value of the mean drag coefficient has been observed to increase as the aspect ratio of the cylinder increases. Presence of tip in drag curve indicates that the tip effect plays an important role in the aerodynamics of even low aspect ratio cylinders. It may be noted that during the tests each of the above mentioned model was exposed to a different flow conditions i. e. different depth of immersion within the same boundary layer in the wind tunnel. The effect of mean velocity profile is minimised by normalising the drag coefficient by local flow velocity. The lower the aspect ratio of the cylinder, the more intense shear and turbulence it experienced due to the boundary layer characteristics. The pattern of variation of mean drag coefficient along the height has been observed to be almost similar in all cases of AR, although in case of very low aspect ratio of 1.71, the reversal of the curve is not seen.

It has been observed (see Fig. 2) that the mean drag coefficient is relatively more within the top one diameter by 5% to 10% in the top open case when compared to a cylinder with the top closed case, for all aspect ratios studied here. However, the opposite trend has been observed in the lower 60% height of the cylinder, i. e. the drag value is higher near the base in case of the top closed model compared to the top open case. This counteracting influence of the top being open and the top being closed could be due to the formation of different types of secondary flows along the span in these two different end conditions.

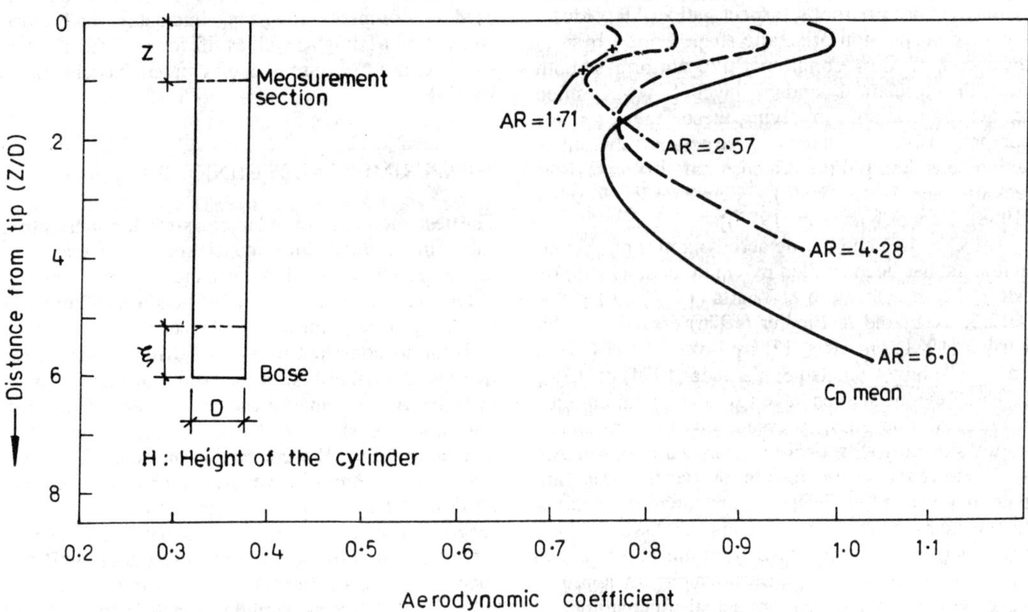

Figure 1. Variation of mean drag coefficient for different aspect ratios (top open case).

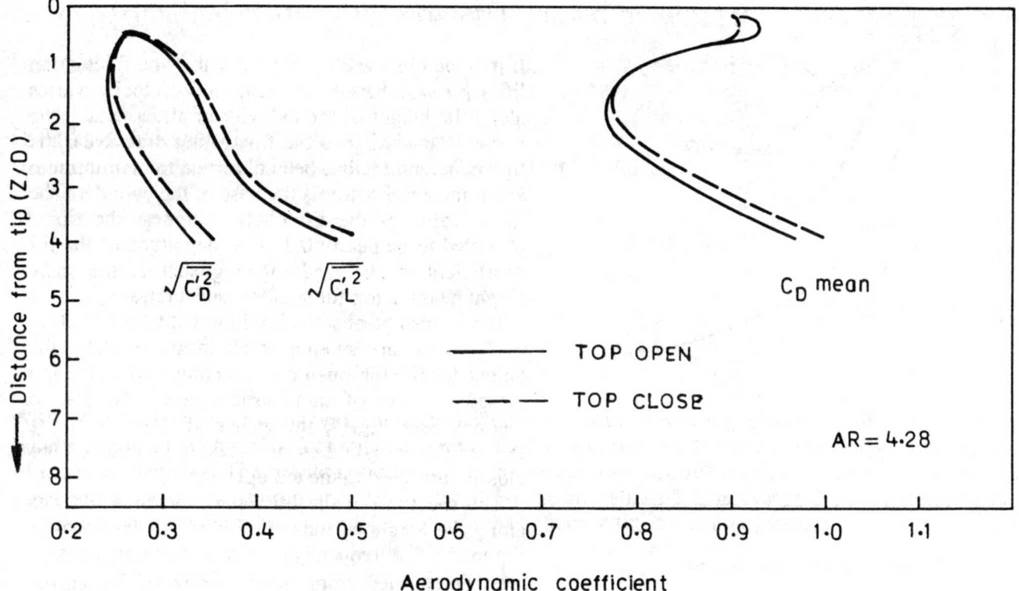

Figure 2. Comparison of aerodynamic coefficient for different end conditions of cylinder. (AR = 4.28).

It may also be emphasized that the observation reported in ESDU (No. 81017) that the mean drag would be constant along the height for cylinders shorter than AR = 5 has not been observed in the present study except for a cylinder of very low aspect ratio of 1.71. This is probably because of the fact that ESDU (No. 81017) values are based on the measurements of Okamoto & Yagita (1973) in uniform, low-turbulent flow while the present study has been carried out in thick turbulent boundary layer flows.

3.2 *Fluctuating drag*

The variation of fluctuating drag C_D *r.m.s* based on local flow velocity along the height for cylinders of aspect ratios 1.71, 2.57, 4.28 and 6 is shown in Figure 3. It has been observed that the pattern of variation of C_D*r.m.s.* along the height is almost similar in all cases of aspect ratios. The value reduces towards the base of the cylinders, the reduction being quite pronounced in case of aspect ratio higher than about 2.5. From Figure 2, it may be observed that the values of C_D *r.m.s* along the height of cylinders are greater in case of top closed compared to the top open case for all aspect ratios.

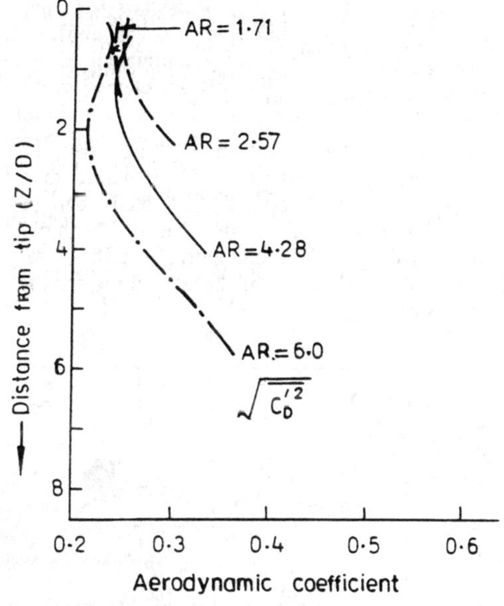

Figure 3. Comparison of C_D*r.m.s* for cylinders of various aspect ratios.

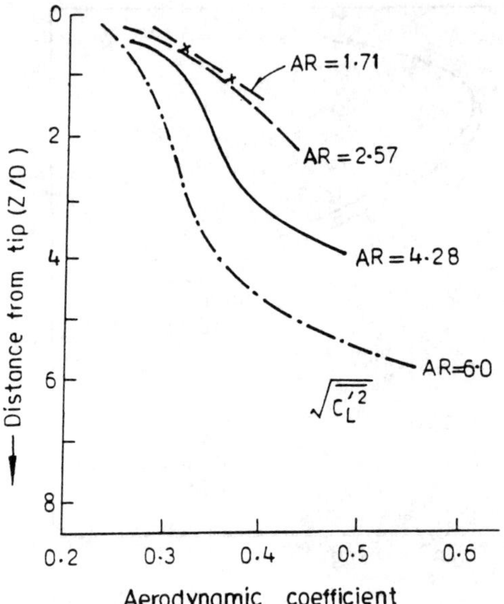

Figure 4. Comparison of $C_L r.m.s$ for cylinders of various aspect ratios.

It may been observed from this Figure that the magnitude of the fluctuating drag coefficient at any location (i. e. at given Z/D below the tip) of the cylinder is greater in magnitude at the same elevation for a lower aspect ratio cylinder.

From simultaneous measurement of pressures, the cross - correlation coefficient of fluctuating drag forces, ρ_{ij} between two vertically separated locations i and j along the span of cylinders has also been obtained. This drag correlation matrix is utilized for calculating the peak value of the drag force for the design of structures.

3.3 *Fluctuating lift*

It may be observed in Figure 4 that the fluctuating lift C_L *r.m.s.* based on local flow velocity varies along the height of the cylinder in all cases and the variation is similar to the fluctuating drag. Near the tip, its magnitude has been observed to be minimum and it increases towards the base of the cylinder. The *r.m.s.* value of the lift coefficient near the tip is observed to be nearly 0.3. The magnitude of the lift coefficient is observed to be greater at the same height near the top for smaller aspect ratios.

It may also be observed in Figure 2 that the values of C_L *r.m.s.* are smaller at all locations along the height for the top open case as compared to the top closed cylinder of same aspect ratio. Similar to $C_D r.m.s.$ (see Fig. 4) the values of $C_L r.m.s.$ along the height of cylinders are greater in case of top closed compared to the top open cylinder.

The spectrum of the lift force at various sections along the height of the cylinder has been shown in Figure 5. The frequency of vortex shedding, f has been determined from these spectra of lift corresponding to the spectral peak. The Strouhal frequency, S(H) is then obtained from Equation 1.

$$S(H) = f D / U_{(H)} \tag{1}$$

where D is the diameter of the cylinder and $U_{(H)}$ is the mean wind speed at the top of the model. Based on the nature of energy distribution in spectra, three regions along the span of a cylinder can be defined. Typical spectra for AR = 6 at various locations on cylinder at ξ /H = 0.9 above the base (i.e. near the top), at ξ/H = 0.5 (i.e. mid section) and at ξ/H = 0.1 (i.e. near the base) are shown in Figure 5. Near the tip, low intensity broad band spectrum of lift has been observed. This confirms the absence of regular vortex shedding in this zone. It has been observed

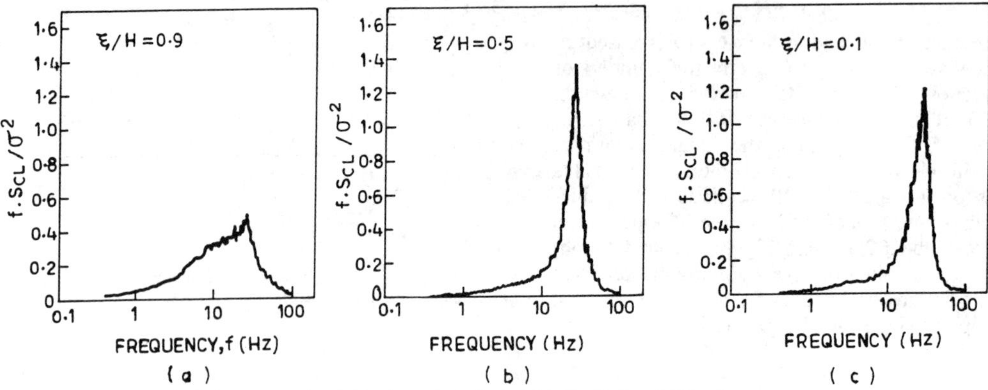

Figure 5. Spectra of lift forces at various elevations along the height of cylinder of AR = 6.

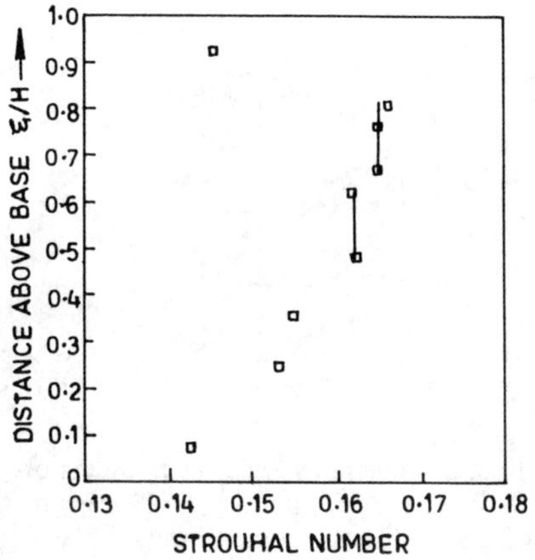

Figure 6. Strouhal number variation along the height of cylinder of AR = 6.

that the separation line near the tip is not a straight line and the wake becomes narrow at certain distance from the tip (see Holscher 1993). In fact it is shown by Tyack (1989) that the free shear layer is influenced by three– dimensional disturbances in the wake. However, this region does not extend deep. The spectrum at $\xi/H = 0.76$ (for cylinder having AR = 6) i. e. 1.4 D below tip, the spectrum contains significant energy and is narrow banded centred around the Strouhal frequency, S(H) of 0.154. Similarly, the narrow banded spectrum has also been observed at 1.29 D below the tip in all other aspect ratios including for AR = 1.71. In the central portion of the cylinder, the vortex shedding peak becomes narrower and the spectral magnitude at the peak increases near mid-section (i.e. $\xi/H = 0.5$) below which the magnitude of the spectral peak gradually reduces. This trend persists but with multiple peaks (see Fig. 5c) up to a level close to 0.58 D above the base and applies to all cylinders studied here. Below this level on the cylinder the data could not be observed. This suggests that the extent of the bottom portion where horse shoe vortices form is limited to the bottom 0.5D of cylinder. Etzold (1976) based on flow measurements has shown that there exists a bottom portion in which secondary flow dominates whose direction was observed towards base; this region extends 0.15D for AR = 1 and varies up to 0.75D for $5 \leq AR < 8$. The presence of secondary flow might be the reason of multiple peaks observed in the bottom portion of the cylinder in the present study (see Fig. 5a).

The variation of vortex shedding frequency, f normalised as Strouhal number, S(H) along the span of cylinders has been drawn and for AR = 6 is shown in Figure 6. The Strouhal frequency has been observed to vary along the height of cylinders of all aspect ratios in an almost similar fashion. A very low S(H) of 0.14 near the tip (within top 20% of cylinder height) has been observed. In the central portion of cylinders, the largest value of S(H) occurs immediately after the top zone somewhere a level near $\xi/H = 0.8$ which gradually decreases in magnitude towards the base. Near the base, the value of S(H) has been observed to be lowest and is marginally lower than the values near the top. The maximum value of S(H) has been observed near a level $\xi/H = 0.75$ above the base and, the maximum value of S(H) increases as the aspect ratio of cylinder increases. The cellular pattern of vortex shedding (region over the span of cylinder in which frequency of vortex shedding remains constant) has not been observed for cylinders shorter than 6. For AR = 6, there seems to exist cellular shedding of vortices in two cells within a region between $\xi/H = 0.4$ to $\xi/H = 0.8$. It may be noted that for AR = 8.5, two such cells of vortex shedding have also been observed (Garg & Niemann 1995).

4 ESTIMATION OF PEAK VALUE OF DRAG

Normally the drag coefficient is defined as average of local drag coefficients obtained at various locations of the cylinder. However, for design purpose the S.F. and B.M. should be computed taking into account the variation of C_D along the span of the cylinder. Therefore, a moment coefficient, $Cm_{(H)}$ for estimating the B.M. at a section Z_i along the cylinder can be defined as :

$$C_{m(H)j} = (1/H)\sum \frac{(\text{Drag force at i-th section})\ (Z_i)}{\frac{1}{2}\rho\ U^2_{(H)}\ A} \qquad (2)$$

Here, the coefficient $C_{m(H)}$ gives B.M. at a section Z_i when multiplied by $\frac{1}{2}\rho\ U^2_{(H)}\ AH$, and A is the exposed area of the structure, $U_{(H)}$ is the mean velocity of wind at the top (denoted as H) of the structure.

The maximum along wind load on the structure at elevation Z may be written as

$$W_{max\ (z)} = W_{(z)} + w_{max\ (z)} \qquad (3)$$

Where $W_{(z)}$ is the mean load and $w_{max(z)}$ is the maximum fluctuating load. The gust response factor (grf), $G_{(z)}$ may then be defined as

$$w_{max(z)} = G_{(z)}\ W_{(z)} \qquad (4)$$

where

Table 1. Comparison of gust response factor (for AR = 6, top open)

Source	Present study					ACI	CICIND
Peak gust factor	3.0	3.5	3.824	4.0	4.5	-	3.824
$C_{m\ max\ (H)}$	0.688	0.734	0.763	0.779	0.825	-	-
Gust response factor (grf)	1.67	1.78	1.851	1.891	2.0	1.787	1.794

$$G_{(z)} = [W_{(max)z} / W_{(z)}] - 1 \quad (5)$$

Peak value of the base bending moment coefficient, $C_{m\ (H)}$ considering the gust effect has been determined by using experimental data and the relation reported by Kasperski & Niemann (1992) as

$$C_{m(H)}\ max = C_{m(H)} + \rho_{ij}\ \sigma_{CDi}\ Z_i\ G_P \quad (6)$$

where G_P is the peak factor ranging from 3.0 to 4.5, Z_i : the lever arm being the distance from the base up to the centre of the equivalent sectional static force, σ_{CD} : standard deviation of the fluctuating drag, and ρ_{ij} : cross correlation coefficient between the drag forces at elevations *i* and *j*.

The peak value of $C_{m(H)}$ and the gust response factor have been computed for AR = 6 by utilizing the experimental values of σ_{CD} and ρ_{ij} based on Equation (2), (5) and (6) and, the values are given in Table 1. For the geometric scale of 1:350, the experimental model (AR = 6) used herein can be assumed to represent a full scale structure having outer diameter as 24.5 m, height as 147 m with the first natural frequency as 0.816 Hz. The peak factors could not be obtained in the experiment. However, a peak factor of 3.824 based on CICIND (1984) has been determined. Further the gust response factor has been computed based on recommendations of ACI-307 (1995) and CICIND (1984) and the comparison with the experimental values (on model) is presented in Table 1. It may be noted that the estimated value of the gust response factor is slightly under-estimated by 3.5% by these two standards compared to the value obtained in the present study for the same peak factor of 3.824.

5. CONCLUSIONS

From the analysis of experimental data from stationary circular cylinders of different aspect ratios namely 1.71, 2.57, 4.28 and 6 the following conclusions can be drawn.

1. The tip effect has been observed on cylinders of aspect ratio as low as 2.57, contrary to the recommendations of ESDU (81017) which limits *AR* as 5 for occurrence of tip effect.
2. The value of the maximum mean drag coefficient observed near the tip is found to increase as the aspect ratio of the cylinder increases.
3. The local mean drag coefficient is observed to be more within one diameter distance by 5% to 10% in case of top open cylinder compared to the closed top one.
4. The span wise distribution of the fluctuating drag along the height is similar to that of mean drag for the cylinders of aspect ratio greater than 2.5. The value of $C_D\ r.m.s.$ is found to be relatively more at the same elevation for a lower aspect ratio cylinder.
5. The span wise distribution of the fluctuating lift along the height is similar to that of mean drag. The value of $C_L r.m.s.$ is found to be relatively more at the same elevation for a lower aspect ratio cylinder.
6. Shedding of vortices has been observed on cylinders of low aspect ratios also. The Strouhal frequency, S(H) varies along the span of the cantilevered cylinders. The maximum value of S(H) increases for cylinders of greater AR.
7. Spectral analysis of lift forces reveals distinct three zones over the span of cylinders. Cellular vortex shedding could occur on cylinders of AR = 6.
8. Various standards may under-estimate (by order of 3.5% in the case of AR=6) the value of gust response factor of drag forces.

ACKNOWLEDGMENT

Garg gratefully acknowledge the support provided by CSIR-DAAD for the fellowship at RUB where the experiments have been carried out.

REFERENCES

ACI -307-95, 1995. Standard practice for the design and Construction of reinforced concrete chimneys. *American Concrete Institute*, Detroit.

Basu, R.I. 1986. Aerodynamic forces on structure of circular cross-section. *J.WEIA*. 24: 33-59.

CICIND 1984. Model code for concrete chimneys, Part A: The Shell. *Intn. Committee on Industrial Chimneys.*

ESDU, Data Item No. 81017, 1981. Mean forces, pressure and moments for circular cylindrical Structure, finite length cylinder in uniform and shear flow. *ESDU.* London, U.K.

Etzold, F & Fiedler, H. 1976, The near-wake structure of a cantilevered cylinder in a cross-flow. *Z. Flugwiss.* 24 (2) : 77-82.

Farivar, D. J. 1984. Turbulent uniform flow around cylinders of finite length. *AIAA J.* 19:275-281.

Fox, T.A. & West, G. S. 1993. Fluid - induced loading of cantilevered circular cylinders in a low - turbulence uniform flow. Part 1 : Mean loading with aspect ratios in the range of 4 to 30. *J. Fluid and Structures.* 7 : 1-4.

Garg, R. K. & Niemann, H.-J. 1995. Three-dimensionality of aerodynamic forces on circular cylindrical structures. *Proc. 9-ICWE.* 1:129-140, New Delhi.

Garg, R.K., Lou, J.X. & Kasperski, M. 1997. Some features of modeling spectral characteristics of flow in boundary layer wind tunnels. *J.WEIA.* 72:1-12.

Gould, R.W.E, Raymer, W.G. & Ponsford, P.J. 1968. Wind tunnel tests on chimneys of circular section at high Reynolds numbers. *In Proc. Symp. On Wind Effects on Building & Structures. Loughborough University of Technology*, Loughborough, U.K.

Holmes, J. D. & Lewis, R. E. 1987. Optimization of dynamic pressure measurement systems I : Single point measurement". *J.Wind Engineering & Industrial Aerodynamics (J.WEIA.)* 25: 219-273.

Hölscher,N.1993. Ein multivariator ansatz fur die aerodynamischer uebertragung function der winddrucke in atmospherischer grenzchicts anstromung. *Ph,D. Thesis, Ruhr University Bochum, Germany.*

Kasperski, M. & Niemann, H.J. 1992. The L.R.C. (load response aorrelation) method : A general method of estimating unfavorable wind load distributions in linear & non linear structural behaviour. *J.WEIA.* 43: 1753-1764.

Niemann, H.-J. & Hölsher, N. 1988. A review of recent experiments on the flow past circular cylinders. *Proc. BBAA-1.* Kyoto. 381-393.

Okamoto, T. & Yagita, M. 1973. The experimental investigation on the flow past a circular cylinder of finite length placed normal to the plane surface in a uniform stream. *Bull. J.S.M.E.* 16(95): 805-814.

Tyack,M. 1989. Development of visualisation techniques and their application in investigating flow around a cylinder of finite height. *B.Eng, Thesis, The university of queensland.*

Uematsu, Y. & Yamada, M. 1994. Aerodynamic forces on circular cylinders of finite height. *J.WEIA.* 51: 249-265.

Wind Engineering into the 21st Century, Larsen, Larose & Livesey (eds) © 1999 Balkema, Rotterdam, ISBN 90 5809 059 0

Pressure distributions on three circular cylinders in different equilateral-triangular arrangements at subcritical Reynolds Number

Zhifu Gu & Tianfeng Sun
Department of Mechanics and Engineering Science, Peking University, Beijing, People's Republic of China

ABSTRACT: Pressure distributions on three circular cylinders in different equilateral-triangular arrangements were investigated by wind tunnel test at a subcritical Reynolds number of 5.5×10^4. The spacing ratios of their centers were in the range $1.7\leq N/d\leq5.0$. It has been shown that the angles of incident flow strongly influence the pressure distributions on any one of the cylinders. Due to the interference of the shear layers and/or wakes, the most complex features of pressure distributions can be found on the cylinder located in downstream. As to the levels of interference regarding to the effect of spacing ratio, the results show mainly four different effect regions, i.e. effect of small, medium, transition and larger spacing can be identified. In small spacing, further more, three basic types of flow pattern can be classified due to various angles of incident flow. These are flow patterns of proximity effect, shear layer reattachment effect and wake effect.

1 INRODUCTION

In many cases of engineering practices, bodies often appear in the form of group, e.g. groups of buildings, chimneys, stacks, chemical reaction-towers, offshore platforms etc. Due to the mutual interference, the aerodynamic characteristics, such as pressure distributions and vortex-shedding patterns on each member of group are completely different from those of isolated one.

Circular cylinder is a typical bluff body and is one of the structural components mostly employed. Numerous investigations have been made of the flow past two circular cylinders, which is the simplest case of group, in the last two decades. An extensive review was given by Zdravkovich (1977) and by Ohya, et al. (1989). At subcritical Reynolds number, many interesting and unexpected phenomena were found, such as the biased bistable flow which occurred at specific ratios between the two circular cylinders in side-by-side arrangement and the fluid forces have a discontinuity for the two cylinders in tandem and staggered arrangements at certain spacing ratios. More detailed investigations of discontinuities of pressure distributions on two circular cylinders in staggered arrangement were recently given by Gu & Sun (1989,1996).

Nevertheless, three circular cylinders in equilateral-triangular arrangement is the typical case of cylinder group. It is extremely interesting to know whether the aerodynamics characteristics of two circular cylinders are still existed or some new features may happen due to the effect of the third one.

Gerhardt & Kramer (1981) invested the interference effects for groups of three or four stacks. Eastop & Turner (1982) studied three cylinders positioned in a straight line, either normal to or in the flow direction. Price & Paidoussis (1984) studied the more general case of three cylinders in a staggered arrangement. They found, in general, the effect of cylinder displacement on the fluid forces for cylinder in a group of three is very similar to that obtained with one cylinder in a group of two. It has been shown that, within a reasonable degree of accuracy, the force coefficients on one of three cylinders may be calculated using a principle of superposition from the force coefficients on one cylinder in a group of two. Sayers (1987) investigated the drag and lift coefficients occurring on cylinder in a group of three equispaced cylinders. He concluded that the angles of orientation to the free stream will strongly influence the force coefficients acting on any one of the cylinders and large reversals in lift magnitude and direction occur at certain inclination angles. Lam & Cheung (1988) studied phenomena of vortex shedding and flow interference of three cylinders in different equilateral arrangements, using a dye injection technique flow visualization at Reynolds numbers 2.1×10^3 and 3.5×10^3. Flow patterns were shot at 10° intervals from 0° to 60°. On the basis of wide and narrow wakes observed behind cylinders, they concluded that the 'bistable' flow characteristic exists at $\beta=0°$ for $N/d=1.27\sim2.29$ depends very much on the starting conditions. More recently, Yunseok & Changkoon (1993) reported the results of the close relationship between aerodynamic responses and pressure distributions on three circular cylinders.

This paper describes the pressure distributions on three equal circular cylinders arranged in an equilateral-triangular manner at different angles of incidence flow and spacing ratios subjected to a cross flow. Particular attention is paid to the classification of flow patterns change and the mechanism of interference between cylinders.

2. EXPERIMENTAL APPARATUS AND DATA REDUCTION

The experiments were conducted in a closed-return low-speed wind tunnel at Peking University. The tunnel has a test section 1.2m wide, 1m high and 8m long. The maximum speed is 36m/s with a turbulence intensity around 0.4%. Two turntables are mounted on the floor of the test section and they are located at the beginning of the test section 1.5m and 5.5m respectively. The first turn-table is being used for experiments in uniform smooth flow and the second one for the experiments in simulated atmosphere boundary layer with the aid of triangle spires, barriers and roughness elements mounted at the beginning of the test section.

Each cylinder of the group was a 640mm long aluminum tube of 48mm external diameter, with a machine-finished surface. Pressure taps were installed circumferentially every 10° around the cylinders at the mid-span. The cylinders with circular end plates of 800mm in diameter at the both ends, were mounted vertically in upper and lower adjustable guide-slots. The lower guide-slots fixed on the first turn-table in the test section. The whole assembly was then located at a distance of 1.5m downstream of the tunnel throat.

The measurement system of the surface pressure consisted of pressure transducers (PDCR–23d), a set of Scanivalve (SGM–48), three DC amplifiers (6M72), an A/D converter and a personal computer (IBM PC/XT). Three individual transducers were used for each cylinder such that the pressure signals at the same azimuth angle θ (see Figure1) of three cylinders under test were recorded simultaneously. A typical run took about 60 seconds to complete the data logging of all pressure taps for three cylinders. All tests were carried out at the wind speed of 18m/s with a Reynolds number 5.5×10^4 that is based on the diameter of single cylinder.

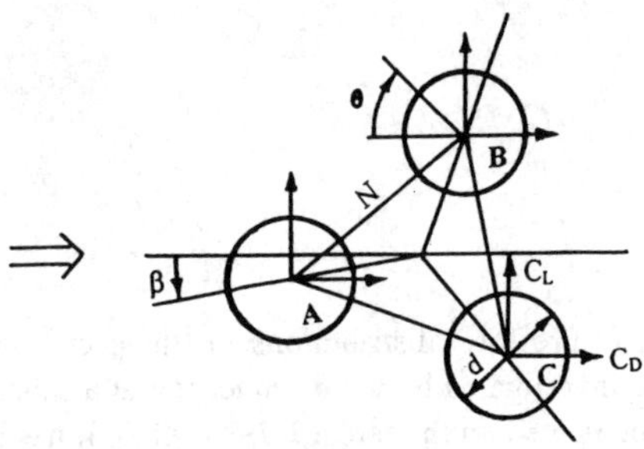

Figure 1. Scheme drawing of the arrangement of the three cylinders and the definition of force coefficients.

Figure 1 is a schematic diagram showing the arrangement of the three cylinders and definition of force coefficients. Here d is the diameter of the cylinder, N the distance between the centers of two cylinders, N/d the spacing ratio between the two cylinders, and β the angle of incident flow. As frequent reference will be made to the individual cylinders, the upstream and the downstream cylinders are labeled cylinder A, B and C respectively in later discussion. Pressure coefficient $C_p(\theta)$ is defined as $C_p(\theta)=[p(\theta)-p_\infty]/0.5\rho V_\infty^2$ where θ is the azimuth angle measured from the wind direction, positive clockwise, p(θ) is the pressure measured on the surface of cylinder, and p_∞, ρ and V_∞ are respectively the pressure, density and velocity of oncoming flow. The drag and lift coefficients, denoted by C_D and C_L respectively, are defined conventionally and are obtained by integrating appropriately the pressure distribution around the circumference of the cylinder at mid span.

No correction for the blockage effects has been made to the pressure data presented in this paper.

3. RESULTS AND DISCUSSION

In order to perceive the effect of group interference, the pressure distribution on single cylinder was measured first and was then compared with the pressure distribution for each of the three cylinders.

The time-mean pressure distribution on a single cylinder at Reynolds number of 5.5×10^4 is shown in Figure 2(a). The pressure distribution on both sides of the cylinder are quite symmetrical and it shows laminar separations at θ=±80°. It is a typical pressure distribution on circular cylinder at subcritical Reynolds number. In the experiments of the three cylinders, seven spacing ratios, i.e. N/d=1.7, 2.0, 2.2, 2.5, 3.0, 4.0 and 5.0, were carried out combined with angles of incident flow β varied from 0° to 60°. To achieve the goal of this research, the increment of β was not fixed, and it was carefully adjusted to the appropriate degree which was depended on the changes of pressure distribution on cylinders during the test.

As for the different levels of interference regarding to the influence of spacing ratio, mainly four different effect regions can be identified. These are effect of small, transition, medium and large spacing.

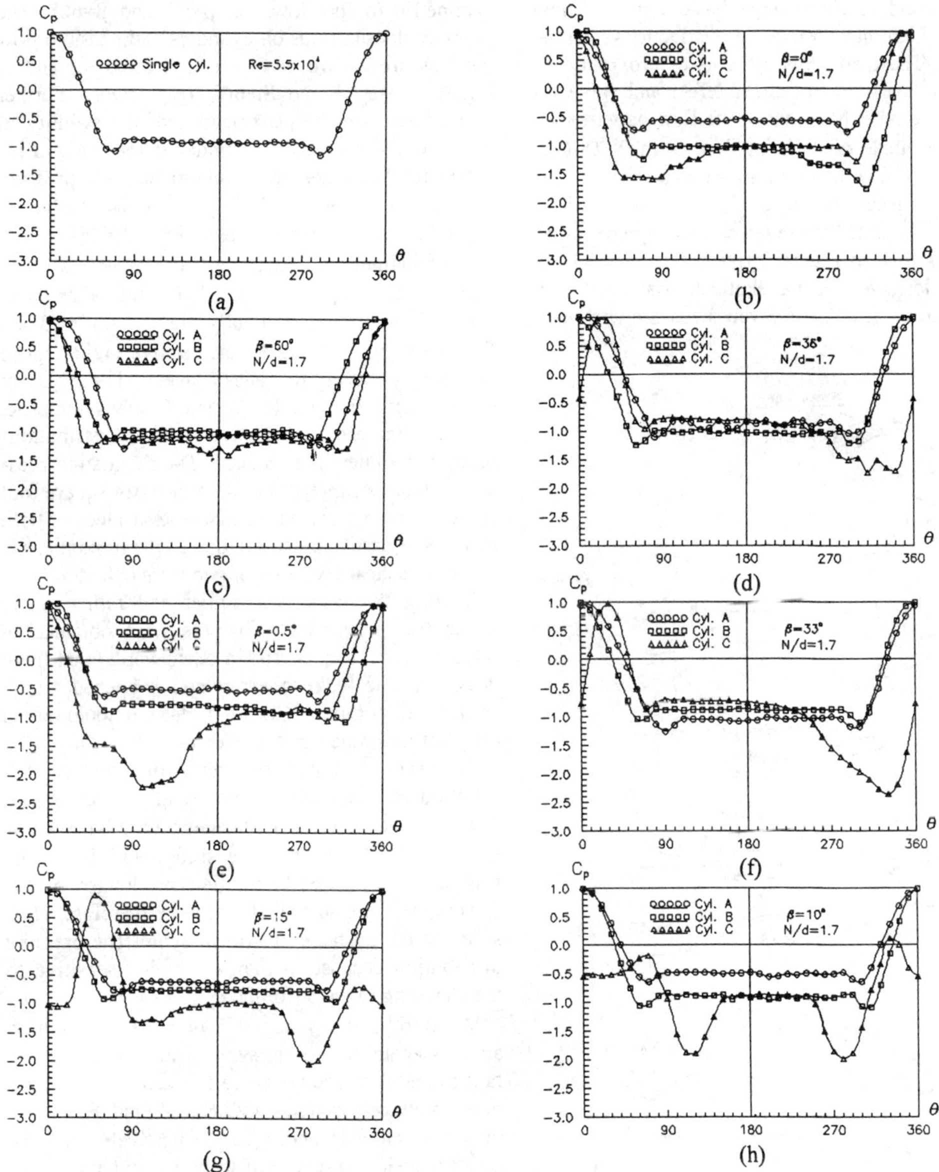

Figure 2. Time-mean pressure distributions of single cylinder and the three cylinders at seven typical angles of incident flow (N/d=1.7).

3.1 *Effect of small spacing: 1.7≤N/d≤2.0*

For small spacing ratio, such as N/d=1.7, the interference between three cylinders is very strong and complex. As the relative positions of cylinder vary with the angles of incident flow, the characteristic of interference vary significantly. However, the separated shear layers may interact directly at certain incident angles of flow. According to the pressure distributions on downstream cylinder varies with the angle of incident flow, three basic types of flow pattern, which have essential distinctions can be classified. These are flow patterns of proximity effect, shear layers reattachment effect and wake effect. On the other hand, when the flow pattern varies with the angle of incident flow from 0° to 60°, seven steps of flow pattern may be identified.

The time-mean pressure distributions on three cylinders at N/d=1.7 are given at seven typical cases of incident flow in Figure 2(b)~(h). In order to explain the flow phenomena, sketched drawing of the possible flow patterns of seven steps are shown in Figure 3.

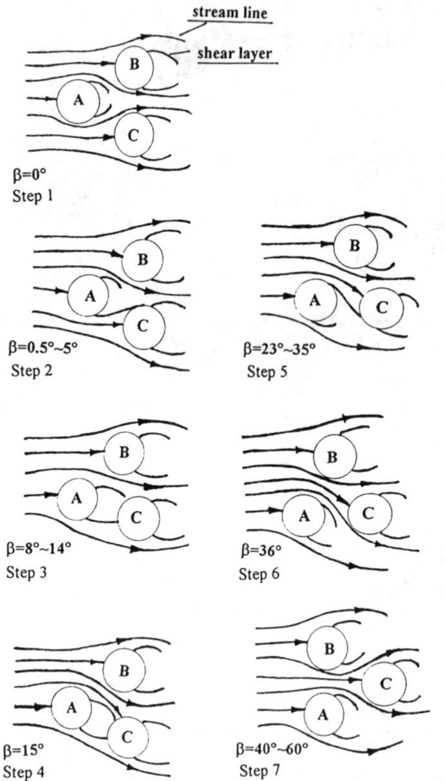

Figure 3. Sketch drawing of the possible flow patterns of seven-steps at small spacing (1.7≤N/d≤2.0)

3.1.1 *Patterns of proximity effect*

The characteristic of this type of effect is that the separated shear layers from each cylinder, as well as their wakes are not interaction directly, at least in the near wake of cylinders. Due to difference in angles of incidence flow, two types of pressure distribution, (A)symmetric and (B) asymmetric, can be subdivided.

(A)Symmetric distribution: This type of pressure pattern appears as the orientation of the group is symmetric to the flow, i.e. β=0°, and β=60°. The pressure distributions on cylinders and possible flow patterns are shown in Figure 2 (b) and (c) and in Figure 3 Step 1 and Step 7, respectively. Further more, based on the performance of the stability of pressure distributions on cylinder as the angle of incident flow changes, the stable or unstable pressure pattern may be identified. At β=0°, due to the blockage effect of downstream cylinders, the base pressure of cylinder A reduces to −0.5, which is much less than the single case of −1. On the other hand, the minimum pressure on gap-side of cylinder B and C increase to −1.5 due to the relative higher speed gap-flow between the shear layers. The pressure distributions on cylinder B and C are symmetric, however, the areas of positive pressure distributions on both cylinders are reduced. The fluctuating pressure, which is expressed as its root-mean-square (not show in this paper), on the gap-side surfaces of both cylinder B and C are strong due to the proximity effect of the shear layers separated from cylinder A.

During the tests, the similar and rather stable symmetric pressure distributions were obtained in different runs. It suggests, in other words, that there is no bistable wake phenomenon happened which mentioned by Lam & Cheung (1988). It could be the effect of difference in Reynolds number.

However, this type of symmetric pressure distribution only appears at very small region around β=0°. As the wind angle changes a little bit, such as β=0.5° which will be discussed detail later, the symmetric pressure distribution no longer exists. This type of pressure distribution is, therefore, classified to the unstable or critical symmetric pressure distribution because of they are highly sensitive to the changing of angles of incident flow.

At β=60°, the symmetric pressure distribution appears again due to the symmetric geometrical arrangement of a group of three cylinders. The stagnation points on upstream cylinder A and B turn to their the gap-side about 10° respectively, and the area of positive pressure distribution extends wider about 5° than that of single cylinder. The positive pressure area on cylinder C, on the other hand, becomes much narrower. The base pressure of the three cylinders are almost the same but higher than that of single one. The fluctuating pressure is stronger than that of single cylinder and the values of fluctuating pressure coefficients are greater than 0.3 on most regions of the cylinders surface. These features are significantly different from the case of β=0°. The symmetric pressure distribution maintains in a rather wide area of angle of incident flow and it could be observed up to β=40°. In the other words, this type of pressure distribution is not sensitive to the changing in angles of incident flow. It, thus, may be attributed to the stable symmetric pressure distribution.

(B)Asymmetric distribution. At a certain region adjacent to the stable symmetric pressure distribution area, e.g. β=36° see Figure 2 (d) and Figure 3 and Step 6, the flow around each cylinder keeps their individual complete wake structure. However, there is still no direct interference between the separated shear layers, as well as the wakes, at least in the near wake of cylinders. It is only the serious effect of proximity, which causes the pressure distribution asymmetrically. The interference between cylinders, in this case, causes the stream lines squeezed and distorted.

3.1.2 *Patterns of shear layer reattachment effect*

The characteristic of this type of effect is that the shear layer separated from upstream cylinder acts directly on one side of downstream cylinder.

At the boundary regions of unstable or stable symmetric pressure distributions, such as at β=0.5° or at about β=33°, the shear layer separated from cylinder A reattaches on gap-side or outer side of cylinder C respectively. Due to the reattachment of the relative high velocity and vortex in the shear layer, the pressure distributions change dramatically on cylinder C (see Figure 2(e) and (f), also see Figure 3 Step 3 and Step 5). The large area of strong fluctuating suction pressure is formed, and the peak value of fluctuating-pressure coefficient may reach 0.4. The rather strong fluctuating lateral force results, which inward the centerline of group in the case of β=0.5° or outward in the case of β=33°. During the test, a probe of hot wire was placed at the near surface region where the strong suction pressure presented. The signal of velocity fluctuating was sampled and processed later. A dominant peak value could be detected in the power spectra, which is corresponding to a Strouhal number of 0.4. It is suggested that the appearance of the great suction on one side of cylinder C is associated with some regular vortex shedding with rather high frequency. It is similar with the case of two circular cylinders in staggered arrangement (e.g. see Gu (1996,1998)).

Because of the changes in orientation of incident flow, the shear layer separated from cylinder A is some away from cylinder B, the pressure distribution on cylinder B approaches to that of single one case.

3.1.3 *Patterns of wake effect*

At the angles of incident flow between the two cases of shear-layer reattachment effect regions which are mentioned above, the interference on cylinder C may be classified as effect of wake. The characteristic of this type of effect is that cylinder C is partially or entirely submerged in the wake of cylinder A, which corresponding to the pressure distribution patterns on cylinder C with single-peak (e.g. β=15°) or double-peak (e.g. β=10°) respectively.

At β=10°, particularly, the double-peak pressure distribution pattern on cylinder C is almost symmetric. It is very similar to the case of downstream cylinder of two circular cylinders in tandem arrangement, except the symmetric line shifts about 30°. The angle between the incident flow and the centerline of cylinder A and C is 20°. It could be interpreted that it is the proximity effect of cylinder B that causes the wake of cylinder A to decline about 20°. The pressure distributions of single-peak and double-peak patterns on cylinder C and the possible flow patterns are shown in Figure 2(g) and (h), and in Figure 3 Step 4 and Step 3, respectively. It should be mentioned that, according to Yunseok & Changkoon (1993), six-steps of flow pattern were determined. However, the Step 4 in this paper, which corresponds to the pressure distribution of single-peak on cylinder C could be add. At this stage, the free stream could reach on the surface of cylinder C directly as it is partially submerged in the wake of cylinder A.

Figure 4 shows the drag and lift coefficients of cylinders as the group of cylinders is rotated through 60° about its axis. The curves of drag and lift force coefficients, particularly the curves of lift force coefficients of cylinder C reveal the process of the four different interference regions.

At small spacing region (e.g. N/d=1.7, see Figure 4(a)), the values of drag coefficient C_D of cylinder A and B are almost the same and increase slowly from 0.8 at β=0° to 1.2 at β=60°. The values of C_D for cylinder C decreases from 1.0 at β=0° to its minimum value of −0.2 at β=25° and then increases. At β=60°, it is slightly less than that of cylinder A and B.

The lift force coefficients C_L on both cylinder A and B are small. As $\beta>20°$, the C_L of cylinder B increases slightly as the angle of incident flow increase, the reverse is true for cylinder A, and they reach the values of ±0.15 at $\beta=60°$ respectively. On the other hand, C_L on cylinder C changes significantly, particularly at small β, such as $\beta=0.5°$, C_L may reach a value of +0.8, but it drops to −0.2 abruptly at $\beta=7\sim8°$, and then it maintains the same value up to $\beta=14°$. It drops again and reaches its minimum value of −1.3 at $\beta=25°$, and then recovers gradually up to zero at $\beta=60°$.

In summery, the C_D of cylinder A and B have no significant difference and also have no significant changes with various angles of incident flow. The significant changes in drag and lift forces on cylinder C is caused by different effects of interference. It is interesting that both the minimum drag and lift

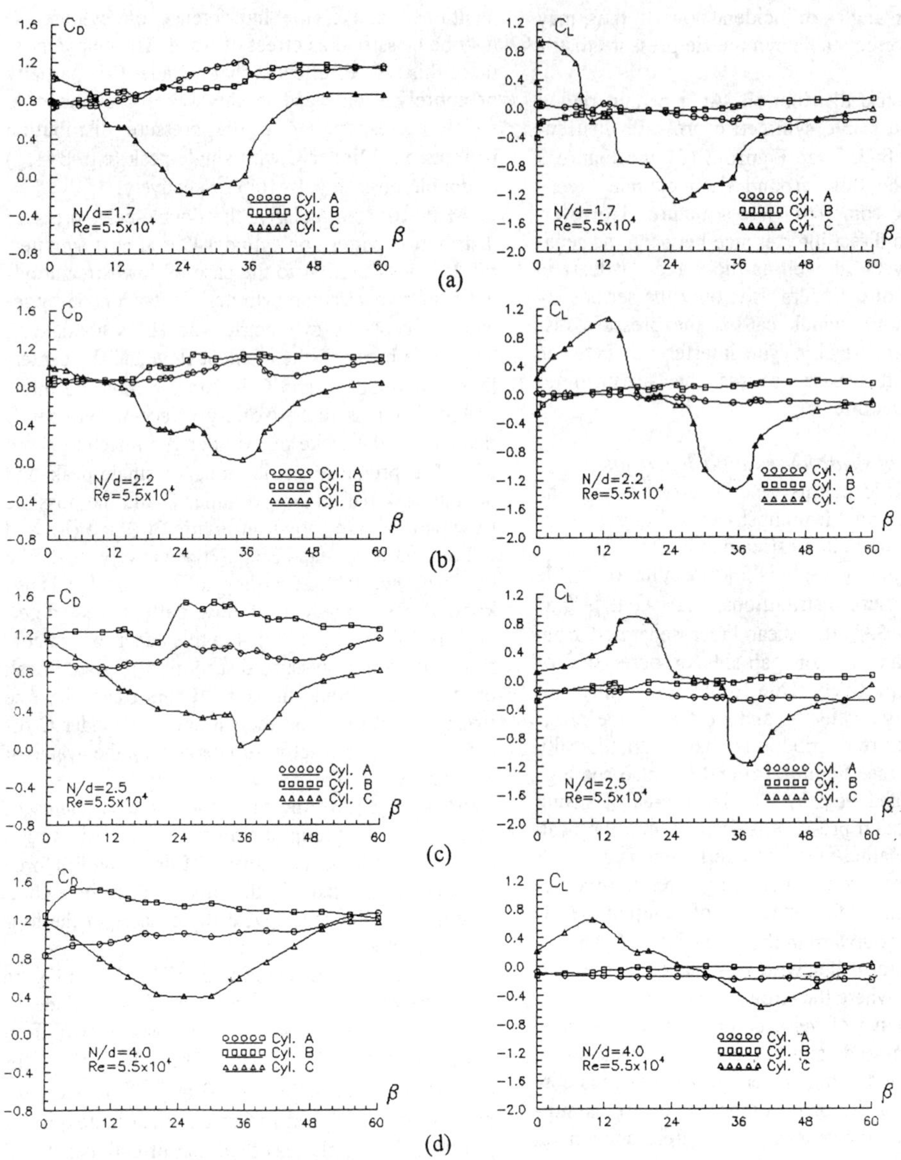

Figure 4. Drag and lift force coefficients C_D and C_L on each of the three cylinders changing with the angles of incident flow for the four typical spacing ratios.

forces (its absolute value) on cylinder C occur almost at the same angle β=25°, thus, a significant outward force results. At small angle of β (β<10°), a significant inward resultant force, which is combined with rather strong lift (C_L=0.8) and drag (C_D =1.0) forces, also results. It is shown that the effect of shear layer reattachment is the most serious case of interference between three cylinders.

3.2 *Effect of medium spacing: 2.5≤N/d≤3.0*

As the spacing ratios increases to a certain value, such as N/d=2.5, the pressure distributions on cylinders show rather different features compared with the case of small spacing. At the medium spacing ratio, the shear layer separated from cylinder A no longer acts on cylinder C directly and, thus, the effect of shear-layer reattachment pressure distribution no longer exists. However, the suction area, which is weaker than the case of shear layer reattachment in small spacing, on outer side of cylinder C still presents, which is caused by the gap-flow of relative high velocity adjacent to the separated shear layer of cylinder A.

The symmetric type of pressure patterns also can be expected to happen at β=0° and β=60°, however, they reveal that the effect of proximity reduces greatly. In the region of wake effect, due to the increment of spacing, the wake behind upstream cylinder could be developed more sufficiently, the wake influence on cylinder C is stronger than the case in small spacing in respect to the fluctuating pressure.

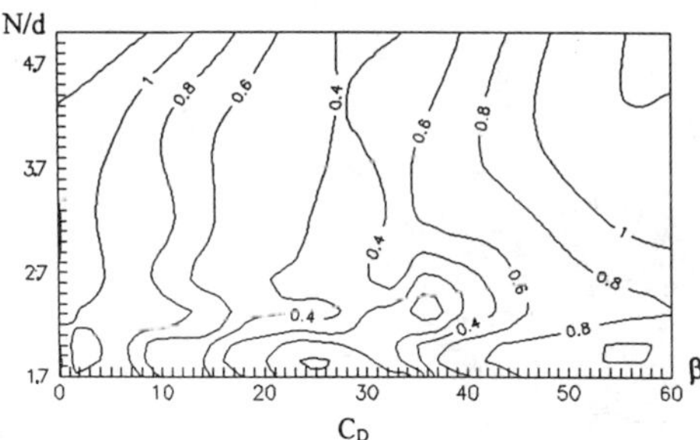

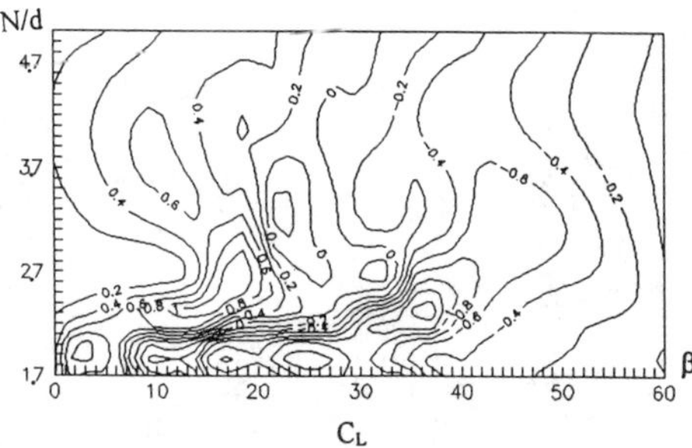

Figure 5. Contours of C_D and C_L of cylinder C with various N/d and β.

The C_D of cylinder A and B are quite different (Figure 4(c)). The values of C_D of cylinder B are larger about 0.4 than those of cylinder A regardless of changing with the angles of incident flow until at the vicinity of β=60°, where, their difference is reduced. As the spacing increases, the tendency of reducing in their difference shifts to smaller angle of incident flow.

The curve of C_D of cylinder C is similar with those in small spacing, however, the minimum C_D of minus value no longer exists. The C_L of cylinder A and B are also similar with those of small spacing, however, the C_L of cylinder C changes and the location of strong positive lift force shifts to β>14°. The variation of lift force on cylinder C with the angles of incident flow forms a curve of one peak and one valley with a small gently area between them. The location of the gently area coincides with the location of the bottom area of the drag curve.

3.3 *Effect of transition spacing: N/d=2.2*

Between the small spacing and the medium spacing are the transition spacing (N/d=2.2). Due to the limited length of the paper, the changes of pressure distributions on cylinders are not shown and discussed here. As to the total effect of interference, it is shown that the drag curves of the three cylinders remains unchanging with those of in small spacing (Figure 4(b). Whereas, the lift curves, particularly C_L of cylinder C behaves the transition process between the curves of small spacing (N/d=2.0) and medium spacing (N/d=2.5).

3.4 *Effect of large spacing: N/d≥4.0*

As the spacing ratio increases up to or larger than 4, the interference on the time-mean pressure distribution is reduced further, but the reverse is true for the fluctuating pressure on cylinders. It could be explained that the regular vortex shedding behind cylinder A and B are re-established, therefore, the most interference on cylinder C is caused by the effect of wake.

The curves of drag force are similar with those at medium spacing but more smoothly (Figure 4(d)). At the vicinity of β=60°, the values of C_D of three cylinders are more approach each other. The C_L curves of cylinder A and B are rather symmetric. The

C_L curve of cylinder C changes gently and just likes one period of sine wave.

The contours of C_D and C_L of cylinder C are presented in Figure 5. It is shown that, within the medium and large spacing region, both the drag and lift coefficients of their minimum values (the absolute value) are located at the same area of β=20°~30°.

4. CONCLUSIONS

The main results of this study may be summarized as follows:

1) The angles of incident flow strongly influence the pressure distributions on any one of the cylinders.

2) Due to the interference of the shear layers and/or wakes, the most complex features of pressure distributions can be found on the cylinder located in downstream.

3) As for the different levels of interference regarding to the influence of spacing ratio, mainly four different effect regions can be identified. These are effect of small, transition, medium and large spacing.

4) In small spacing, three basic types of flow patterns between three cylinders can be classified due to various angles of incident of flow. These are flow patterns of proximity effect, shear layer reattachment effect and wake effect.

5) The effect of shear layer reattachment on the downstream cylinder, which is the most serious interference case between cylinders, the significant lateral forces result either an inward or outward directed which depend on the orientations of incident flow. This phenomenon is associated with some kinds of regular vortex shedding.

ACKNOWLEDGMENTS

The work described in this paper was supported by National Education Committee Foundation of China.

REFERENCES

Eastop, T.D. & Turner, J.B. 1982. Air flow around three cylinders at various pitch-to-diameter ratios for both a longitudinal and a transverse arrangement, *Trans. Inst. Chem. Eng.*, 60 (1982) 359-363.

Gerhardt, H.J. & Kramer, C.1981. Interference effects for groups of stacks, *J. Wind Engineering and Industrial Aerodynamics.*, 8 (1981) 195-202.

Gu, Z.F. et al. 1996. Patterns of flow past two circular cylinders in staggered arrangements, *Proceedings of 4th Asia Flow Visualization Symposium*, Beijing, May 15-19 (1996).

Gu, Z.F. & Sun, T.F. 1998. On interference of two circular cylinders in staggered arrangement at high subcritical Reynolds number, *J. Wind Engineering and Industrial Aerodynamics.* (1998) in press.

Lam, K. & Cheung, W.C. 1988. Phenomena of vortex shedding and flow interference of three cylinders in different equilateral arrangements, *J. Fluid Mechanics*, 196 (1988) 1-26

Ohya, Y., Okajima, A. & Hayashi, M. 1989. Wake interference and vortex shedding, Chap. 10 in *Encyclopedia of Fluid Mechanics*, Gulf Publishing (1989).

Price, S.J. & Paidoussis, M.P. 1984. The aerodynamic forces acting on group of two and three circular cylinders when subject to a cross-flow, *J. Wind Engineering and Industrial Aerodynamics*, 17 (1984) 329-347.

Sayers, A.T. 1987. Flow interference between three equispaced cylinders when subjected to a cross flow, *J. Wind Engineering and Industrial Aerodynamics*, 26 (1987), pp.1-19.

Yunseok Kim & Changkoon Choi 1993. Relationship between aerodynamic responses and pressure distributions on three circular cylinders, *J. of Wind Engineering*, 56 (1993) 1-10

Zdravkovich, M. M. 1977. Review of flow interference between two circular cylinders in various arrangements, Trans. SAME, *J. of Fluids Engineering*, 99 (1977) 618-633.

Wind Engineering into the 21st Century, Larsen, Larose & Livesey (eds) © 1999 Balkema, Rotterdam, ISBN 90 5809 059 0

Influence of turbulence on the self-excited forces on a rectangular cross section

Fred L. Haan, Jr. & Albin A. Szewczyk
Department of Aerospace and Mechanical Engineering, University of Notre Dame, Ind., USA
Ahsan Kareem
NatHaz Modeling Laboratory, Department of Civil Engineering and Geotechnical Sciences, University of Notre Dame, Ind., USA

ABSTRACT: To understand how turbulence affects the self-excited forces responsible for long-span bridge flutter, pressure distributions were examined over a model of rectangular cross section forced to oscillate in a series of turbulent flows. The two main goals were to analyze the role that turbulence scale plays in the relevant flow mechanisms and to understand how turbulence affects the spanwise correlation of the self-excited forces.

1 INTRODUCTION

1.1 *Motivation*

The complexity of flow-structure interactions requires that wind-tunnel testing be an integral element of the design process for long, suspended-span bridges. This complexity coupled with the fact that only a fraction of the turbulent characteristics of the atmosphere can be matched in wind tunnels requires designers to understand the effects of imperfect turbulence simulation as fully as possible. This study addresses the effects that turbulence has on the "self-excited" or "motion-induced" forces responsible for bridge flutter—more specifically, on the flow mechanisms responsible for these effects. This introduction describes both the motivation for this work and the approach taken to address existing issues.

Shortfalls in experimental turbulence simulations can be quantified by examining a number of aspects of the turbulence. Reynolds numbers, *Re*, in wind tunnel experiments can be as much as three orders of magnitude lower than their atmospheric counterparts. Because the size of the smallest scales present in a turbulent flow decreases as *Re* increases, wind tunnel simulations have significantly less energy at small scales than atmospheric flows have. The smaller extent of the inertial subranges in wind-tunnel flows illustrates this lack of small-scale content.

The largest scales of a turbulent flow also fall short in wind tunnels compared to the atmosphere. While atmospheric integral length scales can be as much as ten times the size of a typical bridge deck width (Simiu & Scanlan 1986), wind-tunnel scales are typically on the order of the model deck size.

In spite of these shortcomings, bridge designers must test models in wind tunnels. Unfortunately, tests in turbulence have resulted in some inconsistent results with respect to the self-excited forces. A few examples are described in the following paragraphs.

Aeroelastic stability studies of a bridge deck section model commonly examine its flutter derivatives (most often following the empirical formulation of Scanlan & Tomko (1971)). Some reports of flutter derivatives in various turbulent flows contradict each other. In one of the first studies of how turbulence affects flutter stability, Scanlan & Lin (1978) identified only a slight increase in the magnitude of each of their section model's flutter derivatives going from smooth to turbulent flow. Huston (1986) attempted to address the scale mismatch between wind tunnels to atmospheric flows by employing active vanes and airfoils to generate gusty flow with as large a length scale as possible. These tests on a section model of the Golden Gate Bridge revealed a significant destabilizing trend in the A_2^* flutter derivative.

Unlike the results of Huston's large-scale gustiness, Larose et al. (1993) reported boundary layer turbulence to have a slightly stabilizing effect for their taut-strip models. In addition, their flutter derivatives did not depend on turbulence intensity alone. Despite having similar turbulence intensities, their grid-generated and boundary layer turbulent flows had opposite effects on the flutter derivatives. While grid turbulence decreased the flutter derivative magnitudes, boundary layer turbulence increased them. This suggests that the turbulence effect is not only a function of turbulence intensity but also of other parameters such as scales and spanwise coherence.

It is important to note here that the above studies of flutter derivatives were not all conducted on similar cross sections. Although turbulence will affect different geometries differently, this does not detract from the need to understand the changes it causes.

As further motivation, several numerical studies have considered turbulence effects on bridge aeroelastic stability by employing stochastic tools. Studies of this type (for example, Bucher & Lin 1988, Li & Lin 1995) commonly use a stochastic dynamic pressure to model the turbulence. This approach assumes that only the largest scales of turbulence have significant effects, and that the basic lift-generation mechanisms are the same as those for smooth flow. Given the uncertainties already mentioned regarding scale effects in these flow-structure interaction problems, further study of scale effects is warranted to aid numerical modeling.

Finally, three-dimensional, full-bridge model tests of long-span bridges have often shown flutter boundaries to be well defined in smooth flow and higher or nonexistent in turbulent flow. One often-cited example is the testing of the Lions' Gate Bridge in Irwin & Schuyler (1977). Data such as these has led many researchers to conclude that results from tests conducted in smooth flow are conservative. The proposed physical explanation for this has been that the self-excited forces have a lower spanwise coherence in turbulent flow. Although a number of researchers have mentioned this hypothesis, an analytical demonstration that spanwise diminution of self-excited forces can actually stabilize a bridge deck was put forth by Scanlan & Jones (1990) and more rigorously by Scanlan (1997).

While spanwise coherence of pressures and buffeting forces on static models in turbulent flows have been measured by a number of researchers (for example, Larose & Mann 1998, Sankaran & Jancauskas 1993), the spanwise behavior of the self-excited forces has never been studied experimentally. And although the stabilizing effect of spanwise coherence loss may be apparent for single-mode flutter scenarios, it is not obvious that this will apply to multi-mode coupled flutter cases (for example, Chen et al 1999, Katsuchi et al 1999, Jain et al 1996). Coherence loss along the span may stabilize a deck by reducing negative damping effects, and yet it may destabilize a deck by reducing favorable damping or favorable coupling effects. This issue will become even more important as bridge spans grow longer and multi-mode flutter becomes more probable. Spanwise correlation was therefore included in this study of turbulence effects.

1.2 *The Current Approach*

To study turbulence effects on the self-excited forces on bridge decks, the existing literature on bluff-body aerodynamics was considered as a starting point. A wind-tunnel model of rectangular cross section was chosen both to take advantage of the large amount of work done with rectangular models and to study the flow mechanisms involved without the geometric complications of actual bridge decks. Subsequent phases of this work will involve actual bridge deck cross sections.

Much of the existing literature concerning the effects of turbulence on the pressure distributions around shapes with rectangular cross section examines streamwise pressure distributions on static models (Haan et al. 1998, Li & Melbourne 1995, Bearman & Morel 1983, Kiya & Sasaki 1983, Hillier & Cherry 1981). While a review of all this work is beyond the scope of this paper, several key aspects of turbulence effects are important to note here. Shear layers that separate from a bluff body generate large negative pressure beneath them. Freestream turbulence enhances the entrainment into these separated shear layers. This works to increase the shear layer's curvature and cause reattachment to occur closer to the leading edge. This contraction of the separated region results in steeper surface pressure recoveries and in mean, rms, and peak pressure distributions having maxima of greater magnitudes that occur nearer the leading edge.

Several researchers have stressed the evidence that fine scale turbulence—on the order of the thickness of the separated shear layer—plays a major role in causing these effects (Gartshore 1973, Melbourne 1979, Tieleman & Akins 1990). In this study, the small-scale content of each flow is reported.

In order to conduct this study of self-excited forces in a context similar to that of the above-mentioned bluff-body studies, measurements were done employing the approach of Matsumoto et al. (1995). Rather than extract flutter derivatives directly from free-vibration data, this method obtains streamwise pressure amplitude and phase distributions from forced oscillation experiments. While these distributions are integrated to provide conventional flutter derivatives, they also provide insight into the mechanisms involved in the turbulence effects. The following derivation comes from this reference.

The amplitude and phase distributions for the surface pressures are defined according to the following equations. For brevity, only the results of torsional oscillations are reported in this paper. For torsional motion described in terms of the angular position, α, in equation (1), each position along the model has a pressure fluctuation at the model oscillation frequency described by equation (2).

$$\alpha(t)=\alpha_0\cos(2\pi f_m t) \tag{1}$$

$$C_p(x,t)=-C_p^*(x)\cos(2\pi f_m t-\psi(x)) \tag{2}$$

where α_0 is the oscillation amplitude, f_m is the model oscillation frequency, C_p is the pressure fluctuation (normalized by the dynamic pressure) at frequency f_m, C_p^* is the amplitude and ψ is the phase of this fluctuation as a function of streamwise position, x. The phase can be articulated as the lag between the maximum angle of attack and the maximum negative pressure at a specific location, x.

Self-excited lift and moment are defined according to equations (3) and (4), respectively. Again, only torsional motion is considered here.

$$L = q(2b)\left[kH_2^* b\frac{\dot{\alpha}}{U} + k^2 H_3^* \alpha\right] \quad (3)$$

$$M = q\left(2b^2\right)\left[kA_2^* b\frac{\dot{\alpha}}{U} + k^2 A_3^* \alpha\right] \quad (4)$$

where q is the dynamic pressure, b is half the deck width, $k=2\pi f_m/U$, U is the mean longitudinal velocity, and α is the angular position of the model (and the dot denotes differentiation with respect to time). Flutter derivatives, A_i^* and H_i^*, can be derived from the pressure distributions using equations (5)-(8).

$$A_2^* = \frac{1}{2k^2}\int_{-1}^{1}\frac{2C_p^*(x)x}{\alpha_0}\sin(\psi(x))dx \quad (5)$$

$$A_3^* = \frac{-1}{2k^2}\int_{-1}^{1}\frac{2C_p^*(x)x}{\alpha_0}\cos(\psi(x))dx \quad (6)$$

$$H_2^* = \frac{1}{2k^2}\int_{-1}^{1}\frac{2C_p^*(x)}{\alpha_0}\sin(\psi(x))dx \quad (7)$$

$$H_3^* = \frac{-1}{2k^2}\int_{1}^{1}\frac{2C_p^*(x)}{\alpha_0}\cos(\psi(x))dx \quad (8)$$

The approach taken for this work, then, consists of examining the streamwise distributions of pressure amplitudes and phases along with the spanwise correlation of the fluctuating lift and moment for a model oscillating in smooth and turbulent flows.

2 EXPERIMENTAL SETUP

2.1 *Turbulent Flow Generation*

All work reported here was conducted in the atmospheric wind tunnel at the University of Notre Dame's Hessert Center for Aerospace Research. This open-return facility has a 1.52 m by 1.52 m cross section and a 14.9 m test section. Turbulence was produced with grids of different sizes at various upstream distances from the model. By manipulating grid size and position, several groups of flows were generated having a range of integral length scales but constant turbulence intensities. Both spectral distribution and small-scale content were monitored to observe the impact of the various scales. Results for flows at two different turbulence intensities, 6% and 12%, are presented here.

All turbulent parameters of the longitudinal and vertical velocity fluctuations were quantified using hot-film "X" probes (TSI Model 1241). Integral scales were estimated using Taylor's approximation and the area under the autocorrelation curves (integrated to the first zero-crossing) as a time scale.

2.2 *Pressure Measurements*

While much of the past work has included only the changes in flutter derivatives resulting from turbulence effects, the focus of this research includes the behavior of pressure distributions over the bridge deck as well. This study employs a model of rectangular cross section with frontal height, D, of 38.1 mm, a streamwise width, B, of 254 mm, and a cross-stream span, L, of 1067 mm. End plates of height $7D$ extended $2.5D$ and $5D$ from the model's leading and trailing edges, respectively.

Twenty-one chordwise rows of pressure taps were distributed along the span as shown in Figure 1. A servomotor mechanism is used to force the model in both pitching and plunging motions with various amplitudes and mean angles of attack. The model, mounted in this mechanism in the wind tunnel, is shown in Figure 2. All experiments reported here were done with a torsional amplitude of ±2° with no plunging motion. A potentiometer was used to measure angular displacements. All tests were run at a constant Reynolds number (Re=1.2x10^4), while reduced velocities were varied by changing the model oscillation frequency.

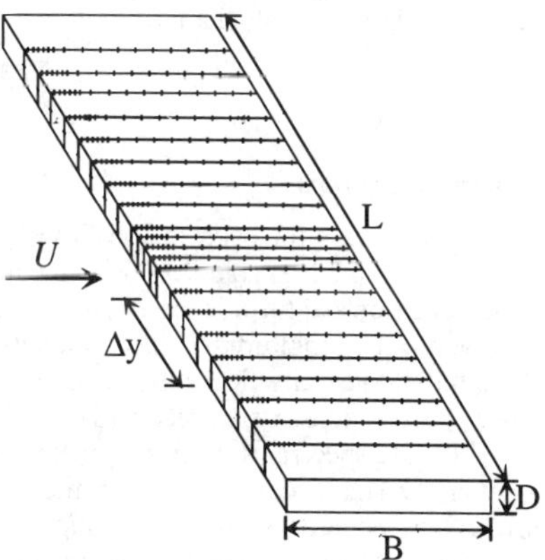

Figure 1. Diagram showing geometric parameters of the model along with the spanwise locations of the streamwise rows of pressure taps.

Sixty-four SenSym pressure transducers were mounted within the model and could be connected to the taps of any two chordwise rows at a given time.

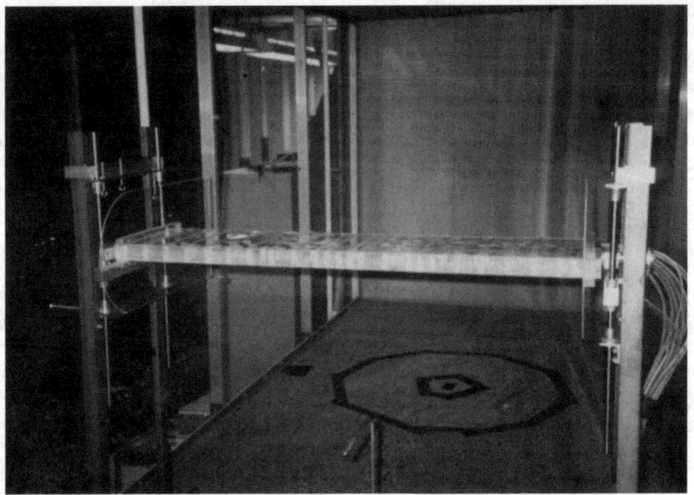

Figure 2. Photograph showing the pressure model mounted in its pitching and plunging mechanism in the wind tunnel.

All pressure signals were corrected for the dynamic effects of the tubing (inner diameter of 1.3 mm and length of 356 mm). Simultaneously sampling all of these transducers enabled observation of both the chordwise and spanwise pressure development. Each reduced velocity condition in each flow was sampled for 49.2 seconds (12 ensembles of 4096 points sampled at 1000 Hz).

The data acquisition system consisted of a G6-180 Gateway PC and 16-bit analog to digital conversion hardware from National Instruments. This hardware included "sample and hold" circuitry for simultaneously acquiring data from all channels.

Table 1. Turbulent Flow Parameters

	I_u	I_w	L_{ux}	S
Smooth Flow				
Case 1	0.4%	0.3%	N/A	N/A
Constant I_u=6%				
Case 2a	6.2%	5.9%	2.0*D*	2.3
Case 2b	5.8%	5.1%	4.8*D*	0.3
Constant I_u=12%				
Case 3a	12.0%	11.4%	1.3*D*	53
Case 3b	12.2%	10.2%	5.3*D*	6.0

3 RESULTS

3.1 *Velocity Measurements*

Turbulence parameters for each of the flows used in this project are listed in Table 1. These include the longitudinal and vertical turbulence intensities (I_u and I_w, respectively), the longitudinal integral length scale of the longitudinal velocity fluctuations (L_{ux}), and the small-scale spectral density parameter (S).

The small-scale spectral density parameter, S, was used to quantify small-scale content in these turbulent flows. This parameter was first suggested by Melbourne (1979) and used subsequently by Tieleman & Akins (1990). It is defined as follows:

$$S = \frac{fG_{uu}(f)}{U^2} \times 10^6 \tag{1}$$

where U is the mean velocity and G_{uu}, the power spectral density function of the longitudinal velocity fluctuations, is evaluated at frequency, $f=aU/D$. This small-scale parameter is essentially a *scale-specific*

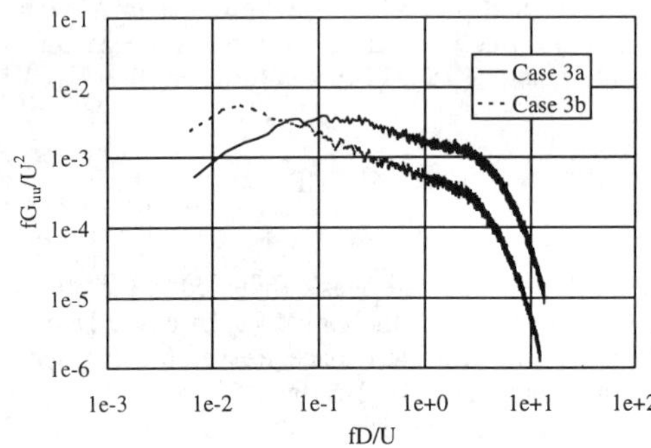

Figure 3. Normalized power spectral densities of longitudinal velocity fluctuations for two flows with equal turbulence intensity, 12%, but different integral length scales, 2.5*D* and 3.9*D*.

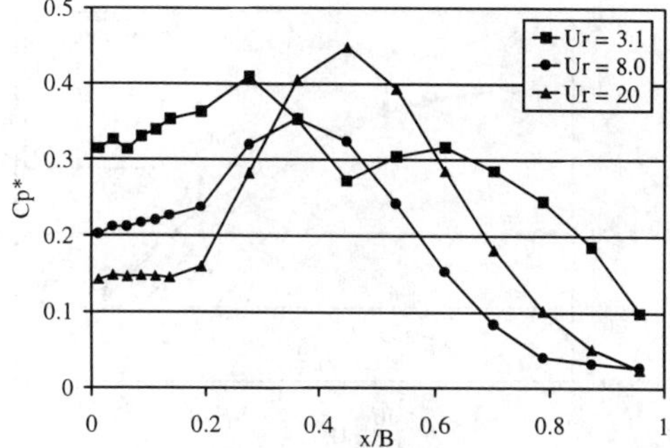

Figure 4. Pressure amplitude distributions for smooth flow.

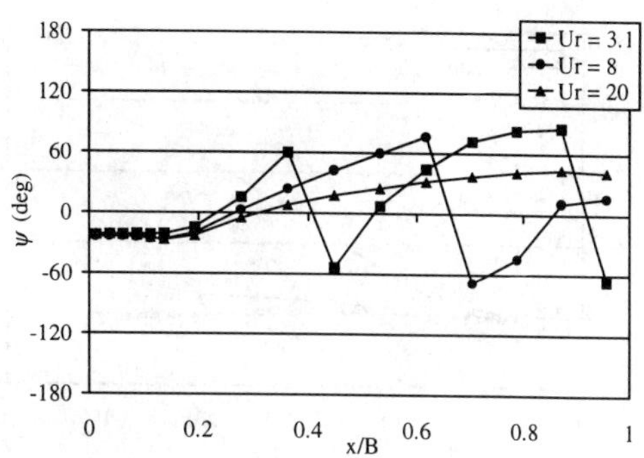

Figure 7. Pressure phase distributions for smooth flow.

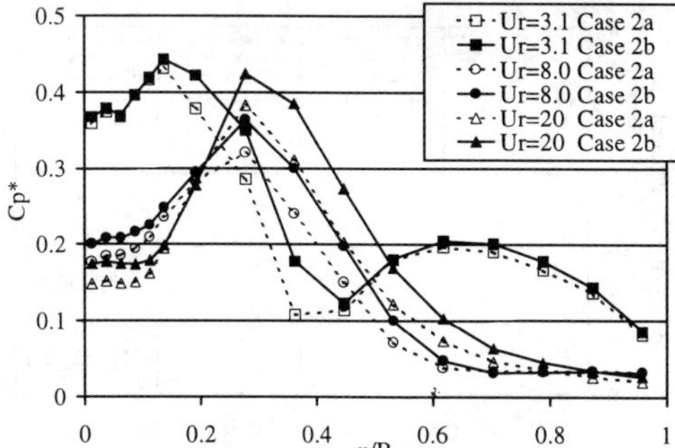

Figure 5. Pressure amplitude distributions for turbulent flows with I_u=6%. Open symbols correspond to the cases with smaller L_{ux}.

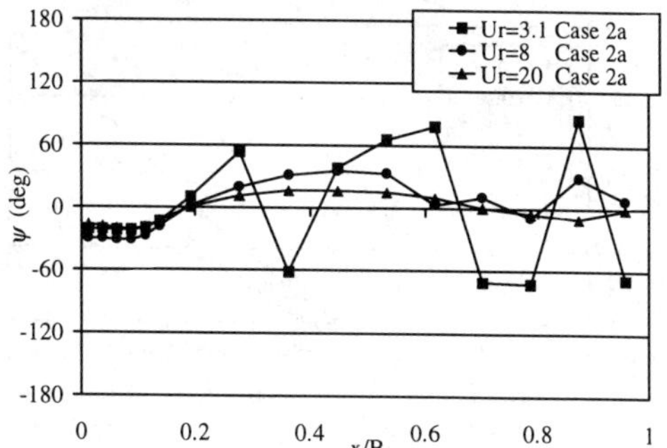

Figure 8. Pressure phase distributions for turbulent flows with I_u=6%.

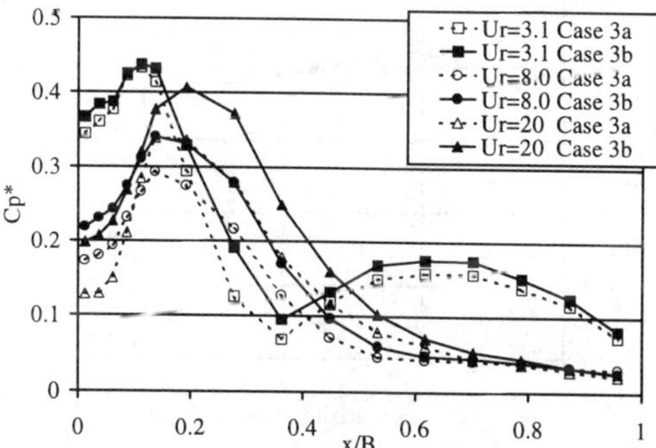

Figure 6. Pressure amplitude distributions for turbulent flows with I_u=12%. Open symbols correspond to the cases with smaller L_{ux}.

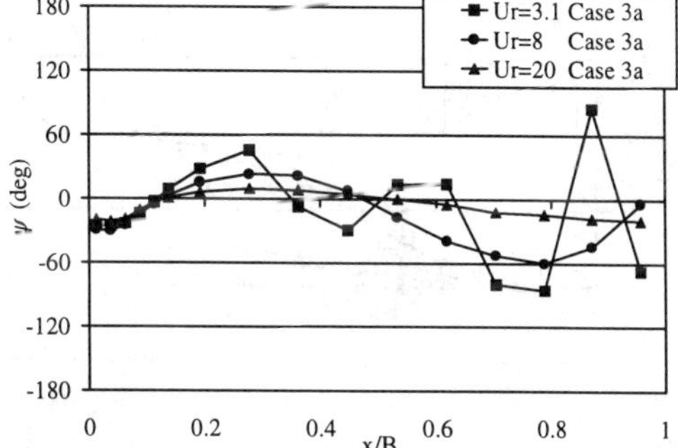

Figure 9. Pressure phase distributions for turbulent flows with I_u=12%.

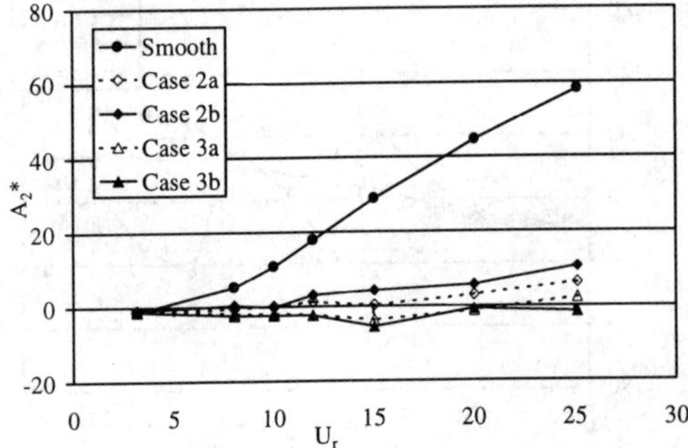

Figure 10. A_2* flutter derivative in smooth and turbulent flow.

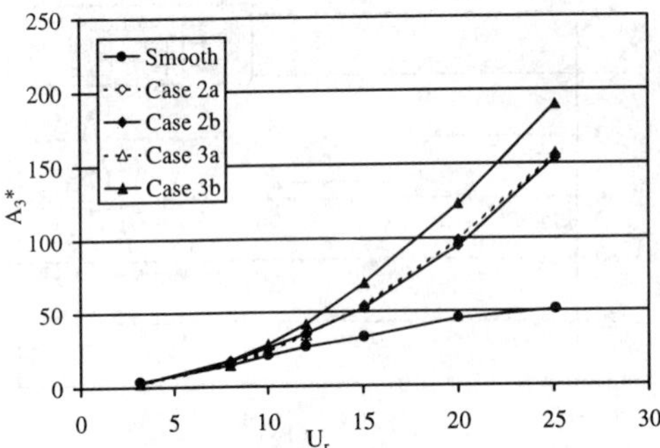

Figure 11. A_3* flutter derivative in smooth and turbulent flow.

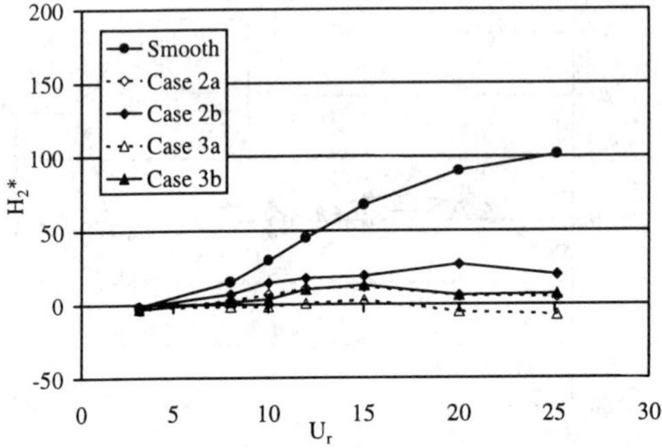

Figure 12. H_2* flutter derivative for smooth and turbulent flow.

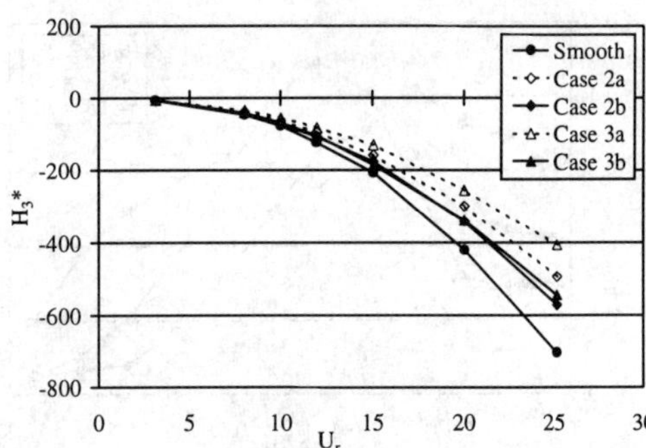

Figure 13. H_3* flutter derivative for smooth and turbulent flow.

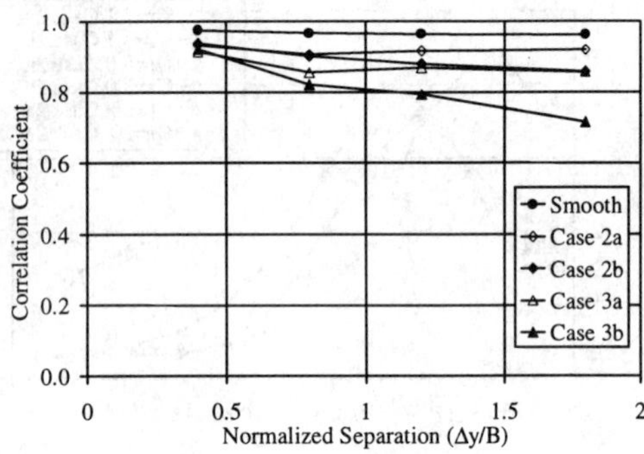

Figure 14. Correlation coefficient for lift force as a function of spanwise separation for U_r=10.

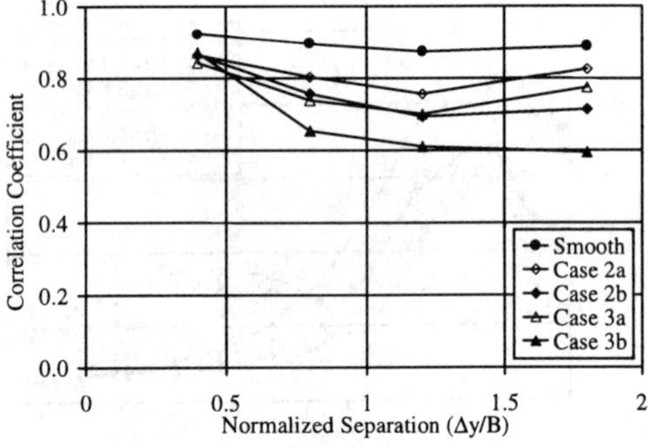

Figure 15. Correlation coefficient for moment as a function of spanwise separation for U_r=10.

turbulence intensity where the choice of a allows one to monitor the energy content at a specified scale. In most cases, including this work, a is set to 10 to monitor scales on the order of the separated shear layer thickness.

Figure 3 shows the normalized power spectral densities of the longitudinal velocity fluctuations for the two cases of 12% turbulence intensity. This plot clearly shows the difficulty in identifying scale effects. One cannot keep the turbulence intensity constant and change the integral length scale without also changing the small-scale content. The small-scale parameter, S, is calculated at fD/U=10. For these flows this is in the dissipation scale range. In the case shown, a 4 times increase in L_{ux} resulted a decrease in S by a factor of 9.

3.2 *Pressure Distributions*

Streamwise distributions of the pressure amplitude and phase were calculated for each flow configuration. Amplitudes, C_p*, for each pressure signal were obtained from the component of the signal's power spectral density at the model oscillation frequency. Phases, ψ, were extracted from the cross spectra between the pressure signals and the torsional displacement signal.

The amplitude distributions for each flow case of Table 1 are shown in Figures 4-6. A basic hump shape that shifts downstream with increasing reduced velocity, U_r, characterizes the smooth flow distribution of Figure 4. This same general shape is present in each of the turbulent cases of Figures 5 and 6 except that it shifts upstream with increasing turbulence intensity. The effect of smaller integral scales (the "a" cases) appears to be to lower amplitudes and to reduce the downstream extent of the hump shape.

Pressure phase distributions for each flow case are shown in Figures 7-9. The smooth flow phase distribution of Figure 7 suggests the presence of a flow structure that elongates as the reduced velocity increases. This structure switches the phase repeatedly between positive and negative values. Similar structures are present in the two turbulent flow cases, but as turbulence intensity increases the downstream extent of the phase switches shifts upstream. Turbulence scale did not change the phase distributions significantly.

3.3 *Flutter Derivatives*

Flutter derivatives were calculated from the amplitude and phase distributions of section 3.2 using equations (5)-(8) and are shown in Figures 10-13.

Each turbulence case had a strong stabilizing effect on A_2*, the flutter derivative associated with torsional damping. Although trends are not particularly clear, pairs having the same turbulence intensity have similar values. The flows with I_u of 12% caused a slightly greater shift from the smooth-flow values.

While A_3* had larger magnitudes in turbulent flow, the magnitudes of both H_2* and H_3* were less than their smooth-flow values. As demonstrated in Chen et al. (1999)—also appearing at this conference—this shift in H_3* is a stabilizing effect in coupled, multi-mode flutter problems. Scale effects were most evident for the H_3* flutter derivative—the flows of smaller integral scales reduced the magnitude of H_3* more than their larger scale counterparts.

3.4 *Spanwise Correlation*

From lift and moment time series integrated from pressures measured at two spanwise positions, correlation coefficient functions were calculated. Correlation coefficients for lift and moment are shown in Figures 14 and 15, respectively, for four different spanwise separations and each flow case. The reduced velocity for both figures was 10.

In each flow studied, the lift forces maintained a greater spanwise correlation than the moments. Case 3b—the flow having the largest turbulence intensity and the largest length scale—reduced the spanwise correlation the most for both lift and moment. In every case, the correlations were strong for spanwise separations much larger than the integral scales of the flows.

The formulations of equations (3) and (4) suggest that the flutter derivatives H_2* and H_3* would follow the correlation of the lift forces in Figure 14 while A_2* and A_3* would follow that of the moments in Figure 15. Further work will involve gaining understanding of the causes of the observed spanwise correlation trends and developing methods to apply the data to full-bridge flutter predictions.

4 CONCLUSIONS

Pressure measurements focusing on the effects of turbulence on the self-excited forces over a rectangular cross section revealed streamwise structure which contracts toward the leading edge with increasing turbulence intensity. The scales of the turbulent flow also alter the shape of this structure. Lift and moment were correlated over spanwise distances significantly larger than any scales present in the flows.

Future work will focus on the physical interpretations of the effects of turbulence on the pressure amplitude and phase distributions and how they relate to the observed changes in the flutter derivatives. How these flutter derivative changes in turbulence can be combined with the spanwise correlation behavior into full-bridge flutter predictions will also be studied. Results of these efforts and greater details of the experiments described in this paper will be presented at the conference.

ACKNOWLEDGEMENTS

The support for this work was provided in part by NSF grants CMS 95-03779 and CMS 95-22145 and an equipment grant from the University of Notre Dame. This support is gratefully acknowledged. The authors also wish to thank Dr. X. Chen for his helpful comments during this project.

REFERENCES

Bearman, P.W. & Morel, T. 1983. Effect of Free Stream Turbulence on the Flow Around Bluff Bodies. *Prog. Aerospace Sci.* 20: 97-123.

Bucher, C.G. & Lin, Y.K. 1988. Stochastic Stability of Bridges Considering Coupled Modes. *Journal of Engineering Mechanics* 114 (12): 2055-2071.

Chen, X., Matsumoto, M., Kareem, A. 1999. Influence of aerodynamic coupling among modes on flutter and buffeting response of long span bridges. *Proc. 10th Intern. Conf. on Wind Engineering*.

Gartshore, I.S. 1973. The effects of free stream turbulence on the drag of rectangular two-dimensional prisms. Boundary Layer Wind Tunnel Laboratory 4-73. University of Western Ontario.

Haan, Jr., F.L., Kareem, A., Szewczyk, A.A. 1998. The effects of turbulence on the pressure distribution around a rectangular prism. *Journal of Wind Engineering and Industrial Aerodynamics* 77 & 78: 381-392.

Hillier, R. & Cherry, N.J. 1981. The Effects of Stream Turbulence on Separation Bubbles. *Journal of Wind Engineering and Industrial Aerodynamics* 8: 49-58.

Huston, D.R. 1986. *The effects of upstream gusting on the aeroelastic behavior of long suspended-span bridges*. Dissertation, Princeton University.

Irwin, H.P.A.H. & Schuyler, G.D. 1977. Experiments on a Full Aeroelastic Model of Lions' Gate Bridge in Smooth and Turbulent Flow. Report LTR-LA-206, Nat. Aero. Estab., Nat. Research Council, Canada.

Jain, A., Jones, N.P., Scanlan, R.H. 1996. Coupled Flutter and Buffeting Analysis of Long-Span Bridges. *Journal of Structural Engineering* 122 (7): 716-725.

Katsuchi, H., Jones, N.P., Scanlan, R.H. 1999. Multi-Mode Coupled Flutter and Buffeting Analysis of the Akashi-Kaikyo Bridge. *Journal of Structural Engineering* 125 (1): 60-70.

Kiya, M. & Sasaki, K. 1983. Free-Stream Turbulence Effects on a Separation Bubble. *Journal of Wind Engineering and Industrial Aerodynamics* 14: 375-386.

Larose, G.L., Davenport, A.G., King, J.P.C. 1993. On the unsteady aerodynamic forces on a bridge deck in turbulent boundary layer flow. In G.C. Hart (ed.), *Conf. Proc. 7th U.S. National Conference on Wind Engineering*.

Larose, G.L. & Mann, J. 1998. Gust Loading on Streamlined Bridge Decks. *Journal of Fluids and Structures* 12:511-536.

Li, Q.C. & Lin, Y.K. 1995. New Stochastic Theory for Bridge Stability in Turbulent Flow II. *Journal of Engineering Mechanics* 121 (1):102-116.

Li, Q.S. & Melbourne, W.H. 1995. An experimental investigation of the effects of free-stream turbulence on streamwise surface pressures in separated and reattaching flows. *Journal of Wind Engineering and Industrial Aerodynamics* 54/55: 313-323.

Matsumoto, M., Kobayashi, Y., Niihara, Y., Shirato, H., Hamasaki, H. 1995. Flutter Mechanism and Its Stabilization of Bluff Bodies. In *Wind Engineering, Papers for the 9th Int. Conf.* New Delhi: Wiley Eastern Limited.

Melbourne, W.H. 1979. Turbulence Effects on Maximum Surface Pressure – A Mechanism and Possibility of Reduction. *Wind Engineering, Proc. of Fifth International Conference*. Pergamon Press.

Sankaran, R. & Jancauskas, E.D. 1993. Measurements of cross-correlation in separated flows around bluff cylinders. *Journal of Wind Engineering and Industrial Aerodynamics* 49: 279-288.

Scanlan, R.H. 1997. Amplitude and Turbulence Effects on Bridge Flutter Derivatives. *Journal of Structural Engineering* 123 (2): 232-236.

Scanlan, R.H. & Jones, N.P. 1990. Aeroelastic Analysis of Cable-Stayed Bridges. *Journal of Structural Engineering* 116 (2): 279-297.

Scanlan, R.H. & Lin, W.H. 1978. Effects of Turbulence on Bridge Flutter Derivatives. *Journal of the Engineering Mechanics Division (ASCE)* 104 (EM4): 719-733.

Scanlan, R.H. & Tomko, J.J. 1971. Airfoil and bridge deck flutter derivatives. *Journal of the Engineering Mechanics Division (ASCE)* 97 (6): 1717-1737.

Simiu, E. & Scanlan, R.H. 1986. *Wind Effects on Structures*. New York: John Wiley & Sons.

Tieleman, H.W. & Akins, R.E. 1990. Effects of Incident Turbulence on Pressure Distributions on Rectangular Prisms. *Journal of Wind Engineering and Industrial Aerodynamics* 36: 579-588.

Wind Engineering into the 21st Century, Larsen, Larose & Livesey (eds) © *1999 Balkema, Rotterdam, ISBN 90 5809 059 0*

Wavelet pattern recognition for bluff body wake characterisation

H. Hangan & G.A. Kopp
The Boundary Layer Wind Tunnel Laboratory, University of Western Ontario, London, Ont., Canada

ABSTRACT: A pattern recognition technique in wavelet space (wavelet pattern recognition - WPR) is introduced and applied to the turbulent near wake field of a circular cylinder and a porous mesh strip. The results of the WPR compare well with results based on a "classical" pattern-recognition method. It is suggested that pattern recognition in wavelet space has enhanced capabilities to detect multiple patterns in bluff body wake flows

1. INTRODUCTION

The overall objective of the proposed work is to develop new tools to identify coherent structures and their role in the turbulence dynamics of bluff body wakes. This work is motivated by the need to: (i) improve the subgrid turbulence models used with LES and (ii) find new solutions in the active control of turbulence.

Previous work (Hangan and Vickery, 1997) focused on the similar and quasi-universal character of the coherent velocity field in bluff body wakes and employed a "variable filtering" spectral method for the eduction of the main wake coherent structures (vortex shedding). Wavelet decomposition is a new approach to understanding and modelling turbulence, based on phase-space structures, allowing the determination of "local" spectra, see Farge (1992).

In the present work a novel pattern recognition in wavelet space ("wavelet pattern recognition" - WPR) technique is presented and applied for the investigation of the turbulent velocity field in the near wake region of two bluff body generators, a (solid) circular cylinder and a porous mesh strip. Results of the WPR method are compared with a "classical" pattern-recognition (PR) technique The two methods and their relative strength is discussed. Results of the WPR method agree well with results via the "classical" pattern-recognition (PR) method (applied in physical space). It is suggested that the advantage of the WPR over the PR methods comes from the fact that, in wavelet space, the templates can be easily identified, thus improving the capability to detect multiple patterns in wakes.

This aspect can be important in developing a better understanding of the complex mechanics of bluff body wake flows with implications in self-induced loading, buffeting and flow control problems all which are related to wind engineering applications.

2. EXPERIMENTAL SET-UP

Two experimental data sets are analysed corresponding to the near wake regions of a (solid) circular cylinder and a low porosity mesh strip. Hot-wire anemometry was previously employed to take simultaneous two-component velocity measurements in the near wake region of the two generators (Kopp and Keffer, 1996).

The Reynolds number for the cylinder flow is approximately 6700 (based on the cylinder diameter, 11.6 mm, and free-stream velocity U_0=9 m/s). The velocity measurements were made with a rake of 8 eight X-wires at several downstream locations in both the near and far regions. Time series of the two velocity components were sampled at a frequency f=5,000 Hz per channel for a total time of 40 seconds. The analysed data corresponds to x/D=16 downstream location and the lateral spacing of the probes was 7 mm. The mesh strip wake has a Reynolds number of 5,000 (based on the nominal mesh strip width W=10 mm) and velocity signals were sampled at a rate of 5,000 Hz/ channel for 40 seconds with a rake of 6 six X-wires at x/W=12 downstream location, and lateral spacing y/W=0.8.

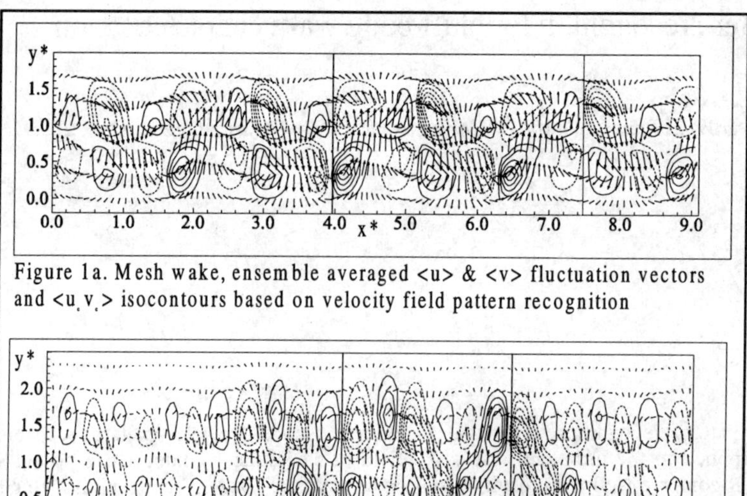

Figure 1a. Mesh wake, ensemble averaged <u> & <v> fluctuation vectors and <u v> isocontours based on velocity field pattern recognition

Figure 1b. Cylinder wake, ensemble averaged <u> & <v> fluctuation vectors and <u v> isocontours based on velocity field pattern recognition

3. ANALYSIS

3.1 Preliminary analysis

The large scale coherent structures associated with the vortex streets of the two wakes are clearly identified in Fig. 1a and Fig. 1b (axis $x^* = \frac{U_0}{f l_0}, y^* = \frac{y}{l_0}$ where l_0 is half velocity wake width), following a classical *pattern recognition* analysis for the mesh-strip and cylinder wake velocity fields, respectively. The pattern-recognition (PR) of Ferre and Giralt (1989) which uses a fluctuating velocity template to trigger the calculation of the ensemble averages, was used. This pattern recognition technique defines an initial estimate of a coherent structure footprint (template) and detects organized motions in the flow as local extrema of the correlation function between the pre-defined template and the prototype, experimental data over a two-dimensional domain. Once these organized events are detected, their ensemble average is used as starting point for new iterations until no significant variation in the ensemble average is detected.

Both vortex streets appear quite regular with little smearing of adjacent vortices, relative the central vortex (which was identified with the PR technique). Here, essentially 100% of the vortices, based on the Strouhal frequency, have been included in the ensemble-average. In other words, all of the street vortices have been classified as belonging to a single type of pattern.

A *continuous wavelet transform*, based on Mexican hat wavelets, was applied to the time series of fluctuating lateral (v) velocity. Wavelet coefficient iso-contours are plotted versus time (sampling points number on the horizontal axis, sampling frequency of 5000 Hz) and versus wavelet scale (the larger the scale the lower the frequency on the vertical axis) in

Fig. 2 a, b for the mesh strip and cylinder, respectively. Contours represent absolute values of the wavelet coefficients coloured by scale and this underlines the patterns and not the energy distribution with scale.

Note that the wavelet coefficient contours in the mesh-strip wake describe extremely regular patterns, as can also be inferred from the velocity time trace also shown. In contrast, the cylinder wake presents relatively irregular patterns, which correlates with the wave form variations observed in the instantaneous lateral velocity signal (top of Fig. 2b). Many different patterns are observable. When the velocity signal displays periodicity (e.g., between points 450-580), the wavelet coefficients are much like those for the mesh-strip wake. Very different patterns are observed when the vortex strength is reduced and a jitter in phase is observed (e.g., near points 720 or 740) correlated with singular patterns in wavelet space. Such irregularities appear much less frequently in the mesh-strip wake.

Identical wavelet patterns (topology and sequence) have been retrieved for the Reynolds stress (uv) signals.

The two flows considered exhibit a very well defined scale interval and the bulk of the energy is at the coherent (vortex street) scales, however, the cylinder wake also contains a large amount of energy at larger scales. This observation is consistent with the "ribs", which link adjacent spanwise (vortex street) vortices, identified by Kiya & Matsumura (1988) at half the Strouhal frequency.

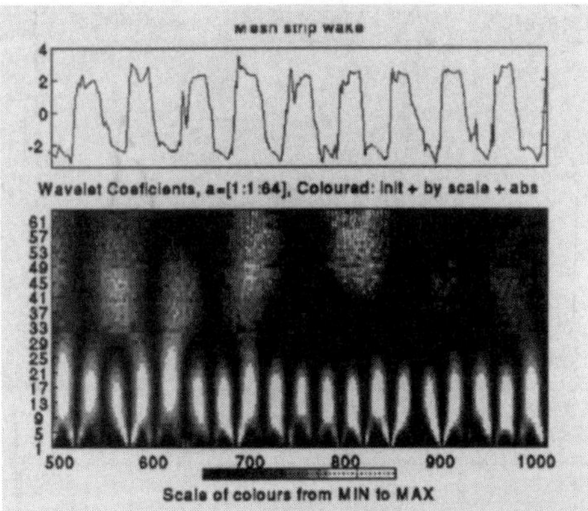

Fig. 2a. Continuous wavelet transform of v velocity near the location of maximumshear, Mesh strip

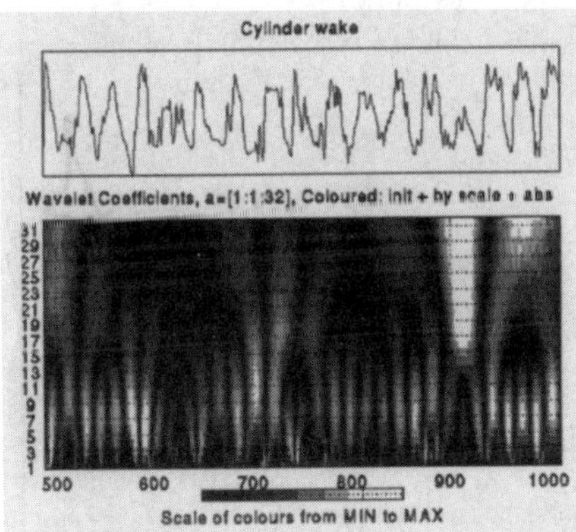

Fig. 2b. Continuous wavelet transform of v velocity near the location of maximumshear, Circular cylinder

3.2. Wavelet pattern recognition

Based on the results and observations from the preliminary analysis a wavelet pattern recognition technique (WPR) has been developed. The method consists of applying the pattern recognition and ensemble average technique developed by Ferre and Giralt (1988) to data in wavelet space, data which is obtained by wavelet transforms of multipoint velocity time series. As for the case of classical pattern recognition (PR) the time coordinate embedded in each time series has been substituted by a spatial downstream coordinate making use of Taylor's hypothesis of frozen turbulence.

Two classes of wavelet coefficients templates have been identified for the two wake flows under consideration. The first wavelet template (A_+) corresponds to the main structure present in the two wake flows, the wake rollers. The second wavelet template (B^+) was chosen to represent the secondary structure observed which is particularly useful for the less regular cylinder wake. This approach allows to define structures which contain energy not only at the vortex shedding scales but also at larger scales. Templates A_+ are presented on the left side of Fig. 3a for the mesh strip and Fig. 3b for the cylinder. These templates correspond to the main wavelet pattern observed during the preliminary analysis (see Fig. 2a and 2b). The second template type (B_+), see Fig. 4 corresponds to the wavelet pattern with energy at larger scales, observed mostly for the cylinder wake (see Fig. 2b).

The design of the initial wavelet templates (both A and B types) focused on the modulus maximum and the position of this maximum, or the maxima lines (Bacry et al., 1993). The *modulus maximum* (xm, a), with xm the spatial location of the maximum and a the scale, can be simply identified by the cancellation of the space derivative of the wavelet transform $\frac{\partial T_\psi}{\partial x}(xm, a) = 0$, where T_ψ is the wavelet transform according to the analysing (Mexican hat) wavelet ψ. The *maxima line* is a connecting line of local maxima (at any scale) which is likely to emanate from a velocity singularity (Arneado et al., 1993).

The wavelet templates have been defined as maxima "bands" rather than maxima lines, see the original templates (upper left-hand side) in Figures 3a for mesh strip and 3b for cylinder. These maxima "bands" have been selected based on preliminary analysis of the wavelet coefficients maps, see Fig. 2a and 2b. The net demarcation between patterns in wavelet space is thought to improve the selection of the initial template, which in physical space (PR method) has to be inferred from previous knowledge of the particular flow field dynamics.

The A type wavelet templates for the cylinder correspond to 50 data points and approx. 250 patterns have been identified based on each of these two templates using an average threshold of 1 r.m.s. of the signal. This corresponds to 53% of the total number of points which is precisely the same "usage rate" found for the initial pattern recognition in physical (u,v) space. The same template (A_+) has been adapted for the mesh strip case, for which the template corresponds to 90 data points. Approximate 205 wavelet patterns have been identified for the A type template. This corresponds to a "usage rate" of 58%

which is larger than the rate found for the initial velocity pattern recognition.

As already mentioned a second class of templates (B_+) has been used for the cylinder wake in order to identify the secondary, larger scale structure which was observed during the preliminary analysis. The B_+ wavelet template for the cylinder wake (see Fig. 4) generated only 41 wavelet patterns which corresponds to an "usage rate" of less than 9%. This low occurrence is expected from the initial wavelet analysis indicating intermittent passage of the larger scale structure. The threshold was set at 2 r.m.s. for the pattern recognition based on template B_+ .

The ensemble average of the events detected by pattern recognition in wavelet space; see lower-left side of Fig. 3a and 3b for A templates for mesh strip and cylinder, respectively, and Fig. 4a for template B_+ (cylinder); are subsequently used to calculate ensemble averages in the physical space. Isocontours of the coherent Reynolds stress are shown on the right hand side of Fig. 3b and Fig. 4a for the cylinder (corresponding to A_+ and B_+ wavelet templates) and Fig. 3a for the mesh strip (corresponding to A templates only). The templates have been created based on the wavelet coefficient map and not on the modulus or square of these coefficients. Single sign templates (even if more significant from the energy point of view) would smear the patterns in the ensemble average process

. The windows in Fig. 1a and Fig. 1b which are ensemble averages of the coherent Reynolds stress for mesh strip and cylinder (based on the classical pattern recognition - PR) compare well with the wavelet pattern recognition (WPR) ensemble averages in Fig. 3a and 3b for the same two cases. This suggests that the proposed wavelet based pattern recognition method (based on A type templates) can be employed for the detection of the main wake coherent structures. The advantage of the WPR over the classical PR methods comes from the fact that, in wavelet space, the templates are very well identified based on the sharp modulus maxima lines and this aspect becomes important in designing templates of different types for the same flow field, for example patterns A and B for the cylinder wake.

The B class wavelet template, which emphasizes on the energy content at larger scales (based on the preliminary wavelet analysis) detects different events compared to wavelet templates of type A. Fig. 4a presents the B_+ initial template and the ensemble average of the events based on this templates and also the ensemble average of the coherent Reynolds stress corresponding to this template. It is observed that the coherent Reynolds stress contours are weak and this fact might be associated to the low number of wavelet events detected based on template B_+. However, note

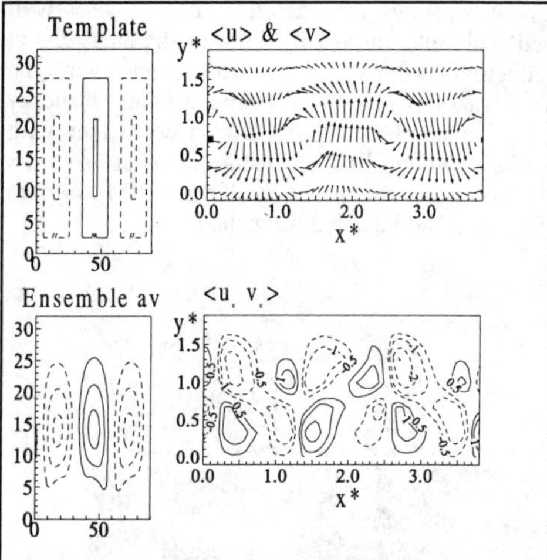

Figure 3a. Mesh wake, template A+, velocity fluctuation vectors and Reynolds stress contours

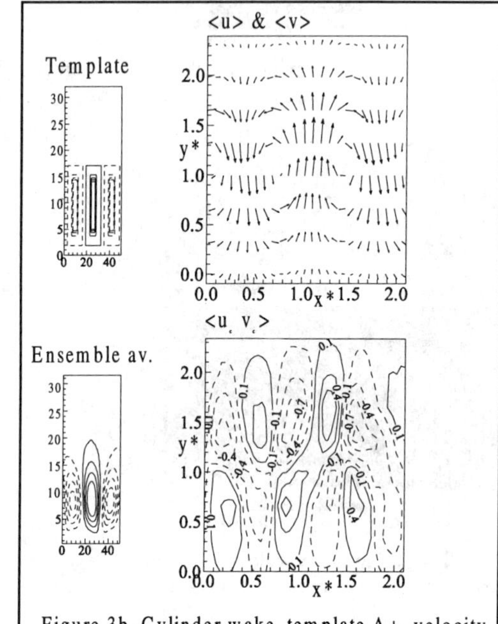

Figure 3b. Cylinder wake, template A+, velocity fluctuation vectors and Reynolds stress contours

that the ensemble averaged velocity field (vectors) shows that the template detected the flow patterns corresponding to the zone between two adjacent rollers which is consistent with the postulated existence of "rib" structures (Kiya and Matsumura, 1988) linking the spanwise wake vortices. Fig. 4b presents some of the events in velocity space corresponding to wavelet pattern B_+, events on which the ensemble averaging process was based. These events present an important scatter which is probably a main cause for the weakness of the coherent Reynolds stress contours in Fig. 4a, but they also present a common pattern - the outflow region between two adjacent rollers.

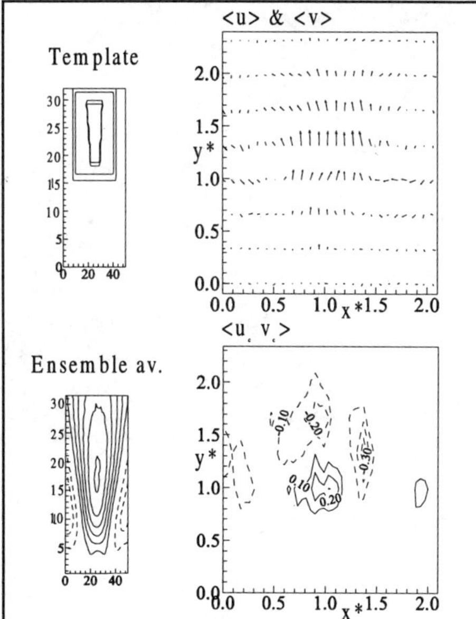

Figure 4a. Cylinder wake, template B+, velocity fluctuation vectors and Reynolds stress contours

The classification of the individual wavelet patterns in order to ascertain their dynamic significance in relation with the physical, turbulence space is presently under investigation. The modulus maxima line calculations will be more rigorously applied in order to increase the robustness of the WPR method.

4. CONCLUSIONS AND FUTURE WORK

A wavelet pattern recognition method is proposed by performing pattern recognition in wavelet space. The method is applied for turbulence analysis of two near bluff body wake flow fields; a circular cylinder and a mesh strip.

Two classes of wavelet templates are designed to identify the main turbulence patterns in the wake. One type of template (*A* type) which correspond to the main coherent structure (vortex shedding), generate ensemble averages for fluctuating velocities and Reynolds stress similar to a classical pattern recognition technique applied directly to the velocity (physical) field. A secondary, (*B* type) wavelet template is also defined and corresponds to a less regular, larger scale pattern in the cylinder wake flow. Based on wavelet pattern recognition it is suggested that this pattern might correspond to a "rib" type sub-structure.

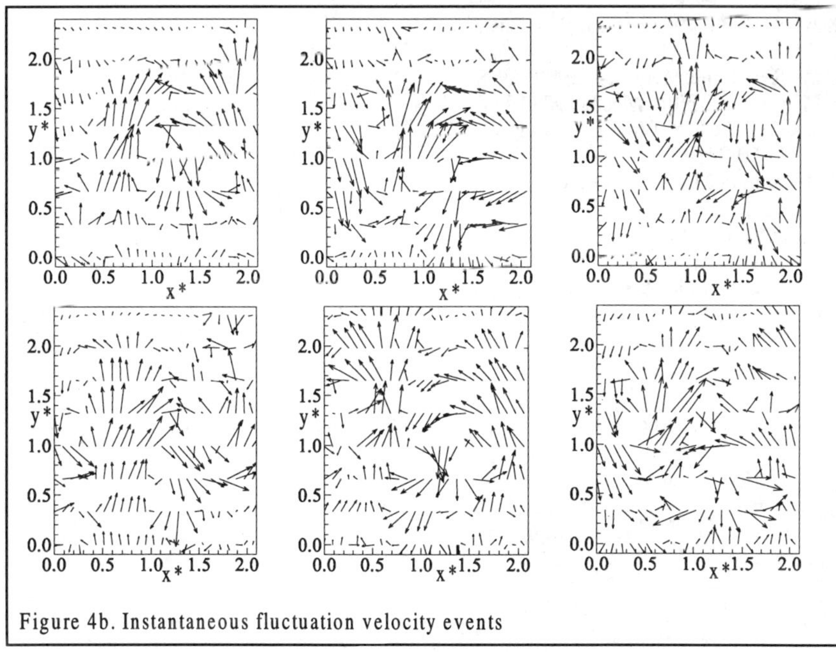

Figure 4b. Instantaneous fluctuation velocity events

The net demarcation of modulus maxima lines of the wavelet coefficients maps allows for a net template definition in wavelet space. This is considered to be an improvement compared to the classical pattern-recognition method for which a certain degree of subjectivity might be associated with the choice of multiple templates (when more than one class of patterns are to be identified).

Further work will investigate the possibilities offered by the wavelet pattern recognition in identifying secondary turbulent structures in other bluff body wake flows with implications in self-induced loading, buffeting and flow control problems..

ACKNOWLEDGEMENTS

This work was supported by the National Sciences and Engineering Research Council of Canada. Thanks are accorded to Professors J.A. Ferre and F. Giralt for making available their PR code. Dr. A. Vernet from the Universitat Rovira I Virgili, Tarragona, Spain helped with the implementation of the wavelet templates into the PR code.

REFERENCES

Arneado A., Bacry E. and Muzy J.F., J.P. Bonnet and M.N. Glauser (eds.), Eddy Structure identification in Free Turbulent Shear Flows, Kluver Academic Publishers, (1993), 153

Bacry E., Muzy J.F. and Arneodo A., J. Stat. Physics, 70 (1993), 635

Farge M., Annu. Rev. Fluid Mech. 24 (1992), 395

Ferre J.A. and Giralt F., J. Fluid Mech. 198 (1989), 27

Hangan H., Vickery B.J., J. of Wind Eng. and Ind. Aerodynamics 72 (1997) 47

Kiya M. and Matsumura M., J. Fluid Mech., 190 (1988), 343

Kopp G.A. and Keffer J.F., Phys. Fluids 8 (1996), 2712

Mallat S. and Hwang W.L., I.E.E.E. Trans. on Inf. Th. 38 (1992), 617

Wind Engineering into the 21st Century, Larsen, Larose & Livesey (eds) © 1999 Balkema, Rotterdam, ISBN 90 5809 059 0

Quasi-steady theory developed with experimental verification

Roger Hoxey & Lynn Short
Silsoe Research Institute, Bedford, UK
Peter Richards
Department of Mechanical Engineering, The University of Auckland, New Zealand

ABSTRACT: Quasi-steady methods seek to attribute variation in surface pressures measured on buildings to the properties of the approach flow. The general theory is well documented e.g. Cook (1990), Letchford et al (1993) and others, where the method has been to use a Taylor series approximation and to truncate the series to the first or second derivatives. This paper seeks to generalise quasi-steady methods and to make appropriate comparisons with experimental data. The experimental building used for this study is a 6m cube which can be rotated through 360° and pitched to 5°. This will enable the derivatives with respect to yaw and pitch to be evaluated and to be used in the simplified quasi-steady methods, but also for higher order polynomial fits to the data to be used in a more complete quasi-steady analysis, which is proposed here.

In previously reported work at international conferences, for example Thomas et al (1995) and Hölscher (1996) the application of quasi-steady or non-linear methods have used spectral estimates to compute the expected pressure spectrum. The approach adopted here is to use the time series information to compute the expected pressure at a tapping point and from this compute the spectra, probability densities and estimates of the peak pressures. This method, therefore has direct application to design loads.

1. QUASI-STEADY THEORY

Letchford *et al* (1993) reviewed the quasi-steady theory and his approach is similar to that of Kawai (1983) and Cook (1990). The instantaneous pressure p(t) at a point on a building surface is expressed in terms of wind velocity V(t) and a pressure coefficient Cp which is a function of the wind direction θ(t) as

$$p(t) = \frac{1}{2}\rho V(t)^2 Cp(\theta(t)). \qquad (1)$$

This equation cannot be evaluated explicitly as it requires information on instantaneous flow properties at the tapping point position in the absence of the building i.e. the wind speed and direction with the building removed. Such instantaneous information is unobtainable in full-scale experiments in an intrinsically nonstationary flow. In practice the wind velocity is measured at some distance beyond the building's sphere of influence, generally upstream and to one side. Using such a reference velocity measurement, the time equivalence on the right and left hand sides of Eq.1 is lost and this restricts the validity of the equation to comparisons of statistical moments and probability and spectral densities. For example, the mean and higher statistical moments of both sides of Equation 1 can be compared as can probability and spectral density functions. In all cases, the error associated with any comparison will be related to the spatial separation of the anemometer from the building, an error that becomes particularly pronounced in single-event statistics such as comparison of peak values. It also follows that cross-correlation of wind dynamic pressure and surface pressure is not implied and is not a test of the quasi-steady theory. In practice, cross-correlations are poor as the transport processes within the flow are complex.

In the first-order or linearised quasi-steady model, the pressure coefficient is approximated by a Taylor polynomial of first order as

$$Cp(\theta) = Cp(\bar{\theta}) + (\theta - \bar{\theta})\frac{\partial Cp}{\partial \theta}. \qquad (2)$$

This ignores second-order terms for the transverse component, first-order terms for the vertical component and all cross product terms.

Analysis of experimental results has shown that linear quasi-steady theory generally underestimates the rms values of pressure as given by Equation 1 when simplified using Equation 2. In the past, this underestimation has conveniently been attributed to the existence of building induced turbulence effects. However, detailed measurements made on the Silsoe Structures Building have shown that the approximations made in the linear quasi-steady model are the main cause of this discrepancy, and that improvements to the theory are required. The method proposed here is to improve the Taylor series representation of Cp(θ) to higher order and to include the vertical flow direction (φ) by using a polynomial representation of Cp(θ,φ) applied to measured time histories. For computational convenience the pressure coefficient is expressed as a power series in θ and φ, viz:-

$$Cp = \sum_m a_m \theta^m + \sum_n b_n \varphi^n \qquad (3)$$

where the values of m and n are selected to give an adequate fit to the measured mean value pressure coefficients. In this simplified form the cross-product terms have been ignored as insufficient experimental evidence is available to determine their value.

2. FULL-SCALE MEASUREMENTS

A 6m cube has been constructed at Silsoe in an exposed position. The building has a plain smooth surface finish and was instrumented with surface tapping points positioned around a centre line section. Simultaneous measurements have been made of 16 pressures and of the simultaneous wind dynamic pressure and direction derived from a sonic anemometer positioned upstream of the building at the roof height. Tapping points are constructed of simple seven millimetre diameter holes (a size sufficient to prevent water blocking the tapping points) and the pressure signals transmitted pneumatically, using 6 mm internal diameter plastic tube to transducers mounted centrally within the cube. Tube lengths of up to 10 metres are used in this system giving a frequency response of 3 dB down at 8 Hz. For the basic data recording, simultaneous measurements of the pressures were made at a rate of 5 Hz together with the three components of wind speed. The records were processed to give mean, fluctuating and spectral properties. Measurements were made with the cube pitched at 0^0, $2½^0$ and 5^0 and at several yaw angles, although natural variations in wind direction contribute to a range of yaw angles. A detailed analyses of mean pressure coefficients has been used to examine the effects of both yaw and pitch and produce numerical information from which the coefficients in Equation 3 have been evaluated. The polynomial fit has been used to enable actual measurements of wind speed and direction to be manipulated to give a full quasi-steady prediction of surface pressure. These estimates are compared with actual measurements in the form of statistical moments, probability densities and power spectral densities.

3. RESULTS

Mean pressure coefficients, derived for a recording period of ten minutes are presented in Figure 2 for a tapping point near the centre of the wall and for a tapping near the centre of the roof. In both cases results are included from symmetrically positioned tapping points and all results are reflected for symmetrical wind directions. In this way the patten over 360° is developed. There are few inconsistent points for the pressure coefficient on the side wall, however for the roof there are patterns which cannot be fully explained and may be associated with approach flow turbulence intensity variations. This will be the subject of further investigation. The maximum, minimum and RMS values for all the data sets have also been evaluated.

A similar set of results is also available for the cube pitched at 2½ and 5° although tapping point symmetry is lost in these cases and wind direction is limited to the cube pitched into the approaching flow. The mean pressure coefficients around the centre section of the cube are shown in Figure 3 for a normal wind direction.

Figure 1. The 6m cube at zero and 5^0 pitch; the cube can be rotated through 360^0.

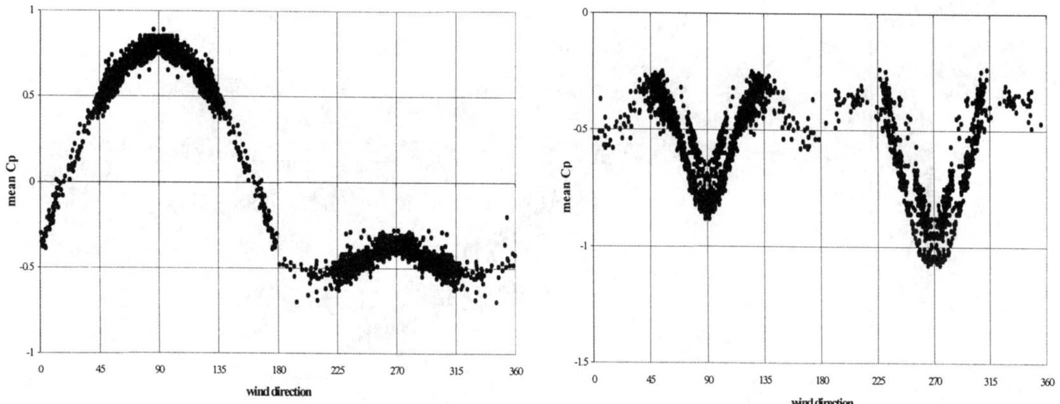

Figure 2. Mean pressure coefficients on the cube at zero pitch for a tapping near the centre of: a) the wall, b) the roof.

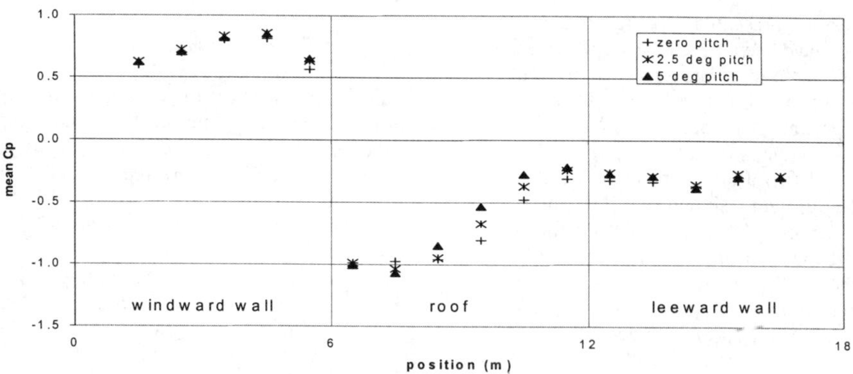

Figure 3. Mean pressure coefficients around the centre section.

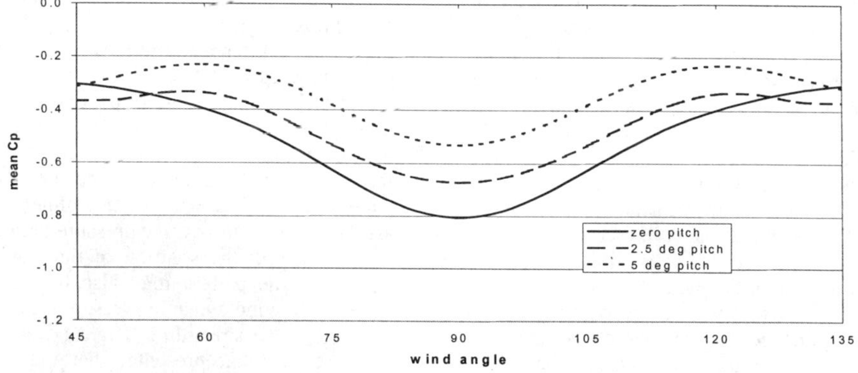

Figure 4. The variation of mean pressure coefficient for a tapping on the roof with angle of pitch.

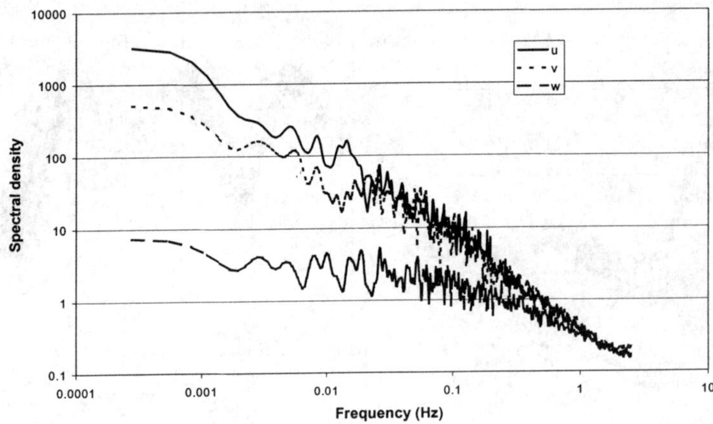

Figure 5. Spectra of the components of velocity.

4. DISCUSSION

From the results presented in Figure 3, three tapping points have been selected as representative of the overall pattern of pressures on the cube. These points are: (i) a point on the windward wall 3.5 m above ground, (ii) a point on the roof 3.5 m from the leading edge (shown as the point at 9.5 m in Figure 3) and (iii) a point on the leeward wall 3.5 m above ground (shown as the point in Figure 3 at 14.5 m). Only the point on the roof shows a significant influence of pitch angle of the cube and hence is the tapping point used to show the influence of the vertical flow direction. There is only space in this report to consider the one case of flow normal to the windward wall.

4.1 *Mean pressure coefficients*

The pattern of mean pressure coefficient on the wall is shown in Figure 2 where the windward wall corresponds to a flow of 90° and the leeward wall to a direction of 270°. The pressure pattern records for 2½ and 5° pitch for the wall taps are indistinguishable from that shown in Figure 2 for zero pitch. For both the windward and leeward wall the flow is symmetrical and hence the odd terms in Equation 3 are zero.

The mean pressure coefficient for the tapping point on the roof, 3.5 m from the leading edge is shown in Figure 2b and the case considered here is for a wind direction of 90°. Again the flow pattern is symmetrical about 90° and hence the odd terms in Equation 3 in θ will be zero. However there is a significant influence of pitch which is shown in more detail in Figure 4, where the mean coefficients have been represented by a least squares fitted curve, for limited wind angles about 90°. The remaining constants in Equation 3 have been evaluated to fit the mean pressure coefficient data.

4.2 *Spectra and admittance functions*

Examples of spectra are presented for flow near normal to the cube. The spectra of the components of the velocity measured upstream of the cube are shown in Figure 5 and are representative of those measured on previous occasions for this site showing a decay of approximately -5/3 at high frequency.

The power spectrum for a tapping point on the windward wall is shown in Figure 6 together with a linear multiplier of the wind dynamic pressure spectrum (mean Cp × q) and also the quasi-steady prediction. The quasi-steady prediction is the use of Equation 3 applied to the wind dynamic pressure with instantaneous measured values of θ and φ. For the windward wall the linear and quasi-steady spectra are indistinguishable as there is little influence of pitch or yaw. The admittance function, defined as the tapping point spectrum divided by the quasi-steady spectrum shows a pattern typical of a windward wall where the values exceed one in the mid frequency range (0.01 to 1 Hz), are near unity at low frequency and fall off below unity at higher frequencies. There are two contributing factors to the admittance exceeding unity: the pattern is indicative of streamline divergence associated with an adverse pressure gradient and is consistent with the solution of the Navier-Stokes equations; the pattern may also be explained by variation in wind dynamic pressure measured at the effective mean streamline height (less than 3 m) compared to that presented here at 6 m cube height.

Similar power spectra and admittance functions are presented for a tapping point on the roof in

Figure 7, here the influence of the pitch terms in Equation 3 are significant and modify the spectrum and the admittance. The terms introduce more energy in the higher frequency region and give a good comparison, in this region, with the measured spectrum for tapping point 12. Two admittance functions are shown in Figure 7; in addition to the tapping point 12 divided by the quasi-steady representation, tapping point 12 is also divided by the linear representation. The distinction here is that the quasi-steady approximation shows values below unity in the mid frequency range between 0.01 and 1 Hz and approximately unity elsewhere which is consistent with streamline convergence associated with the favourable pressure gradient in the flow over the roof. The additional energy in the spectrum for tapping point 12 may have previously been explained by building induced turbulence, however, it is clear from these measurements that the vertical component can account for this increased energy.

The example for the leeward wall is shown in Figure 8 where there is no distinguishable difference between the linear and quasi-steady approach and both agree with the measured pressure which gives in the admittance with a value of near unity over the entire frequency range. Again there is no evidence of building induced turbulence.

4.3 *Probability density functions and statistical moments.*

The probability density functions are shown in Figs. 9 to 11 for the three sample tapping points. In each case the linear and quasi-steady methods are compared to the measured tapping point pressure. Although the same methods were used to process all sets of data, i.e. bin widths and the number of data points, the tapping pressures in each case show a more variable characteristic which may be associated with a point pressure. The tapping point had a diameter of 7 mm, compared to the wind dynamic pressure sensor, the sonic anemometer, which had a measurement volume of a 150 mm cube.

The probability density function for the tapping point on the roof shows that the quasi-steady method can generate positive pressures on the roof which are consistent with the measurements but cannot be explained by the linear approach. The statistics for the three tapping points are presented in Tables 1 to 3 together with the coefficients that had been used in Eq. 3. The statistics for the wind dynamic pressure q and the tapping point pressures are also included for comparison. In each case the coefficient a_0 was selected to give a mean value close to the tapping mean pressure.

The effect of the quasi-steady method is evident in the statistics for the tapping point on the roof where the standard deviation is sensitive to the first order term b_1. The kurtosis is also affected by this term but the skewness only shows a moderate effect. The standard deviation is under predicted in regions of stream line divergence and over predicted in regions of stream line convergence consistent with the pattern observed on the power spectra in Figs 5 to 7. There may also be a contribution to the standard deviation from the size of tapping point compared to the volume of the sonic anemometer.

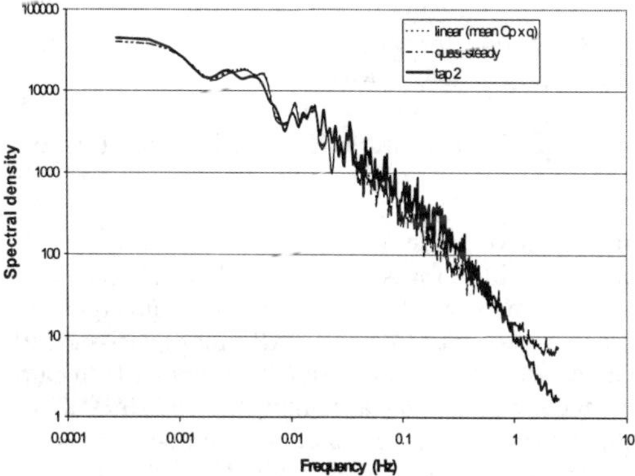

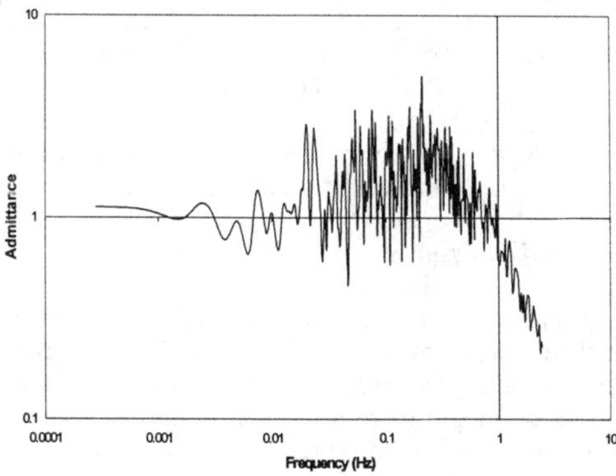

Figure 6. Power spectra and admittance for a tapping point on the windward wall showing the effect of including quasi-steady variations in yaw.

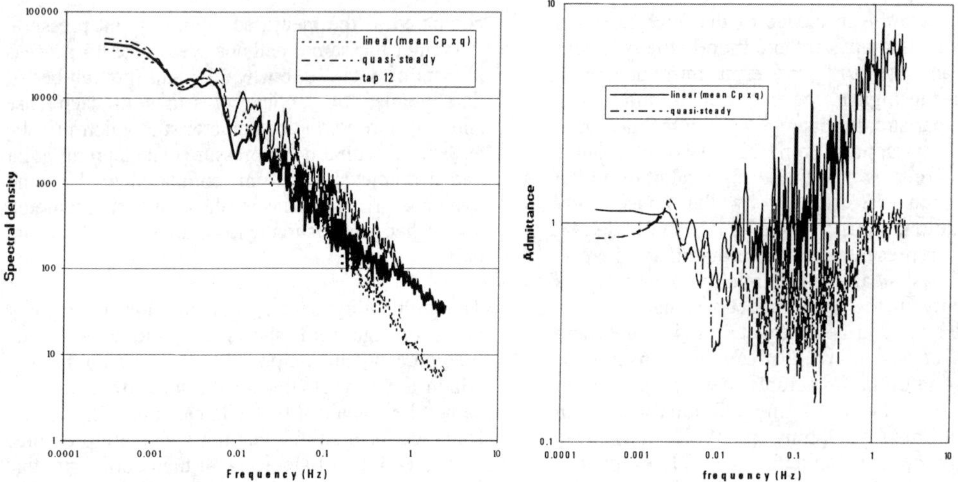

Figure 7. Power spectra and admittance for a tapping point on the roof showing the effect of including quasi-steady variations in yaw and pitch.

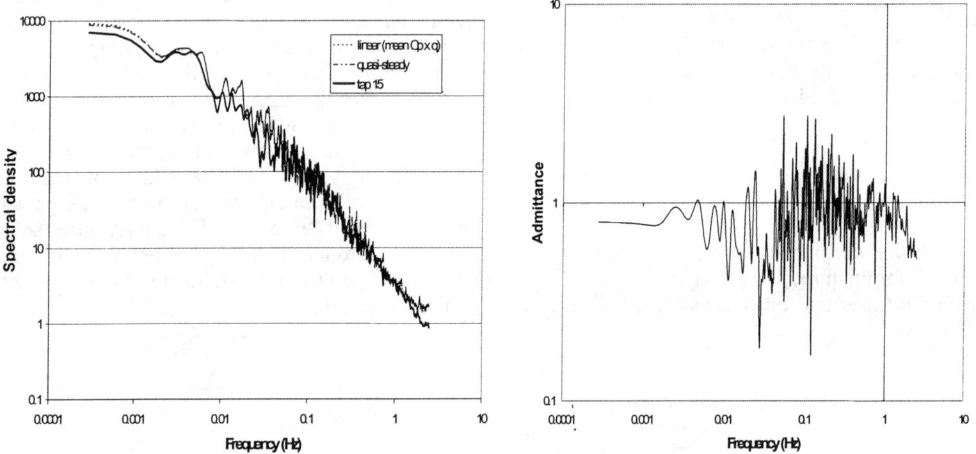

Figure 8. Power spectra and admittance for a tapping point on the leeward wall showing the effect of including quasi-steady variations in yaw.

4.4 *Peak pressures*

The maximum and minimum single occurrence statistics are shown in Tables 1 to 3 where the quasi-steady predictions are compared with the measured tapping pressures. For the windward wall the quasi-steady predictions are relatively insensitive to the coefficients in Eq. 3 with the derived values within the measured extremes. This finding is consistent with streamline divergence. For the tapping on the roof the quasi-steady peaks are sensitive to the coefficients in equation 3 and can exceed the measured peaks. This observation is again consistent with streamline convergence. Finally, for the leeward wall, the quasi-steady values are within the measured extremes which can be accounted for by a combination of streamline straining reducing the standard deviation compared to prediction but with additional fluctuations associated with the small size of the tapping point increasing the standard deviation.

Table 1. Statistics for a tapping point on the windward wall.

		q	p1	p2	p3	p4	p5	p6	tap 2
maximum		161.7	129.3	132.3	132.3	130.0	129.3	129.1	151.7
minimum		7.7	6.1	6.3	6.3	6.4	6.1	6.0	-9.6
mean		53.94	43.15	43.74	43.76	43.55	42.93	42.30	43.57
standard deviation		22.14	17.71	18.07	18.07	17.60	17.68	17.27	19.44
skewness		1.04	1.04	1.03	1.03	1.05	1.03	1.05	1.09
kurtosis		4.25	4.25	4.28	4.28	4.33	4.25	4.35	4.70
coefficients	a_0		0.8	0.83	0.83	0.8	0.8	0.8	
in Eq. 3	a_1		0	0	0	0	0	0	
	a_2		0	-0.6	-0.6	0	0	-0.6	
	a_3		0	0	0	0	0	0	
	a_4		0	0	0.1	0	0	0	
	b_1		0	0	0	0.5	0	0.5	
	b_2		0	0	0	0	-0.6	-0.6	

Table 2. Statistics for a tapping point on the roof.

		q	p1	p2	p3	p4	p5	p6	tap 12
maximum		161.7	-6.1	31.6	36.5	62.0	-5.4	40.4	53.0
minimum		7.7	-129.3	-141.8	-230.9	-245.0	-138.6	-242.1	-187.3
mean		53.94	-43.15	-43.71	-43.29	-44.39	-43.38	-44.06	-43.79
standard deviation		22.14	17.71	19.29	25.15	26.44	18.46	25.85	20.15
skewness		1.04	-1.04	-0.95	-1.08	-1.05	-1.05	-1.09	-0.93
kurtosis		4.25	4.25	4.39	4.95	5.02	4.41	5.05	5.13
coefficients	a_0		-0.8	-0.89	-0.85	-0.95	-0.87	-0.93	
in Eq. 3	a_1		0	0	0	0	0	0	
	a_2		0	2.5	0	2.5	2.5	2.5	
	a_3		0	0	0	0	0	0	
	a_4		0	0	0	0	-4.5	-4.5	
	b_1		0	0	3.2	3.2	0	3.2	
	b_2		0	0	0	0	0	0	

Table 3. Statistics for a tapping point on the leeward wall.

		q	p1	p2	p3	tap 15
maximum		161.7	-3.1	-3.0	-3.0	11.6
minimum		7.7	-64.7	-68.4	-66.7	-76.0
mean		53.94	-21.58	-21.73	-21.70	-21.44
standard deviation		22.14	8.85	8.94	8.92	7.98
skewness		1.04	-1.04	-1.06	-1.05	-1.05
kurtosis		4.25	4.25	4.34	4.32	4.96
coefficients	a_0		-0.4	-0.39	-0.39	
	a_1		0	0	0	
	a_2		0	-0.4	-0.4	
	a_3		0	0	0	
	a_4		0	0	0.18	
	b_1		0	0	0	
	b_2		0	0	0	

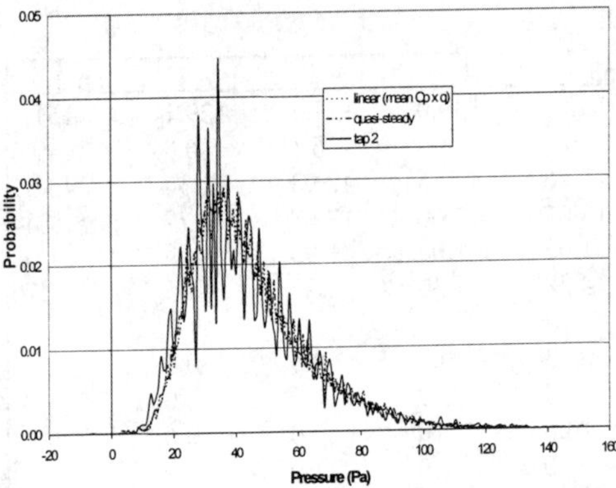

Figure 9. Probability density functions for a tapping point on the windward wall showing the effect of including quasi-steady variations in yaw.

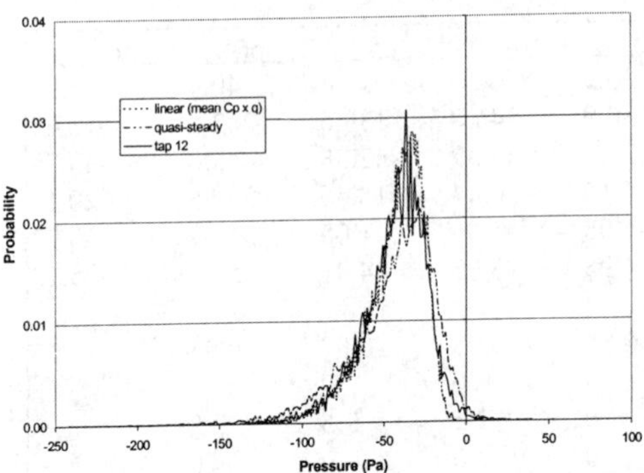

Figure 10. Probability density functions for a tapping point on the roof showing the effect of including quasi-steady variations in yaw and pitch.

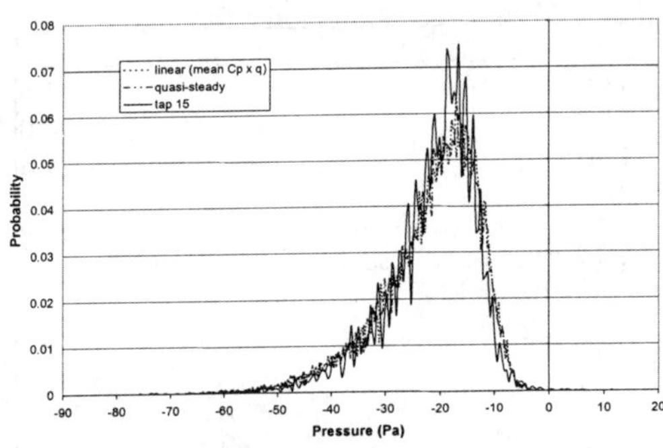

Figure 11. Probability density functions for a tapping point on the leeward wall showing the effect of including quasi-steady variations in yaw.

CONCLUSIONS

1. A more complete quasi-steady method is shown to account for measured pressure fluctuations on a 6 m cube.

2. The influence of pressure gradients, both favourable and advers, on the turbulent flow are apparent in the admittance functions generated from the quasi-steady method.

3. The effect of pressure gradients and the size of tapping compared to the sampling volume of the sonic anemometer are evident in the statistical moments and extremes of the measured and predicted pressure.

4. The quasi-steady method proposed here has direct application to the assessment of design wind loads.

5. Further work is required to quantify the effect of approach flow turbulence variations and spatial variations in static pressure.

REFERENCES

Cook, N.J. (1990), *"The designers guide to wind loading of building structures, Part 2, Static Structures"*, Butterworths, London 1990.

Kawai, H. (1983), "Pressure fluctuations on square prisms - applicability of strip and Quasi-static Theories", *Journal of Wind Engineering and Industrial Aerodynamics*, 13(1983), pp197-208.

Letchford, C.W.; Iverson, R.E. and McDonald, J.R. (1993), "The application of the Quasi-steady theory to full-scale measurements on the Texas Tech Building", *Journal of Wind Engineering and Industrial Aerodynamics*, 48(1993), pp111-132.

Letchford, C.W.; Marwood, R. (1996), *"On the influence of w-component turbulence on roof pressures beneath conical vortices", BBAAIII, Blacksburg, Virginia, USA, July 1996.*

Thomas, G.; Sarkar, P.P.; Mehta, K.C. (1995), *"Identification of admittance functions for wind pressures from full-scale measurements", 9th ICWE, New Delhi, India, Jan 1995.*

Hölscher, N. (1996), *"A non-linear approach for the aerodynamic admittance of wind pressures",BBAAIII, Blacksburg, Virginia, USA, July 1996.*

Wind Engineering into the 21st Century, Larsen, Larose & Livesey (eds) © 1999 Balkema, Rotterdam, ISBN 90 5809 059 0

Aspect ratio effect on the near wake of a free-ended circular cylinder in a cross-flow

Sang-Joon Lee & Cheol-Woo Park
Department of Mechanical Engineering, Advanced Fluids Engineering Research Center, Pohang University of Science and Technology, Korea

ABSTRACT : The free-end effect of a finite circular cylinder in a cross flow on the near wake characteristics has been investigated experimentally. Three finite cylinders having aspect ratio(L/D) of 6, 10 and 13 were mounted on a flat plate. Hot-wire anemometer was employed to measure the wake velocity at the Reynolds number of 20,000. Mean pressure distributions on the cylinder surface were measured and flow near the free-end was visualized to see the flow structure qualitatively. The three dimensionality of the FC wake becomes dominant with decreasing the FC aspect ratio, i.e., the free-end effect is remarkable at smaller aspect ratio. The three dimensionality of the FC wake maybe attributed to the strong entrainment of irrotational fluids caused by the downwash flow at very near the free end.

1 INTRODUCTION

Flow around a circular cylinder has been extensively investigated in the past due to its simple geometry and coherent vortex structure. The circular cylinder is a representative bluff body showing two-dimensional flow characteristics. Recently, the three-dimensionality of the cylinder wake showing different flow structure along the axial direction has been focused. (Szepessy & Bearman 1992) The three-dimensionality has been known to be appeared as the turbulence level of free stream increases. (Slaouti & Gerrard 1981)

In the strict sense, the cylinder wake has three-dimensional flow characteristics over all Reynolds number ranges. In the near wake behind a two-dimensional circular cylinder, however, the three-dimensionality is weak because the vortex shedding is regular and parallel to the cylinder axis in the vortex formation region. Therefore, the cylinder wake has been assumed as a two-dimensional flow in view of the coherent vortex structure in which Strouhal number does not change along the cylinder axis. (Szepessy & Bearman 1992, Baban *et al.* 1989)

The main causes for the three-dimensionality include the non-uniformity of upstream flow, longitudinal vortices and low aspect ratio(L/D) of the cylinder, etc. If there is a free-end at the cylinder end, the free-end condition becomes a significant factor for the three-dimensionality. (Slaouti & Gerrard 1981) Due to the free-end, the flow structures including the vortex formation, vortex shedding pattern and surface pressure distribution have been changed even in the near wake region.

Baban *et al.*(1989) found a strongly fluctuating re-circulation flow in the wake region, in which the drag force fluctuations are also increased. It was attributed to the vortices shed from the end of the cylinder in unsteady pattern. Okamoto & Sunabashiri(1992) found that the wake behind finite cylinders of small aspect ratio(L/D=1-2) was symmetric, but the wake pattern was changed to a three dimensional one when the FC aspect ratio was larger than L/D=4.

However, there are limited studies on the three-dimensionality of the FC wake and some contradictions exist on the basic mechanism, especially, at high Reynolds number flows. Farivar(1981) mentioned that the shedding frequency was disappeared for the aspect ratios smaller than L/D=7.5. On the contrary, Zdravkovich *et al.*(1989) found a regular vortex shedding for the aspect ratio of L/D=2.

The main objective of this study is to investigate the effect of FC aspect ratio on the flow characteristic of near wake, especially, the flow structure near the free end.

2 EXPERIMENTAL APPARATUS AND METHODS

The experiments were carried out in a closed-return type subsonic wind tunnel having a test section of 0.72m wide × 0.6m high × 6m long. Free stream turbulence intensity in the test section was less than 0.08%. Free stream velocity was fixed to 10m/s and corresponding Reynolds number based on the cylinder diameter(D=30mm) was 20000. The schematic diagram of experimental set-up and coordinate system are shown in Fig. 1

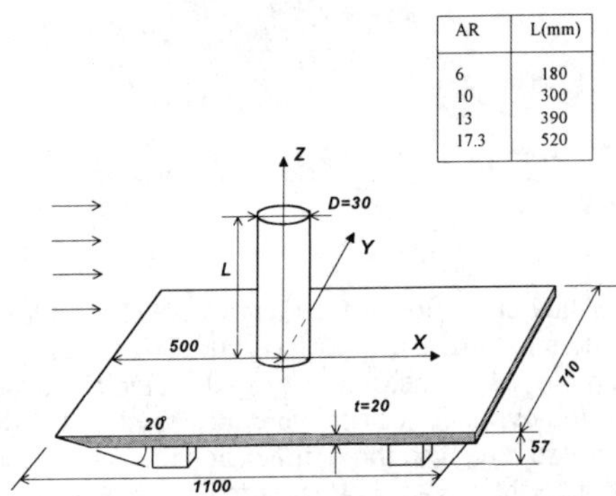

Figure 1. Experimental model and coordinate system (unit : mm)

The FC models were made of stainless steel and their surfaces were polished smoothly with a sandpaper. The finite cylinder was installed vertically on a 15mm thick flat plate whose leading edge was sharp-edged with an angle of 30°. The finite circular cylinder was placed 0.5m downstream from the leading edge of the flat plate.

In this study, three finite cylinders having different aspect ratio(L/D=6, 10, 13) were tested. The aspect ratio was defined as the cylinder height L divided by the cylinder diameter D. For comparison, a two-dimensional circular cylinder with aspect ratio of L/D=17.3 was also included, for which there was no gap between the free-end and test section ceiling.

The wake velocity was measured using an I-type hot-wire probe(DANTEC 55P11) connected to a constant temperature hot-wire anemometer(TSI IFA 100). The hot-wire probe was traversed to the measuring points using a 3-D traverse system having an accuracy of 0.01mm. At each measurement point, 32000 velocity data were acquired at 2KHz sampling rate after low-pass filtering at 800Hz. The schematic diagram of the velocity measurement system is shown in Fig. 2.

In order to measure pressure distribution around the cylinder surface, pressure taps were installed along the longitudinal axis of the cylinder with 5mm intervals. The pressure taps were connected to a micromanometer(FCO-19) and analog pressure signals were digitized using a high precision A/D converter(DT-2838). At each measurement point, 16384 pressure data were acquired at 500Hz sampling rate after low-pass filtering at 200Hz.

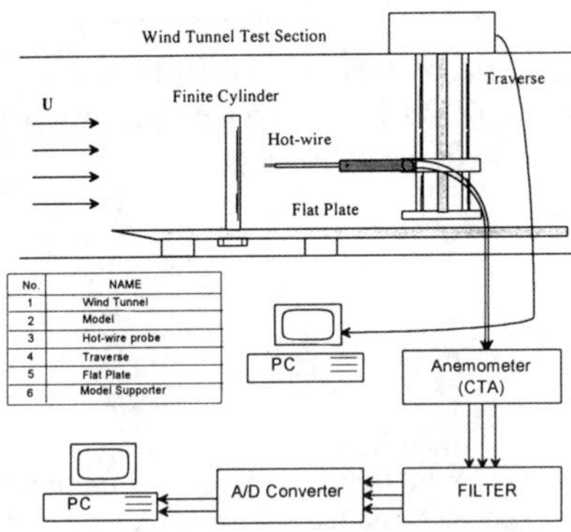

Figure 2. Schematic diagram of velocity measurement system

The pressure difference between the surface pressure p and the static pressure p_0 was divided by the dynamic pressure with free-stream velocity U_0 and air density ρ_a to give the pressure coefficient C_p expressed as:

$$C_p = \frac{p - p_0}{\frac{1}{2}\rho_a U_0}$$

Flow around the FC free-end was visualized using the smoke-wire method to see the flow structure qualitatively. Smoke filaments were generated by heating a fine stainless wire coated with SAFEX liquids. For this flow visualization, the free-stream velocity was set to 2m/s (Re=4000). The flow field near the free-end was illuminated with a thin laser light sheet emitted from a dual-head Nd-Yag laser. Clear flow images were obtained by synchronizing the camera with the laser light sheet.

3 RESULTS AND DISCUSSIONS

3.1 *Vortex shedding frequency*

The velocity signals of wake were analyzed to see the FC aspect ratio effect on the vortex shedding frequency. Fig. 3 shows the power spectral density(PSD) distributions measured at X/D=3, Y/D=2 and Z/L=0.5.

The vortex shedding frequency is decreased as the FC aspect ratio decreases; the peak frequencies were 66Hz(2-D), 60Hz(L/D=13), 53Hz(L/D=10) and 47Hz(L/D=6), respectively. The magnitude of PSD is also decreased with decreasing the aspect ratio. From this, we can see that the coherent structure in the wake is weaken as the aspect ratio decreases. The Strouhal number for the 2-D cylinder is 0.198, which is well agreed with previous results.

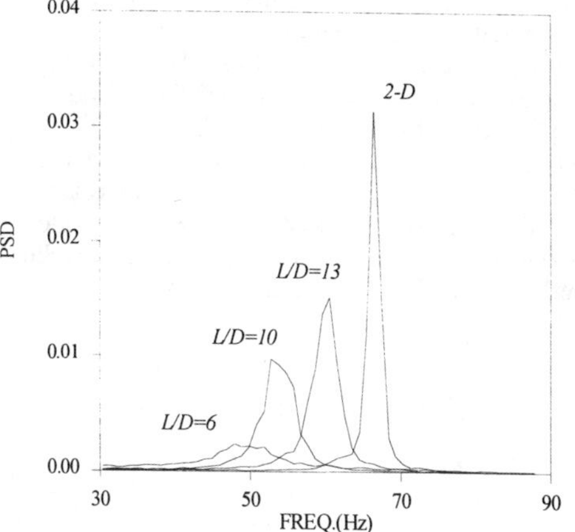

Figure 3. Vortex shedding frequency measured at X/D=3, Y/D=2 and Z/L=0.5

3.2 *Vortex formation region*

The entrainment of ambient fluids into the wake region affects not only on the wake structure but also on the vortex formation behind the finite cylinder. In this study, the end of the vortex formation region was defined as a point that has maximum streamwise velocity fluctuations.

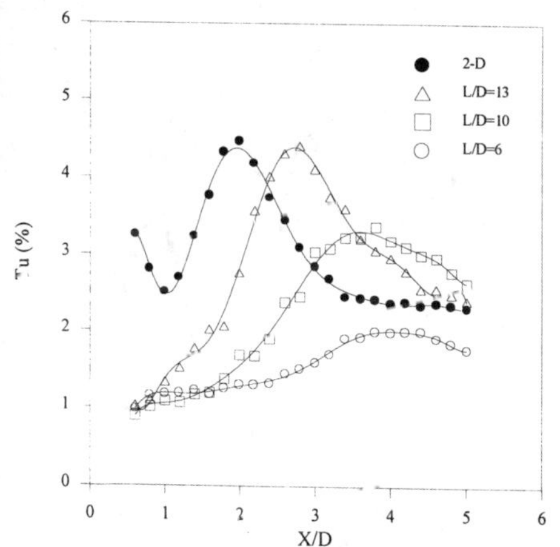

Figure 4. Comparison of vortex formation region measured at Z/L=0.5

Fig. 4 shows the turbulence intensity distributions measured at the middle height(Z/L=0.5) of FC models. The length of the vortex formation region is increased as the aspect ratio decreases. Especially, for the small aspect ratios of L/D=6, 10, the vortex formation length is about two times the two-dimensional cylinder case. This is a peculiar flow characteristic of the finite cylinder wake.

With decreasing the aspect ratio(L/D), the magnitude of turbulence intensity is decreased and the clear peak becomes blunt gradually. Especially, for the case of L/D=6 cylinder, the turbulence intensity distribution around the peak point is nearly flat. From these results, we can see that the free-end of finite cylinder increases the vortex formation region and decreases the velocity fluctuations due to the entrainment of ambient fluids into the wake. And the free-end effect is enhanced as the aspect ratio decreases.

Fig. 5 shows the turbulence intensity distributions of streamwise velocity component measured at the central section(Y=0) of the wake for the L/D=6 cylinder. The turbulence intensity of wake behind the finite cylinder is small and the vortex formation region is relatively large excluding the free-end region, compared to the two-dimensional circular cylinder.

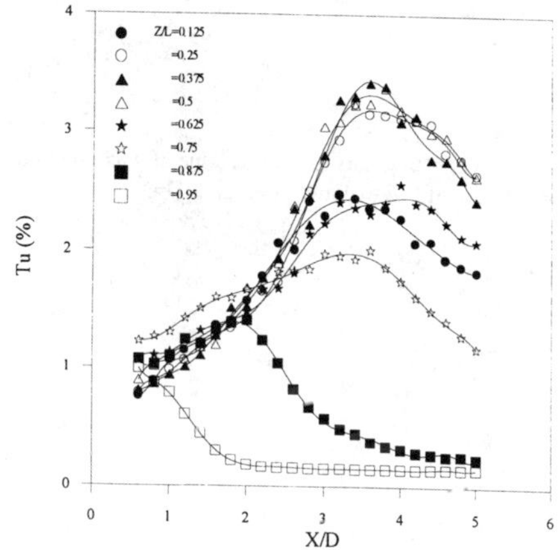

Figure 5. Streamwise turbulent intensity distributions of wake behind the finite cylinder of L/D=10

With approaching to the free-end, the turbulence intensity and the vortex formation region are decreased. This maybe attributed to the fact that the pressure behind the FC free-end is increased due to the interaction between the separated flow from the free-end and vortices shed from each sides of the cylinder. This interaction causes pressure gradient along the longitudinal direction of the cylinder and induces a descending flow from the free-end. Therefore, near the FC free end, the regular vortex shedding is destroyed and the vortex formation region is decreased.

3.3 *Spectral analysis*

Fig. 6 shows the PSD(power spectral density) distribution of wake behind the finite cylinder of L/D=10 at downstream locations of X/D=3,4,5,6. The PSD distributions at Z/L=0.5 show a peak value at the vortex shedding frequency, explaining coherent vortex structure.

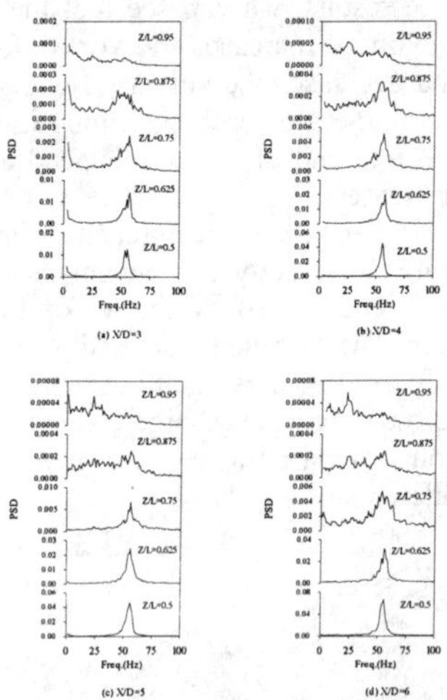

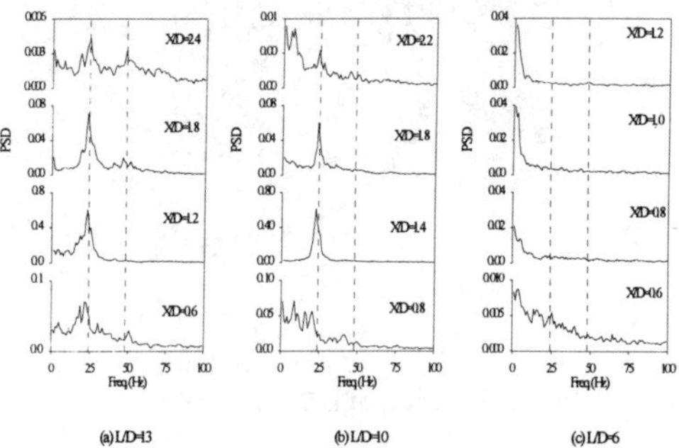

Figure 6. Power spectral density distributions of wake behind the finite cylinder of L/D=10 measured at Y/D=2

Figure 7. Power spectral density distribution along the wake centerline(Y/D=0) at Z/L=0.95

With approaching the FC free-end, the vortex shedding frequency and the PSD value at the peak frequency are slightly decreased. Especially, near the free-end, a peculiar spectral peak is appeared at 24Hz. However, this 24Hz frequency component is appeared only near the FC free-end, where the non-dimensional vertical location Z/L is greater than 0.875. The intermittent frequency characteristics are shown at Z/L=0.875. This results from the fact that the regular vortex shedding and 24Hz frequency component are combined at this location. With going downstream, the 24Hz frequency component becomes clear and the PSD values at 24Hz are increased at Z/L=0.95. Ayoub *et al.*(1982) also found the 24Hz frequency characteristics at the free-end of a finite cylinder at Re=8.5×10^4.

Fig. 7 shows the PSD distributions of streamwise velocity component measured at Z/L=0.95 along the wake centerline(Y/D=0). The 24Hz frequency component is still clear at X/D=2.4 downstream location for the case of L/D=13 cylinder. However, it is difficult to see the 24Hz component at X/D=0.6 location for the FC cylinder of L/D=6. From this, we can see that the 24Hz frequency component is rapidly passed away in the wake centerline as the FC aspect ratio decreases.

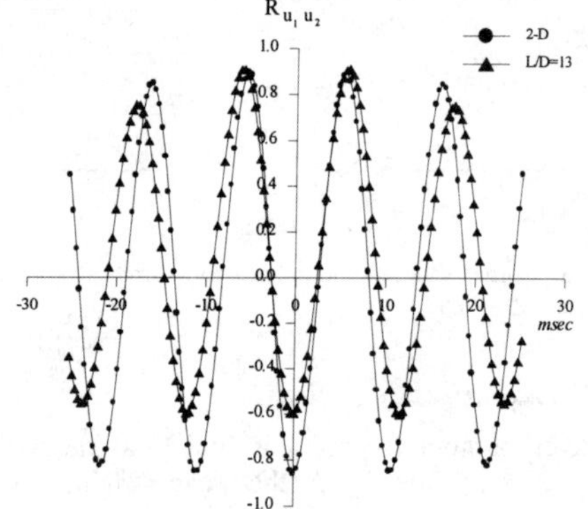

Figure 8. Cross correlation of velocity signals measured from both sides(Y/D=±2) of the finite cylinder at X/D=3, Z/L=0.5

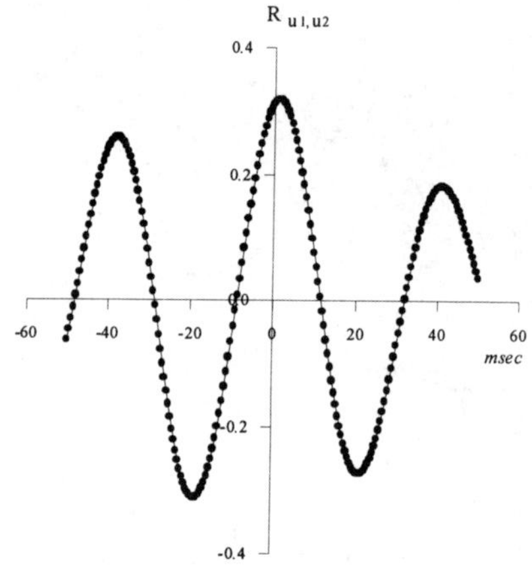

Figure 9. Cross correlation of velocity signals having the 24Hz frequency component at Y/D=±2 for the finite cylinder of L/D=13(X/D=3, Y/D=±2, Z/L=0.95)

The vortex shedding frequencies are changed with respect to the aspect ratio of the finite cylinder, but, the 24Hz frequency component is not affected by the FC aspect ratio at all. This means that the 24Hz component is different from the periodic vortex shedding frequency occurred at a two-dimensional circular cylinder. In order to check this fact, the correlational analysis between vortices shed from both sides of the cylinder was carried out. Two single hot-wire probes were installed in parallel at Y/D=±2 and velocity signals from both probes were measured simultaneously.

Fig. 8 shows the cross-correlation function Ru_1u_2 for the FC models of L/D=13 and 17.3(2-D cylinder) measured at X/D=3, Z/L=0.5. Since the cross-correlation has a local minimum negative value at τ=0(no time-delay), velocity signals measured at both sides of the cylinder are 180° out-of-phase. This indicates that vortices are shed alternatively from both sides of the cylinder in turn. The correlation values for the two-dimensional cylinder are slightly larger than that of the finite cylinder(L/D=13). The time-delay for the finite cylinder is relatively longer than that for the two-dimensional circular cylinder.

Fig. 9 shows the cross-correlation function between the velocity signals measured from both sides(Y/D=±2) of the finite cylinder of L/D=13 at X/D=3. The velocity signals were measured simultaneously at Z/L=0.95, very close to the free-end, and band-pass filtered at 24Hz. The cross-correlation function for the finite cylinder is quite different from that of two-dimensional cylinder. The cross-correlation has a maximum positive value at no time-delay(τ=0). This means that the flows having the 24Hz frequency component on both sides of the cylinder have no phase difference(in-phase).

From this, we can see that the 24Hz frequency component is dominant near the FC free-end and independent of the periodic vortex shedding frequency. This may be closely related with the downwashed shear flow separated from the top of the finite cylinder.

3.4 *Surface pressure*

The surface pressure distributions near the free-end of the finite cylinder of L/D=10 are shown in Fig. 10. This figure shows that, for the finite cylinder, the flow separation from the free-end occurs faster than the two-dimensional cylinder. But, with moving downward along the cylinder axis, the two-dimensionality of surface pressure is gradually recovered. The strong downwash entrainment caused by the free-end reduces the vortex formation region and destroys the periodic vortex shedding near the FC free-end.

The mean pressures on the leeward surface of the finite cylinder are decreased with approaching the free-end. For the finite cylinder, the mean pressures are largely decreased near the free-end because the oncoming flow over the free-end is accelerated. With descending downward from the top along the finite cylinder, however, the mean pressures are increased because the periodic vortex shedding becomes dominant. At Z/L=0.883, the pressure distribution is nearly recovered to that on the two-dimensional cylinder.

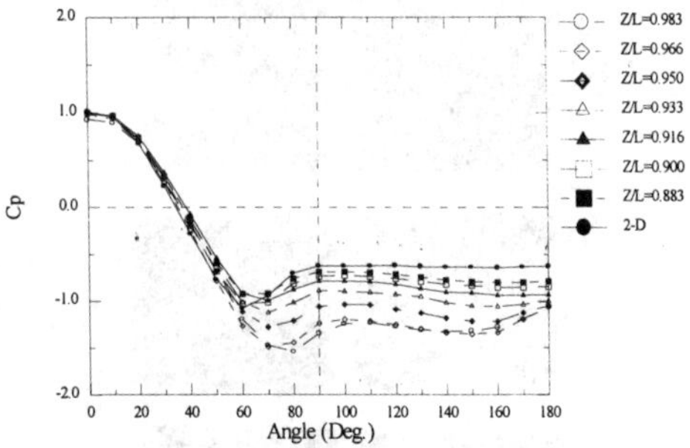

Figure 10. Comparison of pressure coefficient distributions around the free-end of L/D=10 cylinder

3.5 *Flow visualization*

Flow near the free-end of L/D=10 cylinder was visualized using the synchronized smoke-wire technique and some top view results are shown in Fig. 11. A smoke-wire was installed in upstream location(X/D=-2.16) at the same height of the FC free-end(Z/L=1) and the laser light sheet is illuminated horizontally. Flow images near the free-end were photographed with moving the laser light sheet downward from Z/L=1.025 to Z/L=0.962. By investigating the smoke filaments generated in front of the cylinder, the flow structure near the free-end was analyzed qualitatively.

When the smoke filaments come into the horizontal light sheet, they are illuminated whitely as shown in Fig. 11. All these visualized flow are nearly symmetric with respect to the central section(Y/D=0) of the wake. The free-stream ascends in front of the FC free-end and then descends downward just behind the cylinder. The shear flow separated from the free-end descends along the center region of the wake. The vortices in the descending shear layer shown in Fig. 11(d)~(f) seem to be in phase as explained in Fig. 9. The descending downwashed flow entrains large amounts of ambient fluids.

Fig. 12 shows the front view of cross-sectional flow image taken around the free-end of the finite cylinder of L/D=10. The laser light sheet was illuminated vertically at X=0(Y-Z plane) and photographed from a downstream location. A counter-rotating vortex pair is visualized above the FC free-end. Left side vortex rotates in clockwise direction and right one rotates in counter-clockwise. Both vortices have similar size and they are nearly symmetric with respect to the central section(Y/D=0) of the wake as shown in the top views of Fig. 11.

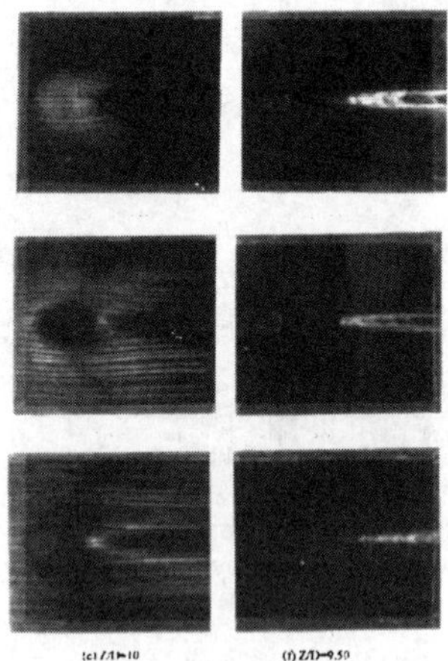

Figure 11. Top view of flow visualization near the free-end of L/D=10 cylinder

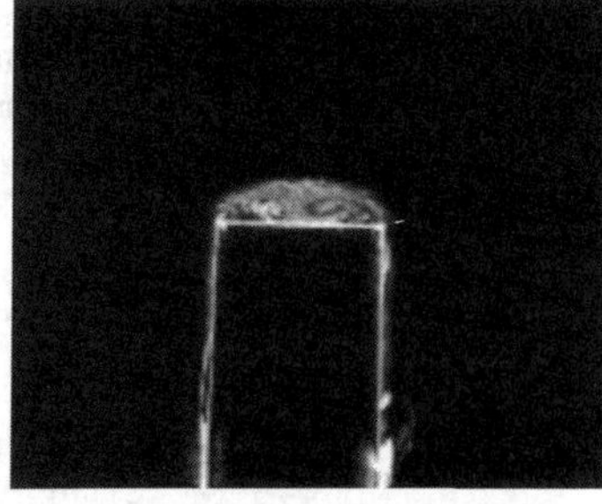

Figure 12. Cross-sectional flow visualization near the free-end of L/D=10 cylinder

The twin-vortex is formed by the roll-up motion of separated shear flow at the edge of the FC free-end. It is continuously separated from top of the cylinder and descends along the central section of the wake. As mentioned above, the 24Hz frequency component seems to be closely related with this twin-vortex.

4 CONCLUSION

The effect of FC(finite cylinder) free-end on the near wake characteristics has been investigated experimentally. The three-dimensionality is attributed to the strong entrainment of ambient irrotational fluids due to the downwash flow at very near the free-end. The downwash flow is concentrated on the center region of the wake. From spectral analysis and cross-correlation of the velocity signals, a peculiar flow structure having 24Hz frequency component exists near the free-end. This 24Hz frequency component seems to be closely related with the counter-rotating twin-vortex formed above the FC free-end. The vortex shedding frequency is decreased and the vortex formation region is increased as the FC aspect ratio decreases. The vortex formation region and periodic vortex shedding are disappeared at near the free-end. This maybe results from the twin-vortex formed above the free-end, strong entrainment and wake interactions. The free-end effect of FC wake becomes dominant with decreasing the FC aspect ratio.

ACKNOWLEDGEMENTS

This work was supported by School of Environmental Engineering, POSTECH and the authors thank for their contribution.

REFERENCES

AYOUB, A. & KARAMCHETI, K. 1982 An experiment on the flow past a finite circular cylinder at high subcritical and supercritical Reynolds numbers. *J. Fluid Mechanics* **118**, 1-26.

BABAN, F., SO, R. M. C. & OTUGEN, M. V. 1989 Unsteady forces on circular cylinders in a cross-flow. *Exp. in Fluids* **7**, 293-302.

FARIVAR, D.1981 Turbulent uniform flow around cylinders of finite length. *J. AIAA* **19**, 275-281.

OKAMOTO, S. & SUNABASHIRI, Y. 1992 Vortex shedding from a circular cylinder of finite length placed on a ground plane. *J. Fluids Eng.* **114**, 512-521.

SLAOUTI, A. & GERRARD, J. H. 1981 An experimental investigation of the end effects on the wake of a circular cylinder towed through water at low Reynolds numbers. *J. Fluid Mechanics* **112**, 297-314.

SZEPESSY, S. & BEARMAN, P. W. 1992 Aspect ratio and end plate effects on vortex shedding from a circular cylinder. *J. Fluid Mechanics* **234**, 191-217.

ZDRAVKOVICH, M. M., BRAND, V. P., MATHEW, G. & WESTON, A. 1989 Flow past short circular cylinders with two free ends. *J. Fluid Mechanics* **203**, 557-575.

Wind Engineering into the 21st Century, Larsen, Larose & Livesey (eds) © 1999 Balkema, Rotterdam, ISBN 90 5809 059 0

The development of separated shear layers of rectangular prisms in turbulent flows

P.C. Lu
Department of Water Resources and Environmental Engineering, Tamkang University, Taiwan

C.M. Cheng
Department of Civil Engineering, Tamkang University, Taiwan

ABSTRACT: Wind tunnel experiments were carried out to study the characteristics of the separated shear layers of rectangular cylinders in two dimensional homogeneous turbulent flows. Measurements of the surface pressures on the lateral sides of the rectangular cylinders and the shear layer velocities were performed simultaneously using pressure taps and split-hot-film, respectively. The results were utilized to characterize the effects of turbulence intensity, turbulence length scale, and the depth-to-width ratio on the development of the separated shear layers. The results show that increases in the turbulence intensity will cause the separated shear layer to thicken and decrease the length of the separation bubble and maximum fluctuating pressure. Changes in the turbulence scale have no significant effects. The reattachment point and position of maximum fluctuating pressure do not coincide, indicating that the reattachment point is not necessarily the position with the greatest pressure values.

1 INTRODUCTION

Under the rapid development of high-rise architecture, wind engineering has become a very important field of research. The forces induced upon high-rise buildings by the wind cannot be neglected. Former research indicates that the effects of such forces are influenced by the bluff body's depth-to-width ratio and the approaching flow field's turbulence intensity (T.I.) and turbulence length scale.

Increases in turbulence intensity cause the separation bubble to shorten and the lateral fluctuating pressure (C'_p) to increase, reaching a maximum at the reattachment zone (Saathoff & Melbourne, 1989; Nakamura & Ozono, 1987).

Small scale turbulence enhances the mixing effect of the flow and increases the growth rate of the separation bubble. Large scale turbulence damages the cross correlation and reduces the vortex strength in the wake, thus diminishing the characteristics of the two-dimensional flow field (Nakamura & Ohya, 1984). Such phenomena is also influenced by the prism's depth-to-width ratio (B/D, where B is the length of the lateral side and D is the length of the side facing the approaching wind). When B/D=0.4, small scale turbulence strengthens regular vortex shedding, causing a lower base pressure. When B/D>0.6, the pressure increases. This pressure reaches a maximum when $L_x/D=3.0$, where L_x is the along-wind turbulence length scale.

As the depth-to-width ratio increases, the mean lateral pressure of a two-dimensional prism in smooth flow decreases (Sant'Anna et al., 1988). However, the pressure recovers near the trailing edge, especially when B/D≥3. At the same time, the fluctuating pressure decreases. In addition, large turbulence length scales have significant effects on the flow structure (Haan Jr. et al., 1998).

The aforementioned researchers studied the separated shear layer using only lateral side prism pressure measurements and not flow field measurements. Those who did actually measure the separated shear layer (Kiya & Sasaki, 1983 and 1985; Ruderich & Fernholz, 1986; Castro & Haque, 1987; Castro & Epik, 1998) utilized long plates. This study was therefore focused on the direct measurements of the separated shear layers and rectangular prism pressure variations in various uniform turbulent flows.

2 EXPERIMENTAL METHOD

A series of experiments were conducted at the Tamkang University Structural Aerodynamics Wind Tunnel Laboratory. The dimensions of this suction open-type wind tunnel are 18.0 m(length) × 2.0 m(width) × 1.5 m(height).

Three models, made of plastic glass, were used in these experiments, with dimensions of 5 cm × 5 cm × 100 cm, 5 cm × 10 cm × 100 cm, and 5 cm × 15 cm × 100 cm, respectively. Two end plates with dimensions of 1.0m × 1.0m, were attached on both ends of the model. The blockage ratio was 2%. The Reynolds number was controlled at 3.3×10^4. Pressure taps were opened along the center of the lateral along-wind sides. Two pressure taps were opened along the width of each model to ensure that the model was placed correctly. Plastic tubings with a length of 110 cm, an outer diameter of 2.5 mm, and an inner diameter of 2.0 mm connected the pressure tappings to a pressure transducer, thus enabling pressure measurements.

The pressure measurements were carried out using a Validyne DP-103-18 pressure transducer and a CD-18, Model MCL-10 signal processor. The pressure exerted upon the taps was transmitted by the tubings to the pressure transducer. The tubing system was calibrated to ascertain the dynamic characteristics of the pneumatic system. Flow field velocity measurements were carried out using a TSI split-hot film 1288, a Model 1125 calibrator, and a 1125R-1 probe rotator. The split-hot film was mounted on a traverse table that moved the split-hot film traversely with a 0.01mm accuracy.

The different turbulence intensities and length scales of the approaching flow fields were generated using square mesh grids, made of wooden bars, mounted upstream prior to the model. Combinations of the bar sizes, the grid sizes, and the distances between the grid and model were used to obtain the different homogeneous turbulent flow fields, as shown in Table 1. The turbulence intensity was found using T.I.=u'/U_0, in which u' is the root mean square of the fluctuating velocity and U_0 is the mean velocity of the approaching flow field. The length scale (L_x) was determined by integrating the corresponding autocorrelation function to its first zero, then multiplying the result by the mean velocity.

Table 1. The longitudinal intensity, u'/U_0, and integral scale, L_x/D, of grid-generated turbulent flows.

Flow pattern	u'/U_0(%)	L_x/D
s	1.1	
a	4.4	1.2
b	4.5	2.0
c	6.1	3.1
d	11.0	2.2
e	10.3	3.5
f	15.2	2.3
g	14.0	3.1

3 RESULTS AND DISCUSSION

3.1 *Lateral mean and fluctuating pressures*

The mean pressure coefficient was defined by $C_p = \overline{(P-P_0)}/0.5\rho U_0^2$, in which P is the instantaneous pressure, P_0 is the static pressure of the Pitot tube inside the wind tunnel, and ρ is the density of the air. Figure 1 shows the mean pressure distribution (C_p) on the lateral sides of all three models in both smooth and turbulent flows (a, b, c, d, e, f, and g). The apparent pressure recovery may be a result of the thickening of the separated shear layer.

The approaching flow separates at the leading edge of the models and curves into a turbulence mixing shear layer, which develops further downstream into a vortex and detaches behind the model. Increasing the turbulence intensity of the flow field alters the formation of the separated shear layer and vortex. As turbulence is induced to the approaching flow, the entrainment of the surrounding flow and instability of the separated shear layer both increase. As a result, the separated shear layer thickens and its central line tends to curve back towards the lateral

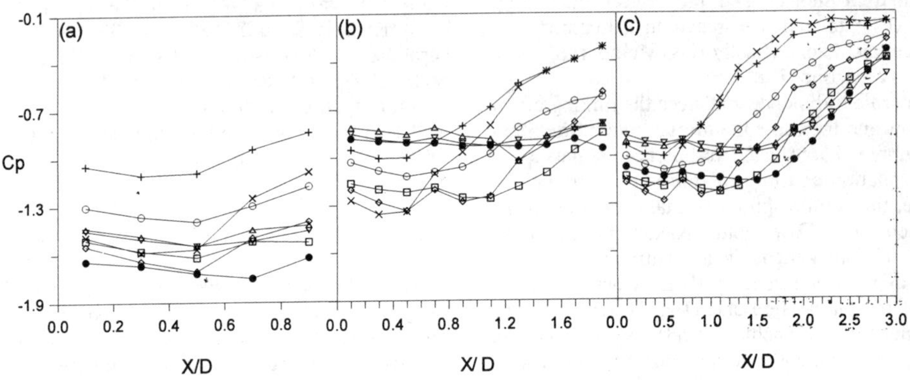

Figure 1. Distributions of mean pressure on the side face. (a) B/D=1 ; (b) 2 ; (c) 3.●,s;△,a; ▽,b;□,c;◇,d;○,e;×,f;+,g.

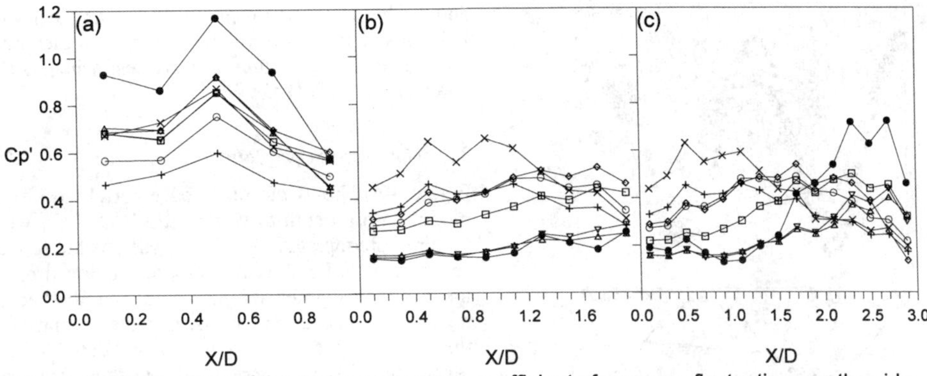

Figure 2. Distributions of the root-mean-square coefficient of pressure fluctuations on the side face. (a)B/D=1;(b)2;(c)3. Symbols as in figure 1.

side, thus causing intermittent reattachment of the shear layer near the trailing edge. Intermittent reattachment in turn affects vortex formation behind the prism. The behavior of the separated shear layer influences the distribution of the lateral mean pressure. Its thickness, curvature, and entrainment of flow also contribute to pressure recovery.

When the turbulence length scale is fixed at approximately 3, (or $L_x/D\approx3$), and the turbulence intensity equals about 5%, 10%, and 15%, as in flow fields c, e, and g, it is clearly shown in Figure 1 that higher turbulence intensity will form thicker separated shear layers, causing greater pressure recovery. This phenomenon can be found in all three models. However, it is less obvious when the turbulence length scale is fixed at approximately 2.

The fluctuating pressure coefficient was defined by $C'_p=p'/0.5\rho U_0^2$, in which p' is the root mean square of the fluctuating pressure. As shown in Figure 2(a), the smooth flow has higher fluctuating pressure (C'_p) values, thus indicating that the turbulence flow fields will disrupt the downstream vortex integrity. In comparing flow fields c, e, and g, greater turbulence intensity creates less fluctuating pressure. This is a result of the turbulence intensity diminishing the energy of the vortex.

In comparing the smooth and turbulent flow fields when B/D=2 and 3, the turbulent flow fields create higher fluctuating pressures than the smooth flow field, as indicated in Figures 2(b) and 2(c). These results are different from those when B/D=1 because the vortices are too weak and no longer dominate the fluctuating pressures. Instead, the approaching flow turbulence determines the fluctuating pressure values. In flow fields b, d, and f the turbulence length scale is approximately 2 and the turbulence intensities are about 5%, 10%, and 15%, respectively (Figure 2(b)). Here the fluctuating pressures along the first half of the lateral side increase with the turbulence intensity. This phenomenon is less obvious further downstream. In flow fields c, e, and g, the same thing occurs at the first half of the lateral sides. However, at the trailing edge, the fluctuating pressure tends to decrease as the turbulence intensity increases.

When B/D=3, as shown in Figure 2(c), the fluctuating pressure increases as the turbulence intensity increases in the front half but decreases near the trailing edge. The turbulence length scale has no significant effects.

Generally, when B/D≥2, the fluctuating pressures on the lateral sides of the prisms are affected simultaneously by the turbulence of the approaching flow and the downstream vortex integrity. Before reattachment occurs, the fluctuating pressure increases with the turbulence intensity, thus indicating dominance of the approaching flow field. However, when reattachment takes place, vortex formation occurs even further downstream. At the same time, higher turbulence intensity disrupts the vortex integrity and creates lower fluctuating pressure, thus indicating dominance of the vortex.

When the turbulence length scale $L_x/D<2$, its effects upon the pressure distribution are not significant (Nakamura & Ozono, 1987). However, large turbulence scales, especially when $L_x/D=3$, weaken the downstream vortex, thus increasing the base pressure (Nakamura & Ohya, 1984).

When the turbulence intensity is approximately 10% and 15% and the turbulence length scale is fixed at approximately 2 and 3, thus including flow fields d, e, f, and g, increases in the turbulence length scales will cause the fluctuating pressure to decrease. The same occurs when B/D=1, 2, and 3.

In Figure 2(a), it is shown that there is a slight pressure recovery near the trailing edge of the model. This does not necessarily mean reattachment has occurred. This can be a result of shear-layer edge direct interaction (Nakamura & Ohya, 1984). In comparison of the pressure distributions in Figures 1

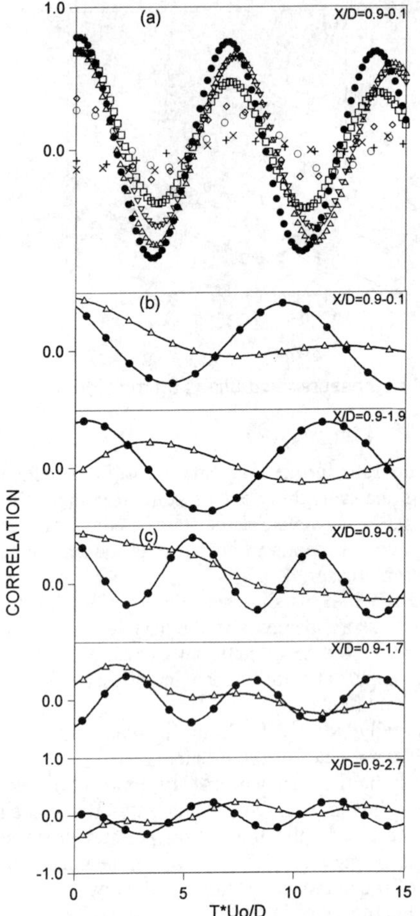

Figure 3. Time-lagged pressure cross-correlation with fixed transducer at X/D=0.9. (a) B/D=1, movable, transducer at X/D=0.1; (b) B/D=2 at X/D=0.1 and 1.9; (c) B/D=3 at X/D=0.1,1.7 and 2.7. Symbols as in figure 1.

and 2, the mean lateral pressures are greatest when B/D=3 and least when B/D=1. On the other hand, the fluctuating pressures are greatest when B/D=1 and least when B/D=3.

3.2 *Time-space correlation*

Figure 3(a) shows the time-space correlation of the model with a depth-to-width ratio B/D=1, in which the pressure tap at X/D=0.9 was used as the reference point. It is obvious that the correlation drops as the turbulence intensity increases, mainly because high turbulence intensity disrupts the formation of the vortex. This also occurs when B/D=2 (Figure 3(b)) and B/D=3 (Figure 3(c)). On the other hand, the turbulence length scale has no significant effect on the correlation, thus proving it also has little effect on the separated shear layer.

In a smooth flow, the depth-to-width ratio has no influence on the time-space correlation, as shown in Figures 3(a), 3(b), and 3(c). However, in a turbulent flow, increases in the depth-to-width ratio result in decreases in correlation. When B/D=2 and 3, the correlation shows stronger periodic characteristics further downstream, indicating the dominance of the vortex shedding.

The peak to peak dimensionless time delay in Figures 3(a), 3(b), and 3(c) are 8.3, 11.0, and 6.3, respectively. The inverse of these values equals the Strouhal number of each prism in a smooth flow. The Strouhal number is 0.12 when B/D=1, 0.09 when B/D=2, and 0.16 when B/D=3 (Okajima, 1982).

3.3 *Velocity profiles*

Figure 4(a) shows the velocity profile of the separated shear layer at X/D=0.3 and 0.7 when B/D=1. The results also indicate that the separated shear layer thickens as the turbulence intensity increases, while

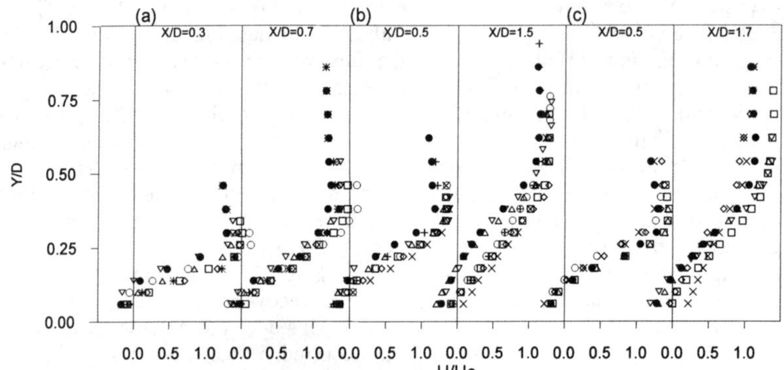

Figure 4. Mean velocity profiles. (a) B/D=1 at X/D=0.3,0.7; (b) B/D=2 at X/D=0.5,1.5; (c) B/D=3 at X/D=0.5,1.7. Symbols as in figure 1.

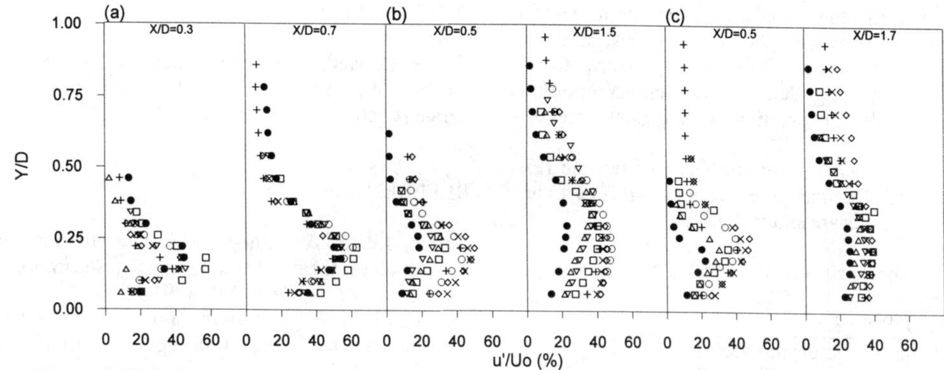

Figure 5. Turbulence intensity profiles. (a) B/D=1 at X/D=0.3,0.7;(b) B/D=2 at X/D=0.5,1.5; (c) B/D=3 at X/D=0.5,1.7. Symbols as in figure 1.

changes in the turbulence scale have no effect. The same can be seen in Figures 4(b) and 4(c), which show that the velocity profiles at X/D=0.5 and 1.5 when B/D=2 and at X/D=0.5 and 1.7 when B/D=3.

The corresponding profiles of the turbulence intensity are shown in Figure 5. The values clearly indicate that as the turbulence intensity of the approaching flow is increased, the turbulence intensity of the separated shear layer also increases. These differences are more evident when B/D=2 and 3.

3.4 *Reattachment of separated shear layers*

After flow separation occurs at the leading edge, the intermittence of the reverse flow is 100% and gradually decreases further downstream (Kiya & Sasaki, 1983). At the time-mean reattachment point, the intermittence of the reverse flow is 50%, and it continues to decrease as the plate becomes longer. Reattachment therefore occurs in a flapping zone. Figure 6 shows the intermittence of the reverse flow at Y=0.06D.

In Figure 6(a), it is obvious that there is a 50% reverse flow at the trailing edge and pressure recovery (Figure 2(a)). However, this is a result of shear-layer edge direct interaction, not reattachment.

When B/D=2, pressure recovery occurs in flow fields c, d, e, f, and g. The positions of maximum fluctuating pressure (Figure 2(b)) and 50% intermittence of the reverse flow (Figure 6(b)) are compared in table 2.

Table 2: The positions of the 50% intermittence of the reverse flow and C'_{pmax} for different flow patterns when B/D=2.

Flow Pattern	50% Intermittence of Reverse Flow	C'_{pmax}
c	1.90D	1.7D
d	1.65D	1.7D
e	1.70D	1.3D
f	1.40D	0.9D
g	1.45D	1.1D

The results showed that increases in the turbulence intensity cause the length of the separation bubble and position of the maximum fluctuating pressure to

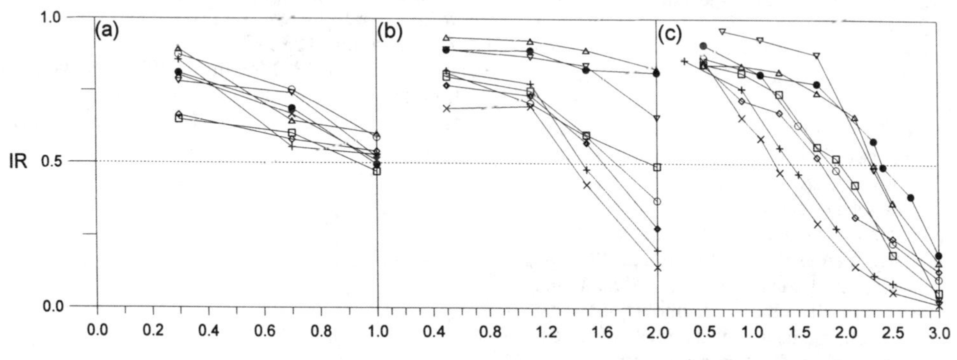

Figure 6. Distributions of the intermittence of reverse flow. (a)B/D=1;(b)2;(c)3. Symbols as in figure 1.

decrease, while the turbulence scale has no significant effects. However, the reattachment point and position of maximum fluctuating pressure do not coincide, indicating that the reattachment point is not necessarily the position with greatest pressure values.

Figure 6(c) shows the intermittence of reverse flow when B/D=3. Reattachment occurs in all flow fields, and the positions are shown below.

Table 3: The positions of the 50% intermittence of the reverse flow and C'_{pmax} for different flow patterns when B/D=3.

Flow Pattern	Intermittence of Reverse Flow 50%	C'_{pmax}
s	2.40D	2.7D
a	2.30D	2.7D
b	2.25D	2.7D
c	1.90D	2.7D
d	1.75D	1.7D
e	1.85D	1.7D
f	1.25D	0.5D
g	1.40D	0.5D

When B/D=2 and 3, the separation bubble shortens as the turbulence intensity increases. Usually, the position of maximum fluctuating pressure is also that of reattachment. However, the positions of reattachment and maximum fluctuating pressures do not completely agree.

4 CONCLUSION

In a two-dimensional flow field, all models with a depth-to-width ratio B/D=1, 2, and 3 have a pressure recovery near the trailing edge of the lateral sides. As the depth-to-width ratio is increased, the negative mean pressure values and fluctuating pressure values both decrease. Without reattachment of the separated shear layer, such as when B/D=1, the pressure distribution on the lateral side is dominated by the downstream vortex. However, before reattachment occurs, the pressure is dominated by the approaching flow field. After reattachment, the pressure is dominated by the downstream vortex, as in the case of B/D=2 and 3. As the turbulence intensity increases, the time-space correlation of the lateral pressures decreases. Changes in the turbulence scale, however, have no significant effects. In two-dimensional flow fields, the separated shear layer thickens as the turbulence intensity increases. In other words, the curvature of the separated shear layer increases. When reattachment occurs, the separation bubble shortens. The point of reattachment, however, does not necessarily have the maximum fluctuating pressure.

ACKNOWLEDGMENT

We acknowledge the grants/financial support from the National Science Council of the Republic of China (contract NO. NSC-86-2621-E-032-007).

REFERENCES

Castro, I.P. & A. Haque 1987. The structure of a turbulent shear layer bounding a separation region. *J. Fluid Mech.* 179:439-468.

Castro, I.P. & E. Epik 1998. Boundary layer development after a separation region. *J. Fluid Mech.* 374:91-116.

Da Matha Sant'Anna, F.A., A. Laneville, J.Y. Trepanier & Z.Y. Lu 1988. Detailed pressure field measurements for some 2-D rectangular cylinders. *J. Wind Eng. Indust. Aerodyn.* 28:241-250.

Haan Jr., F.L., A. Kareem & A.A. Szewczyk 1998. The effects of turbulence on the pressure distribution around a rectangular prism. *J. Wind Eng. Indust. Aerodyn.* 77 & 78:381-392.

Kiya, M. & K. Sasaki 1983. Structure of a turbulent separation bubble. *J. Fluid Mech.* 137:83-113.

Kiya, M. & K. Sasaki 1985. Structure of large scale vortices and unsteady reverse flow in the reattaching zone of a turbulent separation bubble. *J. Fluid Mech.* 154:463-491.

Lee, B.E. 1990. Some observations of the effect of aspect ratio on the influence on the drag of rectangular cylinders. *J. Wind Eng. Indust. Aerodyn.* 33:107-111.

Nakamura, Y. & S. Ozono 1987. The effects of turbulence on a separated and reattaching flow. *J. Fluid Mech.* 178:477-490.

Nakamura, Y. & Y. Ohya 1984. The effects of turbulence on the mean flow past two-dimensional rectangular cylinders. *J. Fluid Mech.* 149:255-273.

Ruderich, R. & H.H. Fernholz 1986. An experimental investigation of a turbulent shear flow with separation, reverse flow and reattachment. *J. Fluid Mech.* 163:283-322.

Saathoff, P.J. & W.H. Melbourne 1989. The generation of peak pressures in separated/reattaching flows. J. Wind Eng. Indust. Aerodyn. 32:121-134.

Wind Engineering into the 21st Century, Larsen, Larose & Livesey (eds) © 1999 Balkema, Rotterdam, ISBN 90 5809 059 0

Clearance interference between structures or structural components

K.F. McNamara & T. Mullarkey
National University of Ireland, Galway, Ireland

ABSTRACT: Structures or their components are sometimes placed in close proximity. When structures or their components vibrate there is a possibility of the clearance gap between the structures being reduced to zero. This may be a 1-dimensional, 2-dimensional or 3-dimensional problem. Solutions to this type of problem have been given for 1-dimensional deterministic situations. Most structures or their components oscillate in a random manner when subject to wind or other natural phenomena. An analysis of the 1-dimensional random vibration problem is considered in this paper. Using analyses in both the time domain and the frequency domain a theoretical estimate for the mean square clearance is obtained. Using this together with appropriate statistical distributions the probability of clearance interference may be obtained.

1 INTRODUCTION

Frequently structures or structural components are placed close enough to each other that the possibility of clashing or clearance interference when they vibrate should be considered. The possibility of structures or structural components hitting off each other may in some cases be only a nuisance but in most cases would be considered a serviceability failure. This kind of behavior may also induce structural degradation. In some cases, such as overhead electrical power conductors, a clearance reduction would lead to a major serviceability failure. The initiative for this study was an analysis of the gap between two water towers that for visual impact rather than space restrictions were built in close proximity as shown in Fig. 1.

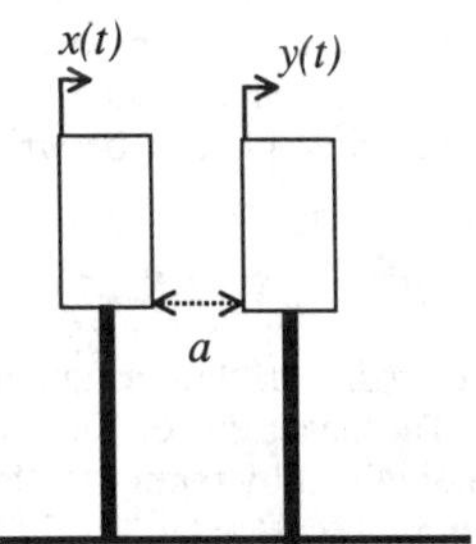

Figure 1. Water towers built in close proximity

Other areas where this type of phenomenon might occur are:

1. Pipe risers in oil production platforms,
2. Clearance between moored structures or ships,
3. Earthquake loading of structures and
4. Clearance between railway cars.

The problem of clearance for electrical power conductors has received much attention (Clapp 1994, Havard et al. 1987, Davenport 1979, Nag 1976). Clearance between railway cars and clearance between moored floating bodies have also been considered (Eickhoff & Scott 1996, DeRabi 1994). Traditional methods made use of deterministic analysis that does not take into account the randomness of the phenomenon. This method may give reasonable results in some situations. In the case of wind loading of structures it is inherently flawed because of the randomness of vibrations observed in reality. Another complicating factor is the phase difference between the two motions. The phase angle is random and may also depend on frequency. An approximation that gives an upper bound on the probability of interference is proposed by assuming that the motions are 180^0 out of phase for all frequencies. If this is assumed to be so then it is shown that the variance obtained will be a maximum.

An analysis is given here which makes use of both time and frequency domains to obtain estimates the statistics of the clearance $z(t) = a - [x(t) - y(t)]$ as shown in Figure 1. The probability of the clearance then being reduced to zero or any other value may be found using standard probability distributions.

2 DETERMINISTIC ANALYSIS

Standard deterministic methods of dealing with this problem assume that the motions represented by $x(t)$ and $y(t)$ are harmonic with amplitudes A_1 and A_2 and phase angles ϕ_1 and ϕ_2 respectively. Then the dynamic clearance $z(t)$ at any time is given by:

$$z(t) = a + A_2 \sin(\omega t + \phi_2) - A_1 \sin(\omega t + \phi_1) \quad (1)$$

where a = undisturbed clearance; and ω = frequency of vibration, which is assumed to be the same for both $x(t)$ and $y(t)$. Letting τ be a shifted time scale such that $\omega\tau = \omega t + \phi_1$, then equation 1 may be rewritten:

$$z(t) = a + A_2 \sin(\omega\tau + \Delta\phi) - A_1 \sin \omega\tau \quad (2)$$

where $\Delta\phi = \phi_2 - \phi_1$.

Clearance interference may be defined as $z(t) \leq 0$ or in certain cases if $z(t)$ is less than some arbitrary value a'. In order to find out if any of these condition occurs it is necessary to find the minimum of $z(t)$. To obtain a minimum (or maximum) differentiate equation 2 with respect to τ and set equal to zero giving the following equation:

$$\frac{\partial z}{\partial \tau} = A_2\omega\cos(\omega\tau + \Delta\phi) - A_1\omega\cos\omega\tau = 0 \quad (3)$$

This implies that:

$$\frac{A_1}{A_2}\cos\omega\tau = \cos(\omega\tau + \Delta\phi) \quad (4)$$

or:

$$\frac{A_1}{A_2}\cos\omega\tau = \cos\omega\tau\cos\Delta\phi - \sin\omega\tau\sin\Delta\phi \quad (5)$$

Rearranging the terms in equation 5 will result in the following relationship:

$$\tan\omega\tau = \frac{\cos\Delta\phi - A_1/A_2}{\sin\Delta\phi} \quad (6)$$

This equation may be solved for $\omega\tau$ and substituting $\omega\tau$ and $\omega\tau + \pi$ into equation 2 the maximum/minimum clearance may be computed exactly.

This method of analysis may in some simple cases give reasonable results but there are major disadvantages with the above theory. Firstly, vibrations of structures due to environmental effects such as wind loading are not deterministic but can only be analyzed correctly by using the theory of random vibration. Secondly, the frequencies of vibration of the two structures may be different. Thirdly the phase angle may also be random. An attempt to overcome some of these difficulties is given in the next section in which the theory of random vibrations is used to obtain an estimate of the variance of the clearance.

3 STOCHASTIC ANALYSIS

The clearance at any time between the two structures shown in figure 1 is given by:

$$z(t) = a - [x(t) - y(t)] \quad (7)$$

This section gives a theoretical method for the evaluation of the variance of $z(t)$ in equation 7. The mean or expected value of $z(t)$ is $E[z(t)] = a$. Knowing the mean and variance of $z(t)$ then probability estimates of clearance reduction or clearance interference may be obtained.

In the following the force spectrum for both structures is given by $S_{II}(f)$. It is assumed here that the force spectrum is the same for both structures. The output spectra and phases for x and y, figure 1, are represented by $S_{xx}(f)$, $S_{yy}(f)$, $\phi(f)$ and $\phi(f)$ respectively. If $z(t)$ and $z(t+\tau)$ represent the clearance between two structures at any arbitrary times t and $t+\tau$, as is given in equation 7, then the auto-correlation function of the clearance between the two structures is given by:

$$R_{zz}(\tau) = \langle z(t)z(t+\tau)\rangle \quad (8)$$

where < > represents time averaging.

The spectrum of the clearance $z(t)$ is given by the Fourier transform of the autocorrelation as follows, (Bendat & Piersol 1971 & 1980):

$$S_{zz}(f) = \int_{-\infty}^{\infty} 2R_{zz}(\tau)e^{-i2\pi f\tau}d\tau \quad (9)$$

An expression for the autocorrelation $R_{zz}(\tau)$ may be obtained by substituting equation 7 for both $z(t)$ and $z(t+\tau)$ into equation 8 giving:

$$R_{zz}(\tau) = \langle[a + y(t) - x(t)][a + y(t+\tau) - x(t+\tau)]\rangle \quad (10)$$

Multiplying out and collecting terms in equation 10 gives the following expression:

$$\begin{aligned} R_{zz}(\tau) = &\langle a^2\rangle + \langle ay(t+\tau)\rangle - \langle ax(t+\tau)\rangle + \langle ay(t)\rangle + \\ &\langle y(t)y(t+\tau)\rangle - \langle y(t)x(t+\tau)\rangle - \langle ax(t)\rangle - \\ &\langle x(t)y(t+\tau)\rangle + \langle x(t)x(t+\tau)\rangle \end{aligned} \quad (11)$$

Recognizing that the time average of both $x(t)$ and $y(t)$ are zero, because they represent vibration about the mean position, then equation 11 may be written as follows:

$$R_{zz}(\tau) = a^2 + R_{xx}(\tau) - R_{xy}(\tau) - R_{yx}(\tau) + R_{yy}(\tau) \quad (12)$$

where R_{xy} and R_{yx} are the cross-correlation functions of x and y respectively. The spectrum of the clearance, $S_{zz}(f)$, may now be obtained by taking the Fourier transform of equation 12 (Bendat & Piersol 1971 & 1980). Substituting $R_{zz}(\tau)$ given by equation 12 into equation 9 yields the following:

$$S_{zz}(f)=2\int_{-\infty}^{\infty}[a^2+R_{xx}(\tau)-R_{xy}(\tau)-R_{yx}(\tau)+R_{yy}(\tau)]e^{-i2\pi f\tau}d\tau$$
$$=2\int_{-\infty}^{\infty}a^2e^{-i2\pi f\tau}d\tau+2\int_{-\infty}^{\infty}R_{xx}(\tau)e^{-i2\pi f\tau}d\tau-2\int_{-\infty}^{\infty}R_{xy}(\tau)e^{-i2\pi f\tau}d\tau-2\int_{-\infty}^{\infty}R_{yx}(\tau)e^{-i2\pi f\tau}d\tau+2\int_{-\infty}^{\infty}R_{yy}(\tau)e^{-i2\pi f\tau}d\tau \quad (13)$$

Equation 13 may be written as follows (Bendat & Piersol 1971 & 1980):

$$S_{zz}(f)=a^2\delta(f)+S_{xx}(f)-S_{xy}(f)-S_{yx}(f)+S_{yy}(f) \quad (14)$$

The cross spectra, $S_{xy}(f)$ and $S_{yx}(f)$ may be written in terms of the frequency response function $H_{xl}(f)$ as follows (Robson 1963, Bendat & Piersol 1971 & 1980):

$$S_{xy}(f) = H^*_{xl}(f)\, H_{yl}(f)\, S_{ll}(f) \quad (15)$$

and

$$S_{yx}(f) = H^*_{yl}(f)\, H_{xl}(f)\, S_{ll}(f) \quad (16)$$

where $H^*_{xl}(f)$ *and* $H^*_{yl}(f)$ are the complex conjugates of the frequency response functions.

Using the following definitions (Bendat & Piersol 1971 & 1980)

$$H(f)=|H(f)|e^{-i\phi(f)};$$

$$H_{xl}(f)=|H_{xl}(f)|e^{-i\phi_{xl}(f)};$$

$$H^*_{xl}(f)=|H_{xl}(f)|e^{i\phi_{xl}(f)};$$

$$H_{yl}(f)=|H_{yl}(f)|e^{-i\phi_{yl}(f)} \text{ and}$$

$$H^*_{yl}(f)=|H_{yl}(f)|e^{i\phi_{yl}(f)}$$

then the cross spectrum $S_{xy}(f)$ may be written:

$$S_{xy}(f)=|H_{xl}(f)|e^{i\phi_{xl}(f)}|H_{yl}(f)|e^{-i\phi_{yl}(f)}S_{ll}(f) \quad (17)$$

or

$$S_{xy}(f)=|H_{xl}(f)||H_{yl}(f)|e^{i[\phi_{xl}(f)-\phi_{yl}(f)]}S_{ll}(f) \quad (18)$$

The product of the frequency response functions is given by (Robson 1963, Bendat & Piersol 1971 & 1980):

$$|H_{xl}(f)||H_{yl}(f)|=\frac{\sqrt{S_{xx}(f)S_{yy}(f)}}{S_{ll}(f)} \quad (19)$$

Using this relationship then equation 18 may be re-written:

$$S_{xy}(f)=\sqrt{S_{xx}(f)S_{yy}(f)}e^{i[\phi_{xl}(f)-\phi_{yl}(f)]} \quad (20)$$

Using similar analysis the cross spectrum $S_{yx}(f)$ may be written:

$$S_{yx}(f)=\sqrt{S_{xx}(f)S_{yy}(f)}e^{i[\phi_{yl}(f)-\phi_{xl}(f)]} \quad (21)$$

Making use of the relationships given by equations 20 and 21 equation 14 may be expressed as follows:

$$S_{zz}(f)=a^2\delta(f)+S_{xx}(f)+S_{yy}(f)-\sqrt{S_{xx}(f)S_{yy}(f)}\{e^{i[\phi_{xl}(f)-\phi_{yl}(f)]}+e^{i[\phi_{yl}(f)-\phi_{xl}(f)]}\} \quad (22)$$

Noting that

$$e^{i[\phi_{xl}(f)-\phi_{yl}(f)]}=e^{-i[\phi_{yl}(f)-\phi_{xl}(f)]}=\cos[\phi_{yl}(f)-\phi_{xl}(f)]-i\sin[\phi_{yl}(f)-\phi_{xl}(f)]=\cos[\phi_{yl}(f)-\phi_{xl}(f)]+i\sin[\phi_{xl}(f)-\phi_{yl}(f) \quad (23)$$

Similarly:

$$e^{i[\phi_{yl}(f)-\phi_{xl}(f)]}=\cos[\phi_{xl}(f)-\phi_{yl}(f)]-i\sin[\phi_{xl}(f)-\phi_{yl}(f)] \quad (24)$$

Substituting equations 23 and 24 into equation 22 and using the fact the cos, is an even function yields:

$$S_{zz}(f)=a^2\delta(f)+S_{xx}(f)+S_{yy}(f)-2\sqrt{S_{xx}(f)S_{yy}(f)}\cos[\phi_{xl}(f)-\phi_{yl}(f)] \quad (25)$$

The mean square value of the clearance $z(t)$ may then be found by integrating the spectrum of $z(t)$ given by equation 25.

$$E[z^2]=\int_0^{\infty}S_{zz}(f)df \quad (26)$$

Substituting the expression for $S_{zz}(f)$ given by equation 25 into equation 26 gives:

$$E[z^2]=a^2+\int_0^{\infty}\{S_{xx}(f)+S_{yy}(f)-2\sqrt{S_{xx}(f)S_{yy}(f)}\cos[\phi_{xl}(f)-\phi_{yl}(f)]\}df \quad (27)$$

The variance of $z(t)$ may then be found by the following :

$$VAR[z]=\sigma^2_{zz}=E[z^2]-E^2[z] \quad (28)$$

Since $E^2[z]$ is equal to a^2 then the variance of $z(t)$ is given by:

$$\sigma_{zz}^2 = \int_0^\infty \{S_{xx}(f) + S_{yy}(f) - 2\sqrt{S_{xx}(f)S_{yy}(f)}$$
$$\cos[\phi_{xl}(f) - \phi_{yl}(f)]\}df \quad (29)$$

Problems arise in the evaluation of the phase factors $\phi_{xl}(f)$ and $\phi_{yl}(f)$. An upper bound on the value of the variance may be obtained by assuming that the phase factors are 180° apart for all frequencies. Using this then the variance given by equation 29 will be a maximum.

4 PROBABILITIES OF CLEARANCE INTERFERENCE

If it is assumed that both $x(t)$ and $y(t)$ are Gaussian variables then $z(t)$ is also a Gaussian variable. The probability distribution of $z(t)$ is then given by:

$$f_z(z)dz = \frac{1}{\sqrt{2\pi}\sigma_{zz}} e^{-\frac{1}{2}\left(\frac{z-a}{\sigma_{zz}}\right)^2} dz \quad (30)$$

The probability of the clearance $z(t)$ being reduced to a value a or less is given by:

$$P[z \le a'] = \int_{-\infty}^{a} \frac{1}{\sqrt{2\pi}\sigma_{zz}} e^{-\frac{1}{2}\left(\frac{z-a}{\sigma_{zz}}\right)^2} dz \quad (31)$$

An estimate of σ_{zz} may be obtained from equation 29. Since the estimate of σ_{zz} given by equation 29 is an upper bound then the probability given by equation 31 will also be an upper limit on the probability. Equation 31 gives the probability of the clearance being reduced to any value a' including a = 0.

5 CONCLUSIONS

A theoretical derivation of the variance of the clearance between two randomly vibrating structures is derived. The expression derived includes phase factors that in some cases may not easily be evaluated. If the phase factors are assumed to be 180° out of phase for all frequencies then an upper limit on the variance may be obtained. Using this value of the variance together with the Gaussian distribution an upper limit on the probability of clearance reduction or clearance interference between randomly vibrating structures may be obtained. It would be useful to investigate these phase factors further, in particular their effects on the variance of the clearance for specific conditions of random vibration. A better estimate of the probability of clearance interference may be obtained by using peak distributions rather than using the parent Gaussian distribution.

REFERENCES

Bendat, J.S. & Piersol, A.G. 1971. Random data: analysis and measurement procedures. New York: Wiley-Interscience.

Bendat, J.S. & Piersol, A.G. 1980. Engineering applications of correlation and spectral analysis. New York: Wiley-Interscience.

Clapp, A.L. 1994. Calculation of horizontal displacement of conductors under wind loading towards buildings and other supporting structures. IEEE Transactions on Industry Applications 30(4): 496-504.

Davenport, A.G. 1979. Gust response factors for transmission line loading. Proc. of Fifth ICWE, Fort Collins CO. July 1979: 899-909.

De Rabi, S. 1994. Development of reliability-based equations for minimum clearance between two moored structures. Proc. of 13th International Conference on Offshore Mechanics and Arctic Engineering, Houston TX. Feb. 27-Mar.3 1994: 289-298.

Eickhoff, B.M. & Scott, G.A. 1996. Structure gauging – maximizing the use of restricted clearance. Proc. of Institution of Mechanical Engineers, Part F: Journal of Rail and Rapid Transit 210(2): 63-71.

Havard, D.G., Pon, C.J. & Watt, G.S. 1987. Clearance for galloping on overhead lines. Proc. of the First International Symposium: Probabilistic Methods Applied to Electric Power Systems, Toronto Ont.: 503-512.

Nag, G.K. 1976. Conductor spacing and clearance to structures for high voltage lines. Journal of the Institution of Engineers (India), Part EL: Electrical Engineering Division 57(EL): 18-24.

Robson, J.D. 1963. An introduction to random vibration. Edinburgh: Edinburgh University Press.

Wind Engineering into the 21st Century, Larsen, Larose & Livesey (eds) © 1999 Balkema, Rotterdam, ISBN 90 5809 059 0

Mechanism of the occurrence for fluctuating forces on a circular cylinder

H. Nishimura
General Building Research Corporation, Osaka, Japan

Y. Taniike
Faculty of Engineering, Osaka City University, Osaka, Japan

ABSTRACT: This paper presents the mechanism of the occurrence for the fluctuating force on a stationary two-dimensional circular cylinder in a smooth flow at a subcritical regime. The characteristic of the fluctuating pressures produced by Karman vortex shedding was searched using the simultaneous multi-pressure measuring system. Furthermore, the fluctuation of the stagnation point and the separation point on the cylinder were clarified by instantaneous pressure distributions. The physical relationship among the fluctuation of the lift force and the fluctuations of their points was discussed to explain the fluctuating wind force induced by the Karman vortex shedding. It was found that the Karman vortex shedding, which periodically oscillates the whole flow around the cylinder, dominates the fluctuation of pressures on the windward body before the separation point as same as the after body.

1 INTRODUCTION

Numerous wind tunnel studies have been carried out to investigate the aerodynamic characteristics of a circular cylinder, which is one of the most simple and basic shape of bluff bodies for the fluid dynamics. Wieselseberger(1921) studied the variation of the drag coefficient of a circular cylinder in wide ranges of the Reynolds number. Fung(1960) determined the fluctuating drag and lift coefficients of a circular cylinder in a super-critical regime. Roshko(1961) found the phenomenon that the periodic vortex shedding appeared in a trans-critical regime, which disappeared in a super-critical regime. Achenbach(1968) and James(1980) also confirmed the Roshko's founding. Bearman(1969) reported that the mean lift force appeared in a critical regime, because of an unbalanced separation from one side of a cylinder, of which the result was confirmed by Schewe(1983). Recently, Zdravkovich(1997) reviewed a large number of studies on characteristics of aerodynamic forces for a circular cylinder (and a group of cylinders). One of interests of many studies subjected to a circular cylinder is the effect of fluctuating forces due to the Karman vortex shedding.

It is well known that the fluctuating pressures induced by the shedding of the Karman vortex on a stationary circular cylinder yield the periodic lift force, especially in a sub-critical Reynolds number.
In this study, we investigated the mechanism of the occurrence for the periodic force on a stationary two-dimensional circular cylinder in a range of the sub-critical Reynolds number, by using the simultaneous multi-pressure measuring system. It was found that the fluctuations of the stagnation point and the separation point on the cylinder are synchronized with the lift fluctuation, both of which have the same frequency as the Strouhal number. We also clarified the relationships among the shedding of the Karman vortex, the movement of the separated shear layer and the occurrence of the fluctuating pressures on a circular cylinder.

2 EXPERIMENTAL ARRANGEMENT

The experiment was performed in the Eiffel type wind tunnel which had the test section of a width of 1.8m and a height of 1.3m. The model had a diameter of B=91mm and a length of 600mm was made of a polyethylene pipe which was polished by #600 emery paper. The model was fixed between two end-plates in order to keep the two dimensionality of the flow around the model. A smooth flow was used as the approaching flow, which had the turbulent intensity of 0.1%. The wind velocity was V=10m/s giving the corresponding Reynolds number of $Re=6.1\times10^4$ that was included in the sub-critical regime. 62 pressure-taps set on the surface of the cylinder in the center plane of the span is shown in Figure 1. Pressures on taps were led to individual pressure transducers through vinyl tubes, owing to acquire simultaneous pressures. The data from each transducer were simultaneously sampled at a rate of

400Hz. Signals from each transducer were corrected in the frequency regime to remove the part of the signal associated with Helmholtz resonance of the transducer/tube cavity. The blokage ratio of the model was 2.3%, and no attempt was made to correct on data for the blokage effect. In order to know the effect of the shedding of Karman vortices, the additional experiment with a splitter plate set behind the cylinder was also carried out.

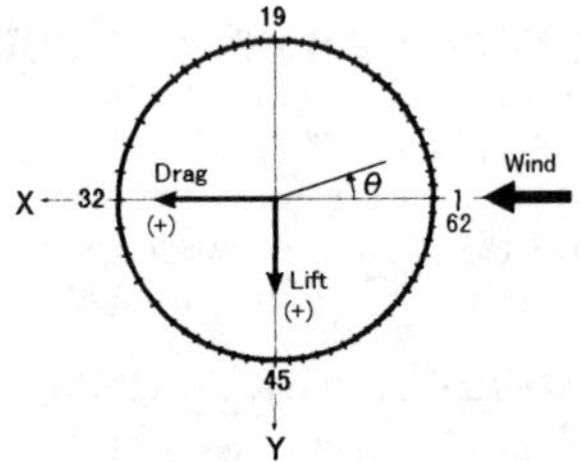

Figure 1 Pressure taps and the definition of wind forces.

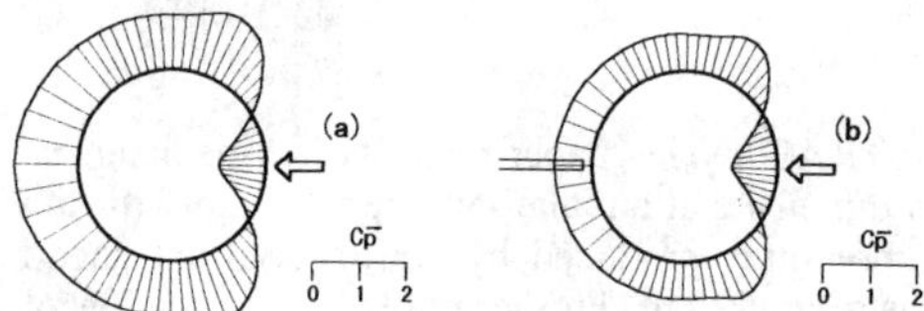

Figure 2 Mean pressure distribution. (a) without a splitter plate, and (b)with a splitter plate.

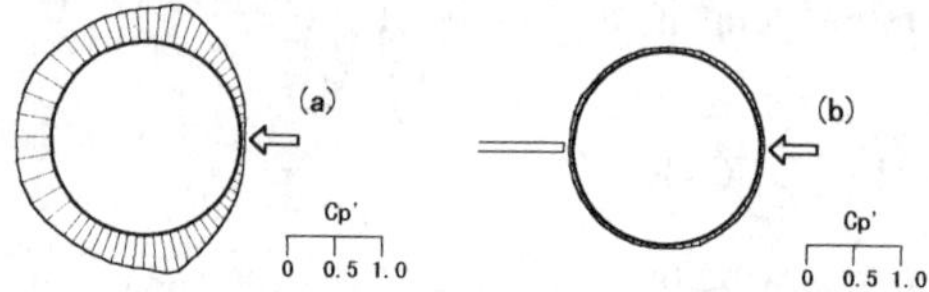

Figure 3 Fluctuating pressure distribution. (a) without a splitter plate, and (b)with a splitter plate.

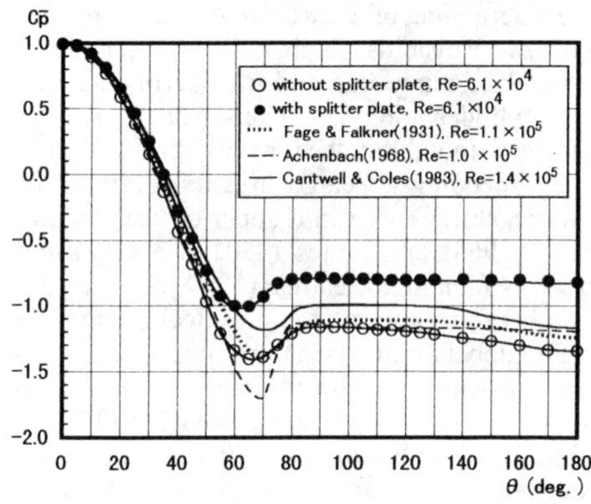

Figure 4 Mean Pressure coefficients, compared with other authors.

3 RESULTS

3.1 *Pressure distribution*

The mean pressure distributions and the r.m.s. fluctuating pressure distributions around the cylinder with and without a splitter plate are shown in Figure 2. When a splitter plate was set behind the cylinder, mean pressures slightly increased in the negative pressure region compared with those without a splitter plate. On the other hand, fluctuating pressures over the whole surface of the cylinder dramatically decreased in the case with the splitter plate, as shown in Figure 3. It suggests that the influence of vortex shedding on the pressures over the surface of the cylinder extended to the windward region before the separation point, because it is considered that the presence of a splitter plate behind a bluff body suppresses the vortex-shedding formation in the wake.

The mean pressure distribution is presented in Figure 4, in which the pressure coefficient $\overline{Cp}$ is plotted against θ, the angle measured from the time-averaging stagnation point, together with the results by Fage & Falkner(1931), Achenbach(1968), and Cantwell & Coles(1983), whose experiments were also carried out in the sub-critical regime. The present result has a good agreement with the results by other researchers. It is derived from Figure 4 that the fitting curve of $\overline{Cp}$ had the minimum value at $\theta=65.4^\circ$ and the reflection point at $\theta_s=76.2^\circ$, which is considered as to be the time-averaging separation angle in accordance with the recommendation by Zdravkovich(1997). The base pressure coefficient was $\overline{Cp}=-1.35$ at $\theta=180^\circ$ which was the minimum value in the wake region of the cylinder.

The fluctuating pressure distribution is showed in Figures 5, in which the r.m.s. pressure coefficient Cp' is plotted together with results by Surry(1972), Kiya et al.(1982), and Norberg(1986). The value of Cp' had the maximum value of 0.43 at $\theta=75^\circ$ which is closed to the time-averaging separation angle.

3.2 *Fluctuating wind forces*

Power spectra of the lift and the drag force fluctuations which were instantaneously integrated pressures over the whole circumference of the cylinder are shown in Figure 6(a). The numbers 1, 2 and 3 above the figure and the corresponding dot lines in the figure indicate the multiple values and their locations of the Strouhal number, $St=fsB/V=0.202$, where *fs* is the peak frequency of the lift spectrum.

It is well known that the spectrum of the lift

fluctuation has an evident sharp peak at the Strouhal number *St* and the spectrum of the drag fluctuation has a peak at 2*St*, respectively. It is noted in the present experiment that the spectrum of the lift fluctuation had a small second peak at 3*St*, which have not been reported yet because the measurement of wind forces using a load-cell type force transmitter could not clearly separate only the force signal from the signal of the natural-frequency component of the instrument mounted the model. The second peak at 3*St* was also seen in the spectrum of the pressure fluctuations in the wake region.

The r.m.s. fluctuating lift and drag force coefficient C_L' and C_D', which were calculated from integrated pressures over the whole surface of the cylinder, showed the values of C_L'=0.56 and C_D'=0.14, respectively.

3.3 *Power spectra of pressures*

Power spectra of pressure fluctuations on some typical points of the cylinder are shown in Figure 7. The power spectrum on the point at θ=0°, which locates on the time-averaging stagnation point, had a small peak at *St*. Whereas the spectrum on the point at θ=5°, which locates on the small angle apart from the stagnation point, showed the evident peak at *St*. It is clear that the influence of the shed Karman vortices appeared not only on the leeward surface but on the windward surface of the cylinder. The fact indicates that the Karman vortex shedding affects to the fluctuation of all over the flow around the cylinder. At θ=90°, the spectrum had the clear peak at *St* and also the small peak at 2*St*. On the other hand, the spectrum of the lift fluctuation has no peak at 2*St* (see Figure 6(a)). It suggests that the shedding Karman vortex affects the pressures not only on the same side as vortex itself but also on the opposite side. The component of the lift fluctuation at 2*St* can be canceled because pressures of both sides simultaneously yielded to the opposite direction for each other. The spectrum at θ=150°, which may be a typical point on the after body in the wake region, showed the peak at 3*St* that was also seen in the spectrum of the lift fluctuation. The spectrum at θ=180° had no clear peak at *St* whereas it had a clear peak at 2*St*, of which configuration was extremely similar to that of the drag fluctuation.

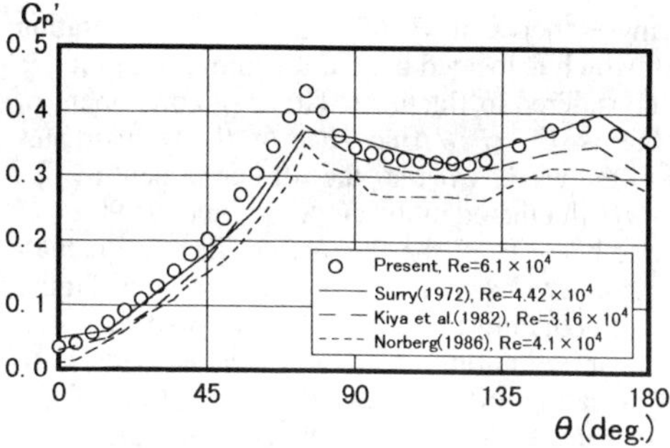

Figure 5 Fluctuating Pressure coefficients, compared with other authors.

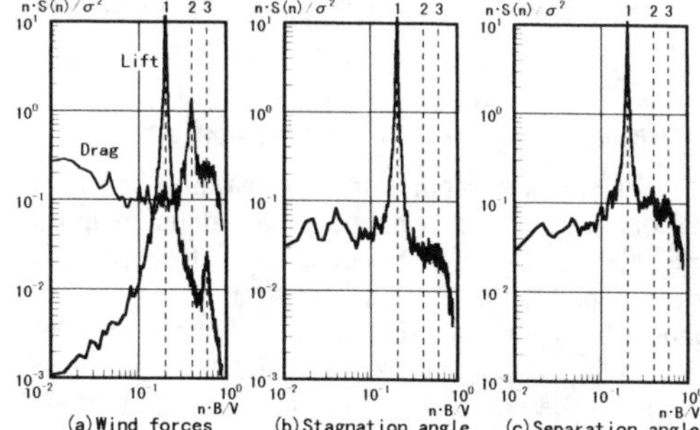

Figure 6 Power spectra of fluctuating (a) wind forces, (b) the stagnation angle, and (c) the separation angle.

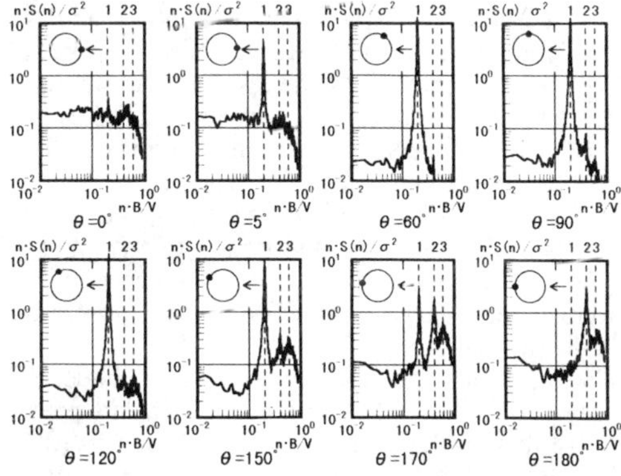

Figure 7 Power spectra of the fluctuating pressures on some typical taps around a circular cylinder.

3.4 *Instantaneous pressures*

Typical time series of instantaneous pressure distributions around the cylinder were obtained by using the conditional sampling technique, which is an ensemble averaging method synchronized the fluctuation of the lift force. Figure 8 shows the instantaneous pressure distributions and their components along the lift and drag directions. Each line in this figure indicates the instantaneous pressure distribution at every 1/400 second in a period of the lift fluctuation. The instantaneous pressure distributions, $\tilde{C}p$, asymmetrically changed with respect to θ=180° at every moment, and the fluctuation especially became larger on the surface of the after body. As minimum pressure of $\tilde{C}p$ remarkably fluctuated at the angles between θ=65° ~70°, the separation point, which is located behind the minimum point, is also considered to fluctuate. The drag component of $\tilde{C}p$, that is, $\tilde{C}p \cdot cos\theta$ fluctuated on the leeward surface in the wake, whereas the lift component of $\tilde{C}p$, $\tilde{C}p \cdot sin\theta$ fluctuated in the angles around θ=90°.

Time histories of the fluctuating lift and fluctuating pressures at several angles were shown in Figure 9. The fluctuating pressure at θ=90° had the sinusoidal configuration, which was synchronized with the fluctuating lift, whereas the fluctuation of the pressure at θ=180° had the twice of frequency of the fluctuating lift. The configuration of the instantaneous pressure distribution continuously changes from the sinusoidal wave with St component to that with 2St component, on which the process 3St component may appeared in their spectra as mentioned before.

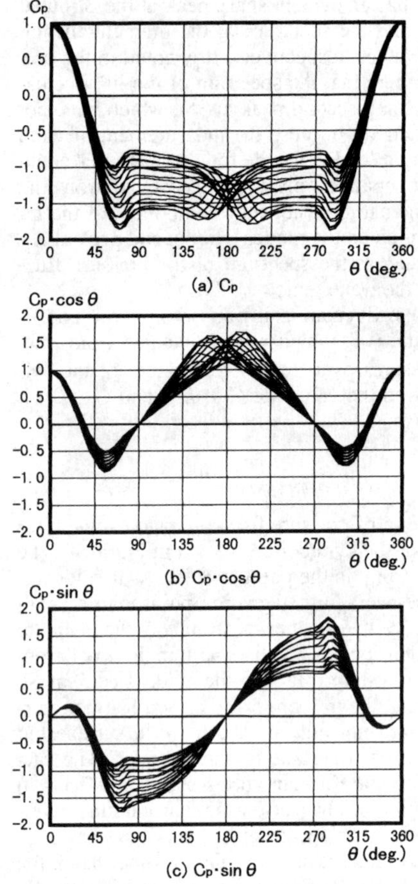

Figure 8 Instantaneous pressure distributions and their components.

3.5 *The stagnation and separation points*

Both of a stagnation point and a separation point on a stationary bluff body have been considered as to be stable the time-averaging meaning. However, as vortices shed from a body depend on the flow around a body, then the separated shear layer and the boundary layer along the windward surface are also affected by the vortex shedding. It is no wonder that the stagnation and separation points on a bluff body with round corners fluctuate by the influence of the wake oscillation, which cause the oscillation of the shear layer and windward flow before the separation point.

It has been reported by Dwyer and McCroskey (1973) that the stagnation and separation angles mutually fluctuate with a reverse phase. Sarpkaya (1979) also presented the same result by the simulation using the discrete vortex method. Figure 10 shows the instantaneous pressure distributions on the windward surface of the cylinder. The instantaneous stagnation angle, θ_{stg}, was decided from the angle which indicated the maximum value of the fitting curve by a trinomial equation. The instantaneous locations of the stagnation point obviously showed a periodic fluctuation.

The separation angle, θ_s, on a circular cylinder is usually defined as the angle that the leeward pressure distribution shows to be uniform, but it is difficult to decide it in practice. In this paper, the separation angle was estimated from the inflection of the curve in the angles from θ=50° to 110°.

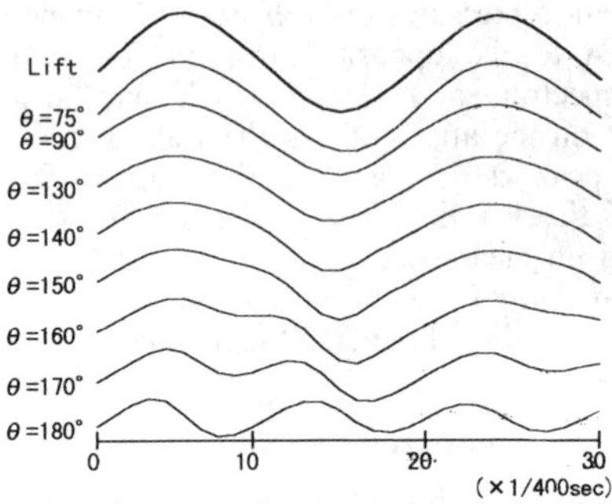

Figure 9 Instantaneous pressure and lift.

The time histories of the instantaneous stagnation angle and the separation angle are shown in Figure 11, together with the time histories of the lift and drag forces. It is an interesting phenomenon that time histories of the fluctuations of the lift force and the stagnation and separation angles were closely resemble with each other. Therefore their spectra were also similar with each other, which had the obvious peak at *St* as shown in Figure 6.

Figure 12 shows the cross-correlations among the fluctuations of the lift force and the stagnation and separation angles. The cross-correlation at time lag

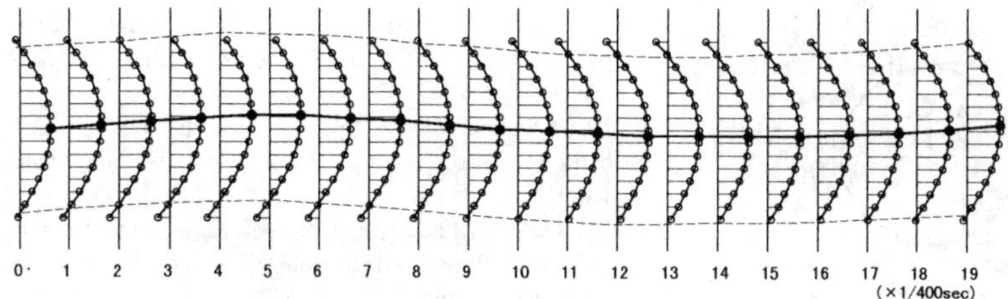

Figure 10 Fluctuation of the stagnation point on the upwind surface of a circular cylinder. •; θ_{stg}, ○; instantaneous pressure on the windward surface.

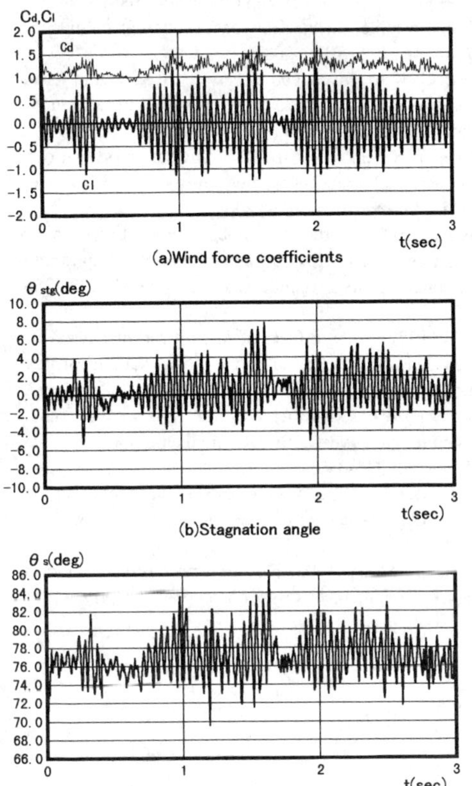

Figure 11 Time histories of (a) wind forces, (b) the stagnation angle, and (c) the separation angle.

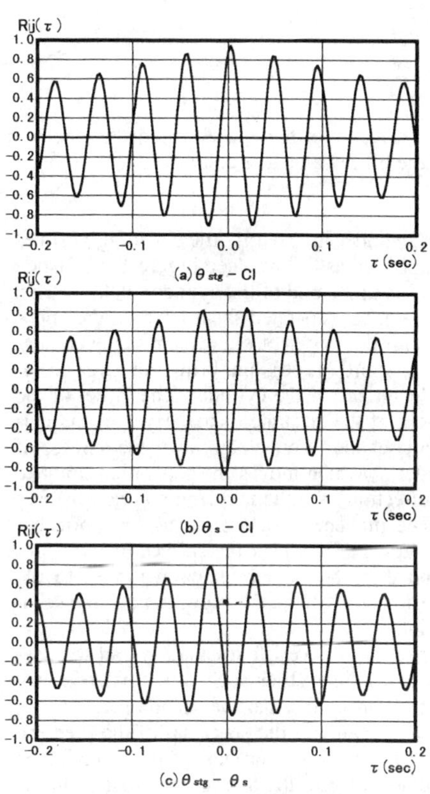

Figure 12 Cross-correlation coefficients between the fluctuation of the lift force and the movements of the stagnation angle and the separation angle.

τ=0 between the stagnation and lift force showed a positive, whereas correlations between the stagnation and separation angles and that between the separation angle and lift force showed negative. The phenomenon suggests that when the separation point moves to the leeward direction and the stagnation point moves toward the reverse direction of the separation point, at this moment the large lift force occurs.

4 CONCLUSIONS

Instantaneous pressure distributions around a two-dimensional static circular cylinder in a smooth flow were clarified by using the simultaneous multi-pressure measurement system. Both of the fluctuations of the stagnation point and the separation point of the cylinder were synchronized with the lift fluctuation.

The mechanism of the occurrence for the fluctuating lift force and the fluctuation of the flow around

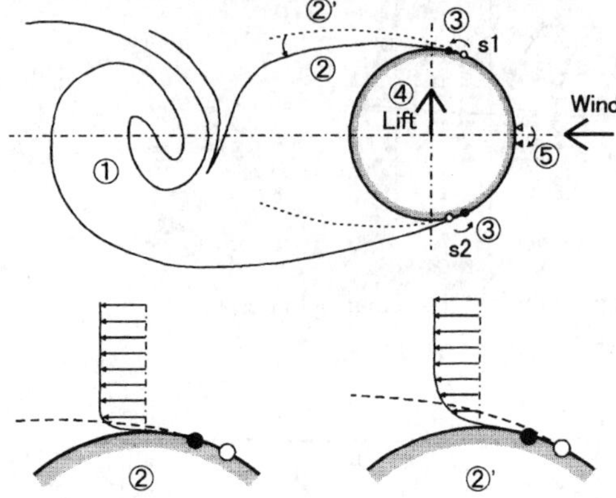

Figure 13 Scheme for the mechanism of the occurrence of the lift force on a circular cylinder.

a circular cylinder induced by the shedding vortices can be explained as follows (see Figure 13); ①when a vortex rolls up behind of the cylinder, ②the vortex takes a part of the separated shear layer on the opposite side into the vortex itself, which has been suggested by Gerrard (1966), and bring the layer closer to the side surface of the cylinder. The effect of the entrainment of the separated shear layer to the cylinder increases the flow velocity near the surface of the cylinder, ③which moves the separation point to leeward direction, ④at the moment, a large lift force occurs. The movement of the separation point entrains the upwind flow near the surface of the cylinder, ⑤and then the stagnation point moves to the opposite direction against the movement of the separation point.

The spectrum of the lift fluctuation had the second peak at 3St, as well as at St which has bee well known. It is considered that the second peak at 3St, which also appeared in the spectrum of the pressure fluctuation on the wake surface of the cylinder, is due to the distortion of the pressure fluctuation in the varying process from the pressure fluctuation on nearby the separation point has the spectrum peak at St to that on the base surface has the spectrum peak at 2St.

REFERENCES

Achenbach, E. 1968. Distribution of local pressure and skin friction around a circular cylinder in cross-flow up to Re=5×10⁶. *J. Fluid Mech.* 34:625-639.

Cantwell, B. & Coles, D. 1983. An experimental study of entrainment and transport in the turbulent near wake of a circular cylinder. *J. Fluid Mech.* 136: 321-374.

Dwyer, H.A. & McCroskey, W. J. 1973. Oscillating flow over a cylinder at large Reynolds number. *J. Fluid Mech.* 61:753-767.

Fage, A. & Falkner, V. M. 1931. The flow around a circular cylinder. *Aero. Res. Counc., Lond., Rep., and Mem.* 1369.

Fung, Y. C. 1960. Fluctuating lift and drag acting on a cylinder in a flow at supercritical Reynolds numbers. *J. Aerospace Sciences*. 27: 801-814.

Gerrard, J. H. 1966. The mechanics of the formation region of vortices behind bluff body. *J. Fluid Mech.* 25: 401-413.

Kiya, M. et al. 1982. A contribution to the free-stream turbulence effect on the flow past a circular cylinder. *J. Fluid Mech.* 115:151-563.

Norberg, C. 1986. Investigation between free-stream turbulence and vortex shedding for a single tube in cross-flow. *J. Wind Eng. and Indust. Aerody.* 23: 501-514.

Sarpkaya, T. & Schoaff, R. L. 1979. Inviscid model of two-dimensional vortex shedding by a circular cylinder. *AIAA Journal*. 17:11793-1200.

Schewe, C. 1983. On the force fluctuations acting on a circular cylinder in cross-flow from subcritical up to transcritical Reynolds numbers. *J. Fluid Mech.* 133: 265-285.

Surry, D. 1972. Some effects of intense turbulence on the aerodynamics of a circular cylinder at subcritical Reynolds number. *J. Fluid Mech.* 52:543- 563.

Roshko, A. 1961. Experiments on the flow past a circular cylinder at a very high Reynolds number. *J. Fluid Mech.* 10: 345-856.

Wieselsberger, C. 1921. Neuere feststellungen uber die gesetze des flüssigkeits und luftwiderstands. *Phys. Z.* 22: 321-328.

Zdravkovich, M. M. 1997. Flow around circular cylinders, vol.1: fundamentals. *Oxford Science Publications.*

Wind Engineering into the 21st Century, Larsen, Larose & Livesey (eds) © 1999 Balkema, Rotterdam, ISBN 90 5809 059 0

Wind response behaviour of elasto-plastic structure with rectangular cross section based on time history analysis considering motion-induced wind forces

Takeshi Ohkuma
Department of Architecture and Building Engineering, Kanagawa University, Yokohama, Japan

Hisao Marukawa & Junko Itoh
Urban Environment Research Center, Izumi Sohken Engineering Company Limited, Sodegaura, Chiba, Japan

ABSTRACT: This paper describes a time history response analysis considering motion-induced wind forces. The target of this paper is elasto-plastic vibration of structures in across-wind direction. First, it compares analytical results for elastic vibration and aeroelastic experimental results to verify the validity of the time history response analysis. Finally, it discusses characteristics of elasto-plastic responses and energy input.

1 INTRODUCTION

Motion-induced wind forces due to the interaction between the vibration of a structure and the stream change around it add to original wind forces. Motion-induced wind forces mainly change the damping effect of a structure. Self-excited vibration with large amplitude as to vortex-induced vibration or galloping occurs when motion-induced wind forces act on a structure as negative damping effect. In AIJ recommendations for loads on buildings motion-induced wind forces are not taken into consideration expect checking the critical wind speed for the aeroelastic instability of buildings. On the other hand the elastic response analysis considering motion-induced wind forces for a few cross sections with a certain vibration mode has been investigated theoretically [Tamura et. al. 1986]. As a result, many characteristics of vibration have been clarified. The elastic response analysis considering motion-induced wind forces has been proposed, by using the results of wind tunnel tests. This is applicable for all cases regardless of vibration modes and cross sections [Ohkuma et. al. 1995, Marukawa et. al. 1992].

With the introduction of new concepts such as performance based objective design or damage control design, the characteristics of elasto-plastic response of buildings, especially buildings with hysteresis-type dampers have been causing concern. One of the authors has discussed the characteristics of the elasto-plastic response of buildings for across-wind vibration [Ohkuma et. al. 1997]. However, these responses were estimated below the resonance wind speed, where the motion-induced wind forces were negligible, because motion-induced wind forces act on buildings as positive damping effect.

In this paper, the method of elasto-plastic response analysis is obtained by the time history response analysis, which is applicable beyond resonant wind speed by utilizing the elastic response analysis method by Ohkuma et al [1995]. According to the results of the responses, the basic characteristics of elasto-plastic vibration are clarified. The two-dimensional structure models of the analysis are a cross section of a completely separated flow type and a cross section of a flow-reattachment type. The analysis is subject to across-wind vibrations.

2 RESPONSE ANALYSIS

2.1 *Structural models for analysis*

The analytical structure models are of a one-mass system based on the following assumptions.

① Damping coefficient is proportional to initial stiffness.

② Restoring force characteristic is of a bi-linear type, and the ratio of second stiffness to first stiffness is 0.5.

③ Displacement of yield point is equivalent to 1/300 for relative story deformation angle of a vertical structure or for deformation ratio to the span of a horizontal structure, if its aspect ratio is 5.

The side ratios ($=D/B$, D: depth, B: breath) of the models are 1 and 2, which are the cross section of completely separated flow type and the cross section of flow-reattachment type respectively. Figure 1 shows the model for analysis. Table 1 shows the analytical parameter of the study.

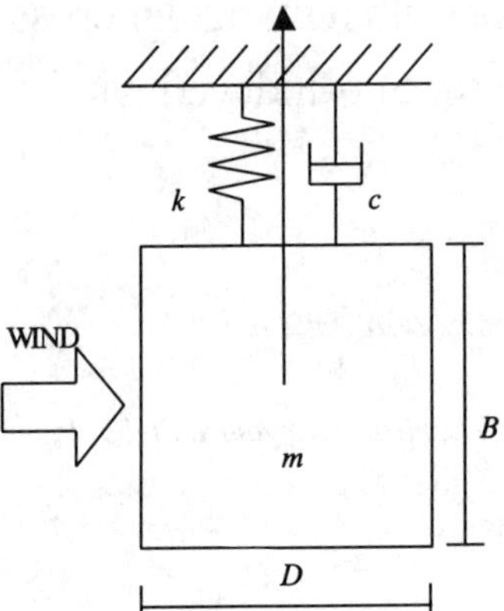

Figure 1 model for analysis

Table 1 analytical parameter

D/B	1	2
S_c^*	15,19,40	3,14,27

$S_c^* = 2\pi h_s \rho_s / \rho$: mass damping parameter
h_s : structural damping constant
ρ_s : density of the structure
ρ : air density

2.2 *Equation of vibration*

The analytical method for wind response is revived from reference 2), expanding elastic time response analysis to elasto-plastic time response analysis. The one-mass motion equation is Equation (1), considering wind-induced force and motion-induced wind forces.

$$m\ddot{y} + c\dot{y} + Q(y) = F(t) + F_c(t) + F_k(t) \tag{1}$$

Where, m: mass, c: structural damping coefficient, $Q(y)$: restoring force, y: response displacement, $F(t)$: fluctuating wind force, $F_c(t)$: out-phase component of motion-induced wind force, $F_k(t)$: in-phase component of motion-induced wind force, d ot: differential with respect to time

In Equation (1), in-phase component of motion-induced wind force is neglected because the aeroelastic stiffness is small enough compared to the structural stiffness. When the response frequency is within the limits of the banded frequency expressed in Equation (3), it is taken into the natural frequency. Equation (4) shows the manner of these in the calculation of the response. The bandwidth may generally be the functions of non-dimensional wind speed, non-dimensional amplitude of response displacement and cross-section shape of a model, etc. However, the bandwidth is treated in proportion to non-dimensional standard deviation of response displacement simply in this paper. The proportional constant γ is 0.5.

$$y = \sum_{n=1}^{N/2} y_n(t) = \sum_{n=1}^{N/2} Y_n \cos(2\pi f_n t + \phi_n) \tag{2}$$

$$f_0(1-\gamma\sigma_y^*) \le f' \le f_0(1+\gamma\sigma_y^*) \tag{3}$$

$$f' \rightarrow f_0, \quad \phi' \rightarrow \phi_0 \tag{4}$$

Where, Y: amplitude of response displacement, f: frequency, f_0: natural frequency, ϕ: phase difference, $\sigma_y(=\sigma_y/B)$: non-dimensional standard deviation of response displacement, f': banded frequency, ϕ': banded phase difference, N: number of data, and subscript n resent the degree of frequency.

The out-of-phase components are divided into the frequency components of negative damping and those of positive damping. The former is evaluated using Equation (5). The latter are evaluated by the effective aerodynamic damping coefficient c_{aj} obtained from Equation (6), in which the natural frequency component and the Strouhal frequency component are considered. The in-phase components of motion-induced wind forces are negligible.

$$F_{ci}(t) = C_{LIi}\left(\frac{\varepsilon Y_i}{B}, \frac{U}{f_i B}\right) qD \cos(2\pi f_i t + \phi_i) \tag{5}$$

$$c_{aj} = \frac{C_{LIj} qD}{2\pi f_j \varepsilon Y_j} \tag{6}$$

Where, C_{LI}: out-of-phase component coefficient obtained from reference [Marukawa et. al. 1992] (Figure 2), U: wind speed, q: velocity pressure, ε : conversion coefficient from non-harmonic vibration to harmonic vibration, and subscript i present component of negative damping.

Based on the above assumptions, vibration Equation (1) can be expressed by Equation (7).

$$m\ddot{y} + (c - \sum_j c_{aj})\dot{y} + Q(y) = F(t) + \sum_i F_{ci}(y) - \sum_j c_{aj}\dot{y}_{r(\neq j)} \tag{7}$$

In this equation, the natural frequency component and the Strouhal number component for positive damping force are indicated by 0 and s for subscript j,r respectively.

2.3 *Equation of energy balance*

The equation of energy balance Equation (8) is multiplied by $\dot{y}dt$ on both sides of Equation (7), and integrated from 0 to T.

$$\begin{aligned} &m\int_0^T \ddot{y}\dot{y}dt + c\int_0^T \dot{y}^2 dt + \int_0^T Q(t)\dot{y}dt = \\ &\int_0^T F(t)\dot{y}dt + \left(\int_0^T \left(\sum_i F_{ci}(t) - \sum_j c_{aj}\dot{y}_{r(\neq j)}\right)\dot{y}dt + \sum_j c_{aj}\int_0^T \dot{y}^2 dt\right) \end{aligned} \tag{8}$$

In Equation (8), the first, second and third terms on the left side represent, kinetic energy, structural damping energy E_{hs} and elastic strain energy plus plastic strain energy E_p respectively. Elastic energy E_e is formed by kinetic energy and elastic strain energy. In the right side, sum of terms represent total input energy E, the first term represent fluctuating wind force energy E_f, second and third terms represent, respectively, negative damping energy and those of positive damping energy. Sum of the second and the third term represents motion-induced wind force energy E_{fc}. Therefore, Equation (8) can be represented by Equation (9).

$$E_{hs} + E_p + E_e = E_f + E_{fc} \tag{9}$$

2.4 *Method of response analysis*

The method of time response analysis is Newmark's β -method. Figure 3 shows a flow chart of the method. The response displacement is expressed by the sum of each frequency component as Equation (2). The response analysis is iteratively carried out until the standard deviation of the response displacement converges, the error between a step and its previous step being equal or less than 1%.

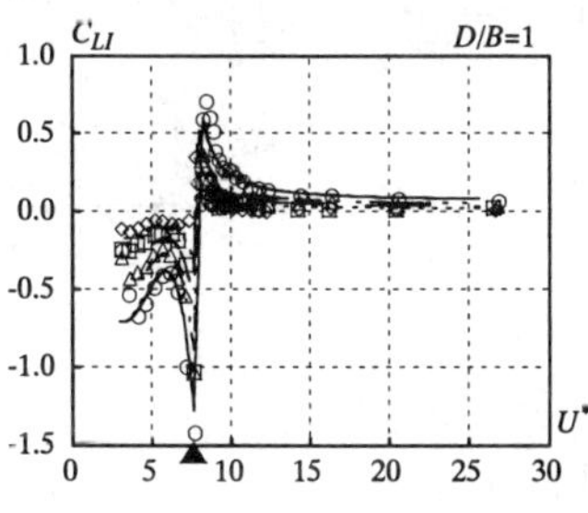

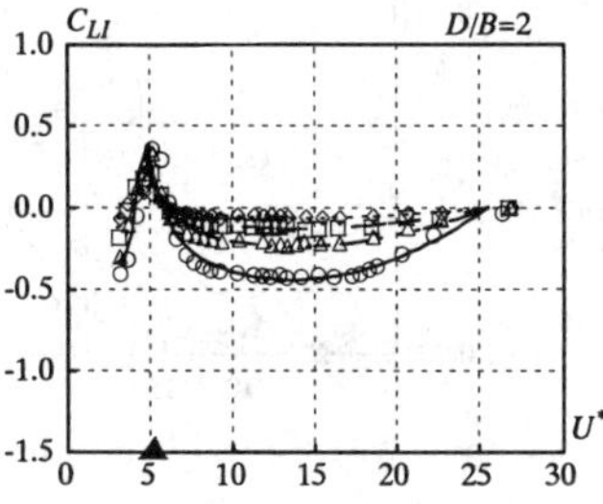

Figure 2 model of out of component coefficient

Evaluation time is 150 times that of the natural period. The values of the response are the ensemble means of 10 values. The response analytical values are shown by non-dimensional value defined in Equation (10).

$$U^* = \frac{U}{f_0 B}, \quad f^* = \frac{fB}{U}, \quad \sigma_y^* = \frac{\sigma_y}{B}, \quad E^* = \frac{E}{(f_0 B)^2 m} \tag{10}$$

Time series of fluctuating wind forces are simulated by the Monte Carlo method using the power spectral density obtained from the wind tunnel test. Figure 4 shows the power spectrum density of the across-wind force.

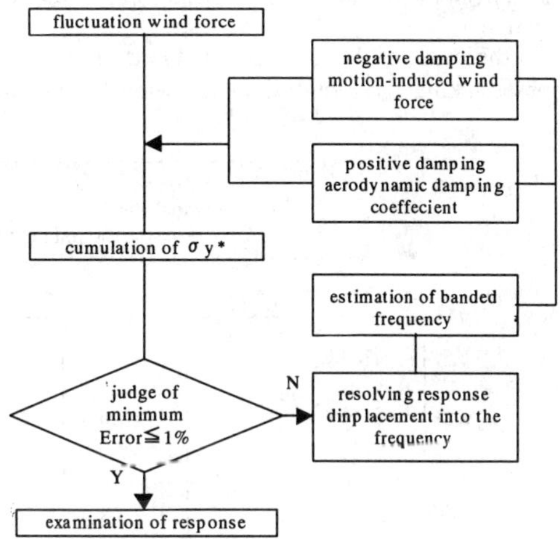

Figure 3 flow chart of method

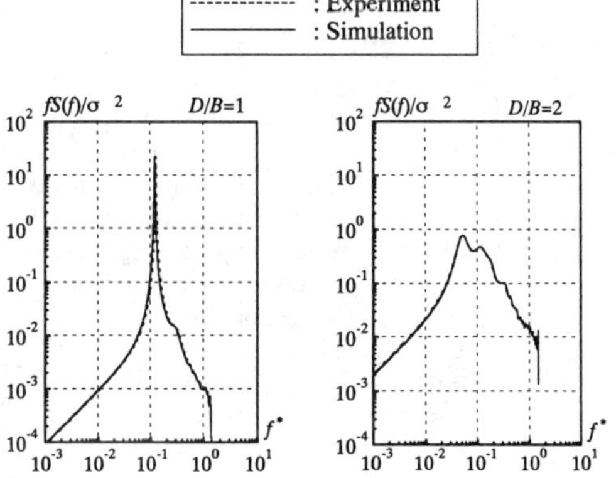

Figure 4 power spectrum density of across-wind force

3 PROOF FOR RELIABILITY OF THE RESPONSE ANALYSIS METHOD

Proof for reliability of the response analysis method, which was discussed at the section 2.4 was examined by means of comparison between the responses by the analysis and ones obtained from the wind tunnel test for the elastic responses, before the analysis for elasto-plastic responses. Figure 5(a) and Figure 5(b) show the comparison between the response obtained from wind tunnel tests and the one obtained by the analysis taking motion-induced wind forces into considering or not. In these figures, mark ▲ shows the reduced resonant wind speed estimated from the power spectrum density of the across-wind force, and mark ↑ shows the wind speed at which the sum value of the critical damping ratio of a structure and the aerodynamic damping ratio estimated from the motion-induced wind force is zero or negative.

In the cases of the side ratio 1, the analytical responses taking the motion-induced wind forces are in good agreement with the responses from the wind tunnel test at the region of wind speed less than $2U_s^*$. In the high wind speed region, the change tendency of the analytical responses according to wind speed is in agreement with the result of the wind tunnel test, but both values are different.

In the cases of the side ratio 2, the analytical responses with the motion-induced wind forces are in good agreement with the responses from the wind tunnel test at the region of wind speed less than $4U_s^*$.

In the both cases of the side ratio 1 and 2, the causes of the difference between the responses obtained from the proposed method and ones from the wind tunnel test in high wind speed region are not clear. One has a problem of modeling the motion-induced wind forces, whose data are lacking in large amplitudes. For γ is treated as a constant value in all wind regions, so the natural frequency components of responses may not be evaluated appropriately. However, except for the high wind speed region, the analysis method proposed in this paper is available.

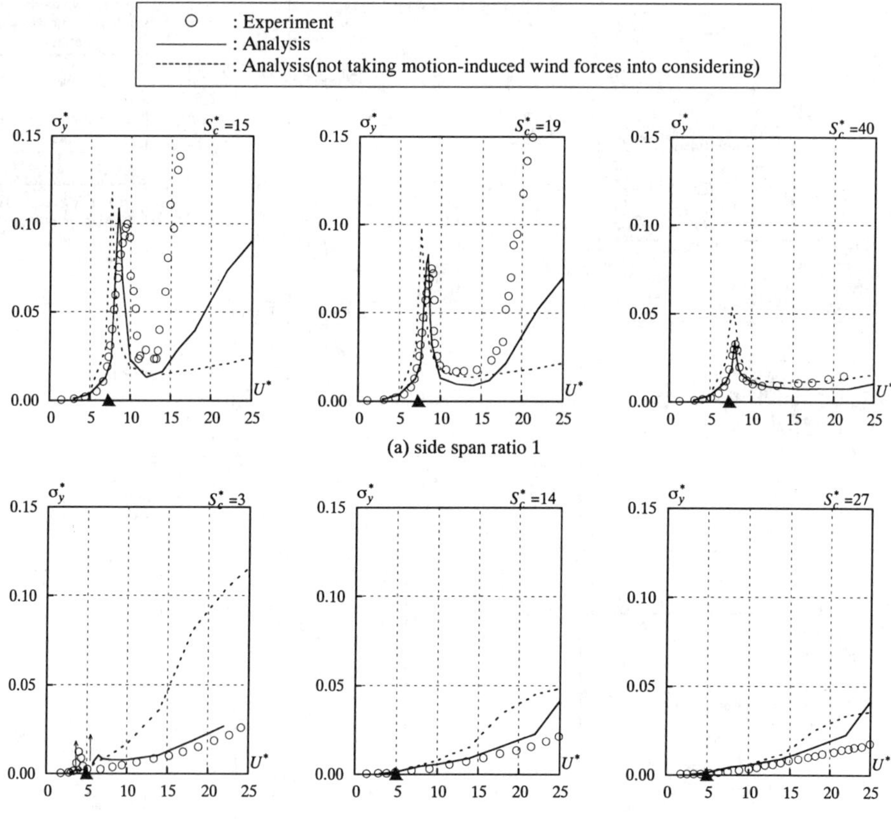

(a) side span ratio 1

(b) side span ratio 2

Figure 5 standard deviations of response displacement

4 RESPONSE ANALYTICAL RESULTS

This chapter describes the response analysis results at the region of wind speed discussed in chapter 3.

4.1 *Characteristics of elasto-plastic response*

Figure 6 shows the standard deviations of the response displacement σ_y^*. In the case of side ratio 1, the elasto-plastic responses σ_y^* are much smaller than those of the elastic responses at the vicinity of the resonance wind speed. The values of the elasto-plastic responses are almost same regardless of mass-damping parameter S_c^*. In the case of side ratio 2, the elasto-plastic responses σ_y^* are slightly smaller than the elastic responses. When S_c^* is less or equal 3, σ_y^* of elasto-plastic response have a peak at resonance wind speed. When S_c^* are greater than 3, σ_y^* of elasto-plastic response increases with a rise in wind speed.

Figure 7 shows the ductility factor μ, which indicates the ratio of maximum response displacement to the displacement of yield point. In the case of side ratio 1, ductility factors have a peak at U_s^*. On the other hand, in the case of side ratio 2, ductility factors increase with a rise in wind speed. In the cases of both, the ductility factors slightly decrease with an increase in S_c^*.

Figure 8 shows the cumulative plastic deformation ratio η, which indicates the ratio of the cumulative value of displacement of yield point. The characteristics of η are the same as those of μ.

Figure 9 shows some examples of power spectrum densities of elasto-plastic response displacement in addition to ones of elastic response displacement. The examples are ones when reduced wind speed U^* is 7.66, which coincides resonance wind speed, and U^* is 14 in the case of side ratio 1 with S_c^* being 15. In the case of side ratio 2 with S_c^*=3, ones are at U^*=14.

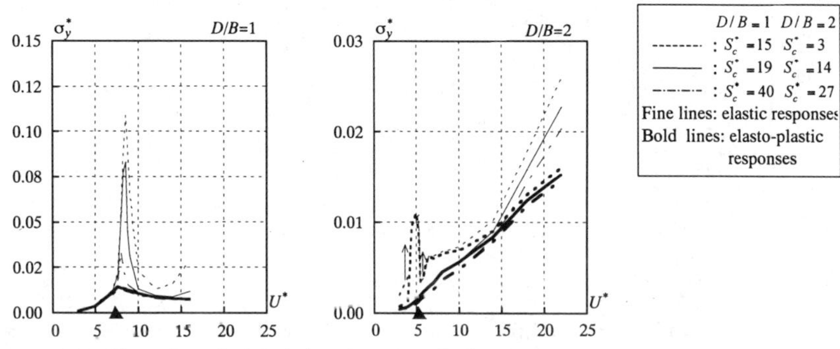

Figure 6 standard deviation of response displacement

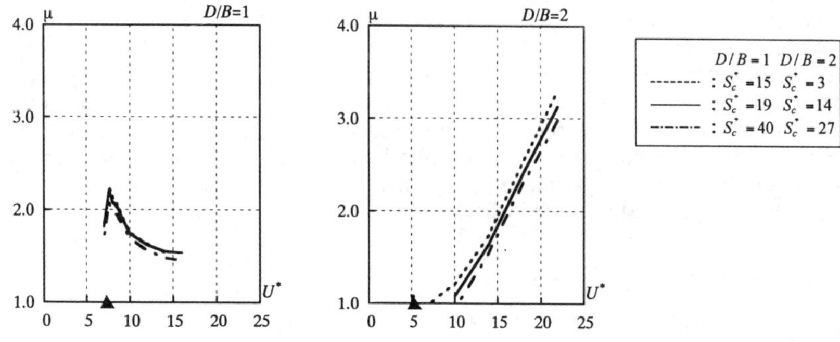

Figure 7 ductility factor

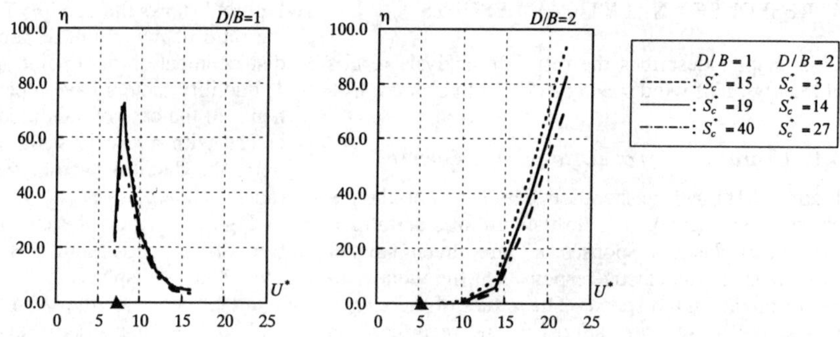

Figure 8 cumulative plastic deformation ratio

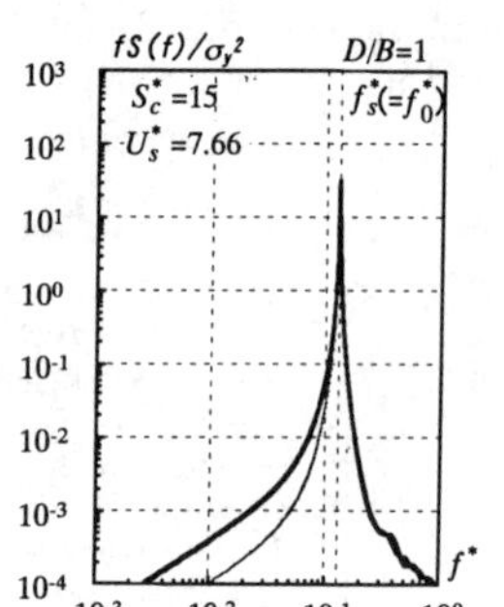

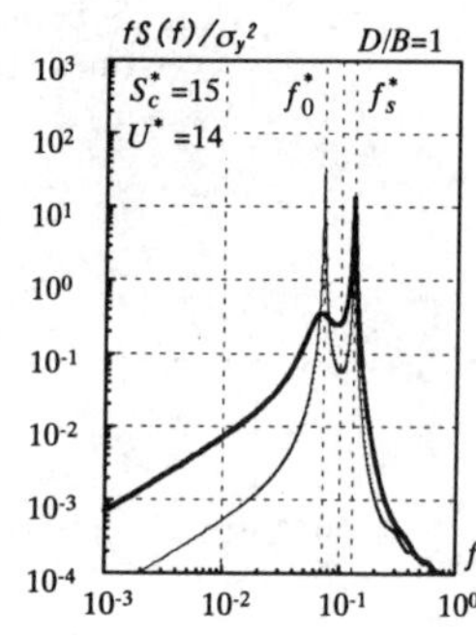

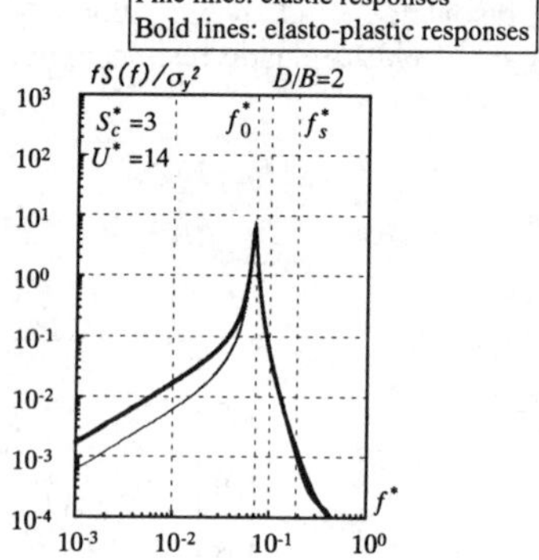

Figure 9 power spectrum density

In the case of side ratio 1 at U^*=7.66, the natural frequency component of elasto-plastic response displacement is prominent, but it is smaller than the natural frequency component of elastic response displacement. At U^*=14, the natural frequency component and the Strouhal number component are prominent in elastic response displacement. However the natural frequency component is considerably smaller than the Strouhal number component.

In the case of side ratio 2 regardless wind speed, the natural frequency component of elasto-plastic response displacement is prominent only, and is smaller than that of elastic response displacement.

4.2 *Characteristics of energy response*

Figure 10 shows plastic strain energy E^*_p. In the case of the side ratio 1, E^*_p have a peak in the vicinity of U^*_s. In the case of side ratio 2, E^*_p increases with a rise in wind speed. In the cases of both, E^*_p slightly decrease with an increase in S^*_c.

Figure 11 shows structural damping energy E^*_{hs}. In the case of side ratio 1, E^*_{hs} have a broad peak in the vicinity of U^*_s, increases with a rise in wind speed. In the case of side ratio 2, E^*_{hs} increases with a rise in wind speed. In the cases of both, E^*_{hs} is smaller than those of elastic responses, increase with an increase in S^*_c.

Figure 12 shows total energy input E^* together with that of elastic response. E^* of elasto-plastic responses are much smaller than that of the elastic response in the case of side ratio 1. E^* of the elasto-plastic response are almost the same or slightly larger than those of elastic response in the case of side ratio 2.

Figure 13 shows fluctuating wind force energy E^*_f together with that of elastic response. E^*_f of elasto-plastic responses are almost the same value as those of elastic response except at the vicinity regions of resonance wind speed.

Figure 14 shows motion-induced wind force energy E^*_{fc} together with that of elastic response. E^*_{fc} of elastic responses have positive values in the vicinity

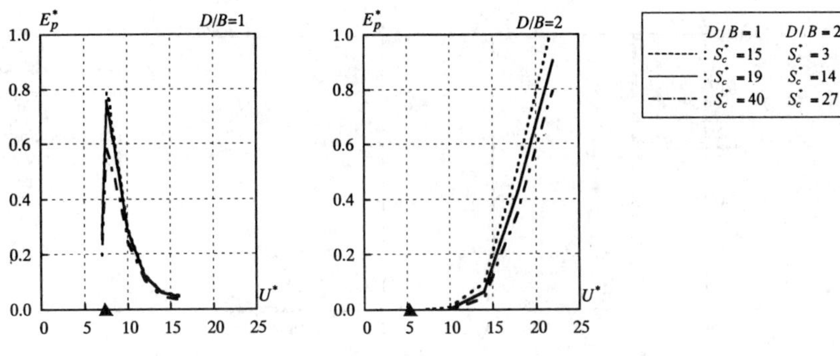

Figure 10 plastic strain energy

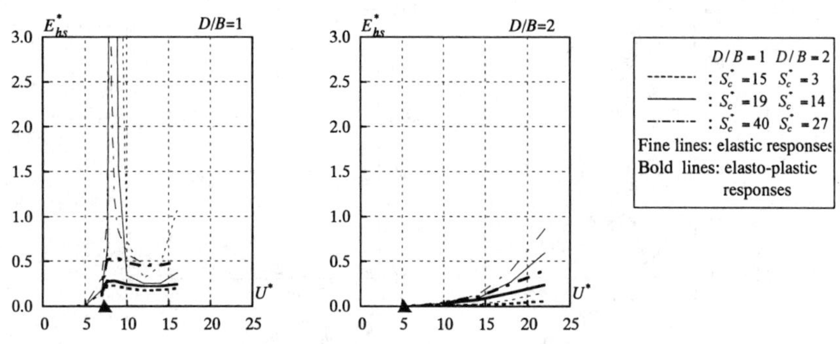

Figure 11 structural damping energy

of the resonant wind velocity. E_{fc} of the elasto-plastic responses become negative with small absolute values in all the analytical wind velocity range in the case of side ratio 1. E_{fc} of the elastic responses and the elasto-plastic responses are negative values, the absolute values of the latter ones are smaller then the former ones in the case of side ratio 2.

5 DISCUSSION

The elasto-plastic responses are smaller than the elastic response because of the hysteretic energy absorption of the structures. According to the response reduction, the effects of the motion-induced wind forces are reduced.

Discussing in detail: In the case of side ratio 1, the elasto-plastic responses are much smaller than the elastic responses. The negative damping effects of the motion-induced wind forces are reduced according to the plastic motion. As a results, the positive damping effects of the motion-induced wind forces are raised in appearance. In the case of side ratio 2, the elasto-plastic responses are almost the same or slightly smaller than the elastic responses. This is caused by the balance between a decrease in response due to plasticity and an increase in response due to the resultant decrease of the motion-induced positive damping force.

6 CONCLUSION

The calculation method of the elasto-plastic responses is presented in this paper. The characteristics of response displacements and energies are clarified by the calculation of the responses using the presented method. The values of response displacements decrease when the properties of a structure become plastic, the decreasing values dependent on the section of a structure. The decreasing values make the effect of the motion-induced wind forces on the response small. As a result, it is important whether the motion-induced wind forces act to the structure as positive damping effect or negative damping effect in elastic responses.

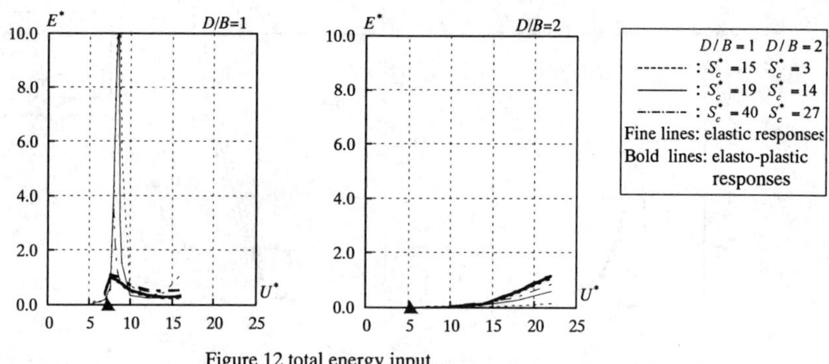

Figure 12 total energy input

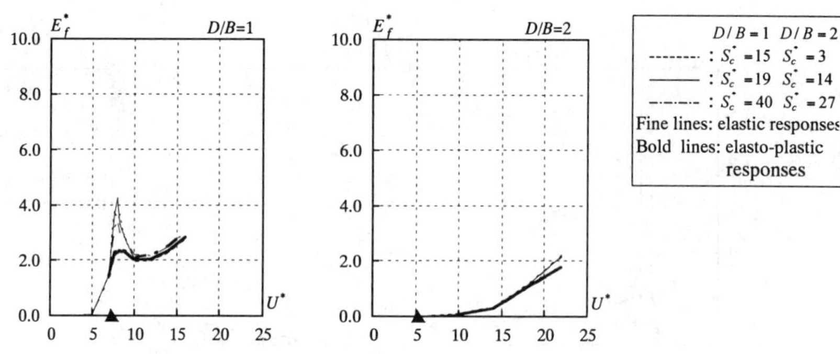

Figure 13 fluctuating wind force energy

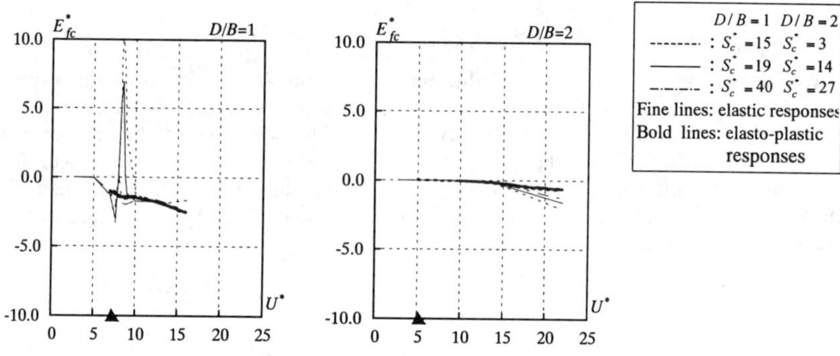

Figure 14 motion-induced wind force

REFERENCE

1) Tamura,Y., Shimada,K., (1986) "A mathematical model for the transverse oscillations of square cylinder", Proceedings of 9th National Symposium on Wind Engineering, pp.163-168, Tokyo, Japan (in Japanese)

2) Ohkuma,T., Mukai,H. and Yamamoto,M. (1995), "Time History Analysis for Across-Wind Aerodynamic Unstable Vibration of Two-Dimensional Rectangular Model", Proceedings of 9th ICWE Vol. II, pp.606-614, New Delhi, India

3) Marukawa,H., Ohkuma,T., Tsurumi,T., Yamamoto,M. and Momomura,Y., (1992) "Vibration-Induced Wind Forces on the Two Dimensional Long after Rectangular Models", Proceedings of 12th National Symposium on Wind Engineering, pp.201-206, Tokyo, Japan (in Japanese)

4) Ohkuma,T., Kurita,T., Ninomiya,M., (1997) "Response Estimation based on Energy Balance for Elasto-Plastic Vibration of Tall Building in Across-wind", Proceedings of ICOSSAR '97, 3, pp.1379-1386, Nov.24-28, Kyoto, Japan

Wind Engineering into the 21st Century, Larsen, Larose & Livesey (eds) © 1999 Balkema, Rotterdam, ISBN 90 5809 059 0

Flow and dispersion in the wakes of three-dimensional porous obstacles in a deep, turbulent boundary layer

A.G. Robins
School of Mechanical and Materials Engineering, University of Surrey, Guildford, UK
P.T. Roberts & L.J. Speirs
Shell Research Limited, Thornton Research Centre, UK

ABSTRACT: This paper gives an overview of an experimental study of flow and dispersion downwind of surface mounted, three dimensional, porous obstacles. Flow visualization was first used to explore the regimes of behavior. This, together with quantitative flow and turbulence measurement, revealed a progressive development from a wake with an attached recirculation region behind a solid plate, to a wake with a detached recirculation region, intermittent recirculation and finally with no recirculation as the porosity became high. These changes in flow structure it turn generated a range of dispersion conditions with significant mean streamline displacements occurring in the decelerating and accelerating regions of the wake and rapid and extensive mixing in regions of mean or intermittent recirculation . These situations offer a severe challenge to the development of dispersion models suitable for routine application at industrial sites.

1 INTRODUCTION

The body of research concerning the wakes of two and three-dimensional, surface mounted porous obstacles, such as fences and wind breaks, is not particularly extensive (e.g. see Gandemer, 1981, Perera, 1981, Raine and Stevenson, 1997). Indeed, the three-dimensional problem, the subject of our work, has been largely neglected, even though it is of considerable industrial relevance. Our interest arose through a need to understand pollutant dispersion near industrial sites that could best be classified as a mixture of solid and porous installations.

2 WIND TUNNEL MODELLING

The experiments were carried out at the University of Surrey in the EnFlo wind tunnel, which has a 20m long working section, 3.5x1.5m in cross-section. Smoke and laser sheet lighting were used for flow visualization studies, hot wire, pulsed hot wire and laser Doppler anemometers for velocity field measurements, and a gas sampling and analysis system for concentration measurement. Passive emissions containing a hydrocarbon tracer were used for all dispersion studies.

Porous plates of size 0.3x0.2m were manufactured from regular grids of rectangular bar, covering a range of porosity from 0 (a solid plate) to 85%. These were mounted either singly or multiply in a nominally 1m deep, turbulent boundary layer in the EnFlo wind tunnel; Figure 1 gives a schematic view of the pates and source arrangement. The boundary layer characteristics were rural in character, with roughness length 0.3mm and friction velocity 0.045Ue, Ue being the free stream velocity.

The effective porosity of the arrays of plates was not the issue, simply the flow characteristics that resulted. The notation used to describe the models is: M(p,n), where 'p' is the percentage porosity of the individual plate and 'n' the number installed.

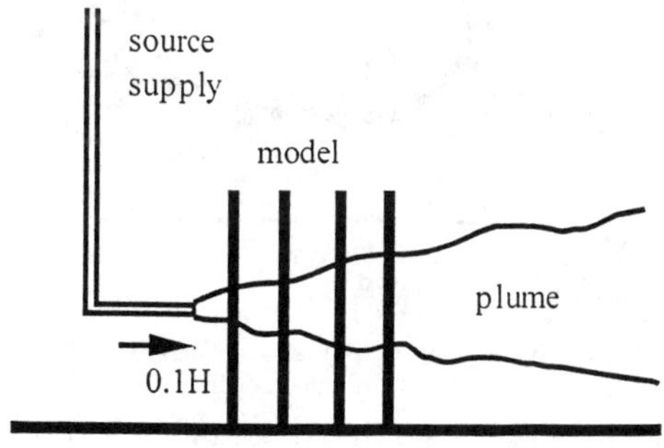

Figure 1. Schematic arrangement of source and model.

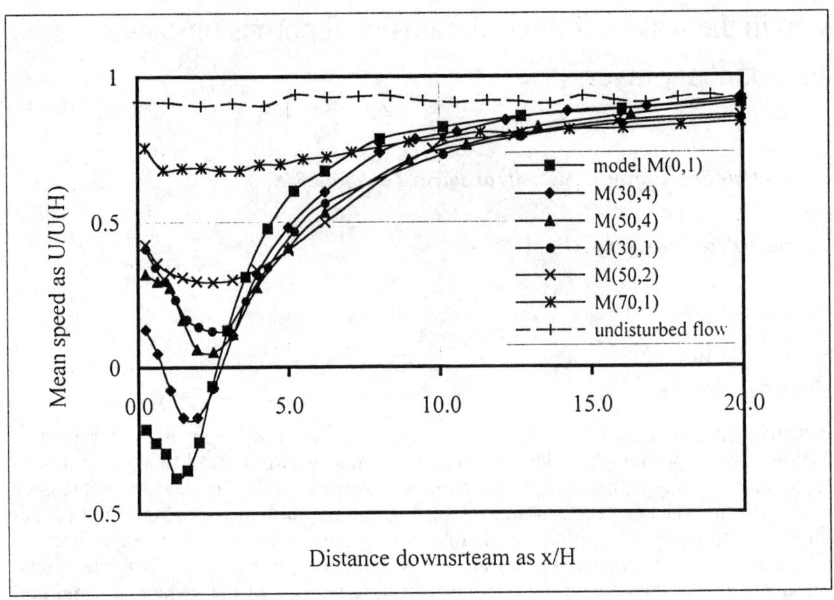

Figure 2. Longitudinal variation of mean vleocity in model wakes at a height of 0.5H; results are non-dimensionalised by the vlocity at the plate heoght, U(H).

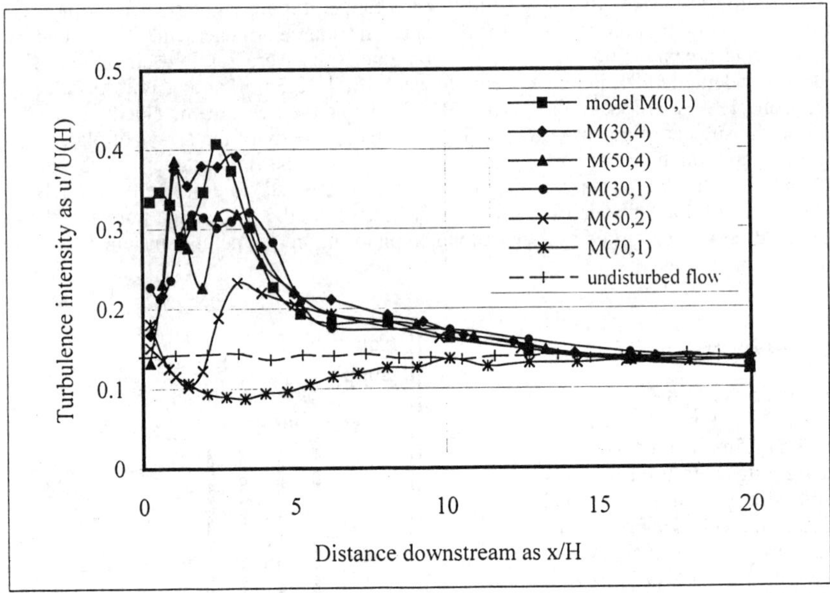

Figure 3. Longitudinal variation of longitudinal turbulence intensity in model wakes at a height of 0.5H; results are non-dimensionalised by the vlocity at the plate heoght, U(H).

3 FLOW VISUALISATION

Initial flow visualization studies revealed a number of regimes of near-wake behavior that can be classified as:

1. recirculating flow 'attached' to the obstacle - the solid plate case,
2. recirculating flow 'detached' from the obstacle - the low porosity case,
3. non-recirculating mean flow with intermittent flow reversal - the intermediate porosity case, and
4. non-recirculating flow - the high porosity case.

The strength of the wakes, as exemplified by the mean velocity deficit, was seen to steadily decrease as the porosity increased. Characteristic configurations were selected on the basis of the flow visualization work and subject to detailed mapping of the mean and fluctuating velocity fields through both the near and far wake regions. Mean concentration fields were also mapped for sources located near the base, mid-height and near the top of the porous plates.

4 VELOCITY FIELD MEASUREMENTS

A series of mean velocity and turbulence intensity profiles at constant height, z=H/2, through the wake is shown in the Figures 2 and 3. Both are made non-dimensional by U(H), the approach flow wind speed at height z=H, where H is the plate height.

The recirculation region behind the solid plate, M(0,1), stretches to about x/H=2.5, where the highest turbulence intensity is seen. Introduction of a small degree of porosity enables bleed flow to pass through the plate causing the recirculation region to be displaced a small distance downstream, as well as reducing its overall length. This is illustrated by the results for model M(30,4). Turbulence intensities are just slightly greater than approach levels ahead of the first mean stagnation point, then high throughout the recirculating region. As the porosity is further increased the first mean stagnation point is displaced further downstream, accompanied by an overall reduction in recirculation region size.

Results for cases M(50,2) and M(30,1) are similar, showing a minimum mean velocity of about 0.1U(H) at about x/H=2.5 and high turbulence levels between x/H=1 and 3. The turbulence intensity in this region is about 0.3U(H), so there is clearly a substantial region of intermittent flow reversal at this height with a distinct possibility of mean flow reversal closer to the surface. There is a small region of shelter between the model and x/H=1, though immediately behind the model turbulence levels are somewhat increased above ambient values due to generation by the plate grid. The mean flow recovery is somewhat slower than for the more solid models, with ambient conditions attained at approximately x/H=20. Ambient turbulence levels are reached a little sooner, at about x/H=15 in all but one case; this however is peculiar to the height of the traverse as more detailed measurements of the vertical distribution show (see below).

The bleed flow strengthens as the blockage is further decreased. For M(50,2) the velocity defect is of order 40 to 50% in the fetch x/H<6, though turbulence levels grow rapidly at about x/H=2.5. The area of shelter is much more extensive behind model M(70,1), though the velocity deficit is only of order 25%. Turbulence intensities are also lower than ambient levels throughout the area of shelter (i.e. for x/H<10).

The results show the existence of three regimes of wake behavior for high, medium and low blockages respectively:

Regime I: near wake flow recirculation for high blockages; attached recirculation for solid plate, otherwise detached; recirculation region bounded by zones of strong shear; e.g. M(0,1), M(30,4),

Regime II: no mean flow recirculation; shelter in the near wake, with bleed flow deceleration; near-wake bounded by zones of strong shear which converge downstream to form a deep and vigorous turbulent mixing region; e.g. M(30,1), M(50,4),

Regime III: persistent mean flow shelter for cases of low blockage; weak shear layers and low levels of wake turbulence; e.g. M(70,1).

Centreline measurements provide only a limited picture of wake behavior and to establish a more comprehensive understanding of the three dimensional flow mechanisms which determine the respective flow regimes, a small subset of model cases were selected:

Regime I - M(30,4)
Regime II - M(50,2)
Regime III - M(70,1).

The solid model, M(0,1), was also used as a reference case.

Detailed mean flow and turbulence mappings were carried out for these cases, followed by dispersion studies. Examples of mean velocity profiles for M(30, 4) are shown in Figure 4. Reversed flow extends vertically to about z/H=1 and laterally to about y/W=0.4 (where W is the total plate width). Excess turbulence levels extend vertically to at least z/H=2 and persist throughout the region investigated (i.e. to x/H=15). The height of the maximum excess rises from around z/H=0.5 at x/H=1 to 1.5H at 15H.

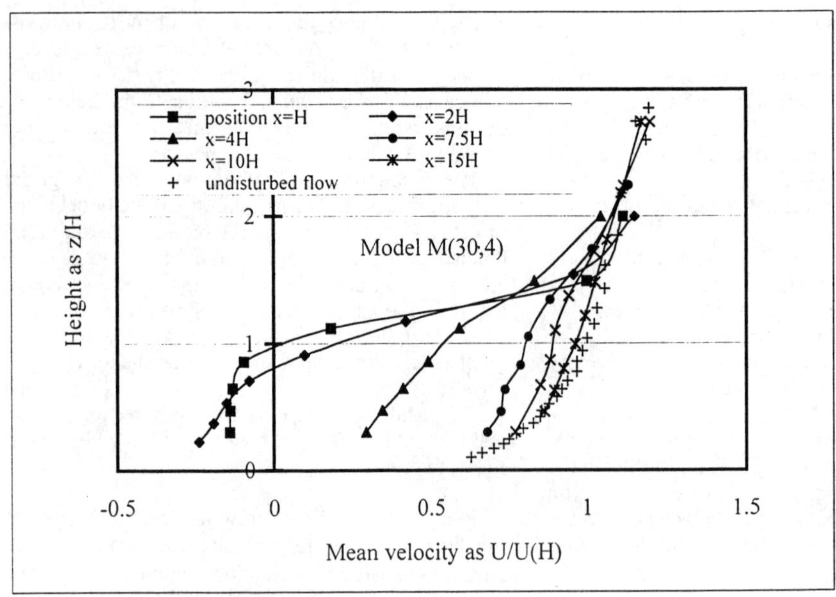

Figure 4. Vertical profile of mean vleocity on centre line of model wakes; results are non-dimensionalised by the vlocity at the plate heoght, U(H).

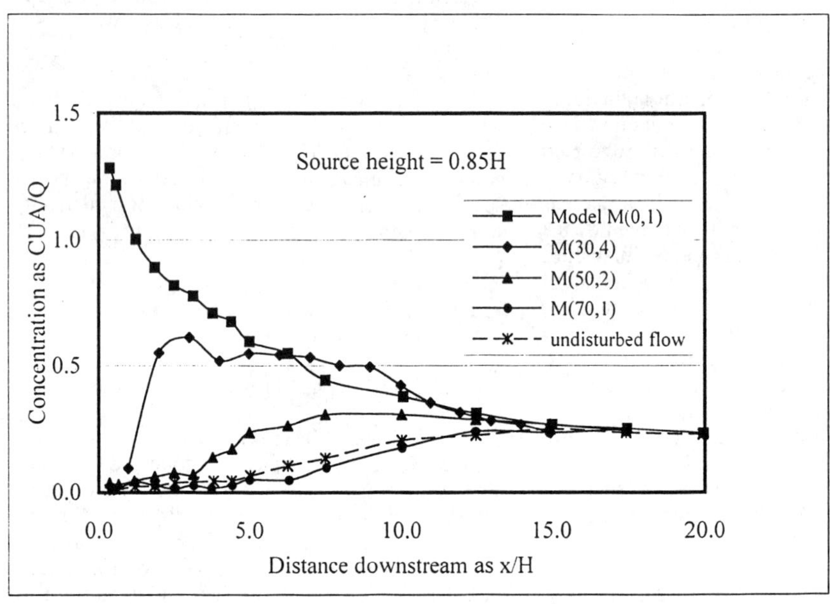

Figure 5. Longitudinal profiles of ground level concentration downwind of models for a source at height 0.85H.

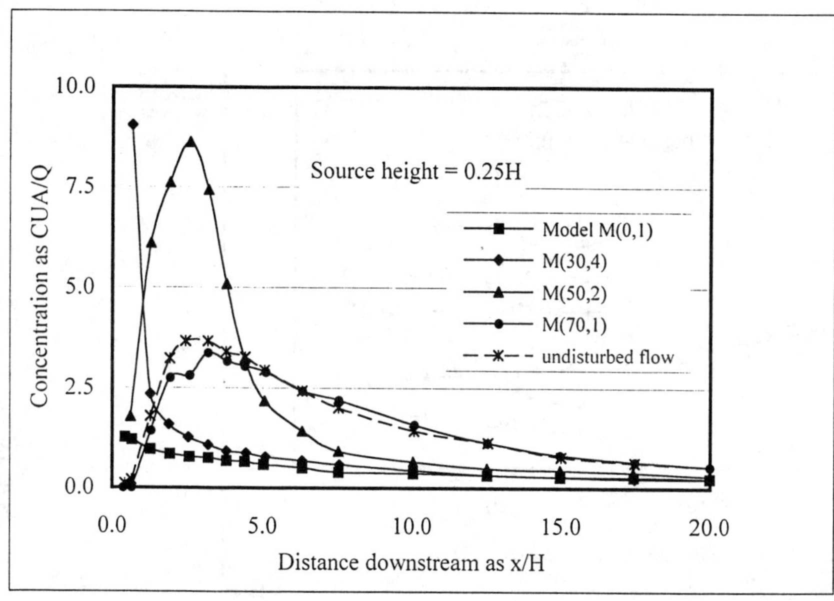

Figure 6. Longitudinal profiles of ground level concentration downwind of models for a source at height 0.25H.

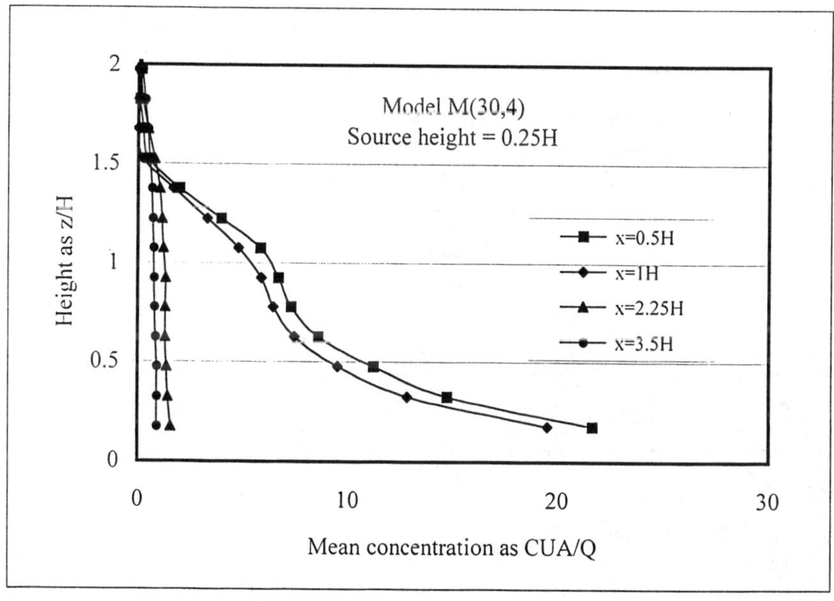

Figure 7. Vertical profiles of ground level concentration downwind of Model M(30,4) for a source at height 0.25H.

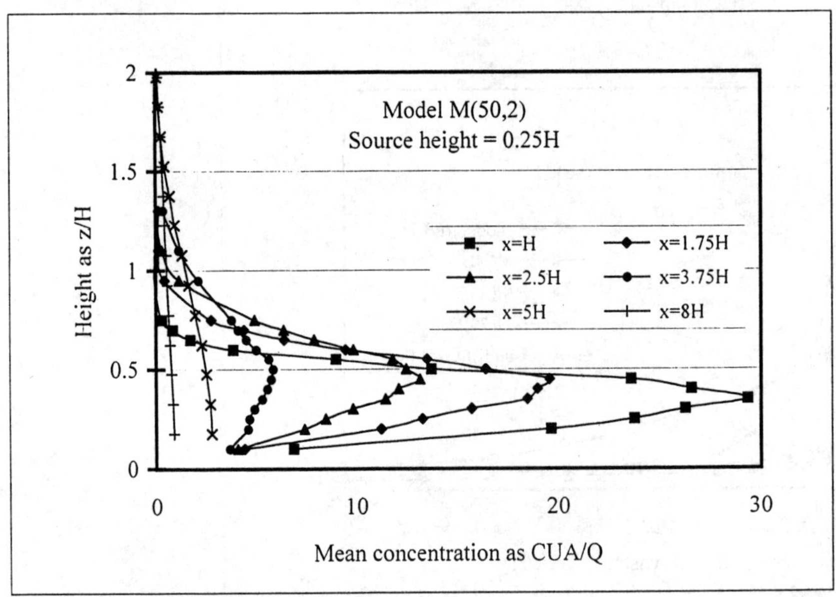

Figure 8. Vertical profiles of ground level concentration downwind of Model M(50,2) for a source at height 0.25H.

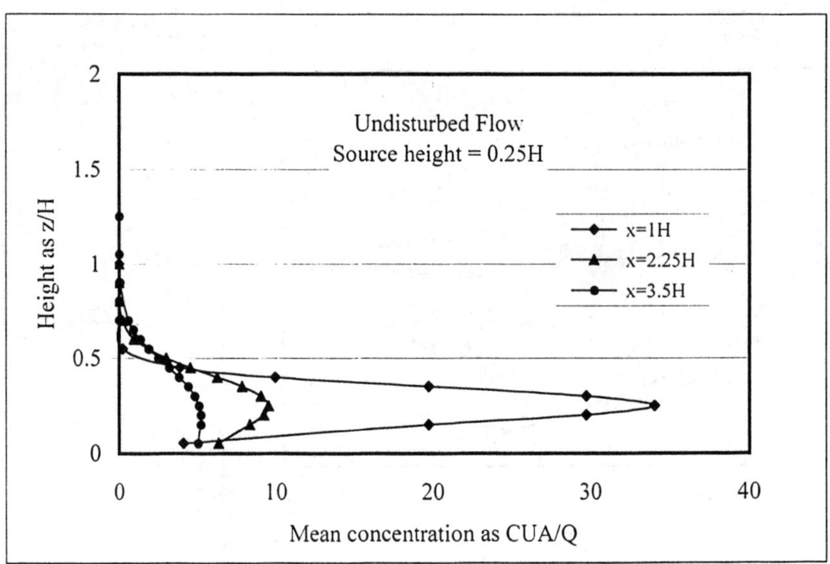

Figure 9. Vertical profiles of ground level concentration in the undisturbed flow for a source at height 0.25H.

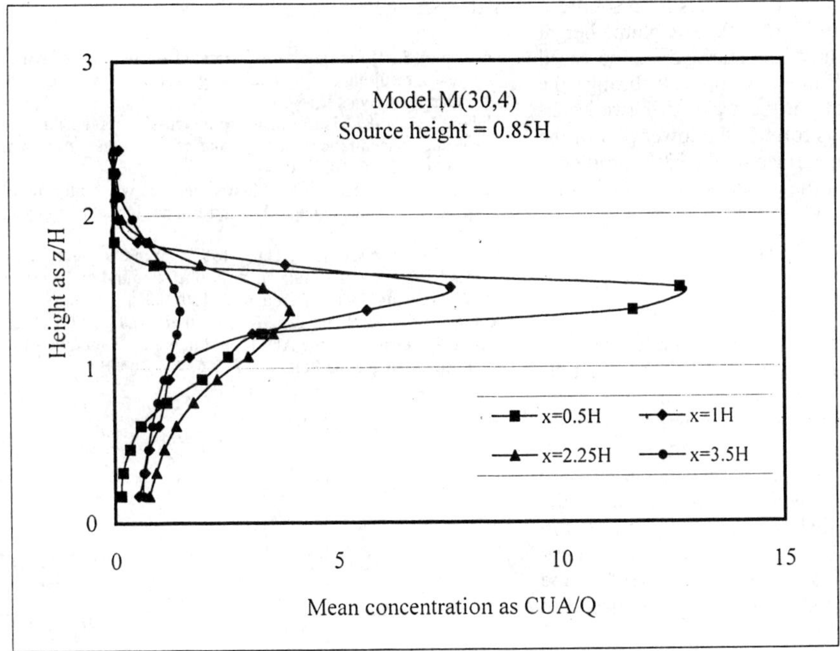

Figure 10. Vertical profiles of ground level concentration downwind of Model M(30,4) for a source at height 0.85H.

5 DISPERSION MEASUREMENTS

Figures 5 and 6 show the variations of ground level, centre line concentration for a range of model porosity (i.e. flow regimes), with source heights of 0.85H and 0.25H respectively. Concentrations are non-dimensionalised as CUA/Q, where U is the reference wind speed at height H, A the projected frontal area of the model and Q the volume emission rate. Results for the solid plate are essentially the same in both cases and results for M(70,1) are approximately the same as those in the undisturbed flow. Enhanced mixing clearly affects the other two cases. The detached recirculation and associated high turbulence levels downwind of M(30,4) mix the elevated plume very rapidly to the surface, which is evidenced for the source at 0.85H by a sudden increase in ground level concentrations at the upstream edge of the recirculation region. The plume from the lower source is mixed to the ground in the low velocity flow behind the plate, producing high ground level concentrations that decay rapidly through the recirculation region and immediately downwind. Results for M(50,2) also show the most significant changes affecting the plume from the lower source height.

These observations are supported by vertical concentration profiles measured downwind from M(30,4) and M(50,2) and shown here as Figures 7 and 8, for a source height of 0.25H. Intense mixing in the detached recirculation region behind M(30,4) deepens the plume very rapidly so that even at x/H=0.5 the concentration field extends to through the full depth of the wake. Flow visualization showed essentially the same picture, with the recirculation lifting material from near the leading stagnation point almost vertically to heights in excess of the plate height. A somewhat simpler pattern is found downwind from M(50,2). The flow deceleration caused mean streamlines to be displaced upwards, carrying the plume with them. High turbulence levels from about x/H=3 the mix the lower part of the plume to the ground. These two figures should be compared with Figure 9, which shows equivalent data for the undisturbed flow (but only for x/H=1, 2.25 and 3.5). Here the plume height remains at the emission height and plume spread is slower.

The deep, detached recirculation downwind from M(30,4) is accompanied by considerable mean streamline displacement, which manifests itself as increases in the plume centre height, as is clearly apparent in the concentration profiles for a source at height 0.85H shown in Figure 10. The plume height has increased to slightly less than 1.5H. Increased mixing also affects plume development through the high turbulence levels in the wake. Skewed concentration distributions form as the lower part of the plume is mixed to the surface by the high turbulence levels associated with the detached recirculation region between x/H=1 and 3.

6 CONCLUSIONS

The differing nature of the wake flows was clearly apparent in the dispersion characteristics. Regions of high turbulence intensity were associated with rapid mixing and plume growth, especially when combined with recirculation. Ahead of such regions the variations in mean longitudinal velocity imply secondary flows which bodily deflected dispersing plumes. The behavior of plumes emitted in the near wake varied strongly with source height, reflecting the properties of the wake (z/H<1), the bounding shear layers (1<z/H<1.5) or the displaced external flow (z/H>1.5).

Existing building wake dispersion models have been developed to account for the effects of near wake recirculation (e.g. Huber, 1987, Robins et al, 1998). Such models are likely to be suitable for Regime I flows since recirculation plays an important role in the dispersion process. It is important to realize, that in the calm bleed flow immediately behind the model, local concentrations are very high. Solid building wake models are therefore only useful beyond the first stagnation point for such cases.

Regime II and III flows show quite different dispersion behavior due to the absence of mean recirculation and presence of a zones of decreased mean flow and turbulence levels. The shape of the concentration distributions suggests that a Gaussian modelling approach may apply. Careful modelling of plume spread and vertical plume displacement is needed, together with the development of a mean flow model.

7 ACKNOWLEDGEMENTS

This work was carried out under the EPSRC Engineering Doctorate Programme in Environmental Technology. Shell Research Ltd. funded the experimental work.

REFERENCES

Gandemer, J., 1981. The aerodynamic characteristics of windbreaks, resulting in empirical design rules. J. Wind Engng. Indust. Aetodyn., 7: 15-36.

Huber, A.H., 1987. Performance of a Gaussian model for centreline concentrations in the wake of buildings. Atmoms. Environ., 22, 6: 1039-1050.

Perera, M.D.A.E.S., 1981. Shelter behind two dimensional solid and porous fences. J. Wind Engng. Indust. Aetodyn., 8: 93-104.

Raine, J.K. & Stevenson, D.C., 1997. Wind protection by model fences in a simulated atmospheric boundary layer. J. Wind Engng. Indust. Aerodyn., 2: 159-180.

Robins, A.G., McHugh, C.A. & Carruthers, D.J., 1997. Testing and evaluating the ADMS building effects module. Int. J. Environment and Pollution, 8, 3-6: 708-717.

Wind Engineering into the 21st Century, Larsen, Larose & Livesey (eds) © 1999 Balkema, Rotterdam, ISBN 90 5809 059 0

Effects of stable stratification on flow and dispersion around a cube

N. Steggel & I. P. Castro
Environmental Flow Research Centre, School of Mechanical and Materials Engineering, University of Surrey, Guildford, UK

ABSTRACT: The first stages of a study on the effects of stable stratification on the flow and dispersion around a cube immersed in simulated atmospheric boundary layers are described. A brief overview of the tunnel arrangements and simulation devices is given and the instrumentation techniques are also discussed. Two boundary layers were used, a neutrally stable layer of depth about 1000mm and a moderately stable layer of depth around 450mm. The stable layer had a Monin-Obukhov depth of around 170mm. With the cube present mean and turbulent velocities were measured together with mean ground level concentrations. In the near-wake region it was found that stability was of little importance and the flow characteristics could be explained by reference solely to the upstream velocities and turbulence levels at the cube height. Further downstream differences in the behaviour of the velocity excesses and turbulence levels were apparent. However mean ground level concentrations were not found to differ greatly between the two layers.

1 INTRODUCTION

1.1 *Background*

There is a considerable literature on the general topic of flows around bluff obstacles submerged in thick turbulent boundary layers. However the overwhelming majority of existing material (e.g. Robins & Castro 1979a, b, Li & Meroney 1983a, b, Akins & Reinhold 1998) is concerned with the case of neutral stability.

A study has therefore been initiated in the EnFlo laboratory with the objective of assessing the effects of stable stratification on the formation and development of wakes downstream of bluff obstacles submerged in turbulent boundary layers. This is of practical consequence since the dispersion characteristics of pollutants released near such obstacles may be significantly changed by the atmospheric conditions.

Some studies have attempted to investigate the influence of stability. Ogawa & Diosey (1980a, b) made comparison between a field and wind tunnel study of the effects of thermal stability on the flow behind a fence. Snyder (1985) examined the effect of stability on the flow and dispersion around a cube although in this case the obstacle was not immersed in a thick boundary layer and only the near wake region was examined. More recently a series of field experiments have been conducted (see Higson et al. 1995 and Mavroidis et al. 1999) in which the effect of stability on the dispersion characteristics in the vicinity of an isolated building has been assessed. They found, for some parameters, a not insignificant effect.

1.2 *EnFlo building wakes project*

Results from the first stages of the present project in which the flow and dispersion characteristics around a 2D fence were studied have recently been published, Steggel & Castro (1998, 1999). These experiments were conducted in the EnFlo stratified flow wind tunnel, in which simulated atmospheric boundary layers can be generated. A 100mm high fence was mounted spanwise within both a neutrally stable boundary layer and a layer of moderate stability.

Mean and turbulent velocities were measured in the first stage, Steggel & Castro (1998). The reattachment length, X_R, was found to be about 12 fence heights in the stable layer and about 8 in the neutral layer. This difference in X_R was explained on the basis of the lower external turbulence levels seen by the separated shear layer in the stable flow, a result which is consistent with the earlier study by Ogawa & Diosey (1980b).

In the second stage, Steggel & Castro (1999), a stack was positioned at a fixed distance upstream of the fence and at several heights. The plume was shown to remain nearer to the ground in the stable layer due to the less severe streamline curvature over the fence in this case. As a result ground level concentrations were considerably higher in the stable layer since a greater fraction of material could be mixed down to ground level. In the present study a cube of 100mm dimensions was used. A series of preliminary measurements has been taken in both the near and far-wake regions and for two cube orientations. In one the cube wall was aligned normal to the oncoming flow and in the other the cube was aligned at 45 degrees to the flow direction. The latter orientation was chosen since it was anticipated that the axial vorticity produced in this case may be weaker and/or may take longer to decay under stably stratified conditions. The origin is taken to be the centrepoint of the cube base.

Simulation of the boundary layers and instrumentation details are discussed in the following section. Measurements of velocity, turbulence and concentration quantities are presented in subsequent sections.

2 EXPERIMENTAL ARRANGEMENTS

2.1 *EnFlo wind tunnel*

Figure 1 indicates the boundary layer simulation method employed in the EnFlo tunnel. The Counihan system comprising barrier wall, vorticity generators and surface roughness was used to generate a neutrally stable boundary layer of depth about 1000mm. A layer of moderate stability whose Monin-Obukhov length scale was around 170mm was generated by cooling the tunnel floor to around 10°C and heating the flow at the entrance to the tunnel working section to 45°C above the cooled floor temperature. An identical roughness arrangement was employed in this case.

2.2 *Instrumentation*

Mean velocity and turbulence measurements were made using a two-component laser Doppler anemometer which, in the case of the stratified flow (in which local temperatures exceeded 55°C) used a specially constructed, air-cooled fibre-optic probe. A cold wire was used to take synchronised temperature measurements such that the fluctuating temperature and heat fluxes could be measured. Heist and Castro (1998) discuss the combined LDA/cold wire measurement technique further.

2.3 *Boundary layer fits and parameters*

Two boundary layers have been simulated in the EnFlo tunnel for use in the current study. These layers correspond to atmospheric conditions of neutral and moderate stability. The neutral layer can be fit to the standard log-profile,

$$\frac{U}{u^*} = \frac{1}{k} ln \frac{z}{z_0}, \tag{1}$$

where u^* is the friction velocity, z_0 the roughness length and k is the Von-Karman constant. It is assumed that the zero-plane displacement is zero due to the particular geometry of the rough surface used in this case.

Fitting the stable boundary layer is not straightforward since four further variables must be determined. These are θ^* the friction temperature, L_{mo} the Monin-Obukhov length scale, β_u and β_T. The last of these are constants required to fit the log-linear expressions known as the Businger-Dyer relationships which can be written:

$$\frac{U}{u^*} = \frac{1}{k}\left\{ ln \frac{z}{z_0} + \frac{\beta_u z}{L_{mo}} \right\}; \tag{2}$$

$$\frac{\Delta\theta}{\theta^*} = \frac{1}{k}\left\{ ln \frac{z}{z_0} + \frac{\beta_T z}{L_{mo}} \right\}. \tag{3}$$

Stull (1993) derives each of the above expressions in his chapter on similarity theory. A full analysis of the boundary layer simulation and fitting process is documented in Steggel (1999). Parameters found in that study have been summarised in Table 1 for each layer. In general the complete datasets

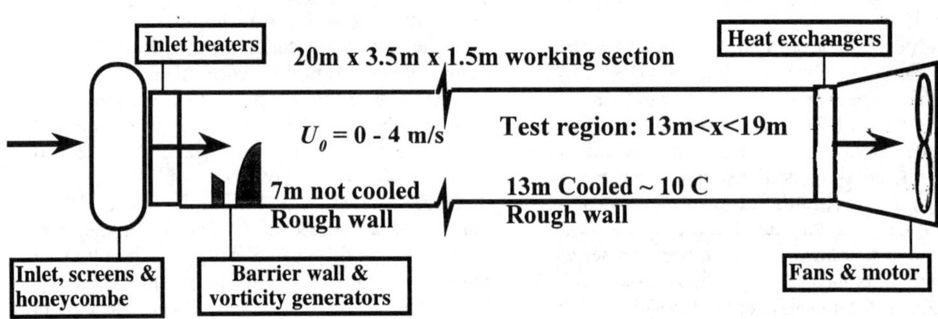

Figure 1. Schematic of boundary layer simulation in the EnFlo tunnel

for these boundary layers were found to be consistent within themselves and with existing atmospheric measurements. A moderately stable boundary layer with $\delta/L_{mo} \approx 2.6$ has been established in the tunnel.

Table 1. Boundary layer parameters and near-wall turbulence levels.

Parameter	Neutral	Stable
δ(mm)	1000	450
z_0(mm)	1.0	0.9
u^*/U_0	0.06	0.036
$\theta^*/\Delta\theta_0$	—	0.033
β_U	—	4.8
β_T	—	6.8
L_{mo}(mm)	—	170
σ_u/u^*	2.0	1.9-2.2
σ_w/u^*	1.3	1.2-1.4
σ_θ/θ^*	—	1.9-2.4
$u^*\theta^*/\sigma_u\sigma_\theta$	—	0.30-0.44

3 LDA MEASUREMENTS

3.1 *Near-wake velocity vectors*

A near-wake mapping of the flowfield (u,v velocities) at half the building height, z/H_b=0.5, was obtained for both boundary layers and both cube orientations. In Figure 2 velocity vectors are shown for the case of neutral flow past a cube orientated at 45°.

The measurements were only taken for y≥0 and have been mirrored although it is clear from the measurements at y=0 that the flow is not exactly symmetric; i.e. there is some cross-flow along the wake centreline. A pair of vortices can clearly be seen and it is possible to pinpoint the centres of these vortices and approximate the zero-velocity point along the wake centreline. In the case of the stable flow (not shown here) the centres of these vortices occurred slightly further downstream and this is probably attributable to the lower upstream turbulence levels seen in that case. This effect was more pronounced for the normally aligned cube but in neither case was the effect as significant as that observed in the fence flow. The general conclusion is that the reattachment position is lengthened by a reduction in upstream turbulence levels and that this effect is more significant for wider obstacles.

3.2 *Wake velocity deficits*

Figure 3 plots velocity excesses against downstream distance for the two cube orientations considered. The excesses have been non-dimensionalised by the upstream velocity at the obstacle height. For clarity a zoom in of the immediate near-wake region is shown in Figure 4. When non-dimensionalised in this manner the curves collapse in the immediate near-wake region, *X*<200mm, according to cube orientation.

Further downstream no collapse is achieved and the excesses appear dependant on both cube orientation and the boundary layer type. In the case of the stable boundary layer and cube aligned at 45 degrees an extremely high velocity excess was observed. This was of the order of 30% of the upstream velocity and persisted along the stretch of measurements made up to *X*=1500mm. The velocity excesses found in the neutral boundary layer are much smaller in comparison. Hayden (1998) submerged a series of buildings (non-cubic) in a neutral boundary layer and found a small velocity excess beyond about ten building heights downstream. This excess generally increased as the building width increased and was greater when the building was aligned at 45 degrees to the flow, a result consistent with the present measurements.

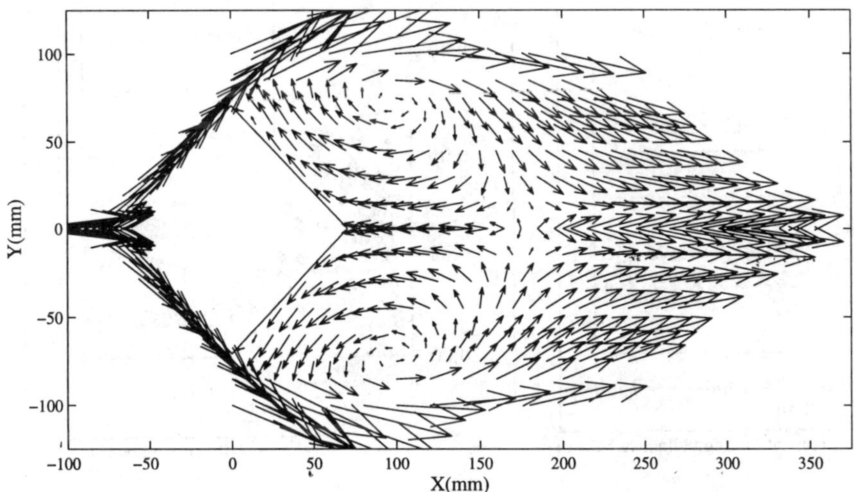

Figure 2. Velocity vectors in near-wake of cube at z/H_b=0.5; neutral boundary layer.

3.3 *Turbulence decay in the wake*

Development of longitudinal and lateral turbulence levels along the wake centreline are shown in Figures 5 and 6. Two additional lines are plotted in Figure 5 indicating the undisturbed longitudinal levels. The turbulence levels have been non-dimensionalised by the upstream velocity at the obstacle height. This would appear to provide a reasonable level of collapse between the two flows in the near wake region.

Further downstream in the stable boundary layer the longitudinal turbulence levels return to roughly their upstream values. However in the neutral boundary layer a large turbulence deficit is observed which does not appear to be dependant on the cube orientation (and is probably a downwash effect). Hayden (1998) also observed a longitudinal turbulence deficit beyond around 5 building heights but data was only presented for flows normal to the building.

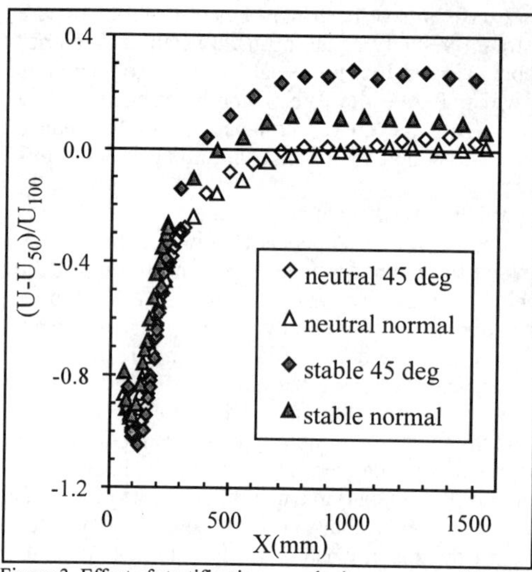

Figure 3. Effect of stratification on velocity excess at z/H_b=0.5.

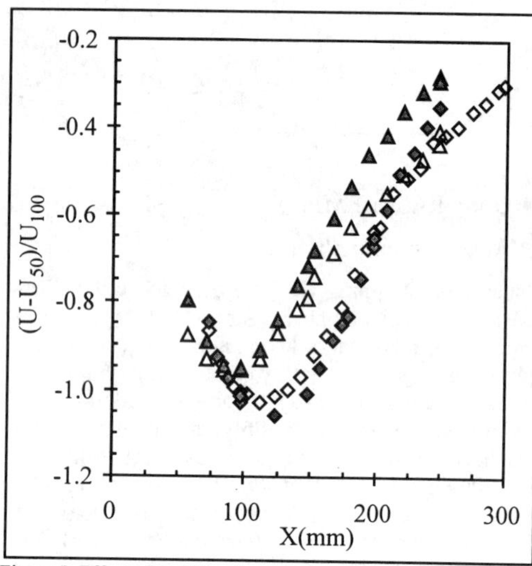

Figure 4. Effect of stratification on near-wake velocity excess.

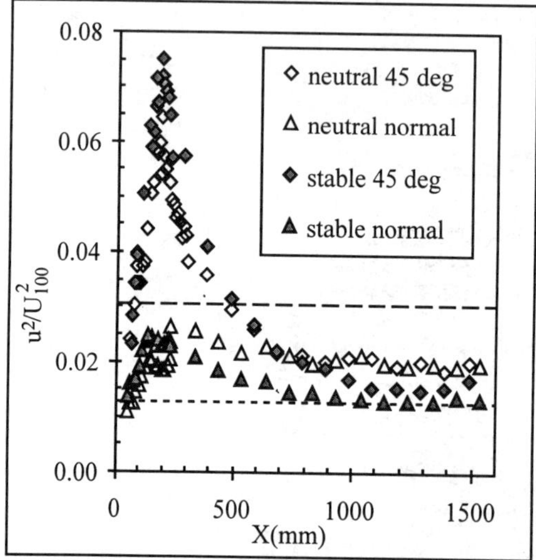

Figure 5. Downstream development of longitudinal turbulence levels, z/H_b=0.5.

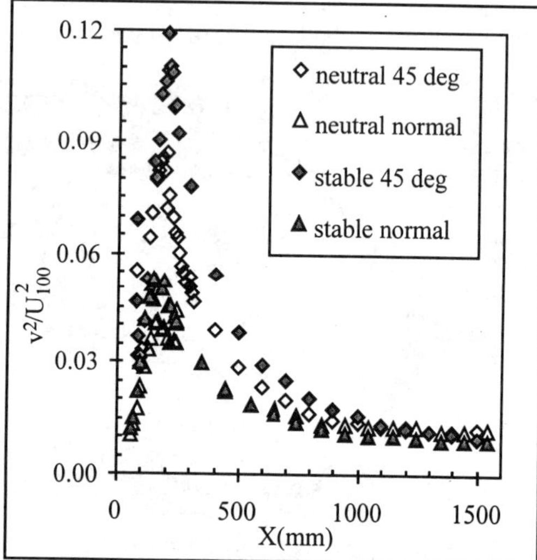

Figure 6. Downstream development of lateral turbulence levels, z/H_b=0.5.

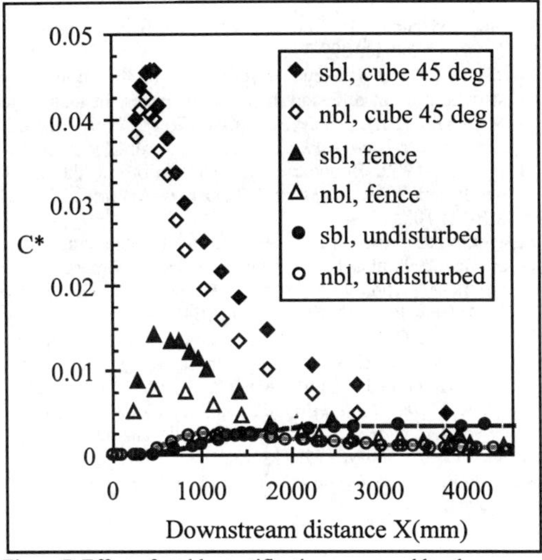

Figure 7. Effect of stable stratification on ground level concentrations, H_S=100mm.

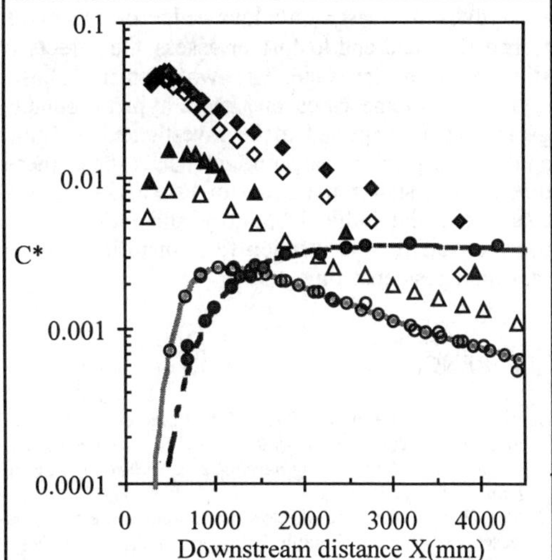

Figure 8. Glc's log-scale.

4 CONCENTRATION MEASUREMENTS

Mean concentration measurements were obtained using a 16 channel gas-sampling system together with a Flame Ionisation Detector (FID). Buoyancy effects had to be considered in the stable layer since there was a 45 degree temperature variation between the surface and the top of the boundary layer. The temperature of the stack release was monitored and maintained within a few degrees of the temperature at the release height as small temperature differences were shown to have an insignificant effect.

Non-dimensional values of ground level concentrations are plotted against downstream distance in Figure 7. An identical curve plotted on a log-scale is shown in Figure 8. The concentrations have been non-dimensionalised using $C^* = CU_{100}A/Q$ where A is the source area, Q the source flowrate, C is the measured concentration (ppm) and U_{100} is the velocity at the obstacle height in the undisturbed boundary layer. In each case the stack height was fixed at 100mm and was positioned 200mm upstream of the obstacle.

In the undisturbed boundary layers it is seen that the plume takes longer to impinge on the ground in the stable case with maximum glc's occurring further downstream. Guassian fits of the undisturbed flows are also indicated on the graphs. With the fence in position the glc's are enhanced in both boundary layers although the maximum observed in the stable case is nearly double that observed in neutral flow. The downwash effect is dominant with the cube in position and glc's are significantly larger than in the fence flow cases. In the stable flow the glc's are still a little larger than in the neutral case but this effect is much less significant than in the fence flow case. Further downstream the differences are more significant but in both boundary layers the glc's appear to be returning to their undisturbed flow values.

5 CONCLUSIONS

Some preliminary investigations of the effects of stable stratification on the flow and dispersion around obstacles have been undertaken. Two obstacles have been studied thus far - a fence and a cube at 45 degrees to the oncoming flow. In both cases it is suggested that the near wake dynamics can be explained by the interaction between the freestream flowfield and the obstacle generated disturbance and that the temperature field is of secondary importance. However in the far-field there appear to be some significant differences, the reasons for which are not yet entirely clear.

Although there are differences in the flowfield the ground level concentrations further downstream do not appear to be much greater than their undisturbed flow values, although in the 45 degree cube case stability seems to have a measurable effect. It is probable, firstly, that glc's will only be significantly affected by stability for wider obstacles but, secondly, that this is due more to the indirect effect of stability on the background velocity and turbulence profiles than on any major changes in the near wake effected directly be the stability.

This project is ongoing and future experiments will concentrate on trying to parameterise the near-wake dynamics based on a knowledge of the undisturbed flowfield and to further assess the effects of stability on the far-wake. The two obstacles chosen represent extreme cases and more typical building geometries are expected to be investigated in future studies. Only mean concentrations have been measured in these stages but a pair of fast FID's are now available in the EnFlo laboratory and these will be used to obtain concentration time histories and assess wake residence times.

REFERENCES

Akins, R.E., and Reinhold, T.A. 1998. Laser Doppler velocimeter measurements of separated shear layers on bluff bodies, *Journal of Wind Engineering and Industrial Aerodynamics*, 74-76, 455-461.

Hayden, P. 1998 Building affected dispersion from elevated releases in neutral and stable boundary layers, *PhD thesis, EnFlo, University of Surrey, UK.*

Heist, D.K. and Castro, I.P. 1998 Combined laser-Doppler and cold wire anemometry for turbulent heat flux measurement, *Experiments in Fluids*, 24, 375-381.

Higson, H.L., Griffiths, R.F., Jones, C.D., and Biltoft, C. 1995. Effect of atmospheric stability on concentration fluctuations and wake retention times for dispersion in the vicinity of an isolated building, *Environmetrics Journal*, 6(6), 571-581.

Huber, A.H. 1988. Video images of smoke dispersion in the near wake of a model building. Part I. Temporal and spatial scales of vortex shedding, *Journal of Wind Engineering and Industrial Aerodynamics*, 31, 189-223.

Li, W.W. and Meroney R.N. 1983a. Gas dispersion near a cubical model building. Part I. Mean concentration measurements, *Journal of Wind Engineering and Industrial Aerodynamics*, 12, 15-33.

Li, W.W. & Meroney R.N. 1983b. Gas dispersion near a cubical model building. Part II. Concentration fluctuation measurements, *Journal of Wind Engineering and Industrial Aerodynamics*, 12, 35-47.

Mavroidis, I., Griffiths, R.F., Jones, C.D., and Biltoft, C. 1999. Experimental investigation of the residence of contaminants in the wake of an obstacle under diferent stability conditions, *Atmospheric Environment*, 33, 939-949.

Ogawa, Y. & Diosey, P.G. 1980a. Surface roughness and thermal stratification effects on the flow behind a two-dimensional fence - I. Field study, *Atmospheric Environment* , 14, 1301-1308.

Ogawa, Y. & Diosey, P.G. 1980b. Surface roughness and thermal stratification effects on the flow behind a two-dimensional fence - II. A wind tunnel study and similarity considerations, *Atmospheric Environment*, 14, 1309-1320.

Ogawa, Y., Oikawa, S., & Uehara, K. 1983b. Field and wind tunnel study of the flow and diffusion around a model cube - I. Flow measurements, *Atmospheric Environment* , 17, 1145-1159.

Ogawa, Y., Oikawa, S., & Uehara, K. 1983b. Field and wind tunnel study of the flow and diffusion around a model cube - II. Nearfield and cube surface flow and concentration patterns, *Atmospheric Environment* , 17, 1161-1171.

Robins, A.G. & Castro, I.P. 1977a A wind tunnel investigation of plume dispersion in the vicinity of a surface mounted cube - I. The flow field, *Atmospheric Environment*, 11, 291-297.

Robins, A.G. & Castro, I.P. 1977b A wind tunnel investigation of plume dispersion in the vicinity of a surface mounted cube - II. The concentration field, *Atmospheric Environment*, 11, 299-311.

Snyder, W.H. 1994. Some observations of the influence of stratification on diffusion in building wakes, In I.Castro & N.Rockliff (eds), *Stably Stratified Flows: Flow and Dispersion over Topography*: 301-324. Oxford: Clarendon.

Steggel, N. 1999. Boundary layer simulation in the EnFlo wind tunnel, *EnFlo Internal Report, University of Surrey*, ME-FD 100.

Steggel, N. & Castro, I.P. 1998. Effects of stable stratification on flow behind a two-dimensional fence submerged in a turbulent boundary layer, In U. Frisch (ed.), *Advances in Turbulence VII*: 487-490. Dordrecht: Kluwer.

Steggel, N. & Castro, I.P. 1999. Effects of stable stratification on dispersion characteristics behind a two-dimensional fence submerged in a turbulent boundary layer, *Proceedings of the 2nd International Conference on Urban Air Quality, Computer Science School of the Technical University of Madrid, 3-5 March 1999*, 12-13.

Stull, R.B. 1993. *Boundary Layer Meteorology*, Dordrecht: Kluwer.

Wind Engineering into the 21st Century, Larsen, Larose & Livesey (eds) © 1999 Balkema, Rotterdam, ISBN 90 5809 059 0

Pressure characteristics for separated flows

H.W.Tieleman & M.R.Hajj
Virginia Polytechnic Institute and State University, Blacksburg, Va., USA
T.A.Reinhold
Clemson University, S.C., USA

ABSTRACT: Flow parameters for eight roughness configurations are presented. For four of these configurations, wind tunnel point pressures from four sets of roof pressure taps are investigated and compared with corresponding WERFL field data. The four sets were strategically chosen to cover areas where extreme suction pressures have been observed. The field pressure records were divided in to three categories, which were based on the level of the lateral turbulence intensity. The probability levels of the field and model peak pressures are discussed in detail. In addition, non conditional and conditional correlation coefficients of the peak pressures were analyzed including the pressure peak duration. Observations are also presented for the convective speed of the pressure waves under the separation bubble and emanating from the leading edge.

1. INTRODUCTION

With the availability of field pressure coefficients from the Wind Engineering Research Facility Laboratory (WERFL) at Texas Tech University (Levitan & Mehta 1992a,b) direct comparison of mean and fluctuating surface pressures between field and wind tunnel model experiments has become feasible. Although for relatively small model scales (1:50) average values of point pressure coefficients have been reproduced adequately, simulation of the fluctuating surface pressures near roof corners, roof edges and roof ridges has proven to be more difficult. The flow phenomena associated with flow separation, which determine the adjacent surface pressures, are affected by the incident turbulence, and in particular those scales which match the thickness of the separated shear layers.

It is generally impossible to satisfy equality of the turbulence Reynolds number for the field and laboratory experiments, with the result that the ratio of large to small scales of the turbulence for field experiments is several orders of magnitude larger than the corresponding ratio for the laboratory flow simulations (Tennekes & Lumley 1972). This means that for atmospheric flows, the frequency extent of the spectral inertial subrange is much larger than that for the flow simulations in the laboratory. As a consequence, the small scale turbulence content as defined by Melbourne may be inadequately modeled. This is especially true if the small scales fall in the frequency range corresponding to the viscous subrange, where spectral densities are appreciably lower than in the inertial subrange (Tieleman & Akins 1996).

If the geometric model scale is based on the ratio of the turbulence integral scale, the small-scale turbulence content of the incident simulation flow, which controls the development of the separated shear layers, will be frequently inadequate. Similarly, if one emphasizes the similarity of the velocity scale associated with the small-scale turbulence parameter, S, geometric scaling of the integral scales can not be maintained. Consequently, simultaneous geometric scaling of the integral scale and simulation of the small scale turbulence does not seem possible, and choices need to be made. Improved matching of the small-scale turbulence parameter can be achieved by using larger models together with increased upstream floor roughness and or small spires. Using larger models and increased floor roughness while relaxing the requirements for strict matching of integral length scales has produced flow simulations which provide a more acceptable laboratory duplication of the peak pressure coefficients in critical areas (Tieleman et al. 1998).

Realizing that simultaneous reproduction of all flow parameters is next to impossible, the final evaluation of the adequacy of any simulation approach must be based on the full-model scale comparison of local pressures and area averaged wind loads. In order to properly assess the validity of the simulation compromises, the following issues must be addressed:

1. Detailed documentation of the distribution of most flow parameters, regardless of whether it is for the field experiments or for laboratory simulations.

2. Reproduction of the flow parameters at those heights where wind loads are being studied.

3. By comparison of flow parameters and pressure coefficients one can evaluate which one of the flow parameters needs more accurate reproduction in order to conduct the best simulation experiments.

4. Meaningful comparison of either flow or pressure measurements requires that record length of the simulation data is matched to that of the field data.

5. Statistical parameters for the pressure records must be analyzed and compared, including covariances and the speed at which pressure waves travel over the building surface.

6. The stochastic nature of the fluctuating wind pressures and area loads must be addressed before a conclusive assessment can be made of the adequacy of the various wind tunnel simulation techniques. One must avoid the comparison of peak pressure coefficients and peak wind loads at different levels of probability (Lieblein 1974).

7. Comparison of the duration of peak wind loads is important since higher structural stresses are allowed in wood for short duration and impact loads.

Area averaged loads and duration of loading can be important for certain structures and materials. Consequently, it is critical that the coherence and duration of loading characteristics of the simulation are formally evaluated and compared with field measurements. This paper provides preliminary results from pressure correlations and duration of peak loading for model and full scale. A companion paper (Yan et al. 1999) presents results of a more detailed study of comparisons between model and field area reduction factors.

2. DESCRIPTION OF THE EXPERIMENTS

The WERFL field laboratory features a 50 m tall, instrumented (6 levels) micro meteorological tower. This tower is located approximately 45 m in the north westerly direction of 280 degrees from the center of the rotatable flat-roof experimental building (9.1 × 13.7 × 4 m). The field data used in this article belong to a large set of 15-minute records of simultaneous flow and pressure observations which were acquired at WERFL between 1990 and 1992. Atmospheric velocity data, speed and direction were first low-pass filtered at 8 Hz and then sampled at a rate of 10 Hz. The pressure data were first filtered at 8 and 10 Hz for the Omega and Valendyne pressure transducers respectively and then sampled at a rate of 40 Hz.

The simulation experiments (scale 1:50) were conducted in the boundary layer wind tunnel of the Wind Load Test Facility (WLTF) at Clemson University. The tunnel has a 21 m long test section with a cross section of 3 × 2 m. Model pressure data were obtained with the use of a transducer and tubing system with a flat response of 5% up to 300 Hz. Flow data were obtained for 8 different simulations at the model location but with the model removed. The arrangement of the upstream spires and surface roughness for configurations #1 through #7 are given in Tieleman et al. 1998. Configuration #8 consists of six 11.5 × 20 cm spires with a 13 cm high trip board, both placed 6.4 m upstream of the model location, and complemented with a row of small spires located as described in configuration #3. The wind tunnel control was set at 800 for each configuration, which corresponds to a free stream speed of approximately 13 m/s. Mean flow and turbulence profiles were based on 45 second long time histories, using a sampling rate of 4000 Hz. The velocity spectra were obtained using time histories of 180 second in duration with the same sampling rate.

Four sets of roof pressures, each consisting of no more than 8 pressure taps were investigated (Fig. 1):

Set A: Consists of eight pressure taps along a line parallel to the long side of the model and at a distance $x/H = 0.663$ from the roof corner for an azimuth angle of 180 degrees.

Set B: Consists of eight pressure taps along the midsection of the building and parallel to its short side for an azimuth angle of 270 degrees.

Set C: Consists of six corner taps which were

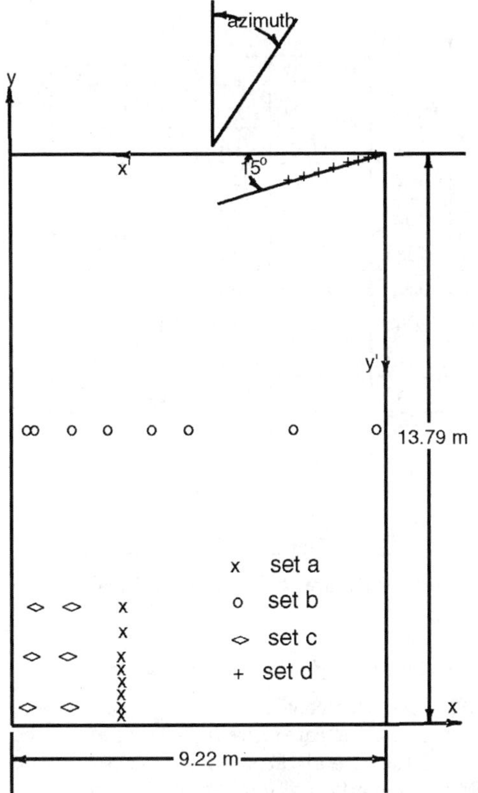

Figure 1. Arrangement of roof pressure taps

investigated for azimuth angles between 200 and 250 degrees.

Set D: Consists of eight pressure taps along a straight line making an angle of 15 degrees with the short side at the opposite corner of Set C for azimuth angles between 20 and 45 degrees.

The mean, rms, and negative peak pressure coefficients observed in the field and laboratory are both based on the dynamic pressure of the mean flow at roof height. The simultaneous pressure time records for each pressure tap group and for each of four simulations (4, 6, 7 & 8) were filtered at 200 Hz and sampled at a rate of 2000 Hz for a record length of 60 seconds. In addition, mean and peak pressure coefficients from these four configurations were obtained for all individual taps at 10 degree azimuth intervals for records of 20 second in duration and repeated 16 times. With field records of 900 seconds in duration, a geometric scale ratio of 1:50 and an approximate mean velocity ratio of $U_m/U_p = 0.9$, the corresponding record length is approximately 20 seconds (i.e., time scale is 45).

As Melbourne (1997) has pointed out, comparison of a single peak pressure coefficients from each record of corresponding length between laboratory and field, still does not satisfy the condition that these coefficients are at the same probability level. Because of the non-stationarity of the field records as a result of widely varying atmospheric conditions the probability level of the field pressure peaks would be difficult to determine. The next best thing may be to repeat the model peak pressure observation a sufficient number of times, and compare its dispersion with that of the field data. Peak-pressure coefficients from many records follow a statistical distribution that will likely be different from that of the data records from which these coefficients are derived. Lieblein (1974) pointed out that the variance of the peak pressure coefficients from individual records approaches its smallest value when the number of records is at least 16, but that actually little improvement is obtained by increasing the sample size beyond 10. The number of field records in consecutive 10-degree sectors varies between 10 and 20. Consequently, the proposed model/full-scale comparison of the sample of field data with a sample of 16 peak pressure coefficients, both derived from records of equivalent duration, should be statistically compatible.

3. DISCUSSION OF RESULTS

3.1 Flow simulations

A summary of roof-height flow data from eight simulations is presented in Table 1. These results complement those presented in Table 2 of Tieleman et al. 1998. The horizontal turbulence intensities are compared with those from WERFL for either S_v/U below or above 20%. Based on these data the simulations 4, 5, 7 & 8 come the closest to duplicating the WERFL results for which the lateral turbulence intensity is less than 20%. New values for the small scale turbulence parameter, S, are presented, replacing the incorrect data from Tieleman et al. 1998. The longitudinal integral scales from simulations 4 through 6 fall below the WERFL scales approximately by a factor of three. Spires close to the model location have a tendency to break up large scale eddies, resulting in still smaller integral scales, but create higher values of S. It is obvious from these results that

Table 1
Flow parameters at roof height

Configuration number	S_u/U (%)	S_v/U (%)	S_w/U (%)	α	U^* (m/s)	z_0* (mm)	$S \times 10^6$	$L_{ux}*$ (m)	$L_{vx}*$ (m)	$L_{wx}*$ (m)
1	7.1	5.1	4.3	0.132	0.554	2.15	11	9.5	1.4	1.4
2	14.3	13.6	7.6	0.109	0.412	0.45	12	34.0	24.5	8.5
3	17.8	17.9	11.6	0.099	0.319	0.12	41	19.5	18.5	4.3
4	18.8	15.7	10.2	0.107	0.291	0.083	28	32.5	16.5	6.0
5	19.6	17.0	11.7	0.108	0.272	0.11	57	30.5	15.0	4.7
6	16.2	12.0	8.6	0.079	0.396	0.202	21	33.0	13.5	5.5
7	17.1	16.0	11.9	0.086	0.253	0.022	133	13.0	4.8	3.1
8	19.3	18.2	12.5	0.085	0.257	0.042	54	17.0	12.0	4.7
WERFL $S_v/U < 20\%$	19.2	16.0	-	-	-	7.0	300	107	90	-
WERFL $S_v/U > 20\%$	22.2	23.1	-	-	-	-	-	185	288	-

*The simulation values of the integral scales and the roughness length are scaled up to full scale.

Table 2
Regression coefficients for peak values of mean and minimum roof pressure coefficients

Sim.	WERFL	C_{pmean}	C_{pmin}
4		0.92	1.03
6	$S_vU < 15\%$	0.94	0.81
7		1.26	1.11
8		0.99	1.08
4		0.98	0.92
6	$15\% < S_v/U < 20\%$	0.99	0.72
7		1.32	0.99
8		1.04	0.97
4		1.00	0.89
6	$S_v/U > 20\%$	1.01	0.69
7		1.35	0.95
8		1.06	0.93

simultaneous simulation of the integral and small scales of the incident turbulent flow has not been achieved, while the turbulence intensities are in reasonable agreement. Consequently, the level of success of these simulations can only be evaluated from the model/field comparison of the mean and fluctuating pressure coefficients.

3.2 Minima of roof pressure coefficients

The pressure coefficients from WERFL were classified in three categories in order to study the effect of the lateral turbulence intensity at roof height. Comparison of minimum values of C_{pmean} and C_{pmin} for the four simulations (4, 6, 7 & 8) with the field results was performed in the following categories: S_v/U less than 15%, S_v/U exceeding 20% and for values in between. Minimum values of these pressure coefficients were obtained for each roof tap (28 total) independently of the azimuth angle for each of the four simulations. The single minimum value for each tap was based on the 16-repeat observations for all azimuth sectors of 10 degrees between 0 and 360 degrees. Similarly, minimum values for the field data were based on all available data for a given tap for each class of lateral turbulence intensity. Regression analysis (Table 2) shows that reproduction of the mean pressure coefficients is quite adequate for simulations 4, 6 & 8. For simulation 7 the minima of these mean coefficients exceed the field results by an average of 30%, which compares favorable with the 27% value cited in Tieleman et al., Table 3. This discrepancy can possibly be attributed to the low values for the turbulence integral scales for this simulation.

Regression analysis of the peak minima shows that simulations 4, 7 and 8 have slopes near unity. Not-withstanding, the results exhibit a small de-

Table 3
Convection velocity ratio V_p/V_H of pressure peaks under a normal separation bubble

Simulation	y = 1m	y = 2m	n*
4	0.64	0.41	10
8	0.81	0.65	7
WERFL	0.90	0.51	13

*number of peaks analyzed.

crease from the low turbulence class to the highest one, with the best comparison observed in the middle class which provides the best agreement with the field observations of the lateral turbulence intensities. For simulation 6 the peak minima fall generally below (0.81) the field data for the low turbulence classification where it should be evaluated. This discrepancy may be attributed to the relatively low content of small-scale turbulence in the incident flow, while at the same time the integral scales are relatively high.

These observations support the hypothesis that mean and fluctuating pressure coefficients vary in some complicated fashion with incident flow parameters, including the turbulence intensities and both integral scale and the small scale turbulence content. These observations are corroborated when one inspects graphical presentations of mean and peak pressure coefficients, for all simulations, as functions of azimuth angle (Figs. 2 & 3). The simulation results, sixteen per angle, are illustrated as vertical bars with the cross bars representing their mean and standard deviation.

3.3 Peak pressure correlations

Before peak wind loads on tributary areas are investigated, it is advisable to study first space correlations of the point pressure coefficients, and specifically the peak pressure coefficients. The correlation of the latter can best be evaluated by employing conditional sampling of the time histories. Only data points whose magnitude exceeds the combined value of the mean pressure coefficient plus twice its standard deviation are retained. The data points not satisfying this condition are replaced with zeros. The area method introduced by Saathoff and Melbourne 1989 is used to calculate the correlation coefficients.

Adequate model/field comparison of the non conditional correlations for set A taps (a = 180 degrees) is shown in Figure 4. The correlation between the 4th tap (50903) and the others of this group does not exhibit symmetry. This demonstrates that the pressures near the leading edge are well correlated, but loose their correlation further downstream. This may result from pressure peaks being dissipated away from the leading edge. Figure 5 depicts correlation coefficients for the data from the field and simulation 7 for pressure tap set C (a = 210 degrees). For this angle of attack a strong vortex hugs the short side of the building, and a negative conditionally sampled correlation as observed for those pressures from taps which are located outside the area of the dominating vortex.

Differences in non conditionally and conditionally sampled time series for set A are presented in Figures 6 & 7. When the condition is assigned to the tap closest to the leading edge little difference can be discerned between the two sets of coefficients (Fig. 6). Both the unconditional and the conditional coefficients decrease from unity to 0.3 or less at the last tap. This trend changes when the condition is assigned to the last tap from the leading edge (Fig. 7). In this case, the condi-

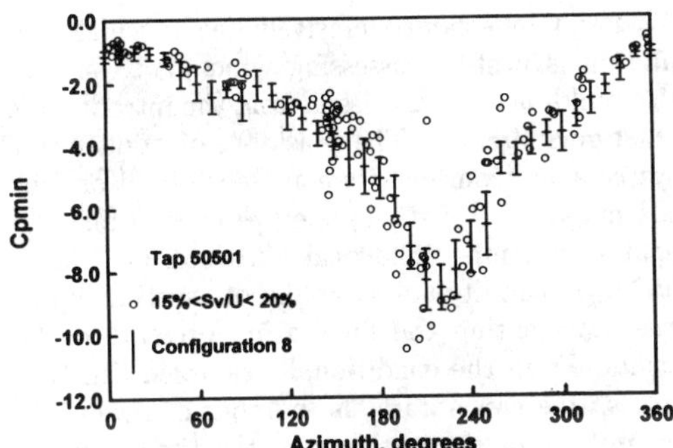

Figure 2. Tap 50501 peak-pressure comparison between simulation 8 and WERFL.

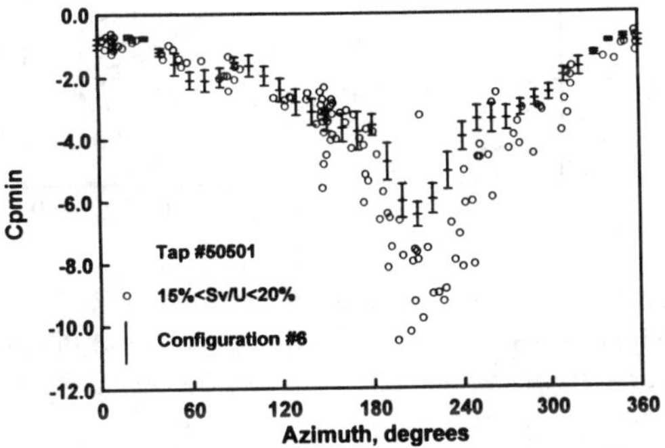

Figure 3. Tap 50501 peak-pressure comparison between simulation 6 and WERFL.

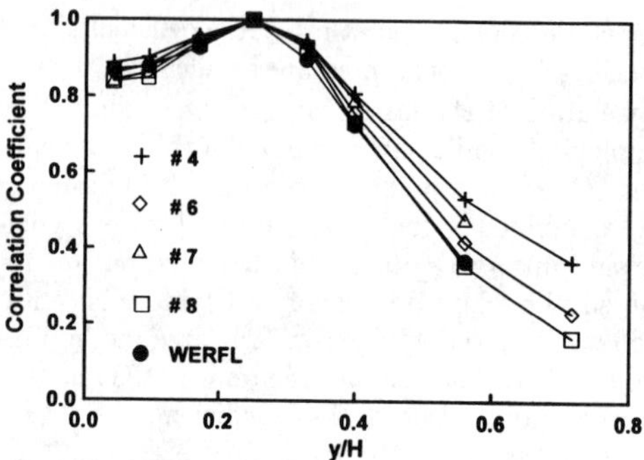

Figure 4. Set A correlation coefficients with tap 50903 as refrence, a = 180 degrees

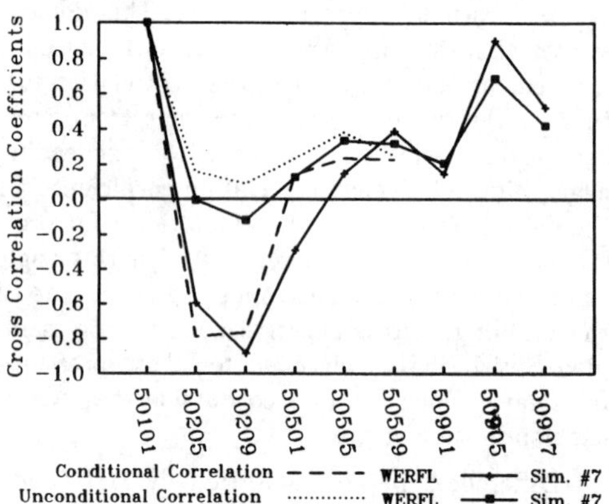

Figure 5. Non conditional and conditional correlation coefficients, Set C, a = 21- degrees, simulation 7.

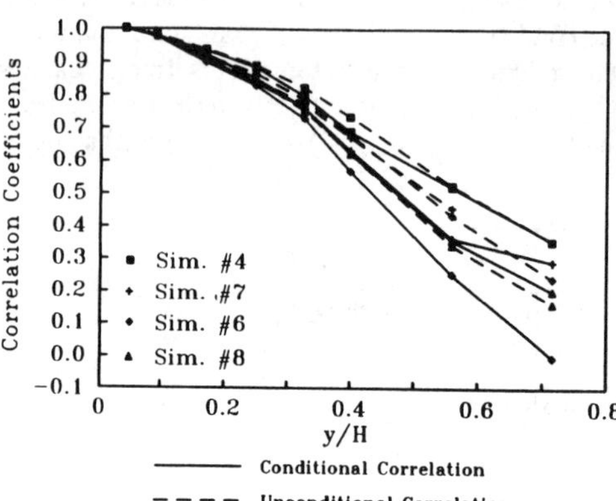

Figure 6. Set A non conditional and conditional correlation coefficients with tap 50900 as reference.

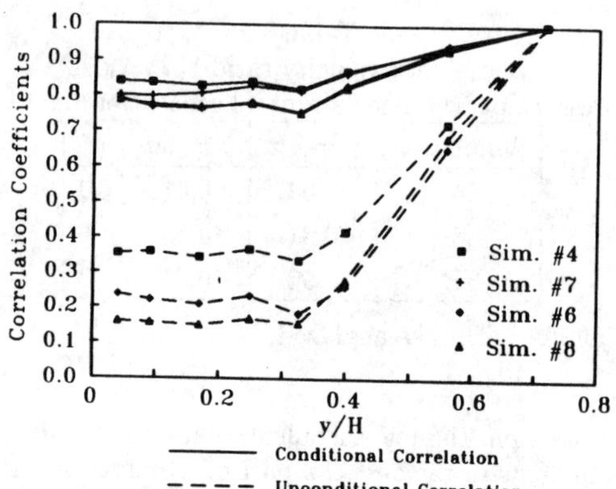

Figure 7. Set A non conditional and conditional correlation coefficients with tap 50907 as reference.

tional correlation coefficients are all of the order of 0.8 or higher. Consequently, the most significant pressure peaks observed at the last tap have their origin near the leading edge, but the peaks which are observed near the leading edge are mostly dissipated before the last tap is reached. From maximum values of the space-time correlations of individual peaks, the convective speed of the pressure wave can be obtained. The average speed of the pressure peaks over the first one meter from the leading edge is often higher than the average speed of the pressure wave over the second meter. Consequently the average speed over the 2-m distance falls below that over the first meter (Table 3). Here, the speed of the pressure wave, V_p, is normalized with the undisturbed velocity at roof height, V_h. The average velocity ratio, V_p/V_h for the field data of 0.51 compares quite well with the 0.5 value quoted by Letchford 1995.

3.4 Pressure-peak duration

Before one can conclusively decide which simulation is best for assessing wind loads on low-rise buildings, the last issue from the introduction must be addressed. The resistance of wooden roof systems and components may depend not only on the magnitude of the applied peak load but also on its duration. The peak duration analysis is based on conditional sampling of the time histories representing the fluctuating pressure coefficients. From the conditionally sampled time histories, one can obtain the number of peaks, the magnitude of the peak value, the time duration that the above condition is satisfied. The average values of the peak pressure coefficients and asso-

Table 4
Area wind loads coefficients
for Set A

Simulation	C_{pmean}	C_{prms}	C_{ppeak}	ΔT^*
#4	-0.82	0.342	-2.26	2.7
#4	-0.76	0.320	-2.46	7.2
#4	-0.83	0.342	-2.66	4.8
#8	-1.10	0.370	-2.93	2.0
#8	-1.04	0.348	-3.02	2.0
#8	-1.11	0.358	-3.17	3.4
WERFL	-1.07	0.289	-3.12	4.6
WERFL	-1.13	0.346	-3.30	4.1
WERFL	-1.17	0.313	-2.82	4.0
WERFL	-1.15	0.307	-2.46	2.2
WERFL	-1.20	0.349	-3.01	7.7

*ΔT - peak duration

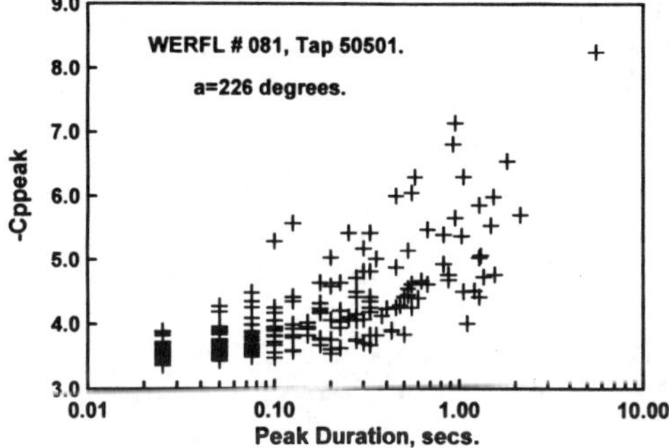

Figure 8. Tap 50501, peak durations for WERFL record #081, $a = 226$ degrees.

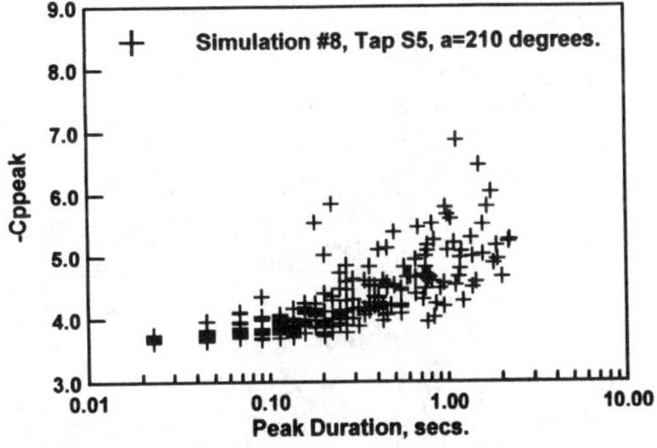

Figure 9. Tap S5, peak duration for simulation 8, $a = 210$ degrees.

ciated time intervals are listed in Table 4 for 6 model records and 5 field records. While more records need to be analyzed to obtain statistically significant results, the observed model and field results compare reasonably well.

With the time scaling coefficient, T_p/T_m, on the average being equal to 45, which also was used to determine the record length for the simulations, model/field comparisons of the observed time intervals is possible. They exhibit a surprising similarity, as the peaks with the largest magnitude have a duration in excess of one second (low frequencies). While the high frequency peaks with short durations, are much smaller in magnitude (Figs. 8 & 9). These observations are generally valid for pressures near roof corners for oblique azimuth angles, and for taps located along the roof edge with the velocity normal to it. For interior roof taps, peak pressures have a relatively low magnitude and their duration rarely exceeds one second.

4. CONCLUSIONS

1. The simultaneous reproduction of the turbulence integral scale and the small scale content of the incident flow could not be achieved.

2. Lateral turbulence intensity has to be carefully reproduced in order to obtain comparable pressure coefficients.

3. Mean and peak pressure coefficients are best reproduced with the flow developed over the two sets of spires of configuration 8.

4. Conditional correlation coefficients of the pressure observations of Set A, show that the pressure peaks migrate down stream from the leading edge but often dissipate after 1 m.

5. Results from space-time correlation of peak pressures in the first meter from the leading edge yield a somewhat higher convection speed than those for the field data.

6. Model/field comparison of the pressure-peak duration is quite good.

7. Future work should include similar analyses for peak wind loads on tributary areas.

REFERENCES

Minimum design loads or buildings and other structures 1995. ASCE 7-95 Standard, American Society of Civil Engineers. New York: ASCE.

Levitan M. L. & K. C. Mehta 1992a. Texas Tech field experiments for wind loads Part I: Building and pressure measuring system. *J. Wind Eng. Ind. Aerodyn.* 41-44: 1565-1576.

Levitan M. L. & K. C. Mehta 1992b. Texas Tech field experiments for wind loads Part II: Meteorological instrumentation and terrain parameters. *J. Wind Eng. Ind. Aerodyn.* 41-44: 1577-1588.

Letchford, C. W. 1995. Simultaneous flow simulation and pressure measurements on the Texas Tech building. *Proc. 9th Int. Conf. Wind Eng.*, New Delhi, India, 9-13 January, 1995. New Delhi: Wiley.

Lieblein, J. 1974. Efficient methods of extreme-value methodology, NBSIR Report, U.S. Department of Commerce, National Institute of Science and Technology.

Melbourne W. H. 1997. Commentary: Status of building aerodynamics research. *J. Wind Eng. Ind. Aerodyn.* 69-71: 135-137.

Saathoff, P. J. & W. H. Melbourne 1989. The generation of peak pressures in separating and reattaching flow. *J. Wind Eng. Ind. Aerodyn.* 32: 121-134.

Tennekes, H. & J. L. Lumley 1972. *A first course in turbulence.* Cambridge, MA: MIT Press.

Tieleman H. W. & R. E. Akins 1996. The effect of incident turbulence on the surface pressures of surface-mounted prisms, *J. Fluids and Struct.* 10: 367-393.

Tieleman, H. W., M. R. Hajj & T. A. Reinhold 1998. Wind tunnel simulation requirements to asses wind loads on low-rise buildings. *J. Wind Eng. Ind. Aerodyn.* 74-76: 675-685.

Yan, S., T. A. Reinhold & H. W. Tieleman 1999. Boundary layer wind tunnel simulation of area averaged wind loads. Proc. 10th Int. Conf. Wind Eng., Copenhagen, Denmark, 21-24 June, 1999.

Wind Engineering into the 21st Century, Larsen, Larose & Livesey (eds) © 1999 Balkema, Rotterdam, ISBN 90 5809 059 0

Flow characteristics in the separated shear layer

P.P.Vlachos & M.R.Hajj
Department of Engineering Science and Mechanics, Virginia Polytechnic Institute and State University, Blacksburg, Va., USA

ABSTRACT: Digital Particle Image Velocimetry (DPIV) is used to measure flow characteristics in the separated flow over a prism. Mean vorticity, rms fluctuations, Reynolds stresses, and mean turbulent kinetic energy are presented. Three different Reynolds numbers (Re = 2500, 9000, and 16500) are considered. The results show that, at the low Reynolds number, the vorticity level is relatively small. As the Reynolds number increases, the level of vorticity increases, reversing flow takes place in a small region near the solid surface and turbulence levels increase throughout the shear layer. In addition, the maximum vorticity inside the separation bubble is strongly related with the regions of maximum Reynolds stresses and turbulent kinetic energy.

1 INTRODUCTION

Flows around low-rise structures are characterized by separation, vortex formation and reattachment regions. Determining the flow parameters in these regions is important since they are directly related to pressure fluctuations and associated wind loads. One generic flow that has been used to model such complex flows is flow separation over prisms. When the flow is normal to the prism it separates at the leading edge and a separated shear layer is formed. To date, there has been success in determining reattachment locations and their dependence on Re number (Kiya and Sasaki, 1983 & 1985) and on free-stream turbulence (Gartshore, 1984) and (Castro and Haque, 1988). The presence of unsteady or low frequency variations in the flow has also been observed (Cherry et al., 1984). Yet, little is known about the details of the flow characteristics in the separation and in reattachment regions.

Difficulties encountered when attempting to make measurements in reversing flows and in boundary layers are the major reason for the lack of data on mean and turbulent flow characteristics in separated regions. For instance, hot-wire anemometry cannot be used in region where instantaneous flow reversal is possible or in regions where the level of turbulence fluctuation is large. Furthermore, the intrusive character of this method can alternate the characteristics of the flow. Trying to get detailed characteristics of the flow using Laser Doppler Velocimetry can be very time consuming and extremely difficult.

The objective of this work is to present the mean velocity field characteristics, r.m.s. of the velocity fluctuations and Reynolds stresses in the separation region over a prism at different Reynolds numbers. Such measurements can be used to explain flow phenomena associated with pressure fluctuations. They also provide quantitative parameters that can be used as a base for comparison with CFD calculations. The velocity field is measured using Digital Particle Image Velocimetry (DPIV) which provides instantaneous two-dimensional measurements of the flow. In addition, the ability to capture a sequence of instances of the flow field adds time as a third dimension. The advantages of DPIV have been discussed extensively by Adrian (1991), Willert and Gharib (1991), and more recently by Adrian (1996).

2 EXPERIMENTAL SETUP

The experiments were carried out in the Engineering Science and Mechanics Water Tunnel facility at Virginia Tech. This tunnel operates in a closed loop with up to 2500 gallons of water. After the settling chamber, a three-way convergence leads to the 0.6m x 0.6m x 1.8m test section made of a clear, acrylic Plexiglas. A 4500 GPM axial flow pump drives the flow at speeds up to 1 m/s, with free stream turbulence levels less than 2%. The prism (L=5.08cm W=3.81cm and H=2.54cm) was mounted on the floor of the test section. The prism was aligned with one side normal to the flow.

For the PIV laser sheet, the light source was a copper vapor 45 Watt pulsing laser that is capable of delivering up to 15 mJ/pulse. Using a set of sheet forming optics, a laser sheet that is 0.002m thick and 0.065m wide is delivered in the test section. The laser sheet plane is positioned in the mid-plane of the prism. The field of view was adjusted to a square region of 0.062m x 0.062m. The laser was synchronized with a high speed CCD camera able to take 1000 images/sec with a resolution of 256x256 pixels. Hollow glass spheres of 15 microns mean diameter were used as flow tracers. For the analysis of the PIV images and the extraction of the velocity vectors, the method of cross-correlation between two sequential frames was implemented using commercial software (AEA VISIFLOW). A total number of 961 vectors were calculated using 50% overlapping. The vectors are placed on a uniform Cartesian grid with dimensions of 31x31. This task was carried out at a rate of one frame/2.5 sec.

The current study was conducted for three free stream velocities corresponding to Reynolds numbers -based on the height of the prism- of Re=2500, Re=9,000 and Re=16500 respectively. The image acquisition frame rate was set at 256 fps (frames per second) for the first two cases (Re = 2500, 9000) and at 512 fps for the third case (Re = 16500). For each of the three cases, 2,048 frames were acquired. The high-speed image acquisition system, provided by the high frequency resolution, is needed to resolve turbulent characteristics of the flow. Based on the spatial resolution of the camera and the field of view, the grid step size is 1.9 mm corresponding to 3.77% of the length of the prism.

3 RESULTS AND DISCUSSION

Particle flow visualizations were carried out in order to provide qualitative information for the character of the flow. The previously described image acquisition setup was used, adjusted to a low frame rate to allow the particle trajectories to show as streaks in the image. For Re=2500 and Re=9000, the frame rate was set at 50 fps while for Re=16500, it was set at 100 fps. Images where acquired for 5 sec in every case. Animation of the images provided visual evidence of the natural unsteady character of the flow. For all three Reynolds numbers, we observed unsteadiness of the free shear layer and instantaneous expansions forming an open wake with no reattachment. When reattachment occurs, the location of the reattachment region is not fixed. Figures (1-a), (1-b) and (1-c) show instants of the flow field for Re=2500, Re=9000 and Re=16500 respectively. For all three cases, the free shear layer and the separation bubble are well defined. As Re number increases a better defined vortical structure is formed. Yet, in all three cases, there is no distinct region of a vortex core.

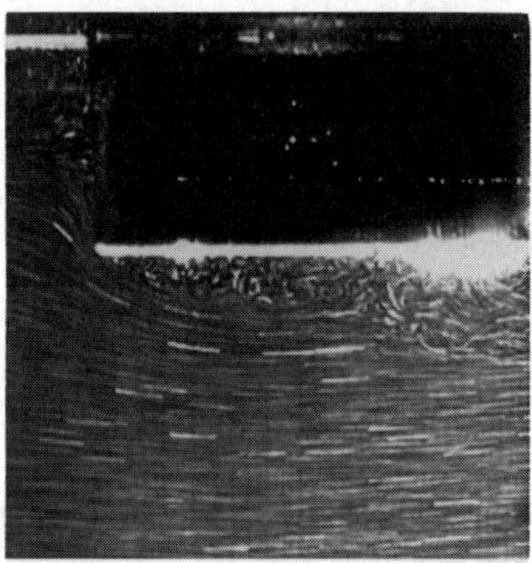

Figure 1-a. Instantaneous image of the flow using particle flow visualization at Re=2500. Frame rate 50 fps

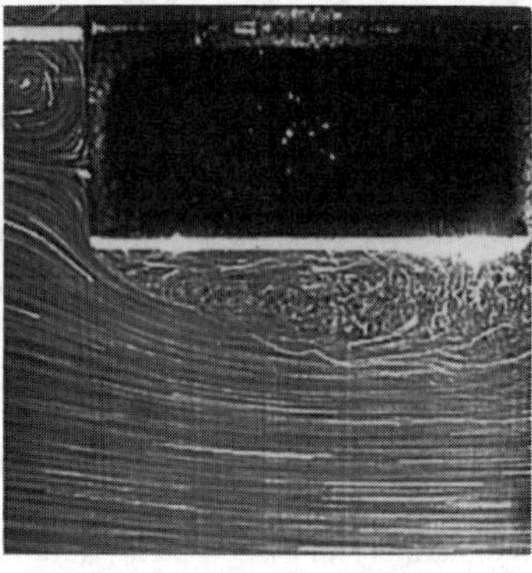

Figure 1-b. Instantaneous image of the flow using particle flow visualization at Re=9000. Frame rate 50 fps

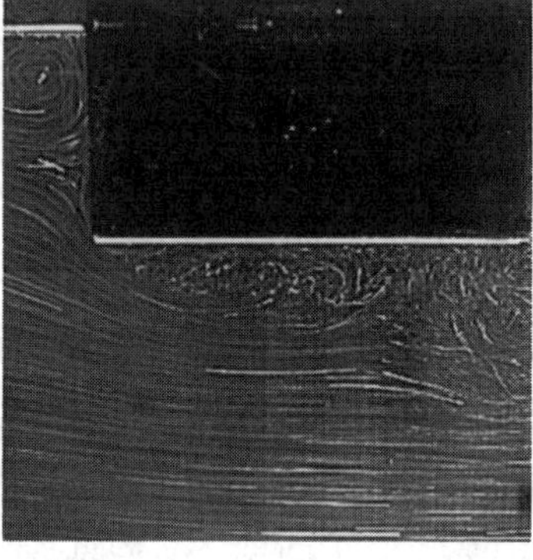

Figure 1-c. Instantaneous image of the flow using particle flow visualization at Re=16500. Frame rate 100 fps

In all the plots that follow, the flow direction is from left to right and the solid surface of the prism is at the x-coordinate of the plot. The leading edge of the prism is located at x=0 and the trailing edge at x=1. Both the x- and y-coordinates are non-dimensionalized with the length of the model. The velocities are non-dimensionalized with the corresponding free stream velocity.

Figures 2-a, 2-b, 2-c show contour plots of the average vorticity over the prism, for the three different Reynolds numbers. In all cases, the plots show a well-defined region of separated flow. Yet, comparison of the three plots reveals significant differences. At Re=2500, the maximum vorticity is near the wall and, throughout the flow field, it is positive. On the other hand at Re=9000 and Re=16500 there is a region of maximum positive vorticity of larger magnitude inside the separation bubble. Moreover, a region of negative vorticity corresponding to the reversing flow is detected near the wall at the higher Reynolds number. This region is small, which implies that the reversing flow component is confined to a relatively small region near the solid surface. For both Re=9000 and Re=16500, the narrow region of negative vorticity spans the whole length of the prism indicating that for the time-averaged flow field there is no reattachment. At Re=2500, the maximum vorticity is located at x=0.45. A second region of very strong vorticity located at x=0.11. At Re=9000, maximum vorticity is measured at x=0.20, while for Re=16500, it is measured at x=0.07.

Figures 3-a, 3-b, 3-c, show the u'_{rms} fluctuations in the separated free shear layer at the three different Reynolds numbers. The maximum level of u_{rms} increases as the Reynolds number increases. The region where u'_{rms} is significant also increases in size with the Reynolds number. For Re=2500 and Re=9000, the maximum values are near the solid surface, inside the separation bubble, while for Re=16500 the maximum value is in the free shear layer. Very large values exist near the edges of the domain. These can be attributed to increased noise levels in the acquired images because of the higher free stream velocity.

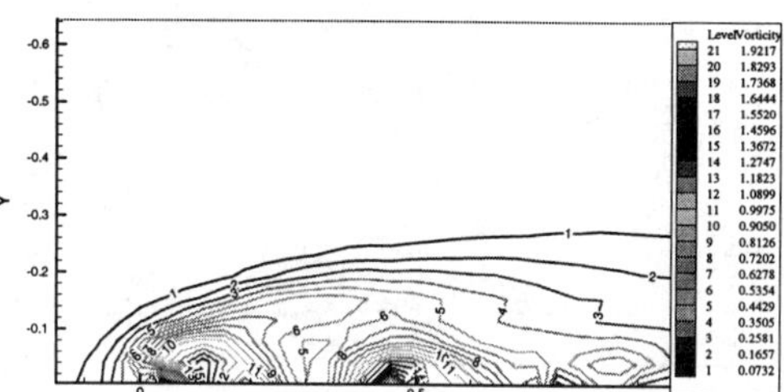

Figures 2-a. Average flow field contours of vorticity at Re=2500

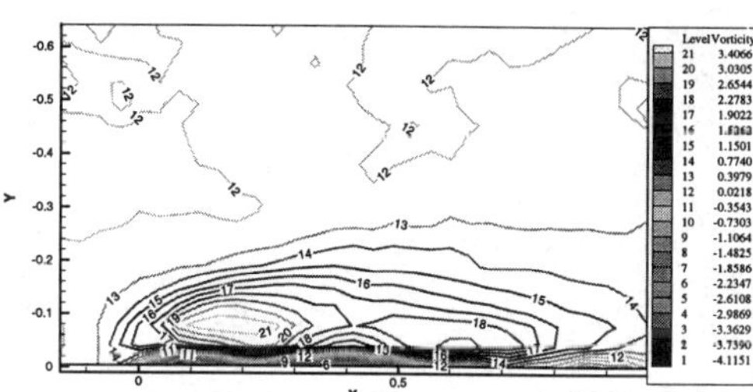

Figure 2-b. Average flow field contours of vorticity at Re=9000

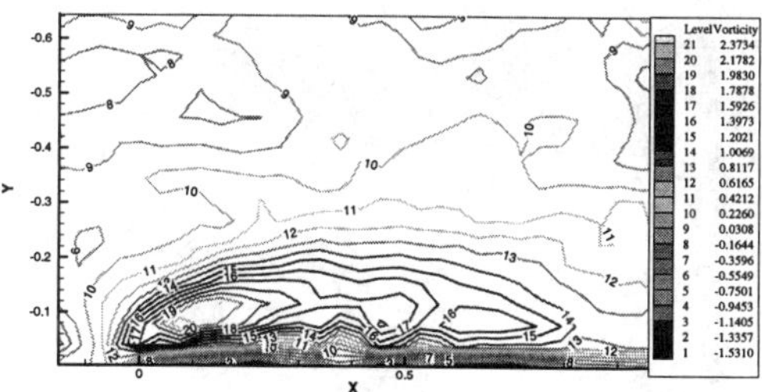

Figure 2-c. Average flow field contours of vorticity at Re=16500

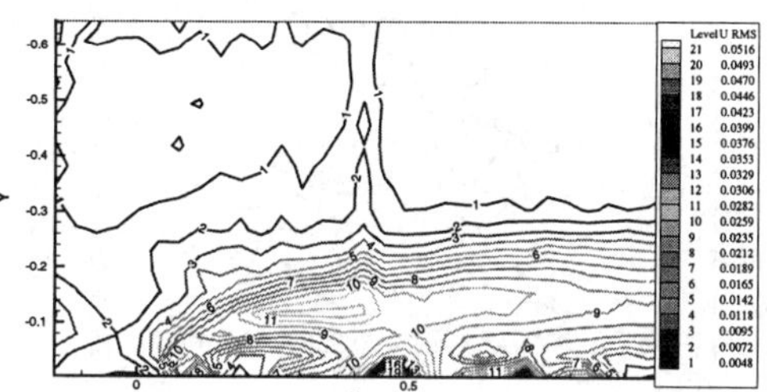

Figure 3-a. Contours of u'_{rms} at Re=2500

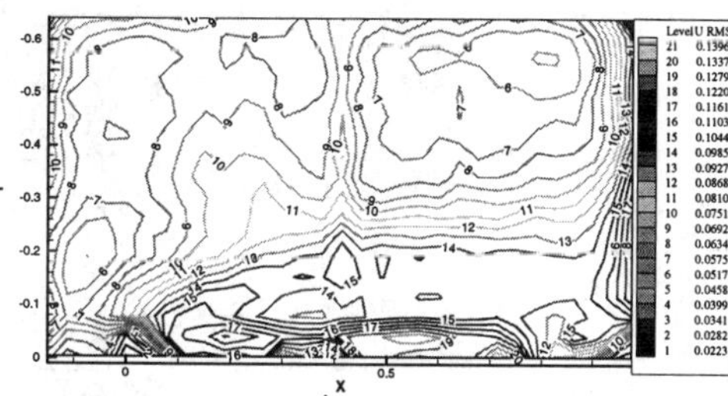

Figure 3-b. Contours of u'_{rms} at Re=9000

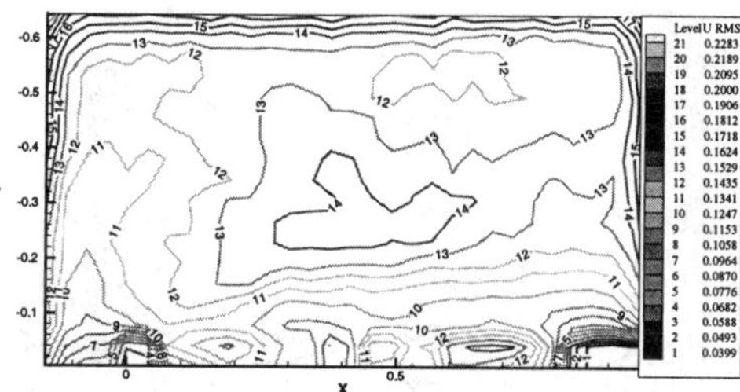

Figure 3-c. Contours of u'_{rms} at Re=16500

The v'_{rms} fluctuations are shown in figures 4-a, 4-b and 4-c. The figures clearly show both the maximum level of v_{rms} and the region where the levels are significant, increase with the Re. For all cases the maximum level of v'_{rms} is smaller than that of u_{rms} at the same Re. We should also note that the values are extremely small near the wall. Large values near the edges of the domain exist at Re=16500. Similarly to the case of u'_{rms}, they can be attributed to noise.

Contours of mean u'v' Reynolds stresses are shown in figures 5-a, 5-b, and 5-c. Contours of mean turbulent kinetic energy (TKE) are shown in figures 6-a, 6-b and 6-c. For the calculation of the mean turbulent kinetic energy, only the contributions of u' and v' are taken into account since the measurements are two-dimensional. The maximum values of u'v' and TKE increase with the increase of the Re number. For Re = 2500, a negative maximum region of u'v' is located at approximately x=0.11 which coincides with the high vorticity values measured at this location. For Re=9000 and 16500, the maximum negative values of u'v' coincide with the position of the maximum vorticity. For Re=9000 we have the maximum at approximately x=0.20, while for Re=16500 at x=0.07. Both, positive and negative maximum values of u'v' are of the same order of magnitude. For Re=2500 the maximum values of TKE occur exactly at x=0.45 where the global maximum of vorticity is present. An extended region of high values is present in the free shear layer. For Re=9000 the corresponding area is larger and well defined within the free shear layer. For Re=16500, high values of TKE are present in the outer field.

4 CONCLUSIONS

DPIV is used to measure flow characteristics in the separated flow over a prism. Quantities including mean vorticity, rms fluctuations, Reynolds stresses,

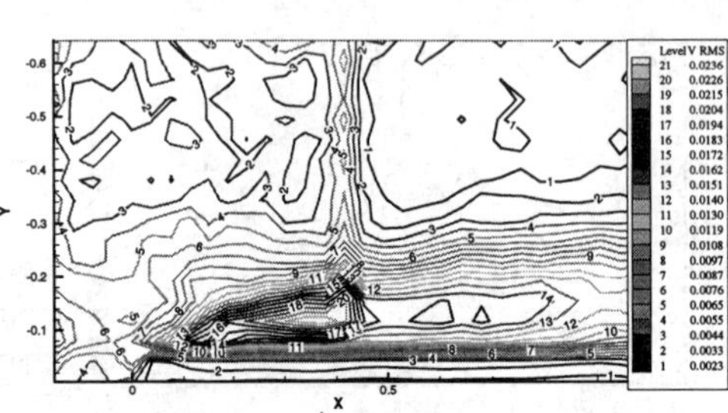

Figure 4-a. Contours of v'_{rms} at Re=2500

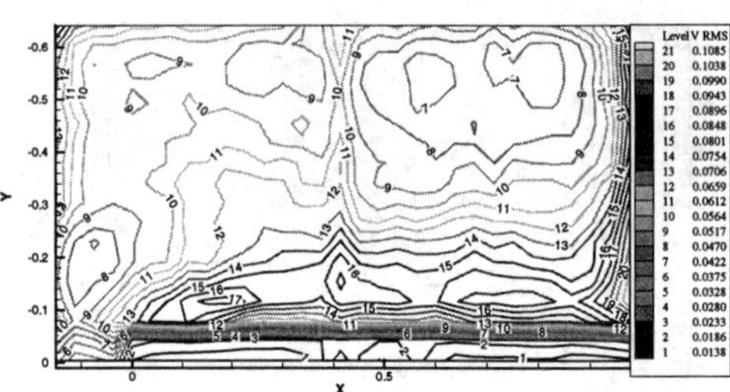

Figure 4-b. Contours of v'_{rms} at Re=9000

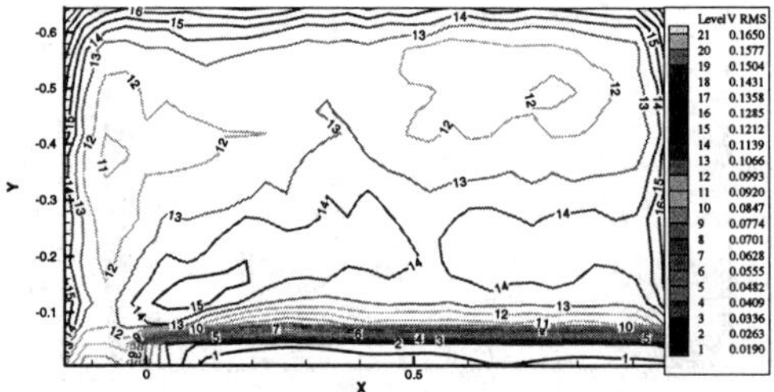

Figure 4-c. Contours of v'_{rms} at Re=16500

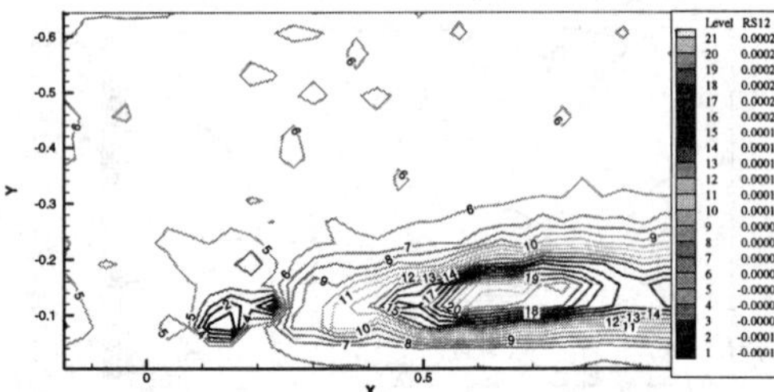

Figure 5-a. Contours of mean u'v' at Re=2500

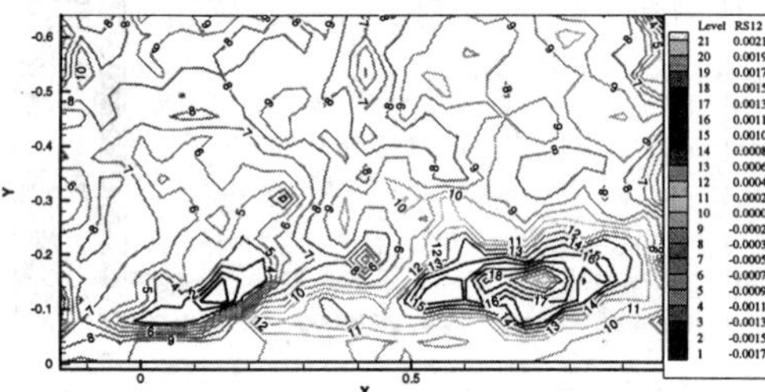

Figure 5-b. Contours of mean u'v' at Re=9000

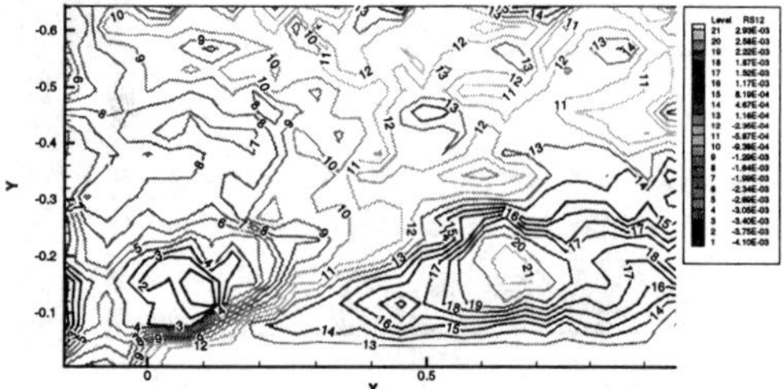

Figure 5-c. Contours of mean u'v' at Re=16500

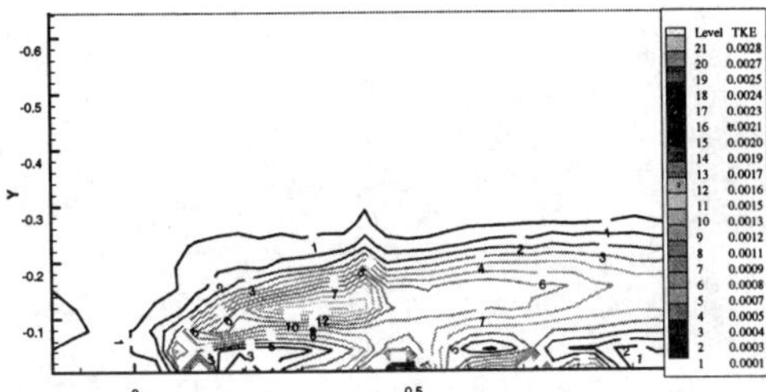

Figure 6-a. Contours of mean turbulent kinetic energy at Re=2500

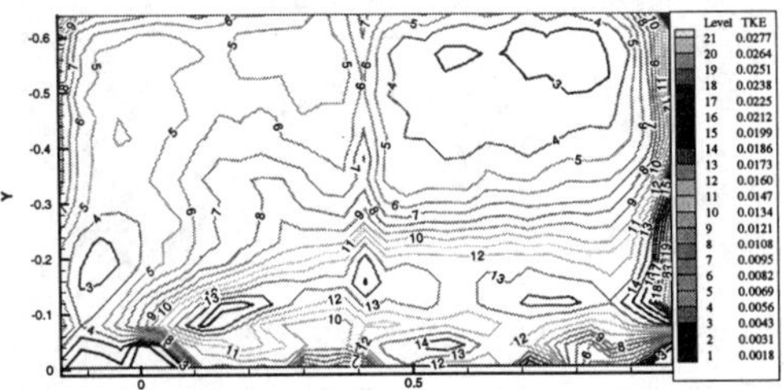

Figure 6-b. Contours of mean turbulent kinetic energy at Re=9000

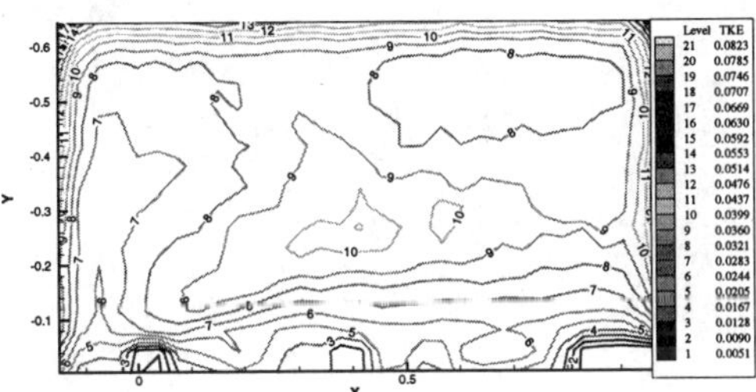

Figure 6-c. Contours of mean turbulent kinetic energy at Re=16500

and mean turbulent kinetic energy are presented. Three different Reynolds numbers are considered. The results show that at the low Reynolds number (Re = 2500), the vorticity level is relatively small. As the Reynolds number increases to 9000 and 16500, the level of vorticity increases and reversing flow takes place in a small region near the solid surface. Moreover, the turbulence levels increase throughout the shear layer. In addition, the vorticity distribution inside the separation bubble is strongly related with the regions of maximum Reynolds stresses and turbulent kinetic energy. This provides evidence that the flow around the shear layer and the vortex is highly turbulent. Large values of velocity fluctuations near the trailing edge of the prism prove that there is not a well-defined reattachment region, so the separated flow forms an open wake.

REFERENCES

Adrian, R. J. 1991. Particle imaging techniques for experimental fluid mechanics. *Ann. Rev. Fluid Mech.*, 1991, 23:261-304.

Adrian, R. J. 1996. Strategies for imaging flow fields with PIV. *AIAA* 96-1988

Castro, I. P. and A. Haque 1988. The structure of a shear layer bounding a separation region. Part 2. Effects of free-stream turbulence. *J. Fluid Mech.* 192:577-595.

Cherry, N.J., Hillier R. and M. E. P. Latour 1984. Unsteady measurements in a separated and reattaching. *J. Fluid Mech.,* 144:13-46.

Gartshore, I.S. 1984. Some effects of upstream turbulence on the unsteady lift forces imposed on prismatic two dimensional bodies. *J. Fluids Eng.* 106:418-424.

Kiya, M. and K. Sasaki 1983. Structure of a turbulent separation bubble. *J. Fluid Mech.* 137:83-113.

Kiya, M. and K. Sasaki 1985. Structure of large-scale vortices and unsteady reverse flow in the reattaching zone of a turbulent separation bubble. *J. Fluid Mech.* 154: 463-491.

Willert, C.E and M. Gharib 1991. Digital Particle Image Velocimetry. *Exp. in Fluids* 10:181-193.

Low-rise buildings

Wind Engineering into the 21st Century, Larsen, Larose & Livesey (eds) © *1999 Balkema, Rotterdam, ISBN 90 5809 059 0*

Numerical evaluation of roofing systems for wind uplift pressures

A. Baskaran, S. M. Zahrai & Y. Chen
National Research Council of Canada, Ottawa, Ont., Canada

ABSTRACT: Wind uplift rating of roofing systems is based on standardized test methods. Currently, there is no consensus on the table size to be used in these testing protocols. The objective of this paper is to study the impact of table size on the roofing system performance. To achieve this purpose, extensive analytical experiments have been conducted to investigate the performance of roofing systems subjected to wind uplift pressures. Analytical results compared well with those obtained from experimental work, benchmarking the numerical modeling. This paper presents some of these comparisons. Required table size has also been suggested for various configurations having TPO (Thermoplastic Polyolefins) as the waterproof roof membrane.

1 INTRODUCTION

1.1 *Need for a numerical model*

Wind resistance rating of roofing systems is based on standardized test methods. In these tests, roof specimens are generally placed in an apparatus whose size (length and width) is normally far less than the size of a real roof. Roofing manufacturers assemble the test specimen with its respective components such as insulation, vapor barrier, etc., on the test frame (see Figure 1). Air pressure is applied until system failure occurs, e.g., membrane tearing, fastener pull out. Although the system configuration (e.g., fastener spacing and membrane width) is the same as that of the field system and the specimen is subjected to the pressures similar to the design requirements, the fastener load and membrane deflection might be different from the field values. This is mainly because the test rig edges offer some resistance to the applied pressure. This effect is greater in narrower tables. If the testing table is sufficiently wide then the roofing system response remains constant or minimum changes can occur. However, the adequate width depends upon the roofing system configuration.

As grouped in Table 1, current test methods consider different table sizes for the performance evaluation of roofing systems. For instance, the FM (Factory Mutual, 1988) tests use table size of 2743 by 1524 mm (9' by 5') or 7315 by 3658 mm (24' by 12') depending on the roofing system. A chamber size of 3048 by 3048 mm (10' by 10') is used by the UL (Underwriters Laboratories, 1991) standard. Present research efforts by a North American roofing consortium, the *Special Interest Group for Dynamic Evaluation of Roofing Systems* (SIGDERS) established at the National Research Council of Canada, have led to the development of a facility making it possible to evaluate roofing systems dynamically (Baskaran and Lei, 1997). A table size of 6100 by 2200 mm (20' by 7.2') is used by SIGDERS.

Careful examination reveals that the table size is important in evaluating roofing systems and it should be selected properly to obtain realistic wind uplift resistance. For example, the use of narrow tables would increase the edge effects on the results particularly for roofing systems having wider membranes, i.e., spacing between fastener rows. On the other hand, using wider tables would make the system response slower. Despite the significance of table dimensions, to the authors' knowledge there still exist no criteria and specific standard to suggest a required table size. A number of parameters may influence the required table size, in particular fastener spacing, F_s, fastener row spacing, F_r, and membrane modulus of elasticity, E.

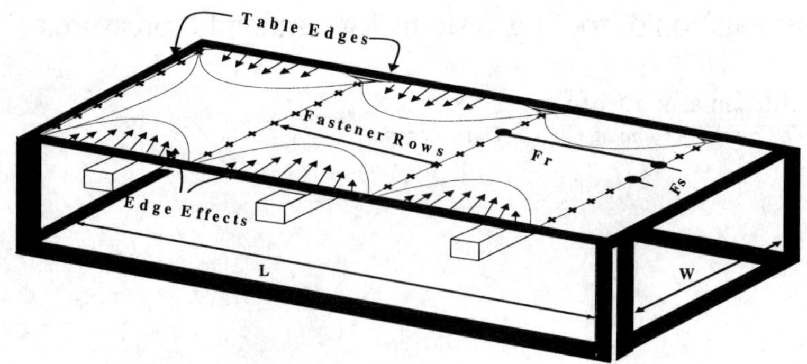

Figure 1. Effect of table size on the roofing system response.

Table 1. Existing table sizes for evaluation of roofing system performance.

No.	Testing Protocol	Table Size mm (ft)	Location	Reference
1	FM 4470 Standard	2700x1500 (9x5)	U.S.A.	FM research 1986
2	Revised FM 4470	7300x3800 (24x12)	U.S.A.	FM Research 1992
3	UL 580 Standard	3000x3000 (10x10)	U.S.A.	UL Inc. 1991
4	UEAtc Standard	6100x1500 (20x5)	Europe	Gerhardt et al 1986
5	BRERWULF	5000x5000 (16.4x16.4)	UK	Cook et al. 1988
6	NT Build 307 Standard	2400x2400 (8x8)	Norway	Paulsen 1989
7	SIGDERS	6100x2200 (20x7.2)	North America	Baskaran and Lei 1997

Therefore, it has been decided to develop a Finite Element (FE) based numerical model for the problem discussed above. This paper discusses the involved steps in the present numerical study as follows:

- Adopt a numerical model to simulate the experimental results;
- Benchmark the model using the experimental data;
- Investigate the effect of table size on the roofing system response; and
- Develop correction factors for tables smaller than the required one.

Numerical techniques can offer flexibility in exploring scenarios that would be too expensive, or in some cases impossible, to set up experimentally. In addition to the economical advantages, the analytical models are generally faster for solving problems where there is a need to investigate the impact of various influencing parameters.

2 ADOPTING A NUMERICAL MODEL

2.1 *General*

ABAQUS version 5.7(1997), a commercially available Finite Element program with non-linear analysis capability was used to carry out all the numerical analyses. The large strains and deformations that occur during the loading of the membrane were accounted through geometrical non-linearity (large deformations theory). Small load increments were considered to accommodate the flexibility of the single-ply membranes. The modeled roofing system has a thermoplastic membranes as the waterproof component. There are two main kinds of thermoplastic roof membrane: Poly[vinyl chloride] (PVC) and Thermoplastic Polyolefin (TPO). In this study, reinforced TPO membrane sheets of varying widths are used. Only a summary of the experimental setup and system details are presented here but details are reported elsewhere (Baskaran et al. 1999).

2.2 *System details*

Figure 2 shows the TPO roof assembly used in the experimental investigation. The roofing system has 1875 mm (6') wide membrane sheets and fastened to the structural deck at every 305 mm (12") apart along the seam, as shown in Figure 2. Two different testing protocols, FM and SIGDERS, were used to evaluate each roofing system. To monitor the system response, appropriate instrumentation was used to measure typical design parameters (i.e., pressure, force, and deflection) at certain locations. Force balances and ultrasonic sensors were respectively used to measure the tensile forces in the fasteners and uplift movements of the membrane.

Typical seam details are also shown in the Figure 2. The seam has an overlap of 127 mm (5"), with the fastener placed 38 mm (1.5") from the edge of the bottom sheet, and 89 mm (3.5") from the edge of the overlapping sheet. The portion of the seam beyond the fastener row was welded with hot air such that a waterproof top surface was obtained. The width of the welded portion varied between 38 and 45 mm (1.5 and 1.75").

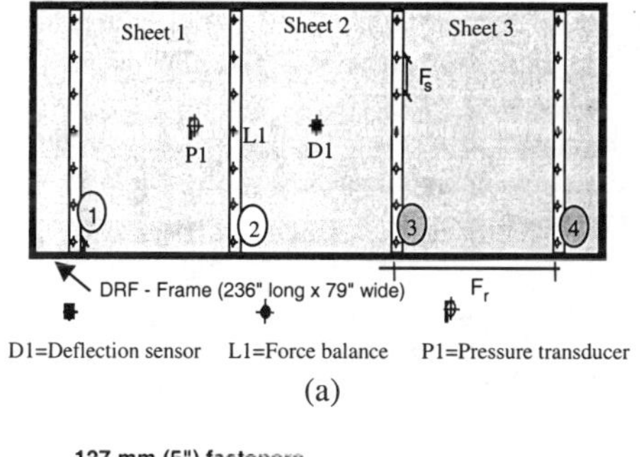

(a)

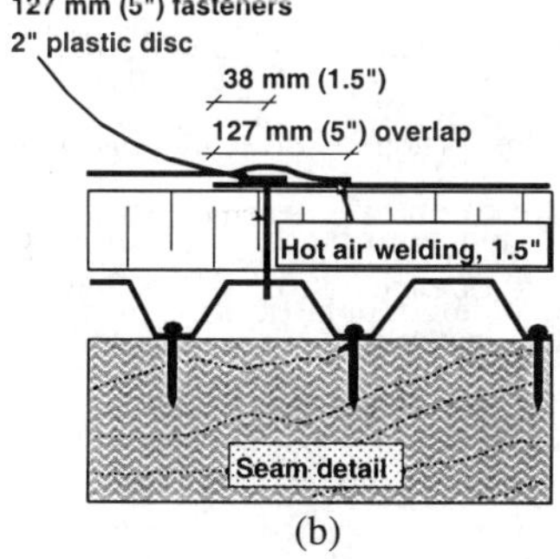

(b)

Figure 2. (a) Roofing system layout for the ex-perimental investigation; (b) longitudinal section of TPO roofing system.

2.3 *FE model*

Figure 3 shows the FE model for the roof systems. The membrane alone was considered in the numerical modeling due to its much greater flexibility compared to other components, insulation and steel deck. In other words, the deflections of the steel deck and insulation were assumed negligible in comparison to the membrane deflection.

A rectangular grid of nodes and shell elements was used in each case to discretize the membrane. The shell elements consisted of 4-nodes and 1.04 mm thick with an equivalent modulus of elasticity of 300 MPa. The modulus of elasticity, the material property of the membrane, was obtained through mechanical tests. Membrane edges were assumed as spring supports and fastener locations were modified to account for the plastic fastener discs. Different material properties were simulated for fastener discs in the seam areas using shell elements. These discs were 3 mm thick with a diameter of 50 mm and a modulus of elasticity of 500 MPa. The geometric non-linear behavior of the membrane was also taken into account using finer meshes particularly around the seam areas and end supports.

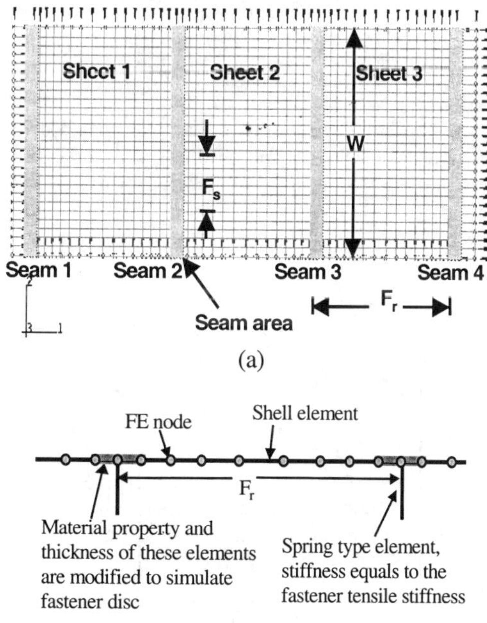

(b)

Figure 3. (a) Roofing system layout for the numerical modeling; (b) seam details considered in the model.

Seam details were modeled by doubling the thickness of the shell element at the seam areas as schematically illustrated in Figure 3, to simulate the spliced region of the membrane. Fixed bar type element were used to simulate fastener attachments with the steel deck. Fasteners were assumed as spring supports with axial stiffness of 20 N/mm.

A typical computed membrane deflected shape is shown in Figure 4 where the membrane ballooning occurs between fastener rows and table edges. Computed fastener forces for three configurations are presented in Figure 5 where the rate of fastener load change is high for a table width less than 3 m. In all cases, with increasing the table width, the fastener forces approach to the tributary loads (i.e., pressure multiplied by the tributary area) for that fastener. Also, note that a larger membrane width (fastener row spacing, F_r) would increase the width of the ideal table.

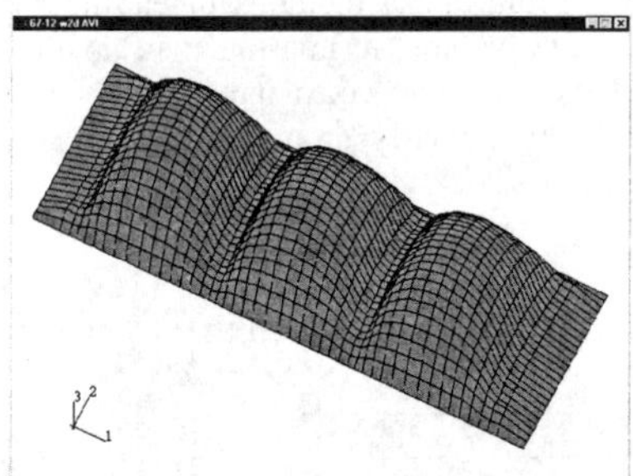

Figure 4. Membrane deflected shape computed by the numerical analysis for 67"/12" configuration (maximum deflection of 117 mm (4.6") due to a suction of 1436 Pa (30 psf)).

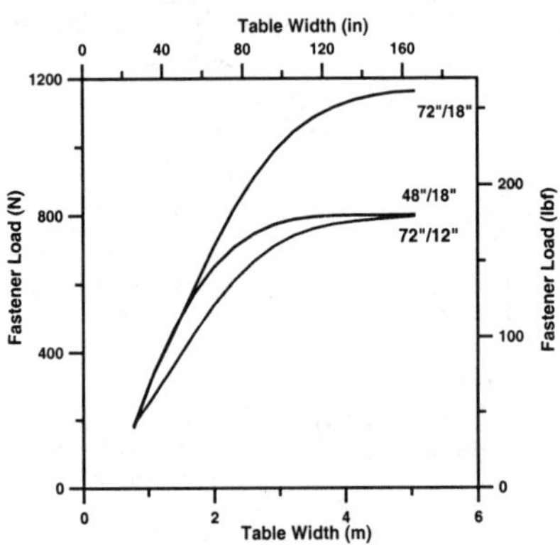

Figure 5. Computed fastener tensile force versus table width for three roofing system configurations.

3 BENCHMARKING THE ADOPTED MODEL

To collect benchmark data, experiments were carried out at the Dynamic Roofing Facility (DRF) applying two (FM and SIGDERS) test protocols. A roofing system with single-ply TPO membrane was initially selected to validate the finite element model. Three F_r/F_s configurations, 1700/305, 1220/460, and 1830/460 mm, (67"/12", 48"/18", and 72"/18") were considered. This testing program provided six sets of data to validate the FE model. Average values of two characteristic parameters, i.e., fastener loads at the center location of the seam, L1, and deflections at the center location of the membrane width, D1, measured from the DRF experiments were compared with the output of the FE analyses. Refer to the Figure 3 for the details of the L1 and D1 locations.

Fastener force comparisons are shown in Figure 6, where the horizontal axis represents the applied suctions on the roof assembly. In the experiments, based on the test protocols (FM or SIGDERS) the required pressures are applied and maintained for specific duration. During the model, simulations were performed at 718 Pa (15 psf) increments. The vertical axis presents fastener forces to represent the roofing systems response for the applied pressure. These comparisons demonstrated that the FE model is a viable tool that can be used to predict the fastener forces of test specimens at any pressure level.

To establish deviations between the two data sets (experiments versus model), the following expression was used:

$$\Delta F = \sum_{i=1}^{N} \left(\frac{F_{FE} - F_{EXP}}{F_{EXP}} \right) \times 100 \tag{1}$$

Where:

F_{FE} is the fastener force obtained from the FE model,

F_{EXP} is the fastener force measured at the DRF,

N is the number of cases (pressure levels) considered for each configuration, and

ΔF is the fastener force deviation between F_{FE}, and F_{EXP}. ΔF with a negative sign (-) means that the model underestimates the roofing system response compared to the experimental approach and vice versa.

Using Equation (1), for the case of F_r/F_s = 67"/12", an under-estimation of 7% by the FE model has been noticed (Figure 6a)

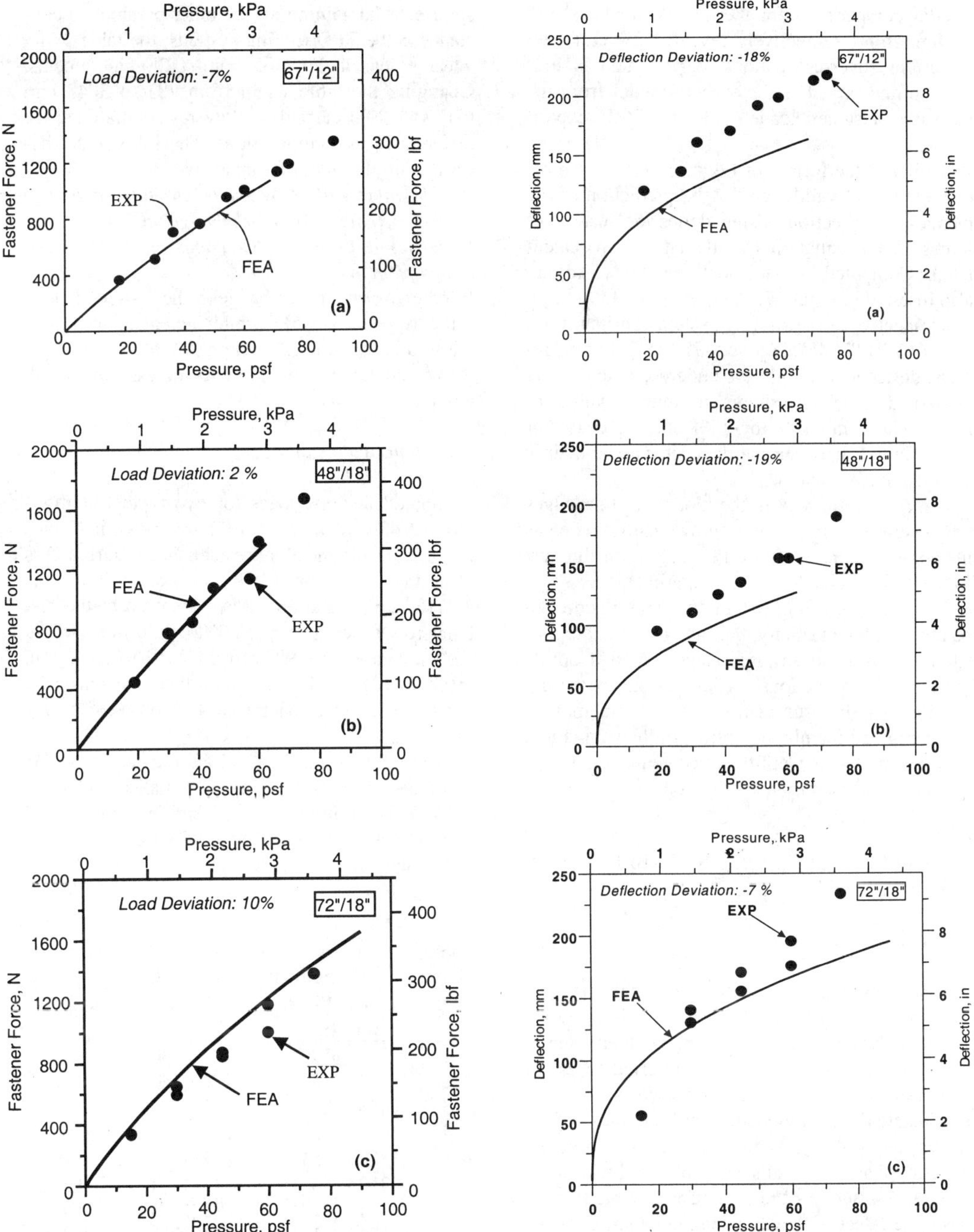

Figure 6. Model Validation – Comparison of the FE simulation results for fastener tensile force with Experimental Data: (a) 67"/12", (b) 48"/18", and (c) 72"/18" configurations.

Figure 7. Model Validation – Comparison of the FE simulation results for membrane deflection with Experimental Data: (a) 67"/12", (b) 48"/18", and (c) 72"/18" configurations.

Similar comparisons for the 48"/18" and 72"/18" configurations respectively revealed 2% and 10% deviations (over-estimations as presented in Figures 6b and 6c) of the analytical model from the measured fastener loads with the DRF experiments.

Similar to the format of Figure 6, Figure 7 presents the model validation for the prediction of the membrane deflection. Using deflection instead of forces in the equation (1), deviations are calculated. Computed displacement deviations (Δd) are also inserted in Figure 7.

Irrespective of the roofing system configurations (F_r/F_s=67"/12", 48"/18" and 72"/18") the membrane deflections are always underestimated by the numerical model. As shown in Figure 7, values of Δd ranged from –7 % to –19% and the cause for the difference between the data sets can be mainly grouped in two factors:

1. Edge Conditions: In the model, all four edges are restrained from any movements, whereas membrane slippage from the edges of the test frame was noticed during the experiments.

2. Membrane Stiffness: In the experiment, depending on the elasticity, the membrane undergoes stretching from its original stage. Also it uplifts for the applied suctions. Thus, the measured deflection is the summation of the membrane stretching and membrane uplift. In the simulation, only the membrane uplift is accounted for in the calculation of membrane deflection.

4 DEVELOPMENT OF TABLE SIZE CORRECTION FACTORS

4.1 *Required table width*

This section focuses on the required table size and development of corresponding correction factors. All three dimensions (i.e., length, width and depth) -as shown in Figure 1- may be referred to as the table size. However, as discussed, during the system installation, components similar to those of the field systems are used. In other words, there is no variation from the field system on the thickness of components such as the insulation and membrane. Also, the effect of the table length is minimum due to the fact that during the system installation on the table, membrane width forms parallel to the table width. Therefore, the present investigation focuses on identifying only the effect of table width on the system response using the validated FE model. Required Table Width (*RTW*) can introduce only minimum changes, if any, in the roofing system response. Maintaining all the other parameters constant for the TPO roofing systems, the table width was increased by 305 mm (12") increments. Changing the table width from 781 to 5048 mm (31" to 199") carried out several simulations, for different F_r/F_s configurations. The following criteria was used to identify the RTW:

"The table with RTW should provide no change in the fastener forces or change in the fastener force should be within 5% compared to those obtained from the smaller table".

In other words, if the change in the fastener tensile force is less than 5% while increasing the table width by 305 mm (12"), the new table width is the RTW and there is no need to increase the width any more.

4.2 *Numerical examples*

Computed fastener loads for two typical configurations (48"/18" and 72"/18") are shown in Tables 2 and 3. Using the above established criterion, it is clear that 2610 mm (103") is the RTW for the 48"/18" configuration. Also, increase in the fastener row spacing from 1220 to 1830 mm (48" to 72") increases the RTW from 2610 to 3219 mm (103" to 127"). Note that both these systems have the same fastener spacing of 457 mm (18") along the seam.

From the established RTW, one can develop correction factors (F_c) for the smaller tables. In Table 2, 2610 mm (103") is identified as the RTW with 748 N as fastener force. Using a 2000 mm (79") table reduces the fastener force to 650 N. To correct this reduction, a magnification factor of

Table 2. Fastener forces for 48"/18" configuration obtained by changing the table width.

No.	Table Width mm (in)	Force N (lbf)	Change (%)
1	5048 (199)	800.6 (180)	0
2	4743 (187)	800.5 (180)	0
3	4438 (175)	800.3 (179.7)	0
4	4134 (163)	800 (179.5)	0
5	3829 (151)	799 (179)	0.1
6	3524 (139)	795 (178)	0.5
7	3219 (127)	788 (177)	0.9
8	2914 (115)	773 (173)	1.9
9	**2610 (103)**	**748 (168)**	**3.2**
10	2305 (91)	708 (159)	5.3
11	2000 (79)	650 (146)	8.2
12	1695 (67)	572 (128)	12
13	1390 (55)	470 (105)	18
14	1086 (43)	341 (77)	27.4
15	781 (31)	186 (42)	45

Table 3. Fastener forces for 72"/18" configuration obtained by changing the table width.

No.	Table Width mm (in)	Force N (lbf)	Change (%)
1	5048 (199)	1162 (261)	0
2	4743 (187)	1159 (260)	0.2
3	4438 (175)	1150 (258)	0.8
4	4134 (163)	1136 (255)	1.2
5	3829 (151)	1116 (250)	1.8
6	3524 (139)	1086 (244)	2.7
7	**3219 (127)**	**1044 (234)**	**3.9**
8	2914 (115)	987 (222)	5.5
9	2610 (103)	912 (205)	7.2
10	2305 (91)	820 (184)	10.1
11	2000 (79)	710 (159)	13.4
12	1695 (67)	588 (132)	17.2
13	1390 (55)	461 (103)	21.6
14	1086 (43)	339 (76)	26
15	781 (31)	240 (54)	29

1.15 (748/650) is needed. Thus, a correction factor of 1 is assigned to the 2610 mm table whereas a correction factor of 1.15 is developed for the 2000 mm table.

For engineering design purposes, the F_c can be useful in two folds:

1: To calculate the design load, the fastener forces should be multiplied by the F_c's.

2: If one assumes, induced forces (F) are directly proportional to the applied pressure (P) via tributary area (P = F.A), then the roofing system's sustained pressures on smaller tables should be divided by the F_c to correct the pressure for the effect of RTW.

At this stage it is worth to mention the benefits of a numerical model. The data in Tables 2 and 3 represent 30 numerical experiments. As discussed in section 1, engineering solutions for problems of this nature can only be obtained through numerical modeling. Experimental approach would not be economically feasible.

Varieties of simulations were performed for the variations of the two influencing factors, namely, fastener row spacing (F_r) and fastener spacing (F_s). Three F_r configurations 1830, 1700 and 1220 mm (72", 67" and 48") with three F_s configurations 152, 305 and 460 mm (6", 12" and 18") are considered. These F_r/F_s combinations represent most of the TPO systems currently available in the roofing industry. However, recently, membrane widths as high as 3048 mm (120") were introduced and F_s of 610 mm (24") were also incorporated in the design of large roof areas. For each configuration, fastener forces are calculated. By applying a procedure similar to that used in Tables 2 and 3, correction factors were developed. The results are presented in the Figures 8, 9 and 10, for the 72", 67" and 48" roof assemblies, respectively.

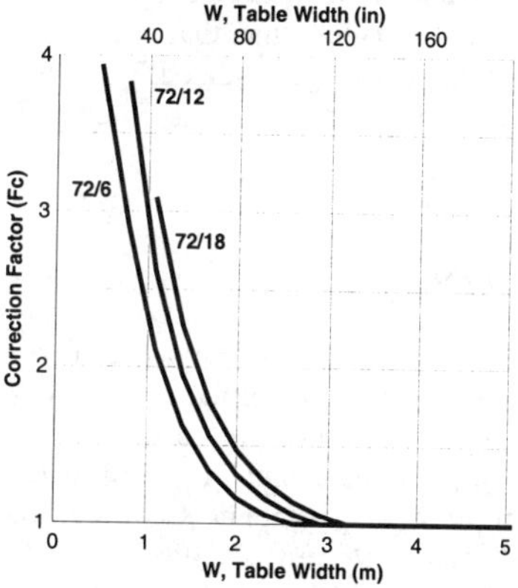

Figure 8. Developed Correction Factors, F_r=72"

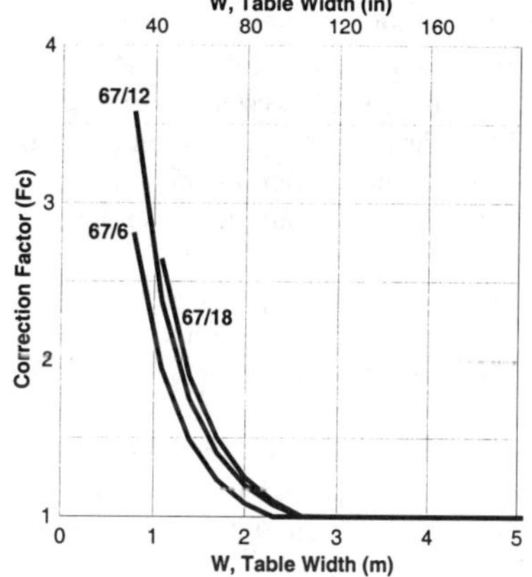

Figure 9. Developed Correction Factors, F_r=67"

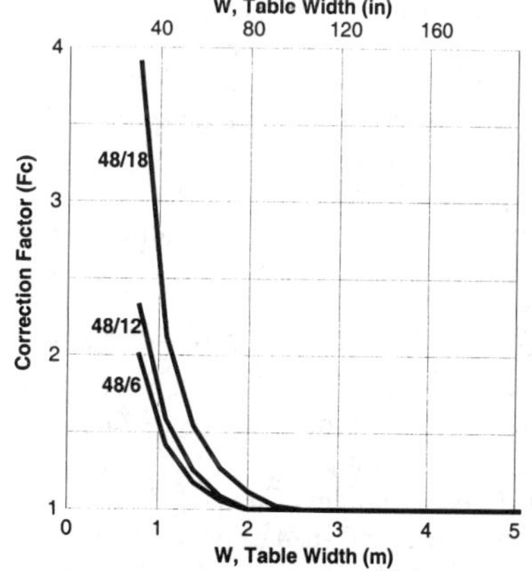

Figure 10. Developed Correction Factors, F_r= 48"

Figures 8 to 10 present an attempt to achieve characteristic curves such that generalized guidelines can be developed for the F_c. The Figures show that the correction factors are higher for the wider membranes and influence of the F_s on the development of F_c is less than that of F_r. Therefore, in the development of generalized F_c, one can assign higher importance factor for F_r compared to F_s.

5 CONCLUSIONS

An analytical model was developed to investigate the wind resistance performance of single-ply roofing systems. Numerical results for various system configurations compared well with those obtained from the experimental studies carried out using the Dynamic Roofing Facility.

The analytical model was also used to investigate the effect of table size on the roofing system performance. Attempts were made to identify the required table width. It was found that an increase in the table width beyond a certain level did not significantly change the system response. This specific limit depends on the system configurations. Systems with wider membranes (fastener row spacing) would generally increase the required table width.

Complementary modeling research is underway to further understand this issue and to investigate the effect of the other thermoplastic membrane, namely, PVC, on the development of the F_c. Only through such modeling efforts will guidelines for table size correction factors (to be applied to the results given by different table sizes) be established.

ACKNOWLEDGMENT

William Lei and Jayne Irwin, Technical Officers, collected the experimental data reported in this paper and their assistance is greatly appreciated.

The SIGDERS consortium is carrying out the presented research. It consists of following members whose support and commitments are greatly appreciated.

Manufacturers: Canadian General Tower Ltd., Carlisle SynTec Inc, GAF Materials Cooperation, Firestone Building Products Co., IKO industries Canada, JPS Elastomerics Corp. - Construction Products Group, Soprema Canada, Vicwest Steel

Building owners: Canada Post Corporation, Department of National Defense, Public Works and Government Services Canada and

Associations: Canadian Roofing Contractors' Association, Industrial Risk Insurers, National Roofing Contractors' Association and Roof Consultants Institute.

REFERENCES

ABAQUS user's manual. 1997. Hibbitt, Karlsson & Sorensen, Inc., Pawtucket, RI.

Baskaran. A, Lei. W & Richardson. C 1999. Dynamic Evaluation of Thermoplastic Roofing Systems for Wind Performance. Journal of Architectural Engineering, ASCE, vol. 5, No 5, pp 16-24.

Baskaran, A. & Chen, Y. 1998. Wind Load Cycle Development for Mechanically Attached Single Ply Roofs. *Journal of Wind Engineering and Industrial Aerodynamics*, Vol. 77 pp. 83-96.

Baskaran, A. & Lei, W. 1997. A New Facility for Dynamic Wind Performance Evaluation of Roofing Systems. *Proceedings of the Fourth International Symposium on Roofing Technology*, NRCA/NIST, Washington, D.C., U.S.A., pp. 168-179.

Cook, N.J., Keevil, A.P. & Stobart, R.K. 1988. BRERWULF- The Big Bad Wolf. *Journal of Wind Engineering and Industrial Aerodynamics*. Vol. 29, pp. 99-107.

Factory Mutual Research, 1986. Approval Standard: Class I Roof Covers (4470), Norwood, Massachusetts, USA.

Gerhardt, H.J. & Kramer, C. 1986. Wind Induced Loading Cycle and Fatigue Testing of Lightweight Roofing Fixations. *Journal of Wind Engineering and Industrial Aerodynamics*, Vol. 23, pp. 237-247.

Paulsen, E.M., 1989. NBI Roof Wind Uplift Strength Test Facility and Load Programs. *Proceedings of the Roof Wind Uplift Testing Workshop*, Oak Ridge, Tennessee, pp. 46-51.

Underwriters Laboratories Inc. 1991. Standard for Wind Uplift Pressure of Roof Assemblies (UL 580). third edition, Nov. 1991.

Wind Engineering into the 21st Century, Larsen, Larose & Livesey (eds) © 1999 Balkema, Rotterdam, ISBN 90 5809 059 0

A 1:50 scale wind-tunnel case study of wind pressure on TTU building

B. Bienkiewicz
Department of Civil Engineering, Colorado State University, Fort Collins, Colo., USA

H.J. Ham
Institute of Industrial Science, University of Tokyo, Japan (Formerly: Department of Civil Engineering, Colorado State University, Fort Collins, Colo., USA)

ABSTRACT: The paper describes a case study of physical modeling of wind flow and wind-induced building pressure, for the Texas Tech University (TTU) test building. Three flow conditions - nominal (prevalent) TTU wind and two bounding flows were simulated in a boundary-layer wind tunnel at Colorado State University. Measurements of the building point pressure were carried out in the three flows and the laboratory results were compared with the field data. It was concluded that the discrepancy between the field and laboratory data, observed for cornering wind at taps near roof edges, were not caused by higher than nominal turbulence intensity of wind during field measurements. The effects of Reynolds number on separated/reattaching shear layer associated with the roof edge region are suggested as a possible cause for the observed discrepancy between the laboratory and field results.

1. INTRODUCTION

One of objectives of the collaborative research effort of the Colorado State University/Texas Tech University Cooperative Program in Wind Engineering (CPWE) was to assess existing and develop new techniques for physical modeling in wind tunnels of wind effects on low-rise structures. Since inception of the CPWE, a number of wind-tunnel studies have been performed at Colorado State University (CSU) to address this objective. The scope of these investigations ranged from refined evaluation of turbulent boundary-layer wind modeled using wind-tunnel techniques previously developed at CSU, development of new flow modeling techniques for studies of low-rise buildings, measurement of point and area-averaged external building pressure, and investigation of building internal pressure. The geometrical scales employed in this effort ranged from 1:100 through 1:25. Flow modeling was focused on wind conditions at the Texas Tech University (TTU) field site and pressure was measured on models of the TTU test building.

Over the years, a number of wind tunnel studies of the TTU building were undertaken at other laboratories (Tieleman et al. 1996, Lin et al. 1995, Xu 1995, Rofail 1994, Jamieson & Carpenter 1993, Okada & Ha 1992, Surry 1991). In general, a good agreement between the laboratory and field data was found. However, the agreement for the peak and RMS point pressure was found to be less satisfactory, at critical locations in the roof corner region, for cornering wind. The discrepancy between the compared data was attributed to the approach flow (lateral turbulence intensity, small and large-scale flow characteristics), Reynolds number effects, pressure tap size, instrumentation limitations (frequency response of the pressure measurement system, size of pressure taps), and others.

This paper presents the results of a recent laboratory modeling of the TTU wind flow and wind-induced building (external) point pressure, performed at CSU, at the geometrical scale of 1:50. It can be considered as Phase II of the ongoing investigation focused on refined modeling of wind flow and wind-induced external pressure on low-rise buildings. The previous effort, denoted herein as Phase I, was reported by Ham and Bienkiewicz (1998). Detailed description of both the phases can be found in Ham (1998).

The paper is organized as follows. First, a brief overview of Phase I is presented. Next, the experimental configuration and the results of Phase II are described in detail. In this effort, two flows - one of higher and one of lower than the TTU nominal turbulence intensity - were modeled. They were subsequently used to evaluate the effects of the flow turbulence intensity on the wind-induced building point (external) pressure. The presented laboratory results are compared with the TTU field data and with wind-tunnel results published by Surry (1991). In addition, other pertinent experimental data are presented. Reasons for observed discrepancy between the laboratory and field results are hypothesized.

2. OVERVIEW OF RESULTS OF PHASE I

The main objectives of the wind-tunnel investigation, denoted herein as Phase I, were: to model prevalent, labeled herein as nominal, wind conditions at the TTU site and to measure wind-induced external point pressure on a model of the TTU test building, at a geometrical scale of 1:50. The ultimate goal was to assess the degree of agreement between the laboratory and field results. This effort was a natural extension of earlier studies of TTU wind flow and external building point pressure carried out at CSU through the CPWE effort. They included 1:100 geometrical scale investigation reported by Cermak & Cochran (1992), for details see Cochran (1992) and 1:25 geometrical scale study by Bienkiewicz & Sun (1992), for details see Sun (1993).

For the present study, the geometrical scale was selected to be 1:50, a reasonable compromise for the flow blockage effects. In addition, wind-tunnel investigations, addressing goals similar to those of the present investigation, were reported elsewhere (Surry 1991, and others) and it was intended to include such results in comparison of the data.

Ham & Bienkiewicz (1998) reported the laboratory results for the nominal TTU wind conditions. A combination of four spires, two barriers and a fetch of the floor roughness led to flow, denoted as CSU B2, which was in a very good agreement with the (nominal) field wind. Figures 1 and 2 depict the laboratory and field mean velocity and turbulence intensity profiles.

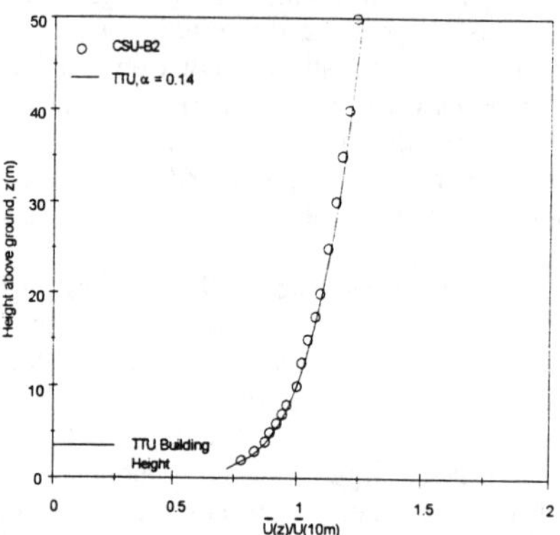

Figure 1. Mean velocity profile.

A comparison of the along-wind velocity spectra is presented in Figure 3, while the main flow parameters are listed in Table 1.

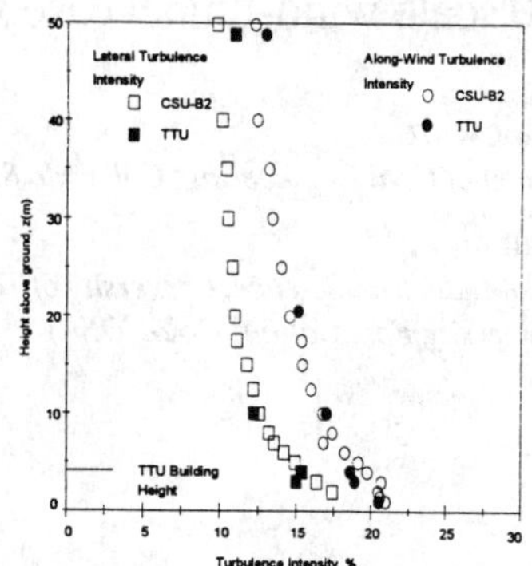

Figure 2. Along-wind and lateral turbulence intensity.

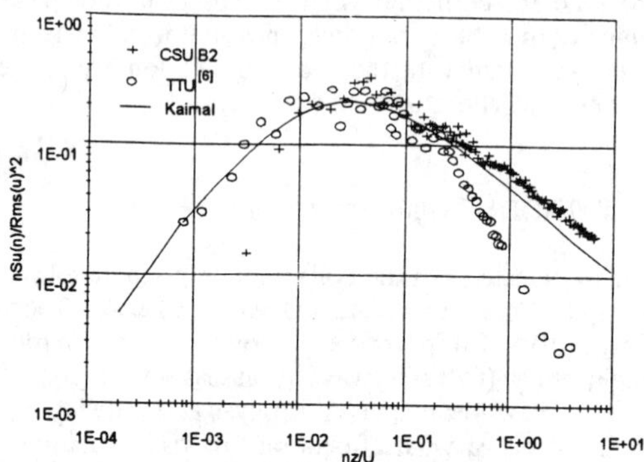

Figure 3. Along-wind velocity spectrum.

Table 1. Flow parameters.

Property	CSU B2	Field (Chock 1988 [1], Tieleman et al. 1996 [2])
Power law exp.	0.14	0.14(0.10-0.17) [1]
Aero. roughness	0.01 m	0.012(0.001-0.038)[1]
Shear velocity	0.74 m/s	0.73(0.54-0.94) [1]
Integr. length scales		
L_{XU}	54 m	76 [2]
L_{XV}	14 m	29 [2]
L_{XW}	8 m	19 [2]

A representative comparison of the laboratory and field wind-induced external point pressure along the mid section of the TTU building is shown in Figure 4.

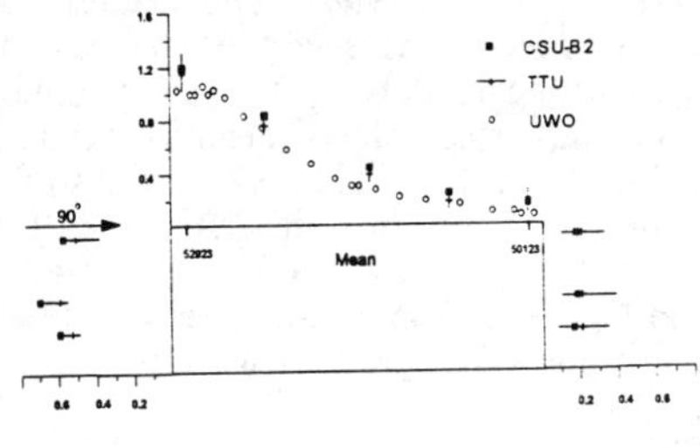

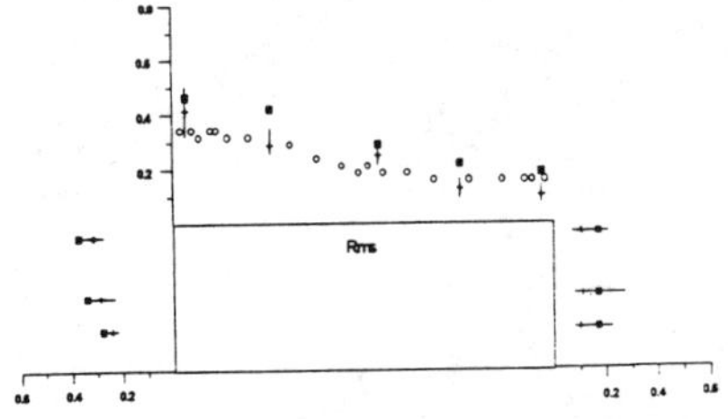

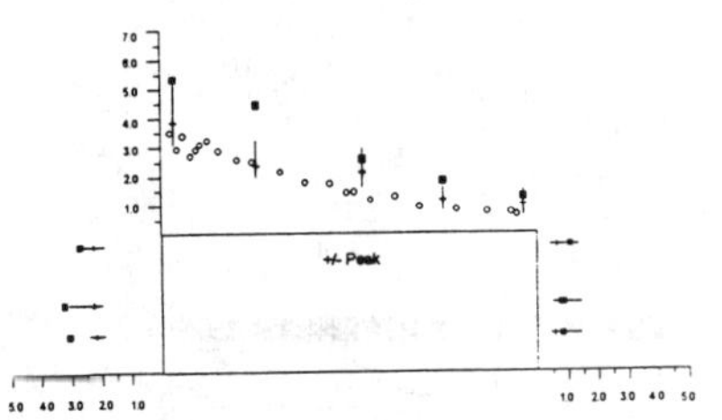

Figure 4. Pressure in mid plane of TTU building.

The plotted pressure coefficient was calculated from measurements at taps in the 90° - 270° plane, see

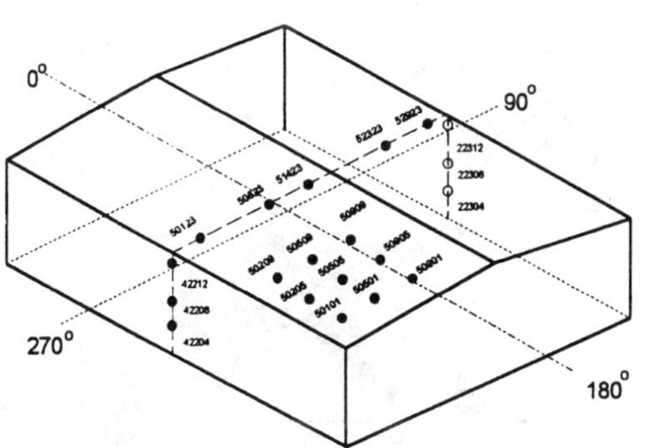

Figure 5. Location of pressure taps.

Figure 5. The data in Figure 4 includes also the results of an earlier study reported by Surry (1991) and denoted as UWO. A very good agreement between the CSU B2 measurements and the field results can be seen for most of the pressure taps. Some discrepancy is noted for the roof pressure.

Detailed measurements of the roof pressure were taken in the roof corner region, at taps schematically marked in Figure 5. Representative results are shown in Figures 6 and 7. Overall, the wind-tunnel and field results were found to be in a very good agreement, as is illustrated in Figure 6. However, a close examination of the data at taps near the roof edges revealed a measurable discrepancy between the laboratory and field peak pressure for cornering wind. This trend is illustrated in Figure 7.

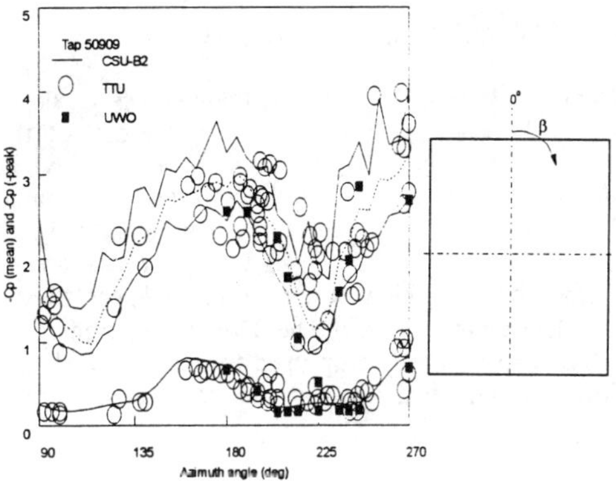

Figure 6. Peak and mean roof pressure at representative tap away from roof edges.

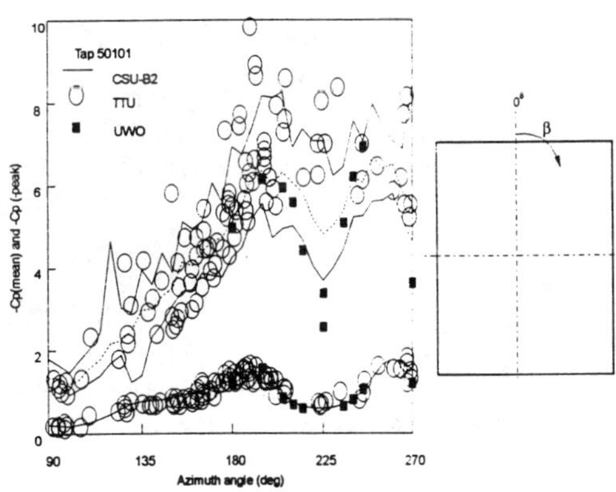

Figure 7. Peak and mean roof pressure at representative tap near roof edges.

Continuous lines in Figures 6 and 7 indicate the highest and lowest peaks (for each wind azimuth) out of ten largest suctions, determined from ten segments of the pressure time series. The dashed line is the mean of the ten considered peak suctions.

Peaks of the field roof suctions near roof edges appeared to be centered around the largest of the considered laboratory peaks, as is illustrated in Figure 7. This observation led to a hypothesis that the field peaks might be associated with wind phenomena of turbulence intensity higher than that of the TTU nominal wind, modeled by the CSU B2 flow. In order to test this postulate, the effects of approach wind flow turbulence on roof point pressure needed to be determined. This goal was addressed during a laboratory investigation denoted herein as Phase II.

3. FLOW MODELING - PHASE II

Wind tunnel modeling of two boundary-layer flows was undertaken as the first step of Phase II of the study. The objective was to generate one flow of turbulence level significantly lower than the nominal flow CSU B2, and one flow of a significantly higher turbulence level. These flows were then considered as the low and high bounds (with respect to the turbulence intensity) of the field wind conditions.

The experimental setup used to generate CSU B2 flow was modified and, after a considerable effort (for details see Ham 1998), two experimental arrangements were established to generate the desired two bounding flows. The setup used to simulate low-turbulence flow (denoted herein as CSU B1 flow) is shown in Figure 8. Details of the arrangement employed to generate the high-turbulence flow (denoted as CSU B3 flow) are depicted in Figure 9. The mean velocity profiles of the generated flows are shown in Figure 10.

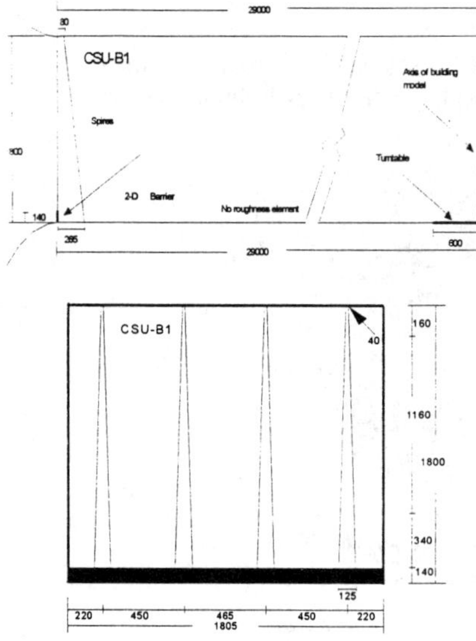

Figure 8. Setup used to generate low-turbulence (CSU B1) flow.

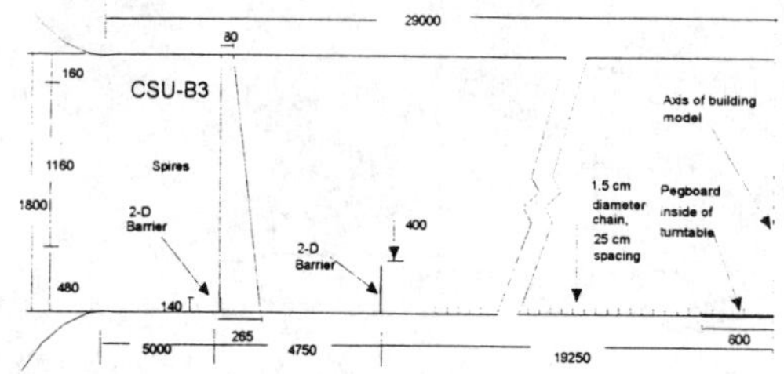

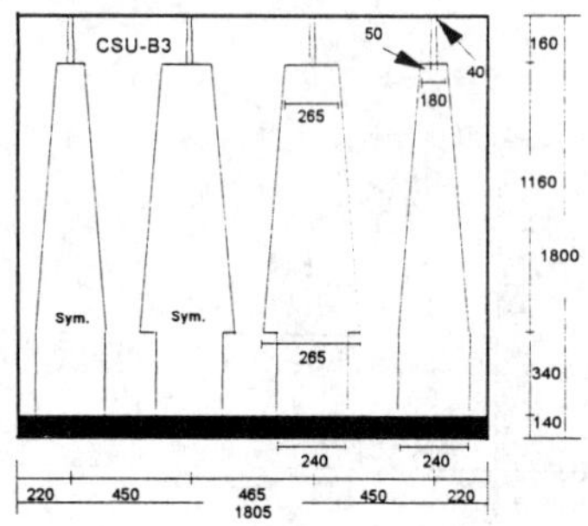

Figure 9. Setup used to generate high-turbulence (CSU B3) flow.

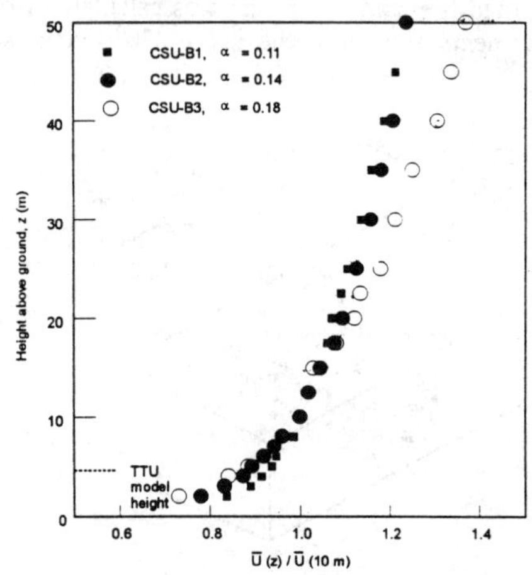

Figure 10. Mean velocity profiles.

The along-wind and lateral turbulence intensities are respectively shown in Figures 11 and 12.

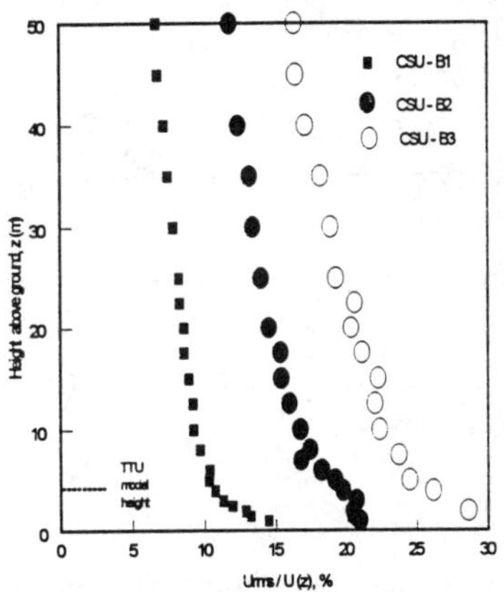

Figure 11. Along-wind turbulence intensity.

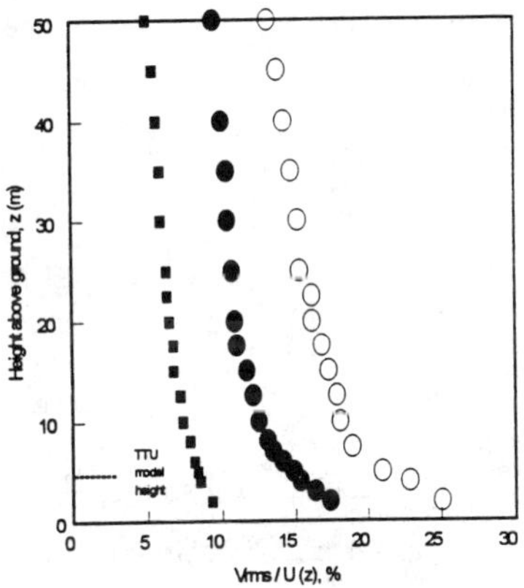

Figure 12. Lateral turbulence intensity.

As a reference, the mean velocity and turbulence intensity profiles obtained in modeled nominal (CSU B2) flow are included in the above figures. It can be seen in Figures 11 and 12 that the generated flows cover a wide range of turbulence intensities. The power spectra of the along-wind and lateral velocity fluctuations, measured in the three flows, are compared in Figures 13 and 14, respectively.

4. COMPARISON OF ROOF PRESSURE

The effects of flow turbulence on point pressure are illustrated in Figures 15 through 18. The building

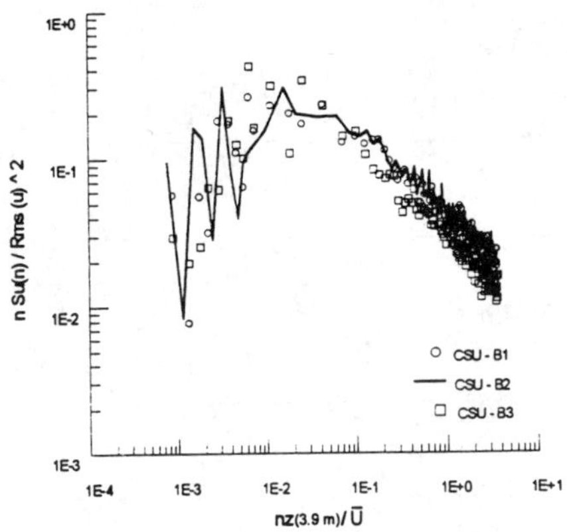

Figure 13. Power spectrum of along-wind velocity at building rooftop elevation.

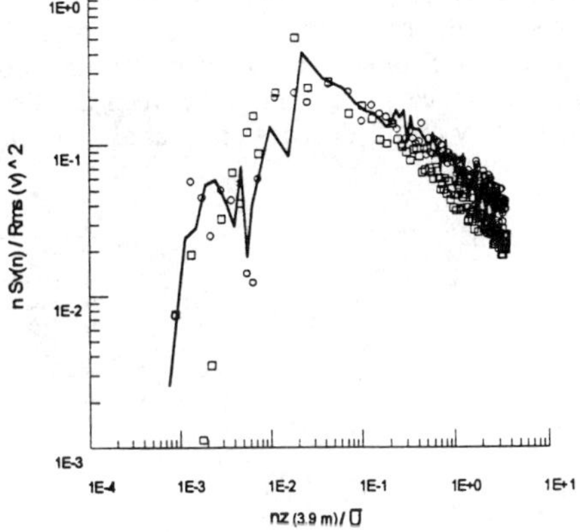

Figure 14. Power spectrum of lateral wind velocity at building rooftop elevation.

mid-plane pressure is shown in Figure 15. It can be seen that the overall best fit of the laboratory and field suctions is obtained in flow CSU B2, modeling the nominal (prevailing) wind conditions at the TTU field site. The effects of lowering the turbulence intensity (replacing flow CSU B2 by CSU B1) are more pronounced than those associated with increasing the turbulence level (replacing flow CSU B2 by CSU B3). Effects of turbulence are also exhibited by the mean pressure, especially at locations near the roof leading edge.

The results presented in Figures 16 and 17 illustrate turbulence effects on point peak pressure at representative (tap) locations in the roof corner region. It can be seen that the field pressure is bracketed by laboratory measurements obtained in the bounding (CSU B1 and CSU B3) flows.

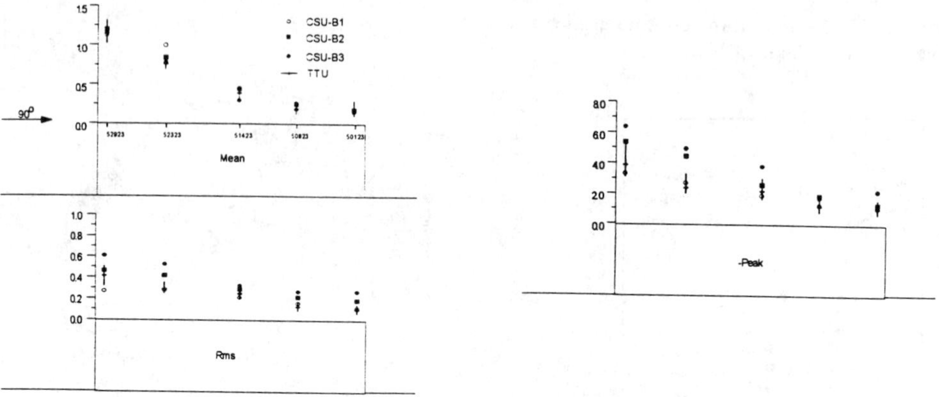

Figure 15. Effects of turbulence on mid plane point pressure.

It also can be seen that, at taps near the roof edges, the cornering wind discrepancy between the field and laboratory data is exhibited by both the peak suctions and the root mean square of the pressure fluctuation. At roof corner taps away from the roof edges, the degree of agreement between the field and laboratory results depends on location. For the pressure rms, these trends are illustrated in Figure 17. Additional evidence is provided in Figure 18. A closer examination of the pressure rms, at the roof corner taps away from the roof edge, revealed departure of the field data from the laboratory results at some locations, for non-cornering wind directions. As an example, this trend is observed at tap 50909, where a

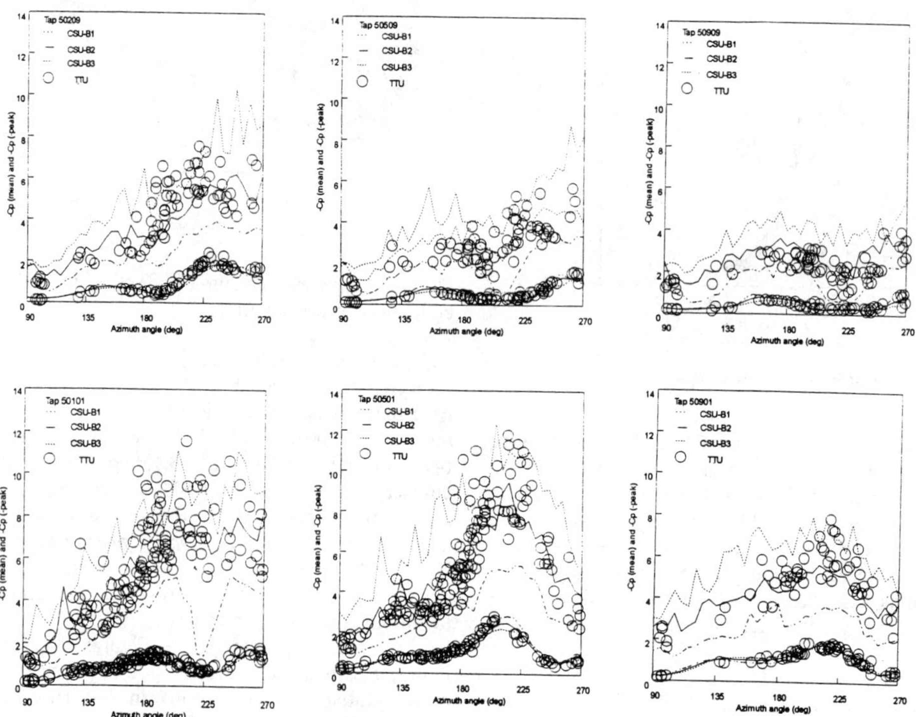

Figure 16. Effects of turbulence on peak suction pressure in roof corner region.

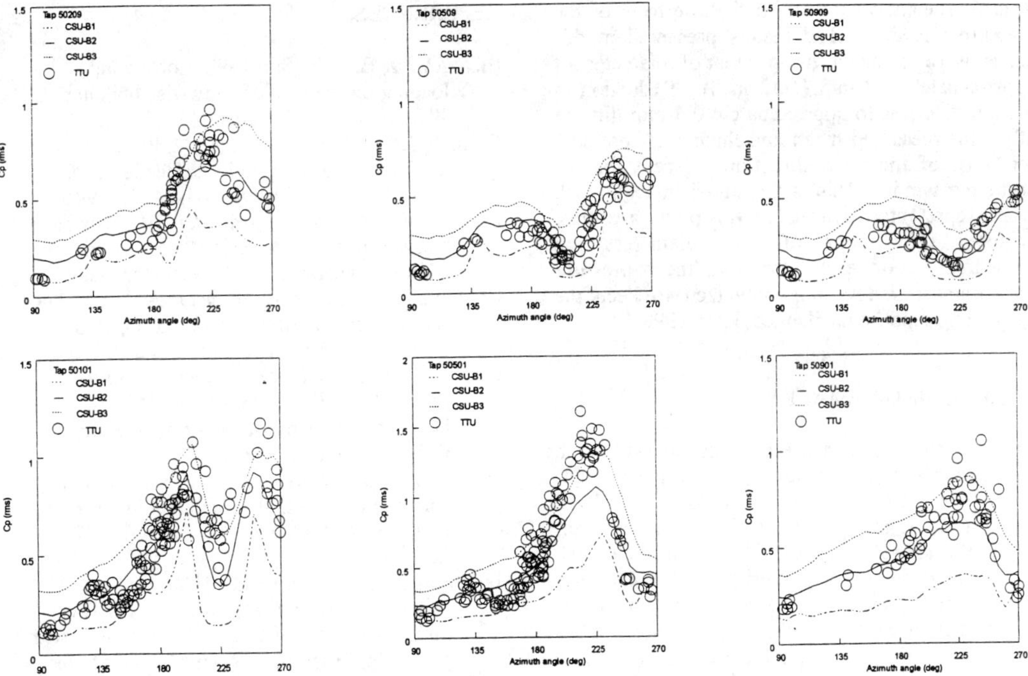

Figure 17. Effects of turbulence on rms of pressure fluctuation in roof corner region.

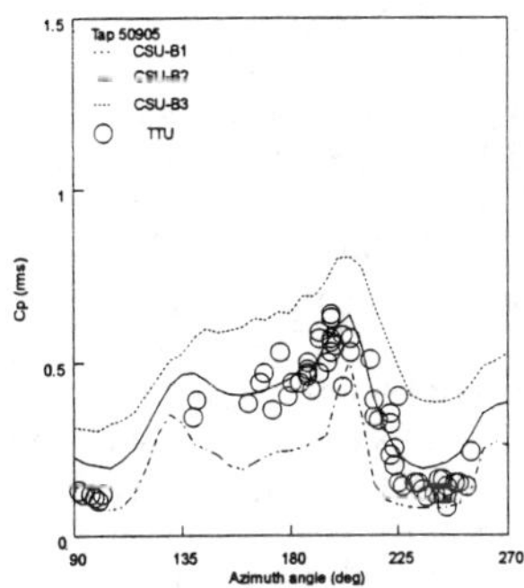

Figure 18. Effects of turbulence on rms roof pressure - tap 50905.

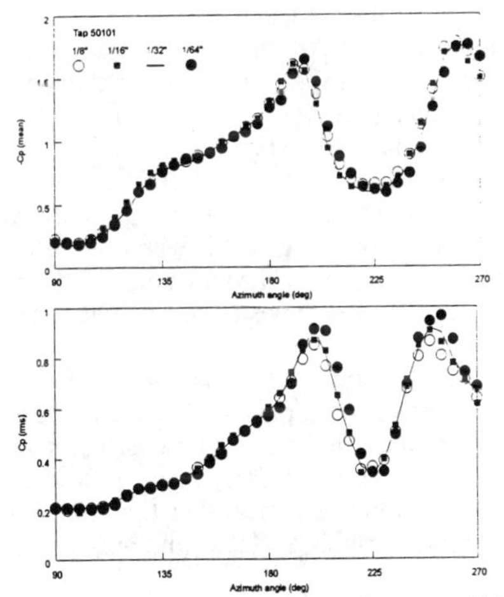

Figure 19. Effects of pressure tap size - tap 50101.

definite departure of the field pressure from the laboratory results in CSU B2 flow can be observed, as the wind azimuth angle of 270° is approached. At this angle, it is plausible to expect that a region in vicinity of this tap might be affected by reattachment of the shear layer separated from the roof longer edge.

5. EFFECTS OF PRESSURE TAP SIZE

A number of instrumentation/experimental optimization issues were addressed as part of this study, see Ham 1998. Topics related to pressure measurement included the frequency response of the pressure measurement system, discussed by

Ham & Bienkiewicz 1998, and the effects of the pressure tap size. The results presented in this paper were obtained using taps of diameter of approximately 0.8 mm (1/32 inch). Reduction in the tap diameter to approximately 0.4 mm did not affect the measured mean and fluctuating pressure for most of the wind directions, except for the cornering winds. This is illustrated in Figure 19, where representative mean and rms point pressures, acquired using taps of four diameters are compared. For each diameter, the pressure measurement system was optimized to keep the frequency response unchanged, Ham 1998.

CONCLUDING REMARKS

The laboratory suction pressure induced at taps in roof corner region was significantly larger in high-turbulence flow than in nominal flow, for all the tested wind azimuths. At locations near the roof edges, this led to an improved laboratory-field peak pressure agreement for cornering wind. However, it also resulted in worsening of the agreement for non-cornering wind azimuths. Similar effects were observed for the pressure standard deviation. This trend seems to suggest that the laboratory-field mismatch of the point pressure near roof edges (at cornering wind) most likely is not caused by higher than nominal (prevalent) level of turbulence during measurement of the compared field data.

Limitations of the instrumentation employed in the wind tunnel study appeared to have minor effects on the measured pressure.

Mismatch between standard deviation of the laboratory and field roof corner pressure was also observed at some locations away from the roof edges, for non-cornering wind azimuths. Precise mapping of the standard deviation the roof corner pressure is desired to investigate this phenomenon in more detail.

The field-laboratory mismatch of the rms of pressure fluctuation points towards possible Reynolds number effects on the structure and dynamics of the separated/reattaching flow shear layer(s) and vortical structures in the roof corner region and the resulting surface wind pressure.

ACKNOWLEDGMENT

This work was performed as a part of the research effort of the CSU/TTU Cooperative Program in Wind Engineering, supported by the National Science Foundation, through engineering grant CMS-9411147.

REFERENCES

Bienkiewicz, B. & Y. Sun 1992. Local wind loading on the roof of a low-rise building. *J. Wind Eng. Ind. Aerodyn.*. 45: 11-24.

Cermak, J.E. & L.S. Cochran 1992. Physical modeling of atmospheric surface layer. *J. Wind Eng. Ind. Aerodyn.* 41-44: 935-946.

Chok, C.V. 1988. *Wind parameters of Texas Tech University field site*, M. S. Thesis, Dept. of Civil Engineering, Texas Tech University.

Cochran, L.S. 1992. *Wind-tunnel modeling of low-rise structures.* Ph. D. Dissertation, Dept. of Civil Engineering, Colorado State University.

Ham, H.J. 1998. *Turbulence effects on wind-induced building pressure.* Ph. D. Dissertation, Dept. of Civil Engineering, Colorado State University.

Ham, H.J. & B. Bienkiewicz 1998. Wind tunnel simulation of TTU flow and building roof pressure. *J. Wind Eng. Ind. Aerodyn.* 77 & 78: 119-133.

Jamieson, N.J. & P. Carpenter 1993. Wind tunnel pressure measurements on the Texas Tech building at scale of 1:25 and comparison with full-scale. Proc. 7USNCWE, University of California at Los Angeles: 303-312.

Lin, J.-X., D. Surry & and H. W. Tieleman 1995. The distribution of pressure near roof corners of flat roof low buildings. *J. Wind Eng. Ind. Aerodyn.* 56: 235-265.

Okada, H. & Y.-C. Ha 1992. Comparison of wind tunnel and full-scale measurement tests on the Texas Tech Building. *J. Wind Eng. Ind. Aerodyn.* 41-44: 1601-1612.

Rofail, A.W. 1994. Full-scale/model-scale comparisons of wind pressure on the TTU building. *Proc. 9ICWE,* New Delhi, India, Wiley Eastern Ltd.:1055-1066.

Sun, Y. 1993. *Wind loading on loose-laid roofing paver system.* Ph. D. Dissertation, Dept. of Civil Engineering, Colorado State University.

Surry, D. 1991. Pressure measurements on the Texas Tech building: Wind tunnel measurements and comparisons with full scale. *J. Wind Eng. Ind. Aerodyn.* 38: 235-247.

Tieleman, H.W., D. Surry & K. C. Mehta 1996. Full/model scale comparison of surface pressures on the Texas Tech experimental building. *J. Wind Eng. Ind. Aerodyn.*, 61: 1-23.

Xu, Y.L. 1995. Model- and full-scale comparison of fatigue-related characteristics of wind pressures on the Texas Tech Building. *J. Wind Eng. Ind. Aerodyn.* 58: 147-173.

Wind Engineering into the 21st Century, Larsen, Larose & Livesey (eds) © 1999 Balkema, Rotterdam, ISBN 90 5809 059 0

Detection of along-wind response problems of mid-rise structures using simulation methods

M. Bontea
Department of Civil Engineering, State University of New York at Buffalo, N.Y., USA & Building Research Institute, INCERC, Bucharest, Romania

M. P. Gaus
Department of Civil Engineering, State University of New York at Buffalo, N.Y., USA

G. Lee
Department of Civil Engineering, State University of New York at Buffalo & Multidisciplinary Center for Earthquake Engineering, State University of New York at Buffalo, N.Y., USA

ABSTRACT: This study involves an investigation of wind loading modeling using simulation methods in the developmental design process for medium-rise structures, which require increased attention to detect possible undesirable dynamic response. Although wind tunnel studies would be desirable for such structures, the economics of their design and construction, frequently discourage such an investment. With recent progress in computational capabilities, the use of deterministic methods to predict the aerodynamic behavior of structures is becoming an alternative to frequency domain methods. One possible approach for interactive design is the use of simulated wind loads, technique which requires the incorporation of proper time history loadings. A study was carried out to evaluate the time-history response of a simple structure, for which ASCE 7-95 provisions may apply. The comparative analysis provides an indication to the designer of the type of results that maybe expected from current code-based procedures, and the results offered by more rigorous dynamic analyses, which can detect undesirable response problems in what concerns the serviceability of the building.

1. INTRODUCTION

Medium-rise buildings are a class of structures, which fall between low-rise (under 4 stories) and high-rise buildings (over 20 stories). The economics of medium-rise buildings are such that many developers do not feel they can make the economic investment needed for specific wind tunnel studies of their aerodynamic behavior. Because the medium-rise structures fall in the transition zone where dynamic response could become a problem, it would be desirable to have methods available during the design process, which could, conveniently and economically, assess the potential of a building to exhibit a dynamic response to wind loading, that may be unacceptable for the structure, occupants and equipment housed in the building.

With recent progress in computational capabilities, the use of deterministic methods to predict the aerodynamic behavior of structures is becoming an alternative to frequency domain methods. The application of simulation methods requires the simulated wind to be as realistic as possible and the method of simulation to be general, convenient and efficient. The advantages of time-domain evaluations are counter-balanced by the difficulties arising from the generation of a family of time-history functions correctly representing the atmospheric turbulence characteristics.

Problems associated with predicting the linear structural response to simulated wind time-histories are presented, and results compared to the building response evaluated using ASCE 7-95 code provisions. SAP2000, the latest version of the SAP series structural analysis programs, has been used for the along-wind 3D linear analysis of the structure.

As the dynamic amplification effects can create different serviceability problems, the maximum along-wind horizontal displacement and acceleration parameters are evaluated, and compared to the acceptable levels of human comfort.

2. SIMPLIFIED MODEL OF THE WIND LOADING

In one dimensional flow, neglecting the components of velocity fluctuations orthogonal to

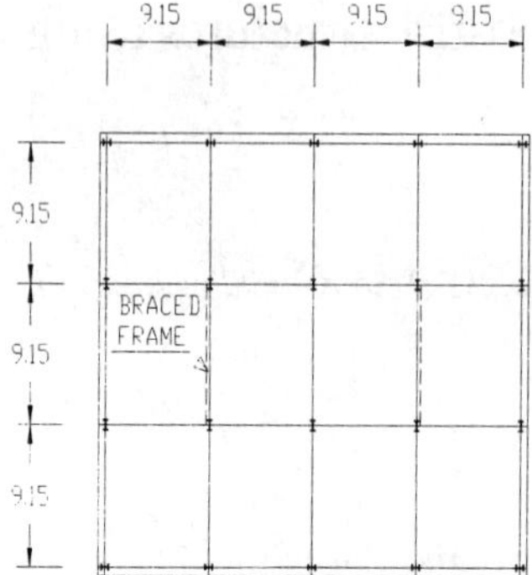

Figure 1. Typical frame plan.

Table 1. Natural frequencies of the structure

Mode No.	Frequency (Hz) E-W direction	N-S direction
1	0.26	0.35
2	0.72	1.36
3	1.20	2.72
4	1.68	4.12
5	2.20	-
6	2.71	-
7	3.26	-
8	3.72	-

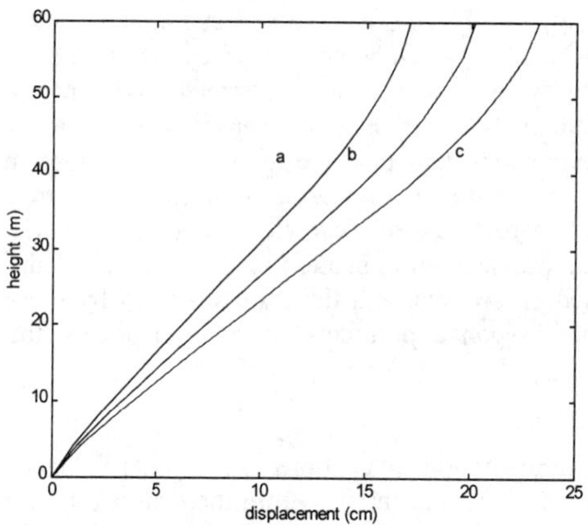

Figure 2. First mode deflection.
a) static component; b) code; c) commentary.

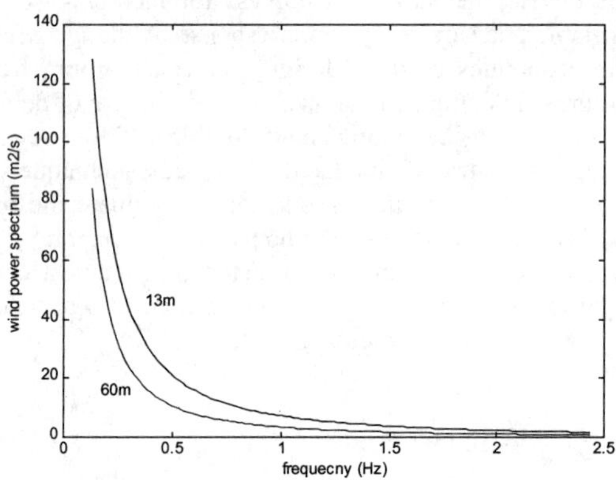

Figure 4. Target wind power spectrum at 13m and 60m height.

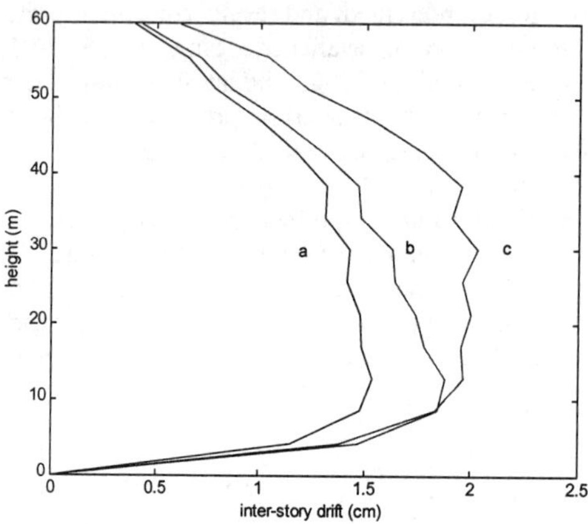

Figure 3. First mode interstory-drift.
a) static component; b) code; c) commentary.

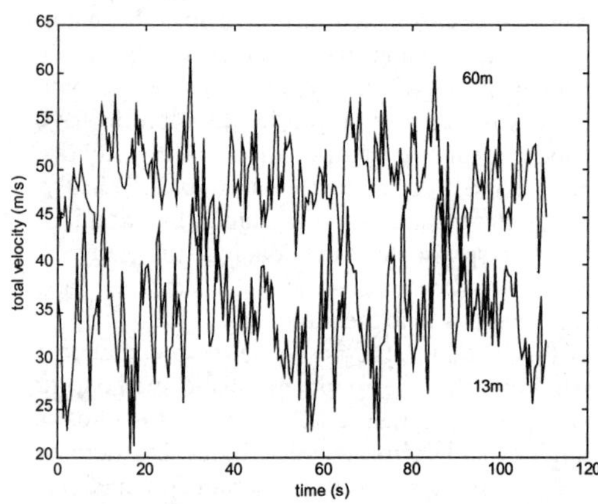

Figure 5. Simulated single time histories at 13m and 60m height.

the mean wind direction, the along-wind aerodynamic forces are generally modeled as the sum of steady and turbulent components:

$$F_D(t) = \overline{F}_D + F(t) \tag{1}$$

with $\overline{F} = (1/2)\rho\overline{U}^2 B^2 C_D$ (2a)

here $\overline{U}$ = wind mean velocity, which is assumed to have a logarithmic variation with height; ρ = air density; C_D = drag coefficient; B = one size of the building; and

$$F(t) = \rho\overline{U}u(t)B^2 C_D \tag{2b}$$

where $u(t)$ = wind fluctuating component which needs to be modeled.

The fluctuating component of the wind, $u(t)$, at a point along a vertical line, can be characterized by its longitudinal turbulence spectrum depending on height. An example of such a spectrum was proposed by Kaimal (1972):

$$S_u(n) = \frac{200u_*^2}{n}\frac{f}{(1+50f)^{5/3}} \tag{3}$$

where n = frequency; and $f = nz/\overline{U}(z)$.

3.SIMULATION OF SINGLE WIND HISTORIES

The actual structure of the turbulent wind is quite complex, and a complete formulation would require both information on the cross correlations for the wind and the local pressure coefficients on small areas on the structure – information generally derived from wind tunnel studies or from computational fluid analysis. Assuming this information is not available, a study was carried out to determine what type of time-history dynamic information for a structure could be developed using single simulated wind records and the spatially averaged pressure coefficients which were developed for Codes such as ASCE 7-95.

Different methods for simulating stationary random wind fields, with specified spectral densities, are proposed in the literature (using series of sine and/or cosine functions; application of linear filters to sequences of random numbers). One of these, the wave superposition approach, is the technique used in this paper. Based on the wind loading model formulation above, the artificial generation of N wind speed time-histories for the windward side of the building is performed using a method developed by Shinozuka & Jan (1972). The along-wind field is simulated by a series of cosine functions with weighted amplitudes:

$$u(t) = \sqrt{2}\sum_{k=1}^{N}[S_u(n_k)\Delta n]^{1/2}\cos(2\pi n_k' t + \phi_k) \tag{4}$$

where N = the number of subintervals $\Delta n = n_{k+1} - n_k$ into which the analyzed frequency band has been subdivided (it contains all the natural frequencies of modes giving significant contributions to the dynamic structural response); $S_u(n)$ = power spectral density of $u(t)$ at a point on the building; n_k = central frequencies of the intervals Δn; $n_k' = n_k + \delta n$ where δn is a small random frequency introduced to avoid periodicity of the simulated process; and ϕ_k = independent random phase-angles. For the leeward side of the building, the other N generated wind time-histories take into consideration the along-wind correlation of velocities at each level as:

$$u(t) = u(t - d/\overline{U}(z)) \tag{5}$$

4.ASCE 7–95 PROVISIONS FOR WIND LOADING AND STRUCTURAL RESPONSE

In the ASCE 7-95 code, the wind design pressures are estimated for flexible structures as:

$$p = qG_f C_p \tag{6}$$

where $q = q_z$ and $q = q_h$ the velocity pressure for windward/leeward side of the building at height z above ground/at mean roof height, respectively; G_f = gust factor evaluated using a complete analysis; and C_p = external pressure coefficient. The velocity pressure is given by $q_z = 0.00256K_z K_{zt} V^2 I$ with V = basic wind speed; I = importance factor based on the building category; K_{zt} = topographic factor; K_z = velocity pressure exposure coefficient for urban and suburban areas given by an exponential law.

The maximum along-wind displacement and acceleration, determined assuming the fundamental vibration mode contribution only are evaluated in

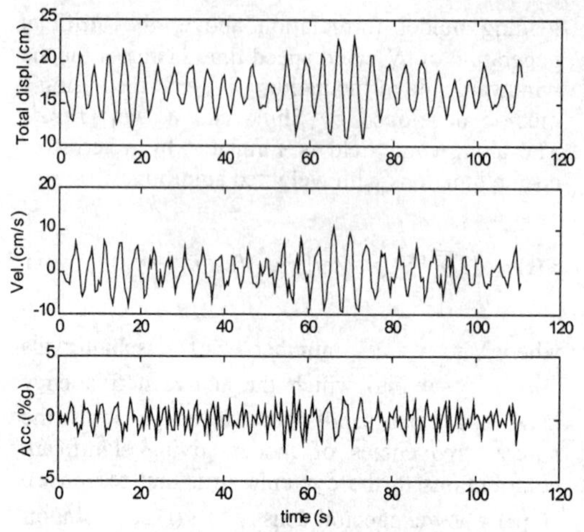

Figure 6. Structural response in time-domain.

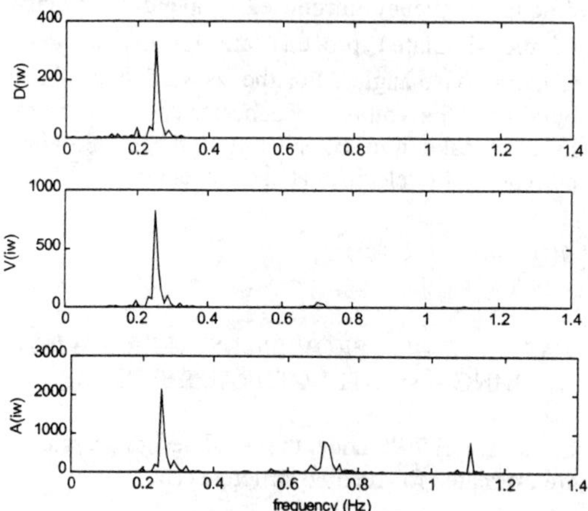

Figure 7. Structural response in frequency domain.

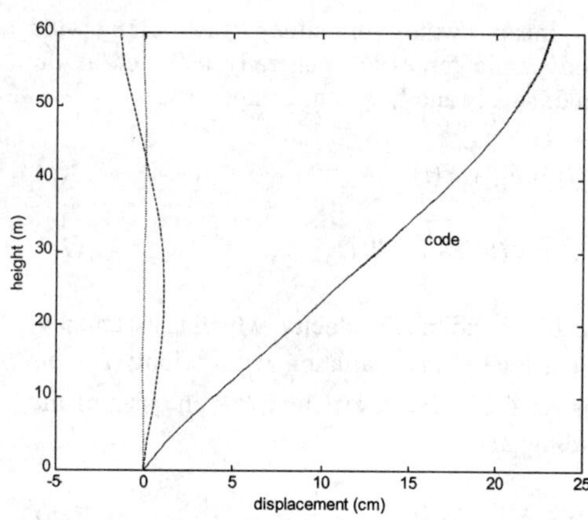

Figure 8. Code predicted and maximum modal displacement at the top: a) 1st mode at t=66.8s; b) 2nd mode at t=58.2s; c) 3rd mode at t=60.3s.

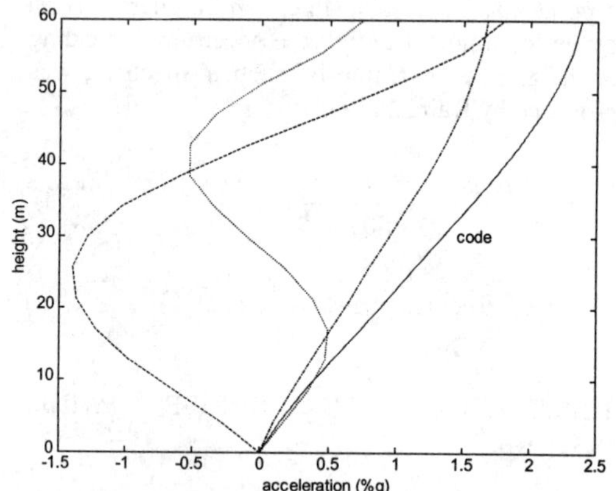

Figure 9. Code predicted and maximum modal acceleration at the top: a) 1st mode at t=72.5s; b) 2nd mode at t=58.2; c) 3rd mode at t=27.8s.

Table 2. Structural response comparative results

Response parameters at the top	ASCE 7-95 code analysis				Simulated time-history analysis				Predicted
	1sr mode	2nd mode	3rd mode	total	1st mode	2nd mode	3rd mode	total	1st mode
$\bar{d}$ (cm)	-	-	-	-	17.03	0.50	0.08	16.59	-
σ_d (cm)	-	-	-	-	2.40	0.27	0.06	2.42	-
$d_{\max}$ (cm)	20.10	0.93	0.19	19.31	23.18	1.43	0.08	22.86	23.15
σ_a (%g)	-	-	-	-	0.68	0.55	0.33	0.92	0.62
$a_{\max}$ (%g)	-	-	-	-	1.68	1.78	0.73	2.71	2.38

Equations 7-8, as:

$$X_{\max}(z)=\frac{\phi(z)\rho bhC_{fx}\hat{V}_{\bar{z}}^{2}}{2m_1(2\pi n_1)^2}KG \qquad (7)$$

with $\varphi(z)=$ fundamental mode shape is $\phi(z)=(z/h)^{\xi}$ with $\xi=$ mode exponent; $\rho=$ air density; $b=$ structure width; $h=$ structure height; $C_{fx}=$ mean along-wind force coefficient; $m_l=$ modal mass, $m_l=\int_0^h \mu(z)\phi^2(z)dz$ with $\mu(z)$, the mass per unit height; $n_1=$ building natural frequency; $G=$ gust effect factor; and

$$\ddot{X}_{\max}(z)=g_{\ddot{x}}\sigma_{\ddot{x}}(z) \qquad (8)$$

respectively, where the acceleration rms value and the peak factor are given by

$$\sigma_{\ddot{x}}(z)=\frac{0.85\phi(z)\rho bhC_{fx}\overline{V}_{\bar{z}}^{2}}{m_1}I_{\bar{z}}KR \qquad (9a)$$

$$g_{\ddot{x}}=\sqrt{2\ln(n_1T)}+0.5772/\sqrt{2\ln(n_1T)} \qquad (9b)$$

with $I_z=$ intensity of turbulence at height $\bar{z}=0.6h$; $R=$ resonant response factor; and $T=$ 3600 sec.

5.COMPARISON OF CODE AND TIME-HISTORY RESULTS

A time history analysis and code analysis were carried out on a medium size office building. The building is a 14-story symmetric steel frame structure with 60m (196ft) height, 29m (94ft) width, 38m (124ft) length and 4.3m (14ft) story height. The structure has two exterior wind force resisting systems (ordinary moment frames) in the along-wind direction (EW), and two interior concentric braced frames in the NS direction (Fig.1). According to the original designers' data, the allowable drift under wind condition is 2/350 of the height of the structure (34cm).

SAP2000, the latest version of the SAP series structural analysis programs, was used for the along-wind 3D linear analysis of the structure. The first natural frequencies are listed in Table 1. The damping coefficient for all calculated eigenmodes had a value of ξ=2%.

Two structural analysis were carried out:

(1) code analysis, where the wind load case consists of concentrated equivalent static forces applied at joints, and suitably evaluated for a flexible structure, based on the current gust factor approach;

(2) time-history analysis, where the wind load case consists of unit forces which form a spatial load vector, applied at joints, and multiplied by time-dependent functions (single wind velocities on the windward and leeward faces of the building).

In each analysis case the modal response has been evaluated for the first three vibration modes (n_1= 0.26Hz; n_2= 0.72Hz; n_3= 1.20Hz): case (1) - displacement; case (2) - displacement and acceleration.

The response of the structure for the code analysis (1) has been evaluated using the following wind and building data: V= 90mph (40m/s); I= 1.0 (building category II); C_p= 0.8 and C_p= 0.5 for the windward and leeward side, respectively; K_{zt}= 1.0; z_g= 366m and α= 7 for area of category B; h= 60m; n_1= 0.26Hz; C_{fx}= 1.3; ρ= 1.2254kg/m^3; K= 0.50; Q= background response factor, Q=0.8295; R= 0.6532 (resonant response factor); g= peak factor, g= 3.5; G_f= 1.035. A comparative evaluation of the gust factor using a simplified/complete analysis for rigid structures gives G_f= 0.8 and G_f= 0.804, respectively.

Based on the code provisions for design wind loads and the response prediction formulation from the Commentary on ASCE 7-95, the along-wind displacement and the interstory-drift in the first mode of vibration are presented in Figures 2-3, together with the structural response associated to the static component of the simulated wind time-histories. The analysis indicates that the commentary prediction values are conservative in terms of maximum displacement and inter-story drift along the height of the structure. The predicted displacement at the top is 36%/15% larger than the static component/ equivalent static value, respectively. The contribution of the 2nd and 3rd mode to the total displacement is insignificant.

The response of the building for the time-history analysis (2) has been evaluated using the following wind and building data: N= 128; T= 110.67s; Δt= 0.434s; $n_{\inf}$= 0.124Hz; $n_{\sup}$= 2.424Hz; Δn= 0.0181Hz; $\overline{U}(10)$= 40m/s; z_0= 0.3m; u_*= 3.71m/s; C_p= 0.8 and C_p= 0.5 for the windward and leeward side, for the mean and fluctuating wind components.

The target wind power spectrum used to simulate the fluctuating velocities at 13m and 60m height is presented in Figure 4. Samples of the generated wind histories at these levels are shown in Figure 5. A different set of 14 single time histories forces has been applied vertically, along the two moment resisting frames, on each side of the building. Due to the software capabilities of SAP2000, the time offset of the wind flow has been considered in the application of time functions at points on the leeward side.

The time-history of the displacement, velocity and acceleration at the top floor and their representation in frequency domain, are presented in Figures 6-7. The different contribution of the 2nd and 3rd mode to the total response is noticed. Figures 8-9 present the maximum modal displacement and acceleration at the top of the building compared to the commentary predicted values. As in the code analysis case, the first mode major contribution to the total response is indicated, together with the reduced contribution of the other two modes to the total displacement. On the other hand, the large contribution of higher modes to the total acceleration is outlined. The level of acceleration at the top floor (60m) is 6% larger on the 2nd mode than on the first mode (1.68%g), while at the sixth level (26m) the modal acceleration is 78% of the maximum value at the top.

Estimates of the rms value of the top floor acceleration were made to check the performance of the building from the comfort criterion point of view, according to which, its value, for a perceptible degree of discomfort, should be in the range of 0.5-1.5%g, at the top floor for a 20min wind flow period. While the code predicted rms along-wind acceleration is 0.62%g for one hour length of time, the respective value for the total and the first three modal responses over 120s is: 0.98%g; 0.68%g; 0.55%g; 0.33%g. Comparative structural response results are presented in Table 2.

6. CONCLUSIONS

Synthetic wind histories, which can excite a medium-rise structure through the range of loadings caused by natural winds, can be directly incorporated into design using current generations of analysis software. The early detection of potential performance problems is desirable in the design analysis of these buildings, usually assumed to do not have significant wind induced vibrations.

Using the SAP2000 package, the case study of a 14-story steel structure, is investigated in time-domain and based on the ASCE 7-95 code provisions. The dynamic response of the structure is determined based on the intrinsic characteristics of the simulated wind, which were used for the evaluation of the structural loading. The time-history analysis provides more detailed information on the performance of the structure, indicating undesirable vibration problems which must be taken into consideration in what concerns the serviceability of the building.

A comparison was made of key behavior parameters (displacement and acceleration levels) to the acceptable occupancy comfort in buildings. The displacement response indicated code values ranging on the conservative side. On the other hand, the contribution of higher modes to the acceleration level was significant, indicating some undesirable problems that these modes may cause for human comfort and equipment performance. The comparative analysis provides an indication to the designer of the type of results that maybe expected from current code-based procedures, and the results offered by more rigorous dynamic analyses.

Further studies are needed to validate this approach for structures having irregular shapes and different structural characteristics which may require more refined wind simulation models and software to conveniently generate such wind time histories for design purposes.

REFERENCES

ASCE 7-95. *Minimum Design Loads for Buildings and Other Structures,* 1996; New York: ASCE Press.

Iannuzzi, A. and Spinelli, P. 1987. Artificial generation and structural response. *J. Struct. Engrg,* Vol. 113, No.12: 2382-2398.

Jesien, W., Stathopoulos, T. & Ha, K.H. 1993. Dynamic along-wind response of buildings: Comparative study. *J.Struct. Engrg.*, Vol.119, No.5: 1498-1515.

Kijewski, T. & Kareem, A. 1998. Dynamic wind effects: A comparative study of provisions in codes and standards with wind tunnel data. In press.

Mehta, K.C., Marshall, R.D. 1998. *Guide to the use of the wind load provisions of ASCE 7-95.* Reston, Va.: ASCE Press.

Pagnini, L.C. & Solari, G. 1998. Serviceability criteria for wind-induced acceleration and damping uncertainties. *J.Wind Engrg.* 74-76: 1067-1078.

Shinozuka, M., and Jan, C.M. 1972. Digital simulation of random process and its application. *J. Sound Vibr., 25(1):* 111-128.

Simiu, E., and Scanlan, R.H. 1996. *Wind effects on structures. Fundamentals and Applications to Design.* New York: J. Wiley & Sons Inc.

Wind Engineering into the 21st Century, Larsen, Larose & Livesey (eds) © 1999 Balkema, Rotterdam, ISBN 90 5809 059 0

Local wind pressures on structures in groups

V. Feranec & T. Feranec
University of Žilina, Faculty of Civil Engineering, Slovak Republic

ABSTRACT: This paper presents the results of the steady local wind pressures on track side structures situated in groups as well as on a single structure in two simulated winds. Wind load for the structures in groups specified by current standards and codes of practice originate from wind studies on isolated structures and give no guide for the assessment of loads under conditions of buffeting. It is well known, however, that the wind pressure distribution on a structure may change drastically when a new structure(s) is built in its neighborhood. A literature review indicates that little information is available on this critical subject. Based on these findings, an experimental study of loads on track side structures in groups has been carried out in the Boundary layer wind tunnel laboratory in the University of Žilina and some results are presented in the paper.

1 INRODUCTION

Wind loading on structures in groups is important field for investigation. Wind loads for the buildings and structures in groups specified by current standards and codes of practice originate from wind studies on isolated structures and give no guide for the assessment of loads under conditions of buffeting. It is well known, however, that the wind pressure distribution on a structures may change drastically when a new structure(s) is built in its neighborhood A literature review indicate that little information is available on this critical subject. In fact, the bulk of information used for code formulation of the wind loads on buildings is based on model tests of free-standing structures. Wind tunnel is factor to have reliable data source. Based on these findings, an experimental study of loads on track side structures in groups has been carried out in the Boundary layer wind tunnel and some results of steady local wind pressures on structures are presented in this paper.

2 WIND TUNNEL EQUIPMENTS AND MEASUREMENTS

Atmospheric boundary-layer wind tunnel in the University of Žilina has a test section 1m wide and 0.85m high, with a fetch length of 3m and wind tunnel length is 5m. A turbulent atmospheric boundary layer in the wind tunnel has been simulated by using wooden grids, tooth mixing devices and turbulence generators followed by smooth terrain or surface roughness, consisting of plastic 7cm cubes in a diamond array at different density.

Wind speed and turbulence at the wind tunnel has been measured with hot-wire anemometers and analyzed with a DISA equipment. Results of the mean wind speed and turbulence intensity profiles for open country, smooth flow, and for urban simulated terrain exposures, turbulent flow, in the wind tunnel are given in Fig. 1. [1]-[2]

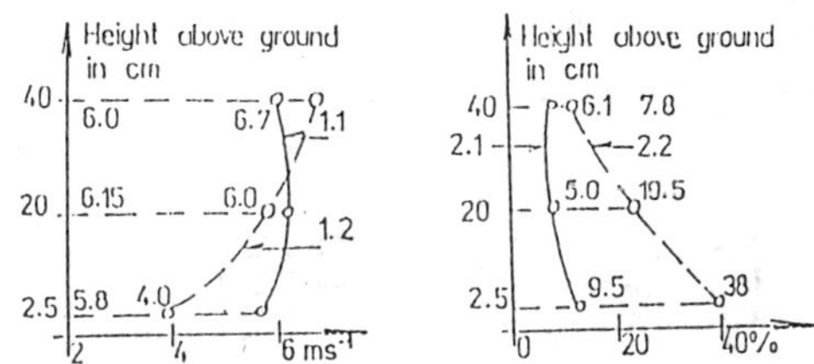

Fig. 1 Mean wind speed and turbulence intensity profiles for open country (——) and urban simulated terrain exposures (- - - -)

Measurement of steady local pressures were carried out with inclined multi-tube manometer. Models of structures were of plax-glass and instrumented with pressure taps. The pressure were converted into mean pressure coefficient form.

3 EXPERIMENTAL RESULTS ON TRACK SIDE STUCTURES

Some wind tunnel results of mean pressure coefficients on models of track-side structure in smooth and turbulent flow are in Fig 2 –12. [1] [2] [3] [5] [8]

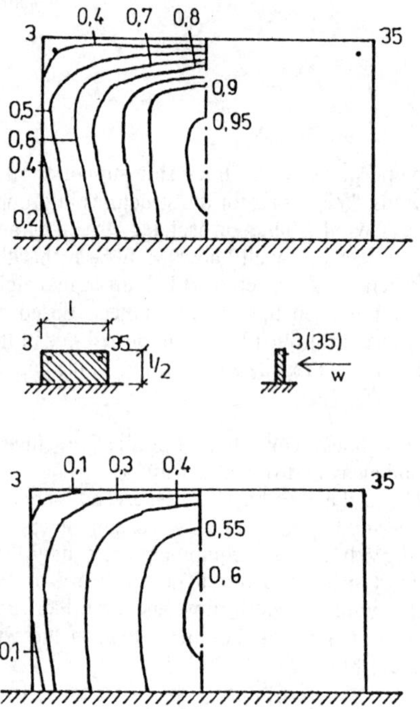

Fig. 2 Steady local wind pressures on a single boundary wall in smooth and turbulent flow

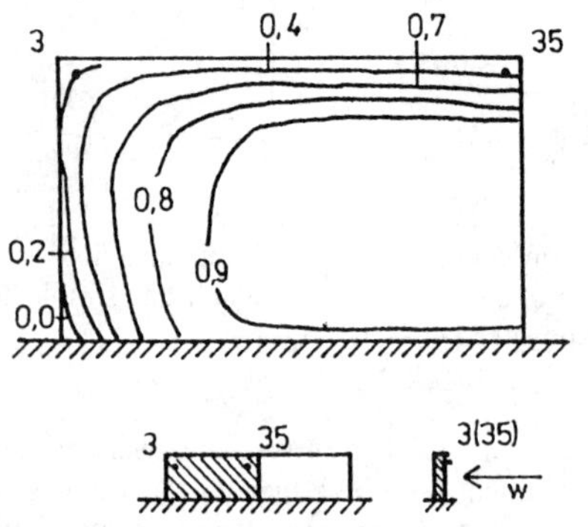

Fig. 3 Steady local wind pressures on solid boundary wall in group in smooth flow

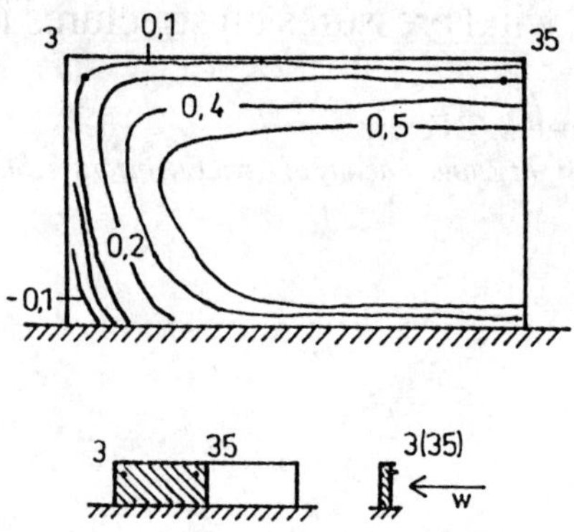

Fig. 4 Steady local wind pressures on solid boundary wall in group in turbulent flow

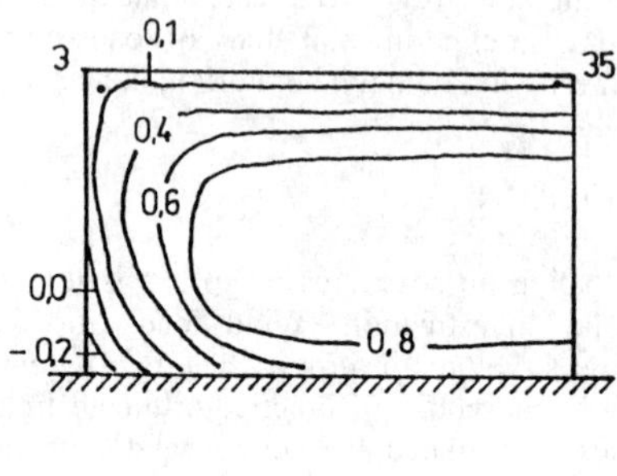

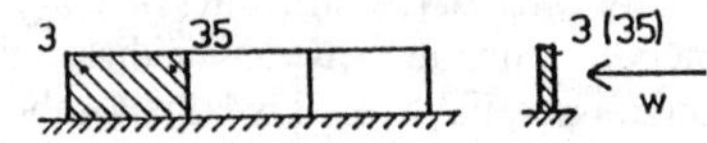

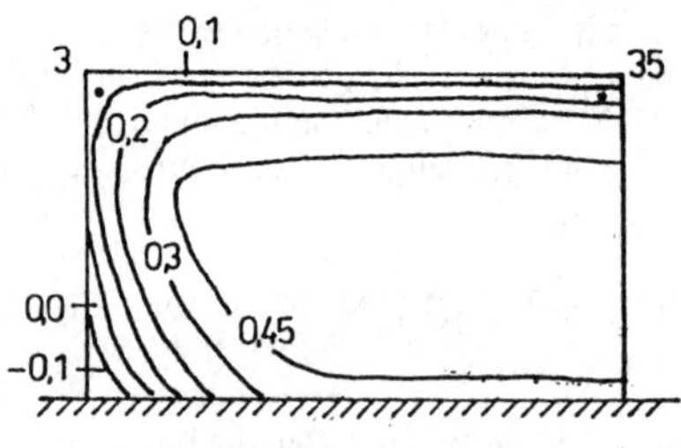

Fig. 5 Steady local wind pressures on boundary wall in group in smooth and turbulent flow

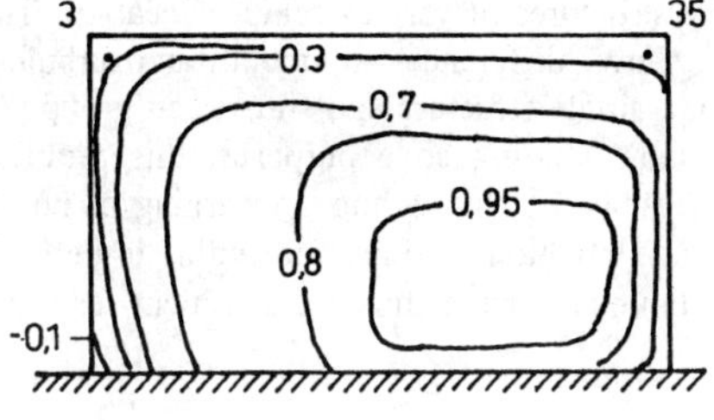

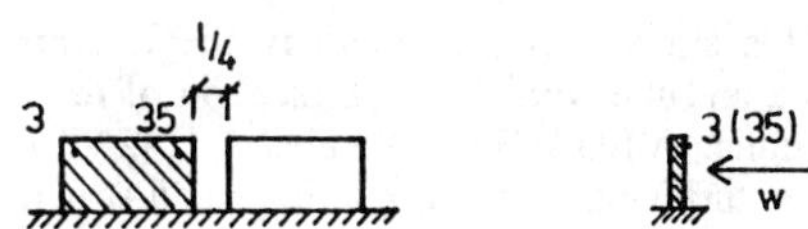

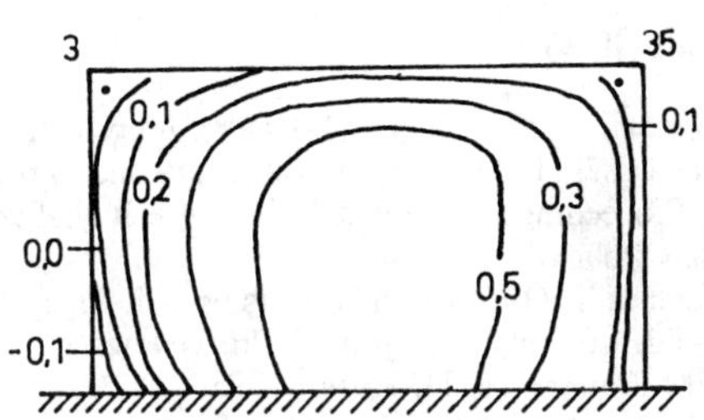

Fig. 6 Steady local wind pressures on solid boundary wall in group in smooth and turbulent flow

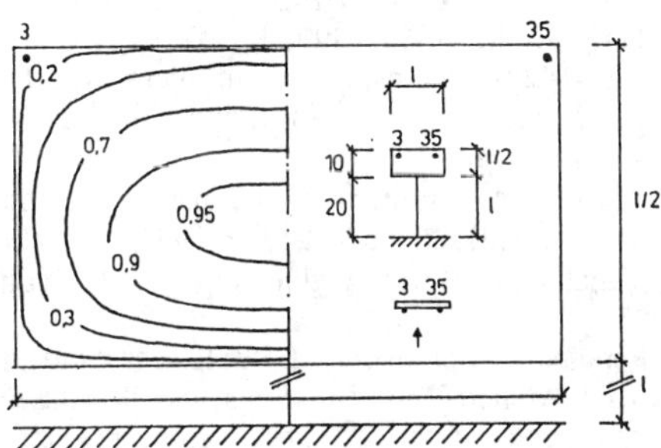

Fig. 7 Steady local wind pressures on a single sign board in smooth flow

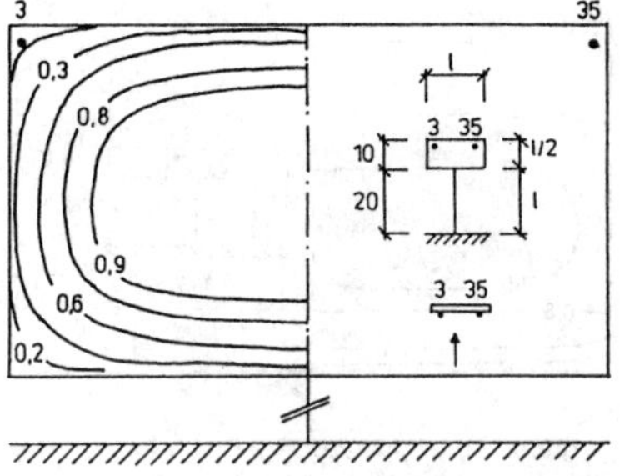

Fig. 8 Steady local wind pressures on a single sign-board in turbulent flow

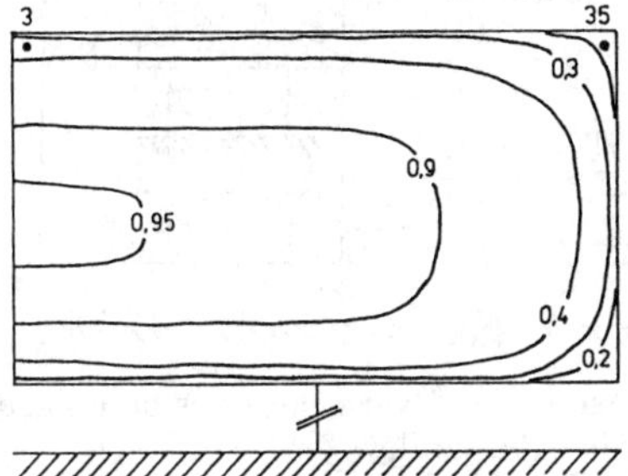

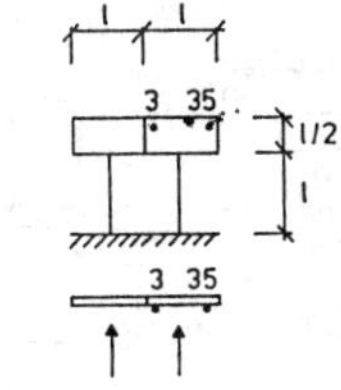

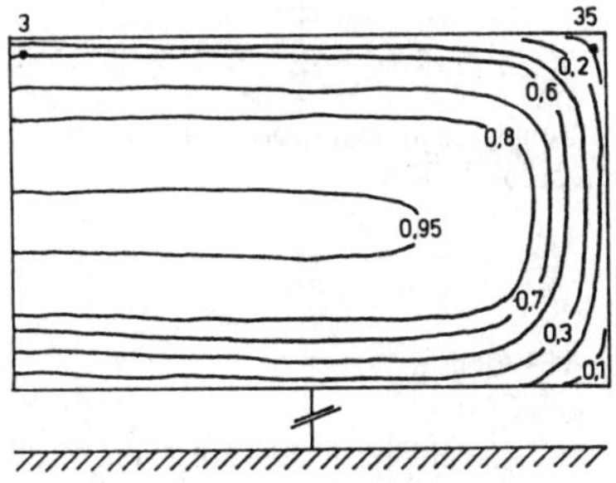

Fig. 9 Steady local wind pressures on signboard in group in smooth and turbulent flow

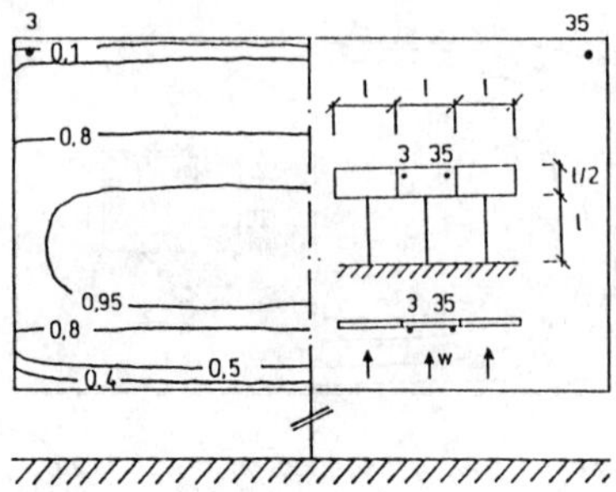

Fig. 10 Steady local wind pressures on a signboard in group in smooth flow

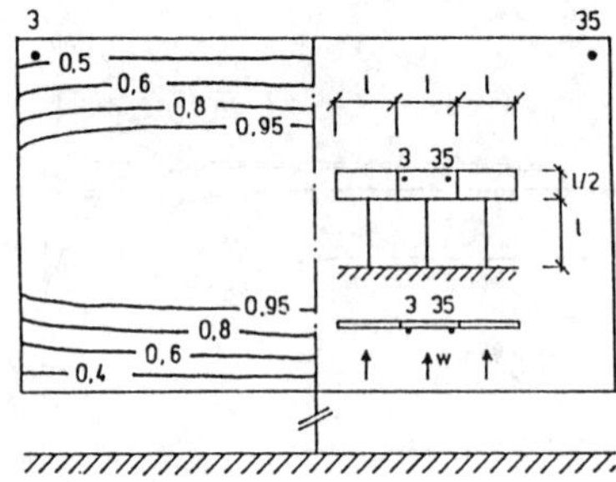

Fig. 11 Steady local wind pressures on a signboard in group in turbulent flow

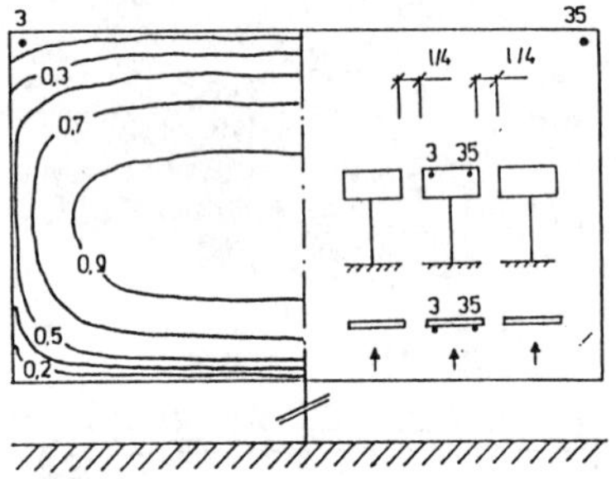

Fig. 12 Steady local wind pressures on signboard in group in smooth flow

CONCLUSIONS

The paper presents the steady pressures on structures in the boundary-layer wind tunnel laboratory. The data shown indicate significant increases of the pressure coefficients under conditions in the presence of a nearby structures at various relative location. The results show differences in smooth and turbulent flow for a single structure and situated in group. At present, for building code purposes, this problem could be treated by providing a warning of possible adverse situations and recommending to perform special boundary wind tunnel experiment tests.[1]-[8]

ACKNOWLEDGEMENTS

This study was supported by the University of Žilina and the Ministry of Education of the Slovak Republic, VEGA Grant Number 1/4142/97 Atmospheric turbulent wind on buildings and structures in groups.

REFERENCES

[1] Feranec V.-Feranec T.: (1994)Proximity Effects on Local Wind Pressures of Buildings and Structures. Proceedings of the EECWE 94, 4-8 July 1994, Warsaw-Poland.

[2] Feranec T. (1995) Wind effects on buildings and track-side structures in groups. Proceedings of the RILEM Conference. TU Košice 1995.

[3] Feranec V.-Feranec T.: (1997) Wind local pressures on buildings and structures situated in groups. Proceedings of the 2nd European & African Conference on Wind Engineering. 2 EACWE 'Palazzo Ducale, Genova, Italy June 22-26, 1997.

[4] Stathopoulos T.: (1984) Adverse wind loads on low buildings due to buffeting. Journal of Str. Eng., Vol. 110, No. 10, October 1984, pp. 2374-2392

[5] Feranec V.: (1997-1999) R. W. VEGA Grant No. 1/4142/97.Atmospheric turbulent wind on buildings and structures in group. University of Žilina . In Slovak.

[6] Larsen A.: Editor, Aerodynamcs of Large Bridges. A. A. Balkema Roterdam 1992.

[7] Tanaka H.: Wind Engineering. DTH, Lyngby, 1992.

[8] Feranec V.-Feranec T.: (1998)Local wind Pressures on Arched-Roof Buildings and Structures in Groups. Proc. of the 2nd East European Conf. on Wind Engineering 7-11 Sept. 1998. Prague, pp. 227-232.

Wind Engineering into the 21st Century, Larsen, Larose & Livesey (eds) © *1999 Balkema, Rotterdam, ISBN 90 5809 059 0*

Wind-induced peak bending moments in low-rise building frames

M. Gioffrè – *Department of Water and Structures Engineering, University of Perugia, Italy*
M. Grigoriu – *School of Civil and Environmental Engineering, Cornell University, Ithaca, N.Y., USA*
M. Kasperski – *Aerodynamik im Bauwesen, Ruhr-Universität Bochum, Germany*
E. Simiu – *Building and Fire Research Laboratory, National Institute of Standards and Technology, Gaithersburgh, Md., USA*

ABSTRACT: We present an investigation into the calculation of peaks of non-Gaussian processes, representative of fluctuating internal forces induced by wind in low-rise building frames. This is aimed at developing a simple procedure which can be easily included in computer programs that use surface pressure databases to obtain peak internal forces. We briefly review the theory applicable to Gaussian processes, and its extension to non-Gaussian processes using translation models. A translation model is calibrated to a record of a one-hour time history of the bending moment at the upwind bent of a low-rise frame, obtained from wind tunnel tests conducted at the University of Bochum. The translation model is then used to obtain estimates of the distribution of the peak bending moment, which is compared to estimates inherent in a Gaussian model. In light of our results the outline of a simple and practical approach for the approximate estimation of non-Gaussian peaks is proposed.

1 INTRODUCTION

Designs based on the occurrence of plastic hinges are increasingly used in engineering practice for low-rise building frames, especially in Germany (Kasperski & Koss 1996). Nevertheless, in the United States and other countries such frames are still predominantly designed by using the allowable stress method or the design strength method. In both methods the calculations require knowledge of the internal forces (i.e., the peak moments, shear forces and axial forces) induced in the elastic frames.

The wind-induced internal forces are stochastic processes with duration τ. For example, in calculating peak along-wind effects on tall buildings according to ASCE Standard 7-95 (American Society of Civil Engineering 1995) τ is assumed to be one hour. The peak effect over the duration τ is a random variable, Y_τ, whose value varies across the ensemble of realizations of the process. If the stochastic process is Gaussian - an assumption that is used for calculating peak along-wind effects on tall buildings - then the expectation of the peak is

$$E[Y_\tau] = \sqrt{2\ln(\nu_0\tau)} + \frac{0.577}{\sqrt{2\ln(\nu_0\tau)}} \quad (1)$$

(Davenport 1964), where ν_0 is the mean zero-upcrossing rate of the process. However, time histories of internal forces in frames are generally not Gaussian, and closed form expressions for their peaks, similar to Equation (1), are not available. An example of such a time history is shown in Figure 1. It represents, in arbitrary non-dimensional units, a one-hour time history of the bending moment at the upwind bent of the two-hinges, center bay frame of a building located in open terrain, with dimensions 11 m (height), 27.5 m (span), and 45 m (length), and a 5 degree pitched roof. The time history was obtained from wind tunnel tests conducted at the Ruhr-University Bochum - see Kasperski et al. (1996) for details.

The purpose of this report is to present a first phase of an investigation into the calculation of peaks of non-Gaussian processes as applied to wind-induced internal forces in low-rise building frames. One of the aims of this investigation is to develop a simple procedure for use in computer programs that use surface pressure databases to obtain peak internal forces (Whalen et al. 1998). Such programs are developed to serve as an alternative procedure for estimating wind loads under the draft provisions of the ASCE Standard 7-98.

We start out by reviewing briefly the theory applicable to Gaussian processes, and its extension for non-Gaussian processes. We apply the theory to and obtain relevant results from the analysis of the time history shown in Figure 1. In light of those results we propose the outline of a simple, practical approach for the approximate estimate of non-Gaussian peaks that can be incorporated

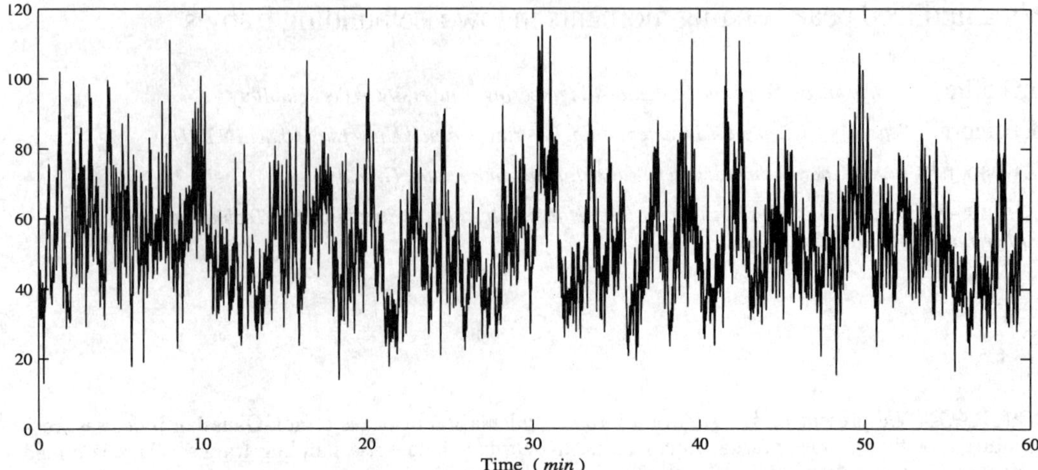

Figure 1. Upwind bending moment record.

in computer programs for calculating peak internal forces in frames.

2 THEORY

As pointed out earlier, the results of our work can be used for the development of standard provisions for wind loads. In view of the need for detailed scrutiny by standard writing groups of the basis for such development, we review briefly in this paper the theory underlying our work.

Consider a stationary stochastic process $X(t)$ and let $\nu(x)$ be the average number of crossing with positive slope of level x of $X(t)$ during the unit time, or the mean x-upcrossing rate. If x is large, it can be assumed that the x-upcrossing of $X(t)$ follows a homogeneous Poisson process with intensity $\nu(x)$. Hence, the probability that the maximum value of $X(t)$ in τ, X_τ, does not exceed x is

$$P(X_\tau \leq x) \approx \exp[-\nu(x)\,\tau] \tag{2}$$

If $X(t)$ is a stationary Gaussian process with mean μ_X, standard deviation σ_X, and standard deviation of $\dot{X}(t) = dX(t)/dt$, $\sigma_{\dot{X}}$, than the mean x-upcrossing rate of $X(t)$ is

$$\nu(x) = \frac{1}{2\pi}\frac{\sigma_{\dot{X}}}{\sigma_X}\exp\left[-\frac{1}{2}\left(\frac{x-\mu_X}{\sigma_X}\right)^2\right] \tag{3}$$

If $X(t)$ is a stationary non-Gaussian process defined by

$$X(t) = g\left[Y(t)\right] = F_X^{-1}\left[\Phi_Y(Y(t))\right] \tag{4}$$

then it is called a translation process (Grigoriu 1995), where $F_X(x)$ is the marginal distribution of $X(t)$ and $\Phi_Y(Y(t))$ is the marginal distribution of the stationary Gaussian process $Y(t)$ with zero mean, unit variance and covariance function $\rho(\tau) = E[Y(t+\tau)Y(\tau)]$. In this case the mean x-upcrossing rate of $X(t)$ is

$$\nu(x) = \frac{\sigma_{\dot{Y}}}{2\pi}\exp\left[-\frac{1}{2}\left(g^{-1}(x)\right)^2\right] \tag{5}$$

in which $\sigma^2_{\dot{Y}}$ is the variance of $\dot{Y}(t) = dY(t)/dt$, and can be calculated from $\sigma_{\dot{Y}} = \eta\,\sigma_{\dot{\tilde{X}}}$. The coefficient η is dependent on the shape of the marginal distribution $F(x)$ and is given by (Grigoriu 1984)

$$\eta = \frac{\sigma_{\dot{Y}}}{\sigma_{\dot{\tilde{X}}}} = \left[\sqrt{\int_{-\infty}^{\infty}\frac{\phi_Y^3(\zeta)}{\{f_{\tilde{X}}[g(\zeta)]\}^2}\,d\zeta}\,\right]^{-1} \tag{6}$$

in which $\tilde{X}(t) = (X(t)-\mu_X)/\sigma_X$ is the standardized non-Gaussian process.

The probability of Equation (2) can be used with $\nu(x)$ given by Equation (3) or Equation (5) to calculate the expectation of the peak during a time interval τ, for both Gaussian and non-Gaussian models. For a Gaussian model Equation (1) gives the expectation. If $X(t)$ is a translation model the expectation of the peak X_τ is

$$E[X_\tau] = \int_{-\infty}^{\infty} x\,dP(X_\tau \leq x)$$

$$\approx \int_{-\infty}^{\infty} x\,\tau\,g^{-1}(x)\frac{f_X(x)}{\phi_Y\left[g^{-1}(x)\right]}\,\nu(x)$$

$$\exp\left[-\nu(x)\,\tau\right]\,dx \tag{7}$$

Table 1. Mean, standard deviation, skewness and kurtosis coefficients.

	μ_X	σ_X	γ_3	γ_4
Estimated	53.1158	15.0458	0.5663	3.3137
$\tilde{X}(t)$	0	1	0.5663	3.3137
Gaussian	0	1	0	3
Gamma	0	1	0.5663	3.4811

where $f_X(x)$ and $\phi_Y(y)$ are the marginal density functions of $X(t)$ and $Y(t)$, respectively. The variance of X_τ is calculated in a similar way

$$\mathrm{Var}[X_\tau] = \int_{-\infty}^{\infty} (x - E[X_\tau])^2 \, dP(X_\tau \le x)$$

$$\approx \int_{-\infty}^{\infty} (x - E[X_\tau])^2 \, \tau \, g^{-1}(x)$$

$$\frac{f_X(x)}{\phi_Y\left[g^{-1}(x)\right]} \, \nu(x) \, \exp\left[-\nu(x)\,\tau\right] \, dx \quad (8)$$

3 ANALYSIS AND RESULTS

3.1 Model calibration

The estimated first four moments (mean, standard deviation, skewness and kurtosis coefficients) of the time series in Figure 1 are reported in the first row of Table 1.

Figures 2 and 3 show the estimated marginal histogram and the spectral density of the bending moment series respectively.

Two models are calibrated on the basis of the normalized time series $\tilde{X}(t) = (X(t) - \mu_X)/\sigma_X$, the Gaussian model and the translation model based on the Gamma distribution function. The Gaussian model is calibrated with the moments in the third row of Table 1 and with the power spectra of Figure 3. The Gamma translation model has marginal density given by

$$f_X(x) = \frac{(x-\gamma)^{\alpha-1} \exp\left[-\frac{x-\gamma}{\beta}\right]}{\beta^\alpha \Gamma(\alpha)}$$

$$(x > \gamma, \ \alpha > 0, \ \beta > 0) \quad (9)$$

where $\alpha = 12.4714$, $\beta = 0.2832$ and $\gamma = -3.5315$ are the parameters estimated by an optimization procedure on the basis of the second row of Table 1. The resulting moments assumed for the Gamma model are listed in the fourth row of Table 1. The spectral density of Figure 3 is also assumed to be the spectrum of the Gamma model.

The Gaussian and Gamma marginal density functions are compared with the histogram of the time series $\tilde{X}(t)$ in Figure 2. The fit of the Gamma density to the data is judged by inspection to be satisfactory.

The standard deviation of the process derivative is calculated for both the Gaussian and the Gamma model using the relation

$$E[\dot{X}(t)^2] = \int_{-\infty}^{\infty} (2\pi f)^2 S_{XX}(f) \, df \quad (10)$$

where $S_{XX}(f)$ is the spectral density of the process considered. The parameter relating $\sigma_{\dot{Y}}$ and $\sigma_{\dot{X}}$ results to be $\eta = 0.9913$.

3.2 Results

Once the models are calibrated the derivation of results is straightforward. Figure 4 shows the obtained expectation and the standard deviation of the peak as functions of the time interval τ for the two models considered.

Figure 5 shows, in addition to the Gaussian and Gamma models of the time series, the corresponding probability density functions of the peak values for calculated for time interval τ corresponding to 10, 20, 30, 45 and 60 minutes. The dotted lines are used for the densities of the Gauusian model, the solid lines are used for the Gamma model.

Table 2 shows, for $\tau = 20$ min and $\tau = 60$ min, the peaks corresponding to various probability of exceedance, quantiles (see Fig. 6), as well as the ratios between the peaks based on the Gamma marginal distribution and those based on the assumption that the marginal distribution is normal.

4 CONCLUSIONS

It is seen from Figures 4 and 5 and Table 2 that estimates of peaks based on the assumption that the time series of the bending moments is Gaussian would be unsafe. Using the largest value obtained in a single wind tunnel test might also be unsafe. This is seen in Figure 1, which shows that, for τ =60 min, the observed peak was about (115 - 53)/15.05 = 4.12 standard deviations larger than the mean, whereas the estimated expected value of the peak is 4.84 standard deviations larger than the mean if the Gamma model is assumed (Fig. 4). In the units of Figure 1, the peak would be 53 +15.05 x 4.84 = 126, that is, about 9.6% percent higher than the observed peak of the recorded time series. We note that our choice of the Gamma model for the marginal distribution of the bending moment is strictly empirical, i.e., it does not rest on fundamental theoretical considerations. A different model resulting in a comparable fit could be proposed. Preliminary estimates suggest that errors in the estimation of the peak associated with

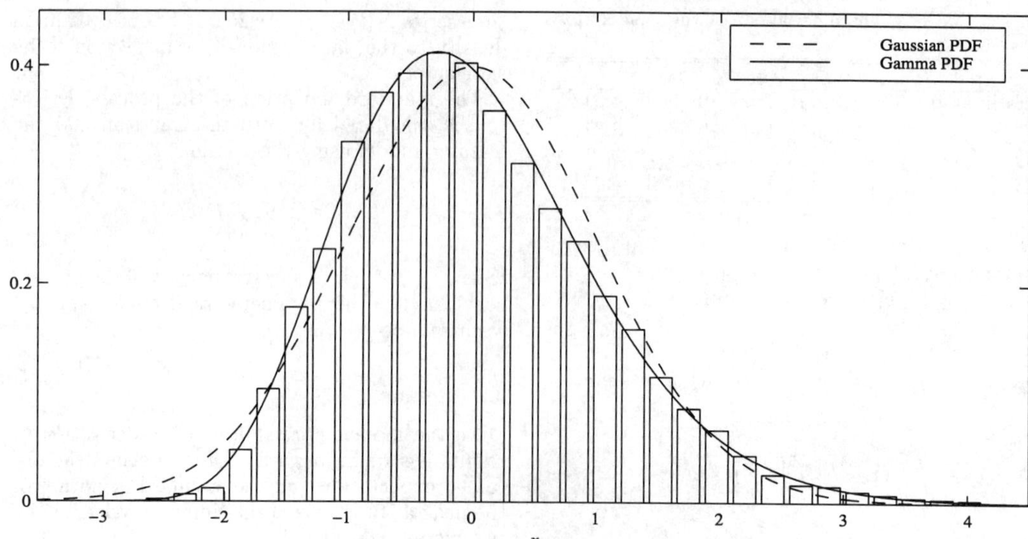

Figure 2. Marginal histogram of the bending moment data compared with Gaussian and Gamma probability density functions, PDF.

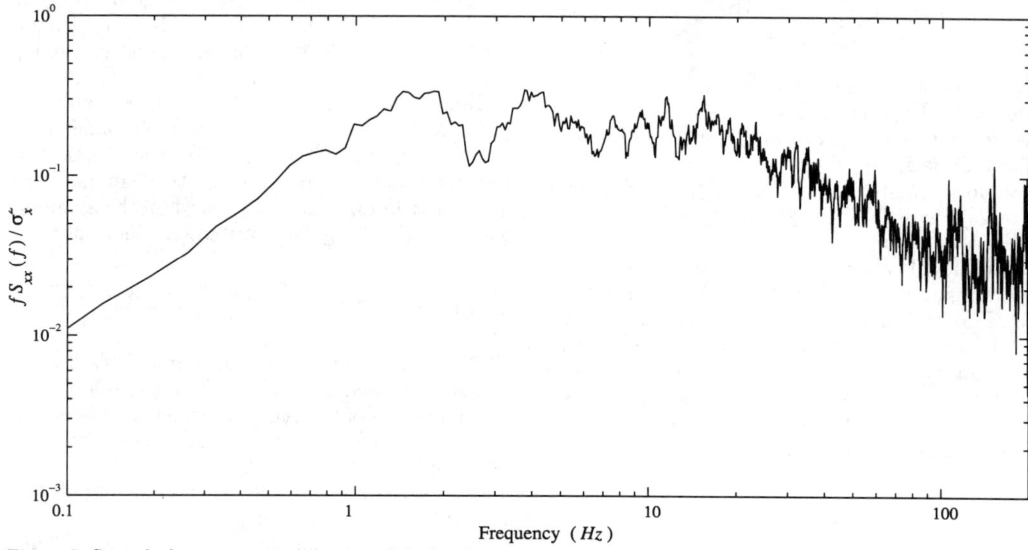

Figure 3. Smoothed power spectral density of the bending moment data.

Table 2. Quantiles $a(\alpha)$ of Gaussian and Gamma models.

	Gaussian model		Gamma model		Ratio	
α	τ=20 min	τ=60 min	τ=20 min	τ=60 min	τ=20 min	τ=60 min
0.500	3.2769	3.5966	4.2092	4.7527	1.28	1.32
0.250	3.5351	3.8333	4.6464	5.1700	1.31	1.35
0.100	3.8087	4.0869	5.1259	5.6312	1.35	1.38
0.050	3.9932	4.2594	5.4590	5.9533	1.37	1.40
0.020	4.2201	4.4728	5.8792	6.3612	1.39	1.42
0.010	4.3824	4.6263	6.1870	6.6611	1.41	1.44
0.001	4.8805	5.1007	7.1700	7.6228	1.47	1.49

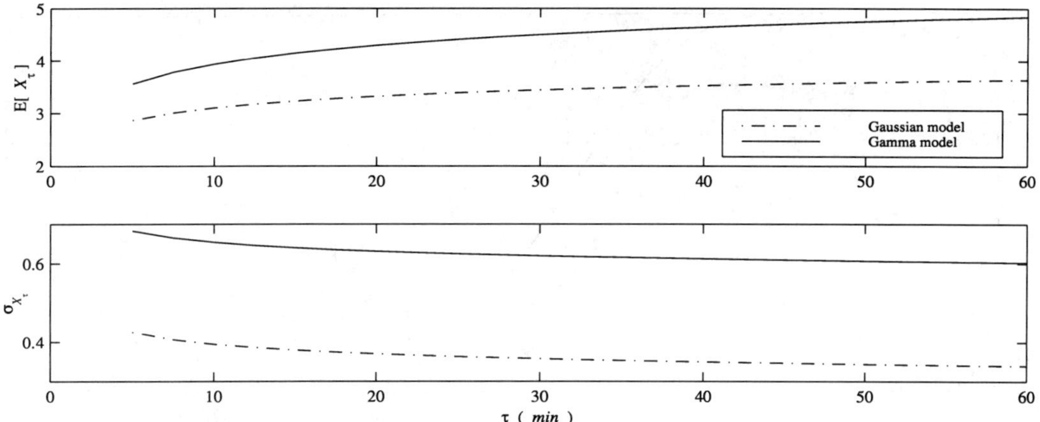

Figure 4. Expected value and standard deviation for peak bending moment.

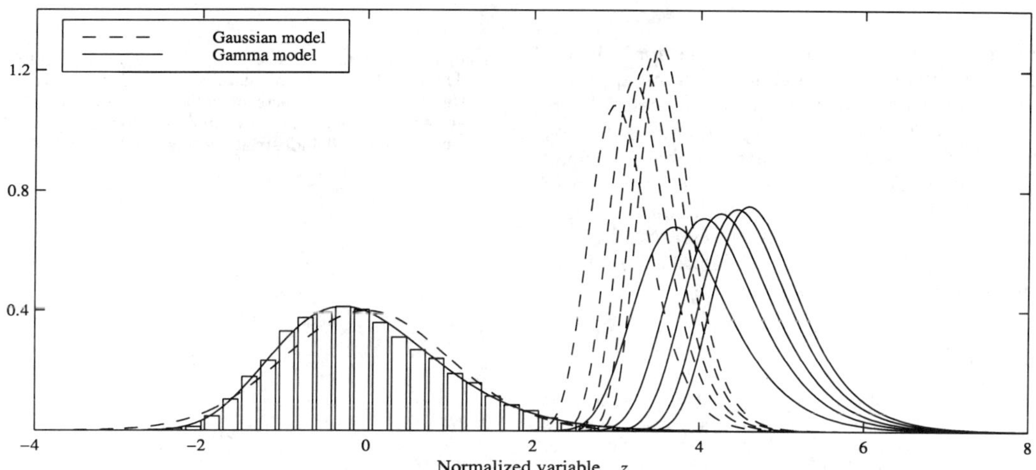

Figure 5. Density functions of the parent distributions and the peak bending moment, for $\tau = 10, 20, 30, 45$ and 60 min.

probabilistic modeling (i.e., with the choice of the probabilistic model for the marginal distribution of the process) are relatively small, except for large percentage points that are of no interest in design practice. This issue will be addressed systematically in a future phase of our work.

An additional remark is that, as seen in Figures 4 and 5, the variability of the peaks is considerably larger for the non-Gaussian than for the Gaussian peaks. For along-wind response of tall buildings, it is standard practice to use for design purposes the expectation of the peaks given by Equation (1). The justification for not selecting a design value larger than the expectation for Gaussian models is that the coefficient of variation (c.o.v.) of the peaks is small (less than 10 percent for τ =60 min - see Figure 4). Therefore, it is permissible to neglect the contribution of the peak's dispersion about the mean to the wind effect's total dispersion, which includes contributions of dispersions of other relevant parameters, e.g. the wind speed, the terrain roughness, and so forth. For some weakly non-Gaussian processes neglecting the dispersion of the peaks in reliability estimates may still be acceptable, but in general this may not be the case.

Based on the procedure illustrated in this paper, calculations may be carried out for non-Gaussian time histories corresponding to a sufficiently large number of sets of standard deviations and coefficients of skewness and kurtosis. Such calculations would yield expectations, standard deviations, and quantiles of the peaks corresponding to each of those sets. This information can be

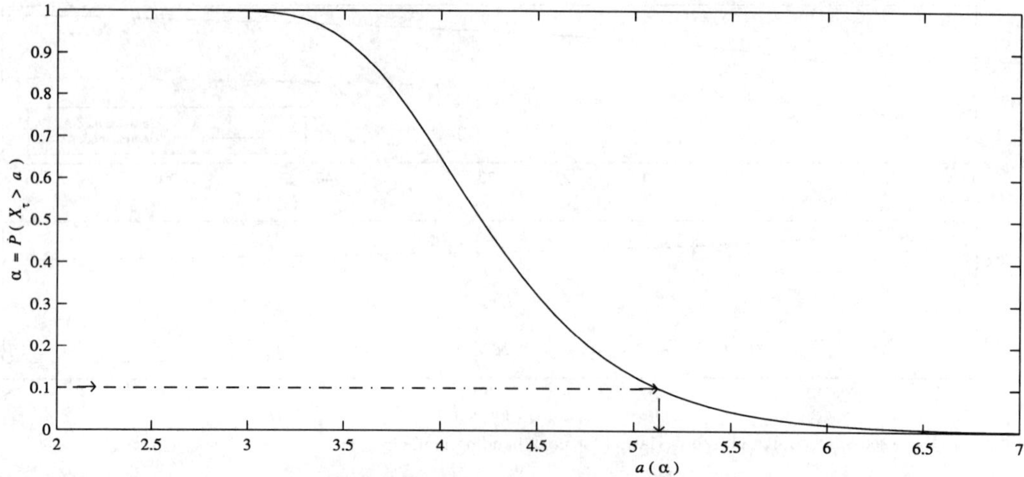

Figure 6. Peak corresponding to a fixed probability of exceedance, quantile, for $\tau = 20$ min.

incorporated in computer programs for calculating wind-induced internal forces from appropriate aerodynamic time history databases. To obtain the peak statistics corresponding to any given time history, subroutines need to be developed that would entail the use of that information in conjunction with calculations of the first four moments of the peaks. Our future research plans include the development of that information and of the requisite computer subroutines.

REFERENCES

American Society of Civil Engineering (1995). *ASCE Standard - ASCE 7-95: Minimum design loads for buildings and other structures.* New York: American Society of Civil Engineering.

Davenport, A. G. (1964). Note on the distribution of the largest value of a random function with application to gust loading. Proceedings, Institution of Civil Engineering 28, 187–196.

Grigoriu, M. (1984). Crossings of non-gaussian translation processes. Journal of Engineering Mechanics 110(4), 610–620.

Grigoriu, M. (1995). *Applied non-Gaussian Processes: Examples, Theory, Simulation, Linear Random Vibration and MATLAB Solutions.* New York:Prentice–Hall: Englewood Cliffs.

Kasperski, M., H. Koss, & J. Sahlmen (1996). Beatrice joint project: wind action on low-rise building. part 1. basic information and first result. Journal of Wind Engineering and Industrial Aerodynamics 64, 101–125.

Kasperski, M. & H. H. Koss (1996, June). Non-linear dynamic analysis of portal frames under wind action. In G. Augusti, C. Borri, & P. Spinelli (eds.), *Proceedings, Structural Dynamics - Eurodyn'96*, Florence, Italy, pp. 275–281. Bakkema, Rotterdam.

Whalen, T., E. Simiu, G. Harris, J. Lin, & D. Surry (1998). The use of aerodynamic databases for the effective estimation of wind effects in main wind-force resisting systems: application to low buildings. Journal of Wind Engineering and Industrial Aerodynamics 77-78, 685–693.

Wind Engineering into the 21st Century, Larsen, Larose & Livesey (eds) © *1999 Balkema, Rotterdam, ISBN 90 5809 059 0*

Area-averaged pressure fluctuations on surfaces at roof corners and gable peaks

N. Hosoya & J. E. Cermak
Fluids Dynamics and Diffusion Laboratory, Department of Civil Engineering, Colorado State University, Fort Collins, Colo., USA

S. M. Dodge
PRC Litton, Camp Springs, Md., USA

ABSTRACT: The paper presents findings of an experimental investigation designed to determine the dependence of area-averaged peak maximum, peak minimum, rms and mean wind pressures on area size and location for surfaces forming roof corners of a flat-roof building and ridge-end areas of a gable roof on a low-rise rectangular building. Data were acquired with the objective of providing information that would be useful in design of roof coverings for these critical areas which are covered by vortices or separation bubbles for certain wind directions.

1 INTRODUCTION

In the United States, public awareness of wind-related damage to low residential structures has clearly increased since the passage of Hurricane Andrew near Miami, Florida in 1992. This hurricane not only caused devastating damage to the local housing structures but also impacted the insurance economy severely. Some of the causes for the structural damages, which are typically initiated by failure of the roof system (Sill and Kozlowsky, 1992), can be directly attributed to the facts that residential structures are often designed and constructed with a limited engineering basis. Wind load effects on small structures and their cladding, particularly near the roof corners, are not well understood. This is because of limited availability of comprehensive field data in addition to difficulty in modeling natural winds in wind tunnels for small structures due to space limitations which prevents adequate growth of turbulence structures (Cermak, 1984). Considerable discrepancies in peak suction pressures exist between field and wind-tunnel observations near the roof corner of a flat roof building as reported by Cochran (1992), and Okada and Ha (1992). For a gable-roof building, on the other hand, good agreement for peak suction was found by Peterka et al. (1998) for a full-scale field installation and a 1:25 scale model despite a notable mismatch in spectral characteristics of the approach wind.

The discrepancies in peak pressure coefficients found on a flat-roof corner have been based on a comparison of point pressures. However, for practical engineering applications, minimum areas of concern would be in the order of 0.1 m^2. This paper presents findings of an investigation on effective pressures (instantaneously area-averaged pressures) over various area sizes near the ridge-end region of a gable roof. The investigation was conducted in a coordinated field and wind-tunnel study on a simple gable roof building with a 5:12 slope. Correlation of the area-averaged pressure fluctuations was of particular interest. Also included are results for roof-corner areas using wind-tunnel data acquired for a flat-roof building by the first author in his doctoral research program which is in progress. This paper is intended to provide information that will be useful in development of design criteria for roof coverings.

2 EXPERIMENTAL CONFIGURATIONS

2.1 *Gable-Roof Building – Field and Wind-Tunnel Tests*

The field test facility of a gable-roof building was designed and is maintained by CPP, Inc. and is located at the eastern foothills of the Rocky Mountains 30 km north of Fort Collins, Colorado. For the prevailing wind directions (W and WNW), the test site is characterized by a typical open-country exposure. Figure 1 illustrates the size and geometry of the rotatable field building. A grid of 48 pressure taps was installed on an area, 0.9 m × 0.9 m, bounded by the ridge line and the roof edge, as shown in the figure. The selected area size corresponded to the minimum design area specified

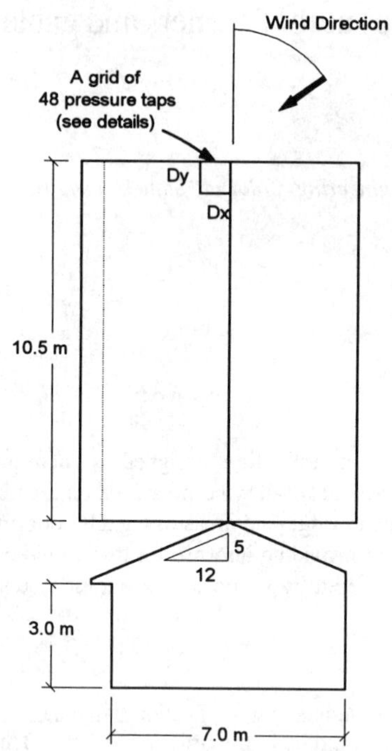

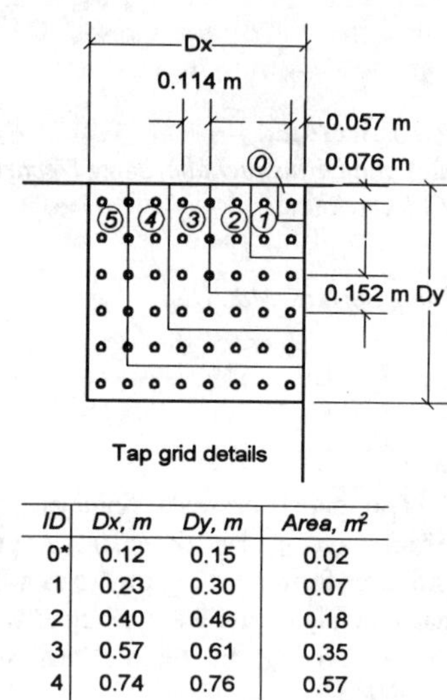

ID	Dx, m	Dy, m	Area, m^2
0*	0.12	0.15	0.02
1	0.23	0.30	0.07
2	0.40	0.46	0.18
3	0.57	0.61	0.35
4	0.74	0.76	0.57
5	0.91	0.91	0.84

* Point pressure near the corner of roof.

Individual tap diameter : 6.35 mm

Figure 1. Gable-roof building and pressure-tap arrangement for the field installation.

by the current wind-load standard, ASCE 7-95. Pressure data were measured simultaneously for all the pressure taps so that effective pressures (area-averaged pressures) over various areas, defined as Areas 0 through 5 in Figure 1, could be obtained. The test facility was fully automated, and a desktop computer inside the building controlled data acquisition from pressure and four wind-velocity sensors mounted on a 10-m tower as well as the orientation of the building relative to approach winds. The system was accessible via a modem for monitoring the test status and downloading data. The tests were setup to obtain pressure data at mean wind speed measured at the eave height exceeding 8.9 m/s (20 mph) for a range of wind angles nominally from 0 to 90 degrees. For each test, the data were sampled at a rate of 30 samples per second for 10 minutes. The field test started in early February 1997, and by the end of the same month, 193 sets of data were obtained. Of the 193 runs, 128 runs were identified valid based on a stationarity test performed on wind speeds and directions at the 0.05 level of significance using the procedure described by Bendat and Piersol (1971).

Pressures on a replica of the field building at a scale of 1:50 were subsequently measured by Dodge (1998) in the Industrial Wind Tunnel at Colorado State University. An atmospheric boundary layer representative of the CPP, Inc. field test site was modeled in the wind tunnel for comparison and validation of the wind-tunnel test. In addition to a traditional boundary-layer simulation technique, a specially designed passive turbulence generator was used to promote a larger scale turbulent flow than otherwise would be obtained. The wind-tunnel data were acquired simultaneously for the 48 taps at a rate of 577 samples per second for 14 seconds, corresponding to 11.5 Hz and 12 minutes at the field test based on the length and nominal wind-speed scales. The tests were conducted for a range of wind directions from 0 to 90 degrees (5° increments). In order to minimize statistical uncertainties, the measurements were repeated 8 times at each wind direction.

2.2 *Flat-Roof Building – Wind-Tunnel Test*

The pressure data for a flat-roof building presented in this paper were obtained by measurements in a wind tunnel at CPP, Inc. The test model was made to represent the flat-roof building at the Wind Engineering Research Field Laboratory of Texas Tech University (TTU) at a scale of 1:50. Figure 2 shows an expanded view of the TTU building model equipped with a total of 239 pressure taps for simultaneous measurement of pressures. The test data from only 9 taps covering the roof corner area as indicated in the figure were used for the present analysis to coincide with the area definition made for the gable-roof building. The approach boundary-layer flow was adjusted to simulate the wind environment at the TTU field site which had characteristics similar to those for the gable building roof tests except for the turbulence length scale which was somewhat smaller for the assumed model scale. The pressure data were sampled for all 239 taps at a rate of 500 Hz for 16 seconds. The measurements were repeated 8 times.

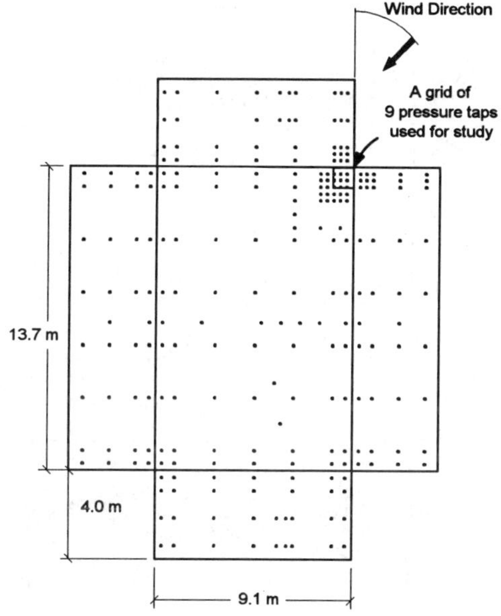

Figure 2. Flat-roof building model — full-scale dimensions of 1:50 scale model of TTU building.

3 RESULTS

3.1 *Peak Area-Averaged Pressures*

Area-averaged pressures over different areas near the ridge end of the gable roof are compared in Figure 3 for the field and wind-tunnel data. The area-averaged pressures were obtained by numerically combining individual point pressures in time series within the areas of interest. The pressure data were then normalized into pressure coefficients using a mean dynamic pressure measured at the eave height of 3 meters (full scale). The figure shows mean, rms, maximum and minimum pressure coefficients as a function of wind direction. For the wind-tunnel data, the coefficients plotted were obtained by averaging the results of repeated measurements. The wind-tunnel results agree reasonably well with the field results especially for areas of 0.18 m^2 and larger sizes, Areas 2 - 5 as defined in Figure 1. For smaller areas, the wind-tunnel data slightly underestimate the peak negative pressures. The most severe peak negative pressure occurs at a wind direction of about 50 degrees, where wind approaches the roof edge at an oblique angle. The peak pressure decreases (becomes smaller in magnitude) with increasing area size — an expected consequence of area averaging. At the critical wind direction, the presence of vortices originating from the roof edge was noted by Peterka et al. (1998) although not as continuous or well formed as over a flat-roof corner.

3.2 *Correlation of Area-Averaged Pressures*

Figure 4 shows examples of time series of point pressures measured at the extreme corners of the largest area examined, Area 5, at the wind direction where magnitude of the peak negative pressure is greatest. The wind-tunnel time was scaled to correspond to the equivalent full-scale value. It is apparent that the time traces of these pressures have similar features for the field and wind-tunnel data for the gable roof. It is interesting to note that for certain durations, individual point pressures exhibit nearly identical patterns. For example, this occurs between 20 and 30 seconds for the field data and between 70 to 90 scaled seconds for the model data. It appears that the influence of ridge-end vortices is intermittent for a gable roof, and the entire ridge-end area is occasionally engulfed by a steady separation bubble lasting for several seconds. This feature is not seen for the flat-roof pressures shown in Figure 4, and the influence of vortices is continuously evident as represented by a series of sharp negative peaks. The collapse of the point pressures for the gable roof suggests a potentially high correlation of pressures over a significant area near the ridge end.

To examine the spatial correlation of pressures further, Area 5 (0.84 m^2) was divided into two parts, and effective (area-averaged) pressures for each part were obtained to estimate the statistical cross-correlation. Definitions of the divided areas are shown in Figure 5. Figure 6 shows the calculated cross-correlation for various wind directions. Correlations for the gable roof are very high (> 0.85) for all wind directions, regardless of the way in which the area is divided. For the flat roof, however, the pattern of the correlation differs significantly depending on the selection of the tributary areas. When the area is divided to include either side of the flat-roof edges, Areas 6 and 7, the correlation varies dramatically with the wind direction. At oblique

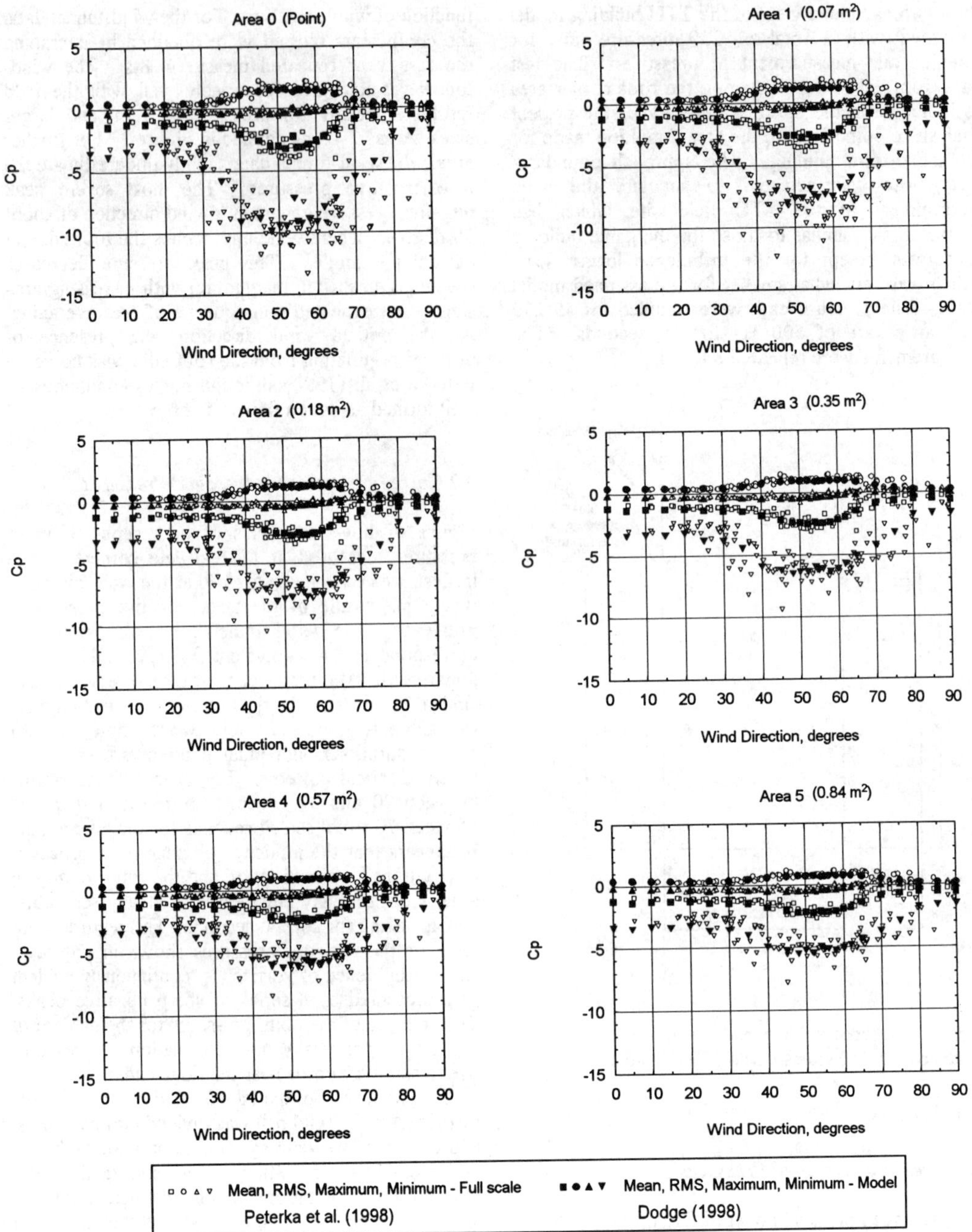

Figure 3. Area-averaged pressure coefficients at ridge end of a gable-roof building.

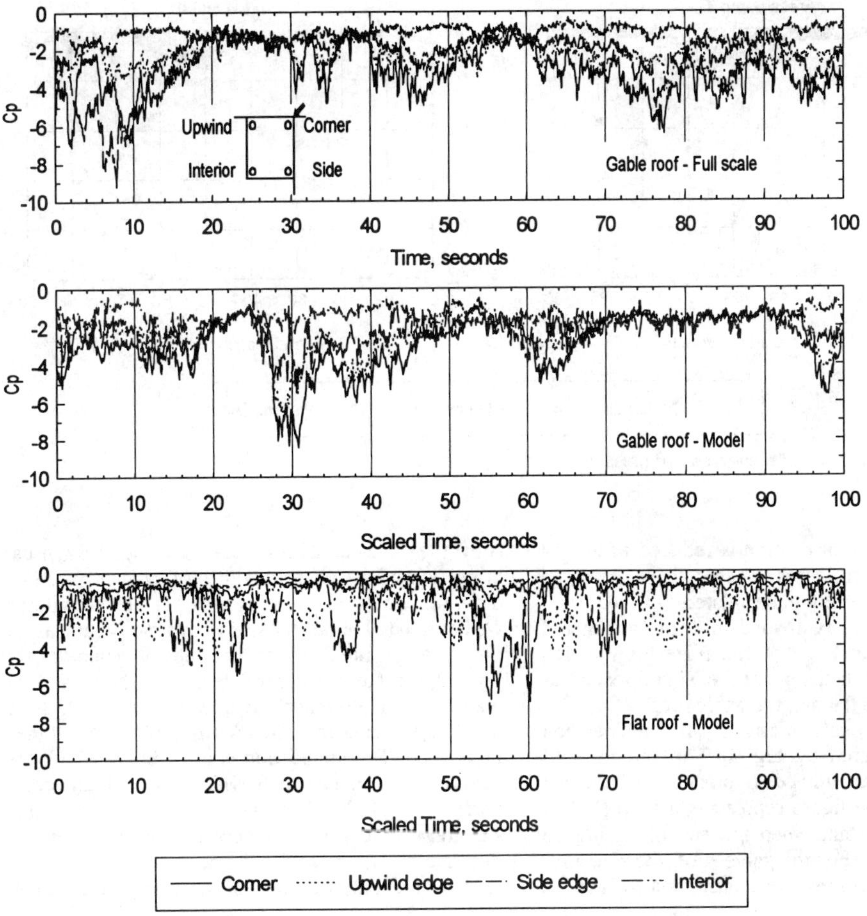

Figure 4. Time series of point pressure coefficients at corners of Area 5 for 50° wind direction.

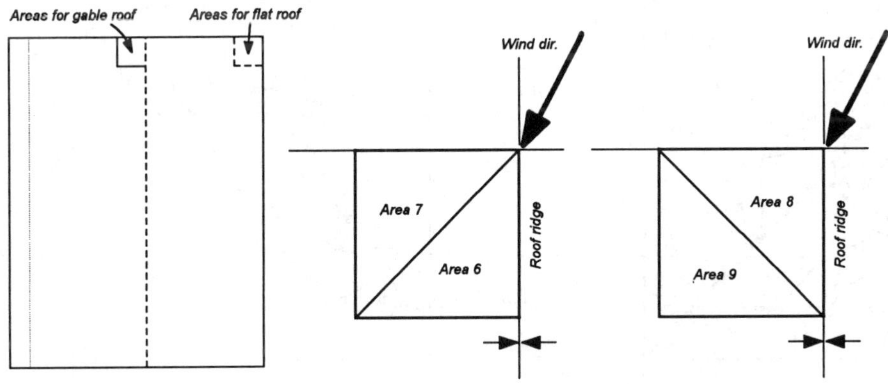

Figure 5. Definition of areas for correlation analysis.

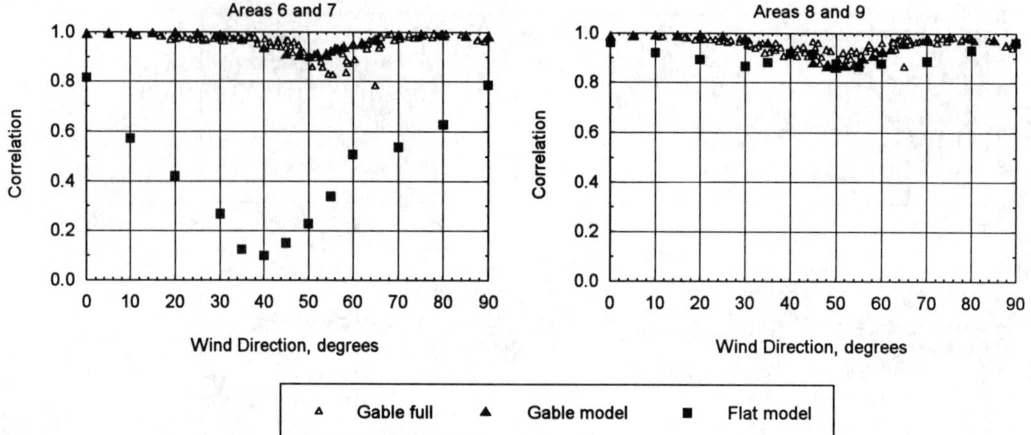

Figure 6. Cross-correlation of area-averaged pressures.

wind directions, the correlation is reduced as low as 0.1, for which the pressures between those areas can be considered practically uncorrelated. On the other hand, when the area is divided into the corner and interior parts, Areas 8 and 9, the correlation is high and insensitive to the change of the wind direction as is seen for both the flat and the gable roof.

A graphical explanation of the correlation mechanism is offered by Figure 7 in the form of contours of *instantaneous* pressure coefficient distributions. The figure represents a snap shot of pressures at an instant when the suction acting on Area 5 is greatest. For the gable roof, the pressures measured in the field test were used, and for the flat roof, the wind-tunnel test data were used. By visual inspection of the contours, lines of symmetry can be drawn as indicated. The pressures can be considered well correlated about the line of symmetry (at least at the instant in time). A pair of lines of symmetry are present for the gable roof, bisecting the total area diagonally. Note that these lines coincide with those dividing the area into Sub-Areas 6, 7, 8, and 9 (see Figure 5). The implication is in agreement with the results given in Figure 6 which shows high cross-correlation of the pressures between the opposing sub-areas. A similar argument can be made to describe the correlation characteristics for the flat roof, where only one line of symmetry is apparent,

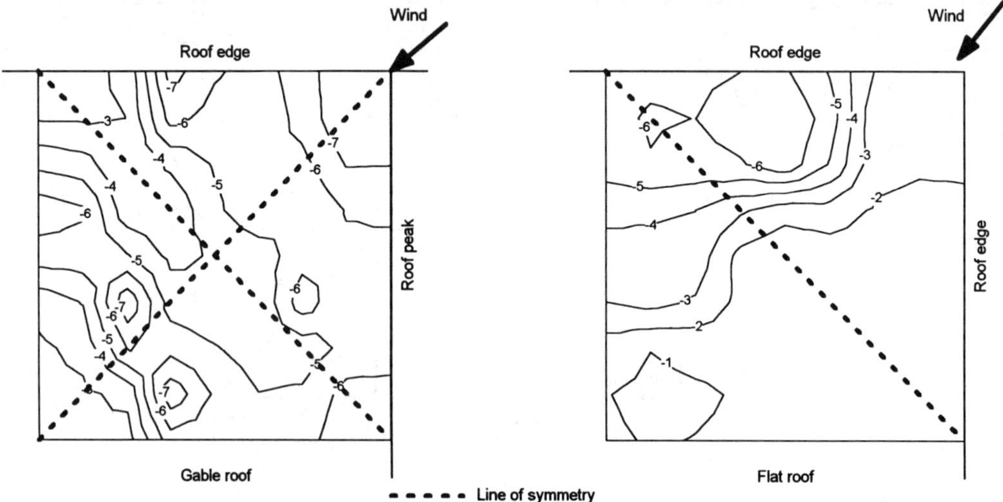

Figure 7. Contour of instantaneous pressure coefficients which minimize total area-averaged pressure.

suggesting a high correlation between Areas 8 and 9, but less for Areas 6 and 7. This observation is consistent with the finding reported by Kawai and Nishimura (1996) who performed an elaborate correlation analysis on a square flat-roof building.

4 CONCLUSIONS

Based on the full-scale field and wind-tunnel measurements of wind-induced pressures near ridge-ends of gable-roof and roof corners of flat-roof buildings, the following conclusions can be made:

1. The magnitude of the extreme minimum area-averaged pressures decreases with increasing area size for the ridge-end region of a gable-roof building.
2. The wind-tunnel test on a 1:50 scale model of a gable-roof building produced area-averaged mean, rms, peak maximum, and peak minimum pressures which are in good agreement with the full-scale field test results.
3. Spatial correlation of pressures near the ridge end of a gable roof is high, and relatively independent of angle of approach wind.
4. There are notable differences in fluctuating pressures under influence of conical vortices between the ridge-end area of the gable-roof and areas at flat-roof corners for the wind direction causing the extreme minimum area-averaged pressure. Over an area comparable to a nominal minimum roof-covering size, the effect of vortices is intermittent for the gable roof, and continuous for the flat roof.

5 ACKNOWLEDGEMENTS

This study was supported by National Science Foundation Grant CMS-9411147 and by CPP, Inc. of Fort Collins, Colorado. Formatting and processing of the paper was accomplished by Gloria Garza.

6 REFERENCES

American Society of Civil Engineers 1995. *Minimum Design Loads for Building and Other Structures.* ASCE 7-95, New York, NY.

Bendat, J.S & Piersol, A.G. 1971. *Random Data: Analysis and Measurement Procedures.* New York, NY: John Wiley & Sons.

Cermak, J.E. 1984. Wind-Simulation Criteria for Wind-Effect Tests. *Journal of Structural Division*, ASCE, 110(2):328-339.

Cochran, L.S. 1992. Wind Tunnel Modeling of Low-Rise Structures. PhD Dissertation, Fluid Dynamics and Diffusion Laboratory, Colorado State University, Fort Collins, CO.

Dodge, S.M. 1998. Validation and Modeling of Wind Pressures on a Gable Roof. MS Thesis, Fluid Dynamics and Diffusion Laboratory, Colorado State University, Fort Collins, CO.

Kawai, H. & Nishimura, G. 1996. Characteristics of Fluctuating Suction and Conical Vortices on a Flat Roof in Oblique Flow. *Journal of Wind Engineering and Industrial Aerodynamics*, 60:211- 225.

Okada, H. & Ha, Y. 1992. Comparison of Wind Tunnel and Full-Scale Pressure Measurement Taps on the Texas Tech Building. *Journal of Wind Engineering and Industrial Aerodynamics*, 41-44:1601-1612.

Peterka, J.A., Hosoya, N., Dodge, S., Cochran, L.S. & Cermak, J.E. 1998. Area Average Peak Pressures in a Gable Roof Vortex Region. *Journal of Wind Engineering and Industrial Aerodynamics*, 77-78:205-215.

Sill, B.L. & Kozlowski, R.T. 1997. Analysis of Storm-Damage Factors for Low-Rise Structures. *Journal of Performance of Constructed Facilities*, 11:168-177.

Wind Engineering into the 21st Century, Larsen, Larose & Livesey (eds) © 1999 Balkema, Rotterdam, ISBN 90 5809 059 0

Improvement of ventilation in the Honolulu Airport ticket lobbies

Anurag Jain & Mukund Srinivasan – *Hart Consultant Group, Santa Monica, Calif., USA*

Gary C. Hart – *Department of Civil and Environmental Engineering, University of California, Los Angeles, Calif., USA*

Richard W. Balcom – *CDS International, Honolulu, Hawaii, USA*

Jon D. Raggett – *West Wind Laboratory, Marina, Calif., USA*

ABSTRACT: This wind study project addresses the comfort requirements for the Ticket Lobbies in the Overseas Terminal (OST) of the Honolulu International Airport (HIA) arising due to the currently existing natural ventilation system. The location of the openings in the terminal combined with the seasonal behavior of wind flow patterns in the Hawaiian Islands leads to uncomfortable situations for passengers and tenants in the ticket lobbies. A set of planned improvements was combined with proposals to improve the natural ventilation in the ticket lobby to provide increased comfort to users of the facility. The project involved modeling of the ticket lobby in a boundary layer atmospheric wind tunnel followed by a statistical analysis of the measured wind flow rates inside the ticket lobbies and wind speed data recorded at the weather station at HIA.

1 INTRODUCTION

The aim of this study is to evaluate the impact to air circulation and cross ventilation of the proposed improvements and additions to the passenger ticket lobbies of the OST at HIA. The ventilation in the ticket lobby areas is exacerbated by the fact that the ventilation louvers along the southern wall of the OST have to be operated in a closed position to avoid dust, smoke and noise being blown through them. This lack of a consistent through passage for the wind to blow from one edge of the terminal to the other creates still spots and lack of air circulation during certain unfavorable conditions. To rectify this lack of adequate air circulation and cross ventilation, improvements are being proposed and their impact on the wind flow in the ticket lobby areas needs to be quantified.

Improvements that were evaluated include the addition of wind scoops on the roof of the OST over the ticketing lobbies and the proposed addition of pyramid-shaped canopies over the second level terraces between the departure drop off curb and ticketing lobbies to provide rain protection. The proposed addition of wind scoops will be in conjunction with strategically placed ceiling fans to supplement natural air flow. Two spatial locations for the placement of the wind scoops were evaluated. One set of wind tunnel tests were conducted with the wind scoops located along the south edge of the roof of the passenger ticketing lobbies and a second set of tests were conducted with the wind scoops located near the center of the roof. In both cases the openings of the wind scoops were oriented toward the south. Two different wind scoop designs were considered. For one set, one building bay wide wind scoops were placed at every other bay. For the other set continuous and larger sized wind scoops were placed on the roof. The impact of the pyramid-shaped canopies to air circulation and cross ventilation in conjunction with the addition of the wind scoops was also evaluated. Another set of testing was conducted to quantify the wind flow rates if the louvers were open.

2 WIND CLIMATOLOGY AT PROJECT SITE

HIA is located on the southern shore of the island of Oahu. Oahu, on which Honolulu is located, is the third largest of the Hawaiian Islands. The Koolau Range, at an average elevation of 2,000 feet parallels the northeastern coast. The Waianae Mountains, somewhat higher in elevation, are parallel to the west coast. HIA, the business and Waikiki districts, and a number of the residential areas of Honolulu lie along the southern coastal plain. The climate of Hawaii is unusually pleasant for the tropics. Its outstanding features are the persistence of the trade winds, the equable temperature, and the general infrequency of severe storms. The prevailing wind throughout the year is the northeasterly trade wind, although its average frequency varies from more than 90 percent during the summer to only 50 percent in January. Because of the trade winds, even the warmest months are usually comfortable. But when the trades diminish or give way to southerly winds, a situation known locally as kona weather, or kona storms, the humidity may become oppressively high. It is this kona weather in conjunction with the closed louvers on the south edge of the OST that create the uncomfortable conditions.

For this wind study, historical wind speed data was collected from Honolulu International Airport. The data was obtained from the National Climatic Data Center in Asheville, North Carolina (1998). Data from 1988 to 1997, a total of 10 years of hourly wind speeds, was used in the statistical analysis.

3 VENTILATION AND COMFORT

3.1 *Ventilation*

For buildings without mechanical cooling and dehumidification, proper ventilation and infiltration airflows are important for providing comfort for occupants. The ticket lobby level of the HIA OST building is naturally ventilated. Natural ventilation is driven by natural and/or artificially produced pressure differentials (ASHRAE 1997). Air exchange can significantly affect energy, air quality and thermal comfort, and can vary with weather conditions, building operations and use. Natural ventilation, usually through operable windows, is dependent on weather and building design but allow occupants to control airborne contaminants and interior air temperature. Natural ventilation can effectively control both temperature and contaminants, particularly in mild climates. Temperature control by natural ventilation is often the only means of providing cooling when mechanical air conditioning is not available.

The arrangement, location and control of ventilation openings should combine the driving forces of wind and temperature to achieve a desired ventilation rate and good distribution of ventilation air through the building. Natural ventilation openings include windows, doors, dormer openings, skylights and specially designed inlet or outlet openings.

Factors arising due to wind forces affect the ventilation rate and include average wind speed, prevailing wind direction, and local obstructions such as nearby buildings, hills, trees and shrubbery. Winds speeds and directional frequency are a function of season. Natural ventilation systems are often designed for wind speeds one-half of the seasonal average (ASHRAE 1997).

Inlets should face directly into prevailing wind. If the inlets are not advantageously placed, wind flow may be less than expected. Desirable outlet locations are on the leeward side of the building directly opposite the inlet, on the roof in low pressure areas created by a flow discontinuity in the wind, on the side adjacent to the windward face where low-pressure areas occur or in a dormer on the leeward side. The inlets should be placed in the exterior high pressure regions; the outlets should be placed in the exterior low-pressure regions.

ASHRAE Standard 62, Ventilation for Acceptable Interior Air Quality prescribes ventilation airflow rates for commercial and institutional buildings. For Transportation Waiting Rooms, Table 2.1 of ASHRAE Standard 62 (ASHRAE 1989) recommends and estimated maximum occupancy of 100 people per 1000 ft^2 of net occupiable space with an indoor air requirement for ventilation of 15 cfm/person. Indoor air quality shall be considered acceptable if this rate of acceptable output air mentioned above is provided for the occupied space.

3.2 *Comfort*

If ventilation and air circulation levels exceed a certain threshold then winds can become discomforting to people experiencing these wind flows. The problem of discomforting winds is a commonplace occurrence in many urban areas with tall buildings and other construction causing local winds to speed up. What constitutes a discomforting wind is very subjective. It depends upon the individual, the temperature, the state of the cloud cover, the state of the weather (rain, snow etc.). The discomfort levels therefore must be quantified specifically for the generally moderate climate of Hawaii.

Wind discomfort is a function of the mean wind speed and some measure of the gustiness of the wind. Discomforting winds are typically measured in terms of an effective wind speed, equal to the mean wind speed, plus some constant multiplied by the standard deviation of the gustiness. All effective wind speeds are defined in terms of a local mean wind speed and local turbulence intensity. The percentage of time that an effective local wind speed exceeds some specified value must be known to evaluate comfort. This can be determined in terms of the percentage of time that a reference wind speed is exceeded at a given location at a given elevation for wind from a specific direction. The effective wind speed for this study is computed using Equation 1

$$u_{eff} = u_{ref} * \mu \, [1 + (3 * TI)] / 1.45 \qquad (1)$$

where u_{eff} is the effective wind speed, u_{ref} is the reference wind speed measured at an elevation of 304.8 feet, μ is the local mean wind speed, and TI is the turbulence intensity. This formula is currently the standard used by the City of San Francisco Planning Department.

3.3 *Ventilation and Comfort Criteria for the Study*

For this study, three levels of flow criteria were used to judge the existing condition and effectiveness of the tested configuration. These criteria are as follows:

1. The first criterion used is the minimum wind speed in the ticket lobby based upon the American Society of Heating, Refrigerating and Air Conditioning Engineers (ASHRAE) criterion for the minimum acceptable ventilation. The required wind speed was computed by dividing the acceptable wind flow rate by an effective tributary area corresponding to each of the points used in the wind tunnel testing. This wind speed was computed to be 0.5 mph. The ASHRAE criterion therefore corresponds to a required exceedance of 100% of the time for wind speeds greater than 0.5 mph. As this may not be feasible, the four different configurations can be evaluated by their deviation from this ventilation criterion.
2. The second criterion is the limiting wind flow inside the lobby that has been set to 300 feet/minute based on conditions of temperature and humidity at HIA. This criterion is a flow/ventilation criterion and is derived from Olgyay (1963).
3. The third criterion is the upper limit for the wind speed inside the ticket lobby based upon Gandemer and Guyot (1981). Hence for the occupant comfort, the wind speeds inside the lobby should not exceed 10 mph more than approximately 15% of the time. This criterion represents an upper limit and is defined as a comfort criterion.

These criteria can be effectively used to compare wind flow rates after mitigation measures are set up.

4 WIND TUNNEL TESTING

4.1 *Wind Tunnel Instrumentation & Description*

A 1:152.4 scale model (2 mm = 1 ft) was made of the area of the OST shown in Figure 1. The aerodynamic effects of the buildings beyond this area was modeled generically with wooden blocks and spires placed upstream of the test model in the wind tunnel. The model was made of plywood and lightweight mat board. This scale model of the OST was tested at the wind tunnel at the West Wind Laboratory. The wind tunnel at the West Wind Laboratory is a 1 m x 4 m open return type atmospheric boundary layer wind tunnel.

For this study, the mean wind speeds were measured with a Sierra Instruments Model 618 Air Velocity meter. Mean and fluctuating wind speeds (for wind tunnel calibration) were measured with a total head tube and Setra System, Inc. 239 Pressure Transducer. Pressures on the building surface were measured with a Setra System, Inc. 239 Pressure Transducer. All analog signals from transducers were digitized on a Keithley MetraByte DAS-8 Analog-to-Digital Converter.

The mean wind velocity and turbulence intensity was measured at a height of 4 – 5 feet above the floor of the ticket lobby level shown in Figure 2. This height corresponds to the region of most concern, the average chest to face level above ground of a human being.

4.2 *Configurations Tested*

A total of seven configurations were tested to quantify air flow rates in the OST ticket lobbies. These are listed below.

1. Existing condition without the addition of winds scoops or canopies
2. Existing condition with addition of wind scoops located at the south edge of roof.
3. Existing condition with addition of wind scoops located at center of roof.
4. Existing condition with addition of canopies only

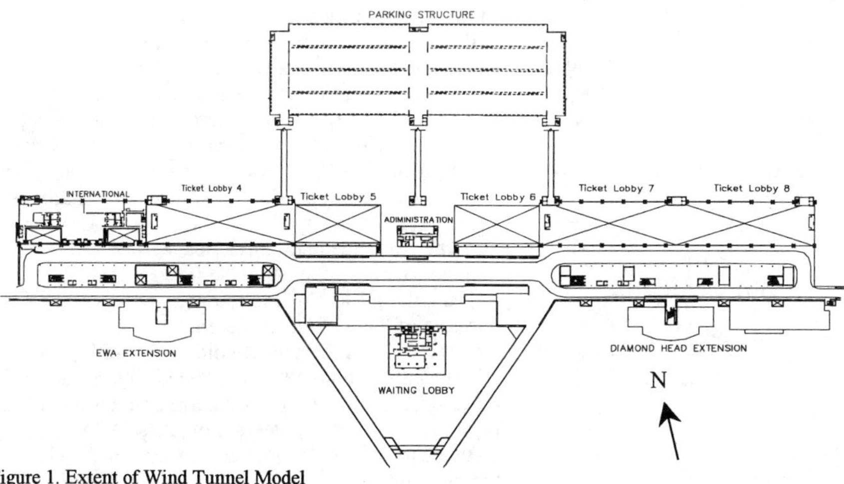

Figure 1. Extent of Wind Tunnel Model

5. Existing condition with addition of canopies and wind scoops located at the south edge of the roof
6. Existing condition with louvers on south wall open
7. Existing condition with continuous wind scoops and canopies

Each of these seven configurations were tested for oncoming winds for the predominant ENE Trade winds and much less frequent SW kona winds. Configuration 2 provided wind flow rates over the entire ticket lobby areas better than in configuration 3 (shown in Figure 3). Configuration 2 was thus selected for testing of Configuration 5. Configuration 6 was tested to determine the improvement in ventilation based on existing capabilities. Configuration 7, shown in Figure 4, illustrates the proportion of the continuous wind scoops relative to the ticket lobby. The wind tunnel results from all seven configurations were integrated into the statistical description of the hourly wind speeds recorded at the HIA weather station.

A significant width (32 feet) of the roof diaphragm would have to be removed to place the wind scoops of configuration 7 compared to 24 feet for configurations 2, 3 and 5. The wind pressure exerted on the surface areas of the wind scoops will induce uplift and shear forces on the roof structure. The combination of increased wind pressure and diaphragm removal caused concern about the load carrying capacity of the roof and rigid frame construction of the Central wing (ticket lobbies 5 and 6), the oldest of the OST. The results of the wind tunnel tests conducted to measure the wind pressures on the wind scoops and roofs of the OST are also presented in this paper.

Figure 2. Cross-section of OST in existing condition

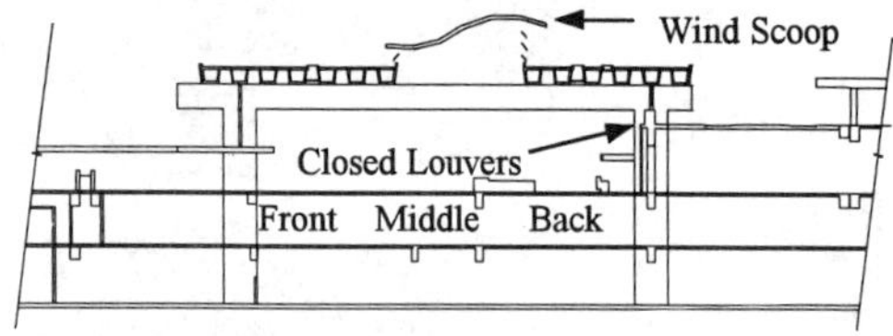

Figure 3. Configuration 3 showing existing condition with wind scoop at roof center

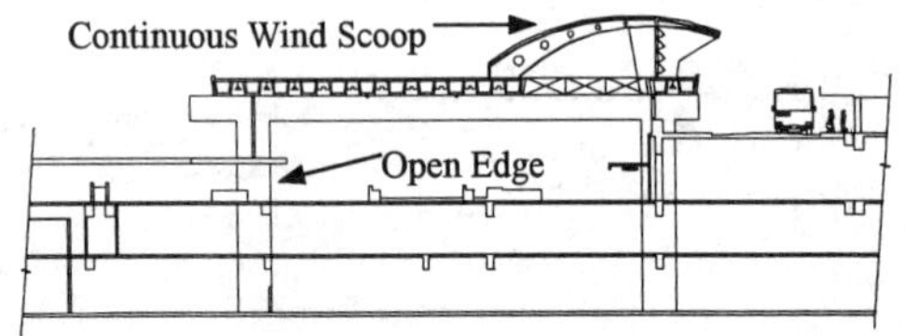

Figure 4. Configuration 7 showing existing condition with continuous wind scoop along south edge of OST roof

5 WIND TUNNEL TESTING RESULTS

5.1 *Ventilation and Comfort*

The amount of air circulation and cross-ventilation were collected as mean wind speed and turbulence intensity at a comprehensive array of points covering ticket lobbies 4 through 8 (Fig. 1). The mean wind speed and turbulence intensity were combined using the previously defined formula of Equation 1. The wind tunnel measurements were obtained as a fraction of the mean wind speed at mid-height of the wind tunnel cross-section. Results of the wind tunnel tests were integrated into a statistical description of wind speeds at the project site to compute the percentage of time that the wind speed exceeds each of the three criteria defined earlier. The exceedance percentage of wind flow exceeding all three criteria were computed at each of the data points for all five lobbies for five different times of day each season. This evaluation did not consider variations in temperature and humidity.

Evaluation of the results for comfortable condition posed a challenge due to the enormity of data. Additional parameters were introduced in the analysis by considering different hours of operation for each airline ticket counters, seasonal change in flight schedules and weather, and spatial variation in the placement of the ticket counters along the lobby width. To consider different times of day in the evaluation of comfort levels required inclusion of temperature and humidity data in the analysis. The amount of air movement required for comfort is a function of ambient temperature and humidity (Olgyay 1963). Historical temperature and humidity data recorded at HIA were used to determine the wind speed required to produce comfortable conditions. This required wind speed for comfort was compared with the experimentally measured wind speed to determine whether adequate wind flow exists or not.

Temperature and humidity data were included for this part of the analysis to determine wind flow rates required for comfortable conditions. Monthly values of humidity, temperature and wind speeds at four times of day were used to determine required wind flow rates. Table 1 gives a sampling of the bioclimatic chart values, for moderate zones in the United States, used for this part of the analysis. Each of the ticket lobbies 4 through 8 were sub-divided into a front (F), middle (M) and back (B) section as illustrated in Figure 3.

Table 1. Wind flow rates required for comfort

Dry Bulb Temp	Relative Humidity	Required Wind Flow Rate
88°F	50%	300 ft/min
87°F	55%	300 ft/min
84°F	60%	300 ft/min
82°F	65%	300 ft/min
80°F	70%	300 ft/min
78°F	75%	300 ft/min
76°F	80%	300 ft/min
73°F	85%	300 ft/min

The monthly annual data for the ten years of record was combined to determine an overall average percentage of time the measured wind flow rates exceeded required. Table 2, 3 and 4 give the average exceedance percentages that the wind speed in the ticket lobbies exceed the bioclimatic chart (Olygyay) criteria at 8:00 am, 2:00 pm and 8:00 pm, respectively. It is to be noted that the hourly wind speed data recorded over the last ten years indicates that Trade winds (Northeast to East) dominate for about 70% of the time Kona winds (Southwest) blow only about 5% of the time and 25% of the time from the remainder of directions. The average exceedance values in Tables 2 – 4 represent values for the whole year but have accounted for wind flow measured from the Trade and Kona directions only. These values for all compass directions should be higher. It should also be noted that the minimum levels of wind flow rates required for adequate ventilation are much lower than wind flow rates required for comfort conditions. The improvement or reduction in wind flow rates provided by each modified configuration over existing can be determined from the values in Tables 2 – 4.

Table 2. Average percentage of time bioclimatic chart (Olgyay 1963) criteria conditions are exceeded at 8:00 am

Lobby	Configuration Number						
	1	2	3	4	5	6	7
4 F	12.4	16.5	11.7	12.6	47.5	36.8	46.8
4 M	11.4	22.1	8.9	10.0	20.2	25.7	21.3
4 B	9.4	8.9	16.0	9.4	17.7	25.5	37.8
5 F	24.3	25.7	37.5	21.5	25.6	30.5	25.1
5 M	10.7	21.8	8.8	14.0	25.7	28.7	28.9
5 B	17.9	9.1	19.0	14.6	23.9	29.5	22.7
6 F	14.9	20.2	24.7	17.0	22.6	28.0	22.8
6 M	9.9	11.7	13.3	9.5	18.1	21.7	20.0
6 B	10.5	12.0	11.2	9.8	11.9	23.0	20.9
7 F	19.6	44.0	44.5	19.6	38.0	30.5	53.6
7 M	11.4	15.3	16.0	11.4	16.7	24.2	25.2
7 B	11.9	13.7	12.5	11.9	12.4	27.7	44.1
8 F	47.2	54.2	51.9	50.1	59.7	60.5	62.1
8 M	24.9	46.9	24.8	18.6	30.7	24.6	25.2
8 B	20.2	34.3	21.5	27.7	39.9	31.7	39.2

Note: F – Front, M – Middle, B – Back

Table 3. Average percentage of time bioclimatic chart (Olgyay 1963) criteria conditions are exceeded at 2:00 pm

Lobby	Configuration Number						
	1	2	3	4	5	6	7
4 F	18.9	28.9	19.7	15.5	67.0	47.2	61.3
4 M	14.6	35.4	10.6	11.2	32.0	21.1	32.9
4 B	11.2	11.0	25.4	9.0	27.9	19.8	51.3
5 F	38.6	43.3	54.9	33.4	43.7	32.0	39.5
5 M	12.6	35.4	10.6	19.1	42.0	29.0	46.4
5 B	28.4	12.8	31.2	23.8	37.2	33.4	34.2
6 F	26.3	31.8	41.4	23.7	35.4	25.1	35.0
6 M	13.1	19.2	23.3	8.7	30.0	14.0	29.2
6 B	14.6	19.0	16.8	13.6	17.7	14.6	32.9
7 F	27.2	60.9	63.0	26.6	53.0	32.7	71.3
7 M	12.5	28.9	25.8	13.0	29.0	16.5	40.2
7 B	15.4	22.2	19.8	14.7	19.5	24.0	60.4
8 F	60.7	72.6	72.1	63.6	72.5	68.3	72.0
8 M	37.6	62.3	37.9	25.0	47.8	48.9	40.0
8 B	27.6	49.2	33.8	42.1	52.3	34.2	52.8

Note: F – Front, M – Middle, B – Back

Table 4. Average percentage of time bioclimatic chart (Olgyay 1963) criteria conditions are exceeded at 8:00 pm

Lobby	Configuration Number						
	1	2	3	4	5	6	7
4 F	23.9	29.3	21.0	23.7	65.3	45.9	63.7
4 M	21.3	35.9	13.3	10.0	20.2	25.6	35.4
4 B	11.2	11.0	16.0	14.7	21.9	25.0	51.1
5 F	40.0	42.7	50.6	35.8	43.2	36.4	41.7
5 M	19.3	35.3	13.3	14.0	25.7	33.6	45.9
5 B	28.4	12.8	19.0	19.9	26.6	35.0	36.2
6 F	28.9	34.3	40.6	28.6	38.3	32.1	37.1
6 M	16.2	20.7	24.9	9.5	18.1	16.6	33.6
6 B	14.6	19.0	11.2	15.1	16.7	17.3	35.0
7 F	33.8	61.5	63.7	32.5	51.3	36.3	70.4
7 M	19.6	28.0	28.6	11.4	16.7	22.5	42.4
7 B	15.4	22.2	12.5	16.9	17.3	30.2	63.2
8 F	64.3	72.4	68.7	66.7	74.1	73.0	75.1
8 M	41.7	63.8	41.6	18.6	30.7	51.5	41.8
8 B	27.6	49.2	21.5	30.8	40.3	40.8	52.9

Note: F – Front, M – Middle, B – Back

Information obtained from these computations of required wind flow rates will lead to a final design solution with the aim of providing an equitable distribution of comfort conditions to occupants and tenants of the ticket lobbies.

5.2 *Wind Pressure on Roof*

A 50-year omni-directional design 3-second average wind speed of 75 mph at 10 m at HIA was used for pressure measurements. This value was determined from the statistical analysis of extreme wind speed data recorded at the HIA weather station and a hurricane-risk analysis wind speed. Oncoming winds from the north and south, normal to the louvers and wind scoops, resulting in maximum wind pressure values on the roof and scoops were used for testing. The parking garage, palm trees in the vicinity of the parking garage and the jetway arms south of the central wing portions of the model were excluded from the testing to avoid any shielding effects on the terminal roofs and wind scoops. Pressure measurements were performed for configurations 6 and 7 of the OST only. Sub-configurations resulting from opening and closing of louvers and scoops during inclement weather were also considered to yield an envelope of pressures. The pressures on the roof and scoops of configurations 6 and 7 are presented in Figures 5 and 6, respectively, for the case of wind from the North and in Figures 7 and 8 for wind from South. Pressure taps were mounted on the inside and outside at each location measured and the time history of the difference in the internal and external pressures was recorded. The peak values of these differential pressure time histories were recorded and are reported in Figures 5 – 8.

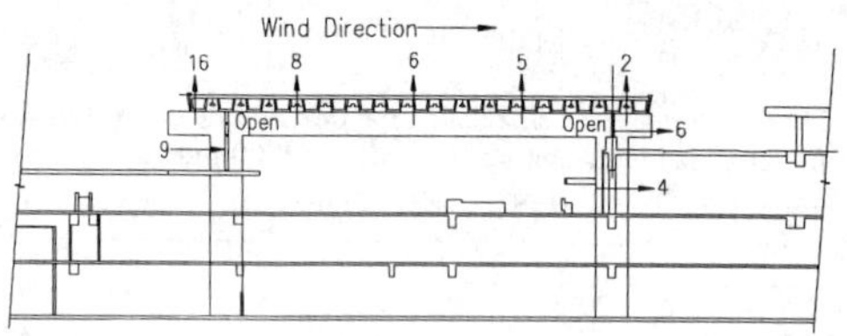

Figure 5. Peak Differential Pressures (psf) on Roof and Scoop

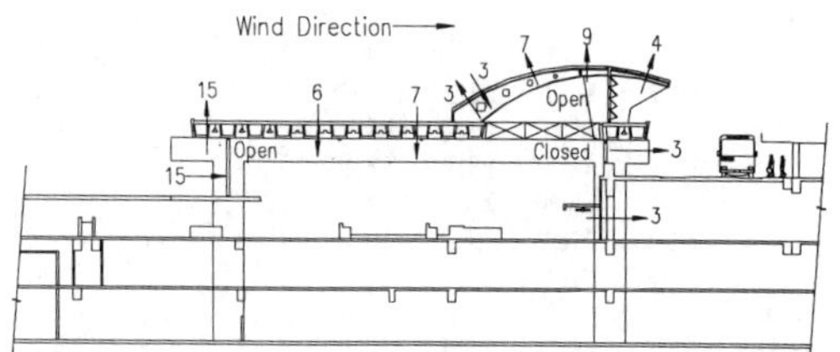

Figure 6. Peak Differential Pressures (psf) on Roof and Scoop

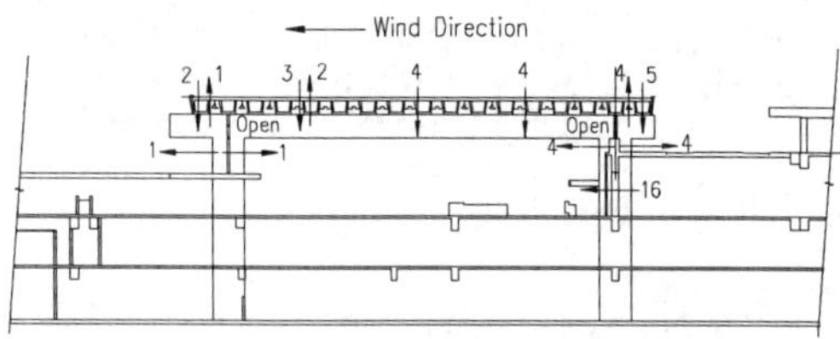

Figure 7. Peak Differential Pressures (psf) on Roof and Scoop

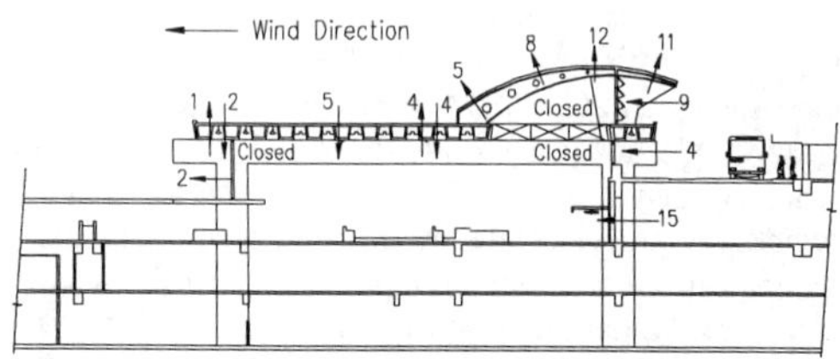

Figure 8. Peak Differential Pressures (psf) on Roof and Scoop

6 CONCLUSIONS

Configuration 7 consistently provides the overall greatest levels of wind flow. Configuration 5 consistently provides the overall second largest levels of wind flow. Configuration 4 causes a decrease in the overall existing flow rate. Configuration 2 results in the third largest overall flow rates followed by Configuration 6 in fourth place. The final choice of a suitable configuration will be made on the relative improvement of the flow condition from the existing conditions and budgetary considerations. Current design work is considering reducing the width of roof openings and reduction in the length of continuous wind scoops of configuration 7 to provide uniform levels of comfort within budget.

Results from the wind pressure data indicate that the measured pressures would require strengthening of the roof framing system to account for the additional loading imposed by wind pressure on the scoops and reduction of the roof diaphragm load carrying capacity.

7 ACKNOWLEDGEMENTS

The authors gratefully acknowledge the support and assistance provided by Mr. Benjamin Cayetano, Governor, State of Hawaii; Mr. Kazu Hayashida, Director, Department of Transportation (DOT), State of Hawaii; Mr. Jerry Matsuda, Airport Administrator – HIA and Mr. Ernest Kurosawa, Engineering Program Manager DOT-Airports Division; in the conduct of this

REFERENCES

1997 ASHRAE Handbook Fundamentals. American Society of Heating, Refrigerating and Air-Conditioning Engineers, Inc., Atlanta, GA.

ANSI/ASHRAE 62-1989. ASHRAE Standard: Ventilation for Acceptable Indoor Air Quality. American Society of Heating, Refrigerating and Air-Conditioning Engineers, Inc., Atlanta, GA.

Gandemer Jacques, and Guyot, Alqin, *LA PROTECTION CONTRE LE VENT*, Centre Scientifique et Technique due Batiment, Paris, 1981.

National Climatic Data Center, Asheville, North Carolina, Local Climatalogical Data for Honolulu International Airport, 1998.

Olgyay, V., *Design With Climate: Bioclimatic Approach to Architectural Regionalism*, Princeton University Press, Princeton, N.J., 1963.

Wind Engineering into the 21st Century, Larsen, Larose & Livesey (eds) © 1999 Balkema, Rotterdam, ISBN 90 5809 059 0

Influence of a scaling mismatch on wind induced action and action effects

M. Kasperski & H. H. Koss
Building Aerodynamics Laboratory, University of Bochum, Germany

ABSTRACT: In the scope of the European research network BEATRICE, an extended comparative wind tunnel study has been performed on wind induced actions and action effects on a typical industrial low-rise building. The partner institutes involved in this study are - in alphabetic order - AIB (Germany), BRE (Great Britain), CSTB (France), DMI (Denmark), NLR (The Netherlands), and TNO (The Netherlands). The study is based on five models with similar geometry but at different geometric scales, i.e. with different heights, tested in the six boundary layer wind tunnels of the partner institutes. Each of them has specific flow properties, e.g. due to different facilities for creating the turbulent boundary layer or due to the construction of the wind tunnel itself. The aim of the study is to answer the question to what extent a change of the boundary layer characteristics is able to influence the results in terms of local pressures, global forces and linear and non-linear response of the structure. Not only the flow properties may influence the results, but also the variation of the geometric scale. To obtain statistically stable estimation, a large sample size of 240 individual events per arrangement have been recorded. Each event corresponds to approx. 10 min in full scale. The paper compares first the simulated boundary layers under the aspect of similarity to data observed in full scale. The attempt is made to describe the functional link of the performed simulations. The scaling mismatch is discussed on a civil engineering background, i.e. the obtained overloading risk of the designed structure is of major interest.

1 INTRODUCTION

The first part of the study, 'Basic information and first results' 1996, describes the test arrangement (acquisition chain and equipment specifications, model size and pressure taps, flow direction, floor mounted taps, etc.), the different boundary layers (profiles of the mean wind velocity and turbulent intensity), the accuracy of the results due to the large sample size and deals about various aspects of analysing the wind induced actions (spectral density functions and statistics of the pressure fluctuations, application of the orthogonal decomposition method). In this part, the authors examine more closely the wind effects on an assumed structure. The used models represent a typical industrial low-rise building with steel frames as the main supporting structure. In figure 1, the basic geometry of the model is sketched with the positions of the pressure taps along the centre bay.

The ratio of height, width and length is fixed for all models at $h/d/l = 0.4/1.0/1.6$. The variation of the geometric scale leads to different eaves heights listed in table 1. For some institutes, a sixth model with a height of 20 mm has been tested, too (model I).

Table 1. Heights of the models in mm

model	II	III	IV	V	VI
height	30	40	50	60	70

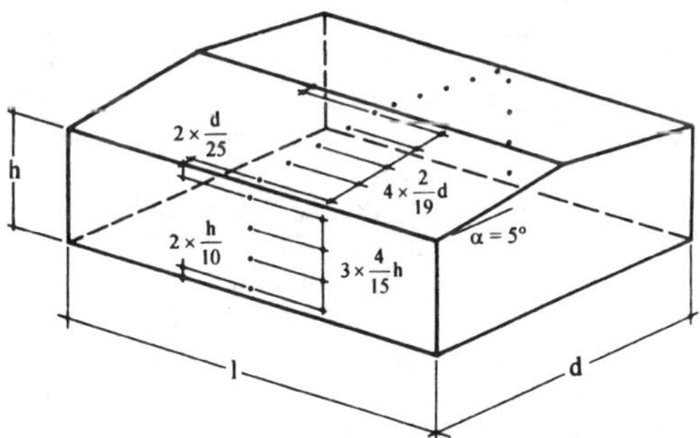

Figure 1. Basic geometry of the model, position of the pressure taps along the centre bay

Figure 2 contains the distribution of the pressure taps along a portal frame in the centre bay. For the study, four taps have been chosen to investigate the pressure fluctuation in typical flow regions:

- stagnation point (tap 3)
- separation at the leading edge (tap 5)
- re-attachment on the downwind part of the roof (tap 12)
- separation and re-circulation on the downwind wall (tap 16)

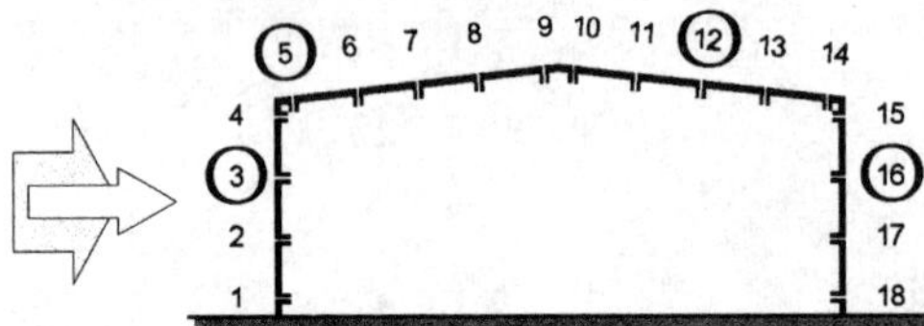

Figure 2. Pressure taps along the frame in the centre bay of the model

The study considers in its initial stage only the flow direction normal to the ridge, i.e. the angle of attack is 0°.

For each model, 240 independent time series of pressure fluctuation along the above pictured frame are monitored and stored on compact disks as input data for the time domain calculation of the structural response.

To maintain the anonymous character of the study, the results obtained at the different partner institutes are only marked with "institute A", "institute B", and so on, not corresponding to the alphabetic order in which they are introduced above. For an easier reading of the paper the same designation as used in part one has been chosen.

2 THE BOUNDARY LAYERS

2.1 *The description*

In Kasperski (1996) the boundary layers have been described by determining the profile of the mean flow velocity $\overline{u}(h)$ and the corresponding distribution of the turbulence intensity over height. The latter was evaluated in its alongwind component $I_u(h)$ and the crosswind component $I_v(h)$.

It was demonstrated that several boundary layers have a distinct non-homogenous character for the mean velocity distribution. The profiles of the turbulence intensities show a fundamental difference for simulations with roughness on the turn table and those with a smooth turn table. For the latter situations, the turbulence intensities remain nearly constant over a larger range of heights.

2.2 *Further comparison*

The initial aim of the study was to compare simulations representing the wind conditions over *open country*. Proceeding from the α-values, determined by fitting the power-law function in the height regions where homogenous flow can be assumed, none of the institutes matches the theoretical requirement of $\alpha = 0.16$ over a height section 3 times the height of the largest model.

To answer the question, if and to what extent the simulations reflect the flow conditions over open country, an independent and neutral criterion is needed. Observations in full scale, compiled and published by ESDU (1986), deliver the appropriate basis.

To assign a terrain category to the simulated flows, ranges of mean velocity are defined. The border of these ranges forms the middle between neighboring ESDU-profiles (Fig. 3). For the assignment of the turbulence intensities analogue ranges are defined by using the ESDU data for $I_u(h)$.

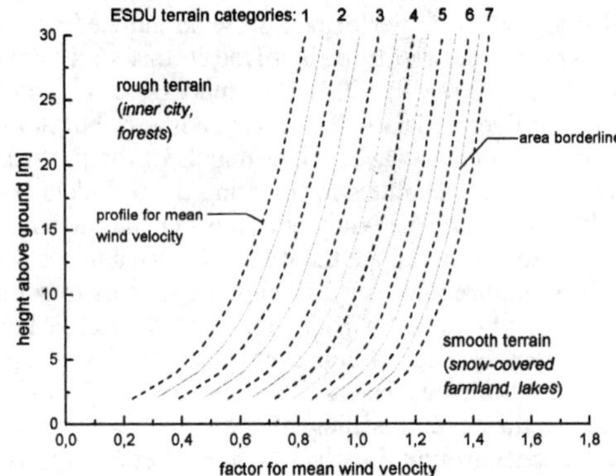

Figure 3. Profiles and areas of mean velocity after ESDU

Because the mean velocities are normalized values, they can be scaled by a factor F_S until the profile shape fits to only one area, i.e. until the profile belongs to one specific type of terrain (Fig. 4). The aimed terrain *open country* corresponds to ESDU category 4.

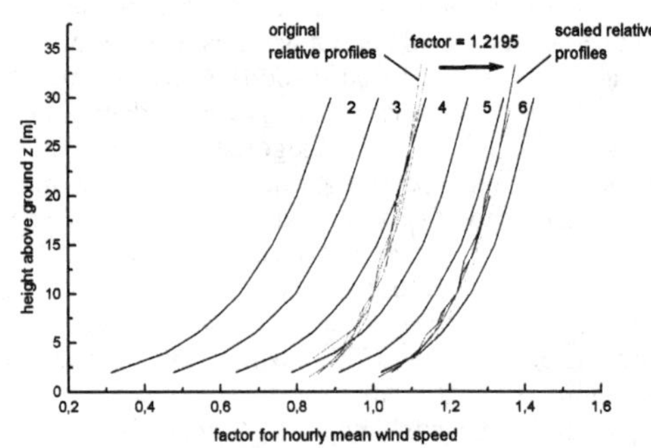

Figure 4. Assignment of the profiles by scaling the curve shape with F_S (institute C).

Due to the different geometric ratios of models and flow, for each simulation not only one profile has

Table 2. Scaling factor F_S

institute	F_S
A	1.0000
B	1.1111
C	1.2195
D	1.0000
E	1.2121
F	1.1905

to be adjusted, but a family of five relative profiles, one for each model. Table 2 contains the factor F_S for each institute.

Herewith, a factor $F_S > 1$ shifts the relative profiles to a more smoother and $F_S < 1$ to a more rougher terrain category.

The following figures (5a - 5f) contain the families of relative turbulence intensity distributions I_u over height h.

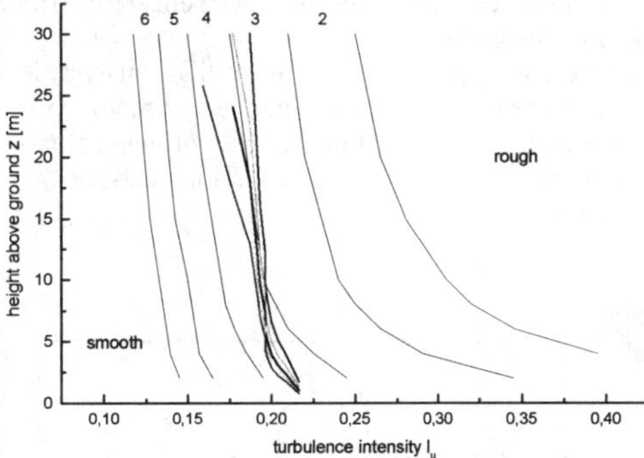

Figure 5a. Family of relative I_u-curves at institute A

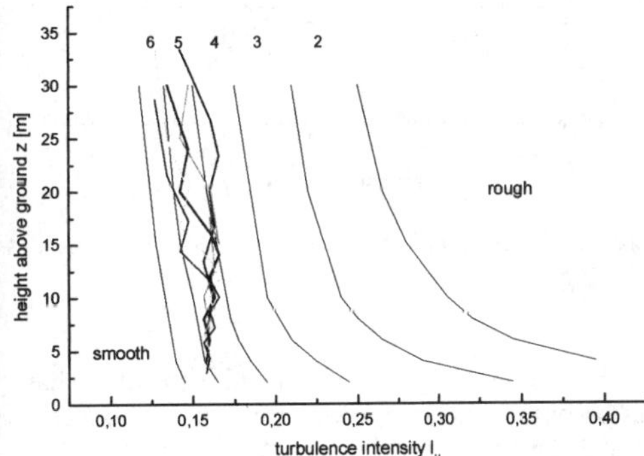

Figure 5b. Family of relative I_u-curves at institute B

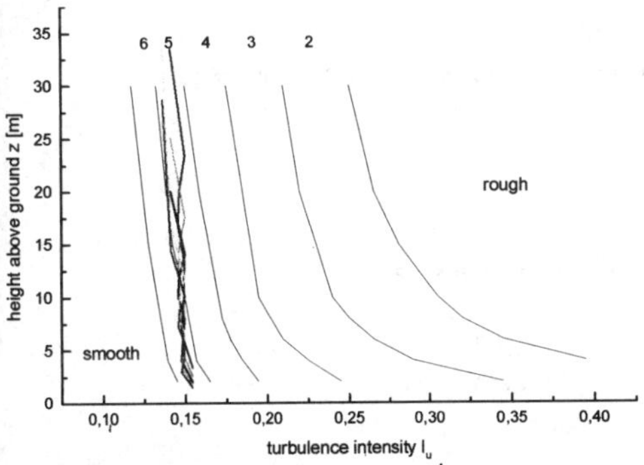

Figure 5c. Family of relative I_u-curves at institute C

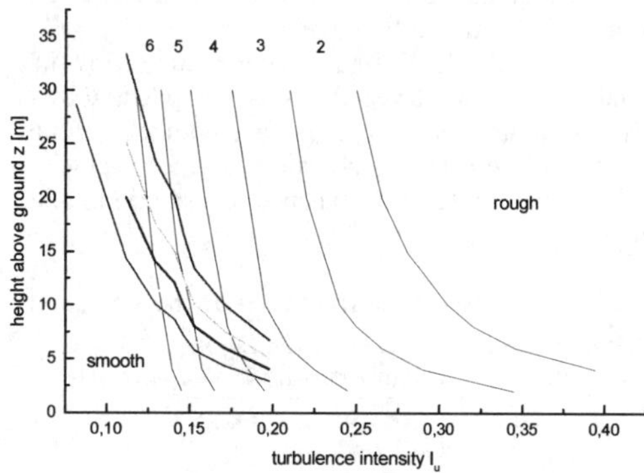

Figure 5d. Family of relative I_u-curves at institute D

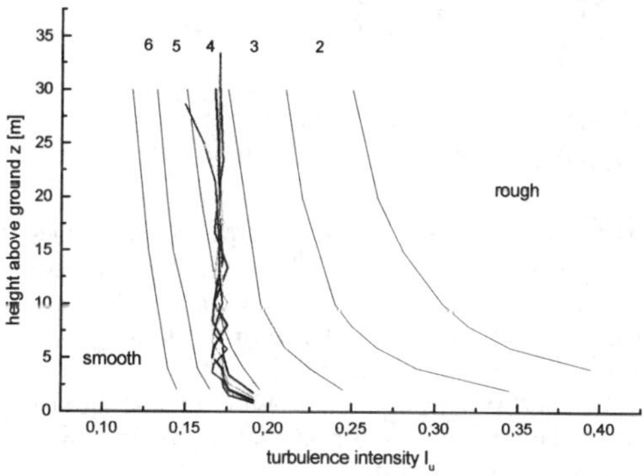

Figure 5e. Family of relative I_u-curves at institute E

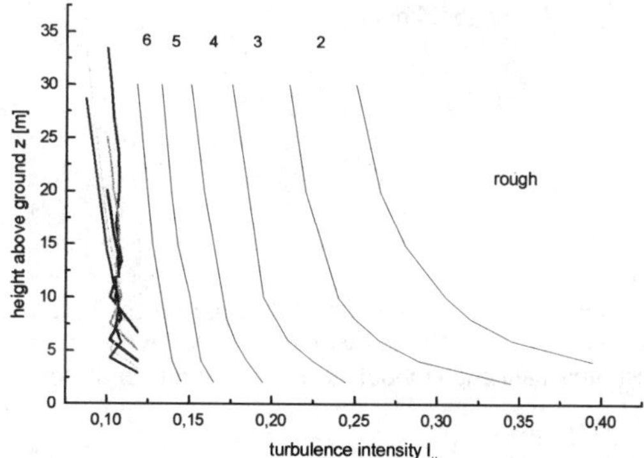

Figure 5f. Family of relative I_u-curves at institute F

The curves of I_u describe already the order of magnitude of the turbulence distribution, hence, no further modification of the shape is required.

The comparison of the turbulence intensity distributions, measured in the experiments, with the ESDU data indicates the difficulty of designing an appropriate boundary layer satisfying all requirements. Two simulations, D and F, have a considerable lack of turbulent energy in a height domain responsible for the flow around the model. This may lead to a large influence on the results and has to be bear in mind for the further evaluation.

In table 3, the ESDU-based interpretation of the simulated boundary layer flows is compiled. Obviously, all simulations represent a smoother terrain than the aimed category 4, some in their velocity profile, some in their turbulence distribution, and some in both.

Table 3. ESDU-based interpretation of the simulated boundary layer flows

institute	velocity profile	I_u-profile
A	5	4
B	5	5
C	6	6-5
D	4	4-7
E	6	5-4
F	6	7

2.3 *The link*

In part one of the study, the spectral density functions (SDF) of the pressure fluctuation were employed, to detect a possible link between the different flows, or to identify the influence of the scaling mismatch in the frequency domain.

In this part of the study, a possible functional link between the simulations for wind actions at typical flow regions is of interest.

For this, peak factors g of the pressure fluctuation at characteristic points in the regarded flow region are introduced as follows:

$$\hat{x} = \bar{x} + \sigma_x \cdot g \tag{1}$$

$$g = \frac{\hat{x} - \bar{x}}{\sigma_x} \tag{2}$$

The statistical evaluation philosophy in this study is based on the 240 individual time series for each test arrangement (model, wind direction, and flow simulation). This means that $\bar{x}$ is the mean of all 240 determined mean values, $\hat{x}$ the mean of all 240 observed maximum values, and σ_x is the mean of 240 rms-values, each estimated within one event.

The parameter g is less susceptible to systematic bias from measurement. The aim is to plot the peak factors versus a characteristic value ξ, which indicates the physical link between the regarded flow simulations. It can be assumed, that following parameters should be contained in ξ: the eaves height h, mean wind velocity $\bar{u}$ at h, the corresponding turbulence intensity $I_u(h)$ and the exponent α of the mean wind speed profile.

Additionally, the information from SDF-analysis in part one (1996) can be taken into account to consider the energetic contribution of the turbulent flow. Figure 6 shows the SDF basic shape for pressure tap 3 (stagnation point).

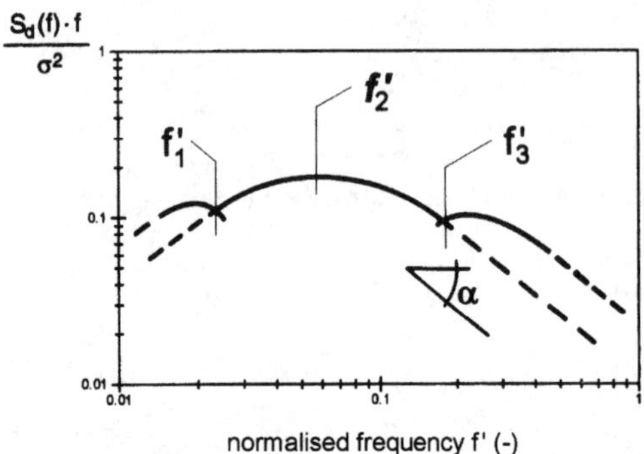

Figure 6. Characteristic SDF of the pressure signal at tap 3 (stagnation point on the upwind wall)

With several variations of trial and error the value ξ can be optimized until a systematic order appears. For tap 3 the characteristic value can be defined as

$$\xi_3 = f_2' \cdot \alpha \cdot \frac{I_u(h)}{h} \tag{3}$$

In which the reduced peak frequency is given by

$$f_2' = f_2 \cdot \frac{h}{\bar{u}(h)} \tag{4}$$

that (3) can be written as

$$\xi_3 = f_2 \cdot \alpha \cdot \frac{I_u(h)}{\bar{u}(h)} \tag{5}$$

In figure 7 the peak values obtained from the different experiments, are plotted over ξ_3. For some institutes, a similar orientation of the peak values can be identified.

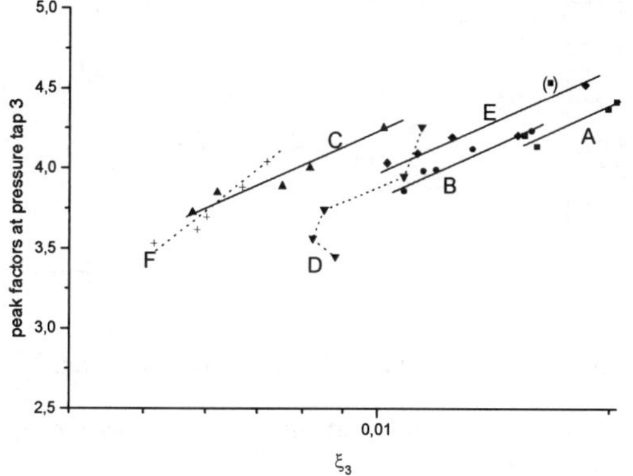

Figure 7. Peak factors at tap 3 over ξ_3

The peak values of the institutes A,B,C, and E are orientated with the same inclination. The simulations D and F can't be included in this group by applying the above defined characteristic value. One reason for this can be the weak turbulent energy, which influences the standard deviation of the process and therefore the orientation of the peak values.

The result in figure 7 suggests that the parallel orientated values of the remaining simulations can be moved to one line. The logarithmic scale requires a factor F_T for the horizontal translation. In figure 8, the correlation between F_T and the previously introduced scaling factor F_S is shown.

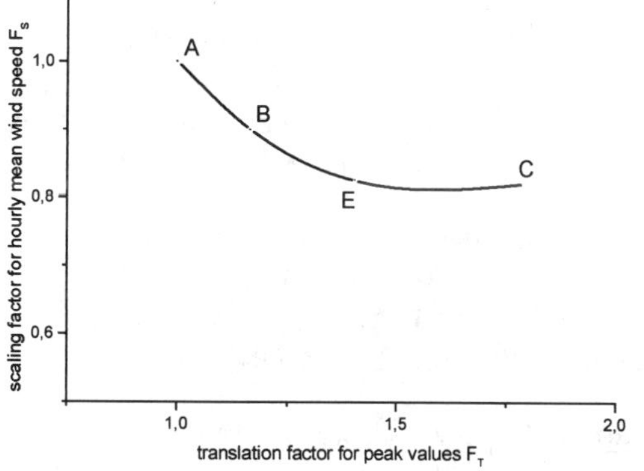

Figure 8. Correlation between F_S and F_T

The regarded family of simulations may deliver a functional correlation of these two factors. To determine a general function, more similar simulations are needed, otherwise the result may be of purely random character.

Using F_T for the characteristic value ξ, equation (5) must be modified as follows:

$$\xi'_3 = f_2 \cdot \alpha \cdot \frac{I_u(h)}{\bar{u}(h)} \cdot F_T \tag{6}$$

Applying ξ'_3, the peak value orientation of the simulations A,B,C, and E at pressure tap 3, show a line-like orientation (Fig. 9).

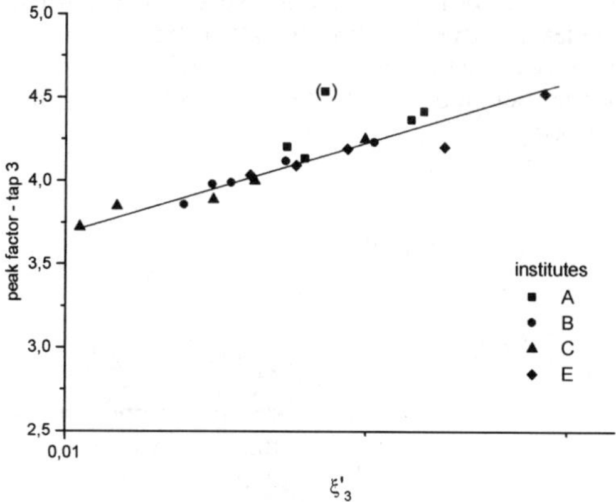

Figure 9. Modified plotting of the peak factors at tap 3 over ξ'_3

The value in parenthesis for the smallest model in simulation A is regarded as an outlier. This is caused by the properties of the lowest flow layer due to a smooth turn table. Again, this indicates the important role of a consistent modelisation of the flow regime, i.e. profile and corresponding turbulence intensity.

2.4 *The influence of the time scale*

Beside the aerodynamic aspects, the sensibility of design values, used for structural engineering, is focussed in this study. This requires a statistical analysis of the wind induced actions. To compare the parameters, one has to ensure that the statistical quality is the same for all examined values. In general, design values are given for global and local actions. *Local action* means for example the wind pressure on single cladding and construction elements. Accordingly, *global action* is the structural response (displacement and internal forces like bending moments), determined by integrating the local actions.

In codes and standards, design values are defined as fractile values of e.g. fluctuating load processes. Using the signals observed at pressure tap 3, 5, 12 and 16, as representative for the four typical flow regions, fractile values are determined assuming gumble distributed extreme values (Eq.7). As basis for this analysis serve the 240 independent events, each delivering one extreme value.

$$F(x) = p(x \le \hat{x}) = e^{-e^{\left(-\gamma - \frac{\pi}{\sqrt{6}} \cdot \frac{x-\bar{x}}{\sigma}\right)}} \tag{7}$$

with $\gamma = 0.5772$

Equation 7 gives the non-exceedence probability of the value $\hat{x}$. Analogue to the peak factor g, the input data for this statistics are the 240 extreme values, each determined within a time series with 4096 samples measured in 2.56 seconds or with an acquisition frequency of 1600 Hz respectively. The corresponding duration in full scale depends on the time scale λ_T:

$$\lambda_T = \frac{\dfrac{L_m}{\bar{u}(h_m)}}{\dfrac{L_{fs}}{\bar{u}(h_{fs})}} = \frac{L_m}{L_{fs}} \cdot \frac{\bar{u}(h_{fs})}{\bar{u}(h_m)} = \lambda_{geom} \cdot \lambda_u^{-1} \tag{8}$$

where L is a geometric dimension of the body, m indicates model size values, fs full scale values, λ_{geom} the geometric scale, and λ_u the velocity scale.

Due to different mean velocities at eaves height in the experiments, the time scale λ_T varies and therewith the corresponding duration in full scale, i.e. the reference period of the observation. Equation (9) can be applied to compensate this effect:

$$p(x \le \hat{x}) = (1 - 0.22)^{600 \cdot \lambda_T / 2.56} \tag{9}$$

The left side of (9) delivers the corrected non-exceedence probability for a 78%-fractile and enters (7) as the given value for $p(x \le \hat{x})$ to determine $\hat{x}$ as the λ_T-compensated 78%-fractile value.

2.5 *Influence of the scaling mismatch*

Acceptable tolerance for the geometric scale of the model are given in terms of following multiples of the geometric scale of the simulated natural wind (Cook 1985):

3 for overall forces and moments (global actions)
2 for local pressures on elements of cladding (local actions

Assuming, that one of the models fits to the geometric scale of the simulated flow, a maximum variation factor of 2.33 was realized within the study. Figure 10 shows the compensated fractile values of the pressure fluctuation at tap 3 for all models (starting with model II on the left) and institutes. Because none of the simulation should be defined as a reference, the comparison of the values is performed within the single institutes.

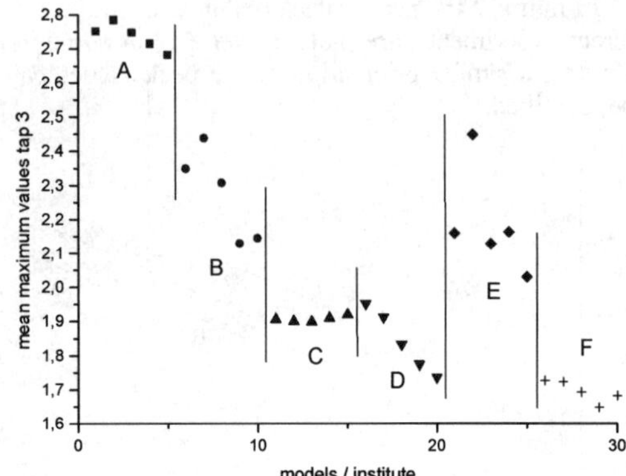

Figure 10. Compensated 78%-fractile values, tap 3

For the examination of local actions only model combinations forming a maximum geometric scale variation of 2 are taken as a basis. For this, the acceptable tolerance of the value scatter is defined as follows:

Supposing that within the regarded combination one value is employed as a design value, the others should not be larger than 10% (conservative but uneconomic design) or should not remain under more than 5% (unsafe side). Successively all values are regarded as design value. The extreme relative deviations of the 78% fractile values are listed in table 4:

Table 4a. Rel. dev. [%]; max. pressure at tap 3

	A	B	C	D	E	F
unsafe > 5%	3.8	14.5	1.1	10.1	20.5	4.7
uneco. > 10%	3.7	12.7	1.1	9.2	17.0	4.5

Table 4b. Rel. dev. [%]; min. pressure at tap 5

	A	B	C	D	E	F
unsafe > 5%	13.3	10.8	9.2	4.8	15.5	93.8
uneco. > 10%	11.8	9.8	8.4	4.6	13.4	51.6

Table 4c. Rel. dev. [%]; max. pressure at tap 12

	A	B	C	D	E	F
unsafe > 5%	25.3	32.3	61.0	51.1	36.0	72.8
uneco. > 10%	20.2	24.4	37.9	33.8	26.5	42.1

Table 4d. Rel. dev. [%]; max. pressure at tap 16

	A	B	C	D	E	F
unsafe > 5%	16.2	30.7	87.2	15.1	14.0	10.9
uneco. > 10%	14.0	23.5	46.6	13.1	12.3	9.8

In some cases, the required tolerance is satisfied but it is to state, that the scaling mismatch influences deeply the result quality.

3 LINEAR STRUCTURAL ANALYSIS

3.1 *The system and wind effects*

For the linear structural analysis a hinged frame is used as static system (Fig. 11). The reactions, limited to the bending moments in the frame corners, are determined by multiplying the wind loads with the influence coefficients of the reactions. Dead load and snow remain unconsidered, because due to linearity they do not influence the result.

Figure 11. Static system

3.2 *Influence of the scaling mismatch*

The estimation of the fractile values of the reactions is equivalent to chapter 2. To estimate the order of magnitude of the scaling mismatch a factor of 3 is taken into account. Hence, all models form the group in which the fractile values can be compared. The acceptable tolerance is again 5% for the unsafe side and 10% for uneconomic design. Tables 5a and 5b contain the relative deviations of the 78%-fractile value of the bending moments in the upwind and downwind knee of the frame.

Table 5a. Relative deviation [%]; bending moment in the upwind frame knee

	A	B	C	D	E	F
unsafe > 5%	11.0	22.3	13.2	10.0	16.0	7.7
uneco. > 10%	9.9	18.2	11.7	9.1	14.5	7.0

Table 5b. Relative deviation [%]; bending moment in the downwind frame knee

	A	B	C	D	E	F
unsafe > 5%	5.8	13.3	43.6	19.2	86.9	24.5
uneco. > 10%	5.5	11.7	30.4	16.1	46.5	19.7

Every simulation leads to an underestimation of the design value by varying the geometric scale, either for one or both bending moments. That indicates, that the global actions are of the same sensitivity than the local actions. Even the simulations A, B, C, and E, found for similar in chap. 2.3, show a large scatter. A comparison in this family leads to following relative deviations (Tab.6):

Table 6. Comparison within institutes A, B, C, and E

	M_{upwind}	$M_{downwind}$
unsafe > 5%	33.9	99.2
uneco. > 10%	28.0	48.1

4 NON-LINEAR STRUCTURAL ANALYSIS

4.1 *Discretization of the system*

The non-linear structural analysis allows the forming of plastic hinges. For this, the program DRAFS (1998) was employed, developed for quasi-static and dynamic structural analysis in time domain. The Finite-Element system for the hinged frame is shown in figure 12.

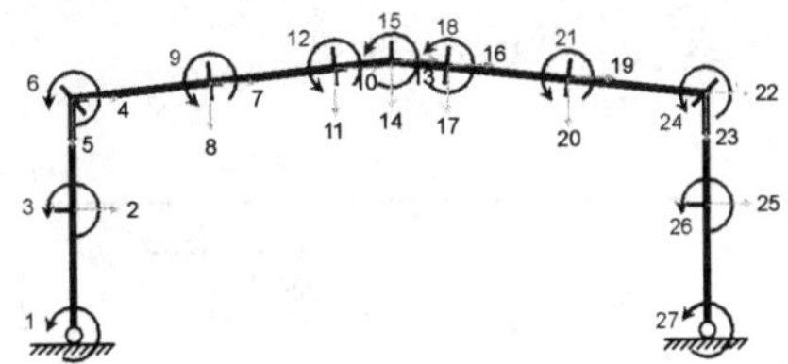

Figure 12. FE-system for time domain analysis

4.2 *Limit states*

For the study, two limit states of the structure are considered: Forming of the first plastic hinge during a storm event, and the occurring of the chain of plastic hinges, and therewith the collapse of the structure.

4.3 *The overloading risk*

In N storm events, the regarded limit state occurs n-times, which can be expressed as the relative frequency f_{rel} of the limit state (Eq. 10). For a sufficient large number of N, which 240 is supposed to be, f_{rel} becomes the probability p.

$$f_{rel} = \frac{n}{N+1} \tag{10}$$

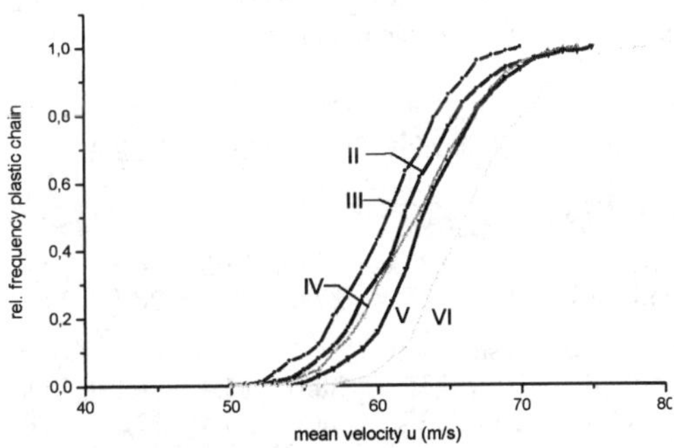

Figure 13. f_{rel} for chain of plastic hinges

The probability p, or f_{rel} respectively, can be determined for various mean wind speeds at eaves height. In figure 13 the probability of obtaining a chain of plastic hinges is plotted versus the mean wind speed.

The analysis is performed for two different types of roof constructions: a lighter roof as it is built in e.g. Australia and a heavier roof which typically is required in regions with a considerable snow load.

The probability $p_{overload}$ to obtain one of the limit states can be estimated by convolution of f_{rel} with an extreme value distribution $p(v)$ of the wind speeds (Eq. 11), sketched in figure 14.

$$p_{overload} = \int_{v=0}^{\infty} p(v) \cdot f_{rel}(v) \cdot dv \qquad (11)$$

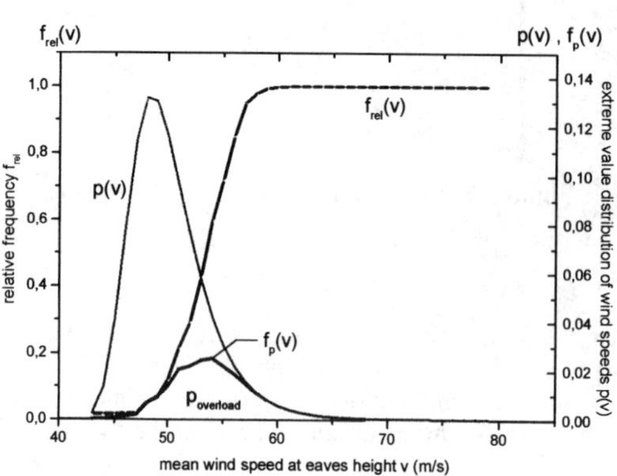

Figure 14. Convolution to estimate $p_{overload}$

To adjust the correct time window of 600 s in full scale, f_{rel} is modified as follows:

$$f_{rel}(600s) = 1 - \left(1 - f_{rel}\right)^{600 \cdot \lambda_T / 2.56} \qquad (12)$$

4.4 *The influence of the scaling mismatch*

If a target overload risk of 5% is introduced for a life time of 50 years, following corresponding mean wind speeds result for the wind climate.

Table 7. Mean wind speeds [m/s] leading to $p_{overload}$ = 5%; limit state: chain of plastic hinges

	A	B	C	D	E	F
light roof construction						
model II	23.9	25.5	27.4	18.5	26.8	29.4
model VI	25.1	27.4	27.5	27.3	28.5	29.7
heavy roof construction						
model II	31.9	35.1	38.4	17.4	36.2	40.8
model VI	33.0	37.2	38.3	30.5	37.2	42.0

The values in table 7 and 8, are the mean wind speeds ($\bar{x}$ in eq. 7), which have to introduced into account, to design the frame with the same structural safety, using the wind load data from the different simulations.

Table 8. Mean wind speeds [m/s] leading to $p_{overload}$ = 5%; limit state: first plastic hinge

	A	B	C	D	E	F
light roof construction						
model II	21.1	22.4	24.1	16.9	23.7	26.0
model VI	22.3	24.2	24.4	25.9	25.0	26.4
heavy roof construction						
model II	24.5	28.6	34.5	14.5	28.7	34.9
model VI	25.0	30.1	30.3	28.2	27.2	38.8

5 CONCLUSIONS

The analysis of the wind induced actions and action effects indicates the enormous influence of a mismatch in the geometric scale on the results.

On the example of the extremes of the local pressures at stagnation, an analysis was performed to transform the results from one simulation to another using a set of different influence parameters like e.g. the shape of the mean velocity profile, the turbulence intensity and mean wind speed at eaves height, a characteristic frequency from the spectral density function of the pressure fluctuation. An additional 'translation' factor is introduced to fit the results from all the institutes and all model heights to one curve. However, the study is far away from delivering a general solution.

The relative deviations within one simulation show that a scaling mismatch may lead to considerable differences (exceeding 20%) in the design values for wind loads and wind load effects, even if the scaling mismatch is smaller than 2 or 3. Differences within one boundary layer are generally smaller than differences between the flow simulations.

The obtained relative deviations are not able to meet the strict demand of 5% to 10% difference to the unsafe of uneconomic side. For the three reasonable simulations of open country (A,B,E), the introduced scaling mismatch of 2.33 leads to differences in the design wind loads of 11, 19 and 17% for the lighter roof and 9, 15 and 20% for the heavier roof construction which is in regard to the safety of the structure too high .

6 REFERENCES

Kasperski, M. & Koss, H. & Sahlmen, J. 1996. BEATRICE Joint Project: Wind action on low-rise buildings. Part 1 - Basic information and first results. *JWEIA Vol. 64, 101-125.* Amsterdam: Elsevier.

ESDU - Engineering Science Data Unit. 1986 *Characteristics of atmospheric turbulence near the ground - Single point data for strong winds.* ITEM 86010

Cook, N.J. 1985. *The designer's guide to wind loading of building structures. Part 1: Background, damage survey, wind dataand structural classification.* London: Butterworths

DRAFS - Dynamic Reliability Analysis of Frame Structures. Vers. 1.3 1998. University of Bochum, Germany.

Wind Engineering into the 21st Century, Larsen, Larose & Livesey (eds) © 1999 Balkema, Rotterdam, ISBN 90 5809 059 0

Wind loading aspects of buildings in groups: Mass-housing developments in South Africa

M. Kasperski & H.-J. Niemann
Aerodynamik im Bauwesen, Ruhr-Universität Bochum, Germany
A. M. Goliger
Division of Building Technology, CSIR, Pretoria, South Africa

ABSTRACT: This is the first in a series of papers describing the results of a comprehensive research programme into the wind loading of low-rise buildings arranged in groups. This programme is aimed in particularly at mass housing developments in South Africa and it has been sponsored by the German Research Foundation - Deutsche Forschungsgemeinschaft (DFG). The main purpose of the research work is to include in the design the effect of a considerably changing overloading over the life time of the structure due to the rapid growth of the townships.

1 INTRODUCTION

Low rise residential buildings, like other structures, are designed for wind effects, on assumption that they are exposed to the action of wind in isolation. The loads are then determined by the details of their geometry. In the presence of neighbouring structures (or other obstructions), the flow and pressure field will often overrule the influence of the shape. Despite of this, most of the research on low-rise structures has been aimed at an idealised case of individual buildings placed in well developed, standard boundary layer flow conditions. The information obtained from such studies formed the basis of various codifications.

Some studies which deal with the effects of grouping of houses have been reported (Hussain & Lee, 1980; Ho et al, 1990). Surry and Lin (1995) suggest that the influence of surrounding structures is very complex and configuration dependent, but in general a reduction in mean roof suction of between 50% and 65% (as opposed to the idealised case of an isolated building) can be expected. This finding indicates not only a potential for significant savings in the costs of the structures but also a challenge and need for a more in-depth comprehensive research which would lead to a certain generalisation and subsequent codification.

The current paper presents the results of an initial stage of a comprehensive study aimed at developing a more realistic, and also economic description of the design wind loads for mass housing developments in South Africa where an ambitious programme of constructing one million houses within 5-year period has been undertaken.

The paper includes selected information on densities of houses, wind exposure of townships and their spatial growth, in time domain. All these have been obtained from an extensive full-scale survey of the existing and proposed large township developments.

A comparative assessment of wind loading stipulations of several codes, in the context of low-rise buildings, is subsequently reported in order to highlight the similarities and discrepancies which are present.

Finally the issue of the life time target probability of failure in respect to low-rise buildings is considered. A comparison of different codes reveals that an internationally agreed value of the target reliability is still missing.

2 MASS HOUSING DEVELOPMENTS IN SOUTH AFRICA

2.1 *General remarks*

At the initial stage of the project a full scale data collection programme was undertaken. The aim of this programme was to assess the relevance of wind damage to the low-rise housing developments in South Africa and to obtain data on the characteristics of those developments in respect to:

- the form (i.e. geometry) of housing units,
- land use in townships (i.e. spatial arrangement of housing units), and
- growth of townships in time domain, affecting the statistics of wind exposure of houses.

In view of the volume of the information involved, the results of the survey will not be reported in the current paper but rather the most relevant issues will be highlighted.

2.2 *Wind damage to low-rise housing developments*

Damage due to strong winds in South Africa is not unusual. Survey of damage suggests that typically more than five large devastating wind events occur each year in which thousands of people could be left homeless or hundreds of houses damaged.

However, it is believed that in many cases the failures are due to faulty design and construction, rather than the wind speeds exceeding the values specified in the code and/or building regulations. Example of such situation is shown in Figure 1 which presents a failure of a large unsupported gable wall. Frequently, the faulty construction results from ignorance of the nature of wind forces e.g. neglecting of negative pressures.

Figure 1. Wind destruction of a house

Interestingly, evaluation of areas subjected to severe winds suggests that in many cases informal housing suffers less damage than formal type of structures. This is due to the form of roofs used in informal construction and also the tradition of placing loads (usually rocks) on the roof sheeting to prevent them from uplift.

Figure 2. Traditional method of preventing the roof uplift

2.3 *Forms of houses*

Figure 3 is a photograph of typical formal housing settlement and Table 1 gives the results of data collection on housing geometry. It can be seen that the area of houses varies between 30 and 50 m² and the roof slope between 3 and 26°. Various other characteristics of houses and their construction methods are included in (DFG 1998) and will not be discussed in the current paper. It is relevant to mention, however, that although both mono- and duo-pitches are used, there is a perception within the low-income communities that duo-pitches are of better standard, and therefore there is a demand (or preference) for this type of roofs. Overhangs are typically used along the side walls and to a lesser extent also along the gable walls as it can be seen in Figure 3.

Figure 3. Typical housing development

Table 1. Characteristics of typical houses

characteristics of units	quantity	range	dimensions commonly encountered
plan area	m²	30 - 56	40
width	m	3.5 - 7.5	6.0
eaves height	m	2.5 - 2.8	2.6
slope of roof	°	3 - 26	10

Interestingly, despite the fact that the roof slope of 10° corresponds to a largest external pressure coefficient stipulated in the South African loading code (SABS 0160-1989), this slope, is in fact, most frequently used in large housing developments.

Finally an interesting concept of housing developments in South Africa, affecting the form and resistance of individual housing units, can be mentioned. It refers to the principle of incremental housing in which a portion (or an empty shell) of a house unit is initially erected and the structure is expected to be completed by its owner (or user). Unfortunately large portion of those structures never become finalised and, therefore, remain in their initial form and unsatisfactory rigidity.

2.4 *Spatial arrangement of houses*

Townships at eight different locations were analysed so to investigate whether there are any similarities regarding the allocation of land for different uses. The distribution of land uses, in turn, determines the layout of townships, and the density of the built-up environment which determines the wind envelope ie. the wind loads of houses.

The use of the land can be broadly described as follows:

- residential, in which vast majority of units are free standing and single storeys,
- streets,
- community facilities (i.e. schools, churches, sport-grounds),
- public open space (parklands), and
- business and industrial, which usually occur in the vicinity of main transportation routes.

The results of the analysis presenting the typical distribution of land use are given in Figure 4, from which it can be seen that most of the land (on average about 80%) is used for residential housing and streets.

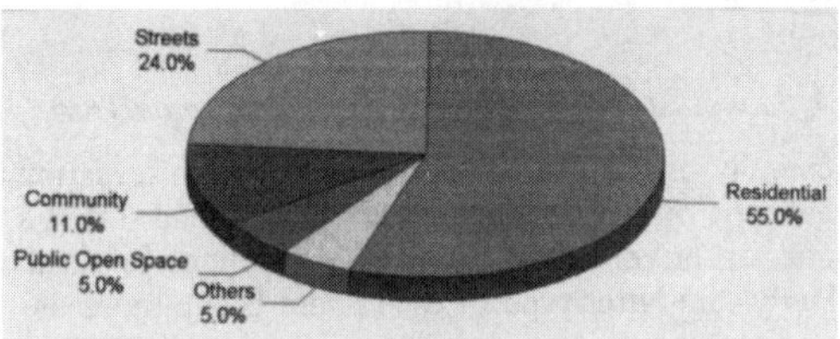

Figure 4. Land use in townships

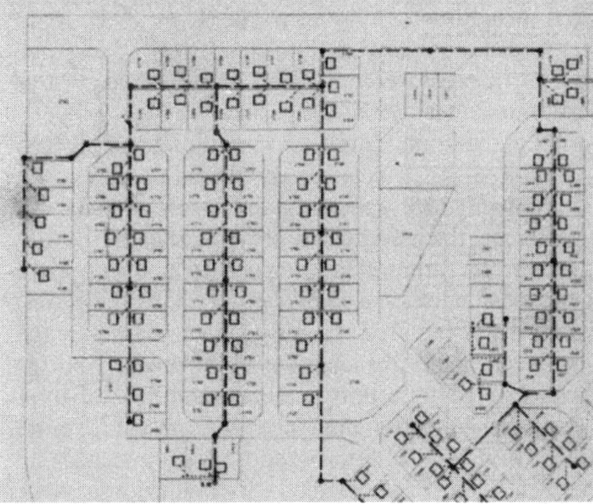

Figure 5. Typical layout of a township

Statistics on density of housing units have been performed. As major result, the average size of stand is obtained. For the wind tunnel study, a size of 250 m² is used.

It was not possible to determine a typical (or preferred) layout of South African townships as these varied substantially depending on several factors i.e. geographic location, composition of the design and development team, availability of funds, etc. One of the layouts which have been analysed is presented in Figure 5.

2.5 *Approach wind conditions and growth of townships*

In general, township developments are located in flat terrain without any prominent features affecting the approach wind flow, although remains of farm infrastructure, i.e. water reservoirs, dams etc. can often be found. Due to the nature of the development of low income townships, tall vegetation within and nearby to the site is invariably removed.

On attempting to obtain the data on township growth, it became apparent that information of this nature is not readily available. Nevertheless, a sample information from six low income housing development areas was analysed.

A characteristic feature of a typical township is that it grows gradually until it becomes restricted by natural obstructions or other limitations (i.e. rivers, wetlands or private ownership of the land). At that stage, the growth stops, but the density of township starts to increase, as a result of additional housing units being built in between the existing ones. At certain stage in the live of a township, and depending on the cause that prevented the growth, it can be overcome and a rapid growth of township, typically within a certain portion or direction, will occur. This refers both to formal and informal housing units. Sample data suggests that between 500 to 1000 dwellings can be typically erected within 3 to 5 years. Such increase translates to approximately a couple of hundred meters of a developed stretch of land. This affects the exposure of the existing houses, in other words the housing units which were at the perimeter of a township (i.e. fully exposed to wind action) become shielded.

It has to be noted, that the amount of data which was available, does not justify to draw any conclusions regarding the effect of the growth of townships in relation to the prevailing wind directions.

3 COMPAPARATIVE ASSESSMENT OF LOADING CODES

3.1 *Description of wind profile at low elevations*

The code models for describing the influence of the terrain roughness on the design wind velocity pressure q_{des} differ considerably. Comparing some leading codes (Austria, Australia, Canada, Germany, South Africa, UK, USA), a range is obtained from 0.625 to 0.842 for ratios of q_{des} (z = 3 m) to q_{des} (z = 10 m). Some codes on the other hand do not allow to reduce the design wind speed for low elevations thus leading to a conservative but uneconomic design.
Partially, this large scatter is due to different concepts to include in the design wind load the gust effects, i.e. the ratios include the mean wind velocity profile and an estimation of the gust factor and the choice on an appropriate gust duration. A bias due to the latter two points can be avoided if the wind tunnel experiments are using directly the mean wind pressure to normalise the pressure coefficients. Then, only the problem remains to choose the appropriate terrain roughness. For the situation in South Africa, it can be assumed that the roughness length z_0 varies only slightly between say 0.01 to 0.03 m. ESDU (1985) then gives the respective ratios for the mean wind speed at 3 m height to the reference wind speed in 10 m height in the reference terrain roughness category with values from 0.905 to 0.78. The corresponding values of the turbulence intensity are 0.185 to 0.205.

3.2 *Wall pressures*

For a standard house (8x6x3m), the comparison of codes in regard to the specified pressure coefficients for walls leads to a range of 0.8 to 0.65 for the upwind wall, i.e. an 'uncertainty' about the appropriate pressure coefficient of more than 20% is obtained. The respective scatter for the downwind is even larger: the specified coefficients range from -0.2 to -0.5. Designing the downwind wall for a load induced by the wake pressure and a positive internal pressure thus leads to load differences from 0.9 to 1.3 or an 'uncertainty' of 30%.

3.3 *Roof pressures*

The separated flow and its effects on the roof pressures are difficult to model in a wind tunnel test. Hence, it is not too surprising that the code models offer a large variability of roof pressure coefficients. In figure 6, an enveloping contour is shown for the pressure coefficients obtained from the investigated leading codes. Obviously, the aerodynamic effects are treated with different philosophies. A lot of codes specify for a wider range of roof slopes two pressure

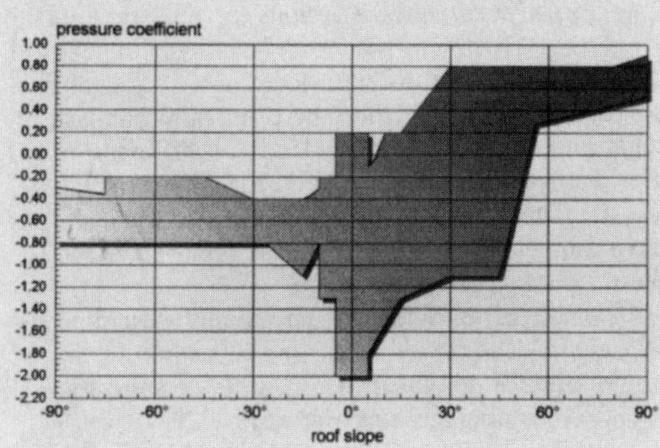

Figure 6: Band width of codified roof pressure coefficients (wind load codes from Austria, Australia, Canada, Germany, South Africa, UK, USA)

coefficients to cover the complexity in the possible effects of the separated flow. However, a general agreement on the appropriate range of possible pressures is missing. Some older codes on the other hand do still use only one coefficient for a specific roof pitch angle.

3.4 *Conclusions drawn from the code comparison*

Generally, the codes will not provide the information what type of pressure coefficient is specified, i.e. if a mean or an extreme pressure coefficient is used. Clearly, the latter type of coefficient is the more appropriate one. In e.g. the Eurocode, a lot of pressure coefficients are of that modern type, using the 78% fractile of the extreme pressures to define the pressure coefficient for the code.

The question may arise, why the presented research programme intends to perform even more wind tunnel studies which will lead to only just one more set of pressure coefficients. The answer to this question includes two major aspects:

- Firstly, with no doubt an internationally agreed method of analysing extreme wind loads is still missing. Strictly speaking, the target fractile depends on the coefficient of variation of the extremes. To estimate this value with an appropriate statistical stability, a larger ensemble of independent runs is required. If only a few runs (e.g. 16) are performed, an adjusting factor should be introduced to take into account the then inherent statistical uncertainties. For the present study, however, this additional source of uneconomic design has to be avoided.
- Secondly, most codes are not able to provide the information on the correlation of extreme effects. For the design of e.g. the gable wall, the simultaneous wind induced actions on the roof (suction) and on the gable wall (pressure) are required. The weighting of these actions will be performed on

the level of stresses in the wall, i.e. specific details of the construction will be taken into account.

4 TARGET RELIABILITY

4.1 *Life expectancy of housing units*

Historically, in view of the severe shortage of suitable accommodation for millions of people in South Africa, certain portion of housing developments have been earmarked as being provisional or temporary, with a life expectancy of 10 to 15 years, after which they were supposed to have been upgraded or replaced. However, in view of the current demand for new units and the growing backlog, that principle has been never been adhered to, and the true life-span of housing units is (and will be) much longer.

In view of that issue, and for the purpose of the current study a standard 50 year life expectancy of structures was adopted.

4.2 *Change in the exposure*

Full-scale investigation (Section 2.4) suggests that the wind exposure of a typical housing unit changes with time. It may initially exist as an isolated structure and with time it will end-up within a dense cluster of similar units, well away from the fringes of a township.

For the purpose of the current study, three typical situations and examples of corresponding time lags have been adopted namely:

- house fully exposed (i.e. flow over flat, open country - terrain category 2), for a period of 0 to 5 years,
- house partially protected in a denser group, group still exposed (i.e. terrain category 2) for a period of 5 to 15 years, and
- house fully protected (i.e. a group in a built-up area - terrain category 3) for a period of 15 to 50 years.

4.3 *Overloading risk due to wind action*

Generally, the design wind load is specified based on a reference wind speed v_{ref} with a specific exceedence probability p_{ex} per year and a partial safety factor γ_w. Comparing some leading codes, a large variety for possible combinations of $p_{ex}(v_{ref})$ and γ_w is obtained. The overloading risk, i.e. the risk of exceeding the design wind load in the life time of the structure is obtained from the exceedence probability of the design wind speed v_{des} which is given as:

$$v_{des} = \sqrt{\gamma_w} \cdot v_{ref} \qquad (1)$$

Table 2: Exceedence probabilities of the reference wind speed per year, partial safety factor for wind loads and basic overloading risk in 50 years life time assuming a Gumbel distribution with a coefficient of variation of 12.5% (typically for storms due to frontal depressions)

	$p(v>v_{ref})$ per year	γ_w	p_{ex} in 50 years life time
Australia	0.001	1.0	0.050
Canada	0.02	1.5	0.047
Germany	0.1	1.5	0.309
Great Britain	0.02	1.4	0.078
Japan	0.01	1.0	0.402
Netherlands	0.08	1.5	0.242
New Zealand	0.001	0.93	0.086
South Africa	0.02	1.35	0.139
Switzerland	0.02	1.5	0.047
USA	0.02	1.3	0.139
Eurocode	0.02	1.5	0.047

To work out the differences in the overloading risks, at this stage, it is reasonable to assume a Gumbel distribution for the extreme wind speeds with e.g. a coefficient of variation of 12.5% which is a typical value for storms due to fully developed frontal depressions. Table 2 gives selected values of p_{ex} and γ_w and the resulting overloading risk. The range of obtained overloading risks varies between 0.05 in the life time of the structure (Eurocode, Australia, Canada, Switzerland) and 0.4 (Japan and Germany). While the latter value with no doubt is too high, it is on the other hand not clear if such a high wind safety standard as it is required in Canada, Australia and the European Union is an appropriate target for a developing country like South Africa. Therefore, at this stage, the study adopts the level of overloading risk which is already obtained with the current South African wind load code, i.e. the target reliability is 14% overloading risk in 50 years life time.

To meet this target, the changing overloading risk per year has to be accumulated over the life time of the structure as follows:

$$p_L(w > w_{des}) = 1 - \prod_{i=k}^{k+L}\left(1 - p_i(w > w_{des})\right)^L \qquad (2)$$

p_i - overloading risk in year i
k - start of exposure
L - life-time of the structure

Figure 7 shows the basic principle of the step-wise changing overloading risk and the accumulation over life time. To meet the target overloading risk of the

South African Standard, an exceedence probability of 1/333 per year is required for the design wind speed for the idealised code situation with an isolated building in open country. Assuming as example a relative shelter effect of 20% for situation 2 (grouped buildings) and 50% for situation 3 (complete township), the required exceedence probability of the design wind speed is reduced to 1/130 which leads to a reduction factor of 0.82 for the design wind speed and 0.67 for the design wind load. As can be seen in the lower part of figure 7, the reduced design wind speed leads to the required overloading risk of 14%, however, most of this risk is accumulated in the first 15 years.

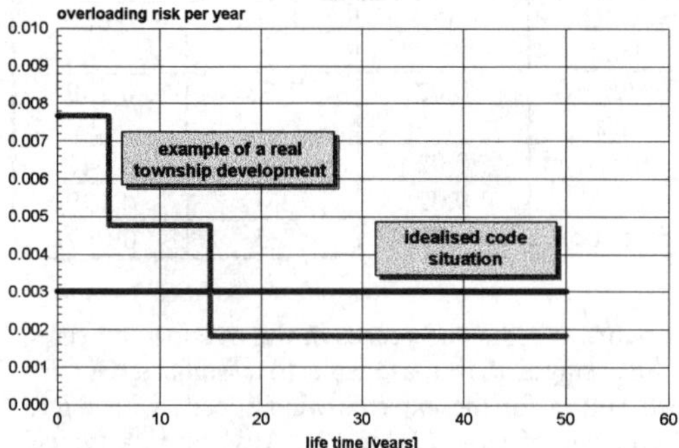

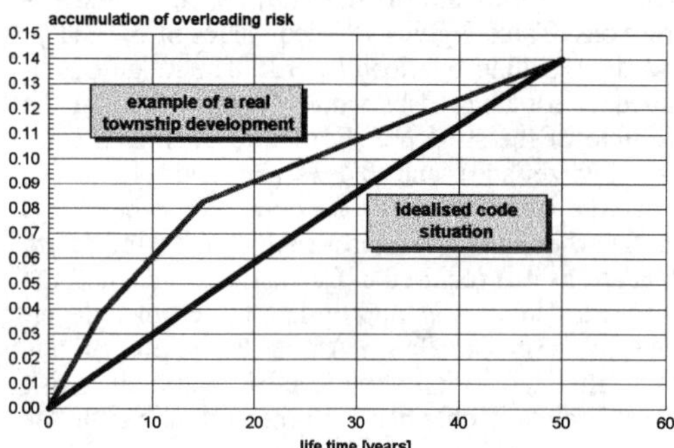

Figure 7: Example of stepwise change of the overloading risk and accumulation over the life time compared to the idealised situation

5 WIND-TUNNEL TESTING

A programme of series wind-tunnel experiments has been developed. For that a typical house unit with a plan of 8x6 metres and eaves-height of 3 metres has been adopted. A duo-pitch roof has been modelled with a variable pitch from 5° to 30° and a step of 5°. A wind-tunnel scale of 1:75 has been used.

Two boundary layer simulations have been modelled which in-turn correspond to two different exposures, as discussed in Section 4.2.

At the time of submission of the current paper the construction of models and modelling of the boundary layers have been completed. Preliminary results of these tests will be included in the conference presentation and will be published soon.

6 CONCLUSIONS

The current paper presents the results of an initial stage of the study into a more realistic and economic description of wind loads for low-rise housing units in groups. Full-scale (i.e. real-life) information has been highlighted and comparative assessment of wind loading stipulation included in various codes is presented. A realistic model for wind exposure and overloading risk has been adopted. A set of comprehensive wind-tunnel measurements is currently in progress and their results will be reported in the future.

7 ACKNOWLEDGEMENTS

Part of this work has been sponsored by the German Science Foundation DFG (Deutsche Forschungsgemeinschaft) and the German Foundation for technical cooperation GTZ (Gesellschaft für Technologische Zusammenarbeit). These financial supports are gratefully acknowledged

8 REFERENCES

DFG 1998. Safety oriented design wind loads for low-rise buildings arranged in groups, with particular consideration of mass housing in South Africa, *Interim Report*, Volume 1, November 1998

ESDU - Engineering Science Data Unit 1985. Characteristics of atmospheric turbulence near the ground - Single point data for strong winds, ITEM 85020

Ho, T., Surry D. & Davenport A. 1990. The variability of low building wind loads due to surrounding obstruction. *Journal of Wind Engineering and Industrial Aerodynamics* 36: 161-170

Hussain, M. & Lee, B. 1980. A wind-tunnel study of the mean pressure forces acting on large groups of low-rise buildings. *Journal of Wind Engineering and Industrial Aerodynamics* 6: 207-225

Surry, D. & Lin, J. 1995. The effect of surrounding and roof corner geometric modifications on roof pressures on lo-rise buildings. *Journal of Wind Engineering and Industrial Aerodynamics*, 58: 113-138

Wind Engineering into the 21st Century, Larsen, Larose & Livesey (eds) © 1999 Balkema, Rotterdam, ISBN 90 5809 059 0

Peak suction and conical vortex on a flat roof

H. Kawai & F. Ohnuma
Tokyo Denki University, Hatoyama, Saitama, Japan

ABSTRACT: Characteristics of peak suction and conical vortex on a flat roof were investigated by wind tunnel tests. Conditional sampling techniques were used to clarify the pressure pattern and its relation to the conical vortex on the roof when very large peak suction was occurring near a windward corner in oblique wind. It is found that a large suction area expands to a half of the roof when the peak suction is occurring, that is almost similar in any wind direction from 0 to 45 degree. The peak suction is induced by a strong conical vortex, which is developed by an approaching gust to the eave from around 25-degree.

1 INRODUCTION

It is well known that large peak suction occasionally occurs along the leading edge close to the roof corner of a low-rise building when flow attacks at an oblique angle to the edge. This suction, which is induced by the conical vortex on a roof, as a result of flow separation from the leading edge, is extremely large at times and often causes severe damage to the roof in strong wind conditions (Metha et al. 1992).

Kawai found that a large suction region extends to the one side of the roof where the peak suction occurs near the roof corner (Kawai & Nishimura 1996, Kawai 1997). This corresponds with Letchford's findings from the field measurement at TTU experimental building that the suction fluctuations at various points on the roof correlate well when the peak suction occurring (Letchford 1995).

The occurrence of the peak suction has a strong relationship with approaching turbulence to the roof. Particularly, the fluctuation of wind direction correlates well with the suction fluctuation (Kawai & Nishimura 1996). Recently, Sarkar found from the field measurement at TTU experimental building that the peak suction was caused by high wind speeds on the occurrence of certain wind angles (Sarkar & Zhao 1998). Kawai also reported that the peak suction may be brought by the gust which attacks at the angle of attack of 25 degree to an eave through the experimental results of simultaneous pressure measurements at multiple points on a flat roof (Kawai & Ohmuma 1998).

This paper investigates the characteristic of the peak suction and the mechanism and the condition of its occurrence by simultaneous measurements of pressure and velocity fluctuations on the roof and walls of a low-rise building with a flat roof in various wind-directions. The main purpose of the paper is to clarify the structure of the flow or the conical vortex when the peak suction is occurring, and to find the cause or the condition of the occurrence of the peak suction. In the analysis of the data, we employed a conditional sampling technique and then were able to describe successfully the clear pattern of the conical vortex on the roof when the peak suction was occurring.

2 OUTLINE OF WIND TUNNEL EXPERIMENT

Simultaneous pressure measurements at various points on the roof and the walls (hereafter referred as Exp1) were carried out in the boundary layer wind tunnel of Maeda Corp. with the working section of 2.3m x 2.0m. A building model with a flat square roof of 120mm x 120mm and 60mm in height was used in the experiment (Figure 1). The model was set in the turbulent boundary layer flow, which simulated the flow over open terrain. The turbulence intensity was 18% at the top of the model. The mean wind speed profile fitted well with the power law of the exponent of 0.15. The velocity of the top of the model was 9.2m/s which was corresponding to the Reynolds number based on the side of the roof of 8.23×10^4. The pressure fluctuations were measured simultaneously at 121 points on the roof and 62 points on the walls by electric pressure scanners of ZOC23B of Scanivalve Corp. The angle of attack was changed from 0 degree (normal to the eave) to

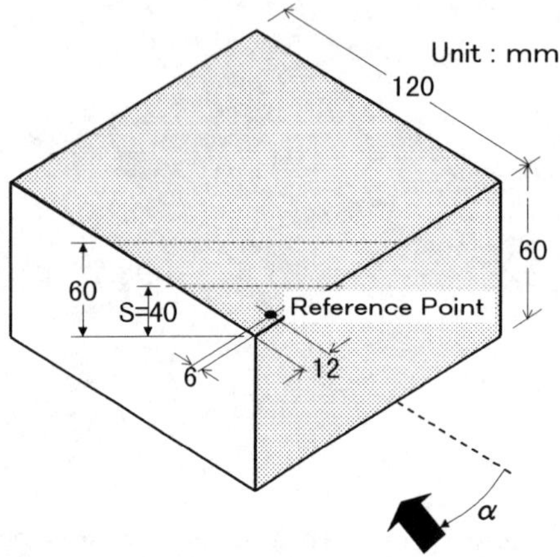

Figure 1. Model and the reference point for conditional sampling

45 degree (diagonal direction) with 5-degree interval.

The simultaneous measurements of velocity over the roof and pressure fluctuation near the roof corner (hereafter referred as Exp2) were carried out in the boundary layer wind tunnel of Tokyo Denki University with the working section of 1.2m x 1.2m. The measurements were to investigate the flow pattern when the peak suction was occurring. The same model was used in the experiment. The longitudinal, lateral and vertical components of the flow in the sections normal to the diagonal centerline at the distance of S=40mm and 60mm from the roof corner shown in Figure 1 were measured by an X-type hot wire probe. A pressure transducer of PD104K of Toyoda Corp was used in the measurement of the pressure fluctuation. The angles of attack were 25 degree and 45 degree. The hot wire probe was set in a line of the mean flow direction both in the angles of attack.

The turbulence intensity of the approaching flow was 22% at the top of the model in Exp2. The mean wind speed profile fitted well with the power law of the exponent of 0.25. The velocity of the top of the model was 7.9m/s which was corresponding to the Reynolds number based on the side of the roof of 7.07×10^4.

3 EXPRIMENTAL RESULTS

20000 pressure signals at various points were simultaneously sampled in every 0.001-second in Exp1. On the other hand, 10000 velocity and pressure signals were simultaneously sampled by every 0.002-second in Exp2. After these signals were normalized by the mean velocity pressure or the mean

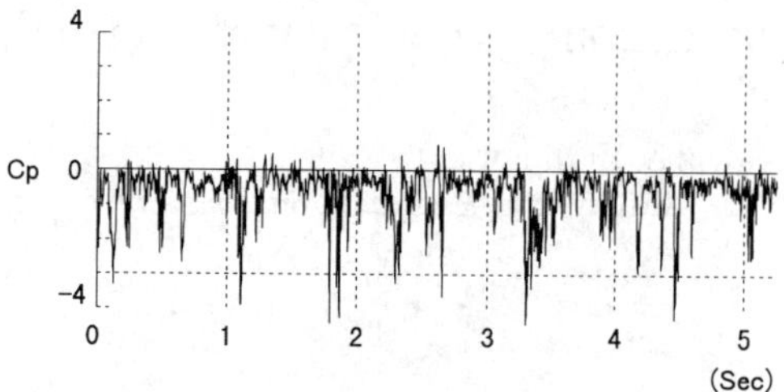

Figure 2. Trace of pressure fluctuation on the reference point

velocity at the top of the model, the signals were resampled under the condition when the instantaneous pressure at a reference point was smaller than a predetermined triggering level. The point on the roof close to the windward corner shown in Figure 1 was chosen as the reference point, because the pressure at the point fluctuates in the spike-like pattern shown in Figure 2 to detect easily the occurrence of the large peak suction. The triggering levels of the pressure coefficient at the reference point are –3.5 for Exp1 and –3.0 for Exp2 respectively.

3.1 *Distribution of mean pressure coefficient*

Figure 3(a) shows the contour maps of the mean pressure coefficients at the angles of attack of 15, 25, 35 and 45 degrees. The shaded zone of the roof shows the domain in which the pressure coefficient is smaller than -0.8. The shaded zone of the windward wall shows the domain in which the pressure coefficient is larger than 0.6.

The large suction zone spreads to the downstream near the windward eave, which is probably induced by the conical vortex on the roof. The zone shrinks gradually to the windward eave as the angle of attack increasing. The maximum mean suction occurs at the point a little apart from the roof corner. The distance between the point and the windward corner increases a little as the angle of attack increasing. The minimum mean pressure coefficient is -2.2 at the angle of attack of 35 degree.

The positive pressure on the windward wall decreases and the point of the maximum positive pressure moves to the windward corner gradually as the angle of attack increasing.

3.2 *Conditional pressure coefficient*

Figure 3(b) shows the contour maps of the conditional pressure coefficients on the occurrence of the large peak suction. The conditional pressure coefficient is the average of the pressure coefficient sampled when the reference pressure coefficient is less than –3.5. The shaded zone of the roof shows the domain in which the pressure coefficient is smaller

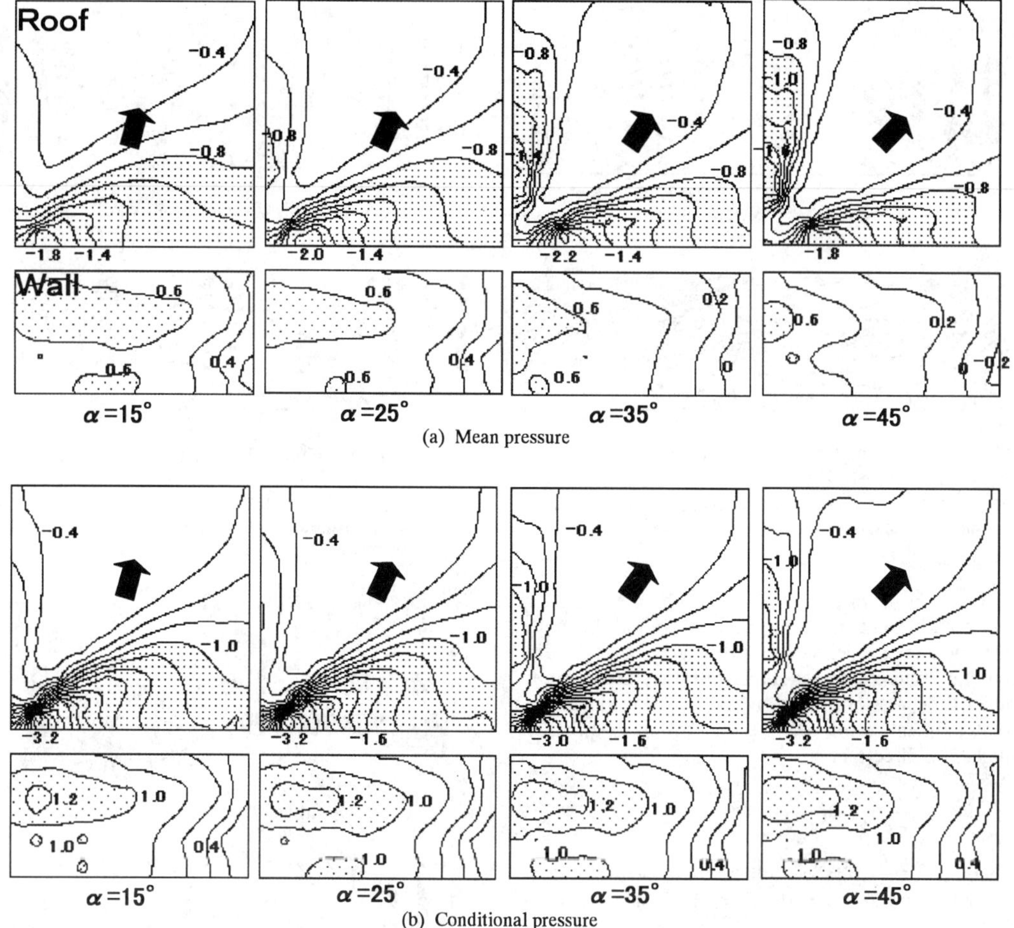

(a) Mean pressure

(b) Conditional pressure

Figure 3. Contour maps of mean and conditional pressure coefficients at various angles of attack

than -1.0. The shaded zone of the windward wall shows the domain in which the pressure coefficient is larger than 1.0.

When the very large peak suction occurs at the reference point, the large suction zone expands to the diagonal centerline of the roof on the one side. At the same time, the large positive pressure acts on a half of the windward wall. The pressure distributions on the side of the roof and on the windward wall are little affected by the angle of the attack. This fact suggests that the peak suction may occur at the same physical condition at the various angles of attack.

3.3 *Structure of conical vortex*

The mean and conditional flow vectors are obtained from the vertical and lateral components of the velocity and are shown in Figure 4 and Figure 5 in the case of S=40mm. The conditional flow vectors are based on the velocity signals when the reference pressure coefficient is smaller than –3.0

In the case of the angle of attack of 45 degree, the pattern of the mean flow vectors is symmetry to the diagonal centerline and a pair of conical vortices is observed on the roof as shown in Figure 4(a). When the peak suction occurs, the velocity increases all over the section and the fluid moves from the right side to the left side in the section as shown in Figure 4(b), where the peak suction occurs at the right side. The vortex at the left side seems to be swept away.

On the other hand, a pair of conical vortices is asymmetry to the centerline at the angle of attack of 25 degree in the mean flow vectors shown in Figure 5(a). The vortex at the right side is more strong and larger than that at the left side. When the peak suction occurs, the vortex at the right side becomes stronger as shown in Figure 5(b).

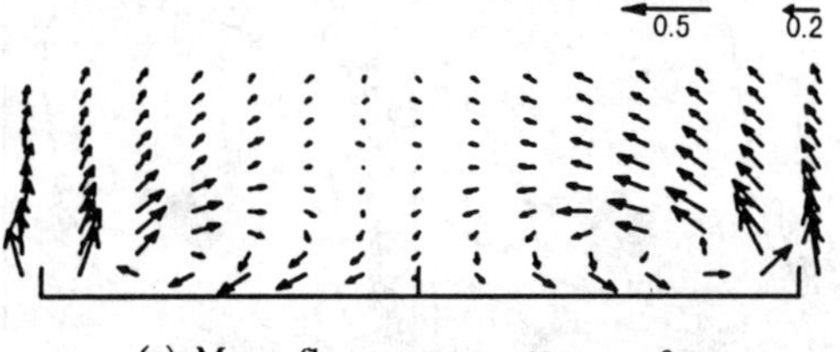

(a) Mean flow vector, s=40 α=45°

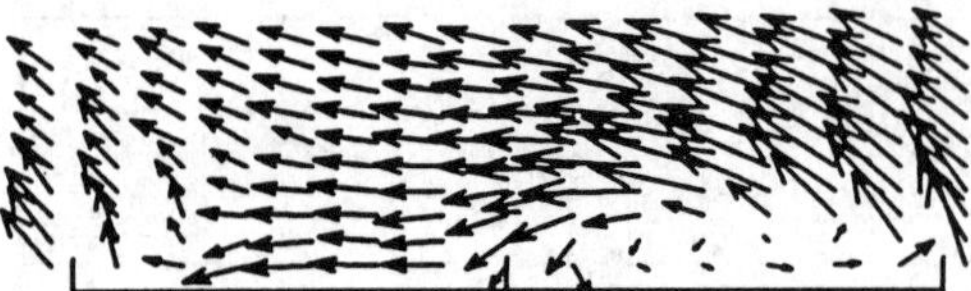

(b) Conditional flow vector, s=40 α=45°

Figure 4. Mean and conditional flow vectors in the section normal to the diagonal line of the roof, α=45°

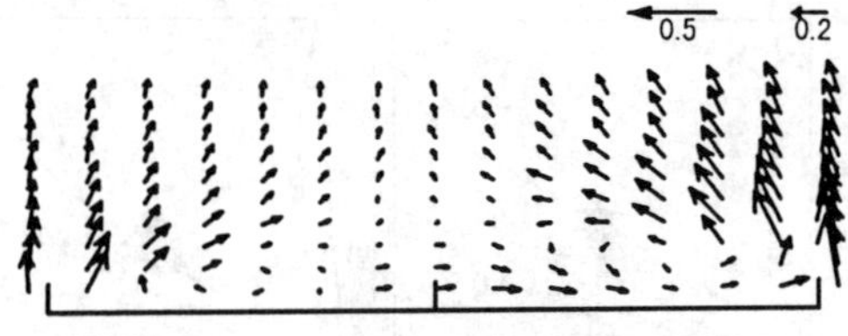

(a) Mean flow vector, s=40 α=25°

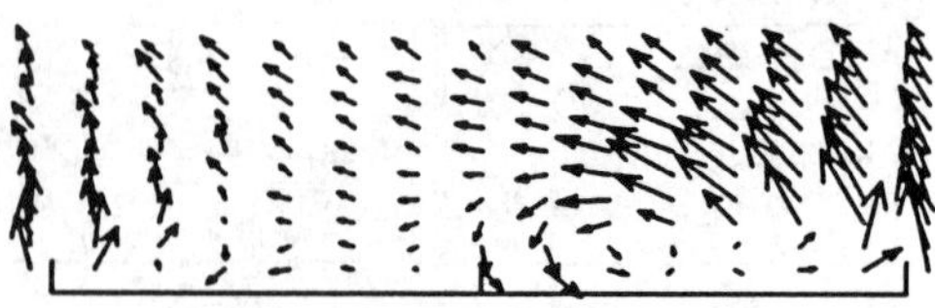

(b) Conditional flow vector, s=40 α=25°

Figure 5. Mean and conditional flow vectors in the section normal to the diagonal line of the roof, α=25°

The flow from the right to the left in the section observed in the conditional flow vectors at the angle of attack of 45 degree is probably brought by the oblique flow to the section. The oblique angle to the mean flow was estimated by the equation (1) as the total momentum in the section becomes vanish in the lateral direction.

$$\theta = \tan^{-1}\left(\frac{\sum qv}{\sum qu}\right), \qquad q = \sqrt{u^2 + v^2 + w^2} \tag{1}$$

where θ is the oblique angle and u, v, w are the longitudinal, lateral and vertical component of the conditional velocity.

The oblique angles were 13.2 degree at the angle of attack of 45 degree and 4.8 degree at the angle of attack of 25 degree. The lateral component of the velocity was corrected by adding $u \sin\theta$ and the conditional flow vectors were redrawn in Figure 6 together with the conditional pressure distributions, which are obtained from contour maps shown in Figure 3. By the correction, the large and strong conical vortex emerged at the right side of the roof. On the other hand, the vortex at the left side is very week and small. The corrected flow vectors and the conditional pressure distributions at the angle of attack of 25 degrees are very similar to that at 45 degree. The suction becomes the maximum around the center of the conical vortex both at the angles of attack of 25 and 45 degrees.

Figure 7 shows the contour maps of the longitudinal component of the velocity at the angle of attack of 45 degree. Figure 7(a) is the mean velocity component, 7(b) is the conditional velocity component and 7(c) is the difference between the conditional and mean components. When the peak suction

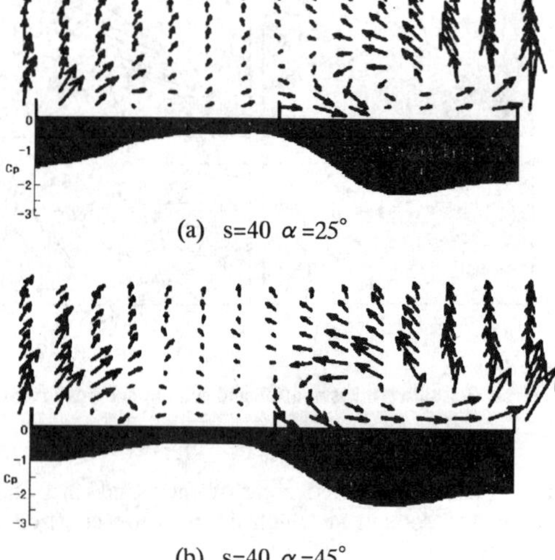

(a) s=40 α=25°

(b) s=40 α=45°

Figure 6. Corrected conditional flow vectors and pressure distributions

occurs at the right side, the longitudinal component of the velocity increases considerably almost over the section particularly near the right side eave except at the center of the conical vortex. In other words, the strong conical vortex is induced when the velocity increases over the roof.

Figure 8 shows the contour maps of the longitudinal component of the velocity at the angle of attack of 25 degree. The contour map of the longitudinal component of the conditional velocity is very

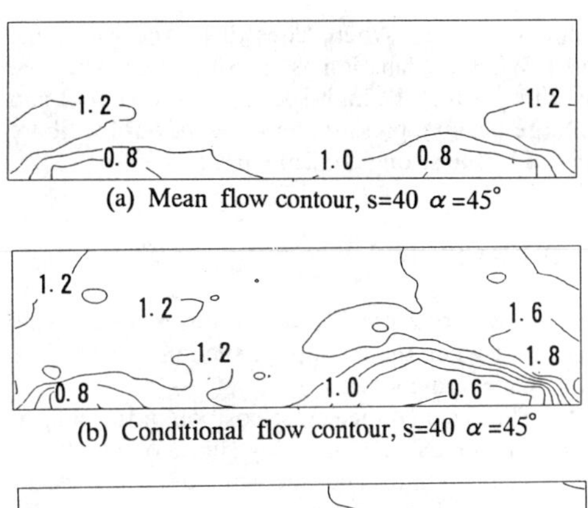

(a) Mean flow contour, s=40 α =45°

(b) Conditional flow contour, s=40 α =45°

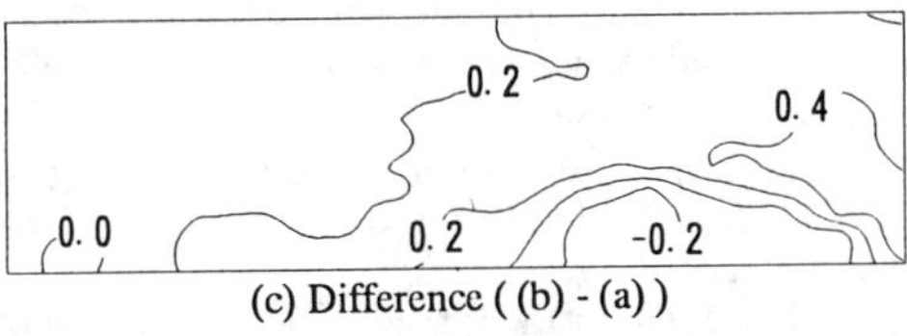

(c) Difference ((b) - (a))

Figure 7. Contour maps of longitudinal component of velocity in the section normal to the diagonal line of the roof, α=45°

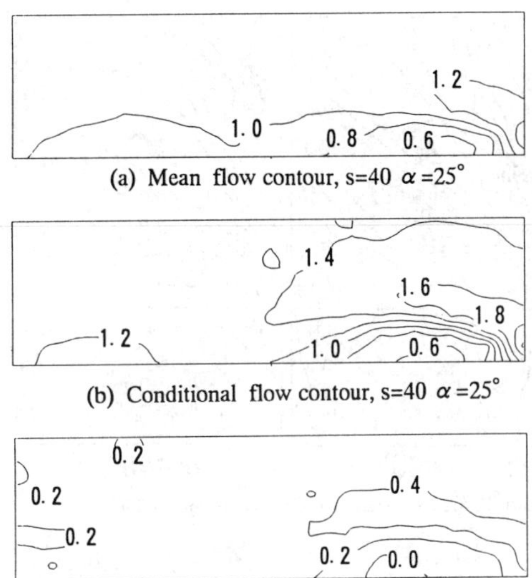

(a) Mean flow contour, s=40 α =25°

(b) Conditional flow contour, s=40 α =25°

(c) Difference ((b) - (a))

Figure 8. Contour maps of longitudinal component of velocity in the section normal to the diagonal line of the roof, α=25°

similar to that at 45 degree in spite of the large difference between the maps of the mean velocity at 25 and 45 degrees. Therefore, the condition of the large and strong conical vortex on the roof at the angles of attack of 25 degree may be as same as that of 45 degrees.

4 DISCUSSION

The experimental data described in the previous section show that the large peak-suction is induced by the strong conical vortex on the one side of the roof, which has very similar at any angles of attack. What is more, the data seems to suggest that the strong conical vortex may be formed on the roof when the high-speed gust attacks to the roof from the certain angle of attack. What angle is the most appropriate for forming the strong conical vortex? Is the high-speed gust always necessary to form the vortex?

4.1 *Angle for the strong conical vortex*

As already discussed in Section 3.2 and 3.2, the contour maps of the conditional pressure coefficients at any angles of attack have the very similar pattern of that of the mean pressure coefficients of the angle attack of 25 degree. Moreover, the pattern of the corrected conditional flow vectors at the angles of attack of 25 and 45 degrees is also similar to that of the mean flow vectors at the angle of attack of 25 degree. These facts strongly suggest that the most appropriate attack angle of the approaching gust to the roof for forming the strong conical vortex may be around 25 degree.

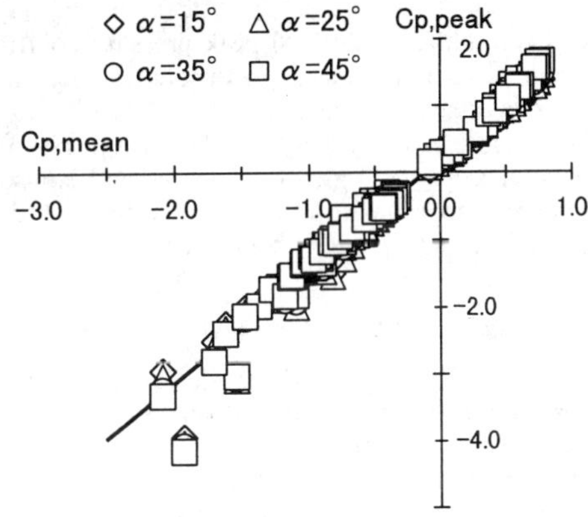

Figure 9. Relation between mean pressure coefficients at the angle of attack of 25° and conditional peak pressure coefficients.

In order to verify this fact, the scatter diagram between the mean pressure coefficients at the angle of attack of 25 degree and the conditional peak pressure coefficients at the angles of attack of 15, 25, 35, 45 degrees is drawn in Figure 10. The lateral axis is the mean pressure coefficient and the vertical axis is

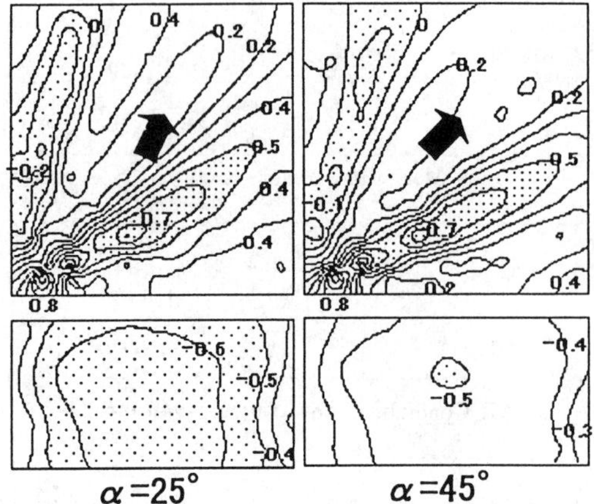

Figure 10. Cross correlation of pressure fluctuations

the conditional peak pressure on the side where the large peak-suction occurs near the windward corner. Except for 2 or 3 points, almost all data including the positive and negative pressure are on the line, which is expressed by the equation (2).

$$C_{p,peak} = 1.65C_{p,mean} + 0.19 \quad (2)$$

where $C_{p,peak}$ is the conditional peak pressure coefficient and $C_{p,mean}$ is the mean pressure coefficient at the angle of attack of 25 degree.

This simple relation supports the hypothesis of the most appropriate angle of 25 degree and is also very variable to estimate the peak pressure coefficient from the mean pressure coefficient.

4.2 *Necessary conditions for the peak pressure*

It is seems to be suggested by the large positive pressure and the large longitudinal velocity when the peak pressure occurs that the strong conical vortex may be formed by the high-speed approaching gust attacking to the roof. Figure 10 shows the cross correlation between the pressure fluctuation at the reference point and the other point. The high positive correlation zone spread to the downwind along the ridge of the large peak-suction zone shown in Figure 3. It is worthy of notice that the correlation with the positive pressure on the windward wall is very high and negative, which is already guessed by the contour maps of the conditional pressure coefficient. The largest negative correlation occurs around the center point of the wall. This fact seems again to support that the high-speed gust attacking to the wall from the angle of attack of 25 degree may induce the strong conical vortex on the roof.

However, the correlation between the pressure fluctuation at the reference point and the longitudinal component of the approaching wind is rather poor as already pointed by Kawai et al. (Kawai & Nishimura 1996). Where does this large gap come from? What explanation is possible for? The one possibility is that the high-speed gust on the roof and the large positive pressure on the windward wall are induced by the strong conical vortex itself.

5 CONCLUSION

1. The strong conical vortex forms on the side of the roof when the large peek suction appear close to the windward corner.
2. The size and the center position of the conical vortex are almost similar in any angles of attack.
3. The strength and the spread of the peak-suction induced by the vortex are also almost similar in any angles of attack.
4. The vortex seems to be formed when the gust attacks to the windward eave from the angle of attack of 25 degree.
5. The pattern of the spread of the peak pressure at any angles of attack is like that of the mean pressure at the angle of attack of 25 degree.
6. The peak pressure coefficients can connect to the mean pressure coefficients at 25 degree by the following equation.

$$C_{p,peak} = 1.65C_{p,mean} + 0.19$$

7. When the peak-suction occurs, the flow-speed over the roof and the pressure on the windward wall increase.

These results seem to support the hypothesis that the peak-suction is induced by the conical vortex, which is formed on the roof when the high-speed gust attacks to the windward wall from the angle of attack of 25 degree.

REFERENCES

Letchford, C.W. 1995, Simultaneous flow visualization and pressure measurements o the Texas Tech building, *Proc. 9th International Conf. Wind Engr.*, Vol.1:524-535.

Kawai, H. & G. Nishimura 1996, Characteristics of fluctuating suction and conical vortices on a flat roof in oblique flow, *J.Wind Eng. Ind. Aerodyn.* 60:211-225.

Kawai, H. 1997, Structure of conical vortices related with suction fluctuation on a flat roof in oblique smooth and turbulent flows, *J.Wind Eng. Ind. Aerodyn.* 69-71:579-588.

Kawai, H. & Ohnuma 1998, Characteristic of peak suction on a flat roof , *Proc. 15th National Symposium on Wind Eng.*:299-304.

Mehta, K.C., M.L. Levintan, R.E. Iverson & J.R. McDonald 1992, Roof corner pressures measured in the field on a low building, *J.Wind Eng. Ind. Aerodyn.* 41-44:181-192.

Sarkar P. & Z. Zhao 1998, Wind flow around buildings, *Cooperative Program in Wind Engineering* Vol.6 Num.1:5-6

Wind Engineering into the 21st Century, Larsen, Larose & Livesey (eds) © *1999 Balkema, Rotterdam, ISBN 90 5809 059 0*

Mean and fluctuating wind loads in built-up areas

H. Kiefer & E. J. Plate
Institute for Water Resources Planning and Rural Engineering, University of Karlsruhe, Germany (Formerly: Institute for Hydrology and Water Resources Planning, Section of Building Aerodynamics)

ABSTRACT: Systematic wind-tunnel experiments were conducted on a series of test buildings embedded in different built-up areas in order to determine the influence of the suroundings on the wind loads. Additionally, full-scale measurements of wind-induced pressures on test buildings in complex surroundings were carried out to verify the wind-tunnel experiments. The results show the variability of the loads due to the influence of the surroundings. From the results of the systematic study the negative pressures on the roof of the full-scale test building were estimated.

1 INRODUCTION

Current design codes are based on wind-tunnel experiments on isolated building models. However, the isolated building is a very rare case in reality. In densely populated areas buildings are grouped together and arranged in characteristic configurations due to their function or location in the urban area. In the design codes the effect of the surrounding roughness is taken into consideration in the parameters of the wind profile. However, the pressure coefficients of the isolated building case are also applied to determine the wind loads in built-up areas. The main reason for this is the lack of data of wind-induced pressures on buildings surrounded by adjacent structures.

The study of Hussain & Lee (1980) leads to the distinction of three flow regimes within roughness fields depending on the spacings and the plan area density of the elements: "Isolated roughness flow" (IRF), "Wake interference flow" (WIF), "Skimming flow" (SF). However, the regularly arranged roughness configurations used by Hussain & Lee are not representative for most of the building arrangements found in reality. Ho et al. (1991) investigated the effect of a typical North American industrial area on the wind loads. Their study was limited to low-rise structures. Hertig & Alexandrou (1995) presented first results of measurements in two models of characteristic urban areas in Switzerland.

However, the above-mentioned lack of data of wind loads on buildings embedded in typical built-up areas still exists. In particular, there are no data from field measurements to verify the findings of wind-tunnel investigations.

For this reason, in the Section of Building Aerodynamics of the former Institute for Hydrology and Water Resources Planning investigations of wind loads on buildings embedded in characteristic built-up areas are carried out. Extensive studies include the analysis of existing building configurations, wind-tunnel experiments on building models in built-up areas, and a full-scale study for verification purposes. For the full-scale experiments two test-buildings on the University campus were instrumented. This paper deals with the results of the systematic wind-tunnel experiments and their application in order to estimate the maximum roof suction of the full-scale test building for oblique wind directions. First results of the full-scale measurements on the roofs of one test building are presented to verify the model tests.

2 EXPERIMENTS

2.1 *Systemtic study (Wind tunnel)*

The wind-tunnel experiments were carried out in one of the boundary layer wind tunnels of the Institute. The profile of mean velocity and turbulence intensity of the modelled approach flow are given in Figure 1. The turbulent length scales indicate a model scale of 1:200 - 1:300. A scale of 1:250 has been chosen for the systematic study. The isolated test buildings of both studies have been exposed to the boundary layer of Figure 1.

A detailed analysis of existing building configurations in German cities led to the definition of 10 subareas with typical building configurations and dimensions. They cover the whole area of a city

(Badde & Plate 1994). Further inspection leads to the distinction of three basic types of building arrangements:

- Type I: Constant building dimensions and distances between the buildings arranged in a homogeneous row pattern
- Type II: Variable building dimensions and distances as in industial areas
- Type III: Buildings are arranged in blocks as in city centers

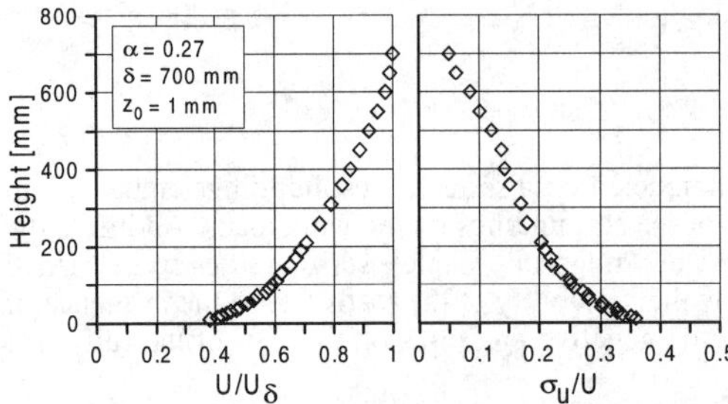

Figure1. Undisturbed approach flow, U_δ = 12 m/s (15 m/s for the specific study)

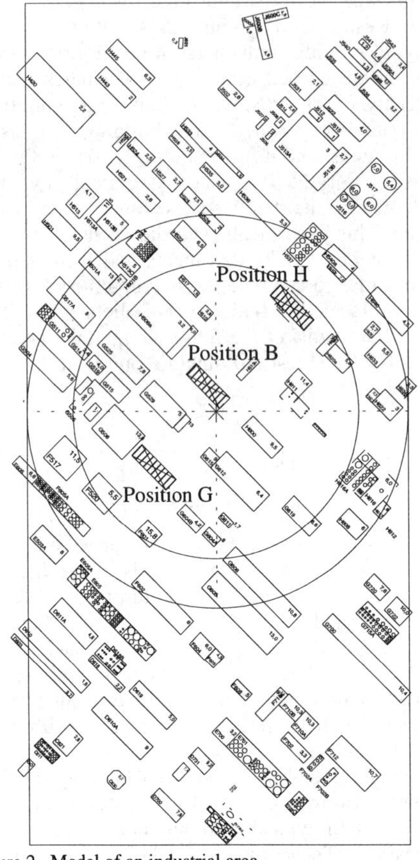

Figure 2. Model of an industrial area

Table 1. Flat-roofed test buildings for the systematic wind-tunnel experiments (full-scale dimensions)

Dimensions	Test buildings					
in full-scale	K1	K2	K3	P1	P2	P3
Height [m]	20	40	80	20	40	80
Length [m]	20	20	20	66	66	66
Width [m]	20	20	20	20	20	20

Mean and fluctuating pressures were measured in the wind tunnel on different flat-roofed building models embedded in these three types of building arrangements. The dimensions of the test buildings are given in Table 1. Additionally, velocity measurements in and above the built-up areas were carried out to investigate the flow development.

The Type I configuration was represented by buildings with identical dimensions as the P1-building of Table 1. The buildings were arranged in a normal grid pattern. The distance between the short side wall was kept constant (t = 1H). Three spacings s between the longer side walls s = 1H, s = 3H, s = 7H, and three different wind directions (0°, 45° 90°) were tested.

An existing industrial area was modelled for the measurements in an arrangement of Type II. Figure 2 shows the wind-tunnel model in plan under an orientation of 45°. Two more orientations (0°, 90°) were modelled for the tests. The arrangement depicted in Figure 2 was used to investigate the influence of the close surroundings on the wind loads of the test buikdings. For this purpose, three different sites of the test buildings (positions B, G, H) located in the inner circle of the model were selected. The measurements were carried out for 8 different orientations of the turnable inner circle of the model. Up to 6 different near field situations for a flow direction perpendicular to the side walls and 12 situations for oblique wind directions were created by this procedure.

Building configuration Type III consists of gable-roofed buildings with varying dimensions. The variation of the heights fits the height distribution of an existing German city (part of the inner city of Frankfurt). The buildings are arranged in a block like shape. The size of the blocks was kept constant. In the first stage of the measurements the test buildings were placed in the middle of an empty block field.

The fetch of all three types varies between 2.2 and 3 meters (550 - 750 m in full-scale). For this reason, the development of the internal boundary layer does not affect the pressures at the test building sites.

The pressure coefficients c_p (Equation 1) of the isolated and the embedded buildings are normalized by the dynamic pressure in building height of the undisturbed approach flow. This implies that the effect of the surroundings is expressed by variation of the pressure coefficients.

$$c_p = (p - p_{ref}) / \frac{1}{2} \rho U_H^2 \quad (1)$$

where $p-p_{ref}$ = pressure difference between the tap and the wind-tunnel side wall; ρ = density of air, U_H = mean velocity in height of the test building.

The mean values and the time series (record length: 60 s and 120 s, sampling frequency: 500 Hz) of the pressures (each run up to 15 taps simultaneously) are recorded. For this purpose, two Scanivalves with MKS Baratron transducers are used for the means, and Setra 237 transducers for the time series with a tubing system optimized for frequency response. The frequency response of the tubing system is almost constant up to ≈ 100 Hz.

2.2 *Specific study (Full-scale and wind tunnel)*

The test buildings for the full-scale experiments called FZU-complex are situated on the University Campus at the northern outskirts of Karlsruhe. The complex consists of two flat-roofed main buildings which are 17 m and 21 m high (Figure 3). The surrounding roughness in the south-west direction consists of the urban area and the University campus. The prevailing wind directions at the site of the test buildings are south-west and north-east. The two test buildings are equipped with 96 pressure taps on side walls and roofs. On the side walls the taps are arranged in vertical rows at edges and centers of the walls. The taps on the roofs are located at the edges and corner regions to record the maximum roof suctions. The wind-induced pressures can be recorded at all taps simultaneously using the Interbus-System of Phoenix Contact for data transmission. A sampling frequency of more than 50 Hz is obtained by this system. A meteorogical tower of 12 m height has been erected on the roof of the higher building. More details of the measurement system are given by Plate & Kiefer (1996). The wind data of the meteorological station at the high-rise physics building (71 m height) not far from the test building site are also included in the analysis of the wind field. These data have been used as reference wind speed for the present results. The velocity in building height was calculated using the profile exponent 0.27 of the undisturbed approach flow of the wind-tunnel experiments. For this reason, the c_p-values are consistent for all investigations.

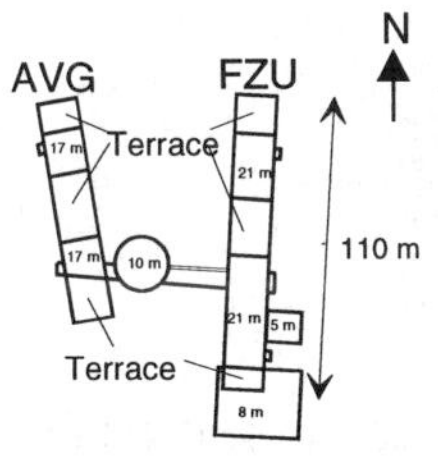

Figure 3. Full-scale test buildings in plan

First data comparing the wind tunnel and full-scale results were given by Kiefer & Plate (1998). Only pressure coefficients of the side walls were presented due to the incomplete instrumentation at this date. Meanwhile, first results of the taps on the roofs became available.

A series of wind-tunnel investigations was also carried out on the scaled model of the test building complex. For the specific study a scale of 1:200 has been used. The surroundings has been modelled up to a distance of 800 m in full-scale for the prevailing wind directions. A detailed model including the vegetation and a simplified version without vegetation have been tested for several wind directions.

The effect of differences in reference pressures in full-scale and model-scale has to be taken into consideration. In full-scale the internal pressure of the test building was used instead of the static pressure. Comparison of the windward and leeward pressure coefficients for the full-scale and model-scale leads to a correction factor of -0.2 for the full-scale data (Kiefer & Plate 1998). This value fits typical internal pressure coefficients well.

2.3 *Parameters of the built-up areas*

The geometric characteristics of the homogeneous row pattern is described by the constant building element spacing. However, additional parameters are necessary for the description of the other built-up areas. The plan area density of the built-up area is not very significant for its aerodynamic behaviour, because most of the buildings are not quadratic in plan and their heights vary. For this reason, the frontal area ratio was defined. This parameter takes into consideration the silhouette areas of the buildings exposed to the flow:

$$\lambda_{ff} = \Sigma A_{silhouette,\, i} / A_{total} \quad (2)$$

where $A_{silhouette}$ = area of the building perpendicular to the flow; A_{total} = total built-up plan area.

Another important parameter is the ratio of the test building height H to the mean height H_{mean} of the surrounding buildings.

$$H_{mean} = \Sigma(H_i\ B_i\ L_i)/\Sigma(L_i\ B_i) \quad (3)$$

where H_i, B_i, L_i = individual building dimensions.

Also important is the amount of variability of the buildings heights. This is described by the ratio σ_H/H_{mean}. These parameters are summarized in Table 2. The relative building height H/H_{mean} of the systematic wind-tunnel investigations varied from 1 to 4. In the case of the FZU-building H/H_{mean} has a value of 1.52. For this reason, the results of the measurements on the test buildings P1 ($H/H_{mean} = 1$) and P2 ($H/H_{mean} = 2$) are presented in section 3, in order to estimate the negative pressures on the roof of the FZU-building in full-scale.

Table 2. Parameters of the built-up areas (full-scale dimensions)

Configurations	Parameters		
	H_{mean}	σ_{H}/H_{mean}	λ_{ff}
	m	-	-
Rows s/H=1 (45°)	20.00	0.00	0.31
Rows s/H=3 (45°)	20.00	0.00	0.17
Rows s/H=7 (45°)	20.00	0.00	0.10
Industrial area (45°)	16.25	0.49	0.20
Block arrangement	19.96	0.21	0.38
Full-scale test site	13.84	0.70	0.17

3 RESULTS AND DISCUSSION

3.1 *Flow regimes*

The results of the P1-building embedded in a row pattern of varying distances between the buildings are well comparable to the study of Hussain & Lee (1980). They attribute to each flow regime a characteristic profile of the $c_{p,mean}$-values along the vertical centerline of the side walls. Their results for a frontal aspect ratio $A_f = 4.0$ are comparable to the geometry of the P1-building ($A_f = 3.3$). Note that the spacings of the present study are defined differently from Hussain & Lee (1980). They defined the distance from building axis to building axis as the building element spacing, whereas in the present study s stands for the empty space between the buildings.

Figure 4 shows the $c_{p,mean}$-values along the vertical centerline of the longer side walls of the P1-building for an approach flow perpendicular to the latter. The spacings s/H = 7 and s/H = 1 are clearly in the range of the "Isolated roughness flow" and the "Skimming flow", the spacing 3H seems to be in the transition between IRF and WIF. In the SF-case (s/H = 1) the profiles of the windward and leeward side are nearly identical due to the stable vortices in the spaces between the buildings. This results in drag coefficients on the buildings which are approximating zero for this flow regime. Figure 5 shows the same profiles for the P2-building. In this case, the height of the test building is more than twice the height of the surrounding buildings. The profile for s/H = 1 is very different in the zone above and below the height of the adjacent buildings. Below the height of the surroundings the largest reduction of the loads occurs. In the zone exceeding the height of the surroundings the $c_{p,mean}$-values are less reduced because for a skimming flow regime the profile of the wind is much smoother than for the other cases.

The homogenous row pattern is described by the constant spacings of the buildings. However, many real existing building arrangements exhibit varying building dimensions and distances between the buildings. The modelled industrial area is a typical example for such an arrangement. Measurements carried out on this model include several different near field situations characteristic for a wide range of existing structures.

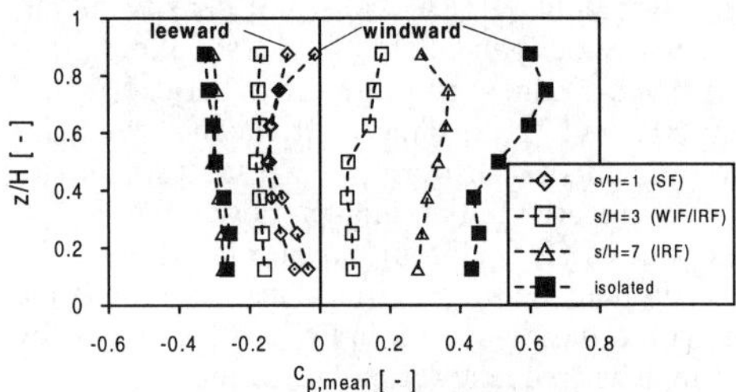

Figure 4. Mean pressure coefficients at the centerline of the windward and leeward wall of the P1 building, $H/H_{mean} = 1$

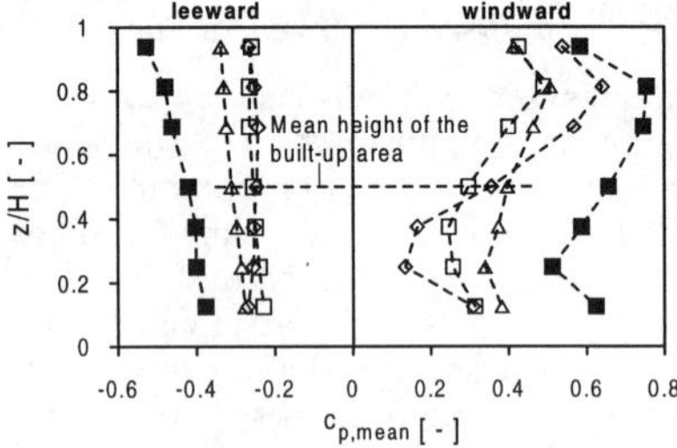

Figure 5. Mean pressure coefficients at the centerline of the windward and leeward wall of the P2 building, $H/H_{mean} = 2$

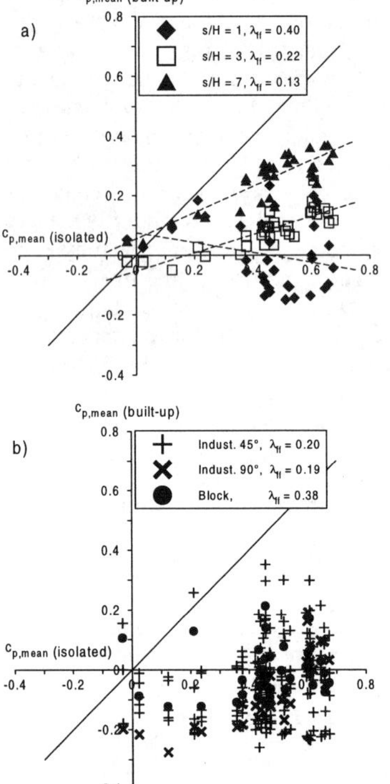

Figure 6. Mean pressure on the windward wall of the P1-building, a) row pattern b) industrial area and block shape

A comparison between the results of the row pattern and the industrial area is useful for an evaluation of the variability of the loads due to varying building spacings in an arrangement. For this reason, the results for the row pattern and the industrial area are plotted against the $c_{p,mean}$-values of the isolated building in Figure 6 a and b. In Figure 6 a the regression lines for the different spacings of the row pattern show a nearly linear relationship between the values of the isolated building and the data for s = 3H and s = 7H. The data for the industrial area under 45° orientation of Figure 6 b include 6 different configurations of the near field. The scatter of these data covers the same range as the three different flow regimes of Figure 6a. This means, that the concept of flow regimes based on building element spacing or plan area density fails when applied to this type of building arrangement. However, Figure 6 shows that the $c_{p,mean}$-values do not exceed a limit of 0.4 in the built-up case. Therefore, we base the estimation of the wind loads on the full-scale test building on envelope curves. The data for these are derived from the measurements with variation of the closest surroundings.

3.2 *Roofs, β = 45°*

The lowest negative pressures on flat roofs occur for oblique wind directions. The prevailing wind direction of 230° at the full-scale test site corresponds to a flow direction of 46° referred to the long building axis. Consequently, the results of the wind-tunnel experiments of the 45° flow direction are comparable to the full-scale data. Therefore, the results for this flow direction will be presented in more detail.

In Figure 7 the $c_{p,mean}$-values for the roof of the P1-building embedded in different built-up areas are plotted against the data of the isolated building. The $c_{p,mean}$-values for the row pattern with a building separation of 7 H are nearly the same as the values of the isolated buildings. In the other cases the negative pressures on the roof are much higher. As expected from the results presented in 3.1 there is a large amount of scatter for the industrial area. The position H under 225° shows the lowest negative pressures due to the lack of large buildings in front of the test site. Unfortunately, the region of lowest negative pressures occurs in a very small area in which no pressure tap is located. In order to evaluate the loads in this area a regression line (dashed line) is added in Figure 7. This line can be interpreted as envelope curve for the worst case of negative pressures on the roofs for the different near field arrangements of the present study.

In Figure 8 the $c_{p,mean}$-values for the P2-building are depicted in the same way as in Figure 7. In this case the data of the row pattern differ from the other building configurations. Lower negative pressures occur for the row pattern. The constant building heights of the row pattern lead to a smoother surface and to the formation of corner vortices. These are absent for the case s/H = 3 where higher pressures are found. The lowest pressures of the different near field situations in the industrial area are of the same order of magnitude. The smallest and the largest building separation of the row arrangement leads to the lowest minimum pressures.

The RMS-values for the roofs of the two buildings are shown in Figures 9 and 10. In general, they show the same characteristics as the mean-values. The RMS-values for the P1-building in the row arrangment s/H = 7 exceed the values of the isolated building. The other cases show reduced fluctuations. The data of the isolated building are not exceeded considerably in case of the P2-building (Figure 10).

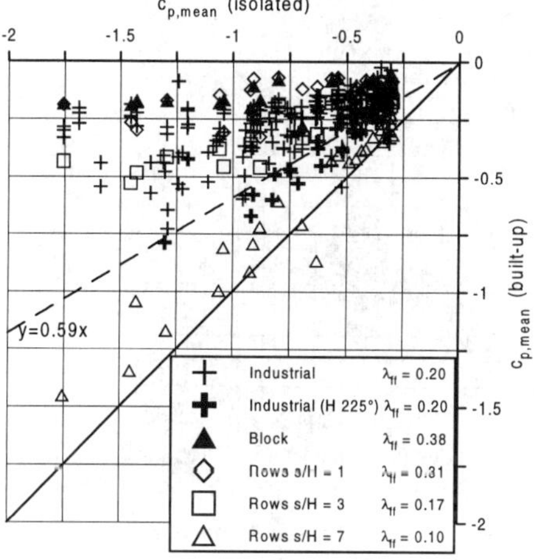

Figure 7. P1-building, roof under 45°, $H/H_{mean} = 1$

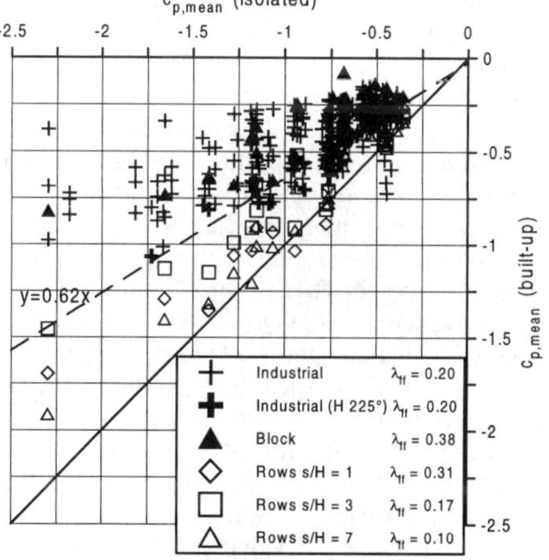

Figure 8. P2-building, roof under 45°, $H/H_{mean} = 2$

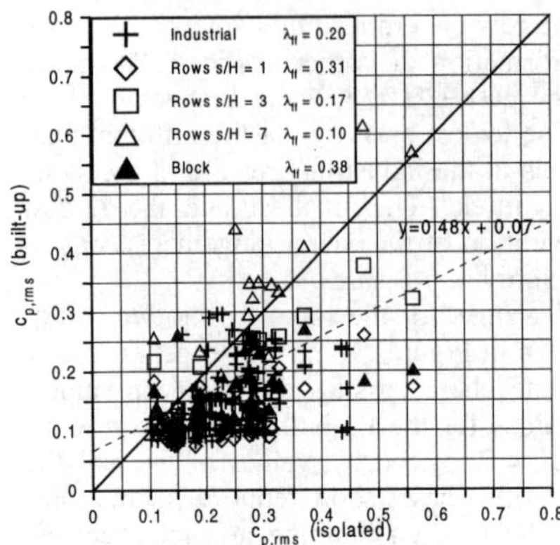

Figure 9. P1-building, roof under 45°, $H/H_{mean} = 1$

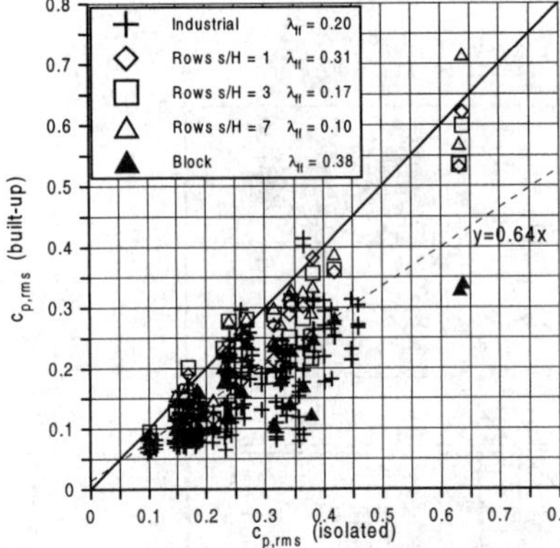

Figure 10. P2-building, roof under 45°, $H/H_{mean} = 2$

3.3 *Comparison with full-scale results*

Full-scale results for the mean pressure coefficients of the side wall taps show a reduction of approximately 50%. The values fit in beween the results of the wind tunnel tests on the simplified and the detailed model. In the frequency domain there is a good agreement, too (Kiefer & Plate 1998).

Locally very low negative pressures occur on a flat roofs for oblique wind directions as shown in Figures 7 and 8. The regression lines of Figures 7-10 were used to estimate reduction factors from the results of the systematic study for the determination of wind loads on the full-scale test building. Due to the relative building height $H/H_{mean} = 1.52$ of the FZU-building the average of $H/H_{mean} = 1$ and $H/H_{mean} = 2$ was used to calculate the reduction factors. In Table 3 the results of the specific wind tunnel measurements, the estimate from section 3.2, and the full-scale results for the minimum local pressures on the roof are summarized. Whereas the mean value of the minimum local pressure is well predicted by the systematic study, the RMS-value is underestimated. The reason for this could be found in the differences in geometry and tap positions of the P1 and the FZU test building. The RMS-value of the FZU in the isolated case is much higher than that of the P1-building. The results of the wind-tunnel experiments on the FZU-model predicts the RMS-values well.

Table 3. Comparison of the maximum local roof suction for oblique wind directions (45°), full-scale - model-scale

		Mean	RMS
Wind tunnel	P1-isolated	-1.76	0.56
	FZU-isolated	-1.97	0.72
	FZU simpl. model	-0.98	0.59
	FZU detail. model	-0.53	0.40
	Estimate from section 3.2	> -1.18	< 0.40
Full-scale	Run 10	-0.91	0.52
	Run 11	-0.85	0.54
	Run12	-0.91	0.49
	Mean of run 10-12	-0.89	0.52

4 CONCLUSIONS

In order to determine the influence of the suroundings on the wind loads, systematic wind tunnel experiments were conducted on a series of test buildings embedded in different built-up areas. The measurements in an industrial area of varying building dimensions show the variability of the pressures caused by different near field configurations. Values typical for all three flow regimes defined by Hussain & Lee (1980) were found for this building layout. Envelope curves based on these results allow to predict the reduction in pressures. This was verified by full-scale experiments on a test building embedded in complex surroundings.

REFERENCES

Badde, O. & Plate, E. J. 1994. Einfluß verschiedener Bebauungsmuster auf die Parameter der Grenzschicht und die windinduzierte Gebäudebelastung, *Abschlußkolloquium SFB 210, University of Karlsruhe*

Hertig, J. A. & Alexandrou, C. 1995. The influence of surroundings on pressure distributions around buildings, *J. E. Cermak et al. (eds.), Wind Climate in Cities*: 293-317, Kluwer Academic Publishers

Ho, T. C. E., Surry, D. & Davenport, A. G. 1991. Variability of low building wind loads due to surroundings, *J. of Wind Eng. and Ind. Aerodyn.* 38: 297-310

Husain, M., Lee & B. E. 1980. An investigation of wind forces on three dimensional roughness elements in a simulated atmospheric boundary layer flow, *Part I - III, BS 55, BS 56, BS 57, Depart. of Build. Science, Faculty of Architect. Studies, University of Sheffield*

Kiefer, H., Plate & E. J. 1998. Wind loads in built-up areas - results from full-scale and wind-tunnel measurements, *Proceedings of the 2nd EECWE, Prague*: 263-270

Plate, E. J. & Kiefer, H. 1996. Modellierung von Windkräften auf Bauwerke in bebautem Gelände, *Research Report, Institute for Hydrology and Water Resources Planning, University of Karlsruhe, Germany*

Wind Engineering into the 21st Century, Larsen, Larose & Livesey (eds) © 1999 Balkema, Rotterdam, ISBN 90 5809 059 0

Some effects of roof shapes on housing wind loads

M. Kumamaru
Japan Science and Technology Corporation, Fukuoka, Japan

N. Tsuru & J. Maeda
Department of Architecture, Faculty of Engineering, Kyushu University, Fukuoka, Japan

A. Miyake
Department of Architecture, Ariake National College of Technology, Omuta, Japan

ABSTRACT: The effects of housing roof shapes on the wind loads in uniform, pulsating and turbulent flows using wind tunnel testing are discussed. Generally the wind load of gable roof shape housing is bigger than that of hipped roof housing. However the existence of ridge tiles reduces the difference. A moderate size of eaves does not influence the drag too much. The same characteristics are seen in pulsating and turbulent flows. The difference between the damage of hipped roof houses and gable houses reported in strong wind disasters can be explained by the above results from the standpoint of the wind loads.

1 INTRODUCTION

Many wind disaster reports (Damage investigation research committee 1993, Kaneto 1992, Okada et al. 1993) have said that the percentage suffering of gable roof houses by strong winds is higher than that of hipped roof houses. The reasons are presumed to be both structural and aerodynamic. The former means that the frame structure of gable is frailer than that of a hipped roof of truss structure. And the latter means that gable housing with a vertical wall is subjected to larger wind pressure than hipped roof housing. We discuss some effects of housing roof shapes, that is, eaves, the pitch of roof, and the ridge tile etc., on the wind loads in uniform, pulsating and turbulent flows using wind tunnel testing.

2 OUTLINE OF WIND TUNNEL TESTING

2.1 *Wind tunnel flows*

We experimented in three kinds of wind tunnel flows as follows: a uniform flow, a pulsating flow and a grid turbulent flow. The wind tunnel used for the tests was the inhaled-Eiffel type. The size of the measurement section is 1.5m×1.5m and the available length is 3m.

The pulsating flow is created by the rotation of shuttlecocks, and is set up on the leeward side of the uniform flow. The wind velocity changes according to the rotation angle and speed of the shuttlecocks almost like a sine wave, we can control the gust factor and the period of pulsation.

The turbulent flow is generated with a square grid frame, which is set up on the windward side of the uniform flow, and 2.2m ahead of a specimen. The frame consists of a bar of 4×4cm in thickness, the mesh length of 14cm, and the mesh number of 10×10.

The wind velocity in each flow is distributed uniformly in the measurement section.

2.2 *Housing models*

The models for testing are shown in Figure 1 and Table 1. We prepared 20 specimens of wooden housing models, and considered roof shapes, roof pitches, and the existence of eaves and ridge. The

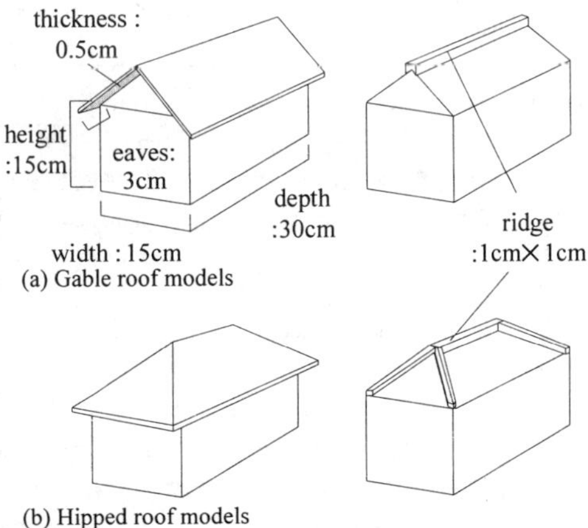

Figure 1. Basic housing models

Table 1. Profiles of housing specimens.

Roof shape	Eaves	Ridge	Pitch of roof	Roof pitch specimen		
				Uniform flow	Pulsating flows	Turbulent flow
Hip	Exist	None	36.9°, 26.6°, 14.0°	All	-	-
	None	Exist		All	36.9°	36.9°
		None		All	36.9°	36.9°
Gable	Exist	None		All	-	-
	None	Exist		All	36.9°	36.9°
		None		All	36.9°	36.9°
Flat	Exist	None	0°	0°	-	-
	None	None		0°	-	-

sizes are as follows. The ratio of the plan section is 2:1; the pitches of roofs are 36.9°, 26.6°, 14.0° and 0°; the length of eaves is 3cm; the section of the ridge is 1cm in height and 1cm in width. All roof pitches in the uniform flow and the roof pitch of 36.9° in the pulsating and turbulent flows are examined.

2.3 *The measurement method*

A site plan of the testing system is illustrated in Figure 2. We measured the drag, which acted on the specimens, using an aerodynamic force balance directly, and the wind velocity using a supersonic anemometer. The experimental data was measured for 15 seconds by a sampling frequency of 200Hz. In the pulsating flows the data was measured 10 times during each period of pulsation.

2.4 *The examination method*

We estimated various drag characteristics in each wind direction by using the drag coefficient defined by Equation 1.

$$C_D = \frac{2F_x}{A\rho U_x^2} \tag{1}$$

where C_D = drag coefficient; F_x = mean drag(kgf); A = the projection area of the model(m^2); U_x = mean wind velocity(m/s); and ρ = air density(kg*s^2/m^4) = 0.125.

In the pulsating flow, we estimated the drag characteristics of the mean and the maximum respectively. At the time, the mean of the maximum of each wave of the data is defined as the maximum drag.

3 TESTING IN UNIFORM FLOW

We measured all models in a uniform flow and understood the drag characteristics of each housing roof shape in every wind direction.

3.1 *Basic characteristics of the uniform flow*

The mean wind velocity is 17m/s, and the intensity of turbulence is 2%. Both the horizontal and per-

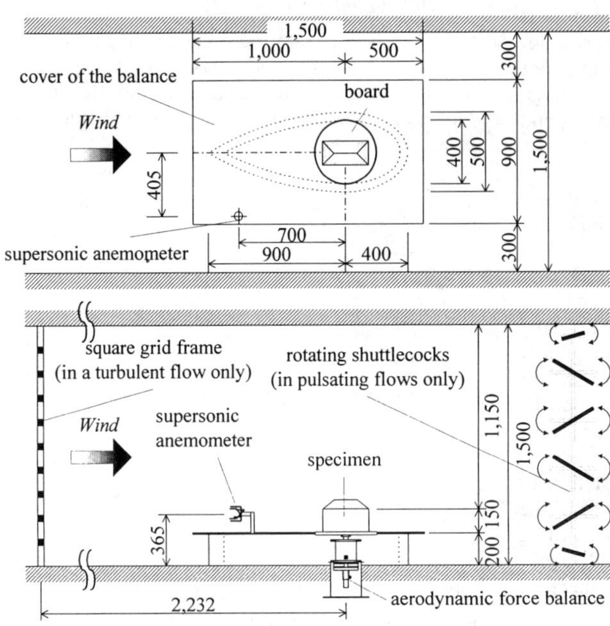

Figure 2. Site plans of the wind tunnel

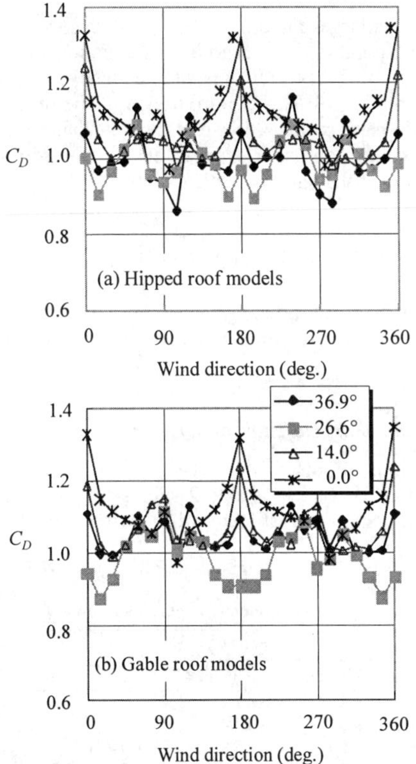

Figure 3. The drag coefficients of models without eaves in each wind direction

pendicular components were almost 0, so it may be assumed that only the component of the mean wind direction contributed to the drag.

3.2 *Effects of wind direction*

An example of the drag characteristics at each wind direction (Kumamaru et al. 1997) is shown in Figure 3, in which pitch 0° means a flat roof. The figure shows the transition of drag coefficients of the hipped and gable roof models without eaves. All of them are symmetrical, which correspond to the plain shape of the models, at every wind direction. However they don't necessarily correspond to the order of the pitch of roof.

3.3 *Effects of roof shape*

The drags at two wind directions from Figure 3 are selected and the effects of the roof shape are shown in Figure 4. W1 and W2 symbolize the wind directions perpendicular to the longer and shorter edges of the model respectively.

In the wind direction of W1, the roof pitch has a larger effect on the drag than the roof shapes such as

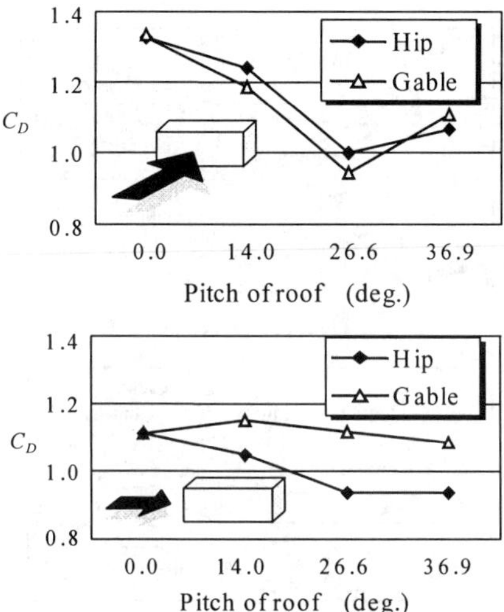

Figure 4. The drag coefficients in the wind directions of W1 (upper) and W2 (lower)

the hip and gable. Both drag coefficients of the hip and gable roof models at each pitch are almost the same, and minimized at the pitch of 26.6°.

On the other hand, in the wind direction of W2, the tendency is the opposite. The drag of the gable roof model is 1.2 times larger than that of the hip roof model. It is thought that the gable shape reduces the effect of roof pitch.

3.4 *Effects of eaves*

Figure 5 shows the drag coefficients in the wind direction of W2. In the figure, the empty symbols represent models without eaves. For the length of eaves in the testing, the existence of eaves reduces the drag in the hip roof models but increases it in gable roof models. However, the effects on the drag characteristics are smaller than those of the ridge described later.

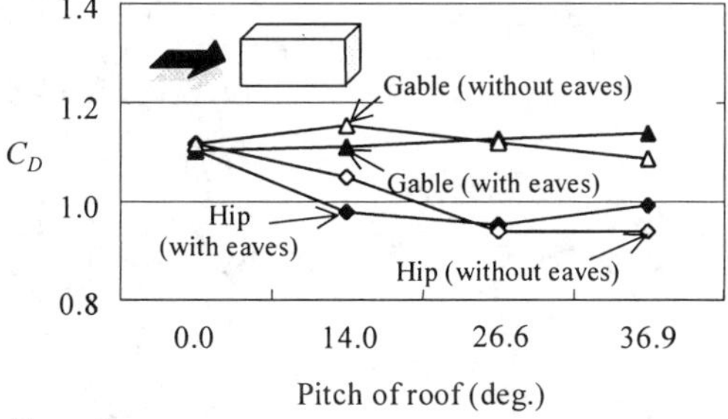

Figure 5.

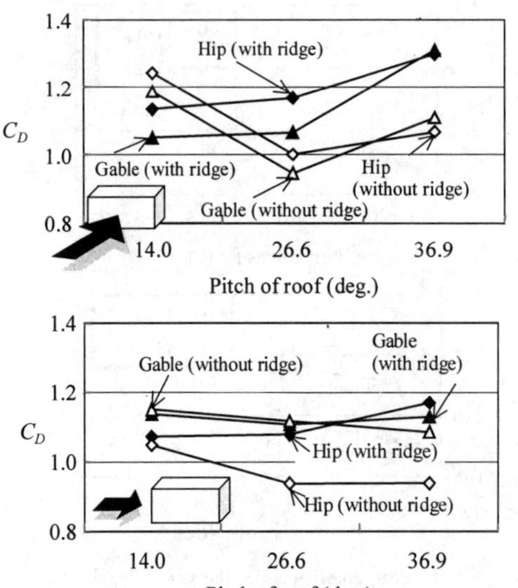

Figure 6. The effects of the ridge in wind directions of W1 (upper) and W2 (lower).

3.5 *Effects of the ridge*

Figure 6 shows the result of comparing the drag coefficients of each roof shape in the above two wind directions. In the figure, the empty symbol means a model without ridge. In the wind direction of W1, the ridge enlarges the drag in both roof shapes of the hip and the gable. However, only the models of roof pitch of 14.0° have an opposite tendency. In the wind direction of W2, the difference of the drag by the roof shape reduces. So it is thought that the existence of a ridge has a larger effect on drag characteristics than the gable's vertical wall.

4 TESTING IN PULSATING FLOWS

The results of the housing models of pitch 36.9° in pulsating flows are compared with the results of the uniform flow.

4.1 *Basic characteristics of pulsating flows*

The periods of pulsation are 1.1, 1.5, 2.3, 4.5 and 9.0 seconds respectively. Table 2 shows the mean, the maximum and the gust factor of the wind velocity in

Table 2. Profiles of wind velocity in pulsating flows.

Period (sec)	Wind velocity (m/s)		Gust factor
	The mean	The max.	
(a) 9.0	8.7	17.7	2.0
(b) 4.5	8.6	17.4	2.0
(c) 2.3	8.5	16.9	2.0
(d) 1.5	8.2	16.5	2.0
(e) 1.1	6.5	14.7	2.3

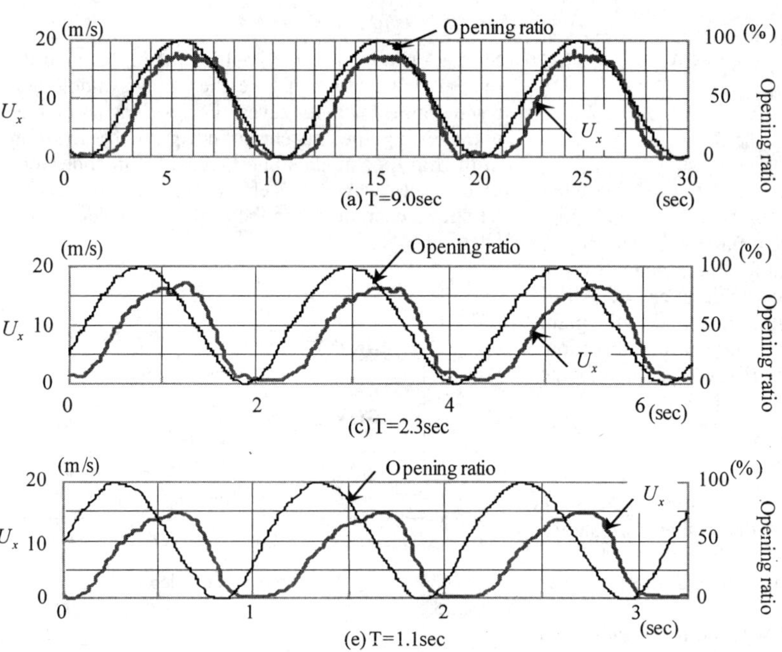

Figure 7. The wave of wind velocity and the opening ratio of the shuttlecocks in each period of pulsation.

the pulsating flows used for the testing. Figure 7 shows examples of the wave of wind velocity and the opening ratio of the shuttlecocks in each period of pulsation. The opening ratio is 0 when the shuttlecocks are closed and 1 when they are open. The maximum wind speed is almost equal to that in the uniform flow in the case of a period of longer than 4.5sec. When the period is shorter than 2.3sec, the wind velocity decelerates gradually. At a period of 1.1sec the wind becomes 5/6 of the velocity in the uniform flow. The wave shape of the flow is almost the same as the sine in the long period of pulsation. But it changes into a shape like the teeth of a saw as the period shortens. The reasons may be considered as follows. When the period of pulsation become fast, the change of wind cannot follow the movement of the shuttlecocks because of the inertia force.

4.2 *Comparisons with the results in the uniform flow*

Figure 8 shows the mean drags, which act on the gable roof model without eaves and ridge, in the pulsating and uniform flows in the wind direction of W1. Though both of the mean winds are equal, the mean drag in the pulsating flow is about 1.5 times as large as that in the uniform flow. We may guess the reason for these differences to be as follows.

The drag $F(t)$ in the pulsating flow is expressed by using the mean wind velocity U and the variable component $u(t)$ by the equation (2).

The mean drag F, which is defined by the expec-

$$F(t)=\frac{1}{2}\rho C_D A(\mathrm{U}+u(t))^2 \qquad (2)$$

tation of $F(t)$, is obtained from the equation (3).

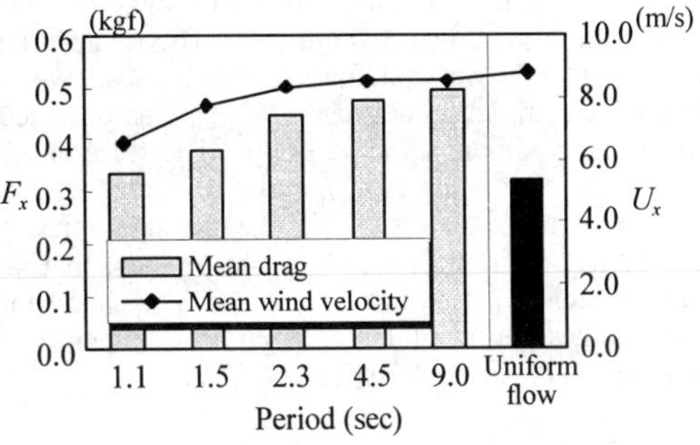

Figure 8. The mean drag and wind velocity in the uniform and pulsating flow

Because the intensity of variation $u(t)$/U is almost

$$\mathrm{F}=\mathrm{E}\left[F(t)\right]$$
$$=\frac{1}{2}\tilde{n}\,\varsigma_D A\mathrm{U}^2\cdot\mathrm{E}\left[1+2u(t)/\mathrm{U}+\left(u(t)/\mathrm{U}\right)^2\right] \qquad (3)$$

zero in the uniform flow, E[ ] of the equation (3) becomes 1. On the other hand, in the case that the intensity of turbulence $u(t)$/U is large, as in the pulsating flow, the second and third terms in the expectation E[ ] should be considered. Then by approximating the pulsating flow to a sine wave, E[ ] becomes 3/2. The result corresponds to the measurement. Therefore the mean drag increases when the variation of the wind velocity grows in the same mean wind.

4.3 *Effects of the periods of pulsation*

The mean drag coefficient C'_D and the maximum

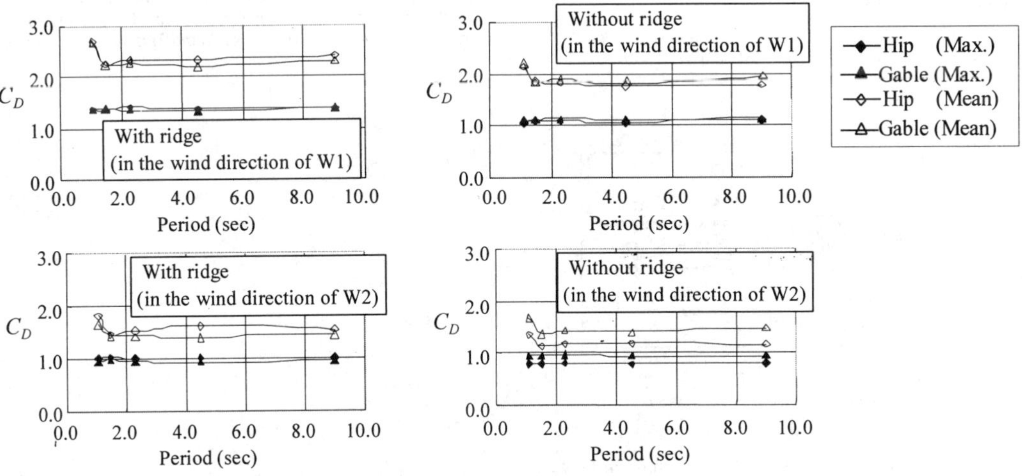

Figure 9. The mean and maximum drag coefficients of each pulsating period

drag coefficient C'_{Dmax} in the flows of some periods of pulsation is shown in Figure 9. The mean drag coefficient C'_D was obtained for each mean wind. The maximum drag coefficient, C'_{Dmax} was obtained for each maximum wind velocity. The differences of the characteristics of C'_D and C'_{Dmax} are seen in the periods of pulsation. While the characteristics of C'_D appear to be influenced by the periods, that is, the coefficient reduces at a period of 1.5sec and increases at a period of 1.1sec again, the tendency is not seen in the characteristics of C'_{Dmax}.

5 TESTING IN THE GRID TURBULENT FLOW

The results of the housing models of pitch 36.9° in the turbulent flow are compared with the results in the uniform flow.

5.1 *Basic characteristics of the turbulent flow*

The mean velocity is 14m/s, the intensity of turbulence is 5.7% and the scale of turbulence is 10-15cm (Matsunaga et al. 1997).

5.2 *Comparisons with the results in the uniform flow*

The drag coefficients in the wind direction of W1 and W2 for every roof shape are shown in Figure 10. In the figure, black and open symbols mean the models with and without ridge respectively. The drag coefficients in the turbulent flow are about 10-20% smaller than in the uniform flow for all wind directions and all roof shapes. Moreover the variation by the roof shapes in turbulent flow is smaller than in the uniform flow.

6 CONCLUSIONS

The results obtained from the experiments in the uniform flow are as follows.

The drag coefficients of the gable roof are 1.2 times as large as those of the hip roof. The presence of ridge increases the drag and reduces the difference between the gable and hip roof shapes. The effects of the ridge on the drag are bigger than the presence of eaves. The results in the other complex flows such as pulsating and turbulent flows are slightly different from the results in the uniform flow. Our discussions in the paper mostly correspond to the damage situation in strong winds (Damage investigation research committee 1993, Kaneto 1992, Okada et al. 1993).

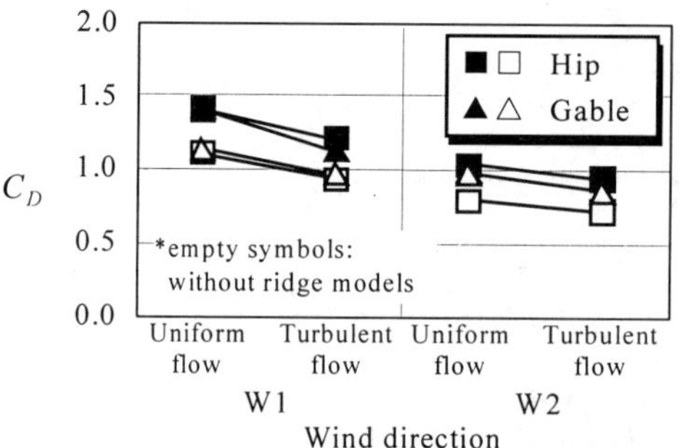

Figure 10. The mean drag coefficients in the uniform flow (left) and turbulent flow (right)

ACKNOWLEDGEMENT

We would like to acknowledge the cooperation and help of Professor Y. Matsufuji, professor of Kyushu University, Japan, Research leader of Core Research for Evolutional Science and Technology of Japan, Science and Technology Corporation.

REFERENCES

Damage investigation research committee 1993, Summary of investigation of housing damage and person's behaviors with passage of the Typhoon 9119, Recommendation for wind load design and construction check of housing, the Building center of Japan, Foundation.

Kaneto, K. 1992, The wind performance of housing in the Northern Kyushu, The Graduation Thesis, Kyushu University.

Okada, H., J. Maeda, et al. 1993, Summary report on Hurricane Andrew, J. of Wind Engineering 56:74-87.

Kumamaru, M., A. Miyake & J. Maeda 1997, Wind tunnel tests on some effects of housing roof shape on its wind, Summaries of technical papers of annual meeting Architectural Institute of Japan 1997, 233-224.

Matsunaga, M., J. Maeda, et al. 1997, Some aerodynamic properties of a transmission tower in turbulent flows – Studies on a new approach for the wind load design of a transmission tower, No1-, J. of Wind Engineering, 71:91-94.

Wind Engineering into the 21st Century, Larsen, Larose & Livesey (eds) © 1999 Balkema, Rotterdam, ISBN 90 5809 059 0

Application of artificial neural network for determination of wind induced pressures on gable roofs

Naveen Kwatra, P.N.Godbole & Prem Krishna
Department of Civil Engineering, University of Roorkee, India

ABSTRACT: Artificial Neural Networks (ANN) have the capability to develop functional relationships between input-output patterns obtained from any source. Thus ANN can be conveniently used to develop a generalised relationship from limited and sometimes inconsistent data. Thus ANN can be applied to tackle the data obtained from the wind tunnel tests on building models with large member of variables. In this paper ANN model has been developed for predicting wind induced pressures in various zones of a Gable Building from limited test data. The procedure can also be extended to account for the interference effects.

1. INTRODUCTION

The usual practice for evaluating wind loads on buildings consists of using codes and standards whose specifications are based on wind tunnel tests performed in the 'stand-alone' configuration. However, buildings seldom exist in isolation being usually surrounded by other buildings. Thus, wind loads on buildings in actual environment differ from those measured on an isolated building. This, apart from findings from the recent work on different facets of flow modelling in the wind tunnel and measurement techniques for pressure fluctuations over building surfaces, has given rise to the need for further study of the subject.

Wind loads on roofs of low buildings are significantly affected by the geometry of the building since it effects the flow pattern around it. This is one of the main reasons which results in the determination of wind loads on low buildings requiring extensive testing, there being a large number of parameters with a wide range of values. To economise on the effort there is a need to explore ways of predicting wind loads from a comparatively reduced test programme. With that objective the application of Artificial Neural Network (ANN) for predicting wind-induced pressures on the basis of data from wind tunnel studies on low buildings has been studied and is reported in this paper. Data used for training and testing of neural network is obtained from wind tunnel tests carried out on the models of gable roof buildings. In the present study Artificial Neural Network is applied for predicting the design wind pressure coefficients for various zones on the building. This has been done also for a particular roof slope while accounting for the interference effect by a similar building on it.

2. NEURAL NETWORK MODEL STRUCTURE

Artificial neural network (ANN) models have been developed by artificial intelligence researchers and are being studied in the hope of simulating 'human like' performance in various fields such as for some complex multivariate and non-linear problems with incomplete or confusing information. The ANN models are composed of many non-linear computational elements (neurons) operating in parallel and arranged in a pattern having extreme similarity with their biological counter part. Neural Network representations are capable of developing functional relationship from discrete values of input-output quantities obtained from computational approaches on experimental results. This generalization property makes it possible to train a network on a representative set of input-output examples and get good results for a new set without training the network on all possible input-output examples.

The Backpropagation algorithm has been used to train the network in the present study. Details about the backpropagation algorithm could be found elsewhere (Rumelhart and McCelland,1986).

3. THE OBJECTIVE

The objective of the study, part of which is reported upon in this paper, is as follows:

1. To examine the efficacy of the use of artificial neural network modelling for predicting the design wind pressure coefficients for different zones in gable buildings based on limited experimental data
2. Application of artificial neural network modelling for predicting the variation of design wind pressure coefficients, due to the interference from a similar building.

4. EXPERIMENTAL PROGRAMME

To cater to the first part of the problem, a gable roof building has been selected with roof slope varying from 10° to 35°. Data used for training of the neural network was obtained experimentally by testing the building models in a boundary layer wind tunnel. The neural network was trained using the roof slope and location of different zones on the building as the input parameter and design wind pressure coefficients on the same as output parameter. The trained network is then expected to predict the design wind pressure coefficients for different zones on the building for the roof slopes not covered in the training set.

Similarly for the second part of the problem the data used for training the neural network is obtained experimentally. The neural network was trained using the x and y co-ordinate of the location of interfering building as input parameter and design wind pressure coefficients on zone 1(refer figure 1) as output parameter. The trained network is then expected to predict the design wind pressure coefficient for the zone 1 for the locations of interfering building not covered in the training set.

The models of a gable roof building 13x7x5 m and roof slope varying from 10° to 35° with the increment of 5° were made at a geometric scale of 1:100. 72 pressure taps were used to cover the whole roof. The tubing system to measure the surface pressure consisted of 500 mm vinyl tube with a 30 mm restrictor at 400 mm from pressure point, and a scanivalve pressure scanner. Pressure measurements were carried out by using Scanivalve ZOC12, a 32-port pressure transducer. The pressure measurement system has a linear response up to 100 Hz. The sampling frequency was kept at 400Hz. 8192 samples of pressure from each port were recorded thus giving a record of approximately 20 seconds. The roof area is divided into different zones as per the Indian Standard Code (IS875-1987, Part- 3), (refer figure 1) with the purpose of making a direct comparison of codal values with those predicted experimentally. Each model was tested for the angle of wind attack 0° to 90° with the increment of 15°.

Experiments were also conducted to find out the interference effect on the building model of roof slope 20° from a similar building. The interfering building was moved in the longitudinal and transverse directions in a regular grid pattern as shown in figure 2. The interfering building was moved longitudinally 40 cm in steps of 5 cm whereas, it was moved upto 30 cm with the same step size transversely. Angle of wind attack was changed from 0° to 90° in the increment of 15° for a every position of interfering building.

Atmospheric surface layer was developed in the wind tunnel over the building models by

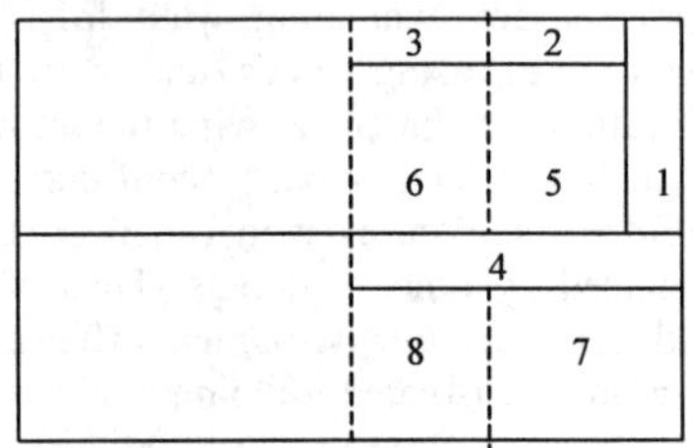

Figure 1. Location of various zones on building roof

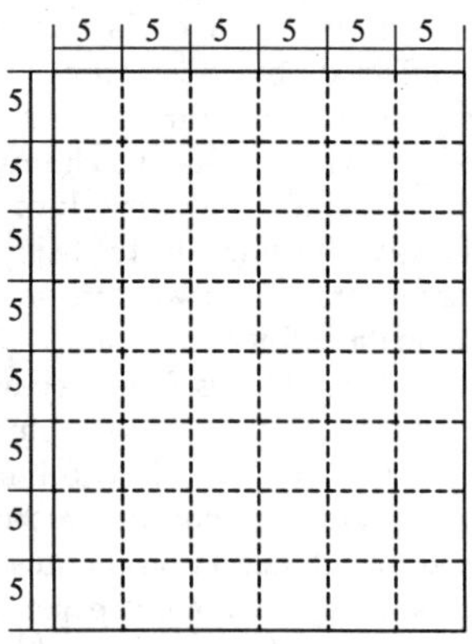

Figure 2. Grid showing the various locations of interfering building. (All dimensions are in cm)

controlling the longitudinal and lateral turbulence intensities and their small scale turbulence content by using the combination of vortex generators, barrier wall and roughening blocks. The target values for these flow parameters were fixed on the basis of the findings of Cermak & Cochran (1992) and Tieleman et al.(1997).

The velocity fluctuations in the wind tunnel were measured by single hot wire probe. The mean velocity and longitudinal turbulence intensity at the eaves height of models were 8.9 m/s and 19% respectively with the velocity profile index α being 0.136. The longitudinal integral scale at the same height was found to be 0.436m. The small scale turbulence content was defined as (Tieleman 1997) $S = (nS_u(n)/Su^2) / (Su/U)^2 \times 10^6$ evaluated at $10U/L_p$ Where n is frequency, $S_u(n)$ spectral density and Su is the standard deviation of the longitudinal velocity, U is the mean velocity and L_p is the characteristic model dimension. In the present study model height is taken as characteristic model dimension. The small scale turbulence content was found to be 216.

5. DATA REDUCTION

The histograms of probability density function of recorded pressure fluctuations show that taps close to the edges and the ridge have significant deviation from Gaussian distribution, as also shown by Xu et al. (1995). Different approaches have been used earlier to transform the observed data to follow the Gaussian distribution, since it is wholly defined by mean and standard deviation. Two Step Power Transformation technique is used in the present work to reduce the observed data into the Gaussian distribution data. In the first step, Box-Cox (1964) transformation is applied to reduce the skewness of observed data to zero or nearly zero. Although the Box-Cox transformation reduces the coefficients of skewness to zero, it is unable to make coefficient of kurtosis equal to 3. Therefore, correction for coefficients is made through modulus transformation A similar transformation has also been suggested by John and Draper (1980) and Gupta et al. (1989). After applying two step power transformation, 99% quantile value of the transformed series is obtained and then it has been back transformed to original time series. This 99% quantile value of observed series of pressure fluctuation (Cpq), is kept as the basis of calculating design pressure coefficients for different zones of the building. Area averaged pressure coefficients were calculated for all the zones of the building for each angle of wind attack. The minimum (worst negative) pressure coefficient from all angle of wind attack has been taken as design pressure coefficient.

6. RESULTS

Two different multilayer feed forward networks were used for both the objectives of the study. Non-linear sigmoidal function was used as activation function. All input and output data were normailsed so that the values remained between 0 to +1. The training session consisted of repeated presentation of training set to the network and reducing the average mean square error of the network output.

For predicting the design wind pressure coefficients for various zones of the building the input parameters of the training data set were zone number and the roof slope. For training of the network Cpq for all zones of roof slopes 10°, 15°, 20° and 30° were taken as output parameter. The network consisted of two hidden layers with twenty neurons in each layer. The network was trained for this data set and average mean square error of the network was reduced to 0.0003. The network performance was tested by checking the output of the network for the same input values as given in training set. Then Cpq for all the zones of the roof slope 25° & 35° were predicted by the network. Figure 3 shows the comparison of experimental values with the ANN predicted values. The maximum error found for the ANN predicted values was around 7%.

To study the application of ANN for predicting the variation of design wind pressure coefficients due to interference from a similar building, zone 1 of the 20° roof slope building was selected. X and Y coordinates of the interfering building were used as input parameters to train the network. Design wind pressure coefficient in zone 1 for particular location of interfering building was taken as output parameter. The neural network consisted of two hidden layers with thirty neurons in each layer. Out of the 63 locations of interfering building, 33 were used for training of the network. The network was trained with this data set and mean square error of the network was reduced to 0.0005. Some typical predictions of ANN are presented and compared with the corresponding experimental values in table 1. It can be seen from the table that ANN predictions are very close to the experimental values.

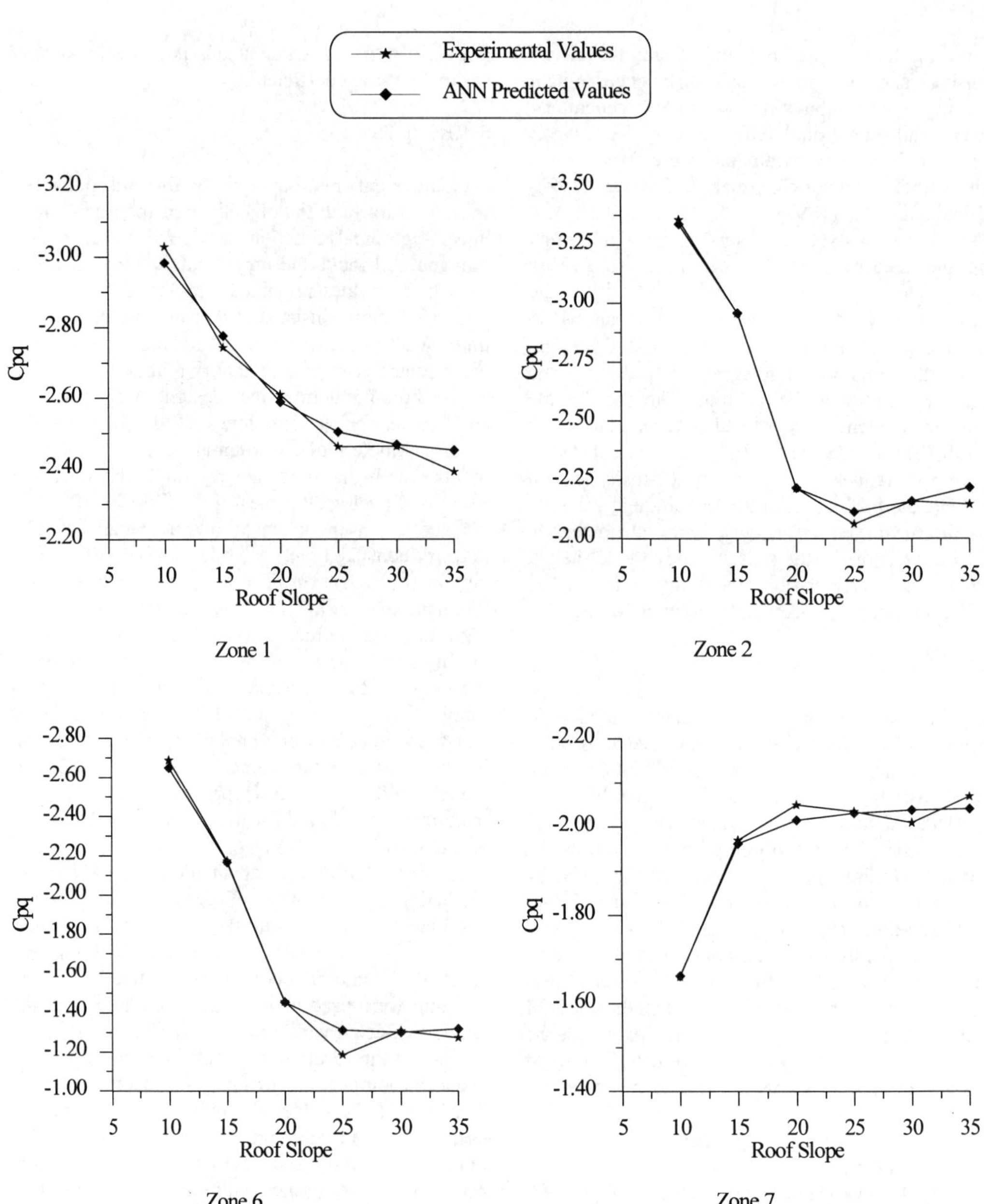

Figure 3. Comparision of Cpq for different zones of the building obtained by experiment and predicted by ANN

Figure 4 shows the comparison of contours obtained by experiment and as predicted by ANN. Values shown in the box is design pressure coefficient for zone 1 when tested in 'stand-alone' case. The predicted contours show close agreement with the experimental values. The maximum error found for the ANN prediction in the interference study was around 6%.

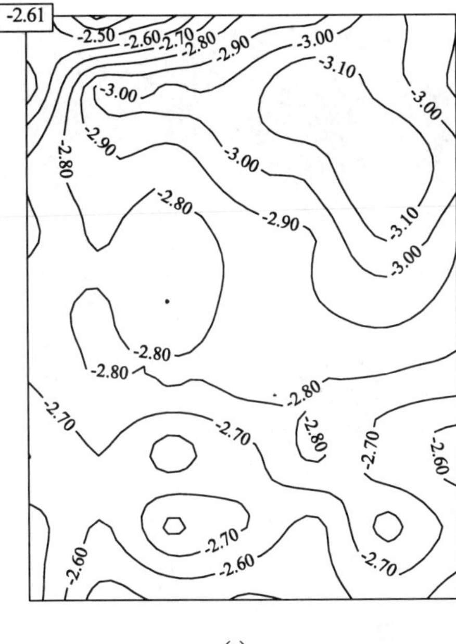

(a)

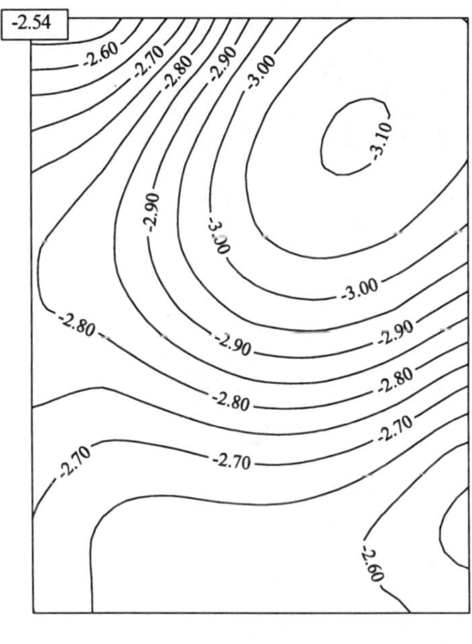

(b)

Figure 4.
(a) Contours of Cpq (Experimental) for zone 1 with the variation of interfering building.
(b) Contours of Cpq (Predicted by ANN) for zone 1 with the variation of intergering building

Table 1

X	Y	Experimental	ANN predictions
0	25	-2.70	-2.76
5	15	-2.817	-2.837
10	0	-2.45	-2.66
15	5	-3.04	-3.01
20	0	-2.52	-2.61

7. CONCLUSIONS

It can be concluded that ANN modelling can be successfully used for determination of wind induced pressure in buildings. It can also be concluded that ANN modelling can also be used to study the interference effect on low buildings and excellent results can be obtained with limited amount of wind tunnel testing. ANN modelling can also be applied for generalising the discrete interference results obtained by experiment. Further research in this regard is still in progress.

8. ACKNOWLEDGEMENT

The work presented in this paper is part of the Ph.D. programme of the by first author under the supervision of second and third author.

9. REFERENCES

Box, G.E.P. & Cox D.R. 1964. An Analysis of Transformation. J. Royal Statistical Soc., 26, 211-252.

Cermak, J.E., & Cocharan, L.S. 1992. Physical Modelling of Atmospheric Surface layer. JWE and Ind. Aerodynamics, 41-44, pp.935-946.

Gupta, D.K., Asthana, B.N. & Bhargava, A.N. 1989. Flood Frequency Analysis by Two Step Power Transformation. J. Instn. of Engrs., India 70, 127-131.

IS:875 (Part 3) 1987. Code of Practice for Design Loads (Other than Earthquake) for Buildings and Structures, part3 Wind Loads, Bureau of Indian Standards, New Delhi.

John, J.A. & Draper, N.R. 1980. Alternative Family of Transformations. J. Appl. Statistics, 29, 190-197.

Tieleman, H.W., Hajj, M.R. & Reinhold, T.A. 1997. Wind Tunnel Simulation Requirements to Assess Wind Loads on Low-Rise Buildings., 2 EACWE, Genova, Italy.

Xu, Y.L., Mehta, K.C. & Reardon, G.F. 1995. Fatigue of Metal Roof Cladding Subject to Wind Loading. Part I: Fatigue-Related Characteristics of Roof Pressures. 9 ICWE, New Delhi, India.

Wind Engineering into the 21st Century, Larsen, Larose & Livesey (eds) © *1999 Balkema, Rotterdam, ISBN 90 5809 059 0*

Detection, analysis and simulation of roof-corner pressure transients

C.L.Pettit
AFRL/VASD, Wright-Patterson, AFB, Ohio, USA

N.P.Jones & R.Ghanem
The Johns Hopkins University, Baltimore, Md., USA

ABSTRACT: Many practical time series, including some pressure signals measured on roof-corners of low-rise buildings in quartering winds, consist of relatively quiescent periods interrupted by transients. In this paper, the dyadic wavelet transform is used to detect these transients in pressure time series. A relatively simple pattern classification scheme is then used detect the underlying structure in these transients. Statistical analysis of the resulting pattern classes yielded a library of signal "building blocks", which are useful for detailed characterization of transients inherent to the signals being analyzed. Nonparametric probability density functions computed for various pattern features and arrival intervals were sampled to generate synthetic time series that mimic the intermittency and transient magnitudes of the original time series.

1 INTRODUCTION

Freestream turbulence and coherent structures may combine with building geometry to produce isolated severe suction peaks. A well-known example of this is the strong suction generated on low-rise building roofs in cornering winds (Lin, et al 1995). The resulting extreme local loads must be accounted for in structural design for wind loads. Thus, the intermittent peak loads need to be quantified and fully understood; however, detailed analysis of these peaks is complicated by their transient nature.

The wavelet transform can be used to account for the nonstationary nature of these processes because it permits the use of analysis functions that are better adapted than sinusoidal functions to analyzing transient features in a signal. Whereas traditional spectral analysis smears the details of transients across a wide frequency range and buries their temporal location in the relative phase angle of the frequency components, the wavelet transform decomposes a time history onto a family of self-similar, localized functions that span the joint time-scale (or time-frequency) plane. Through this approach, transients may be characterized by the magnitude and temporal location of the wavelet coefficients at each scale.

To better quantify the types of transients, as well as their rates of occurrence, that exist in wind pressure measured at a particular location on a structure, the signal should be examined to extract characteristic peaks associated with significant transients. Transients related to a particular flow mechanism are expected to exhibit consistent temporal structure; e.g., the individual vortices behind a circular cylinder are practically indistinguishable when viewed as a grouping of separate entities. This structure can be quantified using pattern recognition techniques, so that a library of "building blocks" for characterizing and simulating intermittent pressure signals can be assembled.

2 THEORY

The primary steps executed by the transient detection and classification algorithm are: (1) denoise the raw signal, (2) perform multi-scale pattern detection, (3) compute a feature vector for each pattern, and (4) use clustering in feature space to detect the underlying structure in the transients. When viewed as a whole, the algorithm conducts conditional sampling based on wavelet coefficients to extract and classify transients. After the transients are classified, a nonparametric probabilistic description of the geometric features of each class is generated using a standard kernel-based density estimation technique. Important features summarized in this manner include each transient's duration and magnitude, as well as the number and location of prominent peaks that dominate the global structure of each transient.

2.1 *Continuous and discrete wavelet transforms*

Only a brief review of wavelet transform theory is presented herein. Daubechies (1992), Kaiser (1994), and Mallat (1998) explain the theory in great detail.

The goal of wavelet analysis is to resolve the local behavior of a signal, $x(t) \in L^2(R)$, onto a set of localized functions called wavelets. These wavelets are derived from a mother wavelet, $h(t)$, through the operations of translation and dilation:

$$h_b^{(a)}(t) = \frac{1}{|a|^{1/2}} h\left(\frac{t-b}{a}\right) \quad (1)$$

where $h \in C$ in general. Translation by the parameter b detects the central location of distinct events in a signal whereas dilation by the parameter a accounts for their extent or scale (in the sense of geographic maps).

The continuous wavelet transform (CWT) of $x(t)$ is

$$\tilde{x}(a,b) = \left\langle h_b^{(a)}, x(t) \right\rangle = \int_{-\infty}^{+\infty} \overline{h_b^{(a)}(t)} x(t) dt, \quad (2)$$

where the overbar denotes complex conjugation. This transform is operationally similar to the Fourier transform; in fact, the scale is roughly inversely proportional to frequency. The CWT can also be inverted to yield a synthesis formula very similar to the inverse Fourier transform; hence, $x(t)$ may be recovered from

$$x(t) = \frac{1}{C_h} \int_0^{+\infty} \int_{-\infty}^{+\infty} \left\langle h_b^{(a)}(t), x(t) \right\rangle h^{*(a)}_{b}(t) db \frac{da}{a^2} \quad (3)$$

where

$$C_h = 2 \int_0^{\infty} \frac{\hat{h}(\omega)}{\omega} d\varpi \quad (4)$$

is essentially a normalization factor that arises in inverting the continuous wavelet transform, and the carat (^) indicates the Fourier transform. The asterisk superscript indicates the family of wavelets that is dual to the original family. Note that (4) indicates that an admissible wavelet must have zero mean value; otherwise C_h will not be finite and the transform will not be invertible. The references cited above provide additional details.

The representation of $x(t)$ in (3) is highly redundant, in that it represents a mapping from the real line to the real plane; that is, $t \in R$ and $(a,b) \in R^2$. A less redundant form of the wavelet transform is used here to improve the computational speed. This form is known as the dyadic wavelet transform (DyWT), in which the scale parameter is sampled at dyadic intervals; i.e., $a_m = 2^m$, where m is an integer. Critical sampling in the time-scale plane is achieved by connecting the scale and translation parameters; the result is the wavelet series, also known as the discrete wavelet transform (DWT), which is useful in signal compression and filtering applications.

2.2 *Multi-scale edge detection*

Coherent transients, which are referred to here as patterns, were detected in a signal by their relatively sharp edges and high local curvatures. These edges were detected in the DyWT as points with relatively high (absolute value) wavelet coefficients contiguous across multiple scales (Mallat 1998). The conceptual basis for this lies in the impulse response of the CWT or DyWT, which (analogously to the Fourier transform) exhibit energy at all scales. Note that the DWT should not be used for pattern detection because it is not translation-invariant (Daubechies 1993).

The signal was first denoised by thresholding the DWT, which is essentially a localized low-pass filtering process (Strang and Nguyen 1996). Then the shorter scales in the DyWT of the denoised signal were examined for localized values that exceeded a pre-determined threshold across contiguous scales. In this way, the boundaries and extrema of individual transients, known as regions-of-sharp-change (rsc) were detected. Finally, empirically-devised tests were applied to the signal segments between each pair of rsc's to determine whether the segments corresponded to true transients or quiescent regions between transients.

2.3 *Pattern classification and recognition*

Looney (1997) provides a good description of the basics of pattern classification and recognition. The concept of pattern is accepted as a primitive from which a foundation of concepts is derived (this concept has already been specialized for the current application to pressure transients). The goal of pattern recognition is to determine whether an object from population P belongs to a well-defined subpopulation S. Hence, recognition is distinct from classification, in which we must decide if multiple patterns constitute an equivalence class. That is, classification is an inductive learning process whereas recognition is deductive reasoning process.

In the current application, patterns are comprised of segments from discrete time signals. Hence, they reside in a relatively high-dimensional vector space, a fact that complicates their comparison. Classification and recognition therefore are implemented in a lower-dimensional feature space: a vector of feature components is computed for each pattern, and the relative similarity of each feature vector to the others is judged via a pre-determined metric (e.g., the Euclidean norm). The choice of these features is application-specific. The goal of feature selection is to summarize the essential nature of each pattern while simultaneously disregarding superfluous aspects, such as background noise.

2.4 *Estimation of Probability Density Functions*

Section 4 discusses the generation of synthetic signals based on the transient classes detected in a given time series. This process depends heavily on sampling from nonparametric estimates of probability density functions (pdf's) for various pattern features. Here, a brief summary of the pdf estimation theory is presented.

The approach is a slight variation on the standard kernel function method, in which a Gaussian kernel function is used to measure the local density of points in the chosen feature space. The feature space is discretized using a regular grid, with **v** denoting a point on this grid. For each grid point, a "mountain function" is constructed (Jang et al. 1997):

$$m(\mathbf{v}) = \sum_{n=1}^{N_p} \exp\left(-\frac{\left\| \mathbf{v} - \mathbf{x}_n \right\|^2}{2\sigma^2} \right) \tag{5}$$

where $\mathbf{x}_n$ is an individual feature vector, N_p is the number of patterns detected (i.e., the number of feature vectors), and σ is an application-specific constant known as the kernel bandwidth, which controls the smoothness of the resulting mountain function. The pdf is then estimated by normalizing the mountain function to a volume of unity.

The choice of a proper kernel bandwidth value involves a certain degree of empiricism. The effect of various choices on the fidelity of the synthetic signals to the original is discussed in Section 4.2.

3 DETECTION OF PRESSURE TRANSIENTS

3.1 *Application to field measurements*

The detection and classification algorithm described above has been applied successfully to signals consisting of artificial transients (Pettit et al. 1998). Next, we describe an application of the algorithm to several intermittent pressure time series measured on the roof-corner of a low-rise building. These data were obtained from the Wind Engineering Research Field Laboratory (WERFL) at Texas Tech University (TTU), a facility that was described by Levitan and Mehta (1992a,b). The work described herein concentrated on pressure measured using a three-by-three grid of transducers on the roof-corner. Because of limited space, the results presented are for a single time series taken from the transducer closest to the roof corner, which was denoted as V1 in the WERFL documentation.

The pressure signal consisted of 36,000 points obtained at a sampling rate of 40 Hz. These data were provided as a pressure coefficient time history, which was computed by dividing each pressure value by the product of the mean atmospheric density and the square of the mean wind speed observed over the entire recording period of fifteen minutes. For the present study, Signal V1 was truncated to 32,768 samples and denoised by thresholding the DWT. This process removed a sizable fraction of the background noise in the signal, but did not noticeably reduce the magnitude or sharpness of the significant transients. Processing the resulting signal using the multi-scale detection algorithm produced a total of 67 transients.

Visual examination of the detected transients indicated some potential for classification based on the number of suction peaks in smoothed versions of the transients. This smoothing was achieved through the following steps:

1. Each pressure coefficient pattern was normalized to a mean absolute value of unity and resampled to a duration of 50 samples. The normalization was implemented to compensate for difference between patterns that had similar shapes but different magnitudes. The resampling yielded patterns with the same duration, so that the relative location of individual peaks in patterns could be stored as separate features. The actual pattern duration was also retained as a separate feature.
2. The normalized and resampled patterns were then approximated using 12-th order Chebyshev polynomials. These approximate patterns generally provided excellent matches to the underlying

global shape of each pattern while simultaneously removing the local peaks that appeared to correspond to noise in the signal.

The patterns were then classified according to the number of suction peaks in the smoothed patterns. Hence, the feature space upon which classification was based was one-dimensional and discrete. Also, as noted above, the duration and normalization factors were retained as separate features for subsequent analysis.

Figure 1 illustrates this pattern-smoothing procedure for two patterns detected in the time series. The first column of plots contains the original time series for each pattern, the second column depicts the resampled patterns, and the third column contains the Chebyshev approximations to the original patterns. From these smoothed patterns, the global suction peaks were counted. For example, the top pattern in Figure 1 was determined to have two suction peaks, and the bottom pattern was characterized as having a single suction peak.

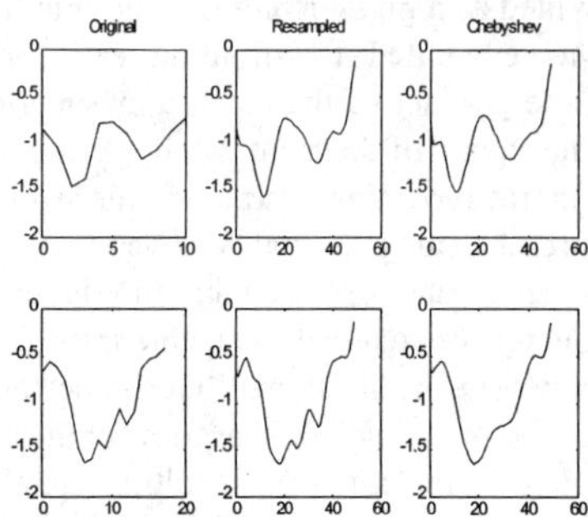

Figure 1. Processing of two patterns detected in Signal V1. The first column contains the two originals patterns, the second column shows the same patterns after resampling to a duration of 50 samples, and the third column depicts the 12th-order Chebyshev approximation of the resampled pattern.

3.2 *Describing the shape of TTU pressure patterns*

After classifying the patterns according to the number of suction peaks, it was found useful to have an economical means of summarizing their basic shape that still maintained important local geometric properties, such as the location and magnitude of each peak. This was achieved by imposing control points at the beginning, end, and local extrema of each smoothed pattern. Each suction peak was assumed to be bordered by a suction "valley" on each side. This is illustrated for a generic two-peak pattern in Figure 2a. For those peaks not bordered by a valley on one side, a control point was imposed as depicted in Figure 2b. Thus, each pattern in a number-of-peak class was assigned the same number of control points.

Because each smoothed pattern consisted of the same number of samples and was scaled to unit mean absolute value, the time index and value of a particular control point, such as that corresponding to the first suction peak in the two-peak class, could be used to judge the relative similarity of the individual patterns within each class. In particular, if this peak should happen to exhibit similar location and values in most of the patterns within each class, this would indicate an additional degree of structure within the number-of-peak class.

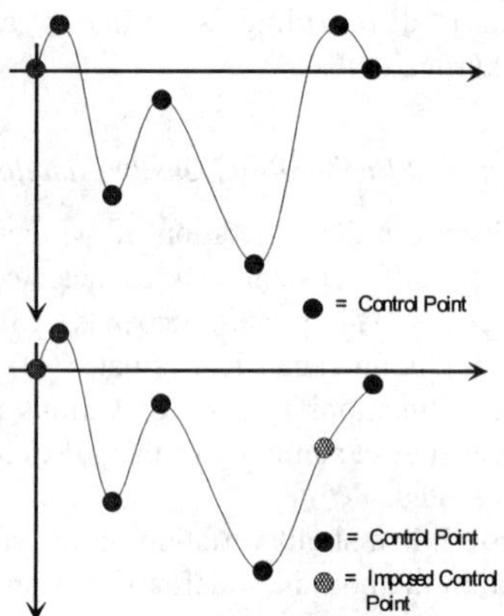

Figure 2. (a) Control point locations for a generic transient pattern. (b) Another generic transient, but with an imposed control point.

3.3 *Results from the detection and classification algorithm*

As noted above, 67 significant transients were detected in Signal V1. The top portion of Figure 3 illustrates the original signal and a pattern indicator function, which was defined to be unity for those time segments that contained a detected transient and zero otherwise. Note that the pattern indicator functions exhibits good visual correlation with most of the visible transients and also indicates that some noise-obscured transients were detected.

The bottom frame of Figure 3 contains a non-parametric estimate of the arrival-interval probability density function (pdf). The dominant primary peak indicates that many of the transients occurred in dense clusters and were separated by relatively short time intervals, whereas the smaller peaks reflect the longer, more quiescent periods between the individual clusters of transients.

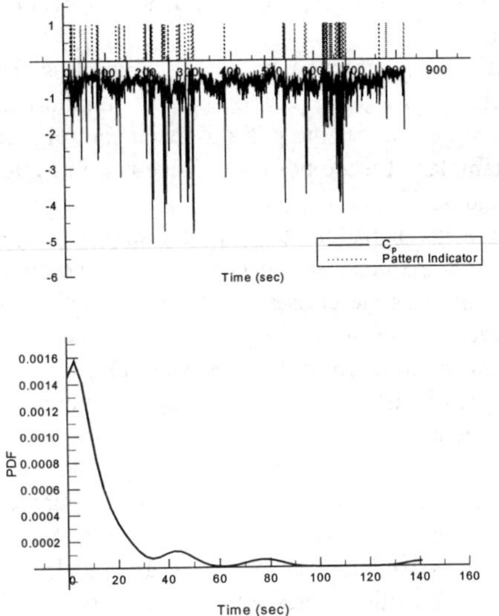

Figure 3. (Top) Signal V1 and its pattern indicator function. (Bottom) Nonparametric arrival interval pdf estimate.

Of the 67 patterns detected in Signal V1, 17 were found to have one suction peak, 33 had two suction peaks, and 17 and three suction peaks. Figure 4 depicts three patterns that were deemed to have a single suction peak. Although each of these patterns exhibits markedly different local behavior from the others, they all share a common underlying single-peak shape. Thus, they were classified together and their unique characteristics, such as their duration, peak value, and peak location, were stored as class-dependent features.

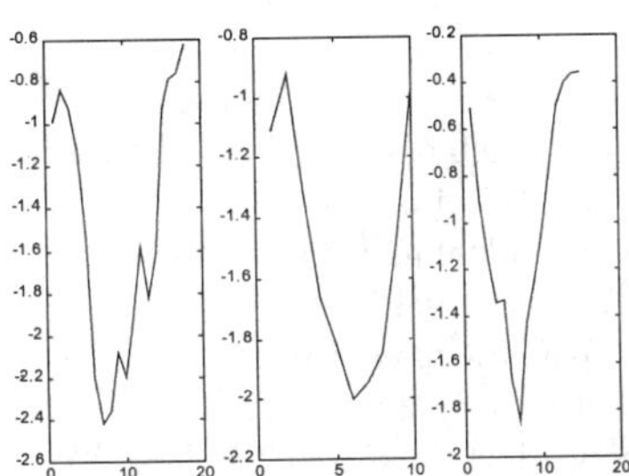

Figure 4. Three single-peak patterns from Signal V1.

Figures 5 and 6 display nonparametric probability density functions (pdf's) associated with several of these class-dependent features. The top frame in Figure 5 shows a contour plot of the joint pdf of duration and magnitude scale factor for the single peak class. The tight cluster of contours reflects the presence of a dominant peak in the pdf. This demonstrates that the majority of the single-peak patterns had similar duration and peak magnitude, and is indicative of additional underlying structure in the single-peak pattern class. That is, most of the patterns in this class have more in common than just a global shape characteristic; they also have similar magnitude and duration.

The second and third frames in Figure 5 contain nonparametric pdf's for the magnitude of the start and end control points, respectively. The relatively large peak in each plot indicates that many of the normalized patterns in the single-peak class had similar beginning and ending values.

The three contour plots in Figure 6 portray the joint time index and value pdf's for the three intermediate control points corresponding to the normalized single-peak patterns in Signal V1. In particular, the center control-point pdf corresponds to the single suction peak. The large peak indicated by the tight clusters simply reflects the structure noted above with regard to the duration and magnitude scale factors.

4 SYNTHETIC SIGNALS

An algorithm was constructed whereby synthetic signals could be assembled based on the patterns extracted from an existing time series, such as the pressure coefficient signal analyzed in the previous section. Synthetic signals are expected to be useful as inputs for computational structure and component modeling, which makes them useful in parameter studies. For example, the effects of increased local suction (e.g., because of higher wind speeds) on roof uplift of a low-rise building might be simulated based on appropriately-scaled versions of characteristic transients observed in the WERFL data (see Section 3). This application is not explored in depth here, but it is the primary motivation for developing the means to simulate intermittent time series based on the patterns detected in a signal.

Instead, the focus of this section is on describing the algorithm used to construct synthetic time series, and comparing example synthetic signals to the original signal on which they are based. Also, the algorithm presented here does not account for simulating the background noise in the original signal; hence, a noise model should be generated separately. One consistent means of doing this is based upon the signal that remains when the detected transients and any underlying trend are removed from the original time series. For Signal V1, this "de-patterned" signal exhibits noise-like visual and spectral characteristics, and has the potential to be modeled reasonably well by standard probability distributions (Pettit 1998).

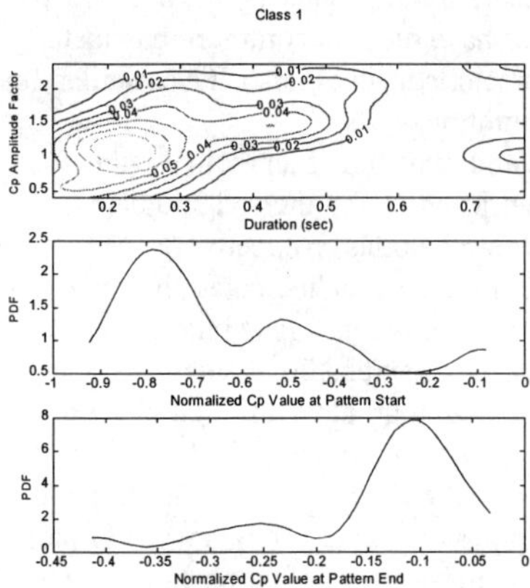

Figure 5. (Top) Joint pdf of pattern duration and normalization factor. (Center and Bottom) Pdf's of start and end control point value.

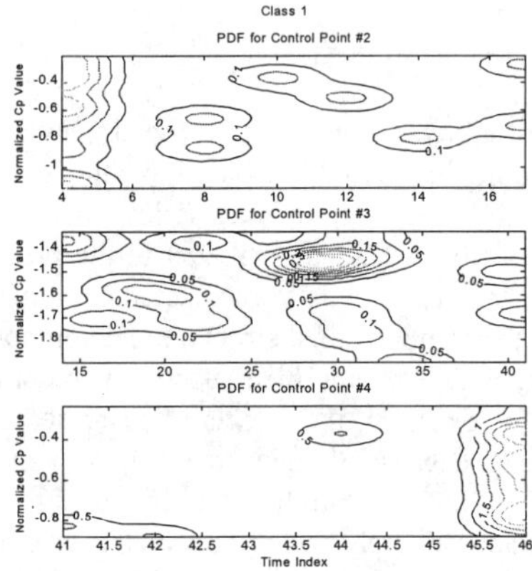

Figure 6. Joint pdf's of intermediate control point time indices and values.

4.1 *Procedure*

A noise-free synthetic time series consists of quiescent periods punctuated by transients from randomly-selected classes, of random duration and magnitude. The bulk of the algorithm is devoted to computing and sampling from the distributions that govern these randomly-selected components. In each case, the associated probability density functions (pdf's) are computed non-parametrically using a Gaussian kernel-based density estimate. The following steps were applied to generate a synthetic signal:

1 Pdf's were estimated for each pattern class. This included the joint pdf for pattern duration and magnitude, as well as pdf's for each control point. Examples of these pdf's are illustrated in Figures 5 and 6.

2 For a given pattern class in the original signal, relative frequencies of occurrence were computed for the possible classes of the next transient observed in the original signal. These relative frequencies were computed by examining each pair of transients in the original signal, where one transient pair consisted of a given transient and the following one in the signal. For each transient pair, the classes of the first and second transients were stored as ordered pairs. After this information was accumulated for the entire signal, a matrix of transitional probabilities was computed for the pattern classes. In this manner, the pattern class for a given arrival was selected based on a one-step memory process in which the pattern class of the most recent transient was assumed to influence the class of the subsequent transient. The pattern classes thus were assumed to constitute a homogeneous Markov chain (Suhir 1997).

3 The transient arrival interval pdf's were estimated based on the time between each pair of arrivals in the original signal. These pdf's were also computed from an assumption of one-step memory, so that the time until the next arrival was assumed to depend on the class of the previous arrival. Note that this assumption seemed appropriate given the assumption in Step 2 that the pattern class of a given arrival was also dependent on that of the previous arrival.

4 A sequence of individual transients was generated in a loop, which ran until the desired signal length was exceeded. Each pass through the loop entailed the following steps to generate an individual transient arrival:
   - The pattern class was selected based on the conditional relative frequencies of occurrence described above (Step 2). Because the TTU patterns were classified according to the number of suction peaks, this would correspond to assigning a particular number of peaks to the synthetic transient in question.
   - The arrival time was sampled from the associated class-conditional nonparametric pdf (see Step 3).
   - Duration and magnitude factors were selected from their joint pdf for the selected class.

- Control point location and value pairs were sampled from the set of joint pdf's (see Step 1). Note that each control point pair was selected independently of the others. A more consistent formulation might entail sampling from a joint pdf for all control points, but this would also require a rather large pattern set of test data to generate a dependable pdf of so many dimensions.
- The "base" transient was constructed from the chosen control points through a two-step process. A piecewise linear interpolation function was fitted through the control points. Then a 15th order Chebyshev approximation of this linear interpolating function was generated. The order of the Chebyshev approximation was chosen empirically.
- The base transient was resampled to obtain the duration selected above. The new transient was then multiplied by the chosen magnitude factor to obtain the desired synthetic transient for this pass through the loop.

The process just described yields a clean synthetic signal. Noise may be added to the synthetic signal at this point, as appropriate to the application.

Table 1. Relevant statistics for the number of arrivals in each of the three sets of synthetic signals. (Note: Avg and SD denote the mean value and standard deviation of Np for the signal set.)

Signal set	σ	Avg	SD	(N_p min, N_p max)	95% C.I. for N_p mean
I	0.01	65.7	11.6	(33, 103)	(64.1, 67.3)
II	0.02	61.3	12.4	(36, 100)	(59.6, 63.1)
III	0.03	58.4	11.2	(33, 92)	(57.3, 60.4)

4.2 *Examples of synthetic time series*

An important factor in the process just described is the kernel-function bandwidth used to estimate the various pdf's, especially those of the class-dependent arrival intervals. Three bandwidth values (listed in the second column of Table 1) were used to estimate the arrival interval pdf's, and the "best" value of the three was judged to be that which produced an average number of arrivals closest to that in the original signal; i.e., $N_p = 67$.

An ensemble of 200 signals was generated for each of the three kernel bandwidths. Relevant statistics are listed in Table 1. These results show that as the kernel bandwidth, σ, was decreased, the mean number of arrivals tended to increase. Furthermore, the 95% confidence intervals for mean number of arrivals in Signal Set I (the right-most column) is the only one that includes the number of arrivals in the original signal (i.e., 67). Hence, of the three bandwidths tried here, $\sigma_I = 0.01$ was the closest to optimal in that it best reproduced the number of arrivals observed in the original signal.

The confidence intervals for the mean number of arrivals in Sets I and III are well separated, so there does exist a clear dependence of the expected number of arrivals on the smoothing parameter. Also, it is interesting that the data in Table 1 indicate a large range of possible numbers of arrivals for each of the smoothing parameter values. Across the entire ensemble of 600 synthetic signals, the lowest and highest numbers of generated arrivals were 33 and 103, respectively.

Figure 7 depicts three synthetic signals, one for each kernel-function bandwidth discussed above, as well as a signal that comprises only the transients detected in the original signal. Each signal also includes the slight linear trend detected in the original signal. This linear trend was computed by fitting a linear regression model to the original signal. Visually, the synthetic signals exhibit a degree of intermittency similar to the original signal, such that in each signal the transients tend to occur in close groups that are well separated in time.

Synthetic pattern magnitudes are also generally similar to those in the original signal, except that a few transients are slightly larger than any found in the original signal. This is because the synthetic transients were constructed by choosing the magnitude scale factor and the control point magnitudes independently. Hence, it is possible that for an individual synthetic transient, these two factors might have combined to produce a larger transient than those observed in the original data. The obvious solution to this problem would be to sample from a joint pdf of magnitude scale factors and control point magnitudes, but this did not seem to be a reasonable approach given the limited number of patterns available in a given transducer signal. Hence, the most practical approach given the limited number of patterns was to sample from each pdf individually.

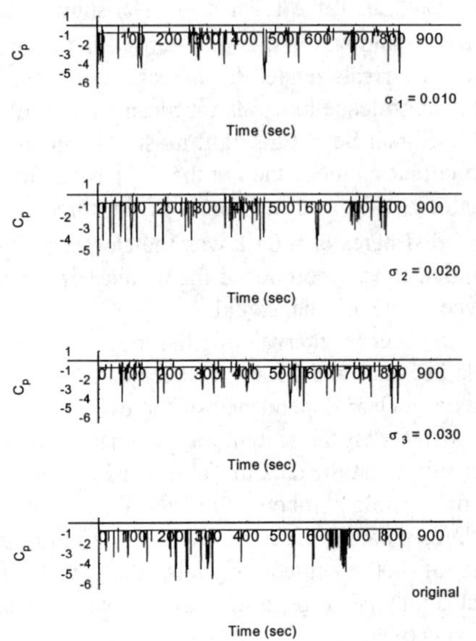

Figure 7. The bottom frame shows the transients in the original signal. The top three frames are examples of synthetic signals based on these transients. One signal is shown for each ot the three kernel function bandwidths described in the text.

5 CONCLUSIONS

A wavelet-based, multi-scale detection and classification algorithm was designed to facilitate the analysis of intermittent, or transient-dominated, time series, with the ultimate goal of perceiving underlying structure in the detected transients.

The joint time-scale localization built into the wavelet transform allows transients to be readily detected by choosing to retain portions of the signal in which the magnitudes of wavelet coefficients exceed a pre-specified threshold across contiguous scales.

The algorithm was applied to an intermittent pressure time series measured on the roof-corner of a low-rise building; sixty-seven transients were detected. These transients were sorted into three classes according to the number of suction peaks exhibited by smoothed versions of the original patterns. Class-dependent probability density functions were estimated nonparametrically for several features that summarize the shape of each transient.

Synthetic signals were assembled by sampling from these nonparametric pdf's. The resulting signals were found to exhibit transient magnitudes and intermittence similar to those of the original signal.

The detection and classigication algorithm should provide quantitative insight into the characteristics of intermittent pressure time series, particularly those associated with roofs of low-rise buildings in quartering winds. In the published literature, there does not appear to have been a concerted effort to examine the statistics of individual pressure transients as distinct entities, even though they are responsible for generating the peak loads on structures subjected to intermittent wind loads; therefore, an immediate application of the algorithm would be to develop a database of wind pressure transients measured on the roofs of low-rise buildings. From this database, sufficient statistical information regarding transient loads could be extracted to improve or fine-tune existing design codes. Furthermore, with appropriate upstream flow measurements, analysis of individual transients could be used to enhance understanding of the mechanics by which these loads are generated.

A related use would be to improve the fidelity of wind-tunnel modeling, especially in the simulation of intermittent turbulent flows. Current wind-tunnel procedure emphasizes the reproduction of turbulence spectra observed in the field, a practice that is predicated on the stationarity of the flow. Unfortunately, atmospheric flows are rarely stationary. An extensive statistical description of the types of transients that should be reproduced in the tunnel could yield more insight into the quality of the wind tunnel modeling process.

REFERENCES

Daubechies, I. 1992. *Ten lectures on wavelets*. SIAM.

Daubechies, I. 1993. Wavelet transforms and orthonormal wavelet bases. *Different perspectives on wavelets* (*Proc. symposia in applied mathematics*), 46: 1-33.

Jang, J.S.R., Sun, C.T., and Mizutani, E. 1997. *Neuro-fuzzy and soft computing*. Prentice-Hall.

Kaiser, G. 1994. *A friendly guide to wavelets*. Birkhauser.

Lin, J.X., Surry, D., and Tieleman, H.W. 1995. The distribution of pressure near roof corners of flat roof low buildings. *Journal of Wind Engineering and Industrial Aerodynamics* 56: 235-265.

Looney, C.G. 1998. *Pattern recognition using neural networks*. Oxford University Press.

Mallat, S. 1998. *A wavelet tour of signal processing*. Academic Press.

Pettit, C.L. 1998. *Wavelet analysis and multi-scale pattern classification in wind engineering*, Ph.D. Dissertation, The Johns Hopkins University.

Pettit, C.L., Jones, N.P., and Ghanem, R. 1998. Analysis of pressure transients via wavelet analysis and pattern classification. *12th ASCE Engineering Mechanics Division Conference*, San Diego.

Suhir, E. 1997. *Applied Probability for Engineers and Scientists*. McGraw-Hill.

Wind Engineering into the 21st Century, Larsen, Larose & Livesey (eds) © 1999 Balkema, Rotterdam, ISBN 90 5809 059 0

Full-scale measurements on a low-rise structure

M.L. Porterfield & N.P. Jones
Department of Civil Engineering, Johns Hopkins University, Baltimore, Md., USA

ABSTRACT: A low-rise structure in Southern Shores, North Carolina has been instrumented to collect wind load and structural resistance data in a variety of wind conditions up to and including hurricane-force winds. The system measures weather parameters (i.e., wind speed and direction, barometric pressure, rainfall, and temperature), wind-induced pressures on the exterior of the structure, internal pressure fluctuations, structural displacements, and strains in structural members. Over two thousand data sets have been collected since the October, 1997 system installation. Much of the data collected to date are from the passage of frontal systems, but two significant wind events, a Northeaster in February, 1998 and Hurricane Bonnie in August, 1998, have been captured. This paper will provide a description of the fundamental components of the measurement system. In addition, meteorological and wind pressure data collected during Bonnie will be presented.

1 INTRODUCTION

Despite significant advances and continued efforts in wind engineering research, the United States continues to experience large property losses in severe winds. As populations continue to migrate toward coastal areas, the United States continues to become more vulnerable to wind-related losses. The 1998 Hurricane season was a relatively gentle reminder of this vulnerability. During that season, seven tropical storms and hurricanes impacted the U.S. coast. None of the storms were worst-case events in either path or strength, yet the United States still experienced considerable losses. Initial estimates put damage totals at approximately $6.5 billion for the 1998 hurricane season (NOAA, 1998).

Low-rise residential and light commercial structures continue to be the most vulnerable to damage in severe winds. In an effort to develop improved means of protecting residential properties against wind damage, the Town of Southern Shores initiated, "*Blue Sky*," a mitigation program involving a cooperative effort between the public and private sectors. As part of *Blue Sky*, a system was developed to obtain full-scale measurements on a low-rise structure in Southern Shores, North Carolina.

Full-scale experiments are necessary to validate wind tunnel and computational models and to study phenomena not easily studied in conventional wind tunnels (e.g., the characteristics of winds in hurricane eye-walls and their effects on low-rise construction). Some comprehensive full-scale field measurement programs have been implemented, most recently, the facility at Texas Tech University's Wind Engineering Research Field Laboratory (Levitan, 1990). These data have been invaluable in improving the understanding of wind-structure interactions. To date, however, none of these programs has reported data in extreme winds such as hurricanes and tropical storms. It is hoped that the Southern Shores measurement program will be able to fill some of these data gaps.

The field site is located in the Town of Southern Shores, on the Outer Banks of North Carolina, about a quarter mile east of the Atlantic Ocean. A location map is provided in Figure 7 later in the paper. In the first year of operation, at least one northeaster and a tropical storm impacted the site. There is a reasonable likelihood that additional and perhaps more severe storms will hit the field site in future seasons. Data on these storms would augment existing field data and perhaps help the wind engineering community better understand the characteristics of these extreme winds (at ground level) and their effects on the constructed environment.

2 THE INSTRUMENTED STRUCTURE

The instrumented structure (Figure 1), the Kern P. Pitts Center, is located adjacent to the Town Hall in Southern Shores. The two-story structure was conceived as a wind hazard training, research, and demonstration facility as part of *Blue Sky*. The structure was designed to resemble low-rise residential construction in the area. The Pitts Center has many interesting architectural features, including a cupola and dormers. In addition, the Pitts Center's structure is composed of three structurally independent sections, each made of a different construction material. The north section of the structure has traditional wood framing. The southwest section is built from concrete masonry units and the southeast section has a galvanized light-gage steel frame.

Since the Pitts Center closely resembles residential construction in the area, data from this structure will be of direct relevance to local construction. Since it was built as a research, training and demonstration facility, some of the barriers associated with instrumenting private homes are avoided. For example, sections of drywall have been left off so the structural frame remains exposed. This allows for easy access to potential instrumentation points. In addition, the owners of the Pitts Center allowed holes to be drilled in the roof and walls of the structure for pressure measurements.

Figure 1: Front View of the Kern P. Pitts Center in Southern Shores, North Carolina. (Note: the front of the structure faces east).

3 OVERVIEW OF INSTRUMENTATION SYSTEM

The instrumentation system was designed to measure wind loads and structural resistance in a variety of wind conditions up to and including hurricane force winds.

3.1 *Meteorological Instrumentation*

Meteorological data are collected at two locations in the vicinity of the structure. Wind speeds are measured along three axes by an ultrasonic anemometer that is located on an anemometry tower 60 feet east of the structure. A propeller-vane anemometer, located on an instrumentation pole above the chimney, measures wind speed and direction. Both anemometers are located 10 meters above ground level. In addition, barometric pressure, rainfall, and temperature are measured with sensors located on the instrumentation pole above the chimney.

3.2 *Pressure Measurements*

The pressure sensors are concentrated in the second floor, southeast corner room. This room was selected for the installation of both the pressure sensors and strain gages for two reasons. First, the structural frame in this room is galvanized steel. Steel is much more readily suitable for strain gage use than wood (which is found in other sections of the structure). Second, there are few obstructions for the wind coming from the southeast direction—it is a fairly open exposure. Other sections of the structure can be easily instrumented as a cabling network was pre-installed throughout the Pitts Center.

Surface pressures are measured at ten exterior points on the structure using differential pressure sensors. Figures 2 to 5 show the locations of the sensors relative to the upper southeast corner of the structure. Six sensors monitor pressures on the walls (three along the south wall, three along the east wall). Additionally, three differential sensors have been installed to measure pressures on the roof and one has been installed to measure pressure on the eave. An additional differential pressure sensor monitors internal pressure in the room.

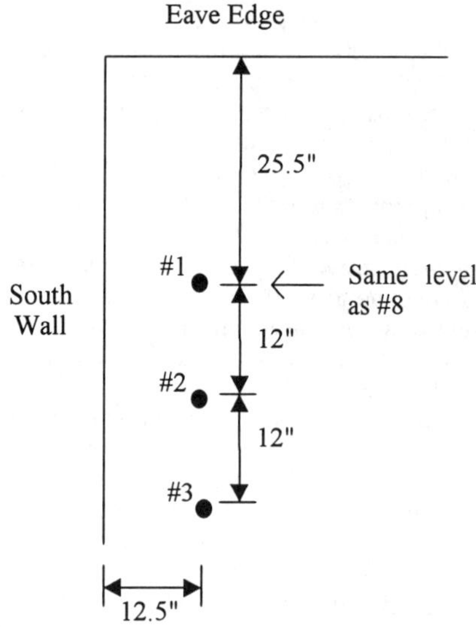

Figure 2: Elevation showing tap locations along east wall.

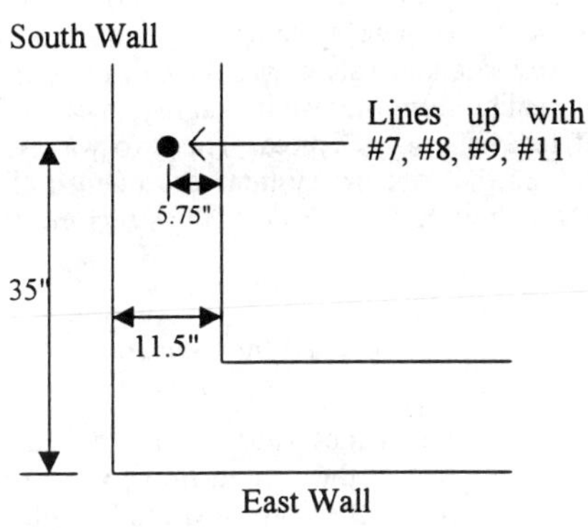

Figure 3: Plan view showing tap locations along eave.

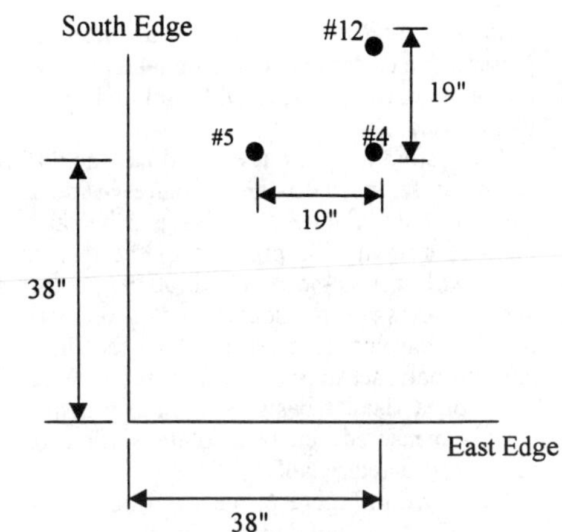

Figure 5: Plan view showing tap locations along roof.

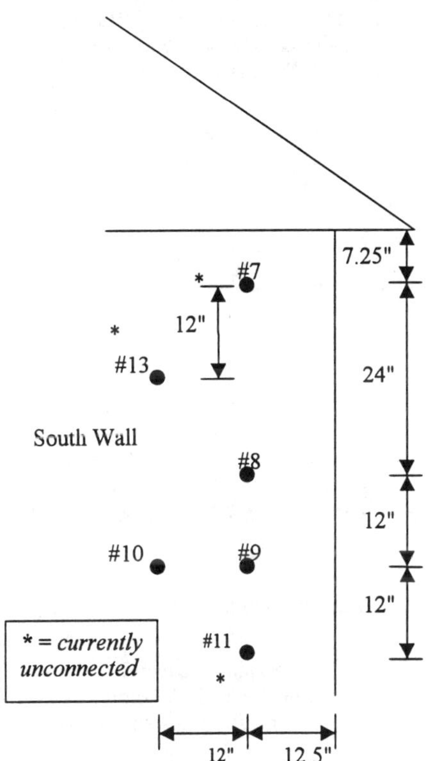

Figure 4: Elevation showing tap locations along south wall.

Since wind-induced pressure fluctuations are small compared to standard atmospheric pressure, and changes in atmospheric pressure can be large in severe winds, it is potentially difficult and costly to find instruments with the accuracy and resolution to measure barometric pressure and its fluctuations as well as the wind-induced pressure components. For example, the stagnation pressure associated with a 100-mph wind (at standard atmospheric conditions) is approximately 0.179 psi. This is small compared to standard atmospheric pressure of 14.7 psi (1.2%).

Most off-the-shelf absolute pressure sensors are not sufficiently accurate to capture the small pressure fluctuations associated with wind. Models that do have the accuracy and resolution to measure small wind-induced fluctuations are expensive, and therefore not a cost-effective means of making multiple pressure measurements. In order to address this problem, differential pressure sensors can be used. Differential sensors are typically relatively inexpensive compared to absolute sensors. The difficulty in using these devices, however, is in establishing a stable reference pressure. In field applications, barometric pressure is often used as the reference pressure. At the Texas Tech Wind Engineering Research Field Laboratory, for example, differential sensors are all referenced to the pressure in a hole in the ground near the structure. This pressure is assumed to be barometric (Levitan, 1991).

At Southern Shores, a Texas-Tech-style reference pressure system was not a feasible solution. The sandy soils would make it difficult to maintain a hole in the ground without considerable expense, and the heavy rain expected in hurricane events would pose challenges to keeping the hole drained.

Also, a semi-portable system was desired so that pressure measurements could be taken at a variety of locations on the structure with minimal disruption to the finished walls.

The system that was developed has the differential sensors referenced to the pressure inside a sealed steel air tank. Figure 5 shows a schematic of the pressure system. The pressure in the tank is monitored with a high-accuracy (0.005% of full scale) absolute pressure transducer and is regularly compared to barometric pressure. To keep the pressure close to barometric, the tank is vented to the atmosphere on a regular basis. Since the pressure in the tank is measured actual absolute surface pressures can be estimated reliably.

As shown in Figure 6, the wind-induced pressure acts through the holes drilled in the exterior of the structure, through a normally open 3-way solenoid valve to the differential sensor. The three-way solenoids are used to control which air supply feeds into the active side of the pressure sensors. During normal operation, the air from outside is in contact with the active side of the gage. While "shunt calibrations" are being performed, however, the solenoid valve closes off exterior air and puts the pressure in the tank on the active side of the gage. This allows the zero offset on the transducers to be calculated since reference pressure is seen on both sides of the gage.

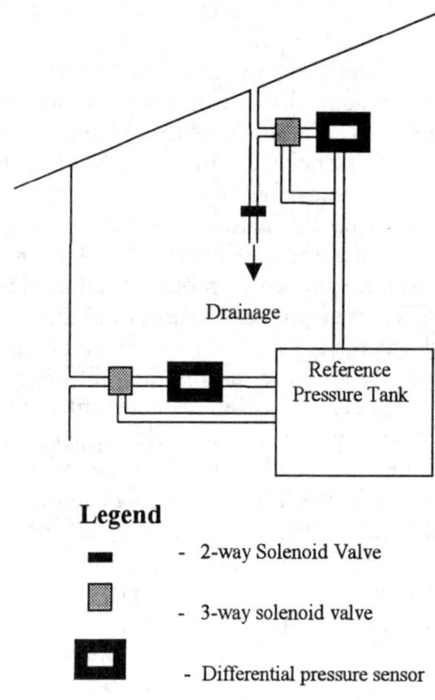

Figure 6: Schematic Diagram of pressure system

Water intrusion was also a concern for the roof taps, so a drainage system using normally closed two-way solenoid valves was developed. During the shunt calibration, the water drainage system for the roof taps is also activated. The two-way solenoid valves in the roof tap systems are opened allowing water to drain to the outside of the structure.

3.3 *Strain and Displacement Measurements*

Twenty strain gages have been installed on the rafters and studs of the second-floor, southeast corner room of the Pitts Center. Currently eight of the strain gages are being supplied with power and used for acquiring data (four on the rafters, four on the studs). The studs and rafters in this section of the structure are steel channels. As shown in Figure 7, a gage was placed on each of the interior flange surfaces to allow the axial and bending strains to be separated. The instrumented rafters and studs are 78" and 96" long, respectively. Gages were placed 60" from the ground on the studs and 49" from the roof crest on the rafters.

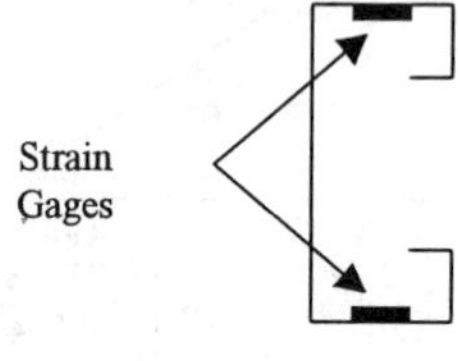

Figure 7: Strain gage locations on channel sections.

A linear potentiometer has been installed along the large, wood-framed north wall of the structure to measure structural displacements.

3.4 *General System Information*

The 31-channel data acquisition system was designed to run unattended. A commercially available remote control and file transfer software package is used with a modem to make system changes and to download data to the remote control and data processing center at the Johns Hopkins University in Baltimore, Maryland. A commercially available data acquisition and process control software program is used to control the data gathering processes. All channels are sampled at 25 Hz; all analog inputs are filtered at 10 Hz using 4-pole Bessel filters.

By default, the system monitors and collects summary statistics on all the channels. When a predetermined trigger threshold (currently set to 3-second gust of 20 mph) is exceeded, the system performs a one-minute shunt calibration on all the differential pressure sensors. The system then records data from all channels at 25 Hz for two five-minute segments. Finally, before returning to the default-monitoring mode, the system performs an additional shunt calibration on the differential sensors.

The system has been installed and operational since October 1997. Since that time, over 2000 data sets have been collected. The remainder of this paper will focus on data and preliminary results from the passage of Hurricane Bonnie.

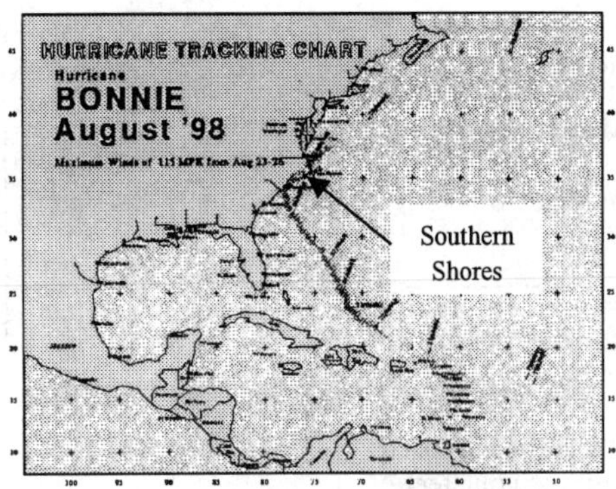

Figure 8: Track of Hurricane Bonnie (Army, 1998)

4 OVERIVEW OF HURRICANE BONNIE

4.1 *Description of the Storm*

Bonnie developed on August 19, 1998 in the eastern Atlantic as the second tropical depression of the 1998 hurricane season. She strengthened into a tropical storm and was named Bonnie on August 20. Bonnie continued to gain strength and was upgraded to a hurricane on August 22. On August 27, after making landfall in North Carolina, Bonnie was downgraded to a tropical storm only to regain hurricane status again on August 28. Bonnie only maintained this status for about 18 hours at which time she was again downgraded to a tropical storm. Meteorologists labeled Bonnie extratropical on August 30 due to her much weakened state.

At her strongest, Bonnie was a large category three hurricane measuring approximately 400 miles across. She first struck the United States coast on August 26 near Cape Fear, North Carolina, just south of Wilmington. According to NOAA, the peak 5-second wind gusts seen in Wilmington were 74 mph, but at Frying Pan Shoals light tower off the North Carolina coast, gusts were recorded up to 104 mph. After impacting the Wilmington area, Bonnie traveled slowly up eastern North Carolina bringing strong winds and large amounts of rain (up to 10 inches in some areas).

Bonnie caused extensive damage in North Carolina. Few structures sustained major structural damage, but the high winds and large amounts of rain toppled numerous trees and hit the tobacco fields very hard. Initial estimates put the damage total between one and two billion dollars in North Carolina alone, with most of these losses in the tobacco industry. Bonnie is directly blamed for the death of a 12-year-old girl in Currituck, NC: a large tree fell on her home, crushing part of the house.

As Bonnie moved past North Carolina and back into the Atlantic, she regained some of the strength she had lost over North Carolina and was able to bring winds greater than 100 mph to the Virginia Beach area. In addition to blowing over trees, Bonnie removed roofs and siding from several structures in this area (USA Today, 1998).

Figure 8 shows the track of Hurricane Bonnie. As indicated on the map, the eye of Bonnie passed near Southern Shores structure site.

Southern Shores is located about 200 miles northeast of Cape Fear, where Bonnie first made landfall. Bonnie had been downgraded to a tropical storm by the time she reached Southern Shores, but she still brought strong winds and rain to the area.

5 DATA RECORDED FROM BONNIE

Between August 25 and August 28, the system was set to trigger continually. An uninterruptable power supply was installed in the system to extend record recording time in the event of a power failure. The system never lost power and ran without interruption for the three days. A total of 693 five-minute data sets were collected over the three-day period.

5.1 *Meteorological Data*

The most severe part of the storm impacted Southern Shores on August 27, 1998. The highest winds were seen around 3:30 p.m. and the lowest barometric pressure was recorded shortly before 9 p.m. Table 1 gives a summary of some of the meteorological data derived over this period. The winds at the site were gusty during the storm and as evidenced in Table 1 averaging time as expected plays an important role in reporting wind gusts.

Table 1: Summary of meteorological data collected during Bonnie (August 27).

Measured Quantity	Value	Approx. Time
Minimum barometric pressure	989.8 mb	8:41 p.m.
Max 5-minute mean wind speed (propeller anemometer)	34.2 mph	3:21 p.m.
Max 5-minute mean horizontal wind speed (sonic anemometer)	40.3 mph	3:11 p.m.
Max 5-minute mean vertical wind speed (sonic anemometer)	3.1 mph	4:15 p.m.
Max 3-second gust speed (propeller anemometer)	51.8 mph	3:21 p.m.
Max 3-second horizontal gust speed (sonic anemometer)	55.3 mph	3:33 p.m.
Max 3-second vertical gust speed (sonic anemometer)	11.5 mph	1:40 p.m.
Max 1-second gust speed (propeller anemometer)	52.9 mph	3:11 p.m.
Max 1-second horizontal gust speed (sonic anemometer)	60.3 mph	3:11 p.m.
Max 1-second vertical gust speed (sonic anemometer)	14.4 mph	1:40 p.m.
Total rainfall	~2 in.	For whole storm

The meteorological data clearly illustrates the passage of Bonnie over the Pitts Center. As seen in Figures 9 and 10, the regions of large temporal barometric pressure gradients (~3 p.m. on August 27 and ~5 a.m. on August 28) are accompanied by the clusters of highest wind gusts. These regions correspond to the passage of the storm's eyewall. Decreased wind speeds accompany the region of minimum barometric pressure: the relatively calm eye of the storm. As indicated on Figure 9, the second cluster of high wind gusts (~5 a.m. on August 28) are considerably lower in magnitude than the first cluster of peaks. This is probably due to the fact that Bonnie continued to degrade as she passed over the Carolinas and by the time the trailing eyewall again passed over the Pitts Center, Bonnie was a considerably weaker storm.

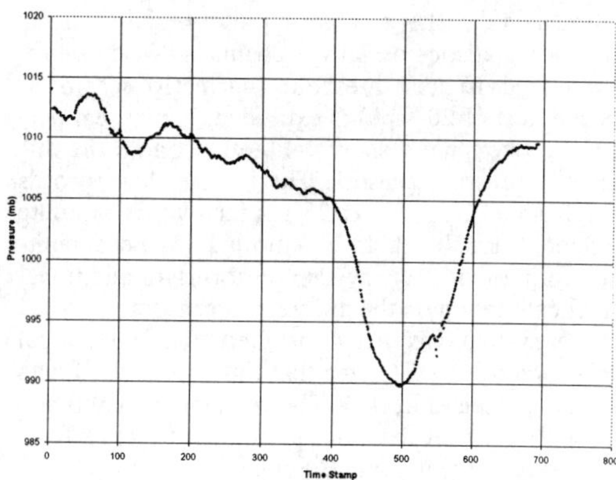

Figure 9: Evolution of barometric pressure.

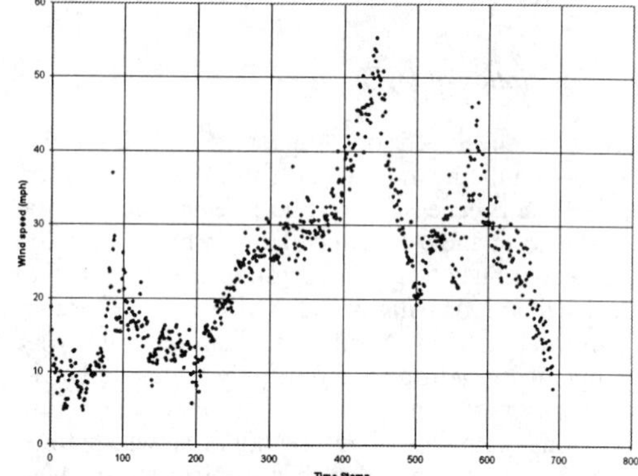

Figure 10: Evolution of 3-sec horizontal wind gusts.

5.1.1 *Pressure Data*

Pressure coefficients were calculated using the standard formula (1)

$$C_p = \frac{p - p_a}{\frac{1}{2}\rho \overline{U}^2} \tag{1}$$

where: p= measured pressure, p_a= atmospheric pressure, ρ= air density, and U = mean wind speed.

For the data presented below, pressure coefficients were calculated for two different choices of U. Both the 5-minute mean and the 3-second horizontal wind speeds were used for U. (The data from the sonic anemometer were used for the calculations below since that anemometer is free of moving parts and can respond more quickly to rapid changes in wind speed.)

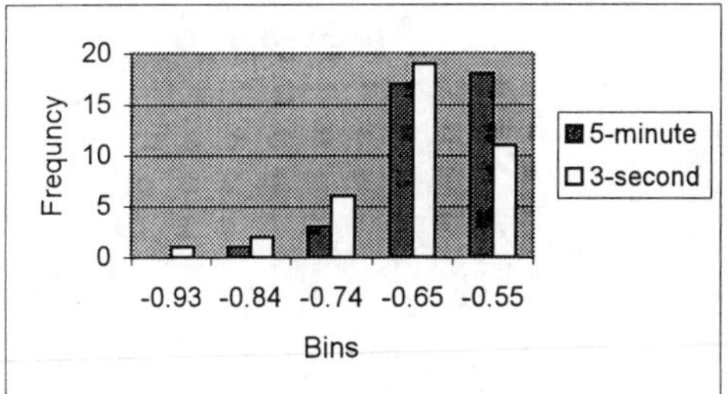

Figure 11: Histogram of mean Cp values for eave tap.

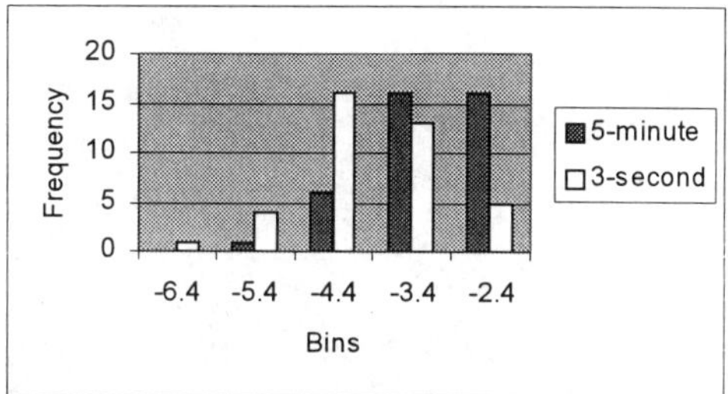

Figure 12: Histogram of peak Cp values for eave tap.

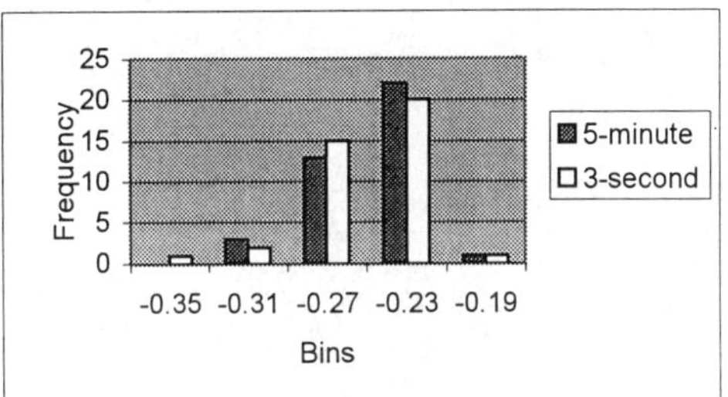

Figure 13: Histogram of coefficient of variation of Cp values for eave tap.

Since some of the records were evidently highly nonstationary and gusty, using the five-minute mean and three-second gust values may overestimate or underestimate instantaneous pressure coefficients. Efforts are currently underway to look more closely at how best the pressure coefficients should be calculated for these types of records so that pressure coefficients are not misrepresented for cases where the wind environment is highly temporally variable, as was most certainly the case here.

Figures 11-13 show histograms of the mean, peak, and coefficients of variation, respectively, of the pressure coefficients calculated for the eave tap. These parameters were calculated for a wind angle of approximately 135 degrees (which bisects the SE corner of the structure). Only records with a minimum three-second gust speed of greater than 20 mph are presented in the figures, resulting in 39 records that meet the criteria. Histograms were chosen to present the data since the distribution of values, not just the mean and peak values, were of interest. Two different choices of U were used: one using the mean wind over the entire 5-minute record and the other using the mean wind based on a moving 3-second window. One of the interesting features highlighted by the histograms is that the distributions of the mean and coefficient of variation for the Cp's calculated are quite similar, but the distribution of the peak Cp values for the two choices of U are substantially different. This indicates clearly that averaging time, or more generally, the precise definition of all the terms in Eq. (1), is extremely important in defining pressure coefficients.

6 CONCLUSIONS

This paper provided a description of the full-scale measurement program being conducted at the Kern P. Pitts Center in Southern Shores, North Carolina, and presented a small subset of the preliminary meteorological and pressure data collected during Hurricane Bonnie. Although Bonnie had been downgraded to a tropical storm by the time she reached the Pitts Center, she brought strong winds to the region. The three-second wind gusts at the site reached nearly 60 mph. Although, in the context of hurricane winds, the recorded wind velocities were not extremely high, this storm provided the opportunity to collect many data sets in a variety of meteorological conditions over a short period of time. Even this very limited subset of data have highlighted some features that will require further analysis and have raised some important questions relative to even simple and generally accepted notions such as the commonly used pressure coefficients.

ACKNOWLEDGEMENTS

The authors wish to acknowledge those who have made this project possible. The Town of Southern Shores provided funding for the design, installation, and maintenance of the initial instrumentation system. The National Science Foundation and the American Society of Civil Engineers (through the O.H. Ammann fellowship) granted money for data analysis and system maintenance.

The authors would like to thank the Town of Southern Shores for their continued support and encouragement. The authors would also like to thank the following individuals from the Johns Hopkins University, Department of Civil Engineering, for their contributions to this research effort: Jack Spangler, Jeff Lowe, Margeurite Jeansonne, Julie McGlosson, Joseph Main, and Steve Kelly.

REFERENCES

Army. 1998. Internet site: http://nab71.nabwc.usace.army.mil/canes/BONNIE.gif

Levitan, M.L., K. C. Mehta, C. V. Chok, & D. L. Millsaps 1990. An overview of Texas Tech's wind engineering field research laboratory. *Journal of Wind Engineering and Industrial Aerodynamics*, 36 : 1037-1046.

Levitan, M.L., K. C. Mehta, & W.P. Vann 1991. Field measurements of pressures on the Texas Tech Building. *Journal of Wind Engineering and Industrial Aerodynamics*, 38: 227-234.

NOAA (National Oceanic and Atmospheric Administration). 1998. Internet site: http://www.nhc.noaa.gov/ftp/pub/forecasts/discussion/MIATWSAT

USAToday. 1998. Internet site: http://www.usatoday.com/weather/huricane/1998/wbonnie.htm

Wind Engineering into the 21st Century, Larsen, Larose & Livesey (eds) © 1999 Balkema, Rotterdam, ISBN 90 5809 059 0

Wind force on canopies attached to low-rise buildings

H. Ueda & H. Hagura
Department of Architecture, Faculty of Engineering, Chiba Institute of Technology, Japan

H. Hamada
Graduate School of Engineering, Chiba Institute of Technology, Japan

ABSTRACT: Large-scale canopies have been attached to the increasing number of buildings. Larger attached canopies are very sensitive to the influence of wind and actually often damaged by wind force considerably. The purpose of the present paper is to describe the characteristics of fluctuating wind pressure and fluctuating wind force acting on a large-scale canopy attached to low-rise buildings with a flat roof. It also describes the characteristics of aerodynamics center that is estimated, based upon fluctuating stress generated in cantilevers supporting a canopy roof.

1 INTRODUCTION

Recently, large-scale canopies have been attached to the increasing number of buildings. Larger canopies are very sensitive to the influence of wind and actually often damaged by wind force considerably.

Although the wind force coefficients for attached canopies on low-rise buildings are contained in the Australian Standard (AS1170.2), other major wind load codes and wind load standards do not include them as was pointed out by Holmes and Kwok (1997). AS1170.2 prescribes the net pressure coefficients for canopies on the wall as well as, awnings or carports adjacent to an enclosed building with a roof slope of 5° or less.

There have been several studies on the fluctuating wind forces acting on canopies attached to eaves of a low-rise building with a gable roof (Cermak et al. 1974, Stathopoulos 1981). On the other hand, wind loads on canopies attached to a building may depend on the position of canopies and the configuration of buildings. Jancauskas and Holmes (1985) investigated the characteristics of fluctuating net uplift force acting on canopy roofs adjacently to an enclosed building with a gable roof, through wind tunnel tests, and proved that the net uplift force acting on canopy roofs depends on the ratios of canopy height to eaves height as well as canopy height to projected length of the canopy.

The purpose of the present paper is to describe the characteristics of fluctuating wind pressures and the fluctuating wind forces acting on a large-scale canopy attached to low-rise buildings with a flat roof. It also presents the characteristics of aerodynamics center that is estimated, based upon fluctuating stress generated in cantilevers supporting a canopy roof.

2 PROCEDURE OF EXPERIMENT

The experiment was performed in a boundary layer wind tunnel at the Department of Architecture, Chiba Institute of Technology. The wind tunnel is an open circuit type with a working section of 6.4 m long, 1.2 m wide, and 0.9 m high.

A surface boundary layer under neutral conditions corresponding to an open terrain was simulated using spires and roughness elements. A mean wind velocity and a longitudinal turbulence intensity profile measured at the center of wind tunnel turntable, without any model on it, are shown in Figure 1. The exponent of the mean wind velocity profile was 0.14, and the value of surface roughness length that was obtained using the logarithmic law was 0.007 cm. A geometric scale of the boundary layer flow was estimated as approximately 1/250 according to the roughness length and the longitudinal turbulence scales. The shape of a flat-roof model with a canopy used in the experiment is shown in Figure 2. A wind velocity at the eaves height of the flat roof model was approximately 10 m/s.

The dimensions of the flat-roof building models are shown in Table 1, and those for the attached canopy models used in this study are shown in Table 2. As shown in Figure 3, the slits of the canopy model made from a brass board had a pair of built-in pneumatic copper tubes with an outer diameter of 1 mm and the length of 18 cm. A copper joint pipe of 8-mm length was placed at one end of each copper pipe. A pressure tap with a diameter of 0.6 mm was arranged at a tip of the upper side of one copper pipe and the lower side of another copper pipe, respectively. The pressure tap was arranged at the center (aa') and gable ends (bb') of the canopy roof. For the canopies of 7-cm and 5-cm length altogether 10 taps (5 on each surface) were uniformly arranged, and for the canopies of 2.5-cm length altogether 6 taps (3 on each surface) were uniformly arranged.

The fluctuating wind pressures acting on each surface of the canopies were measured simultaneously using a pressure measurement system connected to the pneumatic vinyl tubes with 1.6-mm inner diameter and 1.2-m length. The compensation of the tubing system was conducted using the frequency response function of the pneumatic tubing system that was approximated according to the Bergh & Tijdeman theory and the Fast Fourier Transform Method (Ueda et al. 1994). The pressure measurement was carried out for several attacking angles.

Pressure signals were converted for 16384 samples from analog to digital values at a sampling frequency of 500 Hz. A digital low-pass filter with cut-off frequency of 110 Hz was used in analyses of pressure data. The pressure data were converted to pressure coefficient data using the mean dynamic pressure at eaves height. The reference static and dynamic pressure were monitored by a Pitot tube placed at a height of 64 cm above the model roof.

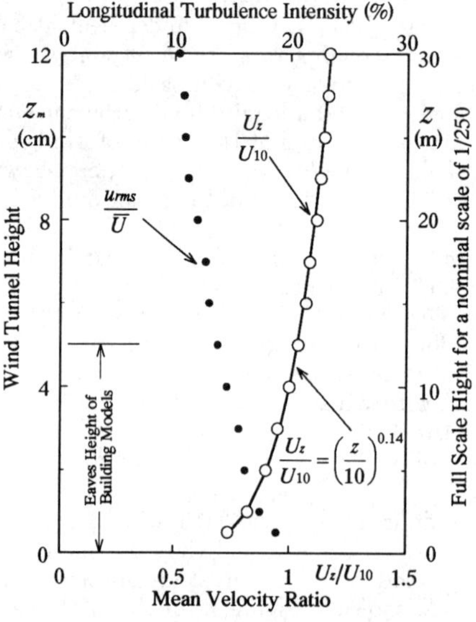

Figure 1. Mean wind velocity and turbulence intensity profiles at model position

3 RESULTS AND DISCUSSION

3.1 *Mean pressure coefficients*

The distributions of mean pressure coefficients of both the upper surface and the lower surface of the center and the gable side of the canopies under the attack angle 0° are illustrated in Figure 4. In this case, span-ratio of the building model $L/H = 1$, wide-ratio $W/H = 10$, length-ratio of the canopy $l/H = 1$.

Table 1. Dimensions of flat roof building models

Parts	Dimensions			
Eaves height ; H (cm)	5			
Ridge direction ; W (cm)	50	37.5	25	12
Span direction ; L (cm)	5	15	25	

Table 2. Dimensions of attached canopy models

Parts	Dimensions			
Thickness ; δ (mm)	1			
Length ; l (cm)	2.5	5	7	
Height ; h (cm)	5	3.75	2.5	1.25
Width ; h (cm)	50	37.5	25	12

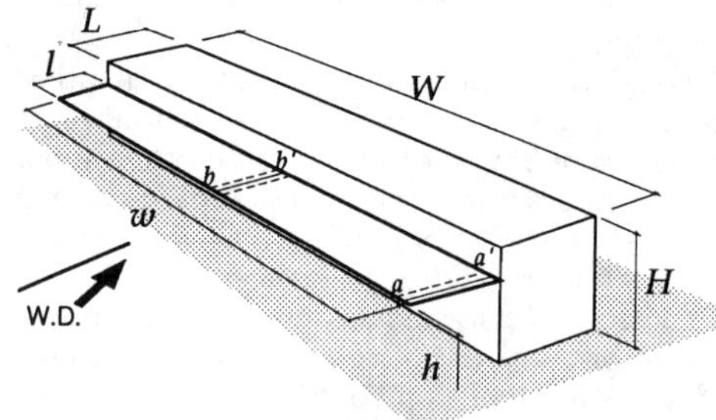

Figure 2. A flat-roof building model with canopy

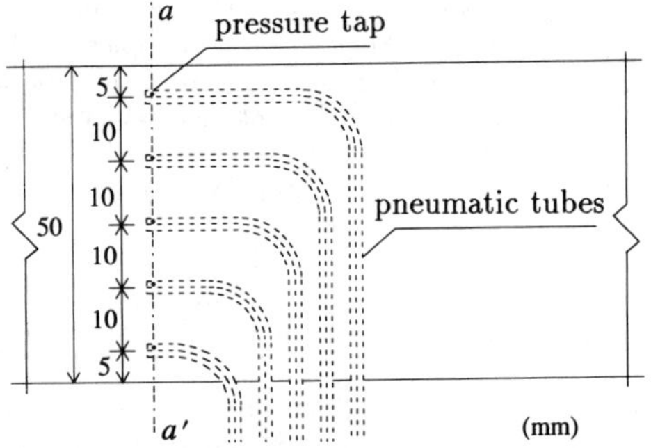

Figure 3. Arrangement of pneumatic tubes on a canopy

The mean pressure coefficient is defined as the average of the instantaneous pressure coefficient over the sampling period. The mean pressure coefficient is plotted against the parameter x/l. Here, x is the distance from the leading edge of the canopy. The mean pressure coefficients of the upper surface of both the canopy center and the gable side recovered with the decrease of the parameter h/H. The pressure recovery resulted from a phenomenon that a flow at the upper side of the canopy was reattached to the building wall, and the tendency was more clear at the canopy center.

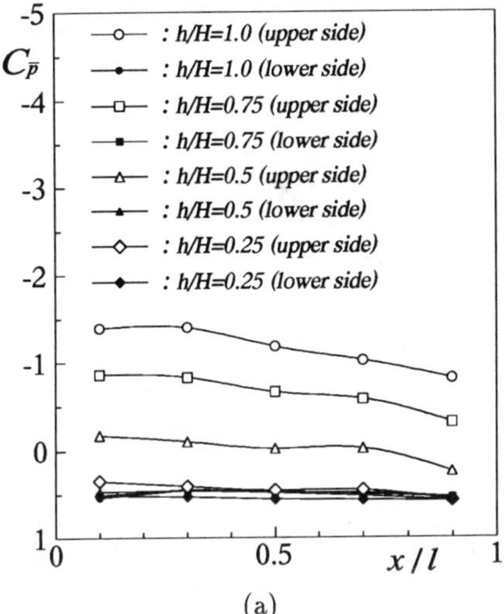

(a)

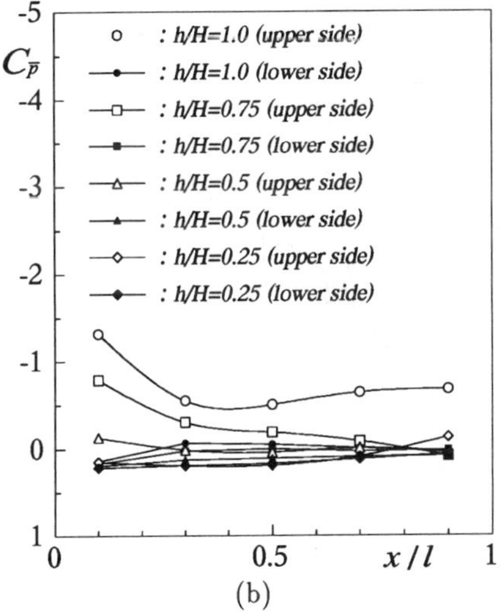

(b)

Figure 4. Distributions of mean pressure coefficients of both upper surface and down surface of canopy under the attack angle 0°: (a) canopy center; (b) gable side

On the other hand, the mean pressure coefficients of the lower surface of both the center and the gable side were not hardly affected by the parameter h/H. This result shows the inflow was stagnated by the building and the canopy beneath the canopy.

3.2 *Wind force coefficients*

Coefficients of wind force acting on the canopies were evaluated based upon the pressure that was measured simultaneously on both the upper surface and lower surface of the canopies. As shown in Figure 5, the properties of the maximum, minimum, mean wind forces acting on the canopies were significantly affected by the attack angles and canopy heights. The mean wind force coefficient $C_{\bar{F}}$ is defined as the mean of instantaneous wind force coefficient measured over the sampling period. The maximum wind force coefficient $C_{\hat{F}}$ is defined as the maximum of the instantaneous wind force coefficient measured over the sampling period. The minimum wind force coefficient $C_{\check{F}}$ is defined as the minimum of the instantaneous wind force coefficient measured over the sampling period. Here, the upward wind force is defined as the negative wind force. The minimum wind force coefficients and the mean wind force coefficients for the canopy center became almost the extreme value in case of the attack angle of 0°. The mean wind force coefficients and the minimum wind force coefficients of gable side became the extreme value in case of the attack angle of about 60°, and the extreme value of the maximum wind force coefficients in this case appeared at the attack angle of 90°. The effects of the attack angles were observed more remarkably at the gable end of the canopy. The minimum wind force coefficient at the gable sides in case of the attack angle of about 60° may have been caused by the effects of conical vortex that occurred at a corner of roof. The minimum wind force coefficient decreased with the decrease of the canopy height. It is considered that the properties of the wind force acting on the canopies are significantly affected by the properties of the pressure on the upper surface of the canopies.

3.3 *Estimate of mean and extreme wind force*

As shown in Figure 6, the mean wind force coefficient and the minimum wind force coefficient at the canopy center under the attack angle of 0° was plotted against the parameter ϕ, that is $\phi = (h/H)^3(w/l)^{0.5}$. These figures show that the mean wind force coefficient and the minimum wind force coefficient at the canopy center in case of the attack angle 0° have good correlation with the parameter ϕ. Accordingly, the relationship between the mean wind force coefficient and the parameter ϕ can be expressed as follows:

$$C_{\bar{F}} \;=\; 0.1\phi^2 - 0.85\phi \qquad (1)$$

where

$$\phi \;=\; (h/H)^3(w/l)^{0.5}$$

$C_{\bar{F}}$: the mean wind force coefficient; h : the attached height of the canopy; H : the eaves height of the building; w : the width of the canopy; and l : the projected length of the canopy.

By using Eq. (1), when the value of the parameter ϕ is the range of $0 \leq \phi < 4.25$, the mean wind force coefficient can be estimated as the range of $-1.8 < C_{\bar{F}} \leq 0$. On the other hand, in case of $\phi \geq 4.25$, the mean wind force coefficient may be estimated as $C_{\bar{F}} = -1.8$. Also, the relationship between the minimum wind force coefficient $C_{\check{F}}$ at tne canopy center in case of the attack angle 0° and the parameter ϕ can be expressed as follows:

$$C_{\check{F}} = 0.1\phi^2 - 1.2\phi - 0.4 \qquad (2)$$

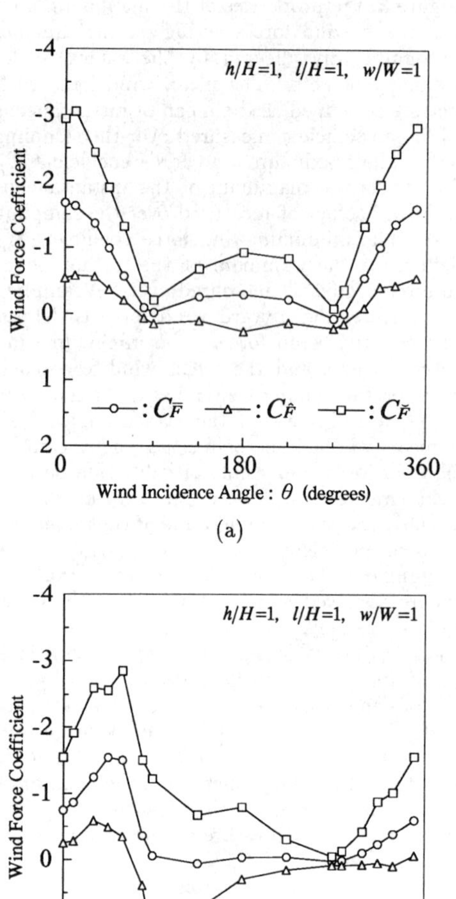

Figure 5. Variation of wind force coefficients with angle of incidence : (a) canopy center ; (b) gable sides

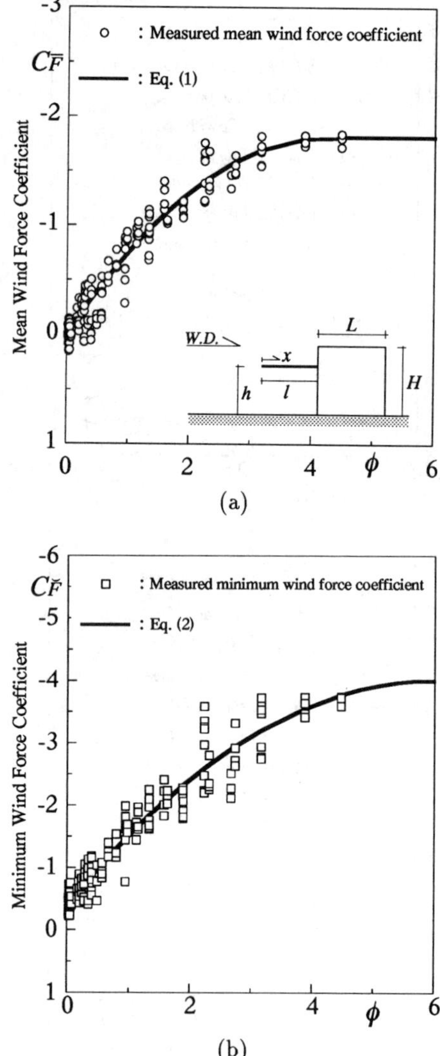

Figure 6. Mean and minimum wind force coefficient at center under attack angle 0° for ϕ : (a) mean force ; (b) minimum force (upward force)

By using Eq. (2), in case of $0 \leq \phi < 6$, $C_{\hat{F}}$ can be estimated as $-4 < C_{\hat{F}} \leq -0.4$. Also in case of $\phi \geq 6$, the minimum wind force coefficient $C_{\hat{F}}$ may be estimated as $C_{\hat{F}} = -4$.

3.4 *Aerodynamics center*

A cantilever beam supporting the attached canopy was assumed for the study. The beam was assumed to be rigid and the stresses were not affected by the dynamic characteristics of the frame structure.

The beam was assumed to exist at the center and at the gable side of the canopy, and each beam was subjected to the wind force acting on the strip with a certain width, as shown in Figure 2. Usually, in case of the cantilever-beams, the maximum moment and the maximum shear force appear at the supporting point of beam. As shown in Figure 7, the maximum moment of this case also appeared at the supporting point of the beam. Here, the maximum moment and the maximum shear force was defined as the maximum of instantaneous stress that occur to the fluctuating wind pressures acting on the beam. An aerodynamics center, which is used to evaluate the wind loads of the canopies, should be decided, based upon both the maximum moment and the maximum shear-force. In this study, it was assumed that both a maximum moment and a maximum shear force simultaneously appeared. Therefore, the aerodynamics center a_c can be defined as follows:

$$a_c = \hat{M}/\hat{Q} \tag{3}$$

where $\hat{M}$: the maximum moment; $\hat{Q}$: the maximum shear force.

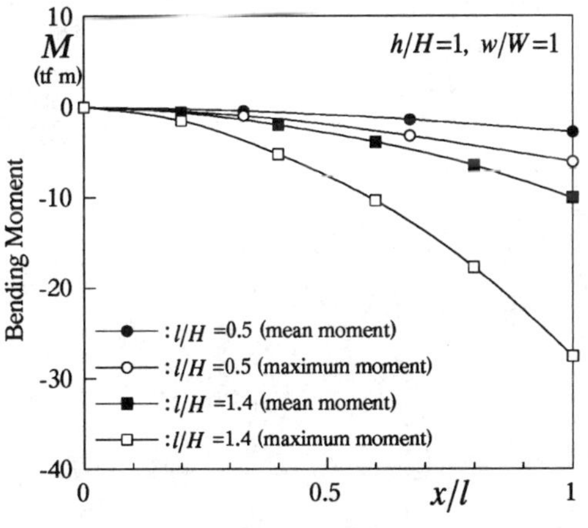

Figure 7. Distributions of bending moment on cantilever-beam of canopy center under attack angle of 0°

In Figure 8, the aerodynamics centerc that were reduced by the projected length of the canopy were plotted against the parameter h/H. The aerodynamics center was not almost affected with the parameter l/H and parameter h/H, and the tendency was more clear at the canopy center. The position of the aerodynamics center under the angle of attack 0° can be estimated from a_c/l=0.4 to 0.5.

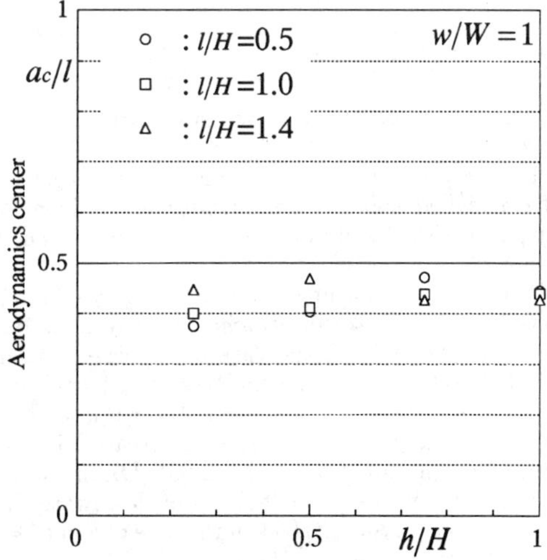

Figure 8. Variation of aerodynamics center for h/H and l/H

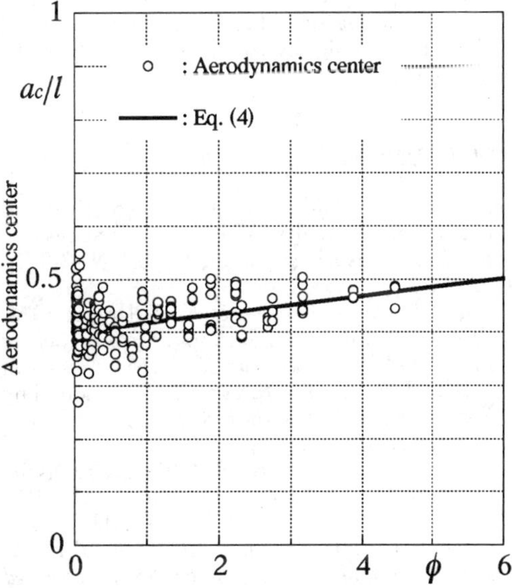

Figure 9. Relationship between a_c/l and ϕ

As shown in Figure 9, the position of the aerodynamics center may be expressed by using the parameter ϕ as follows:

$$a_c/l = 0.017\phi + 0.4, \quad \phi \leq 6 \qquad (4)$$

Where l : the projected length of the canopy.

This fact means that the aerodynamics center of the canopy under the attack angle of 0° is located at the center or slightly toward the tip side of the canopy length.

4 CONCLUSIONS

The results of the tests are summarized as follows:

1. The properties of the wind forces acting on the attached canopy are considerably influenced by properties of wind pressure at the upper side of canopy.
2. Coefficients for upward direction of wind force acting on the center of attached canopy roofs become the greatest in the case of a perpendicular wind direction. The coefficients for upward direction of wind force acting on the gable end of attached canopy roofs become the greatest in the case of a wind direction at about 60°. On the other hand, the wind force coefficients of downward direction become the greatest in the case of a parallel wind direction.
3. The mean wind force coefficient and the maximum wind force coefficient of upward direction acting on the center of attached canopy roofs can be evaluated by the empirical expression.
4. The aerodynamics center of the maximum wind force acting on attached canopy is located at the center or slightly toward the tip side of the canopy length.

REFERENCES

Bergh, H. & H. Tijdeman 1965. Theoretical and experimental results of dynamic response of pressure measuring system, Report NLR-TRF 328,1-19.

Cermak, J. E. , J. Peterka & K. J. Dreher 1974. Wind pressures on a house roof. *Proceedings of the Second U.S.A.-Janpan Research Seminar on Wind effects on Structures* : 125-140. Edited by Hatsuo Ishizaki & Arthur N. L. Chiu : University of Tokyo Press.

Holmes, J. D. & K. C. S. Kwok 1997. Comparisons between the wind load provisions of ISO/DIS 4354, ENV 1991-2-4, ASCE 7-95, AIJ Recommendations and AS1170.2. *International Workshop on Harmonization in Performance-based Building Structural Design in Countries surrounding the Pacific Ocean.* Tsukuba : Japan.

Jancauskas, E. D. & J. D. Holmes 1985. Wind Loads on Attached Canopies, *5th U.S. National Conference on Wind Engineering.* 3A-45-3A-52. Lubbock : Texas.

Stathopoulos, T. 1981. Wind Loads on Eaves of Low Buildings. *Proceedings of the American Society of Civil Engineerings*,Vol.107, No.ST10, October, 1921-1934.

Standard Australia 1989. *Minimumdegin loads on structures. Part 2: Wind loads.* Standards Australia : North Sydney.

Ueda, H., K. Hibi, Y. Tamura & K. Fujii 1994. Multi-channel Simultaneous Fluctuating Pressure Measurement System and its Applications, *Journal of Wind Engineering and Industrial Aerodynamics.* 51:93-104.

Wind Engineering into the 21st Century, Larsen, Larose & Livesey (eds)© 1999 Balkema, Rotterdam, ISBN 90 5809 059 0

Discussion and evaluation of a pilot project for computer-based design of low-rise buildings for wind loads

T.M.Whalen & J.S.Yang
School of Civil Engineering, Purdue University, West Lafayette, Ind., USA

ABSTRACT: Recent research has pursued the development of a new generation of computer-based standard provisions for wind loading, using databases of aerodynamic time histories along with modern computer technologies to improve significantly the description of wind-induced design loads. A prototype computer application for the specification of wind loads on low-rise buildings has been created as part of a pilot project to assess the advantages of this approach, as well as identify areas for further research. Discussion of the procedures and programming of this application is presented. Preliminary research on the effectiveness of this application in improving the design of main wind-force resisting systems of low-rise buildings shows that this approach can identify risk-inconsistencies in designs, non-Gaussian load distributions, and other effects that impact system performance.

1 INTRODUCTION

The impact of computers and related information management technologies on the practice of engineering has been enormous. One area, however, that has not benefited as fully from advances in computer technology is the design standards area. While the ability to analyze building behaviors has increased tremendously, designers rely upon conventional code and standard provisions that still make use of pre-computer age techniques for presenting design information. Such techniques suffer from several drawbacks, such as a limited capacity for storing information in a clear, useable format and a reliance on simplifying formulas and relations that is consistent with the lesser computational capabilities available to previous generations of designers. These drawbacks can reveal themselves in designs that are inefficient, inconsistent with respect to failure risk, or potentially unconservative.

In the specification of design wind loads for structures, code writers have used pseudo-static loads presented in summary tables and plots rather than fluctuating, dynamic wind loads that more accurately represent the true state of loading on the structure under consideration. Additionally, the requirement of calibration with respect to existing practice (needed to preserve safety margins not explicitly represented in the provisions) has resulted in descriptions of loads and building resistance that are overly simplified and, in some instances, obsolete due to changes in building materials, design strategies, and construction methods. Several examples of standard provisions for wind loads that result in risk-inconsistent designs have been identified (Simiu & Stathopoulos 1996, Whalen & Simiu 1998). It has been shown (Whalen et al. 1998) that the use of pseudo-static wind loads specified in ASCE Standard 7-93 (ASCE 1993) in the design of low-rise metal buildings resulted in significant overestimation of the bending moments induced in the supporting portal frames. Moreover, these overestimates varied considerably over a given frame, resulting in uneconomical designs that contributed little to the overall safety of the building. Inconsistencies were also found among the estimates for frames with different structural properties.

These results have motivated developmental work on a next generation of standard provisions for wind loads on structures that makes use of computer-based information storage, transmission, and computation technologies. Previous research has been done (Stathopoulos & Wu 1994) on the use of a knowledge-based expert system for the design of low-rise building roofs, while knowledge-based systems for more general wind design situations have also been investigated (Khanduri et al. 1995). An on-going research effort at the National Institute of Standards and Technology has focussed upon a pilot project that makes use of databases of aerodynamic pressure histories in conjunction with user-supplied building data to calculate realistic wind-induced internal loads in portal frames for low-rise buildings. This procedure has been used to identify risk inconsistencies with respect to internal bending moments in the frames (Whalen et al. 1998), and it has been employed to obtain calculation procedures consistent with the use of a performance criterion for structural collapse, based on the probability of exceeding a collapse wind speed (Simiu & Whalen 1998).

Recently, the authors have extended this procedure into a more advanced application called WiLDE-LRS (*Wi*nd *L*oad *D*esign *E*nvironment for *L*ow-*R*ise *S*tructures). This application, developed using the software package MATLAB®, adopts interactive graphic user interfaces to give a visual, user-friendly design environment. MATLAB scripts are used to perform the required calculations, reducing the time and potential for error associated with traditional code calculations. The application represents a prototype of a computer-based standard provision for wind loading that will ensure realistic, consistent, economical, and safe design in an easy to use system. To demonstrate the potential such a provision has for improving designs, this application was utilized to analyze the behavior of rigid portal frames in low-rise metal buildings under the action of high winds. Attention was paid to the wind-induced load effects (bending moment, shear force, and axial thrust) in the frames. Several features of the induced load effects were studied, including the spatial distribution of peak load effects (both within a frame and across the building), their temporal correlation, and the non-Gaussian nature of their distribution in time. All of these effects can influence significantly the design of the frames, yet little of this information can be obtained using conventional standard provisions. This prototype thus illustrates some of the advantages that may be gained from a computer-based design standard.

2 DISCUSSION OF WILDE

To better understand the analysis procedure employed in this study, a brief introduction to the application WiLDE-LRS is given here. This is not intended to be a comprehensive discussion of features and algorithms used in the application; such details can be obtained from other sources (Yang 1998, Whalen & Yang 1998). Instead, a general overview of the application, with an emphasis on notable features, is provided.

The application is based upon a previously developed MATLAB program frameload used to study wind-induced load effects in single frames. WiLDE-LRS (or, more simply, WiLDE) is an extension of frameload which expands the analysis capabilities, increases its practicality for low-rise building design, and provides a simpler user environment. Three important features characterize WiLDE. The first is its database of wind pressure time histories acting on low-rise buildings. These were obtained from wind tunnel tests performed at the University of Western Ontario's Boundary Layer Wind Tunnel Laboratory on scale models of typical buildings (Lin & Surry 1997). Two different scale models (1:100 and 1:200 scale) were utilized, and various wind exposures, wind speeds, and building geometries were simulated. For any given simulation, time-dependent pressure data were measured on the scale building over a series of wind directions moving around the building in increments of five degrees. Thus, a catalogue of realistic aerodynamic wind pressures is available, allowing for a detailed study of the effects of wind direction on frame behavior. It should be noted that a similar catalogue of wind pressures was used in the development of the pseudo-static pressure coefficients specified in current provisions (Davenport et al. 1987); what distinguishes the computer-based approach is the direct employment of the time-dependent pressure histories in the load specification.

Another significant feature incorporated into WiLDE is the use of graphical user interface (GUI) windows for data entry and task selection. Such windows provide a direct, user-friendly method for specifying relevant building design information, choosing execution options for the application, viewing the results, and modifying inputs. Data entry is facilitated by the use of graphical displays and on-line help menus, while simple data formats improve the accessibility of the application to novice users. The GUI windows also allow for interactive editing of design information, permitting the user to explore the consequences of various design choices quickly and easily. While such capabilities are now commonplace in most engineering software, standard provisions applications have only recently started to make use of them to improve the design calculation process (Schechter et al. 1995). Ease of use is a crucial factor in the acceptance of any design standard. Thus, the simplifications offered by GUI-based applications make them an important part of a computer-based design standard.

The final feature of note for WiLDE is its modular programming approach to the computation of relevant design quantities. In WiLDE, once the building design and wind environment information have been specified, a series of functions performs all of the requisite calculations needed to obtain the induced load effects at user-specified points in the frames. Tributary areas are computed, loads transferred to the girts and purlins and through to the frames are evaluated, and influence coefficients are determined within the application. This reduces the need for designers to employ simplifying assumptions in their evaluation procedures, resulting in a more accurate view of the behavior of their structure. The speed of the calculations is also greatly enhanced, allowing the designer to consider the influence of different frame properties quickly and efficiently. The modular programming approach, additionally, permits easy addition to or modification of these functions. This renders them amenable to a consensus approval process similar to that used to validate current standard provisions. These features, as well as those previously discussed, demonstrate the improvements that may be obtained, without loss of validity or consensus approval, when modern computer technologies are utilized in the design load specification process.

3 EVALUATION OF FRAME PERFORMANCE

As discussed above, rigid portal frames are typically designed using pseudo-static wind pressures that produce enveloping cases of the maximum wind-induced load effects to be resisted. The pressures used are intended to generate conservative values of the frame load effects, independent of the wind direction or exposure category (ASCE, 1993). It is to be expected that frame designs obtained via this procedure will contain excess load carrying capacity due to the conservative nature of the design loads. Moreover, it is not obvious that such large safety margins will actually result in decreased levels of failure risk, since this excess capacity is not necessarily deployed uniformly throughout all locations on all frames. In contrast, a computer-based design standard has the potential to overcome these drawbacks by allowing the designer to investigate the effect of dynamic wind pressures over a variety of directions and exposures, resulting in a more realistic picture of induced load effects in a given frame. In this way, unneeded frame capacity can be identified and eliminated, while still maintaining a uniform level of failure risk.

Other phenomena which can be investigated via the use of a computer-based wind load standard include the temporal distribution of load effects at locations throughout the building, as well as the degree of simultaneity between peak values of these effects. Studying these effects allows the designer to better understand the probabilistic nature of the loads acting on the structure. Such information is needed in order to define properly the ultimate limit states of the structure as well as to provide a consistent framework for serviceability and performance-based design criteria. Increasingly, serviceability and performance-based design criteria are recognized as necessary components of a risk-consistent, multi-hazard approach to the design of high-risk structures and rehabilitation of existing ones (Chock et al., 1998). Thus, availability of probabilistic design information should take on greater importance in the future, and computer-based standard provisions utilizing aerodynamic and climatological databases are ideally suited to provide such information.

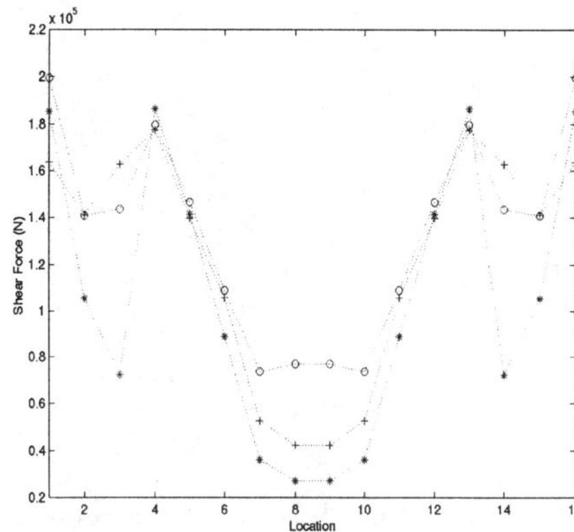

Figure 1. Maximum values of wind-induced shear forces at selected locations in frames designed for building MO32 according to ASCE 7-93 (o) and WiLDE (+ = Frame #2; * = Frame #5 .)

In the sections below, we will study the ability of the application WiLDE to provide the designer with improved information to evaluate the performance of the rigid frames. This will illustrate the potential of the computer-based design standard as well as assist in determining areas of needed research.

3.1 *Identification of Risk-Inconsistent Designs*

To investigate the ability of this computer-based procedure to identify inefficient and risk-inconsistent designs, peak load effect magnitudes as estimated by this procedure were compared to the values associated with use of the wind pressures specified by the ASCE 7-93 Standard. The application of this standard for comparison purposes is based upon its use by the metal building design firm responsible for providing the original design specifications for the frames used in this study. It is recognized that updated versions of this standard could be used to provide improved information; however, the purpose of this effort is to compare the effect of different design pressure specifications (current pseudo-static methods versus modern computer-based ones) on the resulting load effects.

In Figure 1, we show the values of the maximum wind-induced shear force at sixteen positions on frames for building MO32 using the pressures specified by ASCE 7-93 and the aerodynamic databases of WiLDE. Two frames were studied using WiLDE: Frame #2 (near one end of the building) and Frame #5 (in the middle of the building).The shear force values identified by WiLDE are obtained by finding the sample maximum value for each wind direction in the database, then picking the largest value over all of the directional maxima. As expected, the values determined using ASCE 7-93 in most cases bound the sample maxima determined from WiLDE. (The instances where the sample maxima exceed the

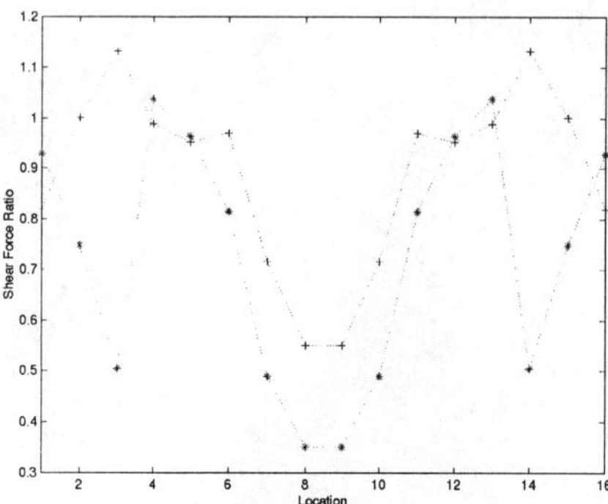

Figure 2. Ratios of sample maximum shear forces obtained using WiLDE to values obtained using ASCE 7-93. (+ = Frame #2, * = Frame #5.)

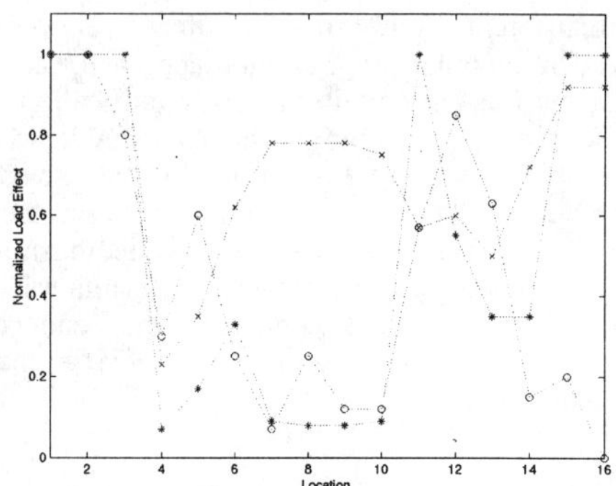

Figure 3. Simultaneity of load effects for Frame #5 of building CS32. (o = instantaneous normalized shear force at bending moment peaks, x = instantaneous normalized axial force at bending moment peaks, * = instantaneous normalized bending moments at axial force peaks.)

ASCE values may be attributed to the fact that load redistribution between frames was not taken into account in the version of WiLDE used in this study.)

One may note that in certain instances the sample maximum values are significantly below the ASCE values, indicating that the design is providing for a large reserve capacity at those locations. It is also apparent that the amount of reserve capacity varies along each frame; locations in the middle of each frame (#7 through #10) have large margins, while locations such as #4 through #6 (on the roof near the eave) have hardly any extra capacity. This is further illustrated by Figure 2, in which ratios of the peak shear forces determined using WiLDE to force values obtained via the ASCE pressure coefficients are plotted for the two frames. Risk inconsistency is revealed by the large fluctuations in the ratios across each frame; design efficiency could be obtained by redesigning the frames in these locations. (Of course, the other load effects must be examined at these locations before performing a redesign.) Similar risk inconsistencies for wind-induced bending moments and axial forces were found using WiLDE for these frames as well as other frames in other buildings. Thus, the computer-based approach gives the designer a rational means to identify and eliminate risk inconsistencies.

3.2 *Simultaneity of Peak Load Effects*

In addition to the magnitudes of the peak load effects experienced, the required resistance of a given frame in a given location of a given building is influenced by the degree to which these peak values occur simultaneously during the wind loading. Due to the additive nature of the load-induced stresses in the frame, highly correlated histories for the load effects would entail high levels of resistance. In contrast, imperfectly correlated load effects would imply a lower likelihood of combined peak stresses, with a corresponding reduced need for resistance.

Information on simultaneity of peak load effects is probabilistic in nature and thus cannot be obtained from traditional wind loading provisions. However, with the aerodynamic pressure histories available to a computer-based standard provision, a designer should be able to assess this effect easily, giving a solid basis for deciding if a reduction in frame capacity is in order. To investigate this, WiLDE was used to provide a simple criterion to evaluate peak load simultaneity for various frames, locations, and wind orientations.

Since the main concern of this investigation was the degree of correlation of extreme load effects, it was decided that traditional measures of cross-correlation of random processes would not be appropriate. This is due to their consideration of the entire history of each process, thus allowing for the possibility that highly correlated peak values could be masked by long regions of less extreme, uncorrelated values. To avoid this possible error, a less sophisticated measure of peak value correlation was adopted. For specified frames and wind orientations, load effect time histories were computed and normalized by the appropriate maximum magnitude at several locations on each frame. When a given normalized load effect reached a value of one, the magnitudes of the other effects were identified and plotted. From this, simultaneity of peak load effects could be roughly identified by noting whether the other normalized load effects were at or close to one in value when a given normalized effect reached a value of one.

In Figure 3, we show plots of these normalized load effects at time instances of peaks for locations in Frame #5 of building CS32 for a wind orientation of 90° (perpendicular to the roof ridgeline). Three cases are shown: instantaneous normalized shear forces when the bending moment is at a peak, instantaneous normalized axial forces at bending

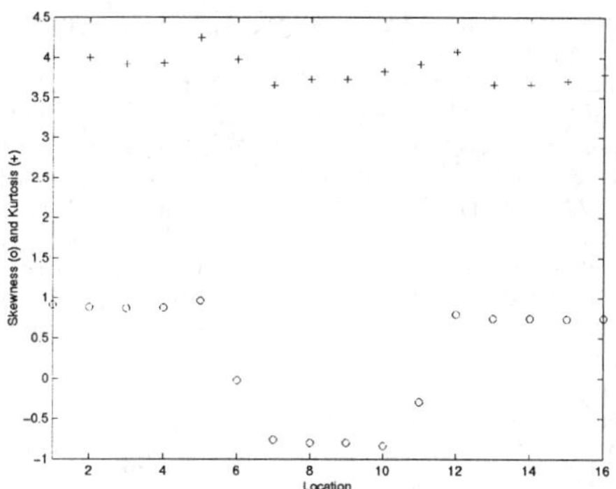

Figure 4. Skewness (o) and kurtosis (+) for bending moment distribution in building MS20 at various locations on frame #2 for 0° wind direction.

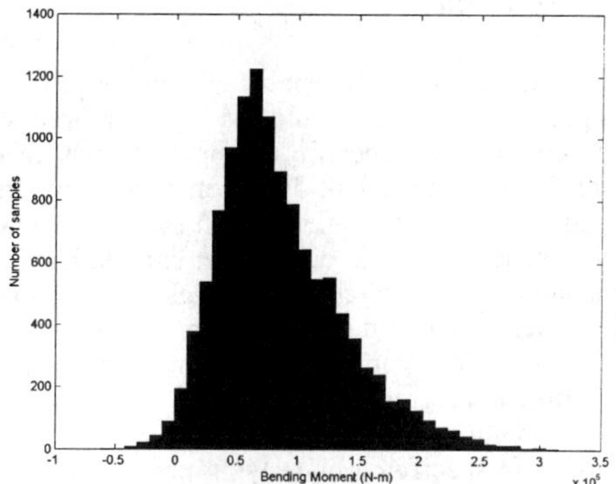

Figure 5. Distribution of bending moment in building MS20 at location #3 on frame #2 for 0° wind direction.

moment peaks, and instantaneous bending moments at peaks of the axial force. The following features may be discerned from this figure: (i) the normalized load effects are all at or near one for locations 1 through 3 (on the windward frame column), indicating highly correlated peaks, (ii) there is a noticeable drop in almost all normalized load effects on the roof of the building (locations 4 through 13) indicative of non-simultaneous peaks, though some locations show significant correlations, (iii) significant simultaneity appears to exist on the leeward column (locations 14 through 16) for the bending moments and axial forces. While not necessarily definitive, such information does provide some indications of nonsimultaneous peak behavior which could be used, for example, to identify locations for which reductions in frame resistance could be justified. Further research is being pursued on the use of this application in the identification of correlated and uncorrelated peak behavior. Nevertheless, this illustrates the sort of probabilistic information that is obtainable from the use of a computer-based provision for wind loads.

3.3 *Non-Gaussian Features of Load Effect Time Histories*

It is clear that the probabilistic nature of a load effect's temporal distribution will influence noticeably its estimated extreme value. In engineering analyses, it is often convenient to assume that a probabilistic quantity has a Gaussian distribution; however, evidence exists that such an assumption is not warranted for wind-induced load effects. Several researchers have documented the presence of non-Gaussian distributions for wind pressures on low-rise buildings, especially at corners, eaves, and other locations where sharp changes in surface geometry occur (Holmes & Best 1981, Kasperski 1992). From this, one can reasonably infer that the wind-induced load effects could show non-Gaussian distribution features that should be taken into account when estimates of peak effects are made. To verify this, WiLDE was used to determine load effect time histories at various locations in the design building MS20, and the statistical properties of these histories were studied.

In Figure 4, we plot statistical information obtained from analyzing a load effect/loading combinations previously identified as critical from a design standpoint. Figure 4 shows the skewness and kurtosis of the induced bending moment at sixteen locations on Frame #2 of building MS20 under an oncoming wind oriented at 0° (parallel to the roof ridgeline). The skewness of a distribution is related to the amount of asymmetry present, while the kurtosis is a measure of the width of a distribution. For a Gaussian distribution, these quantities should equal 0.0 and 3.0 respectively; however, the values calculated for this case reveal significant differences from the Gaussian values. As an additional check on the non-Gaussian behavior, histograms at specific cross sections in each of these frames were calculated. As seen in Figure 5, there is a clear asymmetry in the histogram, with a noticeably longer tail associated with the large magnitudes of the load effect. Similar features were observed for other combinations of building design, frame location, and wind direction (as well as for all three load effects), leading to the conclusion that non-Gaussian load effect distributions are present in low-rise buildings and must be accounted for in the design process. It should be noted that these non-Gaussian features are quite sensitive to the frame location and direction of the oncoming wind. In general, these features are observed for "up-wind" locations on the building, while "down-wind" locations display a more Gaussian behavior. See (Yang 1998) for further details.

4 CONCLUSIONS

As discussed above, current techniques for providing design information for wind loads suffer from serious deficiencies. Specific examples of risk inconsistencies and inefficient design methods have been linked to these deficiencies, which can be associated with potentially unsafe designs on the one hand and uneconomical, inefficient buildings on the other. These deficiencies are inherent in the current medium by which standard provision information is transmitted, resulting mainly from inadequate information content and the use of substantial simplifications in specifying underlying physical phenomena. The use of modern computer and information technologies in standard provisions for wind loads can overcome many of these limitations, providing the designer with easy to use wind load information that is more realistic and better suited for the design situation at hand. The application WiLDE, as part of a pilot project on the development and use of a computer-based standard, has demonstrated some clear advantages over traditional provisions in terms of providing significantly improved descriptions of frame behavior as well as probabilistic information that is currently unavailable to designers. The benefits to the national economy, in terms of both efficient use of resources for buildings and lowered safety risks, are likely to be significant.

REFERENCES

American Society of Civil Engineers 1993. *ASCE Standard 7-93: Minimum Design Loads for Buildings and Other Structures.* New York: ASCE

Chock, G., Boggs, D., & Peterka, J. 1998. A wind and hurricane design framework for multi-hazard performance-based engineering of high-rise structures. In N.K. Srivasta (ed.), *Structural Engineering World Wide 1998*: Paper T139-3. Oxford: Elsevier Science.

Holmes, J.D. & Best, R.J. 1981. An approach to the determination of wind load effects on low-rise buildings. *J. Wind Eng. Ind. Aerodyn.* 7: 273-287.

Kasperski, M. 1992. Extreme wind load distributions for linear and nonlinear design. *Engineering Structures.* 14(1): 27-34.

Khanduri, A.C., Bedard, C., & Stathopoulos, T. 1995. Development of a hybrid KBS for design applications in wind engineering. In *Restructuring: America and Beyond. Proceedings of the 13th ASCE Structures Congress.* Part 2 (of 2): 1427-1430. New York: ASCE

Lin, J. & Surry, D. 1997. *Simultaneous Time Series of Pressures on the Envelope of Two Large Low-Rise Buildings.* BLWT-SS7-1997, The Boundary Layer Wind Tunnel Laboratory, The University of Western Ontario, London , ON, March 1997.

Schechter, M.M., Schechter, E., & Simiu, E. 1995. *Developmental Computer-Based Version of ASCE 7-95 Standard Provisions for Wind Loads.* NIST Technical Note 1415, National Institute of Standards and Technology, Gaithersburg, MD, November 1995.

Simiu, E. & Whalen, T.M. 1998. Aerodynamic time-history and climatological databases for prescriptive and performance-based wind loading provisions. In N.K. Srivasta (ed.), *Structural Engineering World Wide 1998*: Paper T176-2. Oxford: Elsevier Science.

Stathopoulos, T. & Simiu, E. 1996. Codification of wind loads on buildings using bluff body aerodynamics and climatological data bases. *J. Wind Eng. Ind. Aerodyn.* 69-71: 497-506.

Stathopoulos, T. & Wu, H. 1994. Knowledge-based wind loading for envelope design: beyond building codes. *J. Wind Eng. Ind. Aerodyn.* 53(1-2): 177-188.

Whalen, T.M. 1996. *Probabilistic Estimates of Design Load Factors for Wind-Sensitive Structures Using the "Peaks Over Threshold" Approach.* NIST Technical Note 1418, National Institute of Standards and Technology, Gaithersburg, MD, April 1996.

Whalen, T.M. & Simiu, E. 1998. Assessment of wind load factors for hurricane-prone regions. *Structural Safety.* 20: 271-281.

Whalen, T.M., Simiu, E., Harris, G., Lin, J., & Surry, D. 1998. The use of aerodynamic databases for the effective estimation of wind effects in main wind-force resisting systems: application to low buildings. *J. Wind Eng. Ind. Aerodyn.* 77&78: 685-693.

Whalen, T.M. & Yang, J.S. 1998. *A Pilot Project for Computer-Based Design of Low-Rise Buildings for Wind Loads -- The WiLDE-LRS User's Manual.* Structural Engineering Report CE-STR-98-12, Purdue University, West Lafayette, IN.

Yang, J.S. 1998. *A Pilot Project for Computer-Based Design of Low-Rise Building Frames for Wind Loads.* MSCE thesis, Purdue University, West Lafayette, IN. 1998.

Wind Engineering into the 21st Century, Larsen, Larose & Livesey (eds) © *1999 Balkema, Rotterdam, ISBN 90 5809 059 0*

Uncertainty in wind pressure coefficients for low-rise buildings

M.S.de Wit
Delft University of Technology, Subfaculty of Civil Engineering, Netherlands

ABSTRACT: This paper reports on an analysis of the uncertainty in wind pressure (difference) coefficients, which are assessed on the basis of generic knowledge and experimental data rather than with a specific (wind tunnel) experiment. The analysis is carried out on time averaged pressure coefficients in a specific case, concerning a low-rise building in an urban environment. To assess the uncertainty, structured elicitation of expert judgment is employed. Method and results are described here. The acquired uncertainty is compared to an estimate of the uncertainty that could be attributed to wind pressure difference coefficients obtained in a wind tunnel experiment. This comparison serves as input to a discussion whether expert judgment studies could develop as an alternative for wind tunnel tests.

1 INTRODUCTION

For the simulation of natural ventilation flows in buildings, the wind pressure distribution over the building envelope is required. In the design of low-rise buildings, wind tunnel experiments are scarcely employed to measure the wind pressures. Instead, techniques are used, which predominantly rely on inter- or extrapolation of wind pressure coefficients, measured in prior wind tunnel and full-scale experiments. Due to the complexity of the underlying physics, this is a process, which may introduce considerable uncertainty. This uncertainty, however, cannot be quantified with any of the existing techniques.

In this paper, expert judgment is used to assess this uncertainty in a specific case. The case is designed to tentatively explore the effect of the near field on the uncertainty, as well as the effect of the tap position on the building enclosure. The uncertainty is compared with an estimate of the uncertainty in pressure coefficients, obtained in a wind tunnel test. This comparison is the starting point for an evaluation under which provisions an expert judgment study could be an alternative for a wind tunnel test.

2 CASE DEFINITION

The case under study concerns a scale 1:250 wind tunnel model of a four-story office building at the outskirts of Delft, a Dutch town. Figure 1 shows the full scale building dimensions, whereas Figure 2 presents the turntable layout of the building and its immediate surroundings. The building group is immersed in a simulated ‘urban’ boundary layer. The mean velocity profile as it is measured at the center of a void turntable, is shown in Figure 3.

Due to the internal layout of the building, ventilation flows are driven by pressure differences between the long facades (cross-ventilation). Figure 1 shows four marked locations corresponding to the positions of vents in the building façade. This study focuses on two time-averaged pressure differences, i.e. between positions 1a and 1b and between locations 2a and 2b. These pressure differences are related to the mean horizontal wind speed $U(z)$ at reference height z by the pressure difference coefficients ΔC_{p1} and ΔC_{p2}, where ΔC_{pi} is defined as:

$$\Delta C_{p_i} = \frac{p_{ia} - p_{ib}}{\frac{1}{2}\rho U^2(z)} \tag{1}$$

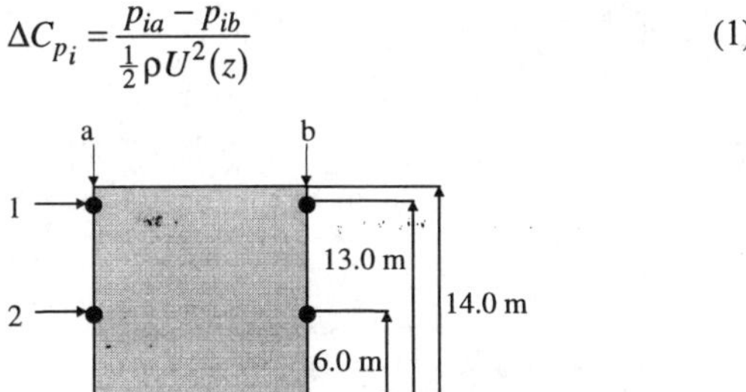

Figure 1 Cross section of the office building, the central building in Figure 2. The full-scale building dimensions are 56 m × 14 m × 14 m. The four pressure taps are located at the centerlines of the long facades. Side a faces the built-up area (north), whereas side b is directed towards the void area (south).

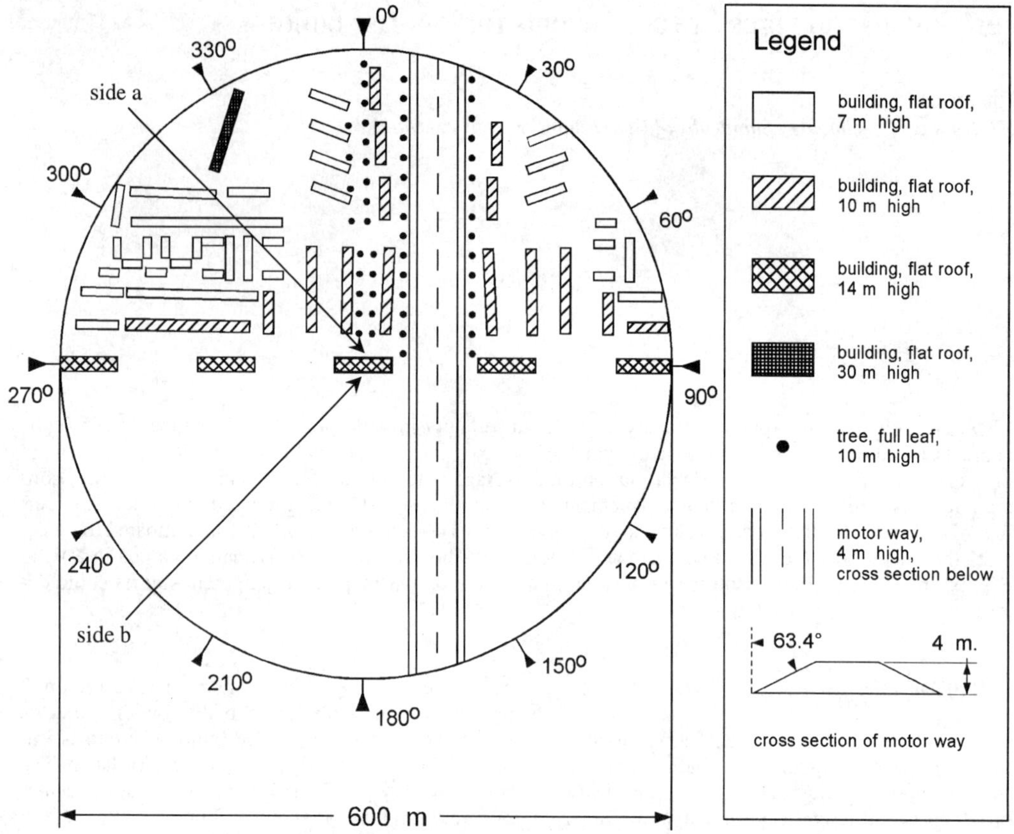

Figure 2 Schematic layout of the turntable model. The dimensions refer to full-scale. The scale of the model is 1:250. The building in the center is the office building under study. Side a and side b are indicated.

with ρ the density of air. In this paper, all coefficients are related to the mean wind speed at building roof height in the unperturbed flow, i.e. upstream of the turntable.

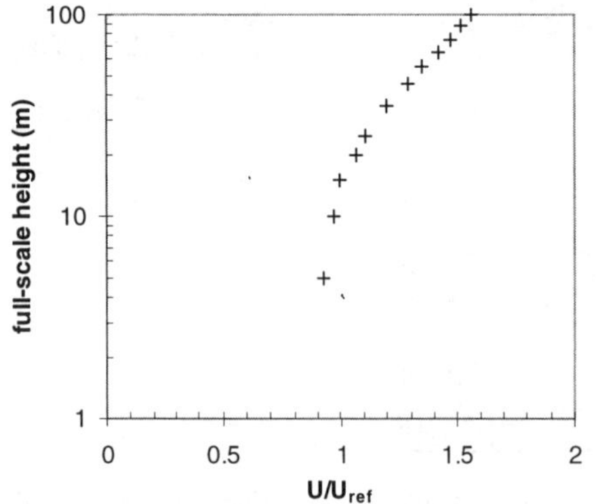

Figure 3 Mean velocity profile in the wind tunnel at the center of a void turntable with 'urban' roughness upstream of the turntable.

3 EXISTING MODELS

A number of models are reported in literature to assist the assessment of time-averaged wind pressure coefficients on the basis of existing experimental data. These models are based on parametric analyses of selected data. Examples of this approach can be found in Allen (1984), Grosso (1992), Grosso et al. (1995), Knoll et al. (1995) and Swami & Chandra (1988). A summary of the key features of these tools can be found in De Wit (1999a).

To obtain a first impression of the uncertainties in pressure difference coefficients derived from existing data, these models were used to estimate the requested pressure difference coefficients for 7 equidistant wind angles between 0° and 180° as indicated in Figure 2. The models by Allen, Knoll et al. and Grosso allow for the assessment of the local coefficients ΔC_{p1} and ΔC_{p2}. Their output is shown in Figure 4 and Figure 5. Liddament, and Swami & Chandra developed models, which calculate façade averaged pressure coefficients. Figure 6 shows the results of Swami & Chandra's model together with

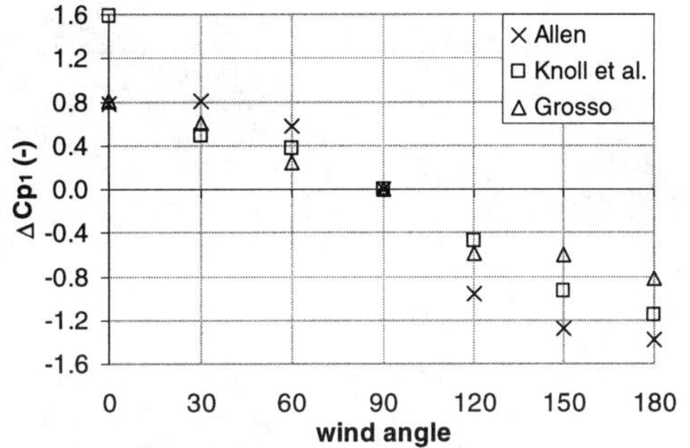

Figure 4 Wind pressure difference coefficients ΔC_{p1}, assessed on the basis of existing data according to three different models.

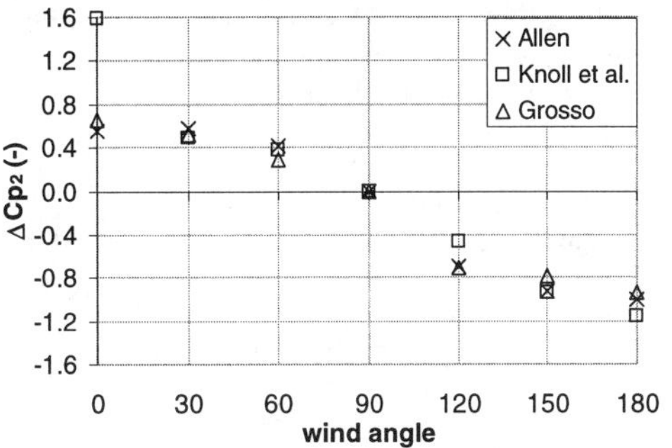

Figure 5 Wind pressure difference coefficients ΔC_{p2}, assessed on the basis of existing data according to three different models.

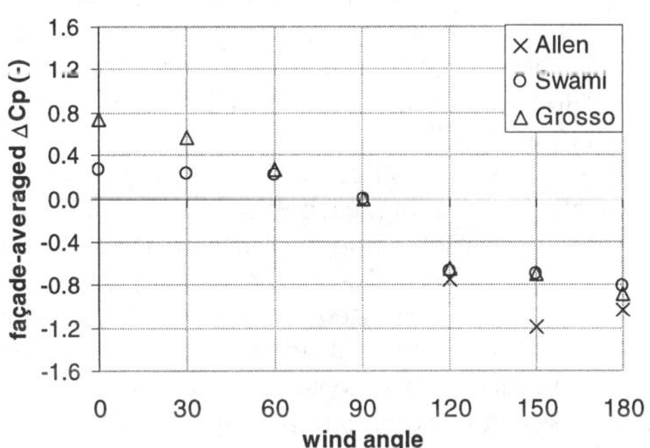

Figure 6 Façade averaged wind pressure difference coefficients, assessed on the basis of existing data according to three different models.

those of Allen and Grosso, which also allow for the assessment of façade averaged coefficients.

Whereas the spread in the outputs from the different models for ΔC_{p1} is considerable, they agree rather well on ΔC_{p2} except for 0°. The degree of consensus on the façade-averaged coefficients strongly depends on the wind angle.

If we would consider adopting the scatter in the model outcomes as a measure of the uncertainty, it is important to contemplate which factors contribute to that scatter and, more importantly, which do not.

First, the model outcomes depend on the choices of the analyst:

- Several models require a characterization of the velocity profile in terms of α or z_o. Which values most adequately represent the profile in Figure 3?
- The case at hand is out of the range of application of some of the models. Are the outcomes still appropriate?
- If the surroundings of the central building have to be classified in a 'shielding' class, which class description best fits the current case?

Moreover, not all uncertainty is captured in the variation of the model output:

- The scatter in the experimental data on which the models are based is eliminated by regression or averaging. Part of this scatter may be measurement error, but part of it results from effects unexplained by the model. Models sharing the same parameters most likely ignore the same effects.
- There is considerable overlap in the data sets underpinning the different models. This overlap introduces a dependency between the model predictions.
- The majority of the data underlying the models that assess the effect of the near field were obtained in (wind tunnel) experiments with regularly arranged near field layouts. The near field in this case is irregular and consists of buildings of different heights.

Although the results of this exercise support the assumption that significant uncertainty exists in wind pressure coefficients, predicted on the basis of existing data, they do not provide a proper basis to assess this uncertainty. Wind-engineering expertise is required, both to provide reliable inputs to the models and to assess the impact of features in the case under study, which are not covered in the existing models.

Hence, the uncertainty in these pressure coefficient assessments can best be quantified by experts in the field, who, acquainted with the complexity of the underlying physics as they are, are best suited to interpolate and extrapolate the data they have available on the subject and assess the uncertainties involved. The next section reports on an experiment in which expert judgment was used to quantify the uncertainties in the wind pressure difference coefficients in the case at hand.

4 EXPERT JUDGMENT

4.1 *Introduction*

In the expert judgment study, the case described in section 2 was presented to six experts, renowned for their expertise in the field of wind pressure measurements on low-rise buildings. For each of the twelve wind angles in Figure 2 they were individually asked to assess the values of ΔC_{p1} and ΔC_{p2}, which would be found if this model would be examined in a wind tunnel with the velocity profile shown in Figure 3. Each expert's assessment of a coefficient did not consist in a 'best estimate', but in a median value plus a central 90% confidence interval expressing his uncertainty. Probability distributions were constructed for each expert from his assessments on the basis of the principle of minimal information (Cooke, 1991).

To implement empirical control, two independent wind tunnel tests were carried out to actually measure the values of the requested wind pressure differences for the twelve wind angles. From a comparison of the experts' assessments with the measured data, a performance score could be calculated for each expert.

Finally, a single (marginal) probability distribution was obtained for each pressure difference coefficient at each wind angle by calculation of a weighted average of the experts' distributions. The weights in this process were based on the experts' performances.

In expert judgment terminology, this resulting distribution for each variable is referred to as the distribution for the decision-maker, or simply the DM.

The most important aspects of the procedure, which was followed to obtain and process the experts' judgments, is outlined in the next section. It is based on Cooke (1991) and Cooke & Goossens (1995). Similar approaches in civil engineering applications can be found in Van Elst (1997) and Ter Haar et al. (1998).

4.2 *Selection of the experts*

A pool of candidates for the expert panel was established by screening recent literature. From the many candidates resulting from the screening, a panel of 7 experts was established on the basis of the following criteria:

- access to relevant knowledge
- recognition in the field
- impartiality with respect to the outcome of the experiment
- familiarity with the concepts of uncertainty
- diversity of background among multiple experts
- willingness to participate

One of the selected experts withdrew in the elicitation stage, as he did not appreciate the concept of subjective probability. Table 1 lists the experts, who participated in the study.

The team of W. de Gids, B. Knoll and H. Phaff worked together and produced one single set of assessments. They will be referred to as a single expert.

Table 1 List of substantive experts in the experiment

W. de Gids, B. Knoll, H. Phaff	Netherlands
J.-A. Hertig	Switzerland
B. Lee	UK
D. Surry	Canada
A. Vrouwenvelder	Netherlands
E. Willemsen	Netherlands

4.3 *Expert training*

None of the experts but one had ever participated as substantive expert in an experiment involving structured elicitation of expert judgment, so they were unacquainted with the motions and underlying concepts of such an experiment. Moreover, acting as a substantive expert entails the assessment subjective quantile values and subjective probabilities, a task the experts were not familiar with.

Hence the experts received a concise training with the aim to:

- provide an overview of the process
- develop confidence
- introduce the experts to the task they must perform
- instill awareness and control of biases
- practice making probabilistic judgments

4.4 *Elicitation*

In the elicitation stage, the core of the experiment, the experts made their judgments available to the analyst. In this experiment, individual meetings with each expert were arranged. The experts were specifically asked not to discuss the experiment with each other. In this way, the diversity of viewpoints would be minimally suppressed.

The elicitation took place in three parts. Prior to the elicitation meeting, each expert prepared his assessments e.g. by looking up relevant literature and making calculations. During the meeting, these assessments were discussed with the analyst, who avoided giving any substantive comments, but merely pursued clarity, consistency and probabilistic soundness in the expert's reasoning. On the basis of the discussion, the expert revised and completed his assessments if necessary.

To ascertain traceability of the results, the elicitation was completed with the writing of the rationale, a report documenting the reasoning underlying the assessments of the substantive expert. The experts' rationales can be found in De Wit (1999b).

4.5 *Wind tunnel tests*

To implement empirical control, two wind tunnel tests were carried out once the experts' assessments had been compiled. The aim of these experiments was to obtain experimental data on which the performances of the experts could be scored.

The first experiment took place in the low speed German-Dutch wind tunnel (DNW-LST), located at the facility of the Dutch Aerospace Laboratory (NLR) in The Netherlands. This is a closed-loop wind tunnel with a closed test section of 8.75m long and a cross-section of 3.00 m × 2.25 m at the position of the turntable. The other test was performed in a boundary layer wind tunnel of the University of Western Ontario in Canada (UWO-BLWT1). This is an open-return tunnel with a test section of 28.7 m length and a cross section of 2.4 m × 2.1 m.

More information on the experiments can be found in De Wit (1999a) and in Willemsen (1998) for the DNW-experiment and in Soerensen (1998) for the experiment in the UWO-tunnel.

4.6 *Analysis of the data*

The analysis of the data focuses on the assessment of the experts' performances and the combination of their judgments to calculate the distributions for the decision-maker, the DM. To accomplish this, the classical model is used as developed by Cooke (1991). This section briefly addresses this model.

Expert performance

An expert's performance w_e is calculated from a comparison of his assessments with measured realizations. It is the product of two measures, calibration C_e and information score I_e:

$$w_e = C_e \, I_e \tag{2}$$

Loosely stated, calibration measures the degree to which the realizations support the expert's assessments. In scoring calibration, an expert is regarded as the statistical hypothesis: 'the realizations are samples, drawn independently from distributions corresponding to the expert's quantile assessments'. The calibration score can be interpreted as the minimum significance level at which this hypothesis would not be rejected. Ergo, a calibration score has a value between 0 and 1 and higher scores are better.

The information score indicates how 'tight' the expert's distributions are. It is calculated per variable as the relative information of the expert's distribution with respect to the uniform distribution on a suitably chosen intrinsic range for that variable. It expresses what we learn from the expert's assessments if we initially believe that the variable is uniformly distributed on the intrinsic range. Information scores are always positive and higher scores are preferred.

The fact that an expert's performance is the product of two scores, calibration and information, suggests that he could compensate a low calibration by being highly informative. This is only possible to a limited extent. An experts performance is dominated by his calibration, as this score may range over five orders of magnitude, whereas the information score rarely varies more than a factor five between experts. Hence, the information score serves to modulate between more or less equally calibrated experts.

Combination of the experts' assessments

For each variable, the experts' assessments must be combined to obtain the DM for that variable. Combination of the experts' assessments according to the classical model uses linear pooling, which means that the DM is a weighted average of the experts' distributions.

Basically, an expert's weight is his normalized performance. The normalization serves to ascertain that the sum of weights for each variable equals 1. However, if an expert's calibration score is below a suitably chosen value of the significance level, the expert, considered as a hypothesis, is 'rejected' and his weight is set to zero. The significance level receives the value that optimizes the performance of the DM, which is calculated in the same way as the performance of individual experts.

It can be shown that this method of scoring experts is 'asymptotically strictly proper', which means that experts on the long run receive the highest weights if they state assessments according to their true beliefs (Cooke, 1991).

5 RESULTS

Figure 7 through Figure 8 show the results of both wind tunnel experiments and the experts' assessments.

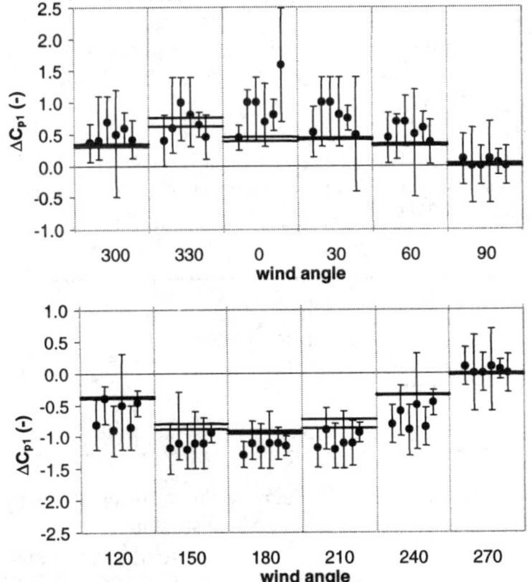

Figure 7 Assessments of the 6 experts for ΔC_{p1}. The dots are their median values, the error bars represent their central 90% confidence intervals. The drawn horizontal lines show the results from the two wind tunnel tests. For each wind angle, the results of experts 1 through 6 are shown from left to right.

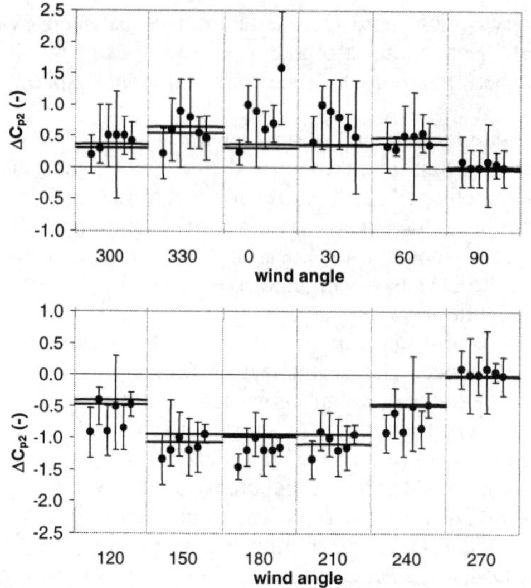

Figure 8 Assessments of the 6 experts for ΔC_{p2}. The dots are their median values, the error bars represent their central 90% confidence intervals. The drawn horizontal lines show the results from the two wind tunnel tests. For each wind angle, the results of experts 1 through 6 are shown from left to right.

Figures 7 and 8 refer to the experts by number. These numbers were randomly attributed to the experts in Table 1 and will be used throughout this paper.

6 ANALYSIS

The experts' performance scores, calculated with the classical model on the basis of 2 realizations for each variable (i.e the wind tunnel results), are listed in Table 2. The last row of the table shows the scores of the DM.

Table 2 Experts' performances, calculated with the classical model and scaled to an effective number of 10 seed variables. The last row shows the score of the optimized DM.

Expert	Calibration C_e	Information I_e	Performance w_e
1	3.0 10^{-2}	0.65	2.0 10^{-2}
2	3.6 10^{-1}	0.55	2.0 10^{-1}
3	1.0 10^{-2}	0.46	4.6 10^{-3}
4	1.0 10^{-3}	0.17	1.7 10^{-4}
5	1.0 10^{-4}	0.81	8.1 10^{-5}
6	3.0 10^{-1}	0.73	2.2 10^{-1}
DM	3.6 10^{-1}	0.55	2.0 10^{-1}

As in Figures 7 and 8, each expert is referred to by his number, which was randomly attributed.

The experts' scores were not directly calculated from their assessments as shown in Figure 7 and Figure 8. From their rationales it became clear that, for each wind angle, the experts first assessed the pressure differences between windward and leeward side. Subsequently they adjusted the sign to obtain the requested pressure differences between side a and side b. Hence, it was considered more appropriate to score the experts on their initial assessments of the pressure difference coefficients between windward and leeward side. In this way, possible systematic tendencies of any of the experts to over- or underestimate would not be masked.

Calibration scores can only be interpreted in relation to the number of variables from which they were calculated. The expert scores in Table 2 are based on 24 variables, i.e. the ΔC_{p1}'s and ΔC_{p2}'s for 12 wind angles. However, the scores have been scaled to an effective number of 10 seed variables, a common number in expert judgment studies, by adjusting the power of the test.

Experts 2 and 6 are almost equally well calibrated. The calibration scores of the other experts are lower by an order of magnitude or more, which is significant.

As explained in section 4.6, in calculating the decision-maker, the significance level is assigned a value, which maximizes the performance of the DM. In this case, maximum performance of the decision-maker was found at a significance level of 3.6 10^{-1}. In other words, all experts except expert 2 received a zero weight, as their calibration scores are lower than the calibration level. Hence, the assessments of the DM with optimized performance are equal to those of expert 2. Figure 9 and Figure 10 show these assessments separately.

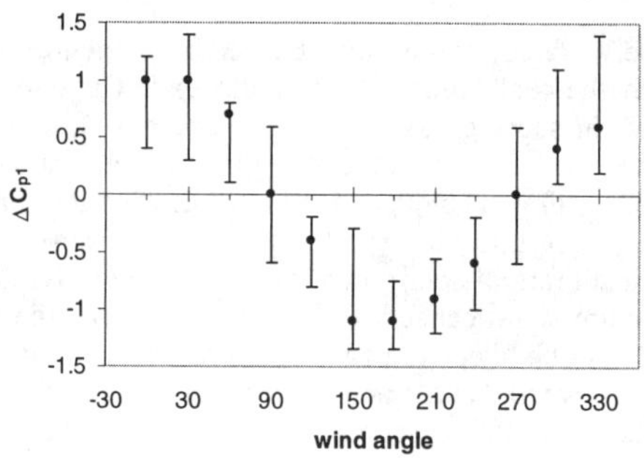

Figure 9 DM's assessments for the ΔC_{p1}'s.

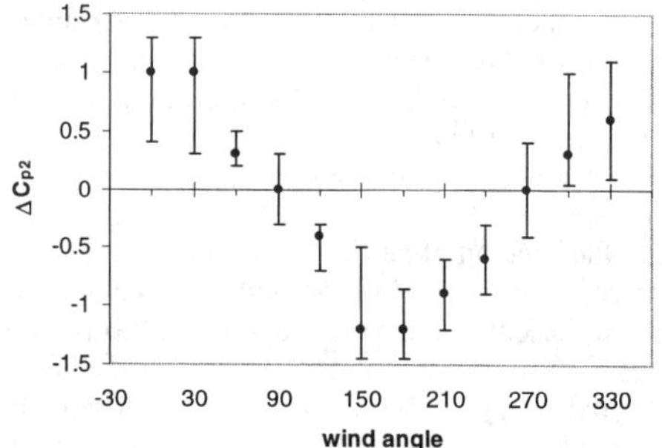

Figure 10 DM's assessments for the ΔC_{p2}'s.

A more detailed analysis of the experts' assessments can be found in De Wit (1999a).

7 DISCUSSION

Figure 9 and Figure 10 show the median values plus a central 90% confidence intervals for the pressure difference coefficients considered in this case study. These values result from an optimal combination of the experts' assessments as explained in section 4.6 and can be interpreted as the uncertainties that should be considered in ΔC_{p1} and ΔC_{p2} when their values are assessed from existing data.

It is interesting to compare these results with the uncertainties in pressure difference coefficients, obtained from a wind tunnel experiment. This reveals the extra information we gain by doing a specific wind tunnel experiment instead of an assessment on the basis of existing data.

As each sound wind tunnel test, compliant with the case description in section 2 yields valid realizations of the wind pressure difference coefficients, the spread in the outcomes of such tests would be a suitable estimator for the uncertainty. In a recently completed study (Hoelscher, 1997), twelve wind tunnel laboratories measured the surface pressures at a floor-mounted cube, corresponding to 50m height in full-scale. They were asked to perform the measurements in a simulated boundary layer flow at neutral stratification corresponding to urban terrain with a profile exponent of $\alpha = 0.22$ +/- 0.02. Apart from a few basic constraints, the participants were free to perform the tasks according to their own judgement and standards.

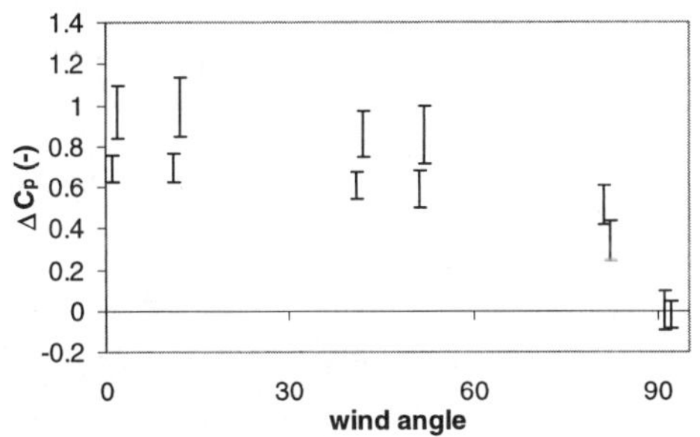

Figure 11 Spread in wind pressure difference coefficients, measured in twelve different wind tunnels on an isolated cube (Hoelscher, 1997). The error bars represent the 90% confidence intervals. The upper error bars are obtained at a level of 0.5 of the cube height, the lower bars were found at 0.93 of the cube height.

Figure 11 shows the 90% confidence intervals for pressure difference coefficients between front and back of the cube, measured at the façade centerline at 0.93H and 0.5H respectively, where H is the obstacle height. These measuring locations are comparable to those of ΔC_{p1} (0.93H) and ΔC_{p2} (0.43H) respectively. Figure 11 shows that the width of the confidence intervals is on average 0.2 and does not significantly depend on the wind angle.

To explore if these results obtained for an isolated cube of 50m full-scale height make sense in the current case, a virtual expert was created. His median values were chosen equal to the outcomes of the UWO wind tunnel test. The 95% and 5% quantile values were set to the median value +/- 0.1 to obtain 90% confidence intervals of width 0.2. Subsequently, the performance of this virtual expert was scored on the results of the test in the DNW tunnel along with the other experts. Table 3 lists the performance score of this virtual expert.

Table 3. Performance of a virtual expert, created on the basis of the data from the UWO-test and scored on the results of the experiment in the DNW-tunnel.

	C_e	I_e	w_e
Virtual expert	$3.8\ 10^{-1}$	1.65	$5.2\ 10^{-1}$

Comparison of the numbers in this table with Table 2 shows that the virtual expert, created on the basis of wind tunnel data, has a calibration score, which is similar to that of the DM, calculated from the expert judgments. However, the information score of the virtual expert is 3 times higher than the information score of the DM. Or, alternatively, the confidence intervals of the virtual expert are 3-6 times narrower than the confidence intervals of the DM. This can be interpreted as the extra information, which is gained by doing a wind tunnel experiment in this case.

Wind pressure coefficients, assessed on the basis of existing data in this study have an uncertainty, which is large compared to both their median values, and to the (estimated) uncertainty in wind tunnel results. This suggests that the output of models and tools, based on a parametric analysis of prior wind tunnel and full-scale data, have little meaning unless the uncertainty in this output is quantified. This study shows that an expert judgment study is an adequate method to perform this quantification. In the form it was implemented here though, it is also a very expensive approach, far more expensive than a wind tunnel study.

The most obvious measure to cut back the costs, i.e. by reducing the number of experts in the panel to e.g. 1 or 2, is not an attractive one at this stage. Indeed, the experts in this study show little agreement in their assessments. This is expressed in their rationales, which underpin their assessments with different and sometimes even conflicting arguments. Moreover, the experts' calibration scores show a large scatter and only two out of six experts receive a fair calibration score.

Thus, at this stage it should be concluded that an expert judgment study is not an attractive alternative for a wind tunnel experiment. However, this tentative conclusion is moderated by the following consideration. If expert judgment studies would be as common as wind tunnel studies are today, experts would be better trained to make subjective probability assessments, which might allow for a reduction of the number of experts in the study. Moreover, experts would probably be better equipped to estimate wind pressure coefficients on the basis of available data, which might lead to a reduction of their uncertainty without a loss of calibration.

Hence, it seems worthwhile to dedicate further research efforts to the exploration of the perspectives of expert judgment in this field.

8 CONCLUSIONS

Expert judgment was successfully employed to quantify the uncertainty in wind pressure difference coefficients for a low-rise building in an urban environment. The uncertainty was measured in the situation that the coefficients are assessed on the basis of existing experimental data and knowledge rather than with a specific (wind tunnel) experiment. The observed uncertainty is large, both compared to the median values of the coefficients, and compared to the (estimated) uncertainty in wind tunnel results. This suggests that point estimates on the basis of existing data do not give useful information. In stead, probability distributions for the coefficients as obtained from an expert judgment study, might be valuable for certain applications, e.g. concerning serviceability limit state evaluations.

However, in the form it was implemented in this study, elicitation and processing of expert judgments is much more expensive than a wind tunnel experiment. Further study is required to investigate to which extent the costs (and possibly the uncertainties) would be reduced if experts would become more familiar with and skilled in the assessment of subjective probabilities.

REFERENCES

Allen, C. 1984. *Wind pressure data requirements for air infiltration calculations.* UK: Report AIC-TN-13-84 of the International Energy Agency-Air Infiltration and Ventilation Centre.

Cooke, R.M. 1991. *Experts in Uncertainty.* New York: Oxford University Press.

Cooke, R.M. & Goossens, L.H.J. 1995. *Procedures guide for uncertainty analysis using structured expert judgment,* Draft report prepared under contract ETNU-CT93-0104-NL for the Commission of European Communities, Directorate general XI, Directorate D (DG 11 CCJH).

De Wit, M.S. 1999a. *Uncertainty analysis in building thermal simulation (in preparation).* Delft: Ph.D.-thesis Delft University of Technology, Subfaculty of Civil Engineering.

De Wit, M.S. 1999b. *Final report of the expert judgment study on wind pressure coefficients, part 2, Rationales,* Delft: Report of Delft University of Technology, Subfaculty of Civil Engineering, B&B group.

Grosso, M. 1992. Wind pressure distribution around buildings: a parametrical model. *Energy and Buildings* 18: 101- 131.

Grosso, M., Marino, D. & Parisi, E. 1995. A wind pressure distribution calculation program for multizone airflow models. *Proc. 3rd Int. Conf. On Building Performance Simulation, Madison, USA.*

Hoelscher, N. 1997. *WTG-Ringversuche in Grenzschichtwindkanälen,* Bochum, November 1997.

Knoll, B., Phaff, J.C. & De Gids, W.F. 1995. Pressure simulation program. *Proc.16th AIVC Conference on Implementing the Results of Ventilation Research, Palm Springs, California, USA, March 1995.*

Soerensen, L.C.1998. *Wind tunnel tests of a low building in a city environment,* Canada: Report Boundary Layer Windtunnel Facility, University of Western Ontario, October 1998.

Swami, M.V. & Chandra, P.E., 1988. Correlations for pressure distributions on buildings and calculation of natural ventilation air flow. *ASHRAE Transactions* 94(1): 243-266.

Ter Haar, T.R., Retief, J.V. & Dunaiski, P.E. 1998. Towards a more rational approach of the serviceability limit states design of industrial steel structures. *Proc. 2nd World Conference on Steel in Construction, paper 283, San Sebastian, Spain.*

Van Elst, N.P. 1997. *Reliability of movable water barriers (in Dutch).* Delft: Delft University Press, WBBM report Series 35.

Willemsen, E. 1998. *Pressure measurements on a 1:250 model for ventilation study TU-Delft (in Dutch),* Netherlands: Report DNW-TR-98.07 of German-Dutch Wind tunnel (DNW).

Wind Engineering into the 21st Century, Larsen, Larose & Livesey (eds) © 1999 Balkema, Rotterdam, ISBN 90 5809 059 0

Understanding the conical-vortex flow on roofs

F.Wu, P.P.Sarkar & K.C.Mehta
Wind Engineering Research Center, Department of Civil Engineering, Texas Tech University, Lubbock, Tex., USA

ABSTRACT: Full-scale synchronized data of wind-flow images, incident wind and roof suction pressures near the roof corner of Texas Tech Experimental Building have shown that a direct relationship exists between incident wind angle of attack, conical vortex structures, and the induced corner high suction pressures. Three-second time-averaging method is applied in the analysis which proves to be a suitable technique for analyzing field data in this experiment. Conical vortex structures generated at the roof corner for oblique wind angles are established as a function of incident wind angle of attack. Suction pressures underneath the conical vortices are found to be directly correlated to its size and location.

1 INTRODUCTION

It is commonly known that high suction pressures occur in the flow separation regions on the roof of a low-rise building, and the damage of roof cladding is inclined to initiate around those regions. Two typical phenomena that are of concern are the separation bubble initiated at the leading roof edge by normal wind and the conical vortices initiated at the leading roof corner by oblique wind. Various full-scale and wind-tunnel model results (Mehta et al. 1992, Cochran & Cermak 1992, Lin et al. 1995) have shown that the highest suction pressures occur near the roof corner underneath the conical vortices. However, the mechanism of generation of high suction pressures at roof corners is yet to be fully understood.

Flow visualization plays an important role in the fundamental research of pressure-generation mechanism. Clarification of separated-flow phenomenon around corners will result in a better understanding of this pressure-generation mechanism and in the improvement of techniques for wind-tunnel simulation which can eventually lead to various effective pressure mitigation techniques. Banks et al. (In press) performed flow visualization studies on conical vortices in a wind tunnel. Other researchers (Kawai & Nishimura 1996, Kawai 1997, Marwood 1997) have also measured the conical vortex structures in wind tunnels. Full-scale flow visualization was initiated by Wagaman (1993) and Letchford (1995) at Texas Tech University to study separation bubble. Thereafter, an extensive study was carried out by Sarkar et al. (1997) and Zhao (1997) to study separated flow on the roof at Texas Tech University.

In the past, studies have revealed a direct relationship between incident wind characteristics, the conical vortex structure, and the induced pressures. Lateral turbulence of the incident wind is considered to play an important role in this process (Tieleman et al. 1997, Letchford & Marwood 1997).

The goal of the present study is to examine various factors that influence the high-suction pressures at roof corners based on full-scale observations. The experiments conducted at Texas Tech Experimental Building that support this study includes simultaneous visualization of the conical vortices along with measurements of the incident wind flow and the roof corner pressures.

2 METHODOLOGY

Visualization of flow patterns can provide us a direct observation and physical understanding of wind flow fields at different regions around buildings. In full-scale experiments, several techniques can be applied to visualize the separated-flow phenomena. Sarkar et al. (1997) explored a few of these techniques that are suitable for full-scale observation. These include tuft-grid, airfoil-grid and smoke-injection methods. While the airfoil-grid method captures a good view of the mean flow, the smoke-injection method is suitable in displaying the dynamic convection or three dimensionality of the flow. In this study, tuft-grid method was used because it could capture the characteristic size of the conical vortices on one plane very effectively. Yarns used as tufts in the tuft-grid method are soft and light such that it can align with the streamline and fluctuate with turbulence within a relatively small dimension of a conical vortex. The airfoil-grid method could have been used instead except it was found not as effective as the tuft-grid method for conical-vortex flow.

The experiment was conducted at the Wind Engineering Research Field Laboratory (WERFL), Texas Tech University. The WERFL facility consists of a 9.1(B) × 13.7(L) × 4.0(H) m (30 × 45 × 13 ft) test building, a 49 m (160 ft) meteorological tower and a pressure measuring and data acquisition system which are fully described by Levitan and Mehta (1992 a,b).

A 2.13 m (7 ft) long by 1.02 m (3.33 ft) high tuft-grid with a mesh size of 10.2 cm (4 in) square was built for this experiment. Bright white yarn pieces with a length of 17.8 cm (7 in) were tied on each node of the steel wire mesh. The grid was placed at different locations close to roof corner normal to the roof diagonal in one case and perpendicular to the short roof edge in another case to visualize a pair or only one line of conical vortices. The results in this paper pertain to the grid placed normal to the short roof edge.

The schematic view of the experimental setup is shown in Figure 1. A 3-component sonic anemometer was selected for its high frequency response. It was mounted at a height of 1.55 m (5.08 ft) above the roof corner and at a distance of 0.91 m (3 ft) upstream of the roof corner to measure the incident wind velocity. The reasons for this particular location are to minimize the influence of the building on the incident wind and to capture the closest reliable upstream wind velocity for correlation with video images and pressures. Since getting reliable wind data is important to this experiment, another separate experiment was conducted with two sonic anemometers, one at the location of Figure 1 and the other at 4 m (13 ft) height near the meteorological tower. This exercise showed that the characteristics of longitudinal and lateral turbulence fluctuations and the mean wind speed values are the same at the two locations after correcting for the difference in heights between the two anemometers. Vortex images were recorded by 8-mm video camera at a speed of 30 frames/second. An LED (Light Emission Diode) was placed between the video camera and the tuft-grid to provide a source of light signal to the camera along with a voltage signal to the data acquisition system; it helped in synchronizing the video images with the wind and pressure data that were simultaneously collected.

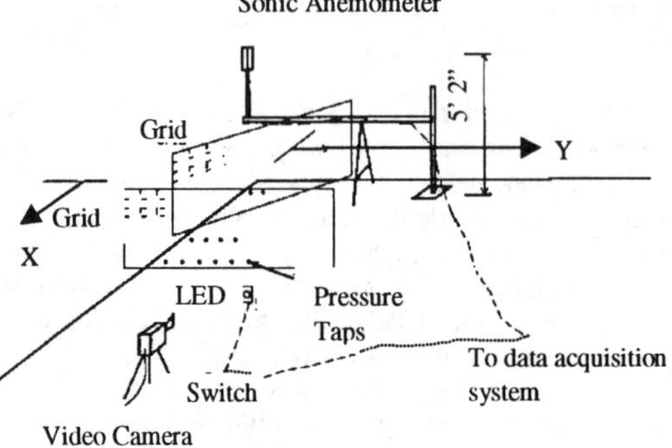

Figure 1. Schematic of synchronized video and wind-pressure data acquisition system

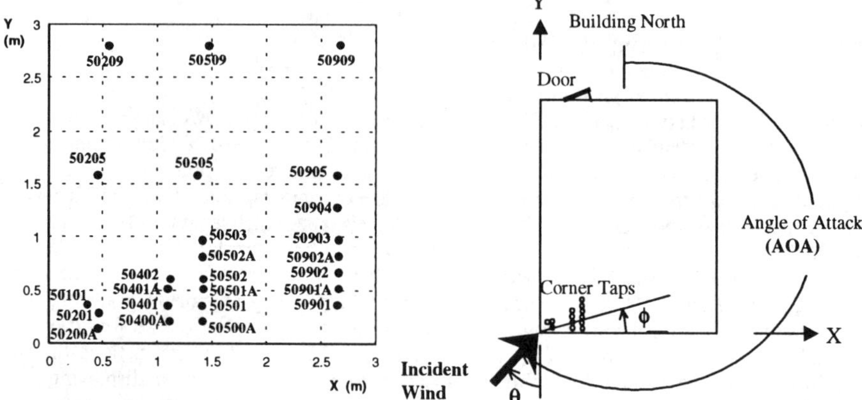

Figure 2. Corner pressure tap locations and definition of coordinates and angles

Corner pressure taps were densely arranged close to the short edge of the roof. Figure 2 shows the layout of all the pressure taps near the WERFL roof corner and also defines the angle of attack (AOA), θ, ϕ and xy coordinate axes. Note the dense array of pressure taps compared to what is described by Levitan and Mehta (1992 a,b).

Both wind and pressure data were collected at a sampling frequency of 30 Hz.

3 RESULTS AND DISCUSSION

Out of 40 runs of data collected, twenty-six runs were validated to be usable. Using the voltage output from the LED that was collected by the data acquisition system and its images recorded by the video camera, each run of wind and pressure data was synchronized with the video images of the vortex. Following this process, a few interesting segments of data were selected for analyses. It needs to be mentioned that viewing the video images frame by frame is a time-consuming and painstaking step in the data processing procedure. Therefore, only 6 runs of data have been analyzed so far and are presented in this paper.

It was observed by Zhao (1997) that the incident wind speed and wind direction change frequently and abruptly during a 15-minute period of data logging in the field. It was noticed by the authors of the present paper that in extreme situations wind direction may change more than 20° in a period of 1/30 second and may vary over a range of more than 100° in a typical 15-minute run. Hence, for this highly turbulent wind, the 15-minute run-average data are less informative for analyzing the conical vortex structures and its effects on roof pressures. The choice of a characteristic time period within a 15-minute run for the current analyses is critical.

It was observed during the analyses that a 3-second period could be appropriately selected as a characteristic time interval within which the wind direction remains steady. A shorter time interval results in an unsteady time average of the data because of fewer data points. A longer time interval incurs fewer segments of data that could be examined for the present purpose because of frequent change in wind directions. Besides, the 3-second gust wind speed is used as the basic wind speed in the design standard for loads in the USA. Therefore, a number of 3-second segments of simultaneous wind and pressure data along with video images were extracted from 6 runs of 15-minute data with the criterion that the variation of wind angles is limited to ±5° about the 3-second mean wind angle. This was carried out to establish the typical vortex structures at various 3-second mean wind angles. It was observed that the overall structure of vortices sustains

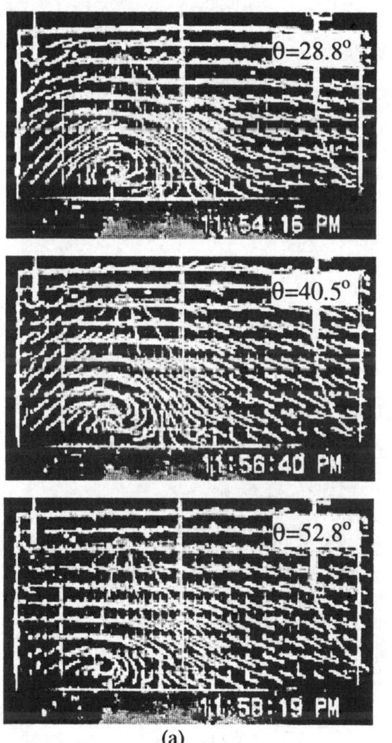

(a)

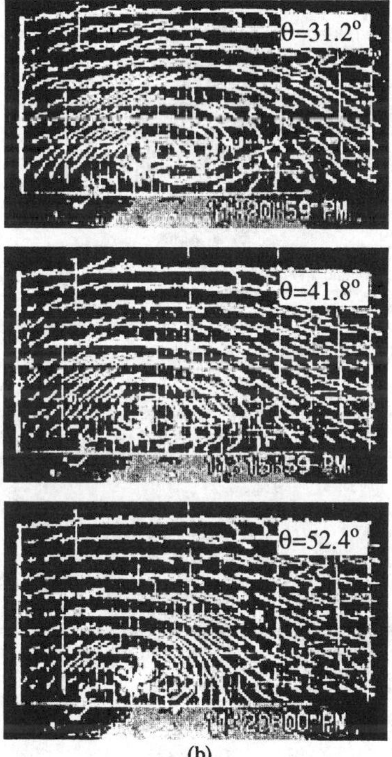

(b)

Figure 3. Several typical vortex structures for different angles of attack: (a) vortices at x=1.43 m , and (b) vortices at x=2.70 m

quite well within this short period and the instantaneous wind speed fluctuations have negligible influence on the overall structure of vortices. This implies that the vortex structures are quite similar for different wind speeds at a relatively steady wind angle. The results of analyses that are presented later also support this observation. Further, when the 3-second mean pressure coefficients were normalized with the corresponding 3-second mean wind speeds, they were found to maintain nearly the same values for a particular 3-second mean wind angle.

The projection of the core or center and the reattachment point of a conical vortex on the roof and the location of a pressure tap are defined by ϕ (Figure 2). φ_h represents the angle that the straight line from the roof corner to the vortex core makes with the roof plane.

3.1 *Typical conical vortex structures*

Vortices above x=1.12 m (3.67 ft), 1.42 m (4.67 ft) and 2.67 m (8.74 ft) lines of taps were visualized in this experiment (Figure 2). To minimize the influence of the tuft-grid on pressures, it was fixed at locations that were slightly offset from each line of taps, specifically at x=1.17 m (3.83 ft), 1.43 m (4.71 ft) and 2.70 m (8.88 ft), respectively.

Several typical vortex images are shown in Figure 3 to illustrate the difference in conical vortex structures as a result of different angles of attack (3-second mean). Figure 4 shows the variation of vortex angular position with different angles of attack. It is quite clear that a good linear relationship exists between ϕ, the horizontal vortex position, and θ (or AOA). Because of the small changes of vortex verti-

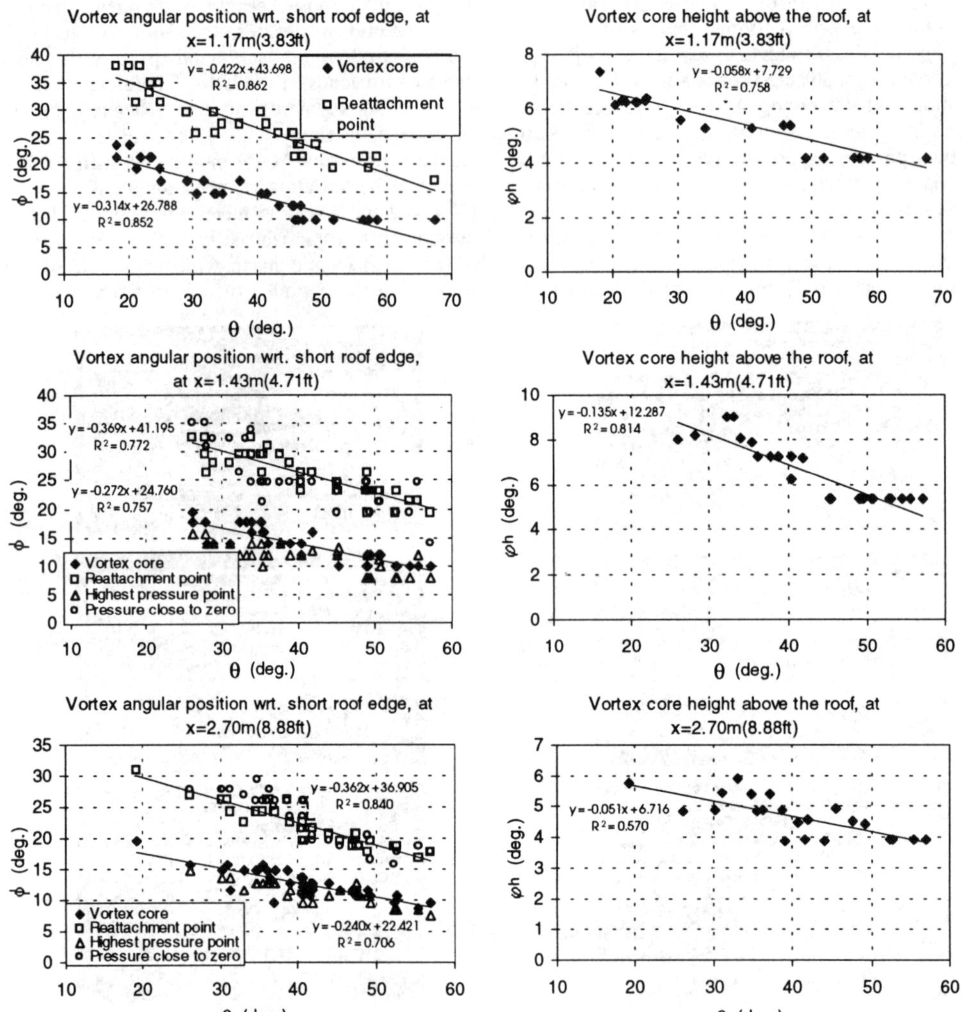

Figure 4. Conical vortex position for different angles of attack

cal position and the relatively poor tuft-grid resolution, the linear relationship between φ_h, the vertical vortex position, and θ is not as good as the others. Nevertheless, the trend that φ_h increases with decreasing values of θ is quite clear. This can be confirmed further with the images of conical vortices in Figure 3; both vortex size and vortex-core height increases with decreasing values of θ.

Linear regression is applied to the data in Figure 4. Equations are listed beside the regression line. New data from the 20 unused data runs will be used in the future to make the regression line more accurate. By applying the regression equations for vortex core position as mentioned in Figure 4, Figure 5 can be obtained which exhibits the projection of the vortex core position on the roof plane corresponding to θ=30°, 40°, and 50°. The lines in Figure 5 are slightly curved towards the short roof edge and are

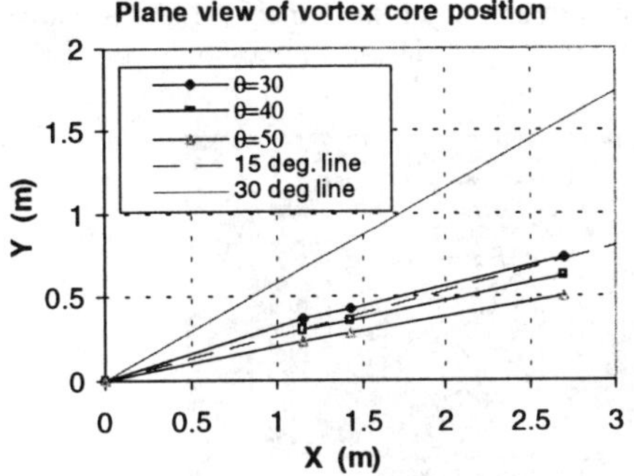

Figure 5. Vortex core position along radial line from roof corner on the roof surface.

quite close to the 14° radial line. This suggests that for nearly quartering winds (~θ=40°), vortex center or core will frequently fluctuate around this line.

3.2 *3-second mean pressure distribution profiles underneath vortices*

Besides the analyses of conical vortex structures, the point pressures underneath them were further analyzed. Several different values of θ were chosen; corresponding pressure distribution profiles of the 4 major lines of taps are presented in Figure 6. It needs to be mentioned that every curved line in Figure 6 represents the mean value of several 3-second segments corresponding to a specific θ.

As shown, 3-second mean technique gives quite a smooth curve for every angle of attack at all locations except at x=0.46 m which has only 2 taps. For each pressure profile, at x=1.42 m and 2.67 m, the highest pressure point and the pressure point close to zero are also plotted in Figure 4 to compare with the vortex structure. In Figure 4, the vortex core almost coincides with the highest pressure point, whereas the flow reattachment point coincides with almost zero pressure point. This shows a direct relationship between vortex structure and the pressures underneath it: peak suction pressure point basically follows the corresponding vortex core position; the suction pressure drops sharply to nearly zero in the region close to and beyond the reattachment point.

From Figure 6, it can be also concluded that high suction pressure usually occurs in the region extending from the apex of roof corner with ϕ bounded between 5° and 25°, and the magnitude of the suc-

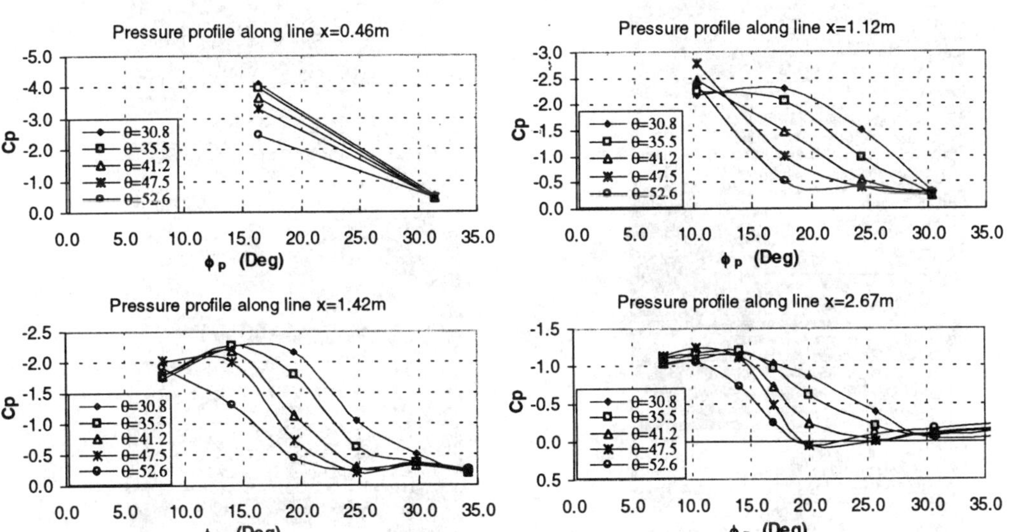

Figure 6. Pressure distribution profiles underneath conical vortices

tion pressure becomes higher near the apex. On each line of pressure taps, the highest pressure point of every profile does not differ much and they are relatively close to each other for a wide range of θ between 30° and 50°. It can be tentatively said that instantaneous peak suction pressure may generate for a wide range of angles of attack at the points close to ϕ=14° line. This may be helpful for future research on understanding the mechanism of peak pressure generation.

3.3 *Conical vortex phenomenon*

Having learned about the influence of θ on the overall structure of the conical vortices, the authors examined the intermittent nature of the vortices within a very short time period (say 1/10 sec.).

Compared to the transient nature of separation bubble that is formed on the roof by a normal wind, conical vortex is a relatively steady separation flow phenomenon. With a quartering wind of sufficient energy (say, in excess of 5 m/s at roof height), the conical vortices are observed to form. However, a close observation of this phenomenon shows that the conical vortex is actually intermittent. It is sealed inside a bubble bounded by a shear layer that attaches at the reattachment point. It grows to a full size (tightening up a little higher and closer towards roof edge) and then convects away from the roof edge by a small amount and eventually dissipates. Almost at the same instant, it re-grows again and the whole process repeats. All this happens in a very short time, approximately 1/3 second. Figure 7 lists 8 images to illustrate this whole process. Time interval between images is 1/30 second.

It can be further scrutinized from Figure 7 that conical vortex core actually shifts away from the roof edge by a very short distance under the shear layer. This shift can be estimated as 0.13 second (4 video frames) to complete. The spatial correlation between pressure taps is also analyzed and is shown in Figure 8a and 8b. These are the cross correlation

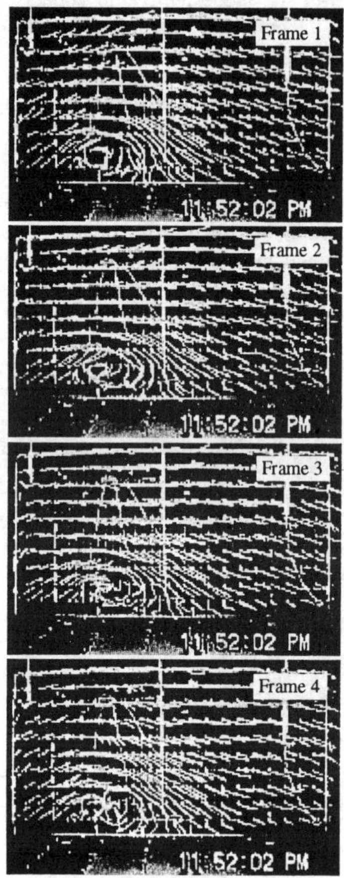

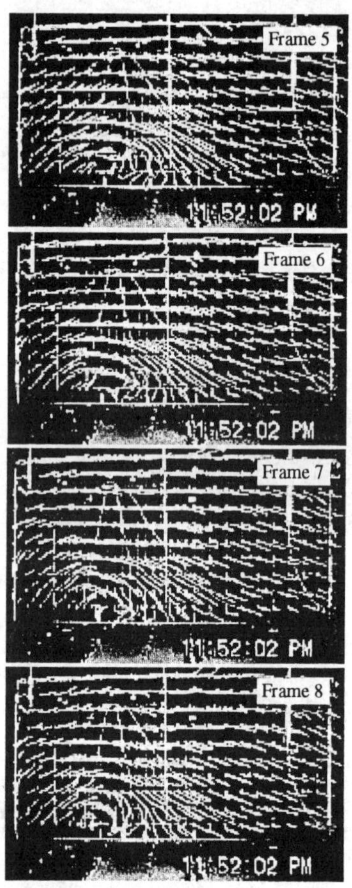

Figure 7. One cycle of conical vortex formation (Interval between frames is 1/30 second)

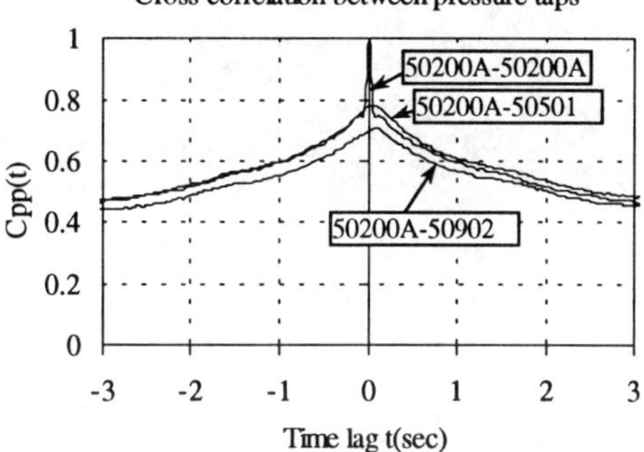

(a)

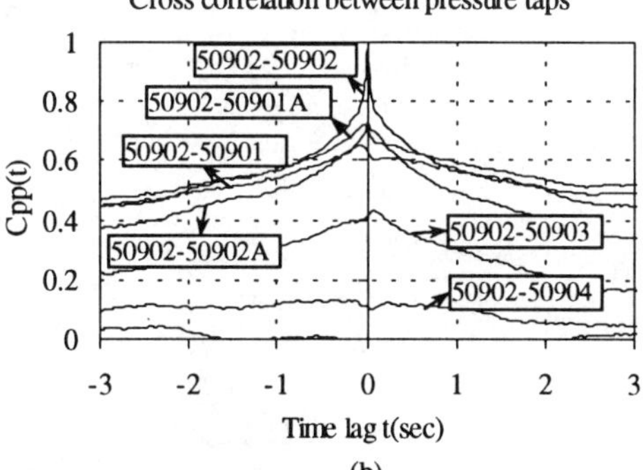

(b)

Figure 8. Cross correlation between corner pressure taps: (a) taps along radial line around 14°, and (b) taps along x=2.67 m line

between taps along ϕ=14° line (Figure 8a) and cross correlation between taps along x=2.67 m line (Figure 8b). From Figure 8b, it can be observed that high correlation exists between pressure tap 50902 and taps from 50901 up to 50903. These taps are located in the vortex prone zone. Pressures for taps beyond 50904 are uncorrelated with those of 50902 because these taps are close to or out of the flow re-attaching zone. The time delay from tap 50901 to tap 50903 is quite small but is a recognizable period of 0.16 second. It may be caused by the shifting movement of the vortex core because its characteristic time shift was estimated as 0.13 second from the video images.

Letchford (1995) recognized the high correlation and negligible time delay between taps 50401, 50501 and 50902. Figure 8a also shows a similar phenomenon. Time delay between 50200A and 50902 is negligible for such a long distance which means vortices along the short roof edge are coherent. Therefore, it implies that high suction pressures near the roof edge occur almost simultaneously; the area-averaged pressures near the roof edges will be higher than those occurring on other areas.

4 CONCLUSIONS

Synchronized wind flow and pressure data along with video images in a full-scale study at Texas Tech University helped to identify the mechanism of corner vortices and the associated pressures. By analyzing this data, typical conical vortex structures that vary with different wind angles of attack are established. It is found that high suction pressures near the roof corner basically follow the core positions of conical vortices. Conical vortex is a relatively stable separation flow. Changing of incident wind angle of attack can break the structure of conical vortex and force it to rebuild to another type of structure.

Although averaging time period is very short (3 seconds) in this study, all the results presented here are still in a mean sense. The mechanism of generating instantaneous peak suction pressure is still not clear and needs further study.

REFERENCES

Banks, D., R.N. Meroney, P.P. Sarkar, Z. Zhao & F. Wu (In press). Flow visualization of conical vortices on flat roofs with simultaneous surface pressure measurement. *J. Wind Eng. Ind. Aerodyn.*

Cochran, L.S. & J.E. Cermak 1992. Full- and model-scale cladding pressures on the Texas Tech University experimental building. *J. Wind Eng. Ind. Aerodyn.* 41-44: 1589-1600.

Kawai, H. & G. Nishimura 1996. Characteristics of fluctuating suction and conical vortices on a flat roof in oblique flow. *J. Wind Eng. Ind. Aerodyn.* 60: 211-225.

Kawai, H. 1997. Structure of conical vortices related with suction fluctuation on a plat roof in oblique smooth and turbulent flows *J. Wind Eng. Ind. Aerodyn.* 69-71: 579-588.

Letchford, C.W. 1995. Simultaneous flow visualization and pressure measurements on the Texas Tech building. *Proceedings of the Ninth International Conference on Wind Engineering, Volume* 1: 524-535.

Letchford, C.W. & R. Marwood 1997. On the influence of v and w component turbulence on roof pressures beneath conical vortices. *J. Wind Eng. Ind. Aerodyn.* 69-71: 597-577.

Levitan, M.L. & K.C. Mehta 1992a. Texas Tech field experiments for wind loads part 1: building and pressure measuring system. *J. Wind Eng. Ind. Aerodyn.* 41-44: 1565-1576.

Levitan, M.L. & K.C. Mehta 1992b. Texas Tech field experiments for wind loads part 2: meteorological instrumentation and terrain parameters. *J. Wind Eng. Ind. Aerodyn.* 41-44: 1577-1588.

Lin, J.-X., D. Surry & H.W. Tieleman 1995. The distribution of pressure near roof corners of flat roof low buildings. *J. Wind Eng. Ind. Aerodyn.* 56: 235-265.

Marwood, R. & C.J. Wood 1997. Conical vortex movement and its effect on roof pressures. *J. Wind Eng. Ind. Aerodyn.* 69-71: 589-595.

Mehta, K.C., M.L. Levitan, R.E. Iverson & J.R. McDonald 1992. Roof corner pressures measured in the field on a low building. *J. Wind Eng. Ind. Aerodyn.* 41-44: 181-192.

Sarkar, P.P., Z. Zhao & K.C. Mehta 1997. Flow visualization and measurement on the roof of the Texas Tech building. *J. Wind Eng. Ind. Aerodyn.* 69-71: 597-606.

Tieleman, H.W., T.A. Reinhold & M.R. Hajj 1997. Importance of turbulence for the prediction of surface pressures on low-rise structures. *J. Wind Eng. Ind. Aerodyn.* 69-71: 519-528.

Wagaman, S. 1993. Full-scale flow visualization over a low rise building. *Master thesis,* Department of Civil Engineering, Texas Tech University.

Zhao, Z. 1997. Wind flow characteristics and their effects on low-rise buildings. *Ph.D. dissertation*, Department of Civil Engineering, Texas Tech University.

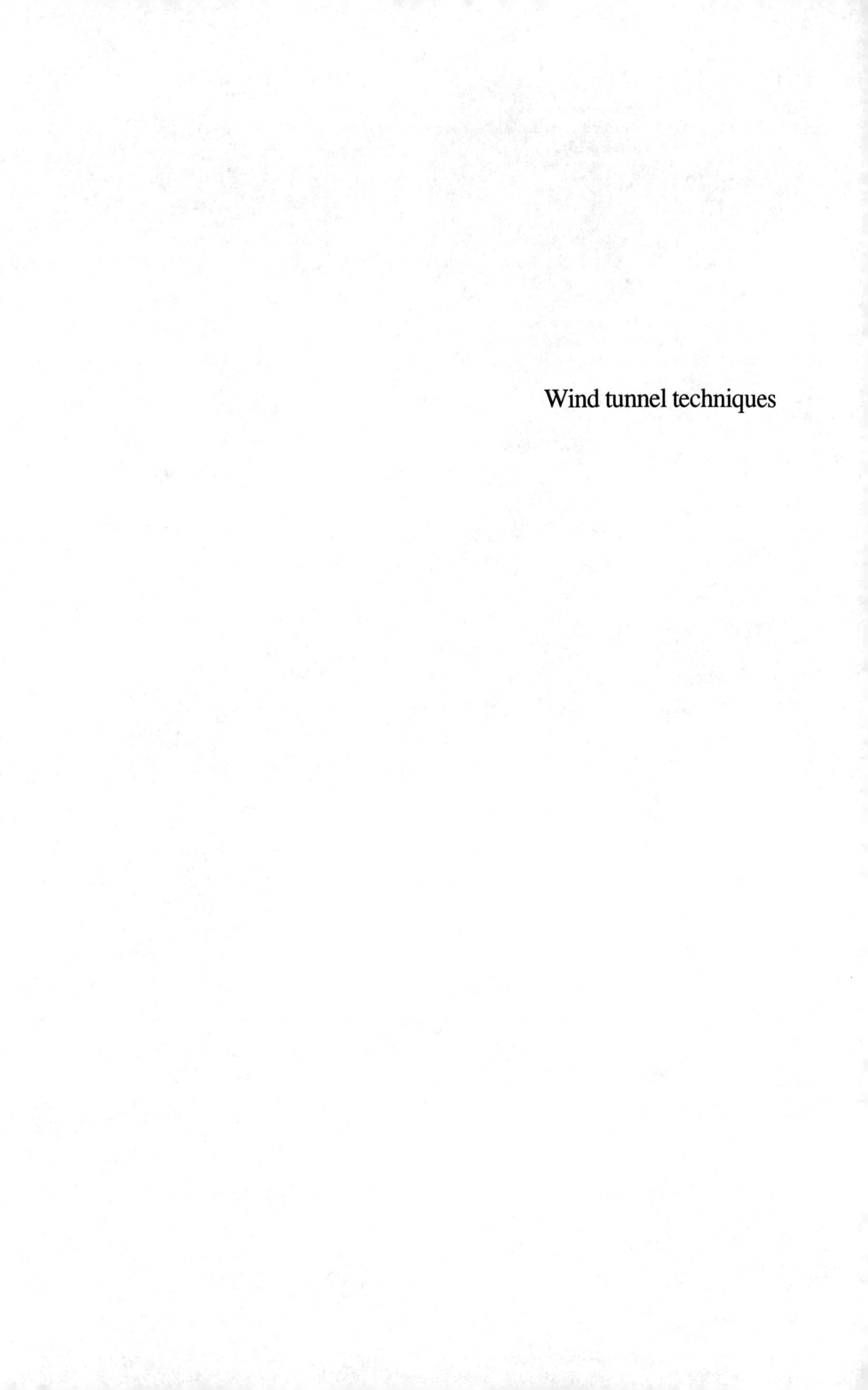

Wind tunnel techniques

Wind Engineering into the 21st Century, Larsen, Larose & Livesey (eds) © *1999 Balkema, Rotterdam, ISBN 90 5809 059 0*

Statistical and spectral comparisons of hot-film and LDV velocity data

M.Chen & T.A.Reinhold
Department of Civil Engineering, Clemson University, S.C., USA

R.E.Akins
Department of Physics and Engineering, Washington and Lee University, Lexington, Va., USA

ABSTRACT: Statistical comparisons of velocity profile data from hot-film anemometer (HFA) and laser Doppler velocimeter (LDV) measurements are made. The extents of discrepancies between the measurements made with the two systems vary with statistical moments. Normal-equally-weighted and sampling-interval-weighted data reduction algorithms are applied to LDV records to investigate the bias effects that might be caused by the LDV's random sampling process on the statistical moments. Spectral analysis of LDV data are performed using four different techniques: the direct unequally spaced Fourier transform method, the Lomb-Scargle periodogram and extirpolation method, the slotted autocorrelation method, and the normal equally spaced fast Fourier transform method. The spectrum of equally spaced HFA data provides a reference to the spectra of LDV data. The major deviation between HFA and LDV spectra lies in the high frequency range. The source of the deviation is discussed and possible approaches to mitigating the deviation are suggested.

1 INTRODUCTION

HFA and LDV are two major tools for obtaining velocity measurements in the wind tunnels with their own advantages and disadvantages. For instance, by using frequency shift techniques LDV can measure flow reversals and large flow fluctuations which are common in the wakes of wind tunnel models. Similar measurements with HFA are not possible unless a pulsed wire system is used. However, LDV obtains unequally spaced samples based on the random passages of the seeding particles through the measuring volume. These random measurement intervals pose difficulties for data post-processing. For example, the normal fast Fourier transform and fast wavelet transform can no longer be used directly. A number of studies have been conducted to assess the LDV sampling scheme and to develop correction factors for bias effects (e.g. Edwards & Jensen 1983), comparative measurements of HFA and LDV data have been performed (e.g. Lau, Morris & Fisher 1979, Lau, Whiffen, Fisher & Smith 1981), and spectral analysis of unequally spaced data have been developed (e.g. Bell 1983, 1986, Gaster & Roberts 1977, Horne & Baliunas 1986, Jones 1972, Lomb 1976, Press & Rybicki 1989, Scargle 1982).

This paper provides comparisons of mean velocities, turbulence intensities, skewness, flatness and power spectral densities obtained for HFA and LDV measurements in simulated turbulent boundary layer flows. Thus the comparisons are specifically tailored to typical wind engineering applications.

Tests were conducted in the boundary layer wind tunnel at Clemson University using both a Thermal Systems Incorporated (TSI) IFA 300 HFA system and a TSI Model L70-2/5 LDV system. The HFA system is a constant temperature anemometer. A single hot-film sensor operated at temperature 250 ^{0}C was used for the HFA measurements. The LDV system has two Model 9253-350 fiberoptic probes with lens focal distance of 350 mm and probe beam diameter of 2.82 mm. One probe carries a pair of green beams and a pair of blue beams, the other transmits a pair of purple beams. Seeding particles were produced by a Martin fog machine evaporating Martin fog fluid, both made by Tracoman, Inc.

2 STATISTICAL COMPARISONS

For the purpose of statistical comparisons, three HFA and three LDV velocity profiles were acquired with the sampling duration of one minute for each location in each profile. The profiles correspond to three different turbulence levels created by changing floor roughness in the test section and spires at the inlet of the wind tunnel test section. HFA data were sampled at a data rate of 4000 points per second and lowpass filtered at 2000 Hz. The LDV data were taken at varying data rates ranging from hundreds to thousands of samples per second depending on the local densities of seeding particles and optical settings of the LDV system. Normal-equally-weighted and sampling-interval-weighted data reduction algorithms were applied to LDV records to investigate the bias effects on the statistical moments. The expressions of the mean velocity, turbulence intensity, skewness and flatness for the sampling-interval-weighted algorithm are those defined in the TSI LDV instruction manual "Find for Windows" (TSI 1996).

Figures 1, 2, 3 and 4 show the comparisons of the mean velocities, turbulence intensities, and flatness respectively. The experimental study reveals that the bias effects on the mean velocity (Fig. 1) and turbulence intensity (Fig. 2) are obvious, and the effects seem to become greater with the increase of turbulence level (from right to left). On average, the normal-equally-weighted mean velocities are 5% larger than the sampling-interval-weighted mean velocities. The latter values are very close to the HFA mean values. In contrast to the mean velocities, the normal-equally-weighted turbulence intensities are a few percent smaller than the sampling-interval-weighted results. Similar to the mean velocities, the latter are generally closer to the HFA measurements. The skewness and flatness of HFA and LDV records exhibit the opposite trend. The normal-equally-weighted skewness and flatness of LDV records agree more closely with those of the HFA data than the sampling-interval-weighted ones. Deviations between the LDV skewness from the two different reduction algorithms are considerable while the differences between the their flatness are much smaller. In addition, Figures 3 & 4 demonstrate that HFA data tend to follow the Gaussian distribution while the shapes of probability distributions of the LDV data are narrower and have longer lower tails than the Gaussian distribution.

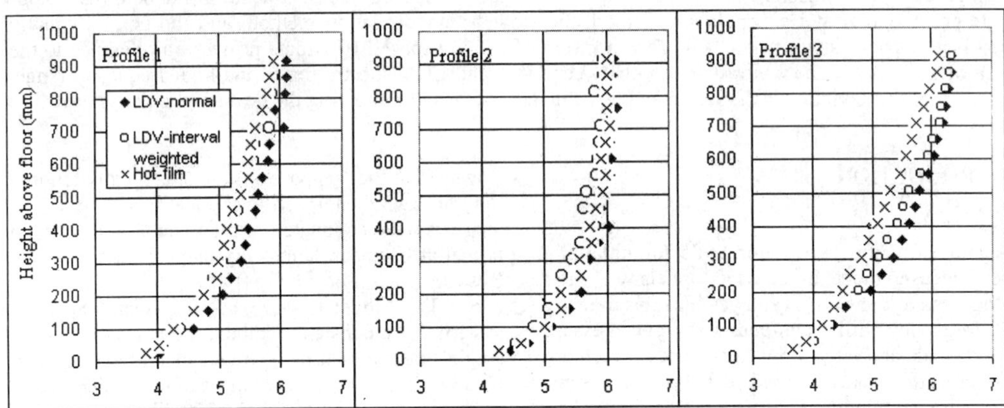

Figure 1. Comparisons of mean velocity. x-axis represents mean velocity (m/s).

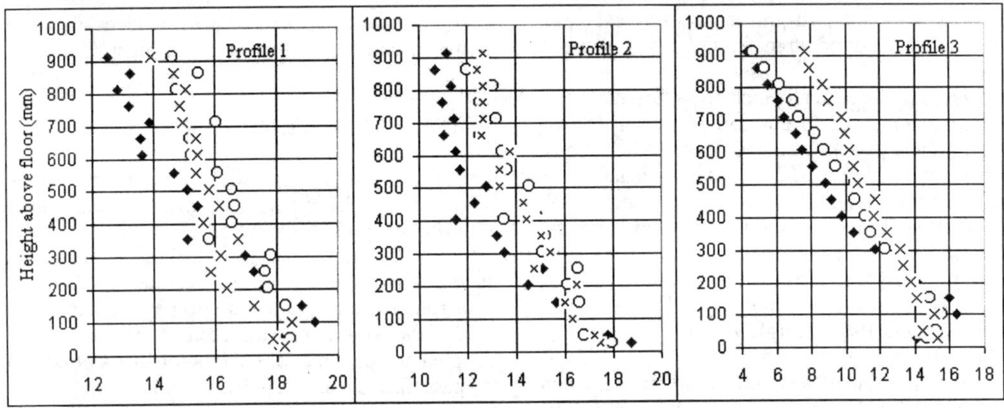

Figure 2. Comparisons of turbulence intensity. x-axis represents turbulence intensity (%); legends are same as Figure 1.

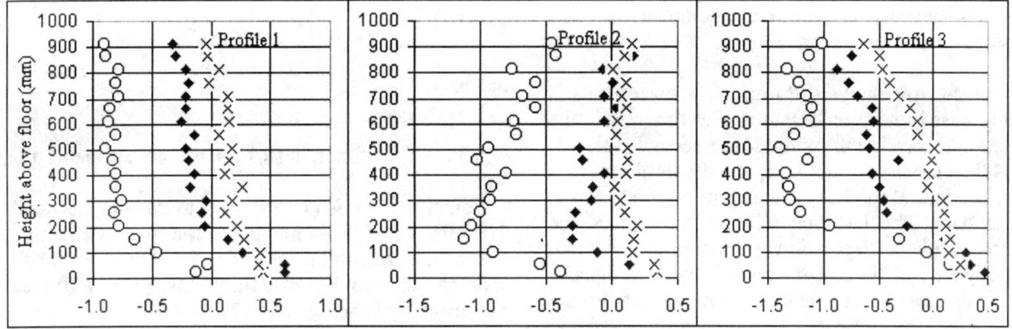

Figure 3. Comparisons of skewness. x-axis represents skewness; legends are same as Figure 1.

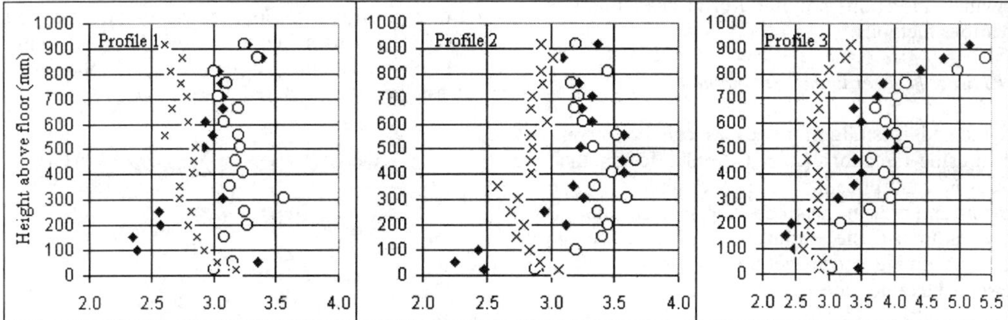

Figure 4. Comparisons of flatness. x-axis represents flatness; legends are same as Figure 1.

3 SPECTRAL ANALYSIS OF LDV DATA

3.1 Power spectral density function of unequally spaced discrete data

The power spectral density function for discrete data can be defined independently in terms of the circular correlation function and the discrete Fourier transform. Assume we have a data set $\{x_r, r = 0,1,2,\ldots, N-1\}$, sampled at instants $\{t_r\}$, the total duration is T, the sampling intervals are $\{\Delta_r = t_{r+1} - t_r\}$. The average interval is $\Delta = T/N$. The two-sided power spectral density function can be defined as

$$S(f_k) = S_k\Delta = \Delta\sum_{r=0}^{N-1} R_r \exp[-i(2\pi kr / N)] \quad (1)$$

or

$$S(f_k) = \frac{1}{T}\left|X(f_k)\right|^2 \quad (2)$$

Where k = 0,1,2, ..., N-1. R_r = circular correlation function, defined as Equation (3); $X(f_k)$ = Fourier transform of $\{x_r\}$, defined in Equation (4); $f_k = k/T$, frequency.

$$R_r = \frac{N-r}{N} RC_r + \frac{r}{N} RC_{N-r} \quad (3)$$

where RC_r = normal correlation function.

$$X(f_k) = \sum_{r=0}^{N-1}(x_r\Delta_r)\exp[-i(2\pi f_k t_r / T] \quad (4)$$

It can be seen that in contrast to the equally spaced discrete data case, the following two inequalities exist for unequally spaced data:

$$\overline{X(f_{N-k})} \neq X(f_k) \quad (5)$$

$$S(2f_c - f_k) \neq S(f_k) \quad (6)$$

where $f_c = 1/(2\Delta)$, the Nyquist frequency. Hence, rigidly speaking, the one-sided power spectral density function with frequency range $[0, f_c]$ is insufficient to represent the overall power spectral density function with frequency range $[-f_c, f_c]$. This is different from the equally spaced data case. Nevertheless, the one-sided power spectral density function is still used in this paper for simplicity.

3.2 Four approaches to spectral analysis of unequally spaced data

Current spectral analysis of unequally spaced data is mainly based on three methods. The first one is the slotted autocorrelation method, the second one employs the classic periodogram, and the third one uses the procedure of the least square fitting of a sinusoid to the data. The last one will be referred as the Lomb-Scargle extirpolation periodogram method. In the second method, data can be treated as equally spaced or retained as unequally spaced. Traditional fast Fourier transform (FFT) routines are available for equally spaced data whereas direct Fourier transform must be employed for unequally spaced data. Thus, four different analyses will be considered. The following subsections provide a brief description of the various methods.

3.2.1 Direct Fourier transform method

This method is straightforward as can be accomplished using Equations (2) & (4) only. The method requires N^2 multiplications. However, this is not too time-consuming using fast computers. In addition, it is normal to use the Barlett segment-and-average scheme to divide a data set into segments, which saves considerable time.

3.2.2 Normal equally spaced data method

This method simply ignores the irregularity of the time steps of a data set. Then Equation (4) can be written as

$$X(f_k) = \Delta \sum_{r=0}^{N-1} x_r \exp[-i(2\pi kr / N)] \tag{7}$$

Equation (2) is again used to calculate the power spectral density function. This is the fastest method.

3.2.3 Slotted autocorrelation method

Bell (1986) gave a complete description of this method. The main idea is to compute the normal autocorrelation function RC_r appearing in Equation (3) using a slotting technique which divides the time lag axis into M equidistant slots of width $\Delta\tau$. All data pairs are grouped according to the time slot in which their time difference falls. The value of RC_r is then estimated as the average of all cross products falling within each slot. That is

$$RC_r(\tau) = \frac{\sum x(t_i)x(t_j)}{H(r\Delta\tau)} \qquad r = 1,2,\ldots,M \tag{8}$$

where $H(r\Delta\tau)$ = the number of cross products falling within the rth slot; t_i, t_j = time instants, must satisfy the relation

$$(r-1/2)\Delta\tau \le |t_j - t_i| < (r+1/2)\Delta\tau, \qquad r = 1,2,\ldots,M$$

instead of

$$(r-1)\Delta\tau \le |t_j - t_i| < r\Delta\tau$$

which, will at times, lead to some unreasonable results.

Equations (3) & (1) are used to calculate R_r and S(f) afterwards. This method is the most ineffective one as the maximum lag index M should be much less than the number of data points, $M \approx N/10$ was suggested by Blackman and Turkey (1958).

3.2.4 Lomb-Scargle periodogram and extirpolation method

The Lomb-Scargle periodogram was first developed by Lomb(1975) using least-squares fitting of sinusoids to a data set and Scargle (1982) by modifying the classic periodogram, redefined as

$$P(\omega) = \frac{1}{2}\left\{ \frac{\left[\sum_{r=0}^{N-1} x_r \cos\omega(t_r - \tau)\right]^2}{\sum_{r=0}^{N-1} \cos^2 \omega(t_r - \tau)} + \frac{\left[\sum_{r=0}^{N-1} x_r \sin\omega(t_r - \tau)\right]^2}{\sum_{r=0}^{N-1} \sin^2 \omega(t_r - \tau)} \right\} \tag{9}$$

where ω = angular frequency; τ = time shift, defined as

$$\tan(2\omega\tau) = \frac{\sum_{r=0}^{N-1} \sin(2\omega t_r)}{\sum_{r=0}^{N-1} \cos(2\omega t_r)} \tag{10}$$

Press & Rybicki (1989) worked out a fast algorithm using an extirpolation technique, which lets this method be used on large data sets. Extirpolation is the reverse of interpolation. One data point is extirpolated as many points (say 4) and spread into a wide enough equally spaced grid to allow the use of the FFT for analysis.

3.3 Comparisons of HFA and LDV data spectra

The data sets used for spectral analysis are different from those used for statistical comparisons in that data were taken in longer sampling duration. Table 1 shows the general descriptions of the data sets.

Table 1. Data description

	Points	Data rate	Mean velocity (m/s)	Turbulence intensity (%)
HFA*	524,288	4000	5.709	14.85
LDV	500,652	2781	6.057	13.00

*HFA data were lowpass filtered 2000 Hz.

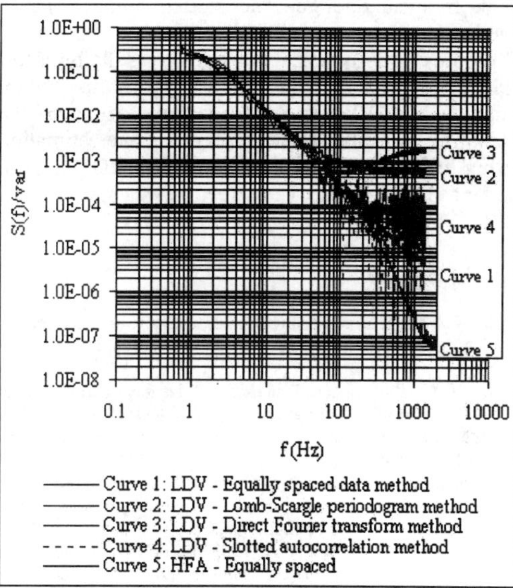

Figure. 5 Power spectra

The spectral plot is shown in Figure 5. The basic Bartlett segment-and-average scheme was used for computation of the spectral density function. Each data segment contains 4096 points without overlap. Data were demeaned with the mean velocities of segments. In Figure 5 var is the global variance of a data set. Results from segments are averaged to give the overall estimation of the spectral density function. The final values are smoothed by averaging five adjacent points except for the first and last two points that are adopted directly.

The power spectra of HFA and LDV data match very closely within the low frequency range. In the high frequency range, the usual trend of attenuation (decreasing spectral density with increasing frequency) that has been well established based on equally spaced data is not observed for the unequally spaced data unless the LDV data are treated as equally spaced data. For LDV data, the normal equally spaced data method yields the best results compared to the other three methods. The direct Fourier transform method produces a rising trend in Figure 5 and the highest variation from the HFA spectrum in the high frequency end. The spectrum, generated by the slotted autocorrelation method using 40,960 points, oscillates violently and contains negative values (not plotted) in the high frequency range. This observation conforms to the conclusions of other authors (e.g. Bell 1983, 1986).

3.4 Source of spectral deviation

It is not hard to understand that all four approaches for the spectral analysis of LDV data involve Fourier transforms. The discrete Fourier transform (DFT) is an approximation of the continuous Fourier transform (CFT); but it is more complicated due to the time domain sampling and truncation required. Furthermore, the DFT of unequally spaced data is even more complex than DFT of equally spaced data. Assume x(t) is a continuous function, its discrete finite data set $\{x_r, r = 0, 1, 2, \ldots, N-1\}$ can be written as

$$x_r = x(t)\Delta_0(t)w(t) \tag{11}$$

where $\Delta_0(t)$ = sampling function, consisting of a series of delta functions as

$$\Delta_0(t) = \sum_{r=-\infty}^{\infty}\delta(t - t_r) \tag{12}$$

w(t) = truncation function, defined as

$$w(t) = \begin{cases} 1, & t \in [0,T] \\ 0, & otherwise \end{cases} \tag{13}$$

Multiplication in the time domain corresponds to convolution in frequency domain. The Fourier transform of $\{x_r\}$ thus is

$$X(f) = [X_c(f) * \Delta_0(f)] * W(f) \tag{14}$$

where $X_c(f)$ = Fourier transform of x(t); W(f) = Fourier transform of w(t), identical for both equally spaced data and unequally spaced data; $\Delta_0(f)$ = Fourier transform of $\Delta_0(t)$. For equally spaced data,

$$\Delta_0(f_k) = \frac{1}{\Delta}\sum_{r=-\infty}^{\infty}\delta(f_k - \frac{r}{\Delta}) \tag{15}$$

which is an equally spaced sequence of impulses. For unequally spaced data,

$$\Delta_0(f_k) = \sum_{r=-\infty}^{\infty}\Delta_r \exp[-i(2\pi f_k t_r)] \tag{16}$$

which does not have a clean analytic expression. It is evident that the spectral deviation of the LDV data from HFA data arises from the irregular sampling of the LDV data. Due to the limited space for this paper, it is not possible to discuss this problem thoroughly. Just examine Figure 6, which is a partial plot of $|FT[\Delta_0(t)w(t)]|^2$ versus frequency f for the first 4096 sampling intervals of the LDV data. The reason that the corresponding curve of equally spaced data is not also shown is that the curve is fairly flat with the magnitude on the order of 10^{-15} except for a large peak at f = 0 Hz. This makes the curve invisible. One can observe from Figure 6 that there are about two turning points: one at $f \approx 20$ Hz, the other at $f \approx 50$ Hz. For frequencies below 20 Hz, the magnitude is very small; within 20 – 50 Hz, the fluctuation level increases steadily; above 50 Hz, the curve varies drastically. Comparing with Figure 5, one will find that the trends of both graphs are very close. This implies that it may be possible to reduce the spectral variation of LDV data in the high frequency range by reducing $|FT[\Delta_0(t)w(t)]|^2$. This can be achieved by using different techniques. For example, the normal equally spaced data method smoothes the sampling intervals completely, while the slotted autocorrelation method smoothes the original data points and drops some points. However, these smoothing methods alter the original data. A better technique might be to find out the relationship between $|FT[\Delta_0(t)w(t)]|^2$ and different sampling rates, then offset the overestimated portions from the spectral density function to recover the true estimates. This is a topic for future research.

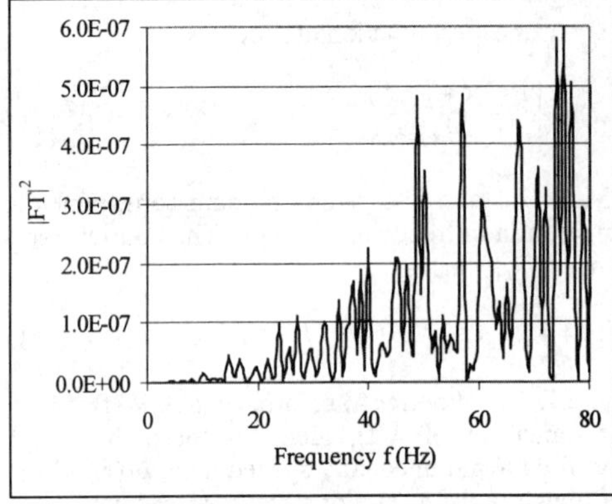

Figure 6. Part of the $|FT[\Delta_0(t)w(t)]|^2$ versus frequency f for the first 4096 sampling intervals of the LDV data

4 CONCLUSIONS

The work has shown that the bias effect caused by the LDV random sampling cannot be ignored. The normal-equally-weighted mean velocities are about 5% larger than the sampling-interval-weighted mean velocities. The higher statistical moments from these two data reduction algorithms have even higher differences for LDV data. Results from sampling-interval-weighted algorithm are generally closer to the HFA measurements. LDV and HFA data have different probability density functions (PDF). HFA's PDFs tend to follow the Gaussian distribution while LDV's PDFs are narrower and more skewed to the low velocity side. The turbulence level has an effect on the differences between HFA and LDV measurements.

LDV data spectral estimates obtained using the normal equally spaced data method provides the best approximation to the spectrum of the HFA data. All four approaches give good spectral estimates in the low frequency end, but create overestimation in the high frequency range. The slotted autocorrelation method is not recommended for further study since it produces highly unstable results over the high frequency range and is the slowest algorithm. Future study is suggested to investigate the effect of the sampling function on the spectral overestimation using both periodic functions and real LDV data in order to find an effective correction scheme.

REFERENCES

Bell, W.A.1983. Spectral analysis algorithms for the laser velocimeter: A comparative study. AIAA Journal 21(5): 714-719.

Bell, W.A. 1986. Spectral analysis of laser velocimeter data with the slotted correlation method. *AIAA/ASME 4th Fluid Mechanics, Plasma Dynamics and Lasers Conference*, 12-14 May 1986. Atlanta, GA.

Blackman, R.B. & Turkey, J.W. 1958. The measurement of power spectra from the point of view of communication engineering. Dover Publications, Inc., New York.

Edwards, R.V. & Jensen, A.S. 1983. Particle-sampling statistics in laser anemometers: sample-and-hold systems and saturable systems. *Journal of Fluid Mechanics* 133: 397-411.

Gaster, M. & Roberts, J.B. 1977. The spectral analysis of randomly sampled records by a direct transform. *Proc. R. Soc. Lond. A*. 354: 27-58.

Horne, J.H. & Baliunas, S.L 1986. A prescription for period analysis of unequally sampled time series. *The Astrophysical Journal* 302: 757-763.

Jones, R.H. 1972. Aliasing with unequally spaced observations. *Journal of Applied Meteorology March*: 245-254.

Lau, J.C., Morris, P.J. & Fisher, M.J. 1979. Turbulence measurements in subsonic and supersonic jets using a laser velocimeter. *Journal of Fluid Mechanics* 93:1-27.

Lau, J.C., Whiffen, M.C & Smith, D.M. 1981. A note on turbulence measurements with a laser velocimeter. *Journal of Fluid Mechanics* 102: 353-366.

Lomb, N.R. 1976. Least-squares frequency analysis of unequally spaced data. *Astrophysics and Space Science* 39: 447-462.

Press, W.H. & Rybicki, G.B. 1989. Fast algorithm for spectral analysis of unevenly sampled data. *The Astrophysical Journal* 338: 277-280.

Scargle, J.D 1982. Studies in astronomical time series analysis. II. Statistical aspects of spectral analysis of unevenly spaced data. *The Astrophysical Journal* 263: 835-853.

TSI Incorporated, 1996. Instruction manual: Find ™ for Windows.

Wind Engineering into the 21st Century, Larsen, Larose & Livesey (eds) © 1999 Balkema, Rotterdam, ISBN 90 5809 059 0

Design and experimental verification of a square-to-square wind tunnel contraction

Fuh-Min Fang, J.C.Chen & Y.T.Hong
Department of Civil Engineering, National Chung-Hsing University, Taiwan

Ta-Ching Liang
Department of Soil and Water Conservation, National Chung-Hsing University, Taiwan

ABSTRACT: Flow characteristics within a square-to-square contraction are investigated numerically and experimentally so as to gain additional insight into the contraction design. Measurements of cross-sectional velocity profiles and longitudinal pressure distributions along the wall centerlines are carried out in the contraction prototype and the results are compared to those predicted by the numerical simulations. Finally, the applicability of the nozzle design method, proposed by Fang (1997), is assessed.

Results show that although the inviscid calculation fails to simulate correctly the nozzle flow near the contracting walls, the general flow behavior of the flow in the regions away from the wall is fairly agreeable with that is predicted. On the other hand, the relative friction losses along the wall centerlines are about 4.3%. After an adjusted way of normalization is made, the measured pressure distributions are in good agreement with the numerical one.

1 INTRODUCTION

A contraction is a major component of a wind or water tunnel. As the flow passes this transitional segment, it is accelerated so that a low-turbulence, uniform flow condition can be achieved in the test section. Generally, the contraction area ratio (CR) is the most dominant factor which affects the extent of flow uniformity, the possibility of flow separation, and downstream turbulence level. Once it is determined, the nozzle shape and length, which also control the flow uniformity and the boundary layer growth along the contracting walls, become the second design choices.

In principle, the problem of contraction design is to search for the optimum shape with minimum nozzle length for a desirable flow quality requirement at the nozzle end. When the length is reduced, the contraction costs less and fits into a smaller space. In addition, the boundary layer will generally be thinner due to the combined effects of decreased length of boundary layer development and increased favorable pressure gradients in the contraction. However, the possibility of flow separation increases.

Typically, a contraction design starts with the selection of a contraction ratio (CR), which is mostly dictated by a consideration of the available space. Once CR is determined, the nozzle shape and length must be chosen so as to satisfy predetermined design criteria, such as exit velocity profile non-uniformity, avoidance of separation, exit boundary layer thickness, and space availability and cost.

By far, the most direct way for engineering contraction design is that suggested by Morel (1975, 1977), who proposed the use of a combination of two matched cubics, each having its apex at one end of the contraction, as the basic shape of 2-D and axisymmetric contractions. When the dimensions of the upstream and downstream sections and the nozzle length are fixed, the location of the matched point becomes the only parameter to determine the entire wall shape. Morel's design procedure starts by prescribing a level of velocity non-uniformity at the contraction exit section, which leads to a corresponding minimum wall pressure near the exit. Based on design charts resulting from potential flow calculations, several possible contraction lengths can be selected corresponding to different locations of the matched point and to different maximum wall pressures near the inlet. After the Stratford criterion (1959) is applied to check for the possibility of separation, the final choice among all the alternatives is the one which leads to the shortest length without the occurrence of separated flows.

For practical design concerns, Fang (1997) extended Morel's goal to a three-dimensional square-to-square contraction case. Considering a level to avoid flow separation, an acceptable flow uniformity at the exit cross section and the minimum nozzle length as the most important design factors, a series of design charts were achieved. Accordingly, a try-

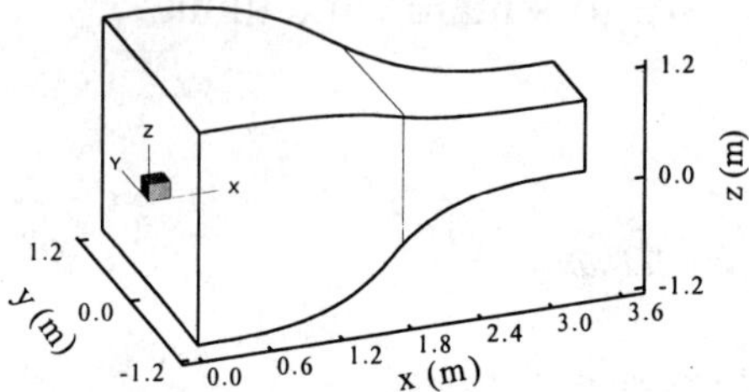

Figure 1. Description of the contraction prototype

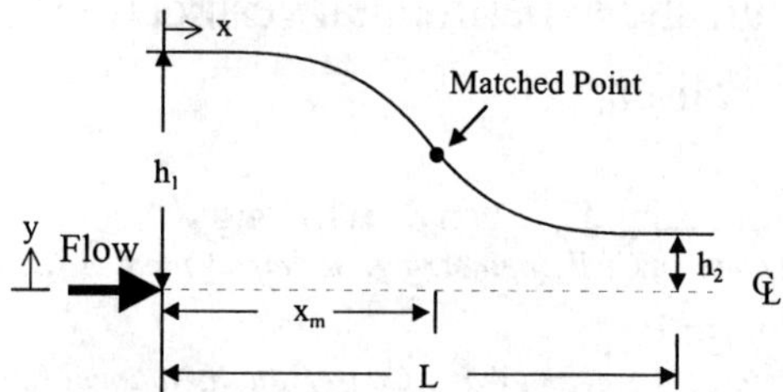

Figure 2. Schematic of the contraction (half plane)

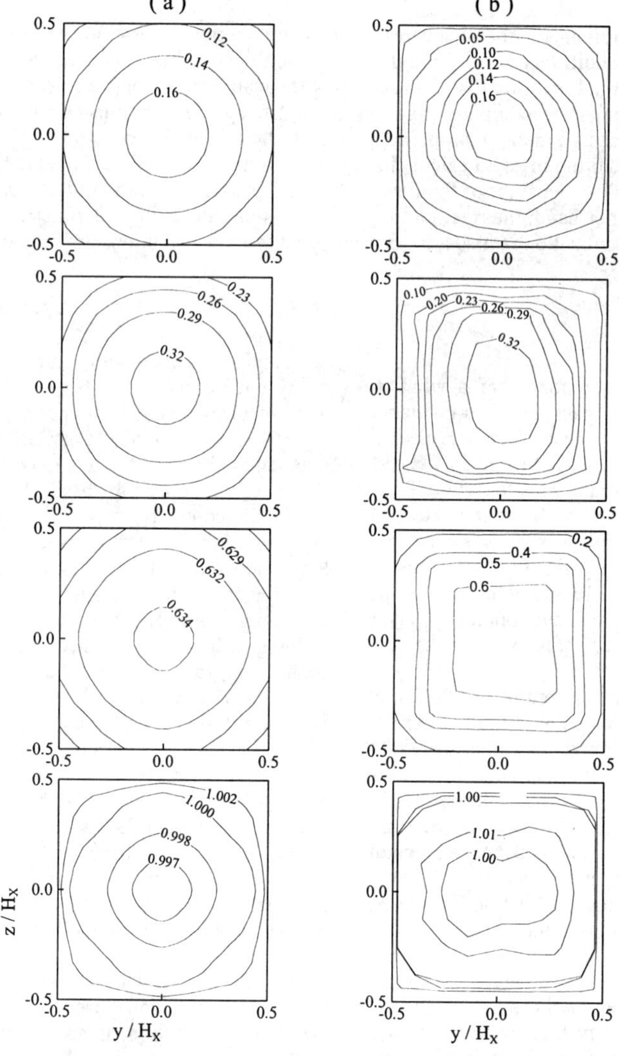

Figure 3. Sectional velocity profiles (x/L=0.32, 0.51, 0.68, and 1.0) (a) calculated (b) measured

and error procedure is proposed to search for an optimum shape with minimum nozzle length for a desirable flow quality requirement.

2 PRESENT PROGRAM

With prescribed design conditions (CR=9), a contraction prototype (Fig. 2) is built in an open-type wind tunnel system according to the design method proposed by Fang (1997). Velocity and pressure measurements were carried out to compare to the numerical predictions and to check the applicability of the design method.

The wind tunnel has a bell-mouth flare at the system entrance. The distances between the edges of the flare and the neighboring boundaries (ceiling, walls and ground) are 1.0, 0.85 and 1.0m, respectively. Next to the flare, a 3-meter-long setting chamber, containing a 10-centimeter-long honeycomb and three screens, is connected. Further downstream, the chamber is followed by the contraction and the test section.

Figure 1 illustrates the geometry of the square-to-square nozzle prototype. The sizes of the inlet and exit cross sections are respectively $2.40m \times 2.40m$ and $0.80m \times 0.80m$. The nozzle length (L) is 3.60m (L/H_1=1.50, H_1 is the width of the entrance section), and the location of the matched point (x_m) is selected as 1.80m ($X_m = x_m/L = 0.50$). The shape of the contraction is a combination of two matched cubics, described as (also see Fig. 2):

$$y\,(\text{or } z) = (h_1 - h_2)\left[1 - \frac{1}{X_m^2}\left(\frac{x}{L}\right)^3\right] + h_2 \;,\quad x < x_m$$

$$\text{or } = \frac{(h_1 - h_2)}{(1 - X_m)^2}\left(1 - \frac{x}{L}\right)^3 + h_2 \;,\quad x > x_m \qquad (1)$$

where h_1 ($=0.5H_1$) and h_2 are respectively the half widths of the inlet and exit sections of the nozzle.

2.1 *Numerical method*

The numerical computations adopt a finite-volume Euler solver (Song & Yuan 1988) to obtain inviscid flow results within the contraction. Under the assumption that the flow density is a function only of pressure (barotropic), for a low-Mach-number flow the equations of motion can be approximated in a conservative form as

$$\frac{\partial G_i}{\partial t} + \nabla \bullet \vec{F_i} = 0\,,\quad i = 1,2,3,4 \qquad (2)$$

where

$$G = \begin{bmatrix} p \\ u \\ v \\ w \end{bmatrix};\quad \vec{F} = \begin{bmatrix} Ku & Kv & Kw \\ u^2 + p/\rho & uv & uw \\ uv & v^2 + p/\rho & vw \\ uw & vw & w^2 + p/\rho \end{bmatrix} \qquad (3)$$

ρ is the density; p, u, v and w are pressure and velocity components, and K is the bulk modulus of elasticity.

The computation proceeds by using a volume integration over a specific control volume ($\forall$). Based on the divergence theorem, one has

$$\frac{\partial G_m}{\partial t} = -\frac{1}{\forall}\int_{\forall} \hat{n} \bullet \vec{F}\, dS \qquad (4)$$

where G_m represents the mean quantity referred to the center of the volume; $\hat{n}$ is the normal vector of the control surface. This equation is used to calculate the change of G_m at a starting time step within an elapsed period (Δt) and thus update the G_m values for the next time step.

In the simulations, appropriate values of pressures and velocities are specified at exterior (phantom) cells outside the boundaries to reflect the correct physical nature of the local boundaries. For the contracting walls, a slip condition is used for the velocity specifications, and the pressures at the phantom cells are given by taking into account the effect of the wall curvature. At the inlet and exit boundaries of the nozzle, on the other hand, the phantom-cell velocities are specified according to the fact that the total head is constant. Finally, the pressures at the entrance are given based on a zero-gradient assumption, and the average pressure at the exit cross-section is taken as the reference pressure of the computational domain.

2.2 *Experimental set-up*

Three rows of pressure taps are set along the centerlines of the top, bottom and one of the side walls of the contraction prototype. All taps are connected to a miromanometer with a resolution of 1/1000 of a centimeter. These allow for accurate evaluations of the longitudinal pressure distributions.

In addition, hot-wire anemometry is used to measure the mean velocity (x-component) profiles at four selected cross-sections within the nozzle (x/L=0.32, 0.51, 0.68 and 1.0). For a typical velocity measurement, signals are recorded within a period of two minutes with a sampling rate of 2 data per second. At each cross section, the velocity is measured at a distance varying from 2cm to 8cm and a total of 144 velocity measurements are conducted.

3 RESULTS

In the study, the measurements and numerical simulations are performed under the condition that the velocity at the nozzle exit section is 15 m/s (u_e). The corresponding Reynolds number, in terms of the velocity and the width of the exit section, is about 8×10^5.

3.1 *Velocity profiles at various cross-sections*

Based on the results of the inviscid calculations, Figures 3a shows the mean velocity profiles at the selected cross-sections (H_x is the local width of the section). Since the flow domain is considered symmetric to the centerline of the nozzle in the computations, the patterns of the calculated results also appear symmetric. At the three upstream sections (x/L<0.68), higher speeds are found at the sectional center regions. At the contraction exit(x/L<0.68), the trend of the cross-sectional velocity variation becomes opposite. That is, larger speeds are near the corners, while smaller speeds are at the center regions.

Compared to the calculated results, Figures 3b depict the corresponding mean velocity profiles from the experiments. In all cases, large transverse velocity gradients, which the inviscid computations fail to predict, are obtained in the near-wall regions due to the existence of boundary layers in the real flow. In the regions away from the wall, on the other hand, the measurement results appear fairly agreeable with those from the calculations.

3.2 *Variations of the cross-sectional non-uniformity*

Figure 4 depicts the calculated cross-sectional velocity non-uniformity (U_n) variation, where U_n represents the ratio (in percentage) of maximum velocity deviation to the local cross-sectional mean velocity. In the simulated flow, the non-uniformity starts with a small value then increases as it goes downstream. As x/L reaches about 0.43, U_n becomes maximum. After that, a minimum value of U_n is obtained as x/L equals to about 0.7. Further downstream, the U_n value again increases then decreases.

In comparison, the experimental data, obtained without the consideration of the near-wall regions, agree well with the result of the calculation (also see Fig. 4).

3.3 *Longitudinal pressure distributions*

By taking the pressure at the most upstream tap as the reference one, Figures 5 show the measured pressure variations along the centerlines of the top, side and bottom walls. In contrast to the corresponding inviscid results, the estimated friction losses along the streamlines are about 5.8 N/m^2, corresponding to 0.59 mm of water head.

To normalize both the measured and calculated results in a more consistent manner, an adjusted dimensionless parameter is used as

$$C_p = \frac{p - p_e}{0.5\rho u_e^2} \frac{(p_i - p_e)_{cal}}{(p_i - p_e)_{meas}} \tag{5}$$

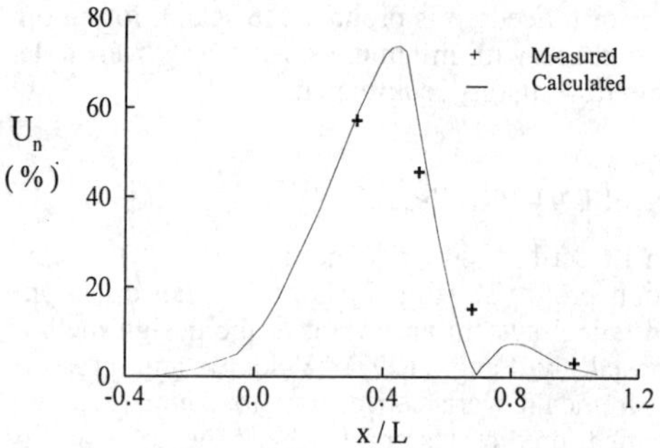

Figure 4. Variations of velocity non-uniformity

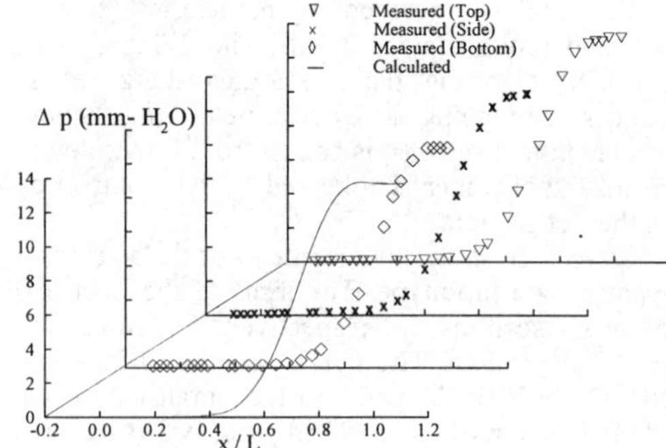

Figure 5. Comparison of pressure distribution along wall centerlines

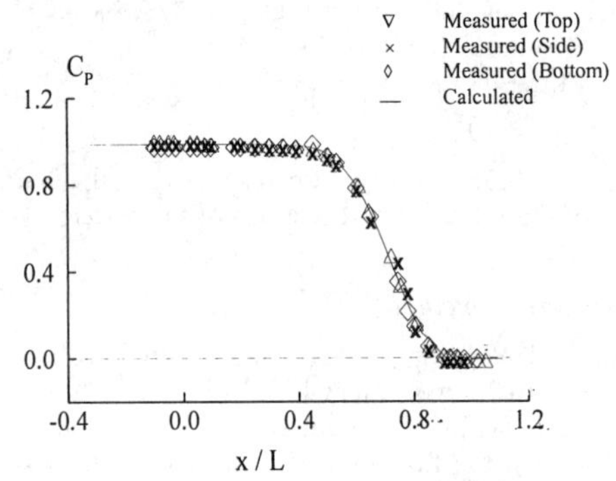

Figure 6. Normalized pressure distributions along wall centerlines

where p_i and p_e, are referred respectively to those along the centerlines at the inlet and exit sections of the contraction; the subscripts "cal" and "meas" stand for the calculated and measured values. Accordingly, all the normalized pressure distributions become almost identical, as shown in Figure 6.

4 DISCUSSIONS

4.1 *On the flow predictions and measurements*

From the viewpoint of nozzle designs, the application of numerical methods in flow analyses always results in errors as the computational domain is truncated into a simplified one. In the simulation, the selected domain of computations contains the portion of the nozzle, preceded and followed by a straight duct with a length of one-half the width of the contraction end section (h_1 and h_2). Further numerical investigations reveal that the relative error produced by this selection is no more than 3%.

The assumption of the simulation is that in the absence of separation, a low-turbulence flow in a major portion of the domain stays as a potential one. Since the boundary layer in the contraction is very thin compared to the size of the local cross-section, the error of the inviscid result is considered small, except at the regions near the walls. Investigations show that the boundary layer thickness along the wall centerlines varies from a few to about one centimeter in the nozzle prototype, indicating that the error introduced by the assumption becomes less significant as it goes downstream.

Moreover, although a honeycomb and screens are set upstream the contraction prototype in the wind tunnel system to pre-condition the inlet flow, the actual flow condition at the nozzle inlet section may still be somewhat different from that in the numerical calculation. This could explain why the actual velocity profiles were not accurately predicted by the numerical methods (Figs. 3). Unfortunately, due to practical limitations, the velocity profile at the nozzle inlet could not possible be measured to support the above statement.

In the experiments, another difficulty encountered is that the application of hot-wire anemometry does not provide enough accuracy to resolve mild velocity variations, especially at the exit section, where the velocity deviates only a few percents. A more appropriate technique or method is therefore needed to solve this problem.

4.2 *Flow behaviors in the contraction*

When a flow passes a nozzle, the increase in mean stream can smooth out spatial irregularities contained in the flow. For any finite-length contraction, there exist two adverse pressure gradients, where the flow might possible separate (Morel, 1975). As the contraction is designed in a way that flow separation is avoided, the majority of the flow should be well described by inviscid analyses (Fang, 1997). Further examinations on the cross-sectional velocity profiles release that the nozzle flow is free of separation, which is to the expectation of the original design.

It is generally known that due to the effects of wall curvature, the velocity profile at any section within the nozzle is always non-uniform. The numerical results in Figures 3a show that higher velocities occur in the core region at the three upstream sections. On the other hand, the trend of the sectional velocity variations becomes opposite at the nozzle exit. Despite of the inconsistency between the predicted and measured velocity profiles in Figures 3, a similar trend is found from the results of measurements (Figures 3b).

The streamwise variation of the cross-sectional velocity non-uniformity shown in Figure 4 is also interesting. Although the inlet velocity profile (at x/L=0) is considered rather uniform (U_n is about 5%), the numerical prediction shows that the value of velocity non-uniformity increases significantly at the beginning and reaches a peak value of about 80% as x/L is equal to 0.43. Further downstream, the value of U_n decays sharply until x/L equals 0.7. The reason why there is a second peak (as x/L is about 0.8) must be related to the change of local wall curvature. More extensive studies would be required in the future to explore the nozzle flow behavior in detail. Nevertheless, the result in Figure 4 indicates that without the consideration of the near-wall regions the experimental data of U_n are consistent with those were predicted.

4.3 *Pressure distributions along the wall centerlines*

By examining the measured pressure distributions along the wall centerlines (Figure 5), it is found that the difference is insignificant, implying that the flow behaviors near the three centerlines are rather similar. This tendency can also be detected from the results in Figures 3b.

Due to the existence of friction, the actual Bernoulli sum along the streamlines decreases. If the wall centerlines can be treated as streamlines of the mean flow, experimental results of the pressure distributions show that the head losses in all cases are about 5.8 N/m^2, approximately 4.3% of the total head.

To have a better comparison between the numerical and experimental results of the centerline pressure distributions, the selection of the normalized pressure C_p, defined by Equation (5), is considered adequate to interpret both sets of data in a more consistent manner. By using this parameter, Figure 6 shows that the measurement results are in good agreement with the numerical one.

4.4 *On the design method*

The design of a wind tunnel system involves a tremendous amount of work due to its complexity. At the stage of the preliminary design, since it is practically not possible to analyze the flow in the entire system, it is generally convenient to divide the system into several segments so as to proceed flow analyses and the design on each individual component. The basic concept of the contraction design method (Fang 1997) as well as that in the present numerical study adopts this approach.

Compared to the results of flow measurements in the nozzle, it has been found that the numerical simulation suffers from certain uncertainties, such as the inconsistency of the actual and numerical flow conditions at the nozzle inlet. Qualitatively, however, the overall flow behavior in the nozzle appears well predicted except in the near-wall region. On the other hand, if the friction loss of the contraction can be accurately estimated, the pressure distributions along the contracting wall can then be achieved based on the numerical results. These would allow for a more extensive check on the possibility of the occurrence of flow separation.

5 CONCLUSIONS

Flow characteristics within a square-to-square contractions were investigated numerically and experimentally so as to gain additional insight into the contraction design. Additionally, by comparing the measurement and the corresponding numerical results of flow behavior in a contraction prototype, the applicability of the nozzle design method proposed by Fang (1997) is assessed.

6 ACKNOWLEDGEMENTS

The study was funded by a grant (NSC 88-2611-E-005-005) from the National Science Council. The computing time was provided by the National Center for High-Performance Computing in Taiwan.

7 REFENENCES

Fang, F. 1997. A design method for contractions with square end sections. *ASME Journal of Fluids Engineering*:454-458.

Morel, T. 1975. Comprehensive design of axisymmetric wind tunnel contractions. *ASME Journal of Fluids Engineering*:225-233.

Morel, T. 1977. Design of two-dimensional wind tunnel contractions. *ASME Journal of Fluids Engineering*:371-378.

Song, C.C.S. & Yuan, M. 1988. A weakly compressible flow model and rapid convergence methods. *ASME Journal of Fluids Engineering*:441-455.

Stratford, B.S. 1959. The prediction of separation of the turbulent boundary layer. *Journal of Fluid Mechanics* 5:1-16.

Wind Engineering into the 21st Century, Larsen, Larose & Livesey (eds) © *1999 Balkema, Rotterdam, ISBN 90 5809 059 0*

Intermittency and coherent structure detection in long time series

X.L.Gilliam & D.A.Smith
Department of Civil Engineering, Texas Tech University, Lubbock, Tex., USA
J.P.Dunyak & J.Kwan
Department of Mathematics and Statistics, Texas Tech University, Lubbock, Tex., USA

ABSTRACT: In this paper, we introduce two statistical techniques to detect coherent structures in long time series: an intermittency rate estimator, and an event counting technique. The intermittency rate estimator calculates the fraction of time occupied by coherent structures, in contrast to incoherent noise. The event counting approach estimates the number of discrete structures that occurred. In both cases, the wavelet transform and modifications of the coherent structure detector (CSD) are used to develop rigorous statistical approaches. The techniques are demonstrated on a seven-hour wind time series collected at the Wind Engineering Research Field Laboratory at Texas Tech.

1 INTRODUCTION

In several earlier papers, the authors developed a rigorous statistical technique, called the Coherent Structures Detector (CSD), for detecting coherent structures in random fluid flows. This technique has been used in various applications such as the detection of coherent gusts (Dunyak et al. 1998); the detection of localized structures in hurricane windfields (Gilliam et al. 1997), and so on. Unlike the standard statistical models, which generally assume that the time series are stationary and incoherent, this model focuses on the identification of intermittent phenomena and coherent structures in non-stationary fields. Two major focus points are used in CSD: that the test statistic used is the popular signal processing tool--the wavelet transform (Bienkiewicz & ham 1997, Gamage & Hagelberg 1993, Gao & Li 1993, Jordan et al. 1994), and the statistical technique is based on one of the oldest methods in non-parametric statistics--the development of a randomized reference distribution (Box et al. 1988). Let Ψ be the analyzing function. Then the wavelet transform at time b and scale a of a given signal x is defined as

$$W_x(b,a) = \frac{1}{\sqrt{a}} \int x(t)\, \Psi^* \left(\frac{t-b}{a}\right) dt .$$

By using a discrete implementation of the continuous wavelet transform, the appropriate wavelet analyzing function is chosen to allow the decomposition of the signal into a frame that reveals the features of the signal (Gilliam 1998). For example, we may use the Haar wavelet analyzing function to detect a sustained change in signal level and we may use the Morlet wavelet to look for concentration in time-frequency of signal energy. In fact, another interpretation is that the wavelet transform is a bank of matched filters between the signal and the scale a and time b version of the analyzing function. To build the randomized reference distribution, the CSD first calculates the Discrete Fourier Transform of the given signal. From this, a large number of exemplars are built by keeping the Fourier transform magnitude of the signal but randomizing the phase. Hence, each exemplar has exactly the same power spectrum as the signal but is known to be incoherent. Each exemplar is then wavelet transformed to provide incoherent transform exemplars. At each scale order statistics are accumulated from the wavelet-transformed exemplars. These order statistics are then used to provide our test criteria-the thresholds. Furthermore, the CSD chooses the thresholds so that, for each scale, the probability of a false alarm anywhere in the time series is small (Dunyak et al. 1998).

However, in practice, many atmospheric measurements are made over long periods of time. In these cases, to avoid a single false alarm anywhere in the long time series, CSD thresholds are pushed so high that many detections are missed. Instead, for long time series, we should seek a large number of correct detections interspersed with an occasional false alarm. Thus, we develop two new approaches that are appropriate for long time series: coherent intermittency rate estimation, and coherent event counting.

The coherent intermittency rate is the fraction of the flow record containing coherent structures (Kwan et al. 1999). Our new statistical technique controls the false alarm probability for each scale and time separately, thereby controlling the false alarm intermittency rate. The statistical test allows two conclusions: that a time series contains intermittent structures at a specific scale, and that the structures are localized to specific portions of the waveform.

In contrast to intermittency rate, event counting identifies discrete, localized coherent structures (Dunyak et al., in press). In the CSD, the probability of a single false alarm anywhere in the time series is controlled. For event counting, the false alarm rate, or number of false alarms per unit time, is controlled instead. Both of the new detection techniques, as well as the CSD, rely on the fundamental distinction between phase relationships in incoherent energy and coherent structures. The techniques are demonstrated on measurements from Texas Tech's Wind Engineering Research Field Laboratory.

2 DEFINITIONS AND HYPOTHESIS

We begin this section with the definition of an incoherent time-series (Dunyak et al. 1998, Gilliam 1998):

DEFINITION 2.1: Consider a discretely sampled real-valued time-series $x=(x_1, x_2,\ldots, x_m)$.Using the notation $\hat{x}$ for the discrete Fourier transform of x, we can write for each frequency component

$\hat{x} = A_i e^{j\theta}$, for $i = 1,2,\ldots,m/2$.

Following standard practice, we call a time-series incoherent if the θ_i are uniform $[0, 2\pi]$ and independent of each other and the A_i. Note that an exemplar is a time-series that has the same spectrum as x but is known to be incoherent.

DEFINITION 2.2: An exemplar of a real signal $x^e(t)$ has Fourier coefficients $\hat{x}_l^e = A_l\, e^{j\theta_l^e}$, where θ_l^e for $l = 1,2,\ldots,N/2$ are uniform $[0, 2\pi]$ and independent of A_l.

Let x be the time-series. Then our test hypothesis is as follows:

H_0: x is incoherent.
H_1: x contains coherent structures.

We use the wavelet transform $W_x(b,a)$ as the test statistics to compare with the scale-dependant threshold $th(a)$ (which is generated from the exemplars) to determine the test hypothesis. Indeed, we choose H_0, if the test statistic is less than the threshold, and we choose H_1 if the test statistic is greater than the threshold. Hence, based on common practice in the signal detection literature (McDonough & Whalen 1995), we define the probability of false alarm (the size of the test) for each time and scale. For example, for the intermittency test that follows,

$$P_{FA} = P(|W_x(b,a)| > th(a) \mid H_0),$$

and the probability of detection is given by

$$P_D = P(|W_x(b,a)| > th(a) \mid H_1) = 1 - \beta .$$

Here β is the type II error of the test.

3 THE INTERMITTENCY RATE

Now we are ready to introduce our first statistical technique for long time series. Our main task is to define intermittency rate calculated from x while controlling the error rate for the entire window. To do so, let us first define the intermittent rate. Using the definition discussed by (Gamage & Hagelberg 1994, 1997), the intermittency rate is the fraction of the time-series containing structures. Hence,

Intermittency rate=(support of structures)/(total record length).

Of course, the intermittency rate is dependent on the type of structures we seek. Then

$$r(a) = \frac{1}{T}\int_0^T I_{|W_x(b,a)|>th(a)}\, db, \tag{1}$$

where $r(a)$ is the intermittent rate; T is the time interval represented by the signal; $W_x(b,a)$ is the wavelet transform of the time-series; $I_{(.)}$ is the indicator function; and $th(a)$ is the threshold used to detect the intermittent coherent structures. Note that a discrete implementation of the wavelet transform is used in practice for the discretely measured time-series. Thus, from Equation (1), the expected value of the intermittency rate is given by:

$$E[(r(a)] = \frac{1}{T}\int_0^T P\{|W_x(b,a)| > th(a)\}\, db. \tag{2}$$

Based on Equation (2), we control the intermittency rate $r(a)$ by controlling the probability

$P\{|W_x(b,a)| > th(a)\}$.

We do this through the process of building our test criteria--the threshold. To build our threshold $th(a)$, we first build a large number of exemplars. As we mention in the introduction section, this is done through the procedure of first calculating the Discrete Fourier Transform of the time-series, then keeping the Fourier transform magnitude but randomizing the phase. Hence, if x is our time series, each exemplar x_i^e has exactly the same power spectrum as the given time-series, but is known to be incoherent. Now, each exemplar is wavelet transformed to provide the wavelet-transformed exemplar $W_i^e(b,a)$. We then use order statistics from $W_i^e(b,a)$ to estimate the threshold $th(a)$. Thus, for each scale a, the threshold $th(a)$ is defined as

$$th(a) = th(a,\gamma,n,N)$$

$$= \{\sup t : (\frac{1}{n}\frac{1}{N}\sum_{i=1}^{n}\sum_{b=1}^{N} I_{|W_i^e(b,a)|>t}) > \gamma\}. \quad (3)$$

Here x is the signal with length N, γ is the probability of the false alarm, n is the number of exemplars, a is the scale, and $th(a)$ is the threshold. With Equation (3), we can establish the performance of our estimation under H_0 as follows:

$$\lim_{n\to\infty} E(r(a) \mid H_0) = \gamma,$$

where $r(a)$ is the intermittency rate, and n is the number of exemplars (Kwan et al. 1998).

Now, we are ready to draw our statistical conclusions. In practices, we conclude that coherent structures are present if $r(a) >> \gamma$. The reason this technique works is that under H_1, the process of making exemplars spreads the coherent energy in the signal over the entire long time-series.

4 COUNTING DISCRETE EVENTS

In this section, we consider the coherent event counting technique. As in the previous section, our goal is to define the false alarm rate and describe a method of calculating the number of false alarms in a long time series. Thus, we first introduce the technique of counting discrete events. Note that there is no unique definition or approach for counting discrete events, but the technique we use here is a useful interpretation for the transient phenomena (coherent events) we seek in wind signals. As we introduced in (Dunyak et al. in press), we consider a time series of length N. The entire wind flow signal x is divided into $\frac{N}{L} = M$ blocks of equal length L (assume N/L is integer). Hence, the wavelet transform W is split up into $\frac{N}{L} = M$ blocks,

$$W_i = \{W_{i,block_m}(b,a)\}_{m=1}^{M} = \{W_{i,m}(b,a)\}_{m=1}^{M}.$$

Now, we are ready to define the event rate:

DEFINITION 4.1: Consider a fixed scale a and threshold th. Let N be the length of time series x with sample spacing Δt. Let W_m be the wavelet coefficient of a particular block. Then, we count an event in a block m if $I_{\max_b |W_m(b,a)|>th}$. Moreover, for the entire time series, the total number of events at scale a for threshold th is defined as

$$K(a) = \sum_{m=1}^{M} I_{\max_b |W_m(b,a)|>th}.$$

Hence, the event rate is given by

$$\gamma(a) = \frac{K(a)}{N\Delta t},$$

where $N\Delta t$ is the time interval represented by the waveform x.

The length L of the blocks plays an important role in counting false alarms. Note that there are many ways to choose the length of the block L. In this paper, we consider the follow conditions as critical:

A. The expected number of events in each block is low so that two events occurring in one block is rare; these two events would be counted as one.

B. L is long enough that a single event rarely crosses over into adjacent blocks. This single event would be counted as two.

Note that these two conditions are opposing, so some middle ground must be found for the choice of L. See (Dunyak et. al. in press) for more details.

We now want to compute the test criteria--the threshold. Again, using the point of view in Definitions 2.1 and 2.2, we build our incoherent exemplars, from which we build our threshold to control the false alarm rate γ at each scale a. Thus each exemplar x_i^e is incoherent but has exactly the same power spectrum as the original measurement. A wavelet transform is applied to each exemplar, yielding $W_i^e(b,a)$. (The transform can be applied in the frequency domain.) Each transformed exemplar is then split into $\frac{N}{L}$ blocks, and order statistics of the $W_{block_m}^e(b,a)$ are accumulated. And each block of length L with a false alarm rate γ will have an expected number of false alarms $\gamma L\Delta t$. For example, in building the test, we generally choose L so that this number is about 0.05. Then the corresponding probability of false alarm α for the block is chosen as $\alpha = \gamma L\Delta t$.

After choosing a false alarm probability α for each block, we now calculate the threshold. For a particular false alarm at a scale a, we define the threshold resulting from n exemplars, each split into M blocks, as

$$th(a, L, n, \gamma) =$$

$$\sup t : \left(\frac{1}{n}\frac{1}{M}\sum_{i=1}^{n}\sum_{m=1}^{M} I_{\max_b |W^e_{i,m}(b,a)|>t} > \alpha\right).$$

At each scale a, this threshold th is used to count events in the original wavelet transform of signal x. More importantly, this threshold is used to display which portions of the original wavelet transform are statistically significant by applying a straightforward thresholding technique

$$|W_x(b,a)| > th(a, L, n, \gamma).$$

Each area of excitation in $W_x(b,a)$ may be assigned a critical value, to act as a p-value from which we are able to indicate statistical significance.

The following expression explains why this threshold $th(a, L, n, \gamma)$ is the correct choice of examining $W_x(b,a)$. Consider a time series x of length N and sampling interval Δt. Consider a desired false alarm rate γ. Then, for each scale a, and n, the number of exemplars, the total number of false alarms $K(th(a, L, n, \gamma))$ is given by

$$\lim_{n\to\infty} \frac{E\left(K(th(a, L, n, \gamma)) \mid H_0\right)}{N\Delta t} = \gamma,$$

see (Dunyak et al. in press) for a detailed proof.

5 ANALYZING WIND DATA

The following example demonstrates both techniques, using seven hours of wind speed data collected at 50-meter height on the meteorological tower at Texas Tech's Wind Engineering Research Field Laboratory. The data was collected during the summer of 1998. Since we are interested in regions of concentrated turbulence in the signal in both techniques, we use the Morlet wavelet as our analyzing function:

$$\Psi(t) = \exp(jct - t^2/2),$$

where $c = 5.4$. In addition, scales from .5 to 2.7 seconds are considered. These scales corresponded to center frequencies of .3 to 1.6 Hz in the scaled Morlet analyzing function.

Figure 1 shows the seven-hour-long wind speed waveform. This was measured during winds created by a low-pressure system. (Our research using wavelets and the Coherent Structure Detector has indicated differences in coherent structures in the boundary layer due to different mesoscale phenomena such as thunderstorm outflows, deep lows, and surface heating-induced mixing.) We now apply our algorithm to the time series. Unfortunately, it is impossible to effectively display the detailed images for the entire seven hours of data. Hence, to illustrate our statistical test, we zoomed in on an interesting portion of the signal, shown in Figure 2(a). For the first example (the intermittent estimation technique), we use a false alarm intermittency rate of $\gamma = 0.001$. Figure 2(b) shows the wavelet transform of Figure 2(a) and Figure 2(c) shows the thresholded-wavelet transform. The information that is below the intermittent CSD threshold had been zeroed out, and only the statistically relevant information remains. Figure 3(a) is the estimated intermittency rate $r(a)$ at each scale, in which case $r(a) >> \gamma$. The estimated intermittency rate is much higher than expected due to false alarms alone, and we have strong statistical evidence of intermittent turbulence.

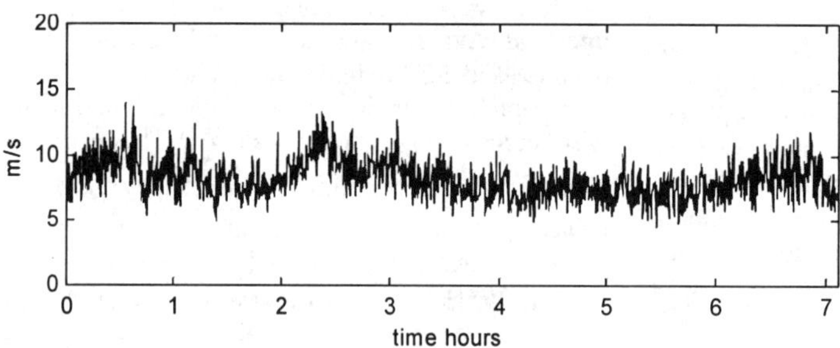

Figure 1. Seven-hour wind time series

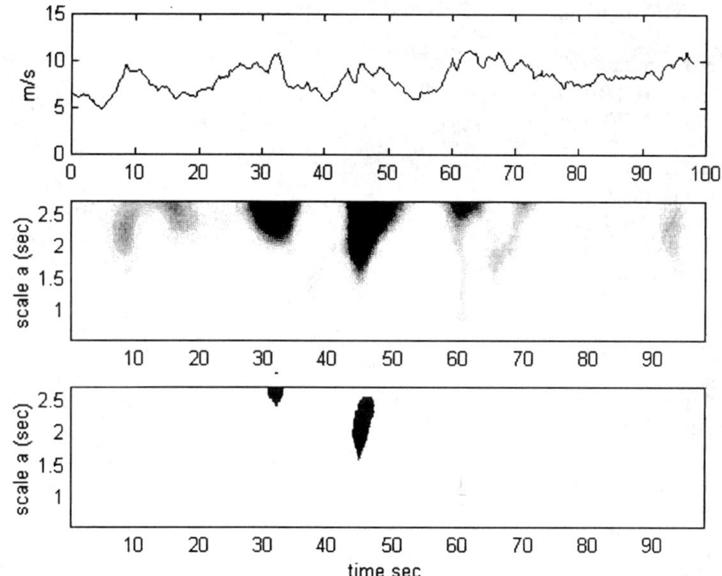

Figure 2. a) Windspeed; b) wavelet transform; c) threshold transform

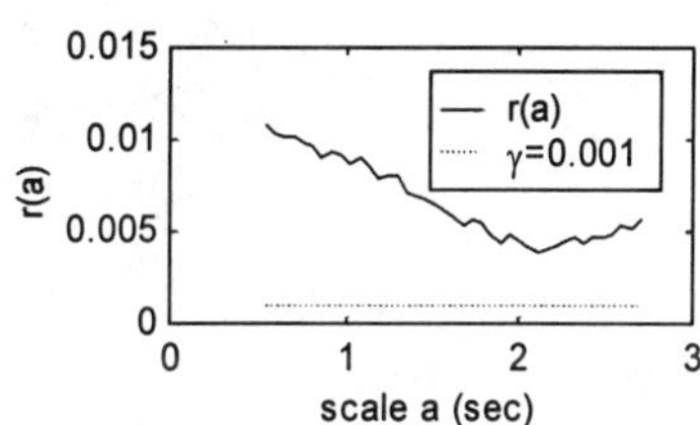

Figure 3(a). The intermittency rate

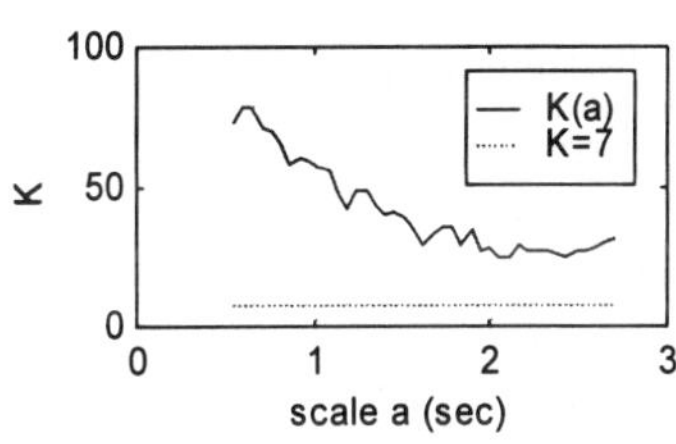

Figure 3(b). The even detections

A similar approach is applied to the discrete event counting approach. This analysis, using a false alarm event rate of one per hour, results in the total number of detections $K(a)$ at each scale shown in Figure 3(b). Note that the total numbers of detections at each scale far exceed the seven expected false alarms in seven hours. See (Dunyak et. al. in press) for details on building a rigorous statistical test from these test statistics. In Figure 3, both intermittency rate estimation and event counting give similar results. This is to be expected in many situations.

6 CONCLUDING REMARKS

In this paper, we enhance the applicability of the coherent structure detector by developing techniques for long time series. The first technique, intermittency rate estimations, provides a measure of the fraction of time a coherent structure is present. The second, event counting, counts coherent structures as discrete events. Both give similar overall results but provide different information about the structures. The two may be combined to develop some ideas about the magnitude, frequency, and characteristic lengths of these localized phenomena.

REFERENCES

Bienkiewicz, B. & H.Ham 1997. Wavelet study of approach-wind velocity and building pressure. *J. Wind Engineering and Industrial Aerodynamics* 69-71:671-683.

Box, G., W.G.Hunter & J.S.Hunter 1988. *Statistics for experimenters, an introduction to design, data analysis, and model building*. John Wiley and Sons, New York.

Dunyak, J., X.Gilliam, R.Peterson & D.Smith 1998. Coherent Gust Detection by Wavelet Transform. *J. Wind Engineering and Industrial Aerodynamics* 77-78: 467-478.

Dunyak, J, X.Gilliam, R.Peterson & D. Smith (in press). Detection of coherent structures in long time series using wavelet transforms. To appear in *Proceedings of Mathematical Theory of Networks and Systems*. July 6-10, 1998, Padova, Italy.

Gao, L. & B. Li 1993. Wavelet analysis of coherent structures at the atmospheric-forest interface. *J. Appl. Meteor*. 32:1717-1724.

Gamage, N. & C.Hagelberg 1993. Detection of microfronts and associated coherent events using localized transforms. *J. Atmos. Sci*. 50-5:750-756.

Gilliam, X. 1998. *Statistical techniques for identification of coherent structures by wavelet analysis*. Ph. D thesis, Texas Tech University, Lubbock, Texas.

Gilliam, X., J.Dunyak, R.Peterson, & D.Smith 1997. Identification of coherent structure by wavelet analysis. *Proceedings of Twelfth Conference on Boundary Layers and Turbulence*, 188-189.

Hagelberg C. & N.Gamage 1994. Applications of structure preserving wavelet decompositions to intermittent turbulence: a case study. *Wavelet in Geophysics*, Academic Press, 45-80.

Hagelberg C. & N.Gamage 1997. Structure-Preserving Wavelet Decompositions of Intermittent Turbulence. *Boundary-Layer Meteorology* 69-71:647-655.

Jordan, D., M.Hajj & H.Tieleman 1994. Wavelet analysis of the relation between atmospheric wind and pressure fluctuations on a low-rise building. *J. Wind Engineering and Industrial Aerodynamics* 70:217-246.

Kumar P. & E. Foufoula-Georgiou 1994. Wavelet Analysis in Geophysics: An introduction. *Wavelet in Geophysics*, Academic Press, 1-44.

Kwan, J., J.Dunyak & X.Gilliam 1999. Estimating intermittency rate using wavelet. *Proceedings of the 13TH Symposium on Boundary Layers and Turbulence*, Dallas, Texas, 468-471.

McDonough, R. & A.Whalen 1995. *Detection of signals in noise*, Academic Press, San Diego, California.

Ross, R. 1996. *Stochastic Processes*, John Wiley and Sons, New York.

Wind Engineering into the 21st Century, Larsen, Larose & Livesey (eds) © 1999 Balkema, Rotterdam, ISBN 90 5809 059 0

History of the aeolian wind tunnel in Denmark

J. D. Iversen
Department of Aerospace Engineering and Engineering Mechanics, Iowa State University, Iowa, USA
K. R. Rasmussen
Institute of Geology, Aarhus University, Denmark

ABSTRACT: The Aeolian Wind Tunnel is here defined as a boundary layer wind tunnel which is used as an aid in the study of the physics of particle transport or to model particle transport past boundary layer obstructions at small scale. The first such experiments in Denmark were carried out by Civilingeniør Martin Jensen in the late 1930's while he was an assistant to Prof. Chr. Nøkkentved in Copenhagen (Laboratoriet for Bygningsstatik). The study of aeolian geomorphology began at the Institute of Geology at Aarhus University in response to a damaging wind storm in 1969. A wind tunnel was built, modeled after the pioneering wind tunnel of R.A. Bagnold in England. A large number of experiments have been conducted in this wind tunnel, as well as a sloping bed addition, and some excellent pioneering work in the areas of particle trajectory, particle size distribution and slope effects have contributed to the store of knowledge in aeolian physics.

1 INTRODUCTION

Wind tunnel investigations of aeolian phenomena have been carried out in Denmark for at least 60 years. Christian Nøkkentved and Martin Jensen began conducting research on snow fences about 60 years ago in a boundary layer wind tunnel.

For about the last 20 years the aeolian wind tunnel at Aarhus University has been in existence. In that time, the contributions to the knowledge of aeolian physics by the researchers at that university have truly been outstanding.

This paper describes the facilities in which the aforementioned researches were conducted and summarizes the results of the research.

2 EARLY WORK WITH SNOW FENCES, SMOKE PLUMES AND DUST TRANSPORT

Christian Nøkkentved (1940) and Martin Jensen (1959) conducted research on the effect of porous screens (snow fences) on the formation of model snow drifts. Their experiments were performed in a wind tunnel with a cross-section of 60 cm x 60 cm, and the model two-dimensional snow fences were 5 cm in height. Their results were among the first to illustrate intermediate drift formations prior to the saturated equilibrium drift corresponding to a "full" snow fence (Figure 1). The modeling particles were beechwood sawdust of specific gravity 0.23. They noted that the threshold wind speed was about 5 m/s and that general drift shape downwind of the snow fence was dependent upon time and independent of wind speed up to about 15 m/s. Jensen and Nøkkentved also measured drift shapes in the field, noting that the shapes were very similar to the intermediate drift shapes formed in the wind tunnel and concluding that none of the drifts measured in the field had reached saturation.

Later, Martin Jensen (1959) performed some pioneering work with the distribution of smoke particles from chimneys, measuring the effect of free-stream wind speed and turbulence and the effect of effluent speed.

Another series of experiments involving the turbulent diffusional transport of coal dust particles due to wind entrainment was carried out some years ago in the 114 cm square open circuit wind tunnel at the Marine Research Laboratory in Lyngby, Denmark (Iversen & V. Jensen 1981).

3 THE AARHUS UNIVERSITY AEOLIAN WIND TUNNEL

The aeolian wind tunnel at the Institute of Geology at Aarhus University was constructed in 1978 and rebuilt in 1986 (Møller 1985, Rasmussen & Mikkelsen 1988, 1991). Aeolian studies became important at the University after strong wind damages occurred in 1969. The wind tunnel (Figure 2) was constructed after Bagnold's design (1941) and important research has been conducted at the Aarhus laboratory since by members of the Statistics and Physics Departments as well as the Institute of Geology. During the last 20 years, the reputation of the sand transport research group has grown such that many foreign scientists from e.g., Scotland, Israel, United States and other countries have spent time at the laboratory.

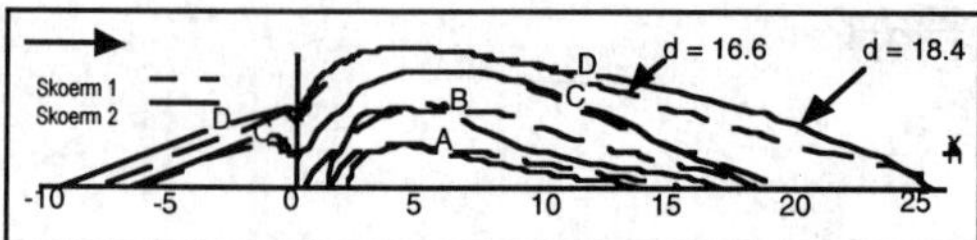

Figure 1. Simulated snowdrift profiles created in a wind tunnel (Nøkkentved 1940, Jensen 1959)

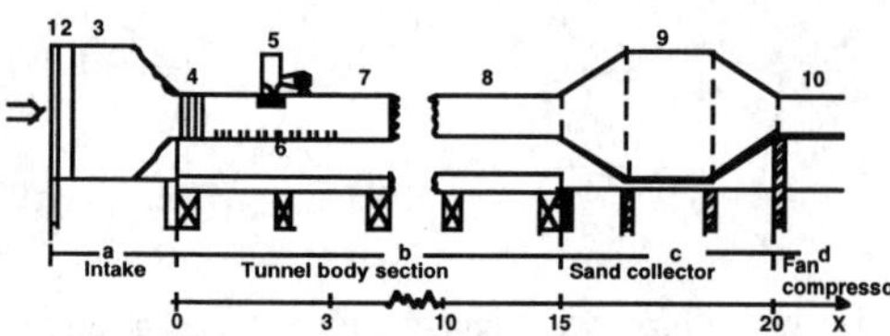

Figure 2. Elevation of the Aarhus wind tunnel (Møller 1985, Rasmussen & Mikkelsen 1988, 1991)

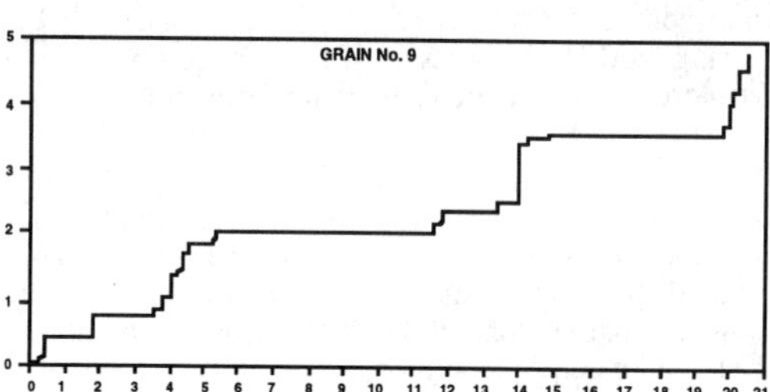

Figure 3. Recorded movement of a radioactive sand grain (Barndorff-Nielsen et al. 1985)

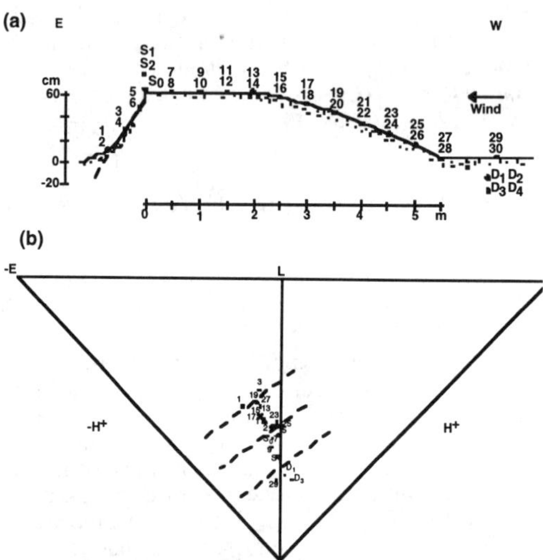

Figure 4. Dune cross-section & hyperbolic shape triangle for particle-size distribution (Barndorff-Nielsen & Christiansen 1985)

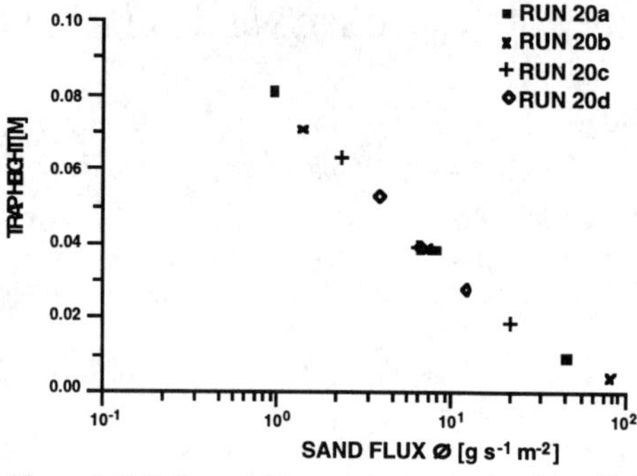

Figure 5. Saltation sand transport rate as a function of height above the surface Rasmussen & Mikkelsen 1988)

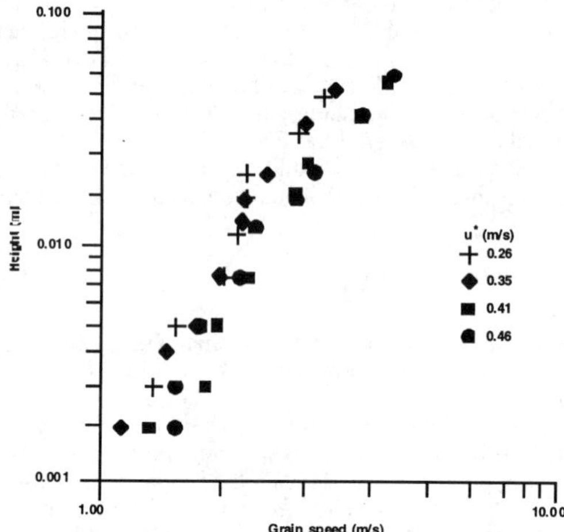

Figure 6. Variation of horizontal grain speed with friction speed within a saltating layer (Rasmussen & Iversen 1996)

4 ADVANCES IN UNDERSTANDING OF SAND TRANSPORT

One of the more interesting and useful series of experiments in the wind tunnel has been the recording of the motion of a single grain during a sand transport experiment by following the motion of a radioactive grain (Barndorff-Nielsen et al 1985). A plot of the motion of one of these grains (longitudinal position versus time) is shown in Figure 3. The results of these experiments clearly showed that smaller grains move with a greater mean speed than larger grains.

Involvement of the Aarhus statisticians has led to more efficient classification of particle grain size distributions and to better understanding of the physical processes involved in the sorting of grains by size that occurs in sand transport (Figure 4, Barndorff-Nielsen & Christiansen 1985). Members of the statistics department have also conducted numerical

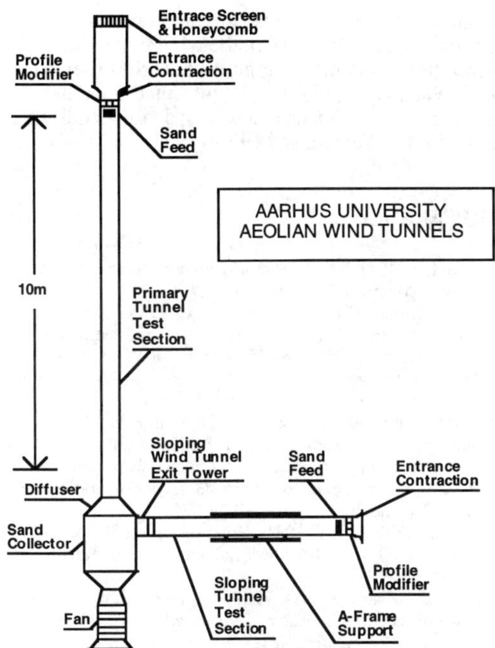

Figure 7. Plan view of the Aarhus wind tunnels, showing both the original Bagnold-type tunnel and the variable slope wind tunnel. (Rasmussen & Iversen 1993)

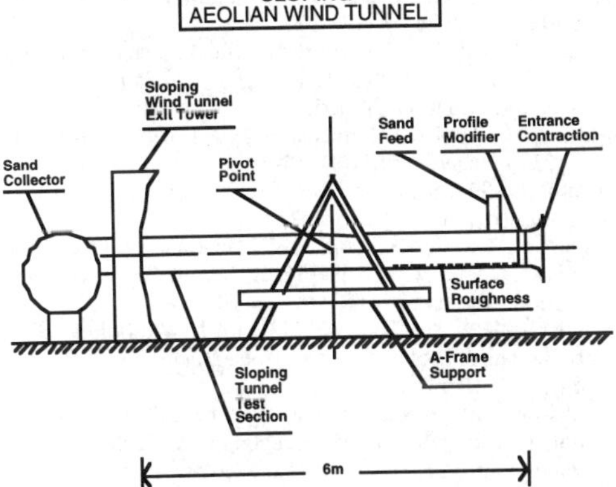

Figure 8. Elevation of the variable slope aeolian wind tunnel at Aarhus University (Rasmussen & Iversen 1993)

studies of particle motion and have significantly contributed to these important studies (Sørensen 1991).

Meanwhile, other improvements in wind tunnel instrumentation have made measurements within the saltation layer possible, significantly contributing to the body of knowledge of saltation physics (Figures 5,6). Model studies at small scale have also been conducted in the wind tunnel (Tsoar et al 1986, Iversen et al 1991).

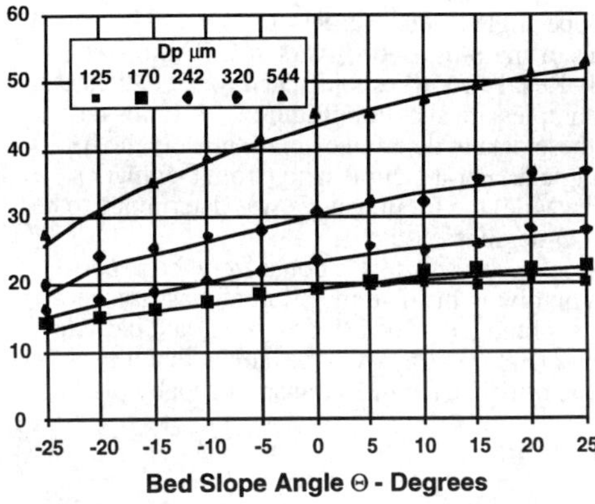

Figure 9. Threshold friction speed versus bed slope angle for 5 sand samples (Iversen & Rasmussen 1996)

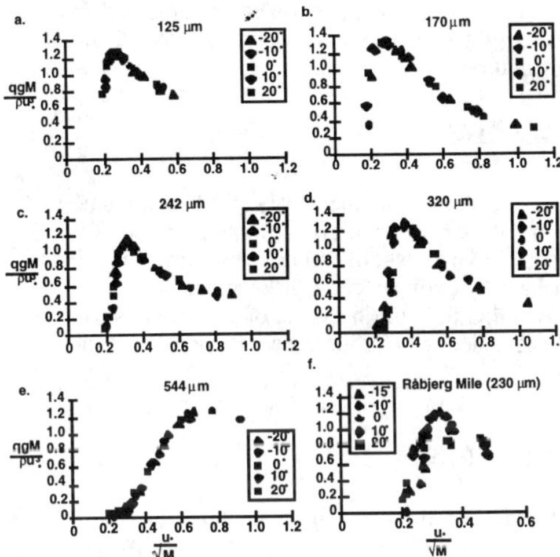

Figure 10. Mass transport of sand as a function of friction speed, mean particle size, and bed slope (Iversen & Rasmussen 1999)

5 THE SLOPING WIND TUNNEL AT AARHUS UNIVERSITY

Using the same driver and sand collector as the primary wind tunnel, a unique addition to the Aarhus laboratory was constructed in the early part of this decade (Rasmussen & Iversen 1993). As shown in the plan view schematic in Figure 7, the new wind tunnel is mounted so that the flow is perpendicular to the flow in the original wind tunnel. The new test section has the capability of rotation about the pivot point as shown in Figure 8, such that the bed slope can be varied in 5 degree increments between plus 25 degrees (uphill) and minus 25 degrees (downhill). This facility has proven to be quite useful. For the first time, it has been possible to make accurate measurements of threshold speed as a function of bed slope angle (Iversen & Rasmussen 1994) as well as rates of mass-transport (Iversen & Rasmussen 1999). Results for threshold speed for five different sand samples are shown in Figure 9. It is possible from these data to determine approximately the friction angle a. For a natural sand (from Denmark's Råbjerg Mile) the friction angle was determined to be about 35 degrees.

Several conclusions have come from the experiments involving the determination of mass transport rates as a function of bed slope, speed, and particle diameter (Figure 10). We have shown that it is possible to normalize the mass transport data by plotting:

$$\frac{qgM}{\rho u_*^3} \text{ vs. } \frac{u_*^2}{\sqrt{M}}$$

where $M = \cos\theta + \sin\theta / \tan\alpha$; q = mass transport per unit width per unit time; g = gravitational acceleration; θ = bed slope angle; ρ = air density; and u_* = friction speed = (shear stress/ air density)$^{1/2}$. Owen (1964) proposed that aerodynamic bed roughness height z_0 in a saltating boundary layer is not a constant but is:

$$z_0 = C\frac{u_*^2}{2g}$$

We have shown that the coefficient C is not a constant but increases rapidly with the increase of wind speed above threshold and reaches a limiting value for large friction speed. Furthermore, we have shown that the limiting value of C decreases as average particle diameter increases (Rasmussen et al 1996).

6 CONCLUSIONS

A number of important fundamental improvements to the understanding of the transport of snow, dust, smoke and sand have been advanced by researchers in Denmark over the past 60 years with the aid of the boundary layer wind tunnel. The existence of the Aarhus University wind tunnel laboratory has provided a valuable tool for understanding the physics of wind-blown sand transport (an important problem on much of the Jutland peninsula in Denmark) and will no doubt continue to play an important role in the future. Recently, work in the wind tunnel has utilized laser technology (Figure 11, T. Sørensen 1998) to restudy and further the information on the shelter effect phenomena that Martin Jensen studied many years earlier (Jensen 1954).

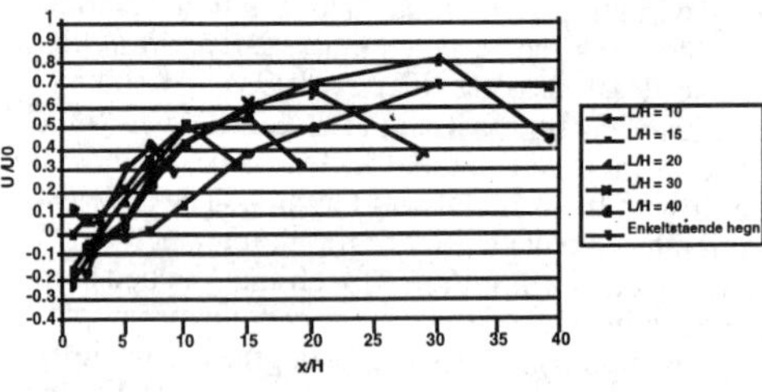

Figure 11. Laser-velocimeter wind speed profiles downwind of a porous fence (Sørensen, T. 1998)

7 REFERENCES

Bagnold, R.A. 1941. The Physics of Blown Sand and Desert Dunes. Methuen, London.

Barndorff-Nielsen, O.E. & Christiansen, C. 1985. The hyperbolic shape triangle and classification of sand sediments. *Proc. intern. workshop on the Physics of Blown Sand, Memoirs #8, Dept. Theoretical Statistics,* Aarhus University, 3: 649-676.

Barndorff-Nielsen, O.E., Jensen, J.L., Nielsen, H.L., Rasmussen, K.R., & Sørensen, M. 1985. Wind tunnel tracer studies of grain progress. *Proc. intern. workshop on the Physics of Blown Sand, Memoirs #8, Dept. Theoretical Statistics,* Aarhus University 2: 243-252.

Iversen, J.D. & Jensen, V. 1981. Wind transportation of dust from coal piles. Skibsteknisk Laboratorium SL Report 81054.

Iversen, J.D. & Rasmussen, K.R. 1994. The effect of surface slope on saltation threshold. *Sedimentology* 41: 721-728.

Iversen, J.D. & Rasmussen, K.R. 1999. The effect of surface slope on saltation mass transport rate. *Sedimentology* (in press).

Iversen , J.D., Wang, W.P., Rasmussen, K.R., Mikkelsen, H.E. & Leach, R.N. 1991. Roughness element effect on local and universal saltation transport. *Acta Mechanica Suppl.* 2: 65-75.

Jensen, M. 1954. Shelter Effect. Copenhagen.

Jensen, M. 1959. Aerodynamik i den Naturlige Vind. Teknisk Forlag, Copenhagen: 173-227.

Møller, J.T. 1985. Wind experiments. Sand transport and desertification in desert lands. El-Baz et al. (eds.), *World Scientific,* London: 348-358.

Nøkkentved, Chr. 1939. Drivedannelse ved sneskærme. *Stads og Havneingeniøren* 31(9): 85-92.

Owen, P.R. 1964. Saltation of uniform grains in air. *Journal Fluid Mechanics* 20: 225-242

Rasmussen, K.R. & Mikkelsen, H.E. 1988. Development of a boundary layer wind tunnel for aeolian studies. Geologisk Institut, Aarhus University, Geoskrifter #27.

Rasmussen, K.R. & Mikkelsen, H.E. 1991. Wind tunnel observations of aeolian transport rates. *Mechanica Acta Suppl.* 1: 135-144.

Rasmussen, K.R., Iversen, J.D. & Rautaheimo, P. 1996. Saltation and wind-flow interaction in a variable slope wind tunnel. *Geomorphology* 17: 19-28.

Rasmussen, K.R. & Iversen, J.D. 1993. A variable slope wind tunnel for testing wind-blown sand. *Proc. 7th U.S. National Conference on Wind Engineering* 2: 543-552.

Rasmussen, K.R. & Iversen J.D. 1996. Grain dynamics and wind-flow in a variable slope wind tunnel. Geologisk Institut, Aarhus University.

Sørensen, M. 1991. An analytic model of wind-blown sand transport. *Acta Mechanica Suppl.* 2: 67-81

Sørensen, T. 1998. Strømning i og Over Systemer af Læhegn, Geologisk Institut, Aarhus University.

Tsoar, H., Rasmussen, K.R., Willetts, B. & Sørensen, M. 1986. Laboratory studies of flow over dunes. *Proc. intern. workshop on the Physics of blown sand, Memoirs #8,* Dept. Theoretical Statistics, Aarhus University.

Wind Engineering into the 21st Century, Larsen, Larose & Livesey (eds) © 1999 Balkema, Rotterdam, ISBN 90 5809 059 0

VELUX A/S wind tunnel

J.O.Jensen
VELUX A/S Wind Tunnel, Østbirk, Denmark

ABSTRACT: VELUX A/S manufactures among other things roof windows. These windows are placed in the part of the building's climate screen most severely exposed to wind and rain, so the need arose within VELUX A/S to simulate weather conditions under controlled circumstances. Therefore, it was decided to develop a wind tunnel for simulation of wind and rain loads on roofs on a scale 1:1.
This abstract contains a description of the VELUX A/S Wind Tunnel and the considerations prior to the construction of this closed-circuit wind tunnel, including description of the rain system, fans, etc. Further, a sketch of the wind tunnel will be presented. In order to optimize the use of the wind tunnel, VELUX A/S is investigating if, by means of Computational Fluid Dynamics (CFD) in connection with initial tests, it will be possible to reach the verified test results even quicker. This process is divided into four phases from the simple case only including the test section to the complicated case including a two-phase flow (rain and wind).

1. INTRODUCTION

VELUX products are typically mounted in the roof, which is the part of the climate screen most exposed to wind and rain. In order to avoid damage to furniture and flooring etc. due to penetration of water, it has always been of great interest to VELUX A/S to be able to predict whether the products could meet the high requirements to tightness regardless of the local climate. Therefore, VELUX A/S decided to develop and build a wind tunnel complying with the following specifications:

2. Test to be carried out on a scale 1:1
3. Wind speed up to 40 m/s
4. Simulation of wind gusts and variable wind directions
5. Simulation of rain
6. Variable roof pitches and room for observation purposes inside the test house.

This abstract is an attempt to describe how the above specifications for a climate wind tunnel were honoured. The calculations and specific results from model tests have not been included, as the abstract would then be too comprehensive.

The final wind tunnel is illustrated in a sketch in section 7.

VELUX A/S has decided to investigate if Computational Fluid Dynamics (CFD) (STAR CD 3.10) can reduce the amount of tests in the wind tunnel. In that way VELUX A/S can perhaps - by means of CFD - relatively quickly and inexpensively identify the problems and then verify the results in the wind tunnel later (section 8).

2. SCALE 1:1

Test on a scale 1:1 has several advantages, as it is not necessary to take model figures into consideration. To a company with production activities it is, however, of greater importance that the tests are carried out with real products. This way, test pieces can be produced very close to the production lines that will be producing the items in large numbers later.

Further, the wind tunnel is also used as a quality control body, which helps ensure that the real products meet the required level of tightness of the prototypes when they are later mass-produced.

The starting point was to enable tests of windows up to 1.4 x 1.4 m. Apart from that, the windows were also to be mounted in a roof surface that opened up for the possibity to test the joint between the window and the roofing material. The size and geometry of the roof surface and the test house were therefore specified as follows:

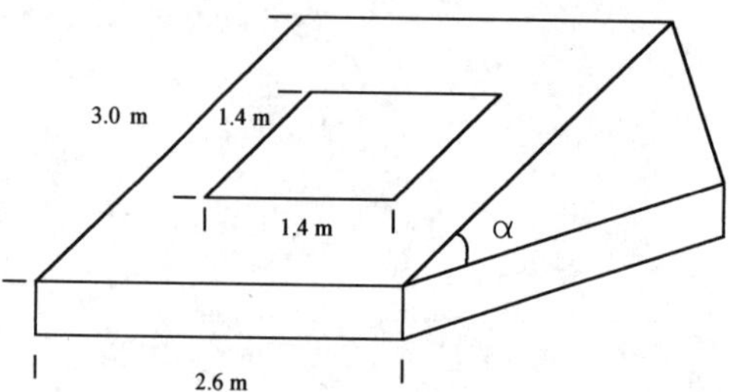

Figure 1: Template of standard test house in the VELUX A/S Wind Tunnel

The construction of the standard house makes it is possible to lift out the whole house by truck. Thus, it is relatively easy to convert the test section for special tests or tests of products from other companies

3. WIND SPEEDS UP TO 40 M/SEC.

VELUX products are sold in many parts of the world and therefore exposed to many different climates. In Japan e.g. typhoons occur with wind and rain simultaneously and wind gusts up to 35 m/sec. or more.

The request of wind speeds up to 40 m/sec. related to the size of the test pieces meant that the inlet jet had to be large, which would have resulted in huge fans. Some types of tests demanded a 4.0 x 4.0 m inlet jet, but did not require wind speeds of 40 m/sec. Therefore, the idea of series-connecting two different inlet jets (contractions) was brought about. We had not previously heard of this solution, but by means of model tests etc. we reached the conclusion that this was a tremendously good solution. The result is that the VELUX A/S Wind Tunnel now has a 4.0 m x 4.0 m (large) downstream followed by a 3.4 x 3.4 m (small) inlet jet.

This has two very obvious advantages: By means of much smaller fans we are able to reach 40 m/sec. in the small inlet jet and at the same time economize on the daily energy consumption in the wind tunnel using the small inlet jet for most assignments. The effect consumption is dependent on the airflow to the third power, so using the small instead of the large one, 62 per cent of the effect consumption is saved.

4. SIMULATION OF WIND GUSTS AND VARIABLE WIND DIRECTIONS

Simulating weather, it is important to be able to simulate wind gusts because penetration of water often occurs in connection with wind gusts.

In principle, the wind speed can be altered in the following ways:

- Opening/closing the damper
- Moving an item abreast of the flow
- Changing the rotation of the fans
- Adjusting the blade angle of the fans (pitch adjustment)

In the VELUX A/S Wind Tunnel, we have chosen to use pitch adjustment of the fans as this is a quicker and relatively simple manner of changing wind speeds. The time of acceleration for changing speed in the VELUX A/S Wind Tunnel is approx. 6 m/sec^2, a little higher at low speeds and a little lower at the highest speeds.

In nature, the wind comes from different directions. Therefore the test house is placed on a huge turning table which can be turned 3.7°/sec.

When the test house is not in use, a huge steel plate is placed in the hole in the floor and on this plate various kinds of test pieces can be installed.

5. SIMULATION OF RAIN

Simulation of rain and heavy shower was one of the most important arguments for building the VELUX A/S Wind Tunnel. Development of the rain system was almost as important as the wind tunnel itself.

It was necessary to use a fluid that has the same physical and chemical properties as normal rain. The most obvious solution would be to establish a huge reservoir for gathering rainwater. But for standard test purposes, VELUX A/S wished to simulate a rain intensity of 120 mm/hr*m^2 which corresponds to 65 l/min. To meet the requirements of the chosen rain system, the reservoir should be extremely large despite the fact that drains have been installed in the test section and in the outlet jet in order to recycle some of the water.

Gathering rainwater involves a great risk that the water would putrefy. Therefore, it was decided to use demineralized water for rain simulation. To produce demineralized water, ordinary drinking water is first sent through an ion exchange system, where the calcium ions are replaced by salt ions. Secondly, the water runs into a reversed osmosis system where it is squeezed through very small-meshed membranes under high pressure and finally 99 per cent clean water is produced. In order not to design the water treatment system to the present consumption, the system is provided with 2 x 2,000 l accumulating tanks. Approximately 50 per cent of the water is gathered, cleaned and recycled.

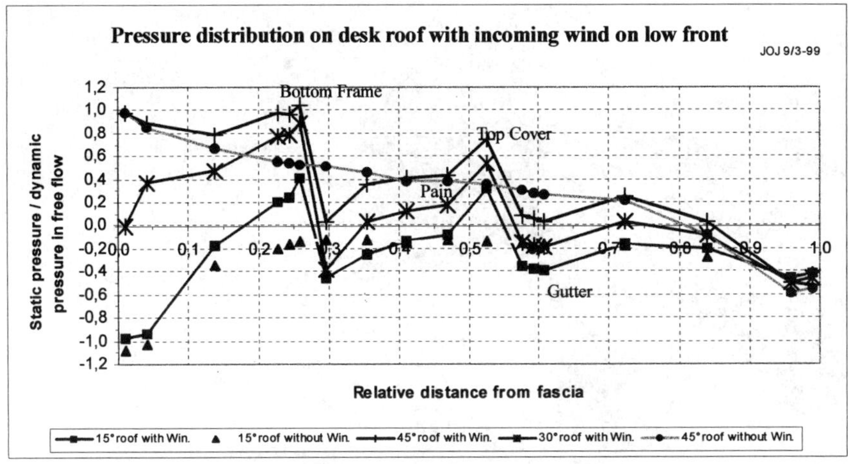

Figure 2: Pressure distribution over a VELUX A/S skylight as a function of the roof pitch.

6. VARIABLE ROOF PITCHES AND OBSERVATION FROM THE INSIDE

VELUX products are usually fitted in roof pitches from 15° to 90°. Some products are used within even narrower roof pitch areas. Thus, it was necessary to be able to change the roof pitch. This is the reason why the standard test house in figure 1 is carried out in such a way that the roof pitch can easily be changed by means of other side and back plates.

Low roof pitches are often characterized as the pitches with the highest risk of water penetration when it is raining and windy at the same time. This is, however, not always the case. On the graph below the pressure distribution on a VELUX skylight is pictured.

From figure 2 it is among other things obvious that the pressure increases when the roof surface is increased. Provided that a ventilating system is incorporated in the product, the risk of penetration of water via the ventilating system is higher in steep roof pitches.

To estimate the penetration of water, it is most efficient if a visual estimation can be made. This means that an observer must be present inside the test house during the test. As mentioned earlier, the standard test house is placed on the turntable above a 2 m big hole with connection to a cellar underneath the test section. This leaves plenty of room for observation.

Normally, the products are tested by the pressure difference between inside and outside that is generated by the wind, but in some situations a higher partial vacuum is required inside the house. This is obtained by establishing a variable suction system from the cellar underneath the test house to the outside.

7. SKETCH OF VELUX A/S WIND TUNNEL

Having thoroughly considered all the matters mentioned above, the VELUX A/S Wind Tunnel was in principle designed as shown in the sketch in figure 3.

As mentioned, this is only a sketch, which means that a number of details have been left out.

The wind tunnel is constructed as a closed-circuit wind tunnel. This has the very obvious advantage that some of the kinetic energy of the wind is recycled and the energy consumption is reduced to approx. 25 per cent compared to an open wind tunnel.

7.1 *Turbines*

For transportation of air in the wind tunnel, 4 pitch-adjusted turbines of 320 Kw each are used. At full load on all four fans, approximately 1.7 million cubic metres of air per hour is recirculated. The channel cross-section is 6.0 m x 6.0 m.

7.2 *Turning Vanes*

In the real wind tunnel, there are 13 turning vanes in each corner. The efficiency of this type of guiding plate is approx. 0.13.

7.3 *Settling Chamber*

The settling chamber consists of a honey comb and a screen.

7.4 *Inlet Jet*

As previously mentioned, the wind tunnel is run by two series-connected inlet jets, first a 4.0 x 4.0 m,

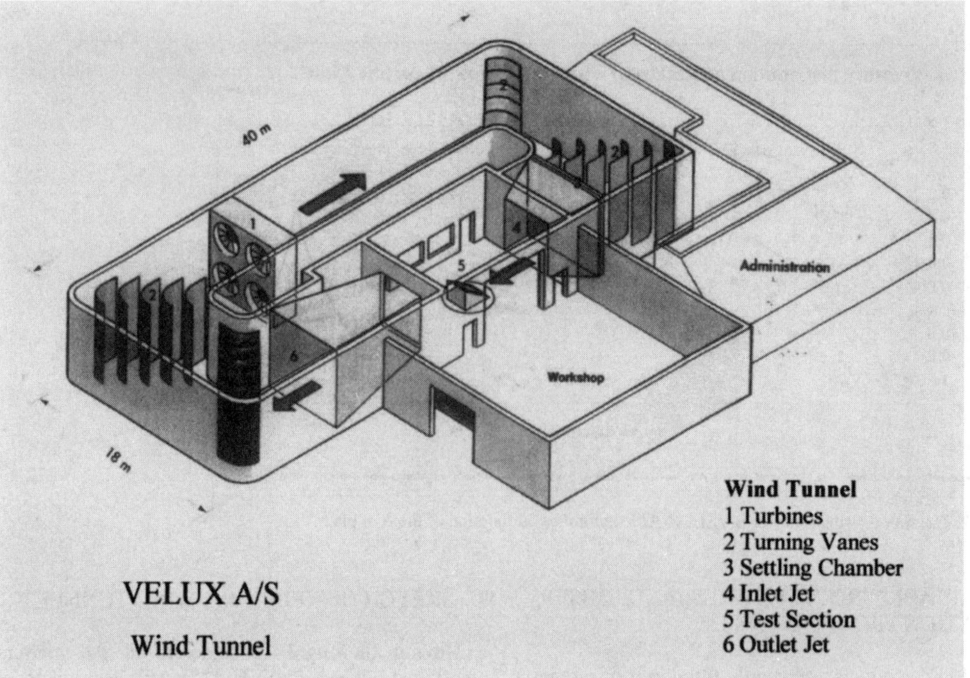

Figure 3: Sketch of the VELUX A/S Wind Tunnel

then a 3.4 x 3.4 m inlet jet (not evident from the chart). Right after the inlet jet, a rain system is placed, which appears from the picture below.

7.5 *Test Section*

The test section is constructed wider and higher than the inlet jet so that the air can move freely around the test piece and no unnatural static pressure is created. Further, the wind directions can be changed by means of the turning plate, on which the test house is placed.

We can test objects as big as 60 per cent blockage of the inlet jet.

7.6 *Outlet Jet*

Approximately 1 m inside the outlet jet, a cooling system is established, which is necessary as all the electrical energy carried to the fans is transformed to

Figure 4: Downstream picture of rain system in the VELUX A/S Wind Tunnel

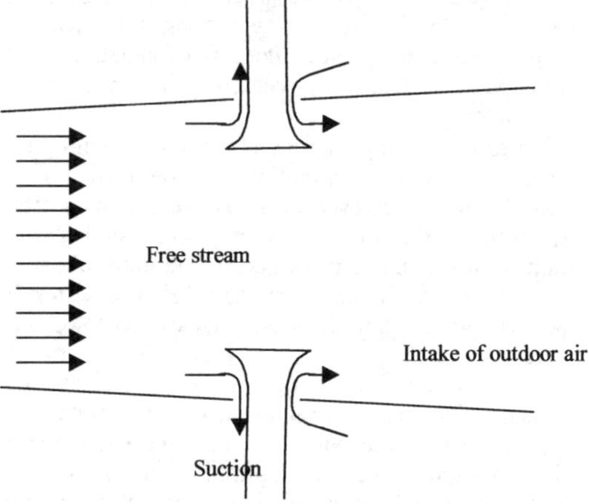

Figure 5: Sketch of the cooling system in the VELUX A/S Wind Tunnel

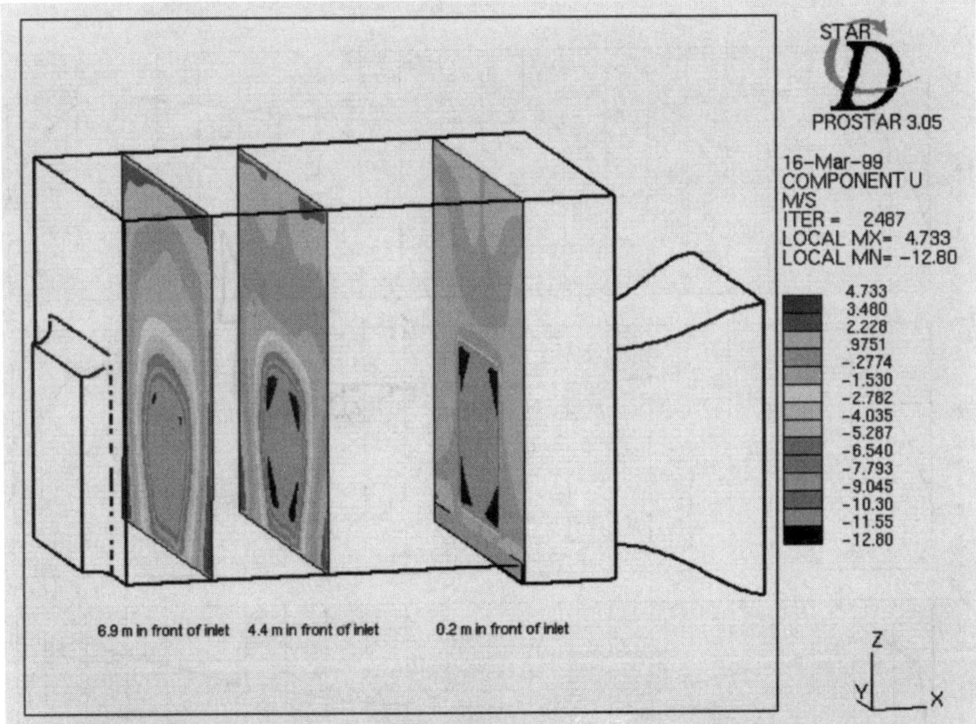

Figure 6. Model of the wind tunnel used for the CFD calculations. The three sections for the comparison with the measurements are shown. The calculated streamwise velocity is shown as greyscale at the three sections.

heat via various losses. As cooling media the outdoor air is used, so the temperature of the outdoor air is crucial to the temperature in the test section. For wind and rain tests the evaporating capacity of the water is also used in huge quantities. Via the cooling system approx. 170,000m^3/h of air can be removed.

During normal operation of the wind tunnel, the temperature can be kept low only by means of the amount of air that is automatically removed, because the inlet of the cooling system has been designed as vertical guiding shovels (see figure 5). They are placed on both sides of the outlet jet from the floor to the ceiling. Being a closed circuit a partial vacuum is built in this area of the wind tunnel. The partial vacuum is also used to take in a corresponding amount of guiding air. This operation, however, does not demand extra energy supply.

For tests with high wind speeds over long time, two suction fans can be activated.

The cooling system also functions as zero for the static pressure in the wind tunnel. It is practical that the zero is as close to the test section as possible as this is the area where the risk of interference with the surroundings is highest.

The space in the outlet jet is slowly expanding resulting in decreasing speed. The result is that the water falls to the floor in the outlet jet from where it is gathered and then returned to the rain system.

The circuit is now closed and the air continues its travel back to the fans.

8. SIMULATION BY MEANS OF COMPUTATIONAL FLUID DYNAMICS (CFD)

Tests in the VELUX A/S Wind Tunnel demand a lot of resources and they are relatively expensive. For the mounting of the products and preparations for testing require staff and so do tests and writing reports on the test results. And as the wind tunnel at the same time is running 11 hrs. every day, it was decided to investigate if simulation by means of CFD could reduce the need for initial tests in the wind tunnel.

Further, CFD is to be used to visualise flow and state numeric values of the pressure in connection with specific problems.

As mentioned in the introduction, we have divided the use of CFD in four phases:

a. Verification of flow in the empty test section

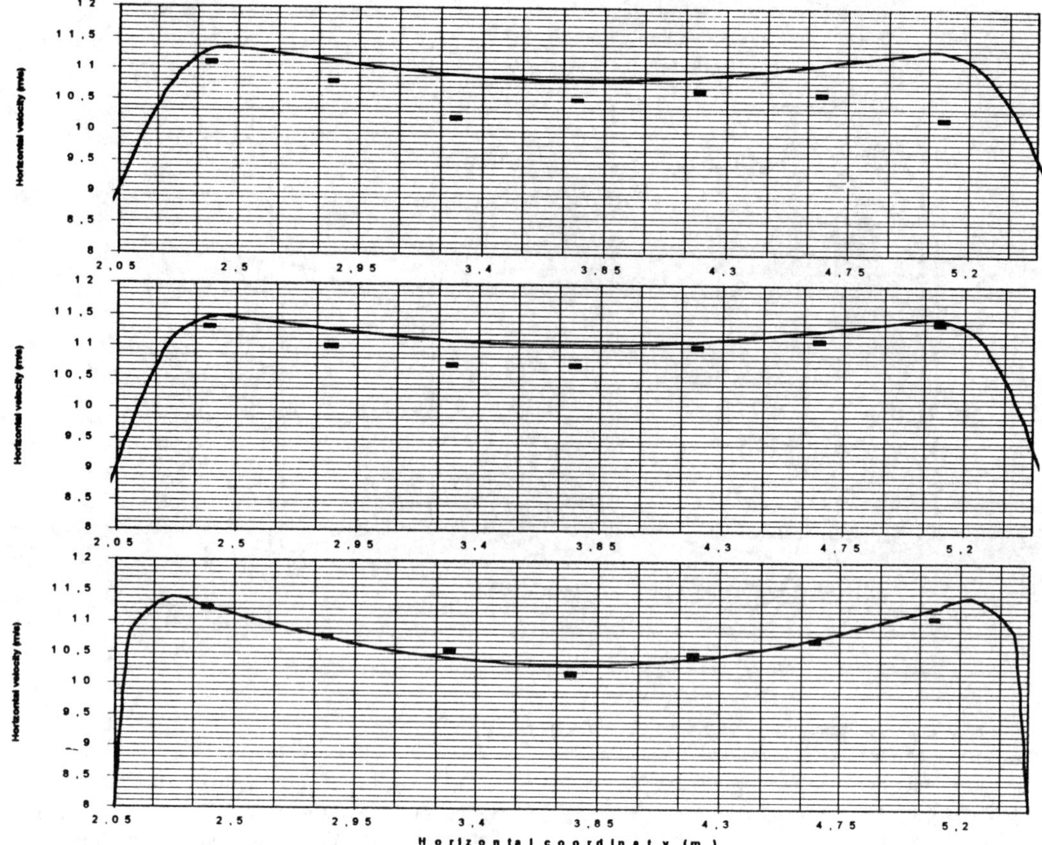

Figure 7. Comparison of measured (squares) and calculated (lines) streamwise velocities at 2.0 m above the floor. Y is the horizontal coordinate perpendicular to the main flow.

b. Verification of flow and pressure conditions on roofs with and without VELUX windows
c. Verification of flow and pressure conditions inside the window construction
d. Simulation of two-phase flow (wind + rain) at the same time in all the above situations.

re a

The precondition for valid results from the CFD simulation is to ensure that the CFD code is calculating correctly in the simplest situation. In this case the test section is empty and the air is flowing like a plume through the test section.

Figure 6 shows three cross-sections of the airflow in the test section. In these sections the speed is measured in the wind tunnel and verified by means of CFD (STAR CD 3.10). The results are from the small inlet jet (3.4 x 3.4 m):

From figure 6 it appears that the downstream velocity profile is equable, with a tendency, though, of higher velocity in the corners of the velocity profile. This phenomenon decreases downstream.

Because of the definition of origo (left corner of figure 6), the downstream speed is stated with a minus.

So measured and simulated speeds are identical, which also appears from figure 7 (page 7), where the horizontal speed 2 m above the floor of each cross-section is stated.

Correlation between measured and simulated speeds gets a little smaller the farther away from the inlet jet the comparison is made.

On figure 7 on "6.9 m downstream", one single measuring lies a bit from the CFD curve. From a symmetric reflection a measuring error has caused this deviation. The comparison of real speeds with simulated speeds is therefore considered satisfactory, and phase b is entered.

re b

If phase b is a success, VELUX A/S has a tool which makes it possible to use the results in practice. The first results with and without roof window mounted in the roof are expected to be ready mid-1999.

re c

Measuring pressure and speed inside the window construction as function of free simulation speed is a complex and difficult task, but at the same time it is useful in connection with the placing and determination of various gaskets etc. Therefore, it would be a huge advantage if VELUX A/S can use CFD in this connection.

re d

Simulation of two-phase flow (wind and rain) simultaneously is an ambitious aim, but considering the development within CFD during the last years VELUX A/S hopes to be able to make valid two-phase simulations in 2 or 3 years.

In order to reduce the amount of initial tests in the wind tunnel, the incorporation of computational fluid dynamics as a primary sorting tool in the tests has been initiated.

The four phases of the CFD project:

a. Verification of flow in the empty test section
b. Verification of flow and pressure conditions on a roof with and without VELUX windows
c. Verification of flow and pressure conditions inside the window construction
d. Simulation of two-phase flow (wind + rain) at the same time in all the above situations.

Phase a has been determined with satisfactory results and phase b is expected determined and evaluated mid-1999. The determination of phase d is not planned for the next 2 or 3 years, but VELUX A/S will be able to benefit from the CFD simulation already after phase b.

9. CONCLUSION

VELUX products comprise roof windows, covers, and various accessories for the roof windows. In order to be able to reduce the development time for these products, VELUX A/S decided to establish a climate wind tunnel, where it was possible to simulate wind and rain loads on a scale 1:1. The wind tunnel was put into operation on 1 September 1992.

In the VELUX A/S Wind Tunnel, it is possible to create wind gusts up to 40 m/sec. with a turbulence intensity of less than 1 per cent.

Wind gusts have influence on the penetration properties of the rain and therefore the possibility for VELUX A/S exists to generate wind gusts with an acceleration of 6 m/sec^2.

The rain system is constructed in such a way that the rain simulation is three-dimensional with an intensity of up to 200 mm/h*m^2. VELUX A/S uses demineralized water as rainwater.

In some cases the test objects are huge and in other cases the test is run at high speed. These requirements are not always present at the same time, so the idea of series-connecting two inlet jets (contraction) arose. Thus, the VELUX A/S Wind Tunnel is equipped with a large inlet jet (4 x 4 m) and a small inlet jet (3.4 x 3.4 m). During daily use of the wind tunnel, the 3.4 x 3.4m inlet jet is used, which reduces the energy consumption significantly compared with the big inlet jet.

A blocking degree of 60 per cent is allowed, measured as vertical projection, which means that objects of 6.9 m^2 can be tested in the small inlet jet and objects of 9.6 m^2 in the large inlet jet.

Wind Engineering into the 21st Century, Larsen, Larose & Livesey (eds)© 1999 Balkema, Rotterdam, ISBN 90 5809 059 0

Turbulence and topographic effects in simulated thunderstorm downdrafts by wind tunnel jet

C.W. Letchford & G. Illidge
Department of Civil Engineering, The University of Queensland, Brisbane, Qld., Australia

ABSTRACT: In this study, a continuous jet impinging on a wall has been used to simulate the downdraft in a thunderstorm. Velocity and turbulence characteristics have been studied at various distances from the stagnation position and over a topographic feature. Comparisons with earlier jet flow studies confirm that speedup over topography is weaker for jet flows than for boundary layer flows, but perhaps not as weak as previously observed.

1 INTRODUCTION

Thunderstorms are responsible for approximately 50% of extreme gust wind speeds in Australia outside of the tropical cyclone region [1] while in the United States it has been reported that about 1/3 of extreme gusts speeds were associated with thunderstorms [2]. Yet, they remain a meteorological phenomenon that has received little attention from Wind Engineers.

Several authors [1,3,4] have drawn the analogy between a downburst and a jet impacting on a flat surface, termed a wall jet. Bakke [5] and Poreh et.al. [6] conducted early experimental studies of wall jets aimed at characterizing the growth in jet thickness and diminution in velocity with radius and height. Proctor [7] compared results from numerical modelling with full-scale data with a degree of success. Later, Holmes et.al. [1,8] undertook physical and numerical modelling of the thunderstorm downburst phenomenon and were able to demonstrate reasonable agreement with the limited full-scale data available. More recently, Wood et.al. [9] conducted further physical and numerical modelling of wall jet flows. They developed an empirical expression for the velocity profile variation with height and distance from impact. Table 1 summarizes the main parameters of these experiments.

These investigations established the usefulness of the wall jet as a tool in developing our understanding of thunderstorm outflows. Holmes & Selvam [1,8] also examined the acceleration of flow in a wall jet as it passes over a topographic feature, in their case a single embankment. Generally it was observed that the speedup in this flow phenomenon was much less than in boundary layer flow.

Table 1. Summary of wall jet experiments

Author	Jet size	Jet speed	Radii studied	Distance to wall
	mm	m/s	mm	mm
Bakke [5]	28.4	34	143 - 303	15
Poreh et.al. [6]	25, 50, 75	52 - 113	458 - 1678	610
Holmes [8]	1150 x 850	~10	1000	1400
Wood et.al. [9]	310	20	165 - 930	165,620, 1550
present	1225 x 820	~9	1000 - 2000	880

Two recent failures of transmission lines in Australia [10,11] occurred on exposed ridgelines during severe thunderstorms. Although Holmes et.al. [1,8] results would indicate a lesser flow acceleration than in boundary layer flow, they examined only the one embankment (height = 40mm, slope = 0.25). The present study grew out of a need for further data on topographic effects in downdraft flows.

2 EXPERIMENTAL PROCEDURE

The tests were conducted using the Department of Civil Engineering's Environmental Wind Tunnel. A jet was formed in the aeronautical section by removing the downstream diffuser and replacing it with a plane wall. The tunnel exit is 1225mm wide and 820mm high and has corner chamfers of 270mm wide by 170mm high, giving a hydraulic diameter of 1.01m. This cross section is 3.6m in length and is proceeded by a 6:1 area contraction and a series of screens. The jet centreline is 1.07m above the laboratory floor and 1.3m below the ceiling above. The wall was constructed of 18mm thick smooth plywood sheeting carefully butt joined and placed 880mm away from the tunnel exit (B). The wall extended floor to ceiling (2.4 high) and 4.7m in width, with jet centreline located 1.8m from one side. A schematic of the arrangement is shown in Figure 1.

The topographical feature studied was the flat-topped embankment with three different slopes; $\phi = 0.2$, 0.4 and 0.6. All hills were 100mm high and were constructed of smooth plywood. The length of the flat top portion on the embankments varied with slope, being 125mm, 185mm and 375mm for slopes of 0.6, 0.4 and 0.2 respectively. The embankments extended the full width of the wall.

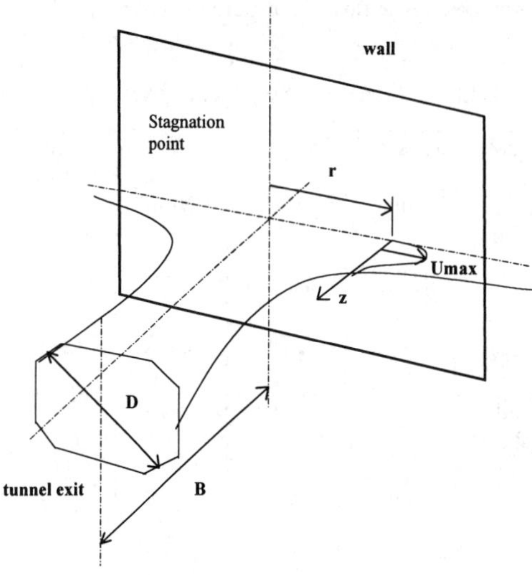

Figure 1. Schematic of experimental arrangement.

Velocity measurements were obtained using DANTEC's streamline system and single hot wire probes mounted parallel to the ground. The hot wire was regularly calibrated against a pitot-static tube located in low turbulence flow. Velocity profiles were obtained above the leading edge of flat top portion of the embankments, on the jet centreline. The data were sampled at 400Hz for 30seconds.

Two jet flow types were studied, a smooth jet, exit level turbulence intensity of 0.6% and mean velocity of 9.3 ± 0.2m/s and a turbulent jet, with ~6% turbulence and a mean velocity varying 8.3 ± 1.0m/s across the jet. The additional turbulence was produced by introducing a grid at the upstream end of the working section of the tunnel.

3 RESULTS AND DISCUSSION

3.1 Wall jet profiles

The mean velocity profiles at various distances from the stagnation point non-dimensionalized by jet diameter are shown in Figure 2 for the smooth jet. Corresponding local longitudinal turbulence intensities (u_{rms}/U) are shown in Figure 3. Figures 4 and 5 show the mean velocity and turbulence intensity profiles at similar radial locations for the turbulent jet.

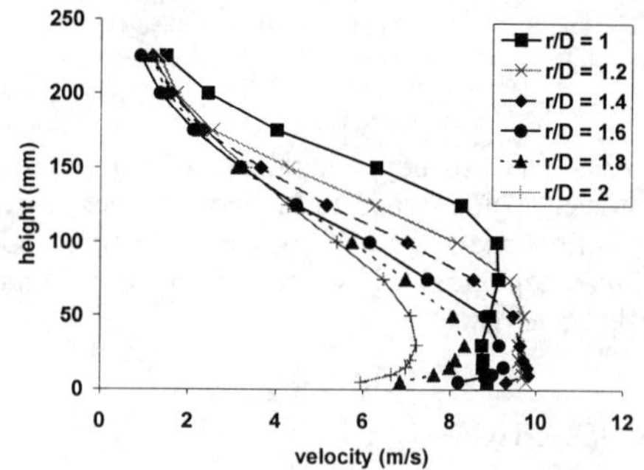

Figure 2. Mean velocity profiles at distance r/D away from the stagnation point for the smooth jet.

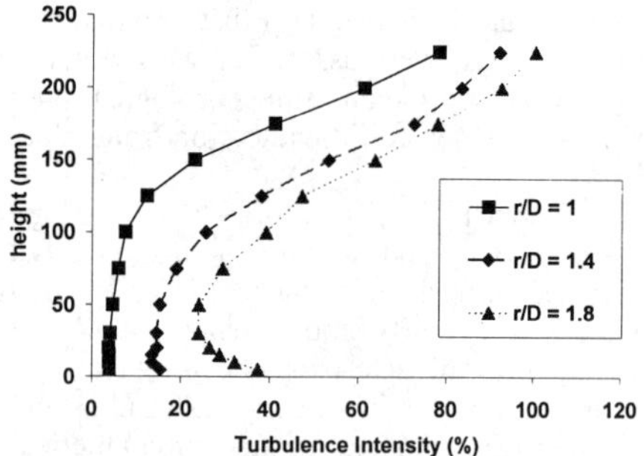

Figure 3, Turbulence intensity profiles at distance r/D away from the stagnation point for the smooth jet.

As the impacting jet spreads out over the ground plane, a characteristic profile develops with a maximum mean speed at a radial position of approximately 1.5 times the jet diameter. At this radial position the velocity profile shows some evidence of boundary layer development up to a height of 50mm and then drops off reaching half the maximum velocity around 150mm. Beyond this radial position the effect of the boundary layer growing on the wall becomes more pronounced, leading to a peak in the velocity profile between 30 and 50mm. When these general flow features are compared with radar measurements of actual thunderstorms [3,4] they indicate a length scale of approximately 1:1000 to 1:2000.

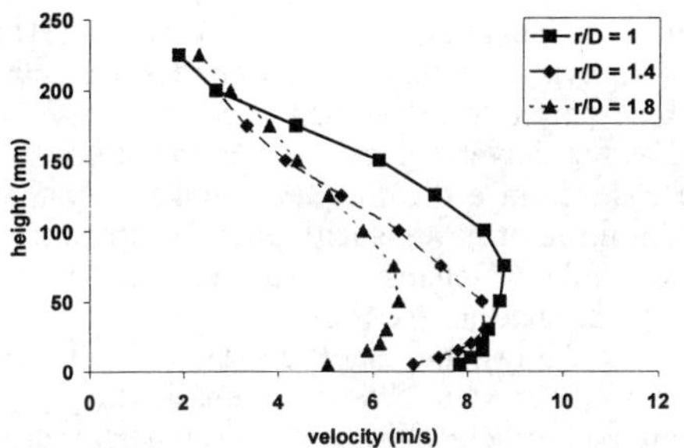

Figure 4. Mean velocity profiles at distance r/D away from the stagnation point for the turbulent jet.

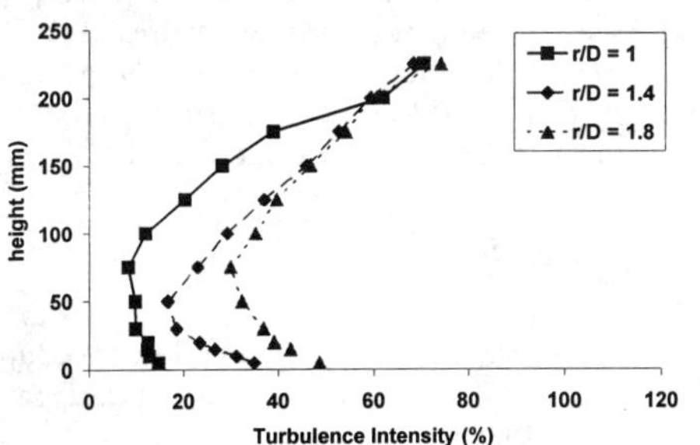

Figure 5. Turbulence intensity profiles at distance r/D away from the stagnation point for the turbulent jet.

When comparing the profiles from the two jet types, that jet with initially higher turbulence leads to higher turbulence levels and lower mean velocities in the associated wall jet flow.

3.2 Gust profiles

In the context of determining wind loads on structures, most codes of practice, eg. [12,13], assume a quasi-steady approach which implies that the peak loading corresponds to the peak or gust wind speed. Thus, it is of interest to determine how the gust wind speed varies with height. In boundary layer flow it has been assumed that the design gust wind speed is related to the mean and turbulence intensity via;

$$\hat{U} = \overline{U} + 3.7I_u$$

where I_u is the turbulence intensity.

Assuming a similar relationship for a downburst, but recognizing the limitations imposed by the trasient nature of a 'real' moving downburst, estimates of quasi-steady gust profiles are presented in Figures 6 & 7.

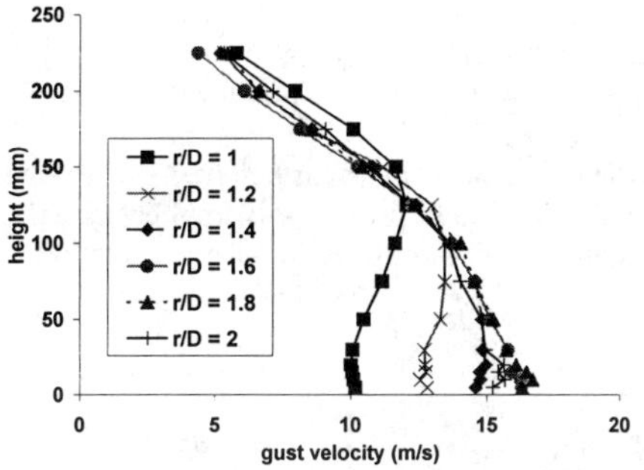

Figure 6. Estimated gust wind speed profiles for smooth jet.

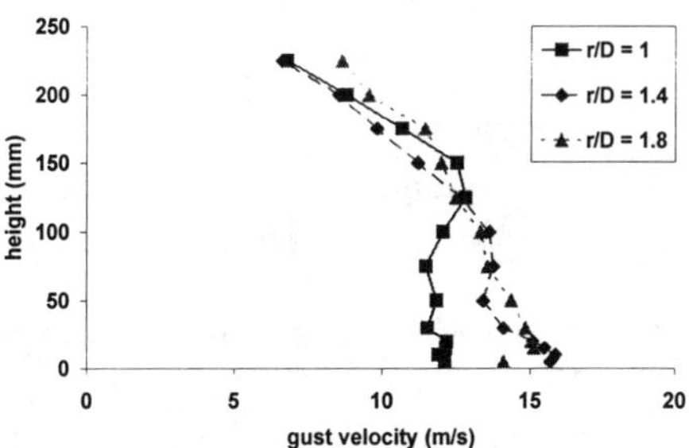

Figure 7. Estimated gust wind speed profiles for turbulent jet.

The gust profiles presented in Figures 6 & 7 indicate that the initial turbulence content of the jet is of little consequence. For both smooth (~0.6%) and turbulent (~6%) jets the maximum 'gust' speed is around 16m/s and occurs at low level (~10mm) between 1.5 and 2 D from the impact point. The 'gust' velocity is approximately constant or slowly decreases with height until about 150mm in each case. The above analysis ignores any of the transient effects present in a 'real' moving downburst.

3.3 Dimensionless velocity profiles

The early wall jet studies [5,6] employed a non-dimensionalizing scheme for the velocity profile in terms of the maximum mean velocity (U_{max}) in each profile and the height (δ) at which half this velocity is reached. The mean velocities of Figure 2 and 4 have been represented in this manner in Figure 8 for the smooth jet and Figure 9 for the turbulent jet. Also plotted in each figure is the expression developed by Wood et.al. [9] for the non-dimensional mean velocity profile:

$$\frac{U}{U_{max}} = 1.55\left(\frac{z}{\delta}\right)^{1/6}\left[1 - erf\left(0.7\frac{z}{\delta}\right)\right]$$

where U is the mean velocity at height z and erf the error function. Wood's profile only applies once the wall jet stabilizes at radii beyond ~1.5 jet diameters, D.

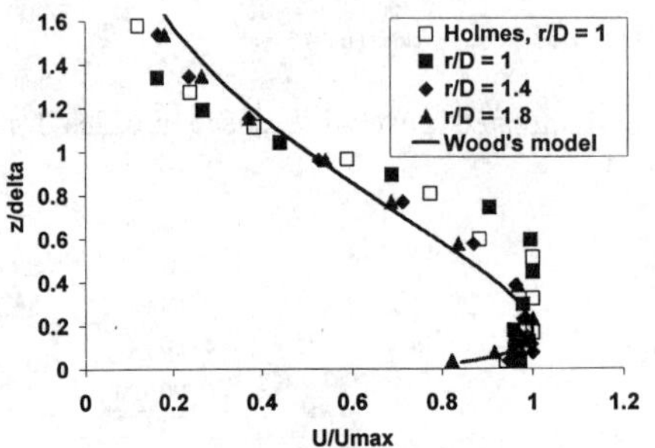

Figure 8. Non-dimensional mean velocity profiles for the smooth jet.

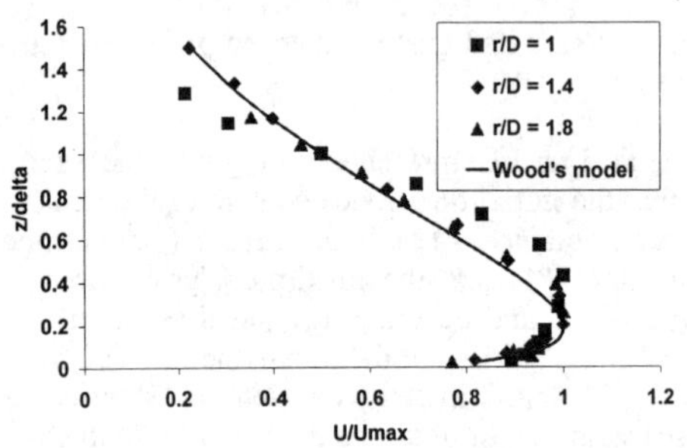

Figure 9. Non-dimensional mean velocity profiles for the turbulent jet.

It is seen from Figures 8 & 9 that there is reasonable agreement with Wood's dimensionless mean velocity profile at radii beyond 1.5 D, ie., once the jets have stabilized. Also, the initial turbulence state of the jet does not appear to greatly effect the non-dimensional profile. Also plotted in Figure 8 is Holmes' profile measured at r = 1000mm from an almost identical jet size (1150 by 850mm) impacting on a plane 1.4m from the exit. (c.f. 880mm here) The agreement for the two radial position's of 1000mm is very good with Holmes having a δ = 157mm and the present study producing 168mm.

Bakke [5], Poreh et.al. [6] and Wood et.al. [9] all give expressions for the growth of δ with radius, although only Wood has made measurements in the vicinity of the maximum mean velocity at r = 1.5D. In a 'real' moving downburst, a structure in it's path would be expected to be exposed to the full range of radii, r, as the downburst moves over the structure. In that case, from a designer's viewpoint only the maximum mean or gust velocity over the structure's height would be of interest. From these results this would be expected to occur once the downburst is within 1.5 to 2 downdraft diameters (D) of the site in question. This is valid should the quasi-steady assumption hold for a travelling jet and although it has been suggested that it does, at least to the first order [4], this remains a subject requiring further study.

It is also of interest to note that whereas Wood [9] and others [5,6] employed axi-symmetric jets in their experiments the irregular shaped jets employed here and by Holmes [8] do not seem to have influenced the mean velocity profiles significantly.

3.4 Topographic effects

It is common practice to specify the flow speedup over topographic features through a topographic multiplier M_t defined as:

$$M_t = \frac{U_{z,hill}}{U_{z,nohill}}$$

Where $V_{z,hill}$ is the velocity at height z with the topographic feature, and $V_{z,nohill}$ the velocity at the same height without the topography.

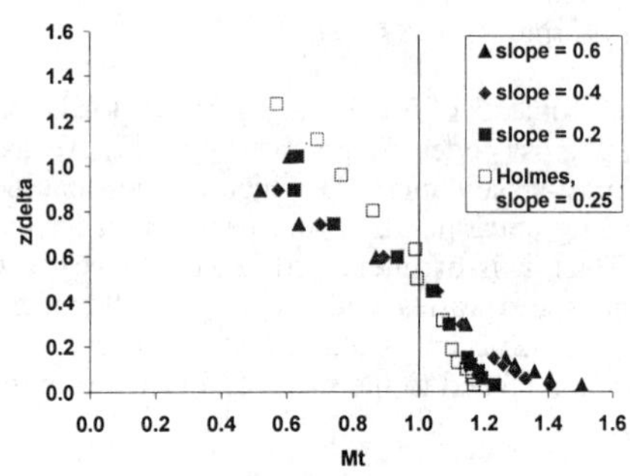

Figure 10. Topographic multipliers at the leading edge of various sloped embankments at r/D = 1.

Figure 10 shows the topographic multipliers for mean velocities above the leading edge of the three different embankment slopes (0.2, 0.4 & 0.6) at the same radial position of r/D = 1. The height has again been non-dimensionalized by δ (=168mm at this radii). Also shown on the figure are the results obtained by Holmes [8] for an embankment of slope 0.25, also at r/D = 1.

The agreement between the present results and Holmes [8] for a similar slope (0.2 & 0.25 respectively) and radial position r/D=1 is encouraging, yet surprising. This is because the ratio, δ/H (jet thickness/hill height) for Holmes was 4 while for the present study it was only 1.7. It could be expected that relative jet thickness would effect the way flow acceleration over the topographic feature took place.

For the shallowest slope (0.2) a maximum topographic multiplier of 1.2 occurs at the closest measurement position to the crest (5mm). When considering the three slopes studied, a relationship for the topographic multiplier at the surface of, $M_t = 1 +$ slope, is suggested with each decreasing almost linearly to unity at $z/\delta = 0.5$. These higher values of M_t for steeper slopes are of particular interest because the transmission line failures reported in [10,11] occurred on steep (slope > 0.25) ridge lines.

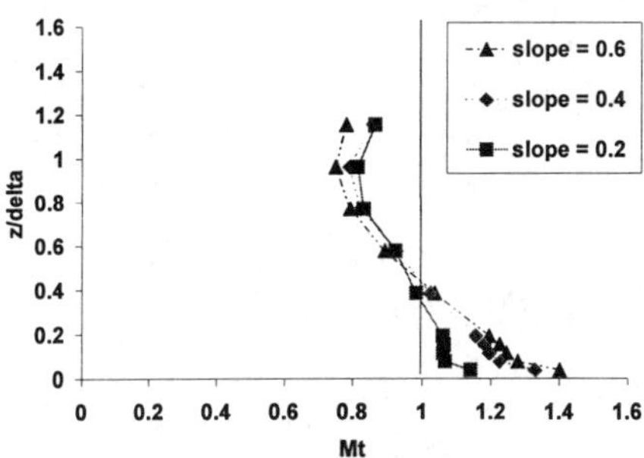

Figure 11. Topographic multipliers at the leading edge of various sloped embankments at r/D – 1.4.

Figure 11 presents the mean topographic multipliers above the leading edge of the embankments at a radial distance of r/D = 1.4. Once again the height has been non-dimensionalized by δ (= 130mm at this radius). It will be recalled from Figure 2 that it is about at this radial distance that the mean velocity has a maximum. The mean topographic multiplier has a surface value approximately given by $M_t = 1 +$ (slope - 0.1) and once again decreases almost linearly with height to unity at $z/\delta = 0.5$.

Comparing Figures 10 & 11 it is seen that these topographic multipliers decrease with increasing radius from stagnation.

In Figure 12 the topographic multipliers for mean wind speed above the leading edge of the embankment of slope 0.4, for two radial distances, r/D = 1 and 1.4 are compared with values recommended by two Australian wind loading codes [12,13]. In this figure the height axis has been non-dimensionalized by the embankment height, H, to facilitate the comparison. The code requirements were developed from speedup in boundary layer flows and it is seen that in all cases they overestimate the speedup involved in wall jet flows as suggested by Holmes [8]. This is likely due to the lack of confinement of the wall jet on its free surface boundary.

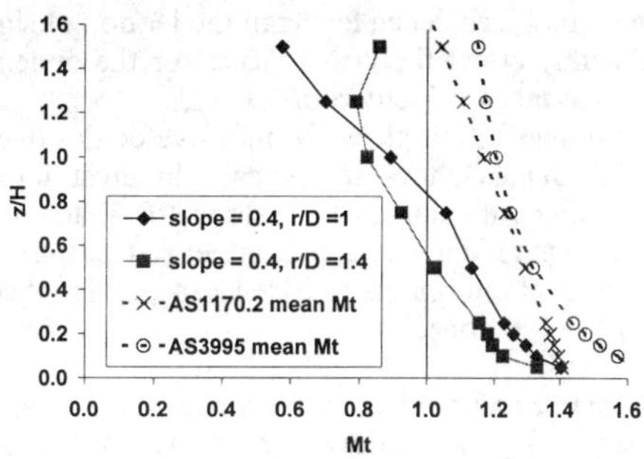

Figure 12. Comparison of topographic multipliers for mean velocity from wall jet for embankment, slope = 0.4 at r/D = 1 and 1.4 with codes [12,13].

4 CONCLUSIONS

Early investigations of wall jets [5,6], the analogy employed for downbursts here, focussed on the stabilized region well away from the point of impact of the jet. Unfortunately, wind engineers need to focus on the maximum velocities in such flows and these occur much closer to the point of impact, in fact within 2 jet diameters. Few investigations have focussed on this area. The present study has drawn on the successful preliminary experiments of Holmes [8] and Wood et.al [9] and conducted a comprehensive study to examine the turbulence characteristics of wall jets close to impact and the flow acceleration over generic topographic features. Reasonable agreement with the earlier work has been found and the results extended for a range of parameters including; initial turbulence of impinging jet, radial position from jet stagnation and slope of topography. These results support the preliminary finding by Holmes that jet flow acceleration over topographic features is less than in boundary layer flow.

The specific conclusions are:

- Wall jets would appear to represent a reasonable tool for studying the behaviour of downbursts.
- The maximum mean and quasi-steady gust velocities in wall jets occur at radii between 1.5 and 2 jet diameters from jet stagnation.
- Mean velocity profiles achieve self-similarity at radii greater than 1.5 jet diameters and Wood's [9] empirical velocity distribution fits these data reasonably well.

- Quasi-steady gust profiles indicate a reasonably constant velocity profile over the lower two-thirds of the wall jet at radii greater than 1.5 jet diameters.
- The initial turbulence levels in the jet do not significantly effect the gust profiles, or the dimensionless mean velocity profiles.
- Topographic multipliers for mean velocities over embankments show an increase in crest level multiplier with slope, ($0.2 \leq$ slope ≤ 0.6) and decrease with radial distance from jet stagnation.
- Crest level multipliers at r/D=1 can be described by $M_t = 1 +$ slope.

The major assumption of the analogy between a stationary wall jet and a downburst is the assumption of quasi-steady behaviour. That is, can the 'real' travelling downburst be replaced by a stationary jet successively displaced from the site in question. The complex flow morphology surrounding downbursts has been addressed in part by Hjelmfelt [4] and this includes leading 'rollers' or horizontal axis vortices during the initial stage of the downburst transient. These vortices could well produce significant velocities and they cannot be reproduced with the quasi-steady model. This discrepancy is being addressed at the University of Queensland at present with the construction of a moving jet wind tunnel facility.

5 ACKNOWLEDGEMENTS

The authors wish to acknowledge the financial support and encouragement of Mr Henry Hawes of Powerlink, Queensland.

6 REFERENCES

[1] R.P. Selvam & J.D. Holmes, Numerical simulation of thunderstorm downdrafts, JWEIA, 41-44 (1992) 2817-2825.

[2] H.C.S. Thom, New distributions of extreme wind speeds in the United States, J. Struct. Div., ASCE, 94 (1969), 1787-1801.

[3] T. Fujita, Tornadoes and downbursts in the context of generalized planetary scales, J. Atmos. Sci., 38 (1981), 1511-1534.

[4] M.R. Hjelmfelt, Structure and life cycle of microburst outflows observed in Colorado, J. App. Meteor., 27 (1988), 900-927.

[5] P. Bakke, An experimental investigation of a wall jet, J. Fluid Mech., 2, (1957) 467-472.

[6] M. Poreh, Y.G. Tsuei & J.E. Cermak, Investigation of a turbulent radial wall jet, J. Appl. Mech., Trans ASME, 34 (1967) 457-463.

[7] F.H. Proctor, Numerical simulations of an isolated microburst. Part 1. Dynamics and structure, J. Atmos. Sci. 45 (1988), 3137-3160.

[8] J.D. Holmes, Physical modelling of thunder storm downdrafts by wind tunnel jet, 2nd AWES Workshop, Monash University, 21-22 February 1992, 29-32, 1992.

[9] G. Wood & K. Kwok, An empirically derived estimate for the mean velocity profile of a thunderstorm downdraft, 7th AWES Workshop, University of Auckland, 28-29 September 1998.

[10] C.W. Letchford, Wind environment experienced by Tarong-Mt. England transmis sion line at Harlin during thunderstorm on 9 February 1990, for Queensland Electricity Commission, Uniquest Report 3032583, 1990.

[11] C.W. Letchford, Investigation of 275kV Transmission line failure near Harlin, 18 March 1998, for Powerlink, Uniquest Report 10764, 1998.

[12] Standards Australia, SAA Loading Code, AS1170 Part 2: Wind Loads, Sydney, 1989.

[13] Standards Australia, Design of steel lattice towers and masts, AS3995, Sydney, 1994.

Wind Engineering into the 21st Century, Larsen, Larose & Livesey (eds) © 1999 Balkema, Rotterdam, ISBN 90 5809 059 0

Simulation of surface-layer velocity fields and their influence on wake

V.P. Mugalev
State Scientific Research Center TsAGI, Moscow, Russia

ABSTRACT: Procedure of stream organization which permits modelling of approximate distribution of surface wind velocity was realized in the wind tunnel. An investigation of effect of varying parameters of boundary layer on flow for a triangular plate at incident angle and on its wake was carried out.

1 INTRODUCTION

In many practical cases one must take into account variability of wind flow conditions and their influence on different objects and their wakes. This relates to streamline of building structures, windfarms, ecology problems, aircrafts take-off and landing in airports, etc.

At the first stage of this work an investigation was carried out, which concerned creation and tests of facility permitting to model a number of mean and fluctuating characteristics of surface boundary layer at neutral stratification in the wind tunnel with a short test section. For this purpose the method of distributed air jets interacting with main flow was used. This method with various modifications was applied in several works (see e.g. Blessmann 1972, 1977, Corke et al. 1975, Gendricson 1979). The method described in the present work has been modified and improved. Then the study of influence of the boundary layer with varying profiles of velocity and turbulence on flow for a triangular wing (plate) at the incident angle and on its vortex wake was carried out. It is supposed that a vortex wake structure may be characterized by the turbulence intensity of longitudinal fluctuation velocity component.

2 TEST PLANT AND MEASUREMENTS

Figure 1 shows a scheme of the plant for modelling given surface boundary layer conditions. The facility works in the wind tunnel T-23 of TsAGI. The plant is a flat platform (length 1200 mm, width 300 mm) with sidewalls. At the beginning of the platform many air jets are blown out through a collector upstream, and then specially selected roughness is located. A uniform

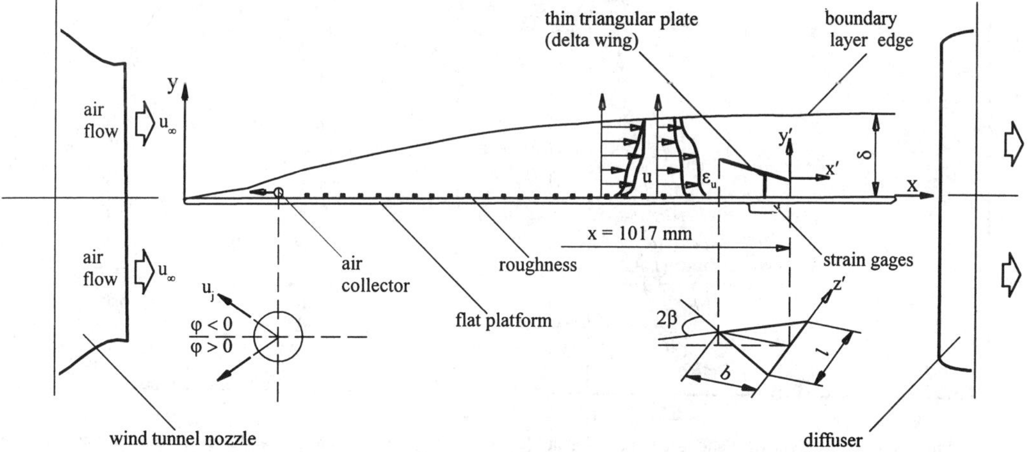

Figure 1. Scheme of the plant for modelling given surface boundary layer conditions in wind tunnel with triangular plate (wing) model.

jet blowing out was achieved over the whole platform width.

A thin triangular plate (wing) with sharp edges located over platform at incidence angle of 19°30′, was used as a generator of stable vortex wake (Belotzerkovski & Nisht 1978, Wiesel 1980). The wing is isosceles triangle with thickness 5 mm, base length l=85 mm, height b=115 mm, sweep angle β=20°18′, aspect ratio λ=1.48. The distance between the platform plane and the nose of the wing is 68 mm, between it and the trailing edge 29.5 mm, between trailing edge and the platform nose 1017 mm. The rest symbols are indicated in Figure 1.

In tests the fields of longitudinal component fluctuation velocity (with hot-wire anemometer of DISA), the fields of mean velocity (with Putot tube with a flat nose at a distance from the tube hole center to the wall of 0,6 mm), the fields of static pressure (with miniature Venturi tube), and also the airforces (with miniature strain gages) were measured.

Now the project of analogous test plant of greater dimensions is in the state of elaboration. It is intended for the wind tunnel T-1 TsAGI with closed test section (flow velocity up to 40 m/s, airstream turbulence less than 0.6%) and expected thickness of the quasi-similar boundary layer up to 0.8 m.

3 EXPERIMENTAL RESULTS AND THEIR ANALYSIS

Below, there are some of the obtained test data. Part of these results is presented in (Mugalev 1993, 1994).

3.1 *Mean velocity and turbulence intensity profiles in boundary layer of facility*

The test results evidenced the predicted evaluations of the plant characteristics and showed possibility of producing thick boundary layer with substantially varying profiles of turbulence intensity $\varepsilon_u = \langle u \rangle^{0.5}/u$ (where u=mean; $\langle u \rangle$ = fluctuated longitudinal velocity components) and mean relative velocity $\bar{u} = u/u_\infty$ (where u_∞ = const = flow velocity over boundary

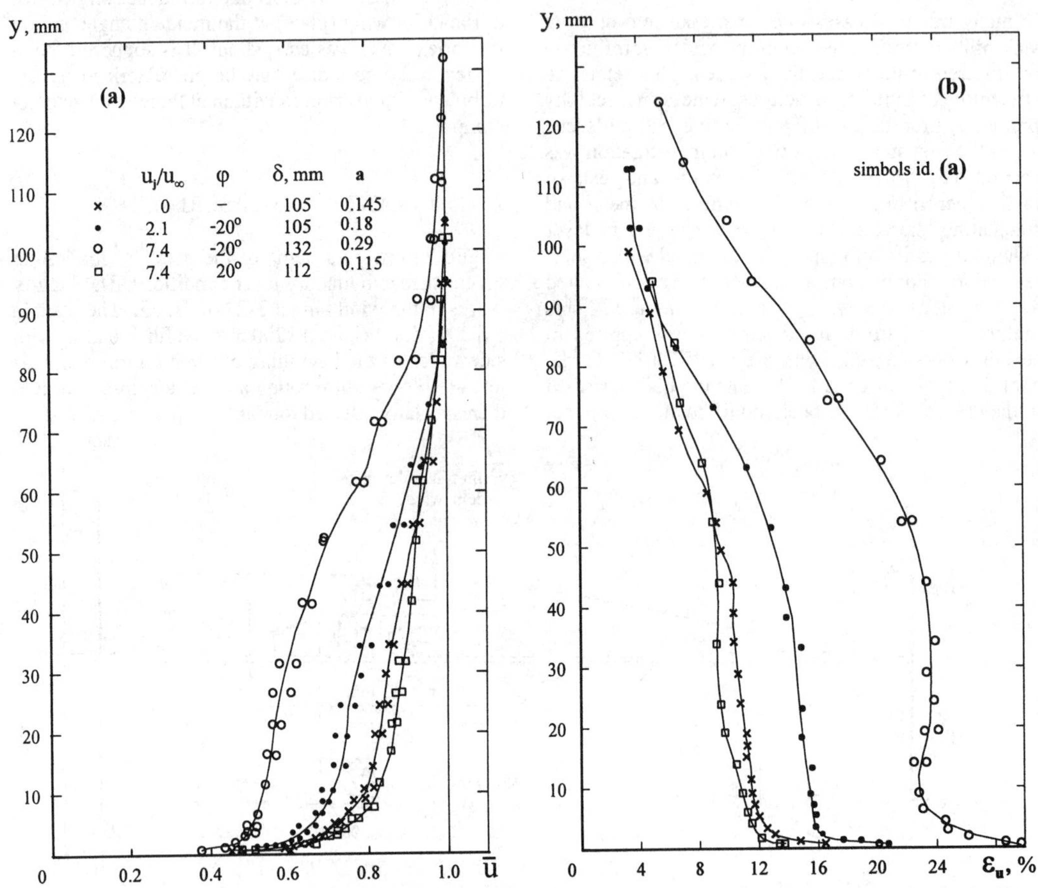

Figure 2. Profiles of mean velocity (a) and turbulence intensity (b) in boundary layer on smooth platform with collector and without wing model (section x=987 mm).

layer). Note that all profile variations can be produced during the tests without wind tunnel stopping.

Figure 2a, b presents the test data on measuring the fields $\overline{u}(y)$ and $\varepsilon_u(y)$ on a smooth platform with collector (but without triangular wing) in the cross section x=987 mm. It is clearly seen that the form of these profiles and the boundary layer thickness (δ) strongly vary according to changes of jets blowing velocity (u_j), angle formed at their collision with flow (φ) and surface roughness. Note that the boundary layer thickness in section x=987 mm is δ_0=28 mm on a smooth platform without collector and wing model. The collector presence on the platform serves also as a factor disturbing the boundary layer in absence of blowing.

It is assumed that the profiles of mean velocity may be described by a power-low:

$$u/u_\infty = (y/d)^a \tag{1}$$

In this case for one experiment set (at varying relative velocity of blowing from 0 to 7.4 and angle $\varphi=\pm 20°$, see Fig.2) the following was obtained: intensity turbulence profiles form difference (over this value magnitude at a fixed height in boundary layer) is almost three times; the mean velocity profile form variation corresponds to varying flow conditions from plane country to borders of forest and low urban buildings (exponent in eq.(1) is $a \approx 0.15 \div 0.3$) (Bysova et al. 1989); the boundary layer thickening (against the thickness boundary layer on smooth platform without collector and wing model) is by almost factor of five.

Note that a set of mean and fluctuation velocity profiles obtained in this work satisfactorily agrees with full-scale measurements in surface boundary layer from the Obninsk meteorological mast (Bysova et al. 1989).

3.2 *Flow for triangular plate and its wake*

Dwell on the results of measuring the turbulence intensity fields in plane (y′, z′) in near wake of trailing edge of triangular wing (plate) at a distance $\overline{x}=x'/b \approx 0.1$ (x′=10 mm; sections y′=0÷14 mm) from the edge downstream.

Figure 3 brings together three versions measuring the wake of flat wing model located above smooth platform: 1) without collector; 2) with collector and without blowing; 3) with blowing at relative velocity air jets $u_j/u_\infty \approx 7.4$ and angle of jet supply φ=-20°. Note that in the first version the wing was outside the boundary layer and was flowed by gradientless low-turbulent flow (in the absence of wing the thickness of boundary layer was δ_0=28 mm, value ε_{u_∞} less 1.4%). In the second and third versions the wing was inside of the boundary layer with substantially different fluctuating and mean velocity profiles (Fig. 2). In this case the frontal area of wing in two latter versions was located in the boundary layer region with the values $\overline{u}(y)$ and $\varepsilon_u(y)$ few varying in height. Analyze the data of Figure 3.

Measurements in a near wake of trailing edge of wing located on smooth surface without collector show that in a number of cross sections (in y′) there exist two rather sharp maxima of the turbulence intensity value (at $\overline{z}_l = z'/l \approx 0.3$ and 0.5). The negative pressure maximum is to be observed also at $\overline{z}_l \approx 0.3$ (the negative pressure coefficient is increased twice in comparison with section $\overline{z}_l \approx 0$). This proves the existence of developed stable vortical structure in a near wake of triangular wing at uniform flow (Wiesel 1980).

A placing on platform of collector in absence of blowing leads to an evident increase of turbulence degree over wingspan. At the same time the character of dependency $\varepsilon_u(y', z')$ qualitatively remains similar to the previous version.

The most intensive effect on turbulence field is produced by air jets blowing through collector. Qualitative variations of profiles $\varepsilon_u(y', z')$ is clearly seen in Figure 3: the sharp maxima disappear and the general level of turbulence degree considerably increases over whole wingspan. Obviously this indicates essential transformation of vortical structure over triangular wing under the effect of substantial variations of velocity and turbulence profiles in the circumfluent of wing flow.

Irrespective of the measurements data considered above for fields $\varepsilon_u(y', z')$ in cross section x′=10 mm, a trajectory of flow points with $(\varepsilon_u)_{max}$ in wake downstream from trailing edge of wing was studied. Measurements were carried out in three cross sections ($\overline{x} \approx 0.1$; 0.6; 1.1; x′=10; 65; 125 mm respectively) at variations of the initial flow conditions on a smooth surface with collector: 1) without blowing; 2) with blowing of air jets at $u=u_j/u_\infty=2.1$; 3) with blowing at $u_j/u_\infty \approx 7.4$. Collected results are presented in Table 1. Here the figures in upper line indicate for each cross section $\overline{x}$ the coordinates and the values $(\varepsilon_u)_{max}$ in wake and the figures in two lower lines - coordinates and values ε_u in middle plane of wake (z′=0) and in flow between wing and vertical sidewall of facility (z′=68 mm). The analysis of data in Table 1 indicates the following.

Increase of blowing through collector leads to reducing the value $(\varepsilon_u)_{max}$ for each given cross section of wake, and this seems to be associated with boundary layer effect on an organized vortex structure.

The turbulence intensity values in middle plane of wing and in region between wing and sidewall of facility in each cross section for given conditions of circumfluent of wing flow are close to each other and slightly differ from the respective values on the profiles $(\varepsilon_u)_{max}$ measured above the platform without wing (Fig.2). This testifies to the fact that maximum disturbances in the flow after wing are concentrated in the so-called vortical plaits in all cases considered. At the same time in cross section $\overline{x} \approx 0.1$ at blowing $u_j/u_\infty \approx 7.4$ the values of $(\varepsilon_u)_{max}$ and ε_u at z′=0 and 68 mm approach each other. Hence blowing at the beginning of platform stimulates destroying the vortical wing wake intensity.

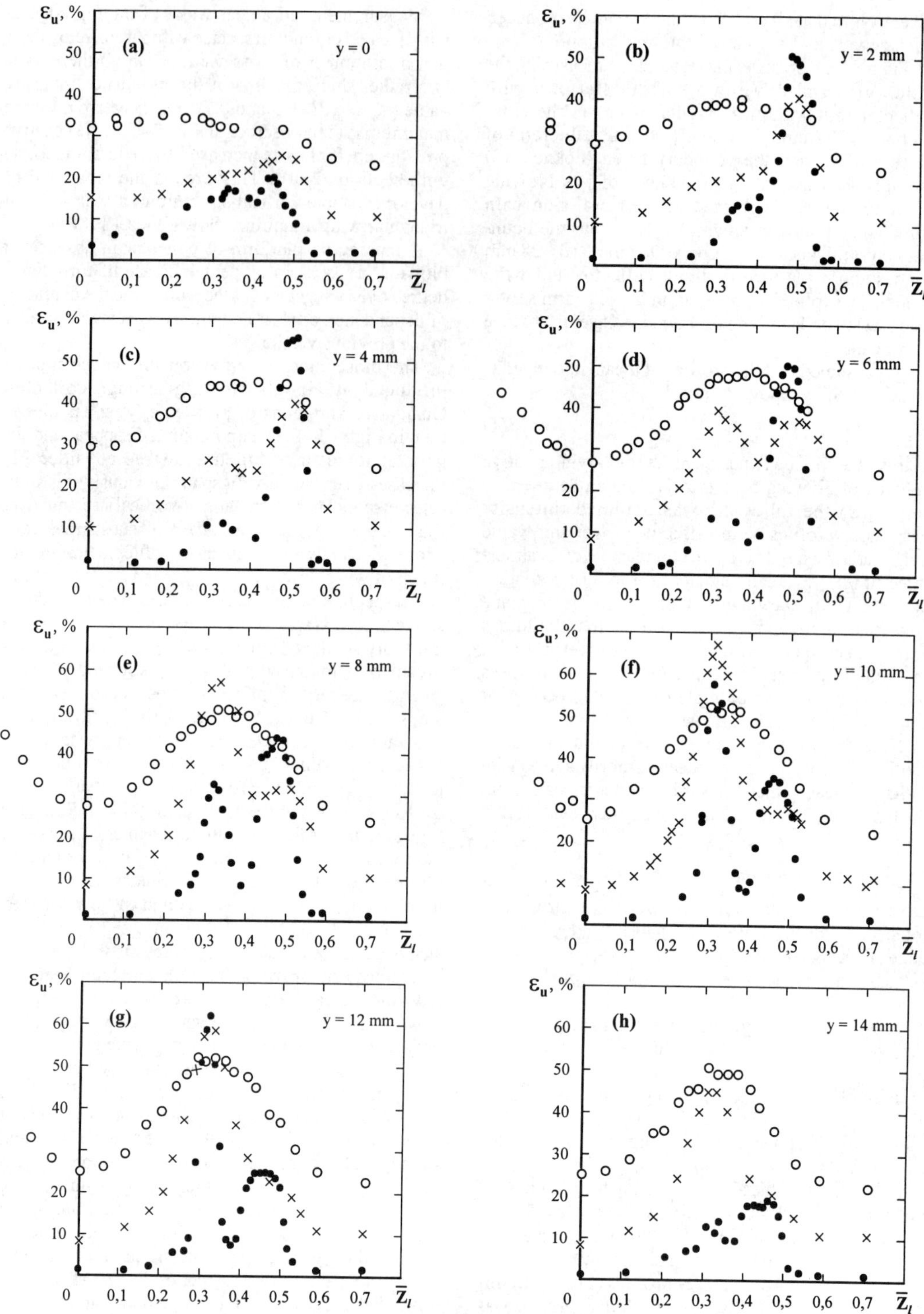

Figure 3. Profiles of turbulence intensity in near wake of trailing edge of triangular wing at $\overline{x} = x'/b \approx 0.1$; $x'=10$ mm; $y'=0\div14$ mm; incidence angle of wing 19°30′ (• wing above smooth platform without collector; × with collector and without blowing; ○ blowing of air jets at $u_j/u_\infty \approx 7.4$; $\varphi = -20°$).

Table 1. Observed data of turbulence intensity in wake of trailing edge of wing downstream at different initial flow conditions (wing above smooth platform with collector).

Section	No blowing ($u_j/u_\infty = 0$)			Blowing ($u_j/u_\infty = 2.1$)			Blowing ($u_j/u_\infty \approx 7.4$)		
	y',mm	z',mm	ε_u,%	y',mm	z',mm	ε_u,%	y',mm	z',mm	ε_u,%
	10	26	66*	8	27	60*	10	24	52*
$\bar{x} = x'/b \approx 0.1$ (x' = 10 mm)	10	0	9	10	0	12	10	0	27
	10	68	10	10	68	13	10	68	22
	5	30	40*	5	28	35*	12	28	33*
$\bar{x} \approx 0.6$ (x' = 65 mm)	5	0	8	5	0	13	5	0	24
	5	68	10	5	68	12	5	68	22
	2	32	32*	5	30	27*	15	30	26*
$\bar{x} \approx 1.1$ (x' = 125 mm)	2	0	7	2	0	11	2	0	21
	2	68	10	2	68	13	2	68	20

* Maximum value.

From analysis of data in Table 1 an important conclusion follows concerning a noticeable evolution of the trajectory of points with maximum value of turbulence intensity over wake length. Closely to trailing edge of wing (at x'=10 mm) the coordinates of $(\varepsilon_u)_{max}$ points are practically independent of conditions circumfluent of wing flow, and this agrees with data in Figure 3f. (Note that the coordinates z' of $(\varepsilon_u)_{max}$ points not much change anywhere, possibly due to two-dimensionality of the flow). At absence of blowing with a larger distance from trailing edge of wing one can see a noticeable lowering of the points $(\varepsilon_u)_{max}$ to platform surface. At blowing with u_j/u_∞=2.1 this lowering is called to a stop and in the cross section which is at the largest distance from wing (x'=125 mm) this tendency becomes the opposite one.

At maximum blowing $u_j/u_\infty \approx 7.4$ in cross sections of wake downstream from trailing edge of wing the point with $(\varepsilon_u)_{max}$ lifts up considerably from platform (for instance, at $\bar{x} \approx 0.6$ coordinates y' more than double and at $\bar{x} \approx 1.1$ increase more than 7 times relatively to cross section $\bar{x} \approx 0.1$). Obviously, this fact is associated both with the effect of variations of the circumfluent of wing flow parameters and with the interaction between the vortex wake and the lower wall of facility.

Finally it should be noted an evident effect of varying characteristics of boundary layer on the aerodynamical parameters of triangular wing (plate) at incident angle. For the test conditions indicated above, the wing stream by flow with considerable increase of the turbulence intensity leads to a significant decrease of drag coefficient and to a noticeable decrease of lift coefficient in comparison with the case of gradientless low-turbulence flow.

4 CONCLUSION

Procedure of approximate modelling a flow of the surface turbulent boundary layer type with considerably varying profiles of fluctuating and mean velocity is realized in the wind tunnel with a short test section.

It is shown that such flow can essentially affect the vortical wake structure and the aerodynamic characteristics of a poorly streamlined object in comparison with the case of uniform flowing stream.

REFERENCES

Belotzerkovski, S.M. & Nisht, M.I. 1978. *Separated and nonseparated flow of thin wings by ideal fluid.* Moscow: Nauka.

Blessmann, J. 1977. The use of cross-jets to simulate wind characteristics. *J. Ind. Aerodyn.* 1(2): 37-47. Amsterdam: Elsevier. (See here reference to Blessman 1972).

Bysova, N.L., Ivanov, V.N. & Garger, E.K. 1989. *Turbulence in boundary layer of atmosphere.* Leningrad: Gidrometeoisdat.

Corke, T.C., Nagib, H.M. & Tan-atichat, J. 1975. Flow near a model of a building in simulated atmospheric surface layers generated by the counter-jet technique. *Proc. 4 th Intern. Conf. on Wind Effects on Build. and Struct.*: 515-526. London.

Gendricson, V.A. 1979. Influence of cross jets system on flow characteristics in wind tunnel. *Turbulent jet flows (collection of reports)*: 183-191.Tallinn.

Mugalev, V.P. 1993. Elaboration of method of approximate modelling a number of parameters of surface wind layer. *Report GosNITs TsAGI.* 12800. Moscow.

Mugalev, V.P. 1994. Experimental investigation of turbulent flow stucture in near wake of triangular wing at incident angle with approximate modelling of surface wind conditions. *Report GosNITs TsAGI.* 12834. Moscow.

Wiesel, E.P. 1980. On effect of triangular wing lengthening on structure of near wortex wake. *Utchenye Zapiski TsAGI.* 5 (11): 119-125. Moscow: TsAGI.

Wind Engineering into the 21st Century, Larsen, Larose & Livesey (eds) © 1999 Balkema, Rotterdam, ISBN 90 5809 059 0

Physical and numerical modelling of thunderstorm downbursts

Graeme S.Wood & Kenny C.S.Kwok
Wind Engineering Services, and Department of Civil Engineering, University of Sydney, N.S.W., Australia

Nicholas A.Motteram & David F.Fletcher
Department of Chemical Engineering, University of Sydney, N.S.W., Australia

ABSTRACT: In this study, a static continuous impinging jet has been used to simulate a thunderstorm downburst. Velocity characteristics have been investigated at various positions in the flow, and over simple topographic features, thereby yielding topographical speed-up factors. A generic empirical equation describing the development of the flow, and comparisons with previous works and conventional boundary layer flow speed-up factors has been completed. Speed-up factors were similar at the crest of the embankment, but dropped off rapidly behind the crest compared to boundary layer flow. Preliminary studies using computational fluid dynamics to predict the flow regime over topographical features has been shown to give reasonable agreement.

1 INTRODUCTION

With thunderstorms causing 50% of extreme wind gusts in Australia outside the tropical cyclone region (Selvan & Holmes, 1992), downbursts of fast flowing air generated in severe thunderstorms have the potential to cause significant damage to structures. It has been previously observed (Selvan & Holmes, 1992) that there is a close relationship between vertical radar gust profiles in thunderstorm downdrafts (Fujita, 1981) and the mean velocity profile above a normal wall jet. Experimentally, density gradient and transient characteristics of the actual downburst are difficult to simulate; therefore, a continuous impinging wall jet has been adopted to investigate the wind profile development over different terrain.

The radially expanding velocity profile from a continuous impinging wall jet has been measured and empirical equations developed to predict the velocity profile Wood & Kwok, (1998). Selvan & Holmes (1992) numerically calculated topographic speed-up factors over a single hill generally finding that the speed-up factors were much less than for developed boundary layer flow. Holmes (1992) measured speed-up factors for a single embankment at a single location, and Letchford & Illidge (1998) measured speed-up factors for various topographic features at a single position and found that the speed-up factors may not have been as low as predicted numerically.

In this current study the effect of thunderstorm downdrafts on topographic speed-up factors is investigated for a single embankment at various radii from the centre of the impact point and for the test surface placed at different distances from the jet outflow. Computational Fluid Dynamics (CFD) modelling is used for comparison with the experimental results.

2 EXPERIMENTAL TECHNIQUES

The downdraft wind tunnel at the Department of Civil Engineering, Fig. 1, was supplied by a 15kW centrifugal blower discharging air into a settling chamber. The settling chamber is 1825mm long and has cross-sectional dimensions of 1235mm high by 620mm wide. The vertical and horizontal contractions have an area ratio of 4:1 and 2:1 respectively. A circular nozzle was connected to the rectangular outlet. The aim of the circular nozzle was to minimise turbulence and to create a symmetrical, steady outflow. The nozzle had a contraction length of 150mm and a diameter of 310mm and circular length of 750mm. The discharge speed at the outflow was approximately $20ms^{-1}$. The vertical test surface was made from 18mm plywood, 2.4m square, mounted on a movable steel frame. Measurements were made with a pair of constant temperature hot-wire anemometers. The reference probe was placed at the centre of the wind tunnel outflow.

Measurements were taken both on the flat surface, Fig. 2, and behind the crest of a flat topped embankment 40mm high with a 2:1 slope, Fig. 3, allowing topographic speed-up factors to be calculated. Measurements were taken for three distances between the wind tunnel outflow and the test surface: 0.5, 2,and 5 times the diameter of the jet. The embankment was tested at three different radii from the centre of the jet: 1, 1.5 and 2 times the diameter of the jet. Mean velocity profiles were obtained above the leading edge of the flat-topped embankment and at several distances behind the leading edge.

A microburst is defined as having a diameter of between 400m and 4km; therefore, a geometric scaling factor of between 1/1300 and 1/13000 is applicable for this series of tests.

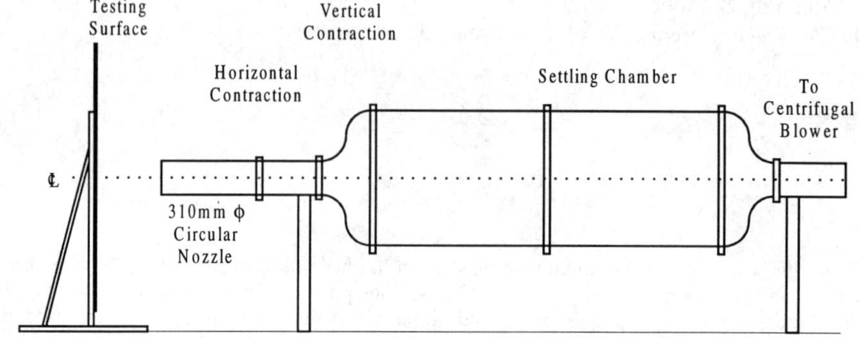

Fig. 1: Schematic Layout of the Thunderstorm Wind Tunnel

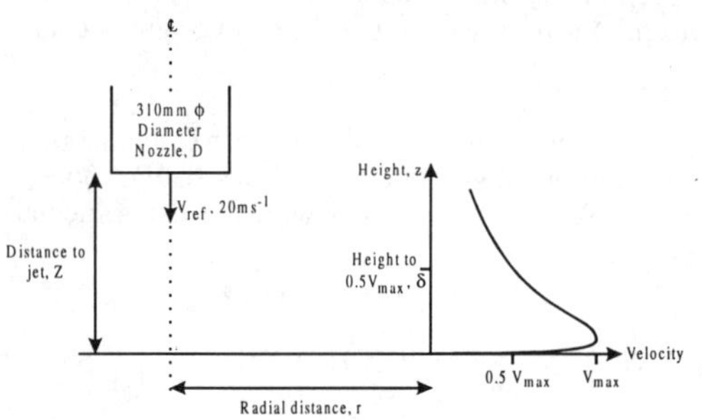

Fig. 2: Definition of Terms for Flat Surface

Fig. 3: Definition of Terms for Embankment

3 CFD ANALYSIS

In order to use the experimental data for full-scale applications it is important to determine any scale effects. This is best done using a model, and therefore CFD was used to make calculations for both the experimental setup and the full-scale geometry. All modelling was performed using the CFX4 package, a commercially-available finite volume CFD code which provides a wide range of models and has undergone extensive validation (CFX4, 1997).

CFD modelling works by splitting the geometry of interest into small volumes over which all variables, such as velocity and pressure, are assumed to be constant. The conservation equations are then transformed to algebraic equations, which relate values at the centroid of each volume to those of the surrounding volumes. These coupled equations are then solved iteratively to produce an approximate solution to the original equations. The accuracy of the solution is checked by making sure that the solution obtained is independent of the number of volumes used.

For this problem the flow-field is determined by conservation of mass and momentum. The only complication is the need to choose an appropriate turbulence model. The geometry used in earlier work by Selvam & Holmes (1992) was used to benchmark the CFX results. In this calculation the flow generated by a jet of width 450m, with a parabolic velocity profile with a maximum speed of $20ms^{-1}$, at a height of 2680m above the ground was simulated. The results obtained in this study agreed very well with those of Selvam & Holmes, as well as with data from laboratory scale and full-scale data. Simulations were also performed to determine the effects of: (i) the choice of turbulence model, (ii) the effect of surface roughness, and (iii) the effect of the inlet velocity.

Two turbulence models were used: the well-known k-ε model, which assumes isotropic turbulence and the Differential Reynolds Stress Model (DSM), which solves conservation equations to determine all of the components of the Reynolds stress tensor. The results from this comparison showed that there were some differences in the shape of the predicted velocity profile. There was also greater similarity between small and large-scale simulations using the DSM model. As the flow will be highly anisotropic, the DSM model was used in all subsequent simulations.

Increasing the surface roughness, which effectively changes the turbulence profile in the vicinity of the wall, caused a more rapid decay of the peak velocity of the wall jet and an increase in the height of the peak velocity above the ground. This was confirmed experimentally by placing carpet over the testing surface, (Pyne, 1993).

For inlet boundary conditions specified at least two jet diameters above the ground, the shape of the predicted velocity profiles was insensitive to the magnitude of the inlet velocity.

4 RESULTS AND DISCUSSION

For a jet diameter, D, mean velocity profile measurements were taken from the centre of the jet impact point up to a distance of 3D, in increments of D and H, where H is the height of the embankment used in the later experiments. The test surface was located at distances of 0.5D, 2D, and 5D from the jet. The development of the mean velocity profile is shown in Fig. 4 for the test surface at 2D from the jet outlet. It can be seen that the height to the peak mean velocity increases as the radial distance increases. Assuming a 1km diameter jet, the height to the peak mean velocity at a distance 3D from the jet impact point is approximately 65m which agrees well with the previous studies.

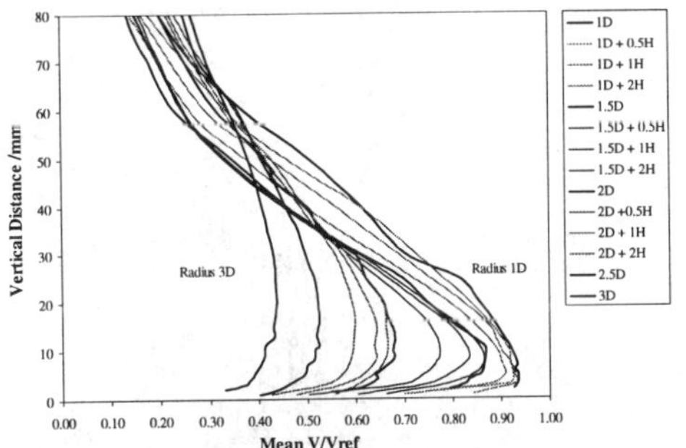

Fig. 4: Radial Development of the Mean Velocity Profile, Surface 2D from Jet

Fig. 5 shows the same velocity profiles normalised with respect to the peak mean velocity at that profile, V_{max}, versus the height normalised with respect to the height where the velocity is equal to half its maximum value, δ. The graph also shows a comparison with previous work. It can be seen that the mean velocity profile stabilises beyond approximately 1.5D, the profiles agree well with previously published data. Similar results were determined for the test surface at 0.5D and 5D. The further the test surface was from the jet outlet, the greater the radius required to develop the stabilised profile. Equation 1 was found to fit the normalised data well, Fig. 5:

$$\frac{V}{V_{max}} = 1.55\left(\frac{z}{\delta}\right)^{1/6}\left[1-erf\left(0.70\frac{z}{\delta}\right)\right] \qquad 1$$

where V is the velocity at height z, and erf is the error function.

The actual mean velocity profile, through the estimation of V_{max} and δ, has been accurately estimated for two different diameters of jet, Wood & Kwok, (1998).

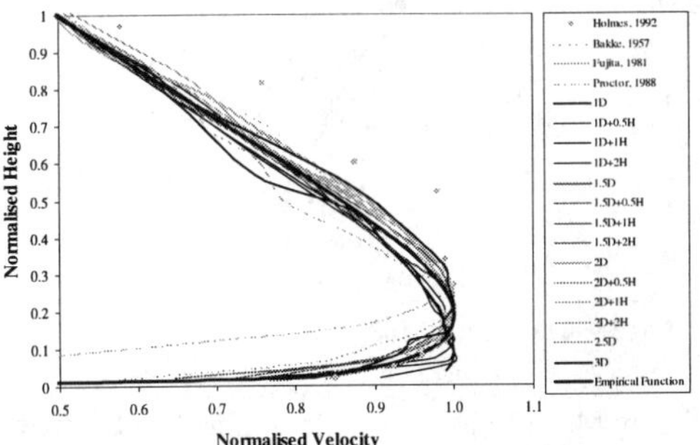

Fig. 5: Normalised Mean Velocity Profiles, Surface 2D from Jet

A comparison between the experimental and CFD vertical mean velocity profiles for the flat board, at a distance 2D from the jet outlet, is shown in Fig. 6. It is evident that the two sets of results do not coincide exactly, but that the ratios of measured mean velocity to the reference mean velocity are of a similar magnitude. It can also be seen that the rate of decrease in the peak velocity ratio is similar in both cases and that the height to the peak velocity ratio increases as the radial distance from the jet impact point increases. Altering the assumed ground roughness in the CFD model could be carried out to improve the closeness of fit between the measured and calculated values, but this was not deemed to be worthwhile. After normalising the data, Fig. 7, it can be seen that the CFD results show a similar trend to the experimental results and that both are predicted well by Eqn. 1.

With the inclusion of a 40mm high embankment and with the crest at a radius of 1.5D, the measured velocity profiles are shown along with those for the flat board, Fig. 8. It can be seen that the peak velocity at the crest of the embankment is significantly larger than that for the flat board, but rapidly drops off behind the crest. Fig. 9 shows the identical results using CFD. Comparing Figs. 8 and 9 it is clear that the trends at the lower level between the two sets of results are similar, especially with the relative change in height to the peak velocity ratio, although the CFD does not predict such a large peak at the

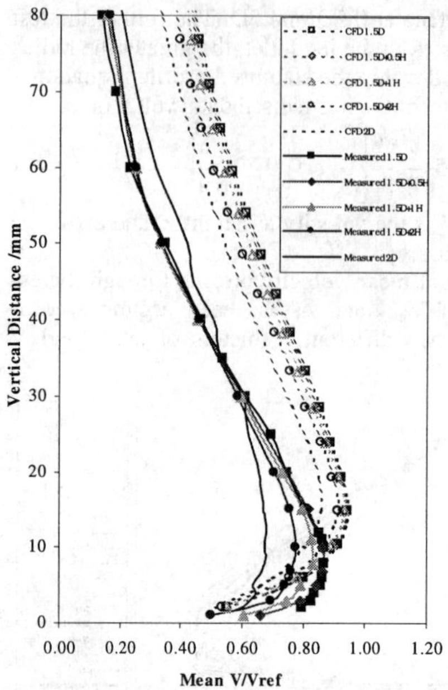

Fig. 6: Mean Velocity Profiles on the Flat Surface, CFD and Wind Tunnel Results, Board at 2D from jet

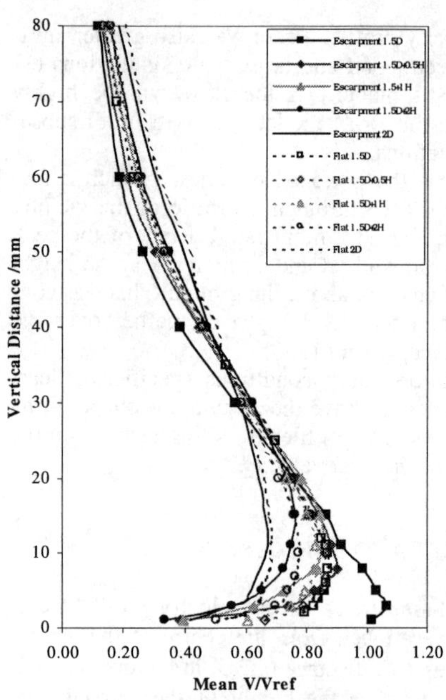

Fig. 8: Wind Tunnel Mean Velocity Profiles for Flat Surface and Embankment, Surface at 2D, Embankment at 1.5D to Crest

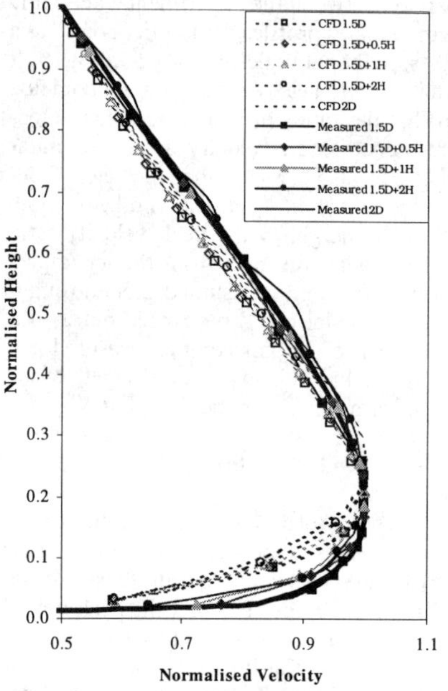

Fig. 7: Normalised Mean Velocity Profiles on the Flat Surface, CFD and Wind Tunnel Results, Board at 2D from jet

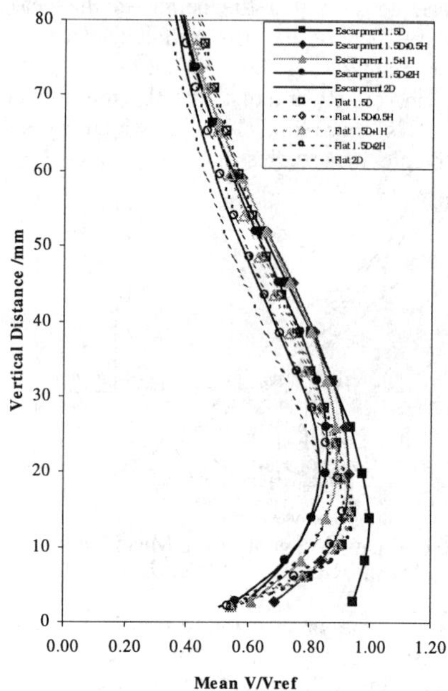

Fig. 9: CFD Mean Velocity Profiles for Flat Surface and Embankment, Surface at 2D, Embankment at 1.5D to Crest

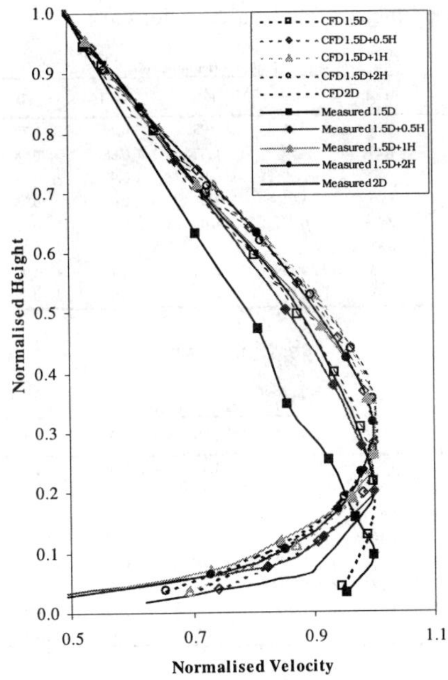

Fig. 10: Normalised Mean Velocity Profiles for Embankment, Surface at 2D

Table 1: Speed-up Factors for an Embankment with a Crest at 1.5D from the Centre of the Impinging Jet, Surface at 2D

	Radial Distance from Centre				
Height /mm	**1.5D**	**1.5D+0.5H**	**1.5D+1H**	**1.5D+2H**	**2D**
3	1.29	0.94	0.83	0.80	0.90
5	1.20	0.98	0.91	0.88	0.97
10	1.11	1.04	1.00	0.95	0.97
20	1.01	1.03	1.07	1.04	0.97
30	0.93	1.00	1.02	1.06	0.97
40	0.84	0.97	1.00	1.03	0.95
50	0.76	0.92	0.96	1.04	0.95
80	0.85	0.85	0.86	0.95	0.92

a. Measured in Wind Tunnel

	Radial Distance from Centre				
Height /mm	**1.5D**	**1.5D+0.5H**	**1.5D+1H**	**1.5D+2H**	**2D**
3	1.56	1.15	1.03	0.97	1.08
5	1.32	1.04	0.95	0.90	1.00
10	1.10	0.97	0.90	0.87	0.94
20	1.07	1.02	0.98	0.96	0.99
30	1.07	1.07	1.06	1.05	1.06
40	1.06	1.09	1.10	1.09	1.07
50	1.03	1.08	1.10	1.10	1.07
80	0.88	0.94	0.98	1.05	1.12

b. Calculated by CFD

crest of the embankment. Normalising the results for the embankment, Fig. 10, it is evident that at the crest of the embankment the developed boundary profile is disrupted by the embankment, but quickly re-establishes itself behind the crest.

Topographical speed-up factors are defined as; mean velocity at a height z above a topographical feature / mean velocity at height z above flat ground, at the same position in the jet flow without the topographic feature. Table 1 compares the speed-up factors for both the CFD and experimental data. Both methods give comparable speed-up factors, which agree well with similar results from Letchford & Illidge (1998). Table 2 shows average, maximum, and minimum speed-up factors at the same relative position behind the crest of the embankment. These tables are formed from the measurements at the nine locations tested: with the crest of the embankment at 1D, 1.5D, and 2D from the centre of the impinging jet, with the test surface 0.5D, 2D, and 5D from the jet outflow. Table 2 also contains computed speed-up factors from the Australian Wind Code, AS1170.2-1989, (1989) for developed boundary layer flow. It is evident that near ground level at the crest of the embankment the speed-up factors can be greater than the code values, but these drop off rapidly behind the crest of the flat-topped embankment and at higher levels. Only at the higher levels remote from the crest are the speed-up factors greater than those given in the code, but the magnitude of the velocity is likely to be less for flow from a thunderstorm.

5 CONCLUSIONS

The measured velocity profiles for the flat surface agreed well with previously published data once the expanding velocity profile had stabilised, beyond approximately 1.5 jet diameters from the centre of the impinging jet. In this stabilised region an generic empirical equation was developed to accurately predict the normalised mean velocity profile.

A CFD analysis using CFX4 with a differential stress model showed similar trends to the measured results for both the flat surface and the embankment.

Average topographic speed-up factors for the embankment were similar to those in the current Australian wind loading code for points above the crest of the embankment, but dropped off rapidly behind and above the top of the embankment. This indicates that the speed-up factors in the current Australian code would be predominantly conservative for a wind profile generated by a thunderstorm downdraft.

Table 2: Topographic Speed-up Factors for Nine Embankment Configurations

Height /mm	Distance Behind Crest				
	Crest	+0.5H	+1H	+2H	+D
3	1.48	1.21	1.09	1.00	1.01
5	1.40	1.18	1.12	0.99	1.01
10	1.25	1.17	1.13	1.06	1.02
20	1.14	1.10	1.10	1.07	1.04
30	1.09	1.06	1.07	1.09	1.08
40	0.98	1.00	1.08	1.07	1.26
50	0.95	1.01	1.08	1.07	1.26
80	0.90	0.92	1.03	1.08	1.24

a. Maximum

Height /mm	Distance Behind Crest				
	Crest	+0.5H	+1H	+2H	+D
3	1.29	0.86	0.73	0.79	0.79
5	1.20	0.97	0.87	0.88	0.88
10	1.10	0.99	0.99	0.92	0.92
20	1.00	1.01	1.01	0.99	0.93
30	0.91	0.97	0.97	0.99	0.92
40	0.83	0.90	0.95	0.98	0.93
50	0.76	0.87	0.84	0.95	0.92
80	0.73	0.75	0.80	0.91	0.92

b. Minimum

Height /mm	Distance Behind Crest				
	Crest	+0.5H	+1H	+2H	+D
3	1.39	1.02	0.89	0.86	0.90
5	1.31	1.08	0.98	0.92	0.95
10	1.18	1.08	1.05	0.99	0.97
20	1.05	1.05	1.05	1.03	1.00
30	0.98	0.99	1.01	1.03	1.01
40	0.90	0.95	0.99	1.02	1.03
50	0.85	0.92	0.95	1.01	1.04
80	0.82	0.85	0.90	0.98	1.03

c. Average

Height /mm	Distance Behind Crest				
	Crest	+0.5H	+1H	+2H	+D
3	1.40	1.36	1.32	1.24	1.09
5	1.39	1.35	1.31	1.23	1.09
10	1.36	1.32	1.29	1.21	1.08
20	1.29	1.26	1.24	1.18	1.07
30	1.23	1.21	1.18	1.14	1.05
40	1.17	1.15	1.13	1.10	1.04
50	1.11	1.09	1.08	1.06	1.02
80	0.92	0.92	0.93	0.95	0.98

d. Code Values

6 ACKNOWLEDGMENT

The authors acknowledge the support given by the Australian Research Council for this work.

7 REFERENCES

Australian Standard, SAA Loading Code, Part 2: Wind Loads, AS 1170.2-1989, *Standards Australia*, 1989.

Bakke, P., An Experimental Investigation of a Wall Jet, *J. Fluid Mech.*, Vol.2, pp.467-472, 1957.

CFX4, CFX4.2 Solver Guide, AEA Technology plc, Harwell, Didcot, Oxon., OX11 0RA, UK, 1997.

Fujita, T.T., Tornadoes and Downbursts in Context of Generalised Planetary Scales, *J. Atmos. Science*, Vol.38, No.8, pp.1151-1534, 1981.

Holmes, J.D., Physical Modelling of Thunderstorm Downdrafts by Wind-Tunnel Jet, *2nd AWES Workshop*, Monash University, pp.29-32, Feb. 1992.

Letchford, C.W., & Illidge, G., Topographic Effects in Simulated Thunderstorm Downdrafts by Wind Tunnel Jet, *7th AWES Workshop*, Auck., 1998.

Proctor, F.H., Numerical Simulations of an Isolated Microburst, Part 1: Dynamics and Structure, *J. Atmos. Science*, Vol.45, No.21, pp.3137-3160, 1988.

Pyne, J., The Modelling of Thunderstorm Downdrafts, *U. Sydney, BE(Hons) Thesis (unpub.)*, 1993.

Selvam, R.P., & Holmes, J.D., Numerical Simulation of Thunderstorm Downdrafts, *J. WEIA*, 41-44, pp.2817-2825, 1992.

Wood, G.S., & Kwok, K.C.S., An Empirically Derived Estimate for the Mean Velocity Profile of a Thunderstorm Downdraft, *7th AWES Workshop*, Auckland, 1998.

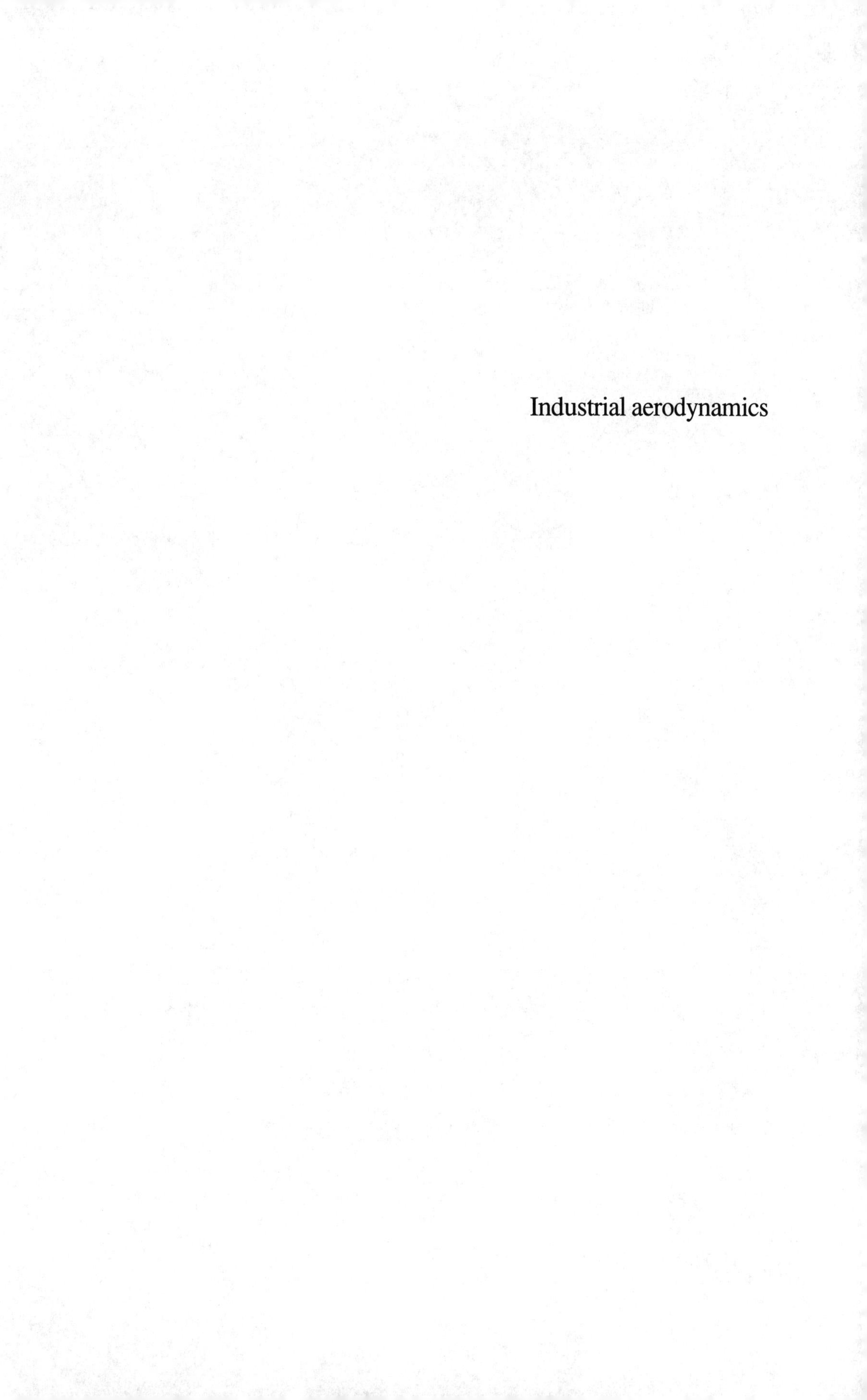

Industrial aerodynamics

Wind Engineering into the 21st Century, Larsen, Larose & Livesey (eds) © 1999 Balkema, Rotterdam, ISBN 90 5809 059 0

Small scale experiments to simulate the smoke extraction from large spaces

H.J.Gerhardt & O.Krüger
I.F.I. Institut für Industrieaerodynamik GmbH, Aachen, Germany

J.Wacker
Wacker Ingenieure, Birkenfeld, Germany

ABSTRACT: The paper is concerned with fire engineering work. It presents results of studies of smoke flow in and from large spaces. When taking into account external wind effects wind tunnel experiments are the only reliable means to optimise the necessary smoke ventilation and air inlet areas and their positions. After giving an introduction into the underlying similarity conditions the paper discusses the proposed methodology using two case studies.

1 INTRODUCTION

Fire safety is an important issue in the design and permit of buildings with large spaces and high occupancy, e. g. train stations, air port terminals, fair buildings, hotel atria, etc. To ensure a proper evacuation of occupants and a quick response of the fire brigades the escape routes – being usually the attack ways for the fire brigades, too – have to be kept smoke free under all environmental influences, e. g. under wind action. It is common fire engineering practice to calculate the thermal effects of fires in large spaces using numerical models based on the energy equation and continuity equation. These so-called multi room fire codes (MRFC) are inappropriate to calculate smoke flows. Here, computational fluid dynamic models (CFD) or experimental methods have to be used to check if the escape routes are unaffected by smoke. For complex room shapes and/or critical fresh air supply flows physical modelling may be the only reliable means of prediction. The first part of the paper will deal with the similarity laws of experimental investigations and develop the limits of the usual practice of simulating the buoyancy effect of fire using low density gas flows at increased temperatures. The limits given are checked by experiments where typical model scales are in the range of 1:5 through 1:20. The second part of the paper will look into the problem of simulating the thermally induced flows in large spaces at model scales of up to 1:200. This will be demonstrated for typical case studies.

2 METHODOLOGY

2.1 Fire and smoke spread

Each fire produces smoke. The smoke volume flow depends on the burning material and in particular on the heat flux of the fire (John 1988). The flow of smoky gases and room air in a large space created by the fire is shown schematically in figure 1. Due to buoyancy effects the smoky gases rise above the fire leading to the so-called plume flow. It is comparable to a jet flow with an imaginary source below the room floor. In the mixing zone of the jet room air is entrained into the jet flow leading to an increase in mass flow and a decrease in smoke temperature with increasing height of the smoke free zone. To keep escape routes free of smoke it is necessary to extract smoke from the fire room, using smoke ventilators. A steady state situation is reached, when the smoke extracted by the smoke ventilators is equal to the smoke in the plume entering the smoke reservoir. Continuity requires an inflow of replacement air, the mass flow of which is equal to the mass flow of the flow vented through the smoke ventilators. For intensive air flow in the room, either due to strong buoyancy flows or due to large momentum inflow of replacement air the risk of strong mixing between replacement air and smoke at the boundary between the smoke reservoir underneath the ceiling and the smoke free area underneath is fairly large. The intensity of the room flow depends strongly on the inflow properties and on the geometry of the room.

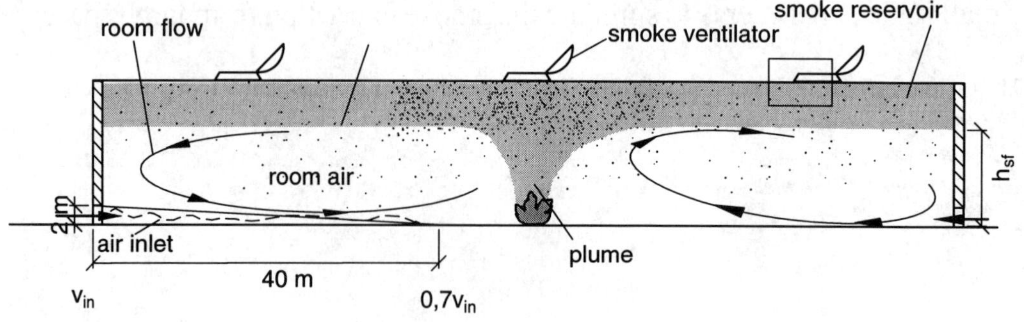

Figure 1: Schematic of fire induced flows in a large space

The strength of the replacement air jet flow is strongly influenced by the external wind pressure distribution at the air inlet and smoke ventilator areas. Commonly used numerical methods to check the stability of the smoke free layer do not account for external wind influences. In particular, for complex building geometries wind tunnel studies are the only means to proof the necessary efficiency of the smoke extraction method proposed. Typical wind tunnel dimensions of 2 m lead to model scales in the range of 1:100 or smaller.

2.2 *Similarity considerations*

Conducting model scale experiments to study smoke venting situations the flow field and the temperature field for model and full scale have to be similar. Since the smoke flow is strongly influenced by buoyancy effects the two fields are coupled. To ensure flow field similarity the ratio of the relevant forces acting on flow particles must be the same for model and full scale. For the present case this requires equality of the Reynolds-number

$$Re = u \frac{L}{\nu} \tag{1}$$

with u = characteristic velocity, L = characteristic length, and ν = kinematic viscosity, and equality of the Archimedes-number

$$Ar = \frac{g \beta \Delta T L}{u^2} \tag{2}$$

with β = thermal expansion coefficient and ΔT = temperature difference for model and full scale. The Reynolds-number gives the ratio of inertia forces to frictional forces, the Archimedes-number the ratio of buoyancy forces to inertia forces.

For real fires and typical characteristic length (= height of room) of 3 m to 15 m and typical temperature differences ΔT = 300 K to 780 K, giving typical thermally induced velocities u = 1 m/s to 5 m/s (Kramer & Thiessen 1986), will give typical Archimedes-numbers Ar = 2.3 to 30.4.

Flows induced by fires in large spaces are usually turbulent. The flow properties may be thought of as consisting of a time averaged and an unsteady, fluctuating part. Reynolds-number and Archimedes-number are commonly obtained by using the time averaged values of velocity and temperature. In addition, the turbulent diffusion parameters – A_τ for the flow field and A_q for the temperature field – have to be considered.

Müllejans (1973) has conducted an in-depth similarity study for thermally induced flows. It distinguished between flows close to and far away from walls. For the wall region the turbulent motion is restricted by the presence of the wall. Similarity requires the equality of Reynolds- and Archimedes-number for model and full scale. In regions without wall influence flow field similarity requires only to satisfy the Archimedes-number. This is due to the fact that the ratio of the turbulent properties of the flow and the temperature field are largely independent of the Reynolds-number and that the parameter describing the turbulent diffusion of the flow field is proportional to the Reynolds-number ($A_\tau \sim Re$). The plume flow for fires in large spaces is outside the wall influence region. Since the plume flow is the driving mechanism of the room flow the similarity condition for model scale experiments may be simplified:

$$\left(\frac{\Delta T}{u^2}\right)_M = M \cdot \left(\frac{\Delta T}{u^2}\right)_G \quad (3)$$

with M = scale factor. Equation (3) shows that the model scale temperature has to be larger than the full scale temperature by the scale factor if in both cases the velocities are equal.

For typical temperatures of smoky gases near the fire similarity cannot be obtained by heating the model gas. We propose instead to simulate the buoyancy effects by using a light gas (helium) for the model smoke flow. Using the density instead of the temperature difference equation (3) may be rewritten:

$$(\Delta\rho \cdot u^2)_M = \frac{1}{M} \cdot (\Delta\rho \cdot u^2)_G \quad (4)$$

The maximum density difference when using helium as model gas is $\Delta\rho$ = 1.09 kg/m^3. For a typical smoke temperature immediately above the fire T_{SG} = 700 K the density difference for the fire is

$$\Delta\rho_{FS} = \rho_{air} - \rho_{SG} = \rho_{air}\left(1 - \frac{T_{air}}{T_{SG}}\right) = 0.71$$

Scale experiments with $u_M = u_{FS}$ and with Archimedes-similarity are possible only for scale factors $M \leq 1.4$. For typical scale factors M = 5 to 20 (and possibly up to 200) the model Archimedes-number is usually smaller than the full scale one. Thus, the stability of the thermal layer at full scale is larger than for the small scale study. Larger stability of the thermal layer will decrease the risk of mixing between smoke and room air. Therefore, model scale experiments with smaller Archimedes-number will provide information on the safe side.

For large scale factors (M > 100) it is advisable to modify the model Archimedes-number by decreasing the plume velocity. However, turbulent flows have still to be ensured. Results using this methods will be presented and discussed in chapter 3.

2.3 *Experimental technique*

Models of buildings or parts of large building complexes have been investigated at typical scales of 1:200 through 1:20. The smoke flow has been investigated qualitatively using helium as a carrier-gas with an aerosol for visualisation purposes. The flow has been documented via video film and photographs. To quantify the smoke density and the concentration of hazardous components of smoke gases the tracer-gas technique has been used, again using helium as carrier gas and adding propane as tracer. Local tracer-gas concentrations have been measured using a flame ionisation detector (FID), either a slow or a fast FID. The decrease in tracer-gas concentration allows also the calculation of local temperatures and hence the evaluation of thermal loads on structural elements.

For both techniques the mass flow rate and thus the smoke exit momentum and the Archimedes-number is controlled by Brooks mass flow regulators. Tests are always conducted without and with wind influence. Usually the yearly mean wind velocity is taken in the wind tunnel tests and eight wind directions ($\Delta\alpha$ = 45°) are investigated. Small scale models are studied in the I.F.I. boundary layer wind tunnel where the atmospheric wind flow is simulated in the usual manner with vortex generators at the beginning of the fetch and roughness elements on the wind tunnel floor. Larger models are tested in the I.F.I. industrial wind tunnel, an open jet wind tunnel (max. nozzle size 4.2 m by 1.6 m), where the wind profile and the turbulence intensity – but not the turbulent eddy size – is modelled using a grid at the nozzle exit plane.

3. RESULTS OF CASE STUDIES

3.1 *Allianz-Kai, Frankfurt*

The Allianz insurance company plans the construction of a new office building in Frankfurt. It consists of a long, 6 storey building which is connected to 4 finger like building parts of roughly the same height, set perpendicular to the long part and to a high rise via an atrium-like space (so-called magistrale). Between high rise and one of the office "fingers" a 2-storey entrance hall is situated which includes a cafeteria. The cafeteria is completely open towards the magistrale. A schematic cross section of the entrance hall and the magistrale is given in Figure 2. For clarity reasons all other building parts have been omitted.

One of the design fires to be considered was a fire in the cafeteria. The Frankfurt fire authority requested that no smoke penetrated into the magistrale for this fire situation since it is considered as a necessary escape route.

An experimental investigation was conducted at a 1:200 scale model of the complete building complex. Both, no wind and wind situations where considered. The necessary smoke ventilator areas in the roof of the entrance hall and the position and size of the air inlets were optimised. For this optimal smoke

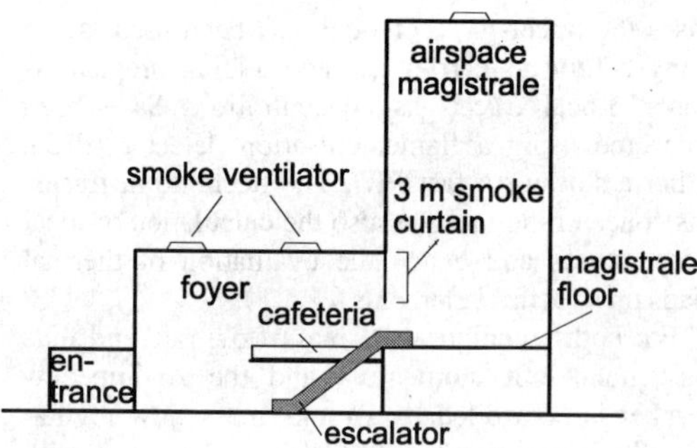

Figure 2: Schematic drawing of cross-section of Allianz-Kai Magistrale and foyer

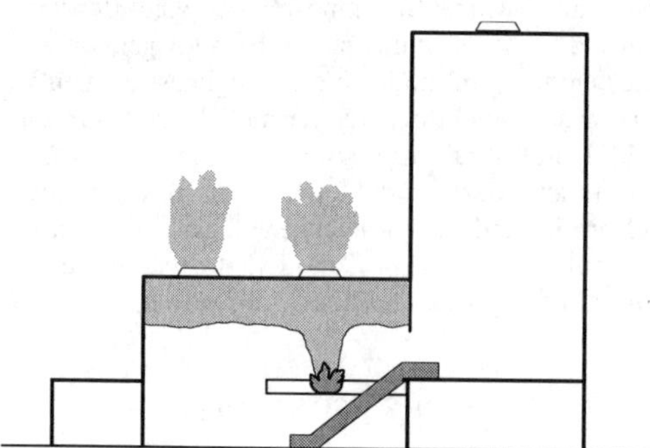

Figure 3: Schematic drawing (taken from photographs) of smoke flow in the foyer of the Allianz-Kai building

extraction situation a 3 m deep smoke curtain, see Figure 2, is necessary to obtain the design aim. Figure 3 shows schematically the smoke flow and smoke reservoir for the optimal situation taken from photographs.

The Archimedes-number for the experimental study using the 1:200 scale model is smaller than for the full scale situation (by a factor of approximately 50). The influence of the discrepancy of the Archimedes-number was checked using a 1:20 scale model and investigating the smoke flow for the no-wind condition. Basically the same smoke flow – as in the 1:200 scale model – was observed. Due to the larger buoyancy effect and therefore a larger stability of the smoke reservoir the height of the smoke layer for the 1:20 scale model is smaller by approximately 30 %. Thus, the smoke curtain specified on the basis of the 1:200 scale experiments is on the safe side.

3.2 *Airshipyard*

The construction of a large shipyard (height 106 m, length 430 m, width 210 m) has recently been started South-east of Berlin. The shipyard will be used to build the large CargoLifter airships. The middle section of the building is a half cylinder, with quarter sections of a sphere on both ends. The spherical end parts consist of sectional doors giving the possibility of almost completely opening the half cylindrical cross section. For ventilation purposes – both, every day venting and smoke venting – ventilators are fitted in the upper portion of the shipyard and replacement air intakes in the lower wall regions. A natural ventilation system was anticipated from the beginning of the design work. Figure 4 gives a schematic drawing of the shipyard showing the smoke ventilator and the air inlets positions.

The smoke extraction efficiency without and with wind action was experimentally and numerically investigated. Due to the size of the building a model scale of 1:200 was necessary. From a fire engineering study the following two fire situations were considered to be critical and to be investigated:

- fire on the ground floor of strength 1.5 MW
- fire in 50 m height (height of airship tail unit) of 10MW

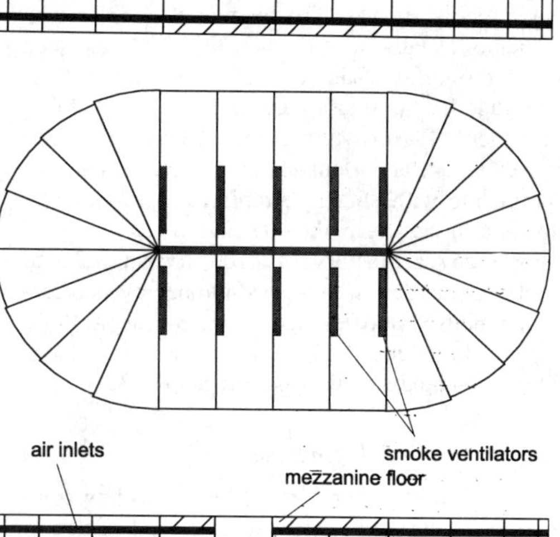

Figure 4: Air let and smoke ventilator areas of airshipyard

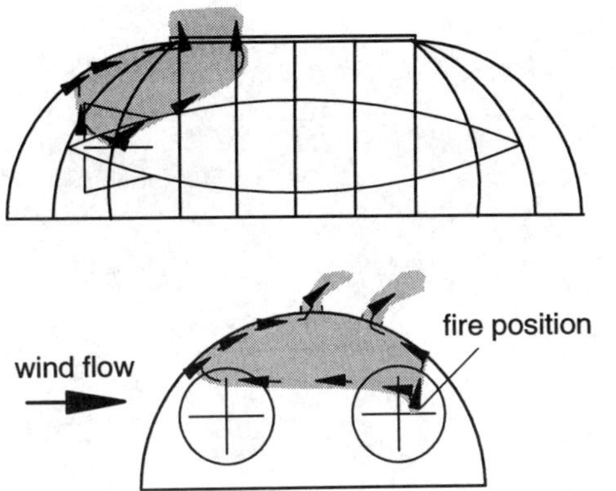

Figure 5: Schematic drawing (taken from photographs of smoke flow in the airshipyard

For the small fire a fire area $A_F = 20\ m^2$ and a fire duration of 30 minutes was considered to be realistic, for the large fire the area and fire duration were $A_F = 20\ m^2$ and 2 hours. For the estimated fire duration a smoke free layer of height 10 m should be ensured without and with external wind influence.

The positions and the size of the smoke ventilation area and air inlet areas were investigated and optimised experimentally; the flow field for the specified smoke ventilators and air intake were checked numerically. The most critical situation is the large fire (10 MW) for which the smoke ventilator and air inlet areas were optimised. The wind tunnel investigation proved that only selected areas should be opened depending on the wind direction. The wind direction dependent control was also elaborated for the large fire situation.

Figure 5 shows a schematic drawing – taken from photographs - of the flow visualisation work for the large fire and wind direction perpendicular to the main axis of the building after the fire lasted for approximately 2 hours (full scale time). Due to the very efficient smoke extraction no smoke is entrained below the fire location.

In the numerical study, using the Phoenix-CFD-Code, the flow field and the temperature field have been calculated. The comparison between experimental and numerical results is best be done by taking the temperature field as basis. A room air temperature of 20 °C was assumed. Thus, the areas with ambient temperature may be considered to be smoke free. Experimental and numerical results agree very good.

CONCLUSIONS

Wind tunnel experiments are the only reliable way to study the smoke flow in and the smoke extraction from large spaces under external wind influence. The flow of smoky gases in large spaces is driven by the thermal effect of the fire and by the pressure difference due to the wind influence between replacement air and smoke extraction openings. Due to the limited size of most boundary layer wind tunnels physical modelling has to be done at typical scales 1:100 or smaller. For those scales the Archimedes-similarity giving the ratio of buoyancy forces to inertia forces will not be satisfying. The results given in the paper show that using small scale models the stability of the smoke reservoir is underestimated leading to smoke flow evaluations on the safe side.

LITERATURE

John, R. 1988. Rauch- und Wärmeabzug bei Bränden in großen Räumen, vfdb 1/88, 21 - 25

Kramer C. & V. Thiessen 1986. Bemessung von maschinellen Abzügen, FLT-Bericht 3/1/64/86

Müllejans, H. 1973. Über die Bedingungen von Modellversuchen in der Klimatechnik, Ki 8/83, Teil 6, S. 63 ff

Wind Engineering into the 21st Century, Larsen, Larose & Livesey (eds) © 1999 Balkema, Rotterdam, ISBN 90 5809 059 0

Internal ventilation study for complex architectural geometry

H. Hangan
The Boundary Layer Wind Tunnel Laboratory, University of Western Ontario, London, Ont., Canada
F. McKenty, L. Gravel & R. Camarero
Centre de Recherche en Calcul Appliqué (CERCA), Montreal, Que., Canada

ABSTRACT: A HVAC efficiency study is carried out for two, geometrically complex halls (atriums). The results of the simulations, in terms of temperature, velocity and air change fields, are used to modify the ventilation systems for the two atriums. The study proves the capability of a commercial Computational Fluid Dynamics (CFD) software to provide useful information to the mechanical engineers in order to assess and improve the HVAC design .

1 INTRODUCTION

Internal ventilation for large and complex architectural spaces is usually a demanding exercise, involving different groups such as architects, mechanical (HVAC) engineers and fluid dynamicists. Each of these groups has specific inputs in the project and so communication between these professionals and their specific tools is very important for the success of the project.

Computational Fluid Dynamics (CFD) is increasingly employed in examining this type of internal flows and assisting in the design of HVAC systems, in order to ensure the thermal and air velocity comfort for the occupants.

A typical CFD ventilation study provides information on the internal flow and air temperature fields as well as the net (or overall) air change rate and ventilation efficiency. The correlation between numerical and physical modelling for internal ventilation studies, both in terms of comparison (calibration) as well as in terms of coupling the two simulations (wind tunnel for external flow, CFD for internal flow) has already been addressed at the University of Tokyo (Kato et al 1992, 1997), for geometrically simple halls.

The present work is based on a BLWTL internal ventilation study for the Miami Center of Performing Arts (Hangan, et al. 1998). The study explores the capability of a commercial CFD code, Star-CD, to provide the temperature, velocity and passive scalar fields for complex architectural internal spaces and to assist the HVAC engineers in improving the internal ventilation systems.

2 METHODOLOGY

CFD simulations are carried out for large internal domains employing complex momentum and heat boundary conditions. The STAR-CD finite volume CFD code is employed for the simulations (STAR-CD v3.0c) which have been carried out on an IBM 595 Risc 6000 station rated at 250 Mflops with 1 GB of RAM.

2.1 *Geometry, Topology and Mesh*

The study is carried out for the Miami Center of Performing Arts consisting of two large halls, a Concert Hall and an Opera House. Both halls present complex internal spaces with two to three tiers, box level and orchestra level, see Fig. 1a and 1b.

The Concert Hall is symmetric, so that the simulation is performed on half of the domain making use of symmetry boundary conditions. The Opera House is non-symmetric and has a complex stage volume. The 3D geometry has been generated based on plane AutoCAD drawings provided by the architects and exported in IGES format, compatible with the CFD software pre-processor (see Fig. 1a, b).

The computational domain is made up of approx. 700,000 grid cells for the Concert Hall and approx. 880,000 cells for the Opera House. Based on preliminary calculations this allows the adequate resolution for the relevant flow and temperature field details. An unstructured, multi-block mesh is employed with higher cell density at the walls and in regions with complicated flow patterns, see Fig. 1a and 1b for the Concert Hall and the Opera House, respectively.

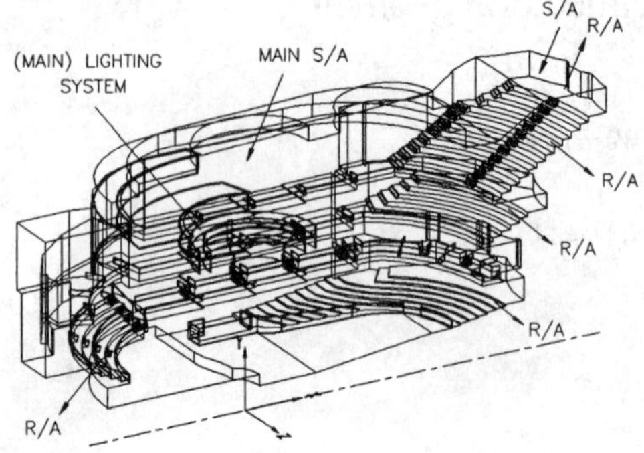

Fig. 1a. Concert Hall, Geometry and HVAC system

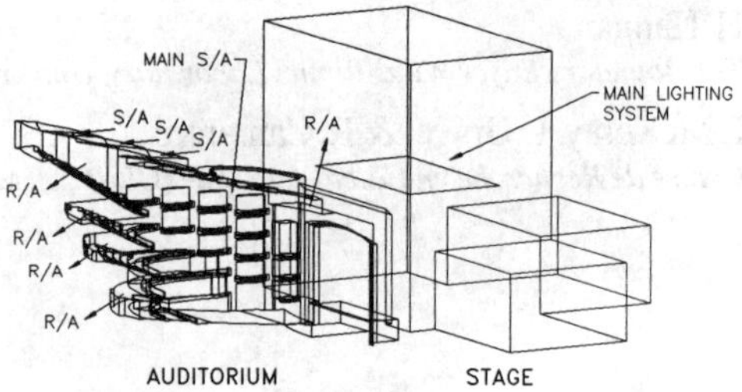

Fig.1b. Opera House, Geometry and HVAC system

The boundary conditions include complex HVAC systems with air supplies (S/A) mainly at the ceiling level and return-air (R/A) at the stage and the occupied auditorium levels. The thermal loads are mainly due to the stage lighting systems and also due to the occupied layer at the auditorium and stage levels. For the Concert Hall the lighting system is situated in a complex confining architectural space (baffles) and this space obstructs the convection of heat. For the Opera House, the stage space above the lighting system is large allowing the heat to be convected upwards by the buoyant flow (Fig. 1a, b).

The HVAC systems and the heating elements are treated as distributed sources. Various occupancies of the halls can be simulated by surface heat fluxes corresponding to the individuals. The walls are considered adiabatic.

2.2 *Solver*

The simulations solve momentum, continuity, enthalpy and scalar equations. A $k-\varepsilon$ turbulence model with wall functions is employed. The computational mesh is refined in areas adjacent to the walls in order to maintain the wall Reynolds number ($y^+ = u_\tau y/\nu$) in the range of y^+=80 - 100, and so insure adequate performance of the turbulent boundary layer wall functions, (STAR-CD v3.0c).

A tracing scalar is used in the simulation of the HVAC system in order to enable better visualizations of the HVAC air dilution in the domain. This scalar called mixture fraction represents the mass fraction (0-1) of newly supplied air in the resulting air mixture and is obtained by performing a transient scalar simulation (1) superimposed on the steady state flow field previously obtained:

$$\frac{\partial\phi}{\partial t} + \frac{\partial}{\partial x_i}(u_i\phi) = \frac{1}{\rho}\frac{\partial}{\partial x_i}(\Gamma\frac{\partial\phi}{\delta x_i}) + S_\phi \qquad (1)$$

where the left hand side scalar convective term is balanced by the right hand side scalar diffusion and source terms.

An additional scalar equation can be included as an option. This additional scalar enables the individuals or block of individuals to be treated as CO_2 sources in an air quality study context. The simulations can also include passive scalars for smoke-dispersion studies and can be related to the external wind tunnel pedestrian wind study in order to assess the impact of external sources of pollutants on the internal domain.

3 RESULTS

The simulations provided flow maps (velocity vectors and contours) and temperature fields in 3D and sectional views corresponding to the main elevations and for selected longitudinal and transverse sections for the two halls. The mixture of fresh (HVAC supplied) air at different time steps is also calculated. These three representations help determining the potential thermal and/or flow problems inside the halls and to correlate these problems with air mixing in the domain. Initially, restricted flow areas have been found where temperatures exceeded prescribed comfort thresholds and these areas corresponded to zones of vortex recirculation which in change have been correlated with poor air exchange.

This complementary approach (temperature, velocity and air change rates) proved to be helpful in improving the preliminary ventilation design in order to obtain quasi-uniformity of the velocity and thermal fields in the occupied layer and provide improved air quality to the halls.

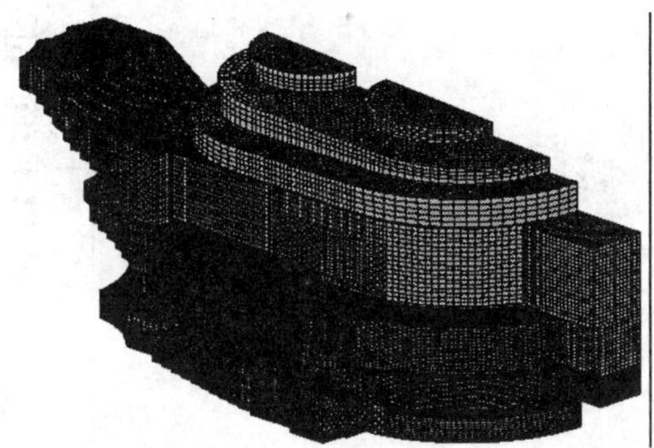

Fig. 2a . Concert Hall, Unstructured grid

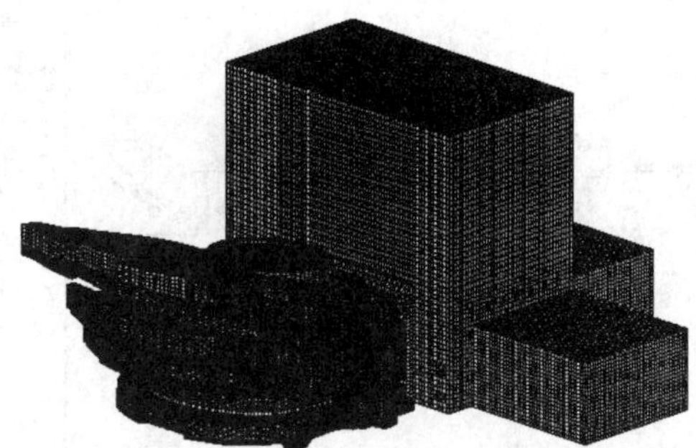

Fig. 2b . Opera House, Unstructured grid

3.1 Concert Hall

The simulation corresponding to the preliminary design of the Concert Hall HVAC system allowed for the identification of some restricted recirculation areas. The supply of air, mainly situated at the ceiling level determined two main vortical systems to develop: one at the front of the auditorium towards the stage and a second one at the level of the auditorium towards the back of the hall. The HVAC supplied air descended at relative high velocity towards the middle of the auditorium and recirculated back (and upwards) in two large vortices, see Fig. 3a.

The recirculation at the front of the auditorium, close to the stage level is enhanced by the presence of the complex stage lighting system, which convected the air upstream towards the ceiling. This frontal vortex resulted into relatively high ambient temperatures above the stage level (see Fig. 4a).

The second recirculation, towards the rear of the auditorium is enhanced by the heat released at the level of the occupied layer, and convected upstream towards the ceiling. This rear auditorium vortex resulted into restricted areas of relative high temperature in the occupied layers at the rear of the auditorium, mainly at the third tier level (Fig. 4a). Secondary recirculation areas have also been identified at the very rear of the third tier level, where air was locally recirculating between the air supplies and air returns in that specific area (Fig. 3a).

Both the main recirculation and the secondary recirculation areas correspond to lower rates of air exchange at various moments in time. Fig 5a shows the mixture of fresh supplied air in the Concert Hall after 15 minutes. These mixture fraction plots of newly supplied air, were obtained by carrying out a transient tracer simulation superimposed on the steady state flow circulation inside the hall. The air mixture plots are an indicator of the exchange rate of air. Fig. 5a can be easily correlated to the velocity and temperature plots (Fig. 3a and 4a). Note that the regions characterized by a lower exchange rate (lower mixture values) corresponding to the recirculation areas previously identified by the velocity plots and which, in change, are related to the higher temperature fields, see Fig. 4a

The information provided by the CFD simulation allowed the HVAC engineers to improve the ventilation system by adequate modifications of the supply and return air inlets and outlets. The main modifications were made at the ceiling level where the air supply (S/A) and air returns (R/A) have been more evenly redistributed on the periphery as well as between the auditorium and the stage parts. The preliminary design air supply was mainly directed towards the auditorium region. These modifications have been re-tested and the results of the second simulation are shown in Fig. 4b (velocity field), Fig. 5b (temperature field) and Fig. 6b (air mixture). These results show the net improvement in air circulation under the revised ventilation system for the Concert Hall, with lower velocity levels (Fig. 3b), more homogeneous and lower temperatures (Fig. 4b) and higher mixing rates (Fig. 5b). Based on the revised air mixture plots it has been determined that the time for a complete air exchange is 35 minutes compared to 45 minutes for the initial (preliminary) HVAC design.

3.2 Opera House

The first simulation of the air circulation for the Opera House showed a generally good air circulation for the HVAC system proposed.

The temperature field (Fig. 6) shows that the large space available above the stage level allowed the hot air, coming from the lighting system area to be convected upwards without restriction to the buoyant flow. The air mixture rates are presented in Fig. 7 after only 5 minutes. Based on the air mixture plots it has been determined that an entire air change (99%) takes place every 45 minutes at the auditorium level.

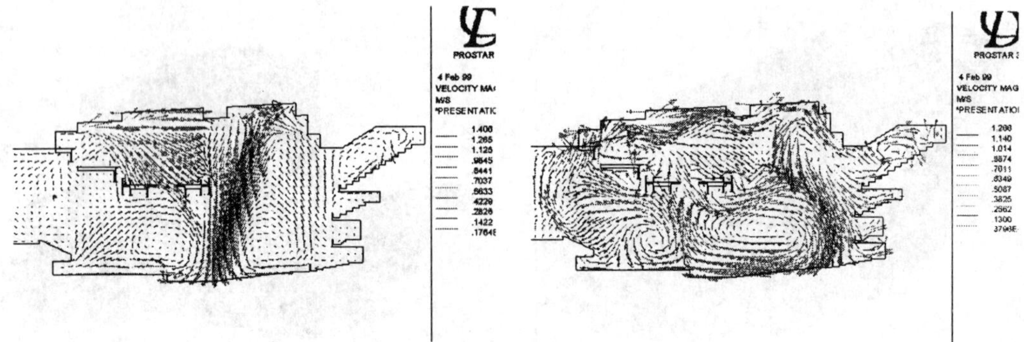

Fig. 3a. Velocity field, Preliminary HVAC system

Fig. 3b. Velocity field, Modified HVAC system

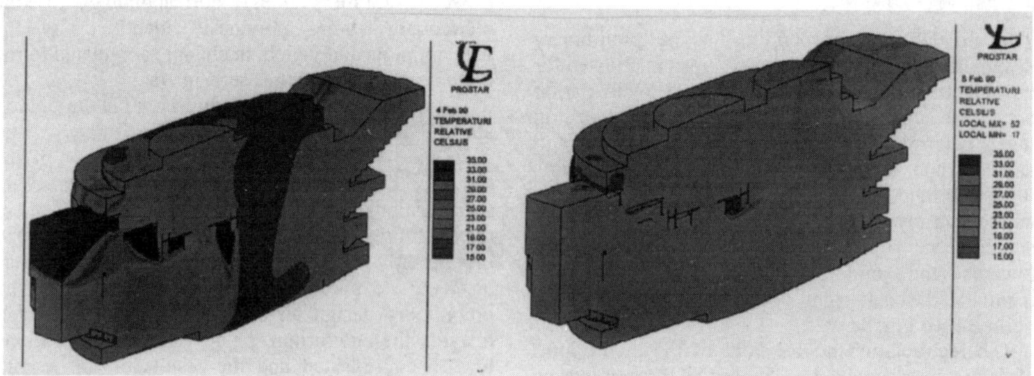

Fig. 4a. Temperature field, Preliminary HVAC system

Fig. 4b. Temperature field, Modified HVAC system

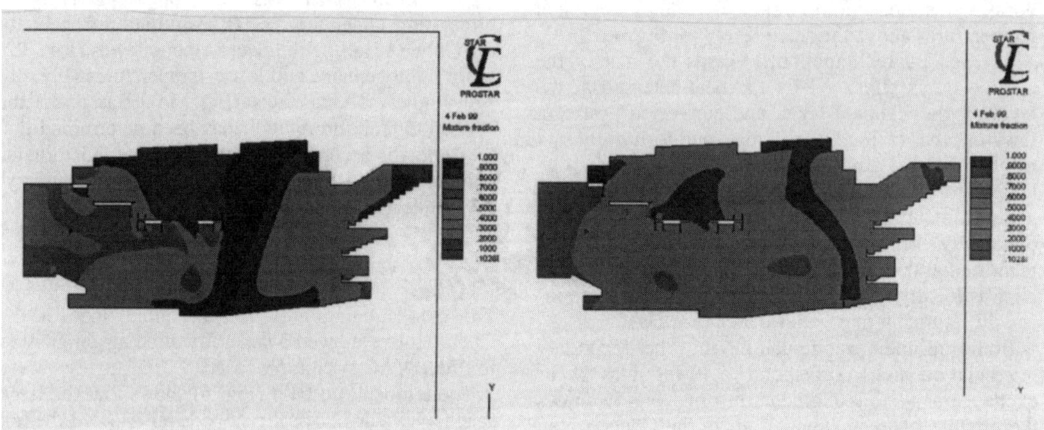

Fig. 5a. Mixture fraction, Preliminary HVAC system

Fig. 5b. Mixture fraction, Modified HVAC system

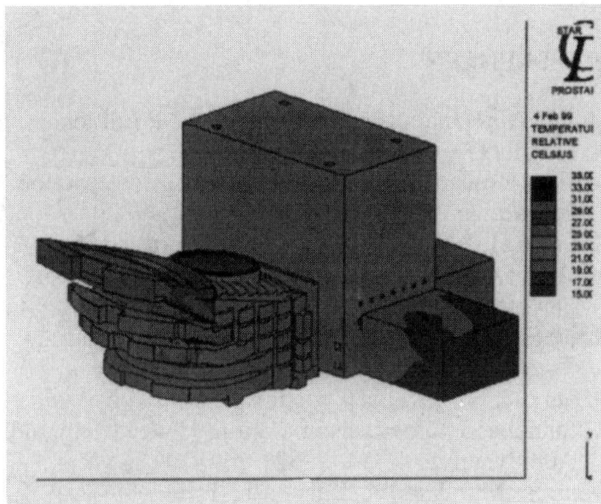

Fig. 6. Opera House, Temperature field

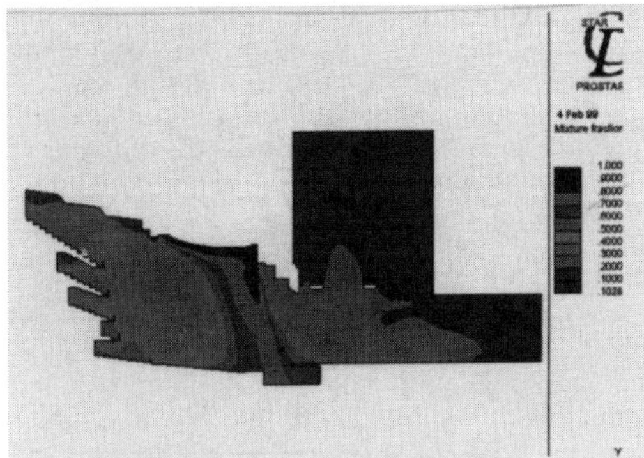

Fig. 7. Opera House, Air mixture, t=5 min

The velocity field (Fig. 8) shows a zone of relatively high velocity at the auditorium level. However this problem is expected to be easily corrected by introducing adequate deflectors for the auditorium air supplies. The HVAC system for the Opera House is presently undergoing final revisions. As was the case of the Concert Hall, the CFD simulations have provided an important validation tool for the ventilation system in terms of air flow and temperature fields. This allows for the optimization of the system in order to provide higher comfort levels which relate to quasi uniform velocity, temperature and exchange rates for the entire internal domain under investigation.

3.3 Comfort Criteria

The air velocity and temperature prediction has been compared with criteria for thermal and air velocity comfort (ASHRAE Handbook 1989). The general acceptability criteria is that at least 80% of the occupants feel comfortable within the environment.

Both thermal and draft (air velocity) comfort criteria have been taken into consideration. Statistical studies revealed correlations between the thermal comfort level and parameters such as temperature, humidity, sex and length of exposure (ASHRAE Handbook 1989). For a 3 (three) hours exposure period and combined occupancy (both male and female), neglecting the humidity factor the following equation defines the "comfort level" *Y*:

$$Y = 0.243t - 6.802 \tag{2}$$

where *t* is the temperature in Celsius and the comfort level *Y* ranges from -3 to +3 (from cold to hot). For both the Concert Hall (final prediction) and the Opera House the average predicted minimum and maximum temperatures of 20 and 28 deg.C (in the occupied layers) correspond to *Y*=-1.9 to *Y*=0

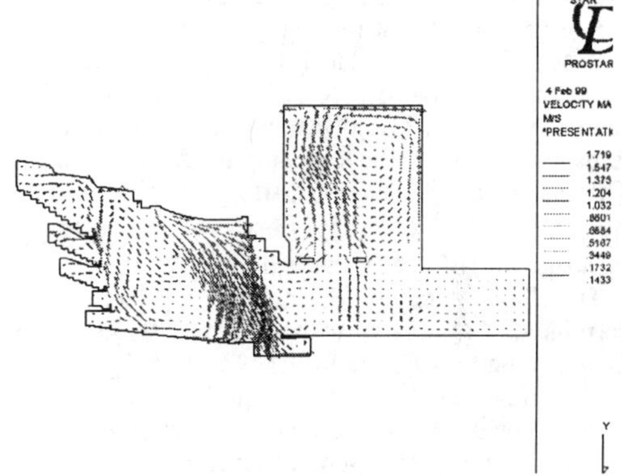

Fig. 8. Opera House, Velocity field

comfort levels range, i.e. slightly cool to perfectly comfortable.

The draft is an undesired local cooling of the human body caused by air flow. The air velocity (draft) comfort criteria is based on the percentage of the population feeling draft when exposed to a certain mean velocity. By correlating this criteria with the general environmental comfort criteria we can interfere that the acceptability is related to a statistical maximum of 20% of people dissatisfied with a certain draft velocity. Based on ASHRAE it has been determined that in average the Opera House provides a "dissatisfaction factor" of 3% corresponding to approx. 0.1m/s drafts. However, locally (see for example the orchestra level, Fig. 8) the air velocity can attend values of approx. 0.6 m/s corresponding to the maximum acceptable 20% dissatisfaction factor and this is subject to further improvements. For the Concert Hall an average dissatisfaction factor of 4% (0.15 m/s) and local factors of up to 17% (0.5 m/s at the orchestra and second tier levels, Fig. 3b) have been interfered.

4 CONCLUSIONS

CFD simulations proved to be an important tool in assessing the ventilation system for two different halls of complex geometry. The temperature, velocity and air mixture fields for a preliminary HVAC design have been analysed and compared with internal environment comfort criteria.

Based on these simulations, the HVAC engineers have been able to adequately modify the ventilation systems. The new design has been re-tested and showed net improvements with respect to the flow, temperature and air mixture fields in the domains.

In order to further assess the reliability of the numerical simulations physical experiments should be performed both in full scale (on the real geometry) (NRC-CNRC 1998) and/or at model scale for generic (simpler) configurations (Kato et al. 1992).

It is envisaged that this type of study can provide further information related to air quality and fire propagation scenarios. Once the steady state flow field is determined, allowances can be made for the transport of any passive scalar in the domain (1). This scalar can be an air quality related parameter such as CO_2 or can relate to the transport (propagation) of smoke in the domain.

The CFD, internal flow simulations can also be correlated ("chained") with external, wind tunnel, flow simulations (Kato et al.1997). Usually, a wind engineering external flow study can provide information related to the pedestrian level winds and pressures on the external building envelope as well as gaseous pollutant concentrations. Once these parameters determined, they can serve as boundary conditions for an internal air quality CFD study, in which external sources of pollutant are transported through the HVAC system inlets into the internal domain. This combined approach between external, wind tunnel (physical) modelling, and internal, numerical modelling expands the area of wind engineering studies which are made available to the construction practitioners.

Other applications such as noise propagation and energy savings might also be envisaged in relation to numerical internal ventilation studies.

ACKNOWLEDGMENTS

The project was sponsored by Cesar-Pelli & Associates. The authors also wish to acknowledge Adapco Inc. and Computational Dynamics for their support in making the successful completition of this project possible.

REFERENCES

ASHRAE Handbook, Fundamentals, I-P Edition, Atlanta GA, ASHRAE, (1989)

Atrium Research Project, NRC-CNRC Construction Innovation, Vol. 3, No.4, (1998)

Hangan H., McKenty F., Gravel L., Camarero R., A CFD internal ventilation study of the Center of Performing Arts, Miami, BLWT-SS51 (1998)

Kato S., Murakami S., Mochida A., Akabayashi S., Tominaga Y., Velocity pressure field of cross ventilation with open windows analysed by wind tunnel and numerical simulation, J. Wind Eng. Ind. Aerodynamin. 41-44 (1992), 2575

Kato S., Murakami S., Takahashi T., Gyobu T., Chained analysis of wind tunnel test and CFD on cross ventilation of large-scale market building, J. Wind Eng. Ind. Aerodynamics 67-68 (1997), 573

STAR-CD v3.0c User Manual (1998)

Wind Engineering into the 21st Century, Larsen, Larose & Livesey (eds) © 1999 Balkema, Rotterdam, ISBN 90 5809 059 0

Direct measurements of the aerodynamic admittance of large ships

G.L.Larose, K.Agdrup & S.V.Larsen
Danish Maritime Institute, Lyngby, Denmark

ABSTRACT: The dynamic actions caused by the gustiness of wind on large ships manoeuvring at low speed can be of importance. Their inclusion in real-time simulation of marine operations is, however, difficult unless knowledge of the unsteady aerodynamic characteristics of a given vessel are readily available. In most of the cases, this information is not available and these actions are either neglected or approximated based on empirical formulations. This paper presents a relatively simple experimental technique to extract the unsteady aerodynamic characteristics of large ships in atmospheric boundary layer flow. The technique is based in the frequency domain and can provide aerodynamic admittance functions for the dominant actions on a vessel for each wind direction.

1 INTRODUCTION

In the study of the manoeuvrability of large ships with important wind area in and out of harbours, the fluctuating loads due to the wind turbulence can play a significant role. During the docking manoeuvres, the ship speed is low and the harbour installations are generally exposed to high winds stretching over large open sea fetches, making conditions for the ship similar to what large buildings in gusty winds would experience. Also, harbours usually have wave breakers but seldom wind sheltering.

The state-of-the-art method applied at the Danish Maritime Institute (DMI) in real-time simulations of navigation in harbours or near major obstacles, such as bridges or wave breakers, is to include the fluctuating wind loads in the definition of the exterior loading conditions on the ship. The fluctuating wind loads are generated by simulating numerically a fluctuating flow field that is combined with a wind load model based on quasi-steady aerodynamic assumptions. The turbulent flow field is generally based on field measurements over a period of time in order to define a wind spectrum representative of the wind conditions at the site. The wind spectrum is then reconverted to the time domain to simulate numerically the fluctuating winds.

Before passing from the frequency domain to the time domain, the wind spectrum is multiplied by the aerodynamic admittance of the ship for the given apparent wind direction in a fashion similar to the application of a filtering function. The aerodynamic admittance represents the unsteady aerodynamic characteristics of the vessel. It expresses the observation that low frequency gusts, i.e. large vortices that could immerse the whole ship, will be more effective as a wind load generator than higher frequency gusts that affect only small parts of the vessel (smaller vortices).

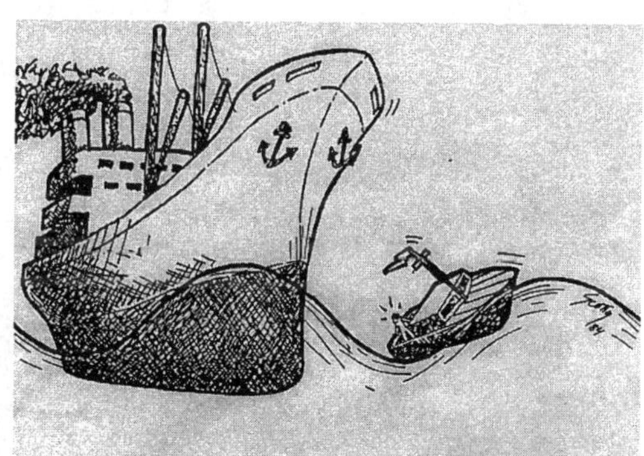

Figure 1. Aerodynamic admittance analogy, after Davenport (unknown source).

This is analogous to the behaviour of ships in waves which is a function of the ratio of the vessel length to the wave length (Figure 1).

The approach is simple and straightforward but representative of the largest part of the dynamic wind loading. Wave loads, current loads and hydrodynamic unsteady forces complete the simulation scenario. However, information on the aerodynamic admittance of ships is scarce and often the calculations are based (see Blendermann 1993) on an empirical expression proposed by Vickery in 1965 and derived from the study of flat plates of varying area normal to a turbulent flow field.

Vickery's admittance $\chi(f)$ is of the form:

$$|\chi(f)|^2 = \left[\frac{1}{1+\left(\frac{2f\sqrt{A}}{U}\right)^{4/3}} \right]^2 \qquad (1)$$

where:

A= projected area of the ship normal to the wind direction;

U= mean wind speed at the point of application of the aerodynamic forces;

f= frequency.

In many cases, this expression has proved to be in relatively satisfactory agreement with direct measurements of along-wind admittance for various bluff bodies including buildings and bridges. For an elongated 3-D body like a ship, comparisons between Equation 1 and experiments are hard to find and seldom published. Nevertheless, this expression remains the standard for marine applications (Blendermann 1993).

2 EXPERIMENTS

To improve the reliability of the numerical simulations of wind loads on large ships such as container vessels, car carriers, cruise liners and oil tankers, DMI has initiated a series of direct measurements of the aerodynamic admittance of ships. The measurement technique evolved from established procedures in bridge aerodynamics (Larose 1997) and bluff-body aerodynamics that are in many ways similar to the original experimental approach put forward by Davenport (1963).

Large ships can be seen as surface mounted prisms with rectangular cross-section and placed horizontally (as opposed to a tall building which is placed vertically). They are 3-D bodies where end effects are of great importance. When studying the static and dynamic wind loading, the approaching flow needs to be simulated with a proper atmospheric turbulent boundary layer having not only a vertical mean wind profile typical of a sea exposure but also scales of turbulence in good proportion with the geometrical scale of the model and adequate intensities of turbulence.

The experiments are performed by measuring time histories of fluctuating wind loads on 1:300 scale ship models (Figure 2) by means of a 6-component force balance to obtain wind load spectra. Since the frequencies of interest are generally very low (eigen periods varying between 5 s and 100 s), the lowest eigen frequency of the model-balance system does not need to be as high as for the high-frequency force balance tests used

Figure 2. View of a 1:300 scale model of a fully-loaded container ship

for tall buildings. In the experiments reported in the next section, the lowest model-balance eigen frequency was of the order of 15 Hz which proved to be sufficiently high. The admittance is obtained by dividing the load spectra by the wind spectra measured at the point of application of the wind forces without the model in place.

2.1 Definitions

The quasi-steady formulation of the wind loads due to the buffeting action of the longitudinal component u of the turbulence is of the form:

$$F_{X,Y}(t) = \frac{\rho\, U\, A_{x,y}}{2} \left(2\, C_{X,Y}\, u(t)\right) \tag{2}$$

where the subscripts x and y mark the heading (x) and beam (y) directions, the subscripts X and Y mark the aerodynamic forces in the x and y directions, ρ is the air density, A is a projected area normal to the x or y directions and C is a steady aerodynamic force coefficient based on the corresponding area.

Assuming that the dynamic wind loading is a stationary random process (Davenport 1963), Equation 2 can be expressed in the frequency domain by the product of spectral density functions, for example for the x force, F_X :

$$S_{F_X}(f) = \left(\frac{\rho U A_x}{2}\right)^2 4 C_X{}^2 S_u(f) \cdot \left|\chi_X(f)\right|^2 \tag{3}$$

where $\chi_X(f)$ is the aerodynamic admittance of the X-force due to the longitudinal component of the wind.

Rigorously, the aerodynamic admittance should be split in two to express the different contributions of the longitudinal (u) and lateral components (v) of the gusts. In practice however, it is not possible to separate the two since the action of the v-component cannot be measured without the u-component being present. Also, the length scales of the lateral gusts (10-30 m) are generally small in comparison with the length of large ships. The contribution of this wind component to the overall dynamic loading is therefore often neglected. This assumption applies for the following analysis.

Rearranging Equation 3 and defining the spectrum of the x and y-force coefficients and healing moment coefficient C_K by:

$$S_{C_{X,Y}}(f) = \frac{S_{F_{X,Y}}(f)}{\left(\frac{\rho U^2 A_{x,y}}{2}\right)^2} \tag{4}$$

$$S_{C_K}(f) = \frac{S_K(f)}{\left(\frac{\rho U^2 A_y H_y}{2}\right)^2} \tag{5}$$

where H_y is the height above water of the centre of the side area.

The aerodynamic admittance functions for the main components of the wind loading on large ships can be expressed by:

$$\left|\chi_X(f)\right|^2 = \frac{U^2 \cdot S_{C_X}(f)}{4 C_X{}^2 S_u(f)} \tag{6}$$

$$\left|\chi_Y(f)\right|^2 = \frac{U^2 \cdot S_{C_Y}(f)}{4 C_Y{}^2 S_u(f)} \tag{7}$$

$$\left|\chi_K(f)\right|^2 = \frac{U^2 \cdot S_{C_K}(f)}{4 C_K{}^2 S_u(f)} \tag{8}$$

All these quantities were measured or calculated using results of the wind-tunnel tests.

2.2 Procedures

A 1:300 scale model of the above-water part of a 295 m long container ship was mounted on a 6-component force balance (force in heading direction x, side force y, vertical force z and corresponding 3 moments) at the centre of the turntable of DMI's Boundary Layer Wind Tunnel 2. Time histories of the forces were recorded with a 100 Hz sampling frequency and a sampling time of 180 seconds. The tests were performed at a model scale free stream mean wind velocity of 16.0 m/s corresponding to approximately 12 m/s at the mean point of application of the wind forces, 55 mm above the wind-tunnel floor (approximately 16 m in full-scale).

The time-varying aerodynamic forces were converted to time-varying coefficients $C_{X,Y,K}$ (*direction, t*), and power-spectral densities of the coefficients were calculated, using a 1024 point fast-Fourier transform with a Hamming window and 50% overlapping. The coefficient of variation of the spectral estimates was of the order of 24 %.

Figure 3 shows a typical power-spectral density of the Y-force coefficient for a wind direction of 90° (beam wind) as a function of reduced frequency based on L_{PP}, the length between perpendiculars of the vessel. A clear resonant peak at 15 Hz can be observed corresponding to the eigen-frequency of the system composed of the 1:300 ship model and the force balance.

A single degree-of-freedom mechanical admittance function was fitted to the resonant peak (light line on Figure 3) using a least square technique. It provided an estimate of the natural frequency and damping of the mechanical system.

The peak was then filtered by simply dividing the spectrum of the force coefficient by the fitted mechanical function (dark line on Figure 3). Note that at 15 Hz the remaining energy of the *u* component of the wind was found to be at least one order of magnitude smaller than at its maximum, around 0.4 - 1 Hz. A typical graphical summary for the analysis of the *X*-force is given in Figure 4.

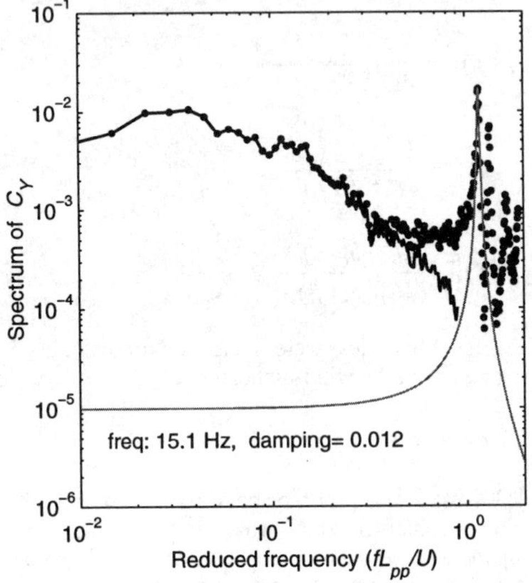

Figure 3. Power spectral density of the *Y*-force coefficient versus reduced frequency for beam winds.

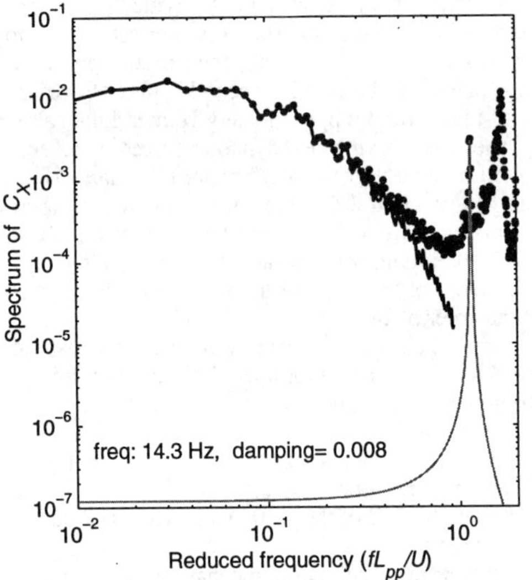

Figure 4. Power spectral density of the *X*-force coefficient versus reduced frequency for head winds.

The aerodynamic admittance functions defined in the previous section (Equations 6 to 8) were calculated by dividing the spectra of force coefficients by the wind spectra of the *u* component measured at the same wind velocity as for the forces, but with the ship model not in place.

Care was taken to simulate adequately the natural atmospheric boundary layer over the sea. The 18-m approaching fetch of the wind tunnel was fitted with vortex-generators and a rough carpet so that a vertical mean wind profile with a power law exponent of 0.11 was generated, with a longitudinal turbulence intensity of 13% at a full-scale equivalent height of 16 m. The wave length of the longitudinal component of the wind turbulence corresponding to the spectral peak in its $fS_u(f)$ representation (Figure 5) was found to be an equivalent full-scale 180 m which is in good agreement with scales of turbulence expected at that height over the sea.

3 RESULTS

The results of the aerodynamic admittance tests conducted for the wind directions 0°, 30°, 60° and 90° are presented in Figures 6 to 9. Table 1 summarises the main characteristics of the container ship studied. The mean static wind force coefficients measured in simulated atmospheric boundary layer are given in Table 2.

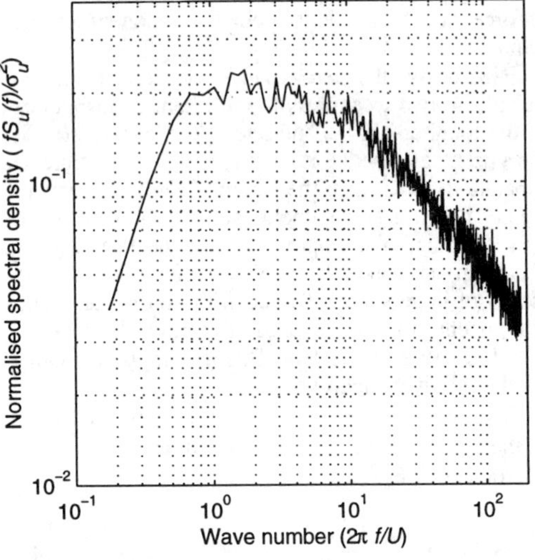

Figure 5. Normalised power spectral density of the *u* component as a function of wave number, 55 mm (16 m in full scale) above the wind-tunnel floor.

Table 1. Main characteristics of the container ship of this study.

	Symbol	Full-scale dimensions
Length between perpendiculars	*Lpp*	282 m
Length over all	*Loa*	295 m
Front area	A_x	1030 m^2
Side area	A_y	7253 m^2
Height above water of side area centre	H_y	9.8 m

Table 2. Mean wind-force coefficients for the container ship of this study.

Wind direction	C_X	C_Y	C_K
0°	-1.04	-0.02	0.01
30°	-1.28	0.69	1.43
60°	-0.64	1.37	2.75
90°	-0.46	1.40	2.77
120°	0.92	1.27	2.54
150°	1.79	0.67	1.34
180°	1.17	0.01	0.03

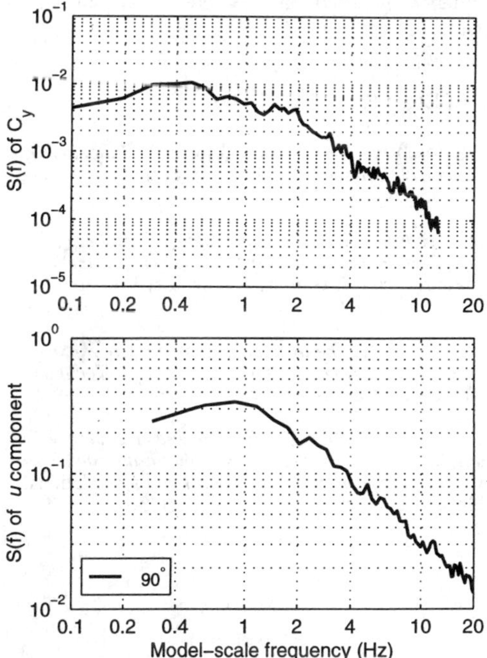

Figure 6. Power spectral densities of the *Y* force coefficients and *u* components of the oncoming flow.

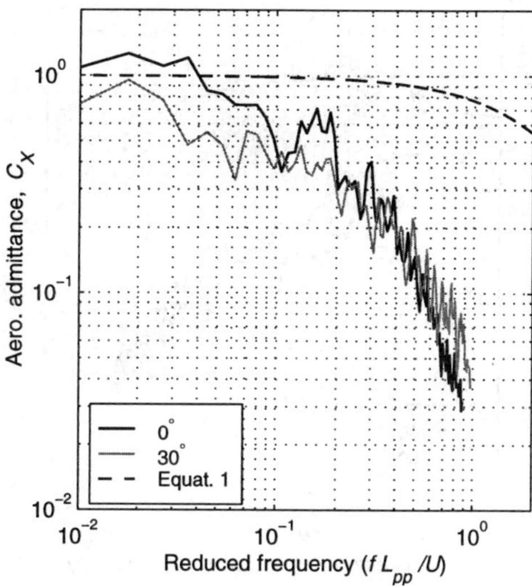

Figure 7. Aerodynamic admittances of the *X* force for two wind directions compared to the empirical expression of Equation 1 (dash line).

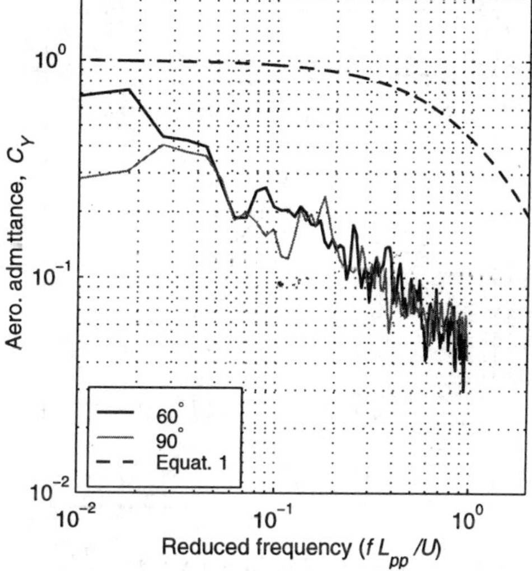

Figure 8. Aerodynamic admittances of the *Y* force for two wind directions compared to the empirical expression of Equation 1 (dash line).

Figure 6 compares the spectrum of the input, the *u* component of the wind, to the spectrum of the output, the *Y*-force coefficient for a wind direction of 90°. The spectral shapes are similar but the spectrum of the y-force coefficient appears to decay faster than the spectrum of the *u* component of the wind fluctuations.

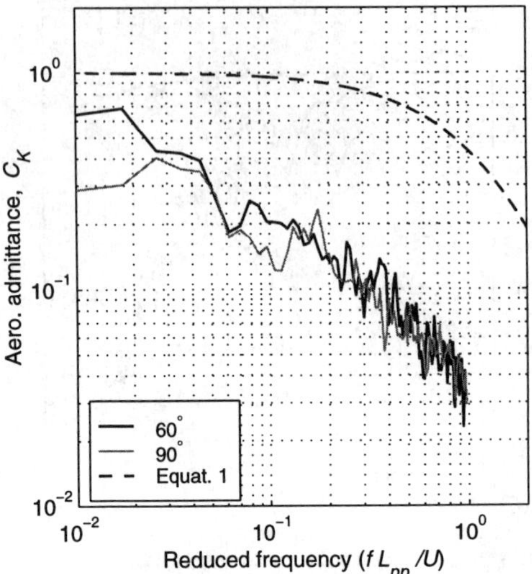

Figure 9. Aerodynamic admittances of the K moment for two wind directions compared to the empirical expression of Equation 1 (dash line).

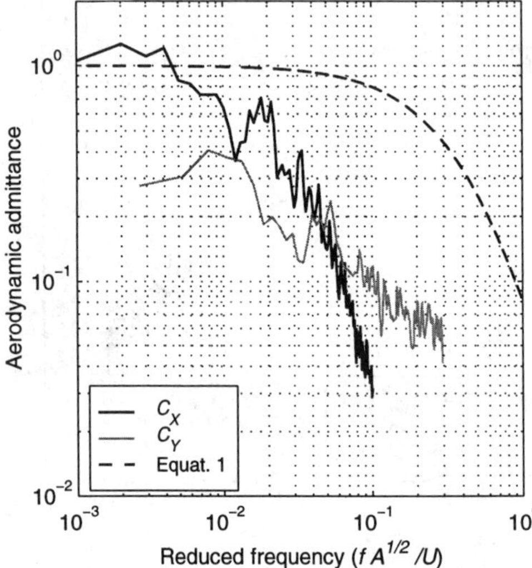

Figure 10. Aerodynamic admittances of the X-and Y-forces for head and beam winds respectively as a function of reduced frequency based on the projected area.

The dashed lines in Figure 7 to 9 represent the aerodynamic admittance calculated using Equation 1, while the solid lines are the results of the present aerodynamic admittance analysis. In all cases at high reduced frequency, the measured admittance was found to be much smaller than the empirical expression. That is to say, the dynamic effects for this ship at low apparent wind speed (10-15 m/s) would be much less important than generally anticipated.

Comparing the admittance curves of Figure 7 to Figure 8, it can be concluded directly that the larger the exposed wind area the smaller the aerodynamic admittance. This is in agreement with the general belief that larger surfaces will be less affected by the wind gustiness than smaller surfaces given the lack of coherence of the wind loading. To illustrate this aspect, Figure 10 reproduces the admittance curves of Figure 7 and 8, but plotted as a function of reduced frequency based on the projected area instead of L_{pp}. The curves fall closer to each other confirming the dependency on the area.

4 CONCLUSION

An experimental technique for the evaluation of the aerodynamic admittance of large ships has been presented. The technique is based on wind tunnel experiments in simulated atmospheric boundary layer flow. Its use has been demonstrated for a 295 m long container ship through tests on a 1:300 geometrical model of the vessel. Results have shown a large degree of conservatism if a conventional empirical expression of the admittance is used in the predictions of the dynamic effects.

REFERENCES

Blendermann W. 1993. “External forces”. *Manoeuvring Technical Manual*, Chapter 3, Seehafen Verlag, Hamburg, pp. 147-154.

Davenport A.G. 1963. “The buffeting of structures by gusts”, in Proc. of Symposium on Wind Effects on Buildings and Structures, Teddington, UK, pp. 357-391.

Larose G.L. 1997. *The dynamic action of gusty winds on long-span bridges.* Ph.D. Thesis, Technical University of Denmark, Lyngby, Denmark.

Vickery B.J. 1965. On the *flow behind a coarse grid and its use as a model of atmospheric turbulence in studies related to wind loads on buildings.* National Physical Laboratory, Aero. Report 1143.

Wind Engineering into the 21st Century, Larsen, Larose & Livesey (eds) © 1999 Balkema, Rotterdam, ISBN 90 5809 059 0

Modelling the failure of trees in high winds

S.E.T.Saunderson & A.H.England
School of Mathematical Sciences, University of Nottingham, UK

C.J.Baker
School of Civil Engineering, University of Birmingham, UK

ABSTRACT: This paper presents an overview of an analytical modelling exercise whose objective was to study the failure of trees in high winds. Four separate, but linked, components are described. The first models the behaviour of a tree root plate under an applied loading at the trunk base. The second and third are dynamic models of the above ground tree system, and are used to determine failure initial failure wind speeds. The final component considers the process of trees tipping over, the final part of the failure process. It is shown that the models predict failure wind speeds that are reasonable, and their use allows the relationship between the many relevant tree, soil and wind parameters to be considered in a coherent manner.

1 INTRODUCTION

The problem of tree failure in high winds is of some concern in a number of ways. The failure of coniferous trees in commercial forests clearly has economic and maintenance consequences. The failure of urban broad leaf trees in windstorms can result in injuries and fatalities and cause significant disruption of transport and other services. In the field of forestry there has been considerable work carried out in recent years – see for example the papers presented in Coutts and Grace (1995). This work has included full scale observations of tree behaviour in high winds (Gardiner, 1995), uprooting tests to measure root strength (Coutts, 1986), wind tunnel tests of the flow over forests, using aeroelastically scaled trees (Marshall, 1998) and, of most relevance to this paper, some limited mathematical modelling work (eg Gardiner, 1992) which is usually based on the concept of treating an individual tree as a simple mass/spring/damper system. As far as urban broad leaf trees are concerned the work has been rather more limited. The third author has been carrying out research into this problem for a number of years, through both full scale experiments (Roodbaraky et al, 1994; Baker, 1997) and fairly simple analytical modeling (Baker, 1995). The latter has involved modelling the tree as two lumped masses (representing the rootball and the crown) connected by a weightless elastic stem (representing the trunk). The response of such a system to a variable wind input has been studied, and predictions made of the maximum bending moment at the stem base for a particular mean wind speed. A comparison of this moment with the stem base strength or the uprooting moment has then been used to determine whether or not tree failure will occur at that wind speed, and thus failure wind speeds can be determined. More recently the author has applied this method, suitably extended to include the effect of rainfall on crop and soil characteristics, to the lodging of cereal crops (Baker et al, 1998). Whilst such modelling has been shown to be able to produce reasonably realistic results for failure wind speeds, it is nonetheless simplistic in many aspects. In particular trees and crops are complex geometric structures that often bear little resemblance to the two mass model that has been used. To remedy this situation, at least for trees, recently more advanced analytical modeling work has been undertaken to investigate the failure of both coniferous and broad leaf trees in high winds (Saunderson, 1997). The purpose was to provide an analytical framework through which the various tree and environmental parameters could be linked, to enable insights to be gained into the problem and to identify those characteristics of trees that can cause them to be particularly susceptible to failure in high winds. The various aspects of this work will be described in some detail in a series of papers that are either in press or in the process of publication (Saunderson et al, 1999, England et al 1999). It is the intention in this paper to give an overview of the model framework that has

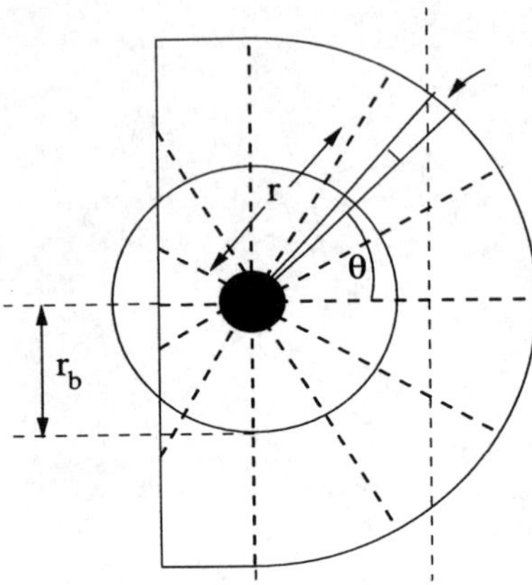

Figure 1 Schematic of the root plate model

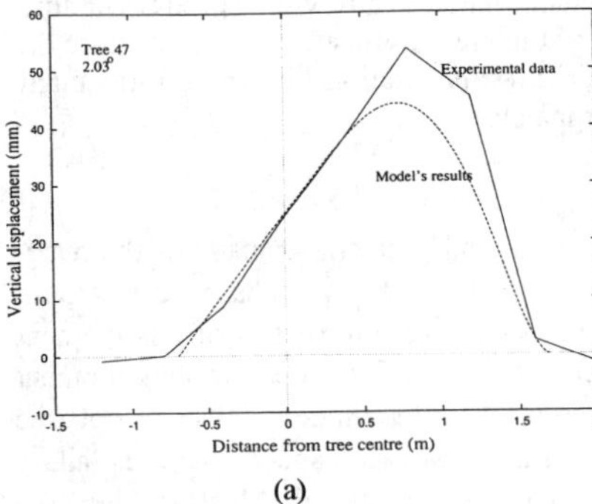

(a)

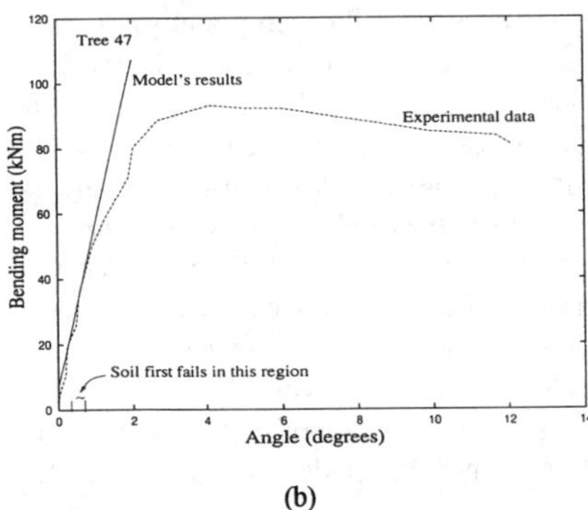

(b)

Figure 2 Results of the uprooting model – a) comparison of model predictions and experimental data for root plate deflection; b) predicted bending moment / deflection relationship

been derived. This framework has three major components as follows.

a) A model of the root / soil system that is used to determine failure criteria for tree uprooting, and also gives information on the flexibility of the tree base for use in (b) below.
b) A model of the initial failure of the tree / root / soil system, either by trunk snapping or by uprooting. This component has two sub components – one describing the behavior of forest trees and another describing the behavior of broad leaf trees.
c) A model of the final stage of tree failure – the tree tipping process.

Each of the components will be briefly discussed in what follows. It will be seen that this approach offers a coherent framework for the consideration of tree failure in high winds that gives considerable physical insight into the problem.

2 THE ROOT / SOIL MODEL

The root system of the tree is assumed to consist of a number of equally spaced roots, which taper from their origin beneath the trunk to a point at the edge of the root plate. A consideration of the bending properties of these roots enables them to be modelled by a circular plate of variable stiffness mounted on a bed of springs, restrained at its edges by a tension due to soil / root friction (Figure 1). The effect of buttressing is allowed for by assuming a rigid portion of the root close to the trunk base (r_b in Figure 1). The root plate is assumed to tip about a point on the leeward side of the trunk (the straight line at the left of Figure 1). The movement of this root plate under various bending moments imposed at the intersection of the root with the trunk is then considered, and a comparison with some experimental data from the uprooting of forestry trees (Coutts, 1986) indicates that it is reasonably realistic (Figure 2a). This model is used to determine the bending moment required for failure of the root / soil system, and also the effective moment / deflection characteristic for the trunk base, which is needed as an input to the above ground components that will be described in what follows (Figure 2b).

3 MODELS FOR INITIAL TREE FAILURE

Two models have been developed for the initial failure of trees under wind loading – one for coniferous forest trees and one for urban broad leaf trees. These will be considered in turn below.

A forest tree is treated as a vertical tapered cantilever, representing the trunk, with a cylindrical mass over the upper portion of the cantilever, representing the crown (Figure 3), under the action of a spatially and temporally varying wind load. The specification of tree geometry, including the degree of taper is taken from the field measurements of Gardiner (1989). The flexibility of the cantilever base is specified by the output of the root model described above. Both mechanical and aerodynamic damping is allowed for. The wind loading is considered to be due to a turbulent fluctuating wind acting over the top portion of the cantilever and canopy, to produce both mean and fluctuating deflections and, more importantly as far as failure is concerned, mean and fluctuating trunk base bending moments. Such a wind field is reasonably consistent with the available full scale data, although recent wind tunnel tests of Marshall (1998) have shown that wind loading on forests trees is primarily due to large scale coherent structures that exist at the top of the canopy due to interaction between trees and the oncoming wind. Such effects have not, to date, been incorporated into the model. The calculation of the mean deflections and moments is fairly straightforward, but the calculation of the unsteady parameters requires a numerical solution of the complex equations of motion. This results in transfer functions that relate the spectrum of the wind fluctuations to the spectra of the tree deflections and trunk base moments. Experimental data is available for the deflection transfer functions, and this enables a check of the models to be made at this stage. A typical comparison is shown in figure 4. It can be seen that there is reasonable agreement, and the first mode of oscillation is well predicted. However at frequencies above about 1Hz the agreement breaks down. This may be due to real physical phenomena – an inadequacy in the model, or tree clashing affecting the fullscale results, or due to the inability of the deflection transducers to operate properly at such frequencies. Nonetheless the degree of agreement is encouraging. These spectra and their moments can then be integrated to obtain, together with the mean values, the extreme values of the deflections and base bending moments for various above canopy wind speeds. By comparing these values with the moments required for uprooting or tree snapping, these values can then be used to determine the wind speeds for which the uprooting bending moment is exceeded, or for which the trunk fails by snapping. In other words the initial failure wind speeds can be obtained. Typical results for a number of Sitka Spruce for which high quality geometric data exists are given in table 1 below. Full geometric and mechanical characteristics for these trees are given in Saunderson et al (1998).

It can be seen that for the above trees the failure wind speed for stem breakage is much less than that for uprooting. The former seem too low, whilst the latter seem reasonable. However it must be stressed that these values are crucially dependent upon the assumed values of uprooting moment and trunk breakage stress, which are amongst the least well specified of the parameters that are required for the calculation.

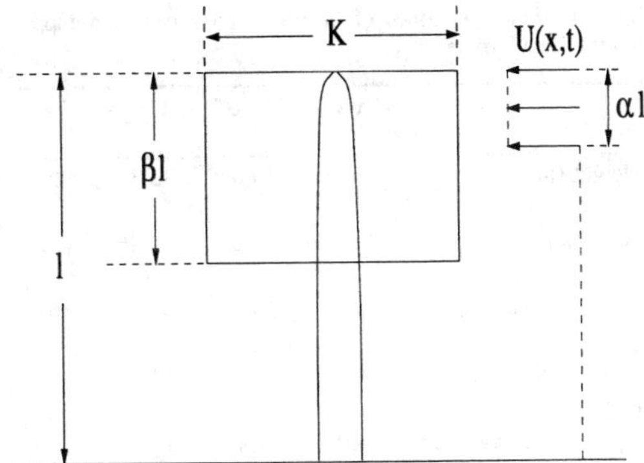

Figure 3 Schematic of the model for a forest tree

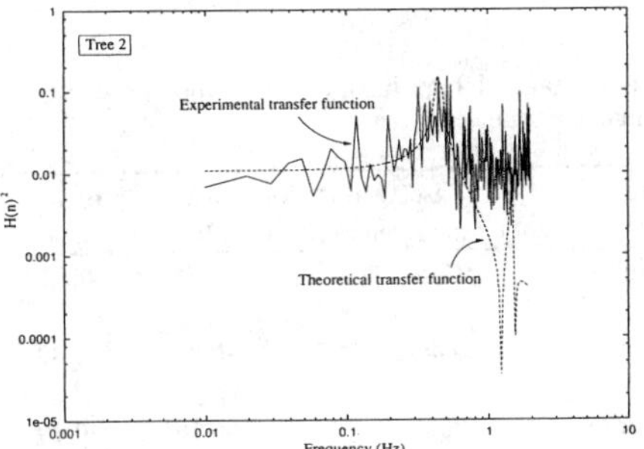

Figure 4 Comparison of predicted and measured deflection transfer functions

This method also enables a parametric investigation of the problem to be carried out to investigate the sensitivity of the failure wind speeds. Some of the results of such an investigation for uprooting wind speed are shown in Table 2. It can be seen from there that the results are surprisingly insensitive to most of the parameters that are considered, with the exception of the uprooting strength.

The component for broad leaf trees is similar to the above except that the tree model is necessarily more complex, and involves the consideration of the dynamic behavior of branched systems. The individual elements of a tree are considered as uniform (non-tapering) beams constrained by connections to other beams at either end, and arranged in a shape that is typical of the fan shape of trees (Figure 5). The dynamic equations of motion are solved for each beam under the action of a spatially and temporally varying wind load as before. Saunderson (1997) shows that these equations of motion and the associated boundary conditions can be represented in a relatively straightforward matrix form. The structure

Table 1 Predicted values of failure wind speeds for experimental Sitka Spruce

	Tree 1	Tree 2	Tree 3	Tree 4
Height (m)	13.3	13.0	13.3	9.9
Base radius (m)	0.09	0.088	0.075	0.05
Uprooting moment (kNm)	50	50	50	50
Stem breakage stress (MPa)	34	34	34	34
Uprooting mean wind-speed (m/s)	14.0	14.1	13.6	20.7
Stem breakage mean windspeed (m/s)	9.1	8.78	6.74	5.54

Table 2 Percentage change in uprooting wind speed for +/- 10% change in parameters

Parameter	Base value of parameter	base +10%	base - 10%
Tree height	10m	+1.0	-0.62
Base radius	0.06m	-0.90	+1.40
Trunk density	900kg/m^3	-0.19	+0.19
Canopy density	0.4kg/m^3	-0.24	+0.25
Canopy radius	1m	+0.20	-0.26
Youngs modulus	6GPa	-0.26	+0.31
Uprooting strength	50kNm	+4.90	-5.10

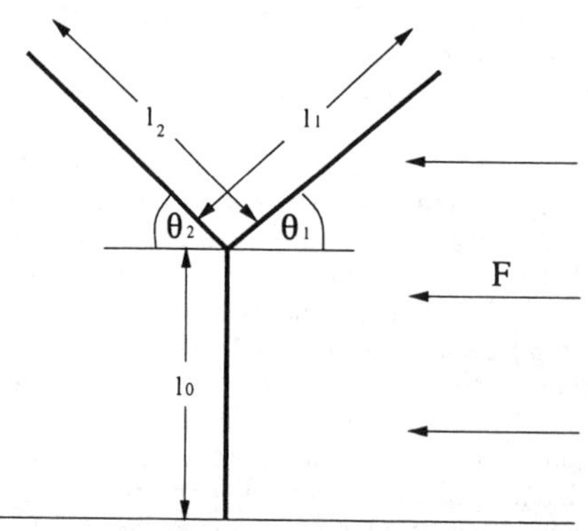

Figure 5 Fan shape model for broad leaf trees

of the matrix that relates the predicted bending moments and displacements to the applied forces can be broken down into a series of sub matrices, each of which represents different facets of the problem – ground boundary conditions, matching conditions between elements, free end boundary conditions etc. As such this matrix is easily generalised and can be extended to quite complex tree geometries. This model has been verified as far as possible by comparing the measured data for the natural frequencies of broad leaf trees with the values predicted using the dynamic model (Baker, 1997). Figure 6 shows the results of this comparison for a large number of lime trees, which are essentially trunk and two branch systems, in both in leaf and out of leaf conditions. Agreement can be seen to be encouraging.

The same procedure can then be carried out as for forest trees and failure wind speeds can be calculated. The results of such a procedure are shown in Table 3 for a typical lime tree in summer and winter condi-

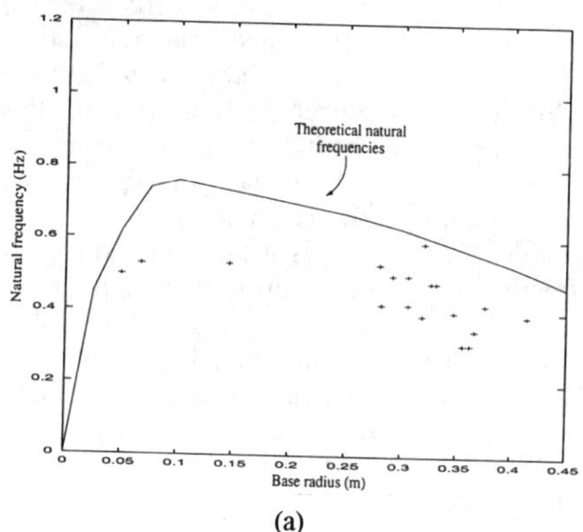

(a)

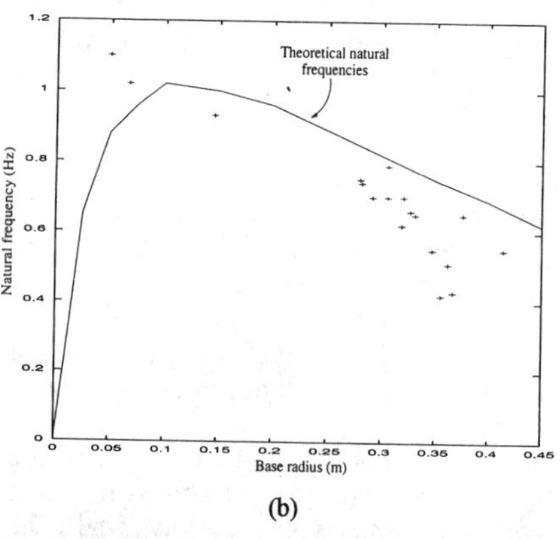

(b)

Figure 6 Comparison of measured and predicted natural frequencies for lime trees (a) in leaf (summer), (b) out of leaf (winter)

Table 3 Predicted value of failure wind speeds for lime trees

	Summer conditions	Winter conditions
Trunk length (m)	2.8	2.8
Trunk radius (m)	0.20	0.20
Trunk density (kg/m3)	657	657
Branch length (m)	8.4	8.4
Branch radius (m)	0.15	0.15
Branch density (kg/m^3)	723	657
Youg's modulus (Gpa)	2.5	4.0
Uprooting moment (kNm)	500	500
Stem breakage stress (Mpa)	54	54
Uprooting mean windspeed (m/s)	9.79	11.21
Stem breakage mean wind-speed (m/s)	8.21	9.24

Table 4 Percentage change in uprooting wind speed for +/- 10% change in parameters for lime trees.

Parameter	Base value of parameter	base +10%	base -10%
Trunk height	2.8m	-6.5	+7.1
Trunk radius	0.20m	+9.1	-8.6
Trunk density	657kg/m^3	-1.0	+1.0
Branch length	8.4m	-15.6	+15.6
Branch radius	0.15m	-3.6	+2.9
Branch density	657kg/m^3	-5.9	+5.7
Youngs modulus	4GPa	+3.1	-3.2
Uprooting strength	400kNm	+4.90	-5.10

tions (in leaf or out of leaf). The stem breakage values again seem lower than would be expected. Similarly table 4 shows the results of a parametric analysis carried out to determine the sensitivity of the calculations for uprooting wind speeds to changes in the tree parameters. In this case it can be seen that the results are rather more sensitive to changes in parameters than are the results for the Sitka Spruce

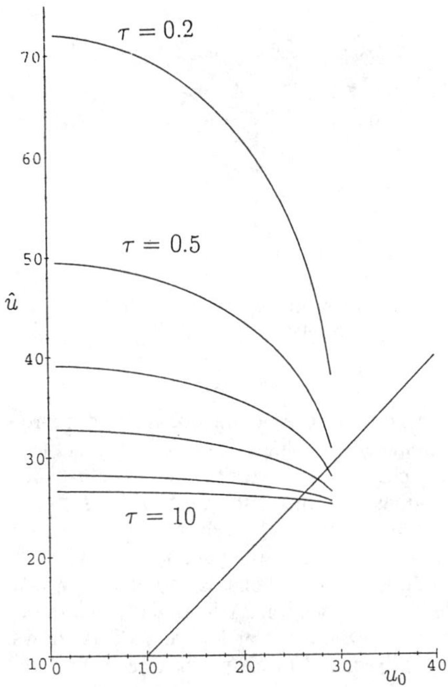

Figure 7 Relationship between mean wind speed, gust speed and averaging time for a Sitka Spruce to blow over completely (x axis is mean velocity in m/s, y axis is gust velocity in m/s and parameter is gust length)

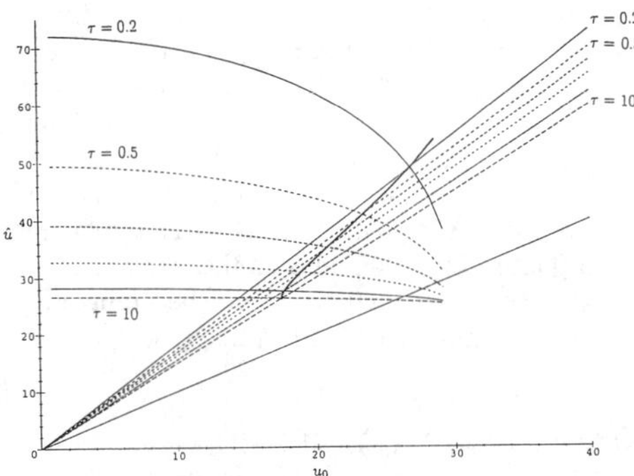

Figure 8 Results of Figure 7 with relationship between mean velocity, gust velocity and averaging time added

4 THE TREE TIPPING MODEL

The final component of the overall model describes the tipping of the tree once initial failure occurs. It considers the dynamic behaviour of the tree at large angles of deflection, under the action of an sustained

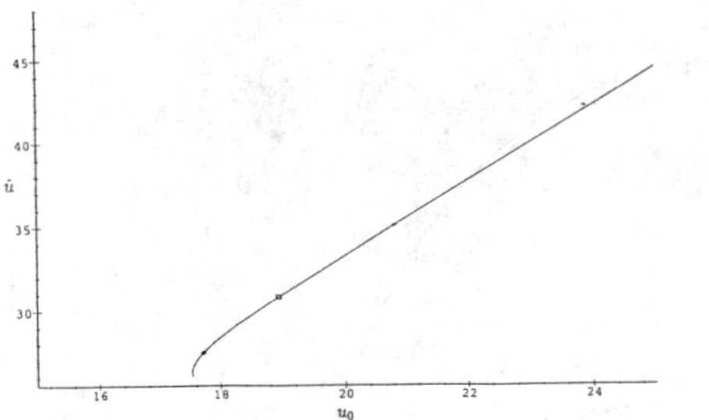

Figure 9 Relationship between failure mean wind speed and failure gust speed for Sitka spruce

average wind force and an impulsive moment representing extreme gust loading, and a resisting moment due to the behavior of the root / soil system under large deflections – from Coutts (1986). This model has been applied to the relatively simple geometric case of the Sitka Spruce (see Figure 1). A basic input for this model is the base bending moment at which initial failure occurs (section 2). Such a procedure results in curves of the form of Figure 7. This shows the gust speed required to cause the tree to fully tip over as a function of the average wind speed and the time period of the extreme gust.

Now extreme wind statistics enable a relationship between extreme wind speed of a certain time period and mean wind speed to be established (eg the well known formula of Wieringa, 1973). Whilst accepting that the accuracy of such a formula above forest canopies must be somewhat dubious, its use does enable the relationship between gust speed and mean wind speed to be expressed in terms of averaging time (Figure 8). The points where the curves for a particular averaging time cross define a curve of mean wind speed and gust wind speed at the failure condition (Figure 9). It can be seen that the minimum mean wind speed at which trees will be completely blown over is at around 17.5m/s in this case.

5 DISCUSSION AND CONCLUSIONS

From what has been presented in the previous sections it can be seen that the different models seem to adequately model the effect of wind on tree / soil systems. Clearly they are in many ways imperfect with some major assumptions involved, but nonetheless they provide a framework through which the complex interactions between a large number of parameters can begin to be understood. In particular they allow the sensitivity of failure wind speeds on different tree or soil characteristics to be determined. It appears that the calculations are most sensitive to the values assumed for the uprooting strength or for the trunk strength, which unfortunately are amongst the least well specified of the parameters involved. The results thus suggest that further research is required in this area.

REFERENCES

Baker C J 1995. The development of a theoretical model for the windthrow of plants. *Journal of Theoretical Biology* 175, 355-372

Baker C J 1997..Measurements of the natural frequency of trees. *Journal of Experimental Botany* 48, 310, 1125-1132

Baker C J, P Berry, J Spink, R Sylvester-Bradley, W Clare, R Scott, J Griffin. A method for the assessment of the risk of wheat lodging. *Journal of Theoretical Biology*

Coutts M 1986. Components of tree stability in Sitka Spruce on peaty gley soil. *Forestry* 59, 2, 173-197

Coutts M, Grace J eds 1994. *Wind and trees.* Cambridge University Press

England A H, Baker C J, Saunderson S E T 1999. A dynamic analysis of the windthrow of trees. submitted to *Forestry*

Gardiner B A 1989. *Mechanical characteristics of Sitka Spruce*. Forestry Commission Occasional Paper 24, 1-11

Gardiner B A 1992. Mathematical modelling of the static and dynamic characteristics of plantation trees, in *Mathematical modelling of forest ecosystems* ed J Franke and A Roede 40-61, Sauerlanders Verlag

Gardiner B A 1994. The interactions of wind and tree movement in forest canopies, in Coutts M, Grace J eds 1994. *Wind and trees*. 41-60, Cambridge University Press

Lavers G 1983. *The strength properties of timber in metric units.* Building Research Establishment, Watford, UK, HMSO

Marshall *B Wind flow structure and wind forces in forests.* DPhil thesis, University of Oxford

Roodbaraky H J, Baker C J, Dawson A R, Wright C J 1994. Experimental observations of the aerodynamic characteristics of urban trees. *Journal of Wind Engineering and Industrial Aerodynamics* 52, 171-184

Saunderson S E T 1997. *The aerodynamic behavior of trees in high winds*. PhD Thesis University of Nottingham

Saunderson S E T, England A H, Baker C J 1999. A dynamic model of the behaviour of Sitka Spruce in high winds" submitted to *Journal of Theoretical Biology*

Wieringa J 1973. Gust factors over open water and built up country. *Boundary Layer Meteorology* 3, 424-441

Wind Engineering into the 21st Century, Larsen, Larose & Livesey (eds) © 1999 Balkema, Rotterdam, ISBN 90 5809 059 0

Experimental measurements and computations of the wind induced ventilation of a cubic structure

M.P.Straw
School of Civil Engineering, University of Nottingham, UK

A.P.Robertson
Silsoe Research Institute, Bedford, UK

C.J.Baker
School of Civil Engineering, University of Birmingham, UK

ABSTRACT: This paper presents the results of an experimental, theoretical and computational investigation of the wind driven ventilation through a 6m cube with openings on the windward and leeward faces. Measurements were made of the surface pressures coefficients and mean and total ventilation rates through the cube. These measurements were then compared with a number of methods for the prediction of mean and fluctuating ventilation rates.

1 INTRODUCTION

The wind driven ventilation through buildings is commonly considered to consist of two components – a mean component driven by the mean pressure field at the ventilation openings, and a fluctuating component driven by the fluctuating pressures and unsteady flows around the openings. The second component can be similarly considered to consist of a number of distinct phenomena (Malinowski, 1971). The first mechanism is pulsation flow, caused by a body of fluid being driven perpendicular to the opening by the difference between the external and internal pressures. The second is known as eddy penetration, and is caused by fluid transfer due to eddies in the unstable shear layer that result from the external turbulent flow.

A number of methods exist for calculating the ventilation due to these components. Mean ventilation rates can be calculated using simple zonal methods based on orifice flows (BSI, 1991). The fundamental equation that is used to obtain the discharge through the orifice Q is the simple discharge coefficient relationship

$$Q = Au_R C_d \sqrt{\Delta C_p} \qquad (1)$$

where ΔC_p is the mean pressure difference across the orifice, A is the orifice area, u_R is the reference velocity and C_d is the orifice discharge coefficient. The discharge coefficient is conventionally taken as 0.61, which is the value for an orifice with flow parallel to its axis. The effect of flow across the orifice (i.e. in its plane) is not usually considered (but see Vickery and Bloxham, 1992). In recent years CFD packages have become a more popular tool for the prediction of mean ventilation flows (Awbi, 1989).

Determination of the fluctuating component of ventilation rates is considerably more complex. The usual approach is to derive the momentum equation for a slug of fluid that is forced in and out of the orifice by the difference between the internal and external pressures – see Holmes (1979), Vickery and Bloxham (1992), Haghighat et al (1991) and Chaplin (1997) amongst others. This non linear equation is then either solved numerically or linearised using one of a number of approaches to produce an equation that is analytically tractable. The equation is then combined with the continuity equation to give second order differential equations relating the discharge through the opening and the internal pressure coefficient to the external pressure coefficient at the opening. Frequency domain approaches are then used to calculate the relationship between the external pressure spectrum, the internal pressure spectrum and the ventilation rate spectrum. Baker (1998, 1999) shows that the use of the technique of orthogonal decomposition of the surface pressure field leads to an elegant solution for these parameters. The ventilation rate spectrum is given by

$$S_{QQ} = \frac{\left(\omega\rho u_R{}^2/2B\right)^2\left(\Pi_1^2 S_{T1} + \Pi_{21}^2 S_{T2}.......\right)}{\left(\left(1-\frac{\omega^2}{\omega_n^2}\right)^2 + \left(\frac{2c\omega}{\omega_n}\right)^2\right)} \quad (2)$$

where ω is an angular frequency, ω_n is the natural frequency of the system $(= (ABN/\rho L)^{0.5})$ and c is the damping of the system $(= K/2(\rho LABN)^{0.5})$; ρ is the density of air, u_R is a reference velocity, K is a coefficient of linearisation, A is the orifice area, L is the effective orifice length (actual orifice length $+0.89\sqrt{A}$) and N is the number of orifices; $B = \gamma P_R/V$ where γ is the ratio of specific heat, P_R is a reference pressure and V is the volume of the enclosure; $\Pi_i = \Sigma_j P_i(x_j)/N$ where $P_i(x_j)$ is the eigenvector of mode i at point j, the position of the openings on the surface; S_{Ti} is the spectrum of mode i. Similarly the internal pressure coefficient spectrum within such an enclosure is given by

$$S_{cp} = \frac{\left(\Pi_1^2 S_{T1} + \Pi_{21}^2 S_{T2}.......\right)}{\left(\left(1-\frac{\omega^2}{\omega_n^2}\right)^2 + \left(\frac{2c\omega}{\omega_n}\right)^2\right)} \quad (3)$$

In deriving the above equations the linearisation method of Haghighat et al (1991) has been used, although other approaches would be equally valid. From such spectra the rms values of ventilation rate and internal pressure coefficient can thus be found. The particular utility of this method is that, when applied to the case with more than one opening, it obviates the need for the calculation of the cross spectra of pressures between orifices that would otherwise be required. This becomes particularly significant where the number of openings becomes large. This being said, in what follows we will consider only the two opening case, for which the main use of equations 2 and 3 lies in their relative simplicity. Finally it needs to be pointed out that all such frequency domain methods implicitly calculate the pulsation flow ventilation mechanism - no methods of calculating ventilation due to eddy penetration are known to the authors.

Some experimental data is available against which methods such as those outlined above can be checked – e.g. at wind tunnel scale the work of Holmes (1979), Liu and Rhee (1986), Vickery and Bloxham (1992) and Rao (1993); and at full scale the work of Bienfait et al (1991), Haghighat et al (1992), Rao (1993), and Furbringer and van der Maas (1995).

This paper presents the results of a large scale experiment of wind driven ventilation and uses the experimental data to calculate ventilation rates by a variety of methods. A simple geometric arrangement has been chosen that, it is hoped, will allow a fundamental understanding of the ventilation flow mechanisms to be gained. The experiments will also provide a simple test case for the different methods of calculating mean and unsteady ventilation rates.

2 THE EXPERIMENTS

The experiments were carried out on a 6m cube constructed on an exposed site at Silsoe Research Institute. The cube was built on a turntable such that it could be rotated to any angle relative to the prevailing wind direction. Two 1m square openings were cut with their axes on the vertical centreline 0.5m above the centre on opposite sides of the cube (i.e. the bottom of the openings were at a height of 3.0m above the ground). In general, the experiments were carried out with the faces containing the openings positioned normal to the wind i.e. with the openings on the windward and the leeward sides. Other experiments have been carried out with the wind parallel to the faces containing the openings. However, these are not considered here.

Reference wind conditions were measured using a sonic anemometer mounted at cube height 2m to the side of the cube and 18m upstream. The atmospheric boundary layer has been measured in the past and shown to be a typical rural boundary layer with a surface roughness length of approximately 0.01m.

The following measurements were carried out.

a) Pressure measurements on the external surfaces of the cube (with the openings closed – the sealed case) and within the cube (with the openings exposed – the open case), using pressure tappings and probes connected to pressure transducers sampled at 5Hz.
b) Three dimensional velocity measurements within the cube using a sonic anemometer, sampled at 20Hz. Measurements were made within the cube along the cube centreline, around the openings and also across a number of planes perpendicular to the wind direction.
c) Tracer gas measurements using carbon monoxide (CO) sampled with a GFC Ambient CO analyser at a frequency of 0.1 Hz. A constant injection method was utilised with the tracer gas being released at nine equally spaced points within the structure. The sampling point was varied in order to ensure that adequate mixing was taking place. The tracer gas measurements provided the total

effective wind driven ventilation i.e. the sum of the mean ventilation and the net unsteady outflow from the cube. Assuming a sinusoidal fluctuation, the net unsteady outflow can be shown to be given by the rms discharge multiplied by the factor $\sqrt{2}/\pi$. Leakage tests were performed for the sealed case.

Table 1. Flow characteristics.

	Mean	Standard deviation
Reference velocity – sealed case m/s	11.78	2.47
Reference velocity – open case m/s	6.70	1.44
External pressure coeff. – upstream opening	0.87	0.326
External pressure coeff. – downstream opening	-0.390	0.135
Internal pressure coefficient – open case	0.081	0.203
Upstream opening - non dimensional discharge	0.884 (measured) 0.67 (corrected)	0.136 (measured) 0.103 (corrected)
Downstream opening - non dimensional discharge	0.681 (measured) 0.71 (corrected)	0.103 (measured) 0.107 (corrected)
Total non-dimensional outflow – tracer gas	0.787	

These experiments were part of a much larger experimental programme to investigate the flow, pressure and pollutant dispersion characteristics around the cube, some of the early results from which are included in Hoxey et al (1999).

For the purposes of the present investigation the results were analysed to determine the mean and unsteady pressure and velocity characteristics at the orifice positions and within the cube. Table 1 shows the values of the mean and rms reference velocities for the sealed and open cases, together with the values of the mean and standard deviations of external and internal pressure coefficients and the mean and standard deviations of the dimensionless flow rates at the openings. The actual measured values of the flow rates were non-dimensionalised with the reference velocity and the orifice area. The total dimensionless flow rate out of the cube, as measured by the tracer gas experiments, is also shown. Leakage from the cube has been allowed for in arriving at this figure. Note that the flow rate values are given as measured (which was at a point 0.2m into the cube from the plane of the orifice) and also corrected to the mean value at the plane of the orifice. This correction was carried out using the results of the CFD calculations (see below) by multiplying by the ratio of the calculated mean velocity at the orifice plane to the calculated velocity at the measurement point. When this process has been carried out it can be seen that the discharges at the two orifices are close to each other as expected. The different results can be seen to be reasonably consistent, with a mean value of 0.69 (+/-0.02) and an rms value of 0.105 (+/-0.002). These figures would suggest that the overall outflow from the cube should be 0.74 (mean + rms x ($\sqrt{2}/\pi$)) which is reasonably close to the measured value of 0.787. It appears, that for this configuration of the cube, the mean ventilation rate accounts for approximately 90% of the total ventilation.

The reference velocity spectrum and internal and external pressure coefficient spectra are shown in figure 1. The spectra are plotted as spectral density / variance for each of the quantities. The velocity spectrum (figure 1(a)) is broadly as expected with a slope close to the value of –5/3. In general the pressure spectra (figures 1(b)-(d)) contain less energy at the lower frequencies, with rather more at the higher frequencies. There is an indication of a peak in the internal pressure spectrum at around 2Hz (which is more apparent in the normalised plot of figure 2). This will be seen to be of some significance in what follows.

The measured velocity distribution through the cube, on the centre-line directly between openings, is shown in figure 3 together with CFD predictions. The latter will be discussed below. At this point it is sufficient to state that the curves are similar in form, although the experimental values are consistently above the predicted values. It can be seen that the velocity at the windward opening reduces rapidly with a recovery towards the leeward opening which was acting as an outlet. It would be expected that the velocity at the openings should be equal for mass conservation. However, as mentioned above it was not possible to measure velocities in the plane of the openings, with the nearest point being 0.2 m from the plane (hence the use of corrected values).

To enable the method of Baker (1998) outlined in section 1, to be used to calculate the unsteady ventilation rate, a proper orthogonal decomposition was carried out on the surface pressure coefficients around the centre line of the cube. Figure 4 shows the distribution of the first three eigenvectors that are obtained in the analysis and the proportion of fluctuating energy associated with each mode. Note that in these plots tappings, 1 to 5 are on the windward face of the cube, tappings 6 to 11 on the cube roof and tappings 12 to 16 on the leeside of the cube. The centre of the orifices correspond to the positions of

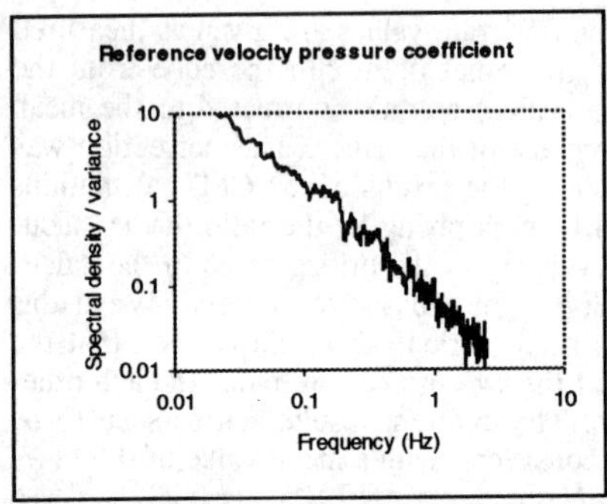

Figure 1(a). Streamwise reference velocity spectrum.

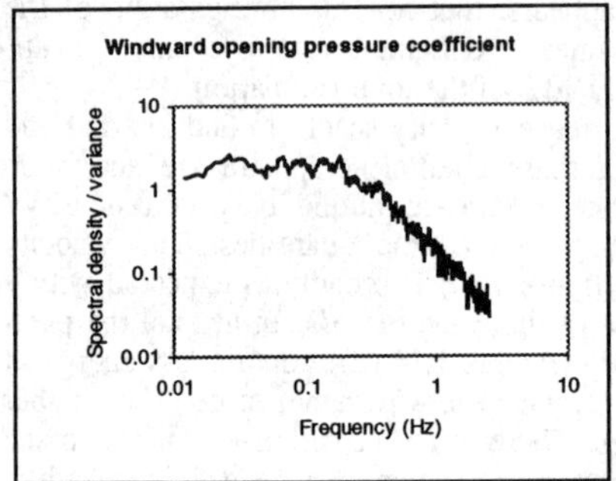

Figure 1(b). Windward opening pressure coefficient spectrum.

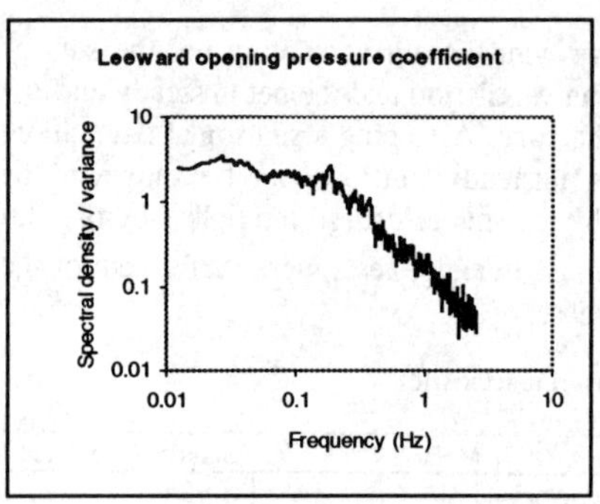

Figure 1(c). Leeward opening pressure coefficient spectrum.

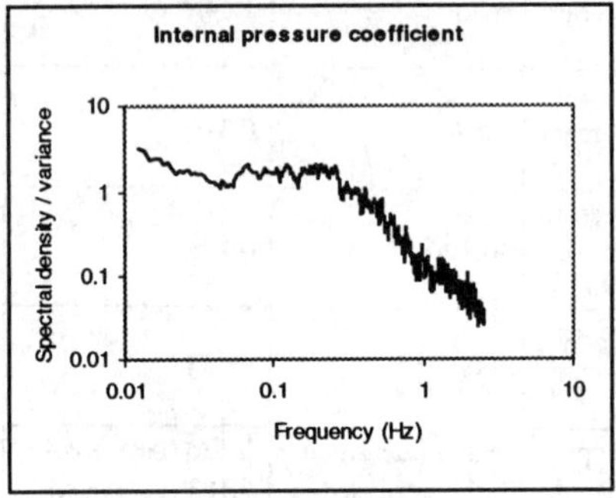

Figure 1(d). Internal pressure coefficient spectrum.

tappings 3 and 14. The mode shapes are broadly as expected, with the shape of the first mode mirroring that of the external pressure distribution (see Baker, 1999). As is usual in such analysis, nearly all of the energy is contained within the first few modes, suggesting that the series in equations 2 and 3 can be truncated after a few terms with little loss of accuracy. The modal spectra for the first three modes are shown in figure 5, again plotted as spectral density / variance against frequency.

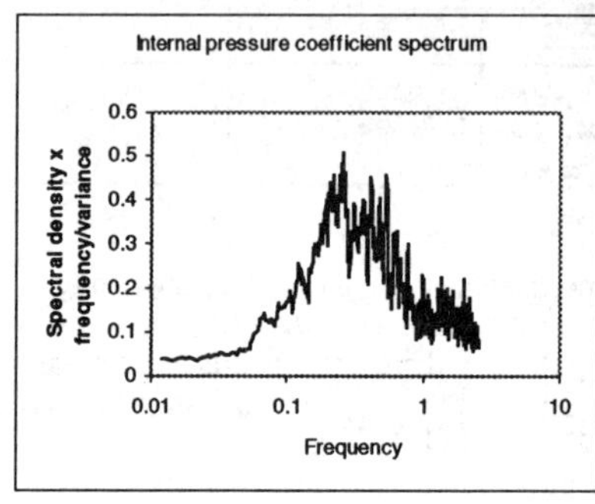

Figure 2. Internal pressure spectrum – plot of spectral density x frequency / variance against frequency.

3 CALCULATION OF VENTILATION RATES

The mean ventilation rates were calculated using two methods as follows;

a) Using the discharge coefficient formula (equation 1) for the measured mean pressure coefficients, and a standard value of the discharge coefficient of 0.61. The pressure coefficient increment (ΔC_p) was taken as that measured between tappings 3 and 14 for the sealed cube case. The effective area was taken as 0.7071m (i.e. the area of a single opening divided by $\sqrt{2}$ as two openings in series were considered). The values were normalised with the reference velocity for the sealed case and the area of a single opening.
b) Using a CFD solution of the flow through the cube (using CFX-F3D). Simulations utilised the RNG k-ε turbulence model with CCCT differencing. A fully independent grid utilising $2\text{x}10^5$ cells produced a fully converged solution of the external flow field and the consequent internal airflow pattern. Both external and internal flow

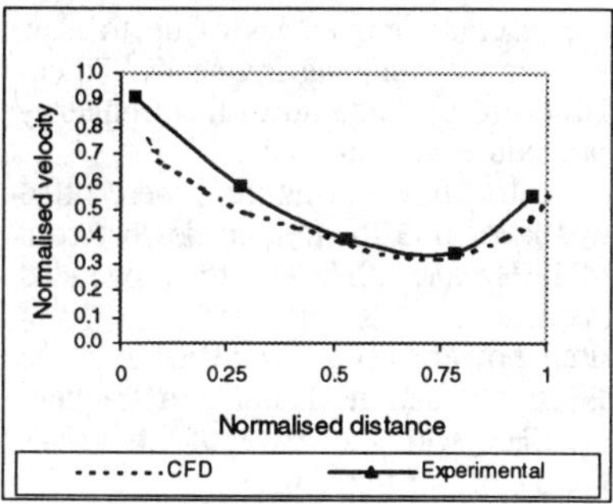

Figure 3. Velocity distribution through the cube.

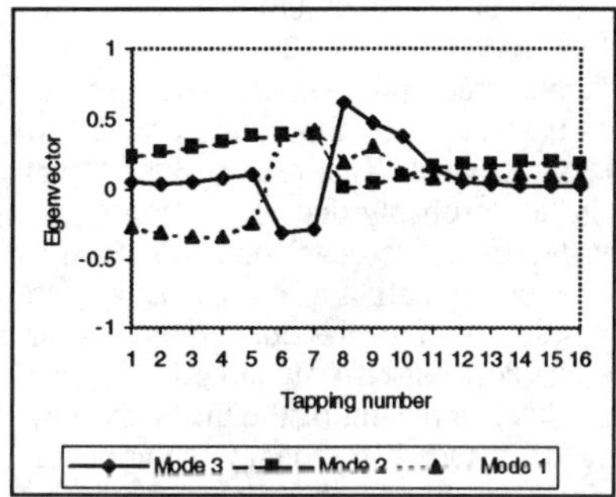

Figure 4(a). Eigenvector distribution of first three modes.

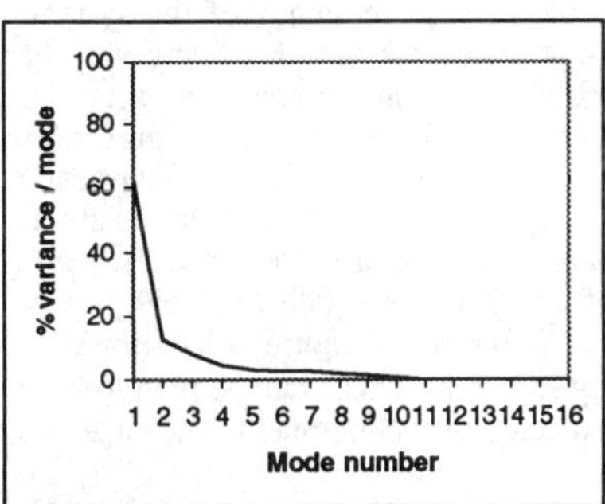

Figure 4(b). Energy contained in individual modes.

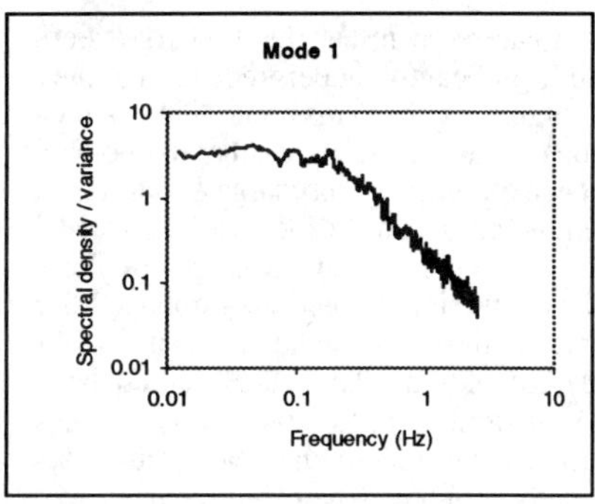

Figure 5(a). Power spectrum of mode 1.

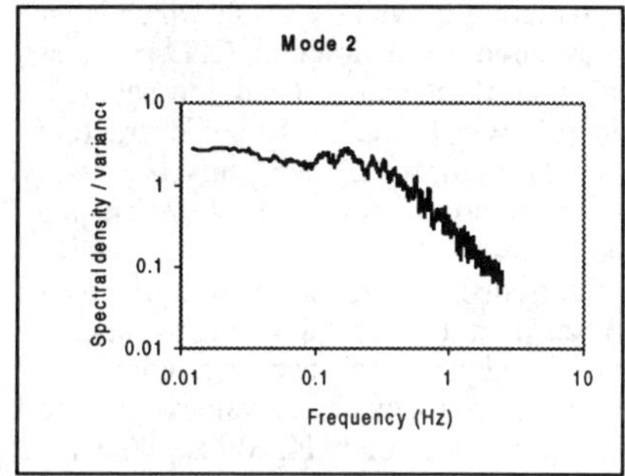

Figure 5(b). Power spectrum of mode 2.

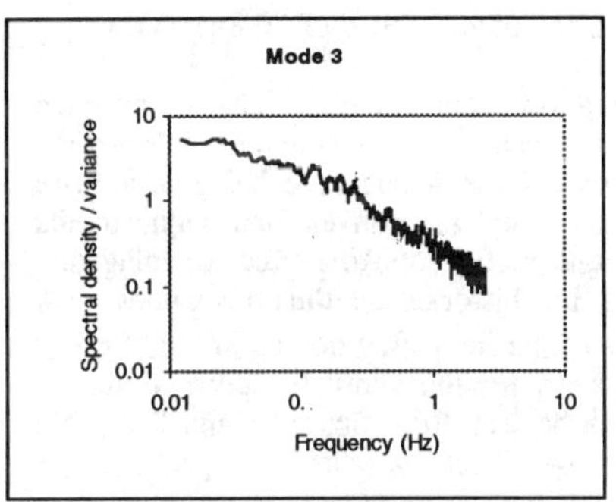

Figure 5(c). Power spectrum of mode 3.

fields were simulated simultaneously. This required a domain of sufficient size that would not affect the external flow field around the cube. The domain size utilised was 5 cube heights upstream, above and to the side of the cube, and 10 cube heights downstream. Data from the site of the cube was used to develop a mean boundary layer profile in terms of both velocity and turbulent kinetic energy profiles with a roughness height of 0.01m.

Table 2 Measured and calculated non dimensional mean discharge rates.

Measured values	0.67/0.71
Discharge coefficient method	0.49
CFD calculations	0.648

From table 2 it can be seen that the discharge coefficient method significantly underpredicts the mean discharge through the cube. This could be for a number of reasons – inaccuracies in the methods of measuring the experimental discharges (but this is unlikely given the consistency of the various results); the assumed (ideal) value of the discharge coefficient being too low (some unpublished data obtained during the course of the study indicated that the discharge coefficients for similar orifices could be as high as 0.75), or (and perhaps most likely) the pressure coefficients measured in the sealed cube case were significantly different when the openings were present. Detailed pressure measurements around the orifice would be required to determine whether this was the case.

The CFD prediction provides a result which is far closer to the measured ventilation rate. CFD provides comprehensive data of the external and internal pressure fields and airflow patterns for the cubic structure making it a useful tool for ventilation studies.

The unsteady ventilation rates were also calculated in two ways as follows

a) Using the frequency domain method of Baker (1998) based on the method of Haghighat et al and the technique of proper orthogonal decomposition, to predict rms values (due to pulsation flow); the following values were assumed for the parameters in equations 2 and 3 - $u_R = 6.7 m/s$, $A = 1m^2$, $L = 1.09m$ (based on an actual orifice length of 0.2m), $V = 216m^3$, $\gamma = 1.4$, $P_R = 100{,}000Pa$ These give values of ω_n and c of 30.8*r/s* and 0.142 respectively.
b) Using a time domain quasi-steady approach with the measured pressure coefficient time histories for the sealed case being used with equation 1 and the conventional value of the discharge coefficients to predict opening discharge time histories and thus rms values

The results of the frequency domain method give a natural frequency for the ventilated cube system of 4.9Hz. It can be seen from figures 1 and 2 that the internal pressure spectrum shows a peak, corresponding to this natural frequency, at a frequency of 2Hz. This difference is likely to be caused by leakage from the cube and the flexibility of the side of the cube, causing a change in the effective bulk modulus of the flow. It can be shown that this is equal to the product $P_R\gamma$ in equations 2 and 3. Vickery and Bloxham (1992) point out that that the effect of building flexibility can reduce the effective value of this parameter to as low as 20% of the normal value. This is consistent with the observed shift in the natural frequency from its predicted value. To allow for this effect the measured natural frequency will be used in what follows. With the natural frequency at this value, it is likely that some of the unsteady ventilation will take place at frequencies of up to, say, 10Hz i.e. higher than the sampling frequency. To enable the calculations to be made up to this frequency the following procedure was adopted.

a) The modal spectra shown in figure 2 were fitted with a power law curve for frequencies between 0.25Hz and 2.5Hz and 10Hz and thus extended to higher frequencies.
b) The measured power spectra were used in the calculations using equations 1 and 2 at frequencies below 2.5Hz, with the extrapolated values being used at the higher frequencies.

The predicted values from both methods were compared with the measured rms discharge rates and the results are shown in table 3. Three different values are given for the frequency domain approach, corresponding to upper frequency limits of 2.5, 5 and 10Hz. It can be seen that the frequency domain method significantly underpredicts the rms ventilation, whilst the quasi-steady method underpredicts to a significantly lesser extent. This seems at first surprising, but is in fact probably due to the configuration under consideration. The quasi-steady method is predicting the rms of ventilation of a reasonably high velocity flow passing through the cube, which is far from being the classical pulsed flow situation. It can be expected that this component of the unsteady flow will have a broad banded frequency component. Pulsed ventilation flow itself, which is predicted by the frequency domain method, assumes that slugs of fluid are forced in and out of the cube by the unsteady surface pressure. This type of ventilation can only exist near the natural frequency of the system, since at low frequencies the length of any slug of fluid must become large in comparison with the length of the cube. This is illustrated in figure 6(a) which shows the spectra of the pulsation flow ventilation predicted by the frequency domain method. The pulsation ventilation spectrum peaks at the natural frequency of the system, and falls off rapidly away from the natural frequency. Figure 6(b) shows an internal velocity spectrum, which can be seen to have a significant low frequency component. Thus it seems

Table 3. Measured and calculated values of fluctuating ventilation rate.

Measured value	0.103/ 0.107
Frequency domain method – 2.5Hz	0.0173
Frequency domain method – 5Hz	0.0178
Frequency domain method – 10Hz	0.0178
Time domain method (quasi-steady)	0.076

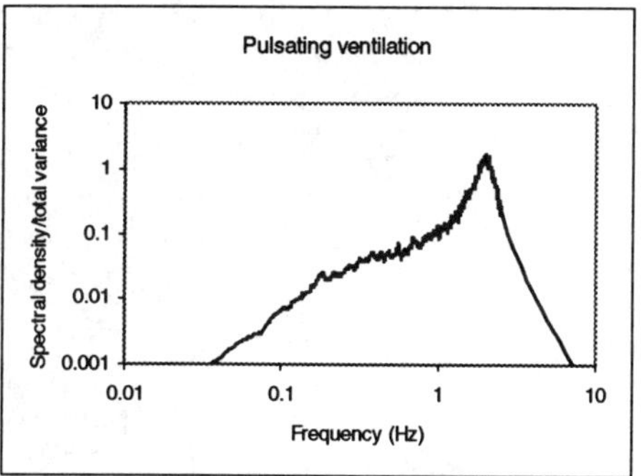

Figure 6(a). Predicted pulsating unsteady ventilation rate spectrum (frequency domain method).

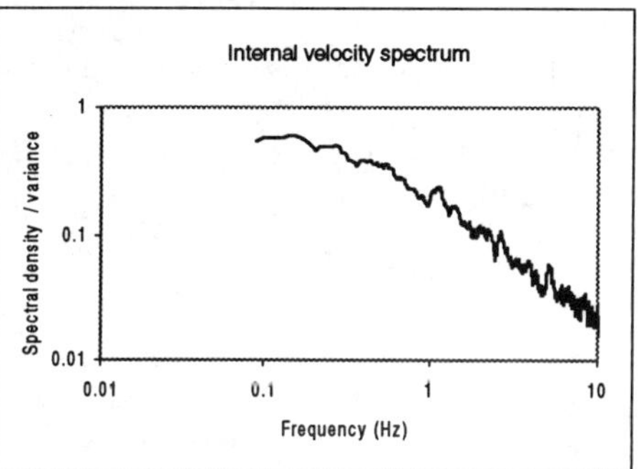

Figure 6(b). Measured internal velocity spectrum.

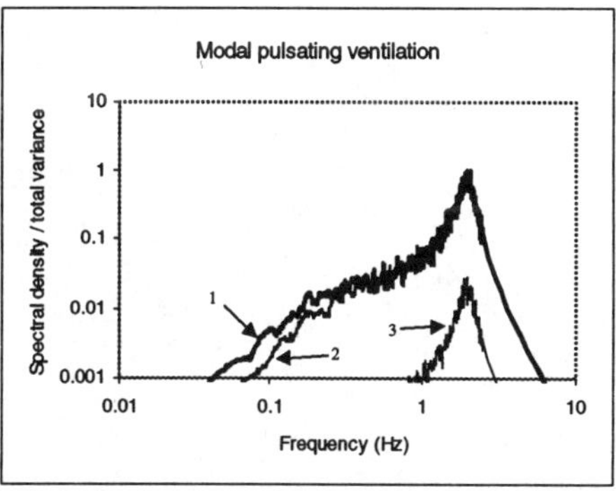

Figure 7. Predicted modal pulsating ventilation rate spectra (3 modes).

that approximately 75% of the fluctuating ventilation is due to this broad banded ventilation, approximately 15% is due to the pulsed flow mechanism, with the remaining 10% unaccounted for - but possibly caused by the eddy penetration mechanism. Two further points of interest emerge. Firstly the value of the upper frequency cutoff has little effect on the frequency domain calculations, suggesting a sharp drop in the spectra above the natural frequency, which figure 6(a) shows to be the case. Secondly it can be seen that for this configuration the pulsation ventilation is dominated by modes 1 and 2 identified in the POD analysis (as shown in figure 7). It should be noted that the modal spectra shown in figure 7 are normalised by the variance of the total predicted pulsation driven ventilation in order to show the relative contributions of each mode.

4 CONCLUSIONS

A large scale experiment has been described for assessing the wind driven ventilation through a cubic structure. The measured mean and rms values indicated that for the configuration under consideration (with openings on the windward and leeward faces of the cube) the mean ventilation rate was an order of magnitude greater than the fluctuating ventilation rate. Taken together, the measured mean and rms ventilation rates were consistent with the results of the tracer gas tests.

A number of methods have been used to predict the mean and unsteady ventilation rates. The discharge coefficient method used to calculate the mean rate significantly underpredicted the measurements. CFD provided a much improved result with the further advantage of producing data of both the internal and external pressure and velocity fields.

The methods used to predict the unsteady ventilation rates suggested that a broad banded quasi-steady fluctuating mechanism was responsible for most of the ventilation (around 75% of the total), whilst the pulsed flow mechanism was only responsible for around 15% of the total ventilation.

REFERENCES

Awbi, H. B. 1989. Application of computational fluid dynamics in room ventilation, *Building and Environment* 24, 1, 73-84.

Baker, C. J. 1998. The use of the technique of orthogonal modal decomposition in understanding the unsteady wind driven ventilation through buildings, *Proceedings of the United Kingdom Wind Engineering Conference*, Bristol.

Baker, C. J. 1999. Aspects of the use of the technique of orthogonal decomposition of surface pressure fields, Proc 10th International Conference on Wind Engineering, Copenhagen.

Bienfait, D., Phaff, H., Vandaele, L., van der Maas, J., Walker, R. 1991. Single sided ventilation, *Proc 12th AIVC Conference*, Ottawa, Canada, 24-27.

BSI (1991) *Code of practice for ventilation – principles and designing for natural ventilation*, British Standard 5925.

Chaplin, G. *1997. Turbulent wind interactions with ventilated structures*, PhD thesis, University of Nottingham, United Kingdom.

Furbringer, J., van der Maas, J. 1995. Suitable algorithms for calculating air renewal rate by pulsating air flow through a large single opening, *Building and Environment* 30, 493-503.

Haghighat, F., Rao, J., Fazio, P. 1991. The influence of turbulent wind on air change rates – a modelling approach, *Building and Environment* 26, 2, 95-109.

Haghighat, F., Rao, J., Riberon, J. 1992. Modelling fluctuating airflow through large openings – ventilation for energy efficiency and optimum indoor air quality, *Proc 13 AIVC Conference*, Nice, France 77-85.

Holmes, J. D. 1979. Mean and fluctuating pressure induced by wind, *Proc 5th International Conference on Wind Engineering*, Fort Collins, Colorado 435-450.

Hoxey, R. P., Richards, P. J., Short, J, L,. 1999. Quasi-steady theory developed with experimental verification, *Proc 10th International Conference on Wind Engineering*, Copenhagen, Denmark.

Liu, H., Rhee, K. H. 1986. Helmholz oscillation in building models, *Journal of Wind Engineering and Industrial Aerodynamics* 24 95-115.

Malinowski, H. K. 1971. Wind effect on the air movement inside buildings *Proc 3rd International Conference on ind on buildings and structures*, Tokyo 125-134.

Rao, J. 1993. *Assessment of the effect of mean and fluctuating wind induced pressures on air infiltration and ventilation in buildings – a system theoretic approach* PhD thesis, Concordia University, Montreal, Canada.

Vickery, B. J., Bloxham, C. 1992. Internal pressure dynamics with a dominant opening, *Journal of Wind Engineering and Industrial Aerodynamics* 41, 193-204.

Wind Engineering into the 21st Century, Larsen, Larose & Livesey (eds) © 1999 Balkema, Rotterdam, ISBN 90 5809 059 0

Effects of the wind on thermal sensation in a house with natural ventilation

J.G.Tsutsumi
University of the Ryukyus, Okinawa, Japan

ABSTRACT: Full-scale measurement of the indoor thermal environmental factors, air temperature, air flow speed and so on, in a single unit house and the wind direction and speed over a roof was carried out to get the relation between indoor thermal sensation and the wind. This house is ventilated naturally in the daytime through opening windows. Indoor air flow speed measured at the center of each room is roughly proportional to the wind speed over the roof in any wind direction even if window screens were put up. Indoor thermal sensation was expressed by the ASHRAE's SET*. The differences between the air temperature and the SET* are almost proportional to the indoor air flow speed, which means the over roof wind can relieve the indoor thermal sensation. These results show that cross-ventilation is an effective passive cooling method.

1 INTRODUCTION

Air flow is an essential factor to form thermal comfort sensation. Especially, sensible air flow is thought to be a most effective factor to mitigate hot and humid sensation (Ishii et al. 1990). Indoor air flow induced by natural ventilation has long since been utilized to relieve thermal comfort of occupants in houses in summer (Tsutsumi et al. 1995). Natural ventilation through opening windows that makes sensible indoor air flow is called cross-ventilation in this paper.

Thermal effect of cross-ventilation is empirically well-known qualitatively, however it is not so clear quantitatively. The authors obtained quantitative data of the thermal effect of cross-ventilation through full-scale measurement in apartment houses (Tsutsumi et al. 1996). It was rather easy to get the relation between the wind and the indoor air flow, because the shape of apartment buildings and the plan of dwelling units were generally simpler than those of single unit houses. However, the effects of window screens and shifts of the wind direction on thermal sensation by cross-ventilation were also unknown in this measurement.

Though thermal sensation is the most important purpose of cross-ventilation, it is very difficult to make the similar thermal conditions in sale models. Then, full-scale measurement was adopted to get exact data on thermal conditions. A single unit house which has windows on all the walls was selected as the measurement stage. Characteristics of cross-ventilation and its effect on thermal sensation in the single unit house are reported in this paper.

2 MEASUREMENT HOUSE

The measurement house is located in the southern suburban area of Fukuoka city, Japan. It is newly developed residential area, and there are still some vacant lots left without houses. The house is surrounded by roads on the three sides and the other side faces a vacant lot. Therefore, the house looks standing alone in a wide area. This area is located on a gently inclined plane, halfway up a small hill, and faces northwest.

It is a 2-story house, and the plans of these floors are shown in Figure 1. There are a living room, a guest room on the southwest side, a kitchen and a bathroom on the northeast side of the first floor, and 3 bed rooms on the second floor. These rooms have their identification number, from A to E, as shown in Figure 1, and H means the entrance hall. The measurement points were set at the centers of these rooms and the hall.

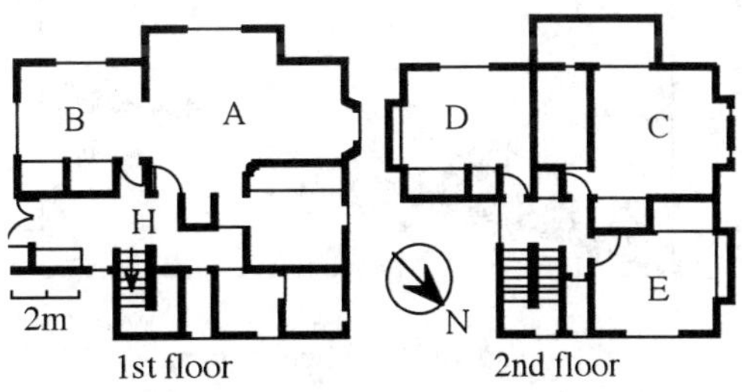

Figure 1. Plans of measurement house.

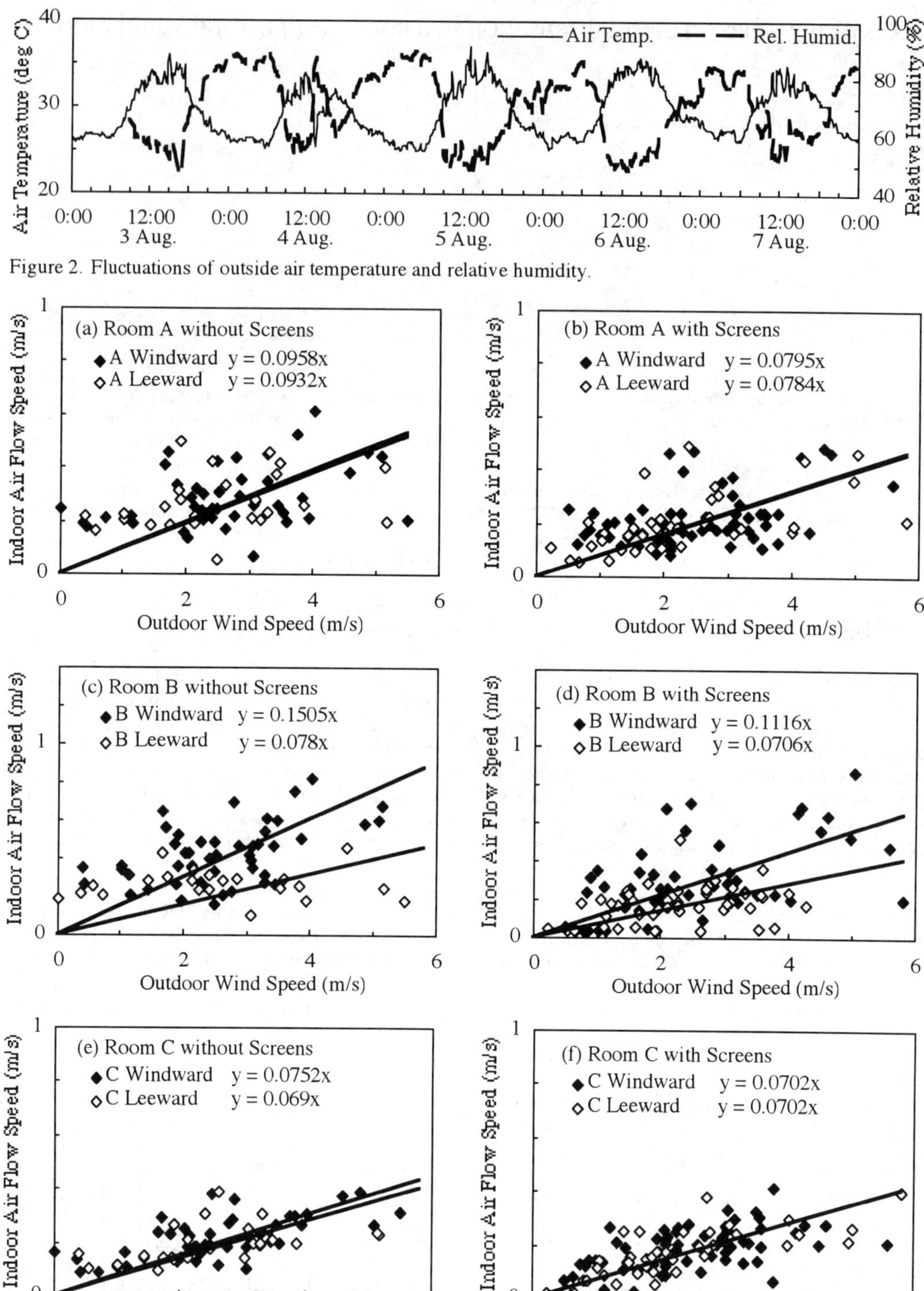

Figure 2. Fluctuations of outside air temperature and relative humidity.

Figure 4. Correlations between over roof wind speed and indoor air flow speed (part 1).

3 MEASUREMENT METHOD

Air temperature, humidity, radiation and air flow were measured as indoor thermal environmental factors at the measurement points. Air temperature was measured by T-type thermo-couple. Relative humidity was measured by electric resistance sensor. A globe thermometer which was a copper globe of 15cm in diameter painted black with T-type thermo-couple was used to measure radiation. Air flow speed was measured by no-directivity thermistor sensor. These sensors were set on a stand at each point at a height of 1.2m from the floor level. Air flow speed was measured at intervals of 5 seconds and the other factors were 10 minutes.

In addition to the indoor thermal environmental factors, the wind direction and speed were measured by a 3-cup anemometer and an arrow-shaped wind vane. They were set on a stand above the roof, and the height of these sensors was 10m from the ground level. Out door air temperature and relative humidity were also measured around the house. These sensors were almost the same as indoor sensors, but covered and ventilated to avoid the effect of solar radiation and rain. They were set at a height of 1.2m from the ground level at a distance of 2m from the outside surface of the wall in the back yard. These data were recorded at intervals of 10 minutes.

4 MEASUREMENT RESULTS

4.1 *Weather conditions*

The measurement was carried out 5 days, from Aug. 3 to Aug. 7, 1998. The outdoor air temperature and relative humidity measured at the measurement site in the measurement term are shown in Figure 2. The wind direction and speed data measured above the roof are also shown in Figure 3. The beginning part of the relative humidity data are missed, because the measurement instrument was not working in this period. Wind direction data were expressed as 16 directions, clockwise from NNE to N.

All these days were almost fine except Aug. 4. It rained a little in the daytime on Aug. 4, and it was partially cloudy on 3 and 5. Wind direction was very changeable in this term except 6. The wind speed data are almost from 1m/s to 5m/s, and stable on 6. The weather conditions of this term on the whole look good as typical summer days.

4.2 *Cross-ventilation ratio*

The correlations between the wind speed above the roof and the indoor air flow speed are shown in Figure 4. The indoor air flow speed data are mean values of 10 minutes, however the over roof wind speed data are instantaneous values. These data are classified into 2 groups by wind direction except H. Windward means the data when one of the windows of the room faces windward in the range of 45 degrees. Leeward means the data of the other wind directions. Since the entrance hall H has no windows itself, the data are not classified. There are 2 graphs for each point, one is of the data without the window screens and the other is with the window screens. Linear regression coefficients by least squares in the graphs indicate the ratios of the indoor air flow speed to the over roof wind speed, which are called cross-ventilation ratio.

The correlations between the over roof wind speed data and the indoor air flow speed data are on the whole found evidently in these graphs, though the correlations are not so clear in some rooms without window screens and the entrance hall H. The data with screens show clearer correlation. When screens are not used, outside wind blows into indoor space directly and makes a cross-ventilation path in a room, which means the air flow speed on the path is high and low in the other area. When screens are used, the screens diffuse the inlet wind and make the air flow distribution in a room flatter.

The cross-ventilation ratios range on the whole from 0.05 to 0.15. The cross-ventilation ratios with the window screens are smaller than those without screens except the leeward data of Room C and D. The measurement point C and D are thought to be located out of the cross-ventilation path. However, the differences between the cross-ventilation ratios with screens and those without screens of Room C and D are small.

The cross-ventilation ratios of the windward data are larger than those of the leeward data except Room A and C. The cross-ventilation ratios of both the windward and leeward data in Room A and C are very close each other, which means that the cross-ventilation ratios at A and C are almost the same in any wind directions. Room A is a living-dining room connected with the kitchen. It has rather large area and the distances from the measurement point to the windows are longer than other rooms. It has also large opening areas on the partitions between Room A and B. Room C has 3 windows, but 2 windows on the northwest wall are very small and can not open widely, and the other window on the southwest wall is surrounded by a verandah with a solid fence. Such characteristics of the rooms and the windows are thought to be causes which make these results.

4.3 *Effects of the wind on thermal sensation*

Thermal effect of the wind on indoor climate is to be described about the data on Aug. 6., because this is a best day as a summer fine day in the measurement term. Fluctuations of indoor air temperature on the 1st floor and that on the 2nd floor on Aug. 6 are shown in two graphs of Figure 5, respectively to compare with outside air temperature.

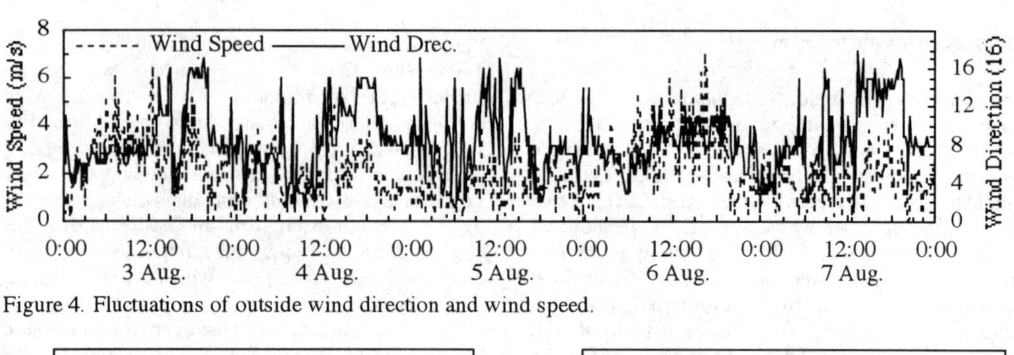

Figure 4. Fluctuations of outside wind direction and wind speed.

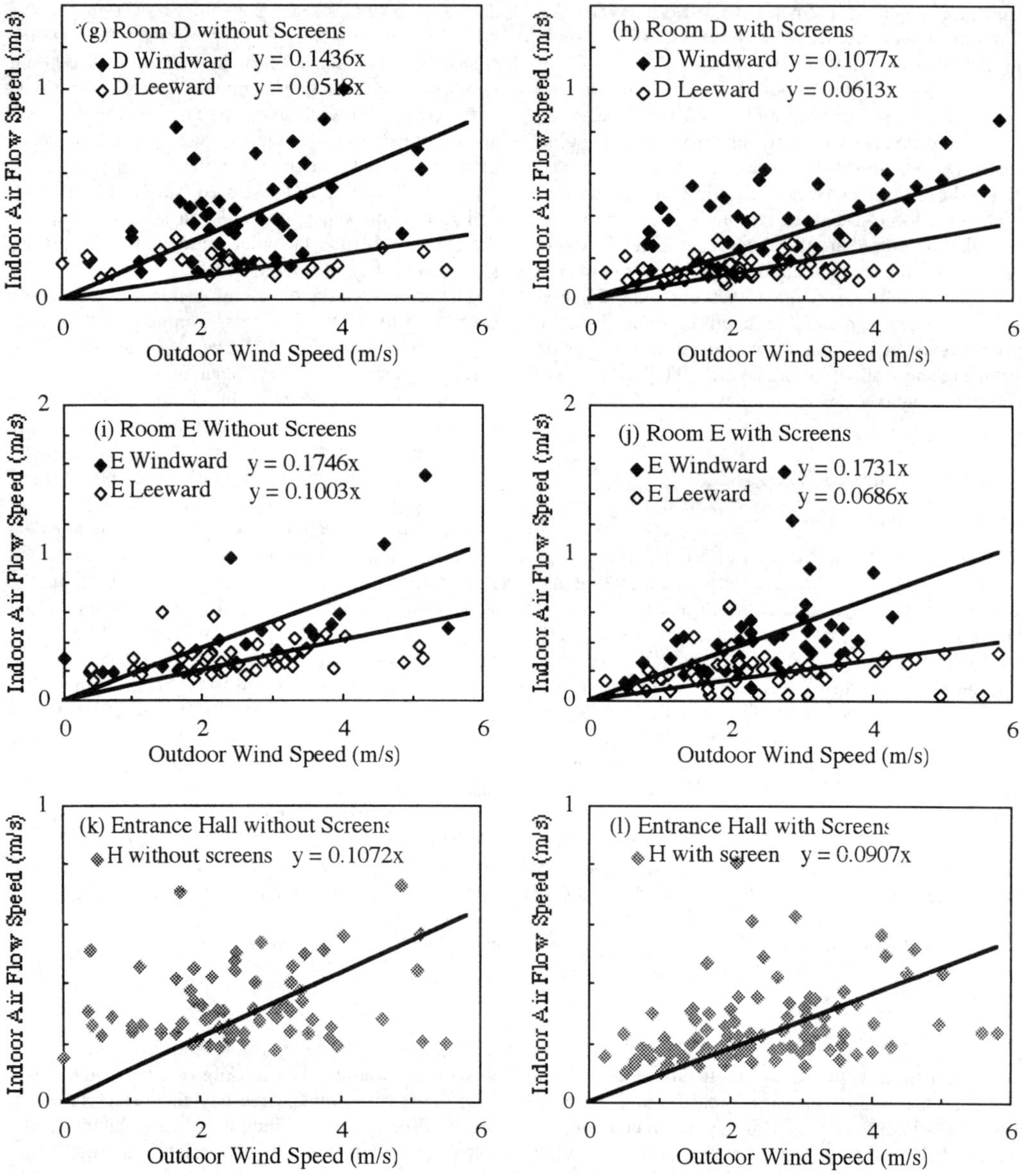

Figure 4. Correlations between over roof wind speed and indoor air flow speed (part 2).

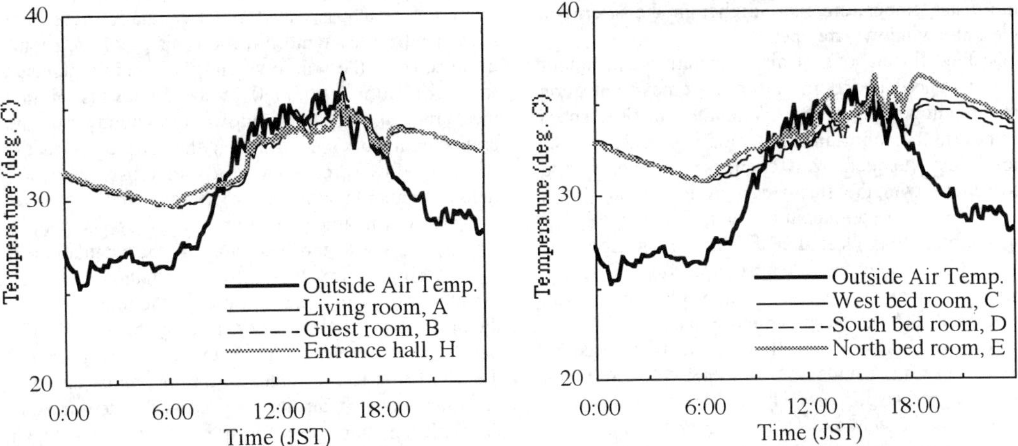

Figure 5. Fluctuations of outside air temperature and indoor air temperature Aug. 6.

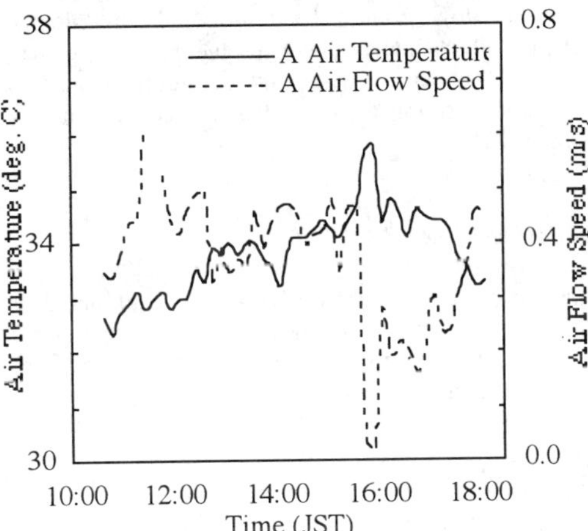

Figure 6. Fluctuations of air temperature and air flow speed in room A on Aug. 6.

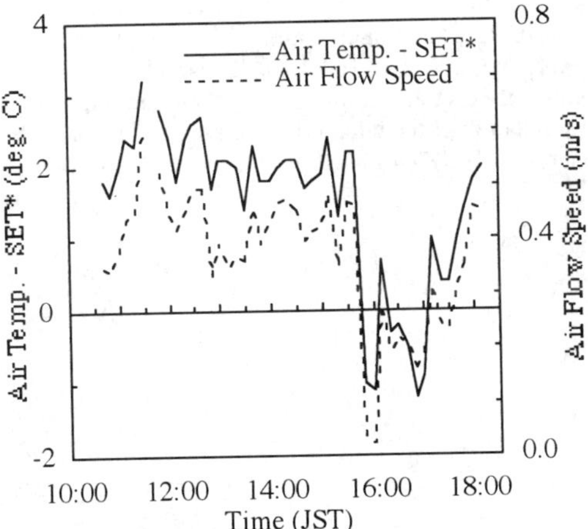

Figure 7. Fluctuations of difference between air temperature and SET* and air flow speed in room A.

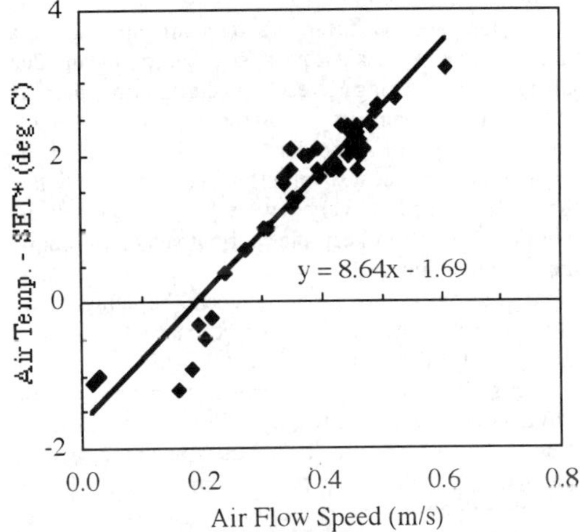

Figure 8. Correlation between air flow speed and temperature difference (air temp. - SET*) in room A.

The outside air temperature is much lower than indoor air temperature in the nighttime, then it rises after the dawn and it is almost the same or a little higher than indoor air temperature in the daytime. The 1st floor air temperature rises more slowly than the outside temperature, and they cross around 8:30. Then indoor air is cooler than outdoor for a while, however when the windows are opened at 10:30, indoor air temperature jumps up by 1 degree. The 1st floor air temperature is a little lower than the outside temperature after that.

The 2nd floor air temperature is higher than the 1st floor all day long. Therefore, the temperature change by opening windows is not found clearly. Both the 1st and the 2nd floor air temperature rise up again just after the windows are closed at 18:30. This is the effect of solar radiation. These data indicate that the indoor air temperature rises higher in the afternoon unless the windows are opened.

Detailed fluctuations of air temperature and indoor air flow speed in Room A for the cross-ventilation time are shown in Figure 6. When the air flow speed decreases, the room air temperature increases. The room air temperature changes frequently in accordance with the fluctuation of indoor air flow, though the air temperature data include diurnal fluctuation. It is clear that the indoor air flow has some effects on the indoor air temperature. It is not only in Room A, but the same phenomena are recognized at all the measurement points.

Moreover, thermal conditions are to be figured by thermal sensation indexes. ASHRAE's SET* is adopted for this purpose. It is calculated under the conditions of 0.4clo and 1.1Met. Fluctuations of the indoor air flow speed and the difference between air temperature and the SET* in Room A in the cross-ventilation time are indicated in Figure 7.

The temperature difference data are shown on a similar curve to the air flow speed, which means that air flow has strong effect on the indoor thermal sensation. The temperature difference data are almost positive except for some data. The SET* is over the air temperature because of effect of radiation when the air flow speed is very low, which means that the indoor air flow can keep the thermal sensation below the air temperature.

To indicate the relation between the temperature difference and the air flow speed mentioned above more clearly, the correlation between them is shown in Figure 8. The positive correlation between these two data is very clear. If the air flow speed is below 0.2m/s, the temperature difference is negative, that is, the SET* is higher than the air temperature. These data show that 0.2m/s is a critical air flow speed for these calculation conditions. If the values of clothing and metabolic ratio are changed, the critical value will be changed. If the indoor air flow speed increases 0.1m/s, the SET* decreases 1 degree. These results are also found at the other measurement points.

5 CONCLUSIONS

Full-scale measurement about cross-ventilation in a single unit house was carried out to get the relation between indoor thermal sensation and the wind. The measurement items are indoor thermal environmental factors, air temperature, relative humidity, air flow speed and globe temperature, and outdoor weather conditions, outside air temperature and relative humidity, over roof wind direction and speed.

The relation between the over roof wind speed and indoor air flow speed measured at 6 points in the house can be roughly regarded as direct proportion. The ratio of the indoor air flow speed to the over roof wind speed, that is called cross-ventilation ratio, are influenced by the wind direction, shape of the room, distance from the windows and the window screens. It ranges from 0.05 to 0.15 as the results of this measurement. If the windows face windward, the cross-ventilation ratio is larger than the other cases. If the window screens are used, the cross-ventilation ratio is smaller than the cases without screens.

Cross-ventilation sometimes raises up the indoor air temperature at the beginning of the ventilation in the morning. However, cross-ventilation can keep the indoor air temperature lower in the afternoon. If the thermal sensation is expressed by the ASHRAE's SET*, the difference between the air temperature and the SET* in a room fluctuates on the similar curve to the fluctuation of air flow speed. The temperature difference is almost proportional to the indoor air flow speed. The critical value of sensible air flow speed is thought to be 0.2m/s. 0.1m/s increase of the indoor air flow reduces the indoor SET* by 1degree Celsius, which means that 1m/s increase of the over roof wind speed decreases indoor thermal sensation by 1degree, if the cross-ventilation ratio is roughly estimated at 0.1.

ACKNOWLEDGEMENTS

This work was funded by YKKAP Co. Ltd. and SxL Co. Ltd. The full-scale measurement in this paper was in part supported by Professor T. Katayama of Kyushu University, Japan.

REFERENCES

Ishii, A. et al. 1990. An experimental study on human sensation to airflow in naturally ventilated rooms. *Environmental International*. 16:61-70.

Tsutsumi, J. et al. 1995. Thermal comfort around a Japanese traditional farm house. *Design of Amenity*. 206-216. Fukuoka: Kyushu Univ. Press.

Tsutsumi, J. et al. 1996. Investigation and numerical simulation of the wind effects on thermal comfort in a house. *J. Wind Eng. Ind. Aerodyn*. 60:267-280.

Wind energy

Wind Engineering into the 21st Century, Larsen, Larose & Livesey (eds) © Balkema, Rotterdam, ISBN 90 5809 059 0

Technical and economic evaluation of the Northern European offshore wind resource

T.T.Cockerill & R.Harrison
Renewable Energy Centre, School of Sciences, University of Sunderland, UK

M.Kühn, W.Bierbooms & G.J.W.van Bussel
Institute for Wind Energy, Delft University of Technology, Netherlands

ABSTRACT: A model encapsulating the major technical and economic features of bottom mounted offshore wind farms is described. The model has been coupled with a Geographical Information System to facilitate evaluation of the offshore wind resource across wide regions. Starting with a limited quantity of environmental information, contour maps showing areas suitable for offshore wind farm construction, and the cost of the energy that would result can be produced. The system has been used to assess the Northern European offshore resource.

1 INTRODUCTION

While there can be no doubt that Northern European offshore wind resources are "truly enormous" (Garrad, Nath et al. 1994), there is considerably less consensus with respect to how much of that resource it is feasible to exploit. Current concerns within the wind industry focus on the technical and economic barriers to offshore wind farm development.

As part of a CEC Joule III funded project (Kühn et al. 1998), two computer based tools have been developed to try to assess these constraints in a coherent manner. Work has dealt exclusively with "bottom-mounted" technologies, which are generally regarded as more realistic for medium term development than floating concepts.

2 METHODOLOGY

2.1 *Problem analysis.*

An offshore wind energy converter system (OWECS) has five major designed "components":

- Turbine
- Support structure
- Number and topographical arrangement of the turbines in a farm
- Connection of the OWECS to the power delivery point
- Hardware and operational strategy employed for maintenance.

The cost of the energy produced by the OWECS can be calculated through a standard discounting calculation:

$$L.P.C. = \frac{I}{aE} + \frac{T.O.M}{E} \qquad (1)$$

where LPC = levelised energy production cost; I = total investment cost associated with constructing the wind farm and E = annual energy production.

The annuity factor, a, is calculated from

$$a = \frac{1 - \left(\frac{1}{(1-d)} \right)^{l}}{d} \qquad (2)$$

where d = discount rate and l = economic lifetime (in years), and $T.O.M.$ = total annualised operation and maintenance (O&M) cost. For all calculations described in this paper an economic lifetime of 20 years was assumed with a discount rate of 5%.

Energy costs will depend in a complex way on both the design of the OWECS components and the characteristics of its location such as the distance from the shore, the water depth and the wind climate. Even though this investigation is confined to bottom mounted concepts, there are a huge range of converter systems that could be constructed at any particular location. The optimum OWECS configuration, in terms of producing energy at minimum cost, will differ between locations.

2.2 *Approach*

The objective of the authors was to produce a series of maps showing locations suitable for OWECS development, together with an estimate of the minimum cost at which energy could be produced at each location. A means of estimating energy cost as a function of OWECS design and environmental conditions was required. To this end a computer based engineering cost model for offshore wind farms, described in more detail in section 3, was developed.

The output from the model was coupled with an electronic mapping system, known as a Geographical Information System (GIS) allowing calculations to be performed for entire regions. When provided with electronic maps of environmental conditions, the system can rapidly calculate an upper-bound on the rated OWECS capacity that could be installed to meet a given constraint on the energy cost. Application of the system to data for Northern Europe produced some illuminating results.

It is important to emphasise that the interest of the authors lay as much in developing the methodology and analysis tools as in assessing the Northern European offshore resource. The tools are general in nature and could be used to evaluate the offshore wind resources of almost any region.

2.3 *Selection of base cases*

Attempting to estimate the cost associated with every conceivable OWECS is clearly impractical. Attention was focused therefore on two "base case" OWECS designs. One of these was a "near-future" design that could be constructed with existing commercial wind energy technology, while the other was a more "forward-looking" concept which would require technology that is currently experimental. The designs were optimised for minimum energy cost using the cost model for a set of environmental conditions representative of those found in Northern Europe. Relationships between changes in the environmental parameters and the cost of energy from the optimised base case farms were also established.

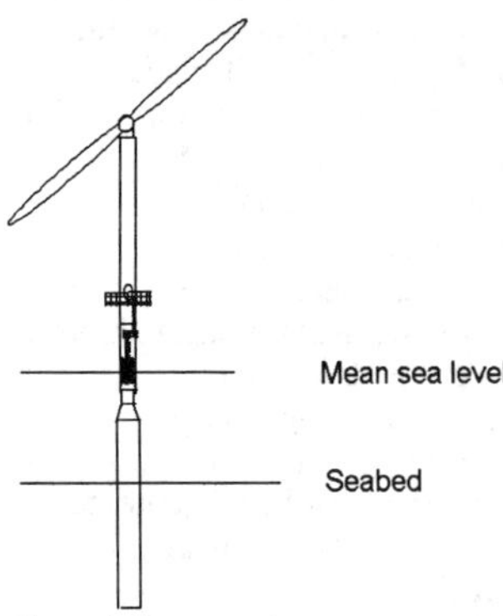

Figure 1: Base case bottom mounted OWEC concept.

The base case concepts were produced during a pilot study that ran in parallel to the development of the cost model. A range of options for all the major subsystems were considered and the most promising selected for development into a wind farm proposal. This work is not the main focus of the paper however and the reader is referred to the Opti-OWECS final report (Ferguson et al. 1998) for further details.

At the time of writing the largest commercially available wind turbines have a rated capacity of 1.5MW. The near-future base case used, therefore, 3 bladed turbines of this capacity based on a design developed by Micon. Each machine would be mounted on a monopile support structure, as shown in figure 1, in a farm consisting of 100 turbine units. Table 1 shows further details.

Table 1. Specification of the "near-future" OWECS.

Feature	Description
Turbine	
Rated capacity	1.5 MW
Rotor diameter	64 m
No. of blades	3
Unit cost	1,000,000 ECU
Energy production	Site dependent
Support structure	
Concept	Monopile
Nacelle height	70m above sea level
Pile height	Local water depth + 10 m
Unit cost	Site dependent
Grid connection	
Type	AC with undersea cables
Cost	Site dependent
Farm layout	
No. of turbines	100
Turbine spacing	640 m
Array efficiency	0.96
Geometry	10 x 10 square
O. & M.	
Annual cost	Site dependent
Turbine availability	Site dependent

It is a common assumption that multi-megawatt capacity wind turbines will be required in order to make offshore wind farms economic. In order to test this, the "forward-looking" base case adopted a 2 bladed 4MW turbine. As no machines of this size are commercially available, the turbine design was developed by partners Kvaerner-Turbin as a paper study based on their existing 3 MW Nasudden II concept (Ferguson et al. 1998). Again, the whole OWECS comprised 100 turbine units mounted on monopile support structures (Table 2).

3 THE OWECS ENGINEERING COST MODEL

3.1 *The OWECS cost model*

The cost model simulates the early stages of the design process for an offshore wind farm, producing estimates for the costs of the main components, the O&M cost and the annual energy production such that the levelised energy production cost can be calculated. Where choices of components are available, the model selects the most economic. Sensitivities of OWECS costs to changes in design concept and environmental conditions were estimated using the model.

Table 2. Details of the "forward-looking" OWECS.

Feature	Description
Turbine	
Rated capacity	4 MW
Rotor diameter	90 m
No. of blades	2
Unit cost	2,550,000 ECU
Energy production	Site dependent
Support structure	
Concept	Monopile
Nacelle height	80m above sea level
Pile height	Local water depth + 10 m
Unit cost	Site dependent
Grid connection	
Type	AC with undersea cables
Cost	Site dependent
Farm layout	
No. of turbines	100
Turbine spacing	900 m
Array efficiency	0.96
Geometry	10 x 10 square
O & M	
Annual cost	Site dependent
Turbine availability	Site dependent

The model runs within the Microsoft Excel 5 spreadsheet system, using its powerful macro language "Visual Basic for Applications". This approach allowed the use of proper programming techniques, avoiding the complications inevitable in large spreadsheets with many interlinked formulae, but still maintaining a familiar spreadsheet interface for users.

3.2 *Validation of the Model*

As far as possible individual components of the model were validated against real data. By way of example, the output of the support structure routines was compared to a selection of designs produced by experienced offshore engineers.

Validation of the model as a whole is difficult. The model simulates aspects of a system for which no physical prototype exists and thus there is no obvious benchmark case against which it can be sensibly evaluated. While a handful of offshore windfarms are in operation, they are of small size with installed capacities below 10 MW. The model, in contrast, has been tailored to studies of much larger farms with installed capacities in excess of approximately 50 MW. Since the engineering and economics of small and large OWECS show very different characteristics, employing existing offshore farms as a benchmark would not be useful.

To obtain an indication of the performance of the model, it was used to re-compute a number of paper based studies available in the literature. Even this is not a completely satisfactory solution as many of the "benchmark studies" embody assumptions at odds with those adopted for the model. It was judged unrealistic to expect the model to reproduce the energy costs in the benchmark studies to a high degree of precision. Instead the validation was defined a priori to be successful if the model was able to rank the OWECS by energy cost in the same order as would be obtained using the original cost data.

The model was able to rank the majority of the recomputed OWECS correctly, although there were a couple of mis-ranked results. Table 3 compares the results for two of the studies recomputed.

Table 3: Extracts from cost model validation.

Study	Skegness	SK Power
Reference	Kölher 1995	Olsen, F.A. et al.
Installed Capacity	180 MW	60 MW
Original L.P.C.	0.137 ECU/kWh	0.065 ECU/kWh
Modelled L.P.C.	0.114 ECU/kWh	0.065 ECU/kWh

4 THE GIS BASED COST MODEL

4.1 *Geographical Information Systems*

A GIS can store and manipulate spatially varying data, such as water depth, along with information fixing the data geographically, for example, the latitude and longitude. All the calculations described in the paper make use of "raster" format data wherein regions are divided into a regular array of cells, and each cell is allocated a value. The entire area encompassed by the cell is assumed to be represented by the single cell value.

A GIS has the ability to perform mathematical manipulation of the stored data. Given, for example, maps of wind conditions in a region, then knowledge of a turbine's Power-wind speed curve allows calculation of a new map showing the energy that would be produced were it located in each of the cells.

4.2 *Choice of GIS*

The GIS adopted for this study is the GRASS 4.1 system (US Army Corps of Engineers, 1993) originally developed by the US Army for munitions and range calculations. Although slightly outdated now, GRASS is a very powerful package and its engineering, rather than purely geographical, heritage makes it arguably the best suited system for our purposes here. GRASS runs under the Unix operating system, with X-windows, and has the distinct advantage of being available free of charge.

4.3 *Implementing a cost model within the GIS*

By necessity, the GIS implementation of the cost model is simplified. It is based on a series of look up tables relating design and environmental factors to overall costs, which have been established using the main model. The look up tables themselves are embedded within a series of GRASS commands, which in turn, are bound together by several Unix shell scripts..

A "cell by cell" approach to the calculations is adopted. The routines deal with each map cell independently and no account is taken of the possibility that wind farms straddling cells might have more favourable qualities than those located wholly within a single cell. In addition, no account is taken of the effect of any changes across a map cell, nor indeed is it possible to do so with any GIS. As a result some care is necessary in interpreting the results. By way of an example, any sudden changes in depth within a map cell, which are not discernible from the average values, could have a significant influence on the energy cost which will not be reflected in these calculations.

The model processes input maps of environmental and design data to produce estimates of six quantities for each map cell:

- Annual energy production
- Cost of the support structure
- Cost of the grid connection
- Cost of the turbine
- Availability of the wind farm
- Annual operation and maintenance costs

Decommissioning costs, which only make a small contribution to the overall energy cost, were neglected.

4.4 *Annual energy production calculation*

The annual energy production calculation estimates the wind speed at the turbine hub height from a reference wind speed and a reference height that must be provided for each cell. The result is used to establish the annual energy production of a single wind turbine and then multiplied by an array efficiency and the total number of turbines, to estimate the total energy production. A turbine availability factor is accounted for after it has been calculated by a later routine.

4.5 *Investment cost calculation*

4.5.1 *Support Structure Cost*

The cost of the support structure is divided into two parts, one for the material costs and one for the physical construction and installation.

Aside from the influence of the design concept and turbine type already noted, the support structure material costs are a strong function of the hub height and the water depth.

For the range of concepts considered the installation costs are largely a function of the number of machines installed, rather than their size or distance from the shore. The installation cost for a single support structure is determined purely on the basis of the total number of turbines to be installed. By combining this value with the previously evaluated material cost, and then multiplying by the number of turbines, the cost of the support structure is determined.

The characteristics of the sea bed and their influence on the dynamic behaviour of the foundation and support structure do have some effect on the design, and therefore cost. Their main contribution is through the dynamics of the support structure. A design which is quite satisfactory at one location, may be unsatisfactory at another by virtue of differences in soil stiffness causing the natural frequency to move into a band where resonance with the rotating blades might occur. While it would be desirable to account for this within the model, the lack of information on the dynamic characteristics of the sea bed makes it impractical in all of the cases investigated here. Instead a constant soil stiffness has been assumed for all the calculations. Work is continuing to improve this aspect of the calculations.

4.5.2 *Grid Connection Cost*

Grid connection costs are most strongly a function of the number of turbines in the OWECS, their individual power capacity, and the distance of the OWECS from the shore. The calculations detailed here only consider AC power transmission is more economic than DC technology for distances to the shore of less than 50-70 km. Variation of the spacing of turbines within a farm has only a small influence on the grid connection costs and its effect is not included.

An important limitation is that no account is taken of the costs of any land based grid connection equipment. Compilation of the data required to allow such costs to be incorporated was, unfortunately, beyond the scope of the project. Thus, the costs quoted are for delivery of electricity to the nearest beach.

4.5.3 *Turbine cost*

The type of turbine to be assumed is specified before running the model and the turbine cost figures are used directly.

4.6 *Operation and maintenance costs and availability calculation*

Operation and maintenance (O&M) costs make a substantial contribution to OWECS economics and thus it is important to include them. Unfortunately, O&M costs are very difficult to predict, and the relevant quantities closely interlinked.

The major features affecting O&M costs are the distance of the OWECS from shore, the exposure of the site (as indicated by the annual mean wind speed), the size of the OWECS, the reliability of the turbines, the maintenance strategy under which they are operated and the availability that is required. An OWECS operation and maintenance model developed in parallel to this work by the Institute for Wind Energy (vanBussel & Schöntag 1997) has been used to investigate the relationships between the main quantities. For the GIS calculations, analytic functions have been fitted to the results so that wind farm availability and operation and maintenance costs can be estimated easily .

4.7 *Calculation of levelised energy production costs*

Each of the intermediate results is stored within a raster map, with a specific value associated with each cell. The values for each cell are combined to produce an estimate for the levelised energy production cost of an OWECS located there.

5 ASSESSMENT OF NORTHERN EUROPEAN OFFSHORE RESOURCES

5.1 *The GIS Offshore Wind Energy Database from JOULE Project JOUR0072.*

As part of an earlier European Commission funded investigation of offshore wind energy [3.3-1], Garrad-Hassan and Germanischer-Lloyd have developed a GIS database of conditions in European waters. This database has been used, with the kind permission of Garrad Hassan and Germanischer Lloyd, as a basis for the calculations presented here.

The information included is specifically oriented towards offshore wind energy and includes the important environmental parameters of wind speed, water depth and distance to shore. Details of areas unsuitable or unavailable for wind energy developments are also incorporated within the database.

At even the relatively modest resolutions of around 2km employed here, a large number of grid cells are required to cover coastal European waters. To prevent processing from becoming unwieldy, the data was subsetted on a country by country basis. For Britain and Ireland this still results in too much data, and thus here that data was further subdivided into separate North and South sets

5.2 *Calculation procedures*

Before the offshore wind resource of a region can be evaluated, locations in which it would not be possible to build a wind farm must be "masked" from consideration. With regard to man made constraints the rules adopted here were:

- *Traffic Zones*: All cells in which there is a record of a shipping traffic zone were excluded.
- *Undersea pipelines and cables:* All cells in which there is evidence of undersea pipeline or cables were added to the 'mask'.
- *Oil platforms*: A 10 km radius around any oil platforms was excluded.
- *Conservation areas:* Only one conservation area, the Wadden Sea National Park, is noted in the database and this was excluded from consideration.
- *Military areas:* Any areas noted as reserved for military use were excluded.

Natural constraints included:

- Locations with a seabed slope in excess of 5°, which is too steep to build upon
- Locations with a water depth in excess of 35m, which was considered to be the practical limit of the bottom mounted concept adopted for the study.

With these areas excluded from consideration, The GIS was used to process the offshore wind energy database for the waters around several northern EU countries, including UK, Western Ireland, Belgium, Netherlands, Germany and Denmark.

6 RESULTS FOR MEDIUM SCALE (150MW CAPACITY) OFFSHORE FARMS

Figure 2 shows the result of calculations for medium scale OWECS around Germany and Denmark. Figure 3 illustrates the total rated capacity that could be installed in the region as a function of the minimum economic energy cost. The graph divides the data into that for sites less then 5 km from the shore, and that for sites greater 5 km from the shore. This, rather arbitrary, division represents an attempt to

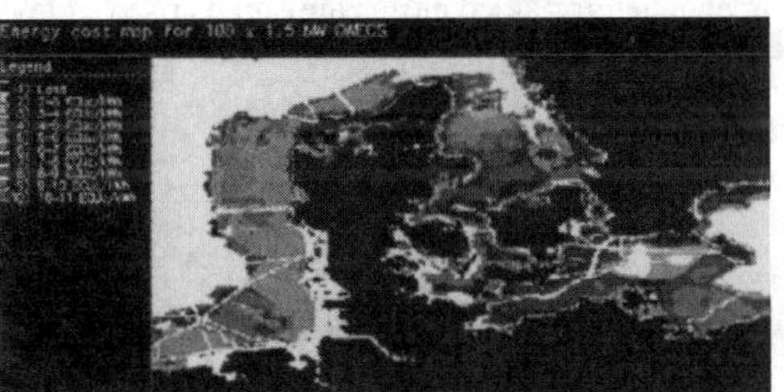

Figure 2: Map of medium scale results for Denmark and Germany.

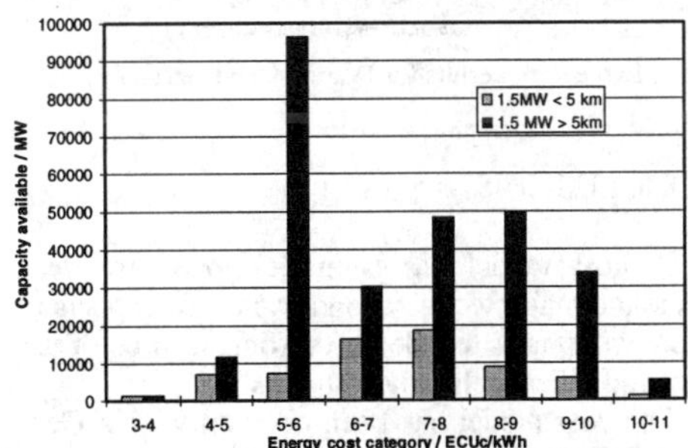

Figure 3: Medium scale results for Denmark and Germany.

take account of the likelihood that public opinion will prevent OWECS developments close to shores.

Calculations were repeated for the other countries. The results are listed in Table 4.

Table 4: Results for medium scale OWECS more than 5 km from coast in Northern EU (Energy costs > 7 ECUc/kWh are excluded).

Region	Maximum Installed Capacity (GW) as function of Energy Cost (ECUc/kWh)				
	3-4	**4-5**	**5-6**	**6-7**	**Total**
Northern GB + W. Ireland	0	4.4	3.6	4	12
Southern GB + W. Ireland	2	22	9	12	45
Belgium + Netherlands	0.5	12	32	3	47.5
Denmark + Germany	0.5	11	97	30	138.5
Northern EU					243

7 RESULTS FOR LARGE SCALE OFFSHORE (400MW CAPACITY) FARMS

The calculations for large scale OWECS were performed almost identically to the medium scale investigation. As large scale OWECS would, by definition, be larger than medium scale developments, they would be more easily visible from the shore. For this region, the buffer zone around shores where it was assumed public opinion would prevent development was extended to 10 km. Details of the results for Denmark and Germany are shown in figure 4. Results obtained for all countries are summarised in table 5.

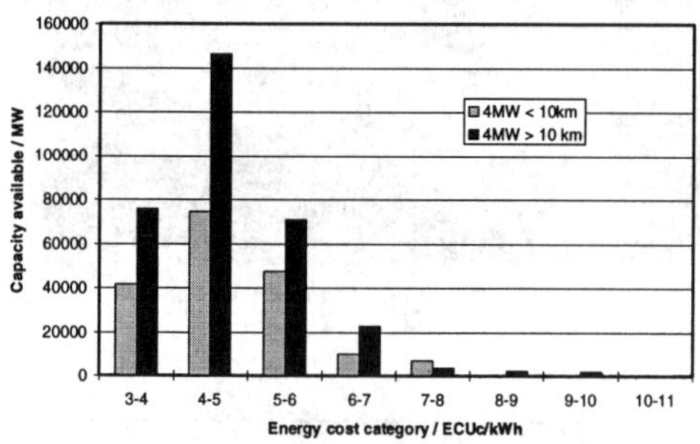

Figure 4: Large scale results for Denmark and Germany.

8 CONCLUSIONS

1. A cost model has been developed that can simulate the major technical and economic aspects of offshore windfarms to a degree sufficient to be of use in pilot and other preliminary studies.

2. The cost model has been coupled with a Geographic Information System, such that the offshore wind potential of whole regions can be rapidly assesses.

3. The combined GIS/cost model has been used to evaluate the offshore resources of the Northern EU in a consistent manner, demonstrating that substantial offshore turbine capacities could be installed to produce power at or below a cost of 4 ECUc/kWh.

4. The major effect of using larger capacity OWEC units was to increase the rated capacity that could be installed to produce energy below any particular cost level. This was only apparent at mediocre sites, where larger OWECS brought about a significant reduction in energy costs. At the best sites, using larger OWECS had almost no influence on the cost.

5. It follows that offshore wind energy developments can begin using currently available technology at the best sites without suffering any economic penalty. Larger scale installations will be required if the Northern European offshore wind resource is to be widely exploited on an economic basis.

Table 5: Results for large scale OWECS more than 10 km from the coast in Northern EU. (Energy costs > 7 ECUc/kWh are excluded).

Region	Maximum Installed Capacity (GW) as function of Energy Cost (ECUc/kWh)				
	3-4	**4-5**	**5-6**	**6-7**	**Total**
Northern GB + W. Ireland	2.5	8	3	4	17.5
Southern GB + W. Ireland	50	42	22	28	142
Belgium + Netherlands	5	50	8	3	66
Denmark + Germany	75	143	71	22	311
Northern EU					536.5

REFERENCES

Ferguson, M. et al. 1998. Opti-OWECS final report volume 4: A typical design solution for an offshore wind energy conversion system. Institute for Wind Energy, Delft University of Technology, Netherlands.

Garrad, A., Nath, C., et al. 1994. Study of offshore wind energy in the EC: Report on CEC JOULE 1 contract JOUR-0072. Verlag Natürliche Energie, Brekendorf, Germany.

Kölher, S 1995. Cost analysis and optimisation of offshore wind energy converter systems. Report IW-95.088R, Institute for Wind Energy, Delft University of Technology, Netherlands.

Kühn, M et al. 1998. Opti-OWECS project final report volume 0: Executive Summary. Institute for Wind Energy, Delft University of Technology, Delft, Netherlands.

Olsen, F.A. et al. 1994. Feasibility study on a large scale wind farm located in the Baltic Sea. In Proceedings of the European Wind Energy Conference 1994, IWEA.

US Army Corps of Engineers 1993. Grass 4.1 users' reference manual. US Army Corps of Engineers, Construction Engineering Research Laboratories, Illinois, USA.

vanBussel, G.J.W. & Schöntag, C 1997. Operation and maintenance aspects of large offshore windfarms. In Watson, R. (ed), Proceedings of the European Wind Energy Conference, Dublin, October 1997: 272 - 275.

Wind Engineering into the 21st Century, Larsen, Larose & Livesey (eds) © Balkema, Rotterdam, ISBN 90 5809 059 0

A dynamic stall model for HAWT performance prediction in yawed flow

Wang Chen & He Dexin
China Aerodynamics Research and Development Center, Mianyang, Sichuan, People's Republic of China

ABSTRACT: In the paper, the Navier-Stokes equation is employed to simulate airfoil steady flow and unsteady flow. The dual time method is applied to calculate unsteady flow. The fgh model offered by Swedish FFA and the generalized dynamic aerodynamic coefficient model brought forward by C.Edward Lan is employed to simulate dynamic aerodynamic coefficient. The parameter of models is identified. The wind performance program is improved by making use of fgh model.

1 INSTRUCTION

Wind turbines operate at all times in an unsteady aerodynamic process. Factors such as atmospheric turbulence, wind shear, skewed flow, and the influence of the turbine tower on the blades, etc. The inflow angle will vary periodically accompanying the turbine rotor rotation in yaw. The variation results in the blades unsteady loading, which plays an important role in the aerodynamic performance of turbine. The horizontal axis wind turbine performance computation is formerly based on traditional blade element and momentum theory. So, the dynamics stall at turbine rotor is ignored. To improve computational accuracy, a semi-empirical algebraic aerodynamic model is employed to predict dynamic aerodynamic in the paper. To develop aerodynamic model, we employed the Navier-Stokes equation to simulate airfoil steady flow and unsteady flow. The dual time method is applied to calculate unsteady flow. The Baldwin-Lomax algebraic model is employed for turbulence model.

2 THE NAIVER-STOKES COMPUTATION OF VISCOUS FLOW FIELD ABOUT AIRFOILS

We use 2D viscous code to calculate steady state as well as unsteady state about airfoil. The code is a finite volume N-S method. It is base on the implicit L-U decomposition by Jameson et al. To increase stability of scheme, the implicit residual smoothing that is modified by using variable coefficients to prevent the loss of flow physics for the unsteady flow is engaged in the calculation. The turbulent model is taken from the Baldmin-Lomax two-layer algebraic eddy model. The dual time method is employed to calculate unsteady flow.

2.1 *Numerical method*

The two-dimensional N-S equation can be written in generalized coordinates (ξ , η) and conservation form as follows:

$$\frac{\partial\hat{Q}}{\partial\tau}+\frac{\partial(\hat{E}+\hat{E}_v)}{\partial\xi}+\frac{\partial(\hat{F}+\hat{F}_v)}{\partial\eta}=0 \tag{1}$$

Where $\hat{Q}$ is the conservative variable, $\hat{E},\hat{F}$ are the inviscid flux vectors, $\hat{E}_v$, $\hat{F}_v$ is the viscid flux. Using the implicit L-U decomposition technique by Jameson et al., Eq (1) are solved by a series of sweeps through the two direction (ξ, η) as follows:

$$(LD^{-1}U)\delta Q=\frac{1}{\omega}RHS \tag{2}$$

And then,

$$L\delta Q^{*}=\frac{1}{\omega}RHS \tag{3}$$

$$\delta Q=\delta Q^{*}-D^{-1}D_{+}\delta Q \tag{4}$$

$$Q^{n+1}=Q^{n}+\delta Q \tag{5}$$

Where, $L=D-(A_{i-1}^{+}+B_{j-1}^{+})$, $U=D+(A_{i+1}^{-}+B_{j+1}^{-})$, $D=\frac{I}{J\Delta t}+(r_A+r_B)I$

And $A^{\pm}=\frac{A\pm\lambda r_A I}{2}$, $B^{\pm}=\frac{B\pm\lambda r_B I}{2}$, $A=\frac{\partial E}{\partial Q}$, $B=\frac{\partial F}{\partial Q}$

r_A, r_B is defined as a value which has to be equal or greater than the spectral radius of A, B respectively. The right of Eq (2):

$$RHS = -\Delta t(\frac{\partial E}{\partial \xi} + \frac{\partial F}{\partial \eta}) + \frac{M_\infty}{\text{Re}}(\frac{\partial E_v}{\partial \xi} + \frac{\partial F_v}{\partial \eta}) \quad (6)$$

2.2 *The Dual time method*

Using the dual time method to simulate unsteady flow is very effective. The unsteady approximate L-U decomposition is:

$$(LD^{-1}U)\delta Q^n = RHS^n \quad (7)$$

First, from t_1 to t_2, using Eq (7) calculate a Q^{n+1} rough value. So,

$$f(Q^{n+1}) = (LD^{-1}U)\delta Q^n - RHS^n$$

Freeze n moment, along virtual moment τ_1, τ_2, there is this relations:

$$f(Q^v) = (LD^{-1}U)(Q^v - Q^n) - RHS^n \quad (8)$$

$$f(Q^{v+1}) = (LD^{-1}U)(Q^{v+1} - Q^n) - RHS^n \quad (9)$$

$$f(Q^{v+1}) = f(Q^v) + (\frac{\partial f}{\partial \tau})^v \Delta\tau + o(\Delta\tau)^2 = f(Q^v) + (\frac{\partial f}{\partial Q})^v (Q^{v+1} - Q^v) + o(\Delta Q^v)^2 \quad (10)$$

Obvious, when $v \to \infty$, $f(Q^{v+1}) = 0$, so set up $f(Q^{v+1}) = 0$
Then,

$$f(Q^v) + (\frac{\partial f}{\partial Q})^v \Delta Q^v = 0 \quad (11)$$

We can get this relation:

$$f(Q^v) + (LD^{-1}U)\Delta Q^v = 0 \quad (12)$$

When v increase, $\Delta Q^v = 0$, so Eq (12) convergence, $f(Q^v) = 0$ is approximate tenable. Setting up:

$$f(Q^v) = \frac{(3Q^v - 4Q^n + Q^{n-1})}{\Delta t} + RHS^n \quad (13)$$

Thus, the unsteady process is changed to solve steady problem. A steady problem is solved every step Δt time.

2.3 *Boundary condition and moving grid*

For the fixed wall, the velocity is not slip, thermoslag, the pressure zero gradient. The Riemann invariant is applied at far field. The turbulent model is taken from the Baldmin-Lomax two-layer algebraic eddy model. The grid is fixed in body and following body moving. It is easily completed by used following relation:

$$\begin{bmatrix} x \\ y \end{bmatrix}_{t+\Delta t} = \begin{bmatrix} \cos\alpha(t) & \sin\alpha(t) \\ -\sin\alpha(t) & \cos\alpha(t) \end{bmatrix} \begin{bmatrix} x \\ y \end{bmatrix} \quad (14)$$

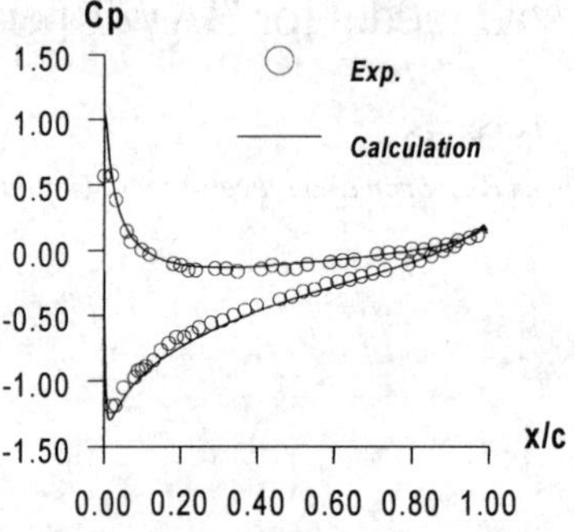

Figure 1. NACA0012 airfoil pressure distribution.

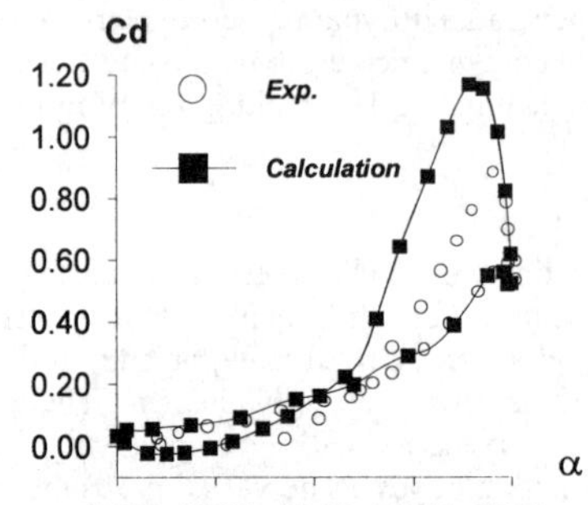

Figure 2. NACA0012 dynamic stall drag hysteresis circle.

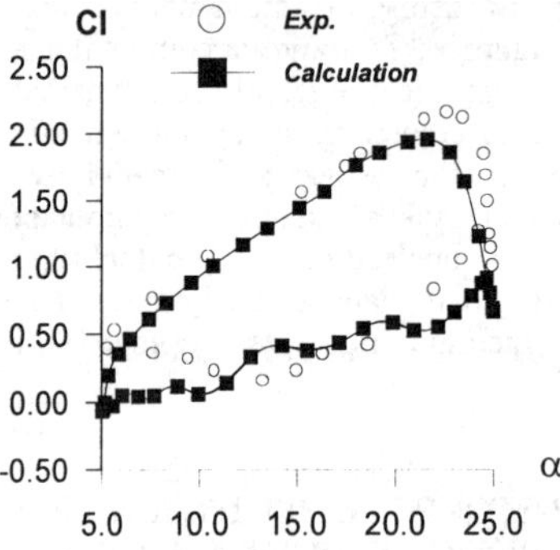

Figure 3. NACA0012 dynamic stall lift hysteresis circle.

2.4 *Numerical result and discussion*

Fig.1 show NACA0012 airfoil steady pressure distribution. The condition is $M_\infty = 0.299$, $Re = 3.45 \times 10^6$, $\alpha = 3.59$. The result of calculation fit exactly experiment. In unsteady process, simulating NACA0012 airfoil forced harmonic oscillation. The attack angle is $\alpha(t)=\alpha_0-\alpha_a\cos(2kM_\infty \bar{t})$. The flow parameter is $\alpha_0=15^0$, $\alpha_a=10^0$, $M_\infty=0.283$, $k=0.151$, $\mathrm{Re}=3.45\times10^6$. Fig.2, Fig.3 and Fig.4 are dynamic stall lift, drag and moment hysteresis circle respectively. Obvious, the result of calculation is not very successful. The B-L model is too simple to fit in heavy separation of flow. So, we can see that during pitching up the separation of flow id suppressed, the simulation closed the experiment data. When pitching down, the separation of flow increased, the result of calculation is poor. An advance turbulent model must be recommended in the future.

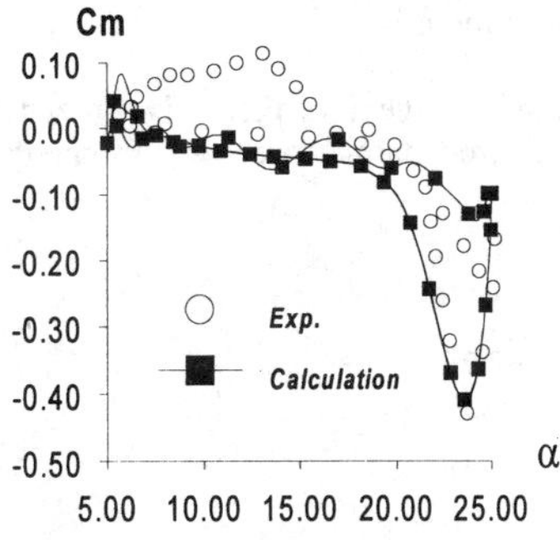

Figure 4. NACA0012 dynamic stall moment hysteresis circle.

3 THE DYNAMIC STALL MODEL IDENTIFICATION

3.1 *The fgh model*

In this paper, the fgh model offered by FFA. The fgh model given as

$$c_n=c_{n,stat}+f\frac{c}{w}\frac{d\alpha}{dt}+g(\frac{c}{w})^2\frac{d^2\alpha}{dt^2}+h\frac{c}{w^2}\frac{dw}{dt} \qquad (15)$$

Where, f, g, h are functions of the angle of attack. For a given set of data, the functions are determined by optimization algorithm such as least-square fitting. The chordwise force and pitching moment equations are described same as the equation for c_n.

3.2 *The generalized dynamic aerodynamic coefficient model*

The GDAC model is brought forward by C.Edward Lan etc. The model given as:

$$c_a=c_0+c_1\alpha+c_2\alpha^2+c_3\alpha|\alpha|+c_4\alpha^3+c_5\dot{\alpha}+c_6\alpha\dot{\alpha}+c_7|\alpha|\dot{\alpha}+c_8|\dot{\alpha}|\dot{\alpha} \qquad (16)$$

Where, c_a represents the aerodynamic coefficients, c_i is coefficients which will be identified. By a given set of data, we can identify aerodynamic coefficients.

3.3 *The generalized dynamic aerodynamic*

Fig.5 and Fig.6 show the result of identification at the reduced frequency k=0.022 with the fgh model and the GDAC model respectively. Fig.7 and Fig.8 show the result at the reduced frequency k=0.044, Fig.9 and Fig10 are at reduced frequency k=0.06. From these results, we can see that the reduced frequency is higher, the result of identification is close the experiment data.

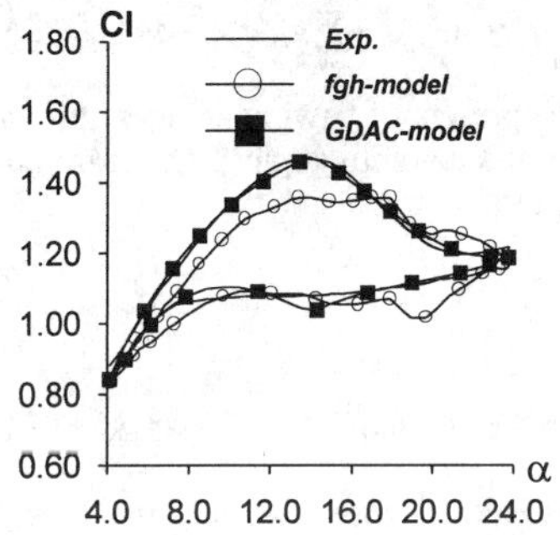

Figure 5. The result of lift model identification (k=0.022).

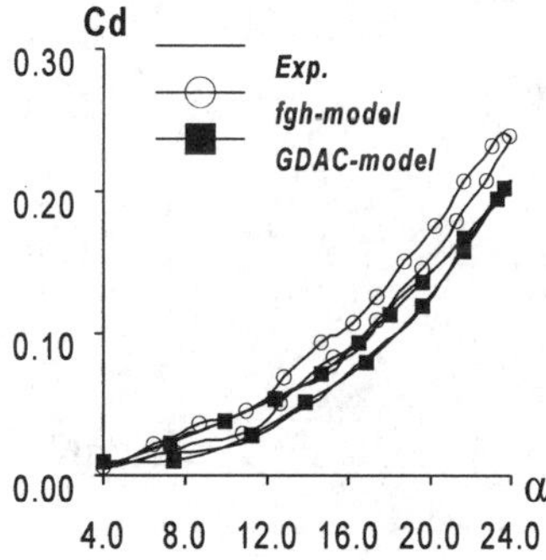

Figure 6. The result of drag model identification (k=0.022).

4 THE HAWT PERFORMANCE COMPUTATION

The classical blade element and momentum theory was employed in HAWT performance. The unsteady aerodynamic at turbine rotor is ignored. So, we lead the unsteady aerodynamic into the classical blade element and momentum theory through employing fgh model.

4.1 *Aerodynamic model*

Blade element and momentum theory is utilized for determining the aerodynamic forces and moments at the blades.

$$\alpha = \phi - \theta - \Delta\alpha \tag{17}$$

$$C_y = C_L \cos\phi + C_D \sin\phi \tag{18}$$

$$C_x = C_L \sin\phi - C_D \cos\phi \tag{19}$$

Where,

$$C_L = C_{L,static} + \sum_{i=0}^{5} f_i^{C_L} \alpha^i (\frac{c}{W})\alpha' + \sum_{i=0}^{5} g_i^{C_L} \alpha^i (\frac{c}{W})^2 \alpha'' \tag{20}$$

$$C_d = C_{d,static} + \sum_{i=0}^{5} f_i^{C_d} \alpha^i (\frac{c}{W})\alpha' + \sum_{i=0}^{5} g_i^{C_d} \alpha^i (\frac{c}{W})^2 \alpha'' \tag{21}$$

Considering the present of yawing angle γ_1,tilt angle ε,con angle δ and azimuth angle β, we can induce a modified expression for ϕ,

$$tg\phi = \frac{W_x}{W_y} \tag{22}$$

$$W_x = W(\cos\gamma \cos\beta + \sin\gamma \sin\beta \sin\varepsilon) - \Omega r \cos\delta(1+b) \tag{23}$$

$$W_y = W(\sin\gamma \cos\delta \cos\varepsilon(1-a) + \sin\delta(\sin\beta \cos\gamma - \sin\gamma \sin\varepsilon \cos\beta)) \tag{24}$$

Where, $\gamma = \pi/2 - \gamma_1$, axial and angular interference factor a and b are given by

$$a = \frac{2s + Ft - \left[Ft^2 + 4sFt(1-Ft)\right]^{0.5}}{2\left(s + Ft^2\right)} \tag{25}$$

$$b = \frac{ss(1-a)}{1 - aFt - ss(1-a)} \tag{26}$$

With,

$$s = \frac{BC \cdot Cy\cos(con)}{\pi r \cdot 8\sin^2(\phi)} \tag{27}$$

$$ss = \frac{BC \cdot Cx}{\pi r \cdot 8Ft\sin(\phi)\cos(\phi)} \tag{28}$$

When axial interference a>0.313,

$$a = 1 + 0.695Sy - \sqrt{(1.815 + 0.483025Sy)Sy} \tag{29}$$

4.2 *Calculation sample*

Rotor description	
Number of blade	3
Rotor radius (m)	2.675

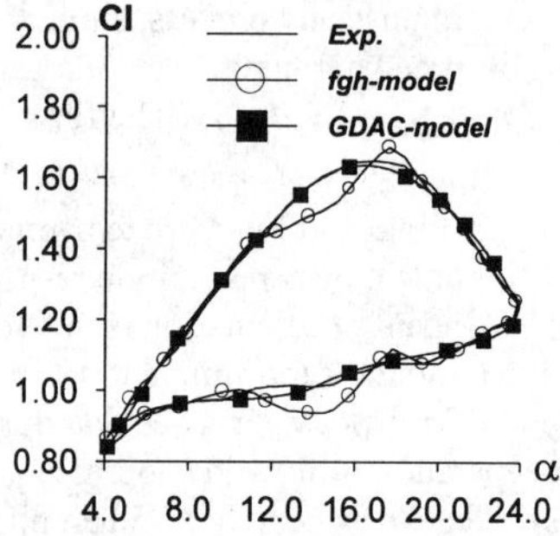

Figure 7. The result of lift model identification (k=0.044).

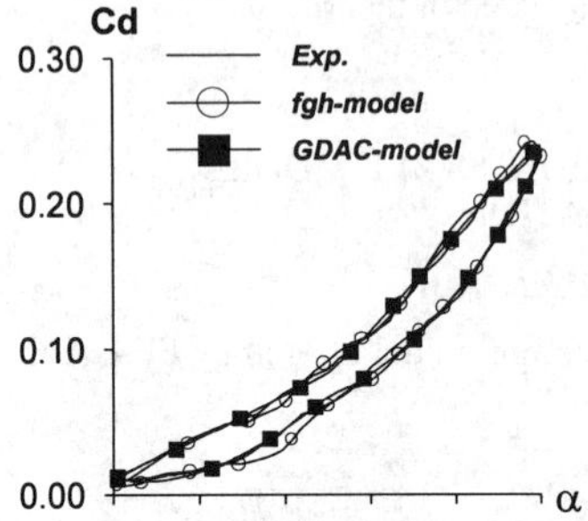

Figure 8. The result of drag model identification (k=0.044).

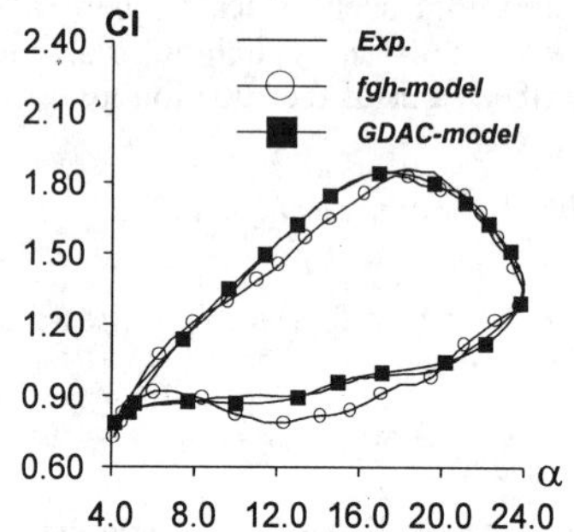

Figure 9. The result of lift model identification (k=0.06).

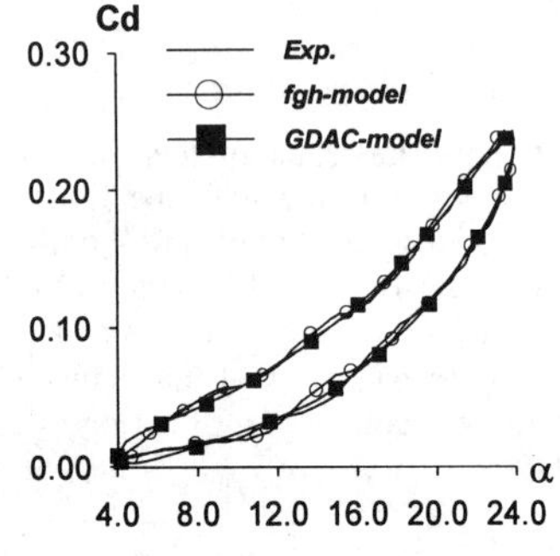

Figure 10. The result of drag model identification (k=0.06).

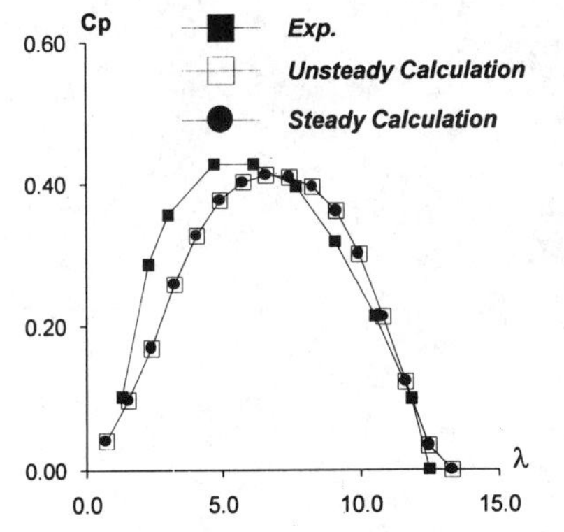

Figure 11. $C_p-\lambda$ calculation and test compared ($\gamma_1 = 0°$).

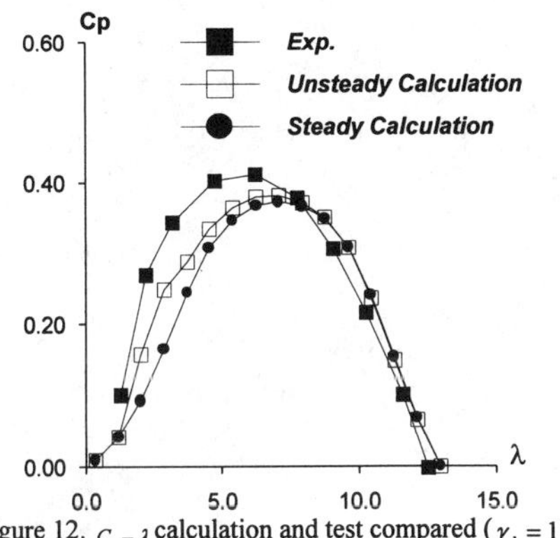

Figure 12. $C_p-\lambda$ calculation and test compared ($\gamma_1 = 15°$).

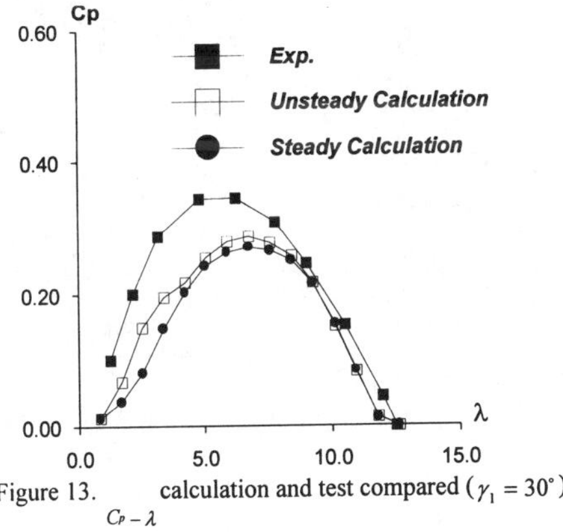

Figure 13. $C_P-\lambda$ calculation and test compared ($\gamma_1 = 30°$).

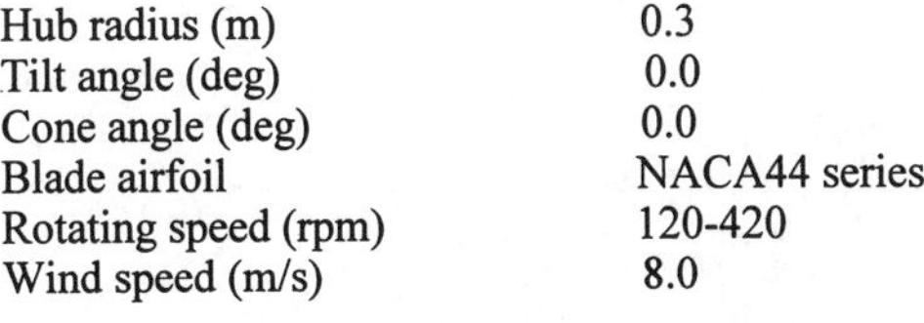

Hub radius (m)	0.3
Tilt angle (deg)	0.0
Cone angle (deg)	0.0
Blade airfoil	NACA44 series
Rotating speed (rpm)	120-420
Wind speed (m/s)	8.0

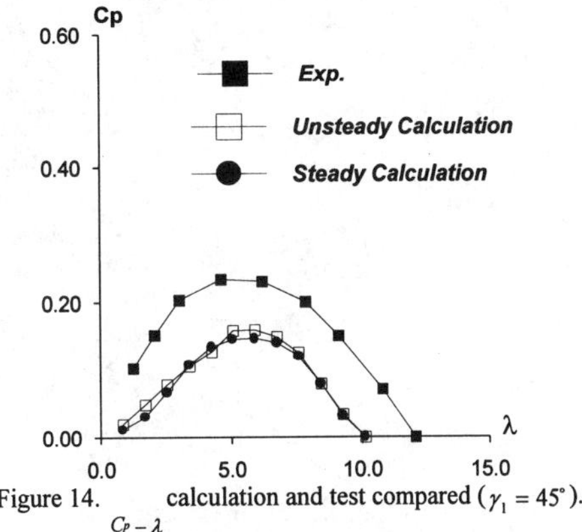

Figure 14. $C_P-\lambda$ calculation and test compared ($\gamma_1 = 45°$).

4.3 *Calculation result*

The rotor aerodynamic performance characteristic curves power vs tip speed ratio with steady model calculation, unsteady model calculation and experiment data is plotted respectively in Fig.11, Fig.12, Fig.13 and Fig.14. The yawing angle γ_1 is $0°, 15°, 30°$, $45°$ respectively. It is apparent that the result of the sample calculation is reasonable. When the unsteady aerodynamic model is taken in the classical blade element and momentum theory, the result of calculation is improved.

5 CONCUSIONS

The dynamic stall affects the aerodynamic performance of HAWT turbine. When the dynamic stall model is introduced in the calculation of performance of HAWT turbinc, the results of calculation close better experiment than classical aerodynamic model. In the future, we will afford to find out the better dynamic stall aerodynamic model, and the three-dimension effects will be considered.

REFERENCES

- Anders björck, Dynamic Stall and Three-dimensional Effects, Report FFA TN 1995-31
- Guo-Feng Lin, C. Edward Lan & Jay M. Brandon, "A Generalized Dynamic Aerodynamic Coefficient Model for Flight Dynamics Applications", AIAA Paper 97-3643.
- Rumsey, C.L. & Anderson, W.K., "Some Numerical and Physical Aspect of Unsteady N-S Computation Over Airfoils Using Dynamic Meshes", AIAA 88-0329, 1988.
- He, D.X., The Theoretical Handbook of Wind Turbine Aerodynamic Design Program, CARDC, 1995.

Wind Engineering into the 21st Century, Larsen, Larose & Livesey (eds) © Balkema, Rotterdam, ISBN 90 5809 059 0

Statistics of extreme loads for wind turbines

P.W. Cheng & W.A.A.M. Bierbooms
Institute for Wind Energy, Delft University of Technology, Netherlands

ABSTRACT: Extreme loading of wind turbines has been treated deterministically as prescribed in the design codes, without taking account the variability of the external loads. In this paper a rational approach to quantify the structural reliability of the wind turbine is presented and a new approach on the simulation of the extreme gusts with constrained simulations is proposed. The result from simulations with deterministic gusts and stochastic gusts are compared and the reliability evaluated. The response distribution at wind speeds beyond the cut out wind speed is obtained with the frequency domain method. The failure probability of the blade in stand still is also determined.

1 INTRODUCTION

In the conventional design practice of wind turbines, the analysis of fatigue loads is performed with stochastic wind field simulations. In contrast, the analysis of extreme wind loads is carried out with deterministic wind signals. With the upcoming generation of more flexible design and the increase in size of the wind turbines, the extreme load becomes a more important issue.
The main objective of this paper is to present a statistical approach to the extreme gust loads of wind turbines and the extreme loads beyond the cut out wind speed. As wind input, deterministic and stochastic gust signals are used to determine the load effects. The analysis of the extreme load in stand still is carried out in the frequency domain. The effects of stochastic gusts on the structural response are compared with the result obtained from simulations with deterministic gusts. The failure probability due to the gust loads is calculated for the blade root. Using the rational approach, it enables the manufacturers to make better estimates of extreme gust loads, achieving a more reliable design of the wind turbine.

2 EXTREME GUST SIGNALS

Extreme gusts that occur in the wind turbulence are usually idealised as a deterministic gust with shapes specified in design codes, e.g. IEC 1997. The choice of the shape and duration of the gust remains rather arbitrary. Recently, efforts have been made in creating gusts with stochastic properties to model the wind gust more realistically (Dragt & Bierbooms 1996). Theoretically, it is possible to generate stochastic gusts with conventional wind field generators. The length of the simulation has to be sufficiently long, as the occurrence probability of extreme gusts is rather small. This is not a very practical approach since it requires data analysis of a lengthy simulation. Moreover, the size and amplitude of the gust generated during the simulation can not be controlled.

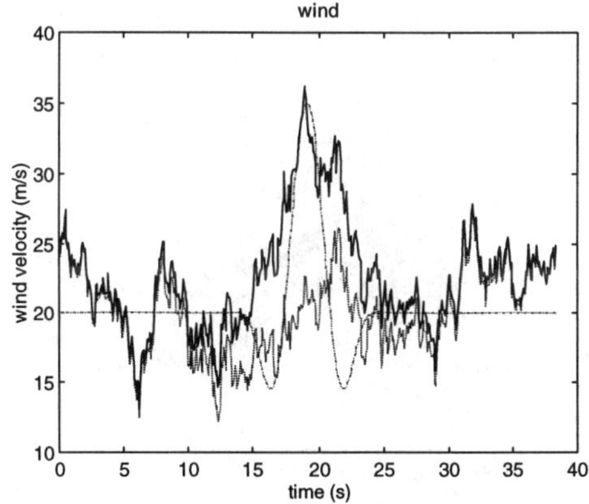

Figure 1 Stochastic and deterministic wind signals. Solid line is constrained gust, the dashed line is the original wind signal and the dash dot line is the IEC gust

2.1 *Constrained wind signals*

The basic idea of constrained simulation is to impose certain constraints on the simulation beforehand. In this case, the constraint is the occurrence of a gust with certain amplitude at a prescribed time point and at a specified location. The constrained wind signals are statistically not distinguishable from random wind signals. For a detailed description of the constrained simulation, refer to Tromans & Vanderschuren 1995.

For a single point in space the constrained wind signals is (Dragt & Bierbooms, 1996)

$$u(t)_{cr} = u(t) + R(t - t_0)(A - u(t_0)) + \frac{\dot{R}(t - t_0)\dot{u}(t_0)}{\dot{R}(t_0)} \quad (1).$$

u(t) is the original wind signal, R(t) is the auto-correlation function, and A is the prescribed gust amplitude. In Figure 1, the original wind signal and the constrained wind signal are shown together with the deterministic gust. The influence of the term with the first time derivative of the auto-correlation function is neglected. Theoretically, it is close to a delta function and contributes only to the constrained gust amplitude at the time point t_0.

As can been seen from Figure 1 the shape of the constrained gust resembles the auto-correlation function. As the gust amplitude A increases, the term with the auto-correlation function in Equation 1 dominates around the time point t_0.

To create a constrained gust in time and space it is necessary to consider the cross-correlation function, which determines the 'deterministic' part of the spatial gust. Subsequently, it is imposed on a random wind field to form part of the constrained spatial gust. The deterministic part of the spatial gust is illustrated in Figure 2. In the following analysis the uniform gust is applied.

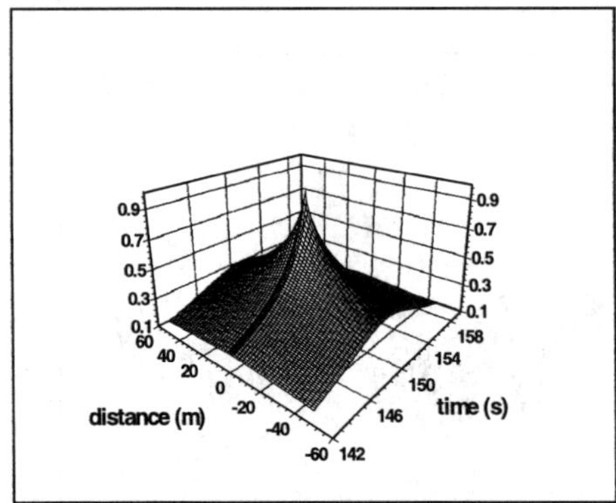

Figure 2 Spatial gust (deterministic part).

3 GUST RESPONSE

3.1 *Wind turbine model*

The effects of the gust response are investigated preliminarily with the prototype wind turbine for offshore application, developed by Kvaerner Turbin (Ferguson et al., 1998). The turbine has two blades with a rotor diameter of 80m. The rated wind speed is 13.7 m/s. The blades are pitch controlled after the rated wind speed. The turbine has a rated power of 3 MW.

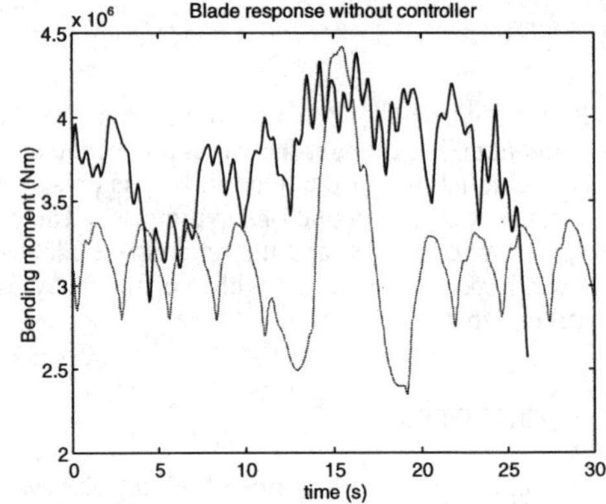

Figure 3 Flap moment at the blade root with a mean wind speed of 13 m/s. Solid line is from constrained simulation and dashed line is with IEC gust.

3.2 *Comparison of gust responses*

The effects of the deterministic and stochastic gust on the bending moment in the flap direction are illustrated in Figure 3. The mean wind speed is 13 m/s, thus the controller is not active. The gust amplitude is 10m/s above the mean wind speed.

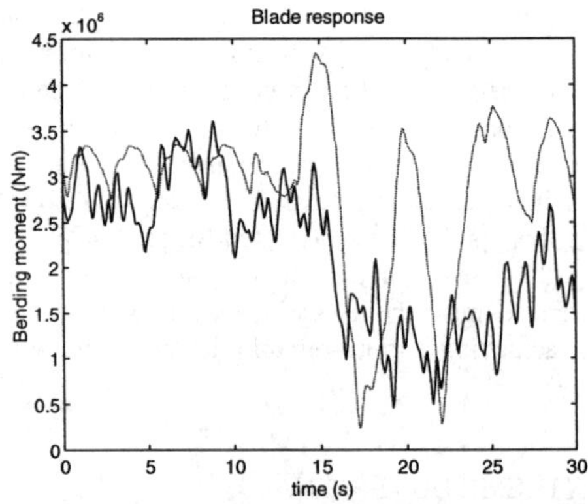

Figure 4 Flap moment at the blade root with a mean wind speed of 14 m/s. Solid line is from constrained simulation and dashed line is with IEC gust.

The stochastic gust has noticeably contributed to the dynamic excitation of the wind turbine, producing an overall higher load on the blade. The dynamic effect of the deterministic gust is rather limited. The response peak occurs simultaneously with the gust peak in case of the deterministic gust. However, the maximum of the bending moment is of comparable magnitude in both cases.

For a mean wind speed of 14 m/s, the response differs considerably. The activation of the controller leads to a limitation of the maximum blade load. As shown in Figure 4, the controller is able to 'filter' the gust excitation. The increase in blade load is negligible. At the same time, the deterministic gust produces a much higher blade load and the effect of the gust peak can be clearly seen. In this case the role of the controller is crucial, since it is unable to react to the IEC gust. For pitch regulated wind turbines, the IEC gust may overestimate the peak bending moment.

3.3 *Statistical variations of the stochastic gusts*

It has to be noted that the stochastic gust signals are different from simulation to simulation. This implies that the structural response is also function of this stochastic variable. Table 1 shows the mean values and standard deviation of the maximums of flap moment for 10 simulations for a mean wins speed of . 13m/s.

Table 1. Mean and standard deviation of the peak bending moment without controller (mean wind speed = 13 m/s)

Gust amplitude above mean m/s	Mean 10^6 Nm	Standard Deviation 10^6 Nm	Max 10^6 Nm
8	4.66	0.14	4.91
10	4.79	0.16	5.09
13	5.62	0.83	6.91
15	6.67	0.81	7.69

It can be seen that for operation without controller the mean of the structural response increases monotonically with the gust amplitude. The standard deviation tends to increase sharply. However, for a statistically meaningful trend interpretation a larger number of simulations is necessary.

In contrast, when the wind turbine operates with active controller the mean of the peak bending moment seems to remain constant, due to the 'filtering' of gust by the controller, see Table 2.

Table 2. Mean and standard deviation of the peak bending moment without controller (mean wind speed = 20 m/s)

Gust amplitude above mean m/s	Mean 10^6 Nm	Standard Deviation 10^6 Nm	Max 10^6 Nm
8	3.55	0.35	3.72
10	3.41	0.33	4.07
12	3.45	0.33	4.15
15	3.29	0.43	3.98

In Table 3 the maximums of the bending moment at the blade root produced by deterministic gusts are listed together with the mean values produced by constrained gusts. Although peak values are of similar order, the values deviate with increasing gust amplitude. It can be seen that the deterministic gusts may underestimate the peak moment for operation without controller.

Table 3. Peak bending moment with the IEC gust and the mean value of the constrained simulations (mean wind speed = 13 m/s).

Gust amplitude above mean m/s	Max IEC gust 10^6 Nm	Mean Constrained 10^6 Nm
8	4.60	4.66
10	4.76	4.79
13	4.91	5.62
15	5.17	6.67

In contrast, above rated wind speed, the IEC consistently delivers a higher estimate of the peak moment. As mentioned before, the controller was not able to react to the IEC gust, thus, enable to reduce the blade load. The characteristic of this particular controller plays a crucial role in determining the peak loads. For pitch regulated wind turbines the short duration gusts are of more relevance. For stall regulated wind turbines, long duration gusts can be relevant as well as the short duration gusts.

Table 4. Peak bending moment with the IEC gust and the mean value of the constrained simulations (mean wind speed = 20 m/s).

Gust amplitude above mean m/s	Max IEC gust 10^6 Nm	Mean Constrained 10^6 Nm
8	4.42	3.55
10	4.18	3.41
12	4.69	3.45
15	4.87	3.29

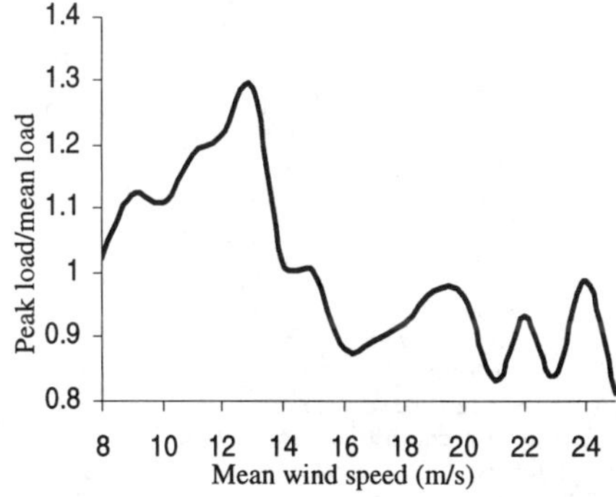

Figure 5 The variation of the peak moment ratio with the mean wind speed. The rated wind speed is 13.7 m/s

One of the characteristics of this wind turbine is that the peak loads are produced by wind speed slightly below the rated wind speed as illustrated in

Figure 5. The fluctuation of the peak loads in the higher wind speed region could be attributed to the inherently stochastic winds. Increasing the number of constrained simulation can reduce this fluctuation.

4 DISTRIBUTION OF THE EXTREME RESPONSE

4.1 *Short term distribution*

The functional dependency between the maximums of the bending moment at the blade root and the gust amplitude is illustrated in the Figure 6a, performed with deterministic gusts with different amplitudes. Since the function is monotonous, the distribution of the bending moment can be transformed from the distribution of the gust amplitude.

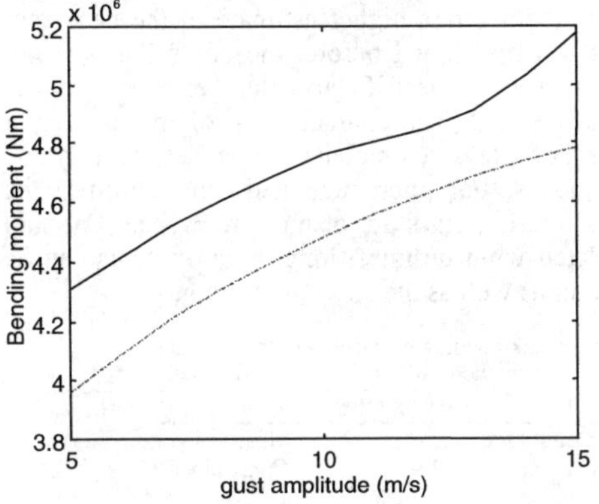

Figure 6a Functional dependency of the peak moment with the gust amplitude for mean wind speed 13 m/s (solid line) and 11m/s (dash dot line)

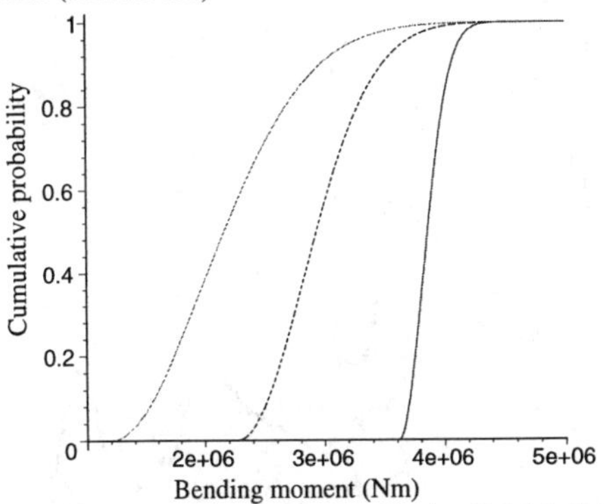

Figure 6b Peak moment distribution fitted to Raleigh distribution. From left to right the respective mean wind speed is 9 m/s, 14m/s, and 25 m/s

Assuming that the wind turbulence is a narrow band process, the maximums follow a Raleigh distribution. The probability of the bending moment is according to the total probability

$$f_m(M) = f_a(A)\frac{da}{dm} \qquad (2).$$

f_m is the probability density function of the peak moment and f_a is the probability density function of the gust amplitude. This transformation has to be done for all mean wind speeds.
Instead of making a direct interpolation of the conditional distributions, it is more practical to interpolate the parameters of the distribution. The distribution of the peak bending moment is fitted to a Raleigh distribution. Another possible choice of distribution is the Weibull distribution, which offers more flexibility.

Figure 6b shows the conditional distribution of the peak bending moment fitted to the Raleigh distribution. As the mean wind speed increases the start value of the distribution increases, however, the dispersion of the distribution decreases. To examine the effect of the distribution, the same distribution is fitted to a Weibull distribution.

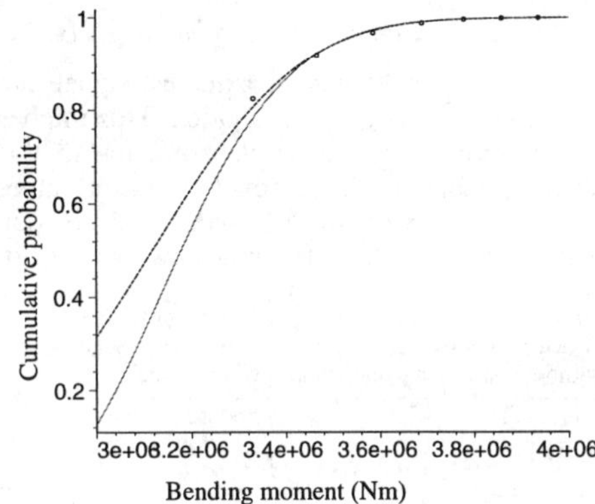

Figure 7 Peak moment distribution fitted to Raleigh distribution (dot line), Weibull distribution (dash line), and the circles are the data points of the original distribution.

From the fitting result showed in Figure 7, it shows that the Weibull distribution tends achieve a quite good agreement due to the extra free parameter. Though in the upper tail region it can be less conservative than the Raleigh distribution. Nevertheless, the choice of the Raleigh distribution can be justified by the fact that the functional dependency between the peak moment and the gust amplitude is approximately linear. Overall, the Raleigh distribution overestimates the original distribution.

Other distribution models may give a better fit of the original distribution. Distribution models with more free parameters have more flexibility to give a better fit (Ronold et al. 1996).

4.2 *Long term distribution*

The conditional distribution of the peak moments contains the information regarding the exceedance probability of the bending moment at the blade root at a given mean wind speed. The distribution of the mean wind speed has to be taken into account to obtain the yearly distribution of the peak moment. It can be determined as follows

$$f_m(M) = \int f_m(M)\Big|_{v_{mean}} f(v_{mean})dv_{mean} \quad (3).$$

The mean wind speed distribution, $f(v_{mean})$, is usually considered as Weibull distributed, for which Raleigh distribution is a special case. The result is the yearly distribution of the maximums of the bending moment. By assuming independence in the yearly distributions, the distribution in T years can be obtained as

$$F_T(m)=(F(m))^T \quad (4).$$

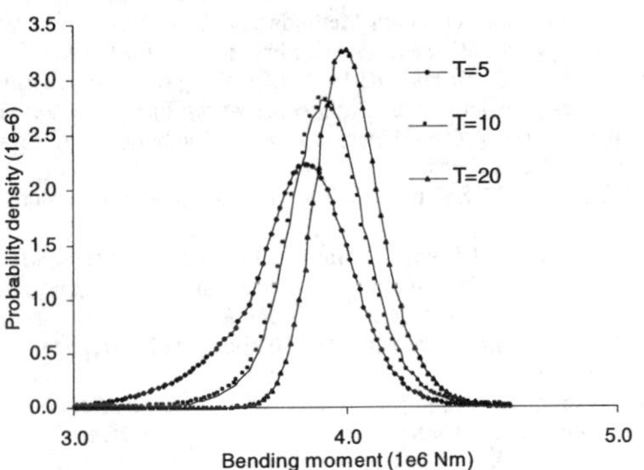

Figure 8 Distribution of the peak moment for different service life T.

As expected the mean value of the distribution increase with the service life T. However, the tails of the distribution gets narrower as the probability density concentrates around the mean value as T increases. This can have important consequence in the failure probability, as the upper tail of the distribution contributes considerably to the failure probability. The probability distribution for different service life is illustrated in Figure 8.

5 STRUCTURAL RESPONSE IN STAND STILL

5.1 *The modal analysis*

The structural response in stand still is obtained with frequency domain method. Modal analysis is performed on the rotor and tower with 5 modal degrees of freedom, Cheng 1998.

The area under the response spectrum is a measure for the variance of the response. Two measures, mean and standard deviation are sufficient to describe the distribution of the blade bending moment. Assuming that wind turbulence is Gaussian and the linearity of the model applies, it follows that the response is also a Gaussian distribution. The two distribution parameters are also function of the mean wind speed.

In contrast to the gust load analysis with operating wind turbines, the emphasis here lies in the extreme loads caused by extreme mean speed. This can be derived from a database of wind records. The Fisher Tippet I (also called Gumbel) is a common choice for the extreme wind speed distribution. The yearly extreme distribution of the bending moment is calculated as stated in the Equation 3 and the distribution for a service life T is determined as stated in Equation 4.

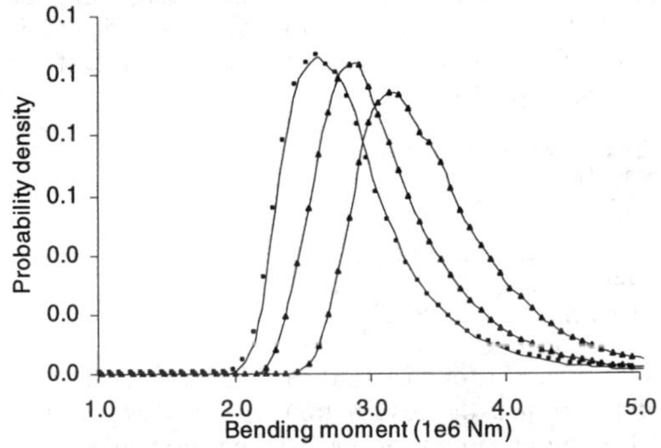

Figure 9 the Distribution of the bending moment close to blade root in stand still. From left to right the curves correspond to a period T of 5, 10 and 20 years

Figure 9 shows the distribution of the bending moment near the blade root for different period T. The increase of T leads to a broader upper tail of the distribution, as the Gumbel distribution has an asymmetry in favour of the upper limit.

The two distributions showed in Figure 8 and Figure 9 are fundamentally different. The first one is concerned with the peak bending moment and weighted with probability of the mean wind speed. The second one is concerned with the distribution of the entire range of the bending moment and weighted with the extreme wind speed distribution

6 FAILURE PROBABILITY

6.1 *Structural strength model*

To calculate the failure probability, it is also necessary to know how the material strength is distributed. The measure for material strength used here is the full plastic moment, under which the stress along the cross section reach the yield stress. The full plastic moment is a random variable that depends on several variables, yield stress and the geometrical variables of the cross section. These stochastic variables are assumed to be Gaussian distributed. The distribution of the full plastic moment is then calculated with a integration procedure (Christensen & Baker, 1982)

By integrating the joint probability density function of the load and resistance along the failure region, the failure probability of the structure is determined. It is performed for the cross section near the blade root.

6.2 *Comparison of the failure probability*

The failure probability from the gust loading is much lower than the failure probability from stand still. Since the strength resistance is the same in both cases, the discrepancy comes from the distribution type of the load. As mentioned before the distribution of gust load has a narrower upper tail, while the upper tail of the load distribution in stand still is broader. This is mainly due to the influence of the Gumbel distribution.

Table 5. Failure probability at a cross section near blade root

Load	Failure probability
Gust load	1.18e-32
Stand still	4.01e-12

Another factor that lead to a lower failure probability is that the distribution of the peak moment takes the maximum of the bending moment associated with certain gust amplitude. While the distribution of the load in stand still take into consideration the whole bending moment range. In another word, all the peaks moment are counted toward the distribution.

The assumption of a narrow band process can be a limiting factor for the load distribution as well. For a broad band process the distribution of the maximums differs from the Raleigh distribution (Cartwright & Longuet-Higgins, 1956)

It has to be noted that, for a very low failure probability, other uncertainty sources, such as the model uncertainty, can influence considerably the outcome. Furthermore, the absolute value of the failure probability is less relevant than the order of it.

7 CONCLUSIONS

A stochastic gust model is proposed for design calculation of the gust load. For pitch regulated wind turbines the IEC gust may overestimate the peak load for operation with pitch control and underestimates the peak load for operation without pitch control.

For a statistically meaningful estimate of the peak gust load, many constrained simulations are necessary.

The failure probability depends strongly on the type of distribution used to integrate the conditional distributions of the blade load.

Secondary peaks in the loads may have significant effects on the failure probability, thus has to be included in the load distribution.

REFERENCES

Bierbooms, W. et. al. 1999. Modelling of extreme gusts for design calculations (New Gust). *Proceeding of the European Wind Energy Conference*: Nice:

Bierbooms, W.& J.B Dragt & H. Cleijne. 1999. Verification of the mean shape of extreme gusts. *Wind Energy*: John Wiley (in press).

Cartwright, D.E. & M.S. Longuet-Higgins 1956. The statistical distribution of the maxima of a random function. *Proceeding of the royal society of London.*A237: 212-232.

Cheng, P.W. 1998. Analyse von Extremlasten von Offshore-Windkraftanlagen durch Methode der Zuverlässigkeitstechnik. *Deutsche Windenergie Konferenz*, Wilhelmshaven.

Dragt, J.B. & W. Bierbooms 1996. Modelling of extreme gusts for design calculations. *Proceeding of the European Union Wind Energy Conference*: 842-845 Göteborg: H.S. Stephens & associates

IEC-TC88. 1993 Safety of Wind Turbine Generator Systems (Draft). IEC.

Ronold, K.O. et al.1996. Calibration of partial safety factors for design of wind turbine rotor blades against fatigue failure in flapwise bending. *Proceeding of the European Union Wind Energy Conference*: 926-933 Göteborg: H.S. Stephens & associates

Thoft Christensen, P. & M.J. Baker 1982. *Structural Reliability Theory and Its Application*. Berlin: Springer Verlag.

Tromans, P.S. & L. Vanderschuren 1995. Response Based Design Method in the North Sea: Application of a New Method. *Proceeding of the 27th Offshore Technology Conference*. Houston.

Wind Engineering into the 21st Century, Larsen, Larose & Livesey (eds) © Balkema, Rotterdam, ISBN 90 5809 059 0

The wind resources in Cape Kaliakra Region

P.T. Ivanov
National Institute of Meteorology and Hydrology, Sofia, Bulgaria

ABSTRACT. The influence of the topography elements (orography, obstacles and roughness) on surface wind climate in cape Kaliakra region and calculation of deferent wind characteristics are solved problems in the paper. Based on ten years synoptical observations (1972-82) in Kaliakra station for the 10 m.a.g.l. the next wind characteristics are determined: prevailing winds, annual, seasonal and daily course of wind speed end wind energy density flow. Influence of the orography elements is expressed by increasing of wind speed and rotation of wind direction. Influence of the obstacles lead to decrease of wind speed. Influence of the roughness in negligible. Numerical modelling of wind speed in different point of the grid and Kaliakra wind atlas are calculated also.

1 INTRODUCTION

The Bulgarian electricity production is based mainly on fuel fossil, nuclear and hydro potential sources. These energy sources are negligible in the country. The renewable energy sources are an alternative of the energy development as an inexhaustible, environmentally sound kind of energy. According to our estimation (Ivanov P. at all,1997) up to 2020 about 8 per cent of primary energy could be received by the renewables (without big water power stations, biggest then 2 MW), the part of these energy may come from the wind. The wind characteristics are also very impotent for generate of wind waves and wind loading on the constructions.

The wind climate estimation is related to the detail investigation of surface wind regime which must be purified from the influence of the topography elements - obstacles, roughness and orography. The WAsP software (Troen Ib., at all, 1989) is one of the suitable instruments which is usually used for such investigation.

The influence of the topography elements on surface wind climate in cape Kaliakra region and calculation of different wind characteristics are solved problems in the paper

2 STATISTICAL PROCESSING OF RAW DATA

The basic supposition in WAsP is: the experimental wind speed data are distributed by the Webull distribution function. Using RAWDATA option 12 monthly raw wind data were imputed and calculated .

The summarised result for investigated period 1972-81 are presented in the paper. The wind speed frequency of total and sector Weibull parameters , the frequency of occurrence, number of observations, number of calms, mean wind speed and mean wind energy density flow are calculated. Fig.1 gives an example of approximated annual wind raw data. Fig. 2 presents the yearly sector wise rose of raw data. As it can be seen the prevailing winds comes from the North and from the South. The seasonal course of prevailing winds are observed during the year. In winter season they are from the North, in spring - from the North-East and in summer they are distributed from the North-West

3 ANNUAL AND DAILY VARIATION OF WIND SPEED AND WIND ENERGY FLOW

Summarising the data for terms of observation it have been obtained the mean wind speeds and wind energy density flow for January and July (Fig.3).

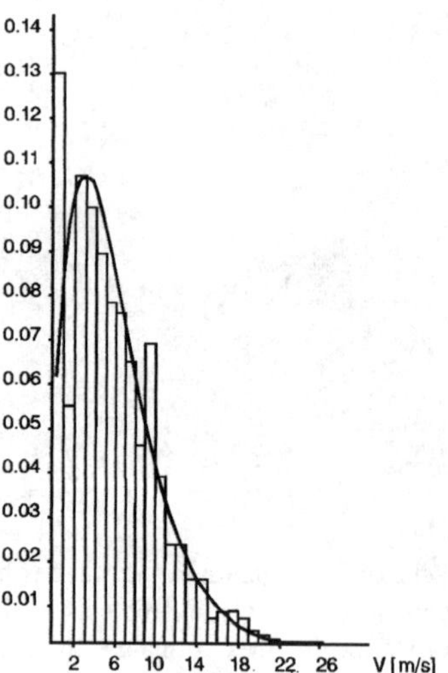

Fig.1. Yearly Weibull distribution function

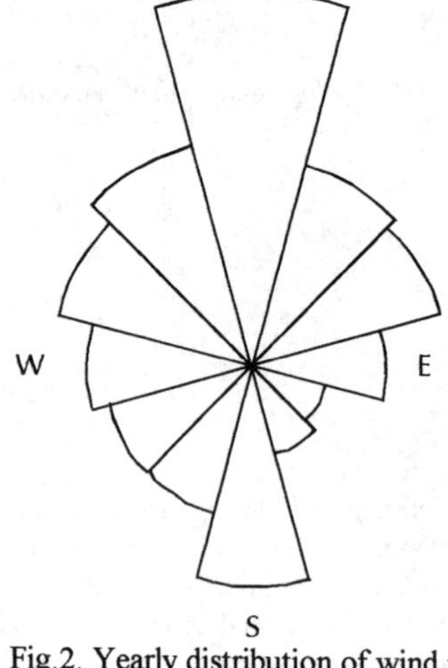

Fig.2. Yearly distribution of wind rose

The smooth lines of daily and yearly variation are obtained using the STATGRAF's quasiqubic splain function. There are observed two maximums and two minimum's in daily variation of the wind speed and wind energy density flow (Fig.3).

The main maximum is observed about 5 o'clock local time (L.T.) and wind speed is changed from 8.3 m/s in January to 5.5 m/s in July. The secondary maximum is observed at 14 L.T. and wind speed is changed from 8 m/s in January to 5.1 m/s in July.

The minimum's are occurred at 11 and 19 h. L.T. in winter season, but in summer they are removed two hours later. As a principle rule wind energy density flow follows the wind speed variation, but in summer there are observed one maximum and one minimum only. This fact is because the wind speed distribution determinate the wind energy density. If the secondary maximum is depends on the air temperature, the main maximum is peculiarity characteristics of the cape Kaliakra. In summers days maximums are better expressed and it could be accepted that a breeze circulation is already prevail.

The annual course of wind speed and wind energy density are presented in Fig.4. The wind

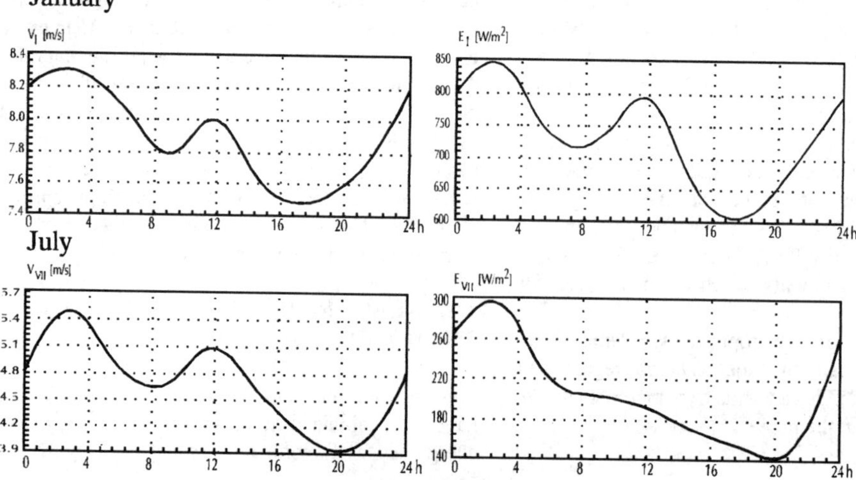

Fig.3. Daily course of wind speed and wind energy density flow for January and July

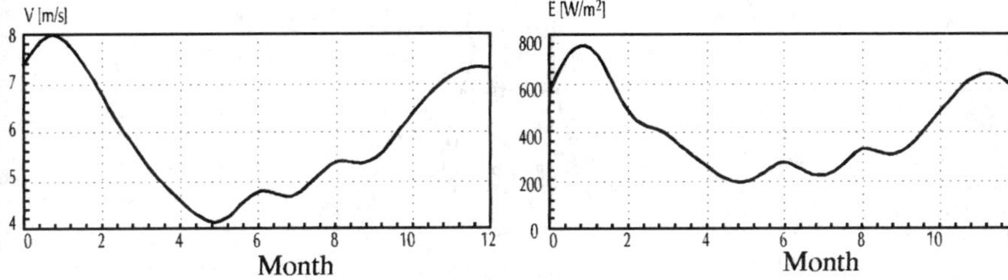

Fig.4. Yearly course of wind speed and wind energy density flow.

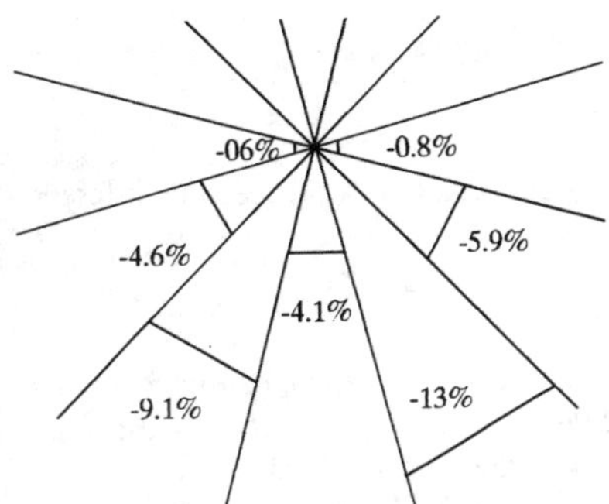

Fig.5. Rose of influence of the obstacles

speed and wind energy density maximums are observed in January with values of 8 m/s and 700 watt/m^2, but the minimum's are in May - 4 m/s and 200 watt/m^2 correspondingly.

4 INVESTIGATION OF THE TOPOGRAPHICAL INFLUENCE ON SURFACE WIND CLIMATE

The basic elements of the topography which perturbed the wind flow are the orography, the shelter behind obstacles and the roughness of the terrain around the station. Fig.5 and Fig.6 gives an influence of the factors mentioned above for every WAsP's sectors for annual wind distribution.

It must be underlined that the topography elements around the station during the investigated period aren't changes.

In accordance with the investigation the influence of the roughness is negligible and it is not mark in the results.

The influence of the obstacles is stronger. So, the existence of the building from the sectors of 150 deg. and 210 deg. decrease wind speed with 13 % and 9 % correspondingly.

Fig.5 is a graphical representation of the obstacles caused decreasing of wind speed.

The orographical influence on wind climate is strongest. It is because of the fact of complicated influence of relief form near the station. The cape Kaliakra is a narrow flat district land just out into the sea with height about 60 m.a.s.l.

Fig.6 gives a scheme of influence of the orography on wind speed in % for each sector. It can be observed an increasing of the wind speed (1.87 times) in sectors 90 and 270 deg. and the bauble a 0.06 in sectors 0 and 180 deg. This fact can be explained by thc wind blowing from the West or from the East when it can examine the cape Kaliakra a swell perpendicular to the flow. The current lines draw close when the wind flow pass above the land and the wind speed increase.

When the wind blows from the North and from the South the flow is divides in two currents which pass parallel to the coast line. So in this case the stream up is too smaller to increase the wind speed. Another sector exist combination of this two effects.

The orography elements changes the wind directions also. For example in sectors 60 and 240 deg. this change is biggest and it is demonstrate in turn clockwise of the wind in 17 deg. from the centre of the sector.

In the sectors 120, 150, 300, 330 deg. this turn is 15 deg. from the counter - clockwise. Like a consequence it must be observed an increasing of the frequency from the North and East in comparison with other stations of the region.

5 NUMERICAL EXPERIMENT OF THE INFLUENCE OF OROGRAPHY ON WIND REGIME

The Wind Atlas Application model allows to do three-dimensional extrapolation of the wind climate.

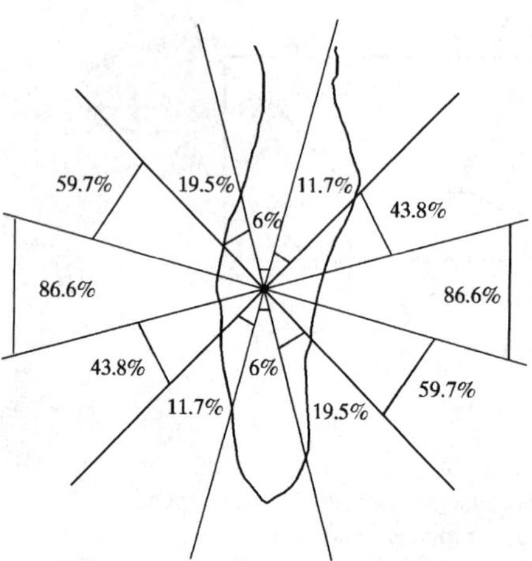

Fig.6. Rose of influence of the orography

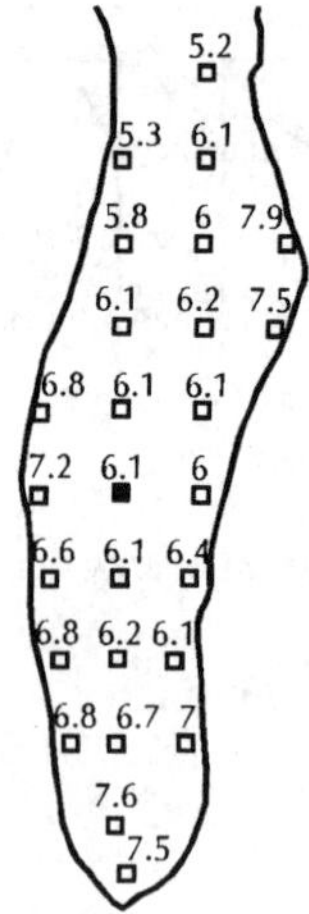

Fig.7. Numerical modelling of wind speed in different points

Using RESURSFILE procedure the mean wind speeds for 10 m.a.g.l. in 26 points of the quadratic grid step in the distance of 50 m were calculated. These way allows to investigate influence of the terrain roughness and orography near the station. The results are presented in Fig. 7 .

It can be seen that there are two areas with higher wind speed than around the station. The first one is situated to the North-East of the station with mean wind speed of 7.9 m/s. The second is situated to the South corner of the cape and the mean wind speed is 7.5 m/s. The wind flow which passed above the land in these areas is not only from the East and the West but from another sectors where the coast line is perpendicular to the flow causing to increase the wind speed.

6 KALIAKRA WIND ATLAS

The basic porpoise of the WAsP is to obtain the "clean" wind climate of the concrete place. Such calculation were obtain for annual wind regime in Kaliakra station (Table 1). As it can be seen frequency occurrence and Weibull's A and k

Table 1. Kaliakra wind characteristics for 10 m.a.g.l. and for different roughness classes and sectors

Sect	0	30	60	90	120	150	180	210	240	270	300	330
Roughness class 0 (Open see with $Zo=3x10^{-4}$ m)												
Freq	21	13	7	3	3	5	12	9	6	4	6	12
A	10.5	6.3	4.3	3.0	3.0	4.2	6.4	6.1	4.5	3.4	5.3	8.6
k	2.01	1.49	1.44	1.37	1.37	1.39	1.67	1.55	1.36	1.23	1.87	2.15
Roughness class 1 (farmland, airport with Zo=0.03-0.5 m)												
Freq	21	11	6	3	3	6	12	8	6	4	7	14
A	7.2	3.9	2.8	2.0	2.2	3.3	4.3	4.0	2.9	2.7	4.3	6.4
k	1.74	1.28	1.22	1.15	1.12	1.22	1.39	1.28	1.14	1.21	1.53	1.82
Roughness class 2 (farmland with Zo=0.10-0.20 m)												
Freq	19	11	6	3	3	7	11	8	5	4	8	14
A	6.1	3.4	2.4	1.8	2.0	3.1	3.8	3.4	2.5	2.5	4.0	5.7
k	1.70	1.2	1.22	1.19	1.11	1.26	1.40	1.27	1.15	1.27	1.53	1.80
Roughness class 3 (suburbs, city with Zo=0.50-1.0 m)												
Freq	18	10	5	3	4	8	11	8	5	5	8	15
A	4.5	2.6	1.9	1.4	1.7	2.6	3.0	2.6	1.9	2.1	3.4	4.6
k	1.63	1.27	1.22	1.22	1.16	1.31	1.40	1.26	1.12	1.34	1.61	1.80

parameters are done for different roughness classes and WAsP's sectors.

7 CONCLUSIONS

The present investigation of the wind climate in Kaliakra station based on the WAsP program allows to make the next basic conclusions:

- the prevailing winds are observed from the North and from the South
- the annual mean wind speed at 10 m.a.g.l. for investigated period is 5.8 m/s and the annual mean wind energy density is 390 w/m^{-2}.
- in the daily course of the wind speed and the wind energy density there are observed two maximums and two minimum's. In summer only one maximum and one minimum of the wind energy density flow there are observed.
- in the yearly variation of the wind speed and wind energy density flow is observed one maximum in January and one minimum in May.
- the shelter behind obstacles around the station lead to decrease the wind speed - mainly in sectors 150 and 210 deg. (1.13 and 1.09 times)
- the influence of the orography is expressed by increasing the wind speed and rotate wind directions. The biggest rotation is observed in sectors 90 and 270 deg. The turning of the wind directions are different for different sectors.
- as a result of the investigation of the wind climate in 26 points in quadratic grid there are defined two areas with higher wind speed then in meteorological station.
- the Kaliakra wind atlas is a compact presentation of the regional wind climate

REFERENCES

Ivanov, P., L.Trifonova (1997). "Renewable Energy Potential and Sustainable Development of Bulgaria". *The European Congress on Renewable Energy Implementation*, Athens, Greece, 446-455.

Troen, Ib and Erik Petersen (1989). "European Wind Atlas". *Riso National Laboratory*, Roskilde, Denmark, 655 p.

Wind Engineering into the 21st Century, Larsen, Larose & Livesey (eds) © Balkema, Rotterdam, ISBN 90 5809 059 0

Estimating winds in a tropical cyclone hitting a wind farm at a coast

Niels Otto Jensen & Jens Carsten Hansen
Wind Energy and Atmospheric Physics Department, Risø National Laboratory, Roskilde, Denmark

ABSTRACT: The characteristics of tropical cyclones are briefly described. The winds encountered are compared to Danish standard design criteria. A description and analysis is presented of a specific severe tropical cyclone which caused significant damage to large wind farms and other structures in the area of Lamba at the coast north of Porbandar in the state of Gujarat, India. Maximum wind speed estimates are made based on available information collected from in-situ inspection and from official sources. An example of the effects on a wind farm site is described and illustrated.

1 INTRODUCTION

A field trip to Gujarat, India was undertaken in the period 3-9 July 1998. The reason for the trip was a reported significant damage to wind farms in the area near Lamba (approx. 21°00′N, 69°23′E) on the Gujarat coast due to a hit by a severe tropical cyclone one month before on 9 June 1998, named "Arabian Sea Tropical Cyclone June 1998" by the Indian Meteorological Department (IMD), Tropical Cyclone Warning Centres in Mumbai and Ahmedabad. The purpose was to try to find out about the circumstances during the cyclone, especially the conditions and time development of wind speed and direction at the Lamba site.

Two field trips were made in Gujarat, on 5 and 6 July 1998, for fact finding and data collection regarding assessment of damage and estimation of surface wind speeds occurring in Gujarat during the cyclone. The field trips were made by car along the main roads 25 and 6 (see Fig. 1). What is reported here is based on our own findings and observations during the travel to the area as well as information gathered from the Indian Meteorological Department (IMD), Mumbai. Before coming to that we first introduce some general information about tropical cyclones.

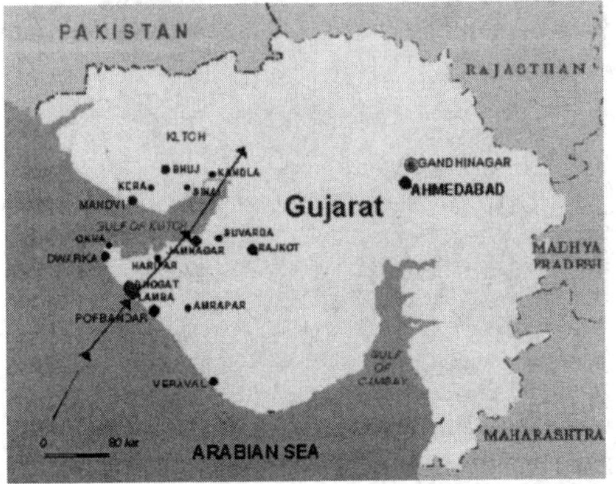

Figure 1 Map of Gujarat, India, indicating location of cyclone hit 9 June 1998, Lamba and Bhogat wind farm sites as well as met stations

2 TROPICAL CYCLONES IN GENERAL

Tropical cyclones develop over the ocean at latitudes typically between 5° and 35° (north or south). At lower latitudes (near the Equator) the Coriolis force is too weak to support the rotation and at higher latitudes the sea surface temperature is too low to support the necessary evaporation. After having developed into a discernible cyclonic state they start moving across the wind (Magnus effect). On the Northern Hemisphere this is to the left (Northwest) according to the general trade wind direction.

A tropical cyclone can produce winds stronger than those occurring in temperate climate zones, i.e. stronger than normal design loads for e.g. wind turbines developed for European conditions. At landfall they reduce in intensity due to lack of latent heat supply from the underlying surface. The ultimate strength of a cyclone depends on the length of time it spends over the ocean before landfall. Regarding the structure of these cyclones such as the spiral bands, the eye wall, etc., Willoughby (1990) reports on a total of 977 instrumented aircraft passes through 19 tropical cyclones occurring over a little more than a decade.

The frequencies of occurrence of tropical cyclones vary depending on the geographical location. World wide there is about 45 tropical cyclones per year with maximum winds above 33 m/s, corresponding to T-4 in the Dvorak system and category 2 on the Saffir-Simpson Scale – see below for definitions. These statistics (Neumann 1993), which cover 22 years (1968-1989), give the numbers for individual regions. Thus North Indian, including the Bay of Bengal as well as the Arabian Sea, has 2.5 per year compared to the north Atlantic with 5 per year. For the Arabian Sea the relative occurrence compared to the Bay of Bengal is even only approximately 1/10 (Mani & Mooley 1983, Table 3.5 p.48), so the number of incidences actually ending up at the Gujarat coast must be judged to be relatively low.

Furthermore their physical extent is relatively small, i.e. some hundreds of kilometers in diameter and the area of real intensive winds is even narrower. The wind maximum is located around the so-called eye-wall. The probability of a given location to be hit by a cyclonic storm of severe hurricane strength is therefore relatively small. Both probability of occurrence and strength should be considered when determining the design wind conditions in tropical cyclone prone areas.

The era of satellites has given the opportunity to discover and follow the development of these storms. The images are also used for quantitative estimation of strength and forecasting of future development. This is done by using the so-called T-scale method introduced by Dvorak (1975). Another scale is the Saffir-Simpson hurricane intensity scale (Simpson and Riehl 1981) used in USA for the Atlantic and Northeast Pacific basins to give an estimate of the potential flooding and damage to property given a hurricane's estimated intensity. We shall return to these scales below.

3 THE PRESENT CYCLONE

A depression formed over the Southeast Arabian Sea on the evening of 4 June at about 500 km Southwest of Goa (see Fig.2). It intensified into a cyclonic storm on 5 June and further into a Severe Cyclonic Storm on the evening of 6 June over the same area. Moving northwards, it intensified into a Very Severe Cyclonic Storm in the afternoon of the 7th at about 700 km south west of Mumbai. While continuing to move northwards, it deepened further with estimated core winds of 39 – 50 m/s (T-4.5 to T-5.5) on the evening of the 8th, 350 km Southwest of Porbandar. The system curved towards the coast of Gujarat and crossed it near Lamba between 7.30 and 8.30 IST of 9 June. It was located near Kandla at 14.30 later the same day. The latest previous very severe tropical cyclone in the Arabian Sea that hit the south Gujarat coast was 16 years back, 5-8 November 1982.

The Indian Express (Financial section) from 12 June 1998 gives the following numbers for the havoc caused by the cyclone: The harbour in Kandla (for geographical location see Fig. 1) incurred a loss of Rs 300 crores (≈70 million USD) in destroyed jetties etc., not accounting for lost cargo and commodities lying in the harbour. 15 large ships were sunk. According to conservative estimates, cited from the same source, The loss to Gujarat Electricity Board is put to Rs 500 crores (≈120 million USD), with a shut down of 4,000 transformers, snapping off 12,000 km of high-tension lines and 7,000 km of low-tension lines, involving some 50,000-80,000 poles and lattice masts blown down. An estimated 200,000

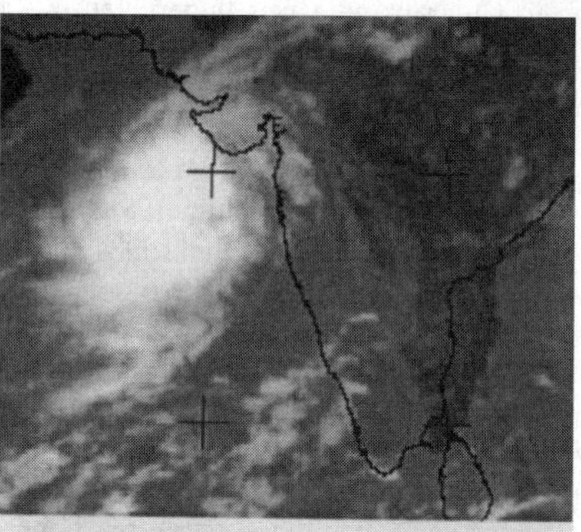

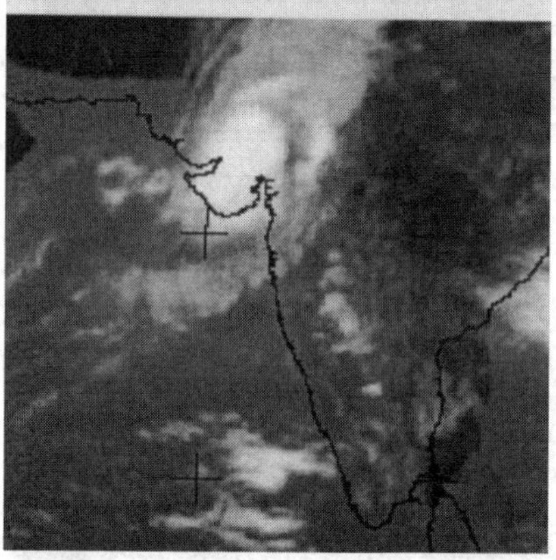

Figure 2 Satellite images of the Arabian Sea Tropical Cyclone June 1998 before and after hit at the Gujarat coast at Lamba 9 June 1998

Figure 3 Damaged wind turbine at the Lamba wind farm site

Figure 4 Damaged factory near Jamnagar

Figure 5 Reloading of tiles for repair works near Lamba

consumers were disconnected for longer periods. The cyclone caused the death of 649 people and much livestock. Houses and other structures of course also suffered (see Figs. 3 and 4). One month after the incidence there was still a lot of trade with roof tiles (Fig. 5). The worst hit area was the coast near Lamba, where a number of wind farms with about 330 large turbines were installed. Of these 125 suffered total destruction (see Fig. 4), 29 were very severely damaged, and 82 had damage that was judged to be possible to repair on site (see Winther-Jensen and Jørgensen, 1999).

The wind measurement systems on the turbines themselves were of no use, since line power to the wind farms were cut well in advance to the arrival of the cyclone. The cyclone obviously had its eye passing right over Lamba. South of Lamba structures were lying away from the coast and north of Lamba they were lying towards the coast. The weather radar in Bhuj was not operational during the cyclone passage due to cyclone induced flooding of the site. Other sources of data are mentioned in the following.

3.1 *Satellite information*

Satellite images are available from the Indian INSAT (used by IMD), the Russian GOMS, and other satellites, e.g. used by the Hawaiian University Tropical Storm Archive and Data Centre. An image from GOMS is shown in Figure 2, at 9 June 11.30 local time, after the cyclone has moved across the coast of Gujarat. Especially interesting in this connection is the reported information regarding the track of the cyclone eye, the classification of strength as a function of time during the cyclone development.

Table 1. T-numbers and corresponding maximum sustained wind speeds in the Dvorak method

T-number	Max.wind speed (m/s)
1	13
2	15
3	23
4	33
5	46
6	59
7	72

At the time of landfall the cyclone was classified by the IMD as T-5 according to the Dvorak method (Dvorak 1975), which means maximum sustained winds of more than 46 m/s, perhaps in the upper end of T-5 thus approaching maximum winds of 59 m/s. However, this method has severe limitations for intensity estimation. It is based on somewhat subjective judgement of the cloud structure (spiral bands). The strength of the satellite analysis is more in the determination of the size, location, and direction of propagation of the cyclone.

3.2 *Met. Station data*

We have had access to two types of station data: from regular meteorological stations with manual reading; and from another set of stations with automatic recording. The locations of the stations are given in Figure 1. The maximum recorded wind speeds are given in Tables 2 and 3.

Table 2. Report of maximum winds from stations operated manually

Station	Max. wind (m/s)	DDHHMM
Bhuj	25	091430
Dwarka	25	090630
Jamnagar (IAF)	50	091100
Kandla Port	44	091200
Okha	17	090930
Porbandar	32	090830
Rajkot	21	091230
Veraval	33	090530

Jamnagar air force base measured a maximum 2-sec gust of 98 knots or 50 m/s with a pressure tube anemometer (Dines). We consider this measurement very reliable. The anemometer is placed on a 3 m high pole on top of a 18 m tall building, inside the camp area.

The station in Dwarka was abandoned during the cyclone. Winds were measured up to the time stated in table 1. The method is a cup-anemometer with an odometer that is read on the hour (HH.30 IST) and again 3 min later.

The same method is used in Porbandar and Dwarka. In Porbandar the anemometer-counter (odometer) broke during the storm. In Lamba the outdoor instruments disappeared and the station was abandoned during the cyclone.

Measurements from Porbandar and Dwarka are of little value in assessing the strength of the cyclone, since they are outside the area of real extreme high wind speeds, but they support the assessment of the size of the cyclone.

Table 3. Report of maximum winds from automatic stations (Second Wind).

Station	Max wind (m/s)	Max. gust (m/s)	gust factor	Time (9 June)
Haripar	42	64	1.52	8 a.m.
Sinai	39	55	1.42	11 a.m.
Kera	29	42	1.42	12 noon
Suvarda	>28	55		9 a.m.–12
Amrapar	33	46	1.40	9 a.m.

The maximum winds in Table 3 are believed to be averages over one hour. The time constant for the gust speeds is not known. The measurements were taken at the height of 20 m.

A problem of comparing wind measurements from different events and even different measurements within the same event is the application of different time-averaging standards. In some cases the averaging period is not even stated or defined, but perhaps just assumed to be common knowledge.

For severe tropical cyclones a standard seems to have developed in which the wind speed given is the largest 1 min. average observed during the entire life of the cyclone. This is in contrast to "normal" micrometeorology where a 10 min average wind, usually at a height of 10 m above ground level, is the standard. Often, however, statements refer to the maximum wind. For a Dines anemometer this would be equivalent to the largest ≈3 sec. gust speed (averaging period defined by the internal time constant of the pressure system).

Under simple boundary-layer flow conditions these different measures can be related. For a Gaussian stochastic process the following relation holds (see e.g. Jensen *et al.*, 1992):

$$<a> = \frac{<u_\tau> - <u>}{<u>} = \frac{\sigma_u}{<u>}\sqrt{2\ln\left[\frac{T}{2\pi}\left(\frac{<u>}{\tau^2 L}\right)^{1/3}\right]}$$

Here $<u_\tau>$ is the mean value of the largest wind speed averaged over the short time τ, $<u>$ is the average wind speed over the longer sampling period T and σ_u and L are the standard deviation and integral length scale of the u-process. With this formula we can e.g. estimate that the gust factor $<u_\tau>/<u>$ is equal to 1.2 (or the amplitude $<a>= 0.2$) given the following conditions: $<u> = 50$ m/s, $\sigma_u/<u>= 0.12$, $T= 1$ min, $\tau = 3$ sec, $L = 100$ m.

Of course the individual values of u_τ or a have a distribution. For measurements over the sea we have found that the frequency distribution of a is very nearly of the Weibull form (Jensen & Kristensen 1989). Then the probability for exceeding a gust of a given amplitude a is

$$F(a) = \exp\left[-\left(\frac{a}{A}\right)^k\right]$$

where the shape parameters $A \approx 0.2$ and $k \approx 3$ are approximately independent of wind speed but may depend on the roughness of the underlying terrain. The physical significance of the parameters is that A is approximately equal to the mean value of the process while the standard deviation is $\approx A/k$. So a "maximum" gust could perhaps be defined as a quite rare but not improbable event, e.g. observed in 1 out of 50 observation periods. For the above example this would correspond to a "maximum" gust of 1.33 times the mean speed. In rougher terrain this value could be higher, mainly through the influence of the turbulence intensity $\sigma_u/<u>$.

Similar reasoning gives a ratio between 1 min and 1 hour average winds (also for $<u>\approx 50$ m/s) of 1.25, so a typical 1 min average wind will be 25% larger than the 1 hour average.

3.3 *Other data*

A third source of "data" that we have tried to use is on site observations of damages, and their correlation to a standard classifications of Hurricanes, e.g. according to specifications of the Saffir-Simpson scale, see Simpson and Riehl, 1981.

Table 4. Brief description of the Saffir-Simpson Hurricane Scale for Categories 1 – 5 (Simpson and Riehl, 1981)

	Wind speed m/s	Storm surge m	Damage Description
1	33-42	1.0-1.7	Some damage to trees, shrubbery, and unanchored mobile homes
2	43-49	1.8-2.6	Considerable damage to shrubbery and tree foliage; some trees blown down. Damage to poorly constructed signs and buildings' roofs. Major damage to mobile homes
3	50-58	2.7-3.8	Foliage torn from trees; large trees and poorly constructed signs blown down. Some damage to roofing materials and structures of buildings. Mobile homes destroyed.
4	59-69	3.9-5.6	Shrubs and trees blown down; all signs and mobile homes down and destructed. Extensive damage to roofing materials, windows and doors. Complete failures of roofs on many small residences.
5	70+	5.7+	Shrubs and trees blown down. Very severe damage to windows and doors. Complete failure of roofs on many residences and industrial buildings. Some complete building failures. Small buildings overturned or blown away.

Two field trips were made in Gujarat on 5 and 6 July 1998 for fact finding and data collection regarding assessment of damage and estimation of wind speeds occurring in during the cyclone. The field trips were made by car along the main roads 25 and 6:

5 July: from Rajkot, via Paddhari, Moti, Jamnagar, Sikka, Khambhalia, Bhatiya, Bhogat to Porbandar

6 July: from Porbandar along the coast past Lamba, Bhogat and Okhamadhi to Dwarka and back.

Damage profiles along the trajectories travelled were made. An example of these observations is given in Table 5. For simplicity and to facilitate assessment of damage from field observations, the multitude of damage observed during the field trips was categorised in a limited number of typical types of damage found. A list of the damage types to be used for the assessment of the tropical cyclone strength is shown in Table 6.

The damage descriptions seem to match Saffir-Simpson scale 2 ("Moderate" hurricane, 43-49 m/s max 1 minute average) near Jamnagar increasing through 3 ("Extensive", 50-58 m/s) to 4 ("Extreme", 59-69 m/s) at the Lamba site at the coast. Please note that these descriptions are made for conditions and structures designed for USA.

Table 5. Observations made on 6 July 1998 driving along road No.6 between Porbandar and Dwarka

Position (km)	Observation
64608	Porbandar: **A** at beach, **K** at airport and naval station
64616	Flooding of road (not due to cyclone)
64618	**I** (60% of exposed plantation), **F**
64622	Village: **L** (partly)
64625	Village: **I**(100%), **L, N**
64633	Village: **G, J, L**(100%), **M, N, O**(few)
64637	Village: **G, J**(incl palms), **L**(100%), **M, N, O**(few)
64641	Village: **G, J**(towards NE), **L**(100%), **M, N, O**(few)
64646	At Harshad: **G, J, L**(100%), **M, N, O**(few)
64651	Open area: **N** or **O** (all)
64654	Lamba site south: **P**
64657	Lamba site: **A, B, C, D, Q**, met station instr. gone
64662	Lamba site north: **C, D, Q**
64665	At road: **D** (towards SSW), **L, M, N, O**
64665	Navadra and Bhogat sites seen from road: **Q**
64670	**D, G, H, J** (towards SSW, opposite tree flagging)
64680	**G, H, J** (towards SSW, opposite tree flagging)
64685	Open area: **N**
64691	Okhamadhi site: **P** (plus other major damage)
64697	Open area: **O** (2/2)
64709	**A** (<50%)
64714	**A** (due to low surface roughness – sea/water)
64718	Dwarka (roads flooded due to recent rain, no obvious serious damage seen)
64722	Dwarka met station abandoned 09Jun98, 06:30. Max wind speed observed was at 05:30: 16.5 m/s. Wind direction at 05:30: NNE; at 11:30: WNW. Average speed for 05:30 to 11:30: 18.6 m/s.

Table 6. Field notes for damage assessment in Gujarat

Damage type	Description
A	LT Concrete power line pole broken at ground level
B	MT steelpipe power line pole (~7 m x Ø0.15m) bent or broken above foundation
C	MT power line steel frame (I-profile) bent at ground level
D	HT power line mast (60 kV or 220 kV) collapsed
E	Telecom relay mast (H=50-70 m) collapsed
F	Shrub and tree branch – leaf rip-off
G	Broken tree branches
H	Broken tree trunk
I	Broken coconut palms
J	Uprooted shrubs and trees
K	Sheds and light roofs blown down
L	Tile roofs blown down
M	Brick walls blown over – broken at ground level
N	Brick buildings seemingly repaired very recently
O	Brick buildings still collapsed
P	Some wind turbine damage
Q	Large number of wind turbines collapsed and damaged

The above estimates are supported by eye-witness reports from Porbandar where the storm tide was over 3 m (this is more than 50 km south of Lamba), and a storm tide at about 4 m, evidenced by seashells and other items that obviously was deposited during the storm (but not windblown) at this height above daily water and some ways inland.

4 SUMMARY

Measurements from Porbandar and Dwarka are of little value in estimating the maximum wind speeds that occurred at the site and data from Lamba itself are completely absent. Furthermore, to our knowledge, no wind farm or no wind turbine has had any wind measurement or data recording system operating during the cyclone, since such systems do not operate during power outages.

Since there is no actual measurements from the center of the cyclone during its passage onshore near Lamba, estimation of wind speeds that occurred has to be based on extrapolation from:

- Satellite image modelling
- Comparison of damage observations with damage descriptions from the literature, e.g. as available from the definition of the Saffir-Simpson scale
- Measurements from 100 km inland at Jamnagar and other stations that are some ways from the site of landfall of the cyclone

Extrapolating from the seaside using satellite derived information, the preliminary estimate of IMD-Mumbai of the maximum 1 minute average wind speed is 56-59 m/s (in the upper end of the T-5 category).

From the field trip and the damage observations, which increase from 2 to 3 to 4 on the Saffir-Simpson scale as the coast is approached, it seems reasonable to estimate a maximum 1 min average wind speed at the Lamba site in the range of 59-69 m/s.

To facilitate comparison between the two cyclone scales used they are shown graphically in Figure 6.

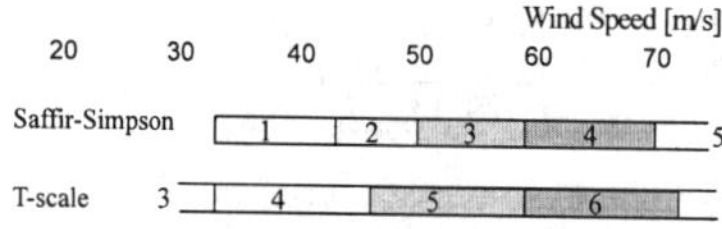

Figure 6 Graphical comparison of the Saffir-Simpson scale and the T-scale bv Dvorak

In extrapolating the data from the surface stations we will use Jamnagar and Haripar which is about half way between Lamba and Jamnagar. Both stations are near the track of the cyclone. From the reasoning above we can convert the gust value from Jamnagar of 50 m/s to a 1 min average by using a gust factor of 1.2 (see reasoning above). This results in 42 m/s. In a similar way we can convert the 1 hour average value at Haripur to a 1 min value by a conversion factor of 1.25, resulting in 53 m/s. Extrapolating the trend to Lamba then gives a 1 min average speed of 64 m/s.

The three estimation methods give fairly consistent results. We conclude that the maximum 1 min speed has been about 62 m/s (the average of the estimates). Finally, from what we have been told it seems that the Arabian Sea Tropical Cyclone June 1998 was as strong and possibly stronger than previous occurrences, maybe the strongest tropical cyclone of the Arabian Sea on record, probably due to its longer life at sea so that a mature state of the cyclone could develop before landfall.

For comparison the design speed from the Danish wind code (DS410, 1978) - i.e. the wind speed over open terrain that on the average is exceeded once in a 50 year period, corrected to 30 m height above ground and for short averaging time - is 40 m/s.

ACKNOWLEDGEMENTS

We would like to thank Deputy Dir. Gen. S. Kumar, IMD, Mumbai; Dr. S. Rangarajan, IITM, Bangalore; Vestas RRB India Ltd. and Vestas A/S, Denmark, for their help and support as well as C.B. Hasager, Risø, for extracting satellite information.

REFERENCES

Dvorak V.F. 1975. Tropical Cyclone Intensity Analysis and Forecasting from Satellite Imagery. *Monthly Weather Review.* 103:420-430.

DS410 1987. Action on building structures, 2. Wind load. Code of practice of the Danish Engineering Society (DS 410.2, translation into English). Teknisk Forlag, Copenhagen. 64pp.

Jensen, N.O. & L. Kristensen 1989. Gust Statistics for the Great Belt Region. Risø-M-2828, Risoe National Laboratory, DK-4000 Roskilde. 21pp.

Jensen, N.O., J. Mann & L. Kristensen 1992. Aspects of the natural wind of relevance to large bridges. In A. Larsen (ed.), *Aerodynamics of Large Bridges*: 25-32. Rotterdam: Balkema.

Mani A. & D.A. Mooley 1983. *Wind Energy Data for India.* Allied publishers Private Ltd. New Delhi. 545pp.

Neumann C.J. 1993. Global Overview. In G.J. Holland (ed.), *Global Guide to Tropical Cyclone Forecasting*: 1.1-1.43. Report No. TCP-31. World Meteorological Organization, Geneva.

Simpson R.H. & H. Riehl 1981. *The Hurricane and its impact.* Louisianna State University Press, Baton Rouge (ISBN 0-8071-0688-7) 398pp.

Willoughby H.E. 1990. Temporal changes of the primary circulation in tropical cyclones. *J. Atmos. Sci.,* 47:242-264

Winther-Jensen M. and E.R. Jørgensen. *When Real Life Wind Speed Exceeds Design Wind Assumptions.* European Wind Energy Conference, Nice, 1-5 March 1999.

Wind Engineering into the 21st Century, Larsen, Larose & Livesey (eds) © Balkema, Rotterdam, ISBN 90 5809 059 0

Short-term prediction of local wind conditions

L. Landberg
Risø National Laboratory, Roskilde, Denmark

ABSTRACT: This paper will describe a prediction system which predicts the expected power output of a number of wind farms. The system is automatic and operates on-line. The paper will quantify the accuracy of the predictions and will also give examples of the performance for specific storm events. An actual implementation of the system will be described and the robustness demonstrated.
Keywords: Short-term prediction, wind farm power output, HIRLAM, WAsP, MOS

1 INTRODUCTION

In many areas of wind engineering predictions of the wind one to two days in advance have become important. These areas include wind energy where electrical utilities with a high wind energy penetration need this information in their scheduling. Transport of large structures as well as other high or low wind dependent engineering situations spring to mind as other possible applications.

A prediction model has been developed to this end and this paper will describe the model in some detail, but the main focus will be on the performance and operational aspects of the implemented model. The on-line implementation was carried out in an EU JOULE-funded project.

2 THE MODEL

This section will describe the model: the idea behind it and the equations used. Furthermore, an actual on-line implementation will be described.

2.1 *The model*

The method for predicting the output of a wind farm is outlined in Figure 1. The idea is to use physical models as 'far' as possible, this is done such that the large-scale flow is modeled by a NWP (Numerical Weather Prediction) model, here HIRLAM (Machenhauer 1988) of the Danish Meteorological Institute (DMI); the wind is transformed to the surface using the geostrophic drag law and the logarithmic profile. As we zoom in on the site more and more detail is required, this detail is provided by the Risø WAsP program (Mortensen et al 1993). WAsP takes local effects (lee from obstacles, effect of roughness and roughness changes and speed-up/down of hills/valleys) into account. To take the shadowing effects of turbines in a wind farm into account the PARK program (Sanderhoff 1993) of Risø is used. Finally, to take any effects not modeled by the physical model and general errors of the method into account two model output statistics (MOS) modules are used. The model is described and analysed in great detail in (Landberg 1999, Landberg & Watson 1994, Landberg 1998).

2.2 *The implementation*

To demonstrate that it is possible, not only to obtain accurate predictions of the wind farm production, but also to use these predictions in an on-line environment, the model was implemented at Risø. The implementation is outlined in Figure 2. The principle is that as soon as the over-all meteorological situation has been predicted by DMI's operational HIRLAM model the forecasts are transmitted via Internet e-mail to Risø. Here a UNIX computer, dedicated to the task, registers the arrival and executes the Risø prediction model, the calculation module then sends the predictions to the output module which generates pages containing the predictions in graphical as well as textual form written in HTML (Hyper Text Mark-up Language). These pages are uploaded to a Web-server accessible via any Web browser (as eg Netscape and Internet Explorer). The address is not public and the site has no links pointing to it, so only utilities with permission can view the predictions. If the information contained on the pages at some point will be con-

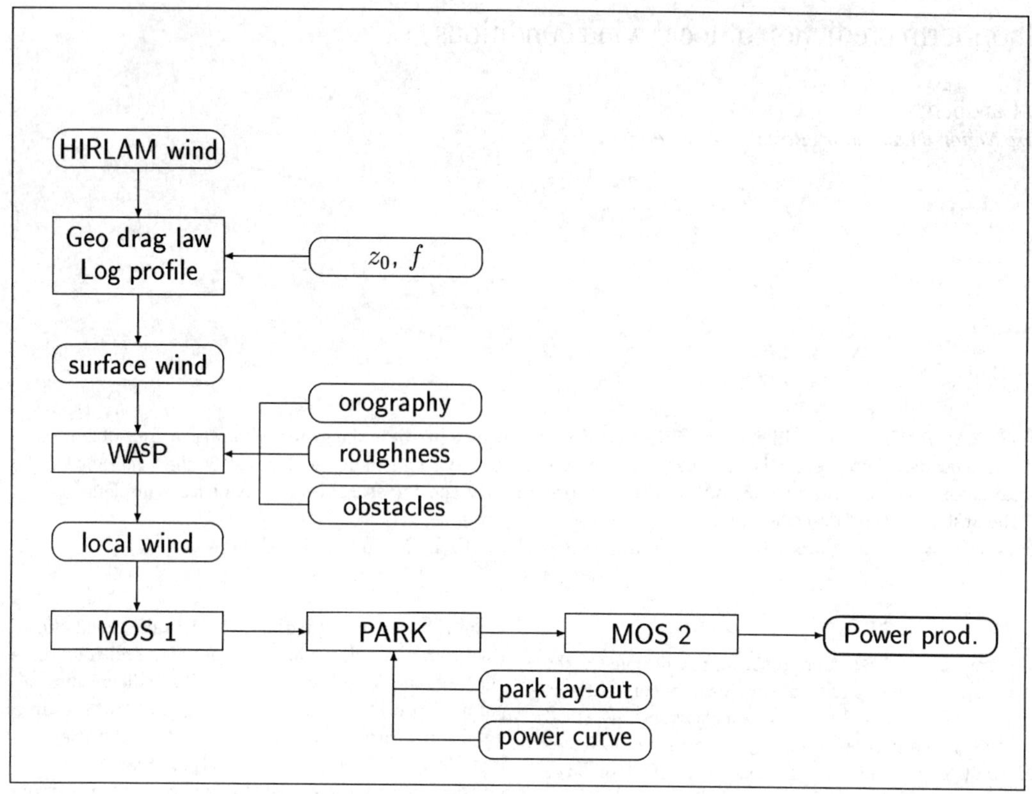

Figure 1: Flow chart describing the idea behind the model. The prediction from HIRLAM enters at the top and is transformed to the local wind moving down the figure, the production (including shadowing effects from the wind farm) is calculated moving along the bottom line. From (Landberg 1999).

sidered requiring a higher level of security, it is very simple to change the site to be password protected.

Predictions are made for the 47 wind farms shown in Figure 3 and the model is run twice a day generating predictions 36 hours ahead each time.

The model is implemented in such a way that a new wind farm can be added in a matter of a few minutes, HIRLAM predictions for the new site are typically available within 24 hours.

Each wind farm can also be changed to represent a region instead of just itself, this is done by scaling the power output with a factor, determined by the total installed capacity of the wind farms in the region.

The on-line predictions are available to the two Danish utilities, and since the focus of the Risø model is on the Elkraft area, the Elkraft utility is the one which has used the predictions in their daily planning and dispatch. This has been going on for seven months now (Feb 99).

3 RESULTS

In this section the quality of the model will be assessed in two ways: the standard way of comparing actual productions to predicted, but also by trying to see how successful the on-line implementation has been. At the end of the section another way of assessing the accuracy and usefulness will be described briefly.

3.1 *Accuracy*

The model has now run for two entire years and it is possible to draw some firm conclusions as to its performance. For independence of the evaluation, the model is evaluated on one year's worth of data taken from the Elkraft SCADA system. The wind farm observations are recorded every hour.

To get an estimate of the skill of the model the predictions are compared to those of the *persistence* model, which is a very simple model stating that

$$P(t+\ell) = P(t) \tag{1}$$

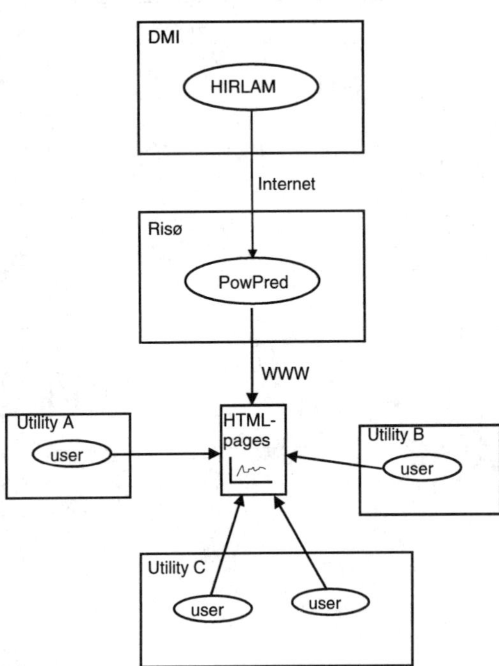

Figure 2: The on-line implementation. See text for details.

where $P(t)$ is the production at time t and ℓ is the look-ahead time. This model could popularly be called the 'what-you-see-is-what-you-get' (WYSIWYG) model! Despite its seeming simplicity, this model describes the flow in the atmosphere rather well, due to the characteristic time scales of weather systems; it is often experienced that the weather in the afternoon is the same as it was in the morning, so for some typical weather situations it is a rather difficult model to beat.

The results of the comparison between the two models for the Nøjsomhedsodde Wind Farm are shown in Figure 4 where the the mean error and the absolute mean error (=|predicted − observed|, a measure of the scatter, giving less weight to outliers than the standard deviation) of the prediction model are compared to those of the persistence model. The following can be seen:

- The prediction model outperforms the persistence model after six hours.
- The mean absolute error of the prediction model is around 15% of the installed capacity.
- The decay of the performance of the prediction model is very gentle, over the 36 hour time-span the error is only increased by less than 10%.
- The mean error of the persistence model is very small. This follows from the definition of the persistence model.

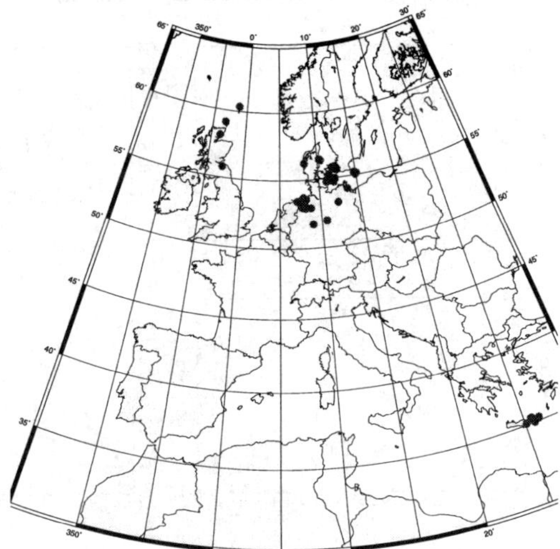

Figure 3: The location of the 47 wind farms. The wind farms are located in Denmark, United Kingdom, Greece and Germany.

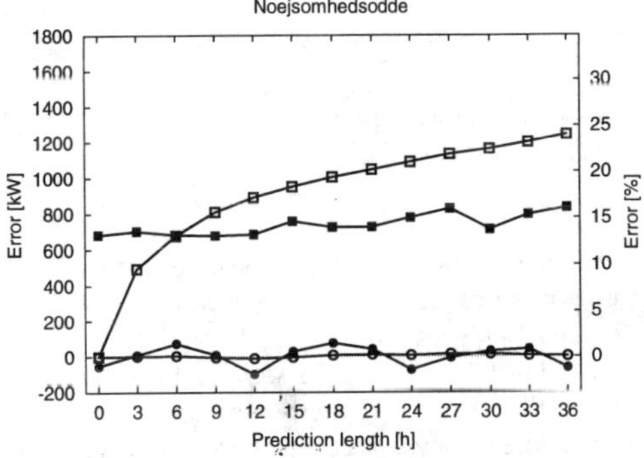

Figure 4: The mean error (circles) and the mean absolute error (squares) for the prediction of the power output of the Nøjsomhedsodde Wind Farm for the prediction model (filled symbols) and the persistence model (open symbols). The left-hand y-axis shows the absolute numbers (in kW) and the right-hand axis the numbers relative to the total installed capacity (in %). One year's worth of data has been used in the comparison. The farm is rated at 5.2 MW.

These predictions are not the best and not the worst examples of the performance, a well-predicted wind farm has a scatter as low as 10% of the installed capacity and a not-so-well-predicted up to 20%.

One very important feature which can not be seen from this figure is how well the model predicts storms.

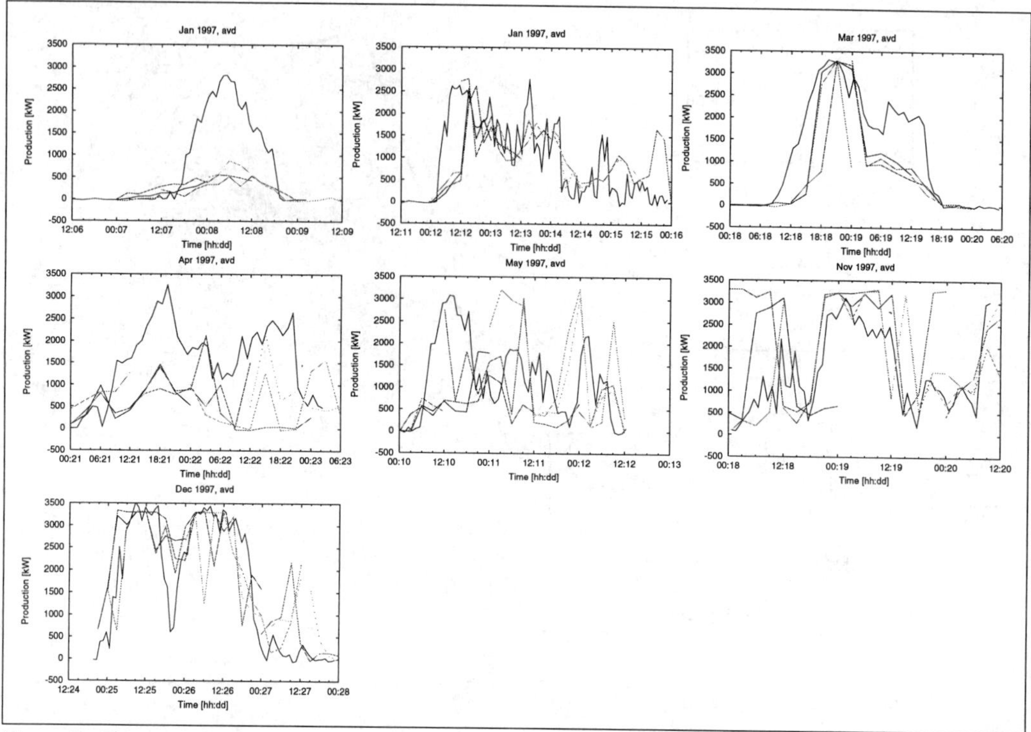

Figure 5: The storms in 1997 as registered at the Avedøre Wind Farm (capacity 3.6 MW). Solid line is the observed production, dashed lines are the predictions.

Here the persistence model has no chance, since it assumes status present. In a previous study (Landberg & Joensen 1998) all storms in 1997 and 1998 in the period 1 Jan 97 to 1 Aug 98 were identified for the Avedøre Wind Farm and the performance of the prediction model was evaluated in qualitative terms. The result can be found in Table 1.

The definition of 'storm' is here taken to mean an event where the production rose from a very low value (ie close to 0 kW) to around 3 000 kW (approx. 80% of the total installed capacity) and then down again to a low value over a period of a few days. Using this selection criteria 7 storms were found in 1997 and 9 in the first seven months of 1998.

These 16 storms are depicted as "thumb-nail images" in Figures 5 and 6 and listed in Table 1.

Studying the data in detail reveals that of the seven storms in 97 two were predicted very well and one was predicted satisfactorily. In 98 five were predicted very well and three satisfactorily. This is very much in line with the results found for the wind speed at the Risø mast: in the first half of 1997 the HIRLAM model had problems, which then seem to have been fixed.

The results shown here differ slightly from the results in (Landberg & Joensen 1998), in that the cases where cut-out (ie the wind is so high that the turbine must stop in order to avoid structural damage) was originally predicted are now removed, since the model has been changed to reflect the fact that cut-out very seldomly is seen in the observations. It is critical to get this right, of course, since missing a cut-out situation will make an enormous difference for the dispatch, by either predicting 100% production (no cut-out) or 0% prediction (cut-out).

3.2 *Track record/reliability*

This section will focus on some operational aspects of the implementation.

The results from the HIRLAM model are typically available after 2 hours, the transit time between DMI and Risø is typically less than a minute and the prediction model runs in less than two minutes for the predictions to be calculated and another two minutes for the web-pages to be generated, the upload is instant. So, typically around two hours after the measurements have been taken at a large number of places around the Northern Hemisphere the predictions are available to the utility.

The system has run operationally for more than two years, with no break-downs. One planned shut-down

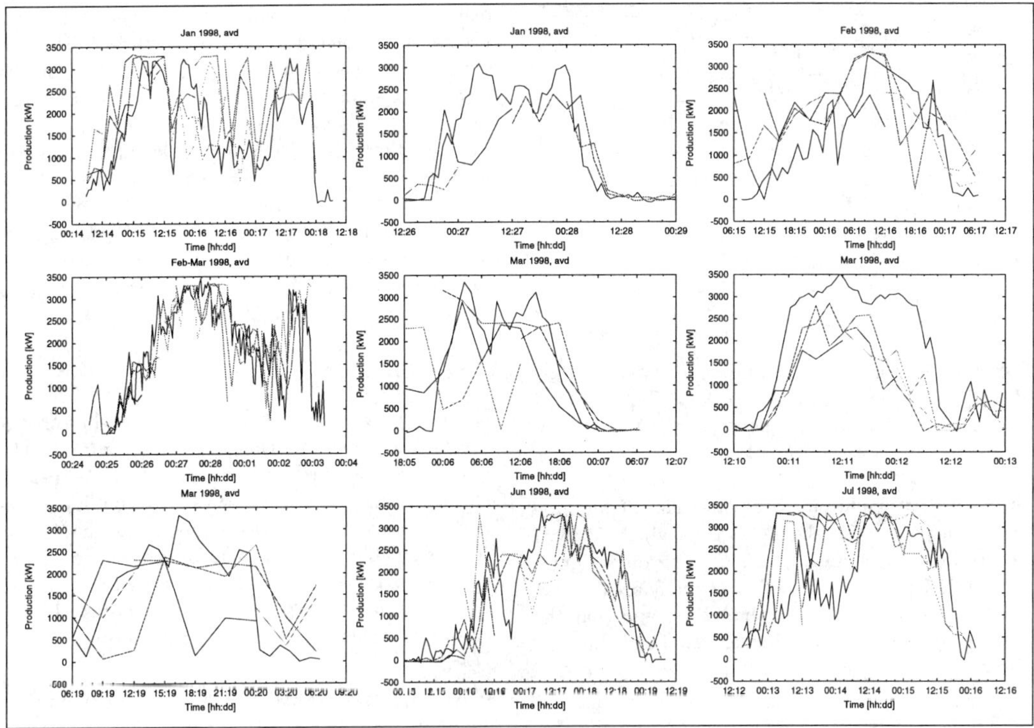

Figure 6: Storms in 1998 (until 1 Aug) as registered at the Avedøre Wind Farm (capacity 3.6 MW). Solid line is the observed production, dashed lines are the predictions.

Table 1: The storms identified in the period 1 Jan 97 to 1 Aug 98. In the column marked 'Quality' the predictions are graded according to the following: − bad, = OK, + good.

1997 Start	End	Quality	Notes	1998 Start	End	Quality	Notes
7/1	9/1	−	level missed	14/1	18/1	+	
12/1	15/1	−	start-up missed	26/1	29/1	+	
18/3	20/3	=	end missed	15/2	18/2	=	
21/4	24/4	−	missed completely	25/2	2/3	+	
10/5	12/5	−	missed completely	6/3	8/3	=	
17/11	22/11	+		10/3	13/3	−	level missed
25/12	27/12	+		19/3	20/3	=	
				15/6	19/6	+	
				14/7	16/7	+	

took place, when the operating system was upgraded. The only problems have occurred when the DMI forecasts have been delayed for some reason. It is of course not possible to do anything about this from the utility side, so the times different from two hours are an indication of how often this occurs.

Figure 7 shows the distribution of the delay time, as can be seen, most of the time the delay is around two hours and after four hours 90% of the predictions have arrived. There are some outliers, and as mentioned earlier all of them are due to the fact that the HIRLAM predictions were delayed.

3.3 *Savings in fossil fuel*

Another very important way of accessing the accuracy from a utility point of view is to see how much conventional fuel and thereby money is saved because of the predictions, other papers are dedicated to this, see (Watson et al 1994, Watson et al 1999). The main con-

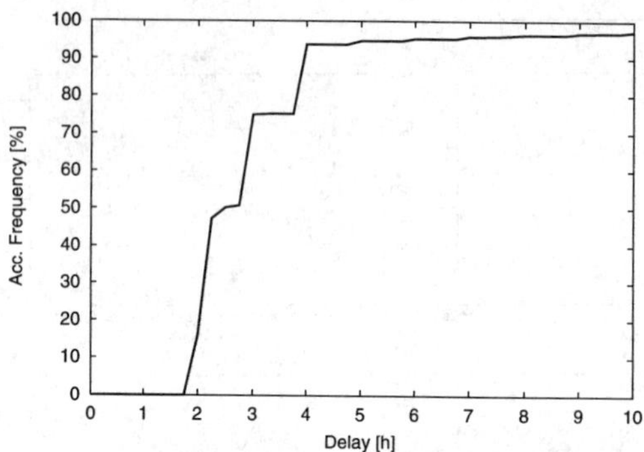

Figure 7: Accumulated distribution of the delay time of the predictions from the E (previously called DKV) version of the HIRLAM model.

clusion is that significant savings are indeed found, the higher the penetration the higher the savings. Typical results are that the use of predictions becomes significant at 25% penetration (note that many areas today have a penetration of more than 20%) and that the savings are more than 25% higher than just using persistence.

4 THE FUTURE

At present the model is used as-is in a JOULE project, predicting the power for 13 wind farms in Germany (location of the wind farms is shown in Figure 3).

For the very near future two model refinements are planned, the first one is a direct improvement of the existing model (described in the following) and the second is a completely new model, where the Risø model described here is combined with the IMM model (Nielsen & Madsen, 1999), this development will be carried out in a project funded by the Danish Energy Agency, with the goal of implementing this model at all the Danish electrical utilities.

It is also hoped that other utilities will be interested, EPRI in the US has already shown interest.

4.1 *Using the 10 m wind as input*

Due to the decrease in grid size and the improvement of the boundary-layer parameterisation of HIRLAM, the level from which the wind has to be taken has changed from when the model was first developed and implemented (see (Joensen et al 1999) for further details). In the original model the wind found in model level 5 (approx. 550 m agl) should be used as the best approximation of the geostrophic wind, now, however, the wind from the lowest level has the smallest scatter when compared to the observed wind at the wind farm site. This means that since the original model had to transform the wind from a geostrophic wind via the geostrophic drag law to the surface the model must be changed, since this transformation has already been done by the HIRLAM model. Referring to Figure 1 the box labeled 'Geo. drag law, Log profile' must therefore be removed. Calculating the mean error and std. dev. of the original model and the model proposed here applied to the Risø mast, the same qualitative picture is found as in Figure 4, however, the predictions using the 10 m wind show a reduction in the standard deviation of typically 20% on the prediction of the wind speed.

5 CONCLUSION

This paper has described an automatic on-line prediction system for wind farm production output. The model was described and analysed and the performance of it was shown to be good. The on-line implementation was also described and it was shown that the model had been operational for two years with very few interruptions. The predictions have been used by the Danish utility Elkraft for seven months. Lastly, new applications of the model were described.

ACKNOWLEDGEMENTS

The on-line implementation of the model has been funded by the Commission of the European Communities under the JOULE programme, contract JOR3–CT95–0008. The on-going project in Germany is also funded by the JOULE programme, contract JOR3–CT98–0272. The development of the original model was as funded by the EFP-programme under the Danish Ministry of Energy and the Environment, contract 1363/94–0005 and the JOULE programme, contract JOUR–0091–MB(C). G Giebel, Risø, is thanked for calculating the delay time of the HIRLAM predictions.

REFERENCES

Joensen, A, G Giebel, L Landberg, H Madsen, 1999: *Model Output Statistics Applied to Wind Power Prediction.* Proceedings from EWEC99, Nice (FR).

Landberg, L, 1998: *A mathematical look at physical power prediction model.* Wind Energy, **1**, 23–28.

Landberg, L and A Joensen, 1998: *A model to predict the power output from wind farms - an update.* In proceedings from BWEA 20, Cardiff(UK), 127–132.

Landberg, L, 1999: *Short-term prediction of the power production from wind farms.* J. Wind Eng. Ind. Aerodyn., **80**, 207–220.

Landberg, L and SJ Watson, 1994: *Short-term prediction of local wind conditions.* Boundary-Layer Meteorol., **70**, 171–195.

Machenhauer, B (ed), 1988: *HIRLAM final report.* HIRLAM Technical Report 5, Danish Meteorological Institute, Copenhagen, Denmark.

Mortensen, NG, L Landberg, I Troen and EL Petersen, 1993: *Wind Atlas Analysis and Application Program (WAsP), User's Guide.* Risø-I-666(EN)(v.2), Risø National Laboratory, Roskilde, Denmark. 133 pp.

Nielsen, TS and H Madsen, 1999: *Experiences with Statistical Methods for Predicting Wind Power.* In proceedings from EWEC99, Nice(FR).

Sanderhoff, P, 1993: *PARK – User's Guide. A PC-program for calculation of wind turbine park performance.* Risø-I-668(EN), Risø National Laboratory, Roskilde, Denmark. 8 pp.

Watson, SJ, L Landberg, and JA Halliday, 1994: *Application of wind speed forecasting to the integration of wind energy into a large power system.* IEE Proc.-Gener. Transm. Distrib., **141**, No. 4, 357–362.

Watson, SJ, G Giebel and A Joensen, 1999: *The Economic Value of Accurate Wind Power Forecasting to Utilities.* Proceedings from EWEC99 Nice (FR).

Wind Engineering into the 21st Century, Larsen, Larose & Livesey (eds) © Balkema, Rotterdam, ISBN 90 5809 059 0

Estimation of offshore wind resources – The influence of the sea fetch

Bernhard Lange
Wind Energy and Atmospheric Physics, Risø National Laboratory, Roskilde, Denmark

Jørge Højstrup
NEG Micon A/S, Randers, Denmark

ABSTRACT: The increasing interest in harvesting offshore wind energy requires reliable tools for the wind resource estimation at these sites. Most commonly used for wind resource predictions on land as well as offshore is the WAsP program. This program has been validated extensively for sites on land and at the coast. However, due to the lack of suitable measurements there is still a need for further validation for offshore sites. New data from ongoing measurements in the Danish Baltic Sea region are available now. The wind resources estimated from these measurements are compared to WAsP-predictions. They are found to agree well. The only deviation found is for two sites with comparable distance to the coast but with a different distribution of land. Here the measurements show slightly different wind resources, which are not predicted by WAsP. Wind speed ratios of several pairs of stations are modelled with WAsP for 12 directional sectors and compared with the measurements. Deviations in the directional wind speed predictions were found to correspond with the length of the sea fetch: For smaller sea fetches WAsP seems to slightly overpredict the wind speed, while for long fetches of more than 30 km an underprediction is found. An analysis of the vertical wind speed profiles at three sites indicates that a combination of different effects is responsible for the correlation between sea fetch and model behaviour.

1 INTRODUCTION

Suitable sites for wind farms on land are scarce in some regions in Europe, while potential areas for offshore sites are huge. Additionally, the wind resource offshore is much better than on land. Therefore the interest in developing offshore sites for wind energy utilisation has been growing in recent years and it is expected that an important part of the future expansion of wind energy utilisation at least in Europe will come from offshore sites.

However, compared to land sites the economic viability of offshore wind farms depends on the compensation of the additional installation cost by a higher energy production. A reliable prediction of the wind resource at offshore sites is therefore crucial for project planning and siting.

The wind resource prediction model WAsP (Mortensen 1993) of the European Wind Atlas (Troen & Petersen 1993) is the standard method for wind resource predictions on land as well as offshore. It has been validated extensively for land conditions. A validation study for coastal stations was performed by intercomparisons of wind measurements at different heights from high meteorological masts close to the sea (Petersen 1993). No significant deviation was found. Only very few measurements are available for a validation of WAsP for offshore sites. A comparison with Vindeby data showed reasonable agreement with a slight overprediction of the wind speed at Vindeby (Barthelmie et al. 1996a).

In Denmark plans are going ahead to install 4000 MW offshore wind turbines until the year 2030. In the current planing phase offshore wind measurements are being made at three prospective wind farm sites. The measurements are located in the confined Danish waters of the Baltic Sea near the islands of Lolland and Falster in distances of about 10 km from the coast. They are used to investigate wind characteristics and estimate wind speeds and power productions for planned wind farms (Barthelmie et al., in press).

The data presently available from these measurements are used in this study together with data from the Vindeby offshore wind farm which is located about 2 km from the coast. Thus the measurements cover the distances most likely encountered in the planning of offshore wind farms.

2 WASP FOR OFFSHORE CONDITIONS

Compared to land conditions the modelling of the wind resource in coastal waters is complicated by a combination of several effects:

The favourable wind resource offshore is mainly caused by the low surface roughness of water areas. Contrary to land conditions, the sea surface roughness is not constant but depends on the wave field present. This in turn is governed by the momentum exchange process between wind and waves which depends on wind speed, water depths, distance from the shore, atmospheric stability, etc.

For fully developed wind waves the dependence on wind speed can be modelled with the Charnock model (Charnock 1955). It describes the sea surface roughness as a function of wind speed, but independent of fetch. In the confined waters around the measurement sites the waves are mainly fetch limited and the length of the upwind sea fetch might also be important for the sea surface roughness (Frandsen et al. 1996, Johnson et al. 1997).

WAsP assumes an average value of 0.2 mm for the sea surface roughness. No dependency on wind speed or fetch is taken into account.

The atmospheric stability is the second parameter, which differs greatly between land and water areas. Its main influence is on the vertical momentum transport, which is reflected in the vertical wind speed profile. It also influences the growth of the internal boundary layer after the land-sea roughness change.

In the WAsP model the atmospheric stability is taken into account as a perturbation of the logarithmic wind speed profile. The mean vertical heat flux and its variability is used to characterise the atmospheric stability. Land and water surfaces are distinguished by different values for these quantities. An interpolation between land and sea areas is used in a transition zone on both sides of the coastal discontinuity.

Roughness changes are described by an internal

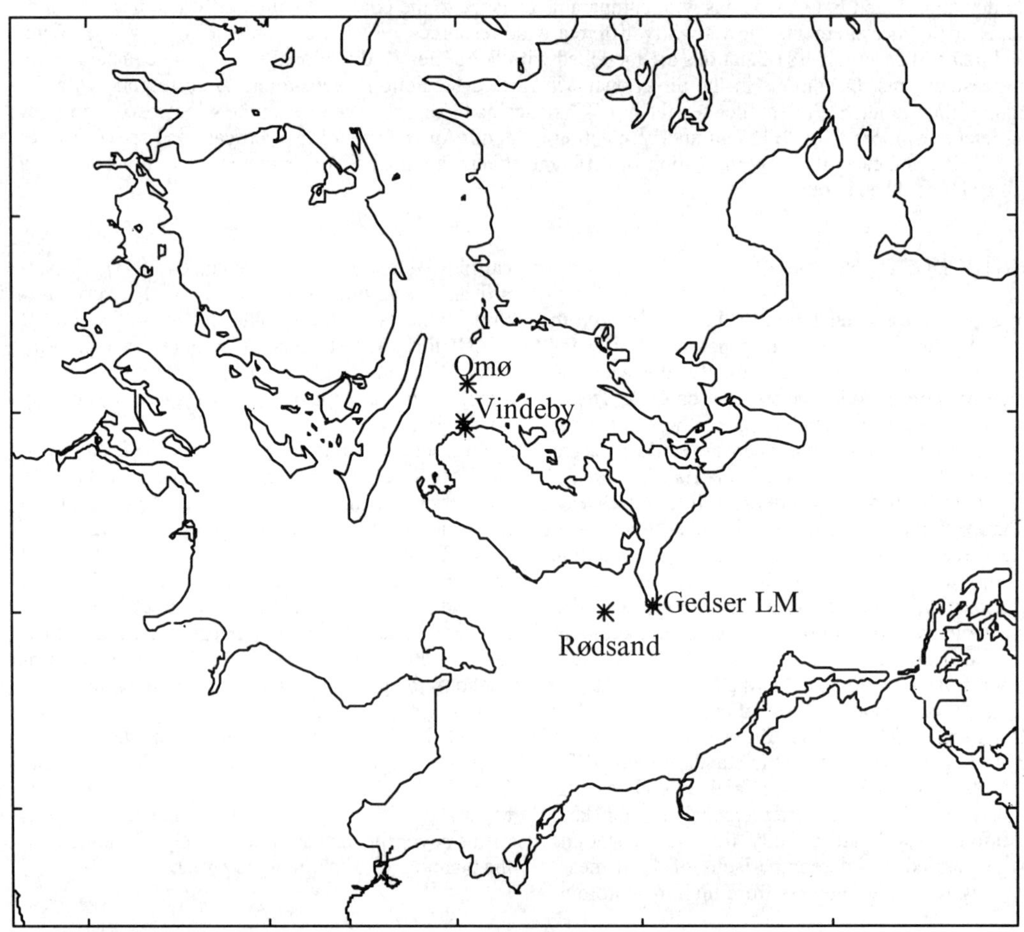

Figure 1: Locations of the measurement sites in the Baltic Sea in the southern part of Denmark

boundary layer (IBL) approach. The IBL model does not depend on the atmospheric stability, i.e. no difference is made between an IBL over land and over water.

A detailed analysis of the influence of the atmospheric stability in the coastal zone has been performed in (Barthelmie 1997) with the Vindeby data. A comparison with WAsP and a discussion of the performance of the atmospheric stability models of WAsP can be found in (Barthelmie et al. 1996b). Modifications of the model parameters of WAsP only lead to small improvements in the prediction accuracy.

The focus of the investigation here is laid on the question if the length of the upstream sea fetch has an influence on the prediction accuracy of the WAsP model. With the new measurements now available, a wide range of sea fetch distances are available for this investigation.

3 MEASUREMENT SITES

Measurements have been performed at six meteorological masts on and around the islands of Lolland and Falster in Denmark. The locations are shown in figure 1. The measurements at Vindeby sea mast west (SMW), Vindeby sea mast south (SMS), Omø and Rødsand are measurements from offshore meteorological masts. The measurements at Vindeby and Gedser Land Masts (LM) are accompanying coastal measurements. All collected data are half hourly averages. Long term meteorological measurements from the station Tystofte are also used.

The Vindeby masts SMW, SMS and LM are situated near the Vindeby offshore wind farm about 1.5 to 2 km off the north coast of Lolland (Fig. 2). The wind farm consists of 11 Bonus 450 kW turbines. For a detailed description of the wind farm see (Barthelmie et al. 1994).

Wind speed measurements at the Vindeby masts are disturbed for some wind direction sectors by wakes of the wind turbines. The measured wind speeds are therefore corrected for the shading effect of the turbines when they are bin-averaged for 30° wind direction sectors. Correction factors for the mean wind speeds are estimated with the wind farm modelling program FCalc (FCalc 1996). Reductions of up to 7% for the SMW and 2% for the LM are predicted.

The sites Omø and Rødsand are offshore sites located in the south-eastern part of Denmark near the island of Lolland. Both sites have a distance of about 10 km to the nearest land. Omø is located to the north of Lolland near the Vindeby site, while Rødsand is situated south-east of Lolland (Fig. 1).

The meteorological station Tystofte is a land measurement located in the southern part of Sealand, in about 5 km distance to the coast and 25 km to the Omø measurement site (Fig. 1).

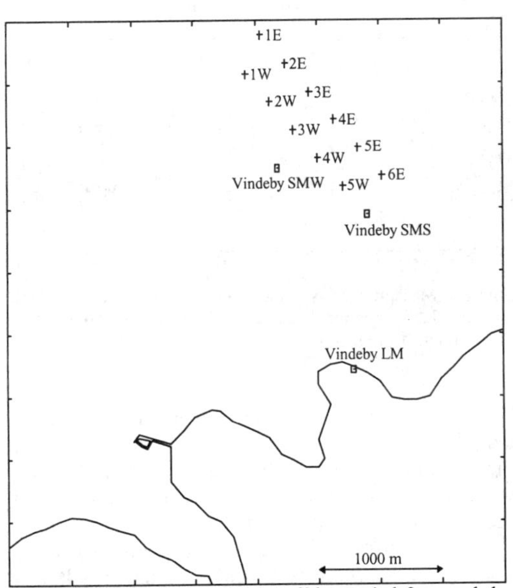

Figure 2: Layout of the Vindeby offshore wind farm and the measurement masts Vindeby SMW, SMS and LM.

4 WIND RESOURCES

The wind resource, i.e. the long term mean wind speed, is estimated from the onsite measurements and compared to WAsP-predictions. Deviations between the wind resource during the measurement period and the long term average were corrected by correlation with a 14 year long time series from the station Tystofte.

It has to be kept in mind, however, that the measurement periods at Omø and Rødsand are only about 9 and 13 months, respectively. Even though they were corrected, this leaves some uncertainty of the estimated mean wind speeds.

The estimation of the long term mean wind speeds from the measurements was done by a combination of different methods using all available measurements (see Barthelmie et al., in press).

WAsP estimations were calculated using two different measurement stations as input:

- The Vindeby land mast, which is a coastal measurement at the Vindeby site. The predictions made with these measurements were corrected for deviations from the long term average.
- The Tystofte station, which is a long term land measurement.

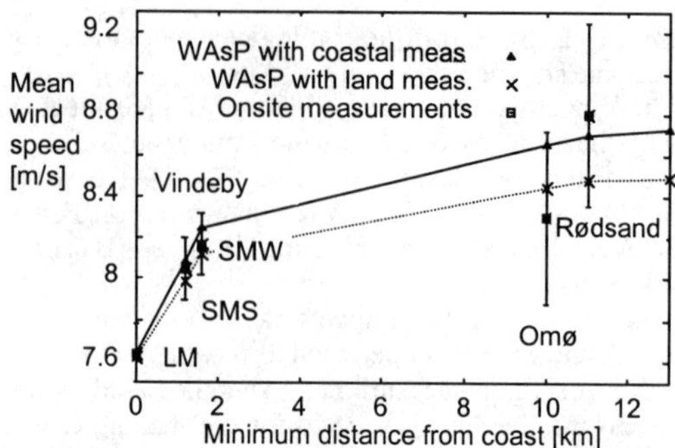

Figure 3: Long term mean wind speeds for a coastal and 4 offshore sites estimated from onsite measurements and predicted by WAsP on the basis of a coastal and a land measurement

Figure 3 shows the mean wind speeds of the sites versus their distances to the nearest coast, i.e. their minimum sea fetch. The points connected with lines are WAsP predictions, the squares are the estimations from the onsite measurements. The error bars indicate the estimated maximum uncertainties of the wind resource estimated from the measurements.

Deviations of the WAsP predictions from the estimations based on the onsite measurements are generally small, up to 5%. They are in the same range as the uncertainties in the wind speed estimations based on the onsite data. As expected, the mean wind speed rises with increasing distance to the coast. The increase in wind speed between the coastal station (Vindeby LM) and the offshore sites in 10 km distance is about 10-15%. The largest increase in wind speed when going offshore from the coast is found for the first kilometres. For the WAsP predictions an additional point is shown which gives the results for the open sea, i.e. the maximum possible value. It can bee seen that the two sites Rødsand and Omø almost reach that limit. WAsP assumes only a very small influence of the land for distances larger that about 10 km.

The WAsP predictions based on the coastal and the land measurements coincide at the site of the coastal station. For stations further away from the coast the predictions differ up to about 3% with the prediction based on the land measurement being lower than that based on the coastal measurement. For the sites near the shore both predictions are very close to the measurement. Differences are only 1-2% and lie within the uncertainty range of the measured data. For the sites Omø and Rødsand, which have larger distances to the shore, the differences are still small (1-5%). However, a considerable difference is found in the measurements of this two sites while WAsP estimates almost equal values.

A possible reason for this is the difference in the fetch situations. Omø is surrounded by land in about 10 km distance in the south, west and north-east directions, while Rødsand has land in this distance only in the northern and north-eastern directions, where the wind speed probability is low. It also has long sea fetches in the most frequent westerly wind directions (Fig. 1).

5 INFLUENCE OF THE SEA FETCH

5.1 Methodology

Ratios of measured wind speeds from two different stations are built for 12 wind directional sectors. These are compared with the respective ratios from WAsP estimated wind speeds. With different fetches for the two stations the influence of the sea fetch can be investigated in its measured and modelled influence on the wind speed.

The WAsP method uses measurements from one site (predictor) to estimate the wind resource at the predicted site. Within the model the measured directional wind speed distributions are described by Weibull functions. For short time series this procedure introduces an error due to deviations of the measured distribution from Weibull curves. To avoid this error the comparison is made for wind speed ratios rather then wind speeds.

An example of the directional dependence of the wind speed ratio is given in figure 4. The ratio is obtained by dividing the measured mean wind speed of Vindeby SMW by that measured at Vindeby LM. It is plotted versus the wind direction measured at Vindeby SMW. No corrections for wind farm shading have been applied here and the directional resolution is 3°.

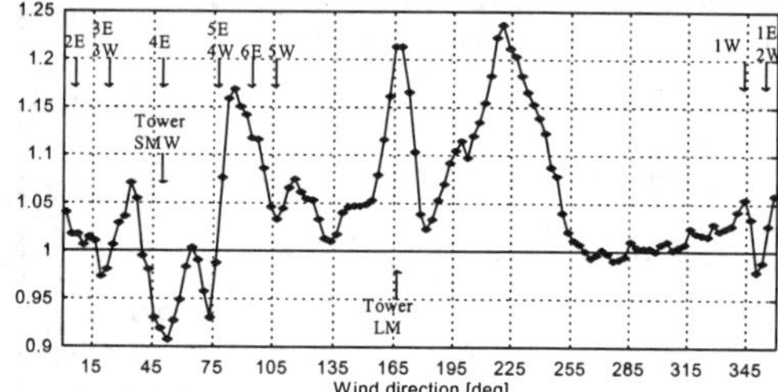

Figure 4: Measured ratios of mean wind speed Vindeby SMW / Vindeby LM; directional resolution 3°; directions of the turbines and measurement masts are shown with arrows

The directions with mast shading are shown and the disturbance can clearly be seen. Also indicated are the directions of the 11 turbines from the Vindeby wind farm. The reduction of the wind speed due to the influence of the wakes is obvious.

For the WAsP estimations of the wind speed ratios, first a common time series of wind speeds and directions is compiled for both stations. The measured data of one site are taken as input to WAsP to derive a wind climatology of the station itself and of the second measurement station. This is used to predict the sectorwise mean wind speeds for both sites.

Subsequently the predictor and predicted station are exchanged and a new ratio calculated. The average ratio is used for comparison with the measurement.

The ratios of the measured data are derived from the same time series as used for the WAsP predictions. The wind speeds of both stations are bin averaged in 30° bins with respect to the wind speed of one station. This is repeated with the wind direction of the other station and the average ratio is used for comparison. Wind direction sectors where the measurement is disturbed by mast interference are omitted. Uncertainties of the measured ratios are estimated as the average standard deviation of the means plus half of the difference of the two ratios calculated.

5.2 Direction dependent wind speed ratios

Examples for measured and WAsP estimated direction dependent wind speed ratios are shown in figures 5, 6 and 7 along with the lengths of the sea fetches of the respective stations. The solid lines show the measured ratios and the dashed lines are WAsP predictions. Measured data with possible mast interference have been omitted.

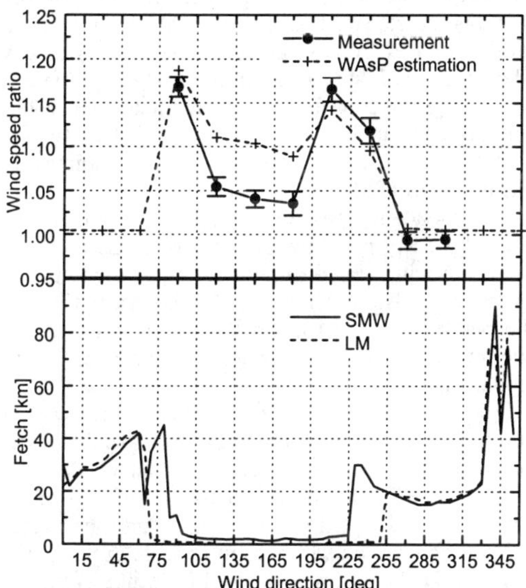

Figure 5: Vindeby SMW / Vindeby LM: Measured and WAsP-predicted wind speed ratios (top) and sea fetches (bottom) versus wind direction

Figure 5 shows the ratios of Vindeby SMW and LM. Vindeby SMW is located 1.6 km off the north coast of Lolland (Fig. 2). Here the behaviour of WAsP for small fetches can be investigated.

The measurements show two distinct maxima of the ratio which are roughly in the direction of the coastline. This situation leads to a large fetch difference since the SMW has long sea fetches while the LM has mainly land fetch. In between this maxima (sectors 120°-180°) SMW has a short sea fetch of 1.6 to 3 km only. For the other directions both stations have similar long sea fetches and the ratio is close to one.

The WAsP-predictions show generally the same directional pattern as the measurements. In most cases the deviations between measurements and WAsP-predictions are small and the general behaviour of the directional wind speed difference between two stations is modelled well. Significant deviations are found only for the case with land fetch for LM and short sea fetch for SMW where WAsP seems to overpredict the difference between the wind speeds on land and at sea.

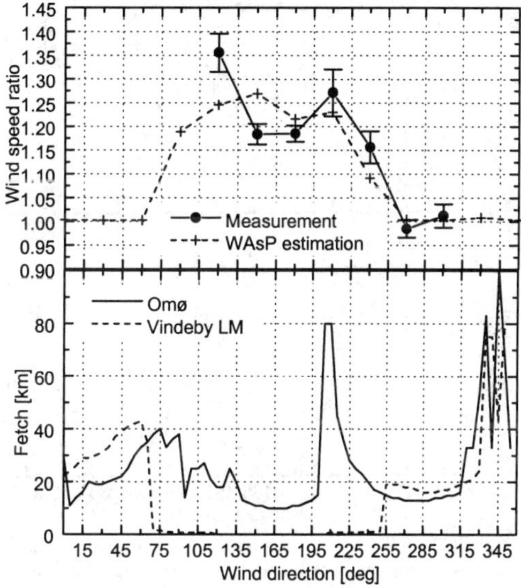

Figure 6: Measured and WAsP-predicted wind speed ratios versus wind direction from Omø / Vindeby LM (top) and sea fetches of Omø and Vindeby LM versus wind direction

Figure 6 show the results for the ratio between Omø and Vindeby LM. The overall situation at these sites is quite similar to the one for Vindeby SMW and LM, only that Omø is located about 10 km further offshore (Fig. 1). 150° and 180° are the sectors with wind from Lolland and a sea fetch of about 10-15 km. Here WAsP again overpredicts the difference between sea and land. The very rapid change of sea fetch length from 15 to more than 80 km at 200° seems to have an influence also on the wind direction sector 180°.

For the very long fetches at Omø at 120° and to a lesser extent at 210° and 240° WAsP underpredicts the difference.

In figure 7 the ratios of the two offshore stations Omø and Rødsand are shown. Rødsand is located 11 km off the south coast of Lolland and the two sites are about 60 km apart from each other (Fig. 1). A comparison of these sites gives the opportunity to study the measured and predicted differences for two offshore stations with very different fetches situations.

The measurements show a minimum in the ratios at wind direction sectors 270° and 300° and a maximum at 330°. This corresponds closely to the very large sea fetches at Rødsand in directions 260° to 290° and at Omø at 330° to 350°. For the other directions the ratio does not deviate much from one.

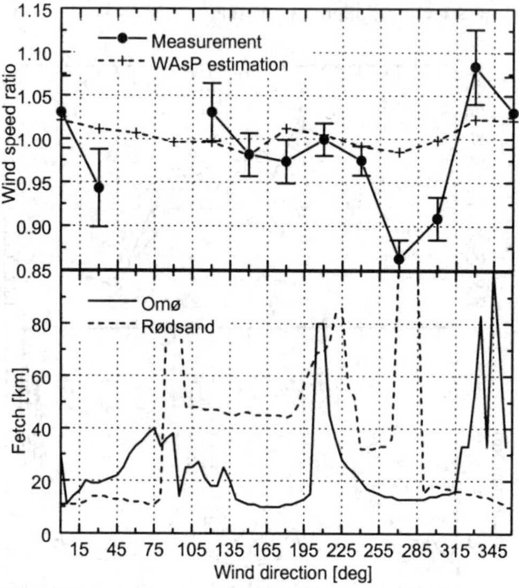

Figure 7: Omø / Rødsand: Measured and WAsP-predicted wind speed ratios (top) and sea fetches (bottom) versus wind direction

The WAsP-predictions show only very small deviations from unity. This leads to significant deviations of 6-12% from the measurements for the two cases mentioned. For the 30° sector a smaller deviation can be seen which can not easily be assigned to a fetch difference. For all other wind directions the deviations are in the order of the measurement uncertainty.

This result explains the measured differences between wind resources at Omø and Rødsand. The very long fetches at Rødsand which also occur in the prevailing wind directions lead to higher wind speeds there compared to Omø. Since WAsP does not take these long fetches into account it underpredicts the wind resource at Rødsand.

6 WIND SPEED PROFILES

To investigate possible causes for deviations of WAsP results, measured and modelled vertical wind speed profiles are compared. Two pairs of stations have been used for this: Vindeby LM and SMW as well as Vindeby LM and Omø. Measurements are available at three heights: 10m, 30m and 48m (46 m at LM). The wind speed measurements have been corrected for the effect of the flow distortion caused by the measurement tower, except for wind directions where the tower was upwind of the anemometer. The correction factors have been estimated after (Højstrup, in press).

WAsP calculations are made using the measured time series at Vindeby SMW and Omø at 48 m height as input. Estimations are made for both sites at all heights.

In most cases the WAsP calculations of the mean wind speeds as well as the estimations of the vertical wind speed profiles agree well with the measurements. Some examples of measured and modelled profiles are shown in figures 8 and 9.

Figure 8 shows the case with wind blowing from land, i.e. Vindeby LM has land fetch, Vindeby SMW has about 1.6 km sea fetch and Omø has about 10 km sea fetch.

The effect of different land and sea roughnesses can clearly be seen as the profiles for the offshore stations are steeper than for the LM. A comparison of the profiles from Vindeby SMW and LM clearly shows a wind speed difference, which decreases with height. At the highest anemometer the difference is only about 3%. This shows that at 1.6 km distance the influence of the land is still dominant at the height of 48 m.

This tendency is also present in the profiles of Vindeby LM and Omø, though to a lesser extent. Here the wind speed difference at the highest anemometer is also much larger. This shows that the influence of the land roughness is much weaker and the internal boundary layer from the land-sea transition is higher.

All measured profiles are to some extent curved towards higher wind speeds at greater heights compared to the logarithmic profile which would be a straight line in the graphs. This is believed to be mainly caused by influences of the atmospheric stability. The large curvature of the profile of the SMW is probably due to the additional influence of the internal boundary layer from the land sea roughness change.

The earlier comparison of the wind speed ratios showed that WAsP overpredicts the difference between LM and the offshore masts at 48 m height (Figs 5, 6). This can also be seen from the profiles in figure 8. However, it also shows that the steepnesses of the predicted and measured profiles do not deviate much. This indicates together with the deviations of the curvatures of the profiles that the simplifications WAsP makes in the description of the sea surface roughness are not the only reason for the deviations. It should be kept in mind, however, that the measurement at Omø / Vindeby LM stems from a relatively short time series. Therefore the atmospheric stability which occurred during the measurement differs from the yearly distribution WAsP assumes.

Figure 9 shows the situation of land fetch for the LM and long sea fetches for both of the offshore masts (240° sector). From the wind speed ratios a slight underprediction of the wind speed difference at 48 m is known, especially for the ratio of Vindeby LM / Omø (Figs 5, 6). This is repeated in the profiles. Again it can be seen that the steepnesses of the offshore profiles compare well between modelled and measured values. For the LM the modelled steepness is slightly too small, which points towards a slight overestimation of the land roughness. As before, the curvatures indicate an influence of the atmospheric stability which is not accurately modelled.

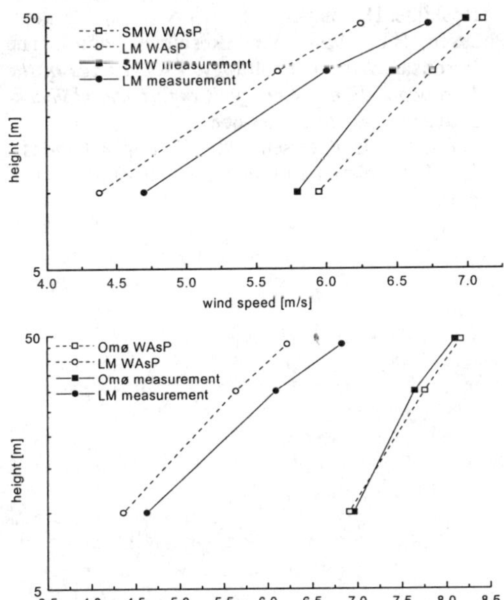

Figure 8: Measured and WAsP-predicted mean wind speeds for wind directions with land fetch situations (150° and 180° sectors) versus measurement heights for the two sites Vindeby LM and SMW (upper graph) and Vindeby LM and Omø (lower graph)

In the examples both over- and underestimation of the difference between land and offshore wind speed are present. In both cases modelled and measured profiles show deviations in curvature and only small differences of steepness. This indicates that a combination of different reasons is responsible for the deviations found from the analysis of the wind speed ratios. It seems likely that the deviations found are not solely caused by a problem of the modelled sea surface roughness. Additionally, the growth of the internal boundary layer and the influence of the atmospheric stability play an important role.

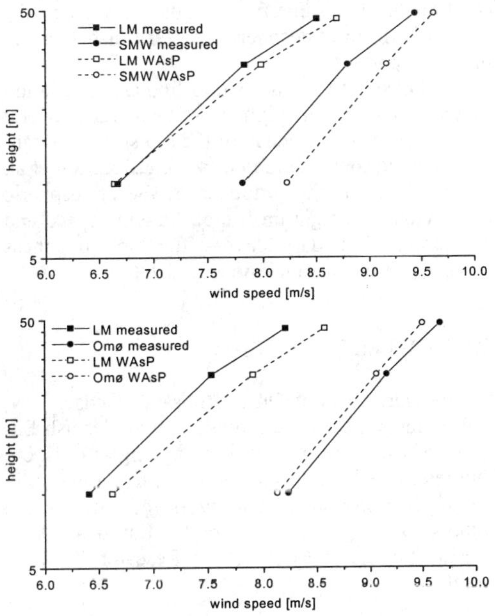

Figure 9: Measured and WAsP-predicted mean wind speeds for wind directions with different fetch situations (240° sector) versus measurement heights for the two sites Vindeby LM and SMW (upper graph) and Vindeby LM and Omø (lower graph)

7 CONCLUSION

The data presently available from ongoing measurements in the Danish Baltic Sea have been analysed and compared with predictions made with the wind resource estimation program WAsP. Measurements of a coastal and an inland station have been used for the predictions. It was found that the predictions of the long term average wind resource are in good agreement with measurements.

The measurement indicates a small difference in wind resources between the stations Rødsand and Omø, which is not predicted by WAsP.

The investigation of the direction dependent wind speed also shows a generally good agreement between WAsP-predictions and measurements. However, for some wind directions significant deviations were found. These deviations show a correlation with the length of the upwind sea fetch. WAsP tends to overpredict the wind speed for situations with short sea fetches and underpredict it for long fetches. This effect explains the measured differences between the Omø and Rødsand sites since at Rødsand the wind comes more frequently from directions with long fetches.

The comparison of measured and predicted height profiles indicates that a combination of different reasons is responsible for the deviations found for some wind directions. The modelled sea surface roughness can not solely be held responsible for the deviations. The growth of the internal boundary layer and the influence of the atmospheric stability also seem to play an important role

A detailed model is needed to find an explanation for these findings. It might have to include several effects like the dependence of the sea surface roughness on wind speed and fetch (see e.g. Johnson et al. 1997) as well as the influence of the atmospheric stability on the height profile of the wind speed and on internal boundary layers due to roughness changes (see e.g. Barthelmie et al. 1996a).

ACKNOWLEDGEMENTS

The measurements at Omø, Rødsand, Gedser Rev, Gedser and Vindeby were funded by the Danish Energistyrelsen in the project 'Kortlægning af Offshore Vindressourcer' and of the European Commission in the JOULE-programs. The work of one of the authors (B. Lange) is funded by the European Commission through a Marie Curie Research Training Grant.

REFERENCES

Barthelmie, R.J., M.S. Courtney, J. Højstrup and P. Sanderhoff 1994. *The Vindeby Project: A Description.* Risø Report R-741(EN). Risø National Laboratory, Roskilde, Denmark

Barthelmie, R.J., M.S. Courtney, J. Højstrup & S.E. Larsen 1996a. Meteorological aspects of offshore wind energy: Observations from the Vindeby wind farm. *Journal of Wind Engineering and Industrial Aerodynamics* Vol. 62: 191-211

Barthelmie, R.J., N.G. Mortensen, L. Landberg & J. Højstrup 1996b. Application of the WAsP Model to Determine the Wind Resource in Non-neutral Conditions in Coastal Areas. *Proceedings of the European Union Wind Energy Conference, Göteborg, Sweden*

Barthelmie, R.J. 1997. The Effect of Atmospheric Stability on the Temporal and Spatial Variability of Offshore Wind Resources. *Proceedings of the OWEMES European Seminar: Offshore Wind Energy in Mediterranean and other European Seas*

Barthelmie, R.J., M.S. Courtney, B. Lange, M. Nielsen, A.M. Sempreviva, J. Svenson, F. Olsen & T. Christensen, in press. Offshore wind resources at Danish measurement sites. *Proceedings of the European Wind Energy Conference and Exhibition 1999, Nice*

Charnock, H. 1955. Wind stress over a water surface. *Quarterly Journal of the Royal Meteorological Society*, Vol. 81: 639-640

FCalc 1996. *Farm Calculation program*. University of Oldenburg, Germany

Frandsen, S. (ed.), L. Chacón, P. Enevoldsen, R. Gómez-Elvira, J. Hernández, J. Højstrup, F. Manuel, K. Thomsen & P. Sørensen 1996. *Measurements on and Modelling of Offshore Wind Farms*. Risø Report R-903 (EN)

Højstrup, J., in press: Vertical Extrapolation of Offshore Windprofiles. *Proceedings of the European Wind Energy Conference and Exhibition 1999, Nice*

Johnson, H.K., H.J. Vested, J. Højstrup, S.E. Larsen & H. Hersbach 1998. On the Dependence of Sea Surface Roughness on Wind Waves. *Journal of Physical Oceanography*. Vol.28: 1702-1716

Mortensen, N.G., L. Landberg, I. Troen & E.L. Petersen 1993. *Wind Atlas Analysis and Application Program (WAsP)*. Risø National Laboratory, Roskilde, Denmark

Petersen, E.L. 1993. Wind Resources Part I, The European Wind Climatology. *Proceedings of the European Wind Energy Conference, Lübeck-Travemünde; 1993*: 663-668

Troen, I. & E.L. Petersen 1989. *European Wind Atlas*. Risø National Laboratory, Roskilde, Denmark

Wind Engineering into the 21st Century, Larsen, Larose & Livesey (eds) © Balkema, Rotterdam, ISBN 90 5809 059 0

A detailed and verified wind resource atlas for Denmark

Niels G. Mortensen, Lars Landberg, Ole Rathmann & Morten N. Nielsen
Risø National Laboratory, Roskilde, Denmark

Per Nielsen
Energy and Environmental Data, Aalborg Ø, Denmark

ABSTRACT: A detailed and reliable wind resource atlas covering the entire land area of Denmark has been established. Key words in the methodology are wind atlas analysis, interpolation of wind atlas data sets, automated generation of digital terrain descriptions and modelling of local wind climates. The atlas contains wind speed and direction distributions, as well as mean energy densities of the wind, for 12 sectors and four heights above ground level: 25, 45, 70 and 100 m. The spatial resolution is 200 meters in the horizontal. The atlas has been verified by comparison with actual wind turbine power productions from over 1200 turbines. More than 80% of these turbines were predicted to within 10%. The atlas is available on CD-ROM and on the Internet.

1 INTRODUCTION

Information on the actual wind resources of a region or an entire country is a prerequisite for the optimal utilisation of wind energy on a large scale. In the past, this information was often given as the wind resource potential, e.g. in the form of a 'wind atlas'. The wind atlas, however, does not provide directly the detailed information required for planning purposes and siting of large wind farms; it only forms the necessary basis for more detailed resource assessments. The purpose of the present work has been to establish a methodology whereby a detailed and reliable 'wind resource atlas' can be constructed for a not-too-complex terrain, and the approach is illustrated and verified by a new wind resource atlas for Denmark.

The methodology can be described briefly as follows: Wind data from existing meteorological stations are first analysed using the wind atlas methodology, in order to determine the regional or site-independent wind climate at each station. The regional wind climate over Denmark can then be described and stations to be included in the analysis selected. Based on these stations, the regional wind climate (wind potential) can be mapped. An interpolation procedure for points between stations is also developed. The wind potential map is verified using measured power production figures from a large number of wind turbines. The actual wind climate and wind resources all over Denmark can now be modelled using the verified wind potential data as well as digital terrain descriptions of the entire country. Finally, these results are used to map the wind resources of Denmark.

2 DANISH WIND-MONITORING STATIONS

The meteorological basis for the wind resource atlas for Denmark is wind atlas data sets for about 24 stations, carefully chosen to represent all regions of Denmark, see Figure 1.

From each station, continuous wind speed and direction measurements in the period 1987–96 have been analysed. The observation intervals range from 10 minutes to 3 hours, but the averaging period for wind speed is 10 min. for all the stations. Wind directions are instantaneous readings or averaged over 10 min.

The wind measurements describe the local or site-specific wind climate at each station – where wind climate is taken here to mean the frequency distributions of wind speed in a number of sectors (here 12) and the frequency distribution of wind direction in the same sectors (the wind rose). The wind atlas data sets, on the other hand, contain the regional or site-independent wind climate estimates, derived from the measurements using the wind atlas methodology.

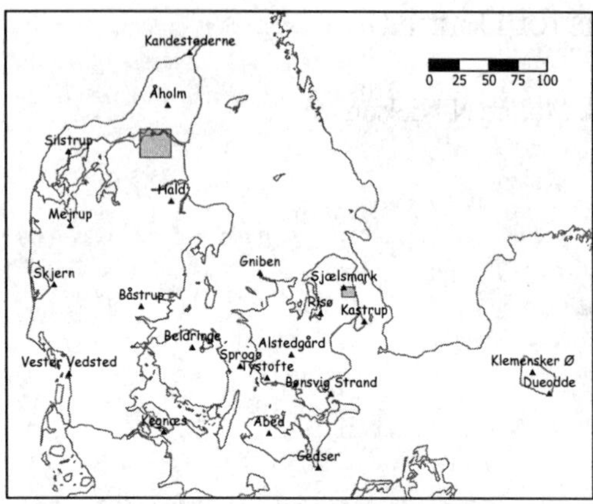

Figure 1: The Danish meteorological stations chosen for the wind atlas analysis. The stations are run by the Danish Meteorological Institute and Risø National Laboratory.

3 THE WIND ATLAS METHODOLOGY

The so-called wind atlas methodology has been applied for the analysis of wind measurements from the meteorological stations. This comprehensive set of models for the horizontal and vertical extrapolation of meteorological data and the estimation of wind resources was developed for the analysis presented in the European Wind Atlas (Troen & Petersen 1989, Petersen et al. 1998a). The actual implementation of the models is known as the *Wind Atlas Analysis and Application Program* – or WAsP for short – which is the software package that has been applied here.

The models are based on the physical principles of flows in the atmospheric boundary layer and they take into account the effect of different surface conditions, sheltering effects due to buildings and other obstacles, and the modification of the wind imposed by the specific variations of the height of ground around the meteorological station in question. Figure 2 illustrates the use of these models on measured wind data to calculate a regional wind climatology.

The figure also illustrates the so-called *application part* of the methodology, following a procedure in which the regional wind climatologies are used as input to the same models to produce site-specific wind climatologies and, given the power curve of a wind turbine, production estimates. For more detailed information on the models and the WAsP-program, the reader is referred to the European Wind Atlas (Troen & Petersen 1989).

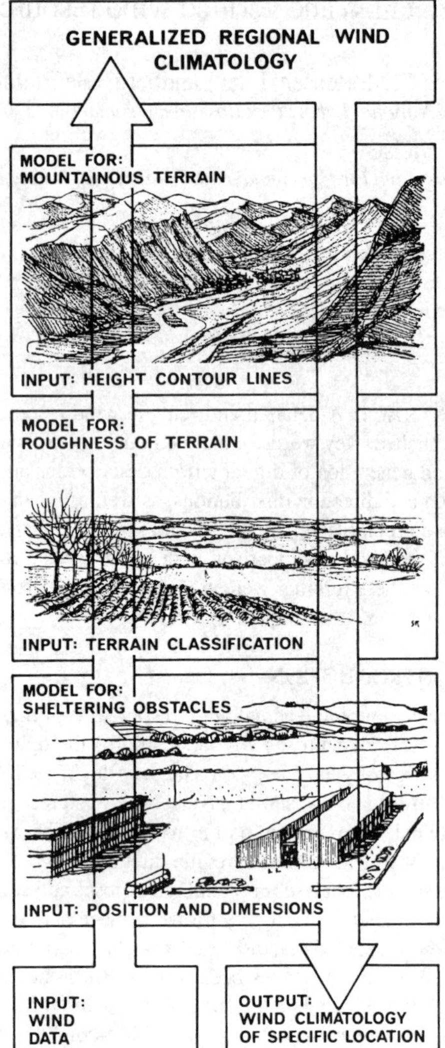

Figure 2: Schematic presentation of the wind atlas methodology. Meteorological models are used to calculate the regional wind climatologies from the raw data. In the reverse process – the application the Wind Atlas – the wind climate at any specific site may be calculated from the regional climatology.

Accurate descriptions of each meteorological station and its surroundings were collected from maps and during field trips. The descriptions include details of the

- terrain roughness, i.e. water areas, forests etc.

- nearby sheltering obstacles such as buildings
- terrain height variations (orography)

Regional wind climatologies were subsequently calculated for the 24 wind atlas stations (Mortensen et al, 1999).

4 REGIONAL WIND CLIMATE OF DENMARK

The collection of wind atlas data sets is a representation of the regional wind climate over Denmark. As an example, Figure 3 shows the variation of the mean wind speed 50 meters above ground level over a flat, uniform surface with a roughness length z_0 of 0.03 m.

One meteorological station only, Bønsvig Strand in Figure 1, was excluded from further analysis because the wind atlas data sets did not compare well with those of the neighbouring stations. Predictions based on data from this station also turned out to be very low when compared to power production data from near-by wind turbines.

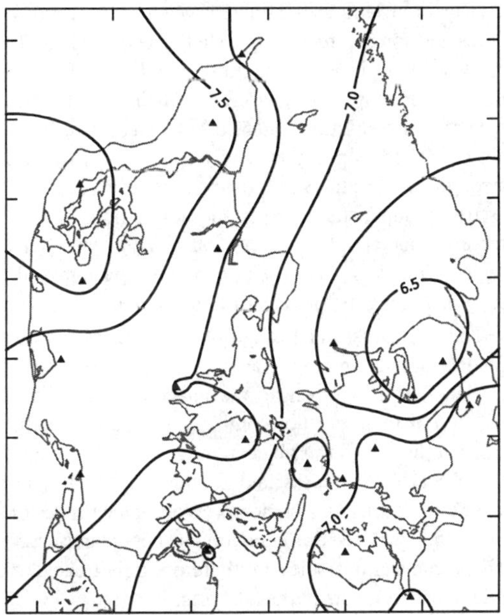

Figure 3: Mean wind speed 50 m a.g.l. in ms^{-1} over a flat, uniform surface with $z_0 = 0.03$ m. The wind speed contour interval is 0.25 ms^{-1}. The distance between tick marks is 50 km.

The mean wind speed shown in Figure 3 vary from about 8 ms^{-1} in the NW part of Denmark to 6 ms^{-1} in NE Zealand. It thus decreases from NW towards SE to a minimum around the Great Belt region and north Zealand, where after it increases towards the Baltic Sea. This picture of the magnitude and variation of the regional wind climate over Denmark is largely supported by similar data from Sweden (Krieg 1992) and Germany (Traup & Kruse 1996).

In order to be able to calculate the regional wind climate (not just the mean wind speed) for any place in Denmark, a continuous 'wind atlas surface' must now be derived by spatial interpolation procedures for both the wind speed and direction distributions.

5 INTERPOLATION OF WIND ATLAS DATA

The starting point of any WAsP prediction is the Wind Atlas file describing the regional wind climate. Mapping the wind resource over a country will involve several regions and in order to avoid discontinuities we have to develop an interpolation routine.

Setting up an interpolation scheme involve arbitrary choices, so perhaps it is worthwhile to review the objectives. We decide that:

- the interpolation must match the statistics of the measuring stations exactly
- the interpolation must be local, i.e. with emphasis on data from nearby stations
- the interpolation must work with an irregular mesh of data
- the interpolation must be as smooth as possible

The interpolation method implicitly assumes that the available data are of high quality and obtained in compatible sample periods, i.e. it does not intend to smooth extreme values or identify global trends in the data set. Corrections of the reference Wind Atlas files are best done from within the WAsP program, i.e. by improved description of the surface roughness and obstacle configuration around the measuring sites.

Wind Atlas files contain wind roses and probability distributions for wind speeds in each sector. The wind speed distributions are approximated by Weibull distributions, and the moments of the wind-speed distributions $\langle u \rangle$ and $\langle u^3 \rangle$ may be found from the Weibull parameters A and k and vice versa. These transformations allow us to interpolate directional distributions and the moments in the total wind-speed distribution independently.

The measuring stations are organised in a grid of triangles with local co-ordinate systems. The first step in the interpolation scheme is to identify the local triangle where the directional distributions and direction-independent $\langle u \rangle$ and $\langle u^3 \rangle$ statistics are known at the corners. The interpolation of $\langle u \rangle$ and $\langle u^3 \rangle$ is based on third-order Bézier polynomia which also require gradients at the measuring points in the corners the triangles. The common gradients ensure continuity between neighbouring Bézier patches and these gradients are settled by an algorithm which minimise the curvature of the interpolation surface. Directional interpolation is made by linear interpolation of Fourier transformations of the three reference distributions relative to a 'typical' wind rose orientation. This orientation is determined as the maximum of the Fourier spline rather than the maximum of the original wind-rose which has a crude directional resolution. Both shape and orientation are smoothly interpolated by this method.

The measuring stations were placed at inland locations and the interpolation grid does not cover all coastal areas. Straightforward extrapolation by the third-order Bézier patches produce large excursions so an alternative method have to be developed. The extrapolation method operates with values and gradients at two reference stations only. The two reference stations and the interpolation point fix a second-order Bézier *curve*. This curve is formally interpreted as a third-order Bézier curve, with along-curve nodal gradients determined by projection. The parameters predicted by the extrapolation routine is a smooth extension of the interpolation surfaces and far-field extrapolation becomes more linear. The Bézier patches are defined only once whereas a new extrapolation has to be defined for every extrapolation point. Therefore extrapolation is computationally less efficient than interpolation.

The interpolation procedures have been implemented in a program which makes it possible to predict the wind atlas data sets for any site in Denmark. For the purpose of calculating the wind resource map, a number of stations in Germany and Sweden, where similar data exist, were included in the final analysis.

6 MODELLING LOCAL WIND CLIMATES

The site-specific wind climate of any place in Denmark can now be estimated using the application part of the WAsP program. The regional wind climate for a given site can be found using the interpolation procedures described above. To account for the influences of local topography we further need digital terrain descriptions of the surrounding terrain. For an area of 43 000 km^2, it was not feasible to establish these 'by hand' – as was done in the analyses of the 24 meteorological stations – so an alternative approach had to be developed.

6.1 *Establishing digital terrain information*

For the flow modelling, we have used a vector map of Denmark prepared by the National Survey and Cadastre. This map database contains height contour lines with 5-m vertical contour intervals for the entire country. The map has been established by hand digitisation as well as vectorisation of scanned contour sheets. The database was reformatted into WAsP map files for direct application with the WAsP orographic flow model.

To account for the effects of the terrain roughness and roughness changes, a digital roughness map had to be established from the land use classification apparent in standard topographical maps. Map sheets with a scale of 1:50 000 were scanned and six layers of land use extracted: water areas, forests, cities and towns, tree groups, shelter belts and single houses. These layers were subsequently transformed into vector maps and analysed with respect to surface roughness. Large water bodies, forests and cities are described as polygons and were simply assigned a roughness length. Small towns and forests, as well as tree groups, shelter belts and single houses were recorded and counted in sub-areas of 1×1 km^2. This registration was then used to assign 'background' roughness lengths for each of the sub-areas.

The polygons and background roughnesses were finally merged into several digital roughness maps in WAsP format covering the entire country. Since the digital maps do not contain information on the size and height of the objects, modelling of the shelter effects from houses and shelter belts had to be abandoned. This is a minor problem, since most of the wind turbines are sited in the open landscape. The sheltering obstacles are therefore only treated as roughness elements in the analyses.

6.2 *Modelling considerations*

In the modelling of site-specific wind climates, two approaches have been tried. One is to employ the wind atlas data sets from one station located centrally in Denmark (here Beldringe on the island of Funen) together with a site-specific correction scheme based on the measured power production figures from a large number (1000+) of wind turbines. This approach is feasible in Denmark, because of the large number of turbine installations. The second, and presumably more physically correct approach, is the one outlined above, where the 'local' regional wind climate is determined from the three nearest met. stations. However, since we verify the methodology against actual wind turbine productions and we further aim at establishing a wind resource map suitable for production estimations, it is not obvious which approach will provide the most accurate results. The experience obtained so far suggests that accurate results may be obtained using the simple approach, whereas the second approach seem to be very sensitive to especially the measured wind roses at the reference stations. The results reported below are based on the simple approach.

7 VERIFICATION OF THE METHODOLOGY

In order to verify the regional wind climate interpolations, the digital terrain maps and the methodology in general, we have compared predictions of wind turbine power productions with actually measured productions. For this purpose, more than 1200 wind turbines all over Denmark were selected. The selection criteria included availability and quality of production data, minimum size of wind turbine (75 kW) and documentation of turbine history and performance. The power curves for the different turbine types were extensively checked and production data were further referenced to a normal wind year. For turbines in wind farms a simplified park efficiency correction was applied, based on the actual efficiencies of a number of wind farms.

Since there are several more or less 'unknowns' in the final modelling and verification process, this was done iteratively and several scenarios were tried. In this process, errors in the different input data files were identified and corrected and the interpretation and transformation of land-use data to roughness maps were calibrated and made consistent.

7.1 *Verification results*

The main results of the verification are: the measured power productions from more than 80% of the 1200+ control turbines fall within ±10% of the predicted productions. About 10% produce less than 90% of the predicted production and about 10% produce more than 110% of the prediction.

For about 20 turbines, the difference between predicted and measured production is larger than 20%. These 'out-liers', which are located randomly all over Denmark, can in many cases be explained by severe local shelter effects not accounted for in the modelling or by errors in the digital terrain height map.

8 WIND RESOURCE MAP OF DENMARK

The wind resources over the land area of Denmark can now be estimated in the following way: the actual wind resource in grid points with a regular spacing of 200 m are calculated by the WAsP program for several heights above the ground surface: 25, 45, 75 and 100 m. Inputs are the wind atlas climatology from Beldringe and the WAsP terrain descriptions derived from existing databases of terrain orography and land-use. For each calculation point, the program uses terrain height information to at least 5 km from the site and a roughness map covering an area with a radius of at least 20 km. The wind climate and power production estimates are corrected as described above. The wind resource atlas (database) thus contains information on the wind speed and direction distributions in 12 sectors, as well as the energy density. As an example, Figure 4 show an energy density map of the municipality of Aalborg in Jutland.

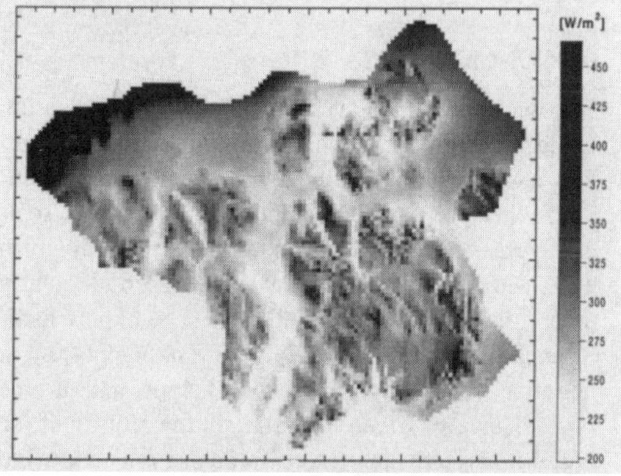

Figure 4: Mean wind energy density in Wm^{-2} 45 m a.g.l. over the southern part of the municipality of Aalborg, Jutland, see Figure 1. Results are shown with a horizontal resolution of 200 m. The distance between tick marks is 1 km.

Given a specific power curve, it is also possible to display directly the estimated power production for an area, an example is given in Figure 5.

The wind resource atlas for Denmark is not published yet as a paper map. The results are stored and distributed on a CD-ROM, together with software for

Figure 5: Expected mean power production from a 600-kW wind turbine with a hub height of 40.5 m in the municipality (full line) of Birkerød, Zealand. Results are shown with a resolution of 100 m. The distance between tick marks is 1 km.

retrieval, display, printing and export of the data to several common file formats. In addition, the CD-ROM contains a number of actual or generalised wind turbine power curves, in order for the user to be able to estimate actual power productions. An Internet version of the resource map is also in the making and will be available at www.emd.dk.

9 CONCLUSIONS

The wind resource atlas for Denmark provides a detailed, reliable and coherent database of the variation and magnitude of the wind resource over Denmark. It can be used by planners on all levels of society to ensure a systematic and efficient planning process, whereby the available sites are used efficiently to optimise the power production. The export facility makes it straightforward to e.g. establish layers in common GIS-systems for certain heights and turbine types. However, given the detail and flexibility of the outputs, also private persons, organisations and wind turbine manufacturers may use the atlas for preliminary siting and power production estimation. Final siting and layout considerations, however, should not be based on the atlas only.

The methodology can be used to establish similar wind resource atlases for other regions where reliable wind climatologies and topographical databases exist, and where the terrain in general is not too complex. In mountainous regions, or regions with a more complex wind climatology, the regional wind climate should be determined using meso-scale models (Frank & Landberg 1997, Petersen et al. 1998b).

ACKNOWLEDGEMENTS

The wind resource map for Denmark is funded by the Danish Energy Agency under the Ministry of Environment and Energy.

Gunnar Jensen, Risø National Laboratory, provided the calibrated and quality-controlled data from the Risø meteorological stations, as well as many observations which were helpful in the analyses of these stations.

REFERENCES

Frank, H.P. and L. Landberg 1997. Modelling the wind climate of Ireland. *Boundary-Layer Meteorol.* **85**, 359–378.

Krieg, R. 1992. Vindatlas för Sverige. In Swedish. Slutrapport på projekt 506 269-2 på uppdrag av NUTEK. SMHI, Norrköping. 26 pp.

Mortensen, N.G., L. Landberg, I. Troen and E.L. Petersen 1993. Wind Atlas Analysis and Application Program (WAsP). Vol. 1: Getting Started, Vol. 2: User's Guide. Risø–I–666(EN). Risø National Laboratory, Roskilde. 133 p.

Mortensen, N.G., L. Landberg, O. Rathmann, G. Jensen and E.L. Petersen 1999. Wind Atlas Analysis of 24 Danish Stations (1987-96). In preparation for publication as Risø-R-1092(EN). Risø National Laboratory, Roskilde.

Petersen, E.L., N.G. Mortensen, L. Landberg, J. Højstrup and H.P. Frank 1998a. Wind Power Meteorology. Part I: Climate and turbulence. *Wind Energy* **1**, 2–22.

Petersen, E.L., N.G. Mortensen, L. Landberg, J. Højstrup and H.P. Frank 1998b. Wind Power Meteorology. Part II: Siting and Models. *Wind Energy* **1**, 55–72.

Traup, S. and B. Kruse 1996. *Wind und Windenergiepotentiale in Deutschland - Winddaten für Windenergienutzer*. In German. Deutscher Wetterdienst, Offenbach am Main. 445 pp.

Troen, I. and E.L. Petersen 1989. *European Wind Atlas*. Risø National Laboratory, Roskilde. ISBN 87–550–1482–8. 656 pp.

Author index *

* Volume 1 pages 1-720; Volume 2 pages 721-1476; Volume 3 pages 1477-2022